Bibliographies on the History
of Science and Technology
Vol. 24

THE HISTORY OF NATURAL HISTORY

Garland Reference Library
of the Humanities
Vol. 991

Bibliographies on the History of Science and Technology

ROBERT MULTHAUF
Smithsonian Institution
Series Editor

THE HISTORY OF METAL MINING AND METALLURGY
An Annotated Bibliography
by Peter M. Molloy

THE HISTORY OF SCIENCE AND TECHNOLOGY IN THE UNITED STATES
A Critical and Selective Bibliography, Vol. II
by Marc Rothenberg

THE HISTORY OF ELECTRICAL TECHNOLOGY
An Annotated Bibliography
by Bernard S. Finn

THE HISTORY OF SCIENCE AND TECHNOLOGY IN RUSSIA
An Annotated Bibliography
by Loren Graham and Paul R. Josephson

THE HISTORY OF AGRICULTURAL SCIENCE AND TECHNOLOGY
An International Annotated Bibliography
by R. Douglas Hurt and Mary Ellen Hurt

THE HISTORY OF NATURAL HISTORY
An Annotated Bibliography
by Gavin Bridson

THE HISTORY OF NATURAL HISTORY

An Annotated Bibliography

Gavin Bridson

GARLAND PUBLISHING, INC.
New York & London / 1994

Library of Congress Cataloging-in-Publication Data

Bridson, Gavin D. R.
The history of natural history : an annotated bibliography / by Gavin Bridson.
p. cm. — (Garland reference library of the humanities ; vol. 991) (Bibliographies on the history of science and technology ; vol. 24)
Includes indexes.
ISBN 0-8240-2319-6
1. Natural history—History—Bibliography. 2. Natural history—Bibliography. I. Title. II. Series. III. Series: Bibliographies on the history of science and technology ; v. 24.
Z7405.H6B75 1994
[QH15]
508'.09—dc20 94-2658
CIP

Printed on acid-free, 250-year-life paper
Manufactured in the United States of America

For Marlene

OUTLINE CONTENTS

OUTLINE CONTENTS

DETAILED CONTENTS

Section Bb. Histories of Naturalists' Societies

Section Bc. Natural History Library Resources

INTRODUCTION

Why Natural History?

At a time when probably the most significant biological research is being pursued at the molecular level, a guide to the history of "natural history" might suggest an interest in a long outmoded branch of life science. Although this term, and the branch of study it describes, has a history stretching back several centuries, it still has meaning and currency today. Natural history has had an uneven history and by the end of the 19th century might have appeared to be fatally waning in the face of growing scientific specialization and professionalism. But by the middle of the present century natural history was still very much alive and in the last three decades has been revived with tremendous vigour.

The Linnean Society of London, the oldest specialist biological society in the world today, was founded in 1788 to honour that peerless 18th-century naturalist Carl Linnaeus. Its royal charter, granted in 1802, described it as "a Society for the Cultivation of the Science of Natural History in all its branches ..." "Natural history" was an unexceptionable term at that time, but as the century progressed the Society's fellowship came to include a growing new class of professional biologists. With this change came some serious attempts to separate the Society's interests into botanical and zoological camps.

One can readily understand why 19th-century naturalists wished to divide their studies into more manageable specialities. Man's attempt to acquire a comprehensive knowledge of the natural world is a massive task and, as the century progressed, an ever growing number of naturalists of every speciality was required to name and describe the vast number of species, to establish their taxonomic status, and to observe and record the bewildering variety of their life histories. New generations of naturalists progressively deepened the level of investigation and consequently focussed on ever narrower segments of nature. The need for greater knowledge about the structure and life processes of individual species led to the growth of a class of indoor researchers, sometimes disparagingly styled "closet naturalists" by their field colleagues. As specialization and professionalization developed, it seemed that the older pursuit of "natural history" might be eclipsed by the growing science of "biology."

The new groups of specialists called themselves protozoologists, pteridologists, lepidopterists, oologists, mammalogists, microscopists, or biologists, rather than naturalists, and began to establish a variety of specialist societies that would better serve their interests. The foundation of the Zoological Society of London in 1826 and the Entomological Society in 1833 marked the beginnings of this movement, one that, incidentally, posed a threat to the continued growth of some general natural history societies.

In the long run, older generalist societies survived these decades of change without too much damage. The Linnean Society of London, for example, by 1988 had completed two centuries of service without ever having altered the wording, at least, of its 1802 mission statement. The resurgence of natural history as a component of biological science in the third quarter of this century gave the Society's interests a new relevance amongst the community of specialized societies and associations. The Linnean Society, generally spoken of as a "biological" society today, now advances into its third century with natural history as one of the three major themes of its biological interest together with the conservation of nature and the study of scientific biology. Today it is largely an association of professional scientists but the current fellowship still includes a healthy proportion of amateur naturalists.

It was almost certainly the foundation and growth of ecology in this century that stimulated a renewed awareness that the many "branches" were but parts of the whole of natural history. Although many of the specialist societies and associations still survive, and indeed thrive, broadly based natural history societies also continue to flourish today and around the world, but especially in "Third World" countries, new "natural history" societies continue to be founded. The total number of such societies must run to several hundreds.

Natural history, then, is clearly alive and well, and anyone looking for an informed statement of its nature and purpose in the mid-20th century should read Marston Bates' *The nature of natural history.*[1] Although the whole study of natural history has seen such a vigourous revival in the second half of this century, the long-standing division of biological science and scientists into botany/botanists and zoology/zoologists still remains entrenched. Some major natural history museums still preserve old botanical/zoological departmentalization behind the scenes even when public policy favours the image of an integrated view of natural history. Most academic institutions still preserve departments and professorial chairs in botany and zoology. Even the Linnean Society of London, the world's oldest surviving biological society, still elects separate Botanical and Zoological Secretaries and publishes a *Botanical journal* and a *Zoological journal*, though it does also publish a *Biological journal*, the successor to its former *Proceedings*.

Rationale for this Guide

In 1989 Ellen Wells of Garland Publishing invited me to prepare a bibliographical guide to the history of natural history. Despite the breadth of the subject, I felt that a guide to the history of the whole field was essential at this stage: no such guide exists at present. The provision of more detailed guides to the history of the various branches of natural history remains an exciting task for future bibliographers.

However, another significant consideration was the existence of another Garland Publishing publication apparently covering very similar territory, that is, Judith A. Overmier's *The history of biology: A selected annotated bibliography*, published in 1989 as vol. 15 of Garland's *Bibliographies in the history of science & technology*. Her 619 entries, listed under alphabetically arranged subject headings, cover most of the significant subdivisions of biology, e.g., Anatomy, Animal Research, Biochemistry, Biogeography, Biology, Botany, Cytology, etc. If a separate guide to natural history was not to look like a poor relation of Overmier's work, it clearly had to be constructed on a wholly different rationale. It also had to direct its efforts at answering the special needs of "natural historians." I was able to exclude natural history of the mediæval world (which lies right outside my own experience or interest!) as it is covered in Claudia Kren's *Medieval science and technology: A selected annotated bibliography*, published in 1985 as vol. 11 of Garland's *Bibliographies in the history of science & technology* (*see* A40). Taking these factors into consideration, I felt able to proceed with Ellen Wells' commission and commenced work in 1990.[2]

---

[1] *See* **M. Bates.** 1990. *The nature of natural history*. With a new preface by H. S. Horn. Princeton, NJ, Princeton Univ. Pr. x+321 pp. *Princeton Science Library*. First published in 1950.

[2] The commission for the present work had previously been accepted by Ronald Wilkinson, well-known for his long-standing interest in the field. Unfortunately, other commitments obliged him to relinquish the project, and I am honoured to chosen to compile this guide in his stead. However, despite his withdrawal, one of those circumstances occurred that is apparently designed to amuse, test, or (save us!) to trick us poor bibliographers in the pursuit of our humble toil. Mr Wilkinson's non-existent book has found an ethereal place in the bibliographical record, reportedly published in

INTRODUCTION

The most cogent reason, in my mind, for producing a guide to the literature of the history of natural history is that so much of it is surprisingly little known, even to working natural history librarians. There are several reasons for this.

Firstly, until recently the secondary historical literature in "book" form was decidedly insignificant when compared with primary natural history — book-length biographies being the chief exception. Most historical literature has appeared as periodical articles, the majority in a wide scatter of primary natural history journals or in some "house" journals such as the Linnean Society's *The Linnean*. There are only a very few historical periodicals dedicated to publishing articles on natural history.

Secondly, almost all of the libraries that acquire this secondary literature are scientific libraries serving scientists and amateur naturalists; historians are a side-issue. The staffs who run these libraries usually have to integrate secondary literature into a system designed to provide a systematic classification of primary literature. For the purposes of classification the biological facet is usually given primary status and, as a result, historical works are integrated with the scientific literature. Historical articles in primary periodicals fare even worse. They rapidly "disappear" into the periodicals stacks, as indeed they must unless offprints can be obtained. Thus, almost the entire corpus of the literature on the history of natural history can become effectively concealed from the readers and, indeed, the librarian.

Thirdly, bibliographical guidance on history of natural history literature is relatively poor compared with that for primary literature. Legée et al. (*see* A19) comes closest to a historical guide but is not widely known and is, in all honesty, somewhat hastily assembled, repetitious, and not designed for reference use. Smit's impressive bibliography (*see* A31) is methodical, informatively and critically annotated, and comprehensive as far as it goes. But its focus is *just* on biology and it describes only the "book" literature, thus excluding a great deal of the periodical literature listed in this guide.

More seriously, current periodical bibliographies provide an inadequate record of historical literature. Many historical articles appearing in current natural history periodicals are, of course, listed in *Kew record* (*see* A168), the *Zoological record* (*see* A170), or even such specialized services as the *Bibliography of systematic mycology*, but these are compiled by and for scientists, and the historical content is pretty insignificant or inadequately indexed. Historians have generally had to rely on the annual *Critical bibliography of the history of science and its cultural influences* (*see* A163) published in *Isis*, but its broad coverage prevents the inclusion of any more than a modest selection of significant natural history titles.

In 1985 the British Museum (Natural History) launched its *Annual bibliography of the history of natural history* (*see* A157) and provided historians for the first time with welcome relief from most previous hindrances to current awareness. At last they had a comprehensive annual list of books and periodical articles, the latter derived from scanning no less than 8,000 periodicals currently being received in the Museum Library and several other large London libraries. (But more of that story later.)

When I commenced this guide in 1990, my principal objective was to provide a retrospective guide to selected literature published before the arrival of the B.M.(N.H.) *Annual bibliography* and thereby to highlight much historical material that probably lies hidden from view in many larger scientific libraries.

1985 — a little rummaging through the 1990 history-of-science reference books will reward the interested ghost hunter!

Author's Credentials

Users of a bibliography that claims to be a "guide" to their field are entitled to know with what authority and experience the compiler presumes to advise them. My principal credentials for attempting this project are thirty years' experience spent on the bibliographical wing of natural history — first as an assistant in the natural history department of the distinguished London antiquarian bookseller Bernard Quaritch, second as Departmental Librarian in the Department of Zoology of the British Museum (Natural History), third as Librarian & Archivist for the Linnean Society of London, and fourth as Bibliographer at the Hunt Institute for Botanical Documentation, Carnegie Mellon University, in Pittsburgh.

This guide reflects my own library oriented experience and undoubtedly differs from one that a historian of science might produce. To those who would see this as a shortcoming I should remark that over many years I provided considerable assistance to historians and biographers in the field and learnt about their needs in the process. But the greater part of my service was directed towards satisfying the needs of working naturalists, especially taxonomists, and particularly those concerned with the historical needs of taxonomy, the historical importance of natural history collections, and the bibliographical problems of nomenclature. It was experience that provided me with a broader view of the history of natural history than an academically trained historian might possess. Because this guide is the work of a British librarian transplanted to America a dozen years ago, it may exhibit a European bias, though I have endeavoured to avoid this.

The work was typed by myself using the time-honoured bi-digital system so familiar to British librarians (described by my fleet-of-finger American colleagues as "hunt and peck"!).[3] Such a disconcerting lack of professionalism might suggest that the text is peppered with errors. In fact I have laboured hard to prevent such unacceptable blemishes and I have been greatly assisted by the careful proof-reading of Adrienne Makowski and her staff at Garland Publishing. I take full responsibility for the typographical errors that remain and ask the reader's forgiveness if any should jar his eye.

Audience

The compiler of any bibliography must keep a clear idea of his ultimate audience if his work is to have any useful purpose. But, in practice, one rarely has a single, clear cut audience in mind. Many people may consult a bibliography for a variety of reasons, and bibliographers try to make their compilations as widely useful as possible. I am no exception to this situation.

Historians of natural history, both amateur and professional, and all naturalists with a curiosity about the history of their subject will obviously make up the primary audience. Professional historians will look for the key scholarly literature and will, I hope, find most works included here. Indeed, well-read historians will find scarcely any "jewels" that are not already known to them but will instead, I fear, shake their heads in dismay when they see many of the older "histories" included here.[4] Perhaps the three background sections on biographical sources, histories of societies, or library and manuscript resources will hold more value for them. It is, rather, the historically oriented naturalists, taxonomists, and curators who will probably form the most significant part of the audience I have in mind. Their needs,

---

[3] This work was compiled on a Macintosh II computer (and a Macintosh Powerbook 170 for field work), using Microsoft Word (versions 4.0 and, later, 5.1). Text designed in "Times" by the author.

[4] For a professional view of what's good and what's not in historical writings, *see* **M. Rothenberg.** 1982. Jewels and trash: An overview of the history of American science. *History of science in America news & views* 2(1): 1-2.

as I saw them over many years, were most influential in guiding my ultimate choice of titles and the arrangement of the guide. Although they are fully appreciative of scholarly histories of ideas in their fields, their practical requirements often call for historical data at a nuts-and-bolts level, hence all the who-the-guys-were lists of field and other workers, precise who-went-where chronicles, accurate where-did-this-come-from documentation of individual specimen histories, and so forth. Because historical and bibliographical research is so often an incidental part of their work, but not of their training, I have included some references to methodology that may be useful. As one of their kind, I naturally envisage that librarians and bibliographers will form part of my audience and have tried to provide for their needs also, using my own experience as my guide.

Sources

Over the course of my library career I always maintained a personal working bibliography of references to the history of natural history gathered from firsthand use of some of the best natural history library resources in the world, namely the British Museum (Natural History), The Linnean Society of London, and the British Library. I naturally turned to this file to provide my first source of materials for this guide. But a personal working bibliography is a somewhat arbitrary tool; one doesn't maintain all its elements with equal emphasis nor does one necessarily preserve consistent citation standards in all entries. I stress this because I was not always able to re-examine all entries that I had previously noted and sometimes had to verify bibliographical details from secondary sources. I hope that they are all in good order.

The library of the Hunt Institute for Botanical Documentation has a remarkable collection of both primary and secondary literature and has been of enormous help in preparing much of this guide. But working in a purely botanical institute separates me from much essential and once very familiar zoological material. When my small personal zoological reference collection failed to provide, I had to turn to the resources of the Carnegie Museum of Natural History — and did so with gratitude. But, for all its worth, their library could not match the great zoological resources I had once lived with in London. I was able to make a short and productive working visit to London in 1992, but my time was severely limited. Therefore, I have been forced to include a number of references that I had noted in my working bibliography but have not seen for some years. In addition to using library resources I also made a bibliographical survey of the entire literature to clarify my view of its overall shape because only when that was achieved could I make a representative selection for a guide such as this. However, because of the difficulties of language and availability, I am certain that my coverage of Russian historical literature is imperfect.

With the whole spectrum of natural history to encompass, I can make no real apology for the large number of works recorded herein; that is the nature of the literature. It is as scattered as the interests, the localities, and the naturalists it chronicles, and much of it is entirely parochial. Out of this very extensive field I have made a generous personal selection. Literature published down to the close of 1992 formed the main survey pool, but a few specially pertinent 1993 titles that came to my notice have been added. In total the four parts of the work comprise some 7,507 references (219 part A; 3,484 in B; 3,508 in C; and 296 in D).

Recent decades have seen the re-publication of many useful older reference works that had long been almost unobtainable on the secondhand book market. Sometimes a re-published version is enhanced by the inclusion of a scholarly introduction and perhaps a new index or other editorial apparatus. In these days of shrinking institutional and personal book budgets, useful savings can be made by acquiring "reprints" instead of struggling to afford the purchase of scarce original editions. Indeed, if the reprint includes additional editorial material it may be distinctly more desirable. I have made a special effort to trace and mention as many re-published texts as I can. They are not always easy to discover, as their publication is usually less publicised than the appearance of new books, and sometimes editions are comparatively small.

Widespread cuts in purchase budgets over recent years have sometimes prevented libraries and scholars from purchasing the latest editions of reference works. It is not uncommon to find that the newest edition shows a far greater increase in price than proportional increase in new content, and one can still satisfactorily use the previous version. In most entries I have noted the existence of the latest prior edition when there is one. Sometimes even a mere comparison of dates and paginations gives an accurate guide to likely overall differences.

Content & Arrangement

My working file had originally been thrown together in a somewhat loosely ordered system that suited the needs of the time. When it came to preparing this guide I had to recast the whole into a structure that reflected some of the peculiarities of natural history, the nature of its literature, and the historical interests and needs of potential readers. The arrangement I chose comprises two major and two minor parts: Part A is a small introductory guide to "Personnel, Organizations, Current Awareness." Part B is a "Background Bibliography" of literature on biography, history of societies, and library resources. The principal portion of the whole guide is Part C, the "Core Bibliography" of the history of natural history and its branches. Finally, Part D is a short guide to selected literature on "Historical and Bibliographical Methods."

The varied needs of the composite audience that I address led me to create the large "background" sections, peripheral to the main concern with the history of natural history, that form such a substantial part the whole. At present there is no thorough bibliographical guide to sources that provide the sort of background information that is so often necessary to the pursuit of research on the history of natural history, i.e., full biographical data on lesser-known naturalists, sources for individual portraits, the history and personnel of natural history societies, literature on natural history library resources of all kinds (but especially retrospective, archival resources), whereabouts of personal papers and scientific notes, and so forth. The "Background Bibliography" is an interim attempt to fill this gap and accounts for its somewhat disproportionate size.

To help users to know what to find and to understand some of the necessary limits imposed on the contents, I shall now elaborate some details of the subject content and rationale of the whole guide, section by section

Part A — Personnel, Organizations, Current Awareness

Previous guides: The present work belongs to a handful of bibliographical guides devoted solely to the history of biology and natural history (e.g., *see* A19 A24 & A31). In the wider domain of the history of science there are several good literature guides and bibliographies that include some biology and natural history, and their broader view of historical literature usefully puts our relatively small field into proper focus (e.g., *see* A8 A11–13 A29 & A30). To supplement the list of these guides I have added a selection of guides on peripheral fields of science and specialized historical interests that may be useful to historians of natural history.

Knowing just who is researching what in the history of natural history cannot easily be ascertained except, perhaps, by asking a few well-placed librarians! There is no helpful register of current research nor any guide to historians classified by their research interests. The nearest approach is the History of Science Society's *Guide* (*see* A49), but its restriction to that Society's membership excludes many of the authors listed in this guide. The Society for the History of Natural History's *Members' handbook* (*see* A50) is better focussed but does not help with research interests. Quite a few societies and other organizations serve the special interests of various branches of the history of science. The History of Science Society's *Guide* (*see* A54) provides the most up-to-date information on them. Literature on the history and functions of individual societies, etc., is limited, as my somewhat patchy selection of titles indicates.

A relatively recent addition to the documentation of current research, forthcoming meetings, and publishing activity is the small but growing number of quickly produced history-of-science newsletters (*see* A74–80). They are replacing the old "notes and news" sections in conventionally produced periodicals — "news" that was usually too late to be of much service. The proper role of the conventional history-of-science periodical is to provide channels for the publication of primary historical research and discussion, material that makes up much of the content of this guide. A list of some general guides to current periodicals (*see* A81–88) is followed by a selection of individual titles (*see* A89–155) based on the periodicals cited in this guide. Bibliographical current awareness in our field is imperfect, especially following the demise of the short-lived *Annual bibliography of the history of natural history* (*see* A157). A list of available sources is provided. Nowadays, the results of much of the formal current research into the history of science can be found in academic dissertations, many of which remain unpublished. Some general bibliographical guides to this elusive field of literature are listed (*see* A172–181) together with a first guide to their availability (*see* A181A).

Lastly, attention is directed to periodic meetings in the history of science. Until comparatively recently there were few regular meetings that served the interests of historians of natural history. The International Congress on the History of Oceanography, first held at Monaco in1966, was the earliest relevant series (*see* A182–185) and the four meetings held so far have presented an impressive quantity of material, including much of natural history interest; a fifth is already planned. Since 1975, the Society for the History of Natural History, despite its modest size and resources, has been a productive organizer of meetings, large and small. Its international biennial meetings have covered a wide spectrum of historical interests (*see* A204–211). An ambitious overseas programme has included a series of meetings in the United States (1981+) (*see* A212–217) and more recently a series of Central European Meetings (1989+), but the latter have not generated any special publications. Of much longer duration is the International Congress on the History of Science which commenced over 50 years ago and has met in a wide range of countries. Their meetings embrace the whole spectrum of science, so that the history of biology and the life sciences form a small part of their deliberations, though they have included some significant contributions over the years.

Part B — Background Bibliography

This part provides references for data in support of Part C, the "Core Bibliography": firstly, biographical data on naturalists; secondly, historical, biographical, and bibliographical information on naturalists' societies; and thirdly, bibliographical guidance on library resources of printed, manuscript, and graphical natural history records.

Section Ba: Biographies of Naturalists. This section surveys biographical information sources of all kinds, commencing with a classified list of "Collected Biographies," including biographical and bio-bibliographical "dictionaries," biographical collections, and some master indexes to general reference works on collected biography. Historians of science might feel that this list is extravagantly large when one considers that all the naturalists who have made really outstanding contributions to the development of the field could probably be embraced in a single reference guide.[5] But for the working naturalist, pursuing a problem that calls for facts about a person only known to him from a specimen label, a nomenclatural citation, an author's acknowledgement, or a casual allusion in print, the search for identity may only be satisfied by some of these more obscure references. Information on the huge number of less significant individuals around the world who have contributed to our present pool of natural history data and to the multitude of specimen collections lies scattered through a large and very varied number of reference works. A unified finding index to all the individuals recorded therein remains a desideratum.

A relatively new movement in historical research on natural history is the development of interest in its social history. One of the historiographical techniques that social historians have brought to the field is prosopography, the biographical study of social groups. Some titles listed herein probably seem to have little more than superficial biographical data, such as the various outdated directories I mention (*see* Ba104–117, 272–293 & 346–374), but are included for their prosopographical value. N.B. One class of collected biography is not included in this section, the many membership registers and records of individual societies, academies and associations. They can be found under the appropriate society heading in Section Bb: Histories of naturalists' societies.

In this picture–oriented age little justification is required for including a subsection on sources of information on "Portraiture." Many of us have experienced surprise upon first seeing the likeness of some bygone naturalist who we thought we already knew quite well. In recent years portraiture has become a major feature of the illustration of historical texts, and authors and publishers maintain a steady demand for more. There are some useful lists and catalogues of naturalists' portraits, and more are in preparation, but documentation is very far from complete at present. I have listed all the reference literature that is known to me.

The second and major part of Section Ba is devoted to "Individual Biographies." The total number of biographical accounts of naturalists, ranging from brief obituary notices to full–scale biographies or autobiographies, must run to many thousands, most of which are probably listed in the various collected biographical works mentioned above. The problem of selecting a modest representative number of individual biographies from this huge roster had to be reduced to a practicable formula. I decided to limit the field to separately published "book-length" biographies and autobiographies. Selection based upon mere bibliographic form might seem to be quite an inappropriate solution but, in practice, book-length biographies are generally confined to naturalists of distinct fame, excellence, novelty, or eccentricity. The resulting selection of just over 550 biographees actually consists of a wide spectrum of interesting personalities, ranging from minor amateurs, such as Thomas Edward, to figures of international standing, like Alexander von Humboldt, and illustrates an immense variety of life stories and intellectual pursuits — for many of them, natural history was but the second or third string of their fiddle.[6] Besides autobiographies and biographies, I have listed personal bibliographies and guides to personal papers. These have special research value, and I have included all that I know, disregarding my "book-length" rule in this case.

Section Bb: Histories of Naturalists' Societies. From the mid-17th century to the present day, co-operative groups of naturalists have played a significant role in encouraging the study and communication of natural history information and the acquisition and curation of natural

---

[5] The *Dictionary of scientific biography*, for example (*see* Ba25), includes just over 1,700 individuals from all periods and cultures under the heading "Life sciences."

[6] One significant shortcoming deserves apology — I found myself unable to make a meaningful selection of biographies of Russian naturalists, of which there are a large number, and have included only a cursory handful. Quite apart from the difficulty of language, there is the problem of distinguishing Soviet propogandist eulogies from objective scientific biographies.

history collections. Some groups of naturalists formed parts of societies with broader scientific interests, but a great many others founded societies entirely concerned with natural history or one of its branches. These societies had enormously varying levels of success, some growing into national academies, others never rising above the level of a local group of naturalists who, perhaps, only met in the field. The more successful societies could boast of considerable achievements. They provided forums for international discussion, accumulated large libraries, sponsored the publication of documents ranging from ephemeral newsletters to some of the greatest sets of scientific journals ever known, formed collections ranging from simple demonstration exhibits to substantial museums, maintained botanical and zoological gardens, and so forth. The world-wide tally of natural history societies and associations in existence today runs to many hundreds, even thousands, some with very long lives, others less than a decade old. With such a diversity of societies playing such a variety of roles one expects to find a mixture of all sorts of literature relating to their histories and activities. Some have histories that are little more than self–congratulatory chronicles, while others exemplify the best in historical scholarship. Some were the dutiful compilation of a reigning officer occasioned by a suitable milestone in the society's history; others were the specially commissioned work of a professional historian. All types are included herein. In selecting societies, associations, and academies for inclusion I cast my net wide enough to include such bodies as geographical societies and national academies, in all of which naturalists once played significant roles and may, indeed, still do.

In addition to listing histories of societies, I have included references to bibliographies and retrospective indexes of their periodical or other publications. Indexes often provide guides to persons and events mentioned in the "proceedings" (the periodic published records of internal business) and, if thoroughly compiled, have obvious historical source value. Their compilation is, of course, an internal matter for the body concerned, and their quality varies enormously. Elaborate retrospective indexes have always been laborious to compile and expensive to print, and limited resources have often prevented the production of the desirable ideal. When I set out to examine or re-examine copies of all the indexes I have cited I was surprised to find how difficult it is to find copies. Cumulative indexes are usually buried away in the periodical stacks, some issued as part of the parent periodical, others (even for the same periodical) sometimes separately published. They are often compiled and published long after the termination of the periodical span they index, and they are often concealed within a bound periodical set without any external indication of their whereabouts. Even some of the largest libraries have on occasion been less than vigilant in acquiring them all, perhaps actually rejecting them sometimes because they are only obtainable as costly extras to an already expensive subscription. So, in the end I was not able to see or re-examine every index that is cited herein.

Most societies and like organizations produced periodic lists of their membership. Although their function was intentionally ephemeral, these lists, if preserved, have prosopographical value for historians. Unfortunately, they are far too numerous to be itemized in this guide and, besides, very few libraries have bothered to keep them.[7] However, some societies compiled and published retrospective records of their membership, and I have listed all that I know of, even including some for medical associations whose earlier membership included naturalists.

The role that societies and the like have played in the development of natural history has received scant attention from most historians, and the literature relating to their individual histories is poorly documented, little-known, and often quite difficult to find. Natural history libraries are often guilty of paying less than usual attention to cataloguing and arranging the

[7] The Linnean Society of London, for example, has regularly issued lists of its Fellows since the 1790s, but which libraries have sets of them, I wonder?

non-scientific literature produced by societies and other associations, often being content with "bundling" them alongside their parent society's periodical publications, hiding them away in uncatalogued "vertical files," or discarding them altogether![8] As a result, the compilation of this section on histories of natural history societies was undoubtedly the most difficult and laborious part of the whole guide. I readily admit that the result is far from perfect, but I hope there is sufficient substance to provide a platform upon which future bibliographers can build a sturdier structure.

Section Bc: Natural history Library Resources. This section forms a guide to library resources chiefly as recorded in published catalogues. Although the needs of the science historian were my first concern when choosing references for this section, I also kept the interests of bibliographers and library historians in mind.

Libraries are almost as significant as specimen collections in the history of natural history. For present-day institutions concerned with the curation and research of important systematics collections the possession of an associated library, with rich retrospective collections, continues to be an indispensable requirement. Research libraries of this kind are also a boon to the historian of natural history because, for practical reasons, significant source literature, instead of being hived-off in some "special collection" as happens with most other sciences, is often shelved with the modern literature on the same topic.

Most of the essential printed, handwritten, and pictorial records of over five hundred years' of natural history observation and research can be found in libraries that are accessible to scholars, but knowing which libraries those are, and their individual specialisms, is the first requirement. This section opens with "Libraries & Library History," a list of the most important guides and directories to public or institutional collections that contain older natural history books and manuscripts, with some emphasis on North American and European resources.

The role that specialist libraries have played in the development of natural history scarcely exists as a separate historical topic. I have put together a list of references on the "History/Description of Individual Libraries," but unfortunately many of the titles listed are only superficial historical sketches (*see* Ba34–158). Literature on the history of personal collecting in this field is little better. A brief list of references on private collecting and some of the individual collectors who assembled notable private natural history libraries, from the superbly systematic Joseph Banks collection to the maniacal conglomeration of Thomas Jefferson Fitzpatrick, must suffice *pro tem* (*see* Bc162–182).

The remaining parts of this section provide lists of published catalogues of books, periodicals and manuscripts. Despite their seemingly large number, it has to be said that in total they provide relatively poor guidance to the wealth of library resources in existence. Many important natural history libraries never published their catalogues at all and most that were published are now decades out of date anyway. There are also other problems and shortcomings that limit the utility of older printed library catalogues. Specially curated materials or special collections acquired as bulk acquisitions were sometimes excluded from the general catalogue with the intent of publishing a separate catalogue that, in the event, never appeared. Libraries sometimes changed policy and disposed of significant older stock.

---

8 Librarians are lightly taken to task more than once in this Introduction. Lest I seem to be taking some unfair digs at them, I should stress that — as one of their kind — my remarks are directed as much against my own past practices as anyone else's! Always doing "the proper thing" often takes more time than one's library users are prepared to grant — they always have first call on a librarian's time. Over many years I was obliged to learn some of the the short cuts and tricks of the trade too! However, to be serious, some of these criticisms are worth consideration for inclusion in any list of "future projects and improvements."

Many printed catalogues regularly excluded useful material, e.g., pamphlets, manuscripts, drawings, etc. In fact, what printed catalogues exist form somewhat of a patchwork, with scattered areas of excellence and reliability. We must await the completion of comprehensive computerized catalogues made available through widespread computer networks. Unfortunately, it will probably be many years before all relevant internal card catalogues are translated into electronic form, and until then the older printed records still have some importance. When available, services such as the United States *National Union Catalog* can overcome some of these problems.

The period from the mid–19th century to the Great War was a fairly active one for the publication of library catalogues in book form, including some significant natural history collections, but activity virtually ceased with the Great War. Increasingly rapid growth in library stocks and rising production costs were largely to blame. One solution was to supplement existing printed catalogues by publishing regular lists of accessions, and a small number of libraries did so.

In the 1960s a new era of library catalogue publication commenced. Conventional, time-consuming, letterpress composition was replaced by direct photo-lithographic reproductions of card catalogues, printed in reduced size with twenty or more cards to a large-format page. Regular supplements were planned for several of these catalogues, but production usually stopped after a couple of volumes or so. The Goliath of this tribe was the colossal 754–volume United States' *National union catalog, pre-1956 imprints* (*see* Bc192) (later re-published in a handy 9,200–microfiche form)! A number of important natural history library catalogues were published in this way. Unwieldy though these library catalogues are, and usually lacking any editorial finesse, they are extremely useful in providing fairly comprehensive information on resources at the time of publication. Unfortunately they were always expensive to acquire and only major reference libraries could afford them.

In selecting material for the "Library Catalogues" section I cast my net very wide as is apparent in the choice of some of the titles included. This was partly because I hoped that the list would perform a dual function: firstly, to provide guidance on the documentation of contemporary library resources, and secondly, to provide data for those who have a special interest in the role of book collecting and library formation in the development of natural history research. These considerations account for my inclusion of obsolete versions of catalogues, of catalogues of since dispersed libraries, and the large selection of lists of personal collections. Most of the latter were dispersed by sale, but some very significant collections were acquired by major institutions and now constitute part of their most valuable resources. I have indicated links between once-private collections and their present institutional homes. A number of smaller local natural history societies maintained libraries in the 19th and earlier part of the 20th century and several published catalogues of their holdings. Few such local libraries exist today owing to the high costs of general maintenance and the considerable funds necessary for sustaining an adequate level of book and periodical acquisitions. Many of these libraries were eventually deposited with larger institutions or dispersed. Outdated and obsolete library catalogues are useful to the historian and bibliographer as sources of information on such things as the library's history, the contemporary availability of literature, the older provenance of identifiable copies of books, the past whereabouts of manuscripts, drawings, etc. The section on "Personal Libraries," mostly consisting of sale catalogues, is also provided for the interest of bibliographers and library historians.

Some libraries outside the mainstream of natural history are included – national libraries, general science, earth sciences, geographical, oceanographical, agricultural, horticultural, medical, and so forth – all of which possess valuable collections of relevant books and manuscripts. I have also kept the mesh fine enough to net such small but relevant items as exhibition catalogues. Within the range of their individual themes, exhibition catalogues may include items acquired since the library catalogue was published, and the items selected for inclusion are almost always more fully described.

"Periodicals Catalogues," or at least title-lists, are much more plentiful than catalogues of books. There have been quite a number of impressive multi-library guides to periodicals resources, largest of which is the 5-volume *Union list of serials in libraries of the United States and Canada* (*see* Bc618). Unfortunately, the *Union list* is now over 30 years behind the times and, with the arrival of on-line computer catalogues, we may not see such union lists in book form again. I have given only a representative selection of periodicals catalogues. Retrospective periodicals holdings consume large areas of storage space, and a number of libraries have already disposed of large sets of older, little–used periodicals in an effort to free shelf space and to reduce conservation costs. With such a constantly changing scene, the printed periodicals catalogue is clearly doomed whereas electronic on-line catalogues give accurate up-to-the-minute availability information. However, like printed book catalogues, older periodicals catalogues still have value to bibliographers and library historians and some distinguished older deposit libraries have preserved their periodicals collections exactly as recorded in older lists.

"Manuscript Resources & Catalogues" lists the literature of manuscript resources, individual repositories, and some individual collections within our field. The significance of this material is eloquently stated by David Allen, well-known as a historian of natural history, whose own experience in using original sources adds authority to his words:

> Manuscripts are the historian's bread: the basic source of sustenance without which his subject can scarcely exist. ... To know what manuscripts exist and where they are is thus crucial to him. ... Natural history has been both fortunate and unfortunate in its legacy of unprinted materials. Fortunate in that, unlike most of science, taxonomy and nomenclature are necessarily historical studies, dependent on dating published works with precision, on checking the claims of earlier collectors, on tracing the sometimes labyrinthine wanderings of type specimens. For these reasons much is preserved that in most other fields is allowed to perish. Fortunate, too, in that its latter-day adherents have remained proud and fond of its ancestry. Yet unfortunate in an over-emphasis on printed texts that stems in part from that very concern with nomenclatural priority and in part from the sheer splendour of an earlier literature which has drawn to it generations of bibliophiles. Preoccupation with the printed surface has led to a general neglect of the untidy, formless world beneath.

Allen's statement was written back in 1979,[9] but sadly the full resources of that "formless world" still remain imperfectly known nearly a decade and a half later. Certainly, awareness of the importance of archival sources is heightened, and there are some ambitious programmes, such as the Australian Science Archives Project's national database (*see* Bc663); Chadwyck-Healey's large-scale microfilming projects in North America, the United Kingdom and France (*see* Bc667–670); and such intensive editorial projects as the publication of the *Correspondence of Charles Darwin* (*see* Ba697), but overall documentation of natural history manuscript resources remains patchy and uneven. In view of this shortfall I have included virtually every published information source of any relevance because, when trying to locate unique natural history records, every reference, however slight, may be significant. In addition, I have provided cross-reference links to any descriptions or catalogues of collections of personal papers listed in "Individual Biographies" in Section Ba whenever possible. It should also be borne in mind that book-length biographies frequently utilize unpublished sources and

---

9 Foreword. In: *Natural history manuscript resources in the British Isles.* London, Mansell; New York, R. R. Bowker, 1980 (*see* Bc666).

their internal bibliographies should signpost locations even if they make little or no effort to describe their nature and content.

Part C — Core Bibliography

Part C, the "Core Bibliography," provides a classified guide to literature on the history of natural history proper. Although the intent of this guide is to cover natural history as a whole, I am obliged to perpetuate the traditional division of the subject into botanical and zoological camps. The specialization and compartmentalization that characterized the development of natural history in the nineteenth and twentieth centuries is paralleled in much of the secondary literature which includes many small histories of individual areas of specialist study or of natural history studies in small geographical areas; major synthetic histories are the exception. This situation, as much as the nature of the subject, dictates the division into separate sections for "History of Botany" and "History of Zoology" in addition to a general section for "History of Natural History." Within each of these three sections I have classified the literature according to the conventional subdisciplines, but I have tried to indicate that these are merely "branches" of natural history by maintaining synonymity of subsection headings throughout. My intention is to help the reader to find historical literature with a common thematic or geographical element across the whole spectrum of natural history. The classification of each of these three sections progresses from the general to the particular and from higher taxonomic groups to lower ones.

In addition to the information contained in narrative histories and biographies, a great deal of source data can be derived from the numerous specialized bibliographies. They record the enormous literary record of centuries of natural history observation, description, and thought. Since there is no recent or comprehensive guide to these bibliographies, it seemed helpful to include a selection in this guide. Accordingly, I have dispersed lists of the principal bibliographies for each branch of the subject, appending them to the relevant subsections in natural history, botany, and zoology.

In order to give a well-rounded picture of the historical literature in various branches of botany and zoology I have included subsections on some topics that are more suggestive of laboratory science than natural history, e.g., morphology, embryology, physiology, etc. However, most of these branches of biological study were initiated by "closet naturalists" long before the era of the professional laboratory scientist and have a place in the history of natural history.

In addition to the historical literature on natural history and its subdivisions, the three parts of the core bibliography include additional subsections on natural history exploration, live and dead plant and animal collections, classification and nomenclature, illustration, and plants and animals in relation to man.

Back in the era when all natural history research was organized from the major centres of culture in Europe and America, knowledge of exotic plants and animals was usually obtained by travelling to distant countries to collect specimens and bringing them home for study. Linnaeus, for instance, sat like a spider at the centre of a radiating web of disciples and other collectors and used the materials they brought back as data for his famous encyclopaedic reference works. Over the next two centuries, collecting expeditions covered the world and varied enormously in objectives, scale, execution, and results. Their importance in the development of natural history needs no further emphasizing here.

For convenience, historical literature on expeditions of all kinds is grouped together in Section Ca as "Natural history Exploration, Voyages & Collecting," regardless of whether the mission was "botanical," "zoological," "ornithological,"or whatever. The descriptive literature about most expeditions shares similar themes of exotic travelogue and adventure, and so forth.

Secondary literature on individual scientific expeditions and voyages is very extensive, and I have confined myself to giving a selection of historical literature concerning some twenty-nine important expeditions spread over two centuries. Primary editions of travel journals are omitted,[10] but subsequent, historically edited, annotated or indexed versions are included.

Living collections of plants and animals have existed for centuries, some being created for entertainment or purely aesthetic reasons, others for the purpose of scientific study. Botanical gardens have been especially plentiful over the centuries and the literature about them is proportionately large. In contrast, menageries and zoological gardens have been much less plentiful, always less enduring, and the historical literature about them is small. General literature relating to the scientific history of each kind, together with works on the individual histories of some ninety-six botanical gardens and seven zoos, is provided in the botany and zoology sections.

Parallel with the practice of maintaining live collections has been the equally ancient, and much less demanding, tradition of making collections of dead plant and animal specimens by both individuals and institutions. Many private collections eventually found their way into formal museums. Some museums grew to an enormous size with collections of immense quantity and variety, historically important provenances, and critical taxonomic and nomenclatorial significance. Over the past two centuries a great deal of historical study has been focussed on private and institutional specimen collections but, until quite recently, the role of museums in the development of natural history was little discussed. On the other hand there is a considerable number of histories of individual natural history museums and their collections, but of very varied quality. Some museums and herbaria have efficiently chronicled their evolution, the origin and growth of their collections, and their literary output, but a great many more have produced only superficial historical sketches, and the histories of others are not yet written. What literature does exist is often hard to locate, and I have almost certainly overlooked some important titles. For natural history, botany, and zoology there are brief lists of general histories of museums, herbaria, and zoos followed by lists of titles relating to selected institutional and private collections.

References to the histories of Education & Research Institutions follow the subsections on museums. In some respects "education and research institutions" might appear to be out of place here. However, some older universities had professorial chairs in "Natural History" (St. Andrews University, for example, still does), most have numbered naturalists among their staff and students, and some have intra-mural natural history field clubs. Many research institutions have been soleley concerned with applied biology, but for some natural history forms part of their research. Agricultural field stations studied the natural history of crops, livestock, and pests in relation to methods of biological control; medical research establishments established relationships between parasites, hosts and vectors; marine stations researched the life histories of various organisms in connection with the protection of submarine structures and the conservation of marine life; biological surveys, at national or regional level, collected field data for the compilation of floras and faunas; and so on. So in order to provide well-rounded coverage of all aspects of natural history I have included some references to this type of institution. However, once again the relevant historical literature is patchy, often very brief, and sometimes more concerned with the trials of administration than with a chronicle of scientific progress. My references could probably be considerably augmented, as I suspect that there are a number of works still unknown to me, but I hope that my present effort will at least form a modest foundation for others to build upon.

---

[10] Comprehensive coverage of the enormous quantity of primary literature on travel is available in bibliographies by Cox and others (*see* Ca800–823).

Biological classification and nomenclature follows next. This has always been major concern of natural history and is discussed in most of the general histories. A number of special historical studies on the subject are listed here. Complementing these are select lists of the great published indexes to botanical and zoological nomenclature. These are, in essence, detailed bibliographies of the earlier taxonomic literature arranged by the names of genera and/or species. They are, of course, intended for the use of taxonomists, but they can also be useful tools for historians as well. A selection of some guides to the etymology of both taxonomic and non-taxonomic terms in natural history concludes this subsection.

The old maxim that "a picture is worth a thousand words" is exemplified in natural history more than in any other science. Some substantial histories of illustration in the various branches of natural history have appeared in recent years. These are complemented by some extensive bibliographies of illustrated works, and, for botany, some impressive indexes to the whereabouts of individual illustrations. Representative lists of these are given for each section.

The "Core Bibliography" concludes with a glimpse of a wider field of interest, the record of man's general relationship with nature, plants, and animals. The exploitation of organisms for sustenance, medicine, industry, and aesthetic pleasure; the extermination of whole species and the continuing search for unknown ones; the pleasures of hunting; the criminal punishment of animals; the faking of "rare" animal specimens; and the folklore and myth associated with plants and animals are some topics treated in an extensive literature for which brief samplings must suffice here.

Part D — Historical & Bibliographical Methods

This concluding part includes some very selective lists of references to methodology and concludes with a few on the history of natural history bibliography.

There are, of course, some good bibliographies on historiographical literature; my selection of works on "Historiography & Historical Methods" is limited to a pot-pourri of works on methods and programmes that have particular relevance for the history of natural history. Experienced historians and biographers will find little that is new to them. However, this section of the guide will probably be useful for working naturalists who occasionally have to deal with historical problems or who decide to research or write on the history of their special interests. There are several entries for programmatic papers that discuss the usual "aims and methods" or "needs and opportunities" and a few dealing with the comparatively recent growth of interest in the social history of natural history.

A small supplementary selection of titles highlights some of the special historical problems that curators and taxonomists encounter when trying to establish precise data about the origin, history, and type status of individual specimens (*see* D38–48). They show how evidence may have to be gleaned from such trifles as the identity of handwriting on specimen labels, the particular kind of pin used to mount an insect, the type and date of manufacture of the paper upon which a dried plant is mounted, etc.

"Archival Sources: Significance and Use" provides a brief list of specialized references, selected from a large general literature, that have particular interest in the context of this guide. It is followed by a handful of titles on handwriting, chiefly concerned with the identification of writing found on specimen labels and other natural history archives (*see* D74–79).

Bibliography has always had a position of special importance in the study of natural history. The great wealth of its literature and the retrospective needs of some important branches of the subject long ago stimulated the compilation of good enumerative bibliographies. Unlike other sciences, though, natural history has international rules governing nomenclatural priority and the validity of publication that have obliged bibliographers to establish exact dates of publication for a great many works. Natural history bibliography therefore includes an extensive literature on the physical makeup and publication history of individual books. This type of research depends on sophisticated analytical techniques which call for a thorough knowledge of the history of the methods of book manufacture and distribution and a recognition of the significance of different kinds of bibliographical evidence. Literature on historical bibliography and bibliographical methods is very extensive, as "Bibliographies of Bibliographical Literature" indicates. I have made a selection of titles on "Bibliographical Methods and Problems," guided by my own experience, that should prove helpful to most users of this guide. A considerable amount of bibliographical work on establishing dates of publication has already been completed, but there is no comprehensive index to it all, especially for zoological literature. The few existing guides to these studies are listed, chief of which is Stafleu & Cowan's *TL2* (*see* D200).

The wide range and considerable number of bibliographies already in existence for various branches of natural history might suggest that little remains to be done. After all, by the beginning of the 19th century naturalists already had such great bibliographies as those of A. von Haller (*see* Cb66), Baron von Münchhausen (*see* Ca100), G. R. Böhmer (*see* Ca92), J. Dryander (*see* Bc486), and J. D. Reuss (*see* Ca102). But, despite the addition of innumerable bibliographies over the intervening years, many areas of the literature are still inadequately documented, and much basic work remains to be done. For anyone wishing to compile one, I have put together a selection of guides to the nature and techniques of systematic bibliography.

The history of historical and bibliographical scholarship in our field barely exists as a separate topic yet. A handful of relevant histories and biographical sketches of some distinguished practitioners conclude this guide.

Indexes & Cross-References

The work is provided with a comprehensive name index and fairly simple geographical and alphabetical subject indexes. Since the bibliography is already arranged according to a fairly elaborate classification, one might question the need for a subject index at all. However, it has been my experience that a classified arrangement is rarely appears as logical or memorable to the user as it was to the compiler! For quick-reference use, the subject index should enable the user to go directly to the required point of the bibliography without the labour of "cycling" through the classified contents list to arrive at the same conclusion. On the other hand the contents list enables the user to understand the "geography" of the work and to formulate a search pattern when making a literature survey, a requirement that can't properly be satisfied by consulting an alphabetical index of terms.

When constructing any subject arrangement, the bibliographer is soon confronted with problems of logical facet-order, cross-classification, acceptable bias in placing entries, misleading titles, and so forth. A partial solution is the time-honoured device of cross-referencing. The reader is not always going to understand the workings of the classifier's mind and should expect signposts to help to guide him around. I have provided numerous cross-references within the body of the work. More could have been provided, but the alphabetical subject index probably obviates the need for many more. I hope I have preserved a helpful balance.

The process of indexing revealed that a some entries were improperly placed and had to be relocated, while some others, upon reconsideration, were entirely withdrawn.[11] Furthermore, some significant works were published too late for inclusion in the original numbered

[11] Entry numbers vacated by these relocations and withdrawals are marked in the text as "Not used."

sequence. These events created some slight irregularities in the entry numberings, the most noticeable of which are the occasionally interpolated "A" and "B" numbers. These must be understood as additional numbers, not subordinate entries.

Acknowledgements

I am indebted to a number of people for various kinds of assistance in compiling this work. Joseph and Nesta Dunn Ewan and Robert P. Multhauf looked over earlier versions of my text and contributed some useful comments and corrections. Gina L. Douglas, Librarian of the Linnean Society of London, provided generous hospitality and extended virtually unlimited library facilities to me when I visited London in 1992. I offer Gina my warmest personal thanks, and I can assure every user of this book that they also can be grateful for her help. Susan Rettger of the Carnegie Museum of Natural History provided a warm welcome at the Museum Library and was generous with assistance. Rex Banks, Head of Library Services, and Ann Datta and Malcolm Beasley of the Department of Library Services at The British Museum (Natural History) provided invaluable help with a number of difficult queries at a rather late stage in my work. I am especially grateful for their patience and expertise in checking bibliographical details for a large number of periodicals indexes and for many photocopies. My colleague Sarah Leroy obtained interlibrary loans and photocopies and searched OCLC files for a number of awkward references. Another colleague, Sharon Tomasic, provided thoughtful and extremely helpful editorial assistance with the introduction. I am, indeed, grateful to the Director, Dr. Robert Kiger, and to all my colleagues at the Hunt Institute for providing the congenial environment that makes this sort of task much more pleasant.

None of these generous helpers bears responsibility for any deficiencies found herein; I alone am to blame for all errors of omission and commission. I shall be very grateful if readers would advise me of any mistakes that they discover. It is well-known to librarians that the better features of reference bibliographies can be immediately appreciated, but insidious weaknesses in detail, such as inconsistencies in classification, misplaced entries, or small but significant errors of fact, require longer use and painful experience to discover. I hope that my efforts will not be found deserving of harsh judgement on that score. I have tried to construct a reference work with the needs of the long-term user in mind rather than the samplings of the quick-reference librarian.

Lastly, to my wife Marlene, the power behind this tiny throne, I owe much gratitude for her patience and cheerful enthusiasm as she kept spurring me onwards — "Closure! When shall we have closure?!" Indeed, the work took longer than I had originally hoped or expected, but I shall feel easier if the finished product is a little better for the extra time and effort.

Finally ...

As a footnote to this introduction I should explain that when I commenced work on this guide it was my intention that readers should be able to rely on the aforementioned *Annual bibliography of the history of natural history* (*see* A157) for a continuing bibliography of the field. During the compilation of this work I was dismayed to learn that this service is about to be terminated after only a few years' existence owing to the United Kingdom government's cost cutting policy. No one has yet offered to take over the project, so once again historians will be forced to accept the old hindrances to bibliographical current awareness.

GDRB | Pittsburgh, PA

PART A —
PERSONNEL, ORGANIZATIONS, CURRENT AWARENESS

PREVIOUS GUIDES

Biology and general science

A1. **Artelt, W.**, ed. 1953. *Index zur Geschichte der Medizin, Naturwissenschaft und Technik, im Auftrag der Deutschen Vereinigung für Geschichte der Medizin, Naturwissenschaft und Technik.* Unter Mitwirkung von J. Steudel (Geschichte der Medizin), W. Hartner (Geschichte der Naturwissenschaften), O. Mahr (Geschichte der Technik). Munich, Urban & Schwarzenberg. vi+398 pp. *Continued* **J. Steudel**, ed. 1966. ___. Vol. 2. Unter Mitwirkung von W. Ricker & C. Nissen. Munich, Urban & Schwarzenberg. viii+312 pp.
Thorough record of the literature of 1945–48 & 1949–52. Continues the bibliographical coverage formerly provided by the *Mitteilungen zur Geschichte der Medizin und der Naturwissenschaft und der Technik*; *see* A137.

A2. **Asmous, V. C.** 1947. Fontes historiae botanicae Rossicae. *Chronica botanica* 11(2): 87–118.
"... an attempt to enumerate all publications which deal, in some way, with the history of botany in Russia."

A3. **Bäumer, A.** 1990. Biologie der Renaissance — Renaissance der Biologie. Eine Bibliographie. *Medizinhistorisches Journal* 25: 305–335.

A4. **Black, G. W., Jr.** 1979. *American science and technology: A bicentennial bibliography.* Carbondale, IL, Southern Illinois Univ. Pr. xi+170 pp.
References to articles in 1,500 periodicals as indexed in the *Biological & agricultural index, Applied science & technology index, Reader's guide, Business periodicals index*, and *Education index*. for 1976.

A5. **Brookman, F. H.** 1983. The history of science in the non-western world: An inventarisation of some important works during the 20th century. *Historia scientiarum* 25: 93–100.

A6. **Browne, J.** 1987. Ch. 13: History of biology. In: H. V. Wyatt, ed. *Information sources in the life sciences*, 3rd ed., pp. 177–186. London, Butterworths. xiv+191 pp. *Butterworths guides to information science.*
The corresponding chapter on this subject in the previous ed. is still useful, i.e.: **Thornton, J. L., & R. I. J. Tully.** 1971. History and biography of biology. In: R. T. Bottle & H. V. Wyatt, eds. *The use of biological literature.* 2nd ed., pp. 311–330. London, Butterworths.

A7. **Carlson, L.** 1982. Bibliography of the history of Australian science. *Historical records of Australian science* 5: 126–131; *Continued* ___. 1984. ___: no. 4, 1983. *Ibid.* 6: 97–107.
See also Gandevia A9 and Mozley A22.

A8. **Corsi, P., & P. Weindling**, eds. 1983. *Information sources in the history of science and medicine.* London, Butterworth Scientific. xvi+531 pp. *Butterworth's guides to information sources.*
Contents include: 1. **P. Corsi.** History of science, history of philosophy and history of theology; 5. **Jordanova, L. J.** The social sciences and history of science and medicine; 8. **P. Corsi.** Guide to bibliographical sources; 9. **P. Weindling.** Periodical literature and societies; 10. ___. Research methods and sources; 17. **D. E. Allen.** Life sciences: Natural history; 18. **M. Pelling.** Experimentalism and the life sciences since 1800; 20. **E. H. Beardsley.** The history of American science and medicine.

A9. **Gandevia, B.** 1957. Natural history in Australia: A list of historical papers. *Victorian naturalist* 73: 131–136.
See also Carlson A7 and Mozley A22.

A10. **Harkányi, K.** 1990. *The natural sciences and American scientists in the revolutionary era: A bibliography.* New York, Greenwood Pr. xiii+511 pp. *Bibliographies and indexes in American history* no. 17.
Covers the period 1760–89. Contents include: Pt. 1. Comprehensive works: Serial publications, Society publications, General bibliographies, Bibliographies & research guides, Histories, Biographies, Natural history, Medicine, Travel literature; Pt. 2. Works of scientists, (pp. 307–431); Appendix: Science collections [books & MSS]; [Classified] List of scientists; Subject index. Part 2, alphabetically arranged by individual scientists, "lists the men and women of sciences, their published monographic works, and studies written about them."

A11. **History of Science Society.** 1971–84. Isis *cumulative bibliography 1913–1965: A bibliography of the history of science formed from* Isis *critical bibliographies 1–90, indexing literature published from 1913 through 1965.* Ed. by M. Whitrow. London, Mansell. 5 vols. Vols. 1 & 2 re-pub. (in microfiche).
Vols. 1–2: Personalities and institutions, 1971; vol. 3: Subjects, 1976 (Contents include: Science – bibliography; literature; sources; Collective biography; Natural history; Biology, microbiology, botany, zoology; Agriculture – plant cultivation & horticulture, forestry.); vols. 4–5: Civilizations and periods [vol. 5 also includes addenda and some corrigenda to vols. 1–3], 1982; vol. 6: Author index, 1984.

A12. ___. 1980–85. Isis *cumulative bibliography 1966–1975: A bibliography of the history of science formed from* Isis *critical bibliographies 91–100, indexing literature published from 1965 through 1974.* Ed. by J. Neu. London, Mansell. 2 vols.
Vol. 1: Personalities and institutions, 1980; vol. 2: Subjects, periods and civilizations, 1985.

A13. ___. 1989. Isis *cumulative bibliography 1976–1985: A bibliography of the history of science formed from* Isis *critical bibliographies 101–110, indexing literature published from 1975 through 1984.* Ed. by J. Neu. London, Mansell. 2 vols.
Vol. 1: Personalities and institutions, 1989; vol. 2: Subjects, periods and civilizations, 1989.

A14. **Hrochová, M., & V. Jankovic.** 1964. Vyberová bibliografia k dejinám prírodnych vied, medicíny a techniky na Slovensku za roky 1918–1938. (A selective bibliography to the history of science, medicine, and technology in Slovakia in the years 1918–1938). *Z dejín vied a techniky na Slovensku* 3: 335–353.

A15. **Hunemörder, C.** 1973. Bibliographie zur Geschichte der Biologie und ihrer wichtigsten Teilgebiete. *Nachrichten aus dem Institut für Geschichte der Naturwissenschaften, Hamburg* 3: 8–22.

A16. **Institut Istorii Estestvoznaniya i Tekhniki, Akademiya Nauk SSSR.** 1947. Knigi i zhurnal'nye stat'i po istorii estestvoznaniya, opublikovannye v SSSR za period s 1939 po 1944 g. vkl. Na russkom i drugikhyazykakh. [Books and articles in periodicals on the history of natural history published in SSSR in Russian and other languages from 1939 to 1944 inclusive]. *Trudy Institut Istorii Estestvoznaniya* 1: 457–535.

A17. **___.** 1949. *Istoriya estestvoznaniya:* [1] *Literatura, opublikovannaya v SSSR, 1917–1947 [History of the natural sciences: ...].* Moscow, Akademii Nauk SSSR. 519 pp.; *Continued* **___.** 1955. *Istoriya estestvoznaniya:* [2]*Literatura, opublikovannaya v SSSR, 1948–1950.* Moscow, Akademii Nauk SSSR. 395 pp.; **___.** 1963. *Istoriya estestvoznaniya:* [3] *Literatura, opublikovannaya v SSSR, 1951–56.* Moscow, Akademii Nauk SSSR. 430 pp.; **___.** 1974. *Istoriya estestvoznaniya:* [4] *Literatura, opublikovannaya v SSSR, 1957–1961.* Moscow, Izd. "Nauka." 547 pp.
Comprehensive bibliographies embracing all branches of the history of natural history. Four later issues, covering literature down to 1980, not seen.

A18. **Jayawardene, S. A.** 1982. *Reference books for the historian of science.* London, Science Museum Library. xiv+229 pp. *Occasional publications* no. 2.
A select classified and annotated handlist.

A19. **Legée, G., M. Guédès, F. Bourdier, & Y. Laissus.** 1975. Introduction bibliographique à l'histoire de la biologie. *Histoire et nature* 3/4: 1–195. *Continued* **___.** n.d. ___*: Supplement.* 13 pp. (Duplicated from typescript).
I. **G. Legée, M. Guédès & F. Bourdier.** La recherche en histoire de la biologie: méthodologie et bibliographie; II. **G. Legée & F. Bourdier.** L'évolution de la biologie; III. **M. Guédès, G. Legée & Y. Laissus.** Bibliographie méthodique <Chap. II. **M. Guédès.** Botanique et physiologie. Microbiologie>.

A20. **Medical Library Association.** 1967. *Medical reference works, 1679–1966: A selected bibliography.* Ed. by J. B. Blake & C. Roos. Chicago, IL. 343 pp.; *Continued* **___.** 1970. ___*: Supplement I, 1967–1968.* By J. S. Richmond. Chicago, IL. 46 pp.; **___.** 1973. ___*: Supplement II, 1969–1972.* By J. S. Richmond. Chicago, IL. [v]+174 pp.; **___.** 1975. ___*: Supplement III, 1973–1974.* By J. S. Richmond. Chicago, IL. [ii]+89 pp.
Extensive, broad ranging and well annotated survey of literature on a specialized area of life sciences history that includes much general material of interest to history of natural history. However, the supplements include very few works on history.

A21. **Miller, G.**, ed. 1965. *Bibliography of the history of medicine of the United States and Canada: 1936–1960.* 2nd ed. Baltimore, MD, Johns Hopkins Univ. Pr. xviii+428 pp.

A22. **Mozley, A.** 1962. A check list of publications on the history of Australian science. *Australian journal of science* 25: 206–214; *Continued* **___.** 1964. Supplement to a check list [etc.]. *Ibid.* 27: 8–15.
See also Carlson A7 and Gandevia A9.

A23. **Neu, J.** 1967. The history of science. In: R. B. Downs & F. B. Jenkins, eds. *Bibliography: Current state and future trends*, pp. 438–454. Urbana, IL, Univ. of Illinois Pr. *Illinois contributions to librarianship* no. 8.
A survey of sources. First pub. as *Library trends* April 1967 issue.

A24. **Overmier, J. A.** 1989. *The history of biology: A selected annotated bibliography.* New York, Garland Publishing. xvii+157 pp. *Bibliographies in the history of science & technology* vol. 15.
619 entries listed under alphabetically arranged subject headings, e.g. Anatomy, Animal Research, Biochemistry, Biogeography, Biology, Botany, Cytology, [etc.].

A25. **Plug, C.** 1990. *Publications on the history of South African science: Annotated bibliography to the end of 1987.* Pretoria, Univ. of South Africa. xii+187 pp.

A26. **Rider, K. J.** 1970. *History of science and technology: A select bibliography for students.* 2nd ed. London, Library Association. 75 pp.
Replaces 1st ed., 1967. 60 pp.

A27. **Rothenberg, M.** 1982. *The history of science and technology in the United States: A critical and selective bibliography.* New York, Garland Publishing. xviii+242 pp. *Bibliographies of the history of science & technology* vol. 2.
Contents include: Ch. II. Special themes: Evolution, pp. 37–43; Learned societies, pp. 47–53; Pollution, ecology, and conservation, pp. 54–58; Ch. IV. The biological sciences: Biology and the biomedical sciences, pp. 137–143; Botany, pp. 143–146; Eugenics and genetics, pp. 147–150; Natural history, pp. 150–155; Paleontology, pp. 155–157; Zoological disciplines, pp. 158–160. *Also* ___. 1982. Jewels and trash: An overview of the history of American science. *History of science in America news & views* 2(1): 1–2.

A28. **Russo, F.** 1959. La documentation en histoire des sciences et des techniques. *Archives internationales d'histoire des sciences* 12: 253–266.

A29. ___. 1969. *Éléments de bibliographie de l'histoire des sciences et des techniques.* 2me éd., refond. et augm. Paris, Hermann & Cie. xv+214 pp.
Contents include: I. La recherche en histoire des sciences et des techniques: Nature, organisation, moyens; II. Sources imprimées générales, toutes époques et toutes disciplines; III. Histoire de toutes les sciences et techniques par époques; IV. Histoire des sciences et des techniques par disciplines (I. Sciences de la vie, pp. 173–179). *Replaces* ___. 1954. *Histoire des sciences et des techniques: Bibliographie.* Paris, Hermann & Cie. 186 pp. *Actualités scientifiques et industrielles* no. 1204.

A30. **Sarton, G.** 1952. *Horus. A guide to the history of science: A first guide for the study of the history of science.* New York, Ronald Pr. Co. xviii+316 pp.
Still a valuable general guide to literature on the history of science. Contents include: Pt. I – Introductory essays; Pt. II – Guide; A. History, B. Science, C. History of science, 14. Chief reference books, 15. Treatises and handbooks, 16. Scientific instruments, 17. History of science in special countries, 18. History of science in special cultural groups, 19. History of special sciences, 20. Journals and serials concerning the history and philosophy of science; D. Organization of the study and teaching of the history of science, 21. National societies devoted to the history of science, ... 24. Institutes, museums, libraries, 25. International congresses, [etc.].

A31. **Smit, P.** 1974. *History of the life sciences: An annotated bibliography.* Amsterdam, Asher. xiv pp. & 1071 cols.
The most important guide to published scholarship, its most serious limitation being the exclusion of periodical articles which form such a significant part of the whole literature. A critically annotated author list of over 4,000 titles with table of contents classified under 130 headings. Contents include: Ch. 1. General references and tools; Ch. II. Historiography of the life and medical sciences; A. Historiography of the life and medical sciences during the ancient and mediaeval periods: Chronological and ethnographical; B. Historiography of the life and medical sciences during the Renaissance and later periods, according to subject; Selected list of biographies, bibliographies, etc. of famous biologists, medical men, etc.

A32. **Weiss, B.** 1990. *Wie finde ich Information zur Geschichte der Naturwissenschaften und der Technik.* 2. ed. Berlin, Verlag Arno Spitz. 454 pp. *Orientierungshilfen* vol. 27.
Replaces ___. 1985. *Wie finde ich Literatur zur Geschichte der Naturwissenschaften und der*

Technik. Berlin, [etc.]. 289 pp. A directory of libraries and organizations and a bibliography of reference sources.

A33. **Wellcome Institute for the History of Medicine.** 1979–80. *Subject catalogue of the history of medicine and related sciences*. Nendeln, K T O Pr. [Kraus International Publications]. 18 vols.
Comprehends all entries in *Current work in the history of medicine* for 1954–77 with other references from the Library's collection and from the MEDLARS data base. Subject section, vols. 1–9 vols, Biographical section, vols. 10–13, & Topographical section, vols. 14–18.

Related subjects

A34. **American Historical Association.** 1961. *Guide to historical literature*. New York, ed. by G. F. Howe et al. Macmillan & Co. xxxv+962 pp.

A35. **Blackwell, R. J.** 1983. *A bibliography of the philosophy of science, 1945–1981*. Westport, CT, Greenwood Pr. 1983. xix+585 pp.
Contents include: §7. Special topics in the philosophy of science: 7.1 Biological sciences—General; 7.2 Biological species; 7.3 Evolution; 7.4 Genetics, pp. 478–516.

A36. **Desmond, R. G. C.** 1990. A bibliography of garden history. *Garden history* 18(1): i–xv.

A37. **Dunbar, G. S.** 1985. *The history of modern geography: An annotated bibliography of selected works*. New York, Garland Publishing. [xvi]+386 pp. *Bibliographies of the history of science & technology* vol. 9.
Over 1,700 entries covering mid C18th to present day.

A38. **Guitard, E. H.**, ed. 1968. *Index des travaux d'histoire de la pharmacie de 1913 à 1963: Repertoire des auteurs et des sujets d'articles et d'ouvrages soit publiés soit analysés dans les revues ou éditions des 51 premières années de la "Societe d'Histoire de la Pharmacie."* Paris, Société d'Histoire de la Pharmacie. 80+103 pp.

A39. **Kinnell, S. K.**, ed. 1987. *Historiography: An annotated bibliography of journal articles, books, and dissertations*. Santa Barbara, CA, Oxford, ABC–Clio. 2 vols.
Contents include: §9. Topics in historiography. Science and technology, pp. 269–278.

A40. **Kren, C.** 1985. *Medieval science and technology: A selected, annotated bibliography*. New York, Garland Publishing. xix+369 pp. *Bibliographies in the history of science & technology* vol. 11.
The present work is limited to post–medieval literature; Kren's bibliography provides good coverage of the preceding period.

A41. **Moore, G.**, ed. 1978. *A guide to the literature of the history of medicine*. Toronto, Univ. of Toronto Library. v+146 pp.

A42. **Morton, L. T.** 1991. *A medical bibliography (Garrison and Morton): An annotated checklist of texts illustrating the history of medicine*. 5th rev. ed. Ed. by J. M. Norman. Aldershot, Scolar Pr; Brookfield, VT, Gower Publishing. xxiv+1,243 pp.
8,927 entries. *Replaces* ___. 1983. ___. 4th ed. London, A. Deutsch. xii+1000 pp. First pub. as **F. H. Garrison & L. T. Morton.** 1943. ___. London, Grafton. viii+412 pp.

A43. **Porter, R.** 1983. *The earth sciences: An annotated bibliography*. New York, Garland Publishing. vii+192 pp. *Bibliographies in the history of science & technology* vol. 3.

A44. **Sarjeant, W. A. S.** 1980. *Geologists and the history of geology: An international bibliography from the origins to 1978*. New York, Arno Pr. 5 vols. *Re-pub.* Melbourne, FL, Krieger Publishing Co., 1986. *Continued* ___. 1987. ___*: Supplement 1979–1984 and additions*. Melbourne, FL, [etc.]. 2 vols.
Contents include: General works (general histories of science, histories of geology, histories of

allied subjects); Historical accounts of societies, museums, and other institutions concerned with geology; ... Index of authors, editors and translators; [etc.].

A45. **Spencer, F.** 1986. *Ecce homo: An annotated bibliographic history of physical anthropology.* Westport, CT, Greenwood Pr. xiii+495 pp. *Bibliographies & indexes in anthropology* no. 2.

A46. **Wright, J. K., & E. T. Platt.** 1947. *Aids to geographical research: Bibliographies, periodicals, atlases, gazetteers and other reference works.* 2nd ed. New York, Columbia Univ. Pr. for the American Geographical Society. xii+331 pp.
An excellent work despite its age.

HISTORIANS OF NATURAL HISTORY

• There is no directory of historians of natural history: The following works must be used:

A47. **Blondel, C., & V. Gourlet.** 1987. *Guide d'histoire des sciences et des techniques en France.* Paris, Cité des Sciences et de l'Industrie; La Villette, Société Française d'Histoire des Sciences et des Techniques. xii+340 pp. *Cahiers d'histoire et de philosophie des sciences*, nouvelle série no. 19.
A directory of contemporary researchers followed by a list of miscellaneous organizations connected with the studies of the history of science.

A48. **Elliott, C. A.**, ed. 1981. Directory of researchers in American science. *History of science in America news & views* 1(3): 22 pp.; *Continued* ___. 1981. Supplement. *Ibid.* 1(4): 11.

A49. **History of Science Society.** 1992. Directory of members. In: *Guide to the history of science*, 8th ed., 1992, pp. 17–168, 269–315. Chicago, IL.
Alphabetical listing; Subject interests; National culture interests; Locations. *Replaces* earlier lists pub. in 1980, 1983 and 1986 eds.

A50. **Society for the History of Natural History.** 1989. *Members' handbook May 1989.* London. 68 pp.
Contents include: Alphabetical list of members; Regional membership lists. *Replaces* ___. 1986. *List of members April 1986.* London. 71 pp.

SOCIETIES & ORGANIZATIONS

Guides

A51. **Anon.** 1983. History of science societies in the U. K. *Newsletter, British Society for the History of Science* 12: 18–26.

A52. **Elliott, C. A.**, ed. 1983. Scientific and technological disciplines: A directory of history centers and projects. *History of science in America news & views* 2 (3): 1–6; 2(4): 12.

A53. ___. 1984. Science documentation and discipline–based centers for the history of science. *A S M news* 50 (1): 2–6.

A54. **History of Science Society.** 1992. Societies and organizations in history of science, technology, and medicine. In: *Guide to the history of science*, 8th ed., 1992, pp. 220–240, 258–268. Chicago, IL.
Geographical arrangement with alphabetical index. *Replaces* earlier lists pub. in 1983 and 1986 eds.

A55. **Sarton, G.** 1952. Ch. 21: National societies devoted to the history of science. Ch. 24: Institutes, museums, libraries. In: *Horus: A guide to the history of science: A first guide for the study of the history of science*, pp. 249–252 and 260–289. New York,

Ronald Pr. Co.

A detailed report of organizations as they were in about 1950 with useful historical notes.

A56. **Verdoorn, F.** 1971. History of science institutions and their universities. *Janus* 58: 278–288.

Individual societies, etc.

Biohistorical Institute of the University of Utrecht.

A57. **Smit, P.** 1968. Das Biohistorische Institut der Reichsuniversität Utrecht. *Medizinhistorisches Journal* 3: 195–205.

A58. **Verdoorn, F.** 1968. Horizontal versus vertical approaches in the history of the life sciences. *Actes du XIe Congrès International d'Histoire des Sciences (Warsaw/Cracow 1965)* 5: 5–9.

Apropos the programme of the Biohistorical Institute of the University of Utrecht.

Center for the History of Entomology.

A59. **Rutschky, C. W., 3rd.** 1979. The Center for the History of Entomology. *Entomological Society of America bulletin* 25 (1): 39–42.

Center for the History of Microbiology.

A60. **Shay, D. E.** 1985. Center for the History of Microbiology. *A S M news* 51 (8): 415–416.

Deutsche Gesellschaft für Geschichte der Medizin, Naturwissenschaft und Technik.

A61. **Winau, R.** 1978. *Deutsche Gesellschaft für Geschichte der Medizin, Naturwissenschaft und Technik 1901–1976.* Wiesbaden, Steiner. 164 pp. *Beiträge zur Geschichte der Wissenschaft und der Technik* no. 15.

History of Science Society.

A62. **History of Science Society.** 1975. Fiftieth anniversary celebration of the Society. *Isis* 66: 443–482.

Founded, Boston, MA, 1924. *Includes* **A. Thackray.** Five phases of prehistory, depicted from diverse documents; **R. P. Multhauf.** The Society and its concerns; **G. Basalla.** Observations on the present status of history of science in the U.S.; **D. J. De Solla Price.** Comment on Basalla's paper; **C. C. Gillispie.** A note on the Society's prizes.

A63. ___. 1992. *Guide to the history of science*, 8th ed., 1992. Chicago, IL. 315 pp.

Includes The History of Science Society [organization, statutes]; Directory of members; Guide to the profession. *Replaces* earlier versions pub. 1980, 1983, 1986.

Hunt Botanical Library (*later* Hunt Institute for Botanical Documentation), Carnegie Mellon University.

A64. **Hunt Botanical Library, Carnegie Mellon University.** 1970. *The Hunt Botanical Library: A decennial report.* By G. H. M. Lawrence. Pittsburgh, PA, Carnegie Mellon Univ. [vii]+47 pp.

Institut für die Geschichte der Naturwissenschaften der Universität Hamburg.

A65. **Sticker, B.** 1970. *Institut für die Geschichte der Naturwissenschaften der Universität Hamburg. Bericht über das erste Jahrzent 1960/1970.* Hamburg. [ ]* pp.

English resumé in *Isis* 63: 393–394, 1972. [*Not examined – pagination not ascertained.]

Internationale Gesellschaft für die Geschichte der Pharmazie.

A66. **Hügel, R.** 1955. *Die Veröffentlichungen der (Internationalen) Gesellschaft für Geschichte der Pharmazie, 1927–1952: Eine Bibliographie.* Eutin. 39 pp. *Veröffentlichungen* n.s. no. 7.

Society for the Bibliography of Natural History (*later* Society for the History of Natural History).

A67. **Society for the History of Natural History.** 1991. *Society for the Bibliography of Natural History 1936 to 1983. Society for the History of Natural History 1983 to 1990. Index to publications 1936–1990.* By J. A. Gibson. London. 137 pp.

A68. **Wheeler, A. C.** 1971. The Society for the Bibliography of Natural History: An account of its formation and activities. *Private library* (2) 4: 65–69.
For a brief account of the Society's infant years *see* **F. Hemming.** 1943. The development of the Society for the Bibliography of Natural History in the first six years of its existence. *Journal of the Society for the Bibliography of Natural History* 1: xi–xvi.

Svenska Linné–Sällskapets.

A69. **Svenska Linné–Sällskapet.** 1987. *Register till årsskrift: 1918–1985.* Stockholm. 24 pp.

A70. **Von Sydow, C.–O.** 1967. Svenska Linné–Sällskapets stiftande och andra första verksamhet. En tillbakablick vid 50–årsminnet. [The founding and first activity of the Swedish Linnaean Society.] *Svenska Linné–Sällskapets årsskrift* 50: 52–91.
English summary, pp. 90–91.

Wellcome Institute for the History of Medicine.

A71. **Skinner, G. M.** 1986. Sir Henry Wellcome's museum for the science of history. *Medical history* 30: 383–418.

A72. **Wellcome Institute for the History of Medicine.** 1988. *Wellcome Institute ...: A brief description.* 3rd ed. London. 30 pp.

A73. **Wellcome Trust.** 1986. *A vision of history: The Wellcome Institute for the History of Medicine: An exhibition to celebrate the 50th anniversary of the Wellcome Trust, 1 September 1986 to 10 April 1987.* London. 72 pp.

NEWSLETTERS

• A selected title–list (current sub-titles may vary):

A74. ***Boletin informativo, Sociedad Latinoamericana de Historia de las Ciencias y la Tecnologia.*** Mexico City, no. 1+, 1983+.

A75. ***History of Australian science newsletter.*** Parkville, Vic., Department of History & Philosophy of Science, Univ. of Melbourne, no. 1+, 1983+.

A76. ***History of science in America: News and views;*** *Newsletter to the Forum for the History of Science in America.* Washington, DC, Forum for the History of Science in America, vol. 1+, 1980+.

A77. ***History of Science Society newsletter,*** no. 1+, 1972+.

A78. ***Newsletter, British Society for the History of Science.*** London, no. 1+, 1979+.

A79. ***Newsletter, International Society for the History, Philosophy and Social Studies of Biology.*** Blacksburg, VA, no. 1+, 1989+.

A80. ***Newsletter, Society for the History of Natural History.*** London, nos. 18–26, 1983–85; no. 31+, 1987+.
Nos. 27–30 contained in *Archives of natural history.* Preceded by *Newsletter, Society for the Bibliography of Natural History.* London. Nos. 1–17, 1979–83; *British newsletter, Society for the Bibliography of Natural History.* London. Nos. 1–7, 1977–78; and *North American newsletter, Society for the Bibliography of Natural History.* N.p. Vol. 1, no. 1, Fall 1977.

PERIODICALS

Guides

A81. **Anon.** 1975. Sovetskie periodicheskie izdaniya po istorii estestvoznaniya i tekhniki. *Voprosy istorii estestvoznaniya i tekhniki* 50: 84–88.

A82. **Boehm, E. H., et al.**, eds. 1981–86. *Historical periodicals directory.* Santa Barbara, CA, Oxford, ABC–Clio Pr. 5 vols.
Vol. 1: **M. S. Ensign**, ed. 1981. *U S and Canada*; vol. 2: **B. H. Pope**, ed. 1981. *Europe West, North, Central and South*; vol. 3: **E. H. Boehm, B. H. Pope, & M. S. Ensign**, eds. 1984. *Europe, East and Southeast; USSR*; vol. 4: *Latin America and the West Indies*; vol. 5: *Australia and New Zealand*. *Replaces* **E. H. Boehm & L. Adolphus.** 1961. *Historical periodicals: An annotated world list of historical and related publications.* Santa Barbara, CA, Clio Pr. 618 pp.

A83. **Caron, P., & M. Jaryc.** 1939. *Liste mondial des périodiques et bibliographies historiques.* Oxford, International Committee of Historical Sciences. 405 pp.

A84. **Goodman, R. E.** 1992. Journals and newsletters in history of science, technology, and medicine. In: History of Science Society. *Guide to the history of science*, 8th ed., 1992, pp. 241–257. Chicago, IL.
Replaces earlier lists pub. in 1980, 1983 and 1986 eds.

A85. **Kramm, H.** 1952–54. *Bibliographie historischer Zeitschriften, 1939–1951.* Marburg, Rasch. 3 vols.

A86. **Numbers, R. L.**, ed. 1990–91. Reviews of journals and serials. *Isis* 81: 281–308, 1990, & *Ibid.* 82: 285–325, 1991.
Retrospective critiques of 26 English language and 47 international history of science journals by various reviewers.

A87. **Sarton, G., & C. F. Mayer.** 1952. Journals and serials concerning the history (and philosophy) of science. In: G. Sarton. *Horus: A guide to the history of science*, pp.194–248. New York, Ronald Pr. Co.

A88. **Weindling, P.** 1983. Bibliography of journals. In: P. Corsi & P. Weindling, eds. *Information sources in the history of science and medicine*, pp. 501–508. London, Butterworth Scientific. *Butterworth's guides to information sources.*

Individual periodicals

• A selected title–list (current sub-titles may vary):

A89. ***Anales de la Sociedad Mexicana de Historia de la Ciencia y de la Tecnnologia.*** Mexico City, vol. 1+, 1969+.

A90. ***Annals of science****; an international review of the history of science and technology from the thirteenth century* [current sub-title]. London, Taylor & Francis, vol. 1+, 1936+.

A91. ***Annaly biologii.*** Moscow, Moskovskoe Obshchestvo Ispytatelei Prirody, Sektsiyo Istorii Estestvoznaniya, vol. 1, 1959.

A92. ***Archiv für Geschichte der Medizin.*** Leipzig, Puschmann–Stiftung an der Universität Leipzig, vols. 1–20, 1907–28.
Continued as *Sudhoffs Archiv für Geschichte der Medizin (und der Naturwissenschaften, zugleich Fortsetzung der Zoologischen Annalen)*; *see* A146.

A93. ***Archiv der Geschichte der Naturwissenschaften.*** Vienna, Gebrüder Hollinek, Heft 1+, 1980/81+.
A periodical that largely comprises a continuation of: **J. Mayerhöfer et al.** 1959–[75?] *Lexikon der Geschichte des Naturwissenschaften: Biographien, Sachwörter und Bibliographien.*

Vols. 1–2(8), Vienna, Gebrüder Hollinek (*see* Ca82). The *Archiv* also includes some original articles and notices of current publications and events.

A94. ***Archives internationales d'histoire des sciences.*** Paris, Union Internationale des Sciences (*later* Rome, Académie Internationale d'Histoire des Sciences), vol. 1+, 1947+.
Also **Anon.** 1967. Tables analytiques, 1947–1967. *Archives internationales d'histoire des sciences* 20: 337–539.

A95. ***Archives of natural history.*** London, Society for the Bibliography (*later* History) of Natural History, vol. 10+, 1981+.
Preceded by *Journal of the Society for the Bibliography of Natural History*; *see* A131. *See also* **J. A. Gibson.** 1991. *Society for the Bibliography of Natural History 1936 to 1983. Society for the History of Natural History 1983 to 1990. Index to publications 1936–1990.* London. 137 pp.

A95A. ***Biology history.*** London, Institute of Biology, History of Biology Group, no. 1+, 1988+.

A96. ***British journal of the history of science.*** Ravenswood, Oxford & London, British Society for the History of Science, vol. 1+, 1962+.

A97. ***Bulletin of the British Museum (Natural History), historical series.*** London, vol. 1+, 1953+.

A98. ***Canadian horticultural history;*** *An interdisciplinary journal.* Hamilton, Ont., Centre for Canadian Historical Horticultural Studies, 1986+.
Also includes historical botanical material.

A99. ***Chronica botanica;*** *[Subtitle varies].* Leiden, Waltham, MA, New York, Chronica Botanica Company *later* Ronald Pr. Co., vols. 1–16(–33), 1935–62.
From vol. 17 onwards this periodical changed to series title for Ronald Pr. publications.

A100. ***Circulaire, Biohistorisch Instituut der Rijksuniversiteit te Utrecht.*** Utrecht.
See *Incidentele mededelingen van het Biohistorisch Instituut der Rijksuniversiteit te Utrecht*, A118.

A101. ***Dejiny ved a techniky.*** Prague, Ceskoslovenská Akademie Ved, vol. 1+, 1968+.

A102. ***Documents pour l'histoire du vocabulaire scientifiques.*** Besançon, Institut de la Langue Française, vol. 1+, 1980+.

A103. ***Endeavour;*** *A quarterly review designed to record the progress of the sciences in the service of mankind.* London, Kynoch Pr. for Imperial Chemical Industries, vols. 1–35, 1942–76; New series, Oxford, Pergamon Pr, vol. 1+, 1977+.
Also pub. in French, German, Italian, and Spanish versions.

A104. ***Environmental review.*** Denver, CO, American Society for Environmental History, vol. 1+, 1976+.

A105. ***Episteme;*** *Rivista critica di storia (delle scienze,) della medicina e della biologia.* Milan, vols. 1–10, 1967–76.
Continued as *History and philosophy of the life sciences*; *see* A115.

A106. ***Folia Mendeliana;*** *Papers relating to Mendel and to the early development of genetics.* Brno, Moravske Zemske Museum, no. 1+, 1966+.

A107. ***Forest & conservation history.*** Durham, NC, Forest History Society, vol. 34+, 1990+.
Preceded by *Journal of forest history*; *see* A127.

A108. ***Forest history.*** St. Paul, MN, New Haven, CT, Santa Cruz, CA, Forest History Foundation Inc. (*later* Forest History Society), vols. 1–17, 1958–74. (Vol. 1 titled *Forest*

history newsletter.)
Continued as *Journal of forest history*; *see* A127.

A109. ***Garden history***; *Journal of the Garden History Society*. Oxford, vol. 1+, 1972+.

A110. ***Gesnerus***; *Vierteljahrsschrift für Geschichte der Medizin und der Naturwissenschaften*. Aarau, Schweizerische Gesellschaft für Geschichte der Medizin und der Naturwissenschaften, vol. 1+, 1943+.
See also **H. Fischer.** 1968. Zum 25. Jahrgang des Gesnerus. *Ibid.* 25: 1–4.

A111. ***Histoire et biologie***; *Cahiers du Cercle d'Étude Historique des Sciences de la Vie*. Paris, Muséum National d'Histoire Naturelle, nos. 1–2,1968–69.
Continued as *Histoire et nature*; *see* A112.

A112. ***Histoire et nature***; *Cahiers de l'Association pour l'Histoire des Sciences de la Nature*. Paris, Muséum National d'Histoire Naturelle, no. 3+ [=n. s. no. 1+], 1973+.
Preceded by *Histoire et biologie*; *see* A111.

A113. ***Historia scientiarum***; *international journal of the History of Science Society of Japan*. Tokyo, Nippon Kagakushi Gakkai [History of Science Society of Japan], no. 19+, 1980+.
Preceded by *Japanese studies in the history of science*; *see* A126.

A114. ***Historical records of Australian science.*** Canberra, A.C.T., Australian Academy of Science, vol. 5+, 1980+.
Preceded by *Records of the Australian Academy of Science*.

A115. ***History and philosophy of the life sciences.*** (*Pubblicazioni della Stazione Zoologica di Napoli*, sect. II). Naples, etc., vol. 1+, 1979+.

A116. ***History of science***; *an annual review of literature, research and teaching*. Cambridge, Heffer, vol. 1+, 1962+.
Also **Anon.** 1987. Cumulative author index to volumes 1–25. *History of science* 25: 427–439.

A117. ***Huntia***; *A yearbook of botanical & horticultural bibliography* (later *A journal of botanical history*). Pittsburgh, PA, Hunt Botanical Library (*later* Hunt Institute for Botanical Documentation), Carnegie Mellon University, vol. 1+, 1964+.

A118. ***Incidentele mededelingen van het Biohistorisch Instituut der Rijksuniversiteit te Utrecht.*** (*Occasional communications from the Utrecht University Biohistorical Institute.*) Utrecht, n. s. no. 1+, 1974+.
Preceded by *Circulaire, Biohistorisch Instituut der Rijksuniversiteit te Utrecht*. Utrecht, nos. ?–86–?, ?–1972–? An elusive serial publication whose publication data has not yet been ascertained.

A119. ***Indian journal of history of science.*** New Delhi, Indian National Science Academy, vol. 1+, 1966+.

A120. ***Isis***; *Revue consacrée a l'histoire de la science, [etc.]. International review devoted to the history of science and civilization*. Ghent, (*later* Washington, DC, etc.), History of Science Society, vol. 1+, 1913+.
Also **G. Sarton.** 1934. Brief table of contents of vols. 1–20 [1913–34]. *Isis* 21: 502–618; **J. Morley, J. Cruise & M. Greenwood.** 1985. *Isis: An index to volumes 44–73, 1953–1982*. Philadelphia, PA. 167 pp. (There is no previous cumulative index. The starting date for this index was chosen to commence the era following George Sarton's editorship). *See also* **C. Rosenberg.** 1987. Isis at seventy-five. *Isis* 78: 515–517.

A121. ***Istoria biologicheskikh nauk.*** Moscow & Leningrad, Akademiya nauk SSSR, Institut Istorii Estestvozneniya i Tekhniki, no. 1+, 1955+.
See Trudy Institut Istorii Estestvoznaniya.

A122. ***Istoriko-biologicheskie issledovaniya.*** Moscow, Akademiya Nauk SSSR, Institut Istorii Estestvoznaniya i Tekhniki, vol. 6+, 1978+.
Preceded by *Iz istorii biologii*; see A124.

A123. ***Iz istorii biologicheskii nauk.*** Moscow, Akademiya Nauk SSSR, Institut Istorii Estestvoznaniya i Tekhniki, vols. 1, 1966.
Previously contained in *Trudy Institut Istorii Estestvoznaniya*; see A148. Continued as *Iz istorii biologii*; *see* A124.

A124. ***Iz istorii biologii.*** Moscow, Akademiya Nauk SSSR, Institut Istorii Estestvoznaniya i Tekhniki, vols. 2–5, 1970–75.
Preceded by *Iz istorii biologicheskii nauk*; *see* A123. Continued as *Istoriko-biologicheskie issledovaniya*; *see* A122.

A125. ***Janus****; revue internationale de l'histoire des sciences, [etc].* Leiden, Foundation Janus, vol. 1+, 1896+.

A126. ***Japanese studies in the history of science.*** Tokyo, Nippon Kagakushi Gakkai [History of Science Society of Japan], vols. 1–18, 1962–79; Supplement, vols. 1–2, 1971.
Continued as *Historia scientiarum*; *see* A113.

A127. ***Journal of forest history.*** Santa Cruz, CA, Forest History Society, vols. 18–33, 1974–89.
Preceded by *Forest history*; *see* A108. Continued as *Forest & conservation history*; *see* A107.

A128. ***Journal of garden history****; An international quarterly.* London, Taylor & Francis, vol. 1+, 1981+.

A129. ***Journal of the history of biology.*** Cambridge, MA, Harvard Univ. Pr, (*later* Dordrecht, D. Reidel Publishing Co.), 1968+.

A130. ***Journal of the history of medicine and allied sciences.*** New York, H. Schuman; London, W. Heinemann, vol. 1+, 1946+.
Also **M. Waserman & C. T. Clausen.** 1977. *Index to the Journal of the History of Medicine and Allied Sciences: Volumes I–XXX, 1946–1975.* New Haven, CT. 131 pp.

A131. ***Journal of the Society for the Bibliography of Natural History.*** London, vols. 1–9, 1936–80.
Continued as *Archives of natural history*; *see* A95. *See also* **J. A. Gibson.** 1991. *Society for the Bibliography of Natural History 1936 to 1983. Society for the History of Natural History 1983 to 1990. Index to publications 1936–1990.* London. 137 pp.

A132. ***Kwartalnik historii nauki i techniki.*** Warsaw, Polska Akademia Nauk, Zaklad Historii Nauki i Techniki, vol. 1+, 1956+.

A133. ***Lychnos****; Lärdomshistoriska samfundets årsbok.* Uppsala & Stockholm, Almqvist & Wiksell, vol. 1+, 1936+.
Also **C. O. von Sydow.** 1965. *Lychnos* 1936–1962. Generalregister. *Ibid.* 1963/1964: 213 pp.; **L. Sellberg.** 1989. *Lychnos* 1963/64–1987. *Ibid.* 1988: 109 pp.

A134. ***Medical history****; a quarterly journal devoted to the history and bibliography of medicine and related sciences.* London, Cambridge University History of Medicine Society [& others] (*later* British Society for the History of Medicine; *later* Wellcome Institute for the History of Medicine), vol. 1+, 1957+.

A135. ***Memorabilia zoologica.*** Wroclaw, Polska Akademia Nauk, Instytut Zoologiczny, vol. 1+, 1958+.
Later volumes ceased to have historical content.

A136. ***Mendel newsletter****; Archival resources for the history of genetics and allied sciences.* Philadelphia, PA, American Philosophical Society, no. 1+, 1973+.

A137. ***Mitteilungen zur Geschichte der Medizin und der Naturwissenschaften (later Mitteilungen zur Geschichte der Medizin, der Naturwissenschaften und der Technik).*** Leipzig, Deutsche Gesellschaft für Geschichte der Medizin, Naturwissenschaft und Technik, vols, 1–40, 1902–42; vol. 41+, 1961+.

The gap in literature coverage between vols. 40 and 41 is filled by the *Index zur Geschichte der Medizin, Naturwissenschaft und Technik*; *see* A1.

A138. ***Osiris***; *Studies on the history and philosophy of science*. Bruges, vols. 1–15, 1936–68.
Continued as *Osiris*. Second series; *see* A139.

A139. ***Osiris***; *A research journal devoted to the history of science and its cultural influences*. Second series. Philadelphia, PA, History of Science Society, vol. 1+, 1985+.
Preceded by *Osiris*; *see* A138.

A140. ***Quipu***; *Revista latinoamericana de historia de las ciencias y la tecnologia*. Mexico City, Sociedad Latinoamericana de Historia de las Ciencias y la Tecnología (SLHCT), vol. 1+, 1984+.

A141. ***Revue d'histoire des sciences***, *et de leurs applications*. Paris, Presses Universitaires de France, Centre International de Synthèse, vol. 1+, 1947+.

A142. ***Scientiarum historia***: *Driemaandelijks tijdschrift voor de geschiedenis van de geneeskunde, wiskunde en natuurwetenschappen*. Antwerp, Ministerie van Nationale Opvoeding en Nederlandse Cultuur, vols. 1–15(2), 1959–73.

A143. ***Studia i materialy z dziejów nauki Polskiej.*** Seria B, historia nauk biologicznych i medycznych. Warsaw, Polska Akademia Nauk, Zaklad Historii Nauki i Techniki, Panstwowe Wydawnictwo Naukowe, vol. 1+, 1957+.

A144. ***Studies in the history of biology.*** Ed. by W. Coleman & C. Limoges. Baltimore, MD, Johns Hopkins Univ. Pr, vols. 1–7, 1977–84.

A145. ***Sudhoffs Archiv***; *Vierteljahrschrift für Geschichte der Medizin und der Naturwissenschaften der Pharmazie und der Mathematik*. Wiesbaden, F. Steiner Verlag, vol. 50+, 1966+. ***Sudhoffs Archiv. Beiheft.*** Wiesbaden, [etc.], vol. 5+, 1966+.
Preceded by *Sudhoffs Archiv für Geschichte der Medizin und der Naturwissenschaften*; *see* A146.

A146. ***Sudhoffs Archiv für Geschichte der Medizin und der Naturwissenschaften.*** Leipzig, Puschmann–Stiftung an der Universität Leipzig, vols. 27–49, 1934–64 [suspended 1943–52]. ***Sudhoffs Archiv ... Beihefte.*** Leipzig, [etc.], vols. 1–4, 1961–64.
Preceded by *Archiv für Geschichte der Medizin*; *see* A92; and *Zoologische Annalen*; *see* A149. Continued as *Sudhoffs Archiv; Zeitschrift für Wissenschaftsgeschichte*; *see* A145.

A147. ***Tijdschrift voor de Geschiedenis der Geneeskunde, Natuurwetenschappen, Wiskunde, en Technick.*** Utrecht, vol. 1+, 1978+.

A148. ***Trudy Institut Istorii Estestvoznaniya*** (from vol. 6+ ***i Tekhniki***). Moscow & Leningrad, vol. 1+, 1947+.
Contents include a separately numbered sub-series: *Istoria biologicheskikh nauk*, e.g. tom. 4 (= vyp. 1), 1955, tom. 14 (= vyp. 2), 1957, tom. 16 (= vyp. 3), 1957, tom. 23 (= vyp. 4), 1959, tom. 24 (= vyp. 5), 1958, tom. 31 (= vyp. 6), 1960, tom. 32 (= vyp. 7), 1960, tom. 36 (= vyp. 8), 1961, tom. 40 (= vyp. 9), 1962, tom. 41 (= vyp. 10), 1961. Continued as *Iz istorii biologicheskii nauk*; *see* A123.

A149. ***Zoologische Annalen***; *Zeitschrift für Geschichte der Zoologie*. Würzburg, Leipzig, A. Stuber's Verlag (C. Kabitzsch), vols. 1–7, 1904–19.
Continued in *Sudhoffs Archiv für Geschichte der Medizin (und der Naturwissenschaften, zugleich Fortsetzung der Zoologischen Annalen)*; *see* A146.

• Various other scientific periodicals also include papers on the history of natural history. In particular, the following titles have published useful material and merit regular scanning:

A150. ***American scientist.*** Burlington, VT, etc., 1943+.

A151. ***Biological journal of the Linnean Society.*** London, vol. 1+, 1969+.
Preceded by *Proceedings of the Linnean Society of London*; *see* A154.

A152. ***Linnean;*** *Newsletter and proceedings of the Linnean Society of London.* London, no. 1+, 1984+.

A153. ***Proceedings of the American Philosophical Society.*** Philadelphia, PA, vol. 1+, 1840+.

A154. ***Proceedings of the Linnean Society of London.*** London, [1839]–1968.
Continued as *Biological journal of the Linnean Society*; *see* A151.

A155. ***Taxon.*** Utrecht, International Association for Plant Taxonomy (IAPT), vol. 1+, 1951+.

PERIODICAL BIBLIOGRAPHIES

• For information on current–awareness bibliographies see:

A156. **Henige, D.** 1986. *Serial bibliographies and abstracts in history: An annotated guide.* Westport, CT, Greenwood Pr. xiv+220 pp. *Bibliographies & indexes in world history* no. 2.
For a detailed guide to older services, some still extant *see* **C. F. Mayer.** 1952. Abstracting and review journals. In: G. Sarton. *Horus: A guide to the history of science*, pp. 105–110. New York, Ronald Pr. Co.

• A selected title–list (current sub-titles may vary):

A157. ***Annual bibliography of the history of natural history.*** London, British Museum (Natural History), vol. 1+, 1985+.
Sadly, the publisher has announced that difficult economic circumstances will force this invaluable service to terminate with no. 9.

A158. ***Bibliografia Italiana di storia della scienza.*** Florence, Olschki, vol. 1+, 1982+. *Biblioteca di bibliografia Italiana* no. 105 [etc.].
Annual, e.g. vol. 5 <1986>, Florence, [etc.], 1991.

A159. ***Bibliographia historiae rerum rusticarum internationalis.*** Budapest, Museum Rerum Rusticarum Hungariae, 1964+.

A160. ***Bibliographie analytique des publications relatives à l'histoire de la science.***
See Critical bibliography of the history of science and its cultural influences; *see* A163.

A161. ***Bibliography of the history of medicine.*** Bethesda, MD, National Library of Medicine, no. 1+, 1965+.
Periodic cumulations: ___. [No. 5.] *1964–1969.* Bethesda [etc.], 1971. vi+1475 pp.; ___. No. 10. *1970–1974.* Bethesda [etc.], 1976. viii+1069+ix–lxxxvii pp.; ___. No. 15. *1985–1989.* Bethesda [etc.], 1981. x+924+xi–lxxii pp.; ___. No. 20. *1980–1984.* Bethesda [etc.], 1990. 1300 pp.; ___. No. 25. *1985–1989.* Bethesda [etc.], 1990. 1454 pp.

A162. ***Bulletin signalétique. 522. Histoire des sciences et des techniques.*** Paris, Centre National de la Recherche Scientifique (CNRS), Centre de Documentation Scientifique et Technique, vol. 23+, 1969+.
Previously included in Bulletin signalétique. C (19–23). Sciences humaines, philosophie.

A163. ***Critical bibliography of the history of science and its cultural influences.*** In: *Isis.* Ghent, (*later* Washington, DC, etc.), History of Science Society, vol. 1+, 1913+.
For cumulations of this annual bibliography *see* History of Science Society. 1971+; *see* A11.

A164. ***Current work in the history of medicine.*** London, Wellcome Historical Medical Library (*later* Wellcome Institute for the History of Medicine), no. 1+, 1954+.
For a cumulation of entries for the the years 1954–77 *see* Wellcome Institute for the History of Medicine. 1979–80; *see* A33.

A165. ***Historical abstracts;*** *bibliography of the world's periodical literature.* Santa Barbara, CA, A[merican] B[ibliographical] C[enter] – Clio Information Services, vol. 1+, 1955+.
Up to 1972 this service covered historical studies of the period 1775 to 1945; since 1973 the period of coverage has been widened to include 1450 to 1964.

A166. ***Index of conference proceedings received by the N L L.*** Boston Spa, National Lending Library for Science & Technology (*later* British Library, Lending Division *later* British Library, Document Supply Centre), 1964+.
Annually cumulated bibliography.

A167. ***International bibliography of historical sciences.*** Paris, International Committee of Historical Sciences – A. Colin (*later* Munich, Saur), vol. 1+, 1930+.

A168. ***Kew record of taxonomic literature relating to vascular plants for 1971*** [+]. London, H.M.S.O. for Royal Botanic Gardens, Kew, 1974+.
Annual bibliography that includes bibliographical, biographical and some historical titles for vascular–plant botany.

A169. ***Recent publications in natural history.*** New York, American Museum of Natural History, 1983+.
Contents include: History and philosophy of science.

A170. ***Zoological record,*** [etc.]. London, Zoological Society of London, vol. 7+, 1870+. (Preceded by ***Records of zoological literature.*** 1864–69. London, J. Van Voorst, vols. 1–6.)
Currently (and for most of the 20th century) this annual service lists historical, biographical and bibliographical literature for each of its sections from "Comprehensive zoology" to "Protozoa."

• There is an online service:

A171. ***HISTLINE: HISTory of medicine onLINE.*** Bethesda, MD, National Library of Medicine.
Commenced in 1969. Online version of the *Bibliography of the history of medicine. See also* **R. E. Skinner.** 1983. Searching the history of science online. *Database* 6(2): 158–165.

GUIDES TO DISSERTATIONS

A172. **British Society for the History of Science.** 1970+. *A list of theses in history of science in British universities, in progress or recently completed. 1970/1971+.* London.

A173. **Fichtner, G.**, ed. 1981+. *Index wissenschaftshistorischer Dissertationen (IWD). Verzeichnis abgeschlossener Dissertationen auf dem Gebiet der Geschichte der Medizin, der Pharmazie, der Naturwissenschaften und der Technik.* Tubingen, Universität Tübingen, Institut für Geschichte der Medizin. In progress.
No. 1: 1970–1980. 1981. 285 pp.

A174. **Fisher, N. W., & W. H. Brock.** 1975. A list of Ph.D. theses in the history of science and related areas in British universities, 1945–74. *British journal for the history of science* 8: 267–278.

A175. **Kuehl, W. F.**, ed. 1965–86. *Dissertations in history. An index to dissertations completed in history departments of United States and Canadian universities.* Lexington, KY, Univ. of Kentucky Pr. (vol. 3: Santa Barbara, CA, & Oxford, ABC–Clio Information Services). 3 vols.
Vol. 1. 1873–1960; vol. 2. 1961–1970; vol. 3. June 1970–June 1980.

A176. **Shcherbakova, A. A.** 1955. Dissertatsii po voprosam istorii biologicheskikh i sel'skokhozyaistvennykh nauk, zashchishchennye v SSSR v 1944–1953gg. *Trudy Institut Istorii Estestvoznaniya* 4: 400–403.

Reports of newly completed dissertations can be found in:

A177. ***Dissertation abstracts*** (*later* ***Dissertation abstracts international***) Ann Arbor, MI, University Microfilms, 1952–69. (Thereafter divided into two parts. Part. B: *The sciences and engineering*. 1966+. Ann Arbor, MI, [etc.], vol. 69+.)
Also *Dissertation Abstracts International retrospective index, vols. I–XXIX [1938–69]*. Ann Arbor, MI, [etc.], 1970. 11 vols. Became international in coverage from vol. 30, 1969, but still emphasizes North American work. Preceded by *Microfilm abstracts*. Ann Arbor, MI, vols. 1–11, 1938–51.

A178. **Fichtner, G.**, ed. 1981+. *Laufende wissenschaftshistorische Dissertationen (LWD). Verzeichnis in Bearbeitung befindlicher Dissertationen auf dem Gebiet der Geschichte der Medizin, der Pharmazie, der Naturwissenschaften und der Technik.* No. 1+. Tubingen, Institute für Geschichte der Medizin. In progress.
No. 1. 1981. 125 pp.

A179. ***Historical research for higher degrees in the United Kingdom.*** London, Institute for Historical Research, vol. 1+, 1967+.

A180. **Jacobs, P. M.** 1976. *History theses 1901–1970: Historical research for higher degrees in the universities of the United Kingdom.* London, Institute for Historical Research, University of London. viii+456 pp.; *Continued* **J. M. Horn.** 1984. *History theses 1971–80 [etc.]*. London, [etc.]. ix+294 pp.
7,633 & 4,416 entries.

A181. ***List of doctoral dissertations in history in progress or recently completed in the United States*** (*later* ***Doctoral disertations in history***). Washington, DC, American Historical Association, Institutional Services Program, vol. 1+, 1976+.

• In addition to these guides, informal notices of newly completed dissertations can be found in other sources such as *History of Science Society newsletter* 1972+, *History of science in America, news & views* 1981+, *Isis*, etc. For information on access to theses see:

A181A. **Borchardt, D. H., & D. Thawley.** 1981. *Guide to the availability of theses.* Munich, K. G. Saur. 443 pp. *International Federation of Library Associations (IFLA) publication* no. 17; *Continued* **G. G. Allen & K. Deubert.** 1984. ___ *II: Non-university institutions.* Munich, K. G. Saur. 124 pp. *IFLA publication* no. 29.

PERIODIC MEETINGS

International Congress on the History of Oceanography.

A182. Premier congrès international d'histoire de l'océanographique, Monaco, 12–17 December 1966. Proceedings pub. in *Bulletin de l'Institut Océanographique, Monaco* numéro spécial 2, 1968; *see* Ca575.

A183. Second international congress on the history of oceanography, *Challenger* expedition centenary, Edinburgh 12–20 September 1972. Proceedings pub. in *Proceedings of the Royal Society of Edinburgh* B 72 & 73, 1972; *see* Ca593.

A184. Third international congress on the history of oceanography, fiftieth anniversary of the founding of Woods Hole Oceanographic Institution, Woods Hole 22–26 September 1980. Proceedings pub. as: **M. Sears & D. Merriman**, eds. 1980. *Oceanography: The past.* New York, Springer–Verlag; *see* Ca596.

A185. Fourth international congress on the history of oceanography, Hamburg 23–29 September 1987. Proceedings pub. as: **W. Lenz & M. Deacon**, eds. 1990. *Ocean sciences: Their history and relation to man. Proceedings [etc.]*. Hamburg, Bundesamt für Seeschiffahrt und Hydrographie; *see* Ca579.

Congrès International d'Histoire des Sciences (International Congress on the History of Science).

A186. **I Congress.** Paris, 20–25 May 1929. [*Actes etc.*]. *Archeion* 2: i–cix, 1929.

A187. **II Congress.** London, 30 June – 4 July 1931. [*Resumés etc.*]. *Archeion* 13 & 14, 1932.

A188. **III Congress.** Oporto & Coimbra, 30 Sept. – 6 Oct. 1934. *Actes, conférences et communications [etc.].* Lisbon, 1936. xlix+463 pp.

A189. **IV Congress.** Prague, 22–27 Sept. 1937. Proceedings not pub. at the time. *See* L. Novy & J. Folta, eds. 1972. The 4th International Congress of the History of Sciences, Prague, 1937. *Acta historiae rerum naturalium nec non technicorum* 6: 3–381.
Historical introduction and publication of the papers read at the congress.

A190. **V Congress.** Lausanne, 30 Sept. – 6 Oct. 1947. *Actes [etc.].* Paris, Hermann & Cie, 1948. 288 pp. *Collection des travaux de l'Académie Internationale d'Histoire des Sciences* no. 2.

A191. **VI Congress.** Amsterdam, 14–21 August 1950. *Actes [etc.].* Paris, Hermann & Cie, 1951–53. 2 vols., 722 pp. *Collection des travaux de l'Académie Internationale d'Histoire des Sciences* no. 6.

A192. **VII Congress.** Jerusalem, 4–12 August 1953. *Actes [etc.].* Paris, Académie Internationale d'Histoire des Sciences; Hermann & Cie, 1954. xii+661 pp. *Collection des travaux de l'Académie Internationale d'Histoire des Sciences* no. 8.

A193. **VIII Congress.** Florence–Milan, 3–9 September 1956. *Actes [etc.].* Vinci, Gruppo Italiano di Storia delle Scienze; Paris, Académie Internationale d'Histoire des Sciences; Hermann & Cie, 1958. 3 vols. *Collection des travaux de l'Académie Internationale d'Histoire des Sciences* no. 9.

A194. **IX Congress.** Barcelona & Madrid, 1–7 September 1959. *Actes [etc.].* Barcelona, Asociacíon para la Historia de la Ciencia Española, 1960. 730 pp. *Collection des travaux de l'Académie Internationale d'Histoire des Sciences* no. 12.

A195. **X Congress.** Ithaca–Philadelphia, 26 August–2 September 1962. *Actes [etc.]. Preceedings [etc.].* Paris, Hermann & Cie, 1965. 2 vols.

A196. **XI Congress.** Warsaw–Cracow, 24–31 August 1965. *Actes [etc.].* Wroclaw, Ossolineum, 1968. 6 vols.

A197. **XII Congress.** Paris, 25–31 August 1968. *Actes [etc.].* Paris, Blanchard, 1970. 12 vols. in 15.
Vol. VIII. *Histoire des sciences naturelles et de la biologie.*

A198. **XIII Congress.** Moscow, 18–24 August 1971. *Proceedings [etc.].* Moscow, "Nauka". 14 vols.

A199. **XIV Congress.** Tokyo–Kyoto, 19–27 August 1974. *Proceedings [etc.].* Tokyo, Science Council of Japan, 1975. 4 vols.
Vol. 3. Biological sciences, etc.

A200. **XV Congress.** Edinburgh, 10–15 August 1977. *Proceedings [etc.].* Ed. by E. G. Forbes. Edinburgh, Edinburgh Univ. Pr. 596 pp.

A201. **XVI Congress.** Bucharest, August 1981.

A202. **XVII Congress.** Berkeley, CA, 31 July–8 August 1985. *Abstracts [etc.].* Berkeley, CA, University of California, Office for the History of Science & Technology, 1985. 2 vols.

A203. **XVIII Congress.** Hamburg–Munich, 1–9 August 1989.

The **XIX Congress** will take place at Zaragoza, Spain from 22nd to 29th August, 1993.

Society for the Bibliography of Natural History (*later* (1983+) Society for the History of Natural History).

A204. **Easter Meeting.** London, 2–4 April, 1975. Session 1: Manuscripts and natural history; Session 2: Natural history and exploration; Session 3: Descriptive bibliography; Session 4: New trends in the history of natural history. Proceedings pub. in *Journal of the Society for the Bibliography of Natural History* 7(4), 1976, pp. 333–537. 15 papers (3 in abstract only).

A205. **2nd Easter Meeting.** London, 13–15 April, 1977. Session 1: Manuscripts and natural history; Session 2: Travel exploration; Session 3: Bibliography; Session 4: History of natural history. Proceedings pub. in *Journal of the Society for the Bibliography of Natural History* 8(4), 1978, pp. 329–536. 18 papers (5 in abstract only).

A206. **3rd Easter Meeting.** London, 3–6 April, 1979: "International conference on the history of museums and collections in natural history." [Session 1]: Collectors and collections in Europe; [Session 2]: Travellers and explorers; [Session 3]: Zoological gardens; [Session 4]: Books in the museum; [Session 5]: North American collectors and collections. Proceedings pub. in *Journal of the Society for the Bibliography of Natural History* 9(4), 1980, pp. 365–670. 25 papers.

A207. **Joint Conference.** (With the Systematics Association,) London 13–16 April 1981: "History in the service of systematics." Proceedings pub. as: Society for the Bibliography of Natural History. 1981. *History in the service of systematics: [etc.]*. London. *Special publication* no. 1; *see* Ca32.

A208. **5th Easter Meeting.** London, 28–31 March, 1983: "Natural history in the early nineteenth century." Proceedings pub. as: Society for the History of Natural History. 1985. *From Linnaeus to Darwin: Commentaries on the history of biology and geology [etc.]*. London. *Special publication* no. 3; *see* Ca73.

A209. **Jubilee symposium.** London, 16–21 March, 1986: "Nature surveyed, natural history – past, present and future." Session 1: (Opening exercises); Sessions 2–4: The naturalist and the community I(–III); Sessions 5–6: Visual presentation of natural history; Session 7: The spoken and written word; Session 8: Natural history in the twenty–first century. Programme & abstracts, 22 pp. 13 papers; some were pub. in *Archives of natural history*; *see* Ca1130.

A210. **6th Easter Conference.** London 21–22 April 1989: "Sources and techniques for biographical and bibliographical research." 10 papers; some were pub. in *Archives of natural history*; *see* Bc692 Bc805(Chalmers) D3 & D37.

A211. **7th International Conference.** London 25–27 April 1990: "Landmarks of natural history." Session 1: Exploring the natural world; Session 2: Ordering the kingdoms of nature; Session 3: Concepts and theories; Session 4: People and places; Session 5: "Tools" for knowledge. 17 papers; some were pub. in *Archives of natural history*.

A212. **8th International Conference.** London 29 April–1 May 1992: "Richard Owen centenary conference." *Programme* [and abstracts]. 15 pp.

North American conferences/meetings:

A213. **1st North American Conference.** Philadelphia, PA, 21–23 October, 1981: "Contributions to the history of North American natural history." Proceedings pub. as: Society for the Bibliography of Natural History. 1983. *Contributions to the history of North American natural history: [etc.]*. London. *Special publication* no. 2; *see* Ca229.

A214. **2nd North American Conference.** San Diego, CA, 9–11 May, 1984: "Exploration of North America west of the Mississippi."

A215. **3rd North American Conference.** Joint meeting with the History of Science Society and the Society for the History of Technology, Raleigh, NC, 9–11 May, 1987: "Natural history in the American South."

A216. **4th North American Meeting.** Joint conference with the National Zoological Park and the Smithsonian Institution, Washington, DC, 13 & 14 October, 1989: "International conference on the history and evolution of modern zoos."

A217. **5th North American Meeting.** Joint meeting with the American Society of Ichthyologists & Herpetologists on the occasion of its 70th anniversary, Charleston, SC, 18–19 June, 1990: "Collection building in ichthyology and herpetology in the 18th, 19th and 20th centuries."

A218. **6th North American Meeting.** Joint meeting with the History of Science Society and the Society for the History of Technology, Madison, WI, 31 October– 3 November, 1991: "Natural history in the Middle West: The nineteenth century."

Central European Meetings:

A219. **1st Central European Meeting.** Hamburg, University Zoological Museum, 5 August 1989.

A220. **2nd Central European Meeting.** Hamburg, Forderkreis der naturwissenschaftlichen Museen Berlins e. V, 5–6 December 1990.

A221. **3rd Central European Meeting.** Jena, Ernst-Haeckel-Haus, 29–30 June 1991. Lecture and tours of the Haeckel Memorial Museum and the Museum of Evolution.

A222. **4th Central European Meeting.** Vienna, Naturhistorisches Museum, 4–8 November 1992. Co-host with the Museum of a meeting on: "Die Erforschung und Erschliessung Amerikas im Spiegel der Naturgeschichte." Proceedings to be published.

PART B — BACKGROUND BIBLIOGRAPHY

Section Ba. BIOGRAPHIES OF NATURALISTS

COLLECTED BIOGRAPHIES

• N.B. Membership registers and records of individual societies, associations, and academies are entered under the heading for the relevant society in the next section, "Histories of naturalists' societies", *see* Bb39–900.

NATURALISTS — General

Ba1. **Abbott, D.**, ed. 1983. *The biographical dictionary of scientists: Biologists.* London, Blond Educational. vi+182 pp.
Ca.200 lives.

Ba2. **Almquist, E.** 1931. *Grosse Biologen: Eine Geschichte der Biologie und ihre Erforscher.* Munich, J. S. Lehmann. 143 pp.
Built round the lives of 23 biologists, viz. G. Mendel, M. J. Schleiden, T. Schwann, A. Meyer, Virchow, A. De Bary, S. Schwendener, O. Brefeld, Koch, K. E. von Baer, C. W. von Nägeli, J. L. Pasteur, A. Jordan, A. Gobineau, W. Harvey, C. Darwin, F. Bacon, J. S. Mill, C. Linnaeus, E. C. Hansen, Aristotle, H. de Vries and T. Smith.

Ba3. **Arnim, M.** 1943–52. *Internationale Personalbibliographie, 1800–1943.* 2. Aufl. Leipzig, Hiersemann. 2 vols.
Lists separately published biographies, articles, entries in biographical dictionaries, etc. Emphasis on German subjects. *Replaces* 1st ed., 1936.

Ba4. ___. 1961–63. ___. Bd. III. *1944–59 und Nachträge zur 2. Auflage [etc.].* By G. Bock & F. Hodes. Stuttgart, Hiersemann. xi+659 pp.; *Continued* ___. 1978–87. *1944–75: Fortgeführt Auflage von Bd. III (1944–1959) mit Nachtragen zur 2. Aufl von Bd, I/II (1800–1943).* By F. Hodes. Stuttgart, Hiersemann. 24 pts.

Ba5. **Asimov, I.** 1982. *Asimov's biographical encyclopedia of science and technology: The lives and achievements of 1510 great scientists from ancient times to the present chronologically arranged.* 2nd rev. ed. New York, Doubleday & Co. xxxv+941 pp., illus.
Replaces 1st ed., Garden City, NY, 1964. x+662 pp.

Ba6. **Barr, E. S.** 1973. *An index to biographical fragments in unspecialized scientific journals.* [Tuscaloosa, AL], Univ. of Alabama Pr. vii+294 pp.
Notices of 7,700 individuals cited in *American journal of science* 1890–1920, *Proceedings of the Royal Society of Edinburgh* 1832–1920, *Proceedings of the Royal Society of London* 1800–1933, *Nature* 1869–1918, *Popular science monthly* 1872–1915 and *Science* 1883–1919. *Replaces* ___. 1965. Nature's "Scientific worthies." *Isis* 56: 354–356. List of portraits of "worthies" pub. in *Nature*, 1873–1938.

Ba7. **Bayle, A. L. J., & A. J. Thillaye**, eds. 1855. *Biographie médicale par ordre chronologique d'après Daniel Leclerc, Eloy, etc Mise dans un nouvel ordre, revue et complétée.* Paris, A. Delahays. 2 vols. *Re-pub.* Amsterdam, Israel, 1967.
2,000 biographical sketches.

Ba8. **Bowker, R. R.** [Pub.]. 1983. *Biographical books, 1876–1949.* New York. lviii+1,768 pp.; ___. 1980. ___, *1950–1980.* New York. xlix+1,557 pp.
Limited to literature published or distributed in the United States.

Ba9. **British Library.** 1988. *International bibliography of biography 1970–1987.* Munich, K. G. Saur. 12 vols.
Biographies published throughout the world covering all spheres of activity and all walks of life. Compiled from MARC records created by the British Library and the Library of Congress and held at the British Library. *Replaces* ___, **Bibliographic Services Division.** 1985. *Bibliography of biography 1970–1984.* London, British Library. 44 microfiches.

Ba10. ___, **Bibliographic Services Division.** 1989+. *Bibliography of biography 1988 [etc.].* London.
Annual index arranged under biographee with author/title index.

Ba11. **Chambers, W. & R.** [Pub.]. 1983. *Chambers biographical encyclopedia of scientists.* Ed. by J. Daintith, S. Mitchell & E. Toothill. Edinburgh. 599 pp.
Replaces ___. 1951. *Chambers's dictionary of scientists.* Ed. by A. V. Howard. Edinburgh. 499 cols.

Ba12. **Cuvier, G. L. C. F. D.**, Baron. 1861. *Recueil des éloges historiques lus dans les séances publiques de l'Institut de France par G. Cuvier.* Nouv. ed. Paris, Firmin Didot Frères, Fils & Co. 3 vols.
Includes L.-J.-M. Daubenton, L.-G. Lemonnier, C.-L. l'Héritier, H.-F. Gilbert, J. Darcet, J. Priestley, J.-M. Cels, M. Adanson, P. M. A. Broussonnnet, P. Lassus, É.-P. Ventenat, C. Bonnet & H. B. de Saussure, A.-F. de Fourcroy, J.-C. Desessarts, H. Cavendish, P.-S. Pallas,

Parmentier & Rumford, G.–A. Olivier, J. Tenon, A.–G. Werner, N. Desmarests, M. de Beauvois, Sir J. Banks, M. Duhamel, R.–J. Haüy, Comte de Berthollet, C.–L. Richard, A. Thouin, Comte de Lacépède, MM. J.–N. Hallé, J.–N. Corvisart & P. Pinel, L.–F.–E. Ramond, L.–A.–G. Bosc, Sir H. Davy, L.–N. Vauquelin, M. de Lamarck, C.–A.–G. Riche, J.–G. Brugières, and J.–V.–M. Fabbroni. *Replaces* ___. 1819–27. ___. Paris, F. G. Levrault. 3 vols. The Paris, 1819–27 edition was *Re-pub.* Brussels, Culture et Civilisation, 1969. 3 vols. *See also* **D. Outram.** 1978. The language of natural power: The "éloges" of Georges Cuvier and the public language of nineteenth century science. *History of science* 16: 153–178. Bibliography of the biographical writing of Cuvier, pp. 165–178.

Ba13. **Daintith, J., S. Mitchell & E. Toothill.** 1981. *A biographical encyclopedia of scientists.* New York, Facts on File. 935 pp.

Ba14. **Debus, A. G.**, ed. 1968. *World who's who in science: A biographical dictionary of notable scientists from antiquity to the present.* Chicago, IL, Marquis Who's Who. xvi+1,855 pp.
Ca. 30,000 entries.

Ba15. **Delpar, H.**, ed. 1980. *The discoverers: An encyclopedia of explorers and exploration.* New York, McGraw–Hill Book Co. viii+471 pp.

Ba16. **Dunmore, J.** 1991. *Who's who in Pacific navigation.* Honolulu, HI, Univ. of Hawaii Pr. xvi+312 pp.
Selected bibliography, pp. 273–300.

Ba17. **Eloy, N. F. J.** 1778. *Dictionnaire historique de la médecine ancienne et moderne, ou mémoires disposés en ordre alphabétique pour servir à l'histoire de cette science, et à celle des médecins, anatomistes, botanistes, chirurgiens et chymistes de toutes nations.* Mons, Hoyois. 4 vols.
First pub. Frankfurt, 1756. Bio-biblio. of over 5,000 persons. For Italian ed. with corrections and additional entries see *Dizionario storico della medecina.* Naples, Gessari, 1761–65. 7 vols.

Ba18. **Embacher, F.** 1882. Die Forschungsreisenden aller Zeiten und Länder. In: *Lexikon der Reisen und Entdeckungen*, Abt. I, pp. 1–302. Leipzig. vi+394 pp. *Re-pub.* Amsterdam, Meridian Publishing Co., 1968.

Ba19. **Figuier, L.** 1870. *Vies des savants illustres depuis l'antiquité jusqu'au dix–neuvième siècle. Savants du XVIIIe siècle.* Paris, Hachette & Cie. 496 pp., illus.
Includes many naturalists.

Ba20. **Fischer, I.**, ed. 1932–33. *Biographisches Lexikon der hervorragenden Aerzte der letzten fünfzig Jahre.* 2. Aufl. Berlin, Urban & Schwarzenberg. 2 vols. *Re-pub.* Munich, [etc.], 1962.
Continuation of: A. Hirsch, ed. 1929–34 & 1935 covering 1880–1930; *see* Ba30.

Ba21. **Freund, H., & A. Berg**, eds. 1963. *Geschichte der Mikroskopie: Leben und Werk grosser Naturforscher.* Band I: *Biologie.* Frankfurt am Main, Umschau Verlag. xv+376 pp., illus. (ports).
Biographies, by various authors, include: E. Abbe, S. von Apáthy, K. E. von Baer, E. S. Becher, G. W. J. Behrens, K. Belar, T. Boveri, R. Chambers, C. E. McClung, L. Dippel, V. von Ebner, G. Haberlandt, P. Harting, O. Hertwig, H. F. von Heurck, R. Ritter von Höhnel, A. Köhler, H. Leitgeb, P. Mayer - A. Dohrn, C. S. Minot, H. von Mohl, H. Molisch, C. W. von Nägeli, M. J. Schleiden, A. Schneider, T. Schwann, S. Schwendener, S. Strugger, J. Ritter von Wiesner, and W. B. Wilson. (Bd. 2: *Medizin.* 1964.)

Ba22. **Fruton, J. S.** 1982. *A bio-bibliography for the history of the biochemical sciences since 1800.* Philadelphia, PA, American Philosophical Society. xiii+885 pp.; *Continued* ___. 1985. *A supplement to a bio-bibliography for the history of the biochemical sciences since 1800.* Philadelphia, PA, American Philosophical Society. vii+262 pp.
Replaces ___. 1974. *Selected bibliography of biographical data for the history of*

biochemistry since 1800. Philadelphia, PA, American Philosophical Society Library. viii+140 pp. *Publications* no. 6.

Ba23. **Garrison, F. H.** 1928. Available sources and future prospects of medical biography. *Bulletin of the New York Academy of Medicine* 1928 (May): 586–607. *Re-pub. in* ___. 1966. *Contributions to the history of medicine from the Bulletin of the New York Academy of Medicine, 1925–1935*, pp. 697–718. New York, Hafner Publishing Co.

Ba24. **Gascoigne, R. M.** 1984. *A historical catalogue of scientists and scientific books from the earliest times to the close of the 19th century*. New York, Garland Publishing. xxii+1,177 pp.

Ba25. **Gillispie, C. C.**, ed. 1970–80. *Dictionary of scientific biography*. New York, Scribner's Sons. 16 vols.; *Continued* **F. L. Holmes**, ed. [1990?] ___*: Supplement II*. New York, [etc.]. Vols. 17–18.

Vol. 16 is an index to people, periodicals, societies, universities, museums, medals, lectureships, prizes, countries, etc. Supplement II adds 20th century scientists to the dictionary. Future supplements are promised. Also *Concise dictionary of scientific biography*. New York, Scribner's Sons. 773 pp. Abridged from the full dictionary. *See also* **C. Scribner Jr.** 1980. Publishing the Dictionary of scientific biography. *Proceedings of the American Philosophical Society* 124: 320–322.

Ba26. **Greenwood, T.** 1886. *Eminent naturalists*. London, Simpkin, Marshall & Co. 202 pp., ports.

Biographical sketches of C. Linnaeus, G. Cuvier, G. Buffon, C. Lyell, R. Murchison, J. Lubbock, T. Edward and L. Agassiz.

Ba27. **Henze, D.** 1965. *Enzyclopaedie der Entdecker und Erforscher der Erde*. Graz, Akademische Druck– und Verlagsanstalt. 2 vols.

Ba28. **Herbert, M. C., & B. McNeil**, eds. 1980. *Biography and genealogy master index: A consolidated index to more than 3,200,000 biographical sketches in over 345 current and retrospective biographical dictionaries*. 2nd ed. Detroit, MI, Gale Research Co. 8 vols. *Gale biographical index series* no. 1; *Continued* ___. 1985. ___*: 1981–85*. Detroit, [etc.]. 5 vols.; ___. 1990. ___*: 1986–90*. Detroit, [etc.]. 3 vols.

Annual supplements, cumulated from time to time. The two cumulations published to date provide ca.2.3 and ca.1.9 million citations respectively. *Replaces* **Gale Research Co.** 1975. *Biographical dictionaries master index*. Detroit, MI. For an even more ambitious, but still far from complete, consolidated index *see* **J. P. Lobies et al.**, eds. 1973+. *Index bio-bibliographicus notorum hominum (IBN)*. Osnabrück, Biblio Verlag. Index to biographies in about 5,000 biographical dictionaries. Pars A: General introduction; pars B: List of evaluated bio-bibliographical works; pars C: "Corpus alphabeticum"; pars D: Supplement; pars E: General index of references.

Ba29. **Herzenberg, C. I.** 1986. *Women scientists from antiquity to the present: An index. An international reference listing and biographical directory of some notable women scientists from ancient to modern times*. West Cornwall, CT, Locust Hill Pr. xxxix+200 pp.

Ba30. **Hirsch, A.**, ed. 1929–34. *Biographisches Lexikon der hervorragenden Aerzte aller Zeiten und Völkern ... unter Spezial–Redaktion von E. Gurlt und A. Wernich, hrsg. von A. Hirsch*. 2. Aufl. Berlin, Urban & Schwarzenberg, 1929–24. 5 vols. *Re-pub.* Munich, Urban & Schwarzenberg, 1962. *Continued* ___. 1935 *Ergänzungsband: Nachträge zu den Bänden I–V*. By W. Haberling & H. Vierodt. Berlin, Urban & Schwarzenberg. 426 pp.

Prominent physicians working before 1880. Continued by: I. Fischer, ed. 1932–33; *see* Ba20.

Ba31. **Huard, P., & M. Wong.** 1965. Bio-bibliographie de quelques médecins naturalistes voyageurs de la marine au début du XIXe siècle. In: Colloque International sur l'Histoire de la Biologie Marine. *Les grandes expéditions scientifiques et la création des* *laboratoires maritimes [etc.]*, pp. 163–217 Paris. *Vie et milieu supplément* no. 19. *Includes* II. Notices bio-bibliographiques, pp. 182–215.

Ba32. **Hyamson, A. M.** 1951. *A dictionary of universal biography of all ages and all peoples.* 2nd ed., entirely re-written London, Routledge. 680 pp. *Re-pub.* Detroit, Gale Research Co., 1981.
Indexes some 110,000 entries in 24 standard biographical dictionaries of its day. *Replaces* 1st ed., 1916. 744 pp.

Ba33. **Inke, G.** 1968. Bibliographie der Bibliographien der Medizin und der in der Medizin angewandten Naturwissenschaften. *Wissenschaftliche Zeitschrift der Martin–Luther Universität Halle–Wittenberg* 1968 Beilage: xiii+658 pp.

Ba34. **Ireland, N. O.** 1962. *Index to scientists of the world from ancient to modern times: Biographies and portraits.* Boston, MA, F. W. Faxon Co. xliii+662 pp.
An index to the biographies of more than 7,475 scientists which appear in 338 English language books published from the 1930s to 1960. Also indicates the presence of portraits.

Ba35. **Krafft, F.**, ed. 1986. *Grosse Naturwissenschaftler: Biographisches Lexikon. Mit einer Bibliographie zur Geschichte der Naturwissenschaften.* 2. Aufl. Dusseldorf, V D I Verlag. 456 pp.
Replaces ___ **& A. Meyer–Abich**, eds. 1970. *Grosse Naturwissenschaftler: Biographisches Lexikon.* Frankfurt am Main, Fischer. 399 pp.

Ba36. **Leistner, O.** 1984–86. *Internationale Bibliographie der Festschriften von den Anfängen bis 1979 mit Sachregister.* 2. erw. Aufl. Osnabrück, Biblio-Verlag. 2 vols. Continued by: ***Internationale Jahresbibliographie der Festschriften:** I J B J.* <1980+> Osnabrück, Biblio-Verlag, 1982+.
Lists Festchriften alphabetically by person honoured. *Replaces* ___. 1976. *Internationale Bibliographie der Festschriften mit Sachregister.* Osnabrück, [etc.]. 893 pp.

Ba37. **McNeil, B., & M. C. Herbert**, eds. 1980. *Historical biographical dictionaries master index: A consolidated index to biographical information concerning historical personages in over 35 of the principal retrospective biographical dictionaries.* Detroit, MI, Gale Research Co. x+1,003 pp. *Gale biographical index series* no. 7.
Selected from the data base used to compile the *Biography and genealogy master index; see* Ba28.

Ba38. **Mayerhöfer, J., et al.** 1959–[75?]. *Lexikon der Geschichte des Naturwissenschaften: Biographien, Sachwörter und Bibliographien.* Vols. 1–2(8). Vienna, Gebrüder Hollinek.
Later continued in: *Archiv der Geschichte der Naturwissenschaften [etc.].* Ed. by R. Hink. Vienna, Gebrüder Hollinek, Heft 1+, 1980/81+. In progress. *See* Ba82.

Ba39. **Milner, R.** 1990. *The encyclopedia of evolution: Humanity's search for its origins.* New York, Oxford, Facts on File. xii+481 pp.
Includes numerous biographical articles.

Ba40. **Morton, L. T., & R. J. Moore.** 1989. *A bibliography of medical and biomedical biography.* Aldershot, Scolar Pr; Brookfield, VT, Gower Publishing. x+208 pp.
1,200 entries. *Replaces* **J. L. Thornton.** 1970. *A select bibliography of medical biography. With an introductory essay on medical biography.* 2nd ed. London, Library Association. 170 pp.

Ba41. **Moskovskoe Obshchestvo Ispytatelei Prirody.** 1940+. *Seriya istoricheskaya. [Historical series].* Moscow. No. [1]+.
Numbered series of biographical sketches of prominent Russian naturalists. Series continued till at least no. 47, 1952.

Ba42. **Neu, J.**, ed. 1980. Isis *cumulative bibliography 1966–1975: A bibliography of the history of science formed from* Isis *critical bibliographies 91–100, indexing literature*

published from 1965 through 1974. London, Mansell.
Vol. 1: *Personalities and institutions*, 1980. xxix+483 pp. Preceded by M. Whitrow, ed. 1971; *see* Ba56.

Ba43. **New York Academy of Medicine.** 1960. *Catalog of biographies in the library of the New York Academy of Medicine.* Boston, MA, G. K. Hall. 165 pp.
3,450 entries for biographies of physicians and scientists, mostly separately published.

Ba44. **Ogilvie, M. B.** 1986. *Women in science; Antiquity through the nineteenth century: A biographical dictionary with annotated bibliography.* Cambridge, MA, M I T Pr. xiii+254 pp.
Bibliography pp. 189–239.

Ba45. **Osborn, H. F.** 1928. *Impressions of great naturalists: Darwin, Wallace, Huxley, Leidy, Cope, Balfour, Roosevelt, and others.* 2nd ed. New York, C. Scribner's Sons. x+294 pp. & 15 port.
Replaces ___. 1924. *Impressions of great naturalists. Reminiscences of Darwin, Huxley, Balfour, Cope and others [etc.].* New York, C. Scribner. xxviii+216 pp. & 12 port.

Ba46. **Pelletier, P. A.**, ed. 1985. *Prominent scientists: An index to collective biographies.* 2nd ed. New York, Neal–Schuman. xxxiii+356 pp.
An index to the biographies of 12,211 scientists which appear in 262 English language books with primary emphasis on works published from 1960 through 1963, thus supplementing Ireland (1962). Includes a list of scientists by field. *Replaces* 1st ed., New York, 1980. xxxviii+311 pp.

Ba47. **Phillips, L. B.** 1889. *The dictionary of biographical reference containing over one hundred thousand names, together with a classed index of the biographical literature of Europe and America.* New ed., rev., corr. ... With a supplement to date by F. Weitenkampf. 3rd ed. London, Sampson Low, Son & Marston. xii+1,038 pp. *Re-pub.* Graz, Akademische Druck- und Verlagsanstalt, 1964; Detroit, MI, Gale Research Co., 1982.
Indexes 40 standard biographical dictionaries of its day and provides many names not included in Hyamson. The "Classed index of the biographical literature ..." *Includes* I. General biographies; II. National biographies; III. Class or particular biographies). *Replaces* 2nd ed., 1881. 1,036 pp.

Ba48. **Plesse, W., & D. Rux**, eds. 1977. *Biographien bedeutender Biologen: Eine Sammlung von Biographien.* Berlin, Volk und Wissen. 384 pp.

Ba49. **Poggendorf, J. C.** 1863+. *Biographisch–literarisches Handwörterbuch zur Geschichte der exakten Wissenschaften.* Leipzig, Barth; Berlin, Akademie Verlag.
Vol. 1–2, down to 1857. 1863; vol. 3, 1858–83. 1898; vol. 4, 1883–1903. 1904; vol. 5, 1904–22. 1926; vol. 6, 1923–31. 1936–40; vol. 7A, 1932–53 (German); vol. 7B, 1932– (latest seen: Teil 8, Lieferung 5, 1989, down to Vlès, F. M. R.). *Vols. 1–6 re-pub.* Ann Arbor, MI, Edwards, 1945. Brief biography followed by personal bibliography for each subject. *See also* **H. Salié.** 1963. Ein Standardwerk zur Geschichte der Naturwissenschaften: Hundert Jahre "Poggendorff." *Forschungen und Fortschritte* 37: 202–205; ___. 1966. Poggendorf and *Poggendorf.* (Translated by R. E. Oesper.) *Isis* 57: 389–392, 1966.

Ba50. **Royal Commonwealth Society.** 1961. *Biography catalogue of the library of the Royal Commonwealth Society.* By D. H. Simpson. London. xxiv+512 pp.

Ba51. **Sackmann, W.** 1985. *Biographische und bibliographische Materialien zur Geschichte der Mikrobiologie und zur bakteriologischen Nomenklatur.* Frankfurt am Main, Lang. 412 pp. *Marburger Schriften zur Medizingeschichte* no. 16.

Ba52. **Selander, S.** 1960. *Linnélärjunger i främmande länder. [Linnaean students in foreign countries].* Stockholm, Bonniers. 90 pp.

Ba53. **Thomson, J. A.** 1932. *The great biologists.* London, Methuen & Co. 173 pp.
A collection of short biographies of about 30 biologists from Aristotle to Lankester.

Ba54. **Waldman, C., & A. Wexler.** 1992. *Who was who in world exploration.* New York, Oxford, Facts on File. viii+712 pp.
Information on over 800 explorers, with index of "Explorers by region of exploration."

Ba55. **Wellcome Institute for the History of Medicine.** 1979–80. *Subject catalogue of the history of medicine and related sciences.* Biographical section. Nendeln, K T O Pr. [Kraus International Publications]. 5 vols.
A cumulation of entries in *Current work in the history of medicine* for 1954–77.

Ba56. **Whitrow, M.**, ed. 1971. Isis *cumulative bibliography 1913–1965: A bibliography of the history of science formed from* Isis *critical bibliographies 1–90, indexing literature published from 1913 through 1965.* London, Mansell.
Vols. 1–2: *Personalities and institutions*, 1971. 2 vols. Continued by J. Neu, ed. 1980; *see* Ba42.

Ba57. **Williams, T. I.** 1982. *A biographical dictionary of scientists.* Assistant editor S. Withers. 3rd ed. London, A. & C. Black. xiv+674 pp.
Replaces 1st ed., 1969. xii+592 pp.; 2nd ed., 1974. xv+641 pp.

Ba58. **Zavatti, S.** 1967. *Dizionnario degli esploratori e delle scoperte geografiche.* Milan, Feltrinnelli. vi+361 pp.
Dictionary list of explorers, with brief bibliographies, followed by a chronological list of expeditions arranged by region.

Ba59. **Zischka, G. A.** 1961. *Allgemeines Gelehrten–Lexikon: Biographisches Handwörterbuch zur Geschichte der Wissenschaften.* Stuttgart, A. Kröners Verlag. 701 pp. *Kröners Taschenausgabe* no. 306.

See also Ca82 & Ca461.

European (& Mediterranean)

Ba60. **Astre, G.** 1966. Catalogue de notices et de mémoires pour l'histoire des naturalistes et des sciences de la nature à Toulouse et dans la région. *Bulletin de la Société d'Histoire Naturelle de Toulouse* 101: 167–188.

Ba61. **Capparoni, P.** 1925–28. *Profili bio-bibliogragrafici di medici e naturalisti celebri Italiani dal sec. XV° al sec. XVIII°.* Roma, Istituto Nazionale Medico Farmacologico "Serono". 2 vols., ports.

Ba62. **Carlo, A. M.** 1932. *Ensayo de una bio-bibliografia de escritores naturales de las islas Canarias (siglos XVI, XVII y XVIII).* Madrid. 717 pp.

Ba63. **Feliksiaka, S.**, ed. 1987. *Slownik biologów polskich. [Dictionary of Polish biologists].* Warsaw, Panstwowe Wydawnictwo Naukowe. 618 pp.

Ba64. **Galton, F.** 1874. *English men of science, their nature and nurture.* 2nd ed. London, Macmillan. *Re-pub.* (with a new introduction by R. S. Cowan) London, F. Cass, 1970. xi+iii+xiv+270 pp. *Social history of science* no. 3.

Ba65. **Institut Istorii Estestvozneniya i Tekhniki, Akademiya Nauk SSSR.** 1949+. *Istoriya estestvoznaniya: Literatura, opublikovannaya v SSSR, 1917–1947[+].* Moscow, Akademii Nauk SSSR.
Periodic bibliographies. See sections "Personalii" in each volume for comprehensive listings of biographical literature on Soviet naturalists.

Ba66. ___. 1958–59. *Biograficheskii slovar deyatelei estestvozneniya i tekhniki. [Biographical dictionary of workers in natural science and technology].* Ed. by A. A. Zvorykin et al. Glavnaya redaktsiya Bol'shoi Sovetskoi Entsiklopedii. Institut Istorii [etc.]. Moscow. 2 vols.
Based on entries in the *Bol'shaya Sovetskaya Entsiklopediya*, 1949–60

Ba67. **Jäggli, M.** 1939. *Naturalisti ticinesi.* Bellinzona, 1939. 95 pp., 16 ports. First pub. in *Bollettino della Società Ticinese di Scienze Naturali* 34: 19–96, 1939.
Lives of about 20 naturalists. *Replaces* ___. 1936. *Naturalisti ticinesi.* Bellinzona, Instituto Editoriale Ticinese. 11 pp. & ports. Estratto dal II volume dell' opera "*Scrittori della Svizzera Italiani.*"

Ba68. **Jessen, J. C.** 1989. *Bibliographie der Autobiographien.* 3: *Selbstzeugnisse, Erinnerungen, Tagebücher und Briefe deutscher Mathematiker, Naturwissenschaftler und Techniker.* Munich, New York, K. G. Saur. 371 pp.

Ba68A. **Lindeboom, G. A.** 1984. *Dutch medical biography: A biographical dictionary of Dutch physicians and surgeons, 1475–1975.* Amsterdam, Rodopi. xxxi pp. & 2,244 cols.

Ba69. **Lindroth, S.**, ed. 1952. *Swedish men of science 1650–1950.* Translated from the Swedish by B. Anderson. Stockholm, Swedish Institute; Uppsala, Almqvist & Wiksell. 295 pp.
Biographies by various authors. *Includes* Lives of O. Rudbeck, N. Rosén von Rosenstein, C. von Linné, C. de Geer, C. P. Thunberg, S. Nilsson, E. Fries, S. Lovén, A. E. Nordenskiöld, H. Nilsson–Ehle.

Ba70. **Lopez Piñero, J. M., et al.** 1983. *Dicionario histórico de la ciencia moderna en España.* Barcelona, Ediciones Península. 2 vols. (1,128 pp.) *Seria universitaria: Historia/ciencia/sociedad* vols. 180–181.
Over 820 biographical sketches of scientists from the late C15th onwards.

Ba71. **Lunn, A. G.**, ed. 1983. *A history of naturalists in North East England.* Newcastle upon Tyne, Univ. of Newcastle upon Tyne, Department of Adult Education. 112+[4] pp.
Articles by various authors.

Ba72. **Meisen, V.**, ed. 1932. *Prominent Danish scientists through the ages with facsimiles from their works.* Copenhagen, Levin & Munksgaard; London, Oxford Univ. Pr. 195 pp. University of Copenhagen 450th anniversary.

Ba73. **Mieli, A.** 1921–23. *Gli scienziati Italiani dall'inizio del medio evo ai nostri giorni. Repertorio bio-bibliografico dei filosofi, matematici, astronomi, fisici, chimici, naturalisti, biologi, medici, geografi Italiani diretto da [etc.].* Rome, Casa editrice Leonardo da Vinci. Vol. 1 (pts. 1–4).

Ba74. **Praeger, R. L.** 1949. *Some Irish naturalists: A biographical note-book.* Dundalk, Dundalgan Pr. 208 pp.
300 individuals briefly covered.

Ba75. **Wagenitz, G.**, ed. 1988. *Göttinger Biologen 1737–1945: Eine biographisch-bibliographische Liste.* Göttingen, Vandenhoeck & Ruprecht. 228 pp. *Göttinger Universitätsschriften* Serie C: *Kataloge*, vol. 2.

See also Bb54 –55 Bb57–58 Bb72 Bb96 Bb110 Bb115 Bb121 Bb123 Bb127 Bb148 Bb156 Bb159 Bb178 Bb202–203 Bb205 Bb263 Bb266 Bb269 Bb311 Bb361 Bb387 Bb409 Bb411 Bb415 Bb431 Bb434 –435 Bb465 Bb472 Bb480 Ca283 Ca1007 Ca1010 & Ca1014–1015.

North American

Ba76. **Bell, W. J., Jr.** 1955. Fifty early American scientists: selected bibliographies. In: *Early American science: Needs and opportunities for study*, pp. 45–80. Williamsburg, VA, Institute of Early American History.
I.e. To about 1820.

Ba77. **Bonta, M. M.** 1991. *Women in the field: America's pioneering women naturalists.* College Station, TX, Texas A & M Univ. xix+299 pp.
Biographies of twenty five naturalists from Colonial to modern times.

Ba78. **Bowker, R. R.** [Pub.]. 1982. *American men and women of science. Editions 1–14. Cumulative index.* New York. 847 pp.
See **J. McK. Catell**; *see* Ba108.

Ba79. **Burgess, R. L.** 1981. Sources of biographical information on American ecologists. *Bulletin of the Ecological Society of America* 62: 236–255.

Ba80. **Clepper, H. E.**, ed. 1971. *Leaders of American conservation.* Sponsored by the National Resources Council of America. New York, Ronald Pr. Co. ix+353 pp.
A biographical dictionary.

Ba81. **Elliott, C. A.** 1979. *Biographical dictionary of American science: The seventeenth through nineteenth centuries.* Westport, CT, Greenwood Pr. xvii+360 pp.
Scientists whose birth dates span the period from 1606 to 1867. "... deliberately designed as a retrospective companion to American men of science (AMS) ... major entries for nearly 600 scientists never included in AMS and minor (or cross–reference) entries for some 300 other scientists who were included in the early editions of AMS." Introduction: "The scientist in American society," pp. 3–9. Five appendices provide an analysis of the the scientists by A. Year of birth, B. Place of birth, C. Education, D. Occupation, and E. Fields of science. *Also* ___. 1982. Models of the American scientist: A look at collective biography. *Isis* 73: 77–93.

Ba82. **___.** 1990. *Biographical index to American science: The seventeenth century to 1920.* Westport, CT, Greenwood Pr. xliii+300 pp. *Bibliographies & indexes in American history* no. 16.
Key to biographical entries in a wide range of reference books and periodicals.

Ba83. **Enros, P. C.** 1985. *A biobibliography of publishing scientists in Ontario between 1914 and 1939.* Thornhill, Ont., HSTC Publications. xxxvi+526 pp. *Research tools in the history of Canadian science & technology* no. 2.

Ba84. **Ewan, J.** 1955. San Francisco as a mecca for nineteenth century naturalists, with a roster of biographical references to visitors and residents. In: California Academy of Sciences. *A century of progress in the natural sciences, 1853–1953*, pp. 1–63. San Francisco, CA.

Ba85. **___.** 1950. *Rocky mountain naturalists: Biographies of nine leading naturalists of the Rocky mountain area supplemented by a roster in biographical dictionary form of natural history collectors, sixteen hundred and eighty two to nineteen hundred and thirty–two.* Denver, CO, Univ. Denver Pr. xiv+358 pp., ports.
I. Avenues of penetration to the Timpanogos and the North Pacific: An historical preview; II. Edwin James; III. John Charles Fremont; IV. Charles Christopher Parry; V. Edward Lee Greene; VI. Thomas Conrad Porter; VII. Harry Norton Patterson; VIII. Marcus Eugene Jones; IX. Eugene Penard; X. Theodore Dru Alison Cockerell; XI. Explanatory notes & bibliography; Roster [etc.].

Ba86. **___, J. & N. D.** 1981. *Biographical dictionary of Rocky Mountain naturalists: A guide to the writings and collections of botanists, zoologists, geologists, artists and photographers.* Utrecht, Bohn, Scheltema & Holkema. xvi+253 pp. *Regnum vegetabile* vol. 107.
Replaces The roster in *Rocky Mountain naturalists*, 1950.

Ba87. **Geiser, S. W.** 1948. *Naturalists of the frontier.* 2nd ed. Dallas, TX, Southern Methodist Univ. 296 pp.

Ba88. **___.** 1958–59. Men of science in Texas, 1820–1880. *Field and laboratory* 26: 86–139, 1958; 27: 20–48, 81–96, 111–160, 163–256,1959.

Ba89. **Holloway, L. M.** 1982. *Medical obituaries: American physicians' biographical notices in selected medical journals before 1907.* New York, Garland Publishing. xiv+513 pp. *Garland reference library of social science* vol. 104.

Ba90. **Judd, W. W.** 1979. *Early naturalists and natural history societies of London, Ontario.* London, Ont., Phelps. 215 pp.

Ba91. **Kaufman, M., S. Galishoff, & T. L. Savitt**, eds. 1984. *Dictionary of American medical biography.* Westport, CT, Greenwood Pr. 2 vols. (xvi+1,027 pp.)
Over 1,000 individuals from the 17th century to the 1970s.

Ba92. **Marquis Who's Who** [Pub.]. 1977. *Who was who in American history: Science and technology. A component of Who's who in American history.* Chicago, IL, Marquis Who's Who. xiii+688 pp.

Ba93. **Meisel, M.** 1924–29. A selected bibliography of the biographies and bibliographies of the principal American naturalists of the pioneer century and of colonial times. (And: Appendix to a selected bibliography [etc.].) In: *A bibliography of American natural history, the pioneer century, 1769–1865, [etc.]*, vol. 1: 156–244, vol. 3: 723–743. Brooklyn, NY, Premier Publishing Co. *Re-pub.* New York, Hafner, 1967.

Ba94. **Oehser, P. H.** 1938. A handlist of American naturalists, based on the Dictionary of American Biography. *American naturalist* 72: 534–546.

Ba95. **Sammons, V. O.** 1990. *Blacks in science and medicine.* New York, Hemisphere. xii+293 pp.

Ba96. **Siegel, P. J., & K. T. Finley.** 1985. *Women in the scientific search: An American bio-bibliography, 1724–1979.* Metuchen, NJ, & London, Scarecrow Pr. xvii+399 pp.
1,517 references classified by interest, including: Biologists (6), Botanists (26), Entomologists (6), Marine scientists (3), Ornithologists (3), and Zoologists (27).

Ba97. **Visher, S. S.** 1951. *Indiana scientists.* Indianapolis, IN, Indiana Academy of Science. 286 pp.

Ba98. **Wild, P.** 1979. *Pioneer conservationists of western America.* Missoula, MT, Mountain Pr. Publishing Co. xxv+246 pp., illus.
Includes 15 biographies.

Ba99. ___. 1986. *Pioneer conservationists of eastern America.* Missoula, MT, Mountain Pr. Publishing Co. xvi+280 pp., illus.
Includes 15 biographies.

See also Bb581 Bb588 Bb591 Bb628 Bb653 & Ca766.

South American

Ba100. **Arachavelata, J.** 1907. Naturalistas en el Uruguay. *Revista histórica de la Universidad Montevideo* 1: 478–506, 828–842.

Ba101. **Looser, G.** 1949. Biografías y bibliografías de naturalistas y antropólogos, principalmente de Chile, publicadas por don Carlos E. Porter. *Revista Chilena de historia y geografía* 113: 186–215.

Asian

Ba102. **Mirza, M. R.** 1986. Some senior naturalists of Pakistan. *Biologia (Lahore)* 32 Supp: 137–153.
Forty-four biographical notices.

Australasian

Ba103. **Moyal, A. M.** 1976. *Scientists in nineteenth-century Australia: A documentary history.* Stanmore, Cassell Australia Ltd. 280 pp.

OLD DIRECTORIES OF SCIENTISTS

• Selected titles, arranged by date of publication:

Ba104. **Gistel, J. von N. F. X.**, ed. 1856. *Die Naturforscher diess– und jenseits der Oceane. Reise– und Correspondenz–Handbuch für Geologen, Geognosten und Mineralogen, Botaniker, Zoologen ganz besonders, Entomologen, Anatomen, rationelle Aerzte, Astronomen, Physiker, Chemiker und Pharmazeuten.* Unter Mitwirkung von dreiundsechzig Naturforschern redigirt durch Johannes Gistel. Straubing, J. Schorner. 372 pp.

Ba105. **Putnam, F. W.**, ed. 1865. *The naturalists' directory: North America and the West Indies.* Salem, MA. 2 pts.; *Continued* ___. 1866. ___*: Appendix.* Salem, MA. 16 pp. ••First pub. in *Proceedings of the Essex Institute.*

Ba106. ***Naturalists' directory*** (or *Naturalist's universal directory*). Salem, Boston, MA, M. E. Cassino. 1877–1881, 1884, 1886–1890, 1894–1954. For the years 1882, 1885, 1892 & 1896 titled: *Scientists' international directory* (or *International scientists' directory*). From 1958–75 titled: *Naturalists' directory (international).* From 1978 onwards titled: *Naturalists' directory and almanac (international).* Baltimore, MD, World Natural History Publications (*later* Gainesville, FL, Flora & Fauna Publications). ••The 43rd ed., (1979–1980), 1978, which was the 100th anniversary edition, included a bibliographical history, pp. 295–298.

Ba107. **Plumb, C. S.** 1889. *A biographical directory of American agricultural scientists.* Knoxville, TN. 100 pp.

Ba108. **Catell, J. McK.** 1906. *American men of science: A biographical directory.* New York, Science Pr. ••Over 4,000 persons. Periodic updates, i.e. **R. R. Bowker** [Pub.]. 1992. *American men and women of science 1992–93.* 18th ed. New York. 8 vols. Vol. 8 is a discipline index. *Also* ___. 1982. *American men and women of science. Editions 1–14. Cumulative index.* New York. 847 pp.

Ba109. **Stephenson, H. H.**, ed. [1912–14]. *Who's who in science (international) 1912* (also *1913, 1914*). London, J. & A. Churchill. 3 vols. *Also* New York, Macmillan Co. ••All published.

Ba110. **Oldenburg, S. F., & E. F. Karsky.** 1923–34. *Nauka i nauchnye rabotniki SSSR. Spravochnik sostavlennyi komissiei "Nauka i Nauchnye Rabotniki SSSR." [Science and scientific workers of the USSR. Science guide book compiled by the commission [etc.].* Moscow, Izd. "Rabotnik Prosveshcheniya." 6 vols. ••Vol. 1: *Nauchnye rabotniki Petrograda. [Scientific workers of Petrograd].* xiii+164 pp.; vol. 2: *Nauchnye uchrezhdeniya Leningrada [Scientific institutions of Leningrad].* 1926. viii+407 pp.; vol. 3: [Not published?]; vol. 4: *Nauchnye rabotniki Moskvy. [Scientific workers of Moscow].* 1930. xx+570+62 pp.; vol. 5: *Nauchnye rabotniki Leningrada. [Scientific workers of Leningrad].* 1934. xx+723 pp.; vol. 6: *Nauchnye rabotniki SSSR bez Moskvy i Leningrada [Scientific workers of USSR excluding those of Moscow and Leningrad].* 1928. viii+810 pp.

Ba111. **Hirsch, G. C.**, ed. 1928. *Index biologorum: Investigatores, laboratoria, periodica.* Editio prima. Berlin, Springer. v+545 pp.

Ba112. **Verdoorn, F. & J. G.**, eds. 1945. Scientific institutions, societies and research workers in the Netherlands Indies. In: P. Honig & F. Verdoorn, eds. *Science and scientists in the Netherlands Indies*, pp. 425–460 <address list, pp. 429–460>. New York, Board for Netherlands Indies, Surinam and Curaçao.

Ba113. **Koetschau Verlag** [Pub.]. 1949. Handbuch der deutschen Wissenschaft. Berlin, Fr. K. Koetschau Verlag. 2 vols. ••Vol. 2: *Biographisches Verzeichnis.* Alphabetically arranged list with a subject index.

Ba114. **Martinez Fontez, E., & J. J. Parodiz.** 1949. *Guias de naturalistes sudamericanos; contiene: Nombres, direcciones y especialidades de profesionales y amateurs en ciencias naturales; instituciones y publicaciones de toda Sud América (incluyendo Centro América, Antillas y México).* Buenos Aires, Talleres Gráficos Lucania. 138 pp.

Ba115. **Scientific Council for Africa South of the Sahara.** 1954. *Systematic botanists and zoologists working in Africa. Botanistes et zoologistes effectuant des études systématique en Afrique.* Bukavu & London, Commission for Technical Co-operation in Africa South of the Sahara. viii+35 pp. *Publication* no. 13.

Ba116. **Telberg, I.** 1964. *Who's who in Soviet science and technology.* 2nd ed. Revised and enlarged by A. Dimitriev & V. G. Telberg. New York, Telberg Book Co. 301 pp.

Ba117. **Turkevich, J. & L. B.** 1968. *Prominent scientists of continental Europe.* New York, Elsevier. 204 pp.

BIBLIOGRAPHIES OF GENERAL BIOGRAPHICAL LITERATURE

Ba118. **Elliott, C. A.** 1990. Collective lives of American scientists: An introductory essay and a bibliography. In: E. Garber, ed. *Beyond history of science: Essays in honor of Robert E. Schofield,* pp. 81–104. Bethlehem, PA, Lehigh Univ. Pr.

Ba119. **Farrell, M. A.** 1979. *Who's whos: An international guide to sources of current biographical information.* New York, New York Metropolitan Reference & Research Library Agency. 102 pp. *METRO miscellaneous publication* no. 22.

Ba120. **Graves, D. J., J. Cargill & B. Alley.** 1986. *Biographical sources: A guide to dictionaries and reference works.* Phoenix, AZ, Oryx Pr. 146 pp.

Ba121. **Kaufman, I. M.** 1955. *Russkie biograficheskie i biobibliograficheskie slovari. [Russian biographical and biobibliographical dictionaries].* 2nd ed. Moscow, Gosudarstvennoe Izdatel'stvo Kul'torno–Prosvetitel'noi Literatury. 751 pp. ••Biologists pp. 141–157; scientific societies pp. 279–300. *Replaces* 1st ed., 1950. 331 pp.

Ba122. **Phillips, L. B.** 1889. Classed index of the biographical literature of Europe and America. In: *The dictionary of biographical reference [etc.].* New ed., rev., corr. ... With a supplement to date by F. Weitenkampf. 3rd ed. London, Sampson Low, Son & Marston. xii+1,038 pp. *Re-pub.* Graz, Akademische Druck- und Verlagsanstalt, 1964; Detroit, MI, Gale Research Co., 1982. ••*Includes* I. General biographies; II. National biographies; III. Class or particular biographies). *Replaces* 2nd ed., 1881. 1,036 pp.

Ba123. **Riches, P. M.** 1934. *Analytical bibliography of universal collected biography, comprising books published in the English tongue in Great Britain and Ireland, America and the British Dominions.* London, Library Association. ix+709 pp. ••Ca.56,000 names.

Ba124. **Slocum, R.,** ed. 1986. *Biographical dictionaries and related works: An international bibliography of more than 16,000 collective biographies, bio-bibliographies, ... bibliographies of biography, biographical indexes, and selected portrait catalogs.* 2nd ed. Detroit, MI, Gale Research Co. 2 vols. ••*Replaces* 1st ed., 1967 & supplement, 1972.

BOTANISTS (& plant collectors) — General

Ba125. **Barnhart, J. H.** 1965. *Biographical notes upon botanists, compiled by John Hendley Barnhart, bibliographer, 1903–1941, and maintained in the New York Botanical garden Library.* Boston, MA, Hall. 3 vols.

Reproduces a typewritten 44,700–card index. New entries were added down to the late 1940s. The original card index is now incorporated with the Hunt Institute for Botanical Documentation's own card file.

Ba126. **Bennigsen, A. von.** 1928. Botanische Ehrennamen bei Laub– und Nadelgehölzen die in Deutschland winterhart sind. *Mitteilungen der Deutschen Dendrologischen Gesellschaft* 39: 471–496.

Ba127. **Bilancioni, G.** 1906. Appendice: Compendiose biografie di botanici. In: *Dizionario di botanica generale, istologia, anatomia, morfologia, fisiologia, biologia vegetale*, pp. 793–906. Milan, U. Hoepli.

Ba128. **Boivin, B.** 1976. *Biography index to* Excerpta botanica, Sectio A, *volumes 1–20, 1959–1973.* Ottawa. 38 pp.

Ba129. **Britten, J., & J. E. Dandy.** 1958. *The Sloane herbarium: An annotated list of the* horti sicci *composing it; with biographical accounts of the principal contributors.* Based on records compiled by the late James Britten. With an introduction by Spencer Savage. Rev. & ed. by J. E. Dandy. London, British Museum (Natural History). 246 pp. & 35 pl.

Part 2, Biographical accounts of [some 400] contributors [etc.], pp. 81–231.

Ba130. **Brown, L. Bush–.** 1964. *Men with green pens: Lives of the great writers on plants in early times, Theophrastus, Dioscorides, Albertus Magnus, Bartholomaeus Anglicus, William Turner.* Philadelphia, PA, Dorrance & Co. [xi]+161 pp.

Ba130A. **Brummitt, R. K., & C. E. Powell**, eds. 1992. *Authors of plant names.* Kew, Royal Botanic Gardens. 732 pp.

Replaces R. D. Meikle, *Draft index of author abbreviations* [etc.]; *see* Ba147.

Ba131. **Clokie, H. N.** 1964. [List of ca. 1,600 collectors.] In: *An account of the herbaria of the Department of Botany in the University of Oxford.* London, Oxford Univ. Pr. 280 pp.

Ba132. **Gould, S. W., & D. C. Noyce.** 1965. *Authors of plant genera.* Volume 2 of the *International plant index.* New York, New York Botanical garden. 336 pp.

A computer prepared list of over 5800 authors who are believed to have named the genera and higher categories of the flowering plants. Reviewed, very unfavourably, by R. C. Rollins, *Rhodora* 68: 35–40, 1966.

Ba133. **Gray, A.** 1889. Biographical sketches. In: *Scientific papers of Asa Gray.* Selected by C. S. Sargent, vol. 2: 283–498. Boston, MA, & New York, Houghton, Mifflin & Co.

I.e. R. Brown & F. Humboldt, A. P. de Candolle, B. D. Greene, C. W. Short, F. Boott, W. J. Hooker, J. Lindley, W. H. Harvey, H. P. Sartwell, C. Dewey, G. A. W.–Arnott, N. B. Ward, M. A. Curtis, H. von Mohl, R. Wight, F. Welwitsch, J. Torrey, W. S. Sullivant, J. Wyman, D. Hanbury, A. Braun, C. Pickering, E. M. Fries, J. Bigelow, J. Carey, T. P. James, J. A. Lowell, C. Darwin, J. Decaisne, G. Engelmann, O. Heer, G. Bentham, A. Fendler, C. Wright, G. W. Clinton, E. Boissier, J. A. C. Rœper, L. Agassiz, & E. Tuckerman.

Ba134. **Harrison, S. G.** 1985. *Index of collectors in the Welsh National Herbarium.* Cardiff, National Museum of Wales. 140 pp.

Ba135. **Hedge, I. C., & J. M. Lamond.** 1970. *Index of collectors in the Edinburgh herbarium.* Edinburgh, Her Majesty's Stationery Office for the Department of Agriculture and Fisheries for Scotland. v+147 pp.

Part II: B. Geographical areas, references and vascular plant collectors; C. Cryptogamic collectors; Part III: Main list of [some 4,500] collectors.

Ba136. **Högrell, B.** 1886. Botanistförteckning med personalnotiser. In: *Botanikens historia in öfversigt*, pp. 220–287. Goteborg, H. L. Bolinder.

Ba137. **Hunt Institute for Botanical Documentation, Carnegie Mellon University.** 1972. *Biographical dictionary of botanists represented in the Hunt Institute*

portrait collection. By T. W. Bossert. Boston, MA, G. K. Hall. viii+451 pp.
Details of some 11,000 botanists.

Ba138. **Jackson, B. D.** 1901. A list of collectors whose plants are in the Herbarium of the Royal Botanic Gardens, Kew to 31st December 1899. *Bulletin of miscellaneous information, Royal Botanic Gardens, Kew* 1901: 1–80.
2,100 collectors listed.

Ba139. ___. 1922. Notes on a catalogue of the Linnaean herbarium. *Proceedings of the Linnean Society of London* 134 supp: 38 pp.
Includes details of ca. 150 collectors.

Ba140. **Lanjouw, J., F. Stafleu, [et al.].** 1954–88. *Index herbariorum.* Part II. *Collectors.* Utrecht, International Bureau for Plant Taxonomy and Nomenclature; Utrecht, A. Oosthoek's Uitgeversmaatschappij; Utrecht, Antwerp, Bohn, Scheltema & Holkema; The Hague, Boston, MA, W. Junk. 7 pts., 1,213 pp. *Regnum vegetabile* vols. 2, 9, 86, 93, 109, 114, 117.
Full names, dates, collecting localities with dates, herbaria where collections are deposited.

Ba141. **Lasègue, A.** 1845. *Musée botanique de M. Benjamin Delessert. Notices sur les collections de plantes et la bibliothèque qui le composent; contenant en outre des documents sur les principaux herbiers d'Europe.* Paris & Leipzig. vi+588 pp. *Re-pub.* (with an introduction by F. A. Stafleu) Lehre, J. Cramer; Codicote, Wheldon & Wesley, 1970. *Historiae naturalis classica* vol. 81.
Includes brief biographical notes on hundreds of collectors.

Ba142. **Lawrence, G. H. M.** 1970. *Plant explorers and their introductions; With a bibliography.* Pittsburgh, PA, Carnegie Mellon Univ. [i]+14 pp.

Ba143. **Lindemann, E.** 1863. Kurze Nachrichten über den Bestand meines Herbariums. *Bulletin de la Société Impériale des Naturalistes de Moscou* 36: 233–253; ___. 1872. Zweiter Bericht über den Bestand meines Herbarium. *Ibid.* 46: 56–101; ___. 1885. Dritter Bericht über den Bestand meines Herbarium. *Ibid.* 61: 37–92.
Includes details of the collectors, their travels, and the collections they contributed.

Ba144. **Lipskii, V. I.** 1908. [Lists of foreign and Russian collectors.] In: *Gerbarii Imperatorskogo S.-Peterburgskago Botanicheskago Sada (1823–1908)*, pp. 48–175. 2nd ed. Yur'yev, K. Mattisena.
Replaces 1st ed., St. Petersburg, 1898. 128+vi pp.

Ba145. ___. 1914. Biografii i literaturnaya deyatel'nost' botanikov i lits, soprikasavshikhsya s imp. Botanicheskim sadom. [Biographies and literary activity of the botanists and the persons connected with the Imperial Botanical Garden] In: A. A. Fischer von Waldheim, ed. *Imperatorskii S.-Peterburgskii Botanicheskii Sad za 200 let ego sushchestvovaniya (1713–1913)*, vol. 3, pp. 1–536. St. Petersburg.

Ba146. **Mears, J. A.** 1981. Guide to plant collectors represented in the Herbarium of the Academy of Natural Sciences of Philadelphia. *Proceedings of the Academy of Natural Sciences of Philadelphia* 133:141–165.
Over 1,250 collectors listed.

Ba147. **Miekle, R. D.** 1980. *Draft index of author abbreviations compiled at the herbarium, Royal Botanic Gardens, Kew.* Kew, Royal Botanic Gardens. v+249 pp.
Replaced by: R. K. Brummitt & C. E. Powell, eds., *Authors of plant names*; *see* Ba130A.

Ba148. **Murray, G., & J. Britten.** 1904. The Department of Botany. In: *The history of the collections contained in the Natural History Departments of the British Museum.* Vol. 1, pp. 77–193. London, British Museum.
Includes a list contributors to the herbarium with biographical details.

Ba149. **Nelmes, E., & W. Cuthbertson.** [1931.] *Curtis's botanical magazine dedications, 1827–1927, portraits and biographical notes.* London, B. Quaritch for the

Royal Horticultural Society. xxi+400 pp. incl. 99 ports.
One hundred dedicatees are described.

Ba150. **Reinikka, M. A.** 1972. Makers of orchid history. In: *A history of the orchid*, pp. 111–305. Coral Gables, FL, Univ. of Miami Pr.
Fifty–one biographies.

Ba151. **Stearn, W. T.** 1983. Some Crocus enthusiasts of the past. *Annales Musei Goulandris* 6: 117–137.
Biographical notes on seven botanists.

Ba152. **Steinberg, C. H.** 1977. The collectors and collections in Herbarium Webb. *Webbia* 32: 1–49.

Ba153. **Urban, I.** 1917. Geschichte des Königlichen Botanischen Museums zu Berlin–Dahlem (1815–1913), nebst Aufzählung seiner Sammlungen. *Beihefte zum botanischen Centralblatt* 34(1): 457 pp. *Also pub. separately* Dresden, 1916.
The history is followed by lists of over 4,000 collectors arranged by major plant groups with a geographical listing.

Ba154. **Wagenitz, G.**, ed. 1982. *Index collectorum principalium herbarii Gottingensis.* Göttingen, Systematisch–Geobotanisches Institut der Georg–August–Universität Göttingen. 214 pp.
A brief history of the herbarium; alphabetical list of collectors with references to biographical literature.

Ba155. **Zander, R., F. Encke, G. Buchheim, & S. Seybold.** 1984. Autoren der Pflanzennamen. In: *Handwörterbuch der Pflanzennamen.* 13. Aufl., pp. 634–764. Stuttgart, E. Ulmer.
A biographical dictionary of plant taxonomists arranged by the author abbreviation used in botanical nomenclature. *Replaces* ___. 1980. Autoren der Pflanzennamen. In: *Handwörterbuch der Pflanzennamen.* 12. Aufl., p.711–839. Stuttgart, Ulmer.

Ba156. [Not used.]

See also Ba386 Cb1032–1033 Cb1040 & Cb1176.

European (& Mediterranean)

Ba157. **Balsamo, F., & M. Geremicca.** 1913. Botanici e botanofili Napoletani, cenni biografici e storici. *Bulletino dell' Orto Botanico della Regia Università di Napoli* 3: 37–74.

Ba158. **Böhm, W.** 1988. *Göttinger Pflanzenbauwissenschaftler: Eine Bibliographie.* Regensberg, Maier–Ruppert. 75 pp.

Ba159. **Bonnet, E.** 1902. Essai d'une bio-bibliographie botanique de la Corse. *Compte rendu de l'Association Française pour l'Avancement des Sciences (Ajaccio, 1901, 2e partie)* 30: 415–434.

Ba160. **Briquet, J.** 1906. *Biographies de botanistes Suisses.* Geneva, H. Kündig. 175 pp. & 5 ports.
I.e. J. Roux, A. de Haller fil., L. Perrot, J.–L. Dupin, C.–I. Fauconnet, F.–S. Alioth.

Ba161. ___. 1940. Biographies des botanistes a Genève de 1500 a 1931. *Berichte der Schweizerischen Botanischen Gesellschaft* 50A: x+494 pp.

Ba162. **Burnat, E.** 1941. Botanistes qui contribué à faire connaître la flore des Alpes–Maritimes. 2nd edn revue et augmentée par F. Cavillier. *Mémoire de l'Association des Naturalistes de Nice et des Alpes–Maritimes* 5: 1–94.

Ba163. **Caspary, R.** 1912. Lebensbeschreibungen ost– und westpreussischer Botaniker. In: *Festschrift zum 50– jährigen Bestehen des Preussischen Botanischen Vereins*, pp. 189–290, ports. Berlin, Friedländer & Sohn.

Ba164. **Christensen, C.** 1924–26. Bibliografi: Kronologisk med korte biografier og systematisk. In: *Den Danske botaniks historie med tilhørende bibliografi*, vol. 2. Copenhagen, Hagerups.
Chronologically arranged author entries, each with a biography. Author index. *Also* **A. Hansen.** 1963. Supplerende data til biografier i Carl Christenen: *Den danske botaniske litteratur.* Nulevende forfattere uden væsentlig botanisk produktion efter 1939. *Dansk botanisk arkiv* 21 (1): 57–59.

Ba165. **Colmeiro y Penido, M.** 1858. Estudios biográficos. Notas biográficas de los botánicos Españoles y Portugueses, y tambien de algunos escritores no botánicos que contribuyeron en la península Hispano–Lusitana y en sus dominios al conocimiento de los vegetales; ordenadas cronológicamente desde el siglo primero hasta el dia. In: *La botánico y los botánicos de la península Hispano–Lusitana*, Parte 2: 141–206. Madrid, Rivadenegra. [xi]+216 pp.

Ba166. **Cox, E. H. M.** 1955. Plant collectors employed by the Royal Horticultural Society 1804–1846. *Journal, Royal Horticultural Society* 80: 263–280.

Ba167. **Daniel, L.** 1894–96. Recherches historiques sur les botanistes mayennais et leurs travaux. *Bulletin de la Société d'Études Scientifiques d'Angers* n.s. 23: 21–53, 1894; *Ibid.* 24: 31–64, 1895; *Ibid.* 25: 85–113, 1896.

Ba168. **Desmond, R. G. C.** 1977. *Dictionary of British and Irish botanists and horticulturists, including plant collectors and botanical artists.* With historical introduction by W. T. Stearn. London, Taylor & Francis. xxvi+747 pp.
New edition in preparation. See *Newsletter, Society for the History of Natural History* no. 31: 7, 1987. *Replaces* **J. Britten & G. S. Boulger.** 1931. *A biographical index of deceased British and Irish botanists.* 2nd ed. rev. & compl. by A. B. Rendle. London, [etc.]. xxii+342 pp.

Ba169. **Direction du Service de l'Écologie et des Espaces Verts de la Ville de Marseille.** 1982. *Les botanistes à Marseille et en provence du 16ème au 19ème siècle: Exposition, Palais de la Bourse de Marseille ... mars et avril 1982.* Marseille. 136 pp. & 12 pl.

Ba170. **Fitting, H.** 1897. Geschichte der hallischen Floristik. *Zeitschrift für Naturwissenschaften* 69: 289–386.
Biographical notes on 8 distinguished botanists.

Ba171. **Georlette, R.** 1949. *Quelques botanistes belges.* Brussels, Office de Publicité. 83 pp. *Collection nationale* no. 95.

Ba172. **Hadfield, M., R. Harling & L. Highton.** 1980. *British gardeners: A biographical dictionary.* London, Zwemmer. 320 pp.
500 biographies.

Ba173. **Hansen, A.** 1963. I. Biografier. In: Den Danske botaniske litteratur: Bibliographia botanica Danica 1940–1959 med tilhørende forfatter–biografier. *Dansk botanisk arkiv* 21 (1): 9–72.
A. Nulevende danske forfattere med væsentlig produktion efter 1939; B. Supplerende data til biografier i Carl Christenen: *Den danske botaniske litteratur.* Nulevende forfattere uden væsentlig botanisk produktion efter 1939; C. Afdøde forfattere. Data om botaniske forfattere afgået ved døden siden 1939.

Ba174. **Hein, W.-G., & H.-D. Schwarz.** 1975–78. *Deutsche Apotheker–Biographie.* Stuttgart, Wissenschaftliche–Verlagsgesellschaft. 2 vols.; *Continued* ___. 1986. ___: *Ergänzungsband.* Stuttgart, [etc.]. 466 pp. *Veröffentlichungen der Internationalen Gesellschaft für Geschichte der Pharmazie e. V.*, N.F. vols. 43, 46 & 55.

Ba175. **Höxtermann, E., et al.** 1985. Kurzbiographien und Porträts von Botanikern in der Geschichte der Berliner Universität. *Wissenschaftliche Zeitschrift der Humboldt Universität zu Berlin, mathematisch–naturwissenschaftliche Reihe* 34: 360–406.

Ba176. **Jansen, P., W. H. Wachter, & H. Engel.** 1939–43. Personalia I (–V). *Nederlandsch kruidkundig archief* 49: 437–452, 1939; *Ibid.* 50: 182–205, 1940; *Ibid.* 51: 340–379, 1941; *Ibid.* 52: 371–415, 1942; *Ibid.* 53: 232–257, 1943.
Biographical notes on over 1,000 Dutch botanists.

Ba177. **Jeffers, R. H.** 1968. Biographical notes. In: *The friends of John Gerard (1545–1612), surgeon and botanist. Index and biographical appendix*, pp. 13–45. Falls Village, CT, The Herb Grower Pr.
Notes on 280 people.

Ba178. **Kent, D. H., & D. E. Allen.** 1984. Collectors. In: *British and Irish herbaria: An index to the location of herbaria of British and Irish vascular plants*, pp. 78–281. London, Botanical Society of the British Isles.
Replaces **D. H. Kent, E. B. Bangerter, & J. E. Lousley.** 1957. Collectors. In: *British herbaria*, pp. 39–87. London, [etc.].

Ba179. **Klástersky, I., A. Hrabetová & J. Duda.** 1970. Botanikové na ceském a Moravskoslezském území od nejstarsích dob. *Zpravy Ceskoslovenské Spolecnosti pro Dejiny ved a Techniky pri Ceskoslovenské Akademii Ved* 14–15(34–35): 1–211.
Includes Floristicky vyzkum Cech (historicky nastin); Floristicky vyzkum Moravy (historicky nastin); Floristicky vyzkum Slezska (historicky nastin); Soupis botaniku Cech, Moravy a Slezska. About 900 botanists noted.

Ba180. **Letacq, A. L.** 1889. Notices sur quelques botanistes ornais et essais sur la bibliographie botanique du département de l'Orne. *Bulletin de la Société Linnéenne de Normandie* (4) 2: 228–291.
Biographies of eleven botanists.

Ba181. **Lipshits, S. Yu.** 1947–52. *Russkie botaniki: Biografo–bibliograficheskii slovar. Botanicorum rossicorum lexicon biographo-bibliographicum.* Moscow, Moskovskoe Obshchestva Ispytatelei Prirody. Vols. 1–4, A–K.
All published.

Ba182. **Losa Quintana, J. M., & M. A. Gimenez.** 1986. Los botanicos Españoles del siglo XIX. 1. Los botanicos de la Facultad de Farmacia de Granada. *Ars pharmacia (Granada)* 27: 363–369.

Ba183. **McCracken, E.** 1971. Some 19th century horticulturists. *Irish booklore* 1: 179–185.

Ba184. **Magnin, A.** 1906–12. Prodrome d'une histoire des botanistes Lyonnais. *Annales de la Société Botanique de Lyon* 31: 1–72, 1 bis –4 bis, 1906; *Ibid.* 32: 1–68, 103–141, 1907; *Ibid.* 35: 13–80, 1911; *Ibid.* 37: 205, 1912.
The first two parts also pub. separately: Lyon, Association Typographique, 1906. [i]+140+4 pp.

Ba185. **Oliver, F. W.**, ed. 1913. *Makers of British botany: A collection of biographies of living botanists.* Cambridge, Cambridge Univ. Pr. vi+322 pp. & ports.
Includes **S. H. Vines.** Robert Morison and John Ray, pp. 8–43, port.; **A. Arber.** Nehemiah Grew, pp. 44–64, port.; **F. Darwin.** Stephen Hales, pp. 65–83, port.; **T. G. Hill.** John Hill, pp. 84–107, port.; **J. F. Farmer.** Robert Brown, pp. 108–125, port.; **F. O. Bower.** Sir William Hooker, pp. 126–150, port.; **G. Henslow.** John Stevens Henslow, pp. 151–163, port.; **F. Keeble.** John Lindley, pp. 164–177, port.; **W. H. Lang.** William Griffith, pp. 178–191, port.; **F. W. Oliver.** Arthur Henfrey, pp. 192–203; **R. L. Praeger.** William Henry Harvey, pp. 204–224, port.; **G. Massee.** Miles Joseph Berkeley, pp. 225–232, port.; **W. B. Bottomley.** Sir Joseph Henry Gilbert, pp. 233–242, port.; **D. H. Scott.** William Crawford Williamson, pp. 243–260, port.; Sir **W. Thiselton–Dyer.** Harry Marshall Ward,

pp. 261–279, port.; **I. B. Balfour.** A sketch of the professors of botany in Edinburgh from 1670 until 1887, pp. 280–301; **F. O. Bower.** Sir Joseph Dalton Hooker, pp. 302–323, port.

Ba186. **Saccardo, P. A.** 1895–1901. Repertorio biografico e bibliografico dei botanici Italiani aggiuntivi gli stranieri che trattarono della flora Italiana and Repertorio biografico e bibliografico supplementare [etc.]. In: *La botanica in Italia: Materiali per la storia di questa scienza*, vol. 1, pp. 7–180, 220–233, vol. 2, pp. 1–132. Venice, C. Ferrari.
Includes geographical indexes. Cenni storici e bibliografici degli orti botanici pubblici e privati d'Italia, vol. 1, pp. 187–216; vol. 2, pp. 133–152. First pub. in *Memorie del R. Istituto Veneto di Scienze, Lettere ed Arte* vol. 25 (4) & vol. 26 (6).

Ba187. **Trochain, J.–L.** 1966. Botanistes du pays Toulousain. *Bulletin de la Société d'Histoire Naturelle de Toulouse* 102: 376–404.

Ba188. **Vernière, A.** 1901. *Les botanistes dans l'Auvergne & Le Velay.* Clermont–Ferrand, Malleval. 47 pp.

Ba189. **Weston, R.** 1773. A catalogue of English authors, who have written on husbandry, gardening, botany, natural history, and subjects relative thereto. Arranged in chronological order. In: *Tracts on practical agriculture and gardening ... To which is added, a chronological catalogue [etc.].* 2nd edn, greatly improved. London, S. Hooper. *Re-pub. as* ___. N.d. *Tracts, on experimental gardening and agriculture; with a catalogue [etc.].* 3rd ed. Leicester, J. Pares.
The catalogue forms a separately paginated section of [i]+136 pp. Actually a chronologically arranged, partly annotated bibliography covering 1480 to 1772 with an author/title index. Few biographical notes.

Ba190. **Zepernick, B., & F. K. Timler.** 1979. Die hauptamtlichen Mitarbeiter. Kurzbiographien. In: *Grundlagen zur 300 jährigen Geschichte des Berliner Botanischen Gartens*, pp. 28–120. Berlin, Botanischer Garten und Botanisches Museum Berlin–Dahlem. 303 pp. *Englera* vol. 1.

See also Bb710 Bb716–717 Cb192 Cb216 Cb341–Cb342 Cb354 Cb765 Cb826 Cb873A Cb887 & Cb1033.

North American

Ba191. **Bailey, L. H.** 1915 (1922). Horticulturists, North American. In: *The standard cyclopedia of horticulture.* Re-issue with corrections, vol. 3, pp. 1563–1603. New York, Macmillan. Various re-issues, e.g. "New" ed. 1937.
160 biographies of American horticulturists by various authors, with portraits.

Ba192. **Barkley, G. A.** 1947. Texas plant collectors as represented in Texas herbaria. *American midland naturalist* 38: 638–670.
List of 1,369 collectors, pp. 645–670.

Ba193. **Barnhart, J. H.** 1909. Some American botanists of former days. *Journal of the New York Botanical Garden* 10: 177–190. Also pub. in *Torreya* 9: 241–257, 1909.

Ba194. ___. 1919. Some fictitious botanists. *Journal of the New York Botanical Garden* 20: 171–180.
Details of 14 fictitious biographies in *Appleton's Cyclopedia of American Biography*, 1887–89.

Ba195. ___. 1921. Biographical notices of persons mentioned in the Schweinitz–Torrey correspondence. *Bulletin of the Torrey Botanical Club* 16: 290–300.
Also **C. L. Shear & N. E. Stevens**, eds. 1921. Correspondence of Schweinitz and Torrey. *Memoirs of the Torrey Botanical Club* 16: 119–289.

Ba196. ___. 1926. Brief sketches of some collectors of specimens in the Barton herbarium. *Bartonia* 9: 35–42.

Ba197. **Clapp, A.** 1935. Kansas plant hunters. *Biennial report, State Board of Agriculture, Kansas* 29: 175–190.
Biographies of eight collectors.

Ba198. **Ewan, J.** 1970. Plant collectors in America: Backgrounds for Linnaeus. In: P. Smit & R. J. Ch. V. ter Laage. *Essays in biohistory*, pp. 19–54. Utrecht, International Association for Plant taxonomy. *Regnum vegetabile* vol. 71.
Includes Roster of collectors and authors concerning America and Linnaeus, pp. 48–50.

Ba199. **Gee, W.** 1918. South Carolina botanists, biography and bibliography. *Bulletin of the University of South Carolina* 72: 3–52.
13 bio-bibliographies.

Ba200. **Geiser, S. W.** 1945. *Horticulture and horticulturists in early Texas*. Dallas, TX, Southern Methodist Univ. Pr. 100 pp.
Pt. 2. Biographies, 1838–1910.

Ba201. **Harshberger, J. W.** 1899. *The botanists of Philadelphia*. Philadelphia, privately pub. xii+457 pp.
Introduction *re-pub. in* **R. L. Stuckey**, ed. 1978. *Development of botany in selected regions of North America before 1900*. New York, Arno Pr.; *see* Cb263.

Ba202. **Humphrey, H. B.** 1961. *Makers of North American botany*. New York, Ronald Pr. xi+265 pp. *Chronica botanica* no. 21.
122 biographies.

Ba203. **Jepson, W. L.** 1928–31. The botanical explorers of California. *Madroño* 1: 167–170, 175–177, 183–185, 188–190, 214–216, 262–270, 1928–29; *Ibid.* 2: 25–29, 83–88, 115–118, 130–133, 156–157, 1931.
Biographies of 12 botanists.

Ba204. **Jones, K. L.** 1977. Our botanical forbears in Michigan. *Michigan academician* 9: 261–283.

Ba205. **Kelly, H. A.** 1914. *Some American medical botanists commemorated in our botanical nomenclature*. Troy, NY, Southworth Co. 216 pp., illus. (incl. many ports.) *Re-pub.* New York, D. Appleton & Co., 1929.
Biographies of: M. S. Sarrazin, J. Mitchell, C. Colden, J. Clayton, J. Bartram, A. Garden, A. Kuhn, M. Marshall, C. Wistar, B. S. Barton, D. Hosack, W. Baldwin, W. Darlington, M. Macbride, J. Bigelow, C. W. Short, J. Torrey, Z. Pitcher, C. Pickering, J. L. Riddell, G. Engelmann, A. W. Chapman, A. Gray, A. W. Saxe, C. C. Parry, E. C. Howe, W. Herbst, G. E. Post, J. T. Rothrock, & H. Hapeman.

Ba206. **Knobloch, I. W.** 1979. *The plant collectors of northern Mexico*. East Lansing, MI, Michigan State Univ., Latin American Studies Center. v+98 pp. *Latin American Studies Center monograph series* no. 17.

Ba207. **Lange, E. F.** 1956. Pioneer botanists of the Pacific northwest. *Oregon historical quarterly* 57: 109–124.

Ba208. **Langman, I. K.** 1955. Ensayo para una bibliografia historico–biografica de la botanica en Mexico. *Memorias, Academia Nacional de Ciencias "Antonia Alzate"* 57: 373–429.

Ba209. **McVaugh, R.** 1972. Index to names of collectors. In: Botanical exploration in Nueva Galicia, Mexico from 1790 to the present time. *Contributions from the University of Michigan Herbarium* 9: 213–316.
Over 300 biographical sketches.

Ba210. **Pennell, F. W.** 1935. Reference–list of collectors. In: The Scrophulariaceae of Eastern temperate North America [pp. 597–624]. *Academy of Natural Sciences of Philadelphia monographs* 1: 650 pp.

Ba211. ___. 1942–43. Botanical collectors of the Philadelphia local area. *Bartonia* 21: 38–57, 64, 1942; *Ibid.* 22: 10–31, 66, 1943.

Ba212. **Reifschneider, O.** 1964. *Biographies of Nevada botanists, 1844–1963.* Reno, NV, Univ. of Nevada Pr. 165 pp.
For notes and corrigenda see review by **J. Ewan** in *Madroño* 19: 30–32, 1967.

Ba213. **Rudolph, E. D.** 1990. Women who studied plants in the pre-twentieth century United States and Canada. *Taxon* 39: 151–205.
Analytical list of 1,454 women actively interested in plants before 1901.

Ba214. **Rydberg, P. A.** 1907. Scandinavians who have contributed to the knowledge of the flora of North America: A memoir prepared for the celebration, at Augustana College and Theological Seminary, of the 200th anniversary of the birth of Linné. *Augustana Library publications* 6: ii+49 pp.

Ba215. **Spaulding, P.** 1908–09. A biographical history of botany at St. Louis, Missouri. *Popular science monthly* 73: 488–499, 1908; *Ibid.* 74: 48–57, 124–133, 240–258, 1909.
Biographies of 20 botanists. *Re-pub. in* **R. L. Stuckey**, ed. 1978. *Development of botany in selected regions of North America before 1900.* New York, Arno Pr.; *see* Cb263.

Ba216. **Stieber, M. T.** 1984. From Edward to Robert Tatnall: Biographical remarks on some contributors to the *Flora of Delaware and the Eastern Shore* with a bibliography of its collectors. *Archives of natural history* 11: 415–429.

See also Cb246 & Cb264.

South & Central American (& Caribbean)

Ba217. **Barkley, F. A., & G. Gutierrez.** 1948. Colectores de plantas en Colombia tal como estan representados en los herbarios de pais. *Revista de la Facultad Nacional de Agronomía, Medellín* 8 (29/30): 85–107.

Ba218. **Ek, R. C.** 1990. *Index of Guyana plant collectors.* Koenigstein, Koeltz Scientiific Books. 85 pp. & map. *Flora of the Guianas, supplementary series* fasc. 1.

Ba219. **Funk, V. A., & S. A. Mori.** 1989. A bibliography of plant collectors in Bolivia. *Smithsonian contributions to botany* 70: iii+20 pp.

Ba220. **Hoehne, F. C.** 1941. Notas bio-bibliográficas de naturalistas botânicos que pretendemos homenagear com a denominação de caminhos e picadas no Jardim Botânico e na Estação Biológica do Alto da Serra. In: Secretaria da Agricultura, Indústria e Comércio de São Paulo, Departamento de Botânica do Estado. *O Jardim Botânico de Sãl Paulo. Precedido de prólogo histórico e notas bio-bibliográficas de naturalistas botânicos que trabalharam para o progresso do conhecimento da flora do Brasil, especialmente no Estado de São Paulo*, pp. 19–246. São Paulo.

Ba220A. **Jimenez, J. de J.** 1985. *Colectores de plantas de la Hispaniola.* Santiago de los Baballeros, UCMM. 196 pp., illus.

Ba221. **Nelson, E. C.** 1990. Early collectors of plants from Honduras and the confusion with British Honduras (Belize). *Taxon* 39: 568–571.

Ba222. **Parodi, L. R.** 1961. Ciento cincuenta años de botanica en la Republica Argentina. *Boletín de la Sociedad Argentina de Botánica* 9: 68 pp.
Includes biographical information on native botanists.

Ba223. **Prance, G. T.** 1971. An index of plant collectors in Brazilian Amazonia. *Acta Amazonica* 1(1): 25–64.

Ba223A. **Renner, S. S.** 1993. A history of botanical exploration in Amazonian Ecuador, 1739–1988. *Smithsonian contributions to botany* 82: 39 pp.
Includes Annotated list of [205] collectors, pp. 7–31.

Ba224. **Urban, I.** 1902. Notae biographicae peregrinatorum Indiae occidentalis botanicorum. In: *Symbolae antillanae seu fundamenta Florae Indiae occidentalis*, vol. 3: 14–158. *This section re-pub.* Amsterdam, Asher, 1964.
Includes Conspectus insularum earumque peregrinatorum botanicorum, pp. 147–158; about 750 biographical notices.

Ba225. ___. 1906. Vitae itineraque collectorum botanicorum, notae collaboratorum biographicae, Flora brasiliensis ratio endendi chronologicam systema, [etc.]. In: C. F. P. von Martius. *Flora brasiliensis: Enumeratio plantarum in Brasilia hactenus detectarum [etc.]*, vol. 1(i), cols.1–211. *Re-pub.* Weinheim, Cramer, 1965.

See also Cb281.

Asian

Ba226. **Borodin, I. P.** 1909. Kollektory i kollektsii po flore Sibiri. [Collectors and collections of Siberian flora]. *Trudy Botanicheskago Muzeya Imperatorskoi Akademii Nauk* 4: iv+245 pp.

Ba227. **Burkill, I. H.** 1927. Botanical collectors, collections and collecting places in the Malay Peninsula: A review of the work so far done towards a knowledge of the plant geography of Malaya. *Gardens' bulletin. Straits Settlements* 4: 113–202.

Ba228. **Fedchenko, B. A.** 1945. Issledovateli flory Irana. [The investigators of the flora of Iran.] *Botanicheskii zhurnal* 30: 31–43.

Ba229. **Lipskii, V. I.** 1905. Botanicheskie kollektsii iz Srednei Azii. Collectiones botanicae Asiae Mediae. In: Flora Srednei Azii ... Flora Asiae mediae seu Turkestaniae Rossicae inclusis chanatis Buchara et Chiwa. Pars III. *Trudy Tiflisskogo Botanicheskogo Sada* 7 (3): 338–841.

Ba230. **Steenis–Kruseman, M. J. van.** 1950. Malaysian plant collectors and collections: Being a cyclopaedia of botanical exploration in Malaysia and a guide to the concerned literature up to the year 1950. In: C. G. G. J. van Steenis, ed. *Flora malesiana.* Series 1, Spermatophyta. Vol. 1. Leyden, Noordhof. cl+639 pp. *Continued* ___. 1958. ___: Supplement 1. In: *Ibid.* Series 1. Vol. 5(4), pp. ccxxxvii–cccxlii. Leyden, Noordhoff; ___. 1974. ___: Supplement 2. In: *Ibid.* Series 1. Vol. 8(1), pp. i–cxv. Leyden, Noordhoff.

Ba231. **Stewart, R. R.** 1967. Plant collectors in West Pakistan and Kashmir. *Pakistan journal of forestry* 17: 337–363.

Ba232. **Wickens, G. E.** 1982. A biographical index of plant collectors in the Arabian peninsula (including Socotra). *Notes from the Royal Botanic Garden, Edinburgh* 55: 632–643.
Studies in the Flora of Arabia III.

See also Cb299 Cb300A.

Australasian

Ba233. **Hall, N.** 1984. *Botanists of Australian acacias: Short biographies of people who have named Australian species, whose names have been given to species, or who have collected type material [etc.].* Melbourne, Vic., Commonwealth Scientific & Industrial Organization. 64 pp.

Ba234. **Hunt Institute for Botanical Documentation, Carnegie Mellon University.** 1974. Hunt Institute collection of biographical sketches relating to Australian botany. By A. L. Karg. *Taxon* 23: 613–618.

Ba235. **Maiden, J. H.** 1907. A century of botanical endeavour in South Australia. *Report, Australian Association for the Advancement of Science* 1907: 1–42.

Ba236. ___. 1908. Records of Victorian botanists. *Victorian naturalist* 25(7): 101–117.

Ba237. ___. 1909? Records of Queensland botanists. *Report, Australian Association for the Advancement of Science* 10: 373–383.

Ba238. ___. 1909. Records of Western Australian botanists. *Journal of the Western Australia Natural History Society* 1909: 5–33.

Ba239. ___. 1909. Records of Tasmanian botanists. *Papers & proceedings of the Royal Society of Tasmania* 1909: 1–21.

Ba240. ___. 1908. Records of Australian botanists — (a) General, (b) New South Wales. *Journal & proceedings of the Royal Society of New South Wales* 42: 60–132; *Continued* ___. 1912. ___: First supplement. *Report, Australian Association for the Advancement of Science* 13: 224–243; ___. 1921. ___: Second supplement. *Journal & proceedings of the Royal Society of New South Wales* 55: 150–169.

Ba241. **Pearn, J.** 1990. *Medicine and botany: An Australian cadaster.* Brisbane, Qld., Amphion Pr. viii+172 pp., ports.
Australian plants and their medical eponyms.

Ba242. **Willis, J. H., et al.** 1986. Australian plants: Collectors and illustrators 1780–1980. *Research notes, Western Australian Herbarium* 12: 1–111.
Over 2,500 names with briefest biographical data.

See also Cb318.

African

Ba243. **Britten, J.** 1920. Some early Cape botanists and collectors. *Journal of the Linnean Society of London, botany* 45: 29–51.
Details of 18 collectors.

Ba244. **Exell, A. W., & G. A. Hayes.** 1968. A list of botanical collectors in the "Flora Zambesiaca" area. *Kirkia* 6: 85–104.

Ba245. **Friis, I.** [1988?] *Botanical collectors in Ethiopia.* 2nd ed. [Copenhagen, Kjøbenhavns Universitet, Botanisk Museum.] 45 pp., map.

Ba246. **Gomes e Sousa, A. de F.** 1940–61. Exploradores e naturalistas da flora de Moçambique. *Moçambique* 6: 111–126, 1940; *Ibid.* 7: 23–31, 1941; *Ibid.* 8: 53–67, 1942; *Ibid.* 9: 39–44, 1943; *Ibid.* 105: 25–33, 1961.

Ba247. **Gunn, M., & L. E. Codd.** 1981. Dictionary of plant collectors. In: *Botanical exploration of southern Africa: An illustrated history of early botanical literature on the Cape flora. Biographical accounts of the leading plant collectors and their activities in southern Africa from the days of the East India Company until modern times,* pp. 77–387. Cape Town, Balkema for the Botanical Research Institute. xv+400 pp.
Introductory volume to the *Flora of Southern Africa. Also* **L. E. Codd & M. Gunn.** 1985. Additional biographical notes on plant collectors in South Africa. *Bothalia* 15: 631–654.

Ba248. **Hepper, N. F., & F. Neate.** 1971. *Plant collectors in West Africa: A biographical index of those who have collected herbarium material of flowering plants and ferns between Cape Verne and Lake Chad, and from the coast to the Sahara at 18°n.* Utrecht, International Bureau for Plant Taxonomy. xvi+96 pp. *Regnum vegetabile* vol. 74.
Details of 1660 names. *Also* **F. N. Hepper.** 1971. A note on "Plant collectors in West Africa" – and a suggested standard format for similar indexes to other parts of Africa. *Mitteilung der Botanischen Staatsammlung München* 10: 510–511.

Ba249. **Letouzey, R.** 1968. Les botanistes au Cameroun. In: A. Aubreville, ed. *Flore du Cameroun*, 7: 1–110.
Histoire de la recherche floristique; Liste des collecteurs; Index géographique.

Ba250. **Verduyn den Boer, J. H.** 1929. *Botanists at the Cape.* I: *Biographical notices of Cape botanists with 21 portraits.* Cape Town, the Author. 71 pp.

See also Cb326.

Cryptogamic botanists, general

Ba251. **Sayre, G.** 1975. Collectors. In: Cryptogamae exsiccatae an annotated bibliography of exsiccatae of algae, lichenes, hepaticae, and musci. V. Unpublished exsiccatae. *Memoirs of the New York Botanical Garden* 19: 277–423.

See also Cb712.

Pteridologists

Ba252. **Underwood, L. M.** 1903–05. The early writers on ferns and their collections. *Torreya* 3: 145–150, 1903; *Ibid.* 4: 49–52, 145–150, 1904; *Ibid.* 5: 37–41, 1905.

Bryologists

Ba253. **Hagen, I. S.** 1897. Norges bryologi i det 18de århundrede. *Kongelige Norske Videnskabers Selskabs skrifter* 1897 (3): 195 pp.

Ba254. **Margadant, W. D., & L. Terken.** 1981. *Index of author abbreviations in the Index Muscorum.* Utrecht, Instituut voor Geschiedenis van de Biologie, R. U. U. 28 pp.
List of 1,100 authors, alphabetically arranged by standard abbreviation.

Ba255. **Reid, A. M. M.** 1987. Pioneer New England bryologists: A prosopography. *Occasional papers, Farlow Herbarium of Cryptogamic Botany* 19: 71 pp.
Data of 33 deceased bryologists.

Ba256. **Sayre, G.** 1977. Authors of names of bryophytes and the present location of their herbaria. *Bryologist* 80: 502–521.
Lists all known authors of taxa in Hepaticae and Musci.

Ba257. **___, C. E. B. Bonner & W. L. Culberson.** 1964. The authorities for the epithets of mosses, hepatics and lichens. *Bryologist* 67: 113–135. *Re-pub.* 1975.
A directory of some 1,150 most frequently cited authorities for the epithets of mosses, hepatics, and lichens with standardized abbreviations of the names for use in bibliographical citations.

Ba258. **Vitt, D. H., S. R. Gradstein & Z. Iwatsuki.** 1985. Bryological collectors. In: *Compendium of bryology: A world listing of herbaria, collectors, bryologists, and current research*, pp. 180–224. Braunschweig, Cramer.

See also Cb604.

Lichenologists

Ba259. **Grummann, V., & O. Klement.** 1974. *Biographisch–bibliographisches Handbuch der Lichenologie.* Lehre, Cramer. 44+839 pp.
3865 lichenologists, down to the early 1960's, includes bibliographies.

Ba260. **Hawksworth, D. L.** 1974. Fungal and lichen authors and their herbaria. In: *Mycologists handbook: An introduction to the principles of taxonomy and nomenclature in the fungi and lichens*, pp. 179–192. Kew, Commonwealth Mycological Institute.

Ba261. **Hertel, H.** 1980. Index collectorum lichenum, herbarii monacensis: Ein Sammler–Verzeichnis des Flechtenherbars der Botanischen Staatssammlung München. *Mitteilungen der Botanischen Staatssammlung München* 16: 333–462.

Ba262. **Laundon, J. R.** 1979. Deceased lichenologists: Their abbreviations and herbaria. *Lichenologist* 11: 1–26.

Ba263. **Sayre, G., C. E. B. Bonner & W. L. Culberson.** 1964. The authorities for the epithets of mosses, hepatics and lichens. *Bryologist* 67: 113–135. *Re-pub.* 1975; *see* Ba257.

Mycologists

Ba264. **Ainsworth, G. C., & P. M. Stockdale.** 1984. Bibliographical notices of deceased medical and veterinary mycologists. *Review of medical and veterinary mycology* 19: 1–13.

Ba265. **Hawksworth, D. L.** 1974. Fungal and lichen authors and their herbaria. In: *Mycologists handbook: An introduction to the principles of taxonomy and nomenclature in the fungi and lichens*, pp. 179–192. Kew, Commonwealth Mycological Institute.

Ba265A. **___.** 1980. Recommended abbreviations for the names of some commonly cited authors of fungi. *Review of plant pathology* 59: 473–480.

Ba265B. **___, B. C. Sutton, & G. C. Ainsworth.** 1983. [Biographical articles on 132 mycologists.] In: *Ainsworth & Bisby's Dictionary of the fungi.* 7th ed. Kew, Commonwealth Mycological Institute. xii+445 pp.

Ba266. **Rogers, D. P.** 1977. *A brief history of mycology in North America.* Amherst, MA, Second International Mycological Congress, Inc. [iii]+65 pp.
A series of biographical sketches.

Ba267. **Stevenson, J. A.** 1960. A list of authors of plant parasite names with recommended abbreviations. In: Index of plant diseases in the United States. *United States Department of Agriculture handbook* 165: 517–547.
Replaces ___. 1944. A preliminary list of authors of plant parasites with recommended abbreviations. *Plant disease reporter* 28: 366–395.

Ba268. **Wright, J. E., & R. J. Lois.** 1949. Lista de siglas de autores empleadas en micologia y fitopatologia. *Lilloa* 21: 225–269.

See also Cb645.

Microbiologists

Ba269. **Bojnansky, V.** 1971. *Bibliography of Czechoslovak plant virologists 1921–1970.* Bratislava, Czechoslovak Academy of Sciences, Slovak Society of Agricultural, Forest and Food Industry Science. 199 pp.

Ba270. **Clark, P. F.** 1961. *Pioneer microbiologists of America.* Madison, WI, Univ. of Wisconsin Pr. xv+369 pp.

Ba271. **Sackmann, W.** 1985. *Biographische und bibliographische Materialien zur Geschichte der Mikrobiologie und zur bakteriologischen Nomenklatur.* Frankfurt am Main, P. Lang. 412 pp. *Marburger Schriften zur Medizingeschichte* no. 16.

See also Ca559.

OLD DIRECTORIES OF BOTANISTS

• Selected titles, arranged by date of publication:

Ba272. **Leggett, W. H.**, ed. 1873. Botanical directory, for North America and the West Indies. *Bulletin of the Torrey Botanical Club* 4 (11): 49–59; ___. 1874. ___: Supplement. *Ibid.* 5 (12): 4 pp.

Ba273. ___. 1878. *Botanical directory for America, 1878.* New York. 16 pp. ••Alphabetical and geographical lists of botanists.

Ba274. **United States Department of Agriculture, Office of Experiment Stations.** 1890. List of botanists of the agricultural experiment stations in the United States, with an outline of the work in botany at the several stations. *Bulletin, Office of Experiment Stations, United States Department of Agriculture* 6: 23 pp.

Ba275. **Engelmann, W.**, [Pub.]. 1891. *Botanisches Adressbuch: Verzeichnis der lebenden Botaniker, sowie der botanischen Anstalten, Gesellschaften un Zeitschriften.* ([and] *Nachtrag*). Hrsg. von Fachmännern. Leipzig. 186+6 pp.

Ba276. **Dörfler, I.** 1896. *Botaniker–Adressbuch: Sammlung von namen und Adressen der lebenden Botaniker aller Länder, der botanischen Gärten und der die Botanik pflegenden Institute, Gesellschaften und periodischen Publikationen.* Vienna, the Author. 292 pp. ••Arranged geographically with an index of persons.

Ba277. ___. 1902. *Botaniker–Adressbuch: Sammlung von namen und Adressen der lebenden Botaniker aller Länder, der botanischen Gärten und der die Botanik pflegenden Institute, Gesellschaften und periodischen Publikationen.* 2. Aufl. Vienna, the Author. x+356 pp. ••Arranged geographically with an index of persons.

Ba278. ___. 1909. *Botaniker–Adressbuch: Sammlung von namen und Adressen der lebenden Botaniker aller Länder, der botanischen Gärten und der die Botanik pflegenden Institute, Gesellschaften und periodischen Publikationen.* 3. Aufl. Vienna, the Author. viii+450 pp. ••Arranged geographically with an index of persons.

Ba279. **Ugrinskii, K. A.** 1912. *Adresnaya kniga botanikov Rossii po dannym 1912 g.* Khar'kov, Izd. Avtora. 31 pp.

Ba280. **Imperial Bureau of Mycology.** 1925 [etc.]. *Mycologists resident in the British Empire ... 1925.* Kew. ••Issued for 1925, 1927, 1938, 1939, [etc?] Later issues published by Commonwealth Mycological Institute.

Ba281. **Zhadovskii, A. E.** 1929. *Adresnaya kniga botanikov SSSR.* Po porucheniyu Gos. Russkogo Botanicheskogo obshchestva Sostavlena A. E. Zhadovskim. *Index botanicorum URSS* [etc.]. Leningrad, izd. Gosudarstvennogo Russkogo Botanicheskkkogo Obshchestva. 178 pp. ••Gives place and date of birth, speciality, etc.

Ba282. **Diels, L., E. D. Merrill & T. F. Chipp**, eds. 1931. *International address book of botanists: Being a directory of individuals and scientific institutions, universities, societies, etc., in all parts of the world interested in the study of botany.* London, Baillière, Tindall & Cox for the Bentham Trustees. xv+605 pp. ••Some 11,000 botanists arranged geographically with an index of persons. Each country's entry includes a list of national societies and a place–arranged list of institutions and local societies.

Ba283. **Polunin, N.** 1950. Botanical research in Scandinavia; 375 contemporary workers and their special interests. *Canadian field naturalist* 64: 105–119.

Ba284. **Fosberg, F. R.** 1957. *List of botanists interested in Pacific basin problems.* Washington, DC, Pacific Science Association. [ii]+64 pp.

Ba285. **Roon, A. C. de.** 1958. *International directory of specialists in plant taxonomy with a census of* *their current interests.* Utrecht, International Bureau for Plant Taxonomy and Nomenclature. 266 pp. *Regnum vegetabile* vol. 13.

Ba286. **Canada Department of Agriculture.** 1959. *A directory of botanists in Canada.* By R. J. Moore. Ottawa. 99 pp.

Ba287. **Torre Florenzano, A.** 1960. *Directorio de botánicos de México (provisional).* San Luis Potosí, Universidad Autónoma Potosina. 85 pp. mimeographed.

Ba288. **Kupriyanova, L. A., & L. A. Aleshina.** 1961. *Adresnaya kniga palinologov Sovetskogo Soyuza.* Leningrad, AN SSSR, Vses. Bot. Obshchestvo. 84 pp.

Ba289. **Pacific Science Information Center.** 1963. *Pacific botanists1963.* Honolulu, HI. [118] pp. *Continued* ___. 1969. *Pacific botanists. Supplement 1969.* Honolulu, HI. 88 pp.

Ba290. **Hattori, S., & H. Matsumura.** 1964. *Address book of Japanese botanists 1964.* Tokyo, Botanical Society of Japan. 118 pp. ••2nd ed. Replaces *Address book of Japanese botanists 1951.*

Ba291. **Canada Department of Agriculture, Research Branch.** 1965. *Canadian botanists, 1965.* Ottawa. ii+55 pp.

Ba292. **Societatea de Stiinte Naturale si Geografie din Republica Socialista Romania.** 1966. *Indexul botanistilor din Republica Socialista Romania.* Ed. by I. Tarjan & G. Magureanu. Bucharest. 66 pp. *Biblioteca naturalistului* no. 2. ••List of 469 botanists with note of their date of birth, address and speciality.

Ba293. **Wornhardt, W. W.** 1968. *International directory of diatomists, with a census of their current interests.* Lehre, J. Cramer. ii+156 pp.

BIBLIOGRAPHIES OF BOTANICAL BIOGRAPHY

Ba294. **Boivin, B.** 1977. A basic bibliography of botanical biography and a proposal for a more elaborate bibliography. *Taxon* 26: 75–105.

ZOOLOGISTS (& anatomists) — General

Ba295. **Bogdanov, A. P.** 1888–92. Materialy dlya istorii nauchnoi i prikladnoi deyatel'nosti v Rossii po zoologii i soprikasayushchimsya s neyu otraslyam znaniya, preimushchestvenno za poslednee tridtsatipyatiletie (1850–1888). (Materials for the history of scientific and applied activities in Russia in zoology and related branches of knowledge especially during the last thirty–five years etc.) *Izvestiya Imperatorskogo Obshchestva Lyubitelei Estestvoznaniya, Antropologii i Etnografii pri Imperatorskom Moskovskom Universitete* 55(3): iv+222 pp. & 12 pl., 1888; *Ibid.* 57(4): 304 pp. & pl. 13–24, 1889; *Ibid.* 70(6): 310 pp. & pl. 25–37, 1891; *Ibid.* 71(bis): 126 pp. & pl. 33–41, 1892.
Biographical notices of about 400 Russian zoologists with portraits.

Ba295A. **Centre National d'Histoire des Sciences, Brussels.** 1990. *Bibliographie de savants belges, 1830–1914.* 1: *Biologie animale.* Sous la direction de H. Elkhadem. Brussels. xix+433 pp.

Ba296. **Clarke, E.** 1981. Biographical sketches. In: M. Neuberger. *The historical development of experimental brain and spinal chord physiology before Flourens.* Translated & ed. with additional material by E. Clarke, pp. 291–374. Baltimore, MD, John Hopkins Univ. Pr.
Sketches of over 350 individuals.

Ba297. **Cleevely, R. J.** 1983. Alphabetical index of collectors. In: *World palaeontological collections*, pp. 38–323. London, British Museum (Natural History); Mansell Publishing Ltd.
List of hundreds of collectors of palaeontological and related material, with brief biographical

notes and literature references. Partly based on: **Sherborn, C. D.** 1940. *Where is the ___ collection? [etc.]*. Cambridge, Cambridge Univ. Pr. 149 pp.

Ba298. **Dewsbury, D. A.**, ed. 1985. *Studying animal behaviour: Autobiographies of the founders.* Chicago, IL, Univ. of Chicago Pr. 519 pp. *Re-pub.* Chicago, [etc.], 1989.
Autobiographical accounts by G. P. Baerends, V. G. Dethier, I. Eibl–Eibesfeldt, J. L. Fuller, D. R. Griffin, H. Hediger, E. H. Hess, R. A. Hinde, J. A. King, P. Leyhausen, K. Z. Lorenz, A. Manning, P. Marler, C. P. Richter, J. M. Smith, J. P. Scott, N. Tinbergen, E. O. Wilson, & V. C. Wynne–Edwards.

Ba299. **Dobson, J.** 1962. *Anatomical eponyms: Being a biographical dictionary of those anatomists whose names have become incorporated into anatomical nomenclature, with definitions of the structures to which their names have been attached and references to the works in which they are described.* 2nd ed. Edinburgh & London, E. & S. Livingstone. [viii]+235 pp.
Replaces 1st ed., London, Baillière, Tindall & Cox, 1946. [x]+280 pp.

Ba300. **Hackethal, S.** 1985. Kurzbiographien und Porträts Berliner Zoologen. *Wissenschaftliche Zeitschrift der Humboldt Universität zu Berlin, mathematisch–naturwissenschaftliche Reihe* 34: 385–406.

Ba301. **Haymaker, W., & F. Schiller**, eds. 1970. *Founders of neurology: One hundred and forty–six biographical sketches by eighty–eight authors.* 2nd ed. Springfield, IL, Thomas. xxi+616 pp., ports.
Replaces **W. Haymaker.** [1953.] *The founders of neurology: One hundred and thirty–three biographical sketches prepared for the Fourth International Neurological Congress in Paris.* With bibliographical and editorial assistance of K. A. Baer. Springfield, IL, Thomas. xxvii+4799 pp., ports.

Ba302. **Huard, P., & M. Montagné.** 1944. Index biographique succinct des anatomistes, morphologistes, anthropologistes et naturalistes. *Travaux de l'Institut Anatomique, Faculté de Médecine de l'Indochine* 9: 1–37. *Also* **P. Huard & M. Montagné.** 1944. Index biographique concernant les sciences biologiques et médicales. *Ibid.* 9: 39–104.

Ba303. **Jardine,** Sir **W.**, ed. 1840. [Forty one biographies of naturalists, each with an engraved portrait.] In: *The naturalist's library.* Edinburgh, Lizars. 40 vols. *Re-pub.* Edinburgh, [etc.]., [1845–46]; London, H. Bohn, 1848; etc.
Lives of U. Aldrovandi, Aristotle, F. de Azara, J. Banks, J. Barclay, T. Bewick, J. Bruce, G. L. L. de Buffon, J. L. Burckhardt, P. Camper, G. L. C. F. D. Cuvier, D. Drury, C. De Geer, C. Gesner, A. von Haller, F. Huber, F. H. A. von Humboldt, J. Hunter, B. G. E. de la V. Lacépède, Lamarck, J. B. P. A. de M. de Latreille, F. Le Vaillant, C. Linnaeus, M. S. Merian, P. S. Pallas, T. Pennant, F. Péron, C. Pliny, S. T. Raffles, J. Ray, G. Rondelet, I. Salviani, R. H. Schomburgk, R. Sibbald, H. Sloane, W. Smellie, J. Swammerdam, J. Walker, A. G. Werner, F. Willughby, & A. Wilson. For details of authorship of these lives *see* **S. Sheets–Pyenson.** 1981. War and peace in natural history publishing: The Naturalist's library, 1833–1843. *Isis* 72: 50–72 (esp. "Table 2. Bibliographic information," pp. 71–71). *A selection re-pub. as* **Anon.** 1840. *Lives of eminent naturalists, with engraved portraits accompanying each.* Edinburgh, W. H. Lizars. Various paging. (I.e. Sloane, Pliny, Werner, Camper, Hunter, & Merian).

Ba304. **Jeudy, R. L.** 1941. *Les principaux anatomistes français du XVIIe siècle.* Paris, Le François. [viii]+106+[2] pp. Thèse médecine, Paris, 1941, no. 255.

Ba305. **Kemper, W. H.** 1905. *The world's anatomists: Concise biographies of anatomic masters, from 300 B.C. to the present time, whose names have adorned the literature of the medical profession.* Revised and enlarged from the original serial publication from the Medical Book News. Philadelphia, PA, Blakiston's Son & Co. xiii+79 pp., illus.

Ba306. **Königliche Friedrich–Wilhelms–Universität, Museum für Naturkunde.** 1896. *Liste der Autoren zoologischer Art– und Gattungs–Namen;*

zusammengestellt von den Zoologen des Museums für Naturkunde in Berlin. Ed. 2. Berlin. 68 pp.
Names of authors with abbreviations. *Replaces* ___. 1888. *Liste der Autoren zoologischer Artbegriffe zusammengestellt für die Zoologische Sammlung des Königlichen Museums für Naturkunde in Berlin.* Berlin. [iv]+87 pp.

Ba307. **Lambrecht, K., & W. & A. Quenstedt.** 1938. *Paleontologi. Catalogus bio-bibliographicus.* The Hague, W. Junk. xx+495 pp. *Fossilium catalogus I. Animalia* pt. 72.

Ba308. **Leake, C. D.**, ed. 1956. *Some founders of physiology: Contributors to the growth of functional biology.* Washington, DC, American Physiological Society. x+122 pp.
Short biographies of great physiologists, prepared for the XXth International Physiological Congress, Brussels, 30 July – 4 August, 1956.

Ba309. **MacGillivray, W.** 1834. *Lives of eminent zoologists, from Aristotle to Linnaeus; with introductory remarks on the study of natural history and occasional observations on the progress of zoology.* 2nd ed. Edinburgh, Oliver & Boyd; London, Simpkin & Marshall. 391 pp., port.
Includes Aristotle, Pliny the Elder, C. Gesner, P. Belon, I. Salviani, G. Rondelet, U. Aldrovandi, J. Jonston, J. Goedart, F. Redi, J. Swammerdam, J. Ray, R. Réaumur, C. Linnaeus Pat., & C. Linnaeus Fil.

Ba310. **O'Connor, W. J.** 1991. *British physiologists, 1885–1914: A biographical dictionary.* Manchester, Manchester Univ. Pr. xvii+582 pp.
Replaces ___. 1988. *Founders of British physiology: A biographical dictionary, 1820–1885.* Manchester, [etc.]. ix+278 pp.

Ba311. **Pilleri, G., & L. Arvy.** 1981. *The precursors in cetology from Guillaume Rondelet to John Anderson.* Berne, Brain Anatomy Institute. 258 pp., ports. *Investigations on cetacea* vol. 12.

Ba312. **Renard–Payen, D.** 1971. *Contribution à la bio-bibliographie de quelques anatomistes du XIXème siècle.* Thèse. Rennes. 97 pp.

Ba313. **Sarjeant, W. A. S.** 1980. *Geologists and the history of geology: An international bibliography.* New York, Arno Pr. 5 vols. *Re-pub.* Melbourne, FL, Krieger Publishing Co., 1986. *Continued* ___. 1987. ___: *Supplement 1979–1984 and additions.* Melbourne, FL, [etc.]. 2 vols.
Includes ... Individual geologists and palaeontologists (over 10,000 references); Index of geologists by nationality and country; Index of geologists by speciality; ... Appendix of women geologists.

Ba314. **Schrader, G. W., & E. Hering**, eds. 1863. *Biographisch–literarisches Lexicon der Thierärzte aller Zeiten und Länder, sowie der Naturforscher, Aerzte, Landwirthe, Stallmeister, u.s.w., welche sich um die Thierheilkunde verdient gemacht haben.* Stuttgart, Ebner & Seubert. x+490 pp. *Re-pub.* Leipzig, Zentralantiquariat der DDR, 1967.

Ba315. **Stirling, W.** 1902. *Some apostles of physiology, being an account of their lives and labours, [labours that have contributed to the advancement of the healing art as well as the prevention of disease].* London, privately pub. iv+129 pp. & 32 pl. *Re-pub.* London, 1966.

Ba316. **Swainson, W.** 1840. Biography of zoologists, and notices of their work. In: *Taxidermy, bibliography and biography*, pp. 98–386. London, Longman, Orme, Brown, Green & Longmans. *Lardner's cabinet cyclopaedia.*

Ba317. **Visser, R. P. W.** 1975. Dutch palaeontologists of the 18th century. *Janus* 62: 125–149.

See also Bb46 Bb111(Ehlers) & Cc266.

Ornithologists

Ba318. **Bahr, P. Manson–.** 1959. Recollections of some famous British ornithologists. *Ibis* 101: 53–64.
Notes on A. Newton, A. Chapman, H. E. Dresser, R. B. Sharpe, W. E. Clarke, & E. G. B. Meade–Waldo.

Ba319. **Gebhardt, L.** 1964. *Die Ornithologen Mitteleuropas.* Giessen, Brühlscher Verlag. 404 pp.
1,250 biographical sketches.

Ba320. **Gladstone, H. S.** 1946. British birds named after persons. *Transactions & journal of proceedings of the Dumfriesshire & Galloway Natural History & Antiquarian Society* (3) 23: 175–189.

Ba321. **Gruson, E. S.** 1972. *Words for birds: A lexicon of North American birds with biographical notes.* New York, Quadrangle Books. xiv+305 pp.

Ba322. **Hume, E. E.** 1942. *Ornithologists of the United States Army Medical Corps. Thirty six biographies.* Baltimore, MD, Johns Hopkins Univ. Pr. xxv+583 pp., 110 illus. *Re-pub.* New York, Arno Pr., 1978.

Ba323. **Mearns, B. & R.** 1988. *Biographies for birdwatchers: The lives of those commemorated in Western Palearctic bird names.* London, Academic Pr. xx+490 pp.

Ba323A. **___.** 1992. *Audubon to Xántus: The lives of those commenorated in North American bird names.* London, Academic Pr. xix+588 pp.
Companion works including over 200 biographies.

Ba324. **Miller, R. F.** 1966. *Biographies of people for whom birds have been named.* Baldwin City, KS, Baldwin Bird Club. 5 vols.

Ba325. **Palmer, T. S.** 1928. Notes on persons whose names appear in the nomenclature of California birds. *Condor* 30: 261–307.

Ba326. **___., et al.** 1954. *Biographies of members of the American Ornithologists' Union. Reprinted from the Auk, 1884–1954.* Washington, DC, privately pub. 630 pp.

Ba327. **Prestwich, A. A.** 1963. *"I name this parrot ..." Brief biographies of men and women in whose honour commemorative names have been given.* [2nd ed.] Edenbridge, Kent, privately pub. [vi]+118+[43] pp.
Replaces 1st ed., London, privately pub., 1958. [v]+86 pp.

Ba328. **Wynne, O. E.** 1969. *Biographical key — names of birds of the world — to authors and those commemorated.* Sandleheath, Hants, privately pub. iv+246 pp.
Provides a guide to at least one source of biographical information for most of the individuals listed. *Replaces* Provisional ed., 1966.

See also Bb826 Bb831 Cc343 & Cc394.

Herpetologists & ichthyologists

Ba329. **Adler, K.**, ed. 1989. *Contributions to the history of herpetology, issued to commemorate the 1 World Congress of Herpetology, Canterbury.* Oxford, OH, Society for the Study of Amphibians & Reptiles. 202 pp., illus. *Contributions to herpetology* no. 5.
Includes **___.** Herpetologists of the past, pp. 5–141 (152 biographies); **J. S. Applegarth.** Index of authors in taxonomic herpetology, pp. 143–178; **R. Altig.** Academic lineages of doctoral degrees in herpetology, pp. 179–202.

Ba330. **McCoy, C. J.** 1989. Index to the biographies of herpetologists and ichthyologists in *Copeia*, 1913–1988. *Smithsonian Herpetological Information Service* 78: 33 pp.

Ba331. **Wright, A. H.** 1949. Scientific and popular writers on American snakes (1517–1944): A check list and short bibliography. *Herpetologica* 5 supp. 1: 1–55.

Entomologists & arthropodologists

Ba332. **Conci, C.** 1975. Repertorio delle biografie e bibliografie degli scrittori e cultori italiani di entomologia. *Memorie della Società Entomologica Italiana* 48: 817–1078.

Ba333. **Gilbert, P.** 1977. *A compendium of the biographical literature on deceased entomologists.* London, British Museum (Natural History). [xiv]+455 pp.; *Continued* **R. Gaedike.** 1985. Berichtigungen und Ergänzungen zu P. Gilbert: *A compendium [etc.].* *Beiträge zur Entomologie* 35: 369–408.
Replaces **M. M. Carpenter.** 1945. Bibliography of biographies of entomologists. *American midland naturalist* 33: 1–116, *Continued* ___. 1953. ___: Supplement. *Ibid.* 50: 257–348; and **J. S. Wade.** 1928. A bibliography of biographies of entomologists, with special reference to North American workers. *Annals of the Entomological Society of America* 21: 489–520.

Ba334. **Hasegawa, H.** 1967. Materials on the lives and contributions of the deceased Japanese entomologists since the Meiji era. *Kontyû* 35 supp: 1–98.
In Japanese.

Ba335. **Lhoste, J.** 1987. *Les entomologistes français: (1750–1950).* Paris, Institut National de la Recherche Agronomique (INRA), Office pour l'Information Eco-Entomologique (OPIE). 355 pp., illus.

Ba336. **Mallis, A.** 1971. *American entomologists.* New Brunswick, NJ, Rutgers Univ. Pr. xvii+549 pp., illus.
211 biographical sketches.

Ba337. **Musgrave, A.** 1932. *Bibliography of Australian entomology 1775–1930, with biographical notes on authors and collectors.* Sydney, Royal Zoological Society of New South Wales. viii+380 pp.
Includes a brief biography for each author listed.

Ba338. **Papavero, N.** 1971–76. *Essays on the history of neotropical dipterology, with special reference to collectors (1750–1905).* São Paulo, Universidad de São Paulo, Museu de Zoologia. 3 vols.

Ba339. **Rohlfien, K.** 1977. Bibliographie entomologischer Bibliographien (1920–1970). *Beiträge zur Entomologie* 27: 313–379.
Includes a large section on personal bibliographies.

Ba340. **Vayssière, P.** 1980. L'entomologie agricole coloniale 1921–1935. Travaux et souvenirs. *Histoire et nature* 16: 120 pp.
Quelques précurseurs (5 people), pp. 13–19; Les entomologistes agricoles coloniaux (17 people), pp. 21–118.

See also Cc538 & Cc583.

Conchologists

Ba341. **Abbott, R. T.** 1973. *American malacologists: A national register of professional and amateur malacologists and private shell collectors and biographies of early American mollusk workers born between 1618 and 1900.* Falls Church, VA. 494 pp.; *Continued* ___. 1975. ___*: Supplement.* Greenville, DE. pp. 495–609.

Ba342. **Barnard, K. H.** 1965. *Personal names in South African conchology.* [Cape Town], Conchological Society of Southern Africa. [i]+53 pp. *Special publication* no. 4.

Ba343. **Dall, W. H.** 1888. Some American conchologists. *Proceedings of the Biological Society of Washington* 4: 95–134, 3 port.

Ba344. **McMillan, N. F.** 1970. Some Irish naturalists, mostly conchological. *Journal of conchology* 27: 197–203.

See also Bb898.

Taxidermists

Ba345. **Herriott, S.**, ed. 1968. *British taxidermists: A historical directory.* Leicester, Leicester Museums & Art Gallery. 36 pp.

OLD DIRECTORIES OF ZOOLOGISTS

• Selected titles, arranged by date of publication:

Ba346. **Meyer, F. A. A.** 1794. Alphabetisches Verzeichniss aller jetzt lebenden Zoologen: Nachrichten von zoologischen Sammlungen. (Verzeichniss der ganzen Welt); Beitrag zu den Methoden Insecten aufzubewahren und zu verschicken. *Zoologische Annalen* 1: 60–132.

Ba347. **Friedländer & Sohn** [Pub.]. 1895. *Zoologisches Adressbuch. Namen und Adressen der lebenden Zoologen, Anatomen, Physiologen und Zoopalæontologen, sowie der technischen Hilfskräfte. Hrsg. im Auftrage der Deutschen Zoologischen Gesellschaft von R. Friedländer & Sohn.* [By W. Junk.] Berlin. viii+470 pp. ••Arranged by country, town, and institution.

Ba348. **Sankt-Peterburkskoe Obshchestva Estestvoispytalei.** 1901. *Adressbuch der Zoologen, Anatomen, Physiologen, Palaeontologen und dgl. Russlands [etc.]. Adresnaya kniga Russkikh zoologhov, [etc.].* Leipzig & St. Petersburg. 147 pp.

Ba349. **Friedländer & Sohn** [Pub.]. 1911. *Zoologisches Adressbuch. Namen und Adressen der lebenden Zoologen, Anatomen, Physiologen und Zoopalæontologen, sowie der künstlerischen und technischen Hülfskräfte.* Hrsg. auf veranlassung der Deutsches Zoologischen Gesellschaft von R. Friedländer & Sohn. 2. vollständig neu bearb. Ausg. Berlin. 1,109 pp. ••Arranged by country, town, and institution.

Ba350. **Union Internationale des Sciences Biologiques.** 1953. *Index des zoologistes.* Paris. viii+429 pp.

Ba351. **Uchida, T.** 1954. List of Japanese zoologists partaking systematic zoology. *Journal of the Faculty of Science, Hokkaido University*, series 6, *zoology* 12 (1/2 suppl.): 33 pp.

Ba352. **Ministerstvo Prosveshcheniya RSFSR.** 1955. *Vydayushchiisya otechestvennye zoologi. [Prominent native zoologists].* By B. N. Mazurmovich. Moscow. 294 pp.

Ba353. **Grabda, E., et al.** 1958. *Parasitology and parasitologists in Poland.* Warsaw, [Polish Parasitological Society]. 118 pp.

Ba354. **Moczar, L.** 1960. A register of Hungarian zoologists. *Allattani kozlemenyek* 47: 195–204.

Ba355. **Blackwelder, R. E. & R. M.** 1961. *Directory of zoological taxonomists of the world.* Carbondale, IL, Southern Illinois Univ. Pr. for the Society of Systematic Zoology. xvii+404 pp.

Ba356. **Strelkov, A. A., & K. B. Yur'ev**, eds. 1961. *Zoologi Sovetskogo Soyuza: Spravochnik.* Moscow/Leningrad, AN SSSR, Zool. Inst. 292 pp. ••*Also* ___. 1962. *Handbook of soviet zoologists.* [English translation.] Washington, DC, U. S. Joint Publications Research Service. 443 pp.

Ba357. **Gallagher, J.** 1962. World list of the rotifer workers. *Transactions of the American Microscopical Society* 81: 282–288.

Ba358. **Polska Akademia Nauk, Komitet Zoologiczny.** 1962. *Informator zoologiczny. Directory of*

Polish zoologists. Warsaw. [viii]+193 pp.

Ba359. **Pacific Science Information Center.** 1968. *Indo-Pacific invertebrate zoologists (other than entomologists).* Honolulu, HI. [176] pp.

Ba360. **Petit, R. E.** 1968. *Directory of conchologists.* [New ed.] Ocean Drive Beach, SC, the Author. 40 pp.

Ba361. **Vsesoyuznoe Paleontologicheskoe Obshchestvo.** 1968. *Paleontologi sovetskogo soyuza: Spravochnik.* Leningrad, Izd. "Nauka." 214 pp.

OLD DIRECTORIES OF ENTOMOLOGISTS & ARACHNOLOGISTS

• Selected titles, arranged by date of publication:

Ba362. **Gistel, J. von N. F. X.** 1834. *Die jetzt lebenden Entomologen, kerffreunde und Kerffreunde und Kerfsammler Europas und der übrigen Continente.* Munich, Jaquet. iv+80 pp.

Ba363. **Silbermann, G.** 1835. *Enumération des entomologistes vivans, suivie de notes sur les collections entomologiques des principaux musées d'histoire naturelle d'Europe, sur les sociétés d'entomologie, sur les recueils périodiques consacrés à l'étude des insectes, et d'une table alphabétique des résidences des entomologistes.* Paris, Roret. 120 pp.

Ba364. **Gistel, J. von N. F. X.**, ed. 1846. *Lexikon der entomologischen Welt, der Carcinologischen und Arachnologischen. Addressenbuch der lebenden Entomologen und Entomophilen etc; der Carcinologen und Arachnologen sammt ihren Schriften, dann der Naturforscher-Akademien und deren Verhandlungen, der zoologischen Ephemeriden, Bibliographien, Biographien und Real-Wörterbücher, der öffentlichen und Privat-Sammlungen der Welt, der Schriften über Sammlungs- und Aufbewahrungsweise der Gliederthiere, mit doppelten Registern und einer Aufzählung aller entomologischen, carcinologischen und arachnologischen Schriftsteller von Aristoteles an bis zur Gegenwart.* Stuttgart, E. Schweizerbart'sche Verlagshandlung. 328 pp. ••*Includes* A: Adressenbuch der Entomologen, pp. 13–81; Supplement, pp. 313–314. Register der Wohnorte und der dahin gehörigen Personalien, pp. 146–154; Supplement, p. 321. Anhang. Aufzählung aller entomologischen Schriftsteller von Aristoteles bis auf die neueste Zeit, pp. 155–193; Supplement, pp. 321–322. B: Adressenbuch der Carcinologen etc., pp. 218–236; Supplement, pp. 322–323. Ortstafel der Carcinologen, pp. 237–239; Supplement, p. 323. Die carcinologischen Schriftsteller des Alterthums, des Mittelalters und der Neuesten Zeit. Nebst der vollständigen Angabe ihrer Werke, pp. 240–261; Supplement, pp. 323–326. C: Adressenbuch der Arachnologen, pp. 291–298. Wohnorte der jezt lebenden Arachnologen, pp. 299. Tafel der arachnologischen Schriftsteller, pp. 300–312. D: Biographien, Gedächtnissreden etc., pp. 108–112; Supplement, pp. 318–320.

Ba365. **Graessner, F.** 1855. *Die Entomologen Europas, Asiens und Amerikas, zum Besten aller Sammler zusammengestellt und mit nöthigen Anmerkungen versehen.* Jena, Mauke. [viii]+96 pp.

Ba366. **Stainton, H. T.** 1856–58. List of British entomologists. [&] Supplemental list of British entomologists. [&] Second supplemental list of British entomologists. *Entomologist's annual* 1856: 11–25; *Ibid.* 1857: 13–26; *Ibid.* 1858: 7–16.

Ba367. **Graessner, F.** 1857. *Die jetzt lebenden Entomologen vorzugsweise Deutschlands und der angrenzenden Länder.* Leipzig, Fleischer. [vi]+144 pp.

Ba368. **Stainton, H. T.** 1859. List of British entomologists. *Entomologist's annual* 1859: 1–54.

Ba369. **Stainton, H. T.** 1860. List of British entomologists. *Entomologist's annual* 1860: 1–65.

Ba370. **Junk, W.** [Pub.]. 1905. *Entomologen-Adressbuch. The entomologists' directory [etc.].* Berlin. vi+295 pp.

Ba371. ___. 1935. *Dipterologi [etc.].* The Hague. [ii]+19 pp.

Ba372. **Anon.** 1960. Liste der ungarischen Entomologen. *Folia entomologica Hungarica* 13: 284–298.

Ba373. **Brennan, J. M., & E. K. Jones.** 1965. *Acarologists of the world: A directory.* Wooster, OH, Ohio Agricultural Experiment Station. [iii]+91 pp.

Ba374. **Pacific Science Information Center.** 1966. *Pacific entomologists, 1966.* Compiled by I. M. Koehler. Honolulu, HI. Unpaged.

ARTISTS (& illustrators)

Ba375. **Burbidge, R. Brinsley–.** 1974. *A dictionary of British flower, fruit, and still life painters.* Leigh–on–Sea, Lewis. 2 vols., incl. 74 pl.
Vol. 1: 1515–1849; vol. 2: 1850–1950.

Ba376. **Desmond, R. G. C.** 1977. *Dictionary of British and Irish botanists ... including ... botanical artists.* London, Taylor & Francis. xxvi+747 pp.

Ba377. **Hardouin–Fugier, E., & E. Grafe.** 1989. *French flower painters of the nineteenth century: A dictionary.* Ed. by P. Mitchell. London, Philip Wilson. 402 pp.

Ba378. **Lewis, F.** 1973. *A dictionary of Dutch and Flemish flower, fruit, and still life painters: 15th to 19th century.* Leigh–on–Sea, Lewis. 83 pp. & 63 pl.

Ba379. ___. 1974. *A dictionary of British bird painters.* Leigh–on–Sea, Lewis. 47 pp. & 48 pl.

Ba380. **Mitchell, P.** 1973. *European flower painters.* London, A. & C. Black. 272 pp. & 51 pl. (col.). Pub. in the U.S.A. as: *Great flower painters.* Woodstock, NY, Overlook Pr., 1973.

Ba381. **Nissen, C.** 1951. *Die botanische Buchillustration: Ihre Geschichte und Bibliographie.* Stuttgart, Hiersemann. 2 vols. *Continued* ___. 1966. ___*: Supplement.* Stuttgart, Hiersemann. vii+97 pp.
The indexes of "Künstler," in Band II. *Bibliographie* and Band III. *Supplement*, provide probably the fullest available list of world botanical artists of all ages and includes skeleton biographical details.

Ba382. ___. 1953. *Die illustrierten Vögelbucher: Ihre Geschichte und Bibliographie.* Stuttgart, Hiersemann. 222 pp.
The index of "Künstler" provides a full list of world ornithological artists of all ages with skeleton biographical details.

Ba383. ___. 1966–78. *Die zoologische Buchillustration. Ihre Bibliographie und Geschichte.* Stuttgart, Hiersemann. 2 vols.
The index of "Künstler," in Band II. *Geschichte*, provide probably the fullest available list of world zoological artists of all ages and includes skeleton biographical details.

Ba384. **Wood, J. C.** 1973. *A dictionary of British animal painters.* Leigh–on–Sea, Lewis. 75 pp. & 40 pl.

See also Bb814 –815.

PORTRAITURE

Portrait indexes & studies

Ba385. **Coats, A. M.** 1965. Notes on some portraits of British botanists and gardeners. *Huntia* 2: 185–215.

Ba386. **Hunt Institute for Botanical Documentation & International Association for Plant Taxonomy.** 1972+. HI–IAPT portraits of botanists. *Taxon* 21+.
Series of reproductions of portraits in *Taxon*; over 100 have been pub. to date, each accompanied by a biographical note. For an introduction to the series *see* **G. S. Daniels & F. A. Stafleu.** 1972. HI–IAPT portraits of botanists. *Taxon* 21: 159–160.

Ba387. **Ipswich Museum.** 1852. *Honorary members of the Ipswich Museum. Published by G. Ransome, F.L.S., Etc., Hon. Sec.* Ipswich. 1 sheet of text & 60 pl.
Lithographed portraits by T. H. Maguire.

Ba388. **Jardine,** Sir **W.**, ed. 1840. [Forty one biographies of naturalists, each with an engraved portrait.] In: *The naturalist's library*. Edinburgh, Lizars. 40 vols. *Re-pub.* Edinburgh, [etc.]., [1845–46]; London, H. Bohn, 1848; etc.
Steel–engraved portraits by Lizars; *see* Ba303.

Ba389. **Junk, W.**, ed. 1926. *Portraits of old botanists*. Berlin, W. Junk. 4 pp. & 25 pls.

Ba390. **Lane, W. C., & N. E. Browne**, eds. 1906. *A. L. A. portrait index: Index to portraits contained in printed books and periodicals*. Washington, DC. lxxv+1601 pp. In the United States Library of Congress.

Ba391. **Literary & Scientific Portrait Club.** [1856?–?] (*"Portraits of the literary and scientific men of the present age ..."*) J. S. Bowerbank (Honorary secretary). London.
Photographic portraits by Maull & Polyblank (*later* Maull & Fox), Gracechurch St. & Piccadilly, London. The Linnean Society of London has a set and other societies probably do also. (Another work entitled *Photographic portraits of living celebrities, executed and published by Maull & Polyblank* was pub. in London from 1856–59 in 40 numbers. There appears to have been some overlap with the above work, some portraits probably occurring in both works.)

Ba392. **Moehsen, I. C. W.** 1771. *Verzeichnis einer Sammlung von Bildnissen, gröstentheils berühmter Aerzte; so wohl in Kupferstichen, schwarzer Kunst und Holzschnitten, als auch in einigen Handzeichnungen*. Berlin, F. W. Birnstiel. 240 pp.

Ba393. **Nelmes, E., & W. Cuthbertson.** [1931.] *Curtis's botanical magazine dedications, 1827–1927, portraits and biographical notes*. London, B. Quaritch for the Royal Horticultural Society. xxi+400 pp. incl. 99 ports.
Ninety nine dedicatees are portrayed, including many botanists.

Ba394. **Sachse, J. D. W.** 1847. *Verzeichnis von Bildnissen von Aertzten und Naturforschern seit den ältesten bis auf unsere Zeiten, mit Biographien*. 1. Hefte: *Abano–Azzoguijoi*. Schwerin, Marcus. vi+93 pp. *Re-pub.* Leipzig, A. Lorentz, 1931.

Ba395. **Wallich, G. C.** 1870. *Eminent men of the day*. Photographed by G. C. Wallich. *Scientific series*. London, J. Van Voorst. 3 ff. & 16 pl.

Portrait collections

Ba396. **American Philosophical Society.** 1961. *A catalogue of portraits and other works of art in the possession of the American Philosophical Society*. Philadelphia, PA. viii+173 pp. (incl. 58 pp. of illus.). *Memoirs of the American Philosophical Society* vol. 54.

Ba397. **Bergianska Botaniska Trädgården, Stockholm.** 1903–05. Illustrerad förteckning öfver bergielunds botaniska trädgårds samling porträtt af botaniska författare;

jämte biografiska notiser. (Catalogus illustratus iconothecae botanicae Horti Bergiani Stockholmiensis anno 1903; notulis biographicis adjectis). By V. B. Wittrock. *Acta Horti Bergiani* 3(2): 198 pp. & 46 pl; *Continued* ___. 1905. ___: Pars II. *Ibid.* 3(3): xciii+245 pp. & 151 pl.
Catalogue of ca. 4,000 portraits arranged by nationality.

Ba398. **Bibliothèque des Conservatoire et Jardin Botaniques de la Ville de Genève.** *See* Hunt Institute for Botanical Documentation. 1987+; *see* Ba401.

Ba399. **British Museum.** 1908–25. *Catalogue of engraved British portraits preserved in the Department of Prints and Drawings in the British Museum.* By F. O'Donoghue (& H. M. Lake). London. 6 vols.
An alphabetical list of individuals with a supplement of group portraits.

Ba400. **Hunt Institute for Botanical Documentation, Carnegie Mellon University.** 1972. *Biographical dictionary of botanists represented in the Hunt Institute portrait collection.* Boston, MA, Hall. viii+451 pp.
Includes some 11,000 botanists.

Ba401. ___. 1987+. *A catalogue of portraits of naturalists, mostly botanists, in the collections of the Hunt Institute, The Linnean Society of London, and the Conservatoire et Jardin Botaniques de la Ville de Genève.* By M. T. Stieber [et al.] Part 1 [etc.]. Pittsburgh, PA.
Pt. 1: Group portraits. 1987; pt. 2: Portraits of individuals, A–D. 1988. For background information *see* **T. W. Bossert.** 1964–65. The portrait collection and program. *Huntia* 1: 174–191, 1964; *Ibid.* 2: 242–278, 1965.

Ba402. **Kungliga Svenska Vetenskapsakademien.** 1971. *Vetenskapsakademiens konstsamlingar: Förteckning över porträtt, medaljer och övriga konstföremal.* By B. S. Nordin–Petterson. Stockholm. 119 pp. *Bidrag till Kungliga Svenska Vetenskapsakademiens historia* 10.

Ba403. **Linnean Society of London.** 1890. [Presidential address, 1889, by W. Carruthers.] *Proceedings, Linnean Society of London* 1888–90: 14–46.
A descriptive catalogue of portraits hanging in the Society's rooms.

Ba404. ___. *See* Hunt Institute for Botanical Documentation. 1987+; *see* Ba401.

Ba405. [**National Library of Medicine.**] **Surgeon–General's Office, United States Army.** 1937. A portrait of physicians; the collection in the Army Medical Library. *Annals of medical history* n.s. 9: 517–532.

Ba406. **National Portrait Gallery, London.** [1986.] *Apples to atoms: Portraits of scientists from Newton to Rutherford.* By W. D. Hackmann. London. 88 pp., illus.
Catalogue of a travelling exhibition.

Ba407. **New York Academy of Medicine.** 1960. *Portrait catalog.* Boston, MA, G. K. Hall. 5 vols. *Continued* ___. 1965. ___*: First supplement, 1959–1965.* Boston, MA, [etc.]. 842 pp.; ___. 1971. ___*: Second supplement, 1965–1971.* Boston, MA, [etc.]. 593 pp.; ___. 1976. ___*: Third supplement, 1971–1975.* Boston, MA, [etc.]. 589 pp.
The main list catalogues 151,792 portraits, both separate and included in books and periodicals.

Ba408. **Reale Istituto Botanico di Padova.** 1899–1902. La iconoteca dei botanici nel R. Istituto Botanico di Padova. By P. A. Saccardo. *Malphigia* 13: 89–123, 1899; *Ibid.* 15: 416–437, 1902.

Ba409. **Royal Botanic Gardens, Kew.** 1906. *Catalogue of portraits of botanists exhibited in the Museums of the Royal Botanic Gardens.* By J. D. Milner. London. vii+105 pp.

Ba410. **Royal College of Physicians of London.** 1952. *Catalogue of engraved portraits in the Royal College of Physicians of London.* By A. H. Driver. London, Royal College of Physicians. 219 pp.

Ba411. ___. 1964. *The Royal College of Physicians of London – portraits.* Ed. by G. Wolstenholme, described by D. Piper. London, J. & A. Churchill. 468 pp., illus.; *Continued* ___. 1977. ___*: Catalogue II.* By G. Wolstenholme & J. F. Kerslake. Amsterdam, New York, Elsevier, Excerpta Medica, North Holland. viii+239 pp.
Replaces ___. 1926. *A descriptive catalogue of the portraits, busts, silver and other objects of interest in the Royal College of Physicians of London.* London. 70 pp.

Ba412. **Royal College of Surgeons of England.** 1960. *A catalogue of the portraits and other paintings, drawings and sculpture in the Royal College of Surgeons of England.* By W. R. LeFanu. Edinburgh, E. & S. Livingstone. xii+119 pp. & 55 pl.
Replaces ___. 1930. *Catalogue of portraits and busts in the Royal College of Surgeons of England with short biographical notices.* Revised to January, 1930. By F. G. Hallett. London. iv+96 pp., illus.

Ba413. **Royal Society of Edinburgh.** 1960. A list of the portraits in the Royal Society of Edinburgh with biographical notes. By D. Guthrie. *Yearbook of the Royal Society of Edinburgh* 1959/60: 5–24.

Ba414. **Royal Society of London.** 1980. *The Royal Society catalogue of portraits.* By N. Robinson. With biographical notes by E. G. Forbes. London. 343 pp., illus.
Replaces ___. 1860. *Descriptive catalogue of the portraits in the possession of the Royal Society.* By C. R. Weld. London, Taylor & Francis. [ii]+75 pp.; **Anon.** 1949. The Society's portraits. *Notes & records of the Royal Society* 7: 97–107; **E. N. de C. Andrade.** 1956. The Royal Society portraits. Endeavour 15: 128–136 & 6 pl; **G. Hamill.** 1969. The Society's portraits and busts. *Notes & records of the Royal Society of London* 24: 156–168.

Ba415. **Royal Society of Medicine.** 1988. *Portraits, paintings and sculptures, Royal Society of Medicine, London.* By A. Sakula. London. xiii+194 pp.

Ba416. **Wellcome Institute for the History of Medicine.** 1973. *Portraits of doctors and scientists in the Wellcome Institute for the History of Medicine.* A catalogue by R. Burgess. London. xxiv+459 pp., illus. *Museum catalogue* no. 3.

• For notes on additional collections *see*

Ba417. **Brooklyn Botanic Garden.** 1920. The Brooklyn Botanic Garden collection of portraits of botanists. By C. S. Gager & A. Gunderson. *Brooklyn Botanic Garden leaflet* ser. 8, no. 6/8: 12 pp.

Ba418. **Deniker, J.** 1895. Sur une collection de portraits des Professeurs du Muséum formée à la Bibliothèque. *Bulletin du Muséum National d'Histoire Naturelle, Paris* 1: 75–76.

Ba419. **Palmer, T. S.** 1944. The Deane collection of portraits of ornithologists — the development of an idea. *Science* 100: 288–290.

Ba420. **Syracuse University.** 1961. The Wolff–Leavenworth collection of engraved portraits at Syracuse University. By L. I. Gardner & L. G. Wells. *Bulletin of the history of medicine* 35: 175–177. ••Collection of over 5,900 portraits of physicians and naturalists.

Ba421. **Van Cleave, H. J.** 1942. [J. Victor Carus collection of photographs of zoologists.] *Bios* 13: 152–153.

MEDALLIC PORTRAITS — STUDIES & COLLECTIONS

Ba422. **College of Physicians of Philadelphia.** 1944. Medical medals in the collection of the College of Physicians of Philadelphia. By E. N. Wade. *Bulletin of the Medical Library Association* 32: 177–213.

Ba423. **Holzmair, E.** 1937. *Katalog der Sammlung Dr. Josef Brettauer. Medicina in nummis [etc.].* Vienna.

383 pp. & 25 pl. Includes descriptions of 1,347 portrait medals.

Ba424. **Johns Hopkins University.** 1964. *Medals relating to medicine and allied sciences in the Numismatic Collection of the Johns Hopkins University: A catalogue.* By S. E. Freeman. Baltimore, MD, Evergreen House Foundation. xx+430 pp. & 32 pl.

Ba425. **Kluyskens, H.** 1859. *Des hommes célèbres dans les sciences et les arts, et des médailles consacrent leur souvenir.* Ghent. 2 vols. ••Over 1,200 items described.

Ba426. **Levinson, A.** 1943. Medallic medallions. *Bulletin of the Medical Library Association* 31: 1–34.

Ba427. **Rudolphi, C. A.** 1862. *C. A. Rudolphi recentiores aevi numismata virorum de rebus medicis et physicis meritorum memoriam servantia denuo edidit emendavit et auxit Carol. Ludov. de Duisburg.* Danzig. 257 pp. Includes medals honouring 611 physicians. *Also* **E. Rüppell.** 1874–76. Beitrag zur Kenntniss der numismatischen Erinnerungen an Ärzte und Naturforscher, die sich nicht in Duisburg's Werk über diesen Gegenstand vorfinden. *Numismatische Zeitschrift* 6/7: 151–232, 1874–75; *Ibid.* 8: 315–338, 1876.

Ba428. **Sarton, G.**, ed. 1926–30. Medallic illustrations of the history of science. *Isis* 8: 333–335, 1926; *Ibid.* 9: 359–362, 420–423, 1927; *Ibid.* 10: 485–488, 1928; *Ibid.* 12: 146–148, 320–322, 1929; *Ibid.* 13: 333–335, 1930; *Ibid.* 14: 215–218, 417–419, 1930.

Ba429. **Storer, H. R.** 1931. *Medicina in nummis: A descriptive list of the coins — medals — jetons relating to medicine, surgery and the allied sciences.* Ed. by M. Storer, M.D. Boston, MA. 1,146 pp.

BIBLIOGRAPHIES OF ICONOGRAPHICAL LITERATURE

Ba430. **Kisch, B.** 1957. Iconographies of medical portraits. *Journal of the history of medicine* 12: 366–387.

INDIVIDUAL BIOGRAPHIES (& bibliographies)

Adanson, M.

Ba431. **Lawrence, G. H. M.**, ed. 1963–64. *Adanson: The bicentennial of Michel Adanson's "Familles des plantes."* Pittsburgh, PA, Hunt Botanical Library. 2 vols. With chapters on his life and aspects of his work by various authors; *Includes* **W. D. Margadant.** 1963. The Adanson collection of botanical books and manuscripts. In: Part 1, pp. 265–368.

Ba432. **Nicholas, J.–P.** 1965. *Adanson et les encyclopédistes.* Paris, Université de Paris. 39 pp. & port.

See also Ba12.

Agassiz, A.

Ba433. **Agassiz, G. R.**, ed. 1913. *Letters and recollections of Alexander Agassiz, with a sketch of his life and work.* Boston, New York, Houghton Mifflin. vii+[ii]+454 pp., 12 pl. & 6 ports. *Also* London, Constable, 1913.

Ba434. **Museum of Comparative Zoology, Harvard University.** 1979. *Letter books of Louis Agassiz and Alexander Agassiz, 1859–1910.* Cambridge, MA. Microfilm edition. 15 vols.

Agassiz, J. L. R.

Ba435. **Agassiz, E. C.**, ed. 1885–86. *Louis Agassiz: His life and correspondence.* Boston, MA, Houghton Mifflin. 2 vols.

Ba436. **Cooper, L.** 1945. *Louis Agassiz as a teacher: Illustrative extracts on his method of instruction.* Rev. & amplified. Ithaca, NY, Comstock Publishing Co. xiii+90 pp.
Replaces 1st ed., 1917.

Ba437. **Herber, E. C.**, ed. 1963. *Correspondence between S. F. Baird and L. Agassiz — two pioneer American naturalists.* Washington, DC, Smithsonian Institution. 237 pp. & 16 pl.

Ba438. **Holder, C. F.** 1893. *Louis Agassiz: His life and work.* New York, G. P. Putnam's Sons. xviii+327 pp., illus. *Leaders in science.*

Ba439. **Lurie, E.** 1960. *Louis Agassiz: A life in science.* Chicago, IL, Univ. of Chicago Pr. xiv+449 pp. & 24 pl.

Ba440. ___. 1974. *Nature and the American mind: Louis Agassiz and the culture of science.* New York, Science History Publications. 124 pp.

Ba441. **Marcou, J.** 1896. *Life, letters, and works of Louis Agassiz, [etc.].* New York, London, Macmillan & Co. 2 vols. *Re-pub.* Westmead, Farnborough, Gregg International Publishers, 1972. 2 vols in 1.

Ba442. **Musée d'Histoire Naturelle, Neuchâtel.** 1983. *Louis Agassiz: Naturaliste romantique et les premières collections du musée ... du 5 mars au 30 octobre 1983.* Ed. by C. Dufour & J.-P. Haenni. Neuchâtel. 64 pp.
Exhibition catalogue.

Ba443. **Museum of Comparative Zoology, Harvard University.** 1979. *Letter books of Louis Agassiz and Alexander Agassiz, 1859–1910.* Cambridge, MA. Microfilm edition. 15 vols.

Ba444. **Robinson, M. L.** 1939. *Runner of the mountain tops: The life of Louis Agassiz.* New York, Randon House. xv+290 pp. & 8 pl.

Ba445. **Schweizerische Naturforschende Gesellschaft.** 1973. *Louis Agassiz, 1807–1873. Vorträge an der Gedenkfeier zum 100. Todestag gehalten an der 153. Jahresversammlung der Schweizerischen Naturforschenden Gesellschaft, Lugano, 20. Oktober.* Berne. 142 pp., illus. *Denkschriften der Schweizerischen Naturforschenden Gesellschaft* vol. 89.

Ba446. **Surdez, M.** 1973. Catalogue des archives de Louis Agassiz (1807–1873). Neuchâtel, Université de Neuchâtel. 202 pp. Also pub. in *Bulletin de la Société Neuchâteloise des Sciences Naturelles* 97: 5–202, 1974.

See also Ba26 & Ba133.

Aldrovandi, U.

Ba447. **Adversi, A.** 1968. *Ulisse Aldrovandi: Bibliofilo, bibliografe e bibliologo del cinquecento.* Milan, A. Giuffre. *Annali, Scuola Speciale per Archivisti e Bibliotecari* vol. 8 (1/2): 85–181.

Ba448. **Aldrovandi, U.** 1907. *Catalogo dei manoscritti di Ulisse Aldrovandi.* By L. Frati, A. Ghigi & A. Sorbelli. Bologna, N. Zanichelli. xxi+287 pp., port.
MSS mostly at University of Bologna.

Ba449. **Baldacci, A., et al.** 1907. *Intorno alla vita e alle opere di Ulisse Aldrovandi.* Bologna, L. Betrami. 223 pp.
Includes **A. Sorbelli.** Contributo alla bibliografia delle opere di Ulisse Aldrovandi, pp. 69–139.

Ba450. **Comitate pel Centenario di Ulisse Aldrovandi.** 1908. *Onoranze a Ulisse Aldrovandi vel terzo centenario dalla sua morte celebrate in Bologna nei giorni XI, XII, XIII giugno MCMVII.* Inola. vi+429 pp. & 14 pl.
A large collection of institutional commemorative addresses.

Ba451. **Fantuzzi, G.** 1974. *Memorie sulla vita e sulle opere di Ulisse Aldrovandi con alcune lettere scelte d'uomini eruditi a lui scritte, e coll'indice delle sue opere mss., che si conservano nella biblioteca dell'Istituto.* Bologna, Lelio dalla Volpe. vi+263 pp.

Ba452. **Mattirolo, O.** 1897. *L'opera botanica di Ulisse Aldrovandi (1549–1605).* Bologna, Fratelli Merlani. xxxi+137 pp.
Includes Elenco dei manoscritti botanici di Ulisse Aldrovandi, pp. 113–117.

Ba453. **Olmi, G.** 1976. *Ulisse Aldrovandi: Scienza e natura nel secondo cinquecento.* Trento, Unicoop. 129 pp. *Quaderni di storia e filosofia della scienza* no. 4.

Ba454. **Pattaro, S. T.** 1981. *Metodo e sistema delle scienze nel pensiero di Ulisse Aldrovandi.* Bologna, Cooperativa Libraria Universitaria Editrice. 251 pp. *Collana di studi epistemologici* no. 3.

Ba455. **Tosi, A.**, ed. 1989. *Ulisse Aldrovandi e la Toscana: Carteggio e testimonianze documentarie.* Florence, Olschki. 472 pp. *Archivio della correspondenza degli scienziati Italiani* no. 5.

See also Ba303.

Alexander, H. G.

Ba456. **Alexander, H. G.** 1974. *Seventy years of bird watching.* Berkhamsted, Poyser. 264 pp. & 8 pp. of illus.

Ali, S.

Ba457. **Ali, S.** 1985. *The fall of a sparrow.* Delhi, New York, Oxford Univ. Pr. xv+265 pp. & 57 pp. of illus.

Allan, P. B. M.

Ba458. **Allan, P. B. M.** 1948. *Moths and memories.* London, Watkins & Doncaster. 316 pp.

Allen, G. M.

Ba459. **Lawrence, M.** 1947. Bibliography of publications of Glover Morrill Allen. *Proceedings of the New England Zoological Club* 24: 1–81.

Allen, J. A.

Ba460. **Allen, J. A.** 1916. *Autobiographical notes and a bibliography of the scientific publications of J. A. Allen.* New York, American Museum of Natural History. xi+215 pp. & port.

Alpini, P.

Ba461. **Ongaro, G.** 1966. I manoscritti di Prospero Alpini [1553–1617]. *Atti e memorie dell'Accademia di Storia dell'Arte Sanitaria* (2) 32: 10–15.

Ba462. **Premuda, L., et al.** 1963. La personalità e l'opera scientifica di Prospero Alpini (Marostica, 16th June 1963). *Acta medicae historiae Patavina* 8/9 (1961/62 & 1962/63): 219 pp.
Proceedings of a symposium with papers by various authors.

Ames, O.

Ba463. **Ames, O.** 1979. *Oakes Ames: Jottings of a Harvard botanist 1874–1950.* Ed. with an introduction by P. A. Plimpton., Cambridge, MA, Botanical Museum of Harvard Univ. [x]+403 pp.
Letters, diaries, and autobiographical notes.

Amici, G. B.

Ba464. **Savelli, R.** 1966. *Giovan Battista Amici, botanico (nel primo centenario della morte).* 2nd ed. Ferrare, Università degli Studi. 41 pp. *Quaderni di storia della scienza e della medicina* vol. 2.

Anderson, N. J.

See Cb759A.

Arber, A.

Ba465. **Arber, M. A.** 1968. List of published works of Agnes Arber, E. A. N. Arber and Ethel Sargant. *Journal of the Society for the Bibliography of Natural History* 4: 372–384; *Ibid.* 5: 70; **R. Schmid & M. A. Arber.** 1976. Supplement to ... *Ibid.* 8: 180–183.

Ba466. **Van Trump, J. D.** 1964. The Agnes Arber collection. *Huntia* 1: 169–171.
Personal papers.

Arnott, G. A. W.–

See Ba133.

Artedi, P.

Ba467. **Lönnberg, E.** 1905. *Peter Artedi: A bicentenary memoir.* Written on behalf of the Swedish Royal Academy of Science. Translated by W. E. Harlock. Uppsala. 44 pp., illus.

Atkins, A.

Ba468. **Schaaf, L. J.** 1985. *Sun gardens: Victorian photograms by Anna Atkins.* Organized by Hans P. Kraus, Jr. New York, Aperture. 104 pp., illus. (some col.).

Audouin, J.–V.

Ba469. **Théodoridès, J.** 1978. *Un zoologiste de l'époque romantique: Jean–Victor Audouin (1797–1841).* Paris, Bibliothèque Nationale. 128 pp. *Mémoires de la Section des Sciences, Comité des Travaux Historiques et Scientifiques* vol. 6.

Audubon, J. J.

Ba470. **Adams, A. B.** 1966. *John James Audubon.* New York, G. P. Putnam's Sons. 510 pp.

Ba471. **Audubon, J. J.** 1930. *The letters of J. J. Audubon, 1826–1840.* Ed. by H. Corning. Boston, MA, Club of Odd Volumes. 2 vols. *Re-pub.* New York, Kraus Reprint Co., 1969.

Ba472. **___, M. R., & E. Coues**, eds. 1897. *Audubon and his journals.* With zoölogical and other notes by E. Coues. New York, Scribners. 2 vols., illus. *Re-pub.* 1900.

Ba473. **Ford, A.** 1988. *John James Audubon: A biography.* New York, Abbeville Pr. 528 pp.

Ba474. **Fries, W. H.** 1973. *The double elephant folio: The story of Audubon's* Birds of America. Chicago, IL, American Library Association xxii+501 pp.
Pt. I, Creation of the double elephant folio of *Birds of America*; Pt. II, Subscribers, sets, and costs; Pt. III, Census and survey of extant complete sets; Appendix, details of other editions, reproductions, the copper plates, etc.

Ba475. **Herrick, F. H.** 1938. *Audubon the naturalist: A history of his life and time.* New York, Appleton–Century. 2 vols. *Re-pub.* New York, Dover Publications, 1968.

Ba476. **Lindsey, A.**, ed. 1985. *The bicentennial of John James Audubon.* Bloomington, IN, Indiana Univ. Pr. 175 pp.
With chapters by six authors.

Ba477. **Low, S. M.** 1988 *An index and guide to Audubon's* Birds of America. New York, Abbeville Pr. 255 pp.

Ba477A. **Princeton University Library.** 1959. The world of John James Audubon: Catalogue of an exhibition in the Princeton University Library. By H. C. Rice, Jr. *Princeton University Library chronicle* 21: 9–88.
Manuscripts, drawings and paintings, annotated books, etc.

Ba477B. **Ripley, S. D.** 1953. Manuscripts of John James Audubon. *Yale University Library gazette* 27: 170–171.
Acquired by Yale University.

Avebury, Lord

See Lubbock, J.

Azara, F. d'

Ba478. **Alvarez López, E.** 1935. *Félix de Azara.* Madrid, M. Aguilar 259 pp. *Biblioteca de la cultura Española.*

See also Ba303.

Babington, C. C.

Ba479. **Babington, C. C.** 1897. *Memorials, journal and botanical correspondence.* Cambridge, Macmillan & Bowes. xcvi+475 pp.

Baer, K. E. von

Ba480. **Akademiya Nauk Estonskoi SSR (Eesti N S V Teaduste Akadeemia), Institut Zoologii & Botaniki.** 1983. *Bibliografiya Baera. Baeri bibliografia. Bibliographie von K. E. V. Baer.* Tallin, Valgus. 109 pp. *Folia Baeriana* vol. 4.

Ba481. **Baer, K. E. von.** 1986. *Autobiography of Dr. Karl Ernst von Baer.* English translation [etc.]. Ed. by J. M. Oppenheimer. Canton, MA, Science History Publications. xiv+389 pp.
Translation of *Nachrichten über Leben und Schriften des ..., mitgetheilt von ihm selbst. Veröffentlichlicht bei Gelegenheit seines fünfzigjährigen Doctor–Jubiläums am 29. August 1864 von der Ritterschaft Esthlands.* 2nd ed. Brunswick, F. Vieweg & Sohn, 1886. xvi+519 pp. *Re-pub.* Hanover/Döhren, 1972.

Ba482. ***Folia Baeriana.*** Tallinn, Akademiya Nauk Estonskoi SSR (Eesti N S V Teaduste Akadeemia), Institut Zoologii & Botaniki, vol. 1+, 1975+.

Ba483. **Knorre, H. von.** 1975. Karl Ernst von Baer (1792–1876): Eine ikonographische Studie. In: K. Mothes & J.–H. Scharf, eds. *Beiträge zur Geschichte der Naturwissenschaften und der Medizin; Festschrift für Georg Uschmann ... zum 60. Geburtstag am 18. Oktober 1873*, pp. 227–268. Halle, Deutsche Akademie der Naturforscher Leopoldina. *Acta historica Leopoldina* no. 9.

Ba484. **Raikov, B. E.** 1968. *Karl Ernst von Baer, 1792–1876: Sein Leben und sein Werk.* Deutsche Übersetzung mit Anmerkungen von H. von Knorre. Leipzig, J. A. Barth. 516 pp. & 20 pl. *Acta historica Leopoldina* no. 5.
Translation of *Karl Ber; Ego zhizn' i trudy.* Moscow, Izd. Akademii Nauk, 1961. 524 pp., illus.

See also Ba2 Ba21 & Ca46.

Baerends, G. P.

See Ba298.

Bailey, F. M.

Ba485. **Kofalk, H.** 1989. *No woman tenderfoot: Florence Merriam Bailey, pioneer naturalist.* College Station, TX, Texas A & M Univ. Pr. xix+225 pp.

Bailey, J. W.

Ba486. **Edgar, R. K.** 1977. An annotated bibliography of the American microscopist and diatomist Jacob Whitman Bailey (1811–1857). *Occasional papers, Farlow Herbarium of Cryptogamic Botany, Harvard University* 1977: 1–26.

Bailey, L. H.

Ba487. **Dorf, P.** 1956. *Liberty Hyde Bailey: An informal biography.* Ithaca, NY, Cornell Univ. Pr. xii+259 pp.

Ba488. **Rodgers, A. D., 3rd.** 1949. *Liberty Hyde Bailey: A story of American plant sciences.* Princeton, NJ, Princeton Univ. Pr. 506 pp. *Re-pub.* New York, Hafner Publishing Co., 1965.

Baird, S. F.

Ba489. **Dall, W. H.** 1915. *Spencer Fullerton Baird 1823–1887: A biography, including selections from his correspondence with Audubon, Agassiz, Dana, and others.* Philadelphia, PA, J. B. Lippincott. Co. xvi+462 pp., illus.

Ba490. **Goode, G. B.** 1883. *Bibliographies of American naturalists.* 1: *The published writings of Spencer Fullerton Baird 1843–1882.* Washington, DC, United States National Museum. xvi+377 pp. *Bulletin of the United States National Museum* 20.

Ba491. **Herber, E. C.**, ed. 1963. *Correspondence between S. F. Baird and L. Agassiz — two pioneer American naturalists.* Washington, DC, Smithsonian Institution. 237 pp. & 16 pl.

Ba491A. **Rivinus, E. F., & E. M. Youssef.** 1992. *Spencer Baird of the Smithsonian.* Washington, DC, Smithsonian Institution Pr. xi+228 pp., illus.

Baldwin, W.

Ba492. **Darlington, W.** 1843. *Reliquiae Baldwinianae: Selections from the correspondence of the late William Baldwin, ... with occasional notes, and a short biographical memoir.* Philadelphia, PA, Kimber & Sharpless. 346 pp. & port. *Re-pub.* ___. 1969. ___. Introduction by J. Ewan. (Four indexes – or persons, plant names, etc. – appended). New York, Hafner Publishing Co. lxix+346 pp.

See also Ba205.

Banister, J.

Ba493. **Ewan, J. & N.** 1970. *John Banister and his Natural History of Virginia, 1678–1692.* Urbana, IL, Univ. of Illinois Pr. xxx+485 pp., incl. 32 pl.
Discusses the role of British botanists Morison, Ray, Plukenet, Bobart, Lister, Lhwyd, Petiver and Sloane, with biographical accounts of each.

Banks, Sir J.

Ba494. **Cameron, H. C.** 1952. *Sir Joseph Banks, K.B., P.R.S. The autocrat of the philosophers.* London, Batchworth Pr. xx+341 pp. & 9 pl. *Re-pub.* Sydney, N.S.W., Angus & Robertson, 1966.

Ba495. **Carter, H. B.** 1987. *Sir Joseph Banks.* London, British Museum (Natural History). xi+671 pp.

Ba496. ___. 1987. *Sir Joseph Banks (1743–1820): A guide bibliographical sources.* Winchester, St. Paul's Bibliographies in association with British Museum (Natural History). 328 pp.

Ba497. **Dawson, W. R.**, ed. 1958. *The Banks letters: A calendar of the manuscript correspondence of Sir Joseph Banks preserved in the British Museum (Natural History) and other collections in Great Britain.* London, British Museum (Natural History).

xiii+965 pp. *Continued* **W. R. Dawson.** 1962. Supplementary letters of Sir Joseph Banks. *Bulletin of the British Museum (Natural History), historical series* 3: 41–70; ___. 1965. ___, 2nd series. *Ibid.* 3: 71–93; **P. M. Teigen.** 1975. ___, 3rd series. *Journal of the Society for the Bibliography of Natural History* 7: 249–257.

Ba498. **Lack, E.** 1985. *Die Abenteuer des Sir Joseph Banks, 1743–1820: Botaniker, Weltreisender und Mäzen.* Vienna, Böhlaus. 104 pp. *Botaniker auf Entdeckungsreisen* no. 1.

Ba499. **Lyte, C.** 1980. *Sir Joseph Banks, 18th century explorer, botanist and entrepreneur.* London, David & Charles. 248 pp.

Ba500. **Maiden, J. H.** 1909. *Sir J. Banks: The "Father of Australia".* Sydney, N.S.W., Gullick. xxiv+244 pp. & 4 pl.

Ba501. **O'Brian, P.** 1987. *Joseph Banks: A life.* London, C. Harvill. 328+[8] pp. *Also* Boston, MA, David R. Godine, 1993.

See also Ba12 Ba303 Ca715 & Ca718–719.

Barclay, J.

See Ba303.

Barton, B. S.

Ba502. **Ewan, J.** 1974. A survey of the Benjamin Smith Barton papers in the library of the American Philosophical Society. *Yearbook of the American Philosophical Society* 1974: 569–571.

Ba503. **MacPhail, I.** 1986. *The Sterling Morton Library bibliographies in botany and horticulture, III: Benjamin Smith Barton and William Paul Crillon Barton.* Lisle, IL, Morton Arboretum. [ix]+39 pp., ports.

See also Ba205.

Bartram, J. & W.

Ba504. **Bartram, J.** 1992. *The correspondence of John Bartram 1734–1777.* Ed. by E. & D. S. Berkeley. Gainesville, FL, Univ. Presses of Florida. xv+809 pp.

Ba505. **Berkeley, E. & D. S.** 1982. *The life and travels of John Bartram: From Lake Ontario to the River St. John.* Tallahassee, FL, Univ. Presses of Florida. xvi+376 pp.

Ba506. **Cutting, R. M.** 1976. *John and William Bartram, William Byrd II and St. John de Crevecoeur: A reference guide.* Boston, MA, G. K. Hall. xxiii+174 pp.

Ba507. **Darlington, W.** 1849. *Memorials of John Bartram and Humphrey Marshall, with notices of their botanical contemporaries.* Philadelphia, PA, Lindsay & Blakiston. xvi+585 pp. & 1 pl. *Re-pub.* New York, Hafner, 1967.
Also **J. Ewan.** 1957. Analytical index to Darlington's Memorials of Bartram and Marshall (1849). *Journal of the Society for the Bibliography of Natural History* 3: 274–280, port.

Ba508. **Earnest, E.** 1940. *John and William Bartram: Botanists and explorers.* Philadelphia, PA, Univ. of Pennsylvania Pr. vii+187 pp. & 2 pl. *Pennsylvania lives.*

Ba509. **Ewan, J.**, ed. 1968. *William Bartram: Botanical and zoological drawings, 1756–1788. Reproduced from the Fothergill Album in the British Museum (Natural History).* Ed. with an introduction and commentary by J. Ewan. Philadelphia, PA. x+180 pp. & 59 pl. *Memoirs, American Philosophical Society* vol. 74.

Ba510. **Fagin, N. B.** 1933. *William Bartram: Interpreter of the American landscape.* Baltimore, MD, Johns Hopkins Univ. Pr. xi+229 pp.

See also Ba205 Ca711–714 & Cb904.

Batalin, A. F.

Ba511. **Manoilenko, K. V.** (Ryazanskaya). 1962. *A. F. Batalin – vydayushchiisya Russkii botanik XIX veka.* Moscow/Leningrad, AN SSSR. 132 pp.

Bates, H. W.

Ba512. **Dickenson, J. P.** 1990. Henry Walter Bates and the study of Latin America in the later nineteenth century: A bibliographic essay. *Inter-American review of bibliography* 40: 570–580.

Ba512A. **Woodcock, G.** 1969. *Henry Walter Bates: Naturalist of the Amazons.* London, Faber & Faber. 269 pp. & 7 pl. *Great travellers.*

Bateson, W.

Ba513. **Bateson, B.** 1928. *William Bateson, F.R.S., naturalist: His essays and addresses together with a short account of his life.* Cambridge, Cambridge Univ. Pr. x+473 pp.

Ba514. **Cock, A. G.** 1977. The William Bateson papers. *Mendel newsletter* 14: 1–4.
At Cambridge University Library.

Ba515. **Coleman, W.** 1968. Bateson papers. *Mendel newsletter* 2: 1–3.

Ba516. **Harvey, R. D.** 1985. The William Bateson letters at the John Innes Institute. *Mendel newsletter* 25: 1–11.

Bauer, F.

Ba517. **Britten, J.** 1909. Ferdinand Bauer's drawings of Australian plants. *Journal of botany* 47: 140–146.

Ba518. **Norst, M. J.** 1988. *Ferdinand Bauer: The Australian natural history drawings.* London, British Museum (Natural History). 120 pp., illus. *Art in natural history series* no. 1.

Bauhin, J. & C.

Ba519. **Hess, J. W.** 1860. *Kaspar Bauhin's, des ersten Professors der Anatomie und Botanik an der Universität Basel, Leben und Charakter. Beitrag zur vierten Säcularfeier der Universität Basel.* Basle, Schweighauser. ii+72 pp. *Beiträge für vaterländische Geschichte* vol. 7.

Ba520. **Legré, L.** 1904. *La botanique en Provence au XVIe siècle. Les deux Bauhin, Jean Henri Cherler, et Valerand Dourez.* Marseille. xi+117 pp.

See also Cb1026.

Beattie, R. K.

See Bc855 for his botanical papers.

Beebe, W.

Ba521. **Berra, T. M.** 1977. *William Beebe: An annotated bibliography.* Hamden, CT, Archon. 157 pp.

Beijerinck, M. W.

Ba522. **Iterson, G. van, Jr., L. E. Den Dooren De Jong, & A. J. Kluyver.** 1940. *Martinus Willem Beijerinck: His life and work.* The Hague, Nijhoff. xi+193 pp. & 13 pl.

Belleval, P. R. de

Ba523. **Planchon, J.-É.** 1869. *Pierre Richer de Belleval, fondateur du Jardin des Plantes de Montpellier.* Montpellier, J. Martel ainé. 72 pp.

Belon, P.

Ba524. **Delaunay, P.** 1926. *Pierre Belon naturaliste.* Le Mans, Monnoyer. [vi]+271 pp.

See also Cb168.

Beneden, E. van

Ba525. **Rabl, K.** 1915. Edouard van Beneden und der gegenwärtige Stand in der Wichtigsten von ihm behandelten Probleme. *Archiv für mikroskopische Anatomie und Entwicklungsgeschichte* 88: 1–470.

Beneden, P. J. van

Ba526. **Kemna, A.** 1897. *P. J. van Beneden, la vie et l'oeuvre d'un zoologiste.* Antwerp, J.–E. Buschmann. 137 pp.

Bentham, G.

Ba527. **Jackson, B. D.** 1906. *George Bentham.* London, Dent. xii+292 pp. *English men of science.*

See also Ba133.

Berkeley, M. J.

Ba528. **Peterson, R. H.** 1980. *"B. & C.": The mycological association of M. J. Berkeley and M. A. Curtis.* Vaduz, Cramer. 120 pp. *Bibliotheca mycologica* vol. 72.

See also Ba185.

Bergius, B.

Ba529. **Kungliga Vetenskapsakademiens Bibliotek, Stockholm.** 1938. *Den Bergianska avskriftssamlingen i Kungl. Vetenskapsakademiens bibliotek. Systematisk förteckning.* By A. Holmberg. Stockholm. xii+137 pp. *Kungl. Bibliotekets handlingar, bilagor* n.f. 4.

Also **A. Holmberg.** 1938. Bengt Bergius och hans avskriftssamling: Randanteckningar till ett register. [Bengt Bergius and his letter collection: Marginal notes for an index]. *Lychnos* 1938: 199–224.

Bernard, C.

Ba530. **Albury, W. R.** 1982. Going beyond the introduction: Claude Bernard scholarship in the 1970s, an essay review. *Bulletin of the history of medicine* 56: 271–275.

Ba531. **Duval, M.** 1881. *L'œuvre de Claude Bernard.* Introduction. Notices par E. Renan, P. Bert et A. Moreau. Table alphabétique et analytique des œuvres complètes de Claude Bernard par R. de la Coudraie. Bibliographie des travaux scientifiques, mémoires, lectures et communications aux académies et sociétés savantes par G. Malloizel. Paris, J. B. Baillière et Fils. viii+333 pp. & port.

Ba532. **Grmek, M. D.** 1967. *Catalogue des manuscrits de Claude Bernard. Avec la bibliographie de ses travaux imprimés et des études sur son œuvre.* Paris, Collège de France et Masson & Cie. 419 pp.

Ba533. **Holmes, F. L.** 1974. *Claude Bernard and animal chemistry: The emergence of a scientist.* Cambridge, MA, Harvard Univ. Pr. xx+541 pp.

Ba534. **Mauriac, P.** 1954. *Claude Bernard.* Paris, Grasset. 236 pp.

Ba535. **Olmsted, J. M. D.** 1938. *Claude Bernard, physiologist.* New York & London, Harper & Bros. xvi+272 pp. & 5 pl.

Ba536. **Schiller, J.** 1967. *Claude Bernard et les problèmes scientifiques de son temps.* Paris, Les Éditions du Cèdre. 237 pp.

Ba537. **Virtanen, R.** 1960. *Claude Bernard and his place in the history of ideas.* Lincoln, NE, Univ. of Nebraska Pr. x+156 pp.

Besler, B.

Ba538. **Baier, H.** 1970. Die Ausgaben des Hortus Eystettensis 1613–1750. *Börsenblatt für den Deutschen Buchhandel, Frankfurter Ausgabe* 95: A273–A280.

Ba539. **Universitätsbibliothek Erlangen–Nürnberg.** 1989. *Hortus Eystettensis: Zur Geschichte eines Gartens und eines Buches. Eine Ausstellung der Universitätsbibliothek, 28. November–16 Dezember 1989.* Ed. by H.–O. Kuenecke. Munich, Schirmer–Mosel. 175 pp. *Schriften der Universitätsbibliothek Erlangen–Nürnberg* no. 20.

Bessey, C. E.

Ba539A. **Overfield, R. A.** 1993. *Science with practice: Charles E. Bessey and the maturing of American botany.* Ames, IA, Iowa State Univ. Pr. xiii+276 pp., illus.

Ba540. **University of Nebraska, Archives & Special Collections.** [1984.] *Charles E. Bessey papers (1865–1915).* Lincoln, NE. 37 reels microfilm & printed guide.

Ba541. ___. 1984. Guide and index to the microfilm edition of Charles E. Bessey papers (1865–1915). Ed. by J. G. Svoboda & P. Churray. *University of Nebraska studies* n.s. 67: 1–198.

Bewick, T.

Ba542. **Bain, I.** 1970. Thomas Bewick; engraver, of Newcastle, 1753–1828. A checklist of his correspondence and other papers. *Private library* (2) 3: 57–77, 124–140.
Another edition: *Re-pub.* (with addenda) [London?], privately pub. 46 pp. & 12 pl.

Ba542A. ___. 1981. *The watercolours and drawings of Thomas Bewick and his workshop apprentices.* Ed. & introduced by I. Bain. London, Gordon Fraser. 2 vols. *Re-pub.* Winchester, Thomas Bewick Birthplace Trust in association with St. Paul's Bibliographies, 1989.
Colour reproductions of over 200 water–colours with the corresponding engraved versions, together with a further 192 drawings and engravings.

Ba543. **Bewick, T.** 1862. *A memoir of Thomas Bewick written by himself. Embellished by numerous engravings by the author for a work on British fishes.* London. xix+344 pp. Another ed., ed. with an introduction by Iain Bain. London, Oxford Univ. Pr., 1975. xliv+258 pp.

Ba544. **Newcastle upon Tyne Public Libraries Committee.** [1904.] *Catalogue of the Bewick collection (Pease bequest)*, by B. Anderton & W. H. Gibson. Newcastle upon Tyne. vi+110 pp.
331 items. Including Books; Albums & Collections; Biographies, criticisms & catalogues; Sale catalogues; Wood–blocks, copper–plates, etc.; Framed engravings and original sketches by Thomas Bewick; Portraits, personal relics, etc.

Ba545. **Roscoe, S.** 1953. *Thomas Bewick: A bibliography raisonné of editions of the* General history of quadrupeds, *the* History of British birds, *and the* Fables of Aesop *issued in his lifetime.* London, Oxford Univ. Pr. xxx+198 pp. *Re-pub.* London, Dawsons, 1973.

Ba546. **Victoria & Albert Museum.** 1953. *Exhibition at the Bethnal Green Museum to commemorate the bicentenary of the birth of Thomas Bewick 1753–1828. Catalogue.* London. 32 pp.
Portraits; Drawings; Books; Impressions from wood blocks; Manuscripts; Personal relics; Works by pupils.

See also Ba303 & Bc331(Mengel).

Bianor, Herman

See Fricquegnon, M. E.

Bigelow, J.

See Ba133 & Ba205.

Blake, E.

Ba547. **Ellwood, C. V., & J. M. V. Harvey.** 1990. The Lady Blake collection: Catalogue of Lady Edith Blake's collection of drawings of Jamaican lepidoptera and plants. *Bulletin of the British Museum (Natural History), historical series* 18: 145–202.

Blanchard, É.

Ba548. **Blanchard, E. de.** 1851. *Notice sur les travaux d'anatomie et de zoologie de M. Émile Blanchard (1835–1850).* Paris, Plon. 20 pp.

Bleeker, P.

Ba549. **Bleeker, P.** 1881. Levensbericht van Dr. P. Bleeker. Geschriften van Dr. P. Bleeker over Ichthyologie, in chronologische volgorde. Register ... [etc.] *Natuurkundig tijdschrift voor Nederlandsch–Indië* 40: 1–158.

Ba550. **Weber, M. C. W., & L. F. De Beaufort.** 1911. Index of the ichthyological papers of P. Bleeker. In: *The fishes of the Indo–Australian Archipelago*, vol. 1. Leiden, L. J. Brill. xi+410 pp. *Re-pub.* 1964.
Includes Bibliography, pp. 1–45; List of his genera & species, pp. 47–410.

Bock, H.

Ba551. **Roth, F. W. E.** 1898. Hieronymus Bock, genannt Tragus. *Botanisches Centralblatt* 74: 265–271, 313–318, 344–347.

Ba552. ___. 1899. Hieronymus Bock, Prediger, Arzt und Botaniker, 1498–1554. In seinem Leben und Wirken dargestellt. Nebst Bibliographie der Schriften Bocks. *Mitteilungen des Historischen Vereins der Pfalz* 23: 25–74. *Also pub. separately* Speier, 1899. 50 pp.

Bodenheimer, F. S.

Ba553. **Bodenheimer, F. S.** 1959. *A biologist in Israel: A book of reminiscences.* Jerusalem, Biological Studies Publisher. 492 pp. & 45 pl.

Boerhaave, H.

Ba554. **Heniger, J.** 1971. Some botanical activities of Herman Boerhaave, Professor of Botany and Director of the Botanic Garden at Leiden. *Janus* 58: 1–78.

Ba555. **Lindeboom, G. A.** 1959. *Bibliographia Boerhaaviana.* Leidon, E. J. Brill. viii+118 pp. *Analecta Boerhaaviana* vol. 1.
Replaces **M. Hertzberger & E. J. Van der Linden.** 1927. *Short–title catalogue of books written and edited by Herman Boerhaave.* Amsterdam, M. Hertzberger. 27 pp., port.

Ba556. ___, ed. 1962–79. *Boerhaave's correspondence.* Leidon, E. J. Brill. 3 vols. *Analecta Boerhaaviana* vols. 3, 5, & 8.

Ba557. ___. 1963. *Iconographia Boerhaavii.* Leiden, Brill. ix+31 pp. & 40 pl. *Analecta Boerhaaviana* vol. 4.

Ba558. ___, ed. 1970. *Boerhaave and his time. Papers read at the international symposium in commemoration of the tercentenary of Boerhaave's birth Leiden 15–16 November 1968.* Leiden, Brill. [ix]+174 pp. *Analecta Boerhaaviana* vol. 6.
13 papers by various authors.

Ba559. ___. 1968. *Herman Boerhaave, the man and his work.* London, Methuen. xx+452 pp. & 34 pl.

Boissier, E.

Ba560. **Burdet, H. M.**, ed. 1985. *Edmond Boissier, botaniste Genèvois 1810–1885–1985.* Geneva, Conservatoire et Jardin Botaniques. 49 pp. *Serie documentaire* no. 17.

Ba561. **Chodat, F., et al.** 1937. *Edmond Boissier, botaniste Genèvois, 1810–1885.* Geneva. ii+76 pp. & 10 pl. *Bulletin de la Société Botanique de Genève* (2) 28.

See also Ba133.

Bolivar, I.

Ba562. **Ruiz, M. C., et al.** 1988. *Ignacio Bolivar y las ciencias naturales en España.* Madrid, Consejo Superior de Investigaciones Cientificas. xvi+205+[4] pp., illus.
Re-publication of a tribute first pub. Madrid, Imprenta Clásica Española, 1921, with the addition of an introduction and appendix by A. Gomis Blanco.

Bolton, J.

Ba563. **Crossland, C.** 1910. *An eighteenth century naturalist: James Bolton, Halifax.* Halifax. 32 pp. First pub. in *Halifax Guardian* . Re-pub. has slight alterations.

Bonaparte, C. L.

Ba564. **Bonaparte, C. L.**, Prince. 1850. *Notice sur les travaux zoologiques de M. Charles Lucien Bonaparte, membre correspondant de l'Institut de France et des Académies de Pétersbourg, Berlin, etc.* Paris, Bachelier. 36 pp. *Revue et magasin de zoologie* 1850.

Bonnet, C.

Ba565. **Anderson, L.** 1982. *Charles Bonnet and the order of the known.* Dordrecht, D. Reidel. xv+159 pp. *Studies in the history of modern science* vol. 11.

Ba565A. **Bibliothèque Publique et Universitaire, Genève.** 1993. *Catalogue de la correspondance de Charles Bonnet. Conservée à la Bibliothèque de Genève.* Geneva, Ville de Genève, Département Municipal des Affaires Culturelles. 134+[5] pp., port. *Manuscrits scientifiques genevois* #1.

Ba566. **Bonnet, C.** 1948. *Mémoires autobiographiques de Charles Bonnet de Genève.* Paris, J. Vrin. 414 pp. *Bibliothèque des textes philosophiques.*

Ba567. **___, G.** 1929. *Charles Bonnet.* Paris, M. Lac. 319 pp.

Ba568. **Riquet, V. A. C. de.** 1859. *Ch. Bonnet, philosophe et naturaliste, sa vie et ses oeuvres.* Paris, A. Vaton. lxiv+436 pp.

Ba569. **Rostand, J.** 1966. *Un grand biologiste: Charles Bonnet, expérimentateur et théoricien.* Paris, Université de Paris. 39 pp. *Les conférences du Palais du Découverte* série D, no. 107.

Bonpland, A.

Ba570. **Bouvier, R., & E. Maynial.** 1949. *Der Botaniker von Malmaison: Aimé Bonpland, ein Freund Alexander von Humboldts.* Neuwied, Lancelot Verlag. 287 pp.

Ba571. **___.** 1950. *Aimé Bonpland, explorateur de l'Amazonie, botaniste de la Malmaison, planteur en Argentine, (1773–1858).* Paris, Société d'Édition d'Énseignement Supérieur. 193 pp. & 4 pl.

Ba572. **Cordier, H.** 1910. *Papiers inédits du naturaliste Aimé Bonpland conservés à Buenos Aires.* Buenos Aires, J. Peuser. 25 pp. *Trabajos del Instituto de Botanica y Farmacologia, Facultad de Ciencias Médicas de Buenos Aires* no. 30.
At the Instituto de Botánica y Farmacología.

Ba573. **Foucault, P.** 1990. *Le pêcheur d'orchides: Aimé Bonpland, 1773–1858.* Paris, Seghers. 333 pp.

Ba574. **Hamy, E.-T.** [1906.] *Aimé Bonpland, médecin et naturaliste: Explorateur de l'Amérique du Sud: Sa vie, son oeuvre, sa correspondence, avec un choix de pièces relatives à sa biographie.* Paris, E. Guilmoto. xcvi+300 pp. & port.

See also Ca761–763.

Boott, F.

See Ba133.

Borlase, W.

Ba575. **Pool, P. A. S.** 1986. *William Borlase.* Truro, Royal Institution of Cornwall. x+314 pp.

Bory de St. Vincent, J.–B.–G.–M.

Ba576. **Bory de St. Vincent, J.–B.–G.–M.** 1908. *Correspondance de Bory de St. Vincent.* Publiée et annotée par P. Lauzun. Agen, Maison d'Édition et d'Imprimerie Moderne. 358 pp., ports.

Ba577. **Lacroix, A.** 1916. *Notice historique sur Bory de Saint–Vincent ... lue dans la séance publique annuelle du 18 decembre 1916.* Paris, Gauthier–Villars. 75 pp., ports.

Borza, A.

Ba578. **Resmerita, I., & I. T. Tarnavschi.** 1976. *Alexandru Borza: Biolog român ctitor al Gradinii Botanice din Cluj–Napoca.* Bucharest, Editura Litera. 160 pp.

Bosc d'Antic, L. A. G.

Ba579. **Greiner, M.**, ed. 1989. *L. A. G. Bosc d'Antic: Un citoyen fraternal au temps de la Révolution Française de 1789.* Paris, Lycée Camille See. 198 pp.
"Un dossier biographique réalisé par la classe de 2ème 4, dans le cadre d'un projet d'action éducative intitulé: "Autour de L. A. G. Bosc, naturliste et girondin."

Bougainville, L. A. de.

See Ca716–717.

Boulenger, G. A.

Ba580. **Boulenger, G. A.** 1921. Liste des publications ichthyologiques et herpetologiques (1877–1920) de G. A. Boulenger. *Annales de la Société Royale Zoologique et Malacologique de Belgique* 52: 11–88.

Bourguignat, J.–R.

Ba581. **Dance, S. P.** 1970. List of J.–R. Bourguignat's publications on molluscs. *Journal of conchology, London* 27: 87–95.

Boveri, T.

Ba582. **Baltzer, F.** 1967. *Theodor Boveri: Life and work of a great biologist, 1862–1915.* Translated from the German by D. Rudnick. Berkeley & Los Angeles, CA, Univ. of California Pr. [xiv]+165 pp. & 6 pl.
Translation of *Theodor Boveri: Leben und Werk.* Stuttgart, Wissenschaftliche Verlagsgesellschaft, 1962. vii+194 pp. & 1 pl.

See also Ba21.

Braun, A.

Ba583. **Mettenius, C.** 1882. *Alexander Braun's Leben nach seinem handschriften Nachlass.* Berlin, Reimer. viii+706 pp. & port.

See also Ba133.

Brehm, A.

Ba584. **Brehm Gesellschaft.** 1929. *Das Brehmbuch. Zum hundertsten Geburtstag von Alfred Brehm.* Berlin, Brehm Verlag. 142 pp., illus.

Ba585. **Schmitz, S.** 1984. *Tiervater Brehm: Seine Reisen, sein Leben, sein Werk.* Munich, Harnack. 280 pp.

Bretschneider, E.

Ba586. **Walravens, H.** 1983. *Emil Bretschneider – russischer Gesandtschaftsarzt: Geograph und Erforscher der chinesischen Botanik. Eine Bibliographie.* Hamburg, Bell Verlag. vi+42 pp. *Han-pao tung-ya shu-chi mu-lu* no. 22.

Breyne, J. P.

Ba587. **Grau, C.** 1968. The scientific correspondence of the Gdansk physician and botanist Johann Philipp Breyne. *Actes du XIe Congrès International d'Histoire des Sciences* 1965 2: 158–161.
In the Landesbibliothek Gotha.

Brightwell, T.

Ba588. **Brightwell, C. L.** 1869. *Memorials of the life of Mr. Brightwell of Norwich.* Norwich, privately pub. 117 pp.

Briquet, J.

Ba589. **Briquet, J. I.** 1900. *Inventaire des publications scientifiques de M. John Briquet.* Geneva. 24 pp.

Ba590. **Crumière–Briquet, V.** 1935. *John Briquet (1870–1931).* Paris, Muséum National d'Histoire Naturelle. 115+[3] pp. & 5 pl.

Brocchi, G.

Ba591. **Berti, G.** 1988. *Un naturalista dall'Ancien Régime alla Restaurazione: Giambattista Brocchi (1772–1826).* Bassano, Verci 156 pp.

Brongniart, A.–T. & A.

Ba592. **Launay, L. de.** 1940. *Une grande famille de savants: Les Brongniarts.* Paris, G. Rapilly et Fils. 208 pp., ports.
i.e. Adolphe–T. & Alexandre Brongniart. Bibliographie d'Alexander Brongniart, pp. 196–208.

Brooks, W. K.

Ba593. **Benson, K. R.** 1983. William Keith Brooks (1848–1908): Primary source material. *Mendel newsletter* 23: 7–11.

Broussonet, A.

Ba594. **Caillé, J.** 1972. *Un savant Montpelliérain: Le Professeur Auguste Broussonet (1761–1807).* Paris, Editions A. Pedone. 183 pp.

See also Ba12

Brown, J. A. Harvie–.

Ba595. **Royal Scottish Museum.** 1983. *Manuscripts in the Royal Scottish Museum.* Pt. 3: *J. A. Harvie–Brown papers.* By J. Pitman. Edinburgh. x+89 pp. *Royal Scottish Museum information series: natural history* no. 9.

Brown, Rachel.

See Cb638.

Brown, Robert.

Ba596. **Mabberley, D.** 1984. *Jupiter botanicus: Robert Brown of the British Museum.* Braunschweig, J. Cramer; London, British Museum (Natural History). 500 pp.

See also Ba185.

Bruce, J.

Ba596A. **Reid, J. M.** 1968. *Traveller extraordinary: The life of James Bruce of Kinnaird.* London, Eyre & Spottiswoode. 320 pp., illus.

See also Ba303.

Bruce, W. S.

Ba597. **Brown, R. N. R.** 1923. *A naturalist at the Poles: The life, work and voyages of Dr. W. S. Bruce, the polar explorer.* With five chapters by W. G. B. Murdoch. London, Seeley, Service & Co. Pp. 11–316, 26 pl., & 3 maps.

Ba598. **Royal Scottish Museum.** 1982. *Manuscripts in the Royal Scottish Museum.* Pt. 2: *William S. Bruce papers and diary of A. Forbes Mackay.* By J. Pitman. Edinburgh. 67 pp. *Royal Scottish Museum information series: natural history* no. 8.

Brunfels, O.

Ba599. **Roth, F. W. E.** 1900. Die Schriften des Otto Brunfels, 1519–1536. *Jahrbuch für Geschichte, Sprache und Literatur Elsass-Lothringens* 16: 257–288.

Ba600. ___. 1900. Otto Brunfels 1489–1534: Ein deutscher Botaniker. *Botanische Zeitung* 58: 191–232.
Includes Beschreibung der botanischen Schriften des Brunfels und ihrer Ausgaben 1530–1537, pp. 225–232.

Ba601. **Sanwald, E.** 1932. *Otto Brunfels 1488–1534: Ein Beitrag zur Geschichte des Humanismus der Reformation.* I. Hälfte. *1488–1524.* Bottrop, W. Postberg. iii+47 pp.

Ba602. **Schmid, A.** 1936. *Zwei seltene Kräuterbücher aus dem vierten Dezennium des sechzehnten Jahrhunderts. Nebst einem bibliographischen Verzeichnis aller bekannten frühen Ausgaben der Brunfelsschen Kräuterbücher als Anhang.* Berne. 20 pp. From *Schweizerisches Gutenbergmuseum* vol. 22.

Buchanan–Hamilton, F.

Ba603. **Prain, D.** 1905. *A sketch of the life of Francis Hamilton (once Buchanan), sometime superintendent of the Honourable Company's Botanic Garden, Calcutta.* Calcutta, Bengal Secretariat Pr. lxxv pp. *Annals, Royal Botanic Gardens, Calcutta* vol. 10 (2).

Buckland, F.

Ba604. **Bompas, G. C.** 1885. *Life of Frank Buckland.* London, Smith & Elder. ix+433 pp.

Ba605. **Burgess, G. H. O.** 1967. *The curious world of Frank Buckland.* London, J. Baker. xii+242 pp. & 15 pl.

Buffon, G. L. L.

Ba606. **Buffon, G. L. L.** 1860. *Correspondence inedité de Buffon, à laquelle ont été réunies les lettres publiées jusqu'à ce jour [etc.].* Ed. by H. Nadault de Buffon. Paris, L. Hachette & Cie. 2 vols. *Re-pub. in* J.–L. Lanessan, ed. 1884–85. *Oeuvres complètes de Buffon*, vols. 13–14. Paris, A. Pilon.

Ba607. **Fellows, O. E., & S. E. Milliken.** 1972. *Buffon.* New York, Twayne. 186 pp. *Twayne's world, authors series: A survey of the world's literature* no. 243.

Ba608. **Flourens, M. J. P.** 1860. *Des manuscrits de Buffon. Avec des fac-similé de Buffon et de ses collaborateurs.* Paris, Garnier. xcv+298 pp. *Re-pub.* Geneva, Slatkine Reprints, 1971. vi+xcvi+300 pp.

Ba609. ___. 1870. *Histoire des travaux et des idées de Buffon.* 3e éd. Paris, Garnier Frères. vii+388 pp.

Ba610. **Genet–Varcin, E., & J. Roger.** 1954. Bibliographie de Buffon. In: J. Piveteau, ed. *Oeuvres philosophiques de Buffon*, pp. 513–575. Paris, Presses Universitaires de France.

Ba611. **Grineveld, P.–M., et al.** 1988. *Buffon, 1788–1988.* Introduction de Jean Dorst. Paris, Imprimerie Nationale. 296 pp.
With chapters by various authors.

Ba612. **Hanks, L.** 1966. *Buffon avant l'histoire naturelle.* Paris, Presses Universitaires de France. 324 pp. *Recherches* vol. 24.

Ba613. **Heim, R., et al.** 1953. *Buffon.* Paris, Publications Françaises. 245 pp. & 25 pl. *Collection des "Grandes naturalistes français."*
With chapters by various authors.

Ba614. **Roger, J.** 1989. *Buffon: Un philosophe au Jardin du Roi.* Paris, Fayard. 645 pp.

Ba615. **Roule, L.** 1924. *Buffon et la description de la nature.* Paris, Flammarion. 248 pp. *L'Histoire de la nature vivante d'après des grands naturalistes Français* no. 1.

See also Ba26 & Ba303.

Bunbury, Sir C. J. F.

Ba616. **Bunbury, C. J. F.** 1895. *Life, letters, and journals of Sir Charles J. F. Bunbury, Bart.* Ed. by his wife, Frances J. Bunbury. [London.] 3 vols.

Ba617. **Lyell, K. M.** 1906. *The life of Sir Charles J. F. Bunbury, Bart.* With an introductory note by Sir J. Hooker. Ed. by Mrs H. Lyell. London, J. Murray. 2 vols.

Burckhardt, J. L.

See Ba303.

Burnat, É.

Ba618. **Burnat, E.** 1922. *Émile Burnat: Autobiographie publiée avec une étude sur le botaniste et son oeuvre, des souvenirs et documents divers.* By J. Briquet & F. Cavillier. Geneva, Conservatoire Botanique. vii+185 pp. & port.

Bute, Earl of

See Stuart, J.

Camper, P.

Ba619. **Daniëls, C. E.** 1880. *Het leven en de verdiensten van Petrus Camper.* Utrecht, J. W. Leeflang. 150 pp., port.

Ba620. **___., & H. C. Rogge.** 1881. *Catalogue de manuscrits de Pierre Camper et de lettres inédites, écrites par lui ou à lui adressées, qui se trouvent dans la Bibliothèque de la Société Néerlandaise pour les Progrès de la Médecine à Amsterdam.* Amsterdam. [iv]+18 pp.

Ba621. **Visser, R. P. W.** 1985. *The zoological work of Petrus Camper (1722–1789).* Amsterdam, Editions Rodopi. [iv]+207 pp. *Nieuwe Nederlandse bijdragen tot de Geschiedenis der Geneeskunde en der Natuurwetenschappen* no. 12.

See also Ba303.

Candolle family

Ba622. **Hochreutiner, B. P. G.** 1898. *Une famille de botanistes. Les Candolles.* Geneva, C. Eggimann & Cie. 43 pp. & 4 port. *Suisse universitaire* ann. 3, nos. 23–25.

Candolle, A. L. P. P. de

Ba623. **Mikulinskii, S. R., L. A. Markova, & B. A. Starostini.** 1980. *Alphonse de Candolle (1806–1893).* Translated from Russian by G. Ludwig. Jena, G. Fischer. 224 pp. *Biographien bedeutender Biologen* vol. 3.

Candolle, A. P. de.

Ba624. **Candolle, A. P. de.** 1862. *Mémoires et souvenirs d'Augustin–Pyramus de Candolle écrits par lui même et publiées par son fils.* Geneva & Paris, Cherbuliez. xvi+599 pp.

Ba625. **De la Rive, A.** 1851. *Notice sur la vie et les écrits de A.–P. de Candolle.* Paris & Geneva, J. Cherbuliez. [iv]+312 pp.

Ba626. **Miège, J.** 1979. Augustin–Pyramus de Candolle: Sa vie, son oeuvre, son action à travers la Société de Physique et d'Histoire Naturelle de Genève. *Memoires de la Société de Physique et d'Histoire Naturelle de Genève* 43(1): 45 pp., illus.

See also Ba133.

Carson, R.

Ba627. **Gartner, C.** 1983. *Rachel Carson.* New York, Frederick Ungar. xiv+161 pp. *Literature and life.*

Ba628. **Sterling, P.** 1970. *Sea and earth: The life of Rachel Carson.* New York, Thomas Y. Crowell. viii+213 pp. *Women of America.*

Carter, H. Gilbert.

Ba629. **Gilmour, J. S. L., & S. M. Walters.** 1975. *Humphrey Gilbert–Carter: A memorial volume.* Cambridge, University Botanic Garden. iv+132 pp.

Cartier, J.

See Cb257.

Carus, C. G.

See Ca46.

Carver, G. W.

Ba630. **Holt, R.** 1943. *George Washington Carver: An American biography.* New York, Doubleday. xi+342 pp., illus.

Castiglioni, L.

See Ca744.

Catesby, M.

Ba631. **Frick, G. F., & R. P. Stearns.** 1961. *Mark Catesby: The colonial Audubon.* Urbana, IL, Univ. of Illinois Pr. x+137 pp. & 16 pl.

Cavanilles, A.

Ba632. **Reyes Prósper, E.** 1917. *Dos noticias históricas del inmortal botánico y sacerdote hispano–valentino Don Antonio José Cavanilles, por Don Antonio Cavanilles y Centi y Don Mariano Lagasca, con anotaciones y los estudios biobibliográficos de Cavanilles y Centi y de la Gasca.* Madrid, "Mateu". 265 pp., ports.

Cesalpino, A.

Ba633. **Viviani, U.** 1922. *Vita ed opere di Andrea Cesalpino (con nuove aggiunte alla prima edizione).* Arezzo, U. Viviani. xi+237 pp., illus. *Collana di pubblicazioni storiche e letterarie Aretine* no. 6.

Chapman, A.

Ba634. **Chapman, A.** 1928. *Retrospect: Reminiscences and impressions of a hunter–naturalist in three continents, 1851–1928.* London, Gurney & Jackson. xix+353 pp. & 53 pl.

Ba635. **___.** 1930. *Memories of four-score years less two, 1851–1920.* With a memoir by G. Bolam. London, Gurney & Jackson. xxviii+257 pp. & 28 pl.

Ba636. **Sawyer, Chas. J.**, (Firm). 1979. *Abel Chapman 'Naturalist'. A collection of his books, drawings and manuscripts.* London. 24 pp. *Catalogue* no. 302.
A bookseller's catalogue; includes 25 lots of material.

Chapman, F. M.

Ba637. **Chapman, F. M.** 1933. *Autobiography of a bird–lover.* New York, D. Appleton–Century Co. xiv+420 pp., illus.

Chase, A.

Ba638. **Stieber, M. T.** 1979. Manuscripts produced and/or annotated by Agnes Chase pertinent to grass collections at the Smithsonian Institution. *Huntia* 3: 117–125.

Cheesman, E.

Ba639. **Cheesman, E.** 1957. *Things worth while.* London, Hutchinson. 330 pp., illus.
Autobiography.

Clark, W.

See Lewis, M.

Clayton, J.

Ba640. **Berkeley, E. & D. S.** 1963. *John Clayton: Pioneer of American botany.* Durham, NC, North Carolina Univ. Pr. x+236 pp.

Ba641. **___.** 1965. *The Reverend John Clayton: A parson with a scientific mind. His scientific writings and other related papers.* Charlottesville, VA, Univ. Pr. of Virginia for the Virginia Historical Society. lxiii+170 pp.

See also Ba205.

Clements, F.

Ba642. **Clements, E. S.** 1960. *Adventures in ecology: Half a million miles ... from mud to macadam.* New York, Pageant Pr. 244 pp., illus.

Clift, W. H.

Ba643. **Austin, F.**, ed. 1983. *The letters of William Home Clift 1803–1832.* Sturminster Newton, Meldon House. ix+115 pp.

Ba644. ___. 1991. *The Clift family correspondence 1792–1846.* Sheffield, Centre for English Cultural Tradition & Language. xxii+265 pp. *CECTAL occasional publication* no. 5.

Ba645. **Dobson, J.** 1954. *William Clift.* London, Heinemann Medical Books. viii+144 pp., illus.

Also see Ba1332.

Cockerell, T. D. A.

Ba646. **Weber, W. A.** 1965. *Theodore Dru Alison Cockerell, 1866–1948.* Boulder, CO, Univ. of Colorado Pr. [iv]+124 pp. & port. *University of Colorado studies, series in bibliography* no. 1.

Cohn, F.

Ba647. **Cohn, P.** 1901. *Ferdinand Cohn: Blätter der Erinnerung.* 2nd ed. Breslau, J. U. Kern's Verlag. x+266 pp. & 3 pl.

Coiter, V.

Ba648. **Herrlinger, R.** 1952. *Volcher Coiter, 1534–1576*. Nuremberg, M. Edelmann. 147 pp., illus. *Beiträge zur Geschichte der medizinischen und naturwissenschaftlichen Abbildung* vol. 1.

Colden, C.

Ba649. **Gray, A.**, ed. 1843. *Selections from the scientific correspondence of Cadwalader Colden with Gronovius, Linnaeus, Collinson, and other naturalists*. Arranged by [etc.]. New Haven, CT, B. L. Hamlen. 51 pp.
First pub. in: *American journal of science & arts* 44: 85–133, 1843.

See also Ba205.

Collinson, P.

Ba650. **James, N. G. Brett–.** [1926.] *The life of Peter Collinson F.R.S., F.S.A.* London, the Author. 292 pp. & 16 pl.

Commerson, P.

Ba651. **Chaumartin, H.** 1967. *Philibert Commerson, médecin naturaliste du roy et compagnon de Bougainville*. Vienne, Ternet–Martin. 32 pp. *Petite histoire de la médecine*.

Ba652. **Laissus, Y.** 1978. Catalogue des manuscrits de Philibert Commerson (1727–1773) conservés à la Bibliothèque Centrale de Muséum National d'Histoire Naturelle (Paris). *Revue d'histoire des sciences* 31: 131–162.

Ba653. **Montessus, F.–B. de.** 1889. *Martyrologe et biographie de Commerson: Médecin–Botaniste et Naturaliste du Roi au XVIIIe siècle*. Chalon–sur–Saône, L. Marceau. vii+225 pp. & port.

Ba654. **Oliver, S. P.** 1909. *The life of Philibert Commerson, D.M. Naturaliste du Roi: An old–world story of French travel and science in the days of Linnaeus*. Ed. by G. F. S. Elliot. London, J. Murray. xvii+242 pp. & 11 pl.

Comstock family

Ba655. **Comstock, A. B.** 1953. *The Comstocks of Cornell*. Ed. by G. W. Herrick & R. G. Smith. Ithaca, NY, Comstock Publ. Associates. 286 pp. & 5 pl.

Conklin, E. G.

Ba656. **Allen, G. E., & D. M. McCullough.** 1968. Notes on source materials: The Edward Grant Conklin papers at Princeton University. *Journal of the history of biology* 1: 325–332.

See also Cc89.

Cook, J.

Ba657. **Beaglehole, J. C.** 1974. *The life of Captain James Cook*. London, A. & C. Black. xi+759 pp., illus. Also pub. as vol. 4 of *The journals of Captain James Cook*. London, Hakluyt Society, 1974.

Ba658. **Hennig, E.** 1952. *James Cook*. Stuttgart, Wissenschaftliche Velagsgesellschaft. viii+141 pp. *Grosse Naturforscher*.

Ba659. **Library of New South Wales.** 1970. *Bibliography of Captain James Cook R.N., F.R.S., circumnavigator*. 2nd ed. Editor K. M. Beddie. Sydney, N.S.W. xvi+894 pp.

See also Ca721.

Cooke, M. C.

Ba660. **English, M. P.** 1987. *Mordecai Cubitt Cooke: Victorian naturalist, mycologist, teacher and eccentric.* Bristol, Biopress Ltd. 357 pp.

Cooper, J. G.

Ba661. **Coan, E. V.** 1982. *James Graham Cooper: Pioneer western naturalist.* Moscow, ID, Univ. Pr. of Idaho. 255 pp., illus. *Northwest naturalist books.*

Cope, E. D.

Ba662. **Osborn, H. F.,** with **H. A. Warren.** 1931. *Cope: Master naturalist. The life and letters of Edward Drinker Cope, with a bibliography of his writings classified by subject. A study of the pioneer and foundation periods of vertebrate palaeontology in America.* Princeton, NJ, Princeton Univ. Pr. xvi+740 pp., illus. Re-pub. New York, Arno Pr., 1978.

See also Cc283.

Correa de Serra, J. F.

Ba663. **Davis, R. B.** 1955. The Abbé Correa in America, 1812–1820: The contributions of the diplomat and natural philosopher to the foundations of our national life. *Transactions of the American Philosophical Society* 45(2): 85–197.

Correns, C.

Ba664. **Saha, M. S.** 1981. The Carl Correns papers. *Mendel newsletter* 21: 1–6.

Coues, E.

Ba665. **Cutright, P. R., & M. J. Broadhead.** 1981. *Elliot Coues: Naturalist and frontier historian.* Chicago, IL, Univ. of Chicago Pr. xv+509 pp.

Coulter, J. M.

Ba666. **Rodgers, A. D., 3rd.** 1944. *John Merle Coulter: Missionary in science.* Princeton, NJ, Princeton Univ. Pr. x+321 pp.

Cranch, J.

Ba667. **Monod, T.** 1970. John Cranch, zoologiste de l'expedition du Congo (1816). *Bulletin of the British Museum (Natural History), historical series* 4: 1–75, & 3 pl.

Crépin, F.

Ba668. **Herring, P.** 1930. *François Crépin.* Copenhagen, P. Haase & Søns Forlag. 156 pp. & port.
Text in Danish.

Croizat, L.

Ba669. **Heads, M., & R. C. Craw.** 1984. Bibliography of the scientific work of Leon Croizat, 1932–1982. *Tuatara* 27: 67–75.

Cuming, H.

See Cc865.

Cunningham, A.

Ba670. **McMinn, W. G.** 1970. *Allan Cunningham, botanist and explorer.* Melbourne, Vic., Melbourne Univ. Pr. ix+147 pp. & 4 pl.

Curtis, J.

Ba671. **Ordish, G.** 1974. *John Curtis and the pioneering of pest control.* Reading, Osprey. vii+121 pp., illus.

Curtis, M. A.

Ba672. **Berkeley, E. & D. S.** 1986. *A Yankee botanist in the Carolinas: The Reverend Moses Ashley Curtis, D.D. (1808–1872).* Berlin & Stuttgart, J. Cramer in der Gebrüder Borntraeger Verlagsbuchhandlung. 242 pp.

See also Ba133 & Ba528.

Curtis, W.

Ba673. **Curtis, W. H.** 1941. *William Curtis, 1746–1799, Fellow of the Linnean Society: Botanist and entomologist. With some notes on his son–in–law Samuel Curtis.* Winchester, Warren. 142 pp.

Cusick, W. C.

See Bc855 for his botanical papers.

Cuvier, G. L. C. F. D.

Ba674. **Coleman, W.** 1964. *Georges Cuvier zoologist: A study in the history of evolution theory.* Cambridge, MA, Harvard Univ. Pr. x+212 pp.

Ba675. **Dehérain, H.** 1908–22. *Catalogue des manuscrits du fonds Cuvier (travaux et correspondence scientifiques) conservés à la bibliothèque de l'Institut de France.* Paris, H. Champion, Hendaye. 2 vols.
First pub. in *Revue des bibliothèques.*

Ba676. **Dujarric de la Rivière, R.**, ed. 1969. *Cuvier, sa vie, son oeuvre. Pages choisies.* Paris, Peyronnet. 116 pp.

Ba677. **Duvernoy, G. L.** 1833. *Notice historique sur les ouvrages et la vie de M. le Baron Cuvier.* Paris, F. G. Levrault. 172 pp.
Includes Ouvrage, mémoires et rapports de M. Cuvier dans l'ordre de leur publication, pp. 155–172.

Ba678. **Flourens, P.** 1841. *Analyse raisonnée des travaux de Georges Cuvier, précedée de son éloge historique.* Paris, Chez Paulin. 287 pp.
Includes Liste des ouvrages de M. Cuvier, pp. 273–287.

Ba679. ___. 1858. *Histoire des travaux de Georges Cuvier.* 3e éd. Paris, Garnier Frères. 295 pp.

Ba680. **Outram, D.** 1976. Scientific biography and the case of Georges Cuvier: With a critical bibliography. *History of science* 14: 101–137.
Bibliography, pp. 119–137.

Ba681. ___. 1980. *The letters of Georges Cuvier: A summary calendar of manuscript and printed materials preserved in Europe, U.S.A. and Australasia.* Chalfont St. Giles, British Society for the History of Science. iv+102 pp. *Monograph* no. 2.

Ba682. ___. 1984. *Georges Cuvier: Vocation, science and authority in post–revolutionary France.* Manchester, Manchester Univ. Pr. viii+299 pp.

Ba683. **Roule, L.** 1933. *Cuvier et la science de la nature.* Paris, Flammarion. 246 pp. *L'Histoire de la nature vivante d'après l'oeuvre des grands naturalistes Français* no. 3.

Ba683A. **Smith, J. C.** 1993? *Georges Cuvier: Annotated bibliography of his published works.* Washington, DC, Smithsonian Institution Press. 352? pp.
Forthcoming publication announced.

See also Ba26 Ba303 Cc55 & Cc462–462A.

Dal Pozzo, Cassiano.

See Ca978.

Dall, W. H.

Ba684. **Bartsch, P., H. A. Rehder, & B. E. Shields.** 1946. A bibliography and short biographical sketch of William Healey Dall. *Smithsonian miscellaneous collections* 104(15): 1–96.

Darlington, W.

Ba685. **Lansing, D. I.** 1985. *That magnificent cestrian, Dr. William Darlington 1782–1863, being a short introductory biography.* Paoli, PA, Serpentine Pr. xi+87 pp.

See also Ba205 & Bc421(Cavanagh).

Darwin, C. R.

• Limitation of space permits no more than the following selection from the large corpus of literature on Darwin's life and ideas:

Ba686. **Allan, M.** 1977. *Darwin and his flowers: The key to natural selection.* London, Faber & Faber. 318 pp., illus. *Also* New York, Taplinger, 1977.

Ba687. **Bell, P. R.**, ed. 1959. *Darwin's biological work, some aspects reconsidered.* Cambridge, Cambridge Univ. Pr. xiii+343 pp. & 2 pl.
Includes **P. R. Bell.** The movement of plants in response to light, pp. 1–49; **J. Challinor.** Palaeontology and evolution, pp. 50–100; **J. B. S. Haldane.** Natural selection, pp. 101–149; **P. Marler.** Developments in the study of animal communication, pp. 150–206; **H. L. K. Whitehouse.** Cross– and self–fertilization in plants, pp. 207–261; **J. S. Wilkie.** Buffon, Lamarck and Darwin: The originality of Darwin's theory of evolution, pp. 262–307.

Ba688. **Bowlby, J.** 1990. *Charles Darwin: A biography.* London, Hutchinson. xiv+511 pp. American ed. as: *Charles Darwin: A new life.* New York, W. W. Norton, 1990.

Ba689. **Bowler, P. J.** 1990. *Charles Darwin: The man and his influence.* Oxford, B. Blackwell. xii+250 pp. *Blackwell science biographies series.*

Ba690. **British Museum (Natural History).** 1910. *Memorials of Charles Darwin: A collection of manuscripts ... books and ... specimens to commemorate the centenary of his birth and the fiftieth anniversary of the publication of "The Origin of Species."* 2nd ed. [By W. G. Ridewood.] London. [v]+50 pp. & 2 pl. *Special guide* no. 4. *Re-pub.* (with an introduction by S. Smith) in *Bulletin of the British Museum (Natural History), historical series* 14(3): 235–298, 1988.

Ba691. **Cambridge University, Christ's College.** 1909. *Darwin centenary: The portraits, prints and writings of Charles Robert Darwin, exhibited at Christ's College, Cambridge.* Cambridge, Cambridge Univ. Pr. [iv]+47 pp.

Ba692. **Cambridge University, Library.** 1960. *Handlist of Darwin papers at the University Library.* By R. V. C. Bailey & J. S. Gosse. Cambridge. 72 pp.

Ba693. **Carroll, P. T.** 1976. *An annotated calendar of the letters of Charles Darwin in the library of the American Philosophical Society.* Wilmington, DE, Scholarly Resources, Inc. xli+235 pp.

Ba694. **Chapman, R. G., & C. T. Duval**, eds. 1982. *Charles Darwin 1809–1882: A centennial commemorative.* Wellington, N.Z., Nova Pacifica. xii+376 pp.
Chapters by R. O'Hanlon, J. R. Durant, J. R. Moore, R. J. Berry, H. E. Gruber, R. T. Keegan & M. Ruse.

Ba695. **Clark, R. W.** 1984. *The survival of Charles Darwin: A biography of a man and an idea.* London, Weidenfeld & Nicholson. x+449 pp. & 16 pp. of illus. *Also* New York, Random House.

Ba696. **Darwin, C. R.** 1958. *The autobiography of Charles Darwin, 1809–1882*. With original omissions restored. Ed. with appendix and notes by his granddaughter Nora Barlow. London, Collins. 253 pp.

Ba697. ___. 1985+. *Correspondence of Charles Darwin.* Ed. by F. Burkhardt, S. Smith, et al. Cambridge, Cambridge Univ. Pr. Vol. 1+.
Latest seen, vol. 8: *1860*, 1993. Each vol. includes a biographical register and index to correspondents. For background information *see* **W. Montgomery.** 1987. Editing the Darwin correspondence: A quantitative perspective. *British journal for the history of science* 20: 13–27.

Ba698. ___. 1990+. *Charles Darwin's marginalia.* Ed. by M. A. Di Gregorio with assistance of N. W. Gill. New York, Garland Publishing. Vol. 1+ (lxii+895 pp. [etc.]). *Research guides in the history of science & technology* no. 783.
Catalogue to and transcription of marginal annotations in Darwin's books. Vol. 1, 1990; [etc.]. To be completed in 3 vols.

Ba699. ___**, F.**, ed. 1887. *The life and letters of Charles Darwin.* London, J. Murray. 3 vols. *Also* New York, D. Appleton & Co., 1896. 2 vols. *Re-pub.* (with a foreword by G. G. Simpson) New York, Basic Books, 1959; *Continued* ___ **& A. C. Seward**, eds. 1903. *More letters of Charles Darwin: A record of his work in a series of hitherto unpublished letters.* London, J. Murray. 2 vols. *Also* New York, D. Appleton & Co., 1903.

Ba700. **De Beer, G. R.**, ed. 1974. *Autobiographies. Charles Darwin, T. H. Huxley.* Ed. with an introduction by [etc.]. Oxford, Oxford Univ. Pr. xxvi+123 pp., illus. *Oxford English memoirs and travels.*

Ba701. **Desmond, A., & J. Moore.** 1991. *Darwin.* London, M. Joseph. xxi+808 pp., illus. *Also* New York, Warner Books, 1992.

Ba702. **Freeman, R. B.** 1977. *The works of Charles Darwin: An annotated bibliographical handlist.* 2nd ed., rev. & enl. Folkestone, Dawson; Hamden, CT, Archon Books. 235 pp.; *Continued* ___. 1986. The works of Charles Darwin, additions and corrections to the second edition. London, privately distributed by the Author. 22 pp.
Replaces 1st ed., London, Dawsons of Pall Mall, 1965. x+81 pp. *Also* **R. B. Freeman.** 1986. Darwin in Chinese. *Archives of natural history* 13: 19–24; **P. J. P. Whitehead.** 1988. Darwin in Chinese: Some additions. *Archives of natural history* 15: 61–62.

Ba703. ___. 1978. *Charles Darwin: A companion.* Folkestone, W. Dawson & Sons; Hamden, CT, Archon Books. 309 pp.

Ba704. ___. 1984. *Darwin's pedigrees.* London, the author. 84 pp.
Includes a reprint of H. F. Burke's "Pedigree of the family of Darwin" 1888, with corrections, additions, and an index of ca. 900 Darwins by Freeman.

Ba705. **Huxley, J. S., & H. B. Kettlewell.** 1965. *Charles Darwin and his world.* New York, Viking Pr. 144 pp.
A pictorial view of Darwin's world.

Ba706. **Manier, E.** 1978. *The young Darwin and his cultural circle: A study of influences which helped shape the language and logic of the first drafts of the theory of natural selection.* Dordrecht, Boston, MA, D. Reidel Publishing Co. ix+242 pp.

Ba707. **Sothebys**, [Firm]. 1992. *Darwin's century: The Jeremy Norman Collection*; *see* Ca499.
Lots 71–193 are Darwin rarities, letters, MSS, etc.

See also Ba133 & Ca777–781.

• Further information on the literature of Darwin studies may be found in the following surveys:

Ba708. **Colp, R., Jr.** 1989. Charles Darwin's past and future biographies. *History of science* 27: 167–197.

Ba709. **Greene, J. C.** 1975. Reflections on the progress of Darwin studies. *Journal of the history of biology* 8: 243–273.

Ba710. **La Vergata, A.** 1985. Images of Darwin: A historiographic overview. In: D. Kohn, ed. *The Darwinian heritage*, pp. 901–972. Princeton, NJ, Princeton Univ. Pr.

Ba711. **Lenoir, T.** 1987. Essay review: The Darwin industry. *Journal of the history of biology* 20: 115–130.

Ba712. **Loewenberg, B. J.** 1965. Darwin and Darwin studies, 1959–1963. *History of science* 4: 15–54.
Bibliography, pp. 47–54.

Ba713. **Oldroyd, D. R.** 1984. How did Darwin arrive at his theory? The secondary literature to 1982. *History of science* 22: 325–374.

Ba714. **Wassersug, R. J., & M. R. Rose.** 1984. A reader's guide and retrospective to the 1982 Darwin centennial. *Quarterly review of biology* 59: 417–437.
Identifies 70 publications and over 100 symposia, lecture series, etc.

Darwin, E.

Ba715. **Darwin, E.** 1981. *The letters of Erasmus Darwin*. Ed. by D. King–Hele. Cambridge, Cambridge Univ. Pr. xxxii+363 pp., illus.

Ba716. **Hele, D. King–.** 1963. *Erasmus Darwin*. London, Macmillan & Co. viii+183 pp.

Ba717. **Krause, E. L.** 1879. *Erasmus Darwin*. Translated by W. S. Dallas. With a preliminary notice by Charles Darwin. London, J. Murray. iv+216 pp. & port. *Re-pub.* Farnborough, Gregg International, 1971.
Translation of *Erasmus Darwin und seine Stellung in der Geschichte der Descendenz–Theorie*. Leipzig, E. Günther, 1880. vi+236 pp.

Ba718. **McNeil, M.** 1987. *Under the banner of science: Erasmus Darwin and his age*. Manchester, Manchester Univ. Pr. viii+307 pp.

Ba719. **Pearson, H.** 1930. *Doctor Darwin*. London, J. M. Dent & Sons. xi+242 pp., illus.

Ba720. **Seward, A.** 1804. *Memoirs of the life of Dr. Darwin, chiefly during his residence at Lichfield. With anecdotes of his friends, and criticisms on his writings*. London, J. Johnson. xiv+430+[2] pp.

See also Cb1177.

Daubenton, L.–J.–M.

Ba721. **Roule, L.** 1925. *Daubenton et l'exploitation de la nature*. Paris, Flammarion. 246 pp. *L'Histoire de la nature vivante d'après des grands naturalistes Français* no. 2.

See also Ba2

Davaine, C.–J.

Ba722. **Théodoridès, J.** 1968. *Un grand médecin et biologiste: Casimir–Joseph Davaine (1812–1882)*. Oxford, Pergamon Pr. 238 pp. & 11 pl. *Analecta medico–historica* vol. 4.

David, A.

Ba723. **Bishop, G.** 1990. *Travels in imperial China: The explorations and discoveries of Pere David*. London, Cassell. 192 pp., illus.

See also Ca795.

Davis, W. T.

Ba724. **Abbott, M.** 1949. *The life of William T. Davis.* Ithaca, NY, Cornell Univ. Pr. xv+321 pp., illus.
Includes List of the published works, pp. 288–315.

Davydov, K. N.

Ba725. **Blyakher, L. Ya.** 1963. *Konstantin Nikolaevitch Davydov.* Moscow, Akademiya Nauk SSSR. 244 pp., illus. *Nauchno–biograficheskaya seriya.*

Deam, C.

Ba726. **Kriebel, R. C.** 1987. *Plain ol' Charlie Deam: Pioneer hoosier botanist.* Lafayette, IN, Purdue Univ. Pr. x+183 pp.

De Bary, A.

Ba727. **Jost, L.** 1930. *Zum hundertsten Geburtstag Anton de Barys.* Jena, G. Fischer. 74 pp. First pub. in *Zeitschrift für Botanik* vol. 24.

Ba728. **Parnes, V. A.** 1972. *Anton de Bari 1831–1888.* Moscow, Izd. "Nauka." 191 pp., illus.
Text in Russian.

See also Ba2

Decaisne, J.

See Ba133.

Delacour, J.

Ba729. **Delacour, J.** 1966. *The living air: The memoirs of an ornithologist.* London, Country Life. 173 pp. & 17 pl.

Delessert, B.

See Cb1027.

Deschamps, L.–A.

Ba730. **Hocquette, M.** 1970. *Louis–Auguste Deschamps 1765–1842: Sa vie – son œuvre.* Saint–Omer, Société des Antiquaires de la Morinie. 76 pp. *Bulletin trimestriel de la Société des Antiquaires de la Morinie, mémoire* 39.

Desfontaines, R.

Ba731. **Chevalier, A.** 1939. *La vie et l'oeuvre de René Desfontaines: Fondateur de l'herbier du Museum.* Paris, Museum National d'Histoire Naturelle. 264 pp. & 8 pl. *Publications* no. 4.

Dethier, V. G.

See Ba298.

De Vries, H.

Ba732. **De Veer, P. H. W. A. M.** 1969. *Leven en werk van Hugo de Vries.* Groningen, Wolters–Noordhoff. viii+252 pp., illus.

See also Ba2

Dick, R.

Ba733. **Smiles, S.** 1878. *Robert Dick, baker of Thurso, geologist and botanist.* London, J. Murray. xx+436 pp., port & illus.

Diderot, D.

Ba734. **Paitre, F.** 1904. *Diderot biologiste [etc.].* Lyon, A. Storck & Cie. vi+106 pp. *Re-pub.* Geneva, Slatkine Reprints, 1971.

Dillenius, J. J.

See Cb1028.

Ditmars, R. L.

Ba735. **Wood, L. N.** 1944. *Raymond L. Ditmars – His exciting career with reptiles, animals and insects.* New York, J. Messner x+272 pp.

Dobbs, A.

Ba736. **Raukun, D. H., & E. C. Nelson**, eds. 1990. *Curious in everything: The career of Arthur Dobbs of Carrickfergus 1689–1765.* Carrickfergus, Carrickfergus & District Historical Society. x+37 pp., illus.

Dobzhansky, T.

Ba737. **Dobzhansky, T.** 1980. *The roving naturalist: Travel letters of Theodosius Dobzhansky.* Ed. by B. Glass. Philadelphia, PA, American Philosophical Society. 327 pp. *Memoirs of the American Philosophical Society* vol. 139.

Ba738. **Fuller, M. B.** 1980. Theodosius Dobzhansky papers. *Mendel newsletter* 18: 1–8.

Dodoens, R.

Ba739. **Meerbeeck, P. J. van.** 1841. *Recherches historiques et critiques sur la vie et les ouvrages de Rembert Dodoens (Dodonæus).* Malines, Hanicq. xiv+340 pp. *Re-pub.* Utrecht, H & S, 1980.
Includes a bibliography, pp. 285–292.

Dohrn, A.

Ba740. **Heuss, T.** 1991. *Anton Dohrn: A life for science.* Ed. by C. Groeben. Translated from the German by L. Dieckmann. Berlin, [etc.], Springer–Verlag. xxxvi+401 pp., port.
Translation of *Anton Dohrn in Neapel.* Stuttgart & Tubingen, R. Winderlich, 1962.

Ba741. **Kühn, A.** 1950. Anton Dohrn und die Zoologie seiner Zeit. *Pubblicazione della Stazione Zoologica di Napoli* [22] 1950 Supp: 205 pp.
Includes Verzeichnis der Schriften A. Dohrns, pp. 186–190.

Ba742. **Simon, H.–R.** 1980. *Anton Dohrn und die Zoologische Station Neapel.* Frankfurt am Main, Erbrich. 164 pp. *Bibliographie et scientia* no. 1.

See also Ba21.

Dombey, J.

Ba743. **Hamy, E.–T.** 1905. *Joseph Dombey, médecin, naturaliste, archéologue, explorateur du Pérou, du Chili et du Brésil (1778–1785). Sa vie, son oeuvre, sa correspondence.* Paris, E. Guilmoto. cx+434 pp. & port.

See also Ca731–735.

Dos Passos, C. F.

Ba744. **Wilkinson, R. S.** 1988. Annotated bibliography of the entomological publications of Cyril F. Dos Passos (1887–1986). *Journal of the Lepidopterists' Society* 42: 168–183.

Douglas, D.

Ba745. **Harvey, A. G.** 1947. *Douglas of the fir.* Cambridge, MA, Harvard Univ. Pr. x+290 pp., illus.

Ba746. **Morwood, W.** 1973. *Traveller in a vanished landscape: The life and times of David Douglas.* London, Gentry Books. [xii]+244 pp. *Re-pub.* New York, C. Potter, 1973.

Drake, D.

Ba747. **Horine, E. F.** 1961. *Daniel Drake (1785–1852): Pioneer physician of the Midwest.* Philadelphia, PA, Univ. of Pennsylvania Pr. 425 pp., port.

Drury, D.

Ba748. **Sherborn, C. D.** 1937. Dru Drury. *Journal of the Society for the Bibliography of Natural History* 1: 109–111.
An index to Drury's letter book containing treanscriptions of his letters to scores of naturalists between 1761 & 1783.

See also Ba303.

Du Bois–Reymond, E.

Ba749. **Akademie der Wissenschaften und der Literatur, Kommission für Geschichte der Medizin und der Naturwissenschaften.** 1981. *Naturwissen und Erkenntniss im 19. Jahrhundert: Emil du Bois–Reymond.* Ed. by G. Mann. Hisdesheim, Gerstenberg. 243 pp.
Papers by various authors given at a meeting held in June 6th–8th, 1979.

Ba750. **Du Bois–Reymond, E.** 1978–81. *Briefe an Emil du Bois–Reymond / Ernst Wilhelm von Brücke.* Hrsg. und bearb. von Hans Brücke ... [etc.] Graz, Akad. Druck– und Verlagsgesellschaft. 2 vols. *Publikationen aus dem Archiv der Universität Graz* vol. 8.

Ba751. **___.** 1985. *Emil du Bois–Reymond (1818–1896), Anton Dohrn (1840–1909): Briefwechsel.* Hrsg. von C. Groeben ... mit K. Hierholzer. Mit einer historischen Einführung von E. Florey. Berlin, Springer–Verlag. xliv+322 pp., ports.

Ba752. **___, Estelle**, ed. 1912. *Reden von Emil du Bois–Reymond.* Mit einer Gedachtnisrede von Julius Rosenthal. 2nd ed. Leipzig, Von Veit. 2 vols.

Ba753. ___, ed. 1982. *Two great scientists of the nineteenth century: Correspondence of Emil du Bois–Reymond and Carl Ludwig.* Collected by Estelle du Bois–Reymond. Foreword, notes and indexes by P. Diepgen. Translated into English by S. L. Ayed. Edited with a foreword by P. F. Cranefield. Baltimore, MD, Johns Hopkins Univ. Pr. xix+183 pp., ports.
Translation of *Zwei grosse Naturforscher des 19. Jahrhunderts: Ein Briefwechsel zwischen [etc.]*. Leipzig, J. A. Barth, 1927.

Ba754. **Rothschuh, K. E., & E. Tutte.** 1975. Emil Du Bois–Reymond (1818–1896): Bibliographie. Originalien und Sekundärliteratur. In: K. Mothes & J.–H. Scharf, eds. *Beiträge zur Geschichte der Naturwissenschaften und der Medizin; Festschrift für Georg Uschmann ... zum 60. Geburtstag am 18. Oktober 1873*, pp. 113–136. Halle, Deutsche Akademie der Naturforscher Leopoldina. *Acta historica Leopoldina* no. 9.

Dufour, L.

Ba755. **Duris, P., & E. Diaz.** 1987. *Petite histoire naturelle de la première moitié du XIXe siècle: Léon Dufour, Correspondant de l'Institut (1780–1865).* Bordeaux, Presses Universitaires de Bordeaux. 308 pp.

Dugès, A.

Ba756. **Smith, H. M. & R. B.** 1969. *Early foundations of mexican herpetology: An annotated and indexed bibliography of the herpetological publications of Alfredo Dugès, 1826–1910.* Urbana, IL, Univ. of Illinois Pr. 85 pp.

Dugmore, A. R.

Ba757. **Thomas, L. J.** 1931. *Rolling stone: The life and adventures of Arthur Radclyffe Dugmore ... with ... illustrations from photographs and paintings by A. R. Dugmore.* Garden City, NY, Doubleday, Doran & Co. viii+311 pp.
Another edition, London, John Long, 1933. 256 pp.

Duhamel, M.

See Ba12.

Dumeril, A. C.

Ba758. **Froissart, L. D.** 1984. *André Constant Dumeril: Médecin et naturaliste: 1774–1860.* Paris, the Author. 268 pp. & 25 pl.

Dumont d'Urville, J. S. C.

Ba759. **Guillon, J.** 1986. *Dumont d'Urville 1790–1842.* Paris, Éditions France–Empire. 355 pp., illus.

Duncan, J.

Ba760. **Jolly, W.** 1883. *The life of John Duncan, Scotch weaver and botanist, with sketches of his friends and notices of the times.* 2nd ed. London, K. Paul, Trench & Co. xix+516 pp.

Dunn, L. C.

Ba761. **Bearman, D.** 1976. L. C. Dunn papers. *Mendel newsletter* 12: 1–5.

Dürer, A.

Ba762. **Koreny, F.** 1988. *Albrecht Dürer and the plant and animal studies of the Renaissance.* Boston, MA, Little, Brown & Co., New York Graphic Society. 278 pp., illus.
Translation of *Albrecht Dürer und die Tier– und Pflanzenstudien der Renaissance*, exhibition held 18th April – 30 June, 1985, Graphische Sammlung Albertina, Vienna, pub. in Munich, Prestel–Verlag, 1985.

Dyche, L. L.

Ba763. **Sharp, B., & P. Sullivan.** 1990. *The dashing Kansan: Lewis Lindsay Dyche. The amazing adventures of a nineteenth–century naturalist and explorer.* Harrow Books in association with the Museum of Natural History, University of Kansas. xiv+222 pp., illus.

Eaton, A.

Ba764. **McAllister, E. M.** 1941. *Amos Eaton: Scientist and educator, 1776–1842.* Philadelphia, PA, Univ. of Pennsylvania Pr. 587 pp.

Edward, T.

Ba765. **Smiles, S.** 1877. *The life of a Scotch naturalist, Thomas Edward, Associate of the Linnean Society [etc.].* 5th ed. London, J. Murray xix+438 pp., 9 pl. & port.

See also Ba26.

Edwards, G.

Ba766. **Anon.** 1776. *Some memoirs of the life and works of George Edwards, Fellow of the Royal and Antiquarian Societies.* London, J. Robson. 1 f.+38 pp. & 5 pl.
The biographical sketch occupies pp. 1–26.

Ba766A. **Mason, A. S.** 1992. *George Edwards: The Bedell and his birds.* London, Royal College of Physicians. viii+71 pp., illus.

Edwards, V. C. Wynne–

See Ba298.

Ehrenberg, C. G.

Ba767. **Laue, M. von.** 1895. *Christian Gottfried Ehrenberg: Ein Vertreter deutscher Naturforschung im neinzehnten Jahrhundert, 1795–1876. Nach seinen Reiseberichten, seinem Briefwechsel mit A. v. Humboldt, v. Chamisso, Darwin, v. Martius u. a., Familienaufzeichnungen, sowie andern handschriftlichen Material.* Berlin, J. Springer. vi+287 pp. & port.
Includes Schriftenverzeichniss, pp. 264–287.

Ehret, G. D.

Ba768. **Calmann, G.** 1977. *Ehret: Flower painter extraordinary.* Oxford, Phaidon Pr. 160 pp. *Also* Boston, MA, New York Graphic Society, 1977.
Includes Ch. 9 "The drawings" (a survey of surviving collections.)

Ehrhart, F.

Ba769. **Alpers, F.** 1905. *Friedrich Ehrhart ... Mitteilungen aus seinem Leben und seinem Schriften. Unter Benutzung von bislang nicht veröffentlichen Urkunden, sowie von Briefen Ehrharts und seiner Witwe.* Leipzig, W. Engelmann. xvi+452 pp.

Eibl–Eibesfeldt, I.

See Ba298.

Eiseley, L.

Ba770. **Christianson, G. E.** 1990. *Fox at wood's edge: A biography of Loren Eiseley.* New York, H. Holt & Co. x+517 pp.

Ba771. **Heidtmann, P.** 1991. *Loren Eiseley: A modern Ishmael.* Hamden CT, Archon. xi+133 pp.

Ellis, J.

Ba772. **Linnean Society of London.** 1948. *Calendar of the Ellis manuscripts. (The correspondence and miscellaneous papers of John Ellis, F.R.S.).* By S. Savage. London. vi+104 pp. *Catalogue of manuscripts in the library of the Linnean Society of London* pt. 4.

Ellis, W. W.

Ba773. **Alexander Turnbull Library.** 1977. The [William W.] Ellis drawings: A symposium. *Turnbull Library record* n.s. 10: 5–37 & 11 pl.
Papers by M. E. Hoare (including an inventory), D. G. Medway (bird drawings), et al.

d'Entrecasteaux, B.

See Ca756–757.

Engelmann, G.

See Ba133 & Ba205.

Engler, A.

Ba774. **Diels, L.** 1931. Zum Gedächtnis von Adolf Engler. *Botanische Jahrbücher für Systematik, Pflanzengeschichte und Pflanzengeographie* 64: lxi pp.

Errera, L.

Ba775. **Fredericq, L., & J. Massart.** 1908. *Notice sur Léo Errera, Membre de l'Académie.* Brussels, Académie Royale de Belgique. 153 pp. First pub. in *Annuaire de l'Académie Royal de Belgique* vol. 74.

Evelyn, J.

Ba776. **Keynes,** Sir **G.** 1968. *John Evelyn: A study in bibliophily with a bibliography of his writings.* 2nd ed. Oxford, Clarendon Pr. xix+313 pp. & 16 pl.
Replaces 1st ed., Cambridge, Cambridge Univ. Pr., 1937. xx+308 pp. & 25 pl.

Ba777. **Ponsonby, A.** 1933. *John Evelyn, Fellow of the Royal Society; author of "Sylva."* London, W. Heinemann. xiii+350 pp.

Fabre, J.-H.

Ba778. **Congrès Jean-Henri Fabre.** 1987. *Jean-Henri Fabre: Anniversaire du jubilé (1910–1985). Actes du Congrès Jean-Henri Fabre, Paris et le Vaucluse, 13–18 mai 1985.* Paris, L. Léopard d'Or. 197 pp., illus.

Ba779. **Coulon, M.** 1924. *Le génie de J. H. Fabre.* Paris, Éditions du Monde Nouveau. xvi+284 pp. *Les emmais du XXe siècle* no. 2.

Ba780. **Cuny, H.** 1967. *Fabre ... choix de textes, bibliographie.* Paris, Seghers. 191 pp., illus. *Savants du monde entier* no. 35.

Ba781. **Delange, Y.** 1981. *Jean-Henri Fabre: Naturaliste total et pédagogue du XIXe siècle.* Paris, J.-C. Lattès. 351 pp., illus. *Re-pub.* Paris, Geneva, Champion-Slatkine, 1986.

Ba782. **Fabre, A.,** Abbé. [1921.] *The life of Jean Henri Fabre, the entomologist.* Translated by B. Miall. London, Hodder & Stoughton. xv+299 pp. *Also* New York, Dodd, Mead & Co., [1921]. xv+398 pp.

Ba783. **Legros, G. V.** 1925. *La vie de J. H. Fabre, naturaliste, suivi du répertoire général analytique des "Souvenirs entomologiques."* Paris, Delagrave. 444 pp.
Also English translation of 1st ed. by B. Miall: ___. 1913. *Fabre, poet of science.* With a preface by J. H. Fabre. London, Unwin. 352 pp. & port. *Re-pub.* New York, Horizon, 1971.

Ba784. **Mauron, M.** 1981. *Jean-Henri Fabre: À la rencontre de l'homme et du poète dans l'oeuvre du savant.* Avignon, A. Barthelemy. 252 pp. & 8 pl. *Collection récits du sud.*

See also Cc266.

Fabricius, J. C.

Ba785. **Fabricius, J. C.** 1928. Autobiographie des Naturforschers Fabricius. Anhang zur Fabricius-Bibliographie. In: J. Schuster, ed. *Linné und Fabricius: Zu ihrem Leben und Werk. Drei Faksimiles zu Linnés 150. Todestag mit einem Nachwort über das natürliche System.* Munich, Verlag der Münchner Drucke. 117+xxix pp., illus. *Münchener Beiträge zur Geschichte und Literatur der Naturwissenschaften und Medizin* Sonderheft 4.

See also Cc866.

Fairchild, D. G.

Ba786. **Fairchild, D. G.** 1938. *The world was my garden: Travels of a plant explorer. ...* assisted by E. & A. Kay. New York, C. Scribner's Sons. xiv+494 pp.

Farrer, R.

Ba787. **Illingworth, J., & J. Routh,** eds. 1991. *Reginald Farrer: Dalesman, planthunter, gardener.* Lancaster, University of Lancaster, Centre for North-West Regional Studies. x+102 pp. *Occasional papers* no. 19.
Contributions by various authors.

Farwell, O. A.

Ba788. **McVaugh, R., S. A. Cain & D. J. Hagenah.** 1953. Farwelliana: An account of the life and botanical work of Oliver Atkins Farwell, 1867–1944. *Bulletin of the Cranbrook Institute of Science* 34: [v]+101 pp.

Fawcett, B.

Ba789. **Desmond, R. G. C.** 1973. Benjamin Fawcett (1808–1893): Printer and engraver of natural history books. In: E. Geck & G. Pressler, eds. *Festschrift für Claus Nissen*, pp. 91–135. Wiesbaden, Pressler.
Includes a detailed bibliography of his natural history productions.

Ba790. **McLean, R.** 1988. *Benjamin Fawcett: Printer and engraver.* London, Scolar Pr. 200 pp., illus.
A detailed bibliography of all his productions.

Fedchenko, A. P.

Ba791. **Leonov, N. I.** 1972. *Aleksei Pavlovich Fedchenko.* Moscow, Izd. "Nauka." 167 pp.

Fischer–Piette, E.

Ba791A. **Backhuys, W.** 1990. *Edouard Fischer–Piette [1899–1988]: Biography, bibliography, new taxa.* Oegstgeest, Universal Book Services. 48 pp.

Fischer von Waldheim, J. G.

Ba792. **Buettner, J. W. E.** 1956. *Fischer von Waldheim: Leven und Wirken des Naturforschers Johann Gotthelf Fischer von Waldheim (1771–bis 1853), [etc.].* Berlin, Akademie Verlag. 83 pp. & 3 pl. *Freiburger Forschungshefte, Kultur & Technik* D15.

Fitch, A.

Ba793. **Barnes, J. K.** 1988. *Asa Fitch and the emergence of American entomology, with an entomological bibliography and a catalog of taxonomic names and type specimens.* Albany, NY, New York State Museum. viii+120 pp. *New York State Museum bulletin* no. 461.

Fitch, W. H.

Ba794. **Hemsley, W. B.** 1915. Walter Hood Fitch, botanical artist, 1817–1892. *Bulletin of miscellaneous information, Royal Botanic Gardens, Kew* 1915: 277–284.
Provides a bibliography of his published illustrations.

Ba794A. **Lewis, J.** 1992. *Walter Hood Fitch: A celebration.* London, HMSO. viii+34 pp. & 73 pp. of illus.

Flower, W. H.

Ba795. **Cornish, C. J.** 1904. *Sir William Henry Flower: A personal memoir.* London, Macmillan & Co. xi+274 pp. & 5 ports.

Ba796. **Lyddeker, R.** 1907. *Sir William Flower.* London, Dent. vii+191 pp., port. *English men of science.*

Forbes, E.

Ba797. **Rehbock, P. F.** 1979. Edward Forbes (1815–1854) — an annotated list of published and unpublished writings. *Journal of the Society for the Bibliography of Natural History* 9: 171–218.

Ba798. **Wilson, G., & A. Geikie.** 1861. *Memoir of Edward Forbes, F.R.S., late Regius Professor of Natural History in the University of Edinburgh.* Cambridge & London, Macmillan. x+589 pp. & port.

Forel, A.

Ba799. **Bibliothèque Cantonale et Universitaire, Lausanne.** 1969. *Inventaire des archives Auguste Forel, 1848–1831 (IS 1925 et 3765).* By S. Unterfichter & O. Pavilion. Lausanne. 91 pp. *Inventaires des fonds manuscrits* no. 5.

Ba800. **Forel, A.** 1937. *Out of my life and work.* Translated by B. Miall. London, Allen & Unwin. 352 pp. *Also* New York, W. W. Norton & Co., [1937].
Translation of *Rückblick auf mein Leben.* [Zurich,] Europa–Verlag, 1935. 295+vii pp.

Ba801. **Meier, R.** 1986. *August Forel, 1848–1931: Arzt, Naturforscher, Sozialreformer. Eine Ausstellung der Universität Zürich, Herbst 1986.* Zurich, Universität Zürich. 159 pp.

Forsskål, P.

Ba802. **Christensen, C.** 1918. *Naturforskeren Pehr Forsskål: Hans Rejse til Ægypten og Arabien 1761–63 og hans botaniske Arbejder og Samlinger [etc.].* Copenhagen, H. Hagerup. [viii]+172 pp. & port.

Ba803. **Schück, H.** [1923.] *Från Linnés tid: Petter Forsskål.* Stockholm, P. A. Norstedt. 491 pp.

Forster, J. G. A.

Ba804. **Fiedler, H.** 1971. *Georg–Forster–Bibliographie, 1767 bis 1970.* Berlin, Akademie–Verlag. 208 pp.

Ba805. **Iredale, T.** 1925. George Forster's paintings. *Australian zoologist* 4: 48–53, & pl. 6–8.

Ba806. **Kersten, K.** 1957. *Der Weltumsegler: Johann Georg Adam Forster 1754–1794.* Frankfurt am Main, Europäische Verlagsanstalt. 400 pp. & 4 pl.

Ba807. **Moleschott, J.** 1862. *Georg Forster, der Naturforscher des Volks.* 2nd ed. Hamm, G. Grote. [viii]+295 pp. & port.

Ba808. **Reintjes, H.** 1953. *Weltreise nach Deutschland: Johann Georg Forsters Leben und Bedeutung.* Dusselforf, Progress–Verlag. 207 pp.

Ba809. **Whitehead, P. J. P.** 1978. The [J. G. A. & J. R.] Forster collection of drawings in the British Museum (Natural History). *Bulletin, British Museum (Natural History), historical series* 6: 25–47.
Drawings by Georg Forster made during Capt. Cook's second circumnavigation.

See also Ca46.

Forster, J. R.

Ba810. **Hoare, M. E.** 1971. Johann Reinhold Forster (1729–1798): Problems and sources of biography. *Journal of the Society for the Bibliography of Natural History* 6: 1–8.

Ba811. ___. 1972. The Forster manuscripts in Berlin. *Journal of Pacific history* 7: 171–173.

Ba812. ___. 1976. *The tactless philosopher: John Reinhold Forster (1729–98).* Melbourne, Vic., Hawthorn Pr. x+419 pp. & 8 pl.

Fossey, D.

See Goodall, J. (etc.)

Foster, M.

Ba813. **Geison, G. L.** 1978. *Michael Foster and the Cambridge school of physiology: The scientific enterprise in late Victorian society.* Princeton, NJ, Princeton Univ. Pr. xix+401 pp.

Fothergill, J.

Ba814. **Fothergill, J.** 1971. *Chain of friendship: Selected letters of Dr. John Fothergill of London, 1735–1780.* With introduction and notes by B. C. Corner and C. C. Booth. Cambridge, MA, Belknap Pr. of Harvard Univ. Pr. xxiv+538 pp.

Ba815. **Fox, R. H.** 1919. *Dr. John Fothergill and his friends: Chapters in eighteenth century life*. London, Macmillan. xxiv+434 pp. & 13 pl.

Ba816. **Lettsom, J. C.** 1786. *Memoirs of John Fothergill, M.D., &c.* 4th ed. London, C. Dilly. [i]+viii+286 pp. & port.

Fountaine, M.

Ba817. **Fountaine, M.** 1980. *Love among the butterflies: The travels and adventures of a Victorian lady, Margaret Fountaine*. Ed. by W. F. Cater. London, Collins. 224 pp., illus.

Ba817A. ___. 1986. *Butterflies and late loves: The further travels and adventures of a Victorian lady*. Ed. by W. F. Cater. London, Collins. 141 pp., illus.

Fourcade, H. G.

Ba817B. **Storrar, C. D.** 1990. *The four faces of Fourcade: A biography of a remarkable forester, land surveyor, botanist and photogrammetrist*. Cape Town, Maskew Miller. ix+198 pp., illus.

Fowler, H. W.

Ba818. **Fowler, H. W., V. T. & M. E. Phillips.** 1965. Writings of Henry Weed Fowler, published from 1897 to 1965. *Proceedings of the Academy of Natural Sciences at Philadelphia* 117: 173–212.
Also **W. F. Smith–Vaniz & R. McC. Peck.** 1991. Contributions of Henry Weed Fowler (1878–1965), with a brief early history of ichthyology at The Academy of Natural Sciences of Philadelphia. *Ibid*. 143: 173–191.

Francis, G. W.

Ba819. **Best, B. J.** 1965. *The life and works of George William Francis*. Adelaide, S.A. [viii]+138+16+[1] pp. & 20 pl.

Ba819A. ___. 1986. *George William Francis, the first director of the Adelaide Botanic Garden: A biography*. [Adelaide, S.A.,] the Author in conjunction with Botanic Gardens of Adelaide. xii+182 pp.

Fricquegnon, M. E.

Ba820. **Rubi, S., & T. Malagarriga.** 1971. *Herman Bianor [i.e. M. E. Fricquegnon], educador y botánico*. Salamanca & Madrid, Instituto Pontifico "San Pio X." 178 pp. *Ediciones "Senderos"* no. 5.

Fries, E.

Ba821. **Eriksson, G.** 1962. *Elias Fries och den romantiska biologien*. Uppsala, Almqvist & Wiksell. ix+487 pp. *Lychnos–Bibliotek* no. 20.

See also Ba69.

Fries, E. M.

See Ba133.

Frisch, K. R. von

Ba822. **Frisch, K. R. von.** 1973. *Erinnerungen eines Biologen*. 3rd ed. Berlin, Springer Verlag. [vii]+196 pp.
For an earlier English version *see* ___. 1967. *A biologist remembers*. Translated by L. Gombrich. Oxford & New York, Pergamon Pr. ix+200 pp. *Monographs in history & philosophy of science* vol. 1.

See also Cc266.

Frohawk, F. W.

Ba823. **Chatfield, J.** 1988. *F. W. Frohawk: His life and work.* Marlborough, Crowood Pr. 184 pp.

Fuchs, L.

Ba824. **Roth, F. W. E.** 1898. Leonhard Fuchs, ein deutscher Botaniker, 1501–1566. *Beihefte zur Botanischen Centralblatt* 8: 161–191.
Includes a bibliography of his works.

Ba825. **Stübler, E.** 1928. *Leonhart Fuchs, Leben und Werk.* Munich, Verlag der Münchner Drucke. vii+133+[3] pp. & 3 pl. *Münchener Beiträge zur Geschichte und Literatur der Naturwissenschaften und Medizin* 13/14.
Includes a bibliography of his works.

Fuertes, L. A.

Ba826. **Boynton, M. F.**, ed 1956. *Louis Agassiz Fuertes: His life briefly told and his correspondence.* New York, Oxford Univ. Pr. 317 pp., illus.

Ba827. **Marcham, F. G.** 1971. *Louis Agassiz Fuertes and the singular beauty of birds. Paintings, drawings, letters, assembled and edited by [etc.].* New York, Harper & Row. xi+220 pp., 88 illus. & 30 col. pl.

Ba828. **Peck, R. McC.** 1982. *A celebration of birds: The life and art of Louis Agassiz Fuertes.* Introduction by Roger T. Peterson. New York, Walker & Co. xiii+178 pp.
Pub. for the Academy of Natural Sciences, Philadelphia. *Includes* Major institutional collections of paintings by Fuertes, pp. 165–168.

Fuller, J. L.

See Ba298.

Fusch, R.

Ba829. **Weil, E.** 1928. *Remaclus Fusch: Arzt, Botaniker, Pharmacologie ca. 1510–1587: Bio-bibliographischer Versuch.* Munich, Münchner Drucke. xii pp. *Münchener Beiträge zur Geschichte und Literatur der Naturwissenschaften und Medizin* Sonderheft no. 5.
Includes Bibliographie der Schriften von Remaclus Fusch, pp. x–xii.

Galdikas, B.

See Goodall, J. (etc.)

Galton, F.

Ba830. **Cowan, R. S.** 1985. *Sir Francis Galton and the study of heredity in the 19th century.* New York, Garland Publishing. xvi+292 pp. *History of hereditarian thought* no. 3.

Ba831. **Galton, F.** [1909.] *Memories of my life.* 3rd ed. London Methuen & Co. viii+339 pp. & 3 ports. *Re-pub.* New York, AMS Pr., 1974.

Ba832. **Pearson, K.** 1914–30. *Life, letters and labours of Francis Galton.* Cambridge, Cambridge Univ. Pr. 3 vols.

Ba833. **University College London, Library.** 1978. *A list of the papers and correspondence of Sir Francis Galton (1822–1911) held in the Manuscripts Room, The Library, University College London.* By M. Merrington & J. Golden. London, Univ. College London, Galton Laboratory. [ii]+iv+89 pp.

Garden, A.

Ba834. **Berkeley, E. & D. S.** 1969. *Dr. Alexander Garden of Charles Town.* Chapel Hill, NC, Univ. of North Carolina Pr. xv+379 pp.

See also Ba205.

Gay, C.

Ba835. **Archivo Nacional, Chile.** 1963. *Catalogo del archivo de Claudio Gay.* Santiago de Chile. 153 pp.

Geddes, P.

Ba836. **University of Strathclyde.** 1983+. *Catalogue of the papers of Sir Patrick Geddes.* Glasgow. In progress.
Vol. 1. 1983. x+307 pp.

Geer, C. de.

See Ba69 & Ba303.

Gegenbaur, C.

Ba837. **Gegenbaur, C.** 1901. *Erlebtes und Erstrebtes von Carl Gegenbaur.* Leipzig, W. Engelmann. vi+69 pp., port.

Geoffroy St. Hilaire, É.

Ba838. **Cahn, T.** 1962. *La vie et l'oeuvre d'Étienne Geoffroy Saint-Hilaire.* Paris, Presses Universitaires de France. [iv]+318 pp.

Ba839. **Geoffroy St. Hilaire, I.** 1847. *Vie, travaux et doctrine scientifique d'Etienne Geoffroy St. Hilaire.* Par son fils. Paris, P. Bertrand; Strasbourg, Veuve Levrault. [viii]+479 pp. *Re-pub.* Brussels, 1968.
Includes a bibliography, pp. 421–471.

Ba840. **Laissus, Y.** 1972. Catalogue des manuscrits d'Étienne Geoffroy St. Hilaire conservés aux Muséum. *Revue d'histoire des sciences et de leurs applications* 25: 365–390.

Ba841. **___.** 1973. Catalogue des manuscrits d'Étienne Geoffroy St. Hilaire conservés aux archives de l'Académie des Sciences de l'Institut de France. *Histoire et nature* 3: 71–89.

Ba842. **Salf, E., P. & L.** 1986. *Un anatomiste et philosophe français, Étienne Geoffroy Saint-Hilaire (1772–1844): Père de la tératologie morphologique et de l'embryologie expérimentale.* Lyon, Faculté de Médecine. 2 vols.
Th. Méd.: Lyon 1: 1986.

See also Cc55.

Gerard, J.

Ba843. **Jeffers, R. H.** 1967. *The friends of John Gerard (1545–1612), surgeon and botanist.* Falls Village, CT, Herb Grower Pr. 99 pp.; *Also* **___.** 1969. ___*: Index.* Falls Village, [etc.]. [11] pp.; **___.** 1969. ___*: Biographical appendix.* Falls Village, [etc.]. 35 pp.

Gesner (Gessner), C.

Ba844. **Fischer, H.** 1966. *Conrad Gesner (26. März 1516 – 13. Dezember 1565): Leben und Werk.* Zurich. 152 pp. *Neujahrsblatt auf das Jahr 1966 als 168. Stück von der Naturforschenden Gesellschaft in Zürich zur Erinnerung an den 400. Todestag, den 13. Dezember 1965, des grossen Zürcher Naturforschers, Universalhistorikers und Arztews, Conrad Gessner.*
Includes **H. Fischer.** Gessner als Bibliograph; **G. Petit.** Gessner, zoologiste; **H. Zoller.** Gessner als Botaniker; etc.

Ba844A. **___, et al.** 1967. *Conrad Gessner 1516–1565: Universalgelehrter, Naturforscher, Arzt.* Mit Beiträgen von H. Fischer, G. Petit, J. Staedtke, R. Steiger, H. Zoller. Zurich, Orell Füssli Verlag. 239 pp.

Ba845. **Hanhart, J.** 1824. *Conrad Gessner: Ein Beytrag zur Geschichte des wissenschaftlichen Strebens und der Glaubensverbesserung im 16ten Jahrhundert. Aus den Quellen geschöpft.* Winterthur, Steiner. xx+355 pp.

Ba846. **Ley, W.** 1929. *Konrad Gesner: Leben und Werk.* Munich, Verlag der Münchener Drucke. viii+154 pp. *Münchener Beiträge zur Geschichte und Literatur der Naturwissenschaften und Medizin* Pt. 15/16.

Ba847. **Société Suisse d'Histoire de la Médecine et des Sciences Naturelles.** 1965. Sonderheft zur Erinnerung an Conrad Gesner 1516–1565. *Gesnerus* 22 (3/4): 115–227.
With chapters on Gesner's scientific interests by various authors.

Ba848. **Wellisch, H. H.** 1984. *Conrad Gesner: A bio-bibliography.* 2nd ed. Zug, Inter Documentation Co. (IDC). xv+145 pp.
Replaces ___. 1975. Conrad Gessner: A bio-bibliography. *Journal of the Society for the Bibliography of Natural History* 7: 151–247.

Ba849. **Zentralbibliothek Zürich.** 1968. Erschliessung des Conrad Gessner–Materials der Zentralbibliothek Zürich. By R. Steiger. *Gesnerus* 25: 29–64.

See also Ba303.

Gilbert, Sir J. H.

See Ba185.

Girard, C.

Ba850. **Goode, G. B.** 1891. *Bibliographies of American naturalists. 5: The published writings of Dr. C. Girard.* Washington, DC, United States National Museum. vi+141 pp. & port. *Bulletin of the United States National Museum* no. 41.

Glover, T.

Ba850A. **Dodge, C. R.** 1888. The life and entomological work of the late Townend Glover, first Entomologist of the U. S. Department of Agriculture. *Bulletin, Division of Entomology, U. S. Department of Agriculture* 18: 68 pp. & 2 pl.

Gmelin, J. G.

Ba851. **Gradmann, R.** 1911. *Johann Georg Gmelin, 1709–1755: Der Erforscher Sibiriens. Ein Gedenkbuch.* Ed. by Otto Gmelin. Munich, O. Gmelin. iii+150 pp.

Goebel, K. von

Ba852. **Bergdolt, E.**, ed. 1942. *Karl von Goebel: Ein deutsches Forscherleben in Briefen aus sechs Jahrzehnten 1870–1937.* 3rd ed. Berlin, Ahnenerbe–Stiftung Verlag. 275 pp. & port.

Goethe, J. W. von

Ba853. **Amrine, F., F. J. Zucker, & H. Wheeler**, eds. 1987. *Goethe and the sciences: A reappraisal.* Dordrecht, Reidel. xvii+442 pp. *Boston studies in the philosophy of science* vol. 97.
Papers by various authors and an annotated bibliography.

Ba853A. **Fischer, H.** 1950. *Goethe's Naturwissenschaft.* Zurich, Artemis Verlag. 91 pp.

Ba853B. **Glaser, H. A.**, ed. 1986. *Goethe und die Natur. Referate des Triestiner Kongresses.* Frankfurt am Main, Lang. 246 pp. *Akten Internationaler Kongresse auf den Gebieten der Ästhetik und der Literaturwissenschaft* no. 1.
Seventeen papers by various authors.

Ba854. **Grünthal, E., & F. Strauss.** 1949. *Abhandlungen zu Goethes Naturwissenschaft.* Berne, P. Haupt. 115 pp. *Berner Beiträge zur Geschichte der Medizin und der Naturwissenschaft* no. 10.

Ba855. **Magnus, R.** 1949. *Goethe as a scientist.* Translated by H. Norden. New York, H. Schuman. xxi+259 pp.
Translation of *Goethe als Naturforscher.* Leipzig, J. A. Barth, 1906. vii+336 pp. & 8 pl.

Ba856. **Schierbeek, A.** 1944. *Goethe als natuuronderzoeker.* Amsterdam, H. Meulenhoff. 184 pp., illus.

Ba857. **Schmid, G.** 1940. *Goethe und die Naturwissenschaften: Eine Bibliographie.* Hrsg. von E. Abderhalden. Halle, Kaiserlich Leopoldinish–Carolinisch Deutschen Akademiae der Naturforscher. xv+620 pp.
Part 1: Goethes naturwissenschaftliches Werk; part 2: Literatur über Goethe; part 3: Goethes Beziehungen zu Naturforschen und Medizinern.

Ba858. **Virchow, R.** 1861. *Goethe als Naturforscher und in besonderer Beziehung auf Schiller.* Berlin. vi+127 pp. *Re-pub.* Wiesbaden, 1971.

See also Ca46.

Goldschmidt, R. B.

Ba859. **Allen, G. E.** 1971. [R. B.] Goldschmidt papers, University of California, Berkeley. *Mendel newsletter* 7: 1–7.

Ba860. **Goldschmidt, R. B.** 1956. *Portraits from memory: Recollections of a zoologist.* Seattle, WA, Univ. of Washington Pr. ix+182 pp. & 8 ports.
For German version *see* **___.** 1959. *Erlebnisse und Begegungen: Aus der grossen Zeit der Zoologie in Deutschland.* Ed. by E. de Lattin. Hamburg & Berlin, P. Parey. 165 pp. & 8 pl.

Ba861. **___.** 1960. *In and out of the ivory tower: The autobiography of Richard B. Goldschmidt.* Seattle, WA, Univ. of Washington Pr. 352 pp.

Goodall, J.

Ba862. **Clewis, B.** 1990. Uncommon devotion: A bibliography of primatologists Jane Goodall, Dian Fossey and Birute Galdikas. *Bulletin of bibliography* 47 (1): 35–45.

Gosse, P. H.

Ba863. **Freeman, R. B., & D. Wertheimer.** 1980. *Philip Henry Gosse: A bibliography.* Folkestone, Dawson. xviii+148 pp.
Replaces **P. Stageman.** 1955. *A bibliography of the first editions of Philip Henry Gosse, F.R.S.* With introductory essays by S. Sitwell & G. Lapage. Cambridge, Golden Head Pr. xi+87 pp. & 6 pl. (For an entertainingly critical review of the introductory text see **D. Tucker** in *Journal of the Society for the Bibliography of Natural History* 4: 223–229, 1957.)

Ba864. **Gosse, E.** 1896. *The naturalist of the sea shore: The life of Philip Henry Gosse.* London, Heinemann. ix+387 pp. & port.

Ba865. **___.** 1907. *Father and son: A study of two temperaments.* London, Heinemann. xi+374 pp. *Re-pub.* Booklover's ed., London, [etc.], 1912.

Gouan, A.

Ba866. **Clap, V.** 1955. *Antoine Gouan, Professeur et botaniste Montpelliérain (1733–1821): Essais et documents inedits.* Montpellier, Imp. P. Déhan. 152 pp. & port.
Includes Œuvres et manuscrits, pp. 15–42.

Gould, J.

Ba867. **Lambourne, M.** 1987. *John Gould — the birdman.* Foreword by D. Attenborough. Milton Keynes, Osberton Productions. 112 pp., 104 col. illus.

Ba868. **McEvey, A. R.** 1968. Collections of John Gould manuscripts and drawings. *La Trobe Library journal* 1 (1): 17–31.

Ba869. ___. [1973.] *John Gould's contribution to British art: A note on its authenticity.* Sydney, Sydney Univ. Pr. ix+28 pp. & 41 illus. on 16 pp. *Art monograph, Australian Academy of Humanities* no. 2.

Ba870. **Sauer, G. C.** 1982. *John Gould: The bird man. A chronology and bibliography.* London, Henry Sotheran Ltd. xxiv+416 pp. & 36 pls.

Ba871. **Sharpe, R. B.** 1893. *An analytical index to the works of the late John Gould, F.R.S. ... With biographical memoir and portrait.* London, H. Sotheran & Co. xlviii+375 pp., port.

Ba872. **Tree, I.** 1991. *The ruling passion of John Gould: A biography of the bird man.* London, Barrie & Jenkins. vii+248 pp., illus. *Also* New York, Grove Weidenfeld, 1992, (subtitled *A biography of the British Audubon*).

Ba873. **Waterhouse, F. H.** 1885. *The dates of publication of some of the zoological works of the late John Gould, F.R.S.* London, R. H. Porter. xi+59 pp.

Gray, A.

Ba874. **Dupree, A. H.** 1959. *Asa Gray, 1810–1888.* Cambridge, MA, Belknap Pr. of Harvard Univ. x+505 pp. & 9 pl. *Re-pub.* 1988.

Ba875. **Goodale, G. L., S. Watson, et al.** 1888. *List of the writings of Dr. Asa Gray ... chronologically arranged, with an index.* New Haven, CT. 67 pp. *American journal of science* ser. 3, vol. 36, appendix.

Ba876. **Gray, J. L.**, ed. 1894. *Letters of Asa Gray.* Boston, MA, Houghton Mifflin & Co. 2 vols. *Re-pub.* New York, Franklin, [1973].

See also Ba205 & Cb256.

Gray, J. E.

Ba877. **Gunther, A. E.** 1975. *A century of zoology at the British Museum through the lives of two keepers 1815–1914.* London, Dawsons. 533 pp.
i.e. J. E. Gray (1800–1875) and A. C. L. G. Günther (1830–1914).

Ba878. ___. 1980. The miscellaneous autobiographical manuscripts of John Edward Gray (1800–1875). *Bulletin of the British Museum (Natural History), historical series* 6: 199–244.

Ba879. **[Saunders, J.]** [1872–75.] *List of the books, memoirs, and miscellaneous papers of Dr. John Edward Gray. With a few historical notes* London, Taylor & Francis. 58 pp.

Gray, T.

Ba880. **Norton, C. E.** 1903. *The poet [Thomas] Gray as a naturalist: With selections from his notes on the Systema Naturae of Linnaeus and facsimiles of his drawings.* Boston, MA, C. E. Goodspeed. 67 pp. & 13 pl.

Gregory, E. L.

Ba881. **Schmid, R.** 1987. Annotated bibliography of works by and about Emily Lovira Gregory (1841–1897). *Bulletin of the Torrey Botanical Club* 114: 319–324.

Grew, N.

Ba882. **LeFanu, W. R.** 1990. *Nehemiah Grew M.D., F.R.S.: A study and bibliography of his writings.* Winchester, St. Paul's Bibliographies; Detroit, MI, Omnigraphics. xvii+182 pp.

See also Ba185.

Griffin, D. R.

See Ba298.

Griffith, W.

See Ba185.

Günther, A. C. L. G.

Ba883. **Gunther, A. E.** 1975. *A century of zoology at the British Museum through the lives of two keepers 1815–1914*. London, Dawsons. 533 pp.
i.e. J. E. Gray (1800–1875) and A. C. L. G. Günther (1830–1914).

Ba884. ___, **R. T.** 1930. Bibliography of the works of Dr. Albert Guenther. *Annals & magazine of natural history* (10) 6 Supp: 233–286.

Haberlandt, G.

See Ba21.

Haeckel, E.

Ba885. **Hemleben, J.**, ed. 1964. *Ernst Haeckel in Selbstzeugnissen und Bilddokumenten*. Reinbek, Rowohlt. 159 pp.

Ba886. **Keitel–Holz, K.** 1984. *Ernst Haeckel: Forscher, Künstler, Mensch: Eine Biographie*. Frankfurt am Main, Fischer. 238 pp.

Ba887. **Klemm, P.** 1966. *Ernst Haeckel: Der Ketzer von Jena. Ein Leben in Berichten, Briefen und Bildern*. Jena, Urania Verlag. 247 pp.

Ba888. **Krausse, E.** 1987. *Ernst Haeckel*. 2. Aufl. Leipzig, B. G. Teubner. 151 pp.

Ba889. **Schmidt, H.**, ed. 1914. *Was wir Ernst Haeckel verdanken. Ein Buch der Verehrung und Dankbarkeit*. Im Auftrag des Deutschen Monistenbundes hrsg. von H. Schmidt [etc.] Leipzig, Verlag Unesma. 2 vols.

Ba890. ___. 1926. *Ernst Haeckel: Leben und Werke*. Berlin, Deutsche Buch–Gemeinschaft g.m.b.h. 446 pp., illus.

Ba891. ___. 1934. *Ernst Haeckel: Denkmal eines grossen Lebens*. Jena, Frommann. 118 pp.

Ba892. **Uschmann, G.** 1961. *Ernst Haeckel: Forscher, Künstler, Mensch*. Briefe ausgewahlt und erlautert von G. Uschmann. 3rd ed. Jena, Urania–Verlag. 275 pp., illus.

Hagenbeck, C.

Ba893. **Hagenbeck, C.** 1909, *Beasts and men, being Carl Hagenbeck's experiences for a half century among wild animals*. Translated by H. S. R. Elliot & A. G. Thacker, with an introduction by P. C–. Mitchell. London, Longmans. xiii+299 pp. & 32 pl.

See also Ca824.

Haldane, J. B. S.

Ba894. **Clark, R. W.** 1969. *J. B. S.: The life and work of J. B. S. Haldane*. London, Hodder & Stoughton. 286 pp., illus.

Ba895. **Dronamraju, K. R.**, ed. 1968. *Haldane and modern biology*. Baltimore, MD, Johns Hopkins Pr. xvi+333 pp.
Contributions by 30 of his colleagues.

Hales, S.

Ba896. **Allan, D. G. C., & R. E. Schofield.** 1980. *Stephen Hales: Scientist and philanthropist*. London, Scolar Pr. 220 pp.

Ba897. **Kennedy, A. E. C.** 1929. *Stephen Hales, D.D., F.R.S.: An eighteenth-century biography.* Cambridge, Cambridge Univ. Pr. xii+256 pp. & 14 pl. *Re-pub.* Ridgewood, NJ, Gregg Pr. Inc., 1965.

See also Ba185.

Haller, A. von

Ba898. **Berner–Haller Symposion.** 1978. *Albrecht von Haller, 1708–1777: Zehn Vorträge gehalten am Berner Haller–Symposion vom 6. bis 8. Oktober 1977.* Basle, Birkhäuser. x+182 pp.

Papers by K. S. Guthke, U. Im Hof, W. Killy, A. Lindt, N. Mani, L. Premuda, M. Stettler, R. Toellner, W. Zimmerli, H. Zoller.

Ba898A. **Biblioteca Nazionale Braidense di Milano.** 1965. *Il fondo Halleriano della Biblioteca Nazionale Braidense di Milano vicende storiche e catalogo dei manoscritti.* By L. P. Vergnano. Milan, Università, Istituto di Storia della Medicina. 129 pp. *Studi e testi, Istituto di Storia della Medicina, Milan* no. 8.

Also ___. 1977. *Libri e manoscritti di Haller: 1777–1977, nel II centenario della morte di Albrecht von Haller.* By M. L. T. Grossi. Milan. 69 pp. Exhibition catalogue. For other Haller literature resources *see* Bc215A.

Ba899. **Chavannes, H.** 1845. *Biographie de Albert de Haller.* 2nd ed. rev. Paris, L.–R. Delay. 322 pp., port.

Replaces ___. 1840. *Albert de Haller. Biographie.* Lausanne, M. Duclaux. 232 pp.

Ba900. **Irsay, S. d'.** 1930. *Albrecht von Haller: Eine Studie zur Geistgeschichte der Aufklärung.* Leipzig, Thieme. [v]+[3]+98 pp. & 3 pl. *Arbeiten aus dem Institut für Geschichte der Medizin an der Universität Leipzig* vol. 1.

Ba901. **Lunsgaard–Hansen von Fischer, S.** 1959. *Verzeichnis der gedruckten Schriften Albrecht von Hallers.* Berne, Haupt. 87 pp. *Berner Beiträge zur Geschichte der Medizin und Naturwissenschaften* no. 18.

Ba902. **Stettler, M., et al.** 1977. Albrecht von Haller 1708–1777. *Actes de la Société Helvétique des Sciences Naturelles, partie scientifique* 157: 181 pp.

Ba903. **Weese, A.** 1909. *Die Bildnisse Albrecht von Hallers.* Berne, A. Francke. 283 pp., illus.

See also Ba303 Cb1029 & Cc207

Hamilton (once Buchanan), F.

See Buchanan–Hamilton, F.

Hanbury, D.

See Ba133.

Hara, H.

Ba904. **Dr Hiroshi Hara Commemoration Committee.** 1991. *Catalogue of the works of Dr Hiroshi Hara.* Tokyo, Yasaka Shobo. 295 pp.

With an introduction by W. T. Stearn.

Harris, T. W.

Ba905. **Scudder, S. H.**, ed. 1869. *Entomological correspondence of Thaddeus William Harris, M.D.* Boston, MA, Boston Society of Natural History. xliii+375 pp. & 4 pl. *Occasional papers* vol. 1.

Memoir and list of publications, pp. 11–47; Correspondence with N. M. Hentz, F. E. Melsheimer, E. Doubleday, E. C. Herrick, J. L. Le Conte, Miss M. H. Morris, etc.

Harrison, R. G.

Ba906. **Wilens, S.** 1980. The Ross Granville Harrison papers. *Mendel newsletter* 19: 1–5.
At Yale University Library.

See also Cc89.

Hartlib, S.

Ba907. **Royal Commission on Historical Manuscripts.** 1980. *Report on the correspondence and papers of Samuel Hartlib (c.1600–1662), educationalist and natural philosopher.* London, for Sheffield Univ. Library. 98 ff.

Harvey, W.

Ba908. **Chauvois, L.** 1957. *William Harvey: His life and times, his discoveries, his methods.* London, Hutchinson Medical Publications. 271 pp.
Translation of *William Harvey: Sa vie et son temps, ses découvertes, sa méthode.* Paris, Société d'Édition d'Enseignement Supérieur, 1957. 253 pp.

Ba909. **Frank, R. G.** 1980. *Harvey and the Oxford physiologists: Scientific ideas and social interaction.* Berkeley, CA, Univ. of California Pr. xviii+368 pp.

Ba910. **Franklin, K. J.** 1961. *William Harvey, Englishman.* London, Macgibbon & Kee. 151 pp.

Ba911. **Keele, K. D.** 1965. *William Harvey, the man, the physician, and the scientist.* London, Nelson. xii+244 pp. *British men of science.*

Ba912. **Keynes,** Sir **G.** 1950. *The portraiture of William Harvey. With a catalogue and reproductions of the pictures.* London, Royal College of Surgeons; Cambridge, Cambridge Univ. Pr. [vi]+42 pp. & 32 pp. of illus. *Thomas Vicary lecture* 1948.
Also ___. 1985. *The portraiture of William Harvey: The Thomas Vicary Lecture 1948.* Historical introduction by D. G. James. London, Keynes Pr. xix+51 pp.

Ba913. ___. 1966. *The life of William Harvey.* Oxford, Clarendon Pr. xviii+483 pp. & 32 pl.

Ba914. ___. 1989. *A bibliography of the writings of Dr. William Harvey 1578–1657.* 3rd ed., rev. by G. Whitteridge & C. English. Winchester, St. Paul's Bibliographies. xvi+136 pp. & 8 pl.
Replaces ___. 1953. ___. 2nd ed., rev. Cambridge, Cambridge Univ. Pr. xiv+79 pp. & 8 pl.

Ba915. **Pagel, W.** 1967. *William Harvey's biological ideas: Selected aspects and historical background.* Basle & New York, S. Karger. 394 pp. & 44 pl.

Ba916. ___. 1976. *New light on William Harvey.* Basle, New York, Karger. vii+189 pp.

Ba917. **Whitteridge, G.** 1971. *William Harvey and the circulation of the blood.* London, Macdonald. xiii+269 pp. *History of science library.*

Ba918. **Zirnstein, G.** 1977. *William Harvey.* Leipzig, B. G. Teubner. 109 pp. *Biographien hervorragender Naturwissenschaftler und Mediziner* no. 28.

See also Ba2.

Harvey, W. H.

Ba919. [**Fisher, L. J.**] 1869. *Memoir of W. H. Harvey, M.D., F.R.S. ... with selections from his journal and correspondence.* London, Bell & Daldy. xvi+372 pp. & port.

Ba920. **Harvey, W. H.** 1988. *The contented botanist: Letters of W. H. Harvey about Australia and the Pacific.* Ed. by S. C. Ducker. Carlton, Vic., Melbourne Univ. Pr. xvi+413 pp. *Miegunyah Press series* no. 3.

See also Ba133 & Ba185.

Hazen, E. L.

See Cb638.

Hediger, H.

See Ba298.

Heer, O.

Ba921. **Malloizel, G.** 1887. *Oswald Heer: Bibliographie et tables iconographiques: Précédé d'une notice bibliographique par R. Zeiller.* Stockholm, F. & G. Beijer. 176 pp.

Ba922. **Mötteli, O.** 1938. *Oswald Heer: Aus dem Leben und Wirken eines schweizerischen Naturforschers.* Zurich, Gute Schriften. 79 pp.

Ba923. **Schröter, C.** 1885–87. *Oswald Heer, Lebensbild eines schweizerischen Naturforschers.* Hrsg. von J. J. Heer. Zurich, F. Schulthess. 2 vols.

See also Ba133.

Heimans, E.

Ba924. **Brouwer, F. I.** 1958. *Leven en werken van E. Heimans en de opbloei der natuurstudie in Nederland in het begin van de twintigste eeuw.* Groningen, J. B. Walters. x+402 pp.

Heinroth, O.

Ba925. **Heinroth, K.** 1971. *Oskar Heinroth: Vater der Verhaltenforschung, 1871–1945.* Stuttgart, Wissenschaftliche Verlagsgesellschaft. 257 pp. *Grosse Naturforscher* vol. 35.

Henfrey, A.

See Ba185.

Henry, A.

Ba926. **Pim, S.** 1984. *The wood and the trees: A biography of Augustine Henry.* 2nd ed., rev. & enl. Kilkenny, Boethius Pr. 252 pp. & [48] pp. of illus.
Replaces 1st ed., London, Macdonald, 1966. 256 pp.

Hensen, V.

Ba927. **Porep, R.** 1970. *Der Physiologe und Planktonforscher Victor Hensen (1835–1924): Sein Leben und sein Werk.* Neumünster, K. Wacholtz Verlag. 147 pp. *Keiler Beiträge zur Geschichte der Medizin und Pharmazie* no. 9.

Henslow, J. S.

Ba928. **Gebbett, J. Russell–.** 1977. *[J. S.] Henslow of Hitcham: Botanist, educationalist and clergyman.* Lavenham, T. Dalton Ltd. 139 pp., illus.

Ba929. **Jenyns, L.** 1862. *Memoir of the Rev. John Stevens Henslow.* London, J. Van Voorst. ix+278 pp.

See also Ba185.

Herold, J. M. D.

Ba930. **Runge, U.** 1983. *Johann Moritz David Herold (1790–1862): Leben und Werk.* Frankfurt am Main, Lang. 299 pp. *Marburger Schriften zur Medizingeschichte* no. 6.

Herrick, C. J.

Ba931. **Roofe, P. G.** 1964. The [C. J.] Herrick papers, an unusual scientific archive. *Isis* 55: 85–87.
Bequeathed to the University of Kansas.

Hertwig, O.

Ba932. **Weissenberg, B.** 1959. *Oscar Hertwig, 1849–1922: Leben und Werk eines deutschen Biologen.* Leipzig, J. A. Barth. 63 pp., illus. *Lebensdarstellungen deutscher Naturforscher* no. 7.

See also Ba21.

Hess, E. H.

See Ba298.

Heurck, H. F. van

Ba933. **Frison, E.** 1959. *Henri Ferdinand van Heurck 1838–1909: Sa vie, son oeuvre.* Leiden, Rijksmuseum voor de Geschiedenis der Natuurwetenschappen. 106 pp., illus. *Communication* no. 111.

See also Ba21.

Hill, J.

Ba934. **Rousseau, G. S.** 1982. *The letters and papers of Sir John Hill 1714–1775.* New York, AMS Pr. lii+208 pp.

See also Ba185.

Hinde, R. A.

See Ba298.

Hodgson, B. H.

Ba935. **Cocker, P. M., & C. Inskipp.** 1988. *A Himalayan ornithologist: The life and work of Brian Houghton Hodgson.* Oxford, Oxford Univ. Pr. 72 pp. & 32 pl.
Includes colour reproductions of some of the paintings prepared for Hodgson's unpublished work on the birds of Nepal, now in the Zoological Society of London.

Ba936. **Gray, J. E. & G. R.** 1846. *Catalogue of the specimens and drawings of mammalia and birds of Nepal and Thibet presented by B. H. Hodgson, Esq. to the British Museum.* London. xi+156 pp.

Ba937. ___. 1863. *Catalogue of the specimens and drawings of mammalia, birds, reptiles and fishes of Nepal and Thibet presented by B. H. Hodgson, Esq. to the British Museum.* 2nd ed. London. xii+90 pp.
The two editions are complementary. Includes list of the unpublished water colour drawings prepared by trained Nepalese artists for Hodgson's uncompleted work on the birds of Nepal.

Ba938. **Low, G. C.**, et al. 1930. A classification of the original watercolour paintings of birds of India by B. H. Hodgson [etc.]. *Proceedings, Zoological Society of London* 1930: 548–625.
In the Zoological Society of London library.

Hofmeister, W.

Ba939. **Geobel, K. von.** 1926. *Wilhelm Hofmeister: The work of a nineteenth century botanist.* With biographical supplement by Frau Professor Ganzmüller (born Hofmeister). Translated into English by H. M. Bower and ed. botanically by F. O. Bower. London, Ray Society. xii+202 pp.
Translation of *Wilhelm Hofmeister: Arbeit und Leben eines Botanikers des 19. Jahrhunderts.* Leipzig, Wissenschaftliche Verlagsgesellschaft, 1924. 177 pp. *Grosse Männer* vol. 8.

Holmes, F. M.

Ba940. **Furley, D. D.** 1989. Notes on the correspondence of E. M. Holmes (1843–1930). *Linnean* 5(3): 23–30.

Holst, E. von.

See Cc266.

Hooke, R.

Ba941. **'Espinasse, M.** 1956. *Robert Hooke.* London, W. Heinemann. xii+192 pp. & 16 pl.

Ba942. **Hunter, M., & S. Schaffer**, eds. 1989. *Robert Hooke: New studies.* Woodbridge, The Boydell Pr. x+310 pp.
Essays by various authors including papers given at a British Society for the History of Science conference held on the 19–21 July 1987.

Ba943. **Keynes,** Sir **G.** 1960. *A bibliography of Dr. Robert Hooke.* Oxford, Clarendon Pr. xix+[v]+115 pp. & 12 pl.

Hooker, J. D.

Ba944. **Allan, M.** 1967. *The Hookers of Kew, 1785–1911.* London, M. Joseph. 273 pp. & 28 pl.

Ba945. **Huxley, L.**, ed. 1918. *Life and letters of Joseph Dalton Hooker based on materials collected and arranged by Lady Hooker.* London, J. Murray. 2 vols.

Ba946. **Turrill, W. B.** 1963. *Joseph Dalton Hooker: Botanist, explorer, and administrator.* London & Edinburgh, T. Nelson & Sons. xi+228 pp. & 25 pl. *British men of science.*

See also Ba185.

Hooker, W. J.

Ba947. **Allan, M.** 1967. *The Hookers of Kew, 1785–1911.* London, M. Joseph. 273 pp. & 28 pl.

Ba948. [**Hooker,** Sir **J. D.**] 1902. A sketch of the life and labours of Sir William Jackson Hooker. *Annals of botany* 16(64): ix+ccxxi.
Appendix C. List of some of Sir W. J. Hooker's chief correspondents (1808–65), pp. clxxxvii–ccxx.

See also Ba133 Ba185 & Ca824.

Hope, J.

Ba949. **Morton, A. G.** 1986. *John Hope, 1725–1786, Scottish botanist.* Edinburgh, Edinburgh Botanic Garden Trust. 47 pp.

Hosack, D.

Ba950. **Robbins, C. C.** 1964. *David Hosack, citizen of New York.* Philadelphia, PA, American Philosophical Society. vii+246 pp. *Memoirs of the American Philosophical Society* vol. 62.

See also Ba205.

Hosking, E. J.

Ba951. **Hosking, E. J.,** with **F. W. Lane.** 1970. *An eye for a bird: The autobiography of a bird photographer.* London, Hutchinson. xviii+302 pp.

Howard, L. O.

Ba952. **Howard, L. O.** 1933. *Fighting the insects — the story of an entomologist.* New York, Macmillan Co. xvii+333 pp.

Huber, F.

See Ba303.

Hübner, J.

Ba953. **Hemming, F.** 1937. *A bibliographical and systematic account of the entomological works of Jacob Hübner and of the supplements thereto by Carl Geyer, Gottfried Franz von Frölich and Gottlieb August Wilhelm Herrich–Schäffer.* London, Royal Entomological Society. 2 vols.

Hudson, W. H.

Ba954. **Payne, J. R.**, ed. 1977. *W. H. Hudson, a bibliography.* London, Dawson; Hamden, CT, Archon Books. xv+248 pp., port.

Ba955. **Roberts, M.** 1926. *W. H. Hudson: A portrait.* London, Nash & Grayson. 320 pp., illus.

Ba956. **Ronner, A. D.** 1986. *W. H. Hudson: The man, the novelist, the naturalist.* New York, AMS Pr. x+127 pp. *AMS studies in modern literature* no. 16.

Ba957. **Shrubsall, D.**, ed. 1982. *Birds of a feather: Unpublished letters of W. H. Hudson.* Bradford on Avon, Moonraker Pr. 108 pp., illus.

Ba958. **Tomalin, R.** 1982. *W. H. Hudson: A biography.* London, Faber & Faber. 314 pp., illus.

Ba959. **Wilson, G. F.** 1922. *A bibliography of the writings of W. H. Hudson.* London, Bookman's Journal. 79 pp. *Re-pub.* Port Washington, NY, Kennikat Pr., 1968; New York, Haskell House Publishers, 1972.

Hultén, E.

Ba960. **Hultén, E.** 1973. *Men roligt har det varit.* Stockholm, Generalstabens Litografiska Anstalts Forlag. 460 pp., illus.

Humboldt, A. von.

• Limitation of space permits no more than the following selection from the large corpus of literature on Humboldt's life and work:

Ba961. **Banse, E.** 1953. *Alexander von Humboldt: Erschliesser einer neuen Welt.* Stuttgart, Wissenschaftliche Verlagsgesellschaft. viii+146 pp., illus. *Grosse Naturforscher* vol. 14.

Ba962. **Beck, H.** 1959–61. *Alexander von Humboldt.* Wiesbaden, F. Steiner. 2 vols.
Vol. 1: Von der Bindungsreise zur Forschungsreise, 1769–1804; vol.2: Vom Reisewerk zum "Kosmos," 1804–1859. Includes a bibliography, vol. 2, pp. 347–356.

Ba963. **Biermann, K. R.**, ed. 1983. *Alexander von Humboldt: Chronologische Übersicht über wichtige Daten seines Lebens.* 2nd ed. Berlin, Akademie–Verlag. 94 pp. *Beiträge zur Alexander–von–Humboldt–Forschung* vol. 1.
Replaces 1st ed., 1968. xvii+86 pp.

Ba964. ___, ed. 1990. *Miscellanea Humboldtiana.* Berlin, Akademie–Verlag. 304 pp. *Beiträge zur Alexander–von–Humboldt–Forschung* vol. 15.
Re-publication of several pieces.

Ba965. **Botting, D.** 1973. *Humboldt and the cosmos.* London, M. Joseph. 295 pp., illus.
For French version *see* ___. 1988. *Humboldt 1769–1859: Un savant démocrate.* Trad. de l'anglais par M. Dupouey. Paris, Belin. 295 pp. *Un savant, une époque.*

Ba966. **Bruhns, K.** 1872. *Alexander von Humboldt: Eine wissenschaftliche Biographie.* 2nd ed. Leipzig, Brockhaus. 3 vols. *Re-pub.* Osnabrück, 1969.
Includes a bibliography of his publications, pp. 485–552 (*see* Ba978 J. Löwenberg). For an earlier English version *see* ___. 1873. *Life of Alexander von Humboldt.* Translated by J. & C. Lassell. London, Longmans. 2 vols. English translation of the first two vols. of the original work.

Ba967. **De Terra, H.** 1958. Studies of the documentation of Alexander von Humboldt. *Proceedings of the American Philosophical Society* 102: 136–141, 560–589.

Ba968. ___. 1968. *Humboldt: The life and times of Alexander von Humboldt, 1769–1859.* 6th ed. New York, A. A. Knopf. xiv+386+ix pp. & 8 pl.

Ba969. **Deutsche Akademie der Wissenschaften zu Berlin.** 1959. *Alexander von Humboldt 14.9.1769 – 6.5.1859: Gedenkschrift zur 100. Wiederkehr seines Todestages, [etc.].* Herausg. von der Alexander von Humboldt–Kommission der Deutschen Akademie der Wissenschaften zu Berlin. Berlin, Akademie–Verlag. [vi]+471 pp. & 29 pl.
Chapters by twelve authors.

Ba970. ___. 1969. *Alexander von Humboldt: Wirkendes Vorbild für Fortschritte und Befreiung der Menschheit.:Festschrift aus Anlass seines 200. Geburtstages.* Hrsg. im Auftrage der Kommission für die Alexander von Humboldt–Ehrungen 1969 der Deutschen Akademie der Wissenschaften zu Berlin. Berlin, Akademie Verlag. xviii+153+31 pp.
Chapters by eight authors.

Ba971. **Heikenroth, H., & I. Deters**, eds. 1986. Alexander–von–Humboldt–Ehrug in der DDR, 3 und 4 Mai 1984 in Berlin. *Abhandlungen der Akademie der Wissenschaften der DDR, Abteilung Mathematik, Naturwissenschaften, Technik* 1985 (2): 121 pp.
Symposium papers, including: **I. Jahn.** Alexander von Humboldt zu biologischen Problemen in seinem Briefwechsel mit Berliner Zoologen, pp. 98–104.

Ba972. **Hein, W. H.**, ed. 1985. *Alexander von Humboldt: Leben und Werk.* Mit Beiträgen von Hanno Beck [et al.]. Ingelheim am Rhein, C. H. Boehringer Sohn; Frankfurt am Main, Wiesbecker Verlag. 334 pp.

Ba973. **Humboldt–Komitee der DDR.** 1959. *Alexander von Humboldt: Bibliographie seiner ab 1860 in deutscher Sprache herausgegeben Werke und der seit 1900 erscheinenen Veröffentlichungen über ihn.* Herausg. von der Deutschen Bücherei. Leipzig. 45 pp.

Ba974. **Ibero–Amerikanisches Institut Preussischer Kulturbesitz.** 1969. *Alexander von Humboldt und seine Welt 1769–1859.* Berlin. 109 pp. & 30 pl.
Catalogue of an exhibition held at Schloß Charlottenburg, Orangerie, Berlin 29th June to 10th August, 1969.

Ba975. **Jahn, I.** 1969. *Dem Leben auf der Spur: Die biologischen Forschungen Alexander von Humboldts.* Leipzig, Urania Verlag. 211 pp. & 16 pl.

Ba976. **Kellner, L.** 1963. *Alexander von Humboldt.* London, Oxford Univ. Pr. [x]+247 pp. & 7 pl.

Ba977. **Lange, F. G.** 1974. *Alexander von Humboldt: Eine Bibliographie der in der Deutschen Demokratischen Republik erscheinen Literatur. Mit einer Übersicht über die Eigentümer von Humboldt–Handschriften in der DDR.* Berlin, Akademie–Verlag. 98 pp. *Beiträge zur Alexander–von–Humboldt–Forschung* vol. 3.

Ba978. **Löwenberg, J.**, ed. 1960. *Alexander von Humboldt: Bibliographische Übersicht seiner Werke, Schriften und zerstreuten Abhandlungen.* Stuttgart, Brockhaus. 68 pp.
Re-pub. of an extract from K. Bruhns 1872 biography; *see* Ba966.

Ba979. **Melken, H.** [1980.] *Alexander von Humboldt: His portraits and their artists, a documentary iconograph.* Berlin, D. Reiner. 179 pp., illus.

Ba980. **Pfeiffer, H.**, ed. 1969. *Alexander von Humboldt: Werk und Weltgeltung.* Munich, R. Piper Verlag. 505 pp., illus.
Chapters by eleven authors. *Includes* **A. Meyer–Abich.** Alexander von Humboldt as a biologist; **C. Troll.** Die Lebensformen der Pflanzen – A. von Humboldts Ideen in der ökologischen Sicht von heute.

Ba981. **Robles, V. A.** 1940. *Alejandro de Humboldt su vida y su obra.* Introducción a la sexta edición castellana del "Ensayo Político Sobre el Reino de la Nueva España." Mexico, D.F., Pedro Robredo. 126 pp., port.

Ba982. **Scurla, H.** 1969. *Alexander von Humboldt: Sein Leben und Wirken.* 7. Aufl. Berlin, Verlag der Nation. 420 pp. & 32 pl.

Ba983. **Uschmann, G.**, ed. 1971. Alexander von Humboldt, (1769–1859): Gedenkfeier der Deutschen Akademie der Naturforscher Leopoldina am 14. September 1969 in Halle (Saale). *Acta historica Leopoldina* 6: 1–59.
Articles by I. P. Gerasimov, E. Neef, A. Watznauer & K. E. Rothschuh.

Ba984. **Zielna, K.** 1989. *Alexander von Humboldt in der polnischen Literatur: Bibliographie.* Berlin, Reimer. 117 pp.

See also Ba303 Ca46 & Ca761–763.

Hume, A. O.

Ba985. **Wedderburn,** Sir **W.** 1913. *A. O. Hume ... 1829 to 1912.* London, Unwin. vi+182 pp.

Hunter, J.

Ba986. **Dobson, J.** 1969. *John Hunter.* Edinburgh & London, E. & S. Livingstone. xvii+361 pp. & 16 pl.

Ba987. **Kobler, J.** 1962. *The reluctant surgeon: A biography of John Hunter.* New York, Doubleday & Co. 450 pp.

Ba988. **LeFanu, W. R.** 1946. *John Hunter: A list of his books.* Cambridge, Cambridge Univ. Pr. for the Royal College of Surgeons. 31 pp.

Ba989. ___. 1978. Natural history drawings collected by John Hunter FRS (1728–1793) at the Royal College of Surgeons of England. *Journal, Society for the Bibliography of Natural History* 8: 329–333.

Ba990. **Oppenheimer, J. M.** 1946. *New aspects of John and William Hunter.* London, W. Heinemann Medical Books. xx+188 pp.

Ba991. **Qvist, G.** 1981. *John Hunter 1728–1793.* London, Heinemann. xvi+216 pp.

See also Ba303 Cc868–870.

Hunter, W.

Ba992. **Fox, R. H.** 1901. *William Hunter, anatomist, physician, obstetrician, (1718–1783), with notices of his friends Cullen, Smellie, Fothergill, and Baillie.* London, H. K. Lewis. vii+75 pp. & 8 pl.

Ba993. **Illingworth, C.** 1967. *The story of William Hunter.* Edinburgh & London, E. & S. Livingstone. viii+134 pp.
Includes a bibliography.

Ba994. ___. 1971. William Hunter's manuscripts and letters: The [University of] Glasgow collection. *Medical history* 15: 181–186.

Ba995. **Oppenheimer, J. M.** 1946. *New aspects of John and William Hunter.* London, W. Heinemann Medical Books. xx+188 pp.

Hurst, C. C.

Ba996. **Hurst, R.** 1975–80. The Hurst collection of genetical letters. *Mendel newsletter* 11: 1–7, 1975; *Ibid.* 12: 9, 1976; ___. 1977. Addition to the C. C. Hurst collection. *Ibid.* 13: 6–7; ___. 1980. Disposition of the Hurst botanical material. *Ibid.* 119: 8–10.
At Cambridge University Library.

Hutchinson, G. E.

Ba997. **Hutchinson, G. E.** 1979. *The kindly fruits of the earth: Recollections of an embryo ecologist.* New Haven, CT, Yale Univ. Pr. 264 pp.

Huxley family

Ba998. **Clark, R. W.** 1968. *The Huxleys.* London, Heinemann. xvi+398 pp. & 8 pl.

Huxley, J. S.

Ba999. **Boothe, N. L.** 1987. The Julian Sorell Huxley papers, Rice University Library, Houston, Texas. *Mendel newsletter* 27: 1–10.

Ba1000. **Huxley, J. S.** 1970. *Memories.* London, Allen & Unwin. 296 pp. & 22 pl.

Ba1001. **Keynes, M., & G. A. Harrison**, eds. 1898. *Evolutionary studies: A centenary celebration of the life of Julian Huxley.* London, Macmillan. 256 pp.

Ba1002. **Rice University, Woodson Research Center, Fondren Library.** 1984. *A guide to the papers of Julian Sorell Huxley.* By S. C. Bates, M. G. Winkler & C. Riquelmy. Houston, TX. 164 pp.

Huxley, T. H.

Ba1003. **Bibby, C.** 1959. *T. H. Huxley, scientist and educator.* London, Watts. xxi+330 pp.

Ba1004. ___. 1972. *Scientist extraordinary: The life and scientific work of Thomas Henry Huxley, 1825–1895.* Oxford, Pergamon Pr. xi+208 pp.

Ba1005. **Clodd, E.** 1902. *Thomas Henry Huxley.* London, Blackwood & Sons. xiii+226 pp.

Ba1005A. **Collie, M.** 1993? *A bibliography of the writings of T. H. Huxley.* Aldershot, Scolar Press. Ca.200 pp., illus.
Forthcoming publication announced.

Ba1006. **Dawson, W. R.** 1946. *The Huxley papers: A descriptive catalogue of the correspondence, manuscripts and miscellaneous papers of the Rt. Hon. Thomas Henry Huxley, P.C., D.C.L., F.R.S. preserved in the Imperial College of Science & Technology.* London, Macmillan & Co. xii+201 pp.

Ba1007. **De Beer, G. R.**, ed. 1974. *Autobiographies. Charles Darwin, T. H. Huxley.* Ed. with an introduction by [etc.]. Oxford, Oxford Univ. Pr. xxvi+123 pp., illus. *Oxford English memoirs and travels.*

Ba1008. **Di Gregorio, M. A.** 1984. *T. H. Huxley's place in natural science.* New Haven, CT, Yale Univ. Pr. xxi+253 pp.
Includes a chronological list of 367 works by Huxley and full list of works about him.

Ba1009. **Huxley, L.** 1900. *Life and letters of Thomas Henry Huxley.* London, Macmillan & Co. 3 vols.

Ba1010. **Imperial College of Science & Technology.** 1968. *Thomas Henry Huxley: A list of his scientific notebooks, drawings and other papers, preserved in the College archives.* By J. Pingree. London. x+94 pp.

Ba1011. ___. 1969. *List of correspondence with Miss Henrietta Anne Heathorn, later Mrs Huxley, 1847–54.* By J. Pingree. London. vii+36 pp.

Ba1012. **Jensen, J. V.** 1991. *Thomas Henry Huxley: Communicating for science.* Newark, DE, Univ. of Delaware Pr. 253 pp.

Ba1013. **MacBride, E. W.** 1934. *Huxley.* London, Duckworth. 143 pp. *Great lives* no. 34.

Ba1014. **Mitchell, P. Chalmers–.** 1901. *Thomas Henry Huxley: A sketch of his life and work.* London, Methuen. xvii+297 pp. & 5 pl.

Ba1015. **Roos, D. A.** 1978. Neglected bibliographical aspects of the works of Thomas Henry Huxley. *Journal of the Society for the Bibliography of Natural History* 8: 401–420.

Ingenhousz, J.

Ba1016. **Reed, H. S.** 1949. Jan Ingenhousz: Plant physiologist. With a history of the discovery of photosynthesis. *Chronica botanica* 11 (5/6): 285–396.

Ba1017. **Van der Pas, P. W.** 1964. The Ingenhousz–Jenner correspondence. *Janus* 51: 202–220.

Ba1018. ___. 1973. [Bibliography of original works.] In: C. C. Gillispie, ed. *Dictionary of scientific biography.* Vol. 7: 14–16. New York, Scribner's Sons.

Ba1019. **Weisner, J.** 1905. *Jan Ingen–Housz: Sein Leben und sein Wirken als Naturforscher und Arzt.* Vienna, C. Konegen. x+252 pp.

Ionides, C. J. P.

Ba1020. **Wykes, A.** 1960. *Snake man: The story of C. J. P. Ionides.* London, Hamish Hamilton. [xi]+225 pp., illus.

Jacquemont, V.

Ba1021. **Brown, A. W., et al.** 1959. *Jacquemont.* Paris, Muséum National d'Histoire Naturelle. 461 pp. & 51 pl. *Les Grands naturalistes Français.*

Ba1022. **Jacquemont, V.** 1867. *Correspondence inédite de Victor Jacquemont avec sa famille et ses amis, 1824–1832.* Précédée d'une notice biographique par V. Jacquemont neveu et d'une introduction par Prosper Mérimée. Paris, M. Lévy. 2 vols.

Jardine, Sir W.

Ba1023. **Royal Scottish Museum.** 1981. *Manuscripts in the Royal Scottish Museum.* Pt. 2: *William Jardine papers.* By J. Pitman. Edinburgh. 46 pp. *Royal Scottish Museum information series: natural history* no. 7.

See also D156.

Jefferies, R.

Ba1024. **Arkell, R.** 1933. *Richard Jefferies.* London, Rich & Cowan. 294 pp.

Ba1025. **Besant,** Sir **W.** 1889. *The eulogy of Richard Jefferies.* New York, Longmans. viii+384 pp.

Ba1026. **Looker, S. J., & C. Porteous.** 1965. *Richard Jefferies, man of the fields: A biography and letters.* London, J. Baker. xix+272 pp.

Ba1027. **Manning, E.** 1984. *Richard Jefferies: A modern appraisal.* Windsor, G. Goldscheider. 162 pp., illus. *The modern naturalists' library.*

Ba1027A. **Miller, G., & H. Mathews.** 1993. *Richard Jefferies: A bibliographical study.* Aldershot, Scolar Press. xxvii+787 pp., illus.

Ba1028. **Thomas, E.** 1909. *Richard Jefferies.* London, Hutchinson. vi+352 pp. *Re-pub.* London, Faber, 1978. x+306 pp.

Jefferson, T.

Ba1029. **Browne, C. A.** 1944. Thomas Jefferson and the scientific trends of his time. *Chronica botanica* 8: 361–423.

Ba1030. **Martin, E. T.** 1952. *Thomas Jefferson: Scientist.* New York, H. Schuman. x+289 pp.

Ba1031. **Miller, C. A.** 1988. *Jefferson and nature: An interpretation.* Baltimore, MD, Johns Hopkins Pr. xii+300 pp.

Johnson, T.

Ba1032. **Kew, H. W., & H. E. Powell.** 1932. *Thomas Johnson, botanist and royalist.* London, Longmans, Green & Co. xi+151 pp.

Johnston, G.

Ba1033. **Johnston, G.** 1892. *Selections from the correspondence of Dr. George Johnston* ... Collected and arranged by his daughter, Mrs Barwell–Carter. Ed. by J. Hardy. Edinburgh, D. Douglas. xxxv+547 pp. & port.

Jones, H.

Ba1034. **Jones, H.** 1976. *The bird paintings of Henry Jones.* London. 108 pp. & 27 pl.
Text by Bruce Campbell.

Ba1035. ___. 1987. *The wildfowl paintings of Henry Jones.* London, Threshold/Harrap. 262 pp., 60 pl. (col.).
Text by Peter Olney. Two selections from the collection of over 1,200 drawings in the Library of the Zoological Society of London.

Jones, M. E.

Ba1036. **Lenz, L. W.** 1986. *Marcus E. Jones: Western geologist, mining engineer and botanist.* Claremont, CA, Rancho Santa Ana Botanic Garden. xvi+486 pp., illus.

Jordan, D. S.

Ba1037. **Benson, K. R.** 1986. David Starr Jordan (1851–1931) papers at Stanford University Archives. *Mendel newsletter* 26: 1–5.

Ba1038. **Hays, A. N.** 1952. *A bibliography of the writings of David Starr Jordan.* Stanford, CA, Stanford Univ. Pr. 195 pp.

Ba1039. **Jordan, D. S.** 1922. *The days of a man: Being memories of a naturalist, teacher and minor prophet of democracy.* Yonkers–on–Hudson, NY, World Book Co. 2 vols.

Ba1040. **Stanford University Archives.** [1969.] *Guide to the microfilm edition of the David Starr Jordan Papers, 1861–1964.* By R. W. Hansen (Project Director), P. J. Palmer (Editor), & C. Stein (Editorial assistant). Sponsored by the National Historical Publications Commission. Stanford, CA, Stanford Univ. Libraries. 31 pp. & 187 reels of microfilm.

Junghuhn, F.

Ba1041. **Schmidt, M. C. P.** 1909. *Franz Junghuhn: Biographische Beiträge zur 100. Wiederkehr seines Geburtstages*, gesammelt und bearbeitet von M. C. P. Schmidt. Leipzig, Dürr. xv+374 pp. & port.

Jussieu family

Ba1042. **Lacroix, A.** 1938. *Notice historique sur les cinq de Jussieu, membres de l'Académie des Sciences (1712–1853). Leur rôle d'animation des recherches d'histoire naturelle dans les colonies françaises. Leurs principaux correspondants.* Lecture ... 21 Décembre, 1936. Paris, Institut de France, Académie des Sciences. 97 pp.

Jussieu, J. de

Ba1043. **Laissus, Y.** 1964. Note sur les manuscrits de Joseph de Jussieu (1704–1779) conservés à la Bibliothèque centrale du Muséum Nationale d'Histoire Naturelle. *Comptes rendus du 89e Congrès National des Sociétés Savantes, Section des Sciences*, pp. 9–16.

Kalm, P.

Ba1044. **Kerkkonen, M.** 1959. *Peter Kalm's North American journey: Its ideological background and results.* Helsinki, Finnish Historical Society. 260 pp.

Ba1045. **Skottsberg, C.** 1950. Pehr Kalm: Minnesteckning. *K. Svenska Vetenskapsakademiens Årsbok* for 1950: 279–365.

See also Ca709–710.

Kamel, G. J.

Ba1046. **Gicklhorn, J. & R.** 1954. *Georg Joseph Kamel, S.J. (1661–1706), Apotheker, Botaniker, Artz, und Naturforscher der Philippineninseln.* Eutin (Holstein), Internationale Gesellschaft für Geschichte und Pharmazie. [123] pp. & 14 pl.

Kaup, J. J.

Ba1047. [**Heldman, G.**] 1955. *Johann Jakob Kaup: Leben und Wirken des ersten Inspektors am Naturalien–Cabinet des Grossherzoglichen Museums 1803–1873.* Darmstadt, the Author. 28 pp. & port.

Keller, B. A.

Ba1048. **Akademiya Nauk SSSR.** 1946. *Boris Aleksandrovich Keller.* Vstupitel'naya stat'ya V. N. Sukacheva i N. V. Turbina. Bibliografiya sostavlena K. V. Ryabova. Moscow/Leningrad. 152 pp. *Materialy k biobibliografii uchenykh SSSR, Seriya biologicheskikh nauk, Botanika* Vol. 2.

Kentmann, J.

Ba1049. **Helm, J.** 1971. *Johannes Kentmann 1518–1574: Ein Sächsischer Arzt und Naturforscher.* Wiesbaden, F. Steiner Verlag. vi+214 pp. *Sudhoff's Archiv, Zeitschrift für Wissenschaftsgeschichte, Beihefte* 13.

Kerner von Marilaum, A.

Ba1050. **Kronfeld, E. M.** 1908. *Anton Kerner von Marilaun: Leben und Arbeit eines deutschen Naturforschers.* Leipzig, C. H. Tauchnitz. xx+392 pp., illus.

Keulemans, G.

Ba1051. **Keulemans, T., & J. Coldewey.** 1982. *Feathers to brush: The Victorian bird artist John Gerrard Keulemans, 1842–1912.* Deventer, the authors. xiii+94 pp. & 24 pl.

King, J. A.

See Ba298.

Kingsley, M.

Ba1052. **Campbell, O. W.** 1957. *Mary Kingsley: A victorian in the jungle.* London, Methuen & Co. 196 pp.

Ba1053. **Frank, K.** 1987. *A voyager out: The life of Mary Kingsley.* London, Hamish Hamilton. 364 pp., illus.

Ba1054. **Gwynn, S.** 1933. *The life of Mary Kingsley.* [2nd ed.] London, Macmillan & Co. vii+285 pp.

Kirby, W.

Ba1055. **Freeman, J.** 1852. *Life of the Rev. William Kirby ... Rector of Barham.* London, Longman. xii+506 pp., 2 pl. & port.

Kirouac, J. L. C.

Ba1056. **Rumilly, R.** 1949. *Le Frère Marie-Victorin et son temps.* Montréal, Les Frères des Écoles Chrétiennes. 459 pp. & 41 pl. (ports.).
I.e. Joseph Louis Conrad Kirouac.

Koch, L.

Ba1057. **Koch, L.** 1955. *Memoirs of a birdman.* London, Phoenix House. 188 pp., illus.

Koch, R.

Ba1058. **Ignatius, A.** 1965. *Robert Koch: Leben und Forsch.* Stuttgart, Frankh. 88 pp. *Kosmos–Bibliothek* no. 248.

Komarov, V. L.

Ba1059. **Akademiya Nauk SSSR.** 1946. *Vladimir Leont'evich Komarov.* Vstupitel'naya stat'ya I. I. Meshchaninova i A. G. Chernova. Moscow/Leningrad. 100 pp. *Materialy k biobibliografii uchenykh SSSR, Seriya biologicheskikh nauk, Botanika* Vol. 1.
Bibliografiya trudov ... Komarov, pp. 39–86.

Ba1060. **Pavlov, N. V.** 1951. *Vladimir Leont'evich Komarov.* Moscow, Izd. AN SSSR. 292 pp.

Kønig, J. G.

Ba1061. **Rendle, A. B.** 1933. John Gerard Koenig. *Journal of botany* 1933: 143–153, 175–187.
Includes a subject index to the 21 vols of his bound manuscripts in the British Museum (Natural History).

Krasheninnikov, S. P.

Ba1062. **Fradkin, N. G.** 1954. *S. P. Krasheninnikov.* Moscow, Gos. izd-vo Geogr. Lit-ry. 48 pp.

Ba1063. **Stepanova, N. N.** 1966. *S. P. Krasheninnikov v Sibiri: Neopublikovannye materialy.* Moscow/Leningrad, Izd. "Nauka." 241 pp. & port.

Krebs, H. A.

Ba1064. **Alton, J., & P. Harper.** 1986. *Catalogue of the papers and correspondence of Sir Hans Adolf Krebs, FRS (1900–1981), deposited in Sheffield University Library.* London, Contemporary Scientific Archives centre. 3 vols.

Ba1065. **Holmes, F. L.** 1991+. *Hans Krebs.* New York, Oxford Univ. Pr. *Monographs on the history and philosophy of biology.* In progress.
Vol. 1: *The formation of a scientific life, 1900–1933.* 1991. xx+491 pp., illus.

Ba1066. **Krebs, H.** 1981. *Reminiscences and recollections.* In collaboration with Anne Martin. Oxford, Clarendon Pr. vii+298 pp., illus.

Krebs, L.

Ba1067. **ffolliott, P., & R. Liversidge.** 1971. *Ludwig Krebs, Cape naturalist to the King of Prussia, 1792–1844.* Cape Town, A. A. Balkema. [xv]+304 pp.

Krukoff, B. A.

Ba1068. **Landrum, L. R.** 1986. *The life and botanical accomplishments of Boris Alexander Krukoff (1898–1983).* New York, New York Botanical Garden. 96 pp.

Kützing, F. T.

Ba1069. **Müller, R. H. W., & R. Zaunick**, eds. 1960. *Friedrich Traugott Kützing, 1807–1893. Aufzeichnungen und Erinnerungen.* Leipzig, J. A. Barth. 300 pp. *Lebensdarstellungen deutscher Naturforscher* no. 8.

Lacépède, B. G. E. de la V.

Ba1070. **Roule, L.** 1932. *Lacépède ... et la sociologie humanitaire selon la nature.* Paris, Flammarion. 244 pp. *L'Histoire de la nature vivante d'après l'oeuvre des grands naturalistes Français* no. 6.

Ba1071. **Villenave, M. G. T. de.** 1826. *Éloge historique de M. le Comte de Lacépède.* Paris, Fournier–Favreux. 76 pp.

See also Ba12 Ba303 & Cc462A.

Laksman, E. G.

Ba1072. **Raskin, N. M., & I. I. Shafronovskii.** 1971. *Erik Gustanovich Laksman: Vydayushchisya puteshestvennikh i naturalist XVIII v.* Leningrad, Izd. "Nauka." 274 pp.

Lam, H. J.

Ba1073. **Jacobs, M.** 1984. *Herman Johannes Lam (1892–1977): The life and work of a Dutch botanist.* Amsterdam, Rodopi. vii+271 pp.

Lamarck, J.–B.

Ba1074. **Barthélemy–Madaule, M.** 1982. Lamarck, the mythical precursor: A study of the relations between science and ideology. Translated by M. H. Shank. Cambridge, MA, M I T Pr. xv+174 pp.
Translation of *Lamarck: ou. le mythe du précurseur.* Paris, Éditions du Soleil, 1979. 185 pp.

Ba1075. **Burkhardt, R. W., Jr.** 1977. *The spirit of system: Lamarck and evolutionary biology.* Cambridge, MA, Harvard Univ. Pr. xi+285 pp.

Ba1076. **Cercle d'Étude Historique des Sciences de la Vie.** 1971. *Colloque international "Lamarck" tenu au Muséum National d'Histoire Naturelle, Paris les 1–2 et 3 juillet 1971* ... Ed. scientifique J. Schiller. Paris, A. Blanchard. [iv]+265 pp.
Includes **Y. Laissus.** Les manuscrits de Lamarck, pp. 191–197 and papers by P.–P. Grassé, P. Omodeo, M. J. S. Hodge, M. Guédès, B. Hoppe, J. Schiller, P. Decourt, G. Legée, S. Szyman, and others.

Ba1077. **Delange, Y.** 1984. *Lamarck: Sa vie, son oeuvre.* [Avignon], Actes Sud. 225 pp., illus.

Ba1078. **Jordanova, L. J.** 1984. *Lamarck.* Oxford, Oxford Univ. Pr. viii+118 pp. *Past masters.*

Ba1079. **Lamarck, J.–B.** 1972. *Inedits de Lamarck, d'après les manuscrits conservés à la Bibliothèque Centrale du Muséum National d'Histoire Naturelle.* Présentés par M. Vachon, G. Rousseau, & Y. Laissus. Paris, Masson & Cie. [ii]+311 pp.

Ba1080. **Landrieu, M.** 1909. *Lamarck, le fondateur du transformisme, sa vie, son oeuvre.* Paris, Société Zoologique de France. xiii+478+[4] pp. & port. *Memoires de la Société Zoologique de France* vol. 21.

Ba1081. **Packard, A. S.** 1901. *Lamarck, the founder of evolution [etc.].* New York, Longmans, Green & Co. xiii+451 pp. *Re-pub.* New York, Arno Pr., 1980.

Ba1082. **Palmer, K. v. W.** 1977. *The unpublished vélins of [J. B. P. A. de M. de] Lamarck: Illustrations of fossils of the Paris Eocene basin.* Ithaca, NY. 67 pp. & 52 pl.

Ba1083. **Polyarkov, I. M.** 1960. The manuscripts of Lamarck. *Russian review of biology* 48: 289–296.
Translation of Russian text in: *Uspekhi sovremennoi biologii* 48: 301–311, 1959.

Ba1084. **Roule, L.** 1927. *Lamarck et l'interprétation de la nature.* Paris, Flammarion. 249 pp. *L'Histoire de la nature vivante d'après l'œuvre des grands naturalistes Français* no. 4.

Ba1085. **Szyfman, L.** 1982. *Jean-Baptiste Lamarck et son époque*. Paris, Masson & Cie (for Fondation Singer Polignac). xxii+448 pp., illus.

Ba1086. **Tschulok, S.** 1937. *Lamarck. Eine kritisch-historische Studie.* Zurich & Leipzig, M. Niehans Verlag. 190 pp.

Ba1087. **Université de Picardie, Centre de Recherche sur l'Histoire des Idées.** 1981. *Lamarck et son temps. Lamarck et notre temps. Colloque international dans le cadre du Centre d'Études et de Recherches Interdisciplinaires de Chantilly 1980 Amiens.* Paris, J. Vrin. 247 pp., illus. *L'Histoire des sciences. Textes et études.*

Ba1088. **Vachon, M., G. Rousseau, & Y. Laissus.** 1968. Liste complète des manuscrits de Lamarck conservés à la Bibliothèque centrale du Muséum National d'Histoire Naturelle, Paris. (Note préliminaire). *Bulletin du Muséum National d'Histoire Naturelle, Paris* 40: 1093–1102.

Ba1089. **Wheeler, W. M., & T. Barbour.** 1933. *The Lamarck manuscripts at Harvard.* Cambridge, MA, Harvard Univ. Pr. xxxi+202 pp. & 4 pl.

See also Ba12 Ba303 Cb1030 & Cc504.

Lambert, A. B.

See Cb1031.

Lang, A.

Ba1090. **Haeckel, E., et al.** 1916. *Aus dem Leben und Wirken von Arnold Lang: Dem Andenken des Freundes und Lehrers gewidmet.* Jena, G. Fischer. iv+285 pp. & port.

Lankester, E.

Ba1091. **English, M. P.** 1992. *Victorian values: The life and times of Dr. E. Lankester, M.D., F.R.S.* Bristol, Biopress. xv+187 pp.

La Pérouse, J. F. de G.

See Ca745–749.

La Peyrouse, P. P. de

Ba1092. **Laissus, Y.** 1972. Philippe Picot de Lapeyrouse (1744–1818), d'après les documents conservés à la Bibliothèque du Muséum National d'Histoire Naturelle. *Bulletin de la Société d'Histoire Naturelle de Toulouse* 108: 365–385.

Latreille, P. A.

Ba1093. **Nussac, L. de.** 1907. *Les débuts d'un savant naturaliste. Le prince d'entomologie: Pierre-André Latreille à Brive de 1762 à 1798.* Paris, Steinheil. vii+264 pp.

Ba1094. **Société Entomologique de France.** 1932. *Livre du centenaire de Latreille.* Paris. 731 pp.

See also Ba303.

Lawes, J. B.

Ba1095. **Dyke, G. V.** 1991. *John Bennet Lawes: The record of his genius.* Taunton, Research Studies Pr. xiv+482 pp. *Research studies in botany & related applied fields* no. 8.

Lawrence, G. N.

Ba1096. **Foster, L. S.** 1892. *Bibliographies of American naturalists. 4: The published writings of G. N. Lawrence, 1844–91.* Washington, DC, United States National Museum. xi+124 pp. *Bulletin of the United States National Museum* no. 40.

Lawson, T.

Ba1097. **Whittaker, E. J.** 1986. *Thomas Lawson, 1630–1691: North country botanist, Quaker and schoolmaster.* York, Sessions Book Trust. xiv+258 pp.

Lea, I.

Ba1098. **Scudder, N. P.** 1885. *Bibliographies of American naturalists. 2: The published writings of Isaac Lea, LL.D.* Washington, DC, United States National Museum. lix+278 pp. & port. *Bulletin of the United States National Museum* no. 23.

Lear, E.

Ba1099. **Hyman, S.** 1980. *Edward Lear's birds.* London, Weidenfeld & Nicholson; New York, Morrow. 96 pp. *Re-pub.* London, Trefoil, 1989.

Ba1100. **Lehmann, J.** 1977. *Edward Lear and his world.* London, Thames & Hudson; New York, Scribner's. 128 pp., illus.

Ba1101. **Noakes, V.** 1968. *Edward Lear: The life of a wanderer.* London, Collins. 359 pp. *Re-pub.* London, Fontana, 1979.

Ba1102. **___.** 1986. *Edward Lear 1812–1888.* New York, H. N. Abrams. 216 pp., illus.

Ba1103. **___.** 1991. *The painter Edward Lear.* Newton Abbot, David & Charles. 96 pp., illus.

Ba1104. **Walker Art Gallery, Liverpool.** 1975. *Edward Lear and Knowsley. An exhibition of watercolours belonging to the Earl of Derby.* Liverpool. 44 pp.
62 items.

L'Écluse, C. de

Ba1105. **Burgenländische Landesregierung.** 1973. *Festschrift anlässlich der 400 jährigen Wiederkehr der wissenschaftlichen Tätigkeit von Carolus Clusius (Charles de l'Écluse) im pannonischen Raum.* Eisenstadt, Burgenländische Landesregierung, Landesarchiv. 309 pp. *Burgenländische Forschungen* Sonderheft 5.
Includes S **Aumüller.** Bibliographie and Ikonographie, and other articles by A. A. Barb, L. Batthyány-Strattmann, E. Grabner, O. F. Guglia, J. Heniger, T. Hortobágyi, E. Horváth, I. Hrabovec, V. Petkovsek, H. Riedl, P. Smit, J. Stannard and G. Traxler.

Ba1106. **Hunger, F. W. T.** 1926. *Clusius–tentoonstellung in het stedelijk museum "De Lakenhal" te Leiden, 19 October – 3 October, 1926. Catalogus.* Leiden, A. W. Sijthoff. 70 pp.
Exhibition catalogue of 257 books, letters, portraits, etc.

Ba1107. **___.** 1927–43. *Charles de l'Escluse (Carolus Clusius): Nederlandsch kruidkundige, 1526–1609 [etc].* The Hague, M. Nijhoff. 2 vols.
Includes a detailed bibliography.

Ba1108. **Morren, C. J. É.** 1875. *Charles de l'Escluse, sa vie et ses oeuvres. 1526–1609.* Liége, privately published. 59 pp.

Ba1109. **Roze, E.** 1899. *Charles de l'Éscluse d'Arras: Le propagateur de la pomme de terre au XVIe siècle: Sa biographie et sa correspondence.* Paris, J. Rothschild. 109 pp., illus.

Ba1110. **Theunisz, J.** 1939. *Carolus Clusius: Het merkwaardige leven van een pionier der wetenschap.* Amsterdam, P. N. van Kampen. 178+[2] pp. & 6 pl.

See also Cb168.

Leconte, John E.

Ba1111. **Rehn, J. A. G.** 1954. The John Eatton Le Conte collection of paintings of insects, arachnids, and myriapods. *Proceedings, American Philosophical Society* 98: 442–448.
8 vols, 4,043 figures, chiefly water colour.

Le Conte, John L.

Ba1111A. **Henshaw, S.** 1878. *The entomological writings of John L. Le Conte*. Ed. by G. Dimmock. Cambridge, MA. [ii]+11 pp. *Dimmock's special bibliography* no. 1.

Le Conte, Jos.

Ba1112. **Le Conte, J.** 1903. *The autobiography of Joseph Le Conte*. Ed. by W. D. Ames. New York, D. Appleton & Co. xvii+337 pp., illus.

Ba1113. **Stephens, L. D.** 1982. *Joseph Le Conte: Gentle prophet of evolution*. Baton Rouge, LA, Louisiana State Univ. Pr. xix+340 pp. *Southern biography series*.

Lees E.

Ba1113A. **Jones M. M.** 1980. *The lookers–out of Worcestershire: A memoir of Edwin Lees, Worcester naturalist, and the history of the first hundred years of the Worcestershire Naturalists' Club*. Worcester. x+216 pp.

Leeuwenhoek, A. van

Ba1114. **Cole, F. J.** 1937. Leeuwenhoek's zoological researches. *Annals of science* 2: 1–46, 185–235.

Ba1115. **Dobell, C.** 1932. *Anthony van Leeuwenhoek and his "little animals": Being some account of the father of protozoology and bacteriology and his multifarious discoveries in these disciplines*. Collected, translated and edited, from his printed works, unpublished manuscripts and contemporary records. Published on the 300th anniversary of his birth. London, Staples Pr. vii+435 pp. & 32 pl. *Re-pub*. New York, Russell & Russell, 1958; New York, Dover Publications, [1960].

Ba1116. **Ford, B. J.** 1991. *The Leeuwenhoek legacy*. Bristol, Biopress; London, Farrand Pr. [xvi]+185 pp.

Ba1117. **Haaxman, P. J.** 1875. *Antony van Leeuwenhoek: De ontdekker der infusorien, 1675–1875*. Leiden, Van Doesburgh. viii+140 pp., illus.

Ba1118. **Leeuwenhoek, A. van.** 1939+. *Alle de brieven van Antoni van Leeuwenhoek. The collected letters of Antoni van Leeuwenhoek*. Ed., illustrated and annotated by a Committee of Dutch scientists. Amsterdam, Lisse, Swets & Zeitlinger. Vol. 1+.
Latest volume, vol. 12: 1696–1699, 1989.

Ba1119. **Palm L. C., & H. A. M. Snelders**, eds. 1982. *Antoni Van Leeuwenhoek, 1632–1723: Studies on the life and work of the Delft scientist commemorating the 350th anniversary of his birthday*. Amsterdam, Rodopi. 209 pp. *Nieuwe Nederlands bijdragen tot de geschiedenis der geneeskunde en der natuurwetenschappen* no. 8.
Chapters by W. H. van Seters, B. C. Damsteegt, J. van Zuylen, P. Bass, J. E. van Bronswijk, G. A. Lindeboom, P. Smit and K. van Berkel.

Ba1120. **Payne, A. S.** 1970. *The cleere observer: A biography of Antoni van Leeuwenhoek*. London, Macmillan. 128 pp.
American ed. pub. as: *Discoverer of the unseen world*. 1966.

Ba1121. **Schierbeek, A.** 1950–51. *Antoni van Leeuwenhoek: Zijn leven und zijn werken*. Lochem, "De Tijdstroom." 2 vols.
For abridged English version *see* **A. Schierbeek.** 1959. *Measuring the invisible world: The life and works of A. van Leeuwenhoek*. With a biographical chapter by M. Rooseboom. London, Abelard–Schuman. 223 pp. & 4 pl.

Ba1122. **Van der Pas, P. W.** 1976. On the portraits of Leeuwenhoek, with a description of an unrecorded portrait. *Aere perennius* 22: 1–9.

Le Moyne de Morgues, J.

Ba1123. **Hulton, P.**, et al. 1977. *The work of Jacques Le Moyne de Morgues: A Huguenot artist in France, Florida and England.* London, British Museum Publications Ltd. in association with the Huguenot Society of London. 2 vols.
Vol. 1: Foreword, catalogue and introductory studies; vol. 2: Plates. The introduction *Includes* I. Jacques Le Moyne de Morgues: A biographical sketch, II. Le Moyne's œuvre, V. **W. T. Stearn.** The plant portraitist and the herbalist, VII. Jacques Le Moyne de Morgues, Peintre.

Lesueur, C. A.

Ba1124. **Bonnemains, J.** 1984. Charles–Alexandre Lesueur en Amerique du Nord (1816–1837): Catalogue des documents du Muséum d'Histoire Naturelle du Havre. *Annales du Muséum du Havre* 29: 1–37, 1984; *Ibid.* 30: 1–105, & 197 pl.

Ba1125. ___. 1986. Les illustrations du livre du Capitaine Nicolas Baudin Expedition de découvertes aux terres Australes (1800–1804). Suite 1: Répertoire des documents rétrouves. *Ibid.* 33: 1–20.

Ba1126. ___. 1986. Origine de la collection Lesueur du Muséum d'Histoire Naturelle. *Ibid.* 38: 1–16.

Ba1127. **Loir, A.** 1920. *Charles–Alexandre Lesueur, artiste et savant français en Amérique de 1816 à 1839.* Le Havre, Muséum d'Histoire Naturelle. 108 pp., illus.

Ba1128. **Vail, R. W. G.** 1938. The American sketchbooks of a French naturalist, 1816–1837: A description of the Charles Alexandre Lesueur collection, with a brief account of the artist. *Proceedings of the American Antiquarian Society* n.s. 48: 49–155.

See also Ca764.

Lettsom, J. C.

Ba1129. **Abraham, J. J.** 1933. *[J. C.] Lettsom: His life, times, friends and descendants.* London, W. Heinemann Medical Books. xx+498 pp., illus.

Levaillant, F.

Ba1130. **Library of Parliament, Cape Town.** 1973. *François Le Vaillant, traveller in South Africa and his collection of 165 water–colour paintings 1781–1784.* Cape Town. 2 vols.
With colour reproductions of all.

Ba1131. **Ogilvie, J. H.** 1962. *A bibliography of Le Vaillant's Voyages and Oiseaux d'Afrique.* Johannesburg, Public Library. vii+72 pp.

See also Ba303.

Lewis, M., & W. Clark

Ba1132. **Bakeless, J.** 1965. *[Meriwether] Lewis and [William] Clark, partners in discovery.* New York, W. Morrow. xii+498 pp.

Ba1133. **Cutright, P. R.** 1969. *Lewis and Clark: Pioneering naturalists.* Urbana, IL, Univ. of Illinois Pr. xvi+506 pp.

See also Ca765–774.

Lewis, G.

Ba1134. **Warner, D. J.** 1979. *Graceanna Lewis: Scientist and humanitarian.* Washington, DC, Smithsonian Institution Pr. for the National Museum of History & Technology. 139 pp.

Leyhausen, P.

See Ba298.

L'Héritier, C. L.

See Ba12.

Lhotsky, J.

Ba1135. **Kruta, V., & T. G. Vallance.** 1977. *Dr. John Lhotsky: The turbulent Australian writer, naturalist and explorer.* Melbourne, Vic., Australia Felix Literary Club. 176 pp., illus.

Lhuyd, E.

Ba1136. **Gunther, R. T.** 1945. *Life and letters of Edward Lhuyd.* London, Dawsons. xv+576 pp. *Early science in Oxford* vol. 14. *Re-pub.* 1968.
See also **M. E. Jahn.** 1971. A note on the correspondence of Edward Lhuyd. *Journal of the Society for the Bibliography of Natural History* 6: 61–62.

Lidforss, B.

Ba1137. **Karlsson, R.** 1983. *Bengt Lidforss och botaniken: En studie av hans populära och vetenskapliga arbeten moty bakgrund av dåtida filosofi och naturvetenskap.* Lund, Liberförlag. 280 pp.

Liebig, J. von.

Ba1138. **Paoloni, C.** 1968. *Justus von Liebig: Eine Bibliographie, sämtlicher Veröffentlichungen mit biographischen Anmerkungen.* Heidelberg, C. Winter. 332 pp.

Lightfoot, J.

Ba1139. **Bowden, J.** 1989. *John Lightfoot: His work and travels. With a biographical introduction and a catalogue of the Lightfoot herbarium.* Kew, Bentham Moxon Trust, Royal Botanic Gardens; Pittsburgh, PA, Hunt Institute for Botanical Documentation. [v]+255 pp.

Lilford, Lord (T. L. Powys)

Ba1140. **Drewitt, C. M.** 1900. *Lord Lilford, Thomas Littleton, fourth Baron: A memoir by his sister.* London, Smith, Elder & Co. xxiii+290 pp., illus.

Liljefors, B.

Ba1141. **Hill, M.** 1987. *The peerless eye, a biography of Bruno Liljefors.* Kingston–upon–Hull, Allen Publishing. 174 pp., 120 illus. (80 col.).

Ba1142. **Russow, K. E.** 1929. *Bruno Liljefors, an appreciation.* Translated from the Swedish by A. Poignant. Stockholm. 190 pp. & 24 pl.

Lillie, F. R.

Ba1143. **Allen, G. E.** 1973. The F. R. Lillie papers in Woods Hole, Massachusetts. *Mendel newsletter* 9: 1–6.

Lindley, J.

See Ba133 & Ba185.

Link, H. F.

Ba1144. **Martius, C. F. P. von.** 1851. *Denkrede auf Heinrich Friedrich Link. Gehalten in der öffentlichen Sitzung der Königlich Bayerischen Akademie der Wissenschaften am 28. März 1851.* Munich. 63 col.
Separatabdruck aus den *Gelehrten Anzeigen* nos. 59–69.

Linnaeus, C.

• Limitation of space permits no more than the following selection from the large corpus of literature on Linnaeus's life and works:

Ba1145. **Ährling, E.** 1878. *Några af de i Sverige befintliga Linnéanska handskrifterna, kritisk skärskådande*. Lund. [ii]+28 pp.

Ba1146. ___. 1885. *Carl von Linnés brefvexling*. Stockholm, P. A. Norstedt & Söner. xviii+111 pp. *Re-pub*. Zug, Inter Documentation Co., 1967. iv+xviii+111 pp.
Also **C. Linnaeus.** [1967?] *Linnaeus's correspondence*. Zug, I D C. 446 microfiche. Reproductions of over 3,000 letters in the collection at the Linnean Society of London.

Ba1147. **Blunt, W.** 1971. *The compleat naturalist: A life of Linnaeus*. By ... with the assistance of W. T. Stearn. London, Collins. 256 pp.

Ba1148. **Bobrov, E. G.** 1970. *Karl Linnei, 1707–1778*. Leningrad, Izd. "Nauka." 287 pp.

Ba1149. ___. 1979. On the works by and on Linnaeus published in Russia and the Soviet Union. *Svenska Linné-Sällskapets årsskrift* 1978: 265–275.

Ba1150. **British Museum (Natural History).** 1907. *Memorials of Linnaeus: A collection of portraits, manuscripts, specimens and books exhibited to commemorate the bi-centenary of his birth*. [By A. B. Rendle.] London. 16 pp. & 2 pl. *Special guide* no. 3.

Ba1151. ___. 1933. *A catalogue of the works of Linnaeus (and publications more immediately relating thereto) preserved in the libraries of the British Museum (Bloomsbury) and the British Museum (Natural History)*. 2nd ed. London. xi+246+(68) pp.; ___. 1936. *An index to the authors (other than Linnaeus) mentioned in the catalogue of the works of Linnaeus [etc.]*. London. 59 pp.; *see* Bc220.

Ba1152. **Broberg, G.**, ed. 1980. *Linnaeus: Progress and prospects in Linnaean research*. Stockholm, Almqvist & Wiksell; Pittsburgh, PA, Hunt Institute for Botanical Documentation. 318 pp. Also pub. as: *Svenska Linné–Sällskapets årsskrift* 1978.
Papers by fifteen authors.

Ba1153. **Bryk, F.** 1923. *Bibliotheca Linnæana*. I. *Die schwedische Linnélitteratur seit 1907*. Stockholm, privately pub. [vii]+52 pp.; ___. 1923. ___. II. *Neue oder wenig bebannte Linnéoriginale*. Stockholm, privately pub. 30 pp. & 1 pl.

Ba1154. ___. 1923. Die Stellung der polnischen Botanik zu Linnés Systeme. [Mit einer Bibliographia Linnæana Polnica]. *Svenska Linné–Sällskapets årsskrift* 6: 35–51.

Ba1155. ___. 1953. Bibliographia Linnaeana ad *Species Plantarum* pertinens. *Taxon* 2: 74–84.

Ba1156. ___. 1954. Bibliographia Linnaeana ad *Genera Plantarum* pertinens. *Taxon* 3: 174–183.

Ba1157. **Carlbring, O., & A. Otterhed.** 1978. *Carl von Linné*. Örebro, Ringströms Boktryckeri. 31 pp. & folded chart.
Chronology of Linnaeus's life and works with a "Beskrivande del" in the form of an alphabetically arranged topical biographical synopsis.

Ba1158. **Drake af Hagelsrum, G.** 1939. *Linnés disputationer: En översikt*. Nässjö. 294 pp.

Ba1159. **Ekelöf, G.** 1913. Förteckning öfver manuskript innehållande anteckningar efter Carl von Linnés Föreläsningar i Zoologi. In: A. J. E. Lönnberg. *Linnés föresläsningar öfver djurriket. Med understöd af Svenska staten för Uppsala Universitet utgifna och försedda ned förklarande anmärkningar af E. Lönnberg*, pp. iv–xiii. Uppsala.
41 MSS described.

Ba1160. **Fries, T. M.** 1903. *Linné: Lefnadsteckning*. Stockholm, Fahlcrantz & Co. 2 vols.
For an abridged English language version by B. D. Jackson. 1923; *see* Ba1170.

Ba1161. **Frängsmyr, T.**, ed. 1983. *Linnaeus: The man and his work*. Berkeley & Los Angeles, CA, Univ. of California Pr. xii+203 pp. & 12 pp. of illus.

Includes **S. Lindroth.** The two faces of Linnaeus, pp. 1–62; **G. Eriksson.** Linnaeus the botanist, pp. 63–109; **T. Frängsmyr.** Linnaeus as a geologist, pp. 110–155; **G. Broberg.** Homo sapiens. Linnaeus's classification of man, pp. 156–194.

Ba1162. **Goerke, H.** 1973. *Linnaeus.* Translated from the German by D. Lindley. New York, C. Scribner's Sons. xi+178 pp.
Translation of *Carl von Linné: Arzt, Naturforscher, Systematiker 1707–1778.* Stuttgart, Wissenschaftliche Verlagsgesellschaft, 1966. 232 pp. & port. *Grosse Naturforscher* vol. 31. *Re-pub.* 1989.

Ba1163. **Gombocz, E.** 1922. Bibliographia Linnæana hungarica. Hrsg von F. Bryk. *Svenska Linné–Sällskapets årsskrift* 5: 90–96.

Ba1164. **Hagberg, K.** 1957. *Carl Linnaeus.* Andra bearbetade upplagen. Stockholm, Natur och Kultur. 237 pp.
Replaces ___. 1939. ___. Stockholm, [etc.]. 324 pp.; For earlier English version *see* ___. 1952. *Carl Linnaeus.* Translated by A. Blair. London, J. Cape. 264 pp.

Ba1165. **Heller, J. L.** 1978. Bibliotheca zoologica Linnaeana. *Svenska Linné–Sällskapets årsskrift* 1978: 240–264.

Ba1166. ___. 1983. *Studies in Linnean method and nomenclature.* Frankfurt am Main, P. Lang. ix+326 pp. *Marburger Schriften zur Medizingeschichte* no. 7.
Re-pub. of several previously published papers.

Ba1167. **Hulth, J. M.** 1907. *Bibliographia Linnaeana. Matériaux pour servir a une bibliographie Linnéenne.* Partie I – Livraison I. Upsala, Almqvist & Wiksell. [iv]+170 pp.
Chronologically arranged annotated bibliography of works by Linnaeus. The proposed 2nd and 3rd parts were never published. *Also* **E. Förberg.** 1923. Företeckning över till Linnésamlingen hörande böcker, sam ej äro upptagna i J. M. Hulth's Bibliographia Linnaeana. *Svenska Linné–Sällskapets årsskrift* 6: 160–165. List of books in the author's Linnaean collection, not mentioned in J. M. Hulth's Bibliographia Linnaeana.

Ba1168. ___. 1921. Uppsala Universitetsbibliotheks förvärv av Linneanska originalmanuskript. In: Uppsala Universitetsbibliotek. *Uppsala Universitets Biblioteks minneskrift 1621–1921,* pp. 407–424. Uppsala, Almqvist & Wiksell.
Acquisitions of original Linnaean manuscripts.

Ba1169. **Hunt Institute for Botanical Documentation, Carnegie Mellon University.** 1976. Linnaean symposium at the Hunt Institute [etc.]. *Taxon* 25: 3–74.
Papers by eight authors.

Ba1170. **Jackson, B. D.** 1923. *Linnaeus (afterwards Carl von Linné): The story of his life.* Adapted from the Swedish of T. M. Fries, and brought down to the present time in the light of recent research. London, H. G. & F. Witherby. xv+416 pp., 7 pl. & 1 map.

Ba1171. **Koninklijk Zoölogisch Genootschap "Natura Artis Magistra" te Amsterdam.** 1878. *Linnaeana in Nederland aanwezig. Tentoongesteld op 10 Januari 1878 in hen Koninklijk Zoölogisch Genootschap "Natura Artis Magistra" te Amsterdam.* By D. J. Coster. Amsterdam, Scheltema & Holkema. 58 pp.
272 items.

Ba1172. **Kungliga Svenska Vetenskapsakademien.** 1909. *Carl von Linnés Bedeutung als Naturforscher und Arzt.* Schilderingen herausgegeben von der Königl. Schwedischen Akademie der Wissenschaften, anlässlich der 200–jährigen Wiederkehr des Geburtstages Linné's. Jena, G. Fischer. xii+575 pp.
Separately paged contributions by various authors, most of which were apparently published separately: **O. E. A. Hjelt.** Carl von Linné als Arzt und Medizinischer Schriftsteller, iv+186 pp.; **E. Lönnberg.** Carl von Linné und die Lehre von den Wirbeltieren, 48 pp.; **C. Aurivillius.** Carl von Linné als Entomolog, 43 pp.; **C. A. M. Lindman.** Carl von Linné als botanischer Forscher und Schriftsteller, iv+188 pp.; **A. G. Nathorst.** Carl von Linné als

Geolog, 86 pp.; **H. Sjögren.** Carl von Linné als Mineralog, 42 pp. Translation of *Carl von Linnés betydelse såsom naturforskare och läkare*. Stockholm, K. Svenska Vetenskaps–Akademien, [1907]. Various paging.

Ba1173. **Lindell, E.** 1932. *Bibliotheca Linnæana – works by or relating to Carolus Linnaeus, his predecessors, contemporaries and pupils, with sequels – from the library of Emil Lindell.* Växjö, privately pub. 123 pp.; *see* Bc508.

Ba1174. **Lindroth, S.** 1966. Two centuries of Linnaean studies. In: T. R. Buckman, ed. *Bibliography and natural history*, pp. 27–45. Lawrence, KS, Univ. of Kansas Libraries.

Ba1175. **Linnean Society of London.** 1940. *Synopsis of the annotations by Linnaeus and contemporaries in his library of printed books.* By S. Savage. London. 20 pp. *Catalogue of manuscripts in the library of the Linnean Society of London* pt. 3.
For a full listing of Linnaeus's library *see* ___. 1925. *Catalogue of the printed books [etc.]*; *see* Bc261

Ba1176. **Linnaeus, C.** 1907–22. *Bref och skrifvelser af och till Carl von Linné.* Utgifna och med upplysande noter försedd af T. M. Fries och J. M. Hulth. Första afdelningen: *Utländska brefväxlingen.* Uppsala & Berlin. 8 vols.

Ba1177. ___. 1916–43. *Bref och skrifvelser af och till Carl von Linné.* Utgifven och med upplysande noter försedd af J. M. Hulth och A. Hj. Uggla. Andra afdelningen: *Utländska brefväxlingen.* Uppsala & Berlin. 2 vols.

Ba1178. ___. 1957. *Vita Caroli Linnaei: Carl von Linnés självbiografier. På uppdrag av Uppsala Universitet.* Ed. by E. Malmeström & A. H. Uggla. Stockholm, Almqvist & Wiksell. 235 pp.
Linnaeus's five autobiographies.

Ba1179. **Pulteney, R.** 1805 *A general view of the writings of Linnaeus.* 2nd ed. with corrections, considerable additions, and memoirs of the author by W. G. Maton. To which is annexed *The diary of Linnaeus* written by himself, and now translated into English, from the Swedish manuscript in the possession of the editor. London, J. Mawman. xv+595 pp.

Ba1180. **Rudolph, G. A., & E. Williams.** 1970. [Catalogue of the K. K. McKenzie collection of] *Linnaeana.* Manhattan, KS. 225 pp. *Kansas State University Library bibliography series* no. 7; see Bc252.

Ba1181. **Smith,** Sir **J. E.**, ed. 1821. *A selection of the correspondence of Linnaeus, and other naturalists, from the original manuscripts.* London, Longman, Hurst, Orme & Brown. 2 vols.

Ba1182. **Stafleu, F. A.** 1971. *Linnaeus and the Linnaeans: The spreading of their ideas in systematic botany, 1735–1789.* Utrecht, Oostheok for the International Association for Plant Taxonomy. xvi+386 pp. *Regnum vegetabile* vol. 79.

Ba1183. **Stearn, W. T.** 1957. An introduction to the *Species Plantarum* and cognate botanical works of Carl Linnaeus. In: C. Linnaeus. *Species plantarum. A facsimile of the first edition 1753.* Vol. 1, pp. 1–176. London, Ray Society.
Includes II. Principal events in the life of Linnaeus; III. Major botanical publications of Linnaeus; IV. Linnaeus's sexual system of classification; V. The *Genera Plantarum* and the typification of Linnaean genera; VI. The *Hortus Cliffortianus*; VII. The *Amoenitates academicae* and the authorship of Linnaean dissertations; VIII. The preparation of the *Species Plantarum* and the introduction of binomial nomenclature; IX. The reception of the *Species Plantarum* in England and its influence on British botany; XI. Pre-Linnaean literature; XII. Linnaean herbaria; XVII. The species–concept of Linnaeus.

Ba1184. ___. 1959. The background of Linnaeus's contributions to the methods and nomenclature of systematic biology. *Systematic zoology* 8: 4–22.

Ba1185. **Strandell, B.** 1966. Linné och Amerika. *Svenska Linné–Sällskapets årsskrift* 49: 85–96.

Linnaean collections at University of Kansas Libraries and Hunt Botanical Library, Pittsburgh, PA.

Ba1186. ***Svenska Linné-Sällskapets årsskrift.*** Stockholm, årgang 1+, 1918+. *Also* ___. 1987. *Register till årsskrift: 1918–1985.* Stockholm. 24 pp.
Includes periodic reports of newly published Linnaean studies, e.g. **C. O. von Sydow.** 1958. Bibliografi över år 1957 nyutgiven litteratur av och om Linné samt dennes lärjungar. *Svenska Linné-Sällskapets årsskrift* 41: 121–132.

Ba1187. **Tournay, R.** 1953. L'oeuvre botanique de Linné et ses diverses éditions. *Bulletin de la Société Royale de Botanique de Belge* 86: 123–133.

Ba1188. **Tullberg, T.** 1907. *Linnéporträtt.* Vid Uppsala universitets minnesfest på tvåhundraårsdagen af Carl von Linnés födelse. Å universitetets vägnar. Stockholm, Aktiebolaget Ljus. [iv]+187 pp. & 21 pl.; *Continued* **I. T. Tullberg Beskow.** 1967. *Linnéporträtt. Supplement till Tycho Tullberg, Linnéporträtt (1907).* Stockholm, Almqvist & Wiksell. 58 pp., pl. XXII–XXXIV. *Acta universitatis upsaliensis, C. Organisation och historia* no. 14. (*Also* **I. T. Beskow & K. B. Tainsh.** 1968. *Register till Linnéporträtt ... af Tycho Tullberg 1907. med Supplement 1967 [etc.].* Uppsala, [Svenska Linné-Sällskapet]. 28 pp.); **M. Tainsh.** 1978–80. *Linnéporträtt. Supplement A (–B).* Stockholm. 26+26 ff.; **K. B Tainsh.** 1984. *Supplement C to Tullberg's "Linneporträtt" 1907.* Stockholm. 18 pp.; **K. B. & M. Tainsh.** 1991. *Linnéporträtt. Supplement 2 till T. Tullberg, Linnéporträtt, Uppsala 1907 med Supplement 1967 ... av I. T. Beskow.* Uppsala. 30 pp.

Ba1189. **United States Department of Agriculture, National Agricultural Library (USDA, NAL).** 1968. *Linnaeana in the collection of the National Agricultural Library.* By M. L. Naftalin. Beltsville, MD. 39 pp. *Library list* no. 89.

Ba1190. **University of Kansas Libraries.** 1957. *A catalog of an exhibition commemorating the 250th anniversary of the birth of Carolus Linnaeus ... and the 200th anniversary of the issue of the Systema naturae, 10th edition, 1758–1958.* By T. R. Buckman. Lawrence, KS, Univ. of Kansas Libraries. 50 pp.
73 items.

Ba1191. **Williams, T.** 1964. *Checklist of Linnaeana [etc.].* Lawrence, KS. x+73 pp. *University of Kansas library series* no. 20; *see* Bc331

Ba1192. **Weinstock, J.**, ed. 1985. *Contemporary perspectives on Linnaeus.* Lanham, MD, Univ. Pr. of America. ix+193 pp.
Papers by nine authors.

See also Ba303 Bb18A Ca708 Ca981 Ca1106 Cb1033–1036 Cc482 Cc523 & Cc871–883.

Lister, M.

Ba1193. **Keynes,** Sir **G.** 1981. *Dr. Martin Lister: A bibliography.* Foxbury Meadow, St. Paul's Bibliographies. ix+52 pp.
Replaces ___. 1979–80. Some uncollected authors, 51. Dr. Martin Lister, F.R.S. *Book collector* 28: 501–520, 1979; *Ibid.* 29: 50–81, 1980.

Lloyd, J. U.

Ba1194. **Simons, C. M.** 1972. *John Uri Lloyd: His life and works 1849–1936, with a history of the Lloyd Library.* Cincinnati, OH, privately pub. [vii]+iii+337 pp. & port.

L'Obel, M. de

Ba1195. **Louis, A.** 1980. *Mathieu de l'Obel 1538–1616. Épisode de l'histoire de la botanique.* Ghent, Story-Scientia. 544 pp.

Ba1196. **Morren, C. J. É.** 1875. *Mathias de l'Obel: Sa vie et ses oeuvres. 1538–1616.* Liège, privately published. 25 pp.

See also Cb164.

Loder, E.

Ba1197. **Pease,** Sir **A. E.** 1923. *Edmund Loder, naturalist, horticulturist, traveller and sportsman: A memoir.* London, J. Murray. x+356 pp. & port.

Lodge, G. E.

Ba1198. **Lodge, G. E.** 1946. *Memoirs of an artist naturalist.* London, Gurney & Jackson. vii+96 pp. & 24 pl.

Ba1199. **Savory, J.**, ed. 1986. *George Lodge – artist naturalist.* London, Croom Helm. 118 pp. & 48 pl. (col.). *Also* Dover, NH, Tanager Books, 1986.

Löfling, P.

Ba1200. **Rydén, S.** 1965. *Pehr Löfling: En Linnélärjunge i Spanien och Venezuela, 1751–1756.* Stockholm. 242 pp. *Bidrag till Kungliga Svenska Vetenskapsakademiens historia* no. 6.

Longfield, C.

Ba1200A. **Hames, J. Hayter–.** 1991. *Madam Dragonfly: The life and times of Cynthia Longfield.* Bishop Auckland, Pentland Pr. [xiv]+208 pp., illus.

Lorenz, K.

Ba1201. **Evans, R. I.** 1975. *Konrad Lorenz, the man and his ideas.* New York, Harcourt Brace Jovanovich. xviii+302 pp.

Ba1202. **Festetics, A.** 1983. *Konrad Lorenz: Aus der Welt des grossen Naturforschers.* Munich, R. Piper. 159 pp., illus.

Ba1203. **Nisbett, A.** 1976. *Konrad Lorenz.* London, J. M. Dent. xii+249 pp. & [16] pp. of illus. *Also* New York, Harcourt Brace Jovanovich, 1977.

Ba1204. **Wuketits, F. M.** 1990. *Konrad Lorenz: Leben und Werk eines grossen Naturforschers.* Munich, Piper Verlag. 286 pp.

See also Ba298 & Cc266.

Lortet family

Ba1205. **Magnin, A.** 1913. *Les Lortets, botanistes Lyonnais particulièrement Clarence, Pierre et Louis Lortet et le botaniste Roffavier.* Lyon, A. Rey. 109 pp. & 1 pl.
First pub. in *Annales de la Société Botanique de Lyon* vol. 37, 1912.

Loudon, J.

Ba1206. **Howe, B.** 1961. *Lady with green fingers: The life of Jane Loudon.* London, Country Life. 184 pp., illus.

Lovén, S.

See Ba69.

Lubbock, Sir J.

Ba1207. **Hutchinson, H. G.** 1914. *Life of Sir John Lubbock, Lord Avebury.* London, Macmillan. 2 vols.

See also Ba26.

Ludwig, C.

Ba1208. **Schröer, H.** 1967. *Carl Ludwig, Begründer der messenden Experimentalphysiologie, 1816–1895.* Stuttgart, Wissenschaftliche Verlagsgesellschaft. 340 pp. *Grosse Naturforscher* vol. 33.

See also Ba753.

Lyonet, P.

Ba1209. **Hublard, E.** 1910. *Le naturaliste hollandais Pierre Lyonet: Sa vie et ses oeuvres (1706–1789) d'après des lettres inédites.* Brussels, Labèque. 159 pp. *Mémoires de la Société des Sciences, des Arts et des Lettres du Hainaut* vol. 61.

Ba1210. **Van Seters, W. H.** 1962. *Pierre Lyonet, 1706–1789, sa vie, ses collections de coquillages et de tableaux, ses recherches entomologiques.* Traduit de Néerlandais par L. Laurent. The Hague, M. Nijhoff. xiv+228 pp. & 28 pl.

McClintock, B.

Ba1211. **Keller, E. F.** 1983. *A feeling for the organism: The life and work of Barbara McClintock.* San Francisco, CA, W. H. Freeman. 235 pp.
For French version *see* **___.** 1988. *L'Intuition du vivant: La vie et l'oeuvre de Barbara McClintock.* Trad. de l'américain par R.–M. Vassallo–Villaneau. Paris, Tiercé. 268 pp.

MacGillivray, W.

Ba1212. **MacGillivray, W.** 1910. *Life of William MacGillivray [etc].* With a scientific appreciation by J. Arthur Thompson. London, J. Murray. xv+222 pp. & 12 pl.

Macintosh, C.

Ba1213. **Coates, H.** 1923. *A Perthshire naturalist: Charles Macintosh of Inver.* London, T. F. Unwin. xx+244 pp., illus. Abridged ed. pub. as: **___.** [1924.] *Charlie Macintosh, post–runner, naturalist, and musician.* London, T. F. Unwin. 142 pp., illus.

M'Intosh, W. C.

Ba1214. **Gunther, A. E.** 1977. *The life of William Carmichael M'Intosh, M.D., F.R.S., of St. Andrews, 1838–1931: A pioneer in marine biology.* Edinburgh, Scottish Academic Pr. for the Univ. of St. Andrews. 214 pp., illus. *St. Andrews University publications* no. 61.

Mackay, A. F.

See Ba598.

Macoun, J.

Ba1215. **Macoun, J.** 1922. *Autobiography of John Macoun, M.A., Canadian explorer and naturalist. Assistant Director of the Geological Survey of Canada 1831–1920, [etc.].* Ottawa, Ottawa Naturalists' Field Club. x+305 pp. & 8 pl.

Ba1216. **Weiser, W. A.** 1989. *The field naturalist: John Macoun, the Geological Survey and natural science.* Toronto, Univ. of Toronto Pr. xiv+253 pp.

Malaspina, A.

See Ca754–755.

Malpighi, M.

Ba1217. **Adelmann, H. D.** 1966. *Marcello Malpighi and the evolution of embryology.* Ithaca, NY, Cornell Univ. Pr. 5 vols.
Vol. 1: The life and works of Marcello Malpighi, pp. 3–726.

Ba1218. **___**, ed. 1975. *Marcello Malpighi: The correspondence.* Ithaca, NY, Cornell Univ. Pr. 5 vols. (lxx+2,227 pp.)

Ba1219. **Cardini, M.** 1927. *La vita e l'opera di Marcello Malpighi.* Rome, Pozzi. 123 pp., port.

Ba1220. **Frati, C.** [1960.] *Bibliografia Malpighiana: Catalogo descrittivo delle opere a stampa di Marcello Malpighi e degli scritti lo riguardano.* London, Dawsons of Pall Mall. [iv]+56 pp.

First pub. as: ___. 1897. Bibliografia delle opere a stampa di Marcello Malpighi e degli scritti che lo riguardano. In: U. Pizzoli, ed. *Marcello Malpighi e l'opera sua: Scritti varii, [etc.]*, pp. 281–338. Milan.

Manning, A.

See Ba298.

Marey, E. J.

Ba1221. **Braun, M.** 1992. *Picturing time: The work of Etienne–Jules Marey (1830–1904).* Chicago, Il, University of Chicago Press. xxii+450 pp.
Bibliography of works by Marey, pp. 425–437.

Ba1221A. **Centre National d'Art et de Culture Georges Pompidou, Paris.** 1977. *É. J. Marey. 1830/1904. La photographie du movement.* Paris. 129 pp.
Illustrated catalogue and exhibition guide; includes bibliography.

Ba1222. **Dagognet, F.** 1992. *Etienne-Jules Marey: A passion for the trace.* Translated by R. Galeta with J. Herman. New York, Zone Books. 204 pp., illus.
Translation of ___. 1987. *Etienne–Jules Marey: La passion de la trace.* Paris, Hazen. 139 pp.

Marler, P.

See Ba298.

Marsh, O. C.

See Cc283.

Marsh, G. P.

Ba1223. **Curtis, J. & W., & F. Lieberman.** 1982. *The world of George Perkins Marsh, America's first conservationist and environmentalist: An illustrated biography.* Woodstock, VT, Countryman Pr. 123 pp.

Marshall, H.

Ba1224. **Darlington, W.** 1849. *Memorials of John Bartram and Humphrey Marshall, with notices of their botanical contemporaries.* Philadelphia, PA, Lindsay & Blakiston. xvi+585 pp. & 1 pl. *Re-pub.* New York, Hafner, 1967.
Also **Ewan, J.** 1957. Analytical index to Darlington's Memorials of Bartram and Marshall (1849). *Journal of the Society for the Bibliography of Natural History* 3: 274–280, port.

Marsili, L. F.

Ba1225. **Longhena, M.** 1930. *Il conte L. F. Marsili, un uomo d'arme e di scienza.* Milan, "Alpes." 346 pp., illus.

Martius, C. F. P. von

Ba1226. **Goebel, K.** 1905. *Zur Erinnerung an K. F. Ph. von Martius.* Munich, Königlich Bayerische Akademie der Wissenschaften. 20 pp.

Ba1227. **Schramm, H.** 1869. *C. F. Ph. v. Martius: Sein Lebens– und Characterbild insbesondere seine Reiseerlebnisse in Brasilien.* Leipzig, L. Denicke. 2 vols.

Mattioli, P. A.

Ba1228. **Fabiani, G.** 1872. *La vita di Pietro Andrea Mattioli raccolta dalle sue opere.* Pubblicata con aggiunte ed annotazioni per cura di L. Banchi. Siena. x+83 pp.
Bibliografia delle opere di P. A. Mattioli, pp. 75–83.

Maulny, L.

Ba1229. **Delaunay, P.** 194? Louis Maulny, naturaliste et collectionneur, 1758–1815. *Bulletin de la Société d'Agriculture, Sciences et Arts de la Sarthe* 60 (1): 132 pp.

Maxwell, M.

Ba1230. **Benson, M.** 1986. *Martha Maxwell: Rocky Mountains naturalist.* Lincoln, NE, Univ. of Nebraska Pr. xix+335 pp. *Women in the West* no. 1.

Meckel, J. F.

Ba1231. **Beneke, R.** 1934. *Johann Friedrich Meckel der Jüngere.* Halle, M. Niemeyer. vii+159 pp., illus. *Beiträge zur Geschichte der Universität Halle–Wittenberg* vol. 3.

Meinertzhagen, R.

Ba1232. **Cocker, M.** 1989. *Richard Meinertzhagen: Soldier, scientist and spy.* London, Secker & Warburg. [xi]+292 pp.

Ba1233. **Lord, J.** 1971. *Duty, honour, Empire: The life and times of Colonel Richard Meinertzhagen.* London, Hutchinson. [xi]+412 pp. & 8 pl.

Mendel, G.

Ba1234. **Correns, C.**, ed. 1905. *Gregor Mendels Briefe an Carl Nägeli 1866–1873: Ein Nachtrag zu den veröffentlichen Bastardierungsversuchen Mendels.* Leipzig, B. G. Teubner. ii+79 pp. *Abhandlungen der Mathematisch–Physischen Classe der Königlich Sächsischen Gesellschaft der Wissenschaften* 29(3).

Ba1235. **Czihak, G.** 1984. *Johann Gregor Mendel (1822–1884): Dokumentierte Biographie und Katalog zur Gesächtnisausstellung anlässlich des hundertsten Todestages, mit Facsimile seines Hauptwerkes:* Verzuche über Pflanzenhybriden. Unter Mithilfe zahlreicher Fachkollegen und Mitarbeiter verfasst von [etc.]. Salzburg, G. Czihak. 120 pp.

Ba1236. ***Folia Mendeliana;*** *Papers relating to Mendel and to the early development of genetics.* Brno, Moravske Zemske Museum, No. 1+, 1966+.

Ba1237. **Iltis, H.** 1932. *Life of Mendel.* Translated by E. & C. Paul. London, Allen & Unwin. 336 pp. & 12 pl. *Re-pub.* 1966.
Translation of *Gregor Johann Mendel, Leben, Werk und Wirkung.* Berlin, Springer, 1924. vii+[3]+426 pp. & 12 pl.

Ba1238. **Jakubicek, M., & J. Kubicek.** 1965. *Bibliographia Mendeliana.* Brno, Universitní Knihovna. 74 pp. *Vyberovy seznam Univ. Knihovny v Brne* no. 101; *Continued* **M. Jakubicek.** 1976. ___*: Supplementum 1970–1974.* Brno, Universitní Knihovna. 56 pp. *Ibid.* no. 152.

Ba1239. **Krízenecky, J.**, Ed. 1965. *Gregor Johann Mendel, 1822–1884: Texte und Quellen zu seinen Wirken und Leben.* Festgabe der Deutschen Akademie der Naturforscher Leopoldina zum Mendel Memorial Symposium 1865–1965, August 1965 in Brünn. Leipzig, J. A. Barth. 198 pp. *Lebensdarstellungen deutscher Naturforscher* no. 11.

Ba1240. **Krumbiegel, I.** 1957. *Gregor Mendel und das Schicksal seiner Entdeckung [etc.].* Stuttgart, Wissenschaftliche Verlagsgesellschaft. [vi]+144 pp. & port. *Grosse Naturforscher* vol. 22.

Ba1241. **Mendel, G.** 1965. *Fundamenta genetica. Revised ed. with a collection of 27 original papers published during the rediscovery era.* Selection and commentary by J. Krízenecky. Prague, Ceskoslovenska Akademie Ved. 400 pp.

Ba1242. **Moravske Zemske Museum.** 1965. *Pamatce Gregora Johanna Mendela k stemu vyroci zverejneni jeho objevu zakladu dedicnosti. Iconographia Mendeliana. To the memory of Gregor Johann Mendel for the centenary of the publication of his discovery of the principle of heredity.* Ed. by L. Marvanova & V. Orel. Brno. 73 pp., illus.

Ba1243. **Orel, V.** 1984. *Mendel.* Oxford, Oxford Univ. Pr. 111 pp.

Ba1244. **___, & A. Matalova**, eds. 1983. *Gregor Mendel and the foundation of genetics.* Brno, Mendelianum of the Moravian Museum. 368 pp.

Proceedings of pt. 1 of the symposium "The past, present and future of genetics," Kuparovice, Czechoslovakia, August 1982. *Includes* **I. Cetl.** The chronology of Mendel's scientific activities, pp. 289–298.

Ba1245. **Sajner, J.** 1967. Entwicklung und Ergebnisse der Gregor Mendel–Forschung. *Medizinhistorisches Journal* 2: 78–91.

Ba1246. **Sosna, M.**, ed. 1966. *G. Mendel memorial symposium, 1865–1965. Proceedings of a symposium held in Brno in August 4–7, 1965*. Prague, Ceskoslovenska Akademie Ved. xxiii+287 pp. & 9 pl.

Ba1247. **Van der Pas, P. W.** 1959. A note on the bibliography of Gregor Mendel. *Medical history* 3: 331–333.
Lists the original publication and reprintings of Mendel's two classic papers.

See also Ba2.

Merian, M. S.

Ba1248. **Lendorff, G.** 1955. *Maria Sibylla Merian, 1647–1717, ihr Leben und ihr Werk, [etc.]*. Basle, Gute Schriften. 64 pp., illus.

Ba1249. **Pöhlmann, O.** 1935. *Maria Sibylla Merian: Roman.* Berlin, Büchergilde Gutenberg. 223 pp., illus.
Fictionalized biography.

Ba1250. **Quednau, W.** 1966. *Maria Sybilla Merian: Der Lebensweg einer grossen Künstlerin und Forscherin.* Gütersloh, S. Mohn. 221 pp.

Ba1251. **Stuldreher–Nienhuis, J.** 1945. *Verborgen paradijzen: Het leven en de werken van Maria Sibylla Merian 1647–1717.* Arnhem, Van Loghum Slaterus. 186 pp. & pl. (col.).

Ba1252. **Ullmann, E.**, ed. [1974.] *Les aquarelles de Leningrad de Maria Sibylla Merian.* Ed. par E. Ullmann; avec la collaboration de H. Ullmann [et al.] Paris. 2 vols.
Facsimile reproductions of drawings in the Akademiya Nauk SSSR, with a separate vol. of commentary.

See also Ba303.

Merriam, C. H.

Ba1253. **Sterling, K. B.** 1977. *Last of the naturalists: The career of C. Hart Merriam.* 2nd ed. New York, Arno Pr. xii+472 pp.

Meyer, F. N.

Ba1254. **Cunningham, I. S.** 1984. *Frank N. Meyer: Plant hunter in Asia.* Ames, IA, Iowa State Univ. Pr. xvii+317 pp., illus.

Michaux, A. & F.

Ba1255. **MacPhail, I.** 1981. *The Sterling Morton Library bibliographies in botany and horticulture, I: André and François Michaux.* Lisle, IL, Morton Arboretum. [vi]+35 pp., port.

Ba1256. ___. 1983. Natural history in Utopia: The works of Thomas Say and François–André Michaux printed at New Harmony, Indiana. In: Society for the Bibliography of Natural History. *Contributions to the history of North American natural history: Papers from the first North American conference of the Society for the Bibliography of Natural History held at the Academy of Natural Sciences, Philadelphia, 21–23 October 1981*, pp. 15–33. London. *Special publication* no. 2.
Bibliography, pp. 25–32.

Ba1257. **Savage, H. Jr. & E. J.** 1987. *André and François Michaux.* Charlottesville, VA, Univ. Pr. of Virginia. ix+435 pp.

Micheli, P. A.

Ba1258. **Ragazzani, S.** 1983. Per una catalogazione degli scritti inediti di Pier Antonio Micheli. *Annali dell' Instituto e Museo di Storia della Scienza di Firenze* 8 (2): 159–172.

Ba1259. **Targioni-Tozzetti, G.** 1858. *Notizie della vita e delle opere di Pier' Antonio Micheli, botanico Fiorentino.* Florence, P. Le Monnier. vi+446 pp.
Includes Catalogus operus Petri Antonii Micheli, pp. 337–360.

Millais, J. G.

Ba1260. **Millais, J. G.** 1919. *Wanderings and memories.* London, Longman, Green & Co. xii+298 pp. & 15 pl. American ed. as *A sportsman's wanderings.* Boston, MA, Houghton Mifflin Co., 1920.

Miller, G. S.

Ba1261. **Miller, G. S., & D. H. Johnnson.** 1954. Bibliography of Gerrit Smith Miller, Jr. *Journal of mammalogy* 35: 329–344.

Miller, P.

Ba1262. **Le Rougetel, H.** 1989. *The Chelsea gardener, Philip Miller and his circle, 1691–1771.* With a contribution by William T. Stearn on the botanical importance of Miller. London, Natural History Museum Publications. 212 pp., illus.

Miquel, F. A. W.

Ba1263. **Stafleu, F. A.** 1966. *F. A. W. Miquel, Netherlands botanist.* Amsterdam. 95 pp. *Wentia* vol. 16.

Mitchill, S. L.

Ba1264. **Hall, C. R.** 1967. *A scientist in the early Republic: Samuel Latham Mitchill, 1764–1831.* New York, Russell & Russell. 162 pp.

Mivart, St. G. J.

Ba1265. **Gruber, J. W.** 1960. *A conscience in conflict: The life of St. George Jackson Mivart.* New York, Columbia Univ. Pr. x+[iv]+266 pp. & port.

Mociño, J. M.

Ba1266. **Palacio, R. G.** 1982. *El peregrinar de las flores Mexicanas, Jose Mariano Mociño y Losada 1757–1822.* Xalapa, Instituto Nacional de Investigaciones sobre Recursos Bióticos. 100 pp. & 16 pl.

Mohl, H. von.

See Ba21 & Ba133.

Molisch, H.

Ba1267. **Molisch, H.** 1934. *Erinnerungen und Welteindrücke eines Naturforschers.* Vienna & Leipzig, E. Haim. 232 pp., illus.

See also Ba21.

Monardes, N. B.

Ba1268. **Guerra, F.** 1961. *Nicolas Bautista Monardes: Su vida y su obra (ca. 1493–1588).* Mexico City, Compania Fundidora de Fierro y Acero de Monterrey. iv+225+[3] pp. & 30 pl. *Publication, Yale University, Department of the History of Medicine* no. 41.

Ba1269. **Marin, F. R.** 1925. *La verdadera biografía del doctor Nicolas de Monardes. Conferencia ... 1913.* Madrid, "Revista de Archivos." 102 pp. *Conferencias y cursos breves organizados por el Ministerio de Instrucción Publica y Bellas Artes.*

Montagu, G.

Ba1270. **Cleevely, R. J.** 1978. Some background to the life and publications of Colonel George Montagu (1753–1815). *Journal of the Society for the Bibliography of Natural History* 8: 445–480.
Manuscript material, pp. 452–456.

More, A. G.

Ba1271. **More, F. M.** 1898. *Life and letters of Alexander Goodman More ... with selections from his zoological and botanical writings.* Ed. by C. B. Moffat. Dublin, Hodges, Figgis & Co. xii+642 pp. & port.

Morgan, T. H.

Ba1272. **Allen, G. E.** 1978. *Thomas Hunt Morgan: The man and his science.* Princeton, NJ, Princeton Univ. Pr. xvii+447 pp.

Ba1273. **Fuller, M. B.** 1981. Thomas Hunt Morgan papers at the American Philosophical Society. *Mendel newsletter* 21: 6–8.

Ba1274. **Glass, B.** 1986. An exciting find: Thomas Hunt Morgan letters. *Mendel newsletter* 26: 6–7.
In the library of the American Philosophical Society.

Ba1275. **Shine, I., & S. Wrobel.** 1976. *Thomas Hunt Morgan: Pioneer of genetics.* Lexington, KY, Univ. Pr. of Kentucky. xv+160 pp.

See also Cc89.

Morison, R.

See Ba185 & Cb1037.

Morris, F. O.

Ba1276. **Morris, M. C. F.** 1897. *Francis Orpen Morris: A memoir.* London, J. C. Nimmo. xi+323 pp. & 3 pl.

Morse, E. S.

Ba1277. **Wayman, D. G.** 1942. *Edward Sylvester Morse: A biography.* Cambridge, MA, Harvard Univ. Pr. xvi+457 pp.

Mougeot, J.–B.

Ba1278. **Laissus, Y.** 1978. Les papiers du Docteur J.–B. Mougeot (1776–1858): Un fonds important pour l'histoire des sciences naturelles en Lorraine. *Comptes rendus du 103 Congrès National des Sociétés Savantes, Section des Sciences,* 5: 233–240.

Mueller, F. von

Ba1279. **Churchill, D. M., T. M. Muir, & D. M. Sikora.** 1978. The published works of Ferdinand J. H. Mueller (1825–1896). *Muelleria* 4: 1–120. *Continued* ___. 1984. ___: Supplement. *Ibid.* 5: 229–248.

Ba1280. **Muir, T. B.** 1984. An index to the new taxa, new combinations and new names published by Ferdinand J. H. Mueller. *Muelleria* 4: 123–168.

Ba1281. **Willis, M.** 1949. *By their fruits: A life of Ferdinand von Mueller, botanist and explorer.* Sydney, Angus & Robertson. xii+187 pp., illus.

Muir, J.

Ba1282. **Clarke, J. M.** 1980. *Life and adventures of John Muir.* [San Diego, CA,] Sierra Club. x+324 pp., illus.

Ba1283. **Cohen, M. P.** 1984. The pathless way: John Muir and the American wilderness. Madison, WI, Univ. of Wisconsin Pr. xvii+408 pp.

Ba1284. **Fox, S.** 1985. *The American conservation movement: John Muir and his legacy.* Madison, WI, Univ. of Wisconsin Pr. xii+436 pp., illus.
First pub. in 1981 as *John Muir and his legacy: The American conservation movement.* Boston, MA, Little, Brown & Co.

Ba1285. **Kimes, W. F. & M. B.** 1986. *John Muir: A reading bibliography, chronologically compiled and annotated by [etc.].* 2nd ed. Fresno, CA, Panorama West Books. xxiv+179 pp.

Ba1286. **Limbaugh, R. H., & K. E. Lewis**, eds. 1986. *The John Muir papers, 1858–1957.* Alexandria, VA, Chadwyck–Healey. 51 reels microfilm, 53 microfiches, & printed guide, 190 pp.

Ba1287. **Stewart, J.** 1975. *Winds in the woods — The story of John Muir.* Philadelphia, PA, Westminster Pr. 126 pp., illus.

Ba1288. **Wolfe, L. M.** 1945. *Son of the wilderness: The life of John Muir.* New York, A. A. Knopf. xvii+364+xvi pp. *Re-pub.* Madison, WI, Univ. of Wisconsin Pr., 1978.

Müller, F.

Ba1289. **Möller, A.** 1915–21. *Fritz Müller: Werke, Briefe und Leben.* Jena, G. Fischer 3 vols. in 4.
Vol. 1: *Gesammelte Schriften*; vol. 2: *Briefe*; vol. 3: *Fritz Müllers Leben.*

Müller, J. P.

Ba1290. **Ebbecke, U.** 1951. *Johannes Müller, der grosse rheinische Physiologe, [etc.].* Hanover, Schmorl & von Seefeld. 191 pp., illus.

Ba1291. **Haberling, W.** 1924. *Johannes Müller: Das Leben des rheinischen Naturforschers auf Grund neuer Quellen und seiner Briefe dargestellt.* Leipzig, Akademische Verlagsgesellschaft. [iv]+500 pp., illus.

Ba1292. **Koller, G.** 1958. *Das Leben des Biologen Johannes Müller, 1801–1858.* Stuttgart, Wissenschaftliche Verlagsgesellschaft. 268 pp. *Grosse Naturforscher* vol. 23.

See also Ca46.

Müller, O. F.

Ba1293. **Anker, J.** 1943. *Otto Friedrich Müller: Et Bidrag til den biologiske Forsknings Historie i det attende Aarhundrede.* Vol. 1: *Barndom, Studenter–og Hovmestertid (1730–67).* Copenhagen. xii+317 pp. *Acta historica scientiarum naturalium et medicinalium* no. 2.
All published.

Ba1294. **___.** 1950. *Otto Friedrich Müller's Zoologia Danica.* Copenhagen, E. Munksgaard. 108 pp. & 9 pl. *Library research monographs, University Library Copenhagen* no. 1.

Mutis, J. C.

Ba1295. **Ferrero, P. M., et al.** 1986. *Actas del simposium CCL aniversario nacimiento de José Celestino Mutis.* Cadiz, Paz Martin Ferrero. 457 pp. *Historia* vol. 3.
Many papers by various authors, including: **F. D. Calonge.** Datos sobre la obra de José Celestino Mutis depositada en el Real Jardín Botánico de Madrid.

Ba1296. **Pinto Escobar, P., & S. Diaz Piedrahita**, eds. 1983. *José Celestino Mutis 1732–1982.* Bogota, Universitad Nacional de Columbia. 284 pp., illus. *Biblioteca Jose Jeronimo Triana* no. 1.
Chapters by six authors. Bibliography, pp. 255–284.

Ba1297. **Gredilla, A. F.** 1982. *Biografía de José Celestino Mutis y sus observaciones sobre las vigilias y suenos de algunas plantas.* Bogota, Plaza & Janes, Academia Colombiana de Historia. 381 pp., illus. *Complemento a la historia extensa de Colombia*

no. 1.
First pub. as: 1911. *Biografía de José Celestino Mutis con relación de su viaje y estudios practicados en el Nuevo Reino de Granada reunidos y anotados por [etc.].* Madrid, Junta para Ampliación de Estudios é Investigaciones Científicas, 1911. 712+[3] pp. & 2 pl.

Ba1298. **Hernandez de Alba, G.** 1947–49. *Archivio epistolar del sabio naturalista don José Celestino Mutis.* Nota biográfica. Bogota, Editorial Kelly. 2 vols.

Ba1299. **Hoyos Sainz, L. de.** 1949. *José Celestino Mutis, naturalista, medico y sacerdota.* Madrid, Editorial Nacional. 259 pp. *Breviarios de la vida española* no. 37.

Ba1300. **Llinas, J. P.** 1982. *Mutis: El hombre y sus suenos.* Bogota, Ediciones Tercer Mundo. 207 pp.

Ba1301. **Mutis, J. C.** 1947–75. *Archivo epistolar del sabio naturalista, Don José Celestino Mutis.* Compilación, prólogo, y notas de G. Hernández de Alba. Bogota, Ministerio de Educacion Nacional (*later* Editorial Kelly). 6 vols.

See also Ca736–743.

Myers, G. S.

Ba1302. **Cox, W. E.** 1988. *Guide to the papers of George Sprague Myers, circa 1903–1986, and undated.* Washington, DC, Smithsonian Institution. vi+86 pp. *Guides to collections, Archives & Special Collections of the Smithsonian Institution* no. 8.

Ba1303. **Myers, G. S.** 1970. Annotated bibliography of the publications of George Sprague Myers (to the end of 1969). *Proceedings of the California Academy of Science* (4) 38: 19–52.

Nägeli, C. W. von

Ba1304. **Cramer, C.** 1896. *Leben und Wirken von Carl Wilhelm von Nägeli.* Zurich, F. Schultess. 91 pp.

See also Ba2 Ba21.

Naumann family

Ba1305. **Leverkühn, P.** 1904. *Biographisches über die drei Naumanns und bibliographisches über ihre Werke nebst den Vorreden zur zweiten Auflage der Naturgeschichte der Vögel Deutschlands.* In: **J. A. Naumann.** *Naturgeschichte der Vögel Mitteleuropas*, vol. 1: xvi–xlvi. Gera–Untermhaus, F. E. Köhler. *Also pub. separately* 32? pp.

Naumann, J. F.

Ba1306. **Naumann, J. F.** 1954. *J. F. Naumanns Briefwechsel mit H. Lichtenstein 1818–1856.* Ed. by E. Stresemann & P. Thomsen. Copenhagen, E. Munksgaard. 74 pp. *Acta historica scientiarum naturalium et medicinalium* vol. 11.

Ba1307. ___. 1969. *Die ornithologische Korrespondenz zwischen Johann Friedrich Naumann und Heinrich Rudolf Schinz in den Jahren 1815 bis 1835.* Ed. by E. Stresemann & L. Baege. Odense, Odense Universitetsforlag. 87 pp., ports. *Acta historica scientiarum naturalium et medicinalium.*

Ba1308. **Thomsen, P.** 1957. *Johan Friedrich Naumann: Der Altmeister der deutschen Vögelkunde; Sein Leben und seine Werke.* Nach des Verfassers Tode bearb. und erganzt von E. Stresemann. Leipzig, J. A. Barth. 212 pp., illus. *Lebensdarstellungen deutscher Naturforscher* no. 6.

Nelson, A.

Ba1309. **Williams, R. I.** 1984. *Aven Nelson of Wyoming.* Boulder CO, Colorado Associated Univ. Pr. xii+407 pp.

Newman, E.

Ba1310. **Newman, T. P.** 1876. *Memoir of the life and works of Edward Newman.* London, J. Van Voorst. 32 pp., port. *Re-pub.* (with a new introduction by E. W. Classey) Faringdon, E. W. Classey Ltd., 1980. 32 pp. *Classica entomologica* no. 6.

Newton, A.

Ba1311. **Wollaston, A. F. R.** 1921. *Life of Alfred Newton, professor of comparative anatomy, Cambridge University, 1866–1907.* London, J. Murray. xv+332 pp. & 5 pl.
LIst of published work, pp. 316–324.

Nice, M. M.

Ba1312. **Nice, Margaret M.** 1979. *Research is a passion with me.* Toronto, Consolidated Amethyst Communications. xii+322 pp., illus.

Nilsson, S.

Ba1313. **Retzius, G.** 1901. *Sven Nilsson: Minnesteckning vid Kongl. Vetenskaps–Akademiens högtidsdag den 30 mars 1901.* Stockholm, Aftonbladets Aktiebolags Tryckeri. 43 pp., port.

See also Ba69.

Nilsson–Ehle, H.

See Ba69.

Nodier, C

Ba1314. **Magnin, A.** 1911. *Charles Nodier, naturaliste: Ses oeuvres d'histoire naturelle publiées et inédites.* Paris, A. Hermann. x+347 pp. & port.

Nordenskiöld, A. E.

See Ba69.

North, M.

Ba1315. **North, M.** 1892. *Recollections of a happy life, being the autobiography of ….* Ed. by her sister, Mrs J. A. Symonds. London, Macmillan. 2 vols.

Ba1316. **___.** 1893. *Some further recollections of a happy life selected from journals, chiefly between the years 1859 and 1869.* Ed. by Mrs J. A. Symonds. London, Macmillan. viii+316 pp.

Ba1317. **___.** 1980. *A vision of Eden: The life and work of Marianne North.* Abridged text by G. Bateman. Exeter. 240 pp.
Includes **B. Moon.** Marianne North, 1830–90, pp. 234–239. Numerous colour reproductions of paintings in the Marianne North Gallery at the Royal Botanic Gardens, Kew.

Ba1318. **Royal Botanic Gardens, Kew.** 1914. *Official guide to the [Marianne] North Gallery.* 6th ed., rev. & augm. London. xxxvi+162 pp.

Nuttall, T.

Ba1319. **Graustein, J. E.** 1967. *Thomas Nuttall, naturalist: Explorations in America, 1808–1841.* Cambridge, MA, Harvard Univ. Pr. xiv+481 pp.

Ba1320. **MacPhail, I.** 1983. *The Sterling Morton Library bibliographies in botany and horticulture, II: Thomas Nuttall.* Lisle, IL, Morton Arboretum. [vi]+35 pp., port.

Oken, L.

Ba1321. **Ecker, A.** 1883. *Lorenz Oken: A biographical sketch, or "in memoriam" of the centenary of his birth, read before the fifty–second meeting of the German Association for the Advancement of Science at Baden–Baden, September 20, 1879. … With explanatory notes [etc.].* From the German by A. Tulk. London, Kegan Paul, Trench & Co.

xxiv+183 pp.
Includes List of Oken's writings, pp. 179–183. Translation of *Lorenz Oken [etc.]*. Stuttgart, Schweizerbart'sche Verlagshandlung, 1880. viii+220 pp.

Ba1322. **Pfannenstihl, M.** 1953. *Lorenz Oken: Sein Leben und Werken.* Freiburg im Breisgau, H. F. Schulz. 24 pp., illus. *Freiburger Universitätsreden* n.s. 14.

See also Ca46.

Opiz, P. M.

Ba1323. **Nemec, B., et al.** 1958. *Philipp Maxmilian Opiz a jeho vyznam pro taxonomii rostlin. Philipp Maxmilian Opiz und seine Bedeutung für die Pflanzentaxonomie.* Prague, Nakladelství Ceskoslovenské Akademie Ved. 271 pp.
Text in German.

Orbigny, A. d'

Ba1324. **Orbigny, A. d'.** 1856. *Notice analytique sur les travaux de géologie, de paléontologie et de zoologie de M. Alcide d'Orbigny, 1823–1856.* Paris. 28 pp.

Ormerod, E. A.

Ba1325. **Wallace, R.**, ed. 1904. *Eleanor Ormerod LL.D. Economic entomologist. Autobiography and correspondence.* London, J. Murray. xx+348 pp., 30 pl., & port. *Also* New York, E. P. Dutton & Co.

Orta, G. da

Ba1326. **Ficalho,** Count **de.** 1886. *Garcia da Orta e o seu tempo.* Lisbon, Impenso Nacional. xii+392 pp.

Osborn, H. F.

Ba1327. **Rainger, R.** 1980. The Henry Fairfield Osborn papers at the American Museum of Natural History. *Mendel newsletter* 18: 8–13.

Ba1328. **___.** 1991. *An agenda for antiquity: Henry Fairfield Osborn and vertebrate paleontology at the American Museum of Natural History, 1890–1935.* Tuscaloosa, AL, Univ. of Alabama Pr. xiii+360 pp. *History of American science & technology series.*

Ba1329. **Ripley, F. E.** 1916–27. *Bibliography of the published writings of H. F. Osborn for the years 1877–1915.* 2nd ed. [New York.] [ii]+74 p.; *Continued* ___. [1927.] ___: Part III. *Chronological arrangement. 1916–27.* [New York.] pp. 75–92.

Osten-Sacken, C. R. von

Ba1330. **Osten-Sacken, C. R. von.** 1903–04. *Record of my life-work in entomology.* Cambridge, MA. Pr. *Re-pub.* (with an appreciation and introductory preface by K. G. V. Smith) Faringdon, Classey, 1978. 10+viii+240 pp., 4 pl. *Classica entomologica* no. 2.

Oudemans, C. A. J. A.

Ba1331. **Moll, J. W.** 1909. *Levensbericht van C. A. J. A. Oudemans.* Amsterdam, J. Müller. 51 pp. *Jaarboek der Koninklijke Akademie van Wetenschappen* vol. 62.

Owen, R.

Ba1332. **Gruber, J. W., & J. C. Thackray.** 1992? *Richard Owen commemoration.* London, The Natural History Museum? 192? pp., illus.
Forthcoming publication announced. Includes a catalogue of the correspondence, manuscripts and drawings of Richard Owen, William Clift and Sir Everard Home contained in the Owen collection at the Natural History Museum, London.

Ba1332A. **Owen, R. S.** 1894. *The life of Richard Owen.* With the scientific portions revised by C. Davies Sherborn. Also an essay on Owen's position in the anatomical

sciences by T. H. Huxley. London, J. Murray. 2 vols.
Bibliography, vol. 2: 333–382. *Also see* A212.

Packard, A. S.

Ba1333. **Henshaw, S.** 1887. The entomological writings of Dr. Alpheus Spring Packard. *Bulletin, Division of Entomology, U. S. Department of Agriculture* 16: 49 pp.

Pallas, P. S.

Ba1334. **Pallas, P. S.** 1967. *A naturalist in Russia: letters from Peter Simon Pallas to Thomas Pennant.* Ed. by C. Urness. Minneapolis, MN, Univ. of Minnesota Pr. vii+189 pp.

Ba1335. **Wendland, F.** 1991. *Peter Simon pallas (1741–1811): Materialen einer Biographie.* Berlin, de Gruyter. 2 vols. *Veroffentlichungen der Historischen Kommission zu Berlin* vol. 80.

Ba1336. **Zaunick, R.** 1925. Peter Simon Pallas (1741–1811), der Begründer der paläarktischen Wirbeltierkunde. Ein bio-bibliographische Studie. *Pallasia* 3: 1–37.

See also Ba12 & Ba303.

Palmer, E.

Ba1337. **McVaugh, R.** 1956. *Edward Palmer: Plant explorer of the American West.* Norman, OK, Univ. of Oklahoma Pr. xvii+430 pp. & 12 pl.

Palmer, T. S.

Ba1338. **United States Library of Congress.** 1958. *Theodore Sherman Palmer: A register of his papers in the Library of Congress.* Washington, DC. 8 ff.
Ornithological papers.

Pander, C. H.

Ba1339. **Raikov, B. E.** 1984. *Christian Heinrich Pander: Ein bedeutender Biologe und Evolutionist, 1794–1865.* Deutsche überzetzung mit Kommentaren und englischen Kurzfassungen von W. E. von Hertzenberg, P. H. von Bitter. Frankfurt on Main, W. Kramer. 144 pp., illus. *Senckenberg–Buch* no. 62.
Translation of *Khristian Pander — vydayushchiisya biolog-evolyutsionist, 1794–1865.* Leningrad, Izd. "Nauka," 1964. 95 pp. *Akademiya Nauk SSSR, nauchno–biograficheskaya seriya.*

Parker, G. H.

Ba1340. **Parker, G. H.** 1946. *The world expands: Recollections of a zoologist.* Cambridge, MA, Harvard Univ. Pr. viii+252 pp., port.

Parker, T. J.

Ba1341. **Parker, T. J.** 1893. *William Kitchen Parker: A biographical sketch.* London, Macmillan. xv+145 pp., port.

Parkinson, S.

Ba1342. **Carr, D. J., et al.** 1983. *Sydney Parkinson: Artist of Cook's "Endeavour" voyage.* London & Canberra, A.C.T., British Museum (Natural History) in association with Australian National Univ. Pr. xvi+300 pp., 253 illus.

Parlatore, F.

Ba1342A. **Parlatore, F.** 1992. *Mie memorie.* Ed. by A. Visconti. Palermo, Sellerio. 482 pp., illus.

Pasteur, L.

Ba1343. **Cuny, H.** 1965. *Louis Pasteur: The man and his theories.* Translated by P. Evans. London, Souvenir Pr. 190 pp., illus. *Also* New York, P. S. Erikswson, 1966. *A profile in science.*
Translation of *Louis Paster ... choix de textes, bibliographie, portraits, facsimiles.* Paris, Seghers, 1963. 223 pp., illus.

Ba1344. **Dagognet, F.** 1967. *Méthodes et doctrines dans l'oeuvre de Pasteur.* Paris, Presses Universitaires de France. vii+263 pp.

Ba1345. **Delaunay, A.** 1967. *Pasteur et la microbiologie.* 3e ed. Paris, Presses Universitaires de France. 126 pp. *"Que sais–je?" Le point des connaissances actuelles* no. 467.

Ba1346. **Descour, L.** 1922. *Pasteur and his work.* Translated by A. F. & B. H. Wedd. London., T. F. Unwin. 256 pp., port.
Translation of *Pasteur et son oeuvre.* Paris, Delagave, 1921. 296 pp., port.

Ba1347. **Dubos, R. J.** 1952. *Louis Pasteur: Free lance of science.* Boston, MA, Little, Brown. xii+418 pp.

Ba1348. ___. 1988. *Pasteur and modern science.* New illustrated ed. Ed. by T. D. Brock. Berlin, Springer–Verlag; Madison, WI, Science Tech Publishers. xxiv+168 pp. *Scientific revolutionaries: A biographical series.*
Replaces Reprints text of 1960 ed. without change.

Ba1349. **Duclaux, É.** 1920. *Pasteur: The history of a mind.* Translated by E. F. Smith & F. Hedges. Philadelphia, PA, W. B. Saunders & Co. xxxii+363 pp., illus.
Translation of *Pasteur: Histoire d'un esprit.* Paris, Masson & Cie, 1896. 400 pp., illus.

Ba1350. **Nicolle, J.** 1961. *Pasteur: A master of scientific enquiry.* London, Hutchinson. 196 pp.

Ba1351. **Prévost, M. L.** 1977. Manuscrits et correspondence de Pasteur à la Bibliothèque Nationale. *Bulletin de la Bibliothèque Nationale* 2: 99–107.

Ba1352. **Vallery–Radot, R.** 1901. *The life of Pasteur.* Translated from the French by Mrs R. L. Devonshire. London, A. Constable. 2 vols. Various eds., e.g. London, Constable, [1948]. xxi+484 pp. *Re-pub.* New York, Dover Publications, 1960.
Translation of *La vie de Pasteur.* Paris, Hachette & Cie, 1900. 692 pp.

Ba1353. ___. 1958. *Louis Pasteur: A great life in brief.* Translated from the French by A. Joseph. New York, A. A. Knopf. vii+199 pp.

See also Ba2

Paterson, .

See Ca730.

Patterson, H. N.

Ba1354. **Patterson, H. N.** 1953. *Afield with plant lovers and collectors: Botanical correspondence of the late Harry N. Patterson with the great botanical collectors and distributors of America from 1870 to 1919.* Ed. by A. L. Kibbe. Carthage, IL. 565 pp., ports.

Pavlov, I. P.

Ba1355. **Akademiya Nauk SSSR.** 1949. *Arkhiv Akademii Nauk SSSR. Rukopisnye materialy I. P. Pavlova v Arkhive Akademii Nauk. [Archives of the Academy of Science of the USSR. Manuscript materials of I. P. Pavlova.].* By G. P. Blok & E. S. Kuliabko. Leningrad. 156 pp. *Trudy arkhiva Akademiya Nauk SSSR* vol. 8.
1. Nauchnye trudy; 2. Otzyvy o trudakh uchenikh; 3. Pis'ma, lichnnye zayavleniya, ankety i pr.;

4. Dokumenty po nauchnoi, nauchn.–organizatsionnoi i administrativnoi deyatel'nosti v Akademii Nauk.

Ba1356. **Cuny, H.** 1964. *Ivan Pavlov: The man and his theories.* Translated by P. Evans. London, Souvenir Pr. 183 pp., illus. *Also* New York, P. S. Eriksson, 1965. *A profile in science.*
Translation of *Ivan Pavlov ... choix de textes, bibliographie, portraits, facsimiles.* Paris, Seghers, 1962. 223 pp., illus.

<u>Pavon, J.</u>

See Ca731–735.

<u>Peale, C. W.</u>

Ba1357. **Miller, L. B.**, ed. 1980. *The collected papers of Charles Willson Peale and his family.* Millwood, NY, Kraus. 449 microfiche.
With a printed: *Guide and index to the microfiche edition.*

Ba1358. ___. 1983. *The selected papers of Charles Willson Peale and his family.* Vol. 1: *Charles Willson Peale: Artist in revolutionary America, 1735–1791.* New Haven, CT, Yale Univ. Pr. liii+673 pp.

Ba1359. ___. 1988. *The selected papers of Charles Willson Peale and his family.* Vol. 2: *Charles Willson Peale: The artist as museum keeper, 1791–1810.* New Haven, CT, Yale Univ. Pr. 2 pts, 678 & 640 pp.

Ba1360. ___, **& D. C. Ward**, eds. 1991. *New perspectives on Charles Willson Peale: A 250th anniversary collection*, pp. 219–235. Pittsburgh, PA, Univ. of Pittsburgh Pr. for the Smithsonian Institution. xviii+317 pp. & 88 pp. of illus.

Ba1361. **Richardson, E. P., B. Hindle, & L. B. Miller.** 1982. *Charles Willson Peale and his world.* New York, H. N. Abrams. 272 pp.

Ba1362. **Sellers, C. C.** 1969. *Charles Willson Peale.* New York, C. Scribner's Sons. xiv+510 pp. & 14 pl.

See also Ca824 Ca982–987.

<u>Peale, T. R.</u>

See Ca786.

<u>Pearson, T. G.</u>

Ba1363. **Orr, O. H., Jr.** 1992. *Saving American birds: T. Gilbert Pearson and the founding of the Audubon movement.* Gainesville, FL, Univ. Pr. of Florida. xii+296 pp., illus.

Ba1363A. **Pearson, T. G.** 1937. *Adventures in bird protection: An autobiography.* New York, D. Appleton–Century Co. xiv+459 pp.

<u>Peattie, D. C. & L. R.</u>

Ba1364. **Scott, W. H. O.** 1989. Donald Culross and Louise Redfield Peattie: A bibliography. *Bulletin of bibliography* 46 (1): 10–30.

<u>Pena, P.</u>

See Cb164.

<u>Pennant, T.</u>

See Ba303.

<u>Péron, F.</u>

Ba1365. **Girard, M.** 1857. *François Péron, naturaliste, voyageur aux terres australes: Sa vie, appréciation de ses travaux, analyse raisonnée de ses recherches sur les animaux* *vertébrés et invertébres d'après ses collections déposées au Muséum d'Histoire Naturelle [etc.]*. Paris, Société d'Émulation de l'Allier. 278 pp.

See also Ba303.

Perrault, C.

Ba1366. **Hallays, A.** 1926. *Les Perraults*. 2nd ed. Paris, Perrin. 309 pp.

Petényi, J. S. von

Ba1367. **Herman, O.** 1891. *J. S. v. Petényi, der Begründer der wissenschaftlichen Ornithologie in Ungarn, 1799 bis 1855*. Ein Lebensbild unter Mitwirkung von J. v. Madarász, St. v. Chernel und G. v. Vastagh. Budapest, Ungarisches Nationalmuseum. 137 pp.

Peterson, R. T.

Ba1368. **Devlin, J. C., & G. Naismith.** 1977. *The world of Roger Tory Peterson: An authorized biography*. New York, Times Books. xxi+266 pp. *Also* Newton Abbot, David & Charles, 1978.

Petiver, J.

Ba1369. **Stearns, R. P.** 1953. James Petiver: Promoter of natural science. *Proceedings of the American Antiquarian Society* 1952: 243–265.

Pettingill, O. S., Jr.

Ba1369A. **Pettingill, O. S., Jr.** 1992. *My way to ornithology*. Norman, OK, Univ. of Oklahoma Pr. xiv+245 pp., illus.

Pfeffer, W.

Ba1370. **Bünning, E.** 1989. *Ahead of his time: Wilhelm Pfeffer, early advances in plant biology*. Translated from the German by H. W. Pfeffer. Ottawa, Carleton Univ. Pr. xii+199 pp. *Carleton science series* no. 1.

Translation of *Wilhelm Pfeffer: Apotheker, Chemiker, Botaniker, Physiologe, 1845–1920*. Stuttgart, Wissenschaftliche Verlagsgesellschaft, 1975. 166 pp. *Grosse Naturforscher* vol. 37.

Philippi, R. A.

Ba1371. **Gotschlich, B.** 1904. *Biografía del Dr. Rodulfo Amando Philippi (1808–1904)*. Santiago de Chile. 184 pp., illus.

Pickering, C.

See Ba133 & Ba205.

Piper, C. V.

See Bc855 for his botanical papers.

Platter, F. & T.

See Cb166 & Cb1038.

Plée, A.

Ba1372. **Thésée, F.** 1989. *Auguste Plée, 1786–1825: Un voyageur naturaliste. Ses travaux et ses tribulations aux Antilles, au Canada, en Colombie*. Paris, Editions Caribéenes. 221 pp.

Plukenet, L.

Ba1373. **Guédès, M.** 1981. Les œuvres de Leonard Plukenet et leurs réimpressions. *Archives of natural history* 10: 67–76.

Plumier, C.

Ba1374. **Urban, I.** 1920. Plumier, Leben und Schriften nebst einen Schlüssel zu einen Blüten–Planzen. *Beihefte zum repertorium specierum novarum regni vegetabilis* 5: 196 pp.

Poivre, P.

Ba1375. **Fels, M. de.** 1968. *Pierre Poivre ou l'amour des épices.* Paris, Hachette. 199 pp.

Ba1376. **Laissus, Y.** 1975. Note sur les manuscrits de Pierre Poivre (1719–1786) conservés à la Bibliothèque du Muséum d'Histoire Naturelle. In: *Comptes rendus du 95e Congrès National des Sociétés Savantes, Sectiondes Sciences,* 2: 161–188.

Pollich, J. A.

Ba1377. **Müller, G. H.** 1990. *Johann Adam Pollich (1741–1780), 'Medicinae Doctor' und Pflanzengeograph der Pflaz.* Bad Dürkheim, Pollichia. 176 pp., illus. *Pollichia–Buch* no. 21.

Popenoe, W.

Ba1378. **Rosengarten, F.** 1991. *Wilson Popenoe: Agricultural explorer, educator, and friend of Latin America.* Lawai, HI, National Tropical Botanical Garden. viii+182 pp., illus.

Ba1379. **Woodger, E. R., & A. Sharpe.** 1983. An inventory of the Wilson Popenoe papers. *Huntia* 5: 23–59.
In the Hunt Institute for Botanical Documentation, Pittsburgh, PA.

Post, G. E.

See Ba205.

Powys, T. L.

See Lilford, Lord

Praeger, R. L.

Ba1380. **Collins, T.** 1985. *Floreat Hibernia: A bio-bibliography of Robert Lloyd Praeger, 1865–1953.* Dublin, Royal Dublin Society. xiv+151 pp. *Historical studies in Irish science & technology* no. 5.

Ba1381. **Praeger, R. L.** 1947. *The way that I went. An Irishman in Ireland.* 3rd ed. Dublin, Hodges, Figgis & Co.; London, Methuen. xiii+416 pp., illus.

Prestele family

Ba1382. **Ravenswaay, C. van.** 1984. *Drawn from nature: The botanical art of Joseph Prestele and his sons.* Washington, DC. 357 pp. incl. 95 pl., & 43 illus.

Pringle, C. G.

Ba1383. **Davis, H. B.** 1936. *Life and work of Cyrus Guernsey Pringle.* Burlington, VT, Univ. of Vermont. [iii]+756+14 pp. & port.

Pulteney, R.

Ba1384. **Jeffers, R. H.** 1960. Richard Pulteney, M.D., F.R.S. (1730–1801), and his correspondents. *Proceedings of the Linnean Society of London* 171: 15–26.
An account of his correspondence held by the Linnean Society of London.

Purkyne, J. E.

Ba1385. **John, H. J.** 1959. *Jan Evangelista Purkyne: Czech scientist and patriot 1787–1869.* Philadelphia, PA, American Philosophical Society. 94 pp. *Memoirs of the American Philosophical Society* vol. 49.

Ba1386. **Kotek, V., & L. Niklicek.** 1987. *Jan Evangelista Purkyne and his place in the history of medicine and natural sciences of the 19th century.* Prague, Academia Praha. 74 pp. & 15 pl.

Ba1387. **Kruta, V.** 1969. *J. E. Purkyne (1787–1869), physiologist. A short account of his contributions to the progress of physiology with a bibliography of his works.* Prague, Academia. 137 pp.

Ba1388. ___, ed. 1969. *Jan Evangelista Purkyne, 1787–1869. Centenary symposium, held at the Carolinium, Prague, 8–10 September 1969.* Brno, Universita Jana Evangelista Purkyne. 287 pp. *Acta Facultatis Universitatis Brunensis* 40.
Papers by ca. 30 authors.

Ba1389. **___, & M. J. P. Flourens.** 1964. *J. E. Purkinje et les débuts de la physiologie de la posture et de l'équilibre.* Paris, Université de Paris. 48 pp. *Les Conférences du Palais de la Découverte* série D, no. 98.

Ba1390. **___, & M. Teich.** 1962. *Jan Evangelista Purkyne.* Translated by S. Kostomlatsky. Prague, State Medical Publishing House. 144 pp. & port.
Translation of *Jan Evangelista Purkyne (1787–1869).* Prague, 1962. 142 pp.

Ba1391. **Purkyne, J. E.** 1964. *K pocatkum vedecke drahy J. E. Purkyne, [etc.] [Beginnings of the scientific career of J. E. Purkyne: Letters with friends from the Prague years 1815–1823, [etc.].* Ed. by V. Kruta. Brno, Lekarska Fakulta, University J. E. Purkyne. 208 pp. & 21 pl. *Acta Facultatis Medicae Universitatis Brunensis* no. 12.

Ba1392. **Zaunick, R., et al.** 1961. Purkyne–Symposium der Deutschen Akademie der Naturforscher Leopoldina in Gemeinschaft mit der Tschechoslowakischen Akademie der Wissenschaften am 31. Oktober und 1. November 1959 in Halle/Saale. *Nova acta leopoldina* 24 (151): 7–230.
Includes Articles by Frankenberger, Kruta, Matousek, Sajner & others.

Quatrefages, J. L. A. de

Ba1393. [**Malard, A. E.**] 1892. Liste des ouvrages et mémoires publiés de 1822 à 1891 par A. de Quatrefages de Bréau. *Nouvelles Archives, Muséum d'Histoire Naturelle, Paris* (3) 4: xlix pp. & port.

Ba1394. ___. 1852. *À la Mémoire de Jean Louis Armand Quatrefages de Bréau. 10 fevrier 1819 – 12 janvier 1892.* Lille. 111+[5] pp., illus.

Raffles, T. S.

Ba1395. **Archer, M., & J. Bastin.** 1978. *The Raffles drawings in the India Office Library, London.* Kuala Lumpur, New York, Oxford Univ. Pr. xiii+106 pp., illus.

Ba1396. **Bastin, J.** 1969. *Sir Thomas Stamford Raffles; with an account of the Raffles–Minto manuscript collection presented to the India Office Library on 17 July 1969 by the Malaysia–Singapore Commercial Association.* Liverpool, Ocean Steam Ship Co. 22 pp., illus.

Ba1397. **Collis, M.** 1966. *Raffles.* London, Faber. 228 pp., illus. *Also* New York, John Day Co., 1968.

Ba1398. **Hahn, E.** 1968. *Raffles of Singapore: A biography.* Kuala Lumpur, Univ. of Malaya Pr. 587 pp., illus.
First pub. Garden City, NY, Doubleday & Co., 1946.

See also Ba303.

Rafinesque, C. S.

Ba1399. **Boewe, C.** 1958. The manuscripts of C. S. Rafinesque (1783–1840). *Proceedings of the American Philosophical Society* 102: 590–595.

See also ___. 1980. Editing Rafinesque's holographs: The case of the short letters. *Filson Club historical quarterly* 54: 37–49.

Ba1400. **Call, R. E.** 1895. *The life and writings of Rafinesque.* Louisville, KY, John P. Morton. xii+227 pp. & port. *Filson Club publications* no. 10.

Ba1401. **Fitzpatrick, T. J.** 1982. *Fitzpatrick's Rafinesque: A sketch of his life with bibliography.* Revised and enlarged by C. Boewe. Weston, MA, M & S Pr. ix+327 pp. & 32 pls.
Replaces ___. 1911. *Rafinesque: A sketch of his life with bibliography.* Des Moines, Historical Department of Iowa. 241 pp.

Ba1402. **Merrill, E. D.** 1949. *Index Rafinesquianis: The plant names published by Rafinesque with reductions, and a consideration of his methods, objectives and attainments.* Jamaica Plain, MA, Arnold Arboretum. ix+296 pp.

Ba1403. **Rafinesque, C. S.** 1836. *A life of travels and researches in North America and South Europe, or outlines of the life, travels and researches of C. S. Rafinesque.* Philadelphia, privately pub. 148 pp. *Re-pub. as* "A life of travels:" Being a verbatim and literatim reprint of the original and only edition. *Chronica botanica* 8 (2): 291–360.

Ba1404. ___. 1978. *Rafinesque: Autobiography and lives.* With an introduction by K. B. Sterling. New York, Arno Pr. Various paging [657] pp. *Biologists and their world.*
Includes Photo–facsimiles of Rafinesque, C. S., *A life of travels ...*, 1836; Call, R. E., *The life and writings of Rafinesque*, 1895; and Fitzpatrick, T. J., *Rafinesque*, 1911.

Rauwolf, L.

Ba1405. **Dannenfeldt, K. H.** 1968. *Leonhard Rauwolf, sixteenth–century physician, botanist, and traveler.* Cambridge, MA, Harvard Univ. Pr. xiv+321 pp. & 4 pl. *Harvard monographs in the history of science.*

See also Cb167.

Ravenel, H. W.

Ba1406. **Haygood, T. M.** 1987. *Henry William Ravenel, 1814–1887: South Carolina scientist in the Civil War era.* Tuscaloosa, AL, Univ. of Alabama Pr. xiii+204 pp. *History of American science & technology series.*

Ray, J.

Ba1407. **Baldwin, S. A.** 1986. *John Ray (1627–1705), Essex naturalist: A summary of his life, work and scientific significance.* Witham, Baldwin's Books. iv+80 pp.

Ba1408. **Gunther, R. W. T.**, ed. 1928. *Further correspondence of John Ray.* London, Ray Society. xxiv+332 pp. *Ray Society publication* vol. 114.

Ba1409. **Keynes, Sir G.** 1976. *John Ray, a bibliography 1660–1970: A descriptive bibliography of the works by John Ray, with introductions, annotations, various indexes and a supplement of new entries, additions and corrections by the author.* Amsterdam, G. T. Van Heusden. xvi+184 pp.
Replaces 1st ed., London, Faber & Faber, 1951. xv+163 pp.

Ba1410. **Lankester, E.**, ed. 1846. *Memorials of John Ray: Consisting of his life by Dr. Derham; biographical and critical notices by Sir J. E. Smith, and Cuvier and Dupetit Thouars. With his itineraries, etc.* London, Ray Society. xii+220 pp.

Ba1411. ___, ed. 1848. *The correspondence of John Ray: Consisting of selections from the philosophical letters published by Dr. Derham, and the original letters of J. Ray in the collection of the British Museum.* London, Ray Society. xvi+502 pp.

Ba1412. **Raven, C. E.** 1950. *John Ray, naturalist, his life and works.* 2nd ed. Cambridge, Cambridge Univ. Pr. xx+506 pp. & port. *Re-pub.* (with a new introduction) Cambridge, [etc.], 1986.

Ba1413. **Thompson, R.** 1974. Some newly discovered letters of John Ray. *Journal of the Society for the Bibliography of Natural History* 7: 111–123.

See also Ba185, Ba303 & Bc331(Kidman & Farley).

Réaumur, R. A. F. de.

Ba1414. **Académie des Sciences, Paris.** 1959. Inventaire de la correspondence de Réaumur conservée aux Archives de l'Académie des Sciences de Paris. *Revue d' histoire des sciences* 1959: 315–326.

Ba1415. **Réaumur, R. A. F. de.** 1962. *La vie et l'oeuvre de Réaumur (1683–1757).* Préface par P.–P. Grassé. Paris, Presses Universitaires de France. vii+188 pp. & 1 pl.

Ba1416. **Torlais, J.** 1961. *Réaumur: Le biologiste, l'entomologiste, l'inventeur, le métallurgiste, le physicien.* Paris, Blanchard. 475 pp.
Replaces ___. 1936. *Un esprit encyclopédique en dehors de "L'Encyclopédie:" Réaumur d'après des documents inédits.* Paris, Desclée de Brouwer 448 pp.

Ba1417. **Trembley, M.**, ed. 1943. *Correspondance inédite entre Réaumur et Abraham Trembley. Comprenant 113 lettres recueillies et annotées par M. Trembley.* Introduction par E. Guyénot. Geneva, Georg & Cie. lvi+432 pp.

Redi, F.

Ba1418. **Anon.** 1964. Dr. Rufus Cole's collection of Redi and his contemporaries. *Academy bookman* 17(2): 2–6.
Presented to the New York Academy of Medicine Library.

Ba1419. **Cardini, M.** [1914.] *Francesco Redi.* Florence, Istituto Micrografico Italiano. 117 pp., illus. *Vite dei medici e naturalisti celebri* no. 2.

Ba1420. **Prandi, D.** 1941. *Bibliografia delle opere di Francesco Redi. Segue la riproduzione della lettera intorno all'invenzione degli occhiali scritta a Paolo Falconieri. Dall'edizione delle "Lettere" 1779–1795.* Reggio Emilia, Nironi & Prandi. viii+38+[8] pp. & 1 pl.

Ba1421. **Viviani, U.** 1924–31. *Vita, opere, iconografia, bibliografia, vocabolario inedito dalle voci aretine e libro inedito dei "Ricordi" di Francesco Redi aretino.* Arezzo, Viviani. 3 vols. *Collana di pubblicazioni storiche e letterarie Aretine* no. 9–11.

Ba1422. [Not used.]

Redouté, P.–J.

Ba1423. **Hunt Botanical Library, Carnegie Mellon University.** 1963. *A catalogue of Redoutéana exhibited at the Hunt Botanical Library, 21 April to 1 August 1963.* Pittsburgh, PA. ix+125 pp.
Includes **F. A. Stafleu.** Redouté – peintre de fleurs, pp. 1–32; **I. MacPhail.** Books illustrated by Redouté, pp. 33–71; **J. V. Brindle.** Redouté's drawings, paintings, and prints [177 items], pp. 73–98; **A. Lawalrée.** Letters and documents by Redouté, pp. 99–117.

Ba1424. **Léger, C.** 1945. *Redouté et son temps.* [Paris,] Éditions de la Galerie Charpentier. 169 pp., & 6 pl. Accompanied by a set of "12 Vélins de Redouté."

Ba1425. **Sitwell, S., R. Madol & W. T. Stearn.** 1954. *Album de Redouté, with twenty-five facsimile color plates from the edition of 1824 and a new Redouté bibliography.* London, Collins. 20 pp. & 25 pl.
Includes **R. Madol & W. T. Stearn**, A Redouté bibliography, pp. 16–20.

Reede tot Drakenstein, H. A. van

Ba1426. **Fournier, M.** 1987–88. Enterprise in botany: Van Reede and his *Hortus Malabaricus* – part I (–II). *Archives of natural history* 14: 123–158, 297–338.

Ba1427. **Heniger, J.** 1986. *Hendrik Adriaan van Reede tot Drakenstein (1636–1691) and Hortus Malabaricus: A contribution to the history of Dutch colonial botany.* Rotterdam, A. A. Balkema. xvi+295 pp.

Ba1428. **Manilal, K. S.**, ed. 1980. *Botany and history of* Hortus Malabaricus. New Delhi, Oxford & I B H. x+237 pp.
Twenty-three essays by various authors.

Renier, S. A.

Ba1429. **Gibin, C., & P.-G. Tiozzi**, eds. 1981. *Stefano Andrea Renier, naturalista e riformatore, Chioggia 1759–Padova 1830.* n.p., Centro Grafico Editoriale. 167 pp., illus.

Richardson, Sir J.

Ba1430. **Curvey, M. F., & R. E. Johnson.** 1969. A bibliography of Sir John Richardson (1787–1865) — printed books. *Journal of the Society for the Bibliography of Natural History* 5: 202–217.

Ba1431. **Johnson, R. E.** 1976. *Sir John Richardson: Arctic explorer, natural historian, naval surgeon.* New York, Crane, Russak. xii+209 pp.

Ba1432. **McIlraith, J.** 1868. *Life of Sir John Richardson.* London, Longman. xi+280 pp.

Richardson, R.

Ba1433. **Turner, D.**, ed. 1835. *Extracts from the literary and scientific correspondence of R. Richardson ... illustrative of the state and progress of botany [etc.].* Yarmouth. lxvi+451 pp., port.

Richter, C. P.

See Ba298.

Ridgway, R.

Ba1434. **Harris, H.** 1928. Robert Ridgway, with a bibliography of his published writings and fifty illustrations. *Condor* 30: 1–119.

Riley, C. V.

Ba1435. **United States Department of Agriculture, National Agricultural Library.** 1990. *The papers of Charles Valentine Riley: A register of his papers in the National Agricultural Library.* Compiled & ed. by J. J. Ho & W. Yuille. Beltsville, MD. xiv+124 pp. *Bibliographies & literature of agriculture* no. 92.

Risso, A.

Ba1436. **Monod, T.** 1931. Inventaire des manuscrits de Risso conservés à la Bibliothèque du Muséum d'Histoire Naturelle. *Archives du Muséum d'Histoire Naturelle, Paris* (6) 7: 102–133, illus.

Ba1437. **___, & J. C. Hureau.** 1977. *Antoine Risso, 1777–1845, volume publié à l'occasion du bicentenaire de sa naissance.* Nice, Centre d'Études Méditerranéennes. 214 pp. & 7 pl. *Annales du Muséum d'Histoire Naturelle de Nice* vol. 5.

Robertson, D.

Ba1438. **Stebbing, T. R. R.** 1891. *The naturalist of Cumbrae, a true story: Being the life of David Robertson.* London, Kegan Paul, Trench & Co. x+398 pp.
Includes Bibliography of Robertson's writings, pp. 385–390.

Rolle, F.

Ba1439. **Martin, G. P. R., & G. Uschmann.** 1969. *Friedrich Rolle, 1827–1887: Ein Vorkampfer neuen biologischen Denkens in Deutschland.* Leipzig, Barth. 151 pp. *Lebensdarstellungen deutscher Naturforscher* no. 14.

Romanes, G. J.

Ba1440. **Romanes, E.** 1896. *The life and letters of George John Romanes.* [By his wife.] 2nd ed. London, Longman. x+360 pp.

Rondelet, G.

Ba1441. **Planchon, J.–É.** 1866. *Rondelet et ses disciples, ou la botanique à Montpellier au XVIme siècle.* Montpellier. 22 pp.
First pub. in: *Montpellier médical* 1866.

See also Ba303.

Rosén von Rosenstein, N.

See Ba69.

Rostand, J.

Ba1442. **Juste, A.** 1971. *La vie et l'oeuvre de Jean Rostand.* Paris, Stock. 172 pp.

Rothschild, L. W., 2nd Baron Rothschild

Ba1443. **Rothschild, M.** 1983. *Dear Lord Rothschild: Birds, butterflies and history.* Glenside, PA, Balaban; London, Hutchinson. xxiii+398 pp.

Rousseau, J.–J.

Ba1444. **Jansen, A.** 1885. *Jean–Jacques Rousseau als Botaniker.* Berlin, G. Reimer. vii+308 pp.

Rouy, G.

Ba1445. **Baranda, J.** 1984. *Georges Rouy (1851–1924), su obra botánica en España.* Madrid, Real Jardín Botánico. 196 pp. *Ruizia* vol. 1.

Rudbeck, O. the elder

Ba1446. **Rudbeck, J.** 1918. *Bibliotheca Rudbeckiana: Beskrivande förteckning över tryckta arbeten, vilka författats eller utgivits av medlemmar av släkten Rudbeckius – Rudbeck samt handla om dem eller deras skrifter. En släkthistoria i elva led från 1600–1900 talen.* Stockholm, Bröderna Lagerström. xii+539 pp.

See also Ba69.

Ruiz Lopez, H.

See Ca731–735.

Rumpf, G. E.

Ba1447. **Ballintijn, G. E.** 1944. *Rumphius, de blinde ziener van Ambon.* Utrecht, W. de Haan. vii+192 pp.

Ba1448. **Koloniaal Museum te Haarlem.** 1902. *Rumphius Gedenboek, 1702–1902.* Amsterdam, J. H. de Bussy. viii+[ii]+221 pp. & 6 pl.
Chapters by various authors including: 4. **J. P. Lotsy.** Over de in Nederland aanwezige botanische handschriften van Rumphius; 14. **G. P. Rouffaer & W. C. Muller.** Eerste proeve van een Rumphius–Bibliographie.

Ba1449. **Wit, H. C. D. de**, ed. 1959. *Rumphius memorial volume.* Baarn, Hollandia. [vi]+v+462 pp. & 27 pl.

Rüppell, E.

Ba1450. **Mertens, R.** 1949. *Eduard Rüppell: Leben und Werk eines Forschungsreisenden.* Frankfurt am Main, W. Kramer. 388 pp., illus.

Rydberg, P. A.

Ba1451. **Tiehm, A., & F. A. Stafleu.** 1990. *Per Axel Rydberg: A biography, bibliography and list of his taxa.* New York, New York Botanical Garden. 75 pp. *Memoirs of the New York Botanical Garden* vol. 58.

Sachs, J.

Ba1452. **Gimmler, H.**, ed. 1984. *Julius Sachs und die Pflanzenphysiologie heute. Festschrift zum 150. Geburtstag des Würzburger Botanikers und Pflanzenphysiologen.* Wurzburg, Verlag der Physikalisch–Medizinischenn Gesellschaft. 216 pp. *Berichte*, Sonderband.

Nine chapters by various authors including: **H. Gimmler.** Tabellarischer Lebenslauf von Sachs und Stammtafel der Sachs–Familie; ___. Bibliographische und biographische Verzeichnisse.

Ba1453. **Pringsheim, E. G.** 1932. *Julius Sachs, der Begründer der neueren Pflanzenphysiologie, 1832–1897.* Jena, G. Fischer. xii+302 pp. & 13 pl.

St. John, H.

See Bc855 for his botanical papers.

Salviani, I.

See Ba303.

Sargent, C. S.

Ba1454. **Ewan, J.**, ed. 1968. Letters of Charles Sprague Sargent to Reginald Somers Cocks, 1908–1926. *Journal of the Arnold Arboretum* 46: 1–44, 122–159, 324–361, 411–439.

Ba1455. **Sutton, S. B.** 1970. *Charles Sprague Sargent and the Arnold Arboretum.* Cambridge, MA, Harvard Univ. Pr. xvii+382 pp.

Sars, M.

Ba1456. **Økland, F.** 1955. *Michael Sars: Et minneskrift.* Oslo, H. J. Dybwad. 221 pp. & port.

Saussure, H. B. de

Ba1457. **Freshfield, D. W., & H. F. Montaignier.** 1920. *The life of Horace Benédict de Saussure.* London, E. Arnold. xii+479 pp., illus.

For French version *see* ___. 1924. *Horace–Bénédict de Saussure.* Ouvrage trad. de l'Anglais par L. Plan. Geneva, Édition Altar. 434 pp., illus.

Savigny, M. J.–C.

Ba1458. **Pallary, P.** 1931. Marie Jules–César Savigny: Sa vie et son oeuvre. *Memoires de l'Institut d'Égypte* 17: 1–111.

Say, T.

Ba1459. **MacPhail, I.** 1983. Natural history in Utopia: The works of Thomas Say and François–André Michaux printed at New Harmony, Indiana. In: Society for the Bibliography of Natural History. *Contributions to the history of North American natural history: Papers from the first North American conference of the Society for the Bibliography of Natural History held at the Academy of Natural Sciences, Philadelphia, 21–23 October 1981*, pp. 15–33. London. *Special publication* no. 2.

Bibliography, pp. 25–32.

Ba1459A. **Stroud, P. T.** 1992. *Thomas Say: New World naturalist.* Philadelphia, PA, Univ. of Pennsylvania Pr. xv+340 pp., illus.

Ba1460. **Summers, G.** 1982. A bibliography of the scientific writings of Thomas Say (1787–1834). *Archives of natural history* 11: 69–81.

Ba1461. **Weiss, H. B., & G. M. Ziegler.** 1931. *Thomas Say, early American naturalist.* Springfield, IL, C. C. Thomas. xv+260 pp. & 19 pl. Re-pub. New York, Arno Pr., 1978.

Schiemann, E.

See Cb196A & Cb197.

Schimper, K. F.

Ba1462. **Götz, H., & A. P. M. Sanders.** 1971. Karl Friedrich Schimper: Eine Lebenskizze nach seinen Schriften und reden – eine Bibliographie. *Janus* 59: 71–93.

Ba1463. **Volger, G. H. O.** 1889. *Leben und Leistungen des Naturforschers Karl Schimper.* Vortrag, gehalten in der ersten Gesammtaltzung der 62. Versammlung Deutscher Naturforscher und Aerzte zu Heidelberg, am ... 1889. Frankfurt am Main. 56 pp.

Schiødte, J. C.

Ba1464. **Gosch, C. C. A.** 1898–1905. *Jørgen Christian Schiødte: Et bidrag til naturvidenskabens historie i Danmark i det nittende aarhundrede.* Copenhagen, Gyldendalsake Boghandels Forlag. 3 vols.

Schlegel, H.

Ba1465. **Schlegel, G.** 1884. Levensschets van Hermann Schlegel. *Jaarboek van de Koninklijke Akademie van Wetenschappen, Amsterdam* 1884: 1–97.
For German version *see* ___. 1886. *H. Schlegel. Lebensbild.* nach dem Holländischen ... hrsg. von H. Köhler. Altenburg. 78 pp. *Mittheilungen aus dem Osterlande* N.F. vol. 3.

Schleiden, M. J.

Ba1466. **Fechner, G. T.** 1856. *Professor Schleiden und der Mond.* Leipzig, A. Gumprecht. xv+427 pp.

Ba1467. **Möbius, M.** 1904. *Matthias Jacob Schleiden: Zu seinem 100. Geburtstage.* Leipzig, W. Engelmann. [iv]+106 pp. & port.

See also Ba2 Ba21.

Schmidt, J.

Ba1468. **Winge, O., & Ä. V. Taning,** eds. 1947. *Naturforskeren Johannes Schmidt, hans liv og ekspeditioner.* Copenhagen, Gyldendal. 187 pp. & 38 pl.

Schmitt, W. L.

Ba1469. **Blackwelder, R. E.** 1979. *The zest for life or, Waldo had a pretty good run: The life of Waldo LaSalle Schmitt.* Lawrence, KS, Allen Pr. xviii+234 pp.

Ba1470. **Cox, W. E.** 1983. *Guide to the papers of Waldo Lasalle Schmitt.* Washington, DC, Smithsonian Institution. iv+94 pp. & 25 ff. *Guides to collections, Archives & Special Collections of the Smithsonian Institution* no. 1.

Schomburgk, R. H.

See Ba303.

Schröter, C.

Ba1471. **Rübel, E.** 1940. *Carl Schröter 1855–1939.* Zürich, Beer. 74 pp. & 6 pl. *Neujahrsblatt zum Besten des Waisenhauses in Zürich* no. 103.

Schwann, T.

Ba1472. **Florkin, M.** 1960. *Naissance et déviation de la théorie cellulaire dans l'oeuvre de Theodore Schwann.* Paris, Hermann. [vi]+236 pp. & 1 pl.

Ba1473. ___. 1961. *Lettres de Théodore Schwann.* Liege, Société Royale des Sciences de Liége. 274 pp.

Ba1474. **Watermann, R.** 1960. *Theodor Schwann. Leben und Werk.* Dusseldorf, L. Schwann. 364 pp.

See also Ba2 Ba21.

Schwarz, E. A.

Ba1475. **Sherman, J. D.**, ed. 1929. The letters of E. A. Schwarz. *Journal of the New York Entomological Society* 37: 182–392.

Schweinfurth, G.

Ba1476. **Guenther, K.**, ed. 1954. *Georg Schweinfurth: Lebensbild eines Afrikasforschers. Briefe von 1857–1925.* Stuttgart, Wissenschaftliche Verlagsgesellschaft. vii+341 pp. & port. *Grosse Naturforscher* vol. 17.

Schwendener, S.

See Ba21.

Sclater, P. L.

Ba1477. **Goode, G. B.** 1896. *The published writings of Philip Lutley Sclater, 1844–96.* Washington, DC, United States National Museum. xix+135 pp. & port. *Bulletin of the United States National Museum* no. 49.

Scott, J. P.

See Ba298.

Scott, Sir P.

Ba1478. **Scott,** Sir **P.** 1961. *The eye of the wind.* London, Hodder & Stoughton. xxii+679 pp. & 35 pl.

Selous, F. C.

Ba1479. **Millais, J. G.** 1919. *Life of Frederick Courtenay Selous, [etc.].* London, Longmans, Green & Co. xi+387 pp., 15 pl. & 1 port.

Ba1480. **Taylor, S.** 1990. *The mighty Nimrod: A life of Frederick Courtenay Selous, African hunter and adventurer, 1851–1917.* London, Collins. xv+312 pp., illus.

Senckenberg, J. C.

Ba1481. **De Bary, A.** 1947. *Johann Christian Senckenberg (1707–1772): Sein Leben auf Grund der Quellen des Archivs der Dr. Senckenbergischen Stiftung.* Frankfurt am Main, W. Kramer. 327 pp.

Ba1482. **Spilger, L.** 1941. Senckenberg als Botaniker und die Flora zu Senckenberg's Zeiten. *Abhandlungen herausgegeben von der Senckenbergischen Naturforschenden Gesellschaft* 458: 175 pp.

Seton, E. T.

See Thompson, E. Seton–.

Shaw, H.

Ba1483. **Faherty, W. B.** 1987. *Henry Shaw: His life and legacies.* Columbia, MO, Univ. of Missouri Pr. xi+228 pp.

Shelford, V. E.

Ba1484. **Croker, R. A.** 1991. *Pioneer ecologist: The life and work of Victor Ernest Shelford, 1877–1968.* Washington, DC, Smithsonian Institution Pr. xviii+222 pp.

Ba1485. **Miller, S. T.** 1977. Victor E. Shelford papers. *Mendel newsletter* 13: 4–6.

Sherff, E. E.

Ba1486. **Sherff, E. E.** 1964. *An annotated list of my botanical writings: A bibliographic list, with notations of observed errata, needed emendations, and additions of taxa described too late for inclusion in the original publications.* Bloomington, IL, by the Author for the Earl Edward Sherff Botanical Library, Illinois Wesleyan Univ. 48 pp.

Shreve, F.

Ba1487. **Bowers, J. E.** 1988. *A sense of place: The life and work of Forrest Shreve.* Tucson, AZ, Univ. of Arizona Pr. xi+195 pp.

Sibbald, R.

See Ba303.

Siebold, P. F. von

Ba1488. **Deutsch–Japanischen Gesellschaft.** 1966. *Philipp Franz von Siebold: Gedenkschrift zur 100. Wiederkehr seines Todestages am 18. Oktober 1966.* Munich, M. Hueber. 139 pp., illus.

Ba1489. **Holthuis, L. B., & T. Sakai.** 1970. *Ph. F. von Siebold and Fauna Japonica: A history of early Japanese zoology.* Tokyo, Academic Pr. of Japan. [xx]+323 pp. & 32 pl.

Simpson, G. G.

Ba1490. **Simpson, G. G.** 1978. *Concession to the improbable: An unconventional autobiography.* New Haven, CT, Yale Univ. Pr. xi+291 pp., illus.

Skryabin, K. I.

Ba1491. **Akademiya Nauk SSSR.** 1947. *Konstantin Ivanovich Skryabin.* Vstupitel'naya stat'ya N. P. Shikhobalovoi. Bibliografiya sostavlena O. V. Isakovoi. Moscow/Leningrad. 91 pp. *Materialy k biobibliografii uchenykh SSSR, Seriya biologicheskikh nauk, Gel'mintologiya* Vol. 1.

Ba1492. **Akademiya Nauk SSSR.** 1976. *Konstantin Ivanovich Skryabin: Zizn'i deyatel'nosti.* Moscow. 254 pp.

Ba1493. [Not used.]

Sloane, Sir H.

Ba1494. **British Museum.** 1904. *Index to the Sloane manuscripts in the British Museum.* By E. J. L. Scott. London. viii+583 pp. *Re-pub.* 1971.

Ba1495. **Brooks, W. E. St. J.** 1954. *Sir Hans Sloane: The great collector and his circle.* London, Batchworth Pr. 234 pp. & 9 pl.

Ba1496. **De Beer, G. R.** 1953. *Sir Hans Sloane and the British Museum.* London, Oxford Univ. Pr. 192 pp. *Re-pub.* New York, Arno Pr., 1975.

Ba1497. **Jacquot, J.** 1953. *Le naturaliste Sir Hans Sloane (1660–1753) et les échanges scientifiques entre la France et l'Angleterre.* Paris, Université de Paris. 25 pp. *Conférence faite au Palais de la Découverte, le 7 Novembre, 1953.*

See also Ba303 Ca824 & Cb1040.

Smellie, W.

See Ba303.

Smith, E. F.

Ba1498. **Rodgers, A. D., 3rd.** 1952. *Erwin Frink Smith: A story of North American plant pathology.* Philadelphia, PA, American Philosophical Society. x+657 pp. *Memoirs of the American Philosophical Society* vol. 31.

Smith, Sir J. E.

Ba1499. **Linnean Society of London.** 1934. *The Smith papers. (The correspondence and miscellaneous papers of Sir James Edward Smith, M.D., F.R.S., First president of the Society).* By W. R. Dawson. London. 114 pp. *Catalogue of manuscripts in the library of the Linnean Society of London* pt. 1.

Ba1500. **Smith, P.**, Lady. 1832. *Memoir and correspondence of the late Sir James Edward Smith, M.D. [etc.].* London, Longman, Rees, Orme, Brown, Green & Longman. 2 vols., port.

Ba1501. **Walker, M.** 1988. *Sir James Edward Smith, 1759–1828, first President of the Linnean Society of London.* London, Linnean Society of London. [vii]+60 pp.
Also **W. T. Stearn.** 1988. James Edward Smith (1759–1828): First President of the Linnean Society and his herbarium. *Botanical journal of the Linnean Society* 96: 199–216.

Smith, J. M.

See Ba298.

Sody, H. J. V.

Ba1502. **Becking, J. H.** 1989. *Henri Jacob Victor Sody (1892–1959). His life and work, a biographical and bibliographical study.* Leiden, E. J. Brill. viii+272 pp.

Solander, D. C.

Ba1503. **Diment, J. A., & A. C. Wheeler.** 1984. Catalogue of the natural history manuscripts and letters by Daniel Solander (1733–1782), or attributed to him, in British collections. *Archives of natural history* 11: 457–488.

Ba1504. **Gisselquist, C., & P. Tingbrand**, eds. 1983. *Daniel Solander, 1733–1782: Naturvetenskapsman och varldsomseglare. [Daniel Solander, 1733–1782: Naturalist and circumnavigator].* Piteå, Kommuns Kulturnamnd. 179 pp. *Piteå Kommuns Kulturnamnd skrifter* no. 1.
Chapters by various authors.

Ba1505. **Rauschenberg, R. A.** 1968. Daniel Carl Solander, naturalist on the "Endeavour." *Transactions of the American Philosophical Society* 58 (8): 66 pp.

Sonnerat, P.

Ba1506. **Ly-Tio-Fane, M.** 1976. *Pierre Sonnerat 1748–1814: An account of his life and work.* Mauritius, the Author. xv+157 pp., illus.

Sowerby family

Ba1507. **Cleevely, R. J.** 1976. A provisional bibliography of natural history works by the Sowerby family. *Journal of the Society for the Bibliography of Natural History* 6: 482–559.

Ba1508. **___.** 1976. The Sowerbys and their publications in the light of the manuscript material in the British Museum (Natural History). *Journal of the Society for the Bibliography of Natural History* 7: 343–368.

Ba1509. **McDonald, J. B.** The Sowerby collection in the British Museum (Natural History): A brief description of its holdings and a history of its acquisition from 1821–1971. *Journal of the Society for the Bibliography of Natural History* 6: 380–401.

Ba1510. **Sowerby, A. de C., A. M. Sowerby, & J. E. Stone.** 1952. *The Sowerby saga: Being a brief account of the origin and genealogy of the Sowerby family and of its history from the earliest times down to the present.* Washington, DC. 3 pts. (v+180 pp.).
Mimeographed production.

Sowerby, A. de C.

Ba1511. **Sowerby, R. R.** 1956. *Sowerby of China, Arthur de Carle Sowerby, [etc.].* Kendal, T. Wilson & Son. [vi]+58 pp. & 5 pl.

Spallanzani, L.

Ba1512. **Bernardi, W., & A. La Vergata**, eds. 1982. *Lazzaro Spallanzani e la biologia del settecento: Teorie, esperimenti, instituzioni scientifiche: Atti del convegno di studi, Reggio Emilia, Modena, Scandiano, Pavia, 23–27 marzo 1981.* Florence, L. S. Olschki. xii+625 pp. & 7 pl. *Biblioteca della "Rivista di storia delle scienze mediche e naturali"* vol. 22.
Chapters by various authors.

Ba1513. **Capparoni, P.** 1941. *Lazzaro Spallanzani.* Turin, Unione Tipografico–Editrice Torinese. 282+[4] pp. & 6 pl. *I grandi italiani* no. 13. *Re-pub. as* ___. 1948. *Spallanzani.* Turin, [etc.]. 310 pp., illus.

Ba1514. **Castellani, C.** 1990. Un'edizione computerizzata di alcuni manoscritti di Lazzaro Spallanzani. *Nuncius* 5: 249–265.

Ba1515. **Museo Spallanzani, Reggio nell' Emilia.** 1939. *Il museo e i manoscritti di L. Spallanzani, esistenti in Reggio d'Emilia. Breve guida e inventario.* [By G. Altana & U. Gualazzini.] Reggio Emilia. 50 pp.

Ba1516. **Prandi, D.** 1951. *Bibliografia della opere di Lazzaro Spallanzani.* Florence, Sansoni Antiquariato. 172 pp., port.

Ba1517. **Rostand, J.** 1951. *Les origines de la biologie expérimentale et l'Abbé Spallanzani.* Paris, Fasquelle. 284 pp.

Ba1518. **Spallanzani, L.** 1958–64. *Lazzaro Spallanzani: Epistolario.* Ed. by B. Biagi. Florence, Sansoni. 5 vols.

Ba1519. ___. 1971. *Lettres à M. l'Abbé Spallanzani ...* Edizione critica condotta sugli originali, introduzione e note di di Carlo Castellani. Milan, Episteme. lxiii+566 pp.

Ba1520. ___. 1984–90. *Edizione nazionale delle opere di Lazzaro Spallanzani.* Parte 1a: *Carteggi.* P. Di Pietro, ed. Modena, E. Mucchi. 12 vols.

See also Ca988.

Sprengel, K. P. J.

Ba1521. **Kaiser, W., & A. Völker.** 1982. *Kurt Sprengel (1766–1833).* Halle, Martin-Luther-Universität Halle–Wittenberg. 115 pp. *Wissenschaftliche Beiträge, Martin–Luther-Universität Halle-Wittenberg* 1982/31.

Ba1522. **Kraus, G.** 1894. *Kurt Sprengel.* Leipzig, W. Engelmann. viii+155 pp. & 2 port. *Der botanische Garten d. Universität Halle* no. 2.

Steller, G. W.

Ba1523. **Ford, C.** 1966. *Where the sea breaks its back. The epic story of a pioneer naturalist [G. W. Steller] and the discovery of Alaska.* Boston, MA, Little, Brown. xi+206 pp. *Also* London, V. Gollancz, 1967.
Apropos the Behring expedition of 1741–42.

Ba1524. **Stejneger, L.** 1936. *Wilhelm Steller: The pioneer of Alaskan natural history.* Cambridge, MA, Harvard Univ. Pr. xxiv+623 pp. & 17 pl. *Re-pub.* Westmead, Farnborough, Gregg International Publishers, 1970.

Stephens, J. F.

Ba1525. **Sherborn, C. D.** 1939. An index to the letters addressed to James Francis Stephens now preserved in the British Museum (N. H.). *Journal of the Society for the Bibliography of Natural History* 1: 253–254.

Sternberg family

Ba1526. **Rogers, K.** 1991. *The Sternberg fossil hunters: A Dinosaur dynasty.* Missoula, MT, Mountain Pr. Pub. Co. 200 pp.
The family of Charles H. and his sons George, Charlie and Levi.

Stoddard, H. L.

Ba1527. **Stoddard, H. L.** 1969. *Memoirs of a naturalist.* Norman, OK, Univ. of Oklahoma Pr. xix+303 pp.

Strickland, H. E.

Ba1528. **Jardine,** Sir **W.** 1858. *Memoirs of Hugh Edwin Strickland.* London, J. Van Voorst. cclxv+xvi+441 pp. & 16 pl.

Stuart, J.

Ba1529. **Coats, A. M.** 1975. *Lord Bute: An illustrated life of John Stuart, Third Earl of Bute, 1713–1792.* Aylesbury, Shire Publications. 48 pp.

Stutchbury, S.

Ba1530. **Branagan, D. F.** 1984. Samuel Stutchbury and his manuscripts. In: M. E. Hoare & L. G. Bell, eds. In search of New Zealand's scientific heritage. *Bulletin of the Royal Society of New Zealand* 21: 123 pp.

Suksdorf, W. N.

See Bc855 for his botanical papers.

Sullivant, W. S.

Ba1531. **Rodgers, A. D., 3rd.** 1940. *"Noble fellow": William Starling Sullivant.* New York, G. P. Putnam's Sons. 361 pp. *Re-pub.* New York, Hafner Publishing Co., 1968.

See also Ba133.

Sumner, F. B.

Ba1532. **Sumner, F. B.** 1945. *The life history of an American naturalist.* Lancaster, PA, J. Cattell Pr. vii+298 pp.

Sutton, G. M.

Ba1533. **Sutton, G. M.** 1980. *Bird student: An autobiography.* Austin TX, Univ. of Texas Pr. 240 pp.

Swainson, W.

Ba1534. **Linnean Society of London.** 1900. An index of William Swainson's correspondence. By A. C. L. G. Günther. *Proceedings of the Linnean Society of London* 1899–1900: 14–61.

Ba1535. **Natusch, S., & G. M. Swainson.** 1987. *William Swainson F.R.S., F.L.S., &c. The anatomy of a nineteenth century naturalist.* 3rd. rev. ed. Wellington, the Author. 184 pp.

Ba1536. **Parkinson, P.** 1984. William Swainson, 1789–1855: Relics in the antipodes. In: M. E. Hoare & L. G. Bell, eds. In search of New Zealand's scientific heritage. *Bulletin of the Royal Society of New Zealand* 21: 123 pp.

Ba1536A. **Swainson, G. M.** 1992? *William Swainson F.R.S., naturalist and artist. Family letters and diaries 1819–1855: Final destiny New Zealand.* Palmertson North?, N.Z., the Author. ? pp.
Forthcoming publication announced.

Swammerdam, J.

Ba1537. **Engel, H.** 1950. Records on Jan Swammerdam in the Amsterdam archives. *Centaurus* 1: 143–155.

Ba1538. **Pöhlmann, O.** 1941. *Jan Swammerdam: Naturforscher und Arzt. Biographischer Roman.* Zurich & Leipzig, Orell Füssli. 226 pp., illus.
Fictionalized biography.

Ba1539. **Schierbeek, A.** 1967. *Jan Swammerdam (12 February 1637 – 17 February 1680): His life and works.* Amsterdam, Swets & Zeitlinger. vi+202 pp., Illus.
Replaces ___. 1946. *Jan Swammerdam (1637–1680), zijn leven en zijn werken (met een hoofdstuk: De genealogie van Swammerdam en verdere Archivalia door Dr H. Engel).* Lochem, De Tijdstroom. 280 pp., illus.

See also Ba303.

Swartz, O.

See Cb759A.

Tamsalu, A.

Ba1540. **Lord, J. B.** 1980. *Aleksander Tamsalu 1891–1960: A botanist in exile.* Hamilton, Ont., Royal Botanic gardens. 126 pp. *Technical bulletin* no. 11.

Taylor, E. H.

Ba1541. **Taylor, E. H.** 1975. *Edward H. Taylor: Recollections of an herpetologist.* Lawrence, KS, Museum of Natural History, Univ. of Kansas. 159 pp. *Monograph* no. 4.
Includes **G. R. Pisani**. The published contributions of E. H. Taylor; contributions by A. B. Leonard and H. M. Smith.

Tegetmeier, W. B.

Ba1542. **Richardson, E. W.** 1916. *A veteran naturalist: Being the life and work of W. B. Tegetmeier.* London, H. F. & G. Witherby. xxiv+232 pp. & 19 pl.

Tenore, M.

Ba1543. **Giacomi, V.** 1961. Ricognizione dell' opera scientifica di Michele Tenore nel primo centenario delle morte (1861–1961). *Delpino* n.s. 3: lxxv pp. & port.
Elenco cronologico ragionato delle pubblicazioni di Michele Tenore, pp. xxxiii–lxxiv.

Thienemann, A.

Ba1544. **Thienemann, A.** 1959. *Erinnerungen und Tagebuchblätter eines Biologen.* Stuttgart, Schweizerbart'sche Verlagsbuchhandlung. [vi]+499 pp. & port.

Thompson, D'A. W.

Ba1545. **Thompson, D'A. R.** 1958. *D'Arcy Wentworth Thompson: The scholar–naturalist, 1860–1948.* London, Oxford Univ. Pr. xi+244 pp. & 9 pl.

Thompson, E. Seton–.

Ba1546. **Anderson, H. A.** 1986. *The Chief: Ernest Thompson Seton and the changing West.* College Station, TX, Texas A & M Univ. Pr. xii+363 pp. & 32 pp. of illus.

Ba1547. **Keller, B.** 1984. *Black wolf: The life of Ernest Thompson Seton.* Topsfield, MA, Salem House; Vancouver, Douglas & McIntyre. 240 pp. & 24 pp. of illus.

Ba1548. **Thompson, E. Seton–.** 1940. *Trail of an artist–naturalist: The autobiography of ... with illustrations by the author.* New York, C. Scribner's Sons. xii+412 pp., illus. *Also* London, Hodder & Stoughton, 1951.

Ba1548A. **Wadland, J. H.** 1978. *Ernest Thompson Seton: Man in nature and the progressive era, 1880–1915.* New York, Arno Pr. xii+528 pp.

Thornton, R. J.

Ba1549. **Grigson, G., & H. Buchanan.** 1951. *Thornton's Temple of Flora with plates faithfully reproduced from the original engravings and the work described by Geoffrey Grigson with bibliographical notes by Handasyde Buchanan.* London, Collins. viii+20 pp. & 36 pl. Reprinted in reduced format London, Collins, 1972. vi+66 pp. (incl. 36 pl.).

Thouin, A.

See Ba12.

Thunberg, C. P.

See Ba69.

Timiryazev, K. A.

Ba1550. **Platanov, G. B.** 1951. *Mirovozzrenie K. A. Timiryazev.* Moscow, AN SSSR. 291 pp.

Tinbergen, N.

Ba1551. **Dawkins, M. S., T. R. Halliday, & R. Dawkins**, eds. 1991. *The Tinbergen legacy.* London, Chapman & Hall. xii+146 pp.
Contributions by 8 authors.

See also Ba298 & Cc266.

Torrey, J.

Ba1552. **Long, C. R.** 1978. Natural history manuscripts and related materials in the archives of the New York Botanical Garden. *Journal of the Society for the Bibliography of Natural History* 8: 343–349.

Ba1552A. **MacPhail, I.** 1992. *The Sterling Morton Library bibliographies in botany and horticulture, IV: John Torrey.* Lisle, IL, Morton Arboretum. [vi]+40 pp., port.

Ba1553. **New York Botanical Garden.** 1973. *Catalog of the manuscript and archival collections and index to the correspondence of John Torrey.* Compiled by Sara Lenley, [et al.] Boston, MA, G. K. Hall. vii+473 pp.

Ba1554. **Robbins, C. C.** 1968. John Torrey (1796–1873): His life and times. *Bulletin of the Torrey Botanical Club* 95: 514–645.

Ba1555. **Rodgers, A. D., 3rd.** 1942. *John Torrey: A story of North American botany.* Princeton, NJ, Princeton Univ. Pr. xii+352 pp. *Re-pub.* New York, Hafner Publishing Co., 1965.

See also Ba133 & Ba205.

Tournefort, J. P. de.

Ba1556. **Heim, R., et al.** 1957. *Tournefort.* Paris, Muséum National d'Histoire Naturelle. 321 pp. & 46 pl. *Les grands naturalistes Français.*
Includes **G. Duprat.** Les manuscrits de Tournefort, pp. 207–238; and other chapters by C. Carrière, J.–P. Dufert, J.–P. Coste, J. Motte, H. Gaussen, H. Humbert, P. Guiral, H. Bianchi, G. Ranson, F. Meyer, R. Dughi, and J.–F. Leroy.

See also Ca707.

Tradescant, J. & J. [elder & younger]

Ba1557. **Allan, M.** 1964. *The Tradescants, their plants, gardens and museum, 1570–1662.* London, M. Joseph. 345 pp. & 27 pl.

Ba1558. **Ross, P. Leith–.** 1984. *The John Tradescants: Gardeners to the rose and lily queen.* London, Peter Owen. 320 pp.

Trelease, W.

Ba1559. **Bay, J. C.** 1945. *William Trelease, 1857–1945. Personal reminiscences.* Chicago, IL, privately pub. 12 pp.

Trembley, A.

Ba1560. **Baker, J. R.** 1952. *Abraham Trembley of Geneva: Scientist and philosopher.* London, E. Arnold. xix+259 pp., illus.

Ba1561. **Trembley, M.**, ed. 1943. *Correspondance inédite entre Réaumur et Abraham Trembley.* Comprenant 113 lettres recueillies et annotées par M. Trembley. Introduction par E. Guyénot. Geneva, Georg & Cie. lvi+432 pp.

Treub, M.

Ba1562. **Zeijlstra, H. H.** 1959. *Melchior Treub, pioneer of a new era in the history of the Malay Archipelago.* Amsterdam, Kon. Inst. voor de Tropen. 129 pp.

Trew, C. J.

Ba1563. **Universitätsbibliothek Erlangen.** 1940. *Die Briefsammlung des Nürnberger Arztes Christoph Jacob Trew (1695–1769) in der Universitätsbibliothek Erlangen.* Bearbeitet von E. Schmidt–Herrling. Erlangen. liv+776 pp. & port. *Katalog der Handschriften der Universitätsbibliothek Erlangen* vol. 5.

Tristram, H. B.

Ba1564. **Bodenheimer, F. S.** 1957. Canon H. B. Tristram 1822–1906: A preliminary bibliography, obituaries and other biographical notes. *Proceedings of the University of Durham Philosophical Society* 13A: 13–22.

Tschudi, J. J. von

Ba1565. **Schazmann, P.–E.** 1956. *Johann Jakob von Tschudi: Forscher, Arzt, Diplomat, [etc.].* Zurich, Verlag Mensch & Arbeit. 199 pp. & 19 pl. *Mensch und Arbeit* no. 40.

Tuckerman, E.

See Ba133.

Turner, W.

Ba1566. **Jones, W. R. D.** 1988. *William Turner: Tudor naturalist, physician, and divine.* London, Routledge. 223 pp.

See also Ba130.

Tyson, E.

Ba1567. **Montagu, M. F. A.** 1943. *Edward Tyson, M.D., F.R.S. (1650–1708) and the rise of comparative anatomy in England. A study in the history of science.* Philadelphia, PA, American Philosophical Society. xxix+488 pp. *Memoirs of the American Philosophical Society* vol. 20.

Unger, F.

Ba1568. **Reyer, A.** 1871. *Leben und Wirken des Naturhistorikers Dr. Franz Unger, Professor der Pflanzen–Anatomie und Physiologie.* Im Auftrage des Vereines der Aerzte in Steiermark. Graz, Leuscher & Lubensky. [iv]+100 pp.

Usinger, R. L.

Ba1568A. **Usinger, R. L.** 1972. *Autobiography of an entomologist.* San Francisco, CA, Pacific Coast Entomological Society. 330 pp., illus.

Valenciennes, A.

Ba1569. **Monod, T.** 1963. Achille Valenciennes et l'histoire naturelle des poissons. *Mémoires de l'Institut Français d'Afrique Noire* 68: 9–45, & 1 pl.

Ba1570. **___, et al.** 1966. Achille Valenciennes (1794–1865). Document inédits. *Archives du Muséum National d'Histoire Naturelle* 9: 9–109, & 3 pl.

Ba1571. **Théodoridès, J.** [1964.] Une amitié de savants au siècle dernier: Alexander von Humboldt et Achille Valenciennes (correspondence inédite.) *Biologie médicale* 1965 (no. hors série): 129 pp.

See also Cc462.

Vavilov, N. I.

Ba1572. **Boiko, V. V., & E. R. Vilenskii.** 1987. *Nikolai Ivanovich Vavilov, 1887–1987: Stranitsy zhizni i deyatel'nosti.* Moscow, Agropromizdat. 187 pp.

Ba1573. **Chernikova, M. S.**, ed. 1973. *Ryadom s N. I. Vavilov: Sbornik vospominanii.* Moscow, Izd. "Sovetskaya Rossiya." 253 pp., illus.

Ba1574. **Goryachev, R. I., & L. M. Zhukova.** 1967. *Nikolai Ivanovich Vavilov ...: Bibliografiya sostavlena [etc.].* Moscow, Izd. "Nauka." 131 pp. *Materialy k biobibliografii uchenykh SSSR, genetika* vol. 1.

Ba1575. **Popovsky, M.** 1984. *The Vavilov affair.* Hamden, CT, Archon Books. viii+216 pp.

Ventenat, É. P.

See Ba12.

Victorin, Frère Marie

See Kirouac, J. L. C.

Vinci, L. da

Ba1576. **Belt, E.** 1968. *Manuscripts of Leonardo da Vinci, their history, with a description of the manuscript editions in facsimile.* Los Angeles, CA, Elmer Belt Library of Vinciana. [iv]+70 pp.

Ba1577. **Emboden, W. A.** 1987. *Leonardo da Vinci on plants and gardens.* Portland, OR, Dioscorides Pr. 234 pp.

Ba1578. **Verga, E.** 1931. *Bibliografia Vinciana 1493–1930.* Bologna. 2 vols. (xii+835 pp.). *Re-pub.* New York, Burt Franklin, 1970. *Burt Franklin bibliography & reference series* no. 367.

Vogt, K. C.

Ba1579. **Vogt, K. C.** 1896. *Aus meinem Leben, Erinnerungen und Rückblicke.* Stuttgart, E. Nägele. vi+202 pp., port.

Ba1580. **Vogt, W.** 1896. *La vie d'un homme.* Paris, Schleicher. 265 pp., ports.

Walker, J.

See Ba303.

Wallace, A. R.

Ba1581. **Brooks, J. L.** 1984. *Just before the origin: Alfred Russel Wallace's theory of evolution.* New York, Columbia Univ. Pr. xiii+284 pp.

Ba1582. **George, W.** 1964. *Biologist philosopher: A study of the life and writings of Alfred Russel Wallace.* London, Abelard–Schuman. xiv+320 pp.

Ba1583. **McKinney, H. L.** 1972. *Wallace and natural selection.* New Haven, CT, Yale Univ. Pr. xix+193 pp.

Ba1584. **Marchant, J.** 1916. *Alfred Russel Wallace: Letters and reminiscences.* London, Cassell & Co. 2 vols. *Re-pub.* New York, Arno Pr., 1975.

Ba1585. **Wallace, A. R.** 1905. *My life: A record of events and opinions.* London, Chapman & Hall. 2 vols. *Re-pub.* New York, AMS Pr., 1974.

Wallace, G. J.

Ba1586. **Wallace, G. J.** 1979. *My world of birds: Memoirs of an ornithologist.* Philadelphia, PA, Dorrance. xii+345 pp.

Ward, F. K–.

Ba1587. **Lyte, C.** 1989. *Frank Kingdon–Ward: The last of the great plant hunters.* London, John Murray. xvi+218 pp.

Ba1588. **Schweinfurth, U., & H. Schweinfurth–Marby.** 1975. *Exploration in the eastern Himalayas and the River Gorge country of southeastern Tibet. Francis (Frank) Kingdon Ward (1885–1958): An annotated bibliography with a map of the areas of his expeditions.* With a preface by Sir G. Taylor. Wiesbaden, Steiner. 114 pp. *Geoecological research* no. 3.

Ward, H. A.

Ba1589. **Ward, R. H.** 1948. *Henry Augustus Ward: Museum builder to America.* Rochester, NY, Rochester Historical Society Publications. xxiv+297 pp., illus. *Publication fund series* vol. 24.

Ward, H. M.

See Ba185.

Ward, N. B.

See Ba133.

Waterton, C.

Ba1590. **Aldington, R.** [1949.] *The strange life of Charles Waterton, 1782–1865.* London, Evans Bros. 200 pp., illus. *Also* New York, Duell, Sloan & Pearse, [1949]. 231 pp., illus.

Ba1591. **Blackburn, J.** 1989. *Charles Waterton, traveller and conservationist.* London, Bodley Head. ix+243 pp., illus.

Ba1592. **Gosse, P.** 1940. *The squire of Walton Hall: The life of Charles Waterton.* London, Cassell & Co. ix+324 pp.

Ba1593. **Hobson, R.** 1866. *Charles Waterton: His home, habits and handiwork. Reminiscences of an intimate and most confiding personal association for nearly thirty years.* London, Whittaker & Co. xl+319 pp.

Ba1594. **Phelps, G.** 1976. *Squire Waterton.* London, E P Publishing. viii+167 pp., illus.

Ba1595. **Waterton, C.** 1955. *Letters of Charles Waterton of Walton Hall, near Wakefield. Naturalist, taxidermist and author of "Wanderings in South America" and "Essays on natural history."* Ed. by R. A. Irwin. London, Rockliff. xii+159 pp. & 7 pl.

Watling, T.

Ba1596. **Hindwood, K. A.** 1969. The [Thomas] "Watling" drawings, with incidental notes on the "Lambert" and the "Latham" drawings. *Proceedings, Royal Zoological Society of New South Wales* 1968–69: 16–32.

Webb, F. B.

Ba1597. **Parlatore, F.** 1856. *Elogio di Filippo Barker Webb.* Florence, Le Monnier. 7+115 pp. & port.

Weismann, A.

Ba1598. **Gaupp, E.** 1917. *August Weismann: Sein Leben und sein Werk.* Jena, G. Fischer. viii+297 pp.

Ba1599. **Sander, K.**, ed. 1985. August Weismann (1834–1914) und die theoretische Biologie des 19. Jahrhunderts: Urkunden, Berichte und Analysen. *Freiburger Universitätsblätter* 24: 20–203.
Eighteen articles by various authors. *Includes* **H. Risler.** August Weismanns Leben und Wirken nach Dokumenten aus seinem Nachlass.

Welwitsch, F.

Ba1600. **Almaça, C.** 1989. Documentos do arquivo histórico do Museu Bocage relativos à exploração histórico–natural angolana do Dr. Frederico Welwitsch (Agosto de 1853–Dezembro de 1860). *Arquivos do Museu Bocage* n.s. 1: 335–347.

Ba1601. **Dolezal, H.** 1959–61. Friedrich Welwitsch: Leben und Werk. I(–II). *Portugaliae acta biologica.* Sér. B 6: 257–323, 1959; *Ibid.* 7: 49–276, 1961.

Ba1602. ___. 1974. *Friedrich Welwitsch vida e obra.* Traduzido e anotado por A. W. Exell & E. J. Mendes. Lisbon, Junta de Investigações Cientificas do Ultramar. 294 pp.

Ba1603. **Edwards, P. I.** 1972. Friedrich Welwitsch, 1806–72: His manuscripts and correspondence in the Departments of Botany and Zoology, British Museum (Natural History). *Biological journal of the Linnean Society* 4: 291–303.

See also Ba133.

Wenderoth, G. W. F.

Ba1604. **Unterhalt–Schüler, I.** 1989. *Georg Wilhelm Franz Wenderoth (1774–1861): Ein Beitrag zur Geschichte der Botanik an der Universität Marburg.* Darmstadt, Selbstverlag der Hessischen Historischen Kommission Darmstadt und der Historischen Kommission für Hessen. xvi+305 pp. *Quellen und Forschungen zur hessischen Geschichte* no. 75.

Werner, A.–G.

See Ba12 & Ba303.

Wetmore, A.

Ba1605. **Cooper, V., & W. E. Cox.** 1990. *Guide to the papers of Alexander Wetmore, circa 1849–1979, and undated.* Washington, DC, Smithsonian Institution, Archives & Special Collections. vi+212 pp., illus. *Guides to collections, Archives & Special Collections of the Smithsonian Institution* no. 11.

Wettstein, R.

Ba1606. **Janchen, E.** 1933. *Richard Wettstein: Sein Leben und Wirken.* Vienna, J. Springer. 195 pp. & port. First pub. in *Oesterreichische botanische Zeitschrift* vol. 82(1/2), 1933.

Wheeler, W. M.

Ba1607. **Evans, M. A. & H. E.** 1970. *William Morton Wheeler, biologist.* Cambridge, MA, Harvard Univ. Pr. xii+363 pp.

White, G.

Ba1608. **Foster, P. G. M.** 1988. *Gilbert White and his records: A scientific biography.* Bromley, C. Helm. xv+240 pp.

Ba1609. **Mabey, R.** 1986. *Gilbert White: A biography of the author of "The natural history of Selborne"*. London, Century. ix+239 pp., illus.

Ba1610. **Martin, E. A.** 1970. *A bibliography of Gilbert White, the naturalist and antiquarian of Selborne. With a biography and a descriptive account of the village of Selborne.* 2nd ed. [Re-pub. with additions.] Folkestone & London, Dawsons of Pall Mall. vii+193+[7] pp.
First pub. London, Halton, 1934. viii+188+[8] pp.

Ba1611. **Prance, C. A.** 1968. Some uncollected authors, XLIII: Gilbert White, 1720–1793. *Book collector* 17: 300–321.

Ba1612. **Rye, A.** 1970. *Gilbert White and his Selborne.* London, W. Kimber. 272 pp.

Ba1613. **Scott, W. S.** 1950. *White of Selborne.* London, Falcon Pr. 260 pp., illus.

Ba1614. **Selborne Society, Library.** 1958. *Selborne Society library ... catalogue of mss., ... devoted to Gilbert White and natural history.* Ealing, Ealing Reference Library. 34 pp.

Ba1615. **White, R. Holt–**, ed. 1901. *The life and letters of Gilbert White of Selborne.* London, J. Murray. 2 vols.

White, M. J.

Ba1616. **McCarthy, G., & O. Manhal.** 1989. *The records of Michael James Denham White (1910–1983).* Parkville, Vic., Australian Science Archives Project. 112 ff.
Papers in the University of Melbourne Archives and the National Insect Collection, CSIRO Division of Entomology.

White, S. A.

Ba1617. **Linn, R.** 1989. *Nature's pilgrim: The life and journeys of Captain S. A. White, naturalist, author and conservationist.* Adelaide, S.A., Government Printer. 172 pp.

Whitman, C. O.

Ba1618. **Maienschein, J.** 1977. [C. O.] Whitman collection, M.B.L. *Mendel newsletter* 13: 1–4.

Wight, R.

See Ba133.

Wikström, J. E.

See Cb759A.

Williamson, W. C.

Ba1619. **Williamson, W. C.** 1896. *Reminiscences of a Yorkshire naturalist.* Ed. by his wife [Annie C. Williamson]. London, G. Redway. xii+228 pp.

See also Ba185.

Willughby (or Willoughby), F.

Ba1620. **Welch, M. A.** 1972. Francis Willoughby, FRS (1635–1672). *Journal of the Society for the Bibliography of Natural History* 6: 71–85.
Includes "Descriptive archival list" of the Willoughby collection and other miscellaneous items

belonging or relating to Willoughby within Lord Middleton's collection of manuscripts (now at the University of Nottingham library), pp. 78–84.

See also Ba303.

Wilson, A.

Ba1621. **Cantwell, R.** 1961. *Alexander Wilson: Naturalist and pioneer: A biography.* Philadelphia, PA, J. B. Lippincott & Co. 319 pp. & 24 pl.

Ba1622. **Hunter, C.**, ed. 1983. *The life and letters of Alexander Wilson.* Philadelphia, PA, American Philosophical Society. xii+456 pp. & 4 pl. *Memoirs of the American Philosophical Society* vol. 154.

Ba1623. **Plate, R.** 1966. *Alexander Wilson, wanderer in the wilderness.* New York, D. McKay & Co. viii+216 pp.

Ba1624. **Stringham, E.** 1958. *Alexander Wilson: A founder of scientific ornithology.* Kerrville, TX, the Author. 29 pp.

See also Ba303.

Wilson, E.

Ba1625. **Roberts, B.** 1967. *Edward Wilson's birds of the Antarctic.* London, Blandford. 191 pp.
Reproduces a selection from the collection of ca.270 water–colour drawings at the Scott Polar Research Institute, Cambridge.

Ba1626. **Seaver, G.** 1933. *Edward Wilson of the Antarctic: Naturalist and friend.* London. xxxiv+299 pp., 2 port. & 25 pl. (6 col.)

Wilson, E. B.

See Cc89.

Wilson, E. H.

Ba1627. **Farrington, E.** 1931. *Ernest H. Wilson, plant hunter: With a list of his most important introductions and where to get them.* Boston, MA, Stratford & Co. xxi+197 pp.

Wilson, E. O.

See Ba298.

Winch, N. J.

Ba1628. **Davis, P., & J. Leathart**, eds. 1988. *The correspondence of Nathaniel John Winch (1768–1838): An index by author.* Newcastle, Hancock Museum. [i]+154 ff.
Held by the Linnean Society of London. *Also* ___. 1986. Nathaniel John Winch (1768–1838), his collections and correspondence. *Linnean* 2(3): 23–32.

Withering, W.

Ba1629. **Peck, T. W, & K. D. Wilkinson.** 1950. *William Withering of Birmingham, M.D, F.R.S, F.L.S.* Bristol, Wright; London, Marshall. viii+239 pp., illus. *Also* Baltimore, MD, Williams & Wilkins, 1950.

Wittrock, V. B.

See Cb759A.

Wolf, J.

Ba1630. **Palmer, A. H.** 1895. *The life of Joseph Wolf, animal painter.* London, Longmans, Green & Co. xviii+328 pp, 53 pl.

Wolff, C. F.

Ba1631. **Uschmann, G.** 1955. *Caspar Friedrich Wolff, ein Pioneer der modernen–Embryologie.* Leipzig, Urania–Verlag. 86 pp., illus.

See also Cc207.

Wollaston, A. F. R.

Ba1632. **Wollaston, A. F. R.** 1933. *Letters and diaries of A. F. R. Wollaston.* Selected and ed. by M. Wollaston. Cambridge, Cambridge Univ. Pr. xv+261 pp. & 4 pl.

Wood, J. G.

Ba1633. **Wood, T.** 1891. *The Rev. J. G. Wood: His life and work.* London, Cassell & Co. xvi+318 pp, port.

Worthington, E. B.

Ba1634. **Worthington, E. B.** 1983. *The ecological century: A personal appraisal.* Oxford, Clarendon Pr. xv+206 pp. *Oxford science publications.*
Including much autobiographical reminiscence.

Wright, S.

Ba1635. **Provine, W. B.** 1986. *Sewall Wright and evolutionary biology.* Chicago, IL, Univ. of Chicago Pr. xvi+546 pp. *Science and its conceptual foundations.*

Wyman, J.

See Ca1046.

Wynne–Edwards, V. C.

See Edwards, V. C. Wynne–

Yerkes, R. M.

Ba1636. **Stark, B. P.** 1985. The Robert Mearns Yerkes papers. *Yale University Library gazette* 59: 162–167.

Section Bb. HISTORIES OF NATURALISTS' SOCIETIES

SOCIETIES (including ASSOCIATIONS & ACADEMIES)

• N.B. Literature on the history of the research institutes that form part of certain academies is entered at Ca1001–1093 (Institutions), Cb1043–1113 (Botanical institutions), and Cc893–962 (Zoological institutions).

General

Bb1. **Cohen, J., C. E. Hansel, & E. F. May.** 1954. Natural history of learned and scientific societies. *Nature* 173: 328–333.

Bb2. **Congrès National des Sociétés Savantes.** 1976. *Les sociétés savantes: Leur histoire. Actes du 100e Congrès National des Sociétés Savantes, Paris, 1975, Section d'Histoire Moderne et Contemporaine [etc.].* Paris, Bibliothèque Nationale. 386 pp.
Includes **Y. Laissus.** Les sociétés savantes et l'avancement des sciences naturelles, les musées d'histoire naturelle, pp. 42–67.

Bb3. **McKie, D.** 1948. Scientific societies to the end of the eighteenth century. *Philosophical magazine* 1948 commemoration no: 133–143.

Bb4. **Sabrosky, C. W.** 1956. Entomological societies. *Bulletin of the Entomological Society of America* 2(22): 1–22.
Worldwide list including reference to their serial publications.

Bb5. **Sparn, E.** 1929. Enumeracion y difusion de las grandes asociaciones de historia natural. *Boletin de la Academia Nacional de Ciencias, Cordoba* 31: 171–184.

Bb6. ___. 1931. Las sociedades cientificas, literarias y tecnicas del mundo con mas de 1000 miembros. Contribucion a la sociedadologia. *Miscelanea de la Academia Nacional de Ciencias, Cordoba* 18: 76 pp.

Bb7. ___. 1931. Las sociedades cientificas, literarias y tecnicas del mundo, de mayor impulso de crocimiento. Contribucion a la sociedadologia. *Ibid.* 19: 23 pp.

Bb8. ___. 1941. Cronologia, diferenciacion, matricula y distribucion geografica de las sociedades botanicas. *Boletin de la Academia Nacional de Ciencias, Cordoba* 35: 135–150.

Bb9. **Sparn, E.** 1943. Cronologia, matricula y distribucion de las actuales sociedades de historia natural general. *Ibid.* 36: 106–139.

Bb10. ___. 1941. In memoriam et ad gloriam. Sociedades de ciencias, letras y artes que llevan los membres de grandes hombres. Contribucion al conocimiento de la nomenclatura de las corporaciones cientificas. *Miscelanea de la Academia Nacional de Ciencias, Cordoba* 25: 82 pp.

Bb11. **Thornton, J. L., & R. I. J. Tully.** 1971. The rise of scientific societies. In: *Scientific books, libraries and collectors: A study of bibliography and the book trade in relation to science.* 3rd rev. ed., pp. 256–276. London, Library Association; *Continued* ___. 1978. ___*: Supplement to the 3rd ed, 1969–1975*, pp. 80–85. London, Library Association.
Replaces 2nd rev. ed., 1962. xiii+406 pp.

Bb12. **Union of International Associations.** 1960. *Les congrès internationaux de 1681 à 1899: Liste complète.* Brussels. 76 pp. *Publication* no. 164; ___. 1964. *Les congrès internationaux de 1900 à 1919: Liste complète.* Brussels. 143 pp. *Publication* no. 188.

European

Bb13. **Allen, D. E.** 1988. The natural history society in Britain through the years. *Archives of natural history* 14: 243–259.

Bb14. ___. 1988. The biological societies of London, 1870–1914: Their interrelations and their responses to change. *Linnean* 4(3): 23–38.

Bb15. **Becker, B. H.** 1874. *Scientific London.* London, H. S. King & Co. viii+340 pp. *Also* New York, D. Appleton & Co., 1875.
Includes historical accounts of learned institutions and societies.

Bb16. **Brown, H.** 1934. *Scientific organizations in seventeenth century France (1620–1680).* Baltimore, MD, Williams & Wilkins. xxii+306 pp. *Re-pub.* New York, Russell & Russell, 1967.

Bb17. **Burkhardt, F.** 1972. England and Scotland: The learned societies. In: T. F. Glick, ed. *The comparative reception of Darwinism*, pp. 32–74. Austin, TX, Univ. of Texas Pr. *Re-pub.* (with a new preface) Chicago, IL, Univ. of Chicago Pr., 1988. xxviii+506 pp.

Bb18. **Drummond, H. J. H.** 1958. Records of medical and scientific societies in Scotland. I: Early medical and scientific societies of North–East Scotland. *Bibliotheck* 1 (2): 31–33; **Finlayson, C. P.** 1958. Records of scientific societies in Scotland. II: Records of scientific and medical societies preserved in the University Library, Edinburgh. *Ibid.* 1 (3): 14–19; ___. 1963. ___: Additions. *Ibid.* 4 (6): 38–39.

Bb18A. **Duris. P.** 1989. Les sociétés linnéennes françaises et le culte de Linné (1780–1830). *Bulletin de la Société Botanique de France* 136: 179–185.

Bb19. **Elliot,** Sir **W.** 1871. Opening address. *Transactions of the Botanical Society of Edinburgh* 11: 1–33, 193–255.
Presidential address, *includes* "... review [of] the results achieved by local societies and field clubs ..." (pp. 11–33), and a "List of field clubs" (pp. 193–255), the latter providing brief historical and bibliographical notes.

Bb20. **Filippov, N. G.** 1986. Dokumenty nauchno–tekhnicheskikh obshchestv Rossi kak istochnikh izucheniya istorii promyshlennosti i tekhniki. [Documents of Russian scientific societies as a source for the history of science and technology]. *Voprosy istorii estestvoznaniya i tekhniki* 1986: 125–131.

Bb21. **Fox, R.** 1980. The savant confronts his peers: Scientific societies in France, 1815–1914. In: R. Fox & G. Weisz, eds. *The organization of science and technology in France: 1808–1914*, pp. 241–282. Cambridge, Cambridge Univ. Pr.

Bb22. **Griffin, F. J.** 1940. The first entomological societies: An early chapter in entomological history in England. *Proceedings of the Royal Entomological Society of London* A 15: 49–68.

Bb23. **Hume, A.** 1853. *The learned societies and printing clubs of the United Kingdom: Being an account of their respective origin, history, objects, and constitution [etc.]*. With a supplement ... by A. I. Evans. London, Longman [etc.]. xxxii+294+72 pp.

Bb24. **Lysaght, A.**, ed. 1959. *Directory of natural history and other field study societies in Great Britain [etc.]*. London, published for the Carnegie United Kingdom Trust by the British Association for the Advancement of Science. xi+217 pp.

Bb25. **McClellan, J. E., 3rd.** 1985. *Science reorganized: Scientific societies in the eighteenth century*. New York, Columbia Univ. Pr. xxxi+413 pp.
Appendix 1. Official scientific societies: 1660–1793, pp. 261–280; appendix 2. Notable private and semi-private scientific societies: 1660–1793, pp. 281–291.

Bb26. **MacLeod, R. M., J. R. Friday, & C. Gregor.** 1975. *The corresponding societies of the British Association for the Advancement of Science 1883–1929: A survey of historical records, archives and publications*. London, Mansell. xxiv+147 pp.

Bb27. **Ornstein, M.** 1938. *The role of scientific societies in the seventeenth century*. 3rd ed. Chicago, IL, Univ. of Chicago Pr. xviii+308 pp. & 14 pl. *Re-pub.* Hamden, CT, Archon Books, 1963; New York, Arno Pr., 1975.
Replaces 2nd ed., 1928.

North American

Bb28. **Allen, F. H.** 1902. Bird clubs in America. *Bird lore* 5: 12–17.

Bb29. **Baker, C. L.** 1960. History of academies of science in America. *Journal of the Tennessee Academy of Science* 35: 224–230.

Bb30. **Bates, R. S.** 1965. *Scientific societies in the United States*. 3rd ed. Cambridge, MA, M I T Pr. xiii+326 pp.
Replaces 2nd ed., New York, Columbia Univ. Pr., 1958. xi+297 pp.

Bb31. **Carnegie Institution of Washington.** 1908. *Handbook of learned societies and instititions. America*. (Ed. by J. D. Thompson.) Washington, DC. viii+592 pp. *Publication* no. 39. *Re-pub.* Detroit, MI, Gale Research Co., 1966.
Covers North and South America. Each entry is arranged under these headings: Name, Address, History, Object, Meetings, Membership, Serial publications, Special publications, etc.

Bb32. **Forbes, S. A.** 1907. History of the former state natural history societies of Illinois. *Science* n.s. 26: 892–898.

Bb33. **Judd, W. W.** 1979. *Early naturalists and natural history societies of London, Ontario*. London, Ont., Phelps. 215 pp.

Bb34. **Kiger, J. C.**, ed. 1982. *Research institutes and learned societies*. Westport, CT, Greenwood Pr. xxv+551 pp. *Greenwood encyclopedia of American institutions* no. 5.
Includes historical sketches of each society by various authors.

Bb35. **Oleson, A., & S. C. Brown**, eds. 1976. *The pursuit of knowledge in the early American republic: American scientific and learned societies from colonial times to the civil war*. Baltimore, MD, Johns Hopkins Univ. Pr. xxv+372 pp.
Eighteen articles by various authors. *Includes* **J. C. Greene.** Science, learning, and utility: Patterns of organization in the early American republic, pp. 1–20; **A. H. Dupree.** The national pattern of American learned societies, pp. 21–32; **B. Hindle.** The underside of the

learned society in New York, 1754–1854, pp. 84–116; **J. Ewan.** The growth of learned and scientific societies in the Southeastern United States to 1860, pp. 208–218; **P. J. Bowler.** The early development of learned societies in Canada, pp. 326–339. References to articles on individual societies are entered in the relevant sections herein.

Bb36. **Skallerup, H. R.** 1963. Bibliography of the histories of American academies of science. *Transactions of the Kansas Academy of Science* 66: 274–281.

Australasian

Bb37. **Hoare, M. E.** 1967. Learned societies in Australia: The foundation years in Victoria, 1850–1860. *Record of the Australian Academy of Science* 1: 7–29.

Bb38. ___. 1969. "All things are queer and opposite": Scientific societies in Tasmania in the 1840's. *Isis* 60: 198–209.

INDIVIDUAL SOCIETIES, etc.

Natural History

International

Commission Internationale pour l'Exploration Scientifique de la Mer Méditerranée

Bb39. **Crovetto, A.** 1968. La Commission Internationale pour l'Exploration Scientifique de la Mer Méditerranée. In: Institut Océanographique, Monaco. Communications, premier Congrès International d'Histoire de l'Océanographie, Monaco, 12–17 décembre 1966. *Bulletin de l'Institut Océanographique, Monaco* numéro spécial 2: 327–335.

Conseil Permanent International pour l'Exploration de la Mer

Bb39A. **Conseil Permanent International pour l'Exploration de la Mer.** 1939. *Index to publications by the International Council for the Exploration of the Sea, 1899–1938.* Copenhagen. 145 pp.

International Union of Biological Sciences

Bb40. **Stafleu, F. A.** 1971. Fifty years of international biology. *Taxon* 20: 141–151.

European

Aargauische Naturforschende Gesellschaft

Bb41. **Aargauische Naturforschende Gesellschaft.** 1986. Aargauische Naturforschende Gesellschaft 1811–1986. Ed. by H. Moor. *Mittheilungen der Aargauischen Naturforschenden Gesellschaft* 31: 500 pp.

Academia (Real) das Sciencias de Lisboa

Bb42. **Academia das Sciencias de Lisboa.** 1926. *Catálogo geral das publicações da Academia das Sciencias de Lisboa (1779–1926), com os preços por que actualmente se vendem no armazém da Academia.* Lisbon. 29 pp.
Replaces ___. 1905. *Catalogo das obras á venda na Typographia da Academia Real das Sciencias, 1779–1904. Annexo á obra as publicações da Academia Real das Sciencias de Lisboa.* By A. A. Girard. Lisbon. xxxiii+118 pp.

Bb43. ___. 1931. *Academia das Sciencias de Lisboa. III jubileu ... 1779–1929.* Lisbon, Coimbra. [iv]+443 pp., illus.

Bb44. ___. 1934. *A infância da Academia (1788–1794). Visita aos arquivos do reino; [etc.].* By A. Baião. Lisbon. 213 pp.

Bb45. ___. 1936. Indice geral das memórias e trabalhos que interessam à Classe de ciências e se encontram nas publicações anteriores: Memórias da Academia real das ciências de

Lisboa, Memórias de matemática e física da Academia real das ciências de Lisboa, História e memórias da Academia real das ciências de Lisboa, Jornal de ciências matemáticas, físicas e naturais, Memórias da Academia real das ciências 1.a classe de ciências matemáticas, físicas e naturais [1779–1924] [etc.]. *Memórias da Academia das Ciências de Lisboa, classe de ciências* 1: 11–61.

Académie (Royale/Impériale/Nationale) de Médecine, Paris

Bb46. **Académie de Médecine, Paris.** 1985. *Index biographique des membres des associés et des correspondants de l'Académie de Médecine: 1820–1984.* [3rd ed.] Paris. xix+287 pp.

Replaces ___. 1972. ___*: 1820–1870.* 2 éd. By G. Nicole–Genty & M. Chapuis. Paris, Éditions Doin. xvi+262 pp.

Académie (Royale/Nationale/Impériale) de Metz (*formerly* Société des Lettres, Sciences, et Arts et d'Agriculture de Metz)

Bb47. **Académie de Metz.** 1908. *Table générale par ordre alphabétique des Mémoires de l'Académie de Metz, [vols. 1–85] 1819–1903. ... Avec une introduction présentant les pièces les plus intéressantes pour l'histoire de la Société royale des sciences et des arts de Metz, 1757–1792 et pour la reconstitution de l'académie en 1819.* Ed. by E. Fleur. Metz. 424 pp.; *Continued* ___. 1932. *Tables des Mémoires ... de [vols. 86–111] 1904 à 1930 ... Avec un répertoire alphabétique des pièces d'archives provenant de la Société Royale des Sciences et des Arts de Metz et de la Société des Philathènes de Metz, 1757–1792.* Nancy. 96 pp.

Académie (Royale) Nationale des Sciences, Belles–Lettres et Arts, Bordeaux

Bb48. **Académie Nationale des Sciences, Belles–Lettres et Arts, Bordeaux.** 1879. *Académie des Sciences, Belles–Lettres et Arts de Bordeaux. Table historique et méthodique, 1712–1875. Documents historiques, 1711–1713. Catalogue des manuscrits de l'ancienne Académie, 1712–1793.* By J. Gères. Bordeaux. 386 pp.

Académie Royale des Sciences, des Lettres et des Beaux–Arts de Belgique

Bb49. **Académie Royale des Sciences, des Lettres et des Beaux–Arts de Belgique.** 1852. *Histoire et bibliographie analytique de l'Académie Royale des Sciences et des Lettres et des Beaux–Arts de Belgique [etc.].* By J. P. Namur. 2e ed. Brussels. xi+164 pp.

Replaces ___. 1838. *Bibliographie académique Belge, ou répertoire systématique et analytique des Mémoires, extraits de mémoires, dissertations, observations, essais et mémoires des prix publiés jusqu'àce jour par l'ancienne et la nouvelle Académie de Bruxelles, [etc.].* Liege. xxv+80 pp.

Bb50. ___. 1858. *Tables générales et analytiques du recueil des Bulletins de l'Académie Royale des Sciences, des Lettres et des Beaux–Arts de Belgique. 1re série, Tomes 1 à 23, (1832 à 1856).* By A. Siret. Brussels. ii+395 pp.; *Continued* ___. 1867. *Tables ... 2 série, Tomes 1 à 20, 1857 à 1866.* By A. Siret. Brussels. 187 pp.; ___. 1883. *Tables ... 2 série, Tomes 21 à 50, 1867 à 1880.* By E. Marchal. Brussels. 380 pp.; ___. 1898. *Tables ... 3 série, Tomes 1 à 30, 1881 à 1895.* By E. Marchal. Brussels. 318 pp.; ___. 1910. *Tables ... 3 série, Tomes 31 à 36, 1896 à 1898.* Brussels. [ ]* pp.; ___. 1919. *Tables générales des Bulletins de la Classe des sciences. 4e série, années 1899 à 1910.* By F. Leurident & J. Perrée. Brussels. 178 pp.; ___. 1919. *Tables générales des Bulletins de la Classe des sciences. 5e série, années 1911 à 1914.* By F. Leuridant. Brussels. 34 pp.
[*Not examined – pagination not ascertained.]

Bb51. ___. 1872. *Centième anniversaire de sa fondation (1772–1872).* Brussels, C. Muquardt. 2 vols.

Vol. 2 (comprising separately paginated reports); *includes* **P. J. van Beneden.** Rapport sur les travaux de zoologie, 224 pp.; **C. J. E. Morren.** Rapport séculaire sur les travaux de botanique et de physiologie végétale (1772–1872), ii+96 pp. *Also pub. separately* ___. 1872. *Mémorandum des travaux de botanique et de physiologie végétale qui ont été publiés par l'Académie Royale des Sciences, des Lettres et des Beaux–Arts de Belgique pendant le premier siècle de son existence (1772–1871). Rapport séculaire.* Bruxelles, F. Hayez. [iii]+ii+96 pp.

Bb52. ___. 1883. Histoire de l'Académie Impériale et Royale des Sciences et Belles–Lettres de Bruxelles, [etc.]. By E. Mailly. *Mémoires couronnés et mémoires des savants étrangers, Académie Royale des Sciences et belles–Lettres de Bruxelles, collection in 8vo* 34 & 35.

Bb53. ___. 1898. *Tables générales des Mémoires de l'Académie Royale des Sciences, des Lettres et des Beaux–Arts de Belgique, 1772–1897.* Brussels. 260 pp.; *Continued* ___. 1919. ___*: Supplément: 1898–1914.* By F. Leuridant. Brussels. 33 pp.; ___. 1965. ___*: Supplément: 1898–1963.* By J. Jeanjot. Brussels. 158 pp.

Bb54. ___. 1909. *Notices biographiques et bibliographiques concernant les membres, les correspondants et les associés, 1907–1909.* 5e éd. Brussels. viii+1,124 pp.
Replaces ___. 1855. *Bibliographie académique ou liste des ouvrages publiés par les membres, correspondants et associés résidents, 1854.* Brussels, M. Hayez. xxiii+254 pp.; ___. 1875. *Notices biographiques et bibliographiques concernant les membres et les correspondants, ainsi que les associés résidents.* [2e éd.] Brussels. lxxxiv+374 pp.; ___. 1887. ___. [3e éd.] Brussels. vii+606 pp.; ___. 1897. ___. [4e éd.] Brussels. viii+832 pp.

Bb55. ___. 1919. *Tables des notices biographiques publiées dans l'Annuaire, 1835–1914.* By F. Leuridant. Brussels, Hayez. [iii]+55 pp.

Bb56. ___. 1922. *L'Académie Royale de Belgique depuis sa fondation (1772–1922).* By P. Pelseneer. Brussels, M. Lamertin. 342 pp.

Bb57. ___. 1964. *Index biographique des membres, correspondants et associés de l'Académie Royale de Belgique de 1769 à 1963.* 2nd ed. By J. Jeanjot. Brussels, Palais de Académies. 298 pp.
Replaces ___. 1948. *Index biographique ... 1769 à 1947.* Brussels. 265 pp.

Bb58. ___. 1968. *Index des lauréats de l'Académie Royale des Sciences, des Lettres et des Beaux–Arts de Belgique, 1769–1967.* By P. Jeanjot. Brussels, Palais des Académies. 178 pp.

Bb59. ___. 1973. *L'Académie Royale des Sciences, des Lettres et des Beaux Arts de Belgique, 1772–1972: Esquisse historique.* By J. Lavalleye. Brussels. 120 pp. *Publications du deuxième centenaire* no. 1.

Académie Royale des Sciences d'Outre–Mer (Koninklijke Academie voor Overzeese Wetenschappen), Brussels

Bb60. **Académie Royale des Sciences d'Outre–Mer, Brussels.** 1982. *Cinquantenaire de l'Académie: 1928–1978. Vijftigjarig bestaan van de Academie.* Brussels. 2 vols.

Académie (Royale/Nationale/Impériale) des Sciences, Arts et Belles–Lettres de Caen

Bb61. **Académie des Sciences, Arts et Belles–Lettres de Caen.** 1884. *Tables chronologique, méthodique et alphabétique des travaux insérés dans les Mémoires ... depuis 1754 jusqu'en 1883 (inclusivement), [etc.].* By A. Gasté. Caen. 151 pp.; *Continued* ___. 1894. *Tables ... Depuis 1884 jusqu'en 1893 (inclusivement). Tables décennales (pour faire suite aux tables de 1754 à 1883) [etc.].* Caen. 35 pp.; ___. 1904. *Tables décennales (1894 à 1903 inclusivement) des Mémoires [etc.].* By H. Prentout. Caen. 24 pp.

Académie des Sciences, Belles Lettres et Arts de Savoie (*formerly* Société (Royale) Académique de Savoie *and* Académie (Royale/Impériale) de Savoie)

Bb62. **Académie des Sciences, Belles Lettres et Arts de Savoie.** 1891 (1892). *Histoire de l'Académie des Sciences, Belles–Lettres et Arts de Savoie de 1820 à 1860.*

Suivie des tables des trente–six premiers volumes des Mémoires et des six premiers volumes des Documents. By L. Pillet. Chambéry. 343 pp.

Bb63. ___. 1903. *État des membres de l'Académie ... depuis sa fondation, 1820, jusqu'au 30 Juin 1903.* Chambéry. lvi+203 pp.
Includes **A.–A. Szerlecki.** Table des noms mentionnés dans les Mémoires et Documents publiés par l'Académie ... de 1825 à 1902; Table des matières contenus dans chacun des volumes des Mémoires [to the end of Sér. 4, tom.9] et des documents publiés par l'Académie ... de 1825 à 1902.

Académie des Sciences, Arts et Belles–Lettres de Dijon

Bb64. **Académie des Sciences, Arts et Belles–Lettres de Dijon.** 1871. *Notes et documents pour servir à l'histoire de l'Académie ... suivis de la table méthodique des travaux renfermés dans les Mémoires de cette académie de 1769–1869.* By P. Milsand. 2nd ed. Dijon. 367 pp.*Mémoires de l'Académie des Sciences, Arts et Belles–Lettres de Dijon* (2) 16.

Bb65. ___. 1915. Tables générale et particulières des travaux contenus dans les Mémoires (1769–1913). By A. Cornereau. *Mémoires de l'Académie des Sciences, Arts et Belles–Lettres de Dijon* (5) 1: 323 pp.

Académie (Royale) des Sciences (de l'Institut de France), Paris

Bb66. **Académie des Sciences, Paris.** 1853. *Table générale des Comptes Rendus des séances de l'Académie des Sciences, publié par MM. le Secrétaires Perpétuels, ... (Tomes 1 à 31, 3 août 1835 à 30 décembre 1850).* Paris. 1,018 pp.; *Continued* ___. 1870. *Table générale ... (Tomes 32 à 61, 6 janvier 1851 à 30 décembre 1865).* Paris. 1,354 pp.; ___. 1888. *Table générale ... (Tomes 62 à 91, 2 janvier 1866 à 27 décembre 1880).* Paris. 1,599 pp.; ___. 1900. *Table générale ... (Tomes 92 à 121, ... 1881 à ... 1895).* Paris, Gauthier Villars. 1,737 pp.; ___. 1927. *Table générale ... (Tomes 122 à 151, ... 1896 à ... 1910).* I: *Auteurs.* Paris, Gauthier Villars. [iii]+590 pp.; ___. 1931. *Table générale ... (Tomes 152 à 181, ... 1911 à ... 1925).* I: *Auteurs.* Paris, Gauthier Villars. [i]+649 pp.; ___. 1947. *Table générale ... (Tomes 182 à 211, ... 1926 à ... 1940).* I: *Auteurs.* Paris, Gauthier Villars. 920 pp.; ___. 1963. *Table générale ... (Tomes 212 à 241, ... 1941 à ... 1955).* I: *Auteurs.* Paris, Gauthier Villars. [iii]+1,097 pp.

Bb67. ___. 1881. *Tables générales des travaux contenus dans les Mémoires de l'Académie ... Première série, Tomes 1 à 14, (An VI, 1815). Seconde série, Tomes 1 à 40, (1816–1878).* Paris. 3 pts. (39+45+50 pp.)

Bb68. ___. 1881. *Tables générales des travaux contenus dans les Mémoires présentés par divers savants à l'Académie ... Première Série, Tomes 1 et 2 (1806–1811). — Seconde Série, Tomes 1 à 25 (1827–1877).* Paris. 3 pts. (15+14+19 pp.)

Bb69. ___. 1937. *Notices et discours, 1924–1936.* Paris. [vi]+753 pp.; *Continued* ___. 1949. ___. Tome 2: *1937–1948.* Paris, Gauthier–Villars. 775 pp.; ___. 1957. ___. Tome 3: *1949–1956.* Paris, Gauthier–Villars. [vi]+731 pp.; ___. 1964. ___. Tome 4: *1957–1962.* Paris, Gauthier–Villars. [iv]+746 pp.; ___. 1972. ___. Tome 5: *1963–1972.* Paris, Gauthier–Villars. [iv]+733 pp.

Bb70. ___. 1966. *Histoire et prestige de l'Académie des Sciences, 1666–1966.* Paris, Musée du Conservatoire National des Arts et Métiers. 262 pp.
Catalogue of the tercentenary exhibition.

Bb71. ___. 1967. *Troisième centenaire, 1666–1966.* Paris, Gauthier–Villars. 2 vols.
Includes Vol. 1: **P. Gauja.** Les origines de l'Académie des sciences de Paris, pp. 1–51; **L. Plantefol.** L'Académie des sciences durant les trois premiers siècles de son existence, ses visages successifs, ses publications, pp. 53–139; Vol. 2: **C. Arambourg & J. Piveteau.** La paléontologie, pp. 116–125; **L. Plantefol.**, Histoire de la botanique, pp. 127–217; **J. Millot.** L'Académie et les découvertes anatomiques, pp. 219–230; **P.–P. Grassé.**

L'Académie des sciences et la théorie de l'évolution, pp. 231–247; **E. Wolff.** Les Précurseurs de l'embryologie expérimentale, pp. 249–264; **E. Fauré-Fremiet.** Les prémices de la biologie moleculaire, pp. 265–291; **L. Gallien.** L'oeuvre des zoologistes française en parasitologie, pp. 293–294.

Bb72. ___. 1979. *Index biographique des membres et correspondants de l'Académie des Sciences du 22 Décembre 1666 au 1er Octobre 1978.* Paris, Garnier. 513 pp.
Replaces Earlier ed., *Index biographique ... 22 Décembre 1666 au 15 Décembre 1967.* Paris, Gauthier-Villars, 1968. xiv+574 pp.

Bb73. **Bertrand, J.** 1869. *L'Académie des sciences et les academiciens de 1666 à 1793.* Paris, Hetzel. iv+435 pp. *Re-pub.* (with an index) Amsterdam, B. M. Israël, 1969. iv+435+[8] pp.

Bb73A. **Crosland, M.** 1992. *Science under control: The French Academy of Sciences 1795–1914.* Cambridge, Cambridge Univ. Pr. xix+454 pp.

Bb74. **Gauja, P.** 1917. *Les fondations de l'Académie des Sciences (1881–1915).* Hendaye, Imp. de l'Observatoire d'Abbadia. [iii]+658 pp.

Bb75. **Godin, M.** 1734. *Table alphabétique des matières contenues dans l'Histoire et les Mémoires de l'Académie Royale des Sciences ... 1–4, 1666–1730.* Paris. 4 vols.; *Continued* **M. P. Demours (& Cotte, –.).** 1747–1809. *Table générale des matières contenues dans l'Histoire et dans les Mémoires de l'Académie Royale des Sciences, depuis l'année 1731 jusqu'à l'année 1740 [–1790] inclusivement, [etc.].* Paris. 6 vols.
Vol. 1: Années 1666–1698; vol. 2: Années 1699–1710; vol. 3: Années 1711–1720; vol. 4: Années 1721–1730; Vol. 5: Années 1731–1740; vol. 6: Années 1741–1750; vol. 7: Années 1751–1760; vol. 8: Années 1761–1770; vol. 9: Années 1771–1780; vol. 10: Années 1781–1790.

Bb76. **Hahn, R.** 1971. *The anatomy of a scientific institution: The Paris Academy of Sciences, 1666–1803.* Berkeley, CA, Univ. of California Pr. xiv+433 pp.
Appendix II: Bibliographical data on Academicians, pp. 330–373.

Bb77. **Hirschfield, J. M.** 1981. *The Académie Royale des Sciences, 1666–1683.* New York, Arno. 222 pp.

Bb78. **Lacroix, A.** 1934. *Notice historique sur les membres et correspondants de l'Académie des Sciences ayant travaillé sur les colonies françaises de la Guyane et des Antilles de la fin du XVIIe siècle au début du XIXe, lue dans la séance publique annuelle du 12 décembre 1932.* Paris, Institut de France, Académie des Sciences. 99 pp. *Mémoires de l'Académie des Sciences de l'Institut de France* 2e sér. vol. 61A.

Bb79. ___. 1936. *Notice historique sur les membres et correspondants de l'Académie des Sciences ayant travaillé sur les colonies françaises des Mascareignes et de Madagascar au XVIIIe siècle et au début du XIXe, lue dans la séance publique annuelle du 17 décembre 1934.* Paris, Institut de France, Académie des Sciences. 118 pp. *Mémoires de l'Académie des Sciences de l'Institut de France* 2e sér. vol. 62A.

Bb80. ___. 1943. *Notice historique sur les membres et correspondants zoologistes et biologistes de l'Académie des Sciences ayant travaillé sur les côtes des colonies françaises de l'Afrique du Nord et du Nord–est depuis le XVIIIe siècle. Lecture faite en la séance annuelle du 20 décembre 1943.* Paris, Institut de France, Académie des Sciences. 42 pp.

Bb81. **Maindron, E.** 1881. *L'ancienne Académie des sciences. Les académiciens 1666–1793.* Paris, B. Tignol. ii+90 pp.

Bb82. ___. 1881. *Les fondations de prix à l'Académie des Sciences. Les lauréats de l'Académie. 1714–1880.* Paris, Gauthier–Villars. [iv]+189 pp.

Bb83. ___. 1888. *L'Académie des Sciences: Histoire de l'Académie, fondation de l'Institut national. Bonaparte Membre de l'Institut National.* Paris, F. Alcan. [vi]+344 pp. & 8 pl.

Bb84. **McClellan, J. E, 3rd.** 1981. The Académie Royale des Sciences, 1699–1793: A statistical portrait. *Isis* 72: 541–567.

Bb85. **Rozier, –., Abbé.** 1775–76. *Nouvelle table des articles contenus dans les volumes de l'Académie Royale des Sciences de Paris depuis 1666 jusqu'en 1770, dans ceux des arts et métiers publiées par cette académie & dans la Collection Académique, etc.* Paris. 4 vols.

Bb86. **Stroup, A.** 1990. *A company of scientists: Botany, patronage, and community at the 17th–century Parisian Royal Academy of Sciences.* Berkeley, CA, Univ. of California Pr. xv+387 pp.
Includes Ch. 3. Botanical research at the Academy, pp. 65–166.

See also Ba12 & Cb527.

Académie des Sciences, Belles–Lettres et Arts de Besançon

Bb87. **Académie des Sciences, Belles–Lettres et Arts de Besançon.** 1954. *L'Académie des Sciences, Belles–Lettres et Arts de Besançon. Deux cents ans de vie comtoise, 1752–1952. Essai de synthèse.* By J. Cousin. Besançon, Ledoux. viii+248 pp.

Bb88. **Cousin, J.** 1959. L'Académie des Sciences, Belles–Lettres et Arts de Besançon au XVIIIe siècle et son oeuvre scientifique. *Revue d'histoire des sciences* 12: 327–344.

Académie des Sciences et des Lettres de Montpellier

Bb89. **Académie des Sciences et des Lettres de Montpellier.** 1892. Index alphabétique des noms d'auteurs contenus dans les 11 volumes de la 1re série [1847/53–1890/92]. Table générale des matières [etc.]. By O. de Coninck. *Mémoires de la section des sciences, Académie des Sciences et des Lettres de Montpellier* 11: 293–324.

Académie (Royale) des Sciences, Belles–Lettres, et Arts, Lyon

Bb90. **Académie des Sciences, Belles–Lettres, et Arts de Lyon.** 1882. *Table des matières contenues dans les Mémoires publiées de 1845 à 1881, suivie d'un catalogue des recueils académiques reçus en échange.* By J. B. Saint–Lager. Lyon. 74 pp.

Bb91. ___. 1900. *Le deuxième centenaire de l'Académie des Sciences … 1700–1900.* Lyon. xxxiv+241 pp. & 9 ports.

Académie (Royale/Impériale) des Sciences, Inscriptions et Belles–Lettres de Toulouse

Bb92. **Académie des Sciences, Inscriptions et Belles–Lettres de Toulouse.** 1854. *Table alphabétique des matières contenues dans les seize premiers tomes des Mémoires de l'Académie impériale des sciences, inscriptions et belles–lettres de Toulouse [1782–1850], suivie de la table générale des auteurs.* Toulouse. 85 pp.; *Continued* ___. [1856.] *Table alphabétique …de la quatrième série des Mémoires … [1851–56] [etc.].* [Toulouse.] 25 pp.; ___. [1862.] *Table alphabétique … de la cinquième série d… (IVe et Ve série) des Mémoires … [1851–62] [etc.].* Toulouse. 66 pp.; ___. [1868.] *Table alphabétique … de la sixième série des Mémoires … de 1863 à 1868 [etc.].* [Toulouse.] 22 pp.; ___. 1879. *Table alphabétique … de la septième série, (1869–78) des Mémoires [etc.].* Toulouse. 48 pp.

Académie de Stanislaus (*formerly* Société Libre (Royale) des Sciences, et Belles–Lettres de Nancy)

Bb93. **Académie de Stanislaus.** 1867. *Table alphabétiques des matières et des noms d'auteurs contenues dans les trois premières séries des Mémoires de l'Academie de Stanislas, (1750–1866).* By M. Simonin. Nancy. 300 pp.

Bb94. ___. 1902. *Table alphabétique des publications de l'Académie de Stanislas (1750–1900).* By J. Favier. Précédée de l'histoire de l'Académie par C. Pfister. Nancy. 225 pp.

Accademia Nazionale dei Lincei (*formerly* Reale Accademia dei Lincei)

Bb95. **Accademia Nazionale dei Lincei.** 1924–53. *Atti degli Accademici pubblicati dall'inizio al 1910 (... dal 1911 al 1924; ... dal 1925 al 1934; ... dal 1935 al 1950).* Rome. 4 pts.

Bb96. ___. 1976. *Biografie e bibliografie degli accademici Lincei.* R. Ridolfi, ed. Rome. lxi+1319 pp.

Bb96A. ___. 1989. *Contributi alla storia della Accademia dei Lincei.* By G. Gabrieli. Rome, Accademia Nazionale dei Lincei. 2 vols. *Storia dell' Accademia dei Lincei, Studi* 1.

Bb97. **Reale Accademia dei Lincei.** 1883. *Breve storia all'Accademia dei Lincei.* By D. Carutti. Rome, R. Accademia. [iv]+259 pp.

(Reale) Accademia delle Scienze Fisiche e Mathematiche, Naples

Bb98. **Reale Accademia delle Scienze Fisiche e Mathematiche, Naples.** 1904. *Indice generale dei lavori pubblicati dal MDCCXXXVII al MDCCCCIII.* Naples. viii+111pp.

Accademia delle Scienze dello Istituto di Bologna

Bb99. **Accademia delle Scienze dell'Istituto di Bologna.** 1855. *Indices generales in Novos Commentarios Academiæ Scientiarum Instituti Bonoiensis [1834–1849].* Bologna. 88 pp.; *Continued* ___. 1864. *Indici generali della collezione publicata dall'Accademia dell Scienze dell'Istituto di Bologna col titulo di Memorie in dodici tomi dal 1850 al 1861.* Bologna. 40 pp.; ___. 1871. *Indici ... seconda serie ... publicata dal 1862 al 1870.* Bologna. 45 pp.; ___. 1880. *Indici ... terza serie ... publicata dal 1871–1879.* Bologna. 49 pp.; ___. 1890. *Indice ... serie quarta ..., 1880–1889.* Bologna. 49 pp.; ___. 1904. *Indice ... serie quinta ..., 1890–1903.* Bologna. 24 pp.; ___. 1915. *Indice ... serie sesta delle Memorie ... Classe di scienze fisiche, 1904–1913.* Bologna. 16 pp.; ___. 1924. *Indice ... serie settima delle Memorie ... Classe di scienze fisiche, 1914–1924.* Bologna. 11 pp.

Bb100. ___ 1881. *Accademia delle Scienze dello Istituto di Bologna dalla sua origine a tutto il MDCCCLXXX.* Bologna. 278 pp.
Includes indexes to the Academy's publications.

(Cesareo–Regia/Imperiale–Reale) Accademia di Scienze, Lettere ed Arti di Padova (Accademia Patavina di Scienze, Lettere ed Arti)

Bb101. **Accademia di Scienze, Lettere ed Arti di Padova.** 1901. *Indice generale per ordine alfabetico di autori e di materie dei lavori letti alla R. Accademia di scienze, lettere ed arti in Padova e pubblicati ne' suoi Atti dall'anno 1779 a tutto l'anno accademico 1899–1900, con brevi notizie biografiche.* By M. Girardi & P. A. Saccardo. Padua, Tip. G. B. Randi. 95 pp.; ___. 1934. *Atti e memorie della Reale Accademia di Scienze, Lettere ed Arti di Padova. Indici ... dall'anno 1900–1901—1929–1930, volumi XVII–XLVI della nuova serie, [etc.].* Padua. 145 pp.

(Reale) Accademia delle Scienze di Torino

Bb102. **Accademia delle Scienze di Torino.** 1883. *Il primo secolo della R. Accademia delle Scienze di Torino. Notizie storiche e bibliografiche, (1783–1883).* Turin. viii+591 pp.
Includes **A. Manno.** Repertorio bibliografico delle pubblicazioni accademiche. 1. Indice per volumi; 2. Indice generale alfabetico ed analitico, pp. 241–588.

Bb103. ___. 1985. I due primi secoli della Accademia Piemontese dal Settecento allo Stato Unitario. Atti del convegno, 10–12 novembre 1983. *Atti della Accademia delle Scienza di Torino, classe di scienze morali, storiche e filologiche* 119 (Supp.): 296 pp.

(Königliche) Akademie Gemeinütziger Wissenschaften, Erfurt

Bb104. **Kiefer, J.** 1988. Angehörige der ehemaligen "Akademie nützlicher (gemeinnütziger) Wissenschaften" zu Erfurt (1754–1945) als Mitglieder der heutigen "Deutschen Akademie der Naturforscher Leopoldina" in Halle. *Beiträge zur Hochschul- und Wissenschaftsgeschichte Erfurts* 21: 85–112.

Bb104A. ___. 1990. Die nationalen und internationalen Tauschbeziehungen der ehemaligen "Akademie nützlicher (gemeinnütziger) Wissenschaften" zu Erfurt (1754–1945) mit gelehrten Gesellschaft, wissenschaftlichen Institutionen und Behörden in den Jahren 1757–1944. *Ibid.* 22: 208–233.

Bb105. **Königliche Akademie Gemeinütziger Wissenschaften, Erfurt.** 1904. Festschrift zur Feier des 150 jährigen Bestehens der Kgl. Akademie. *Jahrbücher der Königlichen Akademie Gemeinnütziger Wissenschaften zu Erfurt* N.F. 30: 652 pp. & 2 pl.

Includes Gesamtregister über die von 1804–1903 in den Sitzungen der Akademie vorgetragenen bzw. in ihren Schriften erscheinen wissenschaftlichen Abhandlungen, pp. 299–371.

Bb106. ___. 1930. *Geschichte der Akademie Gemeinütziger Wissenschaften zu Erfurt, 1754–1929. Festrede aus Anlass der Feier des 175 jährigen Bestehens der Akademie, [etc.].* By J. Biereye. Erfurt. 17 pp.

Bb107. ___. 1931. Verzeichnis der in den Jahrbüchern der Akademie ... Heft 1–50 [1860–1931], gedruckten Abhandlungen und Vorträge. By H. Wiedemann. *Jahrbücher der Akademie Gemeinnütziger Wissenschaften zu Erfurt* 50: 97–118.

Akademie der Wissenschaften, Gottingen (*formerly* (Königliche) Gesellschaft der Wissenschaften, Gottingen)

Bb108. **Akademie der Wissenschaften, Göttingen.** 1951 [1952]. *Festschrift zur Feier des zweihundertjährigen Bestehens der Akademie der Wissenschaften in Göttingen.* Berlin, Springer. 2 vols.

Bb109. **Gesellschaft der Wissenschaften, Göttingen.** 1928. *Autorenverzeichnis zu den Veröffentlichungen der Gesellschaft der Wissenschaften zu Göttingen, 1751–1927.* By M. Arnim. Göttingen, Dieterich. 350 pp.

Bb110. ___. 1928. *Mitgleider-Verzeichnisse der Gesellschaft der Wissenschaften zu Göttingen, 1751–1927.* By M. Arnim. Gottingen. 234 pp.

Bb111. **Königliche Gesellschaft der Wissenschaften, Göttingen.** 1901. *Festschrift zur Feier des hundertfünfzigjährigen Bestehens der Königlichen Gesellschaft der Wissenschaften zu Göttingen. Abhandlungen der mathematisch–physikalischen Klasse.* Berlin. [iv]+668 pp.

Includes **A. Peter.** Geschichte der Gründung und Entwicklung des botanischen Gartens zu Göttingen; **E. Ehlers.** Göttingen Zoologen.

Bb112. **Verlag Dokumentation.** 1971. *Bibliographia academia Germaniae. Abhandlungen und Sitzungsberichte.* Band 1: *Königliche Gesellschaft der Wissenschaften (Akademie der Wissenschaften) Göttingen; Bayerische Akademie der Wissenschaften München (Abhandlungen, Sitzungsberichte, Forschungen zur Deutsche Geschichte).* Ed. by P. Vezényi. Munich. xvi+715 pp.

Akademiya Nauk SSSR (*formerly* Imperatorskaya Akademiya Nauk (Académie Impériale des Sciences de St. Pétersbourg))

Bb113. **Imperatorskaya Akademiya Nauk (Académie Impériale des Sciences de St. Pétersbourg).** 1846. *Registre alphabétique des noms des auteurs, dont les pièces sont insérées dans les différents recueils publiés par l'Académie ... depuis sa fondation jusqu'à l'an 1846.* By P. H. Fuss. [St. Petersburg.] xi+80 pp.

Bb114. **Imperatorskaya Akademiya Nauk.** 1867. *Catalogue des livres publiés en langues étrangères par l'Académie, [etc.]*. St. Petersburg. 124 pp.
Replaces ___. 1854. ___. St. Petersburg. 66 pp.

Bb115. ___. 1872. *Tableau général méthodique et alphabétique des matières contenues dans les publications de l'Académie Impériale des Sciences de St. Pétersbourg depuis sa fondation.* 1–re partie: *Publications en langues étrangères.* (By V. P. Shemiot.) St. Petersburg. xii+489 pp.; *Continued* ___. 1882. ___*: Supplément 1, comprenant les publications en langues étrangères depuis 1871 jusqu'au 1 Novembre 1881.* St. Petersburg. viii+56 pp.
Includes Liste des presidents et des membres de l'Académie depuis sa fondation, pp. 407–488.

Bb116. ___. 1875. *Sistematicheskii i alfavitnyi ukazatel' statei, pomieshchennykh v periodicheskikh izdaniyakh i sbornikakh Imperatorskoi Akademii Nauk, a takzhe sochinenii, izdannykh Akademieyu otdel'no, so vremeni ee osnovaniya po 1872 g. vklyuchitel'no.* Chast' 2. *Sochineniya na russkom yazyke. [Systematical and alphabetical index of the articles published in periodical and non-periodical publications of the Imperial Academy of Sciences, and also separate works published by the Academy from its foundation to the year 1872 inclusive].* Pt. 2. *Publications in Russian language.* (By V. P. Shemiot.) St. Petersburg. x+403 pp.; *Continued* ___. 1882. ___. Pribavlenie 1. *Sochineniya na inostrannykh yazykakh, vyshedshie v svet s 1871 po 1 Noyabrya 1881 g. [Publications in foreign languages, published from 1871 to 1 November 1881.]* St. Petersburg. viii+56 pp.; ___. 1885. ___. Pribavlenie 2. *Sochineniya na russkom yazyke, vyshedshie v svet c 1 yanvarya 1873 po 1 oktyabr 1884 g. [Publications in Russian language, published from 1 January 1873 to 1 October 1884.]* St. Petersburg. [iv]+41 pp.

Bb117. ___. 1876–77. *Catalogue des livres publiés par l'Académie ... 1. Publications en langue russe. 2. Publications en langues étrangères. Katalog, [etc.]*. St. Petersburg. 2 vols.

Bb118. ___. 1879. *Bericht über die Fortschritte, welche die zoologischen Wissenschaften den von der Kaiserlichen Akademie der Wissenschaften zu St. Petersburg von 1831 bis 1879 herausgegebenen Schriften verdanken. Ein Beitrag zur neuern Geschichte der zoologischen Wissenschaftlichen in Russland.* By J. F. Brandt. St. Petersburg. 213 pp. *Beiträge zur Kenntniss des Russischen Reiches und der angräzendern Länder Asiens [etc.]* Folg. 1, Bd. 1.

Bb119. ___. 1885–1900. *Materialy dlya istorii Imperatorskoi Akademii Nauk (1716–1750). [Materials on the history of the Imperial Academy of Sciences].* St. Petersburg. 10 vols.

Bb120. ___. 1888–91. *Katalog izdanii Imperatorskoi Akademii Nauk ... Catalogue des livres pub. par l'Académie Impériale des Sciences ... 1. Publications en langue russe. 2. Publications en langues étrangères.* St. Petersburg. 2 vols.

Bb121. ___. 1908. *Spisok chlenov Imperatorskoi Akademii Nauk, 1725–1907. [List of the members of the Imperial Academy of Sciences, 1725–1907].* By B. L. Modzalevskii. St. Petersburg, Imp. Akad. Nauk. viii+404 pp. *Re-pub.* Leipzig, Zentralantiquariaat der Deutschen Demokratischen Republik, 1971.

Bb122. ___. 1912–16. *Katalog izdanii Akademii Nauk. Catalogue des livres publiés par l'Académie Impériale des Sciences.* Partie1–3. (By I. A. Kubasov.) St. Petersburg. 3 vols.
Partie 1: *Publications périodiques, recueils et séries en langues russe et étrangères. Depuis 1726 jusqu'au 1 Juin 1912*, 1912; partie 2: *Ouvrages publiés séparément en langue russe. Depuis 1726 jusqu'au 1 Juin*, 1915; partie 3: *Ouvrages publiés séparément en langues étrangères. Depuis 1726 jusqu'au 1 Mars 1916*, 1916. The third supplement incorporates material listed in the first two.
Continued ___. 1916. *Pribavlenie k katalogu izdanii Akademii Nauk s 1911 po 1 noyabrya 1916 g.* St. Petersburg. 33 pp.; *Further continued* **Akademiya Nauk SSSR.** 1918. *Pribavlenie k 1, 2 i 3 chastyam kataloga Rossiiskoi Akademii Nauk. (Po 1–e yanvarya 1918 goda).* St. Petersburg. 39 pp.; ___. 1924. *Pribavlenie k katalogu izdanii*

Akademii Nauk. (Po 1 dekabrya 1923 goda). (Ed. by I. A. Kubasov.) St. Petersburg. ii+63 pp.

Bb123. ___. 1915–17. *Imperatorskaya Akademiya Nauk, 1889–1914.* Vol. 3 (pts. 1 & 2): *Materialy dlya biograficheskago slovaria dieistvitel'nykh chlenov Imperatorskoi Akademii Nauk. [Materials for a biographical dictionary of the active members of the Imperial Academy of Sciences].* St. Petersburg. 2 vols. *Re-pub.* Leipzig, Zentralantiquariaat der Deutschen Demokratischen Republik, 1977.

Bb124. ___. 1917. *Imperatorskaya Akademiya Nauk, 1889–1914.* Vol. 1: *Materialy dlya istorii Akademicheskikh uchrezhdenii za 1889–1914 gg. [Materials on the history of the institutions of the Academy for the years 1889–1914].* Petrograd. iii+631 pp.
Includes Biblioteka, pp. 3–46; Botanicheskii muzei, pp. 133–160; Laboratoriya po anatomii i fiziologii rastenii, pp. 161–166; Osobaya zoologicheskaya laboratoriya, pp. 167–176; Zoologicheskii muzei, pp. 177–218; Sevastopol'skaya biologicheskaya stantsiya, pp. 561–587.

Bb125. **Akademiya Nauk SSSR.** 1923–37. *Novye izdaniya Akademii Nauk, vyshedshie v svet [s 1923 – po Aprel' 1937]. [New publications of the Academy of Science, published from 1923 to April 1937].* Leningrad.
A serial bibliography, initially pub. as an annual, then every three months, and from 1927 as a monthly.

Bb126. ___. 1925. *Sistematicheskii ukazatel' izdanii Akademii Nauk Soyuza Sovetskikh Sotsialisticheskikh Respublik, vyshedshikh v svet c 1 Yanvarya 1917 g. po 1 Sentyabrya 1925 g. Catalogue systématique des publications de l'Académie des Sciences de l'URSS parues depuis le 1 Janvier 1917 jusqu'au 1 Septembre 1925.* Leningrad. xiv+127+[3] pp. *Also* ___. 1927. ___. [Another ed.] Leningrad, [etc.]. xii+132 pp. With notes on publications of 1926 & 1927.; ___. [1927.] *Systematisches Verzeichnis über die Veröffentlichungen der Akdemie der Wissenschaften der Union der Sozialistischen Sowjet-Republiken erscheinen vom 1. Januar 1917 bis zum 1. September 1925.* Leipzig. xii+133 pp.

Bb127. ___. 1925. *Spisok deistvitel'nykh chlenov Akademii Nauk Soyuza Sovetskikh Sotsialisticheskikh Respublik, 1725–1925. [List of the active members of the Academy of Sciences of USSR, 1725–1925].* By B. L. Modzalevskii. Leningrad 35 pp.

Bb128. ___. 1927. *Akademii Nauk Soyuza Sovetskikh Sotsialisticheskikh Respublik za desyat' let, 1917–1927. [Academy of Sciences of the USSR for ten years 1917–1927].* Leningrad. [iii]+iii+235 pp.

Bb129. ___. 1935. *Raboty po fiziologii v izdaniyakh Akademii Nauk, 1735–1934.* Moscow/Leningrad. 65 pp.

Bb130. ___. 1936. *Katalog imeyushchikhsya v prodazhe izdanii Akademii Nauk, 1769–1935. Knigi na russkom yazyke. [Catalogue of the publications of the Academy of Sciences from 1769 to 1935 available for sale. Books in Russian language].* (By L. N. Plyushkin & E. K. Geidrich.) Moscow/Leningrad. xiii+240 pp.

Bb131. ___. 1940. *Catalogue des éditions de l'Académie des Sciences en vente, 1760–1938.* Pt. 1. *Livres en langues étrangères.* Moscow. 37 pp.

Bb132. ___. 1945. *220 let Akademiya Nauk SSSR, 1725–1945. Spravochnaya kniga. [200 years of the Academy of Sciences of the USSR, 1725–1945. Commemorative volume.].* Moscow/Leningrad. 328 pp. & 21 pp. of illus.

Bb133. ___. 1945. *Ocherki po istorii Akademii Nauk. Biologicheskie nauki. [Essays on the history of the Academy of Sciences. Biological sciences].* Ed. by L. A. Orbeli & Kh. S. Koshtoyants. Moscow/Leningrad. 66 pp.

Bb134. **Akademiya Nauk SSSR.** 1947. *Izdaniya Akademii Nauk SSSR, 1917–1947.* Vol. 1: *Sobraniya sochinenii. Periodicheskie izdaniya, serii, nauchno–populyarnye knigi. [Periodical publications, series, popular science books].* Moscow/Leningrad. 60 pp.

Bb135. ___. 1957+. *Bibliografiya izdanii Akademii Nauk SSSR. Ezhegodnik, 1956+.* Moscow.
Annual bibliography of AN SSSR publications.

Bb136. ___, **Institut Istorii Estestvoznaniya i Tekhniki.** 1958–64. *Istorii Akademii Nauk SSSR. [History of the Academy of Sciences of the USSR].* Moscow. 2 vols.
Vol. 1: 1724–1803; vol. 2: 1803–1917.

Bb137. **Besterman, T.** 1945. A bibliography of the Bulletin (Izvestiya) of the Academy of Sciences of the USSR. *Journal of documentation* 1: 45–56.

Bb138. **Kol'tsov, A. V.** 1988. *Rol' Akademii Nauk v organizatsii regional'nykh tsentrov SSSR, 1917–1961 gg.* Leningrad, Izd. "Nauka." 261 pp.

Bb139. **Komkov, G. D., B. V. Levshin, & L. K. Semenov**, eds. 1973. *Akademiya Nauk SSSR: Kratkii istoricheskii ocherk.* Moscow, Izd. "Nauka." 520 pp.
For German version *see* ___. 1981. *Geschichte der Akademie der Wissenschaften der UdSSR.* Ed. by C. Grau. Berlin, Akademie Verlag. 745 pp.

Bb140. **Kunik, A. A.** 1865. *Sbornik materialov dlya istorii Imperatorskoi Akademii Nauk v XVIII viekie. [Collection of materials for the history of the Imperial Academy of Sciences in the 18th century].* St. Petersburg. 2 vols., viii+224 pp. & vi+225–530 pp.

Bb141. **Levshin, B. V.**, ed. 1988. *Dokumenty po istorii Akademii Nauk SSSR, 1926–1934 gg.* Leningrad, Izd. "Nauka." 286 pp.

Bb142. **Lipski, A.** 1953. The foundation of the Russian Academy of Sciences. *Isis* 44: 349–354.

Bb143. **Pekarskii, P. P.** 1870–73. *Istoriya Imperatorskoi Akademii Nauk v Peterburgie. [History of the Academy of Sciences of St. Petersburg].* St. Petersburg, Imp. Akademii Nauk. 2 vols.

Bb144. **Schulze, L.** 1985. The Russification of the St. Petersburg Academy of Sciences and Arts in the eighteenth century. *British journal for the history of science* 18: 305–336.

Bb145. **Turkevich, J.** 1963. *Soviet men of science: Academicians and corresponding members of the Academy of Sciences of the USSR.* Princeton, NJ, Van Nostrand. [iii]+441 pp. *Re-pub.* New York, Greenwood Pr., 1976.

Bb146. **Urban, P. K.**, ed. 1971. *Soviet science, 1917–1970.* Part 1: *Academy of Sciences of the USSR.* Metuchen, NJ, Scarecrow Pr. [v]+322 pp.

Bb147. **Vucinich, A.** 1984. *Empire of knowledge: The Academy of Sciences of USSR (1917–1970).* Berkeley, CA, Univ. of California Pr. x+484 pp.

See also Various guides to the archives Bc713 etc.

Allgemeine Schweizerische Gesellschaft für die Gesammten Naturwissenschaften (Société Helvétique (Suisse) des Sciences Naturelles/Società Elvetica delle Scienze Naturali)

Bb148. **Allgemeine Schweizerische Gesellschaft für die Gesammten Naturwissenschaften.** 1915. *1815–1915. Centenaire de la Société Helvétique des Sciences Naturelles. Notices historiques et documents publiés par la Commission historique institutué à l'occasion de la session annuelle de Genève, 12 au 15 Septembre. Jahrhundertfeier des Schweizerischen Naturforschenden Gesellschaft.* Zurich. vi+316 pp. *Neue Denkschriften der Schweizerischen Naturforschenden Gesellschaft (Nouveau mémoires de la Société ...)* vol. 50.

Includes Liste des nécrologies parues dans les publications de la Société Helvétique des Sciences Naturelles dès l'origine de la Société jusqu'à fin 1914, pp. 299–316.

Bb149. ___. 1934. *Bibliographie der Schweizerischen Naturforschenden Gesellschaft. Verhandlungen—"Compte-rendus"—Nekrologie—Denkschriften—"Zeitschrift". Bibliographie de la Société Helvétique ... Bibliografia della Societa Elvetica [etc.].* By F. Custer & C. Linder. Berne. 340 pp.
Covers 1817–1930.

Bb150. ___. 1944. *Bibliographie der Schweizerischen Naturforschenden Gesellschaft. Verhandlungen—Nekrologie—Denkschriften. Bibliographie de la Société Helvétique ... Bibliografia della Societa Elvetica [etc.].* By C. Linder. Berne. 89 pp.
Covers 1931–1940.

Bb151. ___. 1984. *Bibliographie der Publikation der Schweizerischen Naturforschenden Gesellschaft 1941–1974.* By W. Liebi & V. Portner. Berne. 480 pp.

Bataafsch Genootschap der Proefondervindelijke Wijbsbegeerte, Rotterdam

Bb152. **Kuenen, J. P., et al.** 1919. *Gedenboek van het Bataafsch Genootschap der proefondervindelike wijsbegeerte te Rotterdam tot een blijvende herinnering aan het 150-jarig bestaan; uitg. door het bestuur. 1769–17 Augustus–1919. (Het aandeel van Nederland in de ontwikkeling der naturkunde gedurende de laatste 150 jaren).* Rotterdam. ix+341 pp.

Bb153. **Lieburg, M. J. van, & H. A. M. Snelders.** 1989. *"De bevordering en volmaking der proefondervindelijke wijsbegeerte:" De rol van het Bataafsch Genootschap te Rotterdam in de geschiedenis van de natuurwetenschappen, geneeskunde en techniek (1769–1988).* Amsterdam, Rodopi. 270 pp. *Nieuwe Nederlandse bijdragen tot de geschiedenis der geneeskunde en natuurwetenschappen* no. 29.

(Königliche) Bayerische Akademie der Wissenschaften, Munich (*formerly* Königliche Akademie der Wissenschaften, Munich)

Bb154. **Bayerische Akademie der Wissenschaften.** 1913. *Register zu den ersten 50 Jahrgängen der Sitzungsberichte der mathematisch-physikalischen Klasse 1860–1910.* By A. Hilsenbeck. Munich. 152 pp.

Bb155. ___. 1914. *Register zu den Abhandlungen, Denkschriften und Reden der K. Bayerischen Akademie der Wissenschaften 1807–1913.* By A. Hilsenbeck. Munich. [iv]+201 pp.

Bb156. ___. 1959. *Geist und Gestalt: Biographische Beiträge zür Geschichte in Bayerischen Akademie der Wissenschaften vornehmlich im zweiten Jahrhundert ihres Bestehens.* Munich, C. H. Beck. 3 vols.
Vol. 2: *Naturwissenschaften.* (*Includes* **K. von Frisch.** Zoologie; **O. Renner.** Botanik); vol. 3: *Bilder* [portraits].

Bb157. ___. 1978. *Die naturwissenschaftliche Forschung an der Bayerischen Akademie der Wissenschaften im Zeitalter der Aufklärung.* By A. Kraus. Munich. 285 pp. *Abhandlungen, phil.-hist. Klasse* vol. 82.

Bb158. ___. 1983. *Geschichte der Bayerischen Akademie der Wissenschaften, 1759–1807.* Munich, Beck.
1 Bd: Gründungs- und Frühgeschichte, 1759–1769. 2. Bd: Zwischen Stagnation, Aufschwung und Illuminatenkrise, 1769–1786.

Bb159. **Martius, C. F. P. von.** 1866. *Akademische Denkreden.* Leipzig, F. Fleischer. [xii]+619 pp.
44 obituary addresses delivered from 1834 to 1865.

Bb160. **Verlag Dokumentation.** 1971. *Bibliographia academia Germaniae. Abhandlungen und Sitzungsberichte.* Band 1: *Königliche Gesellschaft der Wissenschaften (Akademie der Wissenschaften) Göttingen; Bayerische Akademie der Wissenschaften*

München (Abhandlungen, Sitzungsberichte, Forschungen zur Deutsche Geschichte). Ed. by P. Vezényi. Munich. xvi+715 pp.

Belfast Natural History & Philosophical Society

Bb161. **Deane, A.** 1924. *The Belfast Natural History and Philosophical Society: Centenary volume 1821–1921. A review of the activities of the Society for 100 years with historical notes, and memoirs of many distinguished members [etc.]* Belfast, the Society. viii+212 pp., illus.

Berwickshire Naturalists' Club

Bb162. **Craw, J. H.** 1933. *Berwickshire Naturalists' Club. Centenary volume with index, volumes 1 to 27 (1831–1931).* Berwick–upon–Tweed. 353 pp.

Biological Photographic Association

Bb163. **Gibson, H. L.** 1978–80. The Biological Photographic Association, its half century. *Journal of biological photography* 46: 171–192, 1978; *Ibid.* 47: 82–100, 1979; ___: The era of maturation, 1965–1980. *Ibid.* 48: 41–86, 1980.

British Association for the Advancement of Science

Bb164. **British Association for the Advancement of Science.** 1864. *Index to Reports and Transactions ... from 1831 to 1860 inclusive.* London, J. Murray. 363 pp.

Bb165. **___.** 1893. *Index to the reports and transactions, 1861 to 1890 inclusive.* London, J. Murray. vii+466 pp.

Bb166. **Howarth, O. J. R.** 1931. *The British Association for the Advancement of Science: A retrospect 1831–1931.* Centenary [2nd] ed. London, British Association. vii+330 pp., illus.
Biographical notes, pp. 298–322. *Replaces* 1st ed., London, [etc.], 1922. vii+318 pp., illus.

Bb167. **MacLeod, R. M., & P. Collins**, eds. 1981. *The Parliament of science: The British Association for the Advancement of Science, 1831–1981.* Northwood, Science Reviews. viii+308 pp.

Bb168. **Morrell, J. B., & A. Thackray.** 1981. *Gentlemen of science: Early years of the British Association for the Advancement of Science.* Oxford, Clarendon Pr. xxiii+592 pp. & 16 pp. of illus.

British Ecological Society

Bb169. **British Ecological Society.** 1985. *Catalogue of British Ecological Society archives.* By I. Nottingham. London. 38 pp.
Deposited at the Linnean Society of London library.

Bb170. **Salisbury, E.** 1964. The origin and early years of the British Ecological Society. In: British Ecological Society. *Jubilee symposium*, pp. 13–18.

Bb171. **Sheail, J.** 1987. *Seventy-five years in ecology: The British Ecological Society.* Oxford, Blackwell Scientific. xiv+301 pp.

Cambridge Philosophical Society

Bb172. **Cambridge Philosophical Society.** 1961. *Author index to Proceedings of the Cambridge Philosophical Society, volumes 1 to 50, 1843–1954, and Transactions of the Cambridge Philosophical Society, volumes 1 to 23, 1822–1928.* Cambridge. 106 pp.

Bb173. **Hall, A. R.** 1969. *The Cambridge Philosophical Society: A history, 1819–1969.* Cambridge, Scientific Periodical Library. [vi]+114 pp.

(Královská) Ceská Spolecnosti Nauk v Praze *formerly* Königlich–Böhmische Gesellschaft der Wissenschaften *and* Böhmischer Gelehrte Privatgesellschaft)

Bb174. **Ceské Spolecnosti Nauk v Praze.** 1938. *Index generalis, 1905–1935. Operum a Regia Societate Scientarum Bohemica, annis 1905–1935 editorum. Obecny rejstrík spisu Královské Ceské Spolecnosti Nauk. 1905–1935.* Prague. 69 pp.

Bb175. ___. 1985. *200 let Ceské Spolecnosti Nauk, 1784–1984. [Bicentenary of the Czech Society of Sciences].* Prague, Ustav Ceskoslovenskych a Svetovych Dejin. 573 pp.

Bb176. **Königlich-Böhmische Gesellschaft der Wissenschaften.** 1854. *Systematisch und chronologisch geordnetes Verzeichniss sämmtlicher Werke und Abhandlungen [1771–1852] der Königlich–Böhmische Gesellschaft der Wissenschaften.* By I. J. Hanus. Prague, F. Rohlicek. [ii]+73+[7] pp.

Bb177. ___. 1869. *Repertorium sämmtlicher Schriften der Königlich-Böhmische Gesellschaft der Wissenschaftenvom Jahre 1769 bis 1868. Zur Feier des einhundertjährigen Bestandes der Gesellschaft.* Zusammengestellt von Dr. W. R. Westenweber. Prague. viii+119 pp.

Bb178. ___. 1884. *Generalregister zu den Schriften der Königlich–Böhmische Gesellschaft der Wissenschaften 1784–1884, [etc.].* By J. Wagner. Prague. xvi+160 pp.

Bb178A. ___. 1884. *Die Königlich–Böhmische Gesellschaft der Wissenschaften. Verzeichniss der Mitgleider 1784–1884 [etc.].* Prague. 41 pp.

Bb179. ___. 1884. *Bericht über die mathematischen und naturwissenschaftlichen Publikationen der Königlich–Böhmische Gesellschaft der Wissenschaften während ihres hundertjährigen Bestandes. Ein Beitrag zur Geschichte der Mathematik und der Naturwissenschaften.* By F. J. Studnicka. Prague. 351 pp.

Bb180. ___. 1905. *Obecny rejstrik spisu Královské ceské spolecnosti nauk 1884–1904. Generalregister der Schriften der ... Gesellschaft ... 1884–1904, [etc.].* By J. Wegner. Prague. 106 pp.

Challenger Society

Bb181. **Deacon, M. B.** 1990. British oceanographers and the Challenger Society, 1903 to 1922. In: W. Lenz & M. Deacon, eds. *Ocean sciences: Their history and relation to man: Proceedings of the 4th International Congress on the History of Oceanography Hamburg [etc.],* pp. 34–40. Hamburg, Bundesamt für Seeschiffahrt und Hydrographie. *Deutsche hydrographische Zeitschrift,* Ergänzungsheft, Reihe B, no. 22.

Chester Society of Natural Science, Literature & Art

Bb182. **Chester Society of Natural Science.** 1911. *The formation of the Chester Society of Natural Science, Literature and Art, and an epitome of its subsequent history.* By J. Siddall. Chester. 63 pp., ports.

Bb183. **Williams, E. G.** 1977. *The Chester Society of Natural Science founded by Charles Kingsley in 1871, later becoming the Chester Society of Natural Science, Literature and Art: Its origin and development over one hundred years.* Chester, the Author. 88 pp.

Cotteswold Naturalists' Field Club

Bb184. **Cotteswold Naturalists' Field Club.** 1888. *The origin of the Cotteswold Naturalists' Field Club and an epitome of the proceedings from its foundation [in 1846] to ... 1887.* By W. C. Lucy. Gloucester. vi+129 pp. & 1 pl.

Bb185. ___. 1913. *Index to the Proceedings ... Volumes I.–XVII. 1846–1912.* By R. Austin. [London, J. Wheldon & Co.] 47 pp.

Bb186. **Cotteswold Naturalists' Field Club.** 1946. *The Cotteswold Naturalists' Field Club 1846–1946*. By T. Bainbrigge–Fletcher. Gloucester. 20 pp.

Dansk Naturhistorisk Forening (*formerly* Naturhistorisk Forening i Kjøbenhavn)

Bb187. **Naturhistorisk Forening i Kjøbenhavn.** 1869. Alphabetisk Register til de förste 20 Aargange ... (1849–68). *Videnskabelige meddelelser fra Dansk Naturhistorisk Forening i Kjøbenhavn* Aart II, aarg. 10 : 259–415.; *Continued* ___. 1894. Fortegnelse over samtlige i de 25 aargange af "Videnskabelige Meddelelser fra Dansk Naturhistorisk Forening i Kjøbenhavn" for1869–93 indeholdte Avhabdlinger, ordnede efter Indholdet. Fortegnelse over samtlige i de 25 Aargange ... for 1869–93 meddelte Afbildninger, ordnede systematisk. *Ibid.* 1893: 204–257.

Bb188. ___. 1890. *Festskrift i Anledning af den Naturhistoriske Forenings Bestaaen fra 1833–1883 [etc.]*. Copenhagen. xx+198+53 pp.
Includes Kort oversigt over Foreningens Historie fra 1833–1883, pp. i–xx. Bilag 1. Den naturhistoriske Forenings Medlemmer, [etc.]; Bilag 2. Fortegnelse over foredrag i naturhistorisk forening, [etc.].

Deutsche Akademie der Naturforscher (Leopoldina) (*formerly* Academia Cæsarea Naturæ Curiosorum/Kaiserliche Leopoldinisch–Carolinische Deutsche Akademie der Naturforscher)

Bb189. **Academia Cæsarea Leopoldino–Carolina Germanica Naturæ Curiosorum.** 1860. *Geschichte der Kaiserlichen Leopoldinisch–Carolinischen Deutschen Akademie der Naturforscher während des zweiten Jahrhunderts ihres Bestehens, [etc.]*. By J. D. F. Neigebaur. Jena. 335 pp. & 3 pl.

Bb190. ___. 1889. *Geschichte der Kaiserlichen Leopoldinisch–Carolinischen Deutschen Akademie der Naturforscher während der Jahre 1852–1887, mit einem Rückblick auf die frühere Zeit ihres Bestehens, [etc.]*. By W. Ule. Halle. viii+258 pp.

Bb191. ___. 1894. *Geschichte der Bibliothek und Naturaliensammlung der Kaiserlich Leopoldinisch–Carolinischen Deutschen Akademie der Naturforscher [etc.]*. By O. Grulich. Jena. ix+300 pp. & 1 pl.

Bb192. ___. 1894–99. *Repertorium zu den Acta [Band 1–10 , 1727–54] und Nova Acta [Band 1–8, 1757–91] der Akademie [etc.]*. And *Repertorium zu den Nova Acta [Band 9–63, 1818–1894] der Akademie [etc.]*. By A. Graesel. Leipzig, W. Engelmann. 2 vols.

Bb193. ___. 1937. *Bericht über den Verlauf der Feier der 250. Wiederkehr des Tages der Erhebung der am 1. Jan. 1652 gegründeten Academia naturae curiosorum zur "Sacri romani imperii academia caesareo–leopoldina naturae curiosorum" durch Leopold I. 28. bis 30. Mai 1937*. By E. Abderhalden. Halle. 139 pp., illus.

Bb194. ___. 1937. *Festgabe aus Anlass der 250. Wiederkehr des Tages der Erhebung am 1. Januar 1652 gegründeten privaten Akademie zur Sacri romani imperii academia caesareo–leopoldina naturae curiosorum durch Leopold I. 7. August 1687*. By E. Abderhalden. Halle. 62 pp., illus.

Bb195. **Deutsche Akademie der Naturforscher (Leopoldina).** 1952. *Bericht über den Verlauf der Feier des 300 jährigen Bestehens der Deutschen Akademie der Naturforscher — Leopoldina — am 16. und 17. Februar 1952 in Halle/Saale*. By O. Schlüter. Leipzig. 92 pp. *Nova acta Leopoldina* N.F. vol. 15, Nr. 104.

Bb196. ___. 1977. *Deutsche Akademie der Naturforscher Leopoldina. 1652–1977.* Halle/Saale. 104 pp. *Acta historica Leopoldina, supplementum* no. 1.

Bb197. ___. 1979. *100 Jahre Leopoldina in Halle, 1878–1978: Vortrage zur Festsitzung am 14. Oktober 1978*. Halle/Saale. 64 pp. *Acta historica Leopoldina, supplementum* no. 2.

Bb198. **Duka Zólyomi, N.** 1980. Die Leopoldinische Akademie und die ungarländische Medizin und Naturwissenschaft bis zum Ende des 18. Jahrhunderts. *Acta historica Leopoldina* 13: 51–101.

Bb199. **Reichenbach, E., & G. Uschmann**, eds. 1970. *Nunquam otiosus. Beiträge zur Geschichte der Präsidenten der Deutschen Akademie der Naturforscher Leopoldina.* Festgabe zum 70. Geburtstag des 22. Präsidenten Kurt Mothes. Leipzig, Barth. 349 pp. *Nova acta Leopoldina* n.s. 36 (198).
Includes **W. Artelt.** Vom Akademiegedanken im 17. Jahrhundert; **G. Uschmann & G. Opitz.** Zeittafel zur Akademiegeschichte; and twelve other articles by various authors.

Bb200. **Uschmann, G., & G. Opitz.** 1970. Zeittafel zur Akademie Geschichte. *Nova acta Leopoldina* N.F. 36: 23–28.

Deutsche (*formerly* Königliche Preussische) Akademie der Wissenschaften zu Berlin

Bb201. **Akademie der Wissenschaften der DDR.** 1975. *Die Berliner Akademie der Wissenschaften in der Zeit des Imperialismus.* Teil 1: *Von den neunziger Jahren des 19. Jahrhunderts bis zur Grossen Sozialistischen Oktoberrevolution.* Unter Mitarbeit des Kollektivs der Forschungsstelle verfasst von C. Grau. Leitung der Arbeiten und Gesamtredaktion L. Stern. Berlin, Akademie–Verlag. xviii+276 pp. *Studien zur Geschichte der Akademie der Wissenschaften der DDR*, 2/1.

Bb202. **Deutsche Akademie der Wissenschaften zu Berlin.** 1950. *Die Mitgleider der Deutschen Akademie der Wissenschaften zu Berlin, 1700–1950.* Ed. by E. Amburger. Berlin. x+204 pp.

Bb203. **___.** 1950. *Bildnisse beruhmter Mitgleider der Deutschen Akademie der Wissenschaften zu Berlin.* Hrsg. aus Anlass der 250. Jahresfeier. Berlin, Akademie–Verlag. 111+111 pp., ports.

Bb204. **___.** 1956. *Deutsche Akademie der Wissenschaften zu Berlin, 1946–1956.* Ed. by J. Irmscher & W. Radig. Berlin. vii+447 pp., illus.

Bb205. **___.** 1960. *Deutsche Akademie der Wissenschaften zu Berlin: Biographischer Index der Mitgleider.* By K. R. Biermann & G. Dunken. Berlin, Akademie–Verlag. xii+248 pp. & 10 pl.

Bb206. **___.** 1967. *Von der Brandenburgischen Sozietät der Wissenschaften zur Deutschen Akademie der Wissenschaften zu Berlin.* By W. Hartkopf & G. Dunken. Berlin, Akademie–Verlag. 122 pp., illus.

Bb207. **___.** 1968. *Abhandlungen und Sitzungsberichte der Deutschen Akademie der Wissenschaften (Königl. Preussische Akademie) zu Berlin 1900–1960: Bibliographie.* By P. Vezényi. München–Pullach, Verlag Dokumentation. 264 pp.

Bb208. **___, Hauptbibliothek.** 1966. *Gesamtregister der Abhandlungen, Sitzungsberichte, Jahrbücher, Vorträge und Schriften der Preussischen Akademie der Wissenschaften, 1900–1945.* Ed. by W. Seyffert & I. Weinitschke. Berlin, Akademie–Verlag. xxi+314 pp.

Bb209. **Königliche Preussische Akademie der Wissenschaften zu Berlin.** 1900–01. *Geschichte der Königlich Preussischen Akademie der Wissenschaften zu Berlin.* By A. von Harnack. Berlin, Reichsdruckerei. 3 vols. in 4. *Re-pub.* Hildesheim, Olms, 1970.
Bd. 3: *Gesammtregister über die in den Schriften der Akademie von 1700–1899 ersehienen wissenschaftlichen Abhandlungen und Festreden.* Ed. by O. Köhnke. xiv+588 pp.

Deutsche Wissenschaftliche Kommission für Meeresforschung

Bb210. **Deutsche Wissenschaftliche Kommission für Meeresforschung.** 1962. 60 Jahre Deutsche Wissenschaftliche Kommission für Meeresforschung. By I. Claussen. *Bericht, Deutsche Wissenschaftliche Kommission für Meeresforschung* 17: 7–10.

Bb211. ___. 1962. Rückblick auf die Entwicklung der Deutschen Wissenschaften Kommission für Meeresforschung (DWK) in sechs Jahrzehnten. By K. Bahr. *Ibid.* 17: 34–44.

Devonshire Association for the Advancement of Science

Bb212. **Devonshire Association for the Advancement of Science, Literature and Art.** 1928. *Key to the Transactions of the Devonshire Association, volumes 1–60 [1862–1928].* By R. P. Chope. Plymouth, W. Brendon & Son. 90 pp.

Bb213. ___. 1962. The story of the Devonshire Association 1862–1962. By H. H. Walker. *Report and transactions of the Devonshire Association for the Advancement of Science, Literature and Art* 94: 42–109.

Dublin Naturalists' Field Club

Bb214. **Jackson, P. S. W., C. Moriarty, & J. R. Akeroyd**, eds. 1989. *In the field of the naturalists. Proceedings of a seminar ... held on 27 September 1986 in celebration of 100 years of the Dublin Naturalists' Field Club.* Dublin, Dublin Naturalists' Field Club. viii+72 pp., illus.
Papers by various authors.

Dublin Philosophical Society

Bb215. **Hoppen, K. T.** 1970. *The common scientist in the 17th century: A study of the Dublin Philosophical Society, 1683–1708.* Charlottesville, VA, Univ. of Virginia Pr.; London, Routledge & Kegan Paul. xiv+297 pp.

Bb216. ___, ed. 1983. *The papers of the Dublin Philosophical Society, 1683–1708.* Dublin, Irish Manuscripts Commission. 94 pp. & 20 microfiche (1846 pp.).

Essex Field Club

Bb217. **Essex Field Club.** [1933.] *Essex naturalist. General index to volumes I–XXII, 1887–1930.* By S. J. Barns. Stratford. 217 pp.

Freshwater Biological Association (of the British Empire)

Bb218. **Freshwater Biological Association.** 1988. *List of publications of the Freshwater Biological Association 1929–1978.* By I. Pettman. Ambleside. 133 pp.

Geograficheskoe Obshchestvo SSSR (*formerly* (Imperatorskoe) Russkoe Geograficheskoe Obshchestvo *and* Gosudarstvennoe Russkoe Geograficheskoe Obshchestvo *and* Vsesoyuznoe Geograficheskoe Obshchestvo)

Bb219. **Berg, L. S.** 1946. *Vsesoyuznoe Geograficheskoe Obshchestvo za sto let, 1845–1945. [All-Union Geographical Society for one hundred years, 1845–1945].* Leningrad, Akademii Nauk SSSR. 260 pp., illus.
With a selected list of Society publications.

Bb220. **Semenov, P. P., & A. A. Dostoevskii.** 1896. *Istoriya poluvekovoi deyatel'nosti Imperatorskago Russkago Geograficheskago Obshchestvo 1845–1895. [History of 50 years of activity of the Russian Geographical Society from 1845 to 1895].* St. Petersburg. 3 vols. (1,377 pp.).

See also Guide to the Society's archives Bc761.

Gesellschaft Deutscher Naturforscher und Ärzte

Bb221. **Autrum, H.**, ed. 1987. *Von der Naturforschung zur Naturwissenschaft: Vorträge, gehalten auf Versammlungen der Gesellschaft Deutscher Naturforscher und Ärzte (1822–1958).* Berlin, Springer–Verlag. xii+587 pp.

Bb222. **Engelhardt, D. von.** 1975. Wissenschaftsgeschichte auf den Versammlungen Deutscher Naturforscher und Ärzte im 19. Jahrhundert. In: K. Mothes & J.–H. Scharf, eds. *Beiträge zur Geschichte der Naturwissenschaften und der Medizin; Festschrift für Georg Uschmann ... zum 60. Geburtstag am 18. Oktober 1873*, pp. 337–359. Halle, Deutsche Akademie der Naturforscher Leopoldina. *Acta historica Leopoldina* no. 9.

Bb223. **Gesellschaft Deutscher Naturforscher und Ärzte.** 1958. *Kleines Quellenbuch zur Geschichte der Gesellschaft deutscher Naturforscher und Ärzte. Gedächtnisschrift für die hundertste Tagung der Gesellschaft.* By M. Pfanenstiel. Berlin. 164 pp.

Bb224. **Hoppe, B.** 1966. Meeresbiologie in der ersten Hundert Jahren der Versammlungen Deutscher Naturforscher und Ärzte. *Medizin Historisches Journal* 1: 31–42.

Bb225. **Lampe, H.** 1975. *Die Entwicklung und Differenzierung von Fachabteilungen auf den Versammlungen von 1828 bis 1913. Bibliogr. zur Erfassung d. Sektionsvorträge mit e. Darst. d. Entstehung d. Sektionen u. ihrer Problematik.* Hildesheim, H. A. Gerstenberg. 301 pp. *Schriftenreihe zur Geschichte der Versammlungen Deutscher Naturforschen und Ärzte* vol. 2.

Bb226. **___., & H. Querner**, eds. 1972. *Die Vorträge der allgemeinen Sitzungen auf der 1.–85. Versammlung 1822–1913.* Mit einer Bibliographie der Berichte über die Versammlungen von I. Gärtner. Hildesheim, H. A. Gerstenberg. xi+127 pp. *Schriftenreihe zur Geschichte der Versammlungen deutscher Naturforschen und Ärzte* vol. 1.

Bb227. **Querner, H.** 1975. Darwins Deszendenz– und Selektionslehre auf den deutschen Naturforscher–Versammlungen. In: K. Mothes & J.–H. Scharf, eds. *Beiträge zur Geschichte der Naturwissenschaften und der Medizin; Festschrift für Georg Uschmann ... zum 60. Geburtstag am 18. Oktober 1873*, pp. 439–456. Halle, Deutsche Akademie der Naturforscher Leopoldina. *Acta historica Leopoldina* no. 9.

Bb228. **___., & H. Schipperges**, eds. 1972. *Wege der Naturforschung, 1822–1972, im Spiegel der Versammlungen Deutscher Naturforscher und Ärzte. Im Auftrage der Gesellschaft Deutscher Naturforscher und Ärzte anlässlich ihres 150–jährigen Bestehens.* Berlin, Springer. [vi]+207 pp.
Includes **P. Karlson.** Die Versammlungen der Gesellschaft Deutscher Naturforscher und Ärzte 1920–1960; **G. Rudolph.** Die Physiologie auf den Versammlungen der Deutschen Naturforscher und Ärzte von der Gründung bis zum Jahre1890 unter besonderer Berücksichtigung des ersten Jahrzehnts und der Spannung zwischen Vitalismus und Mechanismus; **H. Querner.** Probleme der Biologie um 1900 auf den Versammlungen der Deutschen Naturforscher und Ärzte; and seven other articles by various authors.

Bb229. **Schipperges, H.**, ed. [1968.] *Die Versammlung Deutscher Naturforscher und Ärzte im 19. Jahrhundert.* Stuttgart, Gentner. 162 pp. *Schriftenreihe der Bezirks–Ärztekammer Nordwürttemberg* no. 12.
Includes **H. Schipperges.** Anspruch und Programm der Naturforschversammlung im 19. Jahrhundert; **W. Schmitt.** Konstituirung der Gesellschaft Deutscher Naturforscher und Ärzte; **H. Querner.** Die Idee der Evolution auf den Naturforscherversammlungen des 19. Jahrhunderts; **H. Rehse.** Die Rolle der Frau auf den Naturforschversammlungen des 19. Jahrhunderts; and five other articles by various authors.

Bb230. **Sudhoff, K.** 1922. *Hundert Jahre deutscher Naturforscher–Versammlungen. Gedächtnisschrift zur Jahrhundert–Tagung der Gesellschaft deutscher Naturforscher und Ärzte, Leipzig, im September 1922 [etc.].* Leipzig. 83 pp. & 8 pl.

Gesellschaft für Naturforschender Freunde zu Berlin

Bb231. **Gesellschaft für Naturforschender Freunde zu Berlin.** 1873. *Festschrift zur Feier hundertjährigen Bestehens der Gesellschaft Naturforschender Freunde zu Berlin[etc.]*. Berlin. vii+264 pp., illus.
Includes **K. B. Reschert.** Festrede gehalten am 9. Jub. 1873.

Bb232. ___. 1973. Bibliographie der Schriften der Gesellschaft Naturforschender Freunde zu Berlin von 1839 bis 1942. *Sitzungsberichte der Schriften der Gesellschaft Naturforschender Freunde zu Berlin* 13(2): 1–132.

Bb233. **Herter, K., & K. Becker**, eds. 1973. Festschrift zum 200 jährigen Bestehen der Gesellschaft. *Sitzungsberichte der Schriften der Gesellschaft Naturforschender Freunde zu Berlin* 13(1): 1–175.
Includes **K. Becker.** Abriss einer Geschichte der Gesellschaft Naturforschender Freunde zu Berlin, pp. 1–58; **K. Herter & R. Bickerich.** Die Mitgleider der Gesellschaft Naturforschender Freunde zu Berlin in den ersten 200 Jahren des Bestehens der Gesellschaft 1773–1972, pp. 59–157.

Hampshire Field Club

Bb234. **Hampshire Field Club & Archaeological Society.** 1932. *Papers and proceedings of the Hampshire Field Club (and Archaeological Society). General index to volumes 1–10 (1885–1931).* By F. W. C. Pepper. Southampton, H. M. Gilbert & Son. 86 pp.

Bb235. ___. 1985. One hundred years of the Hampshire Field Club. By B. Taylor. *Proceedings of the Hampshire Field Club & Archaeological Society* 41: 5–20.

Hollandsche (Hollandse) Matschappij der Wetenschappen te Haarlem (*formerly* Bataafse (*later* Koninklijke) Matschappij der Wetenschappen te Haarlem) (Société Hollandaise des Sciences à Harlem)

Bb236. **Bierens de Haan, J.** 1970. *De Hollandsche Maatschappij der Wetenschappen, 1752–1952*. Haarlem, Tjeenk Willink. vii+461 pp.
With a summary in English.

Bb237. **Hollandsche Matschappij der Wetenschappen te Haarlem.** 1844. Naamlijst van de schrijvers der verhandelingen, waarnemingen, berigten en bijdragen, voorkomende in de Natuurkundige verhandelingen ... Register der onderwerpen, voorkomende in de Natuurkundige verhandelingen ... [vols. 1–24, 1799–1844]. *Natuurkundige verhandelingen van de Hollandsche Matschappy der Wetenschappen te Haarlem* 24 (2): 93–127.

Bb238. ___. 1876. *Société Hollandaise des Sciences à Harlem. Notice historique. Liste des Protecteurs, Présidents ... et membres ... et liste des publications de la Société depuis sa fondation en 1752. Liste des publications des sociétés savantes et des journeaux scientifiques qui se trouvent dans la bibliothèque de la Société, [etc.]*. Haarlem. xxiii+76 pp. & 1 pl.

Bb239. ___. 1902. *Herdenking van het Honderdvijftigjarig bestaan van de Hollandsche Maatschappij der Wetenschappen op 7 Juni 1902*. The Hague, M. Nijhoff. 126 pp.
Complete list of past and present members and a bibliography and index to the society's publications.

Holmesdale Natural History Club

Bb240. **Powell, D., et al.** 1982. *Holmesdale Natural History Club: 125 years 1857–1982*. Reigate, the Museum. 40 pp.

Imperatorskoe Obshchestvo Lyubitelei Estestvoznaniya, Antropologii i Etnografii

Bb241. **Imperatorskoe Obshchestvo Lyubitelei Estestvoznaniya, Antropologii i Etnografii.** 1894. Izdaniya obshchestva s ... 1863 g. no ... 1894 g. — Bibliograficheskii ukazatel. By A. A. Ivanovskii. *Izvestiya Imperatorskago Obshchestva Lyubitelei Estestvoznaniya, Antropologii i Etnografii, [etc.]* 89: 59 pp.; ___. 1913. s 1 avgusta 1894 goda po 15 marta 1913 goda. Bibliograficheskii ukazatel'. By V. V. Bogdanov. *Ibid.* 127: 31 colls.

Bb242. ___. 1914. *Pyatidesyatiletie Imperatorskago Obshchestva Lyubitelei Estestvoznaniya, Antropologii i Etnografii 1863–1913 ... [Fiftieth anniversary of the Imperial Society of Friends of Natural History, Anthropology and Ethnography 1863–1913]. Torzhestvennoe zasedanie 15 Oktyabr' 1913 g. ... i proiskhodivshee odnovremenno chestvovanie 70–letiya zhizni prezidenta ... D. N. Anuchina.* By V. V. Bogdanov. Moscow. 253 pp., ports.

Bb243. [Not used.]

Isle of Man Natural History & Antiquarian Society

Bb244. **Flanagan, C. E.** 1984. Isle of Man Natural History and Antiquarian Society 1879–1979. *Proceedings of the Isle of Man Natural History and Antiquarian Society* 9(2): 169–181.

Isle of Wight Natural History & Archaeological Society

Bb245. **Hutchinson, A. L.** 1970. *Botany, birds, bugs and barrows on the Isle of Wight: The jubilee history of the Isle of Wight Natural History and Archaeological Society, 1919–1969.* Newport, I.O.W. 63 pp.

(Reale) Istituto Lombardo di Scienze e Lettere, Milan (*formerly* (Imperiale Regio) Reale Istituto di Scienze e Lettere)

Bb246. **Reale Istituto Lombardo di Scienze, Milan.** 1891. *Indice generale dei lavori dalla fondazione all' anno 1888 per autori e per materie.* Milan, U. Hoepli. 458 pp.; ___. 1902. *Indice generale dei lavori dal 1889–1900, con le aggiunte e correzioni all' indice generale 1803–1888, [etc.].* Milan. ix+136 pp.; ___. 1930. *Indice generale dei lavori dal 1901 al 1925 [etc.].* Milan, U. Hoepli. v+304 pp.

(Imperiale Regio/Reale) Istituto Veneto di Scienze, Lettere ed Arti

Bb247. **Reale Istituto Veneto di Scienze, Lettere ed Arti.** 1896. *Indice generale dei lavori pubblicati dall'anno accademico 1840–41 di fondazione, al 1893–94 per autori, per nomi e per materie ed elenco delle cariche accademiche.* By A. Carraro. Venice, Tip. Ferrari. 2 vols.; ___. 1939. *Indice generale dei lavori pubblicati dall'anno accademico 1894–95 al 1937–38.* By G. Mioni & C. Mugnaini. Venice. viii+273 pp.

Kettering & District Naturalists' Society & Field Club

Bb248. **Kettering & District Naturalists' Society & Field Club.** 1956. *The first fifty years: A history of the Kettering and District Naturalists' Society and Field Club ... A record of ... work, 1905–1955.* Kettering. 133 pp.

Királyi Magyar Természettudományi Társulat (Königlich–Ungarische Naturwissenschaftliche Gesellschaft)

Bb249. **Királyi Magyar Természettudományi Társulat.** 1884. *Természettudományi közlöny. Névjegyzék és tárgymutató [Name and subject index] a Királyi Magyar Természettudományi Társulat 1841–töl – 1883–ig, megjelent folyóiratához ... Készítette Lengyel I.* Budapest. 80 pp.

Kongelige Danske Videnskabernes Selskab

Bb250. **Kongelige Danske Videnskabernes Selskab.** 1843. *Det Kongelige danske Videnskabernes Selskabs Historie i dets første Aarhundrede 1742–1842.* By C. Molbech. Copenhagen. xvi+618 pp.

Bb251. ___. 1892. *Fortegnelse over de af det Kongelige Danske Videnskabernes Selskab i Tidsrummet 1742–1891 udgivne videnskabelige Arbejder.* By D. Andersen. Copenhagen. x+135 pp.

Bb252. ___. 1930. *Fortegnelse over det Kongelige Danske Videnskabernes Selskabs Publikationer,1742–1930.* By A. Lomholt. Copenhagen. 191 pp.

Bb253. ___. 1942–61. *Det Kongelige Danske Videnskabernes Selskab, 1742–1942. Samlinger til Selskabets Historie.* Ed. by A. Lomholt. Copenhagen. 4 vols.

Bb254. ___. 1957. *Det Kongelige Danske Videnskabernes Selskabs publikationer, 1917–1956, med samlet prisliste, 1742–1956.* Ed. by A. Lomholt. Copenhagen. 98 pp.

Kongelige Norske Videnskabers Selskab (*formerly* Trondhjemske Selskab)

Bb255. **Kongelige Norske Videnskabers Selskab.** 1910. *Utsigt over det Kongelige Norske Videnskabers Selskaps historie 1760–1910.* By J. Richter. Med et tillæg om de naturhistoriske samlinger av O. Nordgaard. Trondheim. 38 pp.

Bb256. ___. 1912. *Fortegnelse over Selskapets Skrifter 1760–1910.* By B. Schmidtnielsen. Trondheim. 80 pp.; *Continued* ___. 1977. *"Skrifter" 1911–1925. An index.* By O. Flo. Trondheim. 20 pp.

Bb257. ___. 1936. *Det Kongelige Norske Videnskabers Selskab. Forhandlinger og Skrifter 1926–1936. Acta et scripta Regiae Societatis Scientiarum Norvegicae 1926–1936. Tabula auctorum et index rerum.* By J. Ansteinsson. Trondheim. 51 pp. Appended to *Kongelige Norske Videnskabers Selskabs skrifter* 1936; *Continued* ___. 1950. ___*: 1937–46.* Trondheim. 44 pp.; ___. 1963. ___*: 1947–1956.* By K. Thalberg. Trondheim. 31 pp.; ___. 1969. ___ *1957–1966. [etc.].* By K. Thalberg. Trondheim. 29 pp.

Bb258. [Not used.]

Koninklijke Akademie van Wetenschappen, Amsterdam

Bb259. **Koninklijke Akademie van Wetenschappen. Amsterdam.** 1884. *Naam– en zaakregister op de Verslagen en mededeelingen der Koninklijke Akademie van Wetenschappen. Afdeeling: Natuurkunde. Tweede reeks. Deel 1–20 [1866–84].* Amsterdam, J. Muller. 29 pp.; *Continued* ___. 1892. *Naam- (en zaak-) register op de Verslagen en mededeelingen ... Afdeeling natuurkunde. Deel I–XXIX [1885–92].* Amsterdam. 55 pp.

Bb260. ___. 1920. *[Naam– en zaak–register op de Verhandelingen der Koninklijke Nederlandsche Akademie van Wetenschappen, Amsterdam. Afdeeling natuurkunde, vols. 1–20, & n.s. vols. 1–21, (1854–1920).]* Amsterdam. [ ]* pp.
[*Not examined – pagination not ascertained.]

Bb261. ___. 1937. *Catalogue of publications of the Royal Academy of Sciences at Amsterdam, Section of Natural Sciences, 1854–1936.* Amsterdam, Noord–Hollandsche Uitgevers Maatschappij. 29 pp.

Bb262. ___. 1943–44. *Naamregister van de Verhandelingen en bijdragen, uitgegeven door de afdeeling Natuurkunde in de jaren 1808–1942.* Amsterdam, Noord–Hollandsche Uitgevers Maatschappij. 3 pts.
Pt. II: Zittingsverslagen; pt. III: Proceedings.

Kungliga Fysiografiska Sällskapet i Lund

Bb263. **Kungliga Fysiografiska Sällskapet i Lund.** 1940. *Kungl. Fysiografiska Sällskapet i Lund, 1772–1940. Historisk overblick och personforteckningar.* By O. Gertz. Lund, H. Ohlssons Boktryckeri. 461 pp.

Kungliga Svenska Vetenskapsakademien

Bb264. **Frängsmyr, T.**, ed. 1989. *Science in Sweden: The Royal Swedish Academy of Sciences, 1739–1989.* Canton, MA, Science History Publications. viii+291 pp. *Bidrag till Kungliga Svenska Vetenskapsakademiens historia* no. 22.
Includes **S. Sörlin.** Scientific travel: The Linnéan tradition; **G. Broberg.** The Swedish Museum of Natural History; **T. Frängsmyr.** Swedish polar exploration; **B. Sundin.** Environmental protection and the national parks.

Bb265. **Kungliga Svenska Vetenskapsakademien.** 1831. *Vetenskapsakademiens skrifter 1739–1825.* By A. J. Ståhl. Stockholm. [iv]+502 pp.; *Continued* ___. 1917. *Kungl. Svenska Vetenskapsakademiens Skrifter, 1826–1917. Register.* By E. Colliander. Stockholm, Almqvist & Wiksell. xxiv+687 pp.; ___. 1965. *Kungl. Svenska Vetenskapsakademien. The Royal Swedish Academy of Science. Publications 1918–1960. Index.* By B.-S. Nordin-Petterson. Stockholm, Almqvist & Wiksell. viii+300 pp.

Bb266. ___. 1915. *Kungl. Svenska Vetenskapsakademien: Personenförteckningar 1739–1915.* Ed. by E. W. Dahlgren. Stockholm. viii+295+[2] pp., illus.

Bb267. ___. 1939. *Kungl. Svenska Vetenskaps Akademien: Forhistoria, grundläggning och första organisation.* By B. Hildebrand. Stockholm, Almqvist & Wiksell. xv+885 pp.

Bb268. ___. 1939. *Kungl. Vetenskapsakademiens äldre skrifter i utländska översättningar och referat.* By A. Holmberg. Uppsala. v+77 pp. & 7 pl. *K. Svenska Vetenskapsakademiens årsbok* 37 bilaga.

Bb269. ___. 1957. *Kungl. Svenska Vetenskapsakademien: Personenförteckningar 1916–1955.* Ed. by C. Skottsberg. Stockholm. ix+123 pp.

Bb270. ___. 1967. *Kungl. Svenska Vetenskapsakademiens historia, 1739–1818.* By S. Lindroth. Stockholm, Almqvist & Wiksell for the Academy. 2 vols. in 3.
Includes Vol. 1: 2. Linné och naturalhistorien, pp. 558–629; Kina och forskningsresorna, pp. 630–672; vol. 2. Linneansk efterklang, pp. 404–485.

Bb271. ___. 1968. *Försäljningskatalog över sällsynta böcker och skrifter utgivna av Kungl. Svenska Vetenskaps Akademien 1739–1967.* Med bihang *Bibliografi över Vetenskaps Akademiens Handlingar för åren 1739–1854.* Materialet sammanställt av P. Thulin. Stockholm, Almqvist & Wiksell. 123 pp.

Kungliga Vetenskaps-Societeten i Upsala (Regia Societas Scientiarum Upsaliensis)

Bb272. **Kungliga Vetenskaps-Societeten i Upsala.** 1889. *Catalogue méthodique des Acta [1740–50] et Nova Acta [and Arsskrift] Regiæ Societatis Scientiarum Upsalienses 1744–1889.* By A. G. S. Josephson. Uppsala, Josephsons Intikvariat. [ii]+33+[2] pp.

Bb273. ___. 1911. *Bicentenaire de la Société Royale des Sciences d'Upsala ...* By N. C. Duner. Uppsala. 62 pp. *Nova acta Regiae Societatis Scientiarum Upsalienses* ser. 4, vol. 3(1).

Bb274. ___. 1977. *Royal Society of Sciences of Uppsala: Catalogue of publications, 1716–1967.* By G. Karlberg. Uppsala. 98 pp. *Nova acta Regiae Societatis Scientiarum Upsalienses* ser. 5, vol. extra ordinum editum.

Kungliga Vetenskaps- och Vitterhets-Samhället i Göteborg

Bb275. **Kungliga Vetenskaps- och Vitterhets-Samhället i Göteborg.** 1928. Företeckning öfver *Göteborgs Kungl. Vetenskaps- och Vitterhets-samhälles ledamöter, 1774–1927.* Ed. by E. Nachmanson. Gothenburg. vi+61 pp. *Handlingar*, fjölden 4, vol.

32, no. 2.
Replaces. ___. 1902. Företeckning öfver *Göteborgs K. Vetenskaps– och Vitterhets–samhälles ledamöter, 1774–1901*. Ed. by S. Aberstén. Gothenburg. vi+61 pp. *Handlingar*, fjölden 4, no. 4.

Bb276. ___. 1978. *Kungliga Vetenskaps– och Vitterhets–Samhället i Göteborg, 1778–1874*. Ed. by N. Eriksson. Gothenburg. 416 pp.; *Continued* ___. 1985. ___, *1875–1953*. Ed. by N. Eriksson. Gothenburg. 364 pp.

Linnean Society of London

Bb277. **Beale, G.** 1991. Early French members of the Linnean Society of London, 1788–1802: Initial encounters. *Studies on Voltaire & the eighteenth century* 292: 129–138.
Also pub. in *Proceedings of the annual meeting, Western Society for French History* 18: 272–282, 1991.

Bb278. **Harris, T. M.** 1971. The minute books of the Linnean Club, from 1811–1955. *Biological journal of the Linnean Society* 3: 343–368.

Bb279. **Linnean Society of London.** 1867. *General index to the Transactions of the Linnean Society of London, vols. 1–25 [1791–1866]*. By G. Bentham. London. 107 pp.; *Continued* ___. 1876. _____ *vols. 26–30 [1868–75]*. London. 65 pp.

Bb280. ___. 1888. *General index to the first twenty volumes of the Journal (Botany) and the botanical portion of the Proceedings, November 1838 to June 1886, of the Linnean Society of London*. London. vii+428 pp.

Bb281. ___. 1896. *General index to the first twenty volumes of the Journal (Zoology), and the zoological portion of the Proceedings, Nov. 1838 to 1890, of the Linnean Society of London*. London. viii+437 pp.

Bb282. ___. 1911. *Catalogue of papers in the Transactions of the Linnean Society from 1791–1905 [etc.]*. London. 27 pp.

Bb283. ___. 1984+. *Linnean: Newsletter and proceeedings of the Linnean Society of London*. London. No. 1+.
Includes articles on various aspects of the Society's history.

Bb284. ___. 1988. *A bicentenary history of the Linnean Society of London*. By A. T. Gage & W. T. Stearn. London, Academic Pr. ix+242 pp., illus.
Replaces ___. 1938. *A history of the Linnean Society of London*. By A. T. Gage. London. ix+175 pp. & 11 pl.

Liverpool Biological Society

Bb285. **Kendall, C. E. Y.** 1928. Index to the Proceedings and transactions ... volumes 1–41, 1887–1927. *Proceedings and transactions of the Liverpool Biological Society* 42 [appendix]: 74 pp.

London Natural History Society

Bb286. **Castell, C. P., et al.** 1958. The Society from 1858–1957. *London naturalist* 37: 2–16.

Bb287. **Fitter, R. S. R.** 1953. Contributions to the bibliography of the natural history of the London area. 2: A subject index of the Society's journals, 1914–51. *Ibid.* 32: 94–113.

Bb288. **Payne, L. G.** 1948–49. The story of our Society. Pt. 1. *Ibid.* 1948: 3–21, 1948; ___. Pt. 2. *Ibid.* 1949: 10–22, 1949.

Marine Biological Association of the United Kingdom

Bb289. **Marine Biological Association of the United Kingdom.** 1887. The history of the foundation of the Marine Biological Association of the United Kingdom. *Journal of the Marine Biological Association* 1: 17–21.

Bb290. ___. 1913. List of publications recording the results of researches carried out under the auspices of the Marine Biological Association of the United Kingdom in their laboratory at Plymouth or on the North Sea cooast from 1886–1913. *Journal of the Marine Biological Association of the United Kingdom* N.S. 10: 143–178.

Medicinisch–Naturwissenschaftliche Gesellschaft zu Jena

Bb291. **Medicinisch–Naturwissenschaftliche Gesellschaft zu Jena.** 1899. *Namen- und Sachenregister zu den Bänden 1–30 [1864–96], nebst den entsprechenden Supplementheften, den Sitzungsberichten der Jenaischen Gesellschaft für Medizin und Naturwissenschaft für die Jahre 1878–86, sowie den Mitteilungen aus dem chemischen Laboratorium der Universität Jena.* Band I und II. By F. Römer. Jena, G. Fischer. iv+250 pp.

Moskovskoe Obshchestvo Ispytatelei Prirody (Société des Sciences Impériale des Naturalistes de Moscou)

Bb292. **Deinega, V. A.** 1936. *Ukazatel' statei, napechatannykh v izdaniyakh Moskovskogo Obshchestvo Ispytatelei Prirody (1805–1934).* Ch. 2. *Table des matières contenues dans les éditions de la Société des Naturalistes de Moscou (1805–1934). Seconde partie.* Moscow–Leningrad, Biomedgiz. 112 pp.; *Continued* ___. 1940. *Ukazatel' statei, ... (1805–1940).* Ch. 3. *Table des matières contenues dans les éditions de la Société des Naturalistes de Moscou (1805–1940). Troisième partie.* Moscow. 87 pp.; **I. D. Strashuna**, ed. 1957. *Ukazatel' statei i otdel'nykh izdanii ... (1940–1954).* Ch. 4. Moscow, Izd. Moskovsk. Universita. 193 pp.

Bb293. **Moskovskoe Obshchestvo Ispytatelei Prirody.** 1940. *Moskovskoe Obshchestvo Ispytatelei Prirody za 135 let ego sushchestvovaniya (1805–1940). Istoricheskii ocherk. Société des Naturalistes de Moscou (1805–1940). Aperçu historique.* By S. Yu. Lipshits. Moscow, Izd. Moskovskoe Obshchestvo Ispytatelei Prirody 131+[4] pp., ports.

Bb294. ___. 1955. 150–letie Moskovskogo Obshchestvo Ispytatelei Prirody i ego rol' v razvitii floristiki, sistematiki i botanicheskoi geografii. [150 years of Moscow's Natural History Society and its role in the development of floristics, systematics and phytogeography]. By S. Yu. Lipshits. *Byulleten' Moskovskogo Obshchestvo Ispytatelei Prirody. Otdel biologicheskii* 60 (5): 5–15; (6): 3–26.

Bb295. **Société Impériale des Naturalistes de Moscou.** 1882. *Table générale et systématique des matières contenues dans les premiers 56 volumes (années 1829–1881) du Bulletin de la Société impériale des naturalistes de Moscou.* By E. Ballion. Moscow, S. Archipoff. 110 pp.

Bb296. **Varsanof'eva, V. A.** 1955. *Moskovskoe Obshchestvo Ispytatelei Prirody i ego zhachenie v razvitii otechestvennoi nauki. [The Moscow Society of Naturalists and its importance in the development of our native science].* Moscow, Izd. Moskovskogo Universiteta. 103 pp. & 2 pl.

La Murithienne, Société Valaisanne des Sciences Naturelles

Bb297. **La Murithienne, Société Valaisanne des Sciences Naturelles.** 1962. Bulletin de la Murithienne ... Répertoire des articles publiés dans les fascicules I à LXXVIII (1861–1961). By A. Donnet. *Bulletin de la Murithienne* 73/80: 89+[2] pp.

Bb298. **Praz, J. C.** 1983. A l'occasion de la publication du centième Bulletin de la Murithienne. *Ibid.* 100: 3–8.

Musealverein für Krain

Bb299. **Musealverein für Krain.** 1898. Repertorium zur 50 jährigen Geschichtschreibung Krains, 1848–1898 ... Enthaltend ein alphabetisches

Autorenverzeichnis und ein Namens–, Orts– und Sachregister zu den "Mittheilungen des historischen Vereines für Krain" (1846 bis 1868), den "Mittheilungen des Musealvereines für Krain" (1866, 1889 bis 1898), den "Jahresheften (1856, 1858 und 1862) des Vereines des Krainisches Landes Museums" und der "Argo" ... (1892 bis 1898), dann auch zu den wissenschaftlichen Aufsätzen ... über Krain ... in "Blätter aus Krain" (1857 bis 1865) und der "Carniolia" (1838 bis 1844). *Mitteilungen des Musealvereins für Krain* 11 Suppl: 76 pp.

Nassauischer Verein für Naturkunde (*formerly* Verein für Naturkunde im Herzogthum Nassau)

Bb300. **Nassauischer Verein für Naturkunde.** 1929. Inhaltsverzeichnis zu Band 1–80 [1844–1929] der Jahrbücher [etc.]. *Jahrbücher des Nassauischen Vereins für Naturkunde* 80: 251–276.

Natural History Society of Edinburgh

Bb301. **Allen, D. E.** 1978. James Edward Smith and the Natural History Society of Edinburgh. *Journal of the Society for the Bibliography of Natural History* 8: 483–493.

Natural History Society of Northumberland, Durham, & Newcastle upon Tyne

Bb302. **Natural History Society of Northumberland, Durham, & Newcastle upon Tyne.** 1929. *History of the Natural History Society of Northumberland, Durham, & Newcastle upon Tyne, 1829–1929.* By T. R. Goddard. Newcastle–upon–Tyne, A. Reid. xvi+195 pp. & 12 pl.

Naturforschende Gesellschaft zu Bamberg (*formerly* Naturforschender Verein zu Bamberg)

Bb303. **Unger, E.** 1984. 150 Jahre Naturforschende Gesellschaft Bamberg. *Bericht der Naturforschende Gesellschaft zu Bamberg* 59: 1–37.

Naturforschende Gesellschaft in Basel

Bb304. **Naturforschende Gesellschaft in Basel.** 1867. *Festschrift herausgegeben von der Naturforschenden Gesellschaft, zur Feier des fünfzigjährigen Bestehens 1867.* Basle. 166 pp. & 1 pl.

Includes **P. Merian.** Geschichte der Naturforschenden Gesellschaft in Basel während der ersten fünfzig Jahre ihres Bestehens.

Bb305. **___.** 1892. *Festschrift zur Feier des fünfundsiebzigjährigen Bestehens der Naturforschenden Gesellschaft in Basel.* Basle. iii≠215 pp.

Bb306. **___.** 1930. *Autoren– und Sachregister der Hefte 1–10, 1834–1852, der Berichte über die Verhandlungen und der Bände 1–40, 1852–1929, der Verhandlungen der Naturforschenden Gesellschaft in Basel.* By C. Walter. Basle, E. Birkhäuser & Cie. 126pp.

Naturforschende Gesellschaft in Bern

Bb307. **Naturforschende Gesellschaft in Bern.** 1854. Alphabetisches Personal– und Sach–Register über die Mitteilungen ... aus den jahren 1843–1854. *Mittheilungen der Naturforschenden Gesellschaft Bern* 1854: 169–184.; *Continued* ___. 1879. Alphabetisches ... –Register ... von Neujahr 1855 bis Neujahr 1880 [etc.]. *Ibid.* 1879: 99–132.; **___.** 1890. Alphabetisches ... –Register ... von Neujahr 1880 bis Neujahr 1891 [etc.]. *Ibid.* 1890: 153–166.; **___.** 1911. Alphabetisches ... –Register ... von 1. Jan. 1891 bis 31. Dez. 1911 [etc.]. *Ibid.* 1911: xxiii pp.

Naturforschende Gesellschaft in Danzig

Bb308. **Eichler, H.** 1984. Die Danziger Naturforschende Gesellschaft, 1743–1789. *Jahrbuch für Geschichte der sozialistischen Länder Europas* 28: 58–78.

Bb309. **Naturforschende Gesellschaft in Danzig.** 1893. *Geschichte der Naturforschenden Gesellschaft in Danzig, 1743–1892.* By E. Schumann. Danzig.

[viii]+149 pp., illus. *Festschrift zur Feier des 150–jährigen Bestehens der Naturforschenden Gesellschaft in Danzig am 2 Januar 1893. Schriften* N.F. vol. 8, No. 2.

Bb310. ___. 1924. *Hauptverzeichnis aller Veröffentlichungen der Naturforschenden Gesellschaft seit ihrer Begründung 1743 bis zum Abschluss des XV. Bandes der neuen Folge ihrer Schriften 1922.* By Prof. Dr. Lakowitz. Danzig. 51 pp.

Naturforschende Gesellschaft zu Emden

Bb311. **Naturforschende Gesellschaft zu Emden.** 1915. *Festschrift der Naturforschenden Gesellschaft zu Emden ... 1814–1914.* Emden. 94+240 pp.
Includes Geschichte der Naturforschenden Gesellschaft zu Emden; Die Sammlungen der Naturforschenden Gesellschaft zu Emden; Verzeichnis sämtlicher Mitgleider seit Bestehen der "Naturforschenden Gesellschaft zu Emden"; Personal Bestand 1814–1914; Grössere Bemerkenswerte veröffentlichungen und Vorträge in den Jahresberichten 1885–1913; [etc.].

Naturforschende Gesellschaft zu Freiburg

Bb312. **Naturforschende Gesellschaft zu Freiburg.** 1921. *Die Naturforschende Gesellschaft zu Freiburg i. Br. in den hundert Jahren ihres Bestehens 1821–1921*, im auftrag des Vorstandes der Gesellschaft dargestellt von L. Neumann. Freiburg, Speyer & Kaerner. 54 pp.

Naturforschende Gesellschaft zu Görlitz

Bb313. **Naturforschende Gesellschaft zu Görlitz.** 1911. Jubiläumsband 1811–1911 der Naturforschenden Gesellschaft zu Görlitz. *Abhandlungen der Naturforschenden Gesellschaft zu Görlitz* vol. 27.
Includes **W. Freise.** Die Naturforschende Gesellschaft zu Görlitz und ihre Präsidenten, vi pp. & 15 ports.

Naturforschende Gesellschaft Graubündens

Bb314. **Naturforschende Gesellschaft Graubündens.** 1917. Verzeichnis der Aufsätze, Mitteilungen und Nekrologe, welche in den Bänden 1–57, Neue Folge [1853/54–1917], der Jahresberichte der Naturforschenden Gesellschaft Graubündens erschienen sind, und Übersicht der Bände und Vereinsjahre. By P. Hitz. *Jahresbericht der Naturforschenden Gesellschaft Graubündens* n.f. 57: xxi–xxxiv.

Naturforschende Gesellschaft zu Halle

Bb315. **Naturforschende Gesellschaft zu Halle.** 1879. *Festschrift zur Feier des hundertjährigen Bestehens der Naturforschende Gesellschaft in Halle a/S.* Halle, M. Niemeyer. xlviii+342 pp. & 12 pl.
Includes **F. Marchand.** Vorbericht über die Naturforschende Gesellschaft zu Halle; Verzeichniss der selbständig erscheinenen Schriften der Gesellschaft [1853–1879].

Naturforschende Gesellschaft Solothurn

Bb316. **Naturforschende Gesellschaft Solothurn.** 1923. *Zum 100 jährigen Jubiläum, 1823–1923.* Solothurn, Gassmann AG. 574 pp., illus.
Includes **J. V. Keller.** Geschichtlicher Überblick über die Entstehung und die Tätigkeit der Naturforschenden Gesellschaft Solothurn.

Naturforschende Gesellschaft in Zurich

Bb317. **Naturforschende Gesellschaft in Zürich.** 1892. *Generalregister der Publikationen (Neujahrsblätter – Mittheilungen – Vierteljahrschrift) ... und Übersicht ihres Tauschverkehres.* Zurich. 92 pp.

Bb318. ___. 1896. *Die Naturforschende Gesellschaft in Zürich 1746–1896.* By E. Rübel. Zurich. x+274 pp. & 6 pl. *Vierteljahrsschrift der Naturforschenden Gesellschaft in Zürich* 41 (Jubelband), Teil I.

Bb319. **Naturforschende Gesellschaft in Zürich.** 1946. Geschichte der Naturforschenden Gesellschaft in Zürich (1746–1946). By E. Rübel. *Ibid.* 91 Beiheft: 123 pp.

Naturforschender Verein in Brünn

Bb320. **Naturforschender Verein in Brünn.** 1912. Die Geschichte des Naturforschenden Vereines in Brünn in den Jahren 1862–1912. By H. Iltis. *Verhandlungen des Naturforschenden Vereines in Brünn* 50: 297–359.
Includes Verzeichniss derjenigen wissenschaftlichen Arbeiten, welche in den bisher erscheinenen 50 Bänden [1862–1911] der Verhandlungen des naturforschenden Vereines in Brünn unter den "Abhandlungen" veröffentlicht worden sind, pp. 343–359.

Naturforscher Verein zu Riga

Bb321. **Naturforscher Verein zu Riga.** 1895. *Festschrift des Naturforscher–Vereins zu Riga in Anlass seines 50–jährigennn Bestehens am 27 März (8 April) 1895*. Riga. vi+296 pp. & 4 pl.
Includes **G. Schweder**, Sen. Rückblick aus die fünfzigjährige Wirksamkeit des Naturforscher Vereins zu Riga.

Bb322. ___. 1932. *Korrespondenzblatt des Naturforscher– Vereins zu Riga. Register zu Bde 1–60 [1845–1930] nebst Verzeichnis der übrigen Druckschriften der vereins.* By A. Meder. Riga. viii+88 pp.

Naturhistorisch–Medicinischer Verein zu Heidelberg

Bb323. **Naturhistorisch–Medicinischer Verein zu Heidelberg.** 1909. Verzeichnis der Veröffentlichen des Naturhistorisch–Medicinischen Vereins ... von 1856–1909. (A.F. Bd. I–VI; N.F. Bd. I–IX der Verhandlungen). Mit Autoren–, Orts–, Sach und Materien–Register, [etc.]. By D. Haberle. *Verhandlungen des Naturhistorisch–Medicinischen Verein zu Heidelberg* N.F. 9: 641–731.

Naturhistorische Gesellschaft zu Hannover

Bb324. **Naturhistorische Gesellschaft zu Hannover.** 1897. *Festschrift. Geschichte und 44.–47. Jahresbericht der Naturhistorischen Gesellschaft zu Hannover.* Hanover. 183+244 pp.
Includes **H. Ude.** Die Geschichte der Naturhistorischen Gesellschaft zu Hannover von 1797 bis 1897, pp. 1–183.

Bb325. ___. 1937. *Festschrift zur 140 Jahrfeier der Gründung der Gesellschaft (11. Dezember 1797).* Ed. by W. Pieper. Hannover. 131 pp. *Jahresbericht der Naturhistorischen Gesellschaft zu Hannover* vols.88 & 89.
Includes **H. Küthmann.** Die Entwicklung der Naturhistorischen Gesellschaft zu Hannover in de Jahren von 1897–1937; **W. Pieper.** Über die Bibliothek der Naturhistorischen Gesellschaft.

Naturhistorische Gesellschaft, Nuremberg

Bb326. **Naturhistorische Gesellschaft, Nuremberg.** [1901.] *Säcular–Feier der Naturhistorische Gesellschaft ... 1801–1901. Festschrift, [etc.].* Nuremberg. xlix+280 pp. & 32 pl.
Includes Einleitung [Die Naturhistorische Gesellschaft Nürnberg (1801–1901)].

Bb327. ___. 1985. *Bibliographie der Naturhistorischen Gesellschaft Nürnberg e. V. die Veröffentlichungen der Naturhistorischen Gesellschaft Nürnberg und ihrer Abteilungen, 1801–1983.* By E. Feist & H. Kohler. Nuremberg. 63 pp.

Naturhistorischer Verein der Preussischen Rheinlande, Westfalens und des Regierungs–Bezirks Osnabrück

Bb328. **Naturhistorischer Verein der Preussischen Rheinlande und Westfalens.** 1885. *Autoren– und Sachregister zu Band. 1–40 (Jahrg. 1844–1883) der Verhandlungen des Naturhistorischen Vereins der preussischen Rheinlande und*

Westphalens, des Correspondenzblattes desselben Vereins und der Sitzungsberichte der Niederrheinischen Gesellschaft für Natur- und Heilkunde zu Bonn. Bonn, C. Georgi. 210 pp.

Naturwissenschaftliche Gesellschaft Isis zu Bautzen

Bb329. **Naturwissenschaftliche Gesellschaft Isis zu Bautzen.** 1896. *Festschrift der Naturwissenschftlichen Gesellschaft Isis in Bautzen zur Feier ihres fünfzigjährigen Bestehens.* Bautzen. 88 pp.
Includes **P. Menzel.** Die Isis in den Jahren 1846 bis 1895.

Bb330. ___. 1921. *1846–1919. Festschrift zur Feier des 75 jährigen Bestehens der Naturwissenschaftlichen Gesellschaft Isis in Bautzen, [etc.].* Bautzen. 98 pp.

Naturwissenschaftliche Gesellschaft Isis zu Dresden

Bb331. **Naturwissenschaftliche Gesellschaft Isis zu Dresden.** 1885. *Festschrift der Naturwissenschaftlichen Gesellschaft Isis in Dresden zur Feier ihres 50 jährigen Bestehens am 14. Mai 1885.* Dresden. iv+178 pp. & 4 pl.

Bb332. ___. 1934. *Festschrift zur Feier des 100 jährigen Bestehens der Naturwissenschaftlichen Gesellschaft Isis in Dresden.* Ed. by A. Schade. Dresden. 222 pp., illus.
Includes **R. Zaunick.** Gründung und Gründer der Naturwissenschaftlichen Gesellschaft Isis in Dresden vor hundert Jahren.

Naturwissenschaftliche Gesellschaft in Winterthur

Bb333. **Naturwissenschaftliche Gesellschaft in Winterthur.** 1935. Generalregister zu Heft. 1–20, ... nach Autoren geordnet. *Mitteilungen der Naturwissenschaftlichen Gesellschaft in Winterthur* 20: 214–219.

Naturwissenschaftliche Gesellschaft Isis zu Meissen

Bb334. **Naturwissenschaftliche Gesellschaft Isis zu Meissen.** 1895. *Festschrift der Naturwissenschaftlichen Gesellschaft Isis zu Meissen zur Feier ihres fünfzigjährigen Bestehens.* Ed. by F. F. Wolf. Meissen. 170 pp.
Includes **F. F. Wolf.** Die "Isis" und ihrer Thätigkeit von 1845–1895.

Naturwissenschaftlicher Verein für Bielefeld

Bb335. **Naturwissenschaftliche Verein für Bielefeld.** 1983. Die Geschichte des "Naturwissenschaftlichen Vereins für Bielefeld und Umgebung e. V." By P. Finke. *Bericht des Naturwissenschaftlichen Vereins für Bielefeld und Umgebung* 26 (1981): 11–159.

Naturwissenschaftlicher Verein zu Bremen

Bb336. **Naturwissenschaftlicher Verein zu Bremen.** 1889. Inhalts-Verzeichnisse zu Band 1 bis 10 [1868–1889] der Abhandlungen. *Abhandlungen herausgegeben vom Naturwissenschaftlichen Vereine zu Bremen* 10: 575–619.; *Continued* ___. 1911. Inhalts-Verzeichnis zu Band 11 bis 20 [1890–1911] der Abhandlungen. *Ibid.* 20: [ ]* pp.
[*Not examined – pagination not ascertained.]

Naturwissenschaftlicher Verein in Carlsruhe

Bb337. **Naturwissenschaftlicher Verein in Carlsruhe.** 1888. Namensverzeichniss von Rednern in den Sitzungen und von Verfassern der Abhandlungen, mit Bezeichung des Gegenstandes. Band 1, 1864 bis Band 10, 1888. *Verhandlungen der Naturwissenschaften Vereins in Carlsruhe* 10: 433–446; *Continued* ___. 1900. Namensverzeichniss ... Band 11, 1896 bis Band 13, 1900. *Ibid.* 13: 637–645; ___. 1907. Namensverzeichniss zu den Bänden 14–20 [1900–06/07]. *Ibid.* 20: 113–119.

Naturwissenschaftlicher Verein für das Fürstentum Lüneberg

Bb338. **Naturwissenschaftlicher Verein für das Fürstentum Lüneberg.** 1901. *Zur Erinnerung an das 50jährige Bestehen des Naturwissenschaftlichen Vereins für das Fürstentum Lüneberg, 1851–1901.* Lüneberg. 101 pp.
Includes **H. Steinworth.** Geschichte des Vereins, 1851–1901.

Naturwissenschaftlicher Verein (*formerly* Gesellschaft) in Hamburg

Bb339. **Naturwissenschaftlicher Verein in Hamburg.** 1887. Festschrift zur Feier des fünfzigjährigen Bestehens des Naturwissenschaftlichen Vereins in Hamburg, 18 November 1887. *Abhandlungen aus dem Gebiete der Naturwissenschaften herausgegeben von dem naturwissenschaftlichen Verein in Hamburg* 10: [523] pp. & 13 pl.
Includes **H. Bolau.** 1837–1887: Zur Geschichte des Naturwissenschaftlichen Vereins in Hamburg. (32) pp.

Naturwissenschaftlicher Verein zu Krefeld

Bb340. **Naturwissenschaftlicher Verein zu Krefeld.** 1908. *Naturwissenschaftlicher Verein zu Krefeld ... gegründet am 1. März 1858. Festschrift zur feier des 50 jährigen Bestehens.* Krefeld. 259 pp.
Includes **G. Boble.** Der Geschichte des Vereins.

Naturwissenschaftlicher Verein für Schleswig–Holstein

Bb341. **Naturwissenschaftlicher Verein für Schleswig–Holstein.** 1904. Register zu Band 1–12 [1873–1902]. By R. Gottschaldt. *Schriften des Naturwissenschaftlichen Vereins für Schleswig–Holstein* 12 [appx.]: 32 pp.

Naturwissenschaftlicher Verein für Steiermark

Bb342. **Naturwissenschaftlicher Verein für Steiermark.** 1884. *Haupt–Repertorium über sämmtliche Vorträge, Abhandlungen und fachwissenschaftliche Notizen, welche sich in den Heften 1 bis einschliesslich 20 (den Jahrgängen 1863 bis einschliesslich 1883) der Mittheilungen des naturwissenschaftlichen Vereines für Steiermark befinden. Nach Materien und den Namen der Autoren geordnet.* Graz. 50 pp. Appended to *Mitteilungen des Naturwissenschaftlichen Vereines für Steiermark* 1883.; ___. 1905. *Haupt–Repertorium ... Heften 21 bis einschliesslich 40 (den Jahrgängen 1884 bis einschliesslich 1903) der Mitteilungen [etc.].* Graz. 39 pp.

Naturwissenschaftlicher Verein zu Zerbst

Bb343. **Naturwissenschaftlicher Verein zu Zerbst.** 1912. *Festschrift zum Feier des fünfzigjährigen Bestehens des Naturwissenschaftlichen Vereins zu Zerbst ... 1862–1912.* Zerbst. 87 pp.
Includes Geschichte des Vereins.

Natuurkundig Gezelschap te Utrecht

Bb344. **Natuurkundig Gezelschap te Utrecht.** 1977. *N G 200 – Natuurkundig Gezelschap te Utrecht 1777–1977.* Ed. by C. Van der Leun. Utrecht. 143 pp.
Chapters by R. Hooykaas, J. G. van Cittert–Eymers, H. A. M. Snelders, N. G. van Kampen & P. H. Kylstra.

Nederlandsche Natuurhistorische Vereniging

Bb345. **Nederlandsche Natuurhistorische Vereniging.** 1941. *Amsterdam natuurhistorisch gezien: Gedenboek uitgegeven ter gelegenheid van het 40–jarig bestaan van de afd. Amsterdam der Nederlandse Natuurhistorische Vereniging, 1901 — 26–Januari–1941.* Amsterdam, Scheltema & Holkema. xxii+314 pp.

Norske Videnskaps–Akademi i Oslo (*formerly* Videnskabs–Selskabet i Christiania)

Bb346. **Norske Videnskaps–Akademi i Oslo.** 1940. *Register til Avhandlinger og skrifter utgitt av det Norske Videnskaps–Akademi i Oslo, 1925–1939.* Oslo. 32 pp.

Bb347. **Videnskabs–Selskabet i Christiania.** 1925. *Register til Forhandlinger og Skrifter, utgit av Videnskapsselskapet i Kristiania, 1858–1924.* By T. Knudsen. Christiania. [ii]+80 pp.

North Staffordshire Field Club

Bb348. **North Staffordshire Field Club.** 1916. *Jubilee volume (1865–1915).* Ed. by S. A. H. Burne, J. T. Stobbs, & H. V. Thompson. Stafford. viii+245 pp., illus.

Northamptonshire Natural History Society & Field Club

Bb349. **Druce, G. C.** 1918. Formation of the Northamptonshire Natural History Society. *Journal of the Northamptonshire Natural History Society and Field Club* 19: 131–142.

Northern Naturalists' Union, Great Britain

Bb350. **Mall, D.** 1984. Sixty years of the Northern Naturalists' Union (1924–84). *Vasculum* 69(3): 17–25.

Oberhessische Gesellschaft (*formerly* Verein) für Natur– und Heilkunde

Bb351. **Oberhessische Gesellschaft für Natur– und Heilkunde.** 1910. *Register zu den Bänden 1–34 (alte Folge, Jahrgang 1849–1904).* By E. Schäfer. Giessen. 88 pp. *Bericht der Oberhessischen Gesellschaft für Natur– und Heilkunde* vol. 34 Appendix.

Obshchestvo Estestvoispytatelei pri Imperatorskom Kazanskom Universitete

Bb352. **Obshchestvo Estestvoispytatelei pri Imperatorskom Kazanskom Universitete.** 1882. *Ukazatel' izdanii ... 1869–1882. Verzeichniss der ... herausgegeben Schriften 1869–1882.* Kazan. 30 pp.

Bb353. **___.** 1894. *Obzor deyatel'nosti Obshchestva ... za pervoe dvadtsatipyatiletie ego sushchestvovaniya, 1869–1894. [Review of the work of the Society of Naturalists at the Imperial Kazan University during the first twenty–five years of its existence].* Kazan. 131+82 pp.

Includes chronological and subject indexes to the Society's *Trudy*.

Obshchestva Ispytatelei Prirody pri Imperatorskom Khar'kovskom Universitete

Bb354. **Obshchestva Ispytatelei Prirody pri Imperatorskom Khar'kovskom Universitete.** 1955. *Sistematicheskii ukazatel' k Trydam Obshchestva Ispytatelei Prirody pri Khar'kovskom Universitete za 1869–1930 gg.* By M. G. Shval'. Khar'kov, Izd. Khar'kovsk Universiteta. 44 pp.

Österreichische Akademie der Wissenschaften (*formerly* (Kaiserliche) Akademie der Wissenschaften in Wien)

Bb355. **Akademie der Wissenschaften in Wien.** 1929. *Generalregister zu den Bänden 1–100, 1850–1926, der Denkschriften der mathematisch–naturwissenschaftliche Klasse.* Vienna. 148 pp.

Bb356. **Kaiserliche Akademie der Wissenschaften in Wien.** 1854. *Sitzungsberichte der Kaiserlichen Akademie der Wissenschaften, Wien. Mathematisch–naturwissenschaftliche Classe: Register zu den ersten X Bänden [1848–53], [etc.].* Vienna. 55 pp.; *Continued* ___. 1856. ___*: Register zu den zweiten 10 Bänden [1853–56], [etc.].* Vienna. 58 pp.; **___.** 1859. ___*: Register ... 21 bis 30 [1856–58], [etc.].* Vienna. 98 pp.; **___.** 1862. ___*: Register ... 31–42, 1858–60, [etc.].* Vienna. 107 pp.; **___.** 1865. ___*: Register ... 43–50, 1861–64, [etc.].* Vienna. 157 pp.; **___.** 1870. ___*: Register ... 51–60, 1865–69, [etc.].* Vienna. 199 pp.; **___.** 1872. ___*: Register ... 61–*

64, 1870–71, [etc.]. Vienna. 98 pp.; ___. 1878. ___*: Register ... 65–75, 1872–77, [etc.]*. Vienna. 107 pp.; ___. 1880. ___*: Register ... 76–80, 1877–79, [etc.]*. Vienna. 63 pp.; ___. 1882. ___*: Register ... 81–85, 1880–82, [etc.]*. Vienna. 90 pp.; ___. 1885. ___*: Register ... 86–90, 1882–84, [etc.]*. Vienna. 81 pp.; ___. 1888. ___*: Register ... 91–96, 1885–87, [etc.]*. Vienna. 88 pp.; ___. 1892. ___*: Register ... 97–100, 1888–91, [etc.]*. Vienna. 110 pp.; ___. 1897. ___*: Register ... 101–105, 1892–96, [etc.]*. Vienna. 144 pp.; ___. 1902. ___*: Register ... 106–110, 1897–1901, [etc.]*. Vienna. 144 pp.; ___. 1907. ___*: Register ... 111–115, 1902–06, [etc.]*. Vienna. 166 pp.; ___. 1912. ___*: Register ... 116–120, 1907–11, [etc.]*. Vienna. 168 pp.; ___. 1917. ___*: Register ... 121–125, 1912–16, [etc.]*. Vienna. 222 pp.; ___. 1922. ___*: Register ... 126–130, 1917–21, [etc.]*. Vienna. 96 pp.; ___. 1927. ___*: Register ... 131–135, 1922–26, [etc.]*. Vienna. 124 pp.

Bb357. ___. 1869. *Verzeichniss sämmtlicher von der Kais. Akademie der Wissenschaften seit ihrer Gründung bis letzten October 1868 veröffentlichten Druckschriften [etc.]*. Vienna, C. Gerold's Sohn. 300 pp.

Bb358. ___. 1897. *Geschichte der Gründung und der Wirksamkeit der Kaiserlichen Akademie der Wissenschaften während der ersten fünfzig Jahre ihres Bestandes*. By A. Hüber. Vienna. [i]+176 pp.

Bb359. ___. 1911. *Register zu den Bänden 1–60 (Jahrgänge 1851–1910) des Almanachs und der Feierlichen Sitzungen der Kaiserlichen Akademie der Wissenschaften*. Vienna, A. Hölder. xii+239 pp.

Bb360. ___. 1915–16. *Verzeichnis der von der Kaiserlichen Akademie der Wissenschaften in Wien herausgegeben oder subventionierten Schriften*. Vienna, A. Hölder. 2 vols.
Vol. 1. 1847 bis 1914; vol. 2. 1915–1926.

Bb361. **Österreichische Akademie der Wissenschaften.** 1947. *Geschichte der Akademie der Wissenschaften in Wien, 1847–1947*. By R. Meister. Vienna, A. Holtzhausen. 411 pp. & 58 pl. *Denkschriften der Gesamtakademie* vol. 1.
Includes III. Übersicht über die Mitgleider, Einrichtungen und Unternehmungen der Akademie, pp. 255–365.

Physikalisch–Medicinische Gesellschaft, Wurzburg

Bb362. **Physikalisch–Medicinische Gesellschaft, Wurzburg.** 1899. *Festschrift zur Feier ihres fünfzighährigen Bestehens*. Wurzburg. 228 pp. & 12 pl.

Physikalisch–Medicinische Sozietät zu Erlangen

Bb363. **Physikalisch–Medicinische Sozietät zu Erlangen.** 1908. *Festschrift der Physikalisch–Medizinischen Sozietät zu Erlangen zur Feier ihres 100–jährigen Bestehens am 27. Juni 1908*. Erlangen, M. Mencke. 124 pp. & 12 pl. *Sitzungsberichte der Physicalisch–Medicinischen Societät zu Erlangen* 40.
Includes **M. Noether.** Geschichte der Physikalisch–Medizinischen Societät zu Erlangen im ersten Jahrhundert ihres Bestehens. 1808–1908.

(Königliche) Physikalisch–Ökonomische Gesellschaft zu Königsberg

Bb364. **Degen, H.** 1960. Vor hundert Jahren: Die Naturforscher und Ärzteversammlung zu Königsberg 1860 nach dem Bericht eines Teilnehmers. *Naturwissenschaftliche Rundschau* 1960 (9): 331–340.

Bb365. **Physikalisch–Ökonomische Gesellschaft zu Königsberg.** 1885. Generalregister zu den publikationen der Physikalisch–Ökonomischen Gesellschaft. 1860–1884. By A. Jentzsch. *Schriften der Physikalisch–Ökonomischen Gesellschaft zu Königsberg* 25: 32 pp.; *Continued* ___. 1911. *General register zu den Jahrg. 26–50, (1885–1909)*. Leipzig & Berlin. [vi]+63 pp.

Bb366. [Not used.]

Pollichia, ein Naturwissenschaftlicher Verein der Rheinpfalz (*formerly* Baierischen Pfalz)

Bb367. **Pollichia, ein Naturwissenschaftlicher Verein der Rheinpfalz.** 1892. *Festschrift zur fünfzigjährigen Stiftungsfeier der Pollichia, Naturwissenschaftlicher Verein der Rheinpfalz.* Durkheim. 65 pp.
Includes Zur Geschichte des Vereins.

Bb368. ___. 1900. *Festschrift zur sechzigjähren stiftungsfeier der Pollichia, naturwissenschaftlichen Vereins der Rheinpfalz.* Dürkheim. 81 pp.
Includes Zur Geschichte des Vereins; Über die Geschichte der Pollichia im letzten Jahrhundert.

Polska Akademija Umiejetnosci, Krakow (*formerly* Akademija Umiejetnosci)

Bb369. **Polska Akademija Umiejetnosci, Krakow.** 1948. *Katalog wydawnictw Polskiej Akademii Umiejetnosci 1873–1947.* By S. Mikucki & R. J. Wojtusiak. Cracow. 2 vols.

Bb370. ___. 1974. *Polska Akademia Umiejetnosci, 1872–1952: Nauki lekarskie, scisle, przyrodnicze i o ziemi. Materialy sesjijubileuszowej, Krakow, 14 XII 1972.* Cracow, Ossolineum. 435 pp.

Poznanskie Towarzystwo Przyjaciól Nauk (*formerly* Towarzystwo Przyjaciól Nauk Poznanskiego *formerly* Naturhistorischer (Naturwissenschaftlicher) Verein der Provinz Posen)

Bb371. **Naturwissenschaftlicher Verein der Provinz Posen.** 1887. *Festschrift zum fünfzigjährigen Jubiläum des naturwissenschaftlichen Vereins der Provinz Posen. 1837 bis 1887.* Poznan. 238 pp.
Includes Prof. Dr. **Magener.** Geschichte des Naturwissenschaftlichen Vereins der Provinz Posen.

Bb372. **Poznanskie Towarzystwo Przyjaciól Nauk.** 1947. *Poznanskie Towarzystwo Przyjaciól Nauk w latach 1927–1947.* By Z. Lisowski. Poznan. 65 pp.

Bb373. ___. 1950. *Katalog wydawnictw Poznanskiego Towarzystwa Przyjaciol Nauk ad roku 1946.* Poznan, Uniwersytet Poznanskie. 61 pp.

Bb374. **Towarzystwo Przyjaciól Nauk Poznanskiego.** 1928. *Towarzystwo Przyjaciól Nauk w Poznaniu w latach 1857–1927.* By A. Wojtkowski. Poznan. 535+xxiv pp.
Includes list of the Society's publications.

Provinciaal Utrechtsche Genootschap van Kusten en Wetenschappen

Bb375. **Provinciaal Utrechtsche Genootschap van Kusten en Wetenschappen.** 1906. *Register op de Aanteekeningen van het verhandelde in de vergarderingen der sectie's van het Provinciaal Utrechtsch Genootschap van Kusten en wetenschappen over de jaren 1845 tot en met 1905. Voorafgegaan door een opgave van de titels der bekroonde en uitgegeven werken en van openingsreden der algemeene vergarderingen.* Utrecht, van Boekhoven. 83 pp.

Quekett Microscopical Club

Bb376. **Cooke, M. C.** 1899. Early memories of the Q. M. C. *Journal of the Quekett Microscopical Club* (2)7: 229–238.

Bb377. **Cunningham, G. J.** 1965. The centenary of the Quekett Club. *Annals of the Royal College of Surgeons of England* 37: 166–168.

Bb378. **Quekett Microscopical Club.** 1882. *General index to the Journal of the Quekett Microscopical Club, first series, vol. 1–6, 1868–1881.* London. 22 pp

Bb379. **Ramsbottom, J. R.** 1931. The Society of Amateur Botanists and the Quekett Microscopical Club. *Journal of the Quekett Microscopical Club* (2) 16: 215–230.

Ray Society

Bb380. **Ray Society.** 1954. *A bibliographical history of the Ray Society.* By R. Curle. London, Ray Society. vi+101 pp.
Supplementary notes appeared in J. Stevens Cox's *Literary repository.*

Real Academia de Ciencias Naturales y Artes, Barcelona

Bb381. **Real Academia de Ciencias Naturales y Artes, Barcelona.** 1895. *Historia de la Real Academia de Ciencias Naturales y Artes. Memoria [etc.].* By J. Balari y Jovany. Barcelona. 203 pp.

Bb382. ___. 1915. *Fiestas científicas celebradas con motivo del CL aniversario de su fundación.* Barcelona. 268+[2] pp. & 15 pl.

Bb383. ___. 1964. *La Real Academia de Ciencias Naturales y Artes en el siglo XVIII.* By J. Iglesies y Fort. Barcelona. 635 pp. *Memorias de la Real Academia de Ciencias y Artes de Barcelona* vol. 36 (1).

Real Sociedad Vascongada de los Amigos del Pais

Bb384. **Real Sociedad Vascongada de los Amigos del Pais.** 1972. *Las ciencias naturales y la Real Sociedad Vascongada de los Amigos del Pais en el siglo XVIII.* By J. Martéinez Ruiz. San Sebastián. 90 pp.

Royal Asiatic Society of Great Britain and Ireland

Bb385. **Royal Asiatic Society of Great Britain & Ireland.** 1923. *Centenary volume of the Royal Asiatic Society of Great Britain and Ireland, 1823–1923.* By F. E. Pargiter. London. xxviii+186 pp.
Includes I. Articles and contributions arranged according to countries: All except India, pp. 4–55; II. Articles and contributions: India, pp. 56–106; III. Authors and their contributions, pp. 107–182. Indexes cover the Journal ... 1834–1922.

Bb386. ___. [1933.] *Index of the contents of the Journal for the decade 1920 to 1929.* London. xii+53 pp.; *Continued* ___. 1940. Index of the contents of the Journal for the decade 1930–1939. *Journal of the Royal Asiatic Society of Great Britain and Ireland* 1940: 259–324.

Royal College of Physicians, London

Bb387. **Royal College of Physicians, London.** 1878–1955 *The roll of the Royal College of Physicians of London; comprising biographical sketches.* Ed. by W. Munk. 2nd ed. London. 3 vols. *Continued* ___. 1955. *Munk's roll. Lives of the Fellows of the Royal College of Physicians of London 1826–1925.* Ed. by G. H. Brown. London. x+637 pp.; ___. 1968. *Lives of the Fellows of the Royal College of Physicians of London. Continued to 1965.* Ed. by R. R. Trail. London. xii+476 pp.
Further supplements continue the record down to the 1980s.

Royal Dublin Society (*formerly* Dublin Society)

Bb388. **Berry, H. F.** 1915. *A history of the Royal Dublin Society, [etc.].* London, Longmans & Co. xv+460 pp.

Bb389. ___**, K. R.** 1986. The Royal Dublin Society and the advancement of popular science in Ireland, 1731–1860. *History of education* 15: 81–88.

Bb390. **Royal Dublin Society.** 1899. *Index to the Scientific Proceedings and Transactions from 1877 to 1898, inclusive Proceedings, vol. I–VIII and of the Transactions vol. I–VI.* Dublin. 43 pp.; *Continued* ___. 1910. *Index to the Scientific Proceedings [vols. 9–11] & Transactions [vols. 7–9] of the Royal Dublin Society from 1898 to 1909 inclusive.* Dublin. 12 pp.

Bb391. ___. 1953. *A bibliography of the publications of the Royal Dublin Society from 1731 to 1951.* 2nd ed., rev. with index. Dublin. 46 pp.
Replaces 1st ed., 1951. 39 pp.

Bb392. ___. 1987. *Nostri plena laboris: An author index to the Royal Dublin Society scientific journals.* By C. Mollan et al. Dublin. 120pp.

Royal Geographical Society

Bb393. **Middleton, D.** 1978. Guide to the publications of the Royal Geographical Society, 1830–1892. *Geographical journal* 144: 99–116.

Bb394. **Royal Geographical Society.** 1844. *General index to the contents of the first ten volumes [1831–41] of the London geographical journal.* By J. R. Jackson. London. iv+216 pp.; *Continued* ___. 1853. *General index to ... the second ten volumes [1841–50] [etc.].* By G. S. Brent. London, J. Murray. 116 pp.; ___. 1867. *General index to the third ten volumes [1851–60] [etc.].* By H. Yule. London, J. Murray. 228 pp.; ___. 1881. *General index to the fourth ten volumes [1861–70] [etc.].* London, J. Murray. 259 pp.; ___. 1884. *General index to the fifth ten volumes [18 71–80] [etc.].* London, J. Murray. 237 pp.

Bb395. ___. 1880. List of papers in the 'Journals' and 'Proceedings' of the Geographical Society [vol. 1–50, 1831–80]. *Journal of the Royal Geographical Society* 50: 153–242.

Bb396. ___. 1886. *Classified index to the maps in the Royal Geographical Society's publications 1830–1883.* Cambridge, MA, Library of Harvard Univ. 43 pp. *Bibliographical contributions* no. 17.

Bb397. ___. 1896. *General index to the Proceedings of the Royal Geographical Society. First series, vols. 1–22, 1855–1878 [etc.].* By M. H. James. London. xxv+367 pp.

Bb398. ___. 1906. *Geographical journal. General index to the first twenty volumes ... 1893–1902.* London, Royal Geographical Society. xxvii+629 pp.; *Continued* ___. 1925. ___ *the second twenty volumes ... 1903–1912.* By L. C. W. Bonacina. London, Royal Geographical Society. xxv+688 pp.; ___. 1930. ___*: Third general index ... volumes XLI to LX, 1913 to 1922.* London, Royal Geographical Society. xx+351 pp.; ___. 1935. ___*: Fourth general index ... volumes LXI to LXXX, 1923–1932.* London, Royal Geographical Society. xxiv+250 pp.; ___. 1951. ___*: Fifth general index ... volumes LXXXI to C, 1933–1942.* London, Royal Geographical Society. xvii+166 pp.; ___. 1959. ___*: Sixth general index ... volumes CI to CXX, 1943–1954.* London, Royal Geographical Society. xiii+130 pp.; ___. 1971. ___*: Seventh general index ... volumes 121 to 130, 1955–1964.* London, Royal Geographical Society. xiii+147 pp.; ___. 1977. ___*: Eighth general index ... volumes 131 to 140, 1965–1974.* London, Royal Geographical Society. xii+202 pp.

Bb399. ___. 1924. *The Royal geographical Society: Its foundation and history, work and publications, charter and bye–laws, house [etc.].* London. 32 pp., illus. *Re-pub.* 1927, 1930, & 1933.

Bb400. ___. 1930. *The record of the Royal Geographical Society, 1830–1930.* By H. R. Mill. Published at the celebration of the Society's centenary, October 1930. London. xvi+288 pp., illus.

Bb401. ___. 1955. *Royal Geographical Society: A record, 1931–55.* By G. R. Crone. London. 36 pp.

See also Synopsis of Society's archives Bc815; Ca637.

Royal Irish Academy

Bb402. **Royal Irish Academy.** 1912. *Index to the serial publications of the Royal Irish Academy (Transactions, Proceedings, Cunningham Memoirs, Todd lecture Series, and Irish*

Manuscript Series) from 1786 to 1906 inclusive. Ed. by R. L. Praeger. Dublin, Hodges, Figgis & Co.; London, Williams & Norgate. x+116 pp.; ___. 1934. *Index to the serial publications of the Royal Irish Academy (Proceedings, Cunningham Memoirs, Todd Lecture Series) from 1907 to 1932.* Dublin, Hodges, Figgis & Co.; London, Williams & Norgate. [ii]+27 pp.; ___. 1959. ___*: 1932–53.* Dublin. 40 pp.; ___. 1974. ___*: 1953 to 1970.* Dublin. 39 pp.

Bb403. **Royal Irish Academy.** 1985. *The Royal Irish Academy: A bicentennial history, 1785–1985.* Ed. by T. O. Raifeartaigh. Dublin. xvi+351 pp., illus.

Royal Microscopical Society *(formerly* Microscopical Society of London)

Bb404. **Fowler, G. H.** 1889. *Index to the Quarterly Journal of Microscopical Science, 1853–1888: To the Transactions of the Microscopical Society, 1844–1852, 1853–1868, and to Proceedings of the Dublin Microscopical Club, 1865–1880.* London, J. & A. Churchill. v+87 pp.

Bb405. **Goodrich, H. L. M. Pixell–.** 1916. *Index to the Quarterly journal of microscopical science, vols. 29–61, inclusive: July, 1888, to July, 1916, inclusive.* London, J. & A. Churchill. 40 pp.

Bb406. **Royal Microscopical Society.** 1895. The President's address: The history of the Royal Microscopical Society. By A. D. Michael. *Journal of the Royal Microscopical Society* 1895: 1–20.

Bb407. ___. 1989. *Royal Microscopical Society: 150th anniversary exhibition catalogue.* By P. J. Evennett. Oxford. 76 pp.
Pub. as a supp. to *Proceedings of the Royal Microscopical Society* 24/25 Sept. 1989.

Bb408. ___. 1989. *God bless the microscope!: A history of the Royal Microscopical Society over 150 years.* By G. L'E. Turner. Oxford. xi+116 pp.

Royal Society of Edinburgh *(formerly* Philosophical Society of Edinburgh)

Bb409. **Bennet, F., et al.** 1983. *The Royal Society of Edinburgh: 100 medical fellows elected 1841–1882.* Edinburgh, University of Edinburgh, History of Medicine & Science Unit. [118] pp. *Scotland's cultural heritage* vol. 4.

Bb410. **Emerson, R. L.** 1979. The Philosophical Society of Edinburgh, 1737–1747. *British journal for the history of science* 12: 154–191; ___. 1981. The Philosophical Society of Edinburgh, 1748–1768. *Ibid.* 14: 133–176; ___. 1985. The Philosophical Society of Edinburgh, 1768–1783. *Ibid.* 18: 255–303.
Appendices list the Membership of the Society: [1]. 1737–1747, pp. 189–191; [2]. 1748–1768, pp. 172–176; [3]. 1768–1783, pp. 298–303.

Bb411. **Jack, G., et al.** 1982. *The Royal Society of Edinburgh: 100 medical fellows elected 1783–1844.* Ed. by S. D. Thorp. Edinburgh, University of Edinburgh, History of Medicine & Science Unit. [116] pp.*Scotland's cultural heritage* vol. 3.

Bb412. **Royal Society of Edinburgh.** 1890. *General index to first thirty-four volumes [of the Transactions] (1783–1888), with history of the institution of the Society, etc.* Edinburgh & London. xi+228 pp.

Bb413. ___. 1910. *General index to the volumes [35–46] issued during the years 1889–1908.* Edinburgh. [iv]+103 pp.

Bb414. ___. 1983. *The Royal Society of Edinburgh (1783–1983): The first two hundred years.* By N. Campbell & R. M. Smellie. Edinburgh, Royal Society of Edinburgh. xvi+186 pp.

Bb415. ___. 1984. *Index of Fellows of the Royal Society of Edinburgh elected November 1783 — July 1883.* By F. Bennet, H. Frew & M. Melrose. Edinburgh, Scotland's Cultural Heritage. 73 pp.

Bb416. **Shapin, S.** 1974. Property, patronage, and the politics of science: The founding of the Royal Society of Edinburgh. *British journal for the history of science* 7: 1–41.

Royal Society of London

Bb417. **Barrow, J.** 1849. *Sketches of the Royal Society and Royal Society Club.* London, J. Murray. vi+212 pp. *Re-pub.* (with a new index) London, F. Cass, 1971. vi+215 pp. *Social history of science* no. 9.

Bb418. **Brasch, F. E.** 1931. *The Royal Society of London and its influence upon scientific thought in the American colonies.* New York, Science Pr. 43 pp., ports.
First pub. in the *Scientific monthly* vol. 33 (Oct./Nov.) 1931.

Bb419. **Clark, A. J.** 1992. The history of the Royal Society: A chronological checklist of books in English. *Notes & records of the Royal Society of London* 46: 335–345.

Bb420. **Crosland, M.** 1983. Explicit qualifications as a criterion for membership of the Royal Society: A historical review. *Ibid.* 37(2): 167–187.

Bb421. **Frick, G. F.** 1976. The Royal Society in America. In: A. Oleson & S. C. Brown, eds. *The pursuit of knowledge in the early American republic: American scientific and learned societies from colonial times to the civil war*, pp. 70–83. Baltimore, MD, Johns Hopkins Univ. Pr.

Bb422. **Gleason, M. L.** 1991. *The Royal Society of London: Years of reform, 1827–1847.* New York, Garland Publishing. x+532 pp. *Harvard dissertations in the history of science.*

Bb423. **Hall, M. B.** 1966. Sources for the history of the Royal Society in the seventeenth century. *History of science* 5: 62–76.
The origins of the Society; Histories of the Society; Manuscript and printed original sources; Biographical information; Interpretations of the Society's activities.

Bb424. ___. 1975. The Royal Society's role in the diffusion of information in the 17th century, I. *Notes & records of the Royal Society of London* 29: 173–192.

Bb425. ___. 1984. *All scientists now: The Royal Society in the nineteenth century.* Cambridge, Cambridge Univ. Pr. xii+261 pp.

Bb426. **Hoppen, K. T.** 1976. The nature of the early Royal Society. *British journal for the history of science* 9: 1–24.

Bb427. **Hunter, M.** 1982. *The Royal Society and its Fellows 1660–1700: The morphology of an early scientific institution.* Chalfont St. Giles, British Society for the History of Science. 266 pp. *Monograph* no. 4.

Bb428. ___. 1989. *Establishing the new science: The experience of the early Royal Society.* Woodbridge, Boydell Pr. xiv+382 pp.

Bb429. **Lyons,** Sir **H.** 1944. *The Royal Society 1660–1940: A history of its administration under its Charters.* Cambridge, Cambridge Univ. Pr. x+354 pp. *Re-pub.* Westport, CT, Greenwood Pr., 1968.

Bb430. **Maty, P. H.** 1787. *A general index to the Philosophical Transactions from the first to the end of the seventieth volume (1665–1780).* London. 801 pp.; *Continued* **Royal Society of London.** 1821. *A continuation to the alpabetical index of the matter contained in the Philosophical Transactions from vol. 71 to vol. 110, the year 1781 to the year 1820 inclusive.* London. 225 pp.; ___. 1833. *A continuation to the ... index of the matter contained in the Transactions from vol. 111 to 120 the year 1821 to the year 1830 inclusive.* London. 101 pp.
Continued in the *Index to the Proceedings ... and Philosophical transactions [etc.]*, 1932+, below.

Bb431. **Royal Society of London.** 1905. Obituary notices of decreased Fellows ... chiefly for the period 1898–1904 with an index to the obituaries published in the "Proceedings" from 1860 to 1899. *Proceedings of the Royal Society of London* 75: 381 pp.

Bb432. ___. 1913. *The celebration of the two hundred and fiftieth anniversary of the Royal Society of London, July 15–19, 1912.* London. [ii]+128 pp., illus.

Bb433. ___. 1913. *Index to the Proceedings ... (old series) vols. 1–75, 1800–1905.* By H. W. Robinson. London, Harrison & Sons. 223 pp.; *Continued* ___. 1932. *Index to the Proceedings of the Royal Society ... (1905–30), and to the Philosophical Transactions ... (1901–30).* By H. W. Robinson. London. [iv]+231 pp.; ___. 1942. *Index to Proceedings and Philosophical Transactions, 1931–1940.* London. 110 pp.; ___. 1951. *Index to Proceedings and Philosophical Transactions and obituary notices, 1941–1950.* London. 56 pp.; ___. 1962. *Decennial index 1951–1960: Index of authors in Proceedings, Philosophical transactions, Obituary notices and Biographical memoirs.* London. [ ]* pp.; ___. 1971. *Decennial index 1961–1970 [etc.].* London. [ ]* pp.; ___. 1981. *Decennial index 1971–1980 [etc.].* London. [ ]* pp.

[*Not examined – pagination not ascertained.]

Bb434. ___. 1932–54. *Obituary notices of Fellows of the Royal Society.* New series. London. Vols. 1–9; *Continued* ___. 1955+. *Biographical memoirs of the Fellows of the Royal Society.* London.

Vol. 26, 1980, pp. 545–565 includes an index to the whole series vols. 1–9 & vols. 1–26, 1932–54; vol. 35, 1990, includes an index to vols. 27–35, 1981–90.

Bb435. ___. 1940. *The record of the Royal Society of London for the Promotion of Natural Knowledge.* 4th ed. London. vi+578 pp. & 16 pl. *Re-pub.* 1993?; ___. 1993? *Supplement to the Record of the Royal Society.* By J. Rowlinson & N. Robinson. London. 316 pp.

Forthcoming publication of supplement and re-publication of 4th ed. announced. *Replaces* 1st ed., London, 1897. vii+224 pp. & 2 pl.; 2nd ed., London, 1901. viii+427 pp.& 11 pl.; 3rd ed., London, 1912. viii+483 pp., 6 pl. & 14 ports.

Bb436. ___. 1938+. *Notes & records of the Royal Society of London.* London. Vol. 1+; ___. 1969. ___*: Index to volumes 1 to 20 (1938–1965).* London. xlviii pp.; *Continued* ___. 1976. Index ... vols. 21 to 30 (1966 to 1975). *Ibid.* 31: 323–353; ___. 1987. Index ... vols 31 to 40 (1976–1986). *Ibid.* 41: 241–262.

Publication suspended 1942–45. Includes articles on various aspects of the Society's history and on the lives and works of its Fellows.

Bb437. ___. 1985. *The archives of the Royal Society.* Pt. I: *Journal books of scientific meetings, 1660–1800.* Bethesda, MD, Univ. Publications of America. Microfilm, 18 reels with printed guide.

Bb438. ___. 1985. *The archives of the Royal Society.* Pt. II: *Council minutes, 1660–1800.* Bethesda, MD, Univ. Publications of America. Microfilm, 3 reels with printed guide.

Bb439. **Stearns, R. P.** 1946. Colonial fellows of the Royal Society of London, 1661–1788. *William & Mary quarterly* (3) 3: 208–268. Also pub. in *Osiris* 8: 73–121, 1948.

Bb440. **Stimson, D.** 1948. *Scientists and amateurs: A history of the Royal Society.* New York, H. Schuman. xiii+270 pp. *Re-pub.* Westport, CT, Greenwood Pr., 1968.

Bb441. **Thomson, T.** 1812. *History of the Royal Society, from its institution to the end of the eighteenth century.* London, R. Baldwin. viii+552+xc pp.

Bb442. **Weld, C. R.** 1848. *A history of the Royal Society, with memoirs of the Presidents.* Compiled from authentic documents. London, J. W. Parker. 2 vols. *Re-pub.* Salem, NH, Ayer, 1975.

See also Guides & catalogues of the Society's archives Bc820–824; Ca829 (Hunter).

Sächsische Akademie *(formerly* Gesellschaft) der Wissenschaften, Leipzig *(formerly* Königlich Sächsische Gesellschaft der Wissenschaften)

Bb443. **Königlich Sächsische Gesellschaft der Wissenschaften, Leipzig.** 1889. *Register zu den Jahrgängen 1846–1885 der Berichte über die Verhandlungen und zu den Bänden 1–12 der Abhandlungen der mathematische–physischen Classe der Königlich Sächsischen Gesellschaft der Wissenschaften zu Leipzig*. Leipzig, S. Hirzel. 86 pp.

Bb444. ___. 1897. *Zur Fünfzigjährigen Jubelfeier der Königlich Sächsischen Gesellschaft der Wissenschaften zu Leipzig am 1. Juli 1896. Reden und Register*. Leipzig. xliv+35+63pp.
Includes Namen– und Sachregister der Abhandlungen und Berichte der mathematisch–physischen Classe der Königlich Sächsischen Gesellschaft der Wissenschaften, 1846–95. (Bibliographische übersicht) etc. 63 pp.

Sankt–Peterburkskoe Obshchestvo Estestvoispytalei (Société Impériale des Naturalistes de St. Petersbourg)

Bb445. **Ginetsinskaya, T. A., & M. A. Zakharova.** 1983. Sankt–Petersburgskoe Obshchestvo Estestvoispytalei i izuchenie fauny Belogo Morya (k 100–letiyu Solovetskoi Biologicheskoi Stantsii). [The St. Petersburg Naturalist's Society and zoological studies in the White Sea (to the centenary of Solovetskii Biological Station]. *Vestnik Leningradskogo Universitetea, ser. biologiya* 21 (4): 100–130.

Schlesische Gesellschaft für Vaterländische Cultur *(formerly* Gesellschaft zur Bedörderung der Naturkunde und Industrie in Schlesien)

Bb446. **Schlesische Gesellschaft für Vaterländische Cultur.** 1853. *Denkschrift zur Feier ihres fünfzigjährigen Bestehens*. Breslau. [iii]+282+[2] pp.
Includes **A. Kahlert.** Geschichte der Schlesischen Gesellschaft für Vaterländische Kultur, pp. 1–46.

Bb447. ___. [1868.] *Verzeichnis der in den Schriften der Gesellschaft von 1804 bis 1863 inclus. enthaltenen Aufsätze*. Breslau. 166 pp.; ___. 1878. *Fortzetzung des Verzeichnisses der in den Schriften der Gesellschaft, von 1864 bis 1878 incl. enthaltenen Aüfsätze, geordnet nach den Verfassern in alphabetischer Folge*. [Breslau] 50 pp.

Bb448. ___. 1878. *General–Sachregister der in den Schriften der Schlesischen Gesellschaft für Vaterländische Cultur von 1804 bis 1876 incl. enthaltenen Aufsätze geordnet in alphabetischer Folge*. Breslau. xii+162 pp.

Bb449. ___. 1904. *Die Schlesische Gesellschaft für Vaterländische Cultur. I: Die Hundertjahrfeier. II: Geschichte der Gesellschaft*. Breslau. 125 (149) pp.

Scottish Marine Biological Association *(formerly* Millport Marine Biological Station *and* Marine Biological Association of the West of Scotland)

Bb450. **Kerr, J. G.** 1949. The Scottish Marine Biological Association. *Notes and records of the Royal Society* 7: 81–96.

Senckenbergische Naturforschende Gesellschaft, Frankfurt am Main

Bb451. **Schöndube, W., & J. Steinbacher.** 1967. Aus der 150–jährigen Geschichte der Senckenbergischen Naturforschenden Gesellschaft zu Frankfurt am Main. *Natur und Museum, Frankfurt am Main* 97: 431–489.

Bb452. ___. 1967. *Die Senckenbergische Naturforschende Gesellschaft zu Frankfurt am Main: Aus ihrer 150–jährigen Geschichte*. Frankfurt am Main, W. Kramer & Co. 67 pp.

Bb453. **Senckenberg–Museum, Frankfurt am Main.** 1967. *Chronik der Senckenbergischen Naturforschenden Gesellschaft 1817–1966*, pp. 169–559. By W.

Kramer. In: *Aus der Geschichte des Senckenberg–Museums* 1–2. Frankfurt am Main, Kramer. 571 pp. *Senckenberg–Buch* 46.

Bb454. **Senckenbergische Naturforschende Gesellschaft.** 1935. *Gesamt–Verzeichnis der Senckenberg–Schriften (1817–1935).* Ed. by R. Richter. Frankfurt am Main. 307 pp.; *Continued* ___. 1940. ___ *(1836–1940).* Ed. by H. T. Reuling. Frankfurt am Main. 142 pp.; ___. 1952. ___. *Zwölfahres–Verzeichnis für 1941–1952.* Frankfurt am Main. 146 pp.; ___. 1961. ___. *Sechsjahres–Verzeichnis für 1953–1960.* Ed. by J. Steinbacher. Frankfurt am Main. 178 pp.; ___. 1973. ___. *Sechsjahres–Verzeichnis für 1967–1972.* Ed. by J. Steinbacher. Frankfurt am Main. 142 pp.

Bb454A. **Ziegler, W., et al.** 1992. 175 Jahre Senckenbergische Naturforschende Gesellschaft. *Natur und Museum* 122: 337–380.

Siebenbürgischer Verein für Naturwissenschaften zu Hermannstadt

Bb455. **Siebenbürgischer Verein für Naturwissenschaften zu Hermannstadt.** 1872. [Index to vols. 1–21, (1849–71).] *Verhandlungen und Mitteilungen des Siebenbürgischen Vereins für Naturwissenschaften zu Hermannstadt* 22: [ ]* pp.; *Continued* ___. 1908. Zusammenstellung der in den Jahrgängen 22 (1872) bis 57 (1907) der Schriften ... veröffentlichten Arbeiten [etc.]. *Ibid.* 58: 169–182.
[*Not examined – pagination not ascertained.]

Bb456. ___. 1896. *Die Siebenbürgische Verein für Naturwissenschaften in Hermannstadt nach seiner Entstehung, seiner Entwicklung und seinem Bestande.* Hermanstadt. 68 pp.

Bb457–458. [Not used.]

Società Adriatica di Scienze Naturali di Trieste

Bb459. **Società Adriatica di Scienze Naturali di Trieste.** 1927. Indici per autori e per materie delle memorie pubblicate nei volumi 1–29 del Bollettino. *Bollettino della Società Adriatica di Scienze Naturali di Trieste* 29: xci–cvii pp.

Società Italiana di Scienze Naturali, Milano

Bb460. **Società Italiana di Scienze Naturali e del Museo Civile di Storia Naturale, Milano.** 1906. *Elenco dei Soci. ... Indice generale per autori e per materie dei lavori pubblicati dalla sua fondazione [1858] a tutto Settembre 1906.* Milan. 132 pp.

Bb461. **Società Italiana di Scienze Naturali.** 1957. *Pubblicazione commemorativa del centenario della Societa: Indice generale per autori ... Elenco dei soci [etc.].* Milan. 207 pp.

Società dei Naturalisti di Modena

Bb462. **Società dei Naturalisti di Modena.** 1882. *Indici generale delle memorie accademiche. Serie prima (1833–58—1882).* Modena. v+235 pp.; *Continued* ___. 1892. *Indice generale ... serie 3, vol. 1 a 10, ann. 16 a 25.* Modena. [ ]* pp.; ___. 1899. *Indice generale ... serie 3, [vol. 11 to 16], ann. 26 a 31.* Modena. [ ]* pp.
[*Not examined – pagination not ascertained.]

Società di Naturalisti in Napoli (*formerly* Circolo degli Aspiranti Naturalisti)

Bb463. **Società di Naturalisti in Napoli.** 1907. *La Società di Naturalisti in Napoli nel XXV anniversario della sua fondazione, 1881–1906.* Naples. 86 pp.

Società Toscana di Scienze Naturali

Bb464. **Tosi, A.** 1986. Collezioni, istituzioni e ricerca scientifica a Pisa fra '800 e '900: La Società Toscana di Scienze Naturali. *Atti della Società Toscana di Scienze Naturali Residente in Pisa* 93 supp: 7–34.

Societas pro Fauna et Flora Fennica (Sällskap pro Fauna et Flora Fennica)

Bb465. **Societas pro Fauna et Flora Fennica.** 1921. Societas pro Fauna et Flora Fennica 1821–1921. By F. Elfving. *Acta Societatis pro Fauna et Flora Fennica* 50: 279 pp. & 3 pl.
Includes a list of ca.150 deceased members.

Bb466. ___. 1939. Index generalis Actorum 1–60 (1875–1937) Societatis pro Fauna et Flora Fennica. By E. Reuter. *Acta Societatis pro Fauna et Flora Fennica* 61: vii+419 pp.; *Continued* ___. 1980. Index generalis serici Acta Societatis ... 61 82 (1938–1980). By G. & O. Biström. *Ibid.* 82 (5): 96 pp.

Bb467. ___. 1944. Index generalis serierum Notiser I–XIV, (1848–1875) et Meddelanden 1–50, (1876–1925) Societatis pro Fauna et Flora Fennica. By E. Reuter. *Acta Societatis pro Fauna et Flora Fennica* 64: xv+676 pp.

Bb468. ___. 1947. Index generalium Memorandorum 1–20 (1927–1945) Societatis pro Fauna et Flora Fennica. By E. Reuter. *Memoranda Societatis pro Fauna et Flora Fennica* 20 supp: xii+238 pp.; *Continued* ___. 1971. Index generalis Memorandorum 21–40 (1945–1964) [etc.]. By L. Fagerström, & Maria & Märta Reuter. *Ibid.* 40 supp: x+195 pp.

Bb469. ___. 1958. Societas pro Fauna et Flora Fennica 1 nov. 1921 – 1 nov. 1946. By A. Palmgren. *Acta Societatis pro Fauna et Flora Fennica* 75: 1–91.

Societas Scientiarum Fennica

Bb470. **Elfving, F.** 1938. Finska Vetenskapssocietetens historia 1838–1938. *Commentationes humanorum litterarum* 10: 310+19+133 pp.
Includes Tables générales des series de publications de Societas Scientiarum Fennica 1838–1938, pp. 1–133.

Bb471. ___, **& G. Michwitz.** 1988. *Finska Vetenskaps–Societetens tredje halvsekel 1938–1987.* Helsinki. 201 pp. *Bidrag till kannedom av Finlands natur och volk* no. 136.

Bb472. **Societas Scientiarum Fennica.** 1912–27. *Tables générales des publications de la Société des Sciences de Finlande, 1838–1910 (1911–1926). Avec des notices biographiques sur les auteurs.* By G. Schaumann. Helsinki. 2 pts.

Societatea de Medici si Naturalisti din Iasi (Société des Médecins et des Naturalistes, Jassy)

Bb473. **Societatea de Medici si Naturalisti din Iasi.** 1954. Societatea de Medici si Naturalisti din Iasi între anii 1830–1850. By I. Lazarescu. *Studii si cercetari stiintifice (Jassy)* 5: 483–508.

Bb474. ___. 1970. La Société des Médecins et des Naturalistes de Iassy, première société scientifique de Roumaine. By V. Rascanu. *Revista medico–chirurgicala a Societatii de Medici si Naturalisti din Iasi* 74: 839–846.

Bb475. ___. 1970. Rapports de la Société des Médecins et des Naturalistes de Iassy avec les hommes de science et les associations similaires de l'étranger. By N. Bumbacesco. *Ibid.* 74: 875–881.

Bb476. ___. 1970. Le rôle de la Société des Médecins et des Naturalistes dans le développement des sciences naturelles en Moldavie. By P. Jitariu. *Ibid.* 74: 883–889.

Bb477. ___. 1970. Personalités importantes dans la direction et l'activité de la Société des Médecins et des Naturalistes. By G. Nastase. *Ibid.* 74: 901–908.

Bb478. ___. 1980. Societatea de Medici si Naturalisti din Iasi (1830–1980). By D. Ciurea. *Ibid.* 84: 587–592.

Bb478A. ___. 1980. Societatea de Medici si Naturalisti din Iasi la a 150-a aniversare (1830–1980). By G. H. Badarau, et al. *Ibid.* 84: 601–611.

Bb478B. **Societatea de Medici si Naturalisti din Iasi.** 1980. Societatea de Medici si Naturalisti si dezvoltarea stiintelor biologicie. By P. Jitariu. *Ibid.* 84: 636–642.

Société de Biologie, Paris

Bb479. **Gley, E.** 1900. La Société de Biologie et l'évolution des sciences biologiques en France de 1849 à 1900. In: *Essais de philosophie et d'histoire de la biologie*, pp. 168–312. Paris, Masson & Cie.

Bb480. **Société de Biologie.** 1899. *Cinquantenaire de la Société de Biologie. Volume jubilaire.* Paris, Masson & Cie. [iii]+740 pp. & 7 pl.
Includes a list of members from 1848 to 1899.

Bb481. **___.** 1923. *Célébration du 75me anniversaire de la fondation de la Société de Biologie par la Société et ses filiales, [etc.].* Paris. 132 pp.

Société Éduenne

Bb482. **Société Éduenne.** 1892. Tables des travaux de la Société Éduenne, 1839–1892. *Mémoires, Société Éduenne des Lettres, Sciences et Arts* n.s. 20: 421–516; *Continued* ___. 1913. 2e table alphabétique des Mémoires ... tomes XXI–XL (1893–1912). *Ibid.* N.S. 41: 359–438.

Société (Royale/Impériale) d'Émulation d'Abbeville

Bb483. **Société d'Émulation d'Abbeville.** 1905. *Table générale des publications de la Société d'Émulation (1797–1904). Suivie de la liste des membres de la Société depuis sa fondation.* By A. Boucher de Crèvecœur. Abbeville. ix+271 pp.

Bb484. **___.** 1949. *Célébration do 150e anniversaire de la Société. Le 3 Octobre 1948.* Abbeville. 32 pp.

Société d'Émulation du Doubs (*formerly* Société Libre d'Émulation du Doubs)

Bb485. **Société d'Émulation du Doubs.** 1907. *Table générale récapitulative des Mémoires ... 1841–1905.* By A. Kirchner. Besançon. vi+127 pp.; **___.** 1931. *Table générale récapitulative des Mémoires ... 1906–1930.* By G. Gazier. Besançon. 38 pp.

Société d'Émulation du Jura

Bb486. **Société d'Émulation du Jura.** 1918. *Table générale récapitulative ... 1818–1917.* By M. Perrod. Lons–le–Saunier. xxiii+128 pp.

Société d'Étude des Sciences Naturelles d'Elbeuf (*formerly* Société d'Enseignement Mutuel des Sciences Naturelles, Elbeuf)

Bb487. **Société d'Étude des Sciences Naturelles d'Elbeuf.** 1981. [Commémoration du centenaire de la Société d'Étude des Sciences Naturelles d'Elbeuf 1881–1981] exposition tenue du 16 au 31 mai 1981 ..., Elbeuf. *Bulletin de la Société d'Étude des Sciences Naturelles et du Musée d'Elbeuf* special no: 93 pp.

Société d'Études Scientifiques d'Angers

Bb488. **Société d'Études Scientifiques d'Angers.** 1882. Table décennale [for vols. 1–10, 1871–80.] *Bulletin de la Société d'Études Scientifiques d'Angers* 11/12: 371–382; *Continued* ___. 1891. Table décennale [for vols. 11–20, 1881–90.] *Ibid.* 2: 243–254; **___.** 1902. Table décennale du Bulletin ... [vols. 21–30] (1891–1900). By T. Surrault. *Ibid.* n.s. 31: 229–248.; **___.** 1911. Table décennale du Bulletin ... [vols. 31–40] (1901–1910). *Ibid.* n.s. 41: 149–166.; ___. 1921. Table décennale du Bulletin ... [vols. 41–50] (1911–1920). *Ibid.* n.s. 51: 57–70.

Société Fribourgeoise des Sciences Naturelles

Bb489. **Société Fribourgeoise des Sciences Naturelles.** 1982. La fondation de la Société Fribourgeoise des Sciences Naturelles. By J. Dubas. *Bulletin de la Société Fribourgeoise des Sciences Naturelles* 71: 1–19.

Société de Géographie de Paris

Bb490. **Société de Géographie de Paris.** 1845. *Bulletin de la Société de Géographie ... Table alphabétique et raisonnée des matières contenues dans les deux premières séries, 1822–1843 (formant 40 volumes), [etc.].* By E. de Froberville. Paris. 251 pp.; *Continued* ___. 1866. *Table ...1844 à 1861 (formant 34 volumes) [etc.].* V. A. Barbié de Bocage. Paris. 189 pp.; ___. [1900.] *Table des matières, séries 5–7, 1861–1899 [etc.].* By P. Lemosof. Paris. viii+235 pp.

Bb491. ___. 1921. *Centenaire de la Société de Géographie, 1821–1921.* Paris. 151 pp., illus.

Bb492. ___. 1975. *Cent–cinquantenaire de la Société de Géographie, 1821–1971.* Paris. 112 pp. *Acta geographica* vol. 20.

See also Inventory of the Society's manuscripts Bc830.

Société Guernesiaise (*formerly* Guernsey Society of Natural Science)

Bb493. **Guernsey Society of Natural Science.** [1934.] *General index to the Transactions of the Society, 1882–1932.* By S. C. Curtis. Guernsey. 82 pp.; *Continued* ___. [1937.] *General index ..., 1933–1936.* By M. A. Brock. Guernsey. 15 pp.

Société d'Histoire Naturelle de la Moselle (*formerly* Société d'Histoire Naturelle du Département de la Moselle *and* Société d'Histoire Naturelle de Metz)

Bb494. **Société d'Histoire Naturelle du Département de la Moselle.** 1870. Table générale des matières contenues dans les 12 volumes de la Ire série [1843–70]. *Bulletin de la Société d'Histoire Naturelle du Département de la Moselle* 12: 155–164.

Société d'Histoire Naturelle de Toulouse

Bb495. **Société d'Histoire Naturelle de Toulouse.** 1887. *But et historique de la société. Table des matières de 1866 à 1886. Liste des sociétés correspondantes françaises et étrangères.* Toulouse. 38 pp.; *Continued* ___. 1900. *Bulletin de la Société d'Histoire Naturelle de Toulouse. Table générale des matières contenues dans les volumes qui ont paru de 1886 à 1899 [etc.].* By J. Laromiguière. Toulouse. 14 pp. ; ___. 1906. ___*: Table générale ... 1900 à 1905 [etc.].* Toulouse. vii pp.

Société Linnéenne de Bordeaux

Bb496. **Société Linnéenne de Bordeaux.** 1901. [Classified index to Actes, vols. 1–50 (*and* Procès-verbaux) de la Société Linnéenne de Bordeaux, 1828–96.] In: *Catalogue de la Bibliothèque.* Fasc. II. Bordeaux. 115 pp. Supplement aux *Actes de la Société Linnéenne de Bordeaux.*

Bb497. ___. 1928. *Centenaire de la reconnaissance d'utilité publique — 1828–1928.* Bordeaux. 118 pp. Appended to *Actes de la Société Linnéenne de Bordeaux* vol. 80.

Bb498–499. [Not used.]

Société Linnéenne de Lyon

Bb500. **Fiasson, J.** 1988. Histoire de la Société Linnéenne de Lyon. *Actes du 112e Congrès National des Sociétés Savantes, Section d'Histoire des Sciences [etc.]*, pp. 129–133.

Bb501. **Société Linnéenne de Lyon.** 1909. Table générale des publications contenues dans les Annales de la Société ... de 1836 à 1908, [etc.]. By A. Bonnet. *Annales de la Société Linnéenne de Lyon* 55 [appx.]: 49 pp.

Bb502. ___. 1924. *Historique de la Société Linnéenne de Lyon pendant le premier siècle de son existence, 1822–1922.* By C. Roux. Lyon. [ii]+54 pp., ports.

Société Linnéenne de Normandie (*formerly* Société Linnéenne du Calvados)

Bb503. **Société Linnéenne de Normandie.** 1926. *Centenaire de la Société, célébré les 31 Mai, 1er et 2 Juin 1924.* Caen. 108 pp.
Includes **A. Bigot.** Notice historique; **P. Bugnon.** Table des travaux originaux publiés de 1824 à 1924 dans les Mémoires de la Société Linnéenne.

Société Linnéenne de Paris

Bb504. **Aymonin, G. G.** 1974. "Adansonia", fêtes champêtres et Linnéens français. *Taxon* 23: 155–162.
Includes brief details of other early Linnaean societies in France.

Bb505. **Beretta, M.** 1991. The Société Linnéenne de Paris (1787–1827). *Svenska Linnésällskapets årsskrift* 1991–1991: 151–175.

Bb506. **Cowan, C. F.** 1969. On the Mémoires de la Société Linnéenne de Paris, 1822–77. *Journal of the Society for the Bibliography of Natural History* 5: 152–154.

Société Linnéenne de Provence

Bb507. **Drouin, J.–M.** 1991. Une société locale de sciences naturelles: La Société Linnéenne de Provence. *Revue d'histoire des sciences* 44: 219–234.

Société Lorraine des Sciences

Bb508. **Société Lorraine des Sciences.** 1980. La biologie dans les publications de la Société Lorraine des Sciences (1828–1978). By G. Percebois. *Bulletin de l'Académie et de la Société Lorraine des Sciences* 19: 39–97.

Société des Naturalistes Luxembourgeois (Gesellschaft Luxemburger Naturfreunde)

Bb509. **Société des Naturalistes Luxembourgeois.** 1915. *Festschrift zur Feier des 25 jahrigen Bestehens, 1890–1915.* Luxembourg. lviii+518 pp. & 48 pl.

Bb510. ___. 1940. *Livre jubilaire publié à l'occasion du cinquantenaire de la Société des Naturalistes Luxembourgeois. 1890–1940.* Luxembourg. 271 pp.
Includes **E. Lahr.** Cinquante années d'activité scientifique.

Bb510A. ___. 1990. Livre du centenaire 1890–1990. La Société des Naturalistes Luxembourgeois du point de vue historique. Table générale des publications de la Société de Botanique du Grand–Duché de Luxembourg (1874–1905) et de la Société des Naturalistes Luxembourgeois (1891–1989). *Bulletin de la Société des Naturalistes Luxembourgeois* 91: 467 pp.

Société des Naturalistes Parisiens

Bb511. **Dupuis, C.** 1954. Les publications scientifiques des Naturalistes Parisiens (1904–1954): Historique et tables. *Cahiers des naturalistes* n.s. 10: 129–164.

Bb512. ___. 1979. Histoire naturelle et naturalistes, dans la France de 1904, année de fondations des Naturalistes Parisiens. *Ibid.* 35 (4): 69–106.

Société Neuchâteloise des Sciences Naturelles (*formerly* Société des Sciences Naturelles de Neuchâtel)

Bb513. **Société Neuchâteloise des Sciences Naturelles.** 1899. *Table des matières des 4 volumes de Mémoires, [1835–74] et des 25 premiers tomes du Bulletin [1842–97]* *comprenant une table analytique des matières et une liste des auteurs.* By J. de Parregaux. Neuchâtel. x+185 pp.

Bb514. ___. 1923. *Table des matières du 5me des Mémoires et du tome 26 au tome 45 du Bulletin [1897/98–1919/21].* By M. Jaquet. Neuchâtel. 62 pp.

Bb515. ___. 1983. Célébration du 150e anniversaire de la Société Neuchâteloise des Sciences Naturelles. By P. A. Siegenthaler. *Bulletin de la Société Neuchâteloise des Sciences Naturelles* (4) 106: 5–22.

Société Philomathique de Paris

Bb516. **Société Philomathique de Paris.** 1888. *Mémoires publiés par la Société Philomathique à l'occasion du centenaire de sa fondation, 1788–1888.* Paris. xvii+158+294 pp.
Includes **M. Berthelot.** Notices sur les origines et sur l'histoire de la Société Philomathique, pp. i–xvii.

Bb517. ___. 1890. *Table générale par noms d'auteurs des articles contenues dans les cinquième, sixième et septième séries des Bulletins, 1836 à 1888.* Paris. vi+55 pp.

Bb518. ___. 1938. *1788–1938. Cent–cinquantième anniversaire de la Société Philomathique de Paris, fondée le 10 Décembre 1788. Reconnue d'utilité publique le 25 Mars 1879.* Paris, Hermann. 30 pp., illus.

Bb519. **Thomas, A.**, ed. 1990. *La Société Philomathique de Paris et deux siècles d'histoire de la science en France. Colloque du bicentenaire de la Société Philomathique de Paris.* Paris, Presses Universitaires de France. vi+154 pp.
Includes **A. Tuilier.** Les débuts de la Société Philomathique de Paris et le contexte idéologique; **J. Roger.** Les sciences naturelles dans les premières décennies de la Société Philomathique de Paris; Liste des membres les plus éminents de la Société Philomathique de Paris de 1788 à 1900.

Société de Physique et d'Histoire Naturelle de Genève

Bb520. **Miège, J.** 1975. Société de Physique et de Sciences Naturelles de Genève; son apport à la botanique genevoise. *Saussurea* 6: 37–41.

Bb521. **Société de Physique et d'Histoire Naturelle de Genève.** 1891. *Centenaire de la fondation de la Société, [etc.].* Geneva. Various paging. *Mémoires de la Société de Physique et d'Histoire Naturelle de Genève* volume supplementaire.

Bb522. ___. 1930. Mémoires de la Société de Physique et d'Histoire Naturelle de Genève. Table générale des matières contenues dans les volumes 1–40 [1821–1927] des Matières [etc.]. By J. Briquet. *Mémoires de la Société de Physique et d'Histoire Naturelle de Genève* 40 (4): 435–500.

Bb523. ___. 1933. Comptes–rendus des séances de la Société ... Table générale des matières des v. 1 à 50, [1884–1933]. *Comptes–rendus des séances de la Société de Physique et d'Histoire Naturelle de Genève* 50 (4): [ ]* pp.
[*Not examined – pagination not ascertained.]

Bb524. ___. 1990. Les origines de la Société de Physique et d'Histoire Naturelle (1790–1822): La science genevoise face au modèle français. By R. Sigrist. *Mémoires de la Société de Physique et d'Histoire Naturelle de Genève* 45 (1): viii+236 pp., illus.

Bb525. ___. 1990. Histoire des sciences de la terre entre 1790 et 1815 vue à travers les documents inédits de la Société de Physique et d'Histoire Naturelle de Genève. Trois grands protagonistes: Marc–Auguste Pictet, Guillaume–Antoine Deluc et Jean Tollot. By A. V. Carozzi. *Ibid.* 45 (2): xxiv+411 pp., illus.

Bb526. ___. 1990. Quelques aspects de la science genevoise vus à travers les activités de la SPHN. By J.–C. Pont [et al.]. *Ibid.* 45 (3): iv+114 pp., illus.

Société Royale Linnéenne de Bruxelles

Bb527. **Société Royale Linnéenne de Bruxelles.** 1900. *La Société Royale Linnéenne de Bruxelles; son origine et son histoire.* By G. de Letter. Brussels. 59 pp.

Bb528. [Not used.]

Société des Sciences et Arts de Vitry–le–François

Bb529. **Société des Sciences et Arts de Vitry–le–François.** 1900. Table par noms d'auteurs et par matières des travaux de la Société de 1861 à 1900 (20 volumes). By Dr. Mougin. *Mémoires, Société des Sciences et Arts de Vitry–le–François* 20: 521–551.

Bb530. ___. 1946. *Tables par noms d'auteurs et par matières des travaux de la Société des Sciences et Arts de Vitry–le–François. De 1861 à 1940.* Châlons–sur–Marne. 46 pp.

Société des Sciences Historiques et Naturelles de la Corse

Bb531. **Société des Sciences Historiques et Naturelles de la Corse.** 1914. Catalogue des publications de la Société des sciences historiques et naturelles de la Corse depuis 1881 jusqu'en 1914. Catalogue chronologique, méthodique et par noms d'auteurs [etc.]. By M. A. Ambrosi. *Bulletin de la Société des Sciences Historiques et Naturelles de la Corse* 34 (361/363): 59 pp.; ___. 1930. Catalogue chronologique, méthodique et par noms d'auteurs des études et documents publiés par le Bulletin ... de 1914 à 1925. Faisant suite au Catalogue dressé en 1914 [etc.]. By M. A. Ambrosi. *Ibid.* 485/489: [ ]* pp.
[*Not examined – pagination not ascertained.]

Société des Sciences Historiques et Naturelles de l'Yonne

Bb532. **Société des Sciences Historiques et Naturelles de l'Yonne.** 1858. Table alphabétique des matières contenues dans les dix premiers volumes [1847–56] du Bulletin [etc.]. By –. Prot. *Bulletin de la Société des Sciences Historiques et Naturelles de l'Yonne* 10: xvi+107 pp.; *Continued* ___. 1875. Tables analytiques ... Première série – deuxième partie [vols. 11–20] 1857–1867. By –. Berthelot. ... partie scientifique. *Ibid.* 20: 156 pp.; ___. 1882. *Tables analytiques. Deuxième série, 1867–1878.* By J. Colin. Auxerre, Paris. 43 pp.; ___. 1910? *Tables analytiques ... 3e sér. vol. I – 4e sér. vol.XIV & sér. gén. vols. XXXIII–LXIV (1879–1910).* Auxerre. [ ]* pp.; ___. 1930? *Tables analytiques ... 4e sér. vol. XV – 5e sér. vol. XIV & sér. gén. vols. LXV–LXXXIV (1911–1930).* Auxerre. [ ]* pp.
[*Not examined – pagination not ascertained.]

Société des Sciences Naturelles de la Charente–Maritime

Bb533. **Société des Sciences Naturelles de la Charente–Maritime.** 1986. *Cent cinquantenaire de la Société des Sciences Naturelles de la Charente–Maritime, 1836–1986.* Ed. by R. Duguy. La Rochelle. 87 pp. *Annales de la Société des Sciences Naturelles de la Charente–Maritime* supplement.

Société (Impériale/Nationale) des Sciences Naturelles de Cherbourg

Bb534. **Société des Sciences Naturelles de Cherbourg.** 1864. Table méthodique des matières contenues dans les 10 volumes composant la 1re série (1852–1864) des Mémoires [etc.]. *Mémoires de la Société Impériale des Sciences Naturelles de Cherbourg* 10: 345–382; *Continued* ___. 1876. Table générale ... 2e série (t. 11 à 20, 1865–1876) des Mémoires [etc.]. By A. Le Jolis. *Mémoires de la Société Nationale des Sciences Naturelles de Cherbourg* 20: 401–429; ___. 1897. Table générale ... 3e série (t. 21 à 30, 1877–1897) des Mémoires [etc.]. *Ibid.* 30: 355–383; ___. 1929. Table générale ... (tomes 31–40 de la 4e série ... publiés de 1900 à 1929) [etc.]. *Mémoires de la Société National des Sciences Naturelles et Mathématiques de Cherbourg* 40: 349–352.

Bb535. ___. 1902. *Cinquantenaire de la Société et jubilé de M. Le Jolis son fondateur. Compte–rendu du banquet et la séance extraordinaire tenus a cette occasion le 29 Décembre*

1901. By L. Corbière. Cherbourg. 96 pp. *Mémoires de la Société National des Sciences Naturelles et Mathématiques de Cherbourg* vol. 32, no. 1.

Société des Sciences Physiques et Naturelles de Bordeaux

Bb536. **Société des Sciences Physiques et Naturelles de Bordeaux.** 1905. *Table générale des matières des publications de la société de 1850–1890.* By J. Chaine & A. Richard. Bordeaux. 145pp.

Société Scientifique de Bruxelles

Bb537. **Société Scientifique de Bruxelles.** 1904. *Revue des questions scientifiques. Table analytique des cinquante premiers volumes, 1877–1901.* Louvain. xii+168 pp.; *Continued* ___. 1922. *Table analytique des volumes de la troisième série (tomes 51–80 — 1902–1921).* Louvain. x+162 pp.; ___. 1937. *Table analytique des volumes de la quatrième série (tomes 31 [i.e. 81]–110 — 1922–1936).* Louvain. 184 pp.

Bb538. ___. 1904. *Annales de la Société Scientifique de Bruxelles. Table analytique des vingt–cinq premiers volumes, 1875–1901, précédée de l'histoire documentaire de la Société scientifique et de la liste générale de ses membres.* Louvain. 250 pp.; *Continued* ___. 1927. *Table analytique des volumes 25 à 46, 1902–1927.* Louvain. 91 pp.

Société Vaudoise des Sciences Naturelles

Bb539. **Société Vaudoise des Sciences Naturelles.** [1870.] Tables des dix premiers volumes [1842–71] du Bulletin [etc.]. By E. Schnetzler. *Bulletin de la Société Vaudoise des Sciences Naturelles* 10: 737–774.; *Continued* ___. [1891.] Tables des volumes 11–20 [1871–85 [etc.]. *Ibid.* 26 Appendix: lxv pp.; ___. [1904.] Table des matières des volumes 21–30 [1885–94] ... Bulletins nos. 92 à 116 [etc.]. *Ibid.* 40 Appendix: l pp.; ___. 1908. Table des matières des vols. 31–40 [1895–1904] ... Bulletins nos. 117 à 151. *Ibid.* 44 Appendix: xlvi pp.; ___. 1916. Table des matières des volumes 41 à 50 [1905–14/15]. By A. Maillefer. *Ibid.* 51 (188): 64 pp.; ___. 1947. Tables des matières des vols. 51–60, 1916–39. *Ibid.* 60 Appendix: [ ]* pp.
[*Not examined – pagination not ascertained.]

Bb540. ___. 1919. *Centenaire de la Société Vaudoise des Sciences Naturelles [1818–1918].* Lausanne. 91 pp.

Bb541. ___. 1969. *Cent cinquantième anniversaire de la Société Vaudoise des Sciences Naturelles: Céremonie commemorative ... à Lausanne, 16 Novembre 1969.* And *La Société Vaudoise des Sciences Naturelles, 1819–1969.* By C. Secrétan. Lausanne. 79 pp.

Society for General Microbiology

Bb542. **Society for General Microbiology.** 1966. The early history of the Society for General Microbiology. By L. A. Allen. *Journal of general microbiology* 42: 171–173.

Srpska Akademija Nauka i Umetnosti, Belgrade (*formerly* Srpska Kraljevska Akademija [etc.])

Bb543. **Srpska Akademija Nauka i Umetnosti.** 1961. *Pregled izdan'a Srpske Akademije Nauka i Umetnosti, 1847–1959.* Belgrade. 273 pp.
List of serial publications and their contents.

Suomalainen Tiedeakatemia (Academia Scientiarum Fennicae)

Bb544. **Suomalainen Tiedeakatemia.** 1946. Suomalainen Tiedeakatemian toimitusen A–sarjan nid. I–LIX hakemisto. Index Annalium Academiæ Scientiarum Fennicæ ser. A. tom. I.–LIX [1909–43.] *Annales Academie Scientiarum Fennicae, Ser. A* 60: 86 pp.

(Cesarsko–Królewskie) Towarzystwo Naukowe Krakowskie (*later re-constituted as* Polska Akademija Umiejetnosci q.v.)

Bb545. **Polska Akademija Umiejetnosci, Krakow.** 1974. *Polska Akademia Umiejetnosci, 1872–1952: Nauki lekarskie, scisle, przyrodnicze i o ziemi. Materialy sesjijubileuszowej, Krakow, 14 XII 1972.* Cracow, Ossolineum. 435 pp.

Towarzystwo Naukowe Warszawskie

Bb546. **Polska Akademia Nauk, Orrodek Rozpowszechniania Wydawnictw Naukowych.** 11966. *Katalog wydawnictw Towarzystwa Naukowego Warszawskiego, Instytu im M. Nenckiego, Kasy im J. Mianowskiego, Biblioteki Kornickiej (pozostalosci z lat 1844–1952).* Ed. by J. Kobuszewska. Warsaw. 204 pp.

Bb547. **Towarzystwo Naukowe Warszawskie.** 1933. *Katalog wydawnictw, 1907–1932.* By H. Szellerowa. Warsaw. 261 pp.

Ural'skoe Obshchestvo Lyubetelei Estestvoznaniya (Société Ouralienne d'Amateurs des Sciences Naturelles)

Bb548. **Ural'skoe Obshchestvo Lyubetelei Estestvoznaniya.** 1930. *Ukazatel' k tomakh 1–40 "Zapisok Ural'skogo Obshchestva Lyubitelei Estestvoznaniya" v Ekaterinburge (Sverdlovske) c 1871 po 1928 god.* By P. P. Makarov. Sverdlovsk. 47 pp.

Verein der Freunde der Naturgeschichte in Mecklenburg

Bb549. **Verein der Freunde der Naturgeschichte in Mecklenburg.** 1856. Systematisches Inhaltsverzeichniss (Alphabetisches Register) zu *Archiv*, Jahrg. 1–10, [1847–56]. By J. Ritter. *Archiv des Vereins der Freunde der Naturgeschichte in Mecklenburg* 10: 137–187; *Continued* ___. 1866. Systematisches Inhaltsverzeichniss zu Archiv XI. bis XX [1857–66]: Autorenregister. *Ibid.* 20: 129–144; ___. 1879. *Systematisches inhaltsverzeichniss zu den Jahrgängen XXI–XXX und alphabetisches Register zu den Jahrgängen XI–XXX , [1857–76.]* By L. Madauss. Neubrandenberg. 62 pp.; ___. 1897. *Systematisches Inhaltsverzeichniss und alphabetisches Register zu Archiv, Jahrg. XXX–L, des Archiv ... nebst Autorenregister zu den Sitzungsberichten der Naturforschenden Gesellschaft zu Rostock.* By E. Geinitz. Güstrow. 45 pp.; ___. 1907. [Index to vols. 51–60, 1897–1906.]* *Ibid.* 61: 14 pp.
[*Not examined – title not ascertained.]

Verein für Natur– und Heilkunde zu Pressburg [= Bratislava] (*formerly* Verein für Naturkunde)

Bb550. **Bokesova–Uherova, M.** 1958. Bratislaver Médizinisch–Naturwissenschaftlicher Verein (1856–1948). *Biologické práce, Slovenskej Akademie Vied, Bratislava* 4(3): 1–86.

Verein für Naturkunde zu Cassel

Bb551. **Verein für Naturkunde zu Cassel.** 1886. *Festschrift des Vereins für Naturkunde zu Cassel zur Feier seines fünfzigjährigen Bestehens.* Cassel, L. Döll. xxvii+279 pp. & 3 pl.
Includes **E. Gerland.** Geschichte des Vereins für Naturkunde. (28 pp.).

Bb552. ___. 1912. *Festschrift des Vereins für Naturkunde zu Cassel zur Feier seines fünfundsiebzigjährigen Bestehens [etc.].* Ed. by B. Schaefer. Cassel. [iii]+xii+341 pp.
Includes **B. Schaefer.** Geschichte des Vereins, (20 pp.); ___. Verzeichnis der Abhandlungen und der wichtigsten Mitteilungen aus den Schriften des vereins für Naturkunde zu Cassel von 1836–1911, (40 pp.).

Verein für Naturkunde zu Zwickau

Bb553. **Verein für Naturkunde zu Zwickau.** 1912. *Festschrift zur Feier des 50 jährigen Bestehens 1862–30 Mai–1912, [etc.].* Zwickau. xxxii+152 pp. & 1 pl. *Jahresbericht des Vereins für Naturkunde zu Zwickau* 1910 & 1911.

Verein für Vaterländische Naturkunde in Württemberg

Bb554. **Verein für Vaterländische Naturkunde in Württemberg.** 1883. General–Register zu den Jahresheften ... Jahrgang 1–39, 1845–83. By E. Koch. *Jahreshefte des Vereins für Vaterländische Naturkunde in Württemberg* 39: 322–381.; *Continued* ___. 1915. General–Register zu den Jahrgängen 40–70, 1884–1914. By J. Eichler. *Ibid.* 71 Appendix: 88 pp.; ___. 1953. Hauptverzeichniss, Jahrgängen 71–106, 1915–50. *Ibid.* 108 Beilage: [ ]* pp.
[*Not examined – pagination not ascertained.]

Verein zür Verbreitung Naturwissenschaftlicher Kenntnisse in Wien

Bb555. **Verein zür Verbreitung Naturwissenschaftlicher Kenntnisse in Wien.** 1910. *Festschrift zur Feier des funfzigjährigen Bestandes des Vereines zur Verbreitung Naturwissenschaftlicher Kenntnisse in Wien am 30. März 1910.* Ed. by F. Toula. Vienna, W. Braumüller. ii+267 pp. & 3 pl.
Includes Chronik des Vereins, 1860–1910.

Wesley Scientific Society, London

Bb556. **Andrews, S.** 1971. The *Wesley Naturalist. History today* 21: 810–817.

Wetterauische Gesellschaft für die Gesamte Naturkunde zu Hanaua. M

Bb557. **Wetterauische Gesellschaft für die Gesamte Naturkunde zu Hanau a. M.** 1908. *Geschichte der Wetterauischen Gesellschaft für die Gesammte Naturkunde. Eine Festgabe zur Feier ihres hundertjährigen Bestehens.* By J. Zingel. Hanau. 215 pp.

Woolhope Naturalists' Field Club

Bb558. **Woolhope Naturalists' Field Club.** 1915. *Index to the Transactions ... 1852 to 1911.* By J. C. Fowler. Hereford. vi+118 pp.; ___. 1939. *Index to the Transactions ... General index 1912–1935 and index of illustrations 1852 to 1935.* By F. Boddington. Hereford. viii+[ii]+139 pp.

Worcestershire Naturalists' Club

Bb559. **Worcestershire Naturalists' Club.** 1980. *The lookers–out of Worcestershire: A memoir of Edwin Lees, Worcester naturalist, and the history of the first hundred years of the Worcestershire Naturalists' Club.* By M. M. Jones. Worcester. x+216 pp.

X Club

Bb560. **Barton, R.** 1990. "An influential set of chaps": The X–Club and Royal Society politics, 1864–85. *British journal for the history of science* 23: 53–81.
Members included G. Busk, J. D. Hooker, J. Lubbock, H. Spencer, T. H. Huxley, etc.

Bb561. **Jensen, J. V.** 1970. The X Club: Fraternity of Victorian scientists. *British journal for the history of science* 5: 63–73.

Bb562. ___. 1972. Interrelationships within the Victorian "X Club." *Dalhousie review* 51: 539–552.

Bb563. **MacLeod, R. M.** 1969. The X Club: A social network of science in late–Victorian England. *Notes & records of the Royal Society of London* 24: 305–322.

Yorkshire Philosophical Society

Bb564. **Yorkshire Philosophical Society.** 1973. *Philosophers and provincials: The Yorkshire Philosophical Society from 1822 to 1844.* By A. D. Orange. York. 76 pp.

(Kaiserlich–Königlich) Zoologisch–Botanische Gesellschaft in Wien (*formerly* Zoologisch–botanischer Verein, Wien)

Bb565. **Blücher, F., & G. Kraatz.** 1876. Entomologisches Inhalts–Verzeichniss zu den Verhandlungen der ... Gesellschaft ... Jahrgang 1–25, 1852–75. *Deutsche entomologische Zeitschrift* 20: 3–30.

Bb566. **Kaiserlich–Königlich Zoologisch–Botanische Gesellschaft in Wien.** 1857. *Personen–, Orts– und Sach–Register der fünf ersten Jahrgänge (1851–55) der Sitzungsberichte und Abhandlungen der Gesellschaft.* By A. Marschall. Vienna. 156 pp.; *Continued* ___. 1862. *Register ... die zweiten fünfjährigen Reihe [1856–60].* By A. Marschall. Vienna. 162 pp.; ___. 1872. Register ... die zweiten zehnjährigen Reihe [Jahrg. 11–20], (1861–70). By A. Marschall. *Verhandlungen der Kaiserlich–Königlichen Zoologisch–Botanischen Gesellschaft in Wien* 22: 484 pp.; ___. 1884. *Personen– Ort– und Sach–Register der dritten zehnjährigen Reihe [Jahrg. 21–30], (1871–80) der Sitzungsberichte und Abhandlungen, [etc.].* By A. Wimmer. Vienna, A. Höldser. 123 pp.; ___. 1895. *Personen– Ort– und Sach–Register der vierten zehnjährigen Reihe [Jahrg. 31–40] (1881–1890) der der Sitzungsberichte und Abhandlungen, [etc.].* By J. A. Knapp. Vienna. 337 pp.

Bb567. ___. 1901. *Botanik und Zoologie in Österreich in den Jahren 1850 bis 1900.* Festschrift herausgegeben von K. K. Zool.–Bot. Gesellschaft in Wien ... anlässlich der Feier ihres fünfzigjährigen Bestandes. Vienna, A. Hölder. ix+620 pp., 5 pl. & 33 port.
Includes **C. Brunner von Wattenwyl.** Geschichte der Kaiserlich–Königlichen Zoologisch–Botanischen Gesellschaft.

Bb568. ___. 1951. *Festschrift anlasslich des 100–jährigen Bestandes der Zoologisch–Botanisch Gesellschaft in Wien.* Ed. by R. Biebl & R. Schubert–Soldern. Vienna. 267 pp. *Verhandlungen der Zoologisch–Botanischen Gesellschaft in Wien* vol. 92.
Includes **J. Gicklhorn.** 1951. 100 Jahre Zoologisch–Botanische Gesellschaft in Wien, pp. 7–24.

North American

Academy of Natural Sciences of Philadelphia

Bb569. **Academy of Natural Sciences of Philadelphia.** 1909. *A short history of the Academy of Natural Sciences at Philadelphia.* By E. J. Nolan. Philadelphia, PA. 38 pp., 3 pl., & 4 ports.

Bb570. ___. 1913. *An index to the scientific contents of the Journal and Proceedings of the Academy ... 1812–1912; published in commemoration of the centenary of the Academy, March 21, 1912.* Ed. by E. J. Nolan. Philadelphia, PA. xiv+1,419 pp.

Bb571. ___. 1948. A brief history of Academy publications. By M. E. Phillips. *Proceedings of the Academy of Natural Sciences of Philadelphia* 50: i–xl.

Bb572. ___. 1967. *Minutes and correspondence of the Academy of Natural Sciences of Philadelphia, 1812–1924.* Philadelphia, PA. 38 reels of microfilm.; With: *Guide to the microfilm publication [etc.].* By V. T. Phillips. Ed. by M. E. Phillips. Philadelphia, PA. 92 pp.
Filmed from the original archives.

Bb573. **Bennett, T. P.** 1983. The history of the Academy of Natural Sciences of Philadelphia. In: Society for the Bibliography of Natural History. *Contributions to the history of North American natural history: Papers from the first North American conference of the Society for the Bibliography of Natural History held at the Academy of Natural Sciences, Philadelphia, 21–23 October 1981*, pp. 1–14. London. *Special publication* no. 2.

Bb574. **Gerstner, P. A.** 1976. The Academy of Natural Sciences of Philadelphia, 1812–1850. In: A. Oleson & S. C. Brown, eds. *The pursuit of knowledge in the early American republic: American scientific and learned societies from colonial times to the civil war*, pp. 174–193. Baltimore, MD, Johns Hopkins Univ. Pr.

Bb575. **Meisel, M.** 1926. Academy of Natural Sciences of Philadelphia (1812). In: *A bibliography of American natural history, the pioneer century, 1769–1865, [etc.]*, vol. 2: 130–218. Brooklyn, NY, Premier Publishing Co. *Re-pub.* New York, Hafner, 1967.
Historical note and bibliography (including list of natural history publications in its *Journal* and *Proceedings* 1817–65).

Bb576. **Porter, C. M.** 1979. The concussion of revolution: Publications and reform at the early Academy of Natural Sciences, Philadelphia, 1812–1842. *Journal of the history of biology* 12: 273–292.

Bb576A. **Rainger, R.** 1992. The rise and decline of a science: Vertebrate paleontology at Philadelphia's Academy of Natural Sciences, 1820–1900. *Proceedings of the American Philosophical Society* 136: 1–32.

See also Ba818; Guide manuscripts & archives Bc710; Ca782 (Rehn).

American Academy of Arts & Sciences

Bb577. **American Academy of Arts & Sciences.** 1882. Centennial celebration of the Academy [etc.]. *Memoirs of the American Academy of Arts & Sciences* n.s. 11: 103 pp.

Bb577A. ___. 1940. *The American Academy of Arts & Sciences, 1780–1940.* Boston, MA 19 pp., illus.

Bb578. **Meisel, M.** 1926. American Academy of Arts & Sciences, Boston. In: *A bibliography of American natural history, the pioneer century, 1769–1865, [etc.]*, vol. 2: 38–56. Brooklyn, NY, Premier Publishing Co. *Re-pub.* New York, Hafner, 1967.
Historical note and bibliography (including list of natural history publications in its *Proceedings* 1780–65).

American Association for the Advancement of Science

Bb579. **American Association for the Advancement of Science.** 1948. *A brief history of the Association from its founding in 1848 to 1948, its present organization and operation, summarized proceedings ... 1940 ... to 1948 [etc.].* Washington, DC. vii+1219 pp.

Bb580. **Kohlstedt, S. G.** 1976. Savants and professionals: The American Association for the Advancement of Science, 1848–1860. In: A. Oleson & S. C. Brown, eds. *The pursuit of knowledge in the early American republic: American scientific and learned societies from colonial times to the civil war*, pp. 299–325. Baltimore, MD, Johns Hopkins Univ. Pr.

Bb581. ___. 1976. *The formation of an American scientific community: The American Association for the Advancement of Science 1848–1860.* Urbana, IL, Univ. of Illinois Pr. xiii+264+[66] pp.
Appendix. Biographical directory of the AAAS, 1848–60. Leadership, [337 biogr. sketches] and Membership [A–Z list].

Bb582. **Meisel, M.** 1929. American Association for the Advancement of Science (1848). In: *A bibliography of American natural history, the pioneer century, 1769–1865, [etc.]*, vol. 3: 50–83. Brooklyn, NY, Premier Publishing Co. *Re-pub.* New York, Hafner, 1967.
Historical note and bibliography (including list of natural history publications in its *Proceedings* 1848–60).

Bb583. **Wolfe, D.** 1970. *Renewing a scientific society: The American Association for the Advancement of Science from World War II to 1970.* Washington, DC, A A A S. ix+337 pp.

American Microscopical Society

Bb584. **American Microscopical Society.** 1890. General index to the Proceedings, including the convention of 1878 to the end of 1890. *Transactions of the American Microscopical Society* 12: 43 pp. *Also pub. separately* Washington, DC, Judd & Detweiler.

Bb585. ___. 1903. Index to volumes 1–25 [1878–1903]. *Ibid.* 25: 187–234.

Bb586. ___. 1979. American Microscopical Society: The first one–hundred years. Part 1 (–2). By G. W. Prescott. *Ibid.* 98: 1–25, 165–194.

Bb587. **Kraus, W. C.** 1905. Twenty–five years of the American Microscopical Society — a retrospect. *Ibid.* 26: 137–144.

American Philosophical Society

Bb588. **American Philosophical Society.** 1865. *List of members of the American Philosophical Society [etc.].* Philadelphia, PA. 56 pp.

Bb589. ___. 1889. *Subject register of papers published in the Transactions [vols. 1–6 old series, vols. 1–16 new series] and Proceedings [Vols. 1–25], [etc.] of the ... Society.* Philadelphia, PA. 75 pp.

Bb590. ___. 1893. 1743–1895. Proceedings commemorative of the hundred and fiftieth anniversary of the foundation of the American Philosophical Society held at Philadelphia for the Promotion of Useful Knowledge, May 22–26, 1893. *Proceedings of the American Philosophical Society* 32 (143): 160 pp., illus.

Bb591. ___. 1900. *Proceedings of the ... Memorial volume I.* Philadelphia, PA. 316 pp.
Biographical notices of past members.

Bb591A. ___. 1912. *General index to the Proceedings ... vols. 1–50, 1838–1911.* Philadelphia, PA. 83 pp.; *Continued* ___. 1937. *Index to vols. 51–75, 1912–1935.* Philadelphia, PA. 71 pp.; ___. 1957. *Proceedings ... Cumulative index to vols. 76–100, 1936–1956.* Philadelphia, PA. 48 pp.; ___. 1982. *Proceedings ... Cumulative index to vols. 101–125, 1957–1981.* Philadelphia, PA. 51 pp.

Bb592. ___. 1927. The record of the celebration of the two hundredth anniversary of the founding of thew American Philosophical Society ... April 27 to April 30, 1927. *Proceedings of the American Philosophical Society* 66: xiii+750 pp.
Includes **C. E. McClung.** The development of biological research in America, pp. 67–78; **J. M. Coulter.** Development of botany in the United States, pp. 309–318; **W. B. Scott.** Development of American palaeontology, pp. 409–430; **H. F. Osborn.** Recent discoveries relating to the origin and antiquity of man, pp. 373–389.

Bb592A. ___. 1940. *Classified index to the publications of the American Philosophical Society: A list of papers, monographs, treatises and books published by the Society, 1769–1940, classified according to subject [etc.].* Philadelphia, PA. v+173 pp.

Bb593. ___. 1942–43. The early history of science and learning in America with especial reference to the work of the American Philosophical Society during the eighteenth and nineteenth centuries. Papers read before the American Philosophical Society Midwinter and Autumn meetings, 1942. *Proceedings of the American Philosophical Society* 86 (1): 1–204, 1942; *Ibid.* 87 (1): 1–119, 1943.
Papers by various authors. *Includes* vol. 86 (1): **B. Willis.** American geology, 1850–1900, pp. 34–44; **M. L.** Wilson. Survey of scientific agriculture, pp. 52–62; **M. L. Fernald.** Some early botanists of the American Philosophical Society, pp. 65–67; **E. D. Merrill.** A generally overlooked Rafinesque paper, pp. 72–90; **P. P. Calvert.** Entomology, scientific and human aspects, pp. 123–129; **G. G. Simpson.** The beginnings of vertebrate paleontology in North America, pp. 130–188; vol. 87 (1): **G. Chinard.** The American Philosophical Society and the world of science, 1768–1800, pp. 1–11; **A. Hrdlicka.**

Contribution to the history of physical anthropology in the United States of America, pp. 61–64; **E. D. Merrill.** Rafinesque's publications from the standpoint of world botany. pp. 110–119.

Bb594. ___. 1955. A brief history of the American Philosophical Society. By E. G. Conklin. *Yearbook of the American Philosophical Society* 1954: 7–28.

Bb595. **Bell, W. J., Jr.** 1964. The American Philosophical Society as a National Academy of Sciences, 1780–1846. *Proceedings of the 10th International Congress of the History of Science* 1: 165–177.

Bb596. **Chinard, G.** 1945. The American Philosophical Society and the early history of forestry in America. *Proceedings of the American Philosophical Society* 89: 444–487.

Bb597. **Glass, B.** 1986. Foreign membership of biological scientists in the American Philosophical Society during the eighteenth and nineteenth centuries. In: R. S. Klein, ed. *Science and society in early America: Essays in honor of Whitfield J. Bell, Jr.*, pp. 345–392. Philadelphia, PA, American Philosophical Society. *Memoirs of the American Philosophical Society* vol. 166.

Bb598. **Meisel, M.** 1926. American Philosophical Society. In: *A bibliography of American natural history, the pioneer century, 1769–1865, [etc.]*, vol. 2: 1–35. Brooklyn, NY, Premier Publishing Co. *Re-pub.* New York, Hafner, 1967.
Historical note and bibliography (including list of natural history publications in its *Transactions* and *Proceedings* 1768–65).

Bb599. **Rosengarten, J. G.** 1907. The early French members of the American Philosophical Society. *Proceedings of the American Philosophical Society* 46: 87–93.

See also Guides to the Society's archives Bc725; Ca782 (Conklin).

American Nature Study Society

Bb600. **Palmer, E. L.** 1957. Fifty years of nature study and the American Nature Study Society. *Nature magazine* 50: 473–480.

American Society of Limnology & Oceanography

Bb601. **Lauff, G. H.** 1963. A history of the American Society of Limnology and Oceanography. In: D. G. Frey. *Limnology in North America*, pp. 667–682. Madison, WI, Univ. of Wisconsin Pr.

American Society of Naturalists

Bb602. **Appel, T. A.** 1988. Organizing biology: The American Society of Naturalists and its "affiliated societies", 1883–1923. In: R. Rainger, K. R. Benson, & J. Maienschein, eds. *The American development of biology*, pp. 87–120. Philadelphia, PA, Univ. of Pennsylvania Pr.

Association of American Geologists and Naturalists

Bb603. **Kohlstedt, S. G.** 1976. Ch. III. The Association of American Geologists & Naturalists. In: *The formation of an American scientific community: The American Association for the Advancement of Science 1848–1860*, pp. 59–77. Urbana, IL, Univ. of Illinois Pr.
Forerunner of the AAAS, 1840–47.

Bb604. **Meisel, M.** 1926. Association of American Geologists and Naturalists (1840). In: *A bibliography of American natural history, the pioneer century, 1769–1865, [etc.]*, vol. 2: 680–698. Brooklyn, NY, Premier Publishing Co. *Re-pub.* New York, Hafner, 1967.
Historical note and bibliography (including list of natural history publications in its *Abstract of the proceedings* and *Reports* 1840–47).

Bb605. [Not used.]

Biological Society of Washington

Bb606. **Biological Society of Washington.** 1924. *Author index to Proceedings of the Biological Society of Washington ..., vols. 1 to 35, 1881–1922.* By F. C. Lincoln. Washington, DC. 98 pp.

Boston Society of Natural History

Bb607. **Boston Society of Natural History.** 1880. *Anniversary memoirs ... published in celebration of the fiftieth anniversary of the Society's foundation. 1830–1880.* Boston, MA. 250 pp. & 32 pl.
Includes **T. T. Bouvé.** Historical sketch of the Boston Society of Natural History [etc.], pp. 14–250.

Bb608. **___.** 1930. *The Boston Society of Natural History, 1830–1930.* By P. R. Creed. Boston, MA, the Society. xii+117 pp.

Bb609. **Kohlstedt, S. G.** 1976. The nineteenth–century amateur tradition: The case of the Boston Society of Natural History. In: G. Holton & W. A. Blanpied, eds. *Science and its public: The changing image*, pp. 173–190. Dordrecht, Reidel. *Boston studies in the philosophy of science* vol. 33.

Bb610. **___.** 1979. From learned society to public museum: The Boston Society of Natural History. In: A. Oleson & J. Voss, eds. *The organization of knowledge in modern America, 1860–1920*, pp. 386–406. Baltimore, MD, Johns Hopkins Pr.

Bb611. **Meisel, M.** 1926. Boston Society of Natural History (1830). In: *A bibliography of American natural history, the pioneer century, 1769–1865, [etc.]*, vol. 2: 457–517. Brooklyn, NY, Premier Publishing Co. *Re-pub.* New York, Hafner, 1967.
Historical note and bibliography (including list of natural history publications in the *Boston journal of natural history*, the Society's *Memoirs* and *Proceedings* 1833–65).

Buffalo Society of Natural Sciences

Bb612. **Buffalo Society of Natural Sciences.** 1907. *The Buffalo Society of Natural Sciences: Historical sketch.* By H. R. Howland. Buffalo, NY. 25 pp. *Bulletin of the Buffalo Society of Natural Sciences* vol. 8, no. 6.

Bb613. **___.** 1938. *Seventy–five years: A history of the Buffalo Society of Natural Sciences, 1861–1936.* Buffalo, NY. 204 pp. & 53 pl. *Bulletin of the Buffalo Society of Natural Sciences* vol. 18.
Some historical information is also contained in: **R. H. Zander et al.** 1986. Featured institution – Collections of the Buffalo Society of Natural Sciences: Buffalo Museum of Science and Tifft Farm Nature Preserve. *Association of Systematics Collections newsletter* 14 (6): 61–65.

Cuvier Natural History Society of Salem

Bb614. **Dexter, R. W.** 1960. The Cuvier Natural History Society of Salem. *Essex Institute historical collections* 96: 149–155; **___.** 1983. More on the Cuvier Natural History Society of Salem. *Ibid.* 119: 95–98.

Ecological Society of America

Bb615. **Burgess, R. L.** 1977. The Ecological Society of America: Historical data and some primary analyses. In: F. N. Egerton, ed. *History of American ecology*, 24 pp. New York, Arno Pr.

Bb616. **Shelford, V. E.** 1938. The organization of the Ecological Society of America, 1914–1919. *Ecology* 19: 164–166.

Elisha Mitchell Scientific Society

Bb617. **Elisha Mitchell Scientific Society.** 1916. General index to vols. 1–31, 1883/84–1916. *Journal of the Elisha Mitchell Scientific Society* 31: 182–213.; *Continued* ___. 1922. Index, vols. 32–37 [1916–22]. *Ibid.* 37: 199–215.; ___. 1929. Index, vols. 38–44 [1922–29]. *Ibid.* 44: 261–296.

Elliott Society of Science & Art (*formerly* Elliott Society of Natural History)

Bb618. **Meisel, M.** 1929. Elliott Society of Natural History, Charleston, S.C. (1853). In: *A bibliography of American natural history, the pioneer century, 1769–1865, [etc.]*, vol. 3: 164–169. Brooklyn, NY, Premier Publishing Co. *Re-pub.* New York, Hafner, 1967.
Historical note and bibliography (including list of natural history publications in its *Proceedings* 1853–60 & *Journal* 1859–60).

Essex Institute, Salem (*formerly* Essex County Natural History Society)

Bb619. **Dexter, R. W.** 1977. The Essex County Natural History Society, 1833–1848. *Essex Institute historical collection* 113: 38–53.

Bb620. ___. 1980. Natural history at the Essex Institute, 1848–1898. *Ibid.* 116: 21–33.

Bb621. **Essex Institute.** 1884. Notes on the condition of zoology, fifty years ago and of today; in connection with the growth of the Essex Institute. By E. S. Morse. *Bulletin of the Essex Institute* 16: 113–121.

Bb622. ___. 1898. The first half–century of the Essex Institute, commemorated at Salem, March 1st and 2nd, 1898. *Ibid.* 30 (1–6): 100 pp. *Also pub. separately* Salem, MA, 100 pp.

Bb623. **Meisel, M.** 1926–29. Essex County Natural History Society, Salem, Mass. (1833). Essex Institute, Salem, Mass., (1848). In: *A bibliography of American natural history, the pioneer century, 1769–1865, [etc.]*, vol. 2: 550–553, 1926; vol. 3: 83–95, 1929. Brooklyn, NY, Premier Publishing Co. *Re-pub.* New York, Hafner, 1967.
Historical note and bibliography (including list of natural history publications in the Society's *Journal* 1836–52, and the Institute's *Proceedings* 1849–65 and *Communications* 1864–65).

Illinois Natural History Society

Bb623A. **Meisel, M.** 1929. Illinois Natural History Society, Bloomington (1858). In: *A bibliography of American natural history, the pioneer century, 1769–1865, [etc.]*, vol. 3: 278–281. Brooklyn, NY, Premier Publishing Co. *Re-pub.* New York, Hafner, 1967.
Historical note and bibliography (including list of natural history publications in its *Transactions* 1858–62).

Linnaean Association of Pennsylvania College, Gettysburg, PA

Bb624. **Meisel, M.** 1926. Linnaean Association of Pennsylvania College, Gettysburg, PA. (1844). In: *A bibliography of American natural history, the pioneer century, 1769–1865, [etc.]*, vol. 2: 730–734. Brooklyn, NY, Premier Publishing Co. *Re-pub.* New York, Hafner, 1967.
Historical note and bibliography (including list of natural history publications in its *Literary record and journal* 1844–49).

Linnaean Society of New England

Bb625. **Bouvé, T. T.** 1880. Historical sketch of the Boston Society of Natural History; with a notice of the Linnaean Society, which preceded it. In: Boston Society of Natural History. *Anniversary memoirs ... published in celebration of the fiftiewth anniversary of the Society's foundation. 1830–1880*, pp. 3–14. Boston, MA.

Literary & Philosophical Society of New York

Bb626. **Harris, J.** 1972. New York's first scientific body: The Literary & Philosophical Society, 1814–1834. *Annals of the New York Academy of Science* 196: 327–337.

Bb627. **Meisel, M.** 1926. Literary and Philosophical Society of New York, New York City (1814). In: *A bibliography of American natural history, the pioneer century, 1769–1865, [etc.]*, vol. 2: 224–227. Brooklyn, NY, Premier Publishing Co. *Re-pub.* New York, Hafner, 1967.
Historical note and bibliography (including list of natural history publications in its *Transactions* 1814–24).

National Academy of Sciences

Bb628. **National Academy of Sciences.** 1877+. *Biographical memoirs of the National Academy of Sciences.* New York, Columbia Univ. Pr. Vol. 1+.

Bb629. ___. 1913. *A history of the first half-century of the National Academy of Sciences: 1863–1913.* Ed. by F. W. True. Washington, DC. xiii+399 pp.

Bb630. ___. 1913. *The semi-centennial anniversary of the National Academy of Sciences, 1863–1913.* Washington, DC. 108 pp.

Bb631. ___. 1926. Publications of the National Academy of Sciences of the United States of America (1915–26). Part I. Index to the first ten volumes of the Proceedings (1915–1924); Part II. List of other publications of the Academy from 1863–1926; Part III. List of publications of the National Research Council from 1916–25. *Proceedings of the National Academy of Sciences of the United States of America* 13 (1): 197 pp.

Bb632. ___. 1966. *History of the Proceedings of the National Academy of Sciences, 1914–1963. Supplemented by an Author index to volumes 1–50.* By E. B. Wilson. Washington, DC. ix+296 pp.

Bb633. ___. 1978. *The National Academy of Sciences: The first hundred years 1863–1963.* By R. C. Cochrane. Washington, DC. xv+694 pp., illus.

National Geographic Society

Bb634. **Pauly, P. J.** 1979. The world and all that is in it: The National Geographic Society, 1888–1918. *American quarterly* 31: 517–532.

Natural History Society of Montreal

Bb635. **Frost, S. B.** 1982. Science education in the 19th century: the Natural History Society of Montreal. *McGill journal of education* 17: 31–48.

New York Academy of Sciences (*formerly* Lyceum of Natural History, New York)

Bb636. **Barnhart, J. H.** 1917. The first hundred years of the New York Academy of Sciences. *Scientific monthly* 5: 463–475.

Bb637. **Fairchild, H. Le R.** 1887. *A history of the New York Academy of Sciences, formerly the Lyceum of Natural History.* New York, the Author. xii+190 pp. & 7 pl.

Bb638. **Meisel, M.** 1926. Lyceum of Natural History of New York (1817). In: *A bibliography of American natural history, the pioneer century, 1769–1865, [etc.]*, vol. 2: 235–273. Brooklyn, NY, Premier Publishing Co. *Re-pub.* New York, Hafner, 1967.
Historical note and bibliography (including list of natural history publications in abstract in American monthly magazine and critical review 1817–19, *American journal of science* 1819–34, and *Annals of the lyceum of natural history of New York* 1824–65).

Bb639. **New York Academy of Sciences.** 1990. Knowledge, culture, and science in the metropolis: The New York Academy of Sciences, 1817–1970. By S. Baatz. *Annals of the New York Academy of Sciences* 584: ix+269 pp.

Bb640. **Nodyne, K. R.** 1970. The founding of the Lyceum of Natural History. *Annals of the New York Academy of Sciences* 172: 139–149.

New York Historical Society

Bb641. **Meisel, M.** 1926. New York Historical Society, New York City (1804). In: *A bibliography of American natural history, the pioneer century, 1769–1865, [etc.]*, vol. 2: 103–106. Brooklyn, NY, Premier Publishing Co. *Re-pub.* New York, Hafner, 1967.
Historical note and bibliography of natural history.

Bb642. **New York Historical Society.** 1905. *The New York Historical Society 1804–1904.* By R. H. Kelby. New York. 160 pp.

Nova Scotian Institute of Science, Halifax

Bb643. **Nova Scotian Institute of Science.** 1953. *Proceedings and transactions of the Nova Scotian Institute of (Natural) Science. Cumulative index to vols. 1–22, 1863/66–1948/50.* By J. G. Aldous. Halifax, N.S. 61 pp.

Ohio (State) Academy of Science

Bb644. **Dexter, R.** 1960. The summer field meetings of the Ohio Academy of Science (1892–1905). *Ohio journal of science* 60: 262–265.

Bb645. **___.** 1962. Conservation and the Ohio Academy of Science — an historical review. *Ibid.* 62: 274–280.

Bb646. **___.** 1966. Milestones: Ohio Academy of Science, 1891–1966. *Ibid.* 66: 353–359.

Ottawa Field Naturalists' Club

Bb647. **Ottawa Field Naturalists' Club.** 1980. *Transactions of the Ottawa Field–Naturalists' Club and The Ottawa naturalist index.* By J. M. Gillett. Ed. by W. J. Cody. Ottawa, Ottawa Field Naturalists' Club. iv+195 pp. *Special publication* no. 2.

Portland Society of Natural History

Bb648. **Meisel, M.** 1926. Portland Society of Natural History (1843). In: *A bibliography of American natural history, the pioneer century, 1769–1865, [etc.]*, vol. 2: 726–729. Brooklyn, NY, Premier Publishing Co. *Re-pub.* New York, Hafner, 1967.
Historical note and bibliography (including list of natural history publications in its *Proceedings* and *Journal* 1862–65).

Royal Canadian Institute *(formerly* Canadian Institute)

Bb649. **Canadian Institute.** 1914. *Canadian Institute. General index to publications, 1852–1912.* By J. Patterson. Toronto. [vi]+516+[2] pp.

Bb650. **Royal Canadian Institute.** 1949. *The Royal Canadian Institute centennial volume, 1849–1949.* Ed. W. S. Wallace. Toronto. ix+232 pp., illus.
Pt. II. A sketch of the history of the Royal Canadian Institute, 1849–1949. Appendix. Officers and councillors of the Royal Canadian Institute, 1849–1949.

Royal Society of Canada

Bb651. **Canadian Library Association.** 1947. *A subject index to the Royal Society of Canada Proceedings and Transactions, third series, vols. I–XXXI, 1907–1937.* Ottawa. 143 pp. *Reference publication* no. 1.

Bb652. **Rousseau, J., M. Gauvreau & C. Morin.** 1939. *Bibliographie des travaux botaniques contenus dans les 'Memoires et comptes rendus de la Société Royale du Canada', de 1882 à 1936 inclusivement.* Montreal, Université de Montréal, Institut Botanique. 117 pp.*Contribution* no. 33.

Bb653. **Royal Society of Canada.** 1894. *Bibliography of the members of the Royal Society of Canada.* By J. G. Bourinot. [Ottawa.] 79 pp.

Bb654. **Royal Society of Canada.** 1908. *General index. Proceedings and Transactions of the Royal Society of Canada. First and second series, 1882–1906.* Ottawa. [vi]+133 pp.

Bb655. ___. 1932. *Fifty years retrospect. Anniversary volume, 1882–1932.* [Ottawa.] xxiv+179 pp.
Essays by various authors including lists of officers, presidents, fellows, etc. and other material on the history of the society.

Bb656. ___. 1943. *Author index. Proceedings and Transactions of the Royal Society of Canada. Sections I–V. Third series: vols. I–XXXV, 1907–1941.* Ottawa. 90 pp.

Bb657. ___. 1966. *Royal Society of Canada. Pioneers of Canadian science. Symposium presented to the Royal Society of Canada in 1964.* Ed. by G. F. G. Stanley. Toronto, Univ. of Toronto Pr. for the Society. xiii+146 pp., illus.

San Diego Society of Natural History

Bb658. **San Diego Society of Natural History.** 1924. *The San Diego Society of Natural History, 1874–1924. A brief history prepared upon the occasion of the fiftieth anniversary of its incorporation.* San Diego, CA. 24 pp.

Sierra Club

Bb659. **Schrepfer, S. R.** 1989. Establishing administrative standing: The Sierra Club and the Forest Service, 1897–1956. *Pacific historical review* 58: 55–81.

Bb660. **Sierra Club.** 1952. *Fifty–seven–year index: Sierra Club bulletin, 1893–1949.* By D. H. Bradley & G. Shochat. San Francisco, CA. 138 pp.; *Continued* ___. *An index to the Sierra Club bulletin, 1950–1975, volumes 35–61.* By E. B. Brazee. Corvallis, OR, Oregon State University Press. x+42 pp. *Oregon State University bibliographic series* no. 16.

Bb661. ___. 1988. *The history of the Sierra Club, 1892–1970.* By M. P. Cohen. San Francisco, CA, Sierra Club Books. xvii+550 pp.

Bb662. **Turner, T.** 1991. *Sierra Club: 100 years of protecting nature.* New York, H. N. Abrams in association with the Sierra Club. 288 pp., illus.

Sociedad Mexicana de Historia Natural

Bb663. **Beltrán, E.** 1964. Reseña historica de la Revista de la Sociedad Mexicana de Historia Natural et su jubileo de plata. *Revista de la Sociedad Mexicana de Historia Natural* 25: 19–28.

Bb664. ___. 1968. El primer centenario de la Sociedad Mexicana de Historia Natural (1868–1968). *Ibid.* 29: 111–169.

Bb665. **Riquelme, I. J.** 1948. Los presidentes de la Sociedad Mexicana de Historia Natural en su 1a época. *Ibid.* 9: 303–329.

Williams College Lyceum of Natural History, Williamstown, MA

Bb666. **Meisel, M.** 1926. Williams College Lyceum of Natural History (1835). In: *A bibliography of American natural history, the pioneer century, 1769–1865, [etc.]*, vol. 2: 596–597. Brooklyn, NY, Premier Publishing Co. *Re-pub.* New York, Hafner, 1967.
Historical note and bibliography (including list of natural history papers 1836–57).

Bb667. **Smallwood, W. M.** 1937. The Williams College Lyceum of Natural History, 1835–1885. *New England quarterly* 10: 553–557.

Yale Natural History Society, New Haven, CT

Bb668. **Meisel, M.** 1926. Yale Natural History Society (1834). In: *A bibliography of American natural history, the pioneer century, 1769–1865, [etc.]*, vol. 2: 563–566. Brooklyn, NY, Premier Publishing Co. *Re-pub.* New York, Hafner, 1967.
Historical note and bibliography (including list of natural history papers 1836–45).

Young Naturalists' Society (Seattle)

Bb669. **Benson, K. R.** 1986. The Young Naturalists' Society and natural history in the Northwest. *American zoologist* 26: 351–361.

Bb670. ___. 1986. The Young Naturalists' Society: From chess to natural history collections. *Pacific northwest quarterly* 77: 82–93.

Asian

(Royal) Asiatic Society of Bengal

Bb671. **Asiatic Society of Bengal.** 1835. Index to the first eighteen volumes of the Asiatic Researches [1788–1833] [etc.]. *Asiatic researches* 18 [appx.]: [iv]+228 pp.; ___. 1856. *Index to volumes 19 & 20 of the Asiatic Researches and to vols 1–23, [1832–54] of the Journal [of the Asiatic Society of Bengal]*. Calcutta. 274 pp.
Appendix C. **H. Piddington.** Index to the Indian geological, mineralogical and palæontological papers in the *Researches* and *Journal of the Asiatic Society*, and in the *Gleanings of science*, pp. 210–252.

Bb672. ___. 1885. *Centenary review of the Asiatic Society of Bengal from 1784 to 1883.* Calcutta. 195+iv+216+ciii+109+xcvii+20 pp.
Pt. I: **R. Mitra.** *History of the society.* 195 pp. (including Index to the papers and contributions to the Asiatic researches and the Journal and Proceedings ... pp. 106–195); Pt. III: **P. N. Bose.** *Natural science.* 109 pp.; Classified index to the scientific papers in the Society's publications from 1788 to 1882 [mathematical and physical science, geology, zoology, botany, geography, [etc.]. xcvii pp.

Bb673. **Blanpied, W.** 1973. Notes for a study on the early scientific work of the Asiatic Society of Bengal. *Japanese studies in the history of science* 12: 121–144.

Bombay Natural History Society

Bb674. **Bombay Natural History Society.** 1901. Journal of the Bombay Natural History Society. General index, vols. 1–13, 1886–1901. *Ibid.* 13: xxix–lxvi; *Continued* ___. 1907. General index, vols. 14–17, 1902–1907. *Ibid.* 17: xli–lxxiv; ___. 1917. General index, vols. 18–24, 1907–1916. *Ibid.* 24: 79 pp.; ___. 1928. *General index to vols. 25–30 [1918–25]*. Madras. 53 pp.; ___. 1938. *General index to vols. 31–36 [1926–33]*. Madras. 70 pp.; ___. 1949. *General index of subjects and authors for vols. XXXVII–XLII [1934–41]*. Madras. 48 pp.

Bb675. ___. 1934. *The Bombay Natural History Society, 1883–1933. Printed in commemoration of the golden jubilee of the Society, 1933.* Bombay, Prince of Wales Museum. 101 pp., illus.
Includes a summary of selected contributions to the Society's *Journal.*

Bb676. ___. 1978–86. Bombay Natural History Society: The founders, the builders and the guardians. Part 1 (–4). By S. Ali. *Journal of the Bombay Natural History Society* 75: 559–569, 1978; *Ibid.* 78: 232–239, 1981; *Ibid.* 79: 38–46, 1982; *Ibid.* 80: 320–330, 1983. *Also* ___. 1986. The *Journal*: Its role in Indian natural history. By S. Ali. *Ibid.* 83 Centenary supp: 1–6, 1986.

Bb677. **Ridley, M., & P. Newton.** 1983. Biology under the Raj: The Bombay Natural History Society, now a hundred years old, began as a club for gentlemen amateurs, both British and Indian. *New scientist* 99 (1376): 857–860.

Deutsche Gesellschaft für Natur– und Völkerkunde Ostasiens

Bb678. **Deutsche Gesellschaft für Natur– und Völkerkunde Ostasiens.** 1902. Festschrift zur Erinnerung an das 25 jährige Stiftungsfest der Deutschen Gesellschaft für Natur– und Völkerkunde Ostasiens am 29. Oktober 1898. Tokyo. 97 pp. *Mitteilungen der Deutschen Gesellschaft für Natur– und Völkerkunde Ostasiens, Supplementband* 6.

Includes General-index zu Band I bis VI der Mitteilungen, pp. 11–75; Mitgleider ... 1873–1898, pp. 77–97.

Bb679. **Deutsche Gesellschaft für Natur- und Völkerkunde Ostasiens.** 1938. *Mitteilungen der Deutschen Gesellschaft für Natur- und Völkerkunde Ostasiens. General-Index der "Mitteilungen" Band I–XXXI (incl. Supplementbände Nr. I–XVI) [1873–1938].* Tokyo, Leipzig. 29 pp.

(Koninklijk) Bataviaasch Genootschap van Kunsten en Wetenschappen

Bb680. **Koninklijk Bataviaasch Genootschap van Kunsten en Wetenschappen.** 1853. Overzigt der Geschiednis van het Bataviaasch Genootschap van Kunsten en Wetenschappen. Van 1778–1853. By P. Bleeker. *Verhandelingen van het Bataviaasch Genootschap van Kunsten en Wetenschappen* 25: 23 pp.

Bb681. ___. 1878. *Het Bataviaasch Genootschap van Kunsten en Wetenschappen gedurende de eerste eeuw zijn bestaan, 1778–1878.* Gedenkboek zamengesteld door den voorzitter van het Genootschap T. H. van Kinderen. Deel 1. Batavia, Ernst. xv+261+lxxxvi pp., illus.

Bb682. ___. 1908. *Register op de artikelen voorkomende in het Tijdschrift voor Indische Taal–, Land– en Volkenkunde en de Verhandelingen van het Bataviaasch Genootschap van Kunsten en Wetenschappen loopende tot het jaar 1907 [vols. 1–56, 1778–1906].* By O. van Hinloopen Labberton. Batavia, Albrecht & Co; The Hague, M. Nijhoff. 276+98 pp.; ___. 1920. *Register op de Verhandelingen van het Bataviaasch Genootschap, deel LVI (5e stuk)–LXI (1907–19) en het Tijdschrift voor Indische taal–, land– en volkenkunde, deel LI–LVII (1909–1919).* Weltevreden, "Papyrus". iv+81+14 pp.; ___. 1942. *Register op de uitgaven van het Koninklijk Bataviaasch genootschap van Kunsten en Wetenschappen, 1920–1940.* Bandoeng. 90 pp.

Royal Asiatic Society, Bombay Branch

Bb683. **Royal Asiatic Society, Bombay Branch.** 1886. *Index to the Transactions of the Literary Society of Bombay vols. 1–3 [1819–23] and to the Journals of the Bombay Branch, Royal Asiatic Society vols. 1–17 [1841–1889. Actually, vols. 1–16, 1841–85].* By G. K. Tiverekar. Bombay, Education Society's Pr. 50+82+xlvii+17 pp.

Royal Asiatic Society, Ceylon Branch

Bb684. **Royal Asiatic Society, Ceylon Branch.** 1895. *Index the Journals and Proceedings of the Ceylon Branch of the Royal Asiatic Society. Volumes 1 to 11, comprising numbers 1 to 41 (1845 to 1890).* By J. F. W. Gore. Colombo. vii+144 pp.

Royal Asiatic Society, (North) China Branch

Bb685. **Royal Asiatic Society, North China Branch.** 1894? A classified index to the articles printed in the Journal of the North–China Branch ... from the foundation of the society1858, to the end of 1893. By J. Haas. *Journal of the North China Branch of the Royal Asiatic Society* n.s. 26: 246–297.

Bb686. ___. 1912. List of articles which have appeared in the Journal of the N. C. B. R. A. S. [vols. 1–42, 1858–1911] classified according to the Dewey decimal index as in use in the library — arranged by Mrs F. W. Ayscough. *Journal of the North China Branch of the Royal Asiatic Society* n.s. 43: 201–216.

Bb687. ___. [1924.] Index to the Journal of the North China Branch of the Royal Asiatic Society from vol. 1 to vol. 54 [1858–1923]. By P. B. Ferguson. *Journal of the North China Branch of the Royal Asiatic Society* 55 Supp: vii+54 pp.

Royal Asiatic Society, Straits (Malayan) Branch

Bb688. **Royal Asiatic Society, Straits Branch.** [1928.] An index to all the Journals nos. 1–86 (1878–1922) of the Straits Branch of the Royal Asiatic Society from its foundation until its change of title to Malayan Branch of the Royal Asiatic Society: And to

Notes and Queries 1 to 4. By C. E. Wurtzburg. *Journal of the Malayan Branch of the Royal Asiatic Society* 5 (4): iii+101+v pp.

Science Society of China, Nanking

Bb689. **Science Society of China.** 1931. *Science Society of China. Its history, organization and activities.* Shanghai. [ii]+31+36+2 pp. & 14 pl.

Australasian, etc.

Australian & New Zealand (*formerly* Australasian) Association for the Advancement of Science

Bb690. **Australian & New Zealand Association for the Advancement of Science.** 1925. *Reports ... index to volumes I to XVI, 1888 to 1923.* Part 1. – Index to authors. Part 2. – Index to subjects. Sydney, N.S.W. 95 pp.

Bb691. **McLeod, R. M.**, ed. 1988. *The commonwealth of science: A N Z A A S and the scientific enterprise in Australia, 1888–1988.* Melbourne, Vic., Oxford Univ. Pr. xvi+417 pp.
Includes **L. Gillbank.** The life sciences: Collections to conservation; **J. M. Powell.** Protracted reconciliation: Society and the environment.

Field Naturalists' Club of Victoria

Bb692. **Field Naturalists' Club of Victoria.** 1980. Centenary: Field Naturalists' Club of Victoria. *Victorian naturalist* 97: 93–134.
Papers by various authors.

Linnean Society of New South Wales

Bb693. **Linnean Society of New South Wales.** 1920. The Society's heritage from the Macleays. By J. J. Fletcher. *Proceedings of the Linnean Society of New South Wales* 45: 567–636; *Continued* ___. 1929. The Society's heritage from the Macleays. Part II: Sir William (John) Macleay, M.L.C., F.L.S. By J. J. Fletcher. *Proceedings of the Linnean Society of New South Wales* 54: 185–272.

Bb694. **Walkom, A. B.** 1925. *The Linnean Society of New South Wales, 1874–1925. Historical notes of its first fifty years (Jubilee publication).* Sydney, N.S.W., A/Asian Medical Pub. Co. 46 pp. & port.

Bb695. ___. 1929. *Proceedings of the Linnean Society of New South Wales: Index, vols. 1–50 [1875–1925].* Sydney, N.S.W. 107 pp.

Royal Society of New South Wales (*formerly* Philosophical Society of New South Wales)

Bb696. **Royal Society of New South Wales.** 1918. A contribution to a history of the Royal Society of New South Wales [etc.]. By J. H. Marden. *Journal and proceedings of the Royal Society of New South Wales* 52: 215–361.

Bb697. ___. 1968. *A century of scientific progress: The centenary volume of the Royal Society of New South Wales; a history of several aspects of Australian scientific development with particular reference to New South Wales.* Sydney, N.S.W. 478 pp.

Royal Society of New Zealand (*formerly* New Zealand Institute)

Bb698. **New Zealand Institute.** 1911. *Index to the Transactions and proceedings of the New Zealand Institute, vols. 1 to 40, 1868–1907. Part I. Authors. Part II. Papers (classified). [etc.].* By –. Riddick. Wellington, NZ. [iv]+88 pp.; ___. 1921. *Index to the Transactions and Proceedings of the New Zealand Institute, vols. 41–51 [1908–19]. Part I. Authors. Part II. Papers (classified). [etc.].* Wellington, NZ. 30 pp.

Bb699. **Royal Society of New Zealand.** 1987. *Science, settlers and scholars: The centennial history of the Royal Society of New Zealand.* By C. A. Fleming. Wellington, N.Z. xiii+353 pp. *Bulletin* vol. 25.

Royal Society of Queensland (*formerly* Philosophical Society of Queensland)

Bb700. **Marks, E. N.** 1959. A history of the Queensland Philosophical Society and the Royal Society of Queensland from 1859–1911. *Proceedings of the Royal Society of Queensland* 71: 17–42, & 2 pl.

Bb701. **Walkom, A. B.** 1914. *Proceedings of the Philosophical Society of Queensland. Index to papers in volumes 1 to 25 [1884–1913] inclusive.* Brisbane, Qld. 18 pp.

Royal Society of South Australia (*formerly* Philosophical Society of Adelaide)

Bb702. **Royal Society of South Australia.** 1907. *Index to the Transactions, proceedings and reports. Volume 1 to 24, 1877–1900.* By R. J. M. Clucas. Adelaide, S.A. 448 pp.; *Continued* ___. 1922. *Index to the Transactions, Proceedings and Reports, vol. 25–44, 1901–20, and to the Memoirs, vol. 1–2, 1899–1912.* Adelaide. 189 pp.

Royal Society of Tasmania (*formerly* Botanical & Horticultural Society of Van Diemen's Land *and* Royal Society of Van Diemen's Land for Horticulture, Botany and the Advancement of Science)

Bb703. **Royal Society of Tasmania.** 1887. *Register of papers published in the Tasmanian Journal and the Papers and Proceedings of the Royal Society of Tasmania from the year 1841 to 1885.* By A. Morton. Hobart, Tas. 58 pp.

Royal Society of Victoria (*formerly* Philosophical Society of Victoria *and* Philosophical Institute of Victoria)

Bb704. **Royal Society of Victoria.** 1934. Author index to the publications of the Royal Society of Victoria together with those of the societies amalgamated with it; namely, of the former Philosophical Institute of Victoria; of the former Philosophical Society of Victoria; of the former Victorian Institute for the Advancement of Science; and of the first Microscopical Society of Victoria (1879–1882); 1855–1934. By F. A. Cudmore. *Proceedings of the Royal Society of Victoria* n.s. 46: 275–319.

Société Royale des Arts et des Sciences de l'Ile Maurice

Bb705. **Société Royale des Arts et des Sciences de l'Ile Maurice.** 1932. *Centenaire de la Société Royale des Arts et des Sciences de l'Ile Maurice 1829–1929.* Port Louis, La Typographie Moderne. [iv]+171 pp. & 10 ports.

Van Diemen's Land Scientific Society

Bb706. **Hoare, M. E.** 1968. Dr. John Henderson and the Van Diemen's Land Scientific Society. *Records of the Australian Academy of Science* 1(3): 7–24.

African

Royal Society of South Africa (*formerly* South African Philosophical Society)

Bb707. **South African Philosophical Society.** 1908. List of contents of the publications of the South African Philosophical Society, [vols. 1–18] 1877–1908. Cape Town. 20 pp. Appended to *Transactions, South African Philosophical Society* vol. 18.

Botanical — General

(Royal) Asiatic Society of Bengal

Bb708. **Das, A. R.** 1969 [1971]. Author and subject catalogue of the botanical and allied articles published in the periodicals of the Asiatic Society of Bengal, 1788–1967. *Journal of the Asiatic Society* 9: 42–62, 1969.

Bombay Natural History Society

Bb709. **Das, A. R.** 1976. Authors' catalogue of the botanical articles published in the *Journal of the Bombay Natural History Society*, (vols. 1–66, 1886–1969). *Journal of the Bombay Natural History Society* 73: 98–120; *Continued* ___. 1983. ___ II (Vol. 67–76, 1970–1979). *Ibid.* 80: 507–516.

Botanical Society (London) 1721–26

Bb710. **Allen, D. E.** 1967. John Martyn's Botanical Society: A biographical analysis of the membership, *Proceedings of the Botanical Society of the British Isles* 6: 305–324. *Also* ___. 1985. John Martyn's Botanical Society: Some further identifications. *Newsletter, Society for the History of Natural History* 24: 9–10.

Bb711. **Edwards, P. I.** 1963. The Botanical Society of London. *Proceedings of the Botanical Society of the British Isles* 5: 117–118.
A note on their archives, now at the British Museum (Natural History).

Botanical Society of America

Bb712. **Tippo, O.** 1956. The early history of the Botanical Society of America. *American journal of botany* 43: 852–858. *Also pub. in* Botanical Society of America. 1958. *Fifty years of botany: Golden jubilee volume of the Botanical Society of America*, pp. 1–13. Ed. W. C. Steere. New York, McGraw Hill Book Co.

Botanical Society of the British Isles (*formerly* Thirsk Natural History Society: Botanical Exchange Club *and* (London) Botanical Exchange Club *and* Botanical Exchange Club of the British Isles)

Bb713. **Allen, D. E.** 1986. *The botanists: A history of the Botanical Society of the British Isles throughout a hundred and fifty years.* Winchester, St. Paul's Bibliographies. xv+232 pp.

Botanical Society of Canada

Bb714. **Connor, J. T. H.** 1986. To promote the cause of science: George Lawson and the Botanical Society of Canada, 1860–1863. *Scientia Canadensis* 10: 3–33.

Botanical Society of Edinburgh

Bb715. **Grieve, S.** 1902. Notes from the archives of the Botanical Society on its origin, history and privileges. *Transactions of the Botanical Society of Edinburgh* 22: 194–220.

Botanical Society of London

Bb716. **Allen, D. E.** 1980. The women members of the Botanical Society of London, 1836–1856. *British journal of the history of science* 13: 240–254.
Appendix: Known women members of the ..., pp. 251–256.

Bb717. ___. 1986. The Botanical Society of London. [and] Appendix 2: The members of the Botanical Society of London. In: *The botanists: A history of the Botanical Society of the British Isles throughout a hundred and fifty years*, pp. 3–65, 202–221. Winchester, St. Paul's Bibliographies.

Botanical Society of Washington

Bb718. **Stevenson, J. A.** 1970. Highlights in the history of the Botanical Society of Washington. *Journal of the Washington Academy of Science* 60: 141–155.

Botanical Society of Western Pennsylvania

Bb719. **Botanical Society of Western Pennsylvania.** 1964. Botanical Society of Western Pennsylvania. *Trillia* 12: 133–167.

Botanischer Verein zu Hamburg

Bb720. **Poppendieck, H.–H.**, ed. 1991. *Botanischer Wanderführer um Hamburg: Zum hundertjährigen Bestehen des Botanischen Vereins zu Hamburg*. Hamburg, Christians. 219 pp.

Botanischer Verein der Provinz Brandenberg (und die Angrenzenden Länder)

Bb721. **Botanischer Verein der Provinz Brandenberg.** 1889. *Register über die Verhandlungen des Botanischen Vereins der Provinz Brandenburg, Band 1–30 (Jahrgang 1859–1888).* By M. Gürke. Berlin. [ii]+60 pp.; **___**. 1910. Verzeichnis der in den Verhandlungen des Botanischen Vereins der Provinz Brandenburg. Band XXXI–L [1889–1908] enthaltenen Arbeiten und Mitteilungen. By C. Shuster. *Ibid.* Beilage zu Bd. 51: 32 pp.

Bb722. **___**. 1909. Die Geschichte des Botanischens Vereins der Provinz Brandenburg 1859–1909. By G. Volkens. *Verhandlungen des Botanischen Vereins der Provinz Brandenburg* 51: (86) pp.

Dansk Botanisk Forening (*formerly* Botanisk Forening i Copenhagen)

Bb723. **Botanisk Forening i Copenhagen.** 1890. *Festskrift , udgivet af den botaniske Forening i Kjøbenhavn i Anledning af dens Halvhundredaarsfest ... 1890, [etc.].* Copenhagen. 296 pp. & 10 pl.

Bb724. **___**. 1904. *Indholdsfortegnelse til Botanisk Tidsskrift ... 1–25. Bind. 1866–1903, Meddelelser fra den Botanisk Forening i København, 1.–2. Bind, 1882–1891, og Festskrift ... 1890.* By M. P. Porsild. Copenhagen, H. Hagerup. 55 pp.

Deutsche Botanische Gesellschaft

Bb725. **Degen, H.** 1974. Die Entstehung der Deutschen Botanischen Gesellschaft. *Naturwissenschaftliche Rundschau* 27: 333–340.

Bb726. **Deutsche Botanische Gesellschaft.** 1906. *Berichte der Deutschen Botanischen Gesellschaft. Registerband zum ersten bis zwanzigsten Jahrgang (Band 1–20) [1882–1902].* Berlin, Gebrüder Borntraeger. iv+305 pp.; *Continued* **___**. 1926. *Berichte der Deutschen Botanischen Gesellschaft. Register zu Band 21–40 [1882–1902].* By B. Leisering. Berlin, Gebrüder Borntraeger. vii+175 pp.

Bb727. **___**. 1908. *Festschrift zur Feier des 25 jährigen Bestehens der Deutschen Botanischen Gesellschaft. Gegründet am 17. September 1882. Gleichzeitig Bd 26 der Berichte der Deutschen Botanischen Gesellschaft.* Berlin, Borntraeger. 406 pp. & 19 pl.

Bb727A. **___**. 1987. Beiträge zur neueren Geschichte der Botanik. *Berichte der Deutschen Botanischen Gesellschaft* 100: xi+441 pp. *Also pub. separately* Stuttgart, Fischer. 1988. Includes several articles on the history of the Society and its publications by H. Lorenzen, R. Meyer, & others.

Edinburgh Society for the Importation of Foreign Seeds & Plants

Bb728. **Emerson, R. L.** 1982. The Edinburgh Society for the Importation of Foreign Seeds and Plants, 1764–1773. *Eighteenth-century life* 7(2): 73–95.

Flora, (Königlich) Sächsische Gesellschaft für Botanik und Gartenbau zu Dresden

Bb729. **Flora, Sächsische Gesellschaft für Botanik und Gartenbau zu Dresden.** 1926. *Festschrift aus Anlass des hunderjährigen Bestehens der Flora, Sächsische Gesellschaft für Botanik und Gartenbau in Dresden.* By W. Dänhardt. Dresden. 278 pp. *Includes* **T. Wittmüss.** Überblick über die geschichtliche Entwicklung der "Flora" von 1826 bis 1926; **F. Tobler.** Veröffentlichungen der "Flora" seit dem Jahre 1841; **G. Dutschmann.** Übersicht der gärtnerischen und botanischen Gesellschaften und Institute, mit denem die "Flora" in Schriftenaustausch steht; Zeittafel über die Amtsdauer der Vorstandsmitglieder der "Flora" von 1826 bis 1926; and other historical papers by various authors.

International Association of Plant Taxonomy

Bb730. **International Association for Plant Taxonomy.** 1972. *Taxon: Journal of the International Association for Plant Taxonomy. Index to vols. 1–20 [1951–71].* Part 1: *Scientific names.* By E. Rouleau. Utrecht, Oosthoek for the International Bureau for Plant Taxonomy & Nomenclature. iv+388 pp.; *Continued* ___. 1973. ___. *Index to vols. 1–20 [1951–71].* Part 2: *Authors and subjects.* By R. M. Lowden. Utrecht, [etc.]. vi+243 pp.; ___. 1985. ___. *Index to vols. 1–30 [1951–81]. Personal names.* By R. M. Lowden. Utrecht, International Bureau for Plant Taxonomy & Nomenclature. iv+816 pp.; ___. 1986. ___. *Index to vols. 21–30 [1972–81]. Authors and subjects.* By R. M. Lowden. Utrecht, [etc.]. [iv]+377 pp.

Bb731. ___. 1982. The origins and early history of the I.A.P.T. By R. S. Cowan & F. A. Stafleu. *Taxon* 31: 415–420.

Kaiserlich–Königlich Gartenbau–Gesellschaft in Wien

Bb732. **Kaiserlich–Königlich Gartenbau–Gesellschaft in Wien.** 1907. *Die K. K. Gartenbau–Gesellschaft in Wien, 1837–1907.* By A. Burgerstein. Vienna. v+128 pp., illus.

Királyi Magyar Természettudományi Társulat (Königlich–Ungarische Naturwissenschaftliche Gesellschaft)

Bb733. **Királyi Magyar Természettudományi Társulat.** 1933. Register der in den Sitzungen der Botanischen Sektion der Königlich–Ungarische Naturwissenschaftliche Gesellschaft vom 8. Januar 1902 an bis zum 8. Juni 1933 gehaltenen Vorträge und der in den Bänden 1–30 (Jahrg. 1902–1933) der Zeitschrift "Növénytani közlemények" (VIII–XXX) erscheinen Abhandlungen. *Botanikai közlemények* 30: 247–306.

Koninklijke Nederlandsche Botanische Vereeniging (Nederlandsche Botanische Vereeniging)

Bb734. **Nederlandsche Botanische Vereeniging.** 1947. Gedenkboek jubileum–uitgave ter gelegenheid van het 100–jarig bestaan van de Nederlandsche Botanische Vereeniging op 15 augustus 1945. By J. T. Henrard. *Nederlandsch kruidkundig archief* 55: 188 pp.

Massachusetts Horticultural Society

Bb735. **Benson, A. E.** 1929. *History of the Massachusetts Horticultural Society.* Norwood, MA, Plimpton Press. ix+553 pp., illus.

Bb736. **Massachusetts Horticultural Society.** 1880. *History of the Massachusetts Horticultural Society: 1829–1878.* By W. C. Strong, et al. Boston, MA. viii+545 pp.

New England Botanical Club

Bb737. **Pease, A. S.** 1951. The New England Botanical Club a half century ago and later. *Rhodora* 53: 97–105.

Norsk Botanisk Forening

Bb738. **Norsk Botanisk Forening.** 1985. The foundation of the Norwegian Botanical Association 50 years ago. By J. Vaage. *Blyttia* 43: 162–168.

Polskie Towarzystwo Botaniczne

Bb739. **Karpowiczowa, L.**, ed. 1972. *50 lat Polskiego Towarzystwa Botanicznego.* Wroclaw, Zaklad Narodowy im. Ossolinskich. 156 pp., illus. *Monografie z dziejów nauki i techniki* vol. 77.

Bb740. **Polskie Towarzystwo Botaniczne.** 1986. 65 lat Polskiego Towarzystwa Botanicznego. *Wiadomosci botaniczne* 30: 217–422.

Preussischer Botanischer Verein

Bb741. **Preussischer Botanischer Verein.** 1912. *Festschrift zum 50 jährigen Bestehen des Preussischen Botanischen Vereins e. V. 1862–1912.* Königsberg. xxviii+360 pp. & 3 pl.

Includes **J. Abromeit.** Geschichte des Preussischen Botanischen Vereins, pp. vii–xxviii.

Regensbergische Botanische Gesellschaft

Bb742. **Bresinsky, A.** 1981. Die Regensburgische Botanische Gesellschaft und die Wurzeln ihres Entstehens im 18. Jahrhundert. In: J. Barthel, ed. *Naturwissenschaftliche Forschung in Regensburgs Geschichte: Vortragsreihe der Universität Regensburg*, pp. 101–130. Regensburg, Mittelbayerische Druckerei- und Verlagsgesellschaft. *Schriftenreihe der Universität Regensburg* no. 4.

Bb743. **Regenburgische Botanische Gesellschaft.** 1890. Geschichte der Kgl. Bayer. Botan. Gesellschaft in Regensburg während ihres 100 jährigen Bestandes vom 14. Mai 1790 bis 14. Mai 1890. By J. Singer. *Denkschriften der Königlich–Baierischen Botanischen Gesellschaft in Regensburg* 6: 1–32.

Bb744. ___. 1940. Zur Geschichte unserer Gesellschaft. By S. Killerman. *Denkschriften der Regensburgischen Botanischen Gesellschaft* 21: i–xiii.

Bb745. ___. 1966. Zur neueren Geschichte der Regensburgischen Botanischen Gesellschaft. By H. Scheuchl. *Denkschriften der Regensburgischen Botanischen Gesellschaft* 26: 7–13.

Bb746. ___. 1984. Die Regensburgische Botanische Gesellschaft: Ihre Entstehung, Entwicklung und Bedeutung. By W. Ilg. *Hoppea* 42: 421 pp.

Bb747. ___. 1990. Geschichte der Botanik in Regensburg: 200 Jahre Regensburgische Botanische Gesellschaft 1790–1990. Katalog zur Ausstellung in den Museen der Stadt Regensburg 14. Juli bis 7. Oktober 1990. By W. Ilg. *Hoppea* 48: 132 pp., illus.

Bb748. ___. 1990. Festschrift: 200 Jahre Regensburgische Botanische Gesellschaft. Teil 1: Zur Geschichte der Gesellschaft von 1790 bis 1990. *Hoppea* 49: 276 pp.

Includes **A. Bresinsky.** Zweihundertjähriges Jubiläum der Regensburgischen Botanischen Gesellschaft: Rückschau und Ausblick, pp. 225–261. Also contains facsimile reprints of several earlier historical accounts including those of J. Singer, S. Killerman, and H. Scheuchl listed herein.

Bb749. ___. 1991. Festschrift: 200 Jahre Regensburgische Botanische Gesellschaft. Teil 2: Wissenschaftliche. *Hoppea* 50: 736 pp.

Includes **A. Bresinsky.** Die Feier des zweihundertjährigen Jubiläums der Regensburgischen Botanischen Gesellschaft, pp. 605–654; Inhaltsverzeichnis der Denkschriften der Regensburgischen Botanischen Gesellschaft von 1815 bis 1990, pp. 721–735 (Updated versions of this list appear in current volumes of *Hoppea*).

Rochester Academy of Science

Bb750. **Rochester Academy of Science.** 1912. Early botanists of Rochester and vicinity and the Botanical Section. By F. Beckwith. *Proceedings of the Rochester Academy of Science* 5 (1910–18): 39–58; *Continued* ___. 1942. ___. Part 2. By A. B. Suydam. *Ibid.* 8 (1941–43): 124–149.

Royal Horticultural Society

Bb751. **Royal Horticultural Society.** 1937. *Index to the Journal and Proceedings 1838–1935; and list of awards, 1859–1935.* London. 1,677 pp.; *Continued* ___. n.d. *Index to Journal and list of awards: Supplement for 1936–1945.* London. 220 pp.; ___. 1957.

Index to the Journal and Proceedings ...and list of awards: Supplement for 1946–1955. London. 1957. 265 pp.; ___. 1967. ___*: Supplement for 1956–1965.* London. 265 pp.

Bb752. ___. 1969. *The story of the Royal Horticultural Society, 1804–1968.* By H. R. Fletcher. London, Oxford Univ. Pr. for the Royal Horticultural Society. xii+564 pp.

Sociedad Botánica de Mexico

Bb753. **Sociedad Botánica de Mexico.** 1981. Fundación y primera década de la Sociedad Botánica de México. By E. Hernández X. *Boletín de la Sociedad Botánica de México* 40: 15–23.

Bb754. ___. 1981. Cronologia de los principales acontecimientos históricos de la Sociedad Botánica de Mexico. By A. Lot & M. Carvajal. *Ibid.* 40: 25–43.

Bb755. ___. 1981. Indice acumulativo del Boletín de la sociedad botánica de Mexico de los numeros 1 (1944) al 40 (1981). By A. B. Cervera. *Ibid.* 40: 73–132.

Società Botanica Italiana

Bb756. **Baccarini, P.** 1904. Notizie intorno ad alcuni documenti della Società Botanica Fiorentina del 1716–1783 ed alla sue vicende. *Annali di botanica* 1: 225–254.

Bb757. **Societa Botanica Italiana.** 1988. *Indici bibliografico dei periodici della Societa Botanica Italiana (1844–1986).* By F. Pedrotti. Florence. 493 pp.

Société Botanique de France

Bb758. **Aymonin, G. G., G. Deysson, & M. Keraudren-Aymonin.** 1975. L'œuvre séculaire de la Société Botanique de France. *Actes du 100e Congrès National des Sociétés Savantes, Section d'Histoire Moderne et Contemporaine [etc.]* pp. 253–265.

Bb759. **Leussink, J. A.** 1985–88. The publication dates of the *Bulletin de la Société Botanique de France* (Vols. 1–25, 1854–1878). *Taxon* 34: 573–606, 1985; ___ (Vols. 26–40, 1879–1893). *Ibid.* 35: 247–261, 1986; ___ (Vols. 41–61, 1894–1914). *Ibid.* 37: 35–62, 1988.

Bb760. **Société Botanique de France.** 1899. *Table générale des articles originaux contennus dans les quarante premiers volumes du Bulletin ... (années 1854–1893])* Paris. 240 pp.

Bb761. ___. 1954. À propos du centenaire de la Société Botanique de France (1854–1954). *Bulletin de la Société Botanique de France* 101 supp: 58 pp.

Includes **R. de Vilmorin.** Images du passé, pp. 1–15, and **F. Pellegrin.** Un siècle de Société Botanique de France, pp. 19–46.

Société Botanique de Genève

Bb762. **Duperrex, A.** 1975. Centenaire de la Société Botanique de Genève 1875–1975. *Saussurea* 6: 17–23.

Société de Botanique de Grand–Duché de Luxembourg

Bb762A. **Société des Naturalistes Luxembourgeois.** 1990. Livre du centenaire 1890–1990. La Société des Naturalistes Luxembourgeois du point de vue historique. Table générale des publications de la Société de Botanique du Grand–Duché de Luxembourg (1874–1905) et de la Société des Naturalistes Luxembourgeois (1891–1989). *Bulletin de la Société des Naturalistes Luxembourgeois* 91: 467 pp.

Société Hallérienne de Botanique

Bb763. **Burdet, H. M.** 1975. La Société Hallérienne de Botanique: 1852–1856. *Saussurea* 6: 25–36.

Société Nationale d'Horticulture de France

Bb764. **Société Nationale d'Horticulture de France.** 1927. *Centenaire de la Société 1827–1927.* Paris. 32 pp.

Société Royale de Botanique de Belgique

Bb765. **Société Royale de Botanique de Belgique.** 1887. *Tables générales du Bulletin de la Société Royale de Botanique de Belgique. Tome I–XXV (années 1862–1887).* By T. Durand. Brussels. 358 pp.; *Continued* ___. 1921. Table générale du Bulletin ... volumes XXVI–XLIX (1887–1912). By P. Van Aerdschot. *Bulletin de la Société Royale de Botanique de Belgique* 50: [ ]* pp.; ___. 1942. *[Bulletin ... Tables ... tome LI–LXXV, 1912–42]**. Brussels. [ ]* pp.; ___. 1974. *Bulletin ... Tables générales des tome 76–100 (1944–1967).* Brussels. 333 pp.
[*Not examined – details not ascertained.]

Society of Amateur Botanists

Bb766. **Ramsbottom, J. R.** 1931. The Society of Amateur Botanists and the Quekett Microscopical Club. *Journal of the Quekett Microscopical Club* (2) 16: 215–230.

Svenska Botaniska Föreningen

Bb767. **Falck, A.** 1870. De botaniska föreningarne i Sverige, ett historiskt utkast. (Die botanischen Gesellschaften Schwedens, ein historischer Entwurf.) *Botanisk tidsskrift* 4: 173–226.

Bb768. **Svenska Botaniska Föreningen.** 1932. *Generalregister till Svensk Botanisk Tidskrift åArg. 1–20 (1907–26).* By T. Arwidsson & G. R. Cedergren. Uppsala. 312 pp.; ___. 1949. ___ *Årg. 21–40 (1927–1946).* By B. Peterson. Uppsala. 308 pp.; ___. 1971. ___ *Årg. 41–60, (1947–1966).* By G. & B. Peterson. Uppsala, Almqvist & Wiksell. 421 pp.

Torrey Botanical Club

Bb769. **Torrey Botanical Club.** 1918. Historical sketch of the Torrey Botanical Club. By J. H. Barnhart. *Memoirs of the Torrey Botanical Club* 17: 12–21.

Bb770. ___. 1955. *Index to the Bulletin of the Torrey Botanical Club, vols.1–75, (1870–1948).* By H. W. Rickett. Lancaster, PA, Science Pr. vii+997 pp.
Preceded by a brief history of the *Bulletin.*

Vserossiiskoe (*later* Vsesoyuznoe) Botanicheskoe Obshchestvo

Bb771. **Lavrenko, E. M., & A. A. Yunatov.** 1965. Pyat'desyat let Vsesoyuznogo Botanichestogo Obshchestvo. [Fifty years' anniversary of the Botanical Society of the USSR]. *Botanicheskii zhurnal* 50: 1205–1247.

Bb772. **___, & O. V. Zalensky.** 1965. 1965. Botaniki–organizatory Russkogo Botanicheskogo Obshchestvo. [The botanist–founders of the Russian Botanical Society]. *Ibid.* 50: 1751–1768.

Bb773. **Sochava, V. B.** 1947. Kratkii obzor deyatel'nosti Vserossiskogo Botanicheskogo Obshchestvo za 30 let (1916–1946). [Brief survey of the activity of the Russian Botanical Society for 30 years]. *Ibid.* 32 (2): 45–59.

Bb774. **Yunatov, A. A.** 1965. K predystorii Vsesoyuznogo Botanicheskogo Obshchestvo. [On the pre-history of the Botanical Society of the USSR]. *Ibid.* 50: 1345–1351.

Wild Flower Society (*formerly* Wild Flower Preservation Society of America)

Bb775. **Wild Flower Society.** 1986. Wild Flower Society 1886–1986. *Wild flower magazine* 407: 63 pp.

Congresses

Bb776. **Rendle, A. B.** 1935. A short history of the International Botanical Congress. *Chronica botanica* 1: 35–40.

Bb777. **Stafleu, F. A.** 1970. A century of botanical congresses. *Proceedings, 11th International Botanical Congress, Washington, 1970*, pp. 9–21.

Bb778. **Verdoorn, F.** 1953. The international plant science congresses, their history and aims. *Proceedings, 7th International Botanical Congress, Stockholm, 1950*, pp. 42–56. Stockholm.

Cryptogamic botany

British Bryological Society (*formerly* Moss Exchange Club)

Bb779. **British Bryological Society.** 1956. *A short account of the Moss Exchange Club and the British Bryological Society.* Compiled from the Annual Reports, [etc.]. By E. Armitage. 2nd ed. Ross, Herefords. 24 pp.

Bb780. **___.** 1983. *The British Bryological Society diamond jubilee.* Proceedings of the Jubilee meeting ... 17–18 September, 1983. Cardiff, The Society. [v]+89 pp. *Special volume* no. 1.
Includes **P. W. Richards.** The British Bryological Society 1923–1983, pp. 3–10; **H. Bischler & S. R. Gradstein.** A causerie on bryological societies, pp. 11–16; ; **E. V. Watson.** The recording activities of the BBS (1923–83) and their impact on advancing knowledge, pp. 17–29; **S. W. Greene.** The British Bryological Society and its future, pp. 65–74.

Bb781. **Foster, W. D.** 1979. The history of the Moss Exchange Club. *Bulletin of the British Bryological Society* 33: 19–26.

Bb782. **Perry, A. R.** 1991. Dates of publication of *Transactions of the British Bryological Society* and *Journal of bryology*. *Journal of bryology* 16: 644–647.

British Mycological Society

Bb783. **British Mycological Society.** 1952. *Transactions of the British Mycological Society: Fifty–year index (1896–1946) : Volumes 1–30.* By G. R. Bisby. Cambridge. [iii]+126 pp.; *Continued* **___.** 195 ? [*Transactions of the British Mycological Society. Index vols. 31–40, (1947–57)*]*. Place? ? pp.; **___.** 1979. *Transactions of the British Mycological Society. Cumulative index 1958–1976; volumes 41–67.* By A. G. Morton & E. M. Grant. Cambridge, [etc.]. 107 pp. *Transactions* ... vol. 73 (3).
[*Not examined – details not ascertained.]

Bb784. **Ramsbottom, J. R.** 1948. The British Mycological Society. *Transactions of the British Mycological Society* 30: 1–12.

Deutsche Gesellschaft für Pilzkunde

Bb785. **Haas, H.** 1973. 50 Jahre Deutsche Gesellschaft für Pilzkunde. *Zeitschrift für Pilzkunde* 39: 9–14.

Nederlandsche Natuurhistorische Vereniging, Bryologische Werkgroep

Bb786. **Sollman, F., A. Touw, & O. van Zanten.** 1979? *[Index to the volumes of Buxbaumia, (1946–1970).*]* Eindhoven, Dutch Bryological Society. 102 pp.
[*Not examined.]

Société Mycologique de France

Bb787. **Société Mycologique de France.** 1934. *Le quarantenaire de la Société Mycologique de France (1884–1924).* By Dr. Guétrot. Paris. [iv]+412 pp.
I. Partie administrative; II. Partie sociale; III. Partie scientifique.

Bb787A. ___. 1984. Le livre d'or du centenaire: Documents historiques sur la mycologie, Société Mycologique de France: 1884–1984. *Bulletin trimestriel de la Société Mycologique de France* 100 (3): CCXXVIII pp., illus.
Many of the mycologists discussed and portrayed were Society members. Photographs of field excursions, see pp. LXVII–LXIX, CCVII–CCIX. For list of contents, *see* Cb663.

Société Mycologique de Genève

Bb788. **Monthoux, O.** 1975. Société Mycologique de Genève: Histoire et activité. *Saussurea* 6: 43–48.

Zoological — General

American Society of Naturalists

Bb789. **American Society of Naturalists.** 1934. Fifty years of the American Society of Naturalists. By E. G. Conklin. *American naturalist* 68: 385–401.

American Society of Parasitologists

Bb790. **American Society of Parasitologists.** 1941. *Journal of parasitology. Index. Volumes 1–25, 1914–39.* Urbana, IL. [iii]+204 pp.; *Continued* ___. 1960. *Index II. Volumes 26–45, 1940–59.* N.p. [v]+332 pp.; ___. 1971. *Index III. Volumes 46–55, 1960–1969. Author — subject — host.* Lawrence, KS. [v]+369 pp.

Bb790A. **Cram, E. B.** 1956. Stepping stones in the history of the American Society of Parasitologists. *Journal of parasitology* 42: 461–473.

Bb791. **Hoogstraal, H.** 1985. The American Society of Parasitologists: Its historical role and modern opportunities. *Ibid.* 71: 685–692.

American Society of Zoologists (*formerly* American Morphological Society, *and* Society of American Zoologists)

Bb792. **American Society of Zoologists.** 1979. The beginnings of the American Society of Zoologists. By C. E. Quinn. *American zoologist* 19: 1247–1249.

Bb793. ___. 1982. Ancestry and beginnings: The early history of the American Society of Zoologists. By C. E. Quinn. *Ibid.* 22: 735–748.

Bb794. ___. 1990. The American Society of Zoologists, 1889–1989: A century oof integrating the biological sciences. By K. R. Benson & C. E. Quinn. *Ibid.* 30: 353–396.

Bb795. **Benson, K. R.** 1991. Epilogue: The development and expansion of the American Society of Zoologists. In: K. R. Benson, R. Rainger, & J. Maienschein, eds. *The expansion of American biology*, pp. 325–335. New Brunswick, NJ, Rutgers Univ. Pr.

Association (*formerly* Institute) for the Study of Animal Behavior

Bb796. **Durant, J. R.** 1986. The making of ethology: The Association for the Study of Animal Behavior, 1936–1986. *Animal behavior* 34: 1601–1616.

Deutsche Zoologische Gesellschaft

Bb797. **Geus, A., & H. Querner.** 1990. *Deutsche Zoologische Gesellschaft, 1890–1990: Dokumentation und Geschichte.* Stuttgart, G. Fischer. ix+187 pp.

Fauna Preservation Society (*formerly* Society for the Preservation of the (Wild) Fauna of the British Empire)

Bb798. **Fauna Preservation Society.** 1978. *The penitent butchers: The Fauna Preservation Society, 1903–1978*. By R. Fitter & P. Scott. London. 48 pp.

Imperatorskoe Obshchestvo Lyubitelei Estestvoznaniya, Antropologii i Etnografii

Bb799. **Imperatorskoe Obshchestvo Lyubitelei Estestvoznaniya, Antropologii i Etnografii.** 1888–90. Letopis zoologicheskikh trudov Obshchestva v pervoe dvatsatipyatilyetie ego sushchestvovaniya (1863–1888 g.) [Chronicles of the zoological works of the Society during the first twenty-five years of its existence (1863–1888)]. By A. P. Bogdanov. *Izvestiya Imperatorskogo Obshchestva Lyubitelei Estestvoznaniya, Antropologii i Etnografii pri Imperatorskom Moskovskom Universitete* 54: 494 pp. & *Ibid.* 59: 538 pp.

Bb800–801. [Not used.]

Koninklijk Zoölogisch Genootschap, Amsterdam

Bb802. **Koninklijk Zoölogisch Genootschap "Natura Artis Magistra" te Amsterdam.** [1898.] *Koninklijk Zoölogisch Genootschap "Natura Artis Magistra," 1838–Mei 1898*. Amsterdam. 64 pp., illus.

National Audubon Society (*formerly* National Association of Audubon Societies for the Protection of Wild Birds & Animals)

Bb803. **Dutcher, W.** 1905. History of the Audubon movement. *Bird lore* 7: 45–57.

Bb804. **Graham, F., Jr., & C. W. Buchheister.** 1990. *The Audubon ark: A history of the National Audubon Society*. New York, A. Knopf. ix+333 pp.

See also Ba1363.

National Wildlife Federation (*formerly* General Wildlife Federation)

Bb805. **Allen, T. B.** 1987. *Guardian of the wild: The story of the National Wildlife Federation, 1936–1986*. Bloomington, IN, Indiana Univ. Pr. 212 pp.

Nederlandsche Dierkundige Vereeniging (Société Néerlandaise de Zoologie)

Bb806. **Nederlandsche Dierkundige Vereeniging.** 1909. *Register op het Tijdschrift der Nederlandsche Dierkundige Vereeniging Serie I: Deel I tot VI; Suppl. Deel I en II. Serie II: Deel I tot X [1875–1908]*. Leiden, E. J. Brill. 21 pp.

Royal Zoological Society of New South Wales

Bb807. **Royal Zoological Society of New South Wales.** [1979?]. *The first one hundred years of the Royal Zoological Society of N.S.W: 1879 to 1979*. By J. H. Prince. [Adelaide]. 81 pp.

Société Zoologique de France

Bb808. **D'Hondt, J.-L.** 1989. *Histoire de la Société Zoologique de France: Son évolution et son rôle dans le dévelopment de la zoologie*. Nancy, Musée de Zoologie de l'Université de la Ville de Nancy. 100 pp. *Revue Française d'aquariologie, herpetologie* 16/3.

Bb809. **Société Zoologique de France.** 1905. *Tables du Bulletin et des Mémoires de la Société Zoologique de France (années 1876 à 1895)*. By F. Secques. Paris. 214 pp.; *Continued* ___. 1919. *Tables ... 1896–1905*. Paris. [ ]* pp.
[*Not examined – pagination not ascertained.]

Bb810. **___.** 1976. Congrès du centenaire de la Société Zoologique de France, Paris, 6–11 Septembre 1976. *Bulletin de la Société Zoologique de France* 101(5) suppl: 765–1133.

Bb811. **Société Zoologique de France.** 1977. La Société Zoologique de France: Ses origines et ses premiers années. By R. Fox. *Ibid.* 101 centenaire supp: 799–812.

Wildlife Society of Southern Africa

Bb812. **Pringle, J. A.** 1982. *The conservationists and the killers: The story of the Wildlife Society of Southern Africa.* Cape Town, Bulpin. 319 pp.

Zoological Society of London

Bb813. **Bastin, J.** 1975. The first prospectus of the Zoological Society of London: New light on the Society's origins. *Journal of the Society for the Bibliography of Natural History* 5: 369–388; *Also* ___. 1973. A further note on the origins of the Zoological Society of London. *Ibid.* 6: 236–241.

Bb814. **Root, N. J., & B. R. Johnson.** 1986. *Proceedings of the Zoological Society of London. An index to the artists 1848–1900.* New York & London, Garland Publ. xxiii+947 pp.

Bb815. ___. 1986. *The Transactions of the Zoological Society of London: An index to the artists 1835–1936.* New York & London, Garland Publishing. xix+444 pp.

Bb816. **Scherren, H.** 1905. *The Zoological Society of London: A sketch of its foundation and development and the story of its farm, museum, gardens, menagerie and library.* London, Cassell. xii+252 pp. & 62 pl.

Bb817. **Zoological Society of London.** 1866. *Proceedings of the Zoological Society of London. Index, 1830–1847.* London. 190 pp.; *Continued* ___. 1863. ___*: Index, 1848–1860.* London. 304 pp.; ___. 1872. ___*: Index, 1861–1870.* London. 481 pp.; ___. 1882. ___*: Index, 1871–1880.* London. 570 pp.; ___. 1892. ___*: Index, 1881–1890.* London. 485 pp.; ___. 1902. ___*: Index, 1891–1900.* London. 595 pp.; ___. 1912. ___*: Index, 1901–1910.* London. 534 pp.; ___. 1922. ___*: Index, 1911–1920.* London. 274 pp.; ___. 1930. ___*: Index, 1921–1928.* London. 180 pp.

Bb818. ___. 1901. *A record of the progress of the Zoological Society of London during the nineteenth century.* Ed. by the Secretary [P. L. Sclater.] London. [vi]+248 pp.

Bb819. ___. 1929. *Centenary history of the Zoological Society of London.* By P. Chalmers–Mitchell. London, Zoological Society. xi+307 pp. & 33 pl.

Bb820. ___. 1937. On the dates of publication of the Society's Proceedings 1859–1926. By F. M. Duncan. With an appendix containing the dates of publication of the 'Proceedings,' 1830–1858, compiled by the late F. H. Waterhouse, and of the 'Transactions,' 1833–1869, by the late Henry Peavot, originally published in P.Z.S. 1893, 1913. *Proceedings of the zoological society of London* 107A(1): 71–84.
Also **C. F. Cowan.** 1973. Proc. Zool. Soc. Lond. publication dates. *Journal of the society for the bibliography of natural history* 6: 293–294.

Bb821. ___. 1976. *The Zoological Society of London 1826–1976 and beyond.* Ed. by ... Lord Zuckerman. London. xviii+353 pp. *Symposia of the Zoological Society of London* no. 40.
Includes **S. Zuckerman.** The Zoological Society of London: Evolution of a constitution, pp. 1–16; **R. Fish & I. Montagu.** The Zoological Society and the British overseas, pp. 17–48; **L. G. Goodwin.** Research, pp. 215–222; **W. W. Bullough & F. Hamilton.** The role of education, pp. 223–231; **R. Fish.** The library and scientific publications of the Zoological Society of London: Part I, pp. 233–252; **M. A. Edwards.** ___: Part II. pp. 253–267; **P. A. Jewell.** The contribution of the Zoological Society of London to field studies and prospects for the future, pp. 269–281.

Zoological Society of San Diego

Bb822. **Shaw, M. B.** 1980. The Zoological Society of San Diego: Its history and development. *Journal of the Society for the Bibliography of Natural History* 9: 565–571.

Ornithological

American Ornithologists Union

Bb823. **American Ornithologists' Union.** 1900. *Index to Bulletin of the Nuttall Ornithological Club, vols. 1–7 – 1876–1883, and to its continuation The Auk, vols. 1–17, 1884–1900.]* Ed. by J. Dwight Jr. New York. viii+426 pp.; *Continued* ___. 1910. *Ten–year index to The Auk, volumes 18–27, 1901–1910.* Ed. by T. S. Palmer & W. W. Cooke. New York. xxviii+250 pp.; ___. 1929. ... *index... vols. 28–37, 1911–20.* Lancaster, PA, for the American Ornithologists' Union. xvii+339 pp.; ___. 1934. ... *index... vols. 38–47, 1921–30.* Ed. by H. S. Swarth. Lancaster, PA. xxiii+328 pp.; ___. 1941. ... *index... vols. 48–57, 1931–40.* Ed. by G. Willett. Lancaster, PA. 374 pp.; ___. 1960. ... *index... vols. 58–67, 1941–50.* Ed. by C. K. Nichols. Lancaster, PA. iv+449 pp.; ___. 1970. ... *index... vols. 68–77, 1951–60.* Ed. by O. J. Austin Jr. Washington, DC. 184 pp.; ___. 1973. ... *index... vols. 78–87, 1961–70.* Ed. by O. J. Austin Jr. Washington, DC. 217 pp.; ___. 1984. ... *index... vols. 88–97, 1971–80.* Ed. by O. J. Austin Jr. Washington, DC. iv+190 pp.; ___. 1991. ... *index... vols. 98–107, 1981–90.* Ed. by J. A. Wiens & A. H. Brush. Washington, DC. [i]+267 pp.

Bb824. **Cannell, P. F.** 1983. An annotated bibliography of the founding of the American Ornithologists' Union. *American birds* 37: 355–357.

Bb825. **Palmer, T. S.** 1933. A brief history of the American Ornithologists' Union. *and* Appendix. A: Meetings and membership of the Union, 1883–1933. Appendix B: Chronology of the ... Union, 1883–1933. In: American Ornithologists' Union. *Fifty years progress of American ornithology, 1883–1933.* On the occasion of its semi–centennial anniversary. New York, November 13–16, 1933, pp. 7–27, 241–249. (Ed. by F. M. Chapman & T. S. Palmer.) Lancaster, PA.

Bb826. **___, et al.** 1954. *Biographies of members of the American Ornithologists' Union. Reprinted from the Auk, 1884–1954.* Washington, DC, privately pub. 630 pp.

British Ornithologists' Club

Bb827. **British Ornithologists' Club.** 1906. Index to the Bulletin of the British Ornithologists Club, vols. I–XV, 1892–1905. Ed. by W. R. Ogilvie–Grant. *Bulletin of the British Ornithologists' Club* 18: 108 pp.; *Continued* ___. 1920. ___ *vols. XVI–XXXIX, 1906–1919.* Ed. by D. Seth–Smith. London, Witherby & Co. [iii]+158 pp.; ___. 1932. ___ vols. XL–LI, 1919–1931. Ed. by G. C. Low. *Ibid.* 3: [vii]+194 pp.; ___. 1963. *Bulletin of the ... Ten year scientific index 1950–1959.* By F. E. Duckett. N.p. 39 pp.

Bb828. ___. 1951. A short history of the Club. By P. Manson–Bahr. *Bulletin of the British Ornithologists' Club* 71: 2–4.

Bb829. ___. 1980. [Various papers.] *Ibid.* 100.
Special issue. *Includes* **R. E. F. Peal.** A short history of the Club and its Bulletin, pp. 4–13; **P. Barclay–Smith.** Recollections of personalities of the Club, pp. 15–22; **D. A. Bannerman.** Reminiscences of the Club, pp. 22–25; **C. E. Godman.** Recollections, pp. 29–34; **N. J. P. Wadley.** Random recollections of the B O C 1950–1960, pp. 36–40.

British Ornithologists' Union

Bb830. **British Ornithologists' Union.** 1900. *General subject–index to 'The Ibis' (first to sixth series), 1859–1894.* Ed. by E. W. Oates. London, Gurney & Jackson. 140 pp.; *Continued* ___. 1900. *General index ... [seventh series] 1895–1900.* London. xlix pp.; ___. 1906. *General index ... [eighth series] 1901–1906.* London, R. H. Porter. xxxvi pp.; ___. 1912. General index ... [ninth series] 1907–1912. *Ibis* (9) 6: 711–738; ___. 1918. General index ... [tenth series] 1913–1918. *Ibis* (10) 6: 773–800.; ___.

1924. General index ... [eleventh series] 1919–1924. *Ibis* (11) 6: 871–892.; ___. 1930. General index ... [twelfth series] 1925–1930. *Ibis* (12) 6: 751–776.

Bb831. **British Ornithologists' Union.** 1908. *Ibis* jubilee supplement. By P. L. Sclater & A. H. Evans. *Ibis* (9)2: 268pp.
Pt 1: Proceedings of the Jubilee Meeting the British Ornithologists' Union; pt. 2: **P. L. Sclater.** A short history of the British Ornithologists' Union; pt. 3: Biographical notices of the original members and principal contributors to the first series of the *Ibis*, and of officers, with portraits; pt. 4: List of members of the British Ornithologists' Union from 1858 to 1908.

Bb832. ___. 1916. *Index of genera and species referred to, and an index to the plates, in 'The Ibis' (seventh, eighth, and ninth series).* Ed. by W. L. Sclater. London. 513 pp.

Bb833. ___. 1918. Index–list of the coloured plates of birds in "The Ibis" 1859 to 1917. By E. Hopkinson. *Ibis* (10) 6: 10–51.

Bb834. ___. 1959. One hundred years of the British Ornithologists' Union. By G. Mountfort. *Ibis* 101: 8–18.

British Trust for Ornithology

Bb835. **Hickling, R.**, ed. 1983. *Enjoying ornithology: A celebration of fifty years of the British Trust for Ornithology 1933–1983.* Calton, T. & A. D. Poyser. 296 pp.
Chapters by various authors, including: **E. M. Nicholson.** The Trust, origins and early days, pp. 15–28; **R. Spencer.** The Trust from 1951 to 1982, pp. 29–53.

Dansk Ornithologisk Forening

Bb836. **Dansk Ornithologisk Forening.** 1927. *Dansk ornithologisk forenings tidsskrift. Register, aarg. 1–20 [1906/07–26].* Copenhagen. 16 pp.; *Continued* ___. 1947. ___. *Register, aarg. 21–40 [1927–46].* Copenhagen. [ ]* pp.; ___. 1967. ___. *Register, aarg. 41–60 [1947–66].* Copenhagen. 40 pp.
[*Not examined – pagination not ascertained.]

Deutsche Ornithologische Gesellschaft (*formerly* Deutscher Ornithologen–Verein, Deutsche Ornithologen–Gesellschaft, Deutsche Ornithologische Gesellschaft zu Berlin *and* Allgemeine Deutsche Ornithologische Gesellschaft)

Bb837. **Deutsche Ornithologen–Gesellschaft.** 1950. Einhundert Jahre Deutsche Ornithologen–Gesellschaft. By E. Stresemann. *Vogelwarte* 15: 209–213.

Deutscher Bund für Vogelschutz e. V

Bb838. **Hannemann, H., & J. M. Simon.** 1987. *Die Chronik eines Naturschutzverbandes von 1899–1984: Bund für Vogelschutz e. V., Deutscher Bund für Vogelschutz e. V.* Wiesbaden, Wirtschaftsverlag. 211 pp. *Schriftenreihe Verbände der Bundesrepublik Deutschland* no. 12.

Bb839. **Simon, J. M., & H. Hannemann.** 1989. *Es begann mit einer Insel ... Deutscher Bund für Vogelschutz e. V.: Die Bilder–Chronik eines Naturschutzverbandes von 1899–1989.* Wiesbaden, Wirtschaftsverlag. 427 pp. *Schriftenreihe Verbände der Bundesrepublik Deutschland* vol. 29.

Nuttall Ornithological Club

Bb840. **Nuttall Ornithological Club.** 1924. *The fiftieth anniversary of the Nuttall Ornithological Club, December 1923.* Cambridge, MA. 27 pp.

Bb841. ___. 1937. *An account of the Nuttall Ornithological Club 1873–1919.* By C. F. Batchelder. Cambridge, MA. 109 pp., & 36 port. *Memoirs* vol. 8.
Re–pub. in: **K. B. Sterling**, ed. 1974. *Contributions to the history of American ornithology.* New York, Arno Pr.; *see* Cc370.

Bb842. ___. 1987. *History of the Nuttall Ornithological Club, 1873–1986.* By W. E. Davis Jr. Cambridge, MA. xi+179 pp. *Memoirs* vol. 11.

(Royal) Australasian Ornithologists' Union

Bb843. **Royal Australasian Ornithologists' Union.** 1951. The first fifty years of the Royal Australasian Ornithologists' Union 1901–1951. By D. J. Dickison. *Emu* 51: 185–284.

Bb844. ___. 1953. *A species index to the "Emu." Volumes 1–50, 1901–1951, together with references to contributors and illustrations.* By A. R. McGill. Melbourne, Royal Australasian Ornithologists Union. vi+[ii]+183 pp.

(Royal) Society for the Protection of Birds

Bb845. **Royal Society for the Protection of Birds.** 1988. *For the love of birds: The story of the Royal Society for the Protection of Birds, 1889–1988.* By T. Samstag. Sandy. 160 pp., illus.

Herpetological & ichthyological

American Society of Ichthyologists & Herpetologists

Bb846. **American Society of Ichthyologists & Herpetologists.** 1984. A chronology of the American Society of Ichthyologists and Herpetologists through 1982. By T. M. Berra. *Special publications, American Society of Ichthyologists & Herpetologists* 2: 1–21.

Bb847. **McCoy, C. J.** 1989. Index to the biographies of herpetologists and ichthyologists in *Copeia*, 1913–1988. *Smithsonian Herpetological Information Service* 78: 33 pp.

Bb848. **Reed, C. F.** 1955. *Index to Copeia, 1913–1954.* Part I – Author index. Lancaster, PA. 106 pp.; ___. 1956. ___. Part II – Subject index. Lancaster, PA. 332 pp.; *Continued* ___. 1965. *Index to Copeia, Supplement 1955–1964.* Lancaster, PA. 329 pp.

Kommission zur Wissenschaftlichen Untersuchung der Deutschen Meere, in Kiel

Bb849. **Kölmel, R.** 1990. The Prussian "Kommission zur Wissenschaftlichen Untersuchung der Deutschen Meere in Kiel". In: W. Lenz & M. Deacon, eds. *Ocean sciences: Their history and relation to man. Proceedings of the 4th International Congress on the History of Oceanography Hamburg [etc.]*, pp. 399–407. Hamburg, Bundesamt für Seeschiffahrt und Hydrographie. *Deutsche hydrographische Zeitschrift*, Ergänzungsheft, Reihe B, no. 22.

Bb850. **Kommission zur Wissenschaftlichen Untersuchung der Deutschen Meere, in Kiel.** 1921. *Festschrift der Preussischen Kommission zur wissenschaftlichen Untersuchung der deutschen Meere zu Kiel aus Anlaß ihres 50 jährigen Bestehens.* Kiel, Lipsius & Tischer. [ii]+194 pp.

Chapters by several authors. *Includes* **J. Reinke.** Botanische Arbeiten der Kommission, pp. 51–75; **K. Brandt.** Die zoologischen Arbeiten der Kieler Kommission 1870 bis 1920, pp. 76–194.

Society for the Study of Amphibians & Reptiles

Bb851. **Society for the Study of Amphibians & Reptiles.** 1983. Society for the Study of Amphibians and Reptiles: A brief history, 1958–1982. *Herpetological circular, Society for the Study of Amphibians & Reptiles* 13: 40–46.

Entomological

Amateur Entomologists' Society

Bb852. **Amateur Entomologists' Society.** 1985. Fifty years of publishing – a bibliography of the Amateur Entomologists' Society. By P. Sokoloff. *Bulletin of the Amateur Entomologists' Society* 44: 146–154.

American Entomological Society (*formerly* Entomological Society of Philadelphia)

Bb853. **American Entomological Society.** [1911.] *A history of the American Entomological Society, Philadelphia, 1859–1909, prepared by request of the Society.* By E. T. Cresson. Philadelphia, PA. 60 pp. & 3 ports.

Bb854. **___.** 1959. The history of the American Entomological Society. By H. W. Allen. *Transactions of the American Entomological Society* 85: 335–371.

Bb855. **Fuester, R. W.** 1984. The Cressons and the American Entomological Society. *Entomological news* 95: 149–154.

Bb856. **Mason, C. E.** 1984. The 125th anniversary of the American Entomological Society. *Ibid.* 95: 125–128.

Bb857. **Meisel, M.** 1929. Entomological Society of Philadelphia (1859). In: *A bibliography of American natural history, the pioneer century, 1769–1865, [etc.]*, vol. 3: 289–299. Brooklyn, NY, Premier Publishing Co. *Re-pub.* New York, Hafner, 1967. Historical note and bibliography (including list of natural history publications in its *Proceedings* 1861–65).

Bb858. **Roback, S. S.** 1984. James A. G. Rehn and the American Entomological Society. *Entomological news* 95: 163–165.

British Entomological & Natural History Society (*formerly* South London Entomological & Natural History Society)

Bb859. **British Entomological & Natural History Society.** 1973. *The new aurelians: A centenary history of the British Entomological & Natural History Society.* By M. J. James, with an account of the collection by A. E. Gardner. London, the Society. 80 pp. & 4 port.

Bb860. **South London Entomological & Natural History Society.** 1955. The history of the "South London" Society. By F. Stanley-Smith. *Proceedings & transactions, South London Entomological & Natural History Society* 1953/54: 55–74.

Brooklyn Entomological Society

Bb861. **Teale, E. W.** 1968. A history of the Brooklyn Entomological Society. *Journal of the New York Entomological Society* 76: 156–164.

City of London Entomological & Natural History Society (*later* London Natural History Society)

Bb862. **Plant, C. W.** 1985. An index to the Transactions of the City of London Entomological and Natural History Society, 1892–1914. *London naturalist* 64: 95–101.

Deutsche Entomologische Gesellschaft (split from Entomologischer Verein q.v. infra in 1880 and re-united in 1914)

Bb863. **Deutsche Entomologische Gesellschaft.** 1887. *Inhalts–Verzeichniss der Deutsche Entomologische Zeitschrift (früher: "Berliner entomologischer Zeitschrift.") Jahrgang. 25–30, (1881–1886).* By F. Blücher. [Berlin.] 17 pp.; *Continued* **___.** 1893. *Inhalts–Verzeichniss der Deutschen entomologischen Zeitschrift, Jahrgang 1887–1892.* By E. Backhaus. Berlin. 31 pp.; **___.** 1900. *Inhalts–Verzeichniss ... Jahrgang 1893–1899.*

By R. Lohde. Berlin. 36 pp.; ___. 1907. *Inhalts–Verzeichniss ... Jahrgang 1900–1906. Gera. Berlin, 1. Dezember 1907.* By F. Müller. [Altenburg.] 40 pp.

Bb864. ___. 1906. *Ein Beitrag zur Geschichte der systematischen Entomologie: Dem Gründer der Deutschen Entomologischen Gesellschaft, des Deutschen Entomologischen National–Museums und des Entomologischen Vereins in Berlin zum 50–jährigen Jubilaeum als entomologischer Präsident und Redakteur und zur Feier des goldenen Doktor–Jubilaeums gewidmet von der Deutschen Entomologischen Gesellschaft. Eine Festgabe für alle Mitglieder.* By G. Kraatz. Berlin. 164 pp. & 5 pl.

Entomological Club

Bb865. **South, R.** 1899. The Entomological Club. *Entomologist* 32: 160–164, 224–226.

Entomological Society of Canada

Bb866. **Connor, J. T. H.** 1982. Of butterfly nets and beetle bottles: The Entomological Society of Canada, 1863–1960. *H S T C bulletin, journal of the history of Canadian sciences, technology, and medicine* 6: 151–171.

Entomological Society of Ontario

Bb867. **Bethune, C. J. S.** 1900. *General index to the thirty annual reports of the Entomological Society of Ontario, 1870–99.* Toronto, Ontario Department of Agriculture. 76 pp.; *Continued* **C. E. Petch.** 1939. *General index to the thirty–eight annual reports of the Entomological Society of Ontario, 1900–1937.* Toronto, T. E. Bowman [printer]. 267 pp.

Entomological Society of Pennsylvania

Bb868. **Entomological Society of Pennsylvania.** 1978. A history of the Entomological Society of Pennsylvania, 1824–44 and 1924–present. By A. G. Wheeler Jr. & K. Valley. *Melsheimer entomological series, Entomological Society of Pennsylvania* 24: 16–26.

Entomological Society of Washington

Bb869. **Entomological Society of Washington.** 1909. The Entomological Society of Washington. By L. O. Howard. *Proceedings of the Entomological Society of Washington* 11: 8–18.

Bb870. ___. 1937. Dr. L. O. Howard and the Entomological Society of Washington. By W. R. Walton & F. C. Bishop. *Ibid.* 39: 121–132.

Bb871. ___. 1984. Vignettes of 100 years of the Entomological Society of Washington. By T. J. Spilman. *Ibid.* 86: 1–10.

(Berliner) Entomologischer Verein (in Berlin) (*See also* Deutsche Entomologische Gesellschaft supra)

Bb872. **Entomologischer Verein.** 1863. *Inhalts–Verzeichniss der Berliner Entomologischen Zeitschrift Jahrgang I–VI – 1857–1862.*] By M. Wahnschaffe. Berlin. 46 pp. Appended to vol. 6; *Continued* ___. 1869. *Inhalts–Verzeichniss (Berliner entomologischer Zeitschrift) Jahrgang VII–XII – 1863–1868.* By M. Wahnschaffe. Berlin. 52 pp. Appended to vol. 12; ___. 1874. *Inhalts–Verzeichniss (Berliner entomologischer Zeitschrift) Jahrgang XIII–XVIII – 1869–1874.* By M. Wahnschaffe. Berlin. 56 pp. Appended to vol. 19.; ___. 1880. *Inhalts–Verzeichniss der Berliner entomologischer Zeitschrift ... Jahrg. 19–24, 1857–80 ... und chronologisches Verzeichniss der Arbeiten der einzelnen Autoren in Jahrg. 1–24, 1857–80, [etc.].* By F. Blücher. Berlin. 32+34 pp.; ___. 1903. *Berliner entomologische Zeitschrift. Inhalts–Verzeichniss der Jahrgänge 25–35, 1881–90.* By W. Roepke. [Berlin.] 70 pp. Appended to vols. 48 & 49.

Bb873. **Hannemann, H.** 1957. *Bericht über die Hunderjahrfeier der Deutschen Entomologischen Gesellschaft, Berlin, 30 September bis 5 Oktober 1956.* Berlin, Akademie Verlag. 295 pp.

Bb874. **Marks, –.** 1960. [Dates of publication, 1884–1939.] *Journal of the New York Entomological Society* 68: 199–207.

Gosudarstvennoe Vserossiiskoe Entomologicheskoe Obshchestvo (*formerly* Russkoe Entomologicheskoe Obshchestvo/Societas Entomologica Rossica)

Bb875. **Bei–Bienko, G. Ya.** 1960. Ocherk deyatel'nosti Vsesoyuznogo Entomologicheskogo Obshchestvo za 100 let (1859–1959 gg.). [An account of the activity of the All–Union Entomological Society for the period 1859–1959]. *Entomologicheskoe obozrenie* 1960 (1): 5–33.
English version in: *Entomological review* 39: 1–19, 1960.

Bb876. **Gosudarstvennoe Vsesoyuznoe Entomologicheskoe Obshchestvo.** 1936. Kratkii ocherk deyatel'nosti Gosudarstvennogo Vserossiiskogo Entomologicheskogo Obshchestvo za 75 let ego sushchestvovaniya (1860–1935 gg.). By E. N. Pavlovskii & N. N. Filip'ev. *Entomologicheskoe obozrenie* 26: 3–14.

Bb877. **Kerzhner, I. M.** 1984. Dates of publication of "Trudy Russkogo Entomologicheskogo Obshchestva" and "Horae Societatis Entomologicae Rossicae." *Ibid.* 63: 849–856.

Bb878. **Russkoe Entomologicheskoe Obshchestvo.** 1899. *Ukazatel' soobshchenii, sdelannikh na obshchikh sobraniyakh Rysskogo Entomologicheskogo Obshchestvo za 35 let ego syshchestvovaniya (1859–1894).* By V. V. Mazarakii. St. Petersburg. 58 pp.

Bb879. ___. 1910. *Tables générales des publications de la Société Entomologique de Russie ainsi que des articles, des synopsis et des formes nouvelles y contenues, 1859–1908.* St. By B. Oshanin. Pétersbourg, Société Entomologique de Russie. 282+48 pp. *Trudy Rysskogo Entomologicheskogo Obshchestvo* 38 Suppl.

Internationaler Entomologischer Verein zu Frankfurt am Main

Bb880. **Internationaler Entomologischer Verein zu Frankfurt am Main.** 1934. *Festschrift zum 50 jährigen Bestehen des Internationalen entomologischen Vereins e. v., Frankfurt am main, 1884–1934.* Frankfurt am Main. 48 pp.

Nederlandsche Entomologische Vereeniging

Bb881. **Nederlandsche Entomologische Vereeniging.** 1869. *Repertorium der acht eerste jaargangen van het tijdschrift [vols. 1–8, 1858–65], [etc.].* By E. A. de Roo van Westmaas. The Hague. 40 pp.; *Continued* ___. 1875. *Repertorium betreffende den 9en tot en met den 16en jaargang, (Nieuwe serie, [= deel 1–8] 1866–73) van het Tijdschrift voor entomologie, uitgegeven door de Nederlandsche Entomologische Vereeniging.* By F. M. Van der Wulp. The Hague. [ ]* pp.; ___. 1882. *Tijdschrift voor entomologie. Repertorium, [ser. 3] deel 17–24 [1874–81], [etc.].* The Hague. [ ]* pp.
[*Not examined – pagination not ascertained.]

Bb882. ___. 1895. *De Nederlandsche Entomologische Vereeniging. Geschiedkundig overzicht, 1845–95.* By F. M. Van Der Wulp. The Hague. 99 pp.

New York Entomological Society

Bb883. **New York Entomological Society.** 1968. A history of the New York Entomological Society. By E. B. Klots. *Journal of the New York Entomological Society* 76: 138–155.

Bb884. **Weiss, H. E.** 1943. The Journal of the New York Entomological Society, 1893–1942. *Ibid.* 51: 285–294.

Polskie Towarzystwo Entomologiczne

Bb885. **Strawinski, K.** 1962. Forty years of Polish Entomologists' Association. *Polskie pismo entomologiczne* B25–26: 7–17.

(Royal) Entomological Society of London

Bb886. **Entomological Society of London.** 1933. *The history of the Entomological Society of London, 1833–1933.* By S. A. Neave & F. J. Griffin. London, Royal Entomological Society. xlvi+224 pp. & 7 pl.

Bb887. **Royal Entomological Society of London.** 1983. The Royal Entomological Society of London 1833–1983. *Antenna* 7: 80–87.

Schweizerische Entomologische Gesellschaft

Bb888. **Schweizerische Entomologische Gesellschaft.** 1983. La Société Entomologique Suisse de 1958 a 1983. By P. Bovey. *Mitteilungen der Schweizerischen Entomologischen Gesellschaft* 56: 3–12.

Società Entomologica Italiana

Bb889. **Società Entomologica Italiana.** 1969. Indice delle pubblicazioni della Società Entomologica Italiana dal 1869 al 1968. By D. Guiglia, M. Magistretti & G. Mattioni. *Memorie della Società Entomologica Italiana* 48: 663–813.

Société Entomologique de Belgique (*formerly* Société Entomologique Belge)

Bb890. **Société Entomologique de Belgique.** 1880. *Assemblée générale extraordinaire convoquée pour la commémoration de la fondation de la Société, 16 Octobre 1880.* Brussels. 14 pp.

Bb891. ___. 1887. *Table générale des Annales de la Société Entomologique de Belgique I–XXX [1857–86] et catalogue des ouvrages périodique de la bibliothèque, 26 Décembre 1887.* By A. Lameere. Brussels & Leipzig. 81 pp.

Société Entomologique de France

Bb892. **Société Entomologique de France.** 1867. *Tables générales alphabétiques et analytiques divisées par ordres d'animaux de toutes les matières contenues dans les trois premières séries des Annales ... suivies de la table des auteurs de ces matières, 1832–1860, [etc.].* By A.–S. Paris. Paris. xiv+739 pp.; *Continued* ___. 1885. *Tables générales de 1861 à 1880 [sér. 4, vol. 1 – sér. 5, vol. 10] inclusivement, [etc.].* By M. E. Lefevre. Paris. 286 pp.; ___. 1895. *Tables générales de 1881 à 1890 [sér. 6, vol. 1 – 10] inclusivement.* By M. E. Lefevre. Paris. 239 pp.

Bb893. ___. 1932. *Société Entomologique de France. Livre du centenaire.* Paris. xii+729 pp., illus., ports.

Society of Aurelians

Bb894. **Allen, D. E.** 1966. Joseph Dandridge and the first Aurelian Society. *Entomologist's record* 78: 89–94.

Conchological

Conchological Society of Great Britain & Ireland

Bb895. **Norris, A.** 1982. The Conchological Society of Great Britain and Ireland: The early years. *Naturalist* 107: 131–134.

Deutsche Malakozoologische Gesellschaft

Bb896. **Zilch, A.** 1967. 150 Jahre Senckenbergische Naturforschende Gesellschaft. 100 Jahre Deutsche Malakozoologische Gesellschaft. *Archiv für Molluskenkunde* 97: 1–5.

Nederlandsche Malacologische Vereniging

Bb897. **Nederlandsche Malacologische Vereniging.** 1984. Jubileunummer: Een halve eeuw Nederlandse Malacologische Vereneging. By J. G. J. Kuiper. *Correspondentieblad van de Nederlandse Malacologische Vereniging* 220: 1554–1601.

Société Malacologique de France

Bb898. **Société Malacologique de France.** 1885–86. *Revue biographique de la Société [etc.]. 1885* (& *1886*). Paris. 2 pts.

Société (Royale) Malacologique de Belgique

Bb899. **Backhuys, W.** 1985. Bibliographical note on the journals of the Société (Royale) Malacologique de Belgique. *Mededelingen van de Werkgroep voor Tertiaire en Kwartaire Géologie* 22: 173–188.

Bb900. **Société Royale Malacologique de Belgique.** 1920. Table des cinquante premiers volumes (1863–1920). *Annales de la Société Malacologique de Belgique* 50 (2): 116 pp.

BIBLIOGRAPHY OF HISTORIES OF SOCIETIES

Bb901. **History of Science Society.** 1971. Isis *cumulative bibliography 1913–1965: A bibliography of the history of science formed from* Isis *critical bibliographies 1–90, indexing literature published from 1913 through 1965*. Ed. by M. Whitrow. Vol. 2. Personalities and institutions. London, Mansell. *Re-pub.* (in microfiche); *Continued* ___. 1980. ___*: 1966–1975: ...* Isis *critical bibliographies 91–100, indexing literature published from 1965 through 1974*. Ed. by J. Neu. Vol. 1: Personalities and institutions. London, [etc.]; ___. 1989. ___*: 1976–1985: ...* Isis *critical bibliographies 101–110, indexing literature published from 1975 through 1984*. Ed. by J. Neu. Vol. 1: Personalities and institutions. London, [etc.].

Section Bc. NATURAL HISTORY LIBRARY RESOURCES

LIBRARIES & LIBRARY HISTORY

General

Bc1. **Ash, L., & W. G. Miller.** 1993. *Subject collections: A guide to special book collections and subject emphases as reported by university, college, public, and special libraries and museums in the United States and Canada.* 7th ed. New Providence, NJ, R. R. Bowker. 2 vols.

A subject guide arranged by alphabetical subject headings. *Replaces* ___. 1985. ___. 6th ed. New York, R. R. Bowker. 2 vols.

Bc2. **Aslib.** 1982–84. *Aslib directory of information sources in the United Kingdom.* 5th ed. London, Association of Special Libraries & Information Bureaux (ASLIB). 2 vols.

Vol. 1. Science and technology. *Replaces* ___. 1977–80. *Aslib directory.* 4th ed. by M. Codlin. London. 2 vols.

Bc3. **Bowker, R. R.** [Pub.]. 1992. *American library directory 1992–93.* 45th ed. New York. 2 vols. (xv+2,477 pp.).
Commenced in 1923, published annually.

Bc4. **Gale Research Co.** 1993. *Directory of special libraries and information centers 1993.* 16th ed. Ed. by J. A. DeMaggio & D. M. Kirby. Detroit, MI. 3 vols.
Over 20,850 libraries in the U.S.A., Canada, etc. Vol. 1. Directory of special libraries and information centers; vol. 2. Geographic and personnel indexes; vol. 3. New special libraries.

Bc5. **Gosudarstvennaya Biblioteka SSSR im. V. I. Lenina.** 1973. *Biblioteki SSSR: Spravochnik. Biblioteki Soyuznykh Respublik (bez RSFSR). [Libraries of the USSR: Handbook. Libraries of the Soviet republics (without the RSFSR)].* Moscow, Izd. "Kniga." 368 pp.

Bc6. ___. 1974. *Biblioteki RSFSR. [Libraries of the RSFSR].* Moscow, Izd. "Kniga." 429 pp.

Bc7. **Lewanski, R. C.** 1978. *Subject collections in European libraries.* 2nd ed. New York, London, Bowker Publishing Co. [xv]+495 pp.
A subject guide to collections arranged under classified headings; omits some important libraries and collections. *Replaces* 1st ed., 1965. xl+789 pp.

Bc8. **Modoc Press Inc.** 1989. *Special collections in college and university libraries.* New York, Macmillan. 639 pp.
Special collections, rare books, and manuscripts in 1,805 institutions. Indexes of personal names, subjects, places, etc.

Bc9. **Saur, K. G.** [Pub.]. 1990. *World guide to special libraries.* 2nd ed. Ed. by B. Bartz et al. Munich, [etc.]. 2 vols. *Handbook of international documentation and information* vol. 17.
Over 32,000 libraries in 160 countries; five categories – General, Humanities, Social Sciences, Medicine & Life Sciences, and Science & Technology.

Bc9A. ___. 1991. *World guide to libraries.* 10th ed. Ed. by B. Bartz, H. Opitz & E. Richter. Munich, [etc.]. xxviii+1,039 pp. *Handbook of international documentation & information* vol. 8.
Over 40,000 libraries in 166 countries arranged by country and type with an alphabetical index. *Replaces* ___. 1989. *World guide to libraries. Internationales Bibliotheks–Handbuch.* 9th ed. Ed. by H. Lengenfelder. Munich [etc.], K. G. Saur. xxxv+1001 pp. The 11th ed. is announced for publication in mid 1993.

Bc10. **Williams, M. I.**, ed. 1985. *A directory of rare book and special collections in the United Kingdom and the Republic of Ireland.* London, Library Association. xiii+664 pp.
Descriptions of individual collections are uneven in treatment and accuracy.

Natural history (& science) libraries

Bc11. **Akademiya Nauk SSSR.** 1986. *Directory of libraries and information centers of the academies of sciences of the socialist countries.* By S. E. Kleshchuk & B. S. Koreneva. Moscow, "Nauka." 120 pp.
Details of 37 major libraries.

Bc12. **Beatty, W. K.** 1969. Biomedical libraries. In: A. Kent & H. Lancour. *Encyclopedia of library and information science*, vol. 2, pp. 554–603. New York, M. Dekker.
Includes Important biomedical libraries, pp. 579–603.

Bc13. **Blanchard, J. R.** 1968. Agricultural libraries and collections. In: A. Kent & H. Lancour. *Encyclopedia of library and information science*, vol. 1: 126–134. New York, M. Dekker.

Bc14. **Boalch, D. H.**, ed. 1960. *World directory of agricultural libraries and documentation centers.* Harpenden, International Association of Agricultural Librarians & Documentalists. 280 pp.
Geographically arranged with a subject index.

Bc15. **Davis, T. J.** 1989. *Latin American research libraries in natural history: A survey.* Washington, DC, American Ornithologists' Union, National Museum of Natural History. [vi]+127 pp.
170 institutions.

Bc16. **DeMaggio, J. A., & D. M. Kirby**, eds. 1992. *Subject directory of special libraries and information centers.* 16th ed. Detroit, MI, Gale Research. 3 vols.
Ca.13,065 entries. Vol. 2: *Computers, enginering, and science libraries.* 725 pp.

Bc17. **Earnshaw, F.** 1971. Some British libraries of interest to biologists. In: R. T. Bottle & H. V. Wyatt, eds. *The use of biological literature,* 2nd ed., pp. 350–357. London, Butterworth.

Bc18. **Edwards, P. I.** 1971. List of libraries in the field of pure and applied biology. *Biological journal of the Linnean Society* 3: 173–188.

Bc19. **Gundry, L. D.** 1970. *Guide to science and history of science resources in the Philadelphia area.* Philadelphia, PA, American Philosophical Society. [iii]+63 pp. *Library publication* no. 3.

Bc19A. **Imperial College of Science & Technology and King's College, London University.** 1986. *Life sciences libraries in and around London.* 6th ed. Ed. by J. Yeadon et al. London. 60 pp.
Replaces 5th ed., 1982.

Bc20. **Poland, U. H.**, ed. 1988. *World directory of biological and medical science libraries.* Munich, New York, K. G. Saur. xi+203 pp. *I F L A publication* vol. 42.
1,371 libraries in 100 countries. Includes a bibliography of national and regional directories of these libraries.

Bc21. **Simpson, P., & D. S. Moulder.** 1980. *Directory of library and information facilities.* Plymouth, United Kingdom Marine & Freshwater Sciences Librarians Group. ii+20 pp.
Information on ten marine & freshwater sciences libraries.

Bc22. **Wyatt, H. V.** 1981. *A directory of information resources in biology in the U. K.* London, British Library. v+83 pp. *British Library research & development report* no. 5606.

Botanical

Bc23. **American Horticultural Society.** 1982. Botanical and horticultural libraries. In: *North American horticulture: A reference guide,* pp. 186–196. B. W. Ellis, ed. New York, Scribner's Sons.

Bc24. **Callery, B. G., E. J. Teeter & M. L. Wolfe.** 1983. *The Council on Botanical and Horticultural Libraries: A directory of member libraries.* Bronx, NY, CBHL. [iii]+60 pp.
An alphabetical list of ca.120 libraries with a subject index.

Bc25. **Evans, P. A., & M. A. Arizmendi**, eds. 1991. *International directory of forestry and forest products libraries.* 3rd ed. Berkeley, CA, University of California Forestry Library. v+108 pp.
Geographical guide to 480 libraries. *Replaces* Earlier eds. 1983 & 1987.

Bc26. **Lawrence, G. H. M.** 1970. Botanical libraries and collections. In: A. Kent & H. Lancour. *Encyclopedia of library and information science*, vol. 3, pp. 104–121. New York, M. Dekker.

Zoological

Bc27. **Gilbert, P., & C. J. Hamilton.** 1990. Libraries holding entomological literature. In: *Entomology: A guide to information sources*, pp. 194–211. 2nd ed. Poole, Mansell.

Bc28. **Kenyon, K. A.** 1984. Zoo/aquarium libraries: A survey. *Special libraries* 75(4): 329–334.

Bc29. **Kisling, V. N.** 1988. American Zoological Park libraries and archives: Historical considerations and their current status. In: E. Mount, ed. *Sci–tech libraries serving zoological gardens*, pp. 49–60. New York, Haworth Pr. A re-issue of *Science & technology libraries* vol. 8, no. 4 as a separate book.
Analysis of a 1987 survey of the status of 117 libraries in the U.S.A.

Bc30. **Mount, E.**, ed. 1988. *Sci–tech libraries serving zoological gardens.* New York, Haworth Pr. xiii+111 pp. Re-issue of *Science & technology libraries* vol. 8, no. 4 as a separate book.

Bc31. ___. 1988. *Sci–tech libraries in museums and aquariums.* New York, Haworth Pr. xiv+204 pp. Re-issue of *Science & technology libraries* vol. 6, nos. 1/2 as a separate book.

Bc32. **Scudder, S. H.** 1880. The entomological libraries of the United States. *Harvard University Library bulletin* 14: 20–23.

Map collections

Bc33. **Cobb, D. A., et al.** 1990. *Guide to U. S. map resources.* 2nd ed. Chicago, IL, American Library Association. xvi+495 pp.
Covers resources in agriculture, biology, forestry, geology, land use, vegetation, zoogeography, etc. *Replaces* 1st ed., 1986. xvi+196 pp.

HISTORY/DESCRIPTION OF INDIVIDUAL LIBRARIES

(Königliche) Akademie Gemeinütziger Wissenschaften, Erfurt

Bc34. **Kiefer, J.** 1990. Die nationalen und internationalen Tauschbeziehungen der ehemaligen "Akademie nützlicher (gemeinnütziger) Wissenschaften" zu Erfurt (1754–1945) mit gelehrten Gesellschaften, wissenschaftlichen Institutionen und Behörden in den Jahren 1757–1944. *Beiträge zur Hochschul- und Wissenschaftsgeschichte Erfurts* 22: 208–233.

Akademiya Nauk SSSR, Leningrad

Bc35. **Akademiya Nauk SSSR, Biblioteka.** 1929. *Biblioteka Akademiya Nauk SSSR, 1728–1929. Kratkii istoricheskii ocherk i putevoditel.* Leningrad. 116 pp.

Bc36. ___. 1959. *Biblioteki Akademii Nauk SSSR: Spravochnik.* By A. I. Chebotarev. Moscow. 323 pp.

Bc37. ___. 1964. *Biblioteka Akademiya Nauk SSSR. 1714–1964. Bibliograficheskii ukazatel'.* E. P. Faidel et al. Leningrad. 307 pp.

Bc38. ___. 1964. *Istoriya Biblioteki Akademiya Nauk SSSR. 1714–1964.* S. P. Luppov et al. Moscow/Leningrad. 599 pp.

Bc39. ___. 1966. *Materialy i soobshcheniya po fondam Otdela rukopisnoi i redkoi knigi Biblioteki Akademii Nauk SSSR*. Ed. by A. I. Kopanev. Moscow/Leningrad. 204 pp.

Bc40. ___. 1977. The libraries of the USSR Academy of Sciences over the past sixty years. By M. V. Varfolomeeva. *Unesco bulletin for libraries* 31: 360–367.

See also Bb124 & Cb611.

Akademiya Nauk SSSR, Botanicheskii Institut im. V. L. Komarova, St. Petersburg (Leningrad)

Bc41. **Lipshits, S. Yu.** 1969. Iz istorii biblioteki Botanicheskogo Instituta imeni V. L. Komarova, Akademii Nauk SSSR. *Botanicheskii zhurnal* 54: 2011–2018.
About the A. K. Razumovsky and F. C. Stephan collections which formed the nucleus of the original library.

See also Cb1044(Lebedev).

American Entomological Society, Philadelphia, PA

Bc42. **Boyd, H. P.** 1984. The library of the American Entomological Society and a brief review of the Society's association with the Academy of Natural Sciences of Philadelphia. *Entomological news* 95: 131–136.

American Museum of Natural History, New York

Bc43. **Root, N. J.** 1977. The rare book and manuscript collection of the American Museum of Natural History library. *Curator* 20: 121–152.

Bc44. ___. 1980. The Library of the American Museum of Natural History. *Journal of the Society for the Bibliography of Natural History* 9: 587–591.

Bc45. ___. 1983. Biography of a museum library. *Curator* 26 (3): 185–198.

Bc46. ___. 1985. American Museum of Natural History library. In: E. Mount, ed. *Sci–tech libraries in museums and aquariums*, pp. 1–7. New York, Haworth Pr. A re-issue of *Science & technology libraries* vol. 6, no. 1/2 as a separate book.

American Philosophical Society, Philadelphia, PA

Bc47. **American Philosophical Society.** 1987. *A new guide to the collections in the library of the American Philosophical Society*. Ed. by J. S. Catlett. Philadelphia, PA. xxv+414 pp. *Memoirs* vol. 665.

Bc48. **Lingelbach, W. E.** 1946. The library of the American Philosophical Society. *William & Mary quarterly* (3) 3: 48–69.
In 1743.

Bc49. ___. 1953. The American Philosophical Society Library from 1942 to 1952, with a survey of its historical background. *Proceedings of the American Philosophical Society* 97: 471–492.
From 1743.

Bc50. **Smith, M. D.** 1976. *Oak from an acorn: A history of the American Philosophical Society library, 1770–1803*. Wilmington, DE, Scholarly Resources. xv+291 pp., illus.

Biblioteca Nacional de Mexico

Bc51. **Moreno, L. R.** 1970. Materiales en la Biblioteca Nacional de Mexico para el estudio de la ciencia novohispana. *Anales de la Sociedad Mexicana de Historia de la Ciencia y de la Tecnologia* 1970 (2): 147–159.

Bibliotheca Bogoriensis, Bogor

Bc52. **Sahertian–Bakhoven, P.** 1974. Bibliotheca Bogoriensis, its history and prospects. *Flora Malesiana bulletin* 27: 2196–2199.

Botanical Research Institute, Pretoria, Mary Gunn Library

Bc53. **Killick, D. J. B.** 1978. The Mary Gunn Library. *Veld & flora* 64(4): 125–128.

British Museum, London

Bc54. **Finch, J. S.** 1942. Sir Hans Sloane's printed books. *Library* (4) 22: 67–72.

Bc55. **Francis,** Sir **F. C.** 1953. The Sloane collection of printed books. *British Museum quarterly* 18 (1): 4–10.

Bc56. **Nickson, M. A. E.** 1979. Sloane's codes: The solution to a mystery. *Factotum* 7: 13–18.

Bc57. **Wood, L.** 1978. Sir Hans Sloane's books. *Factotum* 2: 15–18.

British Museum (Natural History), London

Bc58. **Anon.** 1984. The Botany Library, British Museum (Natural History). *Watsonia* 15: 163–164.

Bc59. **Banks, R. E. R.** 1988. Resources for the history of science in the libraries of the British Museum (Natural History). *British journal for the history of science* 21: 91–97.

Bc60. **British Museum (Natural History).** 1971. A short history of the libraries and list of mss. and original drawings in the British Museum (Natural History). By F. C. Sawyer. *Bulletin of the British Museum (Natural History), Historical* 4: 77–204.

Bc61. **Woodward, B. B.** 1904. The Libraries. In: British Museum (Natural History). *History of the collections contained in the natural history departments in the British Museum*, vol. 1: 1–76. London.

California Academy of Sciences, San Francisco, CA

Bc62. **Pallowick, R. L.** 1985. The California Academy of Sciences library. In: E. Mount, ed. *Sci–tech libraries in museums and aquariums*, pp. 9–17. New York, Haworth Pr. A re-issue of *Science & technology libraries* vol. 6, no. 1/2 as a separate book.

Cambridge University

Bc63. **Butcher, D. W.** 1951. The departmental libraries of the University of Cambridge. *Journal of documentation* 7: 221–243.

Bc63A. **Munby, A. N. L.** 1962. *Cambridge college libraries: Aids for research students.* 2nd ed. Cambridge, W. Heffer. xv+56 pp.
Replaces ___. 1960. ___. Cambridge, [etc.]. xv+55 pp.

Carnegie Mellon University, Hunt Institute for Botanical Documentation (*formerly* Hunt Botanical Library), Pittsburgh, PA

Bc64. **Hunt Botanical Library.** 1961. *The Rachel McMasters Miller Hunt Botanical Library, Carnegie Institute of technology: Its collections, program, & staff.* Pittsburgh, PA. vi+35 pp., illus.

Bc65. **___.** 1970. *The Hunt Botanical Library: A decennial report.* By G. H. M. Lawrence. Pittsburgh, PA, Carnegie Mellon Univ. [vii]+47 pp.

Bc66. **Lawrence, G. H. M.** 1976. A catalogue of Linnaeana. *Taxon* 25: 69–74.
Also **G. H. M. Lawrence & R. W. Kiger.** 1978. Linnaeus and the computer. *Svenska Linné–Sällskapets årsskrift* 1978: 276–295.

Bc67. **Strandell, B.** 1966. Linné och Amerika. *Svenska Linné–Sällskapets årsskrift* 49: 85–96.
Linnaean collections at University of Kansas Libraries and Hunt Botanical Library, Pittsburgh, PA.

Bc68. **___.** 1976. The Linnaean collection. *Taxon* 25: 3–8.

Bc69. ___. 1978. Från invigningen av Linné – biblioteket i Pittsburgh 1973. *Svenska Linné–Sällskapet årsskrift* 1972/74: 163–177.

Canada Agriculture Library, Ottawa

Bc70. **Canada Agriculture, Information Division.** 1971. *Canada Agriculture Library: The first sixty years. Bibliothèque Canada Agriculture: Soixante ans déja!* Ottawa. 15 pp. *Historical series, Canada Agriculture* no. 7.

Carnegie Museum of Natural History, Pittsburgh, PA

Bc71. **Tauber, A. R.** 1981. The library of Carnegie Museum of Natural History. *Carnegie magazine* 55 (8): 25–29.

Cincinnati Zoo

Bc72. **Orendorff, B.** 1988. Cincinnati Zoo library. In: E. Mount, ed. *Sci–tech libraries serving zoological gardens*, pp. 45–47. New York, Haworth Pr. A re-issue of *Science & technology libraries* vol. 8, no. 4 as a separate book.

Commonwealth Forestry Institute, Oxford

Bc73. **Hemmings, E. F.** 1946. The Forest Library at Oxford. *Proceedings of the British Society for International Bibliography* 8: 56–63.

Deutsche Akademie der Landwirtschaftswissenschaften, Berlin, Deutsche Entomologische Institut

Bc74. **Sachtleben, H.** 1958. Die Bibliothek des Deutschen Entomologischen Institutes der Deutschen Akademie der Landwirtschaftswissenschaften zu Berlin. *Entomologische Blätter* 54: 116–118.

Dumbarton Oaks Research Library, Washington, DC

Bc74A. **Thacher, J. S.** 1971. Dumbarton Oaks. VI. The Garden Library. *Harvard Library bulletin* 19: 211–212.

Durham University

Bc75. **Burnett, A. D., & D. M. Knight.** 1975. History of science in Durham libraries. *British journal for the history of science* 8: 94–99. "Collections 5."

Field Museum of Natural History, Chicago, IL

Bc76. **Williams, B. W., & W. P. Fawcett.** 1985. Field Museum of Natural History library. In: E. Mount, ed. *Sci–tech libraries in museums and aquariums*, pp. 27–33. New York, Haworth Pr. A re-issue of *Science & technology libraries* vol. 6, no. 1/2 as a separate book.

Franklin Institute, Philadelphia, PA

Bc77. **Hilker, E. W.** 1967. The Franklin Institute Library. *Pennsylvania Library Association bulletin* 23: 98–104.

Gabinet Zoologiczny Krolewskiego Uniwersytetu Warszawskiego

Bc78. **Laszczkowska, W.** 1974. Z dziejow ksiegozbioru Warszawskiego Gabinetu Zoologicznego (1819–1915). (From the history of the library of the Warsaw Zoological Museum (1819–1915). *Memorabilia zoologica* 26: 72 pp.
Summary in English.

Giardino dei Semplici di Pisa

Bc79. **Sbrana, C.** 1982. Per una recostruzione dell'antica biblioteca del Giardino dei Semplici di Pisa. Nuovi elementi. *Physis* 24: 423–434.

Glasnevin Botanic Garden

Bc80. **McCracken, E.** 1972. The origins of the library at Glasnevin Botanic Garden. *Irish booklore* 2(1): 82–88.

Harvard University, Museum of Comparative Zoology, Cambridge, MA

Bc81. **Jonas, E. S., & S. S. Regen.** 1985. Museum of Comparative Zoology library — The Agassiz Library, Harvard University. In: E. Mount, ed. *Sci–tech libraries in museums and aquariums*, pp. 35–46. New York, Haworth Pr. A re-issue of *Science & technology libraries* vol. 6, no. 1/2 as a separate book.

Bc82. **Work, R. L.** 1952. Ninety years of Professor Agassiz's natural history library. *Harvard library bulletin* 6: 202–218.

Harvard University, Farlow Reference Library of Cryptogamic Botany, Cambridge, MA

Bc83. **Boise, J. R.** 1986. A hundred years in the life of a library. *Newsletter of the Friends of the Farlow* 9: 1–3.

Henry E. Huntington Library, San Marino, CA

Bc83A. **Henry E. Huntington Library.** 1931. Huntington Library collections. *Huntington Library bulletin* 1: 33–106.

Illinois Natural History Survey, Urbana, IL

Bc84. **Meister, C. G.** 1984. Over 125 years of service; The Illinois Natural History Survey Library. *Illinois libraries* 66: 194–195.

Imperatorskii Sanktpeterburghskii Botanicheskii Sad

Bc85. **Fischer von Waldheim, A. A.**, ed. 1913–15. *Imperatorskii S.–Peterburgskii Botanicheskii Sad za 200 let ego sushchestvovaniya (1713–1913). [The Imperial St. Petersburg Botanical Garden during 200 years of its existence (1713–1913].* Yubileinoe izdanie, sostavlennoe chlenami sada pod gl. red. A. A. Fishera–fon–Val'dgeima. St. Petersburg. 3 vols.
Vol. 2 *Includes* **G. A. Nadson.** Biblioteka Imperatorskago S.–Peterburgskago Botanicheskago Sada, pp. 39–95.

Indiana University, Bloomington, IN, Lilly Library

Bc86. **Davidson, H. L.** 1966. The Lilly Library — 76 years. *Special libraries* 57: 391–394.

Bc87. **Indiana University, Lilly Library.** 1985. *The Lilly Library: The first quarter century, 1960–1985*. Bloomington, IN. [vi]+157 pp., illus.

Institut für Geschichte der Medizin und der Naturwissenschaften in Berlin

Bc88. **Geyer, G.** 1940. *Die Entstehung der Bibliothek des Instituts für Geschichte der Medizin und der Naturwissenschaften in Berlin.* Berlin. 15 pp.

Bc89. **Krüger, G.** 1956. Die Bibliothek des Instituts für Geschichte der Medizin und der Naturwissenchaften. *Wissenschaftliche Zeitschrift der Humboldt–Universität zu Berlin, mathematisch–naturwissenschaftliche Reihe* 4 (4) Suppl.

Jardin Botanique de Bruxelles

Bc90. **Aerdschot, P. van.** 1931. Notes sur la bibliothèque collective du Jardin Botanique de l'État à Bruxelles. *Bulletin du Jardin Botanique de l'État* 9: 99–107.

John Crerar Library, Chicago, IL

Bc91. **John Crerar Library.** 1945. *The John Crerar Library, 1895–1944: An historical report prepared under the authority of the Board of Directors by the librarian.* By J. C. Bay. Chicago, IL. x+206+[6] pp., illus.

John G. Shedd Aquarium, Chicago, IL

Bc92. **Powers, J.** 1985. John G. Shedd Aquarium library. In: E. Mount, ed. *Sci–tech libraries in museums and aquariums*, pp. 83–89. New York, Haworth Pr. A re-issue of *Science & technology libraries* vol. 6, no. 1/2 as a separate book.

Bc93. **Roti, D.** 1980. Aquatic Science Center Library, John G. Shedd Aquarium. *Illinois libraries* 62 (3): 264–267.

Kaiserliche Leopoldinisch–Carolinische Deutsche Akademie der Naturforscher, Halle

Bc94. **Kaiserliche Leopoldinisch–Carolinische Deutsche Akademie der Naturforscher.** 1894. *Geschichte der Bibliothek und Naturaliensammlung der Kaiserlichen Leopoldinisch–Carolinischen Deutschen Akademie der Naturforscher [etc].* By O. Grulich. Jena. ix+300 pp. & 1 pl.

Københavns Universitet, Botanisk Centralbibliotek

Bc95. **Christiansen, M. S.** 1974. Botanisk Centralbibliotek: The Central Botanical Library of the University of Copenhagen. *Botanisk tidsskrift* 69: 94–98.

Kungliga Universitetet, Uppsala

Bc96. **Sallander, H.** 1951. The Bibliotheca Walleriana in the Uppsala University library. *Nordisk tidskrift för bok- och biblioteksväsen* 2: 49–74.

Kungliga Vetenskapsakademiens Bibliotek, Stockholm

Bc97. **Holmberg, A.** 1945. Om Bergianska biblioteket och dess uppkomst. *Nordisk tidskrift för bok- och biblioteksväsen* 32: 7–47.

Bc98. **___.** 1946. Bergianska biblioteket i Kungl. Vetenskapsakademiens ägo. *Nordisk tidskrift för bok- och biblioteksväsen* 33: 121–146.

Lincoln Park Zoological Gardens Reference Library, Chicago, IL

Bc99. **Shaw, J. M.** 1984. Lincoln Park Zoo library: An introduction to the library and its special collections. *Illinois libraries* 66: 164–166.

Bc100. **___.** 1988. Lincoln Park Zoological Gardens Reference Library. In: E. Mount, ed. *Sci–tech libraries serving zoological gardens*, pp. 9–20. New York, Haworth Pr. A re-issue of *Science & technology libraries* vol. 8, no. 4 as a separate book.

Linnean Society of London

Bc101. **Barrow, E.** 1987. The James Insch Tea Library. *Linnean* 3(1): 12–24.

Bc102. **Douglas, G. L.** 1988. History of science and technology resources at the Linnean Society of London. *British journal for the history of science* 21: 489–493.

Bc103. **Gage, A. T., & W. T. Stearn.** 1988. Ch. 17: *The library*. In: *A bicentenary history of the Linnean Society of London*, pp. 183–188. London, Academic Pr.

Lloyd Library, Cincinnati, OH

Bc104. **Simons, C. M.** 1954. Lloyd Library: Pharmacy in Cincinnati. *Special libraries* 45: 70–73.

Bc105. **___.** 1964. Centennial of Lloyd Library: A century of contribution to the pharmaceutical, botanical and biological sciences, 1864–1964. *Lloydia* 27: 141–147.

Bc106. **___.** 1972. Lloyd library and museum. In: *John Uri Lloyd: His life and works 1849–1936, with a history of the Lloyd Library*, pp. 263–337. Cincinnati, OH, privately pub. [vii]+iii+337 pp. & port.

McGill University, Montreal, Bibliotheca Osleriana

Bc107. **Teigen, P. M.**, ed. 1983. *Books, manuscripts and the history of medicine: Essays on the fiftieth anniversary of the Osler Library.* New York, Science History Publications.

v+112 pp.

Essays by L. G. Stevenson, C. G. Roland, R. J. Durling, E. Brodman, G. T. Tanselle, & E. J. Freeman on such topics as Osler's bibliographic activities; aspects of current medical bibliography; the development of methods in physical bibliography; and on new directions in medical history in relation to the book.

Marine Biological Laboratory, Woods Hole, MA

Bc108. **Belanger, S. E.** 1975. History of the library of the Marine Biological Laboratory, 1888–1973. *Journal of library history* 10: 255–263.

Minnesota Zoological Garden

Bc109. **Norell, A.** 1988. Minnesota Zoological Garden library. In: E. Mount, ed. *Sci–tech libraries serving zoological gardens*, pp. 39–44. New York, Haworth Pr. A re-issue of *Science & technology libraries* vol. 8, no. 4 as a separate book.

Missouri Botanical Garden, St. Louis, MO

Bc110. **McDermott, J. F.** 1940. The library of Henry Shaw. *Missouri Botanical Garden bulletin* 28: 49–53.

Museo Civico di Storia Naturale di Milano

Bc111. **Schiavone, M.** 1982. *La Biblioteca di Storia Naturale di Milano.* Milan. 24 pp.

Bc112. ___. 1984. La Biblioteca di Storia Naturale di Milano. *Gazzetta del bibliofilo* 3(2): 2–17.

Museu Nacional do Rio de Janeiro

Bc113. **Museu Nacional de Rio de Janeiro.** 1966. *A biblioteca do Museu Nacional de Rio de Janeiro, 1863–1963.* By D. F. F. de Cunha. Rio de Janeiro. 309 pp., illus. *Serie livros* no. 3.

Muséum National d'Histoire Naturelle, Paris

Bc114. **Bultingaire, L.** 1926. Les origines de la collection des Vélins du Muséum et ses premiers peintres. *Archives du Muséum d'Histoire Naturelle* (6) 1: 129–149.

Bc115. **Duprat, G.** 1965. La nouvelle bibliothèque du Muséum d'Histoire Naturelle. *Bulletin des bibliothèques de France* 1965 (1): 11–22.

Bc116. [Not used.]

National Library of Medicine, United States Department of Health, Bethesda, MD

Bc117. **Miles, W. D.** 1982. *A history of the National Library of Medicine. The nation's treasury of medical knowledge.* Bethesda, MD, National Library of Medicine. viii+531 pp. *NIH publication* no. 82–1904.

Bc118. **Smith, K. A.** 1986. The National Library of medicine: From MEDLARS to the sesquicentennial and beyond. *Bulletin of the Medical Library Association* 74: 325–332.

Bc119. **Wasserman, M. J.** 1972. Historical chronology and selected bibliography relating to the National Library of Medicine. *Bulletin of the Medical Library Association* 60: 551–558.

National Zoological Park Branch of the Smithsonian Institution Libraries, Washington, DC

Bc120. **Kenyon, K. A.** 1988. National Zoological Park Branch library. In: E. Mount, ed. *Sci–tech libraries serving zoological gardens*, pp. 1–8. New York, Haworth Pr. A re-issue of *Science & technology libraries* vol. 8, no. 4 as a separate book.

Also ___. 1987. Spotlight on: National Zoological Park Library. *Biofeedback* 12(4): 4–5.

Natural History Museum of Los Angeles County, Los Angeles, CA

Bc121. **Donahue, K. E. S.** 1985. Natural History Museum of Los Angeles County research library. In: E. Mount, ed. *Sci–tech libraries in museums and aquariums*, pp. 75–82. New York, Haworth Pr. A re-issue of *Science & technology libraries* vol. 6, no. 1/2 as a separate book.

Naturhistorische Gesellschaft zu Hannover

See Bb325(Pieper).

Naturwissenschaftlicher Verein zu Regensburg

Bc122. **Hanusch, G.** 1981. Die Bibliothek des Naturwissenschaftlichen Verein, Regensburg. In: H.–J. Genge & M. Pauer, eds. *Wissenschaftliche Bibliotheken in Regensburg: Geschichte und Gegenwart*, pp. 180–205. Wiesbaden, Harrassowitz. 251 pp. *Beiträge zum Buch– und Bibliothekswesen* no. 18.

New York Botanical Garden

Bc123. **Reed, J. F.** 1969. The library of the New York Botanical Garden. *Garden journal of the New York Botanical Garden* 19: 77–88.

New York Public Library

Bc124. **Dickson, S.** 1956. The Arents Collection of books in parts and associated literature. *Bulletin of the New York Public Library* 61: 267–280.

Bc124A. **Williams, S. P., W. V. Jackson, & J. W. Henderson.** 1975. *Guide to the research collections of the New York Public Library.* Chicago, IL, American Library Association. xxxi+336 pp.

New York Zoological Society

Bc125. **Johnson, S. P.** 1988. New York Zoological Society library. In: E. Mount, ed. *Sci–tech libraries serving zoological gardens*, pp. 21–31. New York, Haworth Pr. A re-issue of *Science & technology libraries* vol. 8, no. 4 as a separate book.

Orto Botanico di Palermo, Biblioteca

Bc126. **Azzarello Di Misa, M.** 1988. *Il fondo antico della Biblioteca dell'Orto Botanico di Palermo.* Introduzione di A. M. Ippolito. Palermo, Regione Siciliana. 397 pp. *Sicilia/Biblioteche* no. 9.

Oxford University, Bodleian Library & others

Bc127. **Alston, R. C.** 1988. The Bodleian Library. *Research libraries in O C L C: A quarterly* 28: 4–14.
With details of the numerous special collections.

Bc127A. **Anon.** 1960. Books and manuscripts from the Department of Botany. *Bodleian Library record* 6: 581–583.

Bc128. **Morgan, P.** 1980. *Oxford libraries outside the Bodleian: A guide.* Oxford, Bodleian Library. xxi+264 pp.
Replaces ___. 1973. ___. Oxford, [etc.]. xx+250 pp.

Bc129. **Shaw, D. F.** 1981. *Oxford University science libraries, a guide.* Oxford, Bodleian Library. 83 pp., illus.

Queensland Herbarium Library, Brisbane, Qld.

Bc130. **Everist, S. L.** 1982. History of the Queensland herbarium and botanical library, 1855 to 1976. *Austrobaileya* 1: 429–445.

Regensburgische Botanische Gesellschaft

Bc131. **Genge, H. J.** 1981. Die Bibliothek der Regensburgischen Botanischen Gesellschaft. In: In: H.–J. Genge & M. Pauer, eds. *Wissenschaftliche Bibliotheken in Regensburg: Geschichte und Gegenwart*, pp. 147–179. Wiesbaden, Harrassowitz. 251 pp. *Beiträge zum Buch– und Bibliothekswesen* no. 18.
Also ___. 1980. Zur Geschichte der Bibliothek der Regensburgischen Botanischen Gesellschaft. *Bibliotheksforum Bayern* 8 (1): 12–38.

Royal Botanic Garden, Edinburgh

Bc132. **Mathew, V.** 1986. The RBG and BSE libraries: A fruitful symbiosis. *Transactions of the Botanical Society of Edinburgh* 150th anniversary supp: 5–16.

Bc133. ___. 1987. *The history of the Royal Botanic Garden Library Edinburgh.* Edinburgh. 133 pp., illus. (some col.)

Royal Botanic Gardens, Kew

Bc134. **Davidge, R.** 1960. The library of the Royal Botanic Gardens, Kew. *London librarian* 6 (8): 3–9.

Royal Entomological Society of London

Bc135. **Neave. S. A., & F. J. Griffin.** 1933. Ch. IX: The library. In: *The history of the Entomological Society of London, 1833–1933*, pp. 75–85. London, Entomological Society of London.

Royal Geographical Society, London

Bc135A. **Crone, G. R.** 1955. The library of the Royal Geographical Society. *Geographical magazine* 121: 27–32.

Royal Horticultural Society, London, Lindley Library

Bc136. **Chittenden, F. J.** 1939. The Lindley Library. *Journal of the Royal Horticultural Society* 65: 350–352.

Bc137. **Tjaden, W. L.** 1993. The Lindley Library of the Royal Horticultural Society, 1866–1926. *Archives of natural history* 20: 93–128.
Also ___. 1987. The loss of a library. *Garden, journal of the Royal Horticultural Society* 112: 386–388.

Royal Institution of Great Britain, London

Bc138. **McCabe, I. M., & F. A. J. L. James.** 1984. History of science and technology resources at the Royal Institution of Great Britain. *British journal for the history of science* 17: 205–209. "Collections 10."

Royal Society of London

Bc138A. **Royal Society of London.** 1992. *The Library and Archives of the Royal Society 1660–1990.* By M. B. Hall. London. 81 pp.

Science Museum Library, London

Bc139. **Day, L. R.** 1985. Resources for the history of science in the Science Museum Library. *British journal for the history of science* 18: 71–76. "Collections 11."

Smithsonian Institution of Washington

Bc140. **Adler, C.** 1897. The Smithsonian library. In: Smithsonian Institution. *The Smithsonian Institution, 1846–1896*, pp. 265–302. Ed. by G. B. Goode. Washington, DC.

Bc141. **Churgin, S. J., & R. F. Schallert.** 1980. History of the Smithsonian Institution libraries, with special emphasis of the natural history. *Journal of the Society for the Bibliography of Natural History* 9: 593–606.

Bc142. **Olmert, M.** 1979. A unique library that chronicles dawn of science. *Smithsonian magazine* 10 (7): 128–134.
I.e. the Bern Dibner gift of 11,000 books from the Burndy Library to the Smithsonian Institution.

Sterling Morton Arboretum, Lisle, IL

Bc143. **MacPhail, I.** 1973. Notes on the Morton Arboretum library. *Morton Arboretum quarterly* 9 (4): 56–61.

Bc144. **___.** 1982. Books and gardens. *Ibid.* 18: 33–38.

United States Department of Agriculture, National Agricultural Library, Beltsville, MD

Bc145. **Associates of the National Agricultural Library.** 1974. *The National Agricultural Library: A chronology of its leadership and attainments, 1839–1973.* Beltsville, MD. iv+17 pp.

Bc146. **Fusonie, A.** 1988. The history of the National Agricultural Library. *Agricultural history* 62: 189–207.

Bc147. **Lacy, M. G.** 1921. The library of the U. S. Department of Agriculture and its branches. *Library journal* 46: 493–495.

Bc148. **National Agricultural Library.** 1975+. *Agricultural libraries information notes.* Beltsville, MD. No. 1+.
Includes articles on the library's services and collections.

Bc149. **___.** 1987. *Our agricultural heritage: National Agricultural Library Special Collections.* Beltsville, MD. 18 pp., illus. (col.).

United States Library of Congress, Washington, DC

Bc149A. **Goodrum, C. A., & H. W. Dalrymple.** 1982. *The Library of Congress.* New ed. Boulder, CO, Westview Pr. ix+337 pp. *Westview Library of Federal Departments, Agencies, and Systems.*
Deals with the history, collecting policy, and services. *Replaces* **C. A. Goodrum.** [1974.] ___. New York, Praeger. ix+292 pp. *Praeger library of U. S. Government departments & agencies.*

Università di Napoli, Istituto Botanico, Biblioteca

Bc150. **Guerrieri, G.** 1961. La Biblioteca Tenorea e le tradizioni botaniche napoletane. *Alamanacco dei bibliotecari Italiani* 1961: 97–108.

University College, London

Bc151. **Percival, J., & W. A. Smeaton.** 1978. Library and archive resources in the history of science at University College. *British journal for the history of science* 11: 191–195. "Collections 7."

University of Kansas Libraries, Lawrence, KS

Bc152. **University of Kansas Libraries.** 1964. *A guide to the collections, Department of Special Collections, University of Kansas Libraries.* By T. R. Buckman. Lawrence, KS. 32 pp.
The collections include the Ralph Ellis ornithological and Thomas J. Fitzpatrick natural history collections.

University of Leeds

Bc153. **Wood, P. B., & J. V. Golinski.** 1981. Library and archives resources in the history of science and medicine at the University of Leeds. *British journal for the history of science* 14: 263–281. "Collections 8."

University of Reading, Cole Library of Early Medicine & Zoology

Bc154. **Anon.** 1960. Cole library of Zoology and Early Medicine, University of Reading. *Nature* 188: 1148–1151.

University of Southern California, Hancock Library of Biology & Oceanography, Los Angeles, CA

Bc155. **Halmos, D. M.** 1954. The Hancock Library of Biology and Oceanography. *College & research libraries* 15: 29–32.

University of Texas at Austin, Lundell Rare Book Library

Bc156. **Roddy, R.** 1968. The [Cyrus L.] Lundell Rare Book Library. *Wrightia* 4: 1–7.

Zoological Society of London

Bc157. **Fish, R.** 1976. The library and scientific publications of the Zoological Society of London: Part I. In: Zoological Society of London. *The Zoological Society of London 1826–1976 and beyond.* Ed. by ... Lord Zuckerman, pp. 233–252. London. *Symposia of the Zoological Society of London* no. 40.

Zoological Society of San Diego

Bc158. **Robinson, M. M.** 1988. Zoological Society of San Diego library. In: E. Mount, ed. *Sci–tech libraries serving zoological gardens*, pp. 33–38. New York, Haworth Pr. A re-issue of *Science & technology libraries* vol. 8, no. 4 as a separate book.

Personal collecting

Bc159. **Dickinson, D. C.** 1986. *Dictionary of American book collectors.* New York, Greenwood Pr. xvi+384 pp.
359 individuals including some natural history collectors.

Bc160. **Thornton, J. L., & R. I. J. Tully.** 1971. Private scientific libraries. Scientific libraries of today. In: *Scientific books, libraries and collectors: A study of bibliography and the book trade in relation to science.* 3rd rev. ed., pp. 339–353, 367–380. London, Library Association; *Continued* ___. 1978. ___*: Supplement 1969–75*, pp. 104–109, 113–116. London, Library Association.
Replaces 2nd rev. ed., 1962. xiii+406 pp.

Bc161. **Weimerskirch, P. J.** 1988. Some notable American collectors of science books. *A B bookman's weekly* 1988 (19 Dec.): 2453–2458.

Individual collectors

Ayer, E. A.

Bc162. **Webber, E. L.** 1984. Books, business, and buckskin. *Bulletin, Field Museum of Natural History* 55(7): 5–10, 19–25.
Biographical sketch of E. A. Ayer. Ornithological collection now at the Field Museum of Natural History (*see* Bc451).

Banks, Sir J.

Bc163. **Joppien, R.** 1979. Die Gelehrtenbibliothek des Sir Joseph Banks. In: *Buch und Sammler: private und öffentliche Bibliotheken im 18. Jahrhundert, Colloquium d.*

Arbeitsstelle 18. Jahrhundert, Gesamthochschule Wuppertal, Universität Munster, Düsseldorf, vom 26.–28. September 1977, pp. 115+. Heidelberg, Winter. *Beiträge zur Geschichte der Literatur und Kunst des 18. Jahrhunderts* vol. 3.
Collection now at the British Library (*see* Bc195 & Bc486).

Bergius, B. & P. J.

Bc164. **Holmberg, A.** 1945. Om Bergianska biblioteket och dess uppkomst. *Nordisk tidskrift för bok– och biblioteksväsen* 32: 7–47.

Curtis, W.

Bc165. **Noblett, W.** 1987. William Curtis's botanical library. *Library* (6) 9: 1–22.

Dibner, B.

Bc166. **Dibner, B.** 1977. The Burndy Library in mitosis. *Book collector* 26: 495–523.
Part of the collection now at the Smithsonian Institution (*see* Bc142 & Bc319).

Ellis, R.

Bc167. **Vosper, R.** 1961. A pair of bibliomanes for Kansas: Ralph Ellis and Thomas Jefferson Fitzpatrick. *Papers of the Bibliographical Society of America* 55: 207–225.
Collection now at the University of Kansas Libraries (*see* Bc461).

Fitzpatrick, T. J.

Bc167A. **Wyer, M. G.** 1953. Fitzpatrick as collector. *Books & libraries at the University of Kansas* 4 (November): 2-3.

See also R. Ellis, Bc167.

Gardner, W.

Bc168. **Kenyon, J. R.** 1982. A collector and his books: Dr. Willoughby Gardner and his library of early books. *Nature in Wales* n.s. 1(2): 68–72.
Collection now at the National Museum of Wales (*see* Bc290).

Geer, C. de.

Bc169. **Ehrensvärd, U.** 1968. Leufsta bruks fideikommissbibliotek. *Biblis* 1968: 123–170, illus.
On Carl de Geer's library and collection of natural history drawings.

Hales, S.

Bc170. **Sarton, G.** 1930. Stephen Hales's library. *Isis* 14: 422–423.

Hooke, R.

Bc171. **Feisenberger, H. A.** 1966. The libraries of Newton, Hooke and Boyle. *Notes & records of the Royal Society of London* 21: 42–55.
Robert Hooke's library is described on pp. 46–52.

Jefferson, T

Bc172. **Savin, M. B., & H. J. Abrahams.** 1959. The botanical library of Thomas Jefferson. *Journal of the Elisha Mitchell Scientific Society* 75: 44–52.

Bc173. ___. 1961. The zoology library of Thomas Jefferson. *Ibid.* 77: 98–109.

Keynes, Sir G.

Bc174. **Oates, J. C. T.** 1964. Contemporary collectors XXXIX. Sir Geoffrey Keynes. *Book collector* 13: 473–480, & 6 pl.
Collection now at the Cambridge University Library (*see* Bc223).

Lhuyd, E.

Bc175. **Roberts, B. F.** 1979. Edward Lhuyd's collection of printed books. *Bodleian Library record* 10: 112–127.
Many now at the Bodleian Library, Oxford University.

Martin, H. B.

Bc176. **Cottrell, G. W.** 1963. Contemporary collectors XXXV. Henry Bradley Martin, the ornithological collection. *Book collector* 12: 316–332.

Mathews, G. M.

Bc177. **Mathews, G. M.** 1942. *Birds and books: the story of the Mathews Ornithological Library.* Introduction by T. Iredale. Canberra, A.C.T., Hewitt. 69 pp. & 7 pl.
Collection now at the National Library of Australia (*see* Bc455).

Mead, R.

Bc178. **Russell, K. F.** 1947. The anatomical library of Dr. Richard Mead (1673–1754). *Journal of the history of medicine* 2: 97–109.
Includes a list of anatomical works extracted from the sale catalogue.

Müller, J.

Bc179. **Sturzecher, M.** 1958. Zur Geschichte der Bibliothek des Berliner Naturforschers Johannes Müller. *Libri* 8: 293–302.

Pennant, T.

Bc180. **Rees, E., & G. Walters.** 1970. The library of Thomas Pennant. *Library* (5) 25: 136–149. *Also* **J. Collins.** 1971. The library of Thomas Pennant: Some additions. *Library* (5) 26: 56–58.

Senckenberg, J. C.

Bc181. **Hodes, F., & E. Berninger.** 1968. *Ad Bibliothecam Joh. Christian Senckenberg, Med. Doct. MFrancofort.* Frankfurt am Main. 48 pp. (incl. 20 pl.).

Smith, G. M.

Bc181A. **Merriman, D.** 1952. George Milton Smith: An appreciation of his library. *Yale University Library gazette* 26: 137-146.
Collection of anatomy, medicine, and marine biology with special emphasis on early ichthyology, now at Yale University Library (*see* Bc462).

Thorardson, C. L.

Bc181B. **Bay, J. C.** 1929. Bibliotheca Thorardsoniana: A private collection of scientific and technological literature. *Papers of the Bibliographical Society of America* 23: 1–17, illus. *Also* **R. Hagedorn.** 1950. Bibliotheca Thorardsoniana: The sequel. *Ibid.* 44: 29–54.
Collection now at the University of Wisconsin at Madison Library.

Waller, E.

Bc182. **Lindroth, S.** 1946. *Erik Wallers bibliotek.* Stockholm. 58 pp. & 1 pl.
First pub. in *Svenska bibliotek.* Collection now at the Kungliga Universitet, Uppsala (*see* Bc259).

BIBLIOGRAPHIES OF LIBRARY RESOURCE LITERATURE

Bc183. **Downs, R. B.** 1951. *American library resources: A bibliographical guide.* Chicago, IL, American Library Association. 428 pp.

Bc184. **___, E. C. Downs, & J. W. Heussman.** 1972. *American library resources: A bibliographical guide. Supplement 1961–1970.* Chicago, IL, American Library Association.

viii+244 pp.; *Continued* **R. B. Downs.** 1981. ___: Supplement 1971–1980. Chicago, IL, [etc.]. viii+209 pp.; ___ **et al.** 1981. *American library resources: Cumulative index 1870–1970.* Compiled by C. D. Keller. Chicago, IL, [etc.]. iii+89 pp.

Bc185. ___ **& E. C.** 1973. *British library resources: A bibliographical guide.* Chicago, IL, American Library Association. xvi+332 pp.

LIBRARY CATALOGUES

General — Union catalogues

United Kingdom

Bc186. **Cambridge University.** 1975. *Union catalogue of scientific libraries in the University of Cambridge: Scientific proceedings, 1644–1974.* London, Mansell Publishing. 2 vols.
Ca.25,000 entries for ca.6,000 conferences and similar meetings.

Bc187. ___. 1977. *Union catalogue of scientific libraries in the University of Cambridge: Books published before 1801.* Compiled at the Scientific Periodicals Library, University of Cambridge. London, Mansell Publishing. 9 microfiches.
Ca.5,800 entries with chronological authors and short–titles index, covering 1478–1800.

Ireland

Bc188. **Nelson, E. C.** 1981. *Works of botanical interest published before 1800 held in Irish libraries.* Glasnevin, National Botanic Gardens. 127 pp. *Occasional paper* no. 1.

United States of America

Bc189. **Estok, R., & R. E. Boykin.** 1976. *A union list of oceanographic expeditions, including results of some major cruise reports.* College Station, TX, Texas A & M Univ. Libraries, Center for Marine Resources. 138 pp.
642 entries; holdings of 13 United States' libraries.

Bc189A. **Johnston, S. H., Jr.** 1992. *The Cleveland herbal, botanical, and horticultural collections: A descriptive bibliography of pre-1830 works from the libraries of the Holden Arboretum, the Cleveland Medical Library Association, and the Garden Center of Greater Cleveland.* Kent, OH, Kent State Univ. Pr. [for the Cleveland Herbals Project]. xviii+1,012 pp., illus.
Describes 977 printed books, dating from ca.1477 to 1829, and 27 manuscripts and paintings.

Bc190. **Tucher, A. J.** 1984. *Agriculture in America 1622–1860: Printed works in the collections of The American Philosophical Society, The Historical Society of Pennsylvania, The Library Company of Philadelphia.* New York & London, Garland Publishing. xix+212 pp. *Americana to 1860 [etc.].* vol. II.
An author catalogue of 2,221 works. Appendix I: Subjects and institutions; Appendix II: Chronology.

Bc191. ___. 1985. *Natural history in America 1609–1860: Printed works in the collections of The American Philosophical Society, The Historical Society of Pennsylvania, The Library Company of Philadelphia.* New York & London, Garland Publishing. xix+287 pp. *Americana to 1860 [etc.].* vol. IV.
An author catalogue of 2,390 works. Appendix I: Subjects and institutions; Appendix II: Chronology.

Bc192. **United States National Union Catalog.** 1968–80. *National union catalog, pre-1956 imprints: A cumulative author list representing Library of Congress printed cards and titles reported by other American libraries.* London, Mansell. 685 vols. *Continued*

___. 1980–81. *National union catalog, pre-1956 imprints ... Supplement.* London, Mansell. Vol. 686–754. Microfiche edition, Chicago, IL, American Library Association; London, Bemrose UK Ltd., 1983. 9,200 microfiches; ___. [1970–72.] *National union catalog, 1956 through 1967: A cumulative author list representing Library of Congress printed cards and titles reported by other American libraries.* Totowa, NJ, Rowman & Littlefield. 125 vols.

For background information on the main catalogue see: **D. A. Smith.** 1982. The National Union Catalog, Pre-1956 Imprints. *Book collector* 31: 445–462. The NUC continues to this day (from 1983 on microfiche as: *NUC: Books.* Washington, DC, Library of Congress).

Australia

Bc193. **Burbidge, N. T.** 1979. *Plant taxonomic literature in Australian libraries.* Canberra, A.C.T. [v]+520 pp.

An author union catalogue. *Replaces* ___. 1951. *Select list of publications in systematic botany available in Australia.* Canberra, A.C.T., CSIRO Division of Plant Industry. iii+89 ff. *Divisional report* no. 14.

Individual general catalogues (selected major libraries)

• The catalogues of three great libraries must serve as exemplars of this type of resource:

Bc194. **Bibliothèque Nationale, Paris.** 1897–1981. *Catalogue général des livres imprimés de la Bibliothèque Nationale. Auteurs.* Paris, Imprimerie Nationale. 231 vols. *Re-pub.* (in microfiche) as ___. 1986. *Catalogue général des livres imprimés de la Bibliothèque Nationale. Auteurs1897–1959. Supplement sur fiches.* Paris, Chadwyck–Healey France, 1986; *Continued* ___. 1972–78. *Catalogue général des livres imprimés de la Bibliothèque Nationale. Auteurs – collectivités–auteurs – anonymes, 1960–69.* Paris. 27 vols. *Re-pub.* (in microfiche) Paris, Chadwyck–Healey France, 1986.

The coverage of earlier volumes of the main catalogue is defined by their date of publication; from vols. 189–231 coverage is only taken down to 1959. The 1986 microfiche re-publication should be regarded as a new edition because it reproduces the internally maintained 1.25 million card catalogue which incorporates over 60,000 corrections and amendments made to the original catalogue and cataloguings of items acquired since the publication of the relevant volume of the printed catalogue.

The main catalogue excludes works not assignable to an identifiable personal author. The 27 vol. supplement, (which does include corporate author entries, title entries for collections and anonyma) provides added entries for joint authors, etc, and comprises: Serie 1: Vols. 1–23, *Caractères latins et caractères grecs*; Serie 2: Vol. 1, *Caractères hébraïques*; vols. 2–3, *Caractères cyrilliques*, vol. 4, *Caractères arabes.* For fuller information *see* **A. Bernard.** 1986. *Guide de l'utilisateur des catalogues des livres imprimés de la Bibliothèque Nationale.* Paris, Chadwyck–Healey France. 60 pp.

Bc195 [**British Library** *formerly*] **British Museum.** 1960–66. *General catalogue of printed books ... to 1955.* London. 263 vols.; *Continued* ___. 1968. *General catalogue [etc.] Ten-year supplement 1956–1965.* London. 50 vols; ___. 1971–72. *General catalogue [etc.] Five-year supplement 1966–1970.* London. 26 vols.; ___. 1978. *General catalogue [etc.] Five-year supplement 1971–1975.* London. 13 vols.; ___. [1982.] *General catalogue of printed books, 1976–1982.* London. 402 microfiches.

Also microprint ed. ___. [1960–66] 1967. *General catalogue of printed books to 1955.* Compact ed. New York, Readex Microprint Corp. 263 vols. in 27. *Continued* ___. [1968] 1969. *General catalogue [etc.] Ten-year supplement 1956–1965.* Compact ed. New York, [etc.]. 50 vols. in 5; ___. [1971–72] 1974. *General catalogue [etc.] Five-year supplement 1966–1970.* New York, [etc.]. 26 vols in 3; ___. [1978] 1980. *General catalogue [etc.] Five-year supplement 1971–1975.* New York, [etc.]. 13 vols in 2. Includes books from the Banksian natural history library as itemised in Dryander's catalogue (*see* Bc486) and subsequent additions.

Bc196 ___. 1979–87. *The British Library general catalogue of printed books to 1975.* London, Bingley, (*later* K. G. Saur). 360 vols.; And: ___. 1987–88. ___*: Supplement.* London, New York, K. G. Saur. 6 vols.; *Continued* ___. 1983. *The British Library general catalogue of printed books 1976 to1982.* London, [etc.]. 50 vols.; ___. 1986. ___*: 1982 to1985.* London, [etc.]. 26 vols.; ___. 1988. ___*: 1986 to1987.* London, [etc.]. 22 vols.; ___. 1990. ___*: 1988 to 1989.* London, [etc.]. 28 vols.
A cumulation of the previously pub. *General catalogue* (1960–78) (*see* Bc195), continued by further supplements, with revisions, corrections, and additions. *Also* ___. 1989. *General catalogue ... to 1975.* [Computer file.] London, Saztec Europe; Cambridge, Alexandria, VA, Chadwyck–Healey. 5 laser optical disks, with user guide; addendum on "floppy" diskettes. For background information *see* **A. H. Chaplin.** 1987. *150 years of the General catalogue of printed books in the British Museum.* Aldershot, Scolar Pr. viii+177 pp.

Bc197. ___. 1902–03. *Subject–index of the modern works added to the library of the British Museum in the years 1881–1900.* By G. K. Fortescue. London. 3 vols; *Continued* ___. 1906. ___, *1901–1905.* London; ___. 1911. ___ *,1906–1910.* London; ___. 1918. ___, *1911–1915.* London; ___. 1922. ___, *1916–1920.* London; ___. 1927. ___, *1921–1925.* London; ___. 1933. ___, *1926–1930.* London; ___. 1937. ___, *1931–1935.* London. 2 vols; ___. 1944. ___, *1936–1940.* London. 2 vols; ___. 1953. ___, *1941–1945.* London; ___. 1961. ___, *1946–1950.* London. 4 vols; ___. 1974. ___, *1951–1955.* London. 6 vols; ___. 1965–66. ___, *1956–1960.* London. 6 vols; ___. 1982. *Subject index of modern books acquired, 1961–1970.* London. 12 vols; ___. 1986. ___, *1971–1975.* London. 14 vols; ___. 1986. *The British Library general subject catalogue 1975 to 1985.* London, etc., K. G. Saur. 75 vols.
Forms a continuation of **R. A. Peddie.** 1933. *Subject index of books published before 1880, A–Z.* London, Grafton. xv+745 pp.; ___. 1935. *Subject index of books published up to and including 1880: Second series, A–Z.* London, [etc.]. xv+857 pp.; ___. 1939. *Subject index of books published up to and including 1880: Third series, A–Z.* London, [etc.]. xv+945 pp.; ___. 1948. *Subject index of books published up to and including 1880: New series, A–Z.* London, [etc.]. 872 pp.

Bc198. ___. 1979. *Eighteenth–century British books: A subject catalogue, extracted from the British Museum General catalogue of printed books.* By G. Averley (& others). Folkestone, Wm. Dawson for Univ. of Newcastle upon Tyne, Project for Historical Bibliography. 4 vols. (4,388 pp.)
Vol. 2: Social sciences, pure sciences, technology, the arts.

See also ___. 1991. *Handlist of unpublished finding aids to the London collections of the British Library.* By R. C. Alston. London, British Library. 192 pp.

Bc199. **United States Library of Congress.** 1984–89. *The main catalog of the Library of Congress (MCLC) 1898–1980.* New York, K. G. Saur. 8,794 microfiches.
Catalogue of over 22.8 million entries, the entries for authors, titles, subject series, etc. combined as a dictionary file. *Also* ___. 1975. *The Library of Congress Main reading Room reference collection subject catalog.* By K. A. Gardner. Washington, DC. vi+638 pp. Lists 11,000 monographs & 3,000 serials; ___. 1987. *The tradition of science: Landmarks of western science in the collections of the Library of Congress.* By L. C. Bruno. Washington, DC, Library of Congress. xi+351 pp., illus. Re-pub. as *Landmarks of Western science [etc.].* New York, Oxford, Facts on File, 1989. *Includes* Ch. 2: Botany, pp. 51–71, [bibl.] 304–307; Ch. 3: Zoology, pp. 73–121, [bibl.] 308–312; Ch. 4: Medicine, pp. 123–169, [bibl.] 312–317; Ch. 6: Geology, pp. 195–221, [bibl.] 322–324. Bibliography cites L. of C. shelfmarks.

Bc199A. ___. 1978–79. *Library of Congress shelflist.* Ann Arbor, MI, University Microfilms International. Microfiche.
Accompanied by **L. K. Hamilton**, ed. 1979. *The Library of Congress shelflist: A user's guide to the microfilm edition.* Ann Arbor, MI, [etc.]. 2 vols. Providing a crude subject guide to the library's main collection of 6.8 million titles by LC class.

Natural History — General

Bc200. **Aberdeen University.** (No relevant general catalogue.)
For special list *see* **W. P. D. Wightman.** 1962. An annotated bibliography of the sixteenth–century books relating to the sciences in the Library of the University of Aberdeen. In: *Science and the Renaissance: An introduction to the study of the emergence of the sciences in the sixteenth century*, vol. 2, xix+293 pp. Edinburgh, Oliver & Boyd. *Aberdeen University studies* no. 144.
Over 760 entries.

Bc201. **Academia Nacional de Ciencias, Cordoba, et al.** 1972. *Catalogo colectivo de obras fundamentales de las ciencias exactas y naturales y de historia de estas Ciencias: Biblioteca de la Academia Nacional de Ciencias, Córdoba; Biblioteca Mayor, Universidad Nacional de Córdoba; Biblioteca de la Universidad Católica de Córdoba.* Ed. by T. García Castellanos & G. Olmos. Cordoba, Academia Nacional de Ciencias. 175 pp.

Bc202. **Académie Royale des Sciences, des Lettres et des Beaux–Arts de Belgique.** 1881–90. *Catalogue des livres de la bibliothèque de l'Académie Royale des Sciences, des Lettres et des Beaux–Arts de Belgique.* Brussels. 4 vols.
1re part: Sociétés, établissements, administrations publiques, etc., recueils périodiques, 1881; 2e part: Ouvrages non périodiques. Sciences, 1883 (B. Sciences naturelles, pp. 255–442). *Replaces* ___. 1850. *Catalogue des livres de la Bibliothèque de l'Académie Royale des Sciences, des Lettres et des Beaux–Arts de Belgique.* Brussels. ii+247 pp.

Bc203. **Academy of Natural Sciences of Philadelphia.** 1972. *Catalog of the library.* Boston, MA, G. K. Hall. 16 vols.
Dictionary catalogue to a collection of ca.150,000 vols. *Replaces* ___. 1837. *Catalogue of the library of the Academy of Natural Sciences of Philadelphia.* Philadelphia, J. Dobson. x+300 pp.

Bc204. ___. 1981. *The Wolf Room collection.* Catalogued by Friends of the Library. Philadelphia, PA. 80 pp.
The special collection of pre-Linnaean books and selected rarities of later dates.

Bc205. **Akademiya Nauk SSSR.** (No general catalogue.)
For special list *see* ___. 1976. *Rukopisnye i redkie pechatnye knigi v fondakh biblioteki AN SSSR: Sbornik nauchnykh trudov. [Manuscripts and rare printed books in the collection of the library of the Academy of Sciences of the USSR: Record of scientific works.]* By S. P. Luppov & A. A. Moiseeva. Leningrad. 157 pp.

Bc206. **Allgemeine Schweizerische Gesellschaft für die Gesammten Naturwissenschaften.** 1864. *Verzeichnis der Bibliothek der Schweizerischen Naturforschenden Gesellschaft.* Berne. x+188 pp.; *Continued* ___. 1882. ___*: [Supplement].* Berne. [ ]* pp.
Ca.4,500 entries. *Replaces* ___. 1850. *Verzeichnis der in der Bibliothek der Schweizerischen Naturforschenden Gesellschaft vorhandenen Bücher.* By C. Christener. Berne. [iii]+172 pp.
[*Not examined – pagination not ascertained.]

Bc206A. **American Academy of Arts & Sciences.** 1802. *Catalogue of books in the library of the American Academy of Arts and Sciences, Boston, 1802.* [Boston, MA.] 68 pp.

Bc207. **American Geographical Society.** 1962. *Research catalog of the American Geographical Society.* Boston, MA, G. K. Hall. 15 vols; *Continued* ___. 1972. ___*: First supplement: Regional catalogue.* Boston, MA, [etc.]. 2 vols; ___. 1974. ___*: First supplement: Topical catalogue.* Boston, MA, [etc.]. 2 vols; ___. 1978. ___*: Second supplement.* Boston, MA, [etc.]. 2 vols.
219,000 entries in the main work which is divided thus: Vol. 1. General geography, travel & exploration, mathematical geography, physical geography; vol. 2. Human geography, history of geography, [etc.].; vol. 3. North America; vols. 4–5. United States; vols. 6–7. Mexico &

Central America, Bermuda & West Indies, South America; vols. 8–10. Europe; vol. 11. Africa; vols. 12–13. Asia; vol. 14. Australasia; vol. 15. Polar regions, oceans, tropics.

Bc208. **American Museum of Natural History.** 1977–78. *Research catalog of the library of the American Museum of Natural History.* Boston, MA, G. K. Hall. 25 vols.; *Continued* ___. 1983. *New catalog of the library.* Boston, MA, G. K. Hall. 6 vols.
325,000 vols of books and periodicals covering general natural history and animal sciences. Author catalogue, 13 vols; Classed catalog, 12 vols. The "New catalog" supplements the earlier one with nearly 70,000 entries.

Bc209. **American Philosophical Society.** 1968. *Science and medicine to 1870: Pamphlets in the American Philosophical Society Library.* Ed. by S. J. Crowther & M. Fawcett. Philadelphia, PA. viii+164 pp. *Library publication* no. 1.

Bc210. ___. 1970. *Catalog of books in the American Philosophical Society library.* Westport, CT, Greenwood. 28 vols.
Dictionary catalogue, including manuscripts. *Replaces* ___. 1824. *Catalogue of the library of the American Philosophical Society held at Philadelphia for promoting useful knowledge.* Philadelphia, PA. xv+290 pp.; ___. 1863–84. *Catalogue of the Americ an Philosophical Society library [etc.].* By J. P. Lesley. Philadelphia, PA. 4 pts (1,390 pp.).

Bc211. ___. 1986. *Natural history and the New World, 1524–1770: An annotated bibliography of printed materials in the library of the American Philosophical Society.* Compiled by A, Guerrini. Philadelphia, PA. xi+83 pp. *Library publication* no. 11.

See also **A. J. Tucher.** 1984. *Agriculture in America 1622–1860 [etc.]*; *see* Bc190; ___. 1985. *Natural history in America 1609–1860 [etc.]*; *see* Bc191.

Bc212. **Arctic Institute of North America.** 1968. *Catalogue of the library of the Arctic Institute of North America, Montreal.* Boston, MA, G. K. Hall. 4 vols; *Continued* ___. 1971. ___*: First supplement.* Boston, MA, [etc.]. 902 pp.; ___. 1974. ___*: Second supplement.* Boston, MA, [etc.]. 2 vols.; ___. 1974. ___*: Third supplement.* Boston, MA, [etc.]. 3 vols.
Main work lists their collection of some 30,000 books and pamphlets.

Bc213. **Asiatic Society of Bengal.** 1908–10. *A catalogue of printed books in European languages in the library of the Asiatic Society of Bengal.* Calcutta. 664 pp.

Bc214. **Bataafsch Genootschap der Proefondervindelijke Wijsbegeerte te Rotterdam.** 1899. *Catalogus van de bibliotheek van het Bataafsch Genootschap der Proefondervindelijke Wijsbegeerte te Rotterdam.* Rotterdam, Wijt & Zonen. 215 pp.; *Continued* ___. 1917. ___*: Aanvullende catalogus.* Rotterdam. [ ]* pp.
Replaces ___. 1870. *Catalogus [etc.].* Rotterdam, J. van Baalen & Zonen. 127 pp. [*Not examined – pagination not ascertained.]

Bc215. **Bernice Pauahi Bishop Museum, Honolulu.** 1964. *Dictionary catalog of the library.* Boston, MA, G. K. Hall. 9 vols; *Continued* ___. 1967. ___*: First supplement.* Boston, MA, [etc.]. 676 pp.; ___. 1969. ___*: Second supplement.* Boston, MA, [etc.]. 239 pp.
The second supplement includes the Fuller library of books on the history and exploration of the Pacific Ocean.

Bc215A. **Biblioteca Nazionale Braidense di Milano.** 1983–93. *Catalogo del fondo Haller della Biblioteca Nazionale Braidense di Milano.* M. T. Monti, ed. Milan, Consiglio Nazionale delle Ricerche, Franco Angeli Libri. 10 vols. *Centro di Studi del Pensiero Filosofico del Cinquecento e nel Seicento in Relazione ai Problemi della Scienza, Consiglio Nazionale delle Ricerche*, Serie 2, *Strumenti bibliografici* vols. 6–15.
Parte 1: *Libri.* 4 vols; Parte 2: *Dissertazioni.* 1985. 5 vols; Parte 3: *Libri delle Biblioteche Lombarde.* 1990. xxvi+502 pp. Parts 1 & 2 catalogue 9,000 books and 10,000 dissertations still preserved in the Biblioteca Braidense; part 3 lists books found in other libraries. *Also*

___. 1977. *Libri e manoscritti di Haller: 1777–1977, nel II centenario della morte di Albrecht von Haller*. By M. L. T. Grossi. Milan. 69 pp. Exhibition catalogue. *Also see* Bc737.

Bc215B. **Biblioteche Estense e Universitaria, Modena.** [1984.] *Immagine e natura: L'immagine naturalistica nei codici e libri a stampa delle Biblioteche Estense e Universitaria. Secolo XV–XVII*. Modena; *see* Ca1129.

Bc216. **Bibliothèque Municipale de Toulouse.** 1976. *Catalogue des flores et faunes illustrées, antérieures à 1815, conservées à la Bibliotheque Municipale de Toulouse*. By C. Voisin & E. Cabrol. [Toulouse.] 115 pp., illus.

Bc217. **Bibliothèque Sainte–Genèvieve, Paris.** 1980. *Catalogue des ouvrages imprimés au XVIe siècle: Sciences, techniques, médecine*. Réd. par J. Linet & D. Hillard, avec la collab. de X. Lavagne. Paris, etc., K. G. Saur. xvi+493+[12] pp.

Bc218. **British Library, Science Reference Library.** 1975+. *Science reference Library catalogue*. London. Microfiche.
Author/title and classified catalogues of additions to stock since 1975.

Bc219. **British Museum (Natural History).** 1903–15. *Catalogue of the books, manuscripts, maps and drawings in the British Museum (Natural History)*. London. 5 vols; *Continued* ___. 1922–40. ___*: Supplement*. London. 3 vols. *Re-pub.* Weinheim, J. Cramer; Codicote, Wheldon & Wesley, 1964; New York, Maurizio Martino, [1992]. 8 vols. in 4.
An author catalogue. Periodicals are catalogued under corporate author headings or titles. The first six vols. were mostly the work of B. B. Woodward. *Replaces* ___. 1884, *Catalogue of books in the General Library*. London. [iv]+147 pp.; and: ___. 1884. *Catalogue of books in the Department of Zoology*. 3rd ed. London. [ii]+399 pp. (1st ed. 1880. 124 pp.; 2nd ed, 1882. 192 pp.)

Bc220. ___. 1933. *A catalogue of the works of Linnaeus (and publications more immediately relating thereto) preserved in the libraries of the British Museum (Bloomsbury) and the British Museum (Natural History)*. 2nd ed. London. xi+246+(68) pp.; ___. 1936. *An index to the authors (other than Linnaeus) mentioned in the catalogue of the works of Linnaeus [etc.]*. London. 59 pp.
Largely compiled by B. H. Soulsby and often cited as "Soulsby." The catalogue is arranged under 20 headings, incl. I: Works; II: Systema Naturae; III: Travels; IV: Botanical works; VI: Medical works; VIII: Miscellaneous works; XII: Theses & orations; XIII: Letters; XIV: Life; XV: Herbarium & collections; XVII: Works edited by Linnaeus; XVIII: Works written with co-operation of Linnaeus. *Replaces* ___. 1907. ___. London. 27 pp.

Bc221. ___. 1956–74. *List of accessions to the museum library, British Museum (Natural History) 1955 (–1974)*. London. 20 annual vols.
Bi-monthly lists classified by U D C and arranged in three sections. Section A: Books, pamphlets and special publications; Section B: New periodicals; Section C: Selected papers and monographs from periodicals received. Each annual volume has a name index.
Also ___. 1905. *Guide to an exhibition of old natural history books, illustrating the origin and progress of the study of natural history up to the time of Linnaeus*. [By B. B. Woodward.] London. 27 pp. *Special guide* no. 1; ___. 1909. *Books and portraits illustrating the history of plant classification exhibited in the Department of Botany*. 2nd ed. [By A. B. Rendle.] London. 19 pp. & 4 pl. *Special guide* no. 2; ___. 1907. *Memorials of Linnaeus: A collection of portraits, manuscripts, specimens and books [etc.]*, (*see* Ba1150); ___. 1910. *Memorials of Charles Darwin: A collection of manuscripts ... books and ... specimens [etc.]*, (*see* Bac90); ___. 1979. *Catalogue of an exhibition: Arranged for the International Conference on the history of the Museums and Collections in Natural History, 3–6 April 1979*. London. ii+18 pp.

Bc222. **Buffalo Society of Natural Sciences.** 1972. *Milestones of science: Epochal books in the history of science as represented in the Library of the Buffalo Society of Natural Sciences*. By R. A. Sparrow. Buffalo, NY. xiv+307 pp., illus. *Buffalo Museum*

of Science collection catalog no. 1.
Selection of 198 titles.

Bc223. **Cambridge University Library.** *See* ___. 1977. *Union catalogue [etc.]; see* Bc186.
CUL possesses Sir Geoffrey Keynes's library as itemised in his catalogue (*see* Bc505) and subsequent additions.

Bc224. **Colombo Museum.** 1904. *Catalogue of the Colombo Museum Library. Printed Books*. By G. A. Joseph. Colombo. 170 pp.
Replaces ___. 1894. ___. Pt. 2. *Printed books, [etc.]*. By A. Haly & G. A. Joseph. Colombo. 90 pp.; ___. 1896. ___*: Supplement*. Colombo. 92 pp.; ___.1899. ___*: Supplement no. 2*. Colombo. 59 pp.

Bc225. **Colonial Museum & Geological Survey Department, New Zealand.** 1900. *Catalogue of the Colonial Museum Library*. By J. Hector. 2nd ed. Wellington, N.Z. [v]+156 pp.
Replaces ___. 1890. ___. By J. Hector. Wellington, N.Z. [iii]+56 pp.

Bc226. **Deutsche Akademie der Wissenschaften zu Berlin.** 1874. *Verzeichniss der Bibliothek der Königlichen Akademie der Wissenschaften in Berlin*. Berlin. 98 pp.

Bc227. **Deutsche Geologische Gesellschaft.** 1930. *Katalog der Bibliothek*. By P. Dienst. Berlin, Stuttgart. xi+1,161 pp.
Ca.30,000 entries. *Replaces* ___. 1887. ___ *(Bestand am 1. April 1887)*. Berlin. 166 pp.

Bc228. **Feuille des Jeunes Naturalistes.** 1887–93. *Catalogue de la bibliothèque*. Paris, Lille. 16 pts.

Bc229. **Finska Vetenskapssocietetens.** 1881. *Katalog öfver Finska Vetenskaps–Societetens bibliothek. År 1881*. Hensinki, A. Moberg. 179 pp.

Bc230. **Fondation Teyler, Haarlem.** 1885–1915. *Catalogue de la bibliothèque*. Haarlem, Héritiers Loosjes. 4 vols.
2e livr: Anatomie, physiologie, développement; Hist. nat. universelle; Hist. nat. de différents pays; Anthropologie, ethnologie, by C. Ekama, 1885. Vol. 3: Sciences exactes et naturelles, 1888–1903, by G. C. W. Bohnenseig, 1904. Vol. 4: 1904–1912, by J. J. Verwijnen, 1915. *Replaces* **Teyler's Stichting.** 1826. *Catalogus der bibliotheek van Teyler's Stichting 1826*. By M. van Marum. Haarlem, De Wed. A. Loosjes. xxi+126 pp.; ___. 1832. ___ *1832*. Haarlem, [etc.]. xxii+171 pp.; ___. 1837. ___ *1837*. Haarlem, [etc.]. Various paging; ___. 1848. *Appendix tot den Catalogus ... 1848*. By J. A. van Bemmelen. Haarlem, De Erven Loosjes. Various paging; ___. 1865. *Catalogue de la Bibliothèque du Musée Teyler à Harlem, 1865*. 3rd ed. Ed. by D. Lubach. Haarlem, Heritiers Loosjes. vi+iv+228 pp.

Bc231. **Freshwater Biological Association.** 1979. *Catalogues of the library of the Freshwater Biological Association, United Kingdom*. Boston, MA, G. K. Hall. 6 vols.
Author catalogue of ca.126,000 entries, 5 vols; Subject index of limnology papers published in the British Isles 1957–1975 and index to F B A staff publications 1931–79, 1 vol.

Bc232. **Geological Society of London.** 1881. *Catalogue of the library of the Geological Society of London*. Compiled ... by J. Dallas. London. 618 pp.
Ca.15,000 entries. *Replaces* ___. 1846. *Catalogue of the books and maps in the library of the Geological Society*. London. xv+226 pp.

Bc233. ___. 1895–35. *Geological literature added to the Geological Society's library. July–December 1894 (–1934)*. London. 37 annual vols.

Bc234. **Geologists' Association.** 1927. [Library catalogue – See London University, University College; *see* Bc264.]
Replaces ___. 1879. *Catalogue of the library of the Geologists' Association*. [By B. B. Woodward.] London. [iv]+79 pp. With *Addenda & corrigenda*. 12 pp.

Bc235. **Gesellschaft für Erdkunde zu Berlin.** 1903. *Katalog der Bibliothek der Gesellschaft für Erdkunde zu Berlin.* By P. Dinse. Berlin, Mittler. xxvii+926 pp.
Catalogue divided into two parts, pt. 1. General geographical literature arranged by subjects, pt. 2. Literature of particular regions and seas, arranged by regions; author index. *Replaces* ___. 1888. *Bibliothek. Verzeichniss der Bibliothek [etc.].* Berlin. [ii]+xvi+418 pp.

Bc236. **Gesellschaft für Natur– und Heilkunde in Dresden.** 1905. *Verzeichnis der Bücher–Sammlung [etc.].* Dresden. x+359 pp.
Ca.60,000 titles. *Replaces* ___. 1841. *Personalbestand und Büchersammlung der Gesellschaft für Natur– und Heilkunde in Dresden.* Ed. M. L. Löwe. Dresden. viii+56 pp.; *Continued* ___. 1845–50. ___*: Erster (& zweiter) Nachtrag.* Dresden. viii+57–94 pp. & 95–112 pp.; ___. 1877. *Katalog der Bibliothek der Gesellschaft für Natur– und Heilkunde in Dresden.* Dresden. [ii]+164 pp.

Bc237. **Gesellschaft für Naturforschender Freunde zu Berlin.** 1908. *Verzeichnis d. Bibliothek der Gesellschaft für Naturforschender Freunde zu Berlin.* Berlin. 2 pts. 369 pp.
Replaces ___. 1828. *Verzeichniss der Bücher der Gesellschaft [etc.].* By F. Klug. Berlin. iv+102 pp.

Bc237A. **Henry E. Huntington Library.** (No relevant general catalogue.)
For special list *see* ___. 1937. *Science and the New World (1526–1800): An exhibition to illustrate the scientific contributions of the New World and the spread of scientific ideas in America.* By T. Hornberger. San Marino, CA. 18 pp.

Bc238. **Herzog–August Bibliothek, Wolfenbüttel.** 1976–87. *Verzeichnis medizinischer und naturwissenschaftlisher Drucke, 1472–1830.* Bearbeitet von U. Zachart (& W. Arnold), unter Mitarbeit von U. Zeidler (& G. Kabisch). Band 1–4: Reihe A, *Alphabetischer Index*; Band 5–7: Reihe B, *Chronologischer Index*; Band 8–10: Reihe C: *Ortsindex*; Band 11–13: Reihe D: Systematischer Index. Nendeln, Leichenstein, K[raus] T[hompson] O Pr; Millwood, NY, Kraus International; Munich, Saur.

Bc239. **Historical Society of Pennsylvania.** (No relevant general catalogue.)
For special lists *see* **A. J. Tucher.** 1984. *Agriculture in America 1622–1860 [etc.]; see* Bc190; And: ___. 1985. *Natural history in America 1609–1860 [etc.]*; *see* Bc191.

Bc240. **Hollandsche Maatschappij der Wetenschappen te Haarlem.** 1910. *Catalogus der Bibliotheek van de Hollandsche Maatschappij der Wetenschappen te Haarlem.* By P. C. Molhuijsen. Haarlem, Erven Loosjes. 151 pp.

Bc241. **Hunterian Museum, University of Glasgow.** 1930. *The printed books in the library of the Hunterian Museum in the University of Glasgow. A catalogue.* By M. Ferguson. With a topographical index by D. B. Smith. Glasgow. xxiii+396 pp.

Bc241A. **Illinois State Museum of Natural History.** 1911. Catalogue of the library of the [etc.]. In: *Report on the progress and condition of the Illinois State Museum of Natural History 1909 and 1910*, pt. II, pp. 52–445. By A. R. Crook. Springfield, IL.

Bc242. **Imperatorskaya Mediko–Khirurgicheskaya Akademii v S.–Peterburg.** 1871–76. *Sistematicheskii katalog biblioteki Imperatorskoi Mediko–Khirurgicheskoi Akademii v S.–Peterburg.* Chast 1 (–II). By I. Merklin. St. Petersburg. 2 vols.
Ch. 1, tom. 1 & 2: Katalog knig, broshyur i dissertatsii; Ch. 2, tom. 1 & 2: Glavneishie stat'i zhurnalov, gazet i sbornikov. Includes botany, zoology, anatomy, histology, embryology, comparative anatomy & histology, biology, physiology, etc.

Bc243. **Imperial College of Tropical Agriculture.** 1975. *Catalogue of the Imperial College of Tropical Agriculture, University of the West Indies, Trinidad.* Boston, MA, G. K. Hall. 8 vols.

Bc244. **Indiana University, Bloomington, Lilly Library.** (No general catalogue.)
For special lists *see* ___. 1965. *Discovery: An exhibition of books relating to the age of geographical discovery and exploration: Lilly Library, Indiana University, November 12–13, 1965.* Prepared for the fifth annual meeting of the Society for the History of Discoveries. By R.

B. Reed & C. K. Byrd. Bloomington, IN. [40] pp.; *Lilly Library publication* no. 1.; ___. 1970. *Biology: An exhibition at the Lilly Library in honor of the meetings of The American Institute of Biological Sciences and The American Physiological Society at Indiana University August 23 – September 3, 1970.* Bloomington, IN. 32 pp., illus. *Lilly Library publication* no. 10. 77 items.

Bc245. **Institut für Geschichte der Medizin in Wien, Josephinische Bibliothek.** 1974. *Katalog der Josephinischen Bibliotek des Instituts für Geschichte der Medizin in Wien.* By E. Lesky. Graz, Akademische Druck- und Verlagsanstalt. 733 pp.

Bc246. **Istituto e Museo di Storia della Scienza, Florence.** 1966–68. Il catalogo dell'antica Biblioteca del reale Museo di Fisica e Storia Naturale. By M. L. Bonelli. *Physis* 8: 392–453, 1966; *Ibid.* 9: 425–478, 1967; *Ibid.* 10: 123–172, 347–389, 1968.

Bc247. **John Crerar Library.** 1902. *A list of bibliographies of special subjects, July, 1902.* Chicago, IL. 504 pp.
Natural history, pp. 183–288.

Bc248. ___. 1904. *A list of cyclopedias and dictionaries, with a list of directories.* August, 1904. Chicago, IL. vi+272 pp.
Classified list of 1,260 cyclopedias etc. and 350 directories.

Bc249. ___. 1911. *A list of books on the history of science, January 1911.* Prepared by A. G. S. Josephson. Chicago, IL. x+297 pp.; *Continued* ___. 1917. ___*: Supplement ... Dec. 1916.* Chicago, IL, [etc.]. vi+139 pp. *Re-pub.* New York, Kraus Reprint Corp., 1966–67. 2 vols.; ___. 1941–45. ___*: Second supplement.* Pt. 1. General science. By R. B. Gordon. Chicago, IL. 6 pts. [All published?]
Includes Pt. 1. General science; pt. 6. Geology, paleontology.

Bc250. ___. 1967. *Author–title catalog.* Boston, MA, G. K. Hall. 35 vols.; ___. 1967. *Classified–subject catalog.* Boston, MA, [etc.]. 42 vols.; ___. 1967. *Subject index to the classified–subject catalog.* Boston, MA, [etc.]. 610 pp.
Catalogues of a collection of some 1,100,000 volumes and pamphlets in pure and applied sciences. *Also* ___. 1978. *Science through the ages: A selection of pioneer works in science, technology, medicine in the collections of the John Crerar Library.* An exhibit held at the Cultural Center of the Chicago Public Library, November 1978, through January, 1979. Chicago, IL. [64] pp., illus.; ___. 1984. *Nature disclosed: Books from the collections of the John Crerar Library illustrating the history of science.* By A. Waleson. Chicago, IL, Univ. of Chicago Pr. xi+77 pp.

Bc251. **Kaiserliche Leopoldinisch–Carolinische Deutsche Akademie der Naturforscher.** 1887–1905. *Katalog der Bibliothek der Kaiserlichen Leopoldinisch–Carolinischen Deutschen Akademie der Naturforscher.* Ed. by O. Grulich. Halle. Vols. 1–3(1).
Incomplete — all published.

Bc252. **Kansas State University Library.** (No general catalogue.)
For special list *see* ___. 1970. *Linnaeana.* By G. A. Rudolph & E. Williams. Manhattan, KS. 225 pp. *Kansas State University Library bibliography series* no. 7. The collection formed by K. K. MacKenzie. See also **B. Strandell.** 1966. Linné och Amerika. *Svenska Linné–Sällskapets årsskrift* 49: 85–96. Linnaean collections at University of Kansas Libraries and elsewhere.

Bc253. **Királyi Magyar Természettudományi Társulet.** 1901. *A Kir. Magyar Természettudományi Társulet könyveinek czímjegyzéke összeállította. [Catalogue of the library.]* By A. Ráth. Budapest. [iv] pp. & 588 cols.
Replaces ___. 1877–86. ___. By A. Heller. Budapest. 2 vols. Vol. 2. Az 1877–1885. Végeig szerzett könyvek.

Bc254. **Königlich–Preussische Geologische Landesanstalt und Bergakademie.** 1880. *Katalog der Bibliothek der Königl. Geologischen Landesanstalt und Bergakademie zu*

Berlin. 2nd ed. [With *Nachtrag etc*. 1875–79]. Berlin, Ernst & Korn. xvii+661+49 pp.; *Continued* ___. 1887. *Nachtrag zum Katalog [etc. 1875–86]*. Berlin, [etc.]. [ ]* pp.
Main catalogue has ca.10,000 entries. *Replaces* 1st ed., 1876. [*Not examined – pagination not ascertained.]

Bc255. **Königlich–Württemburg. Landwirtschaftliche Hochschule, Hohenheim.** 1905. *Katalog der Bibliothek der K. Württemb. ... Angeftertigt nach dem Stande vom 1. November 1905*. Pleiningen, F. Find. viii+346 pp.

Bc256. **(Koninklijk) Bataviaasch Genootschap van Kunsten en Wetenschappen.** 1940. *Catalogus der bibliotheek. F–G: Geschiedenis*. Bandoeng. xvi pp. & 474 cols.
Ca.7,500 entries. *Replaces* ___. 1853. *Bibliothecae Societatis Artium Scientiarumque quae Bataviae floret, catalogus systematicus*. By P. Bleeker. 2nd ed. by J. Munnich. Batavia. xviii+156 pp.; ___. 1864. *Catalogus van der bibliotheek van het Bataviaasch Genootschap van Kunsten en Wetenschappen. Bibliothecae societatis artium scientiarumque quae Bataviae floret catalogus systematicus [etc.]* [3rd ed.] By J. A. van der Chijs. Batavia. xx+487 pp.; *Continued* ___. 1872 & 1877. ___*: Eerste (tweede) vervolg–catalogus der bibliotheek, en catalogus der Maleische, Javaansche en Kawi handschriften van het Bataviaasch Genootschap [etc.]*. Batavia, The Hague, M. Nijhoff. 2 vols.

Bc257. **Koninklijke Akademie van Wetenschappen, Amsterdam.** 1858–81. *Catalogus van de boekerij der Koninklijke Akademie der Wetenschappen, gevestigd te Amsterdam*. Amsterdam, F. & J. Müller. 3 vols.; *Also* ___. 1874. ___.1 deel. 1 stuk. New ed. Amsterdam, C. G. Van der Post. xvi+192; ___. 1885. *Register op den Catalogus van de boekerij der Koninklijke Akademie [etc.]*. Amsterdam, J. Müller. 193 pp.; *Continued* ___. 1891. *Catalogus van de boekerij [etc.]. Eerste vervolg. Met register*. Amsterdam, J. Müller. xv+476 pp.
Deel 3 *Includes* T. Natuurlijke geschiedenis. U. Plantenkunde. V. Dierkunde. *Replaces* **Koninklijk Nederlandsch Instituut van Wetenschappen, Letterkunde en Schoone Kunsten te Amsterdam.** 1841. *Catalogus bibliothecae Instituti Regii Neerlandici*. Pars prior. Amsterdam. [ ]* pp.; **Koninklijke Akademie van Wetenschappen, Amsterdam.** 1855. *Catalogus van de boekerij van de Koninklijke Akademie van Wetenschappen gevestigd te Amsterdam. Eerste aflevering*. By W. Vrolik. Amsterdam. viii+83 pp. (All pub.) [*Not examined – pagination not ascertained.]

Bc257A. ___. 1937. *Catalogus der 15de en 16de eeuwsche drukken der Koninklijke Akademie der Weyenschappen, in bruikleen ep de Koninklijke Bibliotheek*. By R. Pennink. The Hague. 30 pp.

Bc258. **Koninklijke Natuurkundige Vereeniging in Nederlandsch–Indië.** 1905. *Catalogus der boekerij van de Koninklijke Natuurkundige Vereeniging in Nederlandsch–Indië*. The Hague, M. Nijhoff. 326 pp.
Replaces ___. 1861. *Catalogus van de bibl. der Koninklijke Natuurkundige Vereeniging in Nederlandsch–Indië. Batavia*. [ ]* pp.; ___. 1876. *Catalogus [etc.]*. By J. Heringa. Batavia. [ ]* pp.; ___. 1884. *Catalogus [etc.]*. [By J. P. van der Stok & C. P. Sluiter.] Batavia, Ernst. vi+393 pp.; ___. 1895. ___*: Supplement–catalogus 1883–1893*. Batavia, The Hague. iv+LXXVI+19 pp. [*Not examined – pagination not ascertained.]

Bc259. **Kungliga Universitetet, Uppsala.** 1955. *Bibliotheca Walleriana: The books illustrating the history of medicine and science collected by Dr. Erik Waller and bequeathed to the library of the Royal University of Uppsala*. A catalogue compiled by Hans Sallander. Stockholm, Almqvist & Wiksell. 2 vols., 55 pl.
Arranged under broad headings, alphabetical within each section, good name index. Vol. 1, *includes* Incunabula; Medicine; etc; vol. 2, *includes* Natural sciences in general; Botany; Zoology; History of science in general; History of natural sciences; Biography: General; Biography: Special; Bibliography; etc.

Bc260. **Library Company of Philadelphia.** (No relevant general catalogue.)
For special lists *see* ___. 1971. *James Logan, 1674–1751, bookman extraordinary.* By E. Wolf 2nd. Philadelphia, PA, Library Company of Philadelphia. 51 pp. An exhibition catalogue of books and manuscripts from his library; ___. 1974. *The library of James Logan of Philadelphia, 1674–1751.* By E. Wolf 2nd. Philadelphia, PA. 578 pp.; **A. J. Tucher.** 1984. *Agriculture in America 1622–1860 [etc.]*; *see* Bc190; ___. 1985. *Natural history in America 1609–1860 [etc.]*; *see* Bc191.

Bc261. **Linnean Society of London.** 1925. *Catalogue of the printed books and pamphlets in the library of the [etc.].* New ed. by S. Savage. London. v+860 pp.
An author catalogue of ca.17,500 entries. Includes the library of Carl Linnaeus (entries for books from Linnaeus's own library are identified by the [Gothic] letter **L**). Supplemented by annual lists of "Additions and donations to the Library ..." in *Proceedings of the Linnean Society of London* down to the list for 1962–63. Lists of acquisitions resumed in *The Linnean* vol. 1+, 1984+. *Replaces* 1st ed. in *Transactions of the Linnean Society* 5: 277–293, 1800, (with periodic supplements appearing in the *Transactions* through vol. 21); ___. 1827. *Catalogue of the library of the Linnean Society of London.* London. 76 pp.; ___. 1866–77. *Catalogue of the natural history library of the Linnean Society.*Pt. I. *Separate works and papers.* ([and] Pt. II. *Transactions of societies, journals, and supplement to part I.* [and] Pt. III. *Additions ... July 1866 to ... June 1876.*) London. 3 vols., ___. 1896. *Catalogue of the library [etc.].* New ed. London. [iii]+727 pp.
Also ___. 1915. A selection of the additions to the library presented by Thomas Henry Riches, Esq., from the library of Alfred Russel Wallace. (1823–1913.) *Proceedings of the Linnean Society of London* 127: 62–70; ___. 1953. [James] Insch Tea library. *Ibid.* 164: 286–292; ___. 1957. Additions and gifts to the library 1955–56 [Including the A. A. Pearson Bequest of books on mycology, each item from his library is identified]. *Ibid.* 168: 31–42.

Bc262. **Linnean Society of New South Wales.** 1886. *Catalogue of the library of the Linnean Society of New South Wales.* Sydney, N.S.W. 74 pp.

Bc263. **London University, University College.** 1887. *Catalogue of books in the medical and biological libraries at University College, London.* With an appendix. London, Taylor & Francis. [ii]+411 pp.; *Continued* ___. 1892. *Catalogue of additions to the general library and to the medical and biological libraries at University College, London, during the years 1887–91.* London. [ ]* pp.
Ca.12,500 titles in main work. [*Not examined – pagination not ascertained.]

Bc264. ___. 1927. *Catalogue of the geological books in the Library of University College, London, including the library of the Geologists' Association, which is incorporated in the College Library.* By W. Bonser. London, E. Stanford. viii+102 pp.
Ca.1,225 entries.

Bc265. **Louisiana State University, McIlhenny Collection.** 1987. *Nature classics: A catalogue of the E. A. McIlhenny natural history collection at Louisiana State University.* Comp. & ed. by A. H. Perrault. Baton Rouge, LA, Friends of Louisiana State Univ. Library. xvi+280 pp. & 16 pls (col.).
Includes **D. M. Lank.** Natural history illustration: An introductory essay, pp. 3–75.

Bc266. **Lyceum of Natural History of New York.** 1830. *Index to the library of the Lyceum of Natural History, of New York.* New York. 72 pp.

Bc267. **McGill University, Montreal, Bibliotheca Osleriana.** 1929. *Bibliotheca Osleriana: A catalogue of books illustrating the history of medicine and science, collected, arranged, and annotated by ... and bequeathed to McGill University,* by Sir William Osler. Oxford, Clarendon Pr. xxxvi+786 pp. *Re-pub.* (with Prologue by L. G. Stevenson) Montreal, McGill—Queen's Univ. Pr., 1969. xli+792 pp.

Bc268. **Madras Government Museum.** 1894. *Catalogue of the Madras Government Museum.* Madras. 243 pp.
Ca.10,000 entries.

Bc269. **Manchester Museum, Owens College.** 1895. *A catalogue of the books and pamphlets in the library [etc.].* By W. E. Hoyle. Manchester. xv+302 pp. *Museum handbook.*

Bc270. **Marine Biological Association of the United Kingdom.** 1977. *Catalogues of the library of the Marine Biological Association of the United Kingdom, Plymouth, England.* Boston, MA, G. K. Hall. 16 vols.
Author catalogue and extensive subject index of their collection of over 13,000 books, 50,000 reprints, and large collection of periodicals.

Bc271. **Marine Biological Laboratory, Woods Hole.** 1971. *Catalog of the library of the Marine Biological Laboratory and the Woods Hole Oceanographic Institution.* Boston, MA, G. K. Hall. 12 vols.
Author catalogue of ca.339,000 entries, 11 vols; Journal catalog, 1 vol. Catalogue of ca.12,000 books, 138 expedition reports, and 30,000 periodical articles.

Bc272. **Museo Civico di Storia Naturale di Milano.** (No general catalogue.)
For special list *see* ___. 1971. *Mostra del libro antico naturalistico: Attraverso una scelta di esemplari della biblioteca del Museo Civico di Storia Naturale di Milano.* Catalogo a cura di C. Conci & M. Schiavone. Milan. 63 pp.

Bc273. **Museo di Fisica e Storia Naturale di Firenze, Biblioteca Medico–Lorense.** 1970. *Museo di Fisica e Storia Naturale di Firenze. Catalogo della Biblioteca Medico–Lorense.* Ed. G. Di Pietro & M. L. R. Bonelli. Florence, Olschki. 578 pp. *Istituto e Museo di Storia della Scienza, Biblioteca* vol. 8.
Books published before 1860.

Bc274. **Museu Paraense Emílio Goeldi.** 1987–88. *Catalogus librorum Musaei Goeldiani.* Ed. R. E. Horch & M. Rotenberg. Belem, Ministerio da Ciencia e Technologia, Conselho Nacional de Desenvolvimento Cientifico e Technologico, Museu [etc.]. 2 vols. *Coleçao catalogos de obras raras.*
Vol. 1: *Cimelia: Catalogo destritivo das obras raras dos seculos XVI, XVII e XVIII.* 1987. 257 pp.; Vol. 2: *Botanica: Catalogo descritivo das obras raras sobre botanica dos seculos XIV e XX.* 1988. 218 pp.

Bc275. **Museum Francisco–Carolinum, Linz.** 1897. *Bibliotheks–Katalog des Museums Francisco–Carolinum in Linz.* By G. Bancalari. Linz. ix+670 pp.

Bc276. **Muséum National d'Histoire Naturelle, Paris, Bibliothèque Centrale.** 1948+. *Liste des acquisitions en 1948 [etc.], Bibliothéque Centrale du Muséum National d'Histoire Naturelle.* Paris. In progress.

Bc277. ___. 1978. *Catalogue des incunables et des ouvrages du 16 siècle.* By R. Loubatières. Paris. 113 pp.

Bc278. **Museum of Practical Geology & Geological Survey.** 1878. *A catalogue of the library of the Museum of Practical Geology and Geological Survey.* Compiled by H. White & T. W. Newton. London. [iv]+602 pp.

Bc279. **National Agricultural Library, United States Department of Agriculture.** 1967–70. *Dictionary catalogue of the National Agricultural Library, 1862–1965.* New York, Rowman & Littlefield. 73 vols; *Continued* ___. 1972–85. *National Agricultural Library catalog, 1966(–1985).* Totowa, NJ, Rowman & Littlefield (*later* Littlefield, Adams & Co.). 20 vols.
Replaces **United States Department of Agriculture, Library.** 1887. *Class list of works in the library of the United States Department of Agriculture.* Washington, DC. 8 vols.

Bc280. ___. 1975. *Heritage of American agriculture: A bibliography of pre-1860 imprints.* By A. Fusonie. Beltsville, MD. 71 pp. *Library list* no. 98.

Bc281. ___. 1967. *Historic books and manuscripts concerning general agriculture in the collection of the National Agricultural Library.* By M. L. Naftalin. Washington, DC. 94 pp. *Library list* no. 86.

Bc282. ___. 1968. *Historic books and manuscripts concerning horticulture and forestry in the collection of the National Agricultural Library.* By M. L. Naftalin. Beltsville, MD. 106 pp. *Library list* no. 90.

Bc283. [**National Library of Medicine, United States Department of Health.**] **Surgeon–General's Office, United States Army.** 1880–1961. *Index–catalogue of the library of the Surgeon General's Office, United States Army.* Washington, DC, Government Printing Office. 61 vols.
5 series: First series, vols. 1–16, 1880–95; Second series, vols. 1–21, 1896–1916; Third series, vols. 1–10, 1918–32; Fourth series, vols. 1–11, 1936–55 (A–Mnuchin only; unfinished); Fifth series, vols. 1–3, 1959–61. For background information *see* **C. F. Mayer.** 1953. The Index–catalogue as a tool of research in medicine and history. In: E. A. Underwood, ed. *Science, medicine and history: Essays ... written in honour of Charles Singer*, vol. 2: 482–493. London, Oxford Univ. Pr.

Bc284. **National Library of Medicine, United States Department of Health** (*formerly* **United States Army Medical Library** (*and*) **Armed Forces Medical Library**). 1949–66. *Catalog cards [*later *Author catalog,* and *Catalog] 1948–65.* Washington, DC, Library of Congress; *Continued* ___. 1966+. *National Library of Medicine current catalog 1966 [etc.].* Washington, DC, Bethesda, MD.
Reproduction of catalogue cards. Five–yearly cumulations were pub. for 1950–54, 1955–59, & 1960–65. Continued by a quarterly list that is cumulated at yearly and five–yearly intervals.

Bc285. ___. 1961. *Early American medical imprints: A guide to works printed in the United States, 1668–1820.* By R. B. Austin. Washington, DC. x+240 pp.

Bc286. ___. 1967. *A catalogue of sixteenth century printed books in the National Library of Medicine.* Compiled by R. J. Durling. Bethesda, MD, U. S. Department of Health, Education, & Welfare. xii+698 pp.; *Continued* ___. 1971. *A catalogue of incunabula and sixteenth century printed books in the National Library of Medicine. First supplement.* Compiled by P. Krivatsy. Bethesda, MD, U. S. Department of Health, Education, & Welfare. v+51 pp. *DHEW publication* no. (NIH) 71–296.

Bc287. ___. 1979. *A short title catalogue of eighteenth century printed books in the National Library of Medicine.* Compiled by J. B. Blake. Bethesda, MD, U. S. Department of Health, Education, & Welfare. [v]+501 pp. *DHEW publication* no. (NIH) 79–104.

Bc288. ___. 1989. *A catalogue of seventeenth century printed books in the National Library of Medicine.* Compiled by P. Krivatsy. Bethesda, MD. xiv+1,315 pp.
Also **E. B. Wells.** 1967. Pictorial resources at the National Library of Medicine. *Medical annals of the District of Columbia* 36: 365–366; **National Library of Medicine.** 1989. *Medicine and the naturalist tradition: A brochure to accompany an exhibition of medically–related natural history books from the collections of the National Library of Medicine.* By K. Boyd. Bethesda, MD. 34 pp. (List of the 54 items in the exhibition, pp. 31–34).

Bc289. **National Maritime Museum.** 1968. *Catalogue of the library.* Vol. 1: *Voyages and travel.* London, Her Majesty's Stationery Office. xi+403 pp.

Bc290. **National Museum of Wales, (Amgueddfa Genedlaethol Cymru).** (No general catalogue.)
For special list *see* ___. 1982. *The Willoughby Gardner Library: A collection of early printed books on natural history.* By J. R. Kenyon. Cardiff. 54 pp.
Books chiefly of the 15th to 18th centuries.

Bc291. **Naturforschende Gesellschaft zu Emden.** 1892. *Verzeichniss der Bücher der Naturforschenden Gesellschaft in Emden.* Emden. 143 pp.
Replaces ___. 1861. ___. Emden. 39 pp.

Bc292. **Naturforschende Gesellschaft, Zurich.** 1885. *Katalog der Bibliothek der Naturforschenden Gesellschaft in Zürich.* 3. Aufl. Zurich. xxvii+438 pp.
Replaces ___. 1815. *Catalogus bibliothecae Societatis physicae Turicensis.* Zurich. [iv]+xvi+284 pp.; *Continued* ___. 1825. ___*: Supplementum I.* Zurich. [ii]+54 pp.; ___. 1829. ___*: Supplementum II.* Zurich. 39 pp.; ___. 1836. ___*: Supplementum III.* Zurich. 106 pp.

Bc293. **Naturforschender Verein in Brünn.** 1875. *Katalog der Bibliothek des Naturforschenden Vereines in Brünn.* (Abgeschlossen im September 1874). Brno. 161+62 pp.; ___. 1880. ___*: I. Supplement–Heft.* (Abgeschlossen im März 1880). Brno. 81+35 pp.

Bc294. **Naturforscher Gesellschaft bei der Universität Jurjew (Dorpat).** 1910. *Katalog' Obshchestva Estestvoispytatelei pri Imperatorskom Yur'evskom Universitet', izdavaemyi pod redaktsiei Bibliotechnoi Kommissii. Katalog der Bibliothek der Naturforscher–Gesellschaft bei der Universität Jurjew (Dorpat).* 2. Teil: *Bücher und Atlasse.* Jurjew. 107 pp.

Bc295. **Naturhistorische Gesellschaft in Colmar.** 1910. *Katalog der Bibliothek der Naturhistorischen Gesellschaft von Colmar i. Els.* 2. Ausg. By E. Rodé. Colmar, Decker. 324 pp.
Replaces ___. 1869. *Bibliothèque de la Société d'Histoire Naturelle de Colmar.* By F. Faudel. Colmar. 103 pp.; ___. 1878. ___. 2. Ausg. By C. Méspalet & F. Faudel. Colmar. [ ]* pp. [*Not examined – pagination not ascertained.]

Bc296. **Naturhistorischer Verein der Preussischen Rheinlande, Westfalens und des Reg.–Bezirks Osnabrück, Bonn.** 1897. *Katalog der Bibliothek des Naturhistorischen Vereins der Preussischen Rheinlande, Westfalens und des Reg.–Bez. Osnabrück. Abgeschlossen im September 1897.* By B. Wenzel. Bonn. vi+323 pp.

Bc297. **Nederlandsche Maatschappij ter Bevordering van Nijverheid, Koloniaal Museum.** 1908. *Koloniaal Museum, bibliothek. Catalogus der boekverzameling.* Nieuwe uitg. Haarlem, Loosjes. viii+208 pp.; *Continued* ___. 1911. *Eerste supplement van den Catalogus der Boekverzameling betreffende de aanwinsten der Boekerij van Juni 1908 tot Mei 1911. Met verbeteringen en aanvullingen van de eerste Uitgave.* Haarlem. viii+70 pp.
Replaces ___. 1891. *Koloniaal Museum. Catalogus [etc.].* 1e gedeelte. Haarlem. 62 pp.; ___. 1901. *Catalogus van de bibliotheek der [etc.].* Haarlem. 71 pp.

Bc298. **New York Public Library.** 1979. *Dictionary catalog of the research libraries of the New York Public Library, 1911–1971.* New York, New York Public Library (printed & distributed by G. K. Hall, Boston, MA). 800 vols.
Also ___. 1971. *Dictionary catalog of the Rare Book Division.* Boston, MA, G. K. Hall. 21 vols; *Continued* ___. 1973. ___*: First supplement.* Boston, MA, [etc.]. 754 pp. These cataloguings are incorporated into the 1979 *Dictionary catalog.*

Bc298A. ___. 1957. *The Arents collection of books in parts ... A complete checklist.* New York. 88 pp.; *Continued* ___. 1964. *The Arents collection of books in parts ... A supplement to the checklist 1957–1963.* New York. 38 pp.
Including natural history books. *Includes* **S. Dickson.** The Arents Collection of books in parts and associated literature: A brief survey, pp. 7–20.

Bc299. **New Zealand Oceanographical Institute.** 1961. Expedition reports and related publications in the library of the New Zealand Oceanographical Institute. By B. N. Krebs. *Miscellaneous publications, New Zealand Oceanographical Institute* 10: 87 pp.

Bc299A. **Newberry Library, Chicago.** 1992. *A checklist of the Newberry Library's printed books in science, medicine, technology, and the pseudosciences ca.1470–1750.* By J. S. Gottlieb. New York, Garland Publishing. xxiv+312 pp., illus.

Bc300. **Oxford University, Radcliffe Library.** 1835. *Catalogue of the works in medicine and natural history contained in the Radcliffe Library.* By J. Kidd. Oxford. vii+330 pp.
Classed catalogue of ca.6,000 works, incl: Voyages and travels, scientific transactions, natural philosophy and zoology, botany, British natural history, chemistry, mineralogy and geology, comparative anatomy and physiology, human anatomy and physiology, and medicine.

Bc301. ___. 1877. *Catalogue of books on natural science in the Radcliffe Library at the Oxford University Museum up to December 1872.* Oxford, for the Radcliffe Trustees. [iv]+566 pp.; *Continued* ___. 1874–1927. *Catalogue of the books added, 1873–1926.* Oxford. 54 pts.

Bc302. **Queen's Borough Public Library, New York.** 1952. *Scientific expeditions.* By E. Terek. New York. xvii+176 pp.

Bc303. **Rothamsted Experimental Station.** 1940. *Catalogue of printed books and pamphlets on agriculture published between 1471 and 1840.* 2nd ed. Compiled by M. S. Aslin. Harpenden. 293 pp.; *Continued* ___. 1949. ___*: Supplement.* Harpenden. 15 pp.
Replaces 1st ed., 1926. 332 pp.

Bc304. **Royal College of Physicians of London.** 1912. *Catalogue of the library.* By J. F. Payne & H. Barlow. London. [v]+1,354 pp.

Bc305. **Royal Empire** [*later* **Commonwealth**] **Society, Library.** 1930–37. *Subject catalogue of the Royal Empire Society, formerly Royal Colonial Institute.* By E. Lewin. London. 4 vols. *Re-pub.* London, 1967.
Vol. 1: The British Empire generally, and Africa; vol. 2: The Commonwealth of Australia, the Dominion of New Zealand, the South Pacific, general voyages and travels, and Arctic and Antarctic regions; vol. 3: The Dominion of Canada and its provinces, the Dominion of Newfoundland, the West Indies, and colonial America; vol. 4: Mediterranean colonies, Middle east, Indian Empire, Burma, Ceylon, British Malaya, East Indian Islands, and the Far East. *Also* ___. 1961. *Biography catalogue of the library of the Royal Commonwealth Society.* By D. H. Simpson. London. xxiv+512 pp. *Replaces* **Royal Colonial Institute.** 1881. *Catalogue of the library of the Royal Colonial Institute to October, 1881.* London. 29 pp.; ___. 1886. *Catalogue [etc.].* London. 179 pp.; ___. 1895. ___. By J. R. Boosé. London. clv+543 pp.; ___. 1901. *First supplementary catalogue of the library.* By J. R. Boosé. London. cclxxiii+793 pp.

Bc306. ___. 1971. *Subject catalogue of the Royal Commonwealth Society.* Boston, MA, G. K. Hall. 7 vols.
Supplements the preceding subject catalogues.

Bc307. **Royal Geographical Society.** 1895. *Catalogue of the library of the Royal Geographical Society containing the titles of all works up to December 1893.* By H. R. Mill. London. viii+833 pp.
Author catalogue with appendices on Collections of voyages and voyages, Government anonymous, and other miscellaneous publications, and Transactions and periodical publications. *Replaces* ___. 1852. *Catalogue of the library ... to May, 1851 [etc.].* London. iv+143 pp.; ___. 1865. *Catalogue of the library ... May 1865.* London. [ii]+541 pp.; ___. 1871. *Supplement to the alphabetical catalogue ... containing the additions from 1865 to December, 1870.* London. [ii]+135 pp.; ___. 1882. *Second supplement to the alphabetical catalogue ... containing the additions from December, 1870, to the end of 1880.* By E. C. Rye. London. [iii]+380 pp.

Bc308. ___. 1871. *Classified catalogue of the library of the Royal Geographical Society to December 1870.* By G. M. Evans. London. [iii]+478 pp.

Bc309. **Royal Institution of Great Britain.** 1857. *A new classified catalogue of the library of the Royal Institution of Great Britain with indexes of authors and subjects.* By B. Vincent. London. xvii+926 pp.; 1882. *A new classified catalogue [etc.].* Vol. 2: *Including the additions from 1857 to 1882.* By B. Vincent. London. xvii+388 pp.; *Continued* ___.

1890. *Additions to the second volume of the catalogue ... 1882–90.* London. 71 pp.
Class III. Sciences and arts. *Replaces* ___. 1809. *A catalogue of the library of the Royal Institution of Great Britain.* By W. Harris. London, T. Payne & R. H. Evans. xvi+482 pp.; ___. 1821. ___ *including a complete list of all the Greek writers* ... By C. Burney ... arranged; with an alphabetical index of authors, by W. Harris. London. xix+507 pp.

Bc310. **Royal Society of Edinburgh.** 1858. *Catalogue of books and maps in the Royal Society of Edinburgh.* Edinburgh, R. Grant & Son. 133 pp.

Bc311. **Royal Society of London.** 1881–83. *Catalogue of the scientific books in the library of the Royal Society.* London. 2 vols.
Pt. I. *Transactions – journals – [etc.].*; [Pt. II.] *General catalogue.* 1,199 pp. *Replaces* ___. 1825. *Catalogue of the library of the Royal Society.* London. [ii]+608 pp.; ___. 1839. *Catalogue of the scientific books in the library of the Royal Society.* [Begun by Sir A. Panizzi.] London. [iii]+776 pp.

Bc312. ___. 1982. *Book catalogue of the library of the Royal Society.* By A. J. Clark. [Frederick, MD,] Univ. Publications of America. 5 vols.

Bc313. **Royal Society of New South Wales.** 1889. *Catalogue of the scientific books in the library of the Royal Society of New South Wales.* Part 1. *General catalogue.* Sydney, N.S.W. 110 pp.

Bc314. **Royal Society of Tasmania.** 1885. *Catalogue of the library of the Royal Society of Tasmania.* By A. Morton. Hobart, Tas. 98 pp.
Replaces ___. 1856. *Catalogue ... on the first day of March, 1856.* Hobart, Tas. [ ]* pp. [*Not examined – pagination not ascertained.]

Bc315. **Science Society of China** (Chung–Kuo K'o Hsüeh Shê). 1939. *Catalogue of the Science Society of China.* Shanghai. xvi+496 pp.

Bc316. **Scott Polar Research Institute, Cambridge.** 1976. *The library catalogue of the Scott Polar Research Institute, Cambridge, England.* Boston, MA, G. K. Hall. 19 vols.

Bc317. **Scripps Institution of Oceanography.** 1970. *Catalog of the Scripps Institution of Oceanography library.* Boston, MA, G. K. Hall. 12 vols; *Continued* ___. 1973. ___*: First supplement, author-title catalog.* Boston, MA, [etc.]. 3 vols.; ___. 1973. ___*: First supplement, subject catalog, shelf list, shelf list of documents and papers.* Boston, MA, [etc.]. 998 pp.
Author–title catalog, 7 vols; Shelf list, 2 vols; Shelf list of documents, reports and translations collection, 1 vol; Subject catalog, 2 vols. Catalogues of the collection of 80,000 bound volumes and 13,000 reports.

Bc318. **Shanghai Science Institute.** 1937. *Classified catalogue of books and periodicals in the Shanghai Science Institute [etc.].* Shanghai. [814] pp.; *Continued* ___. 1939. ___*: Supplement.* Shanghai. [ ]* pp.
Japanese, Chinese and European language sections. [*Not examined – pagination not ascertained.]

Bc319. **Smithsonian Institution.** (No general catalogue.)
For special list *see* ___. 1980. *Heralds of science as represented by two hundred epochal books and pamphlets in the Dibner Library, Smithsonian Institution.* 25th anniversary ed. Norwalk, CT, Burndy Library; Washington, DC, Smithsonian Institution. xiii+96 pp.
Preface and notes by Bern Dibner. *Includes* Botany #18–36; General science #73–99; Zoology #186–200. First pub. 1955.

Bc320. **Societatii de Medici si Naturalisti din Iasi.** 1986. Cartea straina veche in Biblioteca de Medici si naturalisti (sec. XVI–XVIII). By D. Lemny. *Revista medico–chirurgicala a Societatii de Medici si Naturalisti din Iasi* 1986 (Supp.): 369–413.

Bc321. **Société d'Agriculture, Sciences et Arts de la Sarthe.** 1877–81. *Catalogue de la bibliothèque de la Société d'Agriculture, Sciences et Arts de la Sarthe.* Réd. par L. Brière. Le Mans, E. Monoyer. 515 pp.

Bc322. **Société Linnéenne de Bordeaux.** 1894–1901. *Catalogue de la bibliothèque de la Société Linnéenne de Bordeaux.* Bordeaux, J. Durand. 2 vols. Supplement aux *Actes de la Société Linnéenne de Bordeaux.*
Replaces ___. 1869. *Catalogue de la Société Linnéenne de Bordeaux.* Bordeaux. xxx pp.

Bc323. **Société Nationale des Sciences Naturelles et Mathématiques, Cherbourg.** 1881–83. *Catalogue de la bibliothèque de la Société Nationale des Sciences Naturelles et Mathématiques.* By A. Le Jolis. Cherbourg. Pts. I, II, 1–3.
1re partie, 2e ed. Janvier 1881. 84 pp.; 2e partie, 31 Déc. 1872 – Avril 1883. 384 pp.
Replaces **Société Impériale [etc.].** 1870–78. *Catalogue de la bibliothèque de la Société Nationale des Sciences Naturelles et Mathématiques.* By A. Le Jolis. Cherbourg. 2 pts in 3.

Bc324. **Southampton University, Perkins Agricultural Library.** 1961. *Catalogue of the Walter Frank Perkins Agricultural Library.* Southampton, Southampton Univ. Library. 291 pp.
Over 2,000 items.

Bc325. **Természettudományi Múzeum Könyvtára.** (No general catalogue.)
For special list *see* ___. 1968. *Gyüjteményes munták és sorozatok katalógusa. Katalog der Sammelwerke und Serien.* Budapest. [x]+174 pp.
A detailed catalogue of all parts of multi–volume treatises, serial monographs, faunas, floras, etc. with author index.

Bc326. **Towarzystwo Przyjaciól Nauk Poznanskiego.** 1869. *Katalog biblioteki Towarzystwa Przyjaciól Nauk Poznanskiego.* By H. Feldmanowski. Poznan. vi+291 pp.

Bc327. **United States Geological Survey.** 1964. *Catalog of the United States Geological Survey Library, Department of the Interior (Washington, D.C.).* Boston, MA, G. K. Hall. 25 vols.
Ca.500,000 documents in geology, palaeontology and allied sciences.

Bc328. **Universitaria de Padova, Biblioteca.** (No relevant general catalogue.)
For special list *see* **F. M. Scapin.** 1956–57. Catalogo degli incunaboli medici e naturalistici della Biblioteca Universitaria di Padova. *Acta medicæ historiæ patavina* 3: 93–163. *Also pub. separately* Padua, 1956. 69 pp.
101 titles.

Bc329. **Universiteit Amsterdam, Bibliotheek.** (No relevant general catalogue.)
For special list *see* ___. 1922. *Catalogus van de boeken en tijdschriften geplaatst in de wiskundige leeszaal en in de Laboratoria voor natuurkunde, dierkunde, plantekunde, plantenphysiologie en pharmacognosie.* Amsterdam. 22 pp.
Ca.1,500 entries.

Bc330. **University of Illinois at Urbana–Champaign.** 1972. Library. Rare Book Room. *Catalog.* Boston, MA, G. K. Hall. 11 vols. *Continued* ___. 1978. ___*: Supplement.* Boston, MA [etc.]. 2 vols.
For special list *see* ___. 1984. *Geology emerging: A catalog illustrating the history of geology (1500–1850) from a collection in the library of the University [etc.].* By D. C. Ward & A. V. Carozzi. Urbana–Champaign, IL. [xiv]+565 pp., illus.

Bc331. **University of Kansas Libraries.** (No general catalogue.)
For special lists *see* **R. Kidman & E. Farley.** 1955. [John] Ray in Kansas. *Books & libraries at the University of Kansas* 9 (May): 9-10; **University of Kansas Libraries.** 1957. *A catalog of an exhibition commemorating the 250th anniversary of the birth of Carolus Linnaeus ... and the 200th anniversary of the issue of the Systema naturae, 10th edition, 1758–1958.* By T. R. Buckman. Lawrence, KS, Univ. of Kansas Libraries. 50 pp.; ___. 1964. *A checklist of Linneana 1735–1835 in the University of Kansas Libraries.* Compiled by T. Williams. Lawrence, KS. x+73 pp. *University of Kansas library series* no. 20.

Bc332. **University of Oklahoma Libraries.** 1976. *The catalogue of the History of Science collections of the University of Oklahoma Libraries.* By D. H. D. Roller & M. M. Goodman. London, Mansell. 2 vols.
39,000 entries for books and periodicals with subject entries. *Replaces* ___. 1969. *Short–title catalog of the history of science collections.* By D. H. D. Roller. Norman, OK, Oklahoma Univ. Library. 631 pp.

Bc333. **University of Washington, Fisheries–Oceanographic Library.** 1972. *Selected references to literature on marine expeditions, 1700–1960.* Boston, MA, G. K. Hall. iv+517 pp.

Bc333A. **Verein der Freunde der Naturgeschichte in Mecklenburg.** 1896. *Katalog der Bibliothek des Vereins der Freunde der Naturgeschichte in Mecklenberg [etc.].* Güstrow. [iv]+178 pp.

Bc334. **[Wellcome Institute for the History of Medicine] Wellcome Historical Medical Library.** 1962. *A catalogue of printed books in the Wellcome Historical Medical Library.* I: *Books printed before 1641.* London. xv+407 pp.

Bc335. ___. 1966+. *A catalogue of printed books in the Wellcome Historical Medical Library.* II: *Books printed from 1641 to 1850.* London. Vol. 1+. In progress.
Vol. A–E. 1966; vol. F–L. 1976.

Bc336. **Wellcome Institute for the History of Medicine.** 1980. *Subject catalogue of the history of medicine and related sciences.* Munich, Kraus–Thompson. 18 vols.

Bc337. ___. 1983. *An annotated catalogue of medical Americana in the Library of the Wellcome Institute for the History of Medicine: Books and printed documents 1557–1821 from Latin America and the Caribbean Islands and manuscripts from the Americas 1575–1927.* Ed. by R. Price. London. xix+319 pp.

Bc338. **Wetterauische Gesellschaft für die Gesamte Naturkunde zu Hanau a. M.** 1883 *Katalog der Bibliothek der Wetterauische Gesellschaft für die Gesamte Naturkunde zu Hanau.* By H. Eisenach & C. Kirn. Hanau. 185 pp.; *Continued* ___. 1902. *Erster Nachtrag zum Katalog der Bibliothek der Wetterauischen Gesellschaft für die Gesammte Naturkunde zu Hanau.* Hanau. [iii]+124 pp.

Bc338A. **Williams College, Chapin Library.** 1935. *The Chapin Library, Williams College: A short-title list.* By L. E. Osborne. Portland, ME, The Southworth-Anthoensen Pr; London, B. Quaritch Ltd. viii+595 pp.
Includes Books in science, printed after 1500, pp. 521–540 ("Books about birds," pp. 536–540). For special list *see* ___. 1965. *Landmarks of biology 1470–1859: Four centuries of printed editions from Aristotle to Darwin in the Chapin Library.* Williamstown, MA. 24 pp.

Local natural history societies

Bc339. **Birmingham Natural History & Microscopical Society.** 1892. *Catalogue of library.* Birmingham. 60 pp. ••*Replaces* ___. 1877. *Laws and catalogue of the library.* Birmingham. 17 pp.; ___. [1880.] ___. Birmingham. 22 pp.

Bc340. **Boston Society of Natural History.** 1837. Catalogue of the library of the Boston Society of Natural History. *Boston journal of natural history* 1(4): 497–512. *Also pub. separately* 1837. Boston, MA, Freeman & Bolles. 27 pp.

Bc341. **Bristol Naturalists' Society.** 1881. *Catalogue of the books in the library of the [etc.].* Clifton. [ii]+24 pp.

Bc342. **Glasgow Natural History Society.** 1883. *Catalogue of books in the library.* Glasgow. 20 pp.

Bc343. **Hertfordshire Natural History Society & Field Club.** 1885. *Catalogue of the library [etc.].* By J. Hopkinson. Rev. to July, 1885.

London, Watford & Hertford. 52 pp.; *Continued* ___. 1890. ___: *Supplementary catalogue ... Field Club, July, 1885, to December, 1889.* London & Hertford. 53–60 pp.

Bc344. **Natural History & Philosophical Society of Belfast.** 1892. *Catalogue of books in the library [etc.].* Belfast. 39 pp. ••*Replaces* Earlier eds., 1847, 1874.

Bc345. **Naturforschende Gesellschaft in Danzig.** 1914. *Katalog der Bibliothek der Naturforschenden Gesellschaft in Danzig.* 3 Heft E. *Zeitschriften und Gesellschaftschriften.* Danzig. vi+58 pp.

Bc346. **Naturforschende Gesellschaft des Osterlandes.** 1884. *Catalog der Bibliothek der Naturforschenden Gesellschaft des Osterlandes zu Altenburg.* Altenburg, Hiller & Schuster. 123 pp.

Bc347. **Naturforscher Verein zu Riga.** 1908. *Katalog der Bibliothek des Naturforscher–Vereins zu Riga.* 1.A: *Bibliographie und Biographie.* B: *Naturwissenschaftliche Schriften gemischten Inhalts.* C: *Vereins– und periodische Schriften.* Riga. 45 pp.

Bc348. **Naturhistorische Gesellschaft, Nuremberg.** [1890.] *Katalog der Bibliothek der Naturhistorischen Gesellschaft in Nürnberg. Stand am 30. April 1890.* [Nuremberg.] 64 pp.

Bc349. **Naturwissenschaftliche Gesellschaft Isis zu Dresden.** 1861. *Bibliothek–Katalog der Gesellschaft für [etc.].* Dresden. 31 pp.

Bc350. **Naturwissenschaftliche Verein für Sachsen und Thüringen.** 1858. *Geschäftsordnung und Verzeichniss der Bibliothek des Naturwissenschaftlichen Vereines für Sachsen und Thüringen in Halle.* 2. Ausg. Ed. by C. Giebel. Halle. 56 pp.

Bc351. **Naturwissenschaftlicher Verein zu Regensburg.** 1905. *Katalog der Bibliothek des Naturwissenschaftlichen Vereins in Regensburg.* Regensburg. 83 pp. *Bericht* 1903/04.

Bc352. **Physikalisch–Medicinische Gesellschaft zu Würzburg.** 1869. *Verzeichniss der Bibliothek der Physikalisch–Medicinischen Gesellschaft zu Würzburg.* Wurzburg. 127 pp.

Bc353. **Società Italiana di Scienze Naturali.** 1868. *Catalogo della biblioteca.* Milan. 47 pp.

Bc354. **Société Vaudoise des Sciences Naturelles.** 1858. *Catalogue de la bibliothèque.* Ed. by V. Cérésole. Lausanne. 32 pp.

Bc355. **Verein für Naturkunde zu Cassel.** 1875. *Catalog der Bibliothek des Vereins [etc.].* Cassel. [ii]+36 pp.

Bc356. **Verein für Naturkunde, Mannheim.** 1898. *Katalog der Bibliothek des Mannheimer vereins für Naturkunde.* Neu Ausg. By M. Oeser. Mannheim. [iv]+59 pp.

Bc357. **Woolhope Naturalists' Field Club.** 1941. *Catalogue of the library of the Woolhope Naturalists' Field Club.* By F. C. Morgan. Hereford. 92 pp.

Microscopical science

Bc358. **Quekett Microscopical Club.** 1929. *Catalogue of books in the library of the Quekett Microscopical Club.* London. vii+178 pp.
Replaces New ed., London, 1915. 105 pp. Author & subject catalogue.

Bc359. **Royal Microscopical Society.** 1929. *Catalogue of the printed books and pamphlets in the library [etc.].* London. vii+177 pp.
Author catalogue (no subject index). *Replaces* ___. 1888. *Catalogue of the books in the library, [etc.].* London. 69 pp.; ___. 1910. *Catalogue of the library.* London. 141 pp. Subject catalogue, pp. 109–141. Library now largely dispersed. In 1970 the Museum of the History of Science, Oxford, purchased a selection of pre-1851 titles from the Society.

Botanical — General

Bc360. **Aberdeen University.** (No relevant general catalogue.)
For special list *see* **A. M. Davidson.** [1923.] Books on botanical subjects bequeathed to the University by Professor Trail. In: *James William Helenius Trail: A memorial volume*, 56 pp. Aberdeen. *Aberdeen University studies* no. 91.
1,250 titles. *Also* **F. Cusick.** 1978. Early botanical books in Aberdeen University Library. *Aberdeen University review* 47: 121–132.

Bc361. **Bayerische Botanische Gesellschaft zur Erforschung der Heimischen Flora.** 1912. Katalog der Bibliothek der Bayerischen Botanischen Gesellschaft zur Erforschung der heimischen Flora. Tl. II: Einzelschriften und Sonderabdrücke (Sep.) sowie Nachträge zu Tl. I. By A. Toepffer. *Berichte der Bayerischen Botanischen Gesellschaft zur Erforschung der heimischen Flora* 13: 1–52.

Bc362. ___. 1914. Katalog der Bibliothek der Bayerischen Botanischen Gesellschaft zur Erforschung der heimischen Flora. Nachtrag vom 1. Oktober 1911 bis 30. September 1913. By A. Toepffer. *Ibid.* 14: 217–230.

Bc363. **Bibliothèque des Conservatoire et Jardin Botaniques de la Ville de Genève.** 1985. *Ouvrages botaniques anciens. Catalogue des ouvrages prélinnéens de la [etc.]*. By Hervé M. Burdet. Geneva. xxiii+599 pp.

Bc364. **Botanic Gardens, Hong Kong.** 1896. *A classified catalogue of the library of the Botanic Gardens, Hong Kong.* By C. Ford. Hong Kong. [ ] pp.

Bc365. **Botanical Survey of India.** 1967+. *Union catalogue of the books in the libraries of the Botanical Survey of India.* By A. R. Das. Calcutta. Vol. 1+.

Bc366. **Botanischer Verein der Provinz Brandenburg.** 1911. *Katalog der Bibliothek des Botanischen Verein der Provinz Brandenburg.* Im Auftrage des Vereins zusammengestellt von C. Schuster. Dahlem–Steglitz bei Berlin. viii+191 pp. *Verhandlungen des botanischen Vereins der Provinz Brandenberg* Bd. 51, Beilage 1.
I: Periodische Schriften; II: Selbständig erscheinene Schriften [etc.]; III: Karten, Varia; IV: Nachträge. Ca.3,000 titles.

Bc367. **Brooklyn Botanic Garden.** (No general catalogue.)
For special list *see* ___. 1935. Books and manuscripts illustrating the history of botany: An annotated list. By E. P. Chichester & C. S. Gager. *Brooklyn Botanic Garden record* 24: 159–194. Exhibit of 186 books from the library

Bc368. **Carnegie Mellon University, Hunt Institute for Botanical Documentation** (*formerly* **Hunt Botanical Library**). 1958–61. *Catalogue of botanical books in the collection of Rachel McMasters Miller Hunt.* Pittsburgh, PA, Hunt Botanical Library. 2 vols. in 3. *Re-pub.* New York, Maurizio Martino, [1991]. 2 vols.
A chronologically arranged catalogue with author indexes. Vol. 1: *Printed books 1477–1700. With several manuscripts of the 12th, 15th, 16th & 17th centuries.* Compiled by J. Quinby; Vol. 2, part i: *Introduction to printed books 1701–1800.* (*Includes* **G. Dunthorne.** Eighteenth–century botanical prints in color; **J. S. L. Gilmour.** Gardening books of the eighteenth century; **W. T. Stearn.** Botanical gardens and botanical literature in the eighteenth century; **A. Stevenson.** A bibliographical method for the description of botanical books); Vol.2, part 2: *Printed books 1701–1800.* Compiled by A. Stevenson.
Also **Carnegie Institute, Pittsburgh.** 1952. *Botanical books, prints and drawings: From the collection of Mrs. Roy Arthur Hunt. 1952 January third to February seventeenth, Department of Fine Arts, Carnegie Institute.* Pittsburgh, PA, Carnegie Institute. xvi+63+[9] pp., frontis. & 48 pl.
Clements Library, University of Michigan. 1949. *Fifty–five rare books from the botanical library of Mrs. Roy Arthur Hunt.* Loaned for exhibition in honor of the Botanical Society of America at their Summer meeting, Ann Arbor, August, 1949. Described by H. H.

Bartlett. Ann Arbor. [vii]+55 pp.
Also see Adanson collection, Bc520.

Bc369. **Chelsea Physic Garden.** 1956. *Catalogue of the library of the Chelsea Physic Garden.* London, C. F. Roworth. [vi]+35 pp.
Books and manuscripts.

Bc370. **Cleveland Medical Library.** (No general catalogue.)
For special list *see* ___. 1947. The [George Gehring] Marshall Collection of Herbals in the Cleveland Medical Library. By R. B. & M. H. Fisch. *Bulletin of the history of medicine* 21: 224–261.
See also Bc189A.

Bc371. **Conservatoire et Jardin Botanique de la Ville de Genève.** 1985. *Ouvrage botaniques anciens: Catalogue des ouvrages prélinnéens de la Bibliothèque des Conservatoire et Jardin Botaniques de la Ville de Genève.* Dressé et publié sous la direction de H. M. Burdet. Geneva. xxliii+599 pp.
Chronologically arranged catalogue with name and place/printer indexes and numerous illustrations.

Bc372. **Dumbarton Oaks Research Library & Collection.** (No general catalogue.)
For special list *see* ___. 1983. *Hortus librorum: Early botanical books at Dumbarton Oaks.* Catalogue by Laura Ten Eyck Byers. Washington, DC. vi+42 pp.

Bc373. **Edinburgh Museum of Science & Art.** 1897. *List of books, etc., relating to botany and forestry, including the Cleghorn Memorial Library, in the library of the museum.* Edinburgh, H. M. Stationery Office. iv+199 pp.
Ca.3,000 titles. Now at Royal Botanic Garden Library, Edinburgh.

Bc374. **Flora, Sächsische Gesellschaft für Botanik und Gartenbau in Dresden.** 1909. *Verzeichnis der Bibliothek d. "Flora" zu Dresden.* By –. Löbner. Dresden. 51 pp.
Replaces ___. 1841. *Katalog über die Bibliothek der Gesellschaft Flora.* Ed. M. L. Löwe. Dresden. 24 pp.; ___. 1897. *Verzeichnis der Bücher–Sammlung des Gesellschaft [etc.].* Ausg. im März 1897. Dresden. 32 pp. Supplement zu den *Sitzungsberichten und Abhandlungen der Genossenschaft "Flora" zu Dresden.*

Bc375. **Forestry Commission.** 1977. *Library catalogue of books. Revised to December 1976.* N.p. vii+234 pp.
Replaces Earlier eds., 1957 and 1964; ___. 1971. *Library catalogue of books. Revised to October 1970.* London. 176 pp.

Bc376. **Harvard University, Library of Arnold Arboretum.** 1914–17. *Catalogue of the library of the Arnold Arboretum of Harvard University.* Compiled under the direction of C. S. Sargent by E. M. Tucker. Cambridge, MA. 2 vols. *Publications of the Arnold Arboretum* no. 6. *Continued* ___. 1933. ___. Vol. 3. *Supplement 1917–1933.* Compiled by ... E. M. Tucker. Cambridge, MA. [v]+346 pp.
Vol. I: Serial publications – Authors and titles; Vol. II: Subject catalogue with supplement to vol. I; Vol. III: Serial publications – Authors and titles. There is no subject catalogue to the supplement.
Also ___. 1972. *Early botanical books: An exhibit commemorating the centennial of the Arnold Arboretum 1872–1972.* May 21, 1972 – June 9, 1972. Cambridge, MA, The Houghton Library. 52 pp. 60 items dating from 1475 to the mid 18th century.

Bc377. **Horticultural Society of New York.** 1970. *Printed books 1481–1900 in the Horticultural Society of New York.* A listing by E. C. Hall. New York. xiii+279 pp.
Author catalogue of ca.3,000 titles.

Bc378. **Imperatorskii Sanktpeterburghskii Botanicheskii Sad.** 1886. *Catalogus systematicus Bibliothecae Horti Imperialis Botanici Petropolitani. (Sistematicheskii katalog biblioteki S.–Peterburgskogo botanicheskogo sada).* Ed. nov. By F. E. Herder. St.

Petersburg. xi+510 pp. *Continued* ___. 1899. *Catalogus ... 1886–1896.* By J. G. Klinge. St. Petersburg. ii+vii+253 pp.; ___. 1899. *Catalogus ... 1898.* Curavit J. G. Klinge. St. Petersburg. 47 pp.

Classified catalogue of ca.15,000 (ca.4,000 in main supplement) titles (books and periodicals) with name indexes.

Bc379. **Jardin Botanique de Buitenzorg** (Kebun Raja, Bogor). 1894. *Catalogue de la bibliothèque du Jardin Botanique de Buitenzorg.* 2nd ed. By J. J. Brutel de la Rivière. Batavia, Imprimerie de l'État. vii+371+6 pp.; *Continued* ___. 1896–1903? *Catalogus der bibliotheek.* 4 (–9) vervolg. [Batavia.] 6 pts.

Ca.4,000 titles. *Replaces* ___. 1887. *Catalogus der bibliotheek van 's Lands Plantentuin te Buitenzorg.* By W. Burck & F. Westerman. Batavia, Landsdrukkerij. xi+194 pp. & 2 supps., 1889 & 1891. 37 pp. & [ ]* pp. [*Not examined – pagination not ascertained.]

Bc380. **John Innes Institute.** (No general catalogue.)

For special list *see* ___. 1978. *The John Innes special collection: A catalogue of selected books from the library of the John Innes Institute.* By E. Atchison. Norwich. ix+108 pp.

Bc381. **Kaiserlich. Königlich. Gartenbau–Gesellschaft in Wien.** 1911. *Bibliothekskatalog der Kaiserl. Königl. Gartenbau–Gesellschaft in Wien.* Verfasst von A. Vollbracht. Vienna. 67 pp.

Bc382. **Koninklijke Algemeene Vereeniging voor Bloembollencultuur.** 1956. *Catalogus: Bibliotheek.* Haarlem. 262 pp. *Continued* ___. 1960. *Catalogus (eerste supplement): Bibliotheek.* Haarlem. 128 pp.

A classified catalogue with name indexes. Periodicals pp. 113–133 and supplement pp. 58–61.

Bc382A. **Landbouwuniversiteit Wageningen, Bibliotheek.** 1993+. *Catalogus speciale collecties.* Wageningen, Jan–Kopshuis. Part 1+. In progress.

I. *Tuinbouw t/m 1850.* Deel 1. *Tuinbouw Algemeen en Sierteelt.* 1993. vii+84+[9] pp.; Deel. 2 will cover *Groente– en Fruitteelt.*

Bc383. **Liverpool Museum, Botanical Library.** 1992. *Hand–list of the botanical library, Liverpool Museum.* By J. Edmondson. Liverpool, National Museums & Galleries on Merseyside. [vi]+74 pp.

Bc384. **Lloyd Library, Cincinnati.** 1914–18. *Bibliography relating to botany, exclusive of floras. Embracing botanical sections A, B, C, D, E, F, G, H, I, J, W, of the Lloyd Library.* Cincinnati, OH. 2 vols. *Bibliographical contributions from the Lloyd Library* vols. 2–3 (795+337 pp.)

The library's own holdings of these works are highlighted.

Bc385. **Massachusetts Horticultural Society.** 1962. *Dictionary catalog of the library of the Massachusetts Horticultural Society.* Boston, MA, G. K. Hall. 3 vols. ___. 1972. ___*: First supplement.* Boston, MA, [etc.]. 441 pp.

Their collection of ca.31,000 vols. is recorded in the main work. *Replaces* ___. 1918–20. *Catalogue of the library of the ... Part I: Alphabetical list of authors and titles. Part II: Subject catalogue.* Cambridge, MA. vii+587 pp.; ___. 1932. *Selected catalog of the library.* By D. S. Manks. Boston, MA. 59 pp.; ___. 1936. ___. Rev. ed. Boston, MA. 64 pp.; *Continued* ___. 1942. ___*: Supplement.* Boston, [etc.]. 32 pp.

Also **Sotheby Parke Bernet Inc.** 1980. *Fine books and autograph letters ...from the collection of the Massachusetts Horticultural Society ... public auction ... October 1, 1980 [etc.].* New York. [100] pp.

Bc386. **Missouri Botanical Garden.** (No general catalogue.)

For special lists *see* ___. 1896. The Sturtevant PreLinnean Library. By J. C. Bay & C. E. Hutchings. *Report, Missouri Botanical Garden* 7: 119–209. *Continued* ___. 1903. A supplementary catalogue of the Sturtevant PreLinnean Library. By C. E. Hutchings. *Ibid.* 14: 233–316. *Also* ___. 1932. Some incunabula in the Garden library. By E. M. *Missouri Botanical Garden bulletin* 20: 153–159; ___. 1959. *Flower prints of five centuries. An exhibition of flower books from the Library of the Missouri Botanical Garden.* October 30 through November 26, 1959. St. Louis, MO. [27] pp.

Bc387. **National Museum of Wales, (Amgueddfa Genedlaethol Cymru).** (No general catalogue.)
For special list *see* ___. 1942. *Plant illustration from woodcut to process block. Handbook to a temporary exhibition, based on the collection in the Department of Botany*, July – December, 1942, by A. E. Wade. Cardiff. 32 pp.
149 items.

Bc388. **Nederlandsche Botanische Vereeniging.** 1924. *Catalogus van de bibliotheek der Nederlandsche Botanische Vereeniging te Amsterdam.* Groningen, Noordhoff. 211 pp.
A classified catalogue of with a name index. Periodicals pp. 2–12. *Replaces* ___. 1883. *Catalogus der Bibliotheek [etc.].* Nijmegen, H. C. A. Thieme, 1883. 55 pp.

Bc389. **New York Public Library.** (No relevant general catalogue.)
For special list *see* ___. 1965. [Mr. & Mrs Pál] Kelemen gift: Early books on botany. By L. M. Stark. *Bulletin of the New York Public Library* 69: 73–76. A gift of 40 rare books.
___. 1989. *Nature illustrated, flowers, plants, and trees 1550–1900: From the collections of the New York Public Library*, text by B. McTigue. Foreword by E. Perényi. New York, H. N. Abrams. 128 pp., illus. (col.). 99 items.

Bc390. **Pennsylvania Horticultural Society.** 1941. *Catalog of the Library: The Pennsylvania Horticultural Society ... January 1941.* Philadelphia, PA. 161 pp.
Replaces ___. 1840. *Catalogue of the library of the [etc.].* Philadelphia, PA. 24 pp.; ___. 1931. ___. Philadelphia, PA. 88 pp. For special lists *see* ___. 1976. *From seed to flower: Philadelphia 1681–1876. A horticultural point of view.* Philadelphia, PA. 120 pp. "The great majority of the books are from the Society's own collection"; ___. 1978. *Selected books from the library of the Pennsylvania Horticultural Society.* Philadelphia, PA. 96 pp.

Bc391. **Pennsylvania State University, G. Weidman Groff Collection.** 1961. *Report on The G. Weidman Groff Collection: Books, periodical publications and manuscript materials recording observations of the occurrence and use of plants in China and surrounding countries of Southeast Asia (1907–1947), in the University Library, Pennsylvania State University.* Prepared by H. S. Brunner with J. Hsueh–Ming Chen & M. H. Runner. University Park, PA. 510 pp.

Bc392. **Phytopathologisch Laboratorium "Willie Commelin Scholten", Bibliotheek, Baarn.** *See* Rijksuniversiteit te Utrecht, Bibliotheek. 1936. *Catalogus [etc.]*; *see* Bc396.

Bc393. **Pierpont Morgan Library.** (No general catalogue.)
For special lists *see* ___. 1947. *Flowers of ten centuries: An exhibition.* April 28 to July 26, September 15 to November 15. New York. 70 pp. & 12 pl.
187 items, books and drawings.
___. 1980. *Flowers in books and drawings, c. 940 – 1840.* New York. [82] pp. incl. 24 pl. Exhibition held 2 September – 9 November, 1980, with 141 items.

Bc394. **Regensburgische Botanische Gesellschaft.** 1976. Katalog der bibliothek der Regensburgischen Botanischen Gesellschaft. By M. Müllerot. *Hoppea* 34: 1–424.
Replaces ___. 1895–97. *Katalog der bibliothek der Kgl. Botanischen Gesellschaft in Regensburg.* I. *Nichtperiodische Schriften.* II. *Periodische, bezw. Gesellschaften und Nachträge zu I.* By F. Vollmann. Regensburg. 142+441 pp.

Bc395. **Reichsstadtische Bibliothek Lindau.** (No general catalogue.)
For special list *see* ___. 1987. *Botanik, Kräuter und Arzneien in alten Büchern. Kostbarkeiten aus fünf Jahrhunderten im Besitz der ehemals Reichsstädtischen Bibliothek Lindau.* By W. Dobras. Lindau, Kulturamt der Stadt Lindau. 64 pp.

Bc396. **Rijksuniversiteit te Utrecht, Bibliotheek.** 1936. *Catalogus van botanische boekwerken in de Universiteitsbibliotheek, in het Botanisch Laboratorium, in het Phytopathologisch Laboratorium "Willie Commelin Scholten" (te Baarn), zoomede op*

enkele andere plaatsen te Utrecht. Utrecht. [3]+294p.
An author catalogue with a classified index. Omits general works pub. before 1918. *Replaces* **Phytopathologisch Laboratorium "Willie Commelin Scholten", Bibliotheek, Baarn.** 1909. *Catalogus van de bibliotheek 1909.* Amsterdam. 39 pp.

Bc397. **Royal Botanic Gardens, Kew.** 1974. *Author catalogue of the Royal Botanic Gardens Library, Kew, England.* Boston, MA, G. K. Hall. 5 vols.
Replaces ___. 1899. *Catalogue of the library [etc.].* London. viii+790 pp. *Bulletin of miscellaneous information, Royal Botanic Gardens, Kew* Additional series III; and: ___. 1919. ___*: (Supplement.) Additions received or incorporated in the years 1898–1915.* London. 433 pp. *Bulletin of miscellaneous information, Royal Botanic Gardens, Kew* Additional series III–2. Both are divided into four parts, viz. section 1 – General; section 2 – Travels; section 3 – Periodicals; section 4 – Manuscripts.

Bc398. ___. 1974. *Classified catalogue of the Royal Botanic Gardens Library, Kew, England.* Boston, MA, G. K. Hall. 4 vols.
Mostly arranged according to the Dewey Decimal Classification except for systematic works which are arranged according to the Bentham & Hooker botanical classification. With geographic and subject indexes. Supplemented by the periodical: ___. 1978+. *Current awareness list.* Kew, Royal Botanic Gardens. In progress (not published extra–murally for several years).

Bc399. **Royal Botanic Gardens, Peradeniya.** 1902. *Catalogue of the library of the Royal Botanic Gardens, Peradeniya, Ceylon. November, 1900.* By M. Willis. Colombo, H. C. Cottle, Acting Government Printer. 126 pp.
Replaces ___. 1889. *A classified catalogue of the library of the Royal Botanic Gardens, Pérádeniya, Ceylon.* By H. Trimen. Colombo. [iv]+28 pp.

Bc400. **[Royal Horticultural Society.] Horticultural Society of London.** [1859.] *Catalogue of the ... library of the ... Society, [etc.].* London. 50 pp.
Auction sale catalogue.

Bc401. **Royal Horticultural Society, Lindley Library.** 1927. *The Lindley Library. Catalogue of books, pamphlets, manuscripts and drawings.* [By H. R. Hutchinson.] London. viii+488 pp.
An author catalogue. *Replaces* ___. 1898. *Catalogue of the Lindley Library.* [Ed. by H. R. Hutchinson.] London. iv+162 pp.; *Also* ___. 1954. *Exhibition of manuscripts, books, drawings, portraits, medals and congratulatory addresses on the occasion of the Society's 150th anniversary celebrations.* London. 64 pp. & 5 pl. 186 items.; ___. 1991. Brent Elliott's treasures of the Lindley Library. *Garden, journal of the Royal Horticultural Society* 116: [12 articles], Viz: Alphand's *Promenades de Paris,* 116 (1): 30–31; Donald Beaton: The greatest gardening journalist of the early 19th century? 116 (2): 64–67; Memories of old Vienna: Franz Antoine's *Imperial Winter Garden,* 116 (3): 126–127; Conservatory catalogues of yesteryear, 116 (4): 172–174; Artist with an eye for detail: Turpin and Poirot's *Leçons de Flore,* 116 (5): 232–235; The gastronomic works of Edward Ashdown Bunyard, 116 (6): 320–321; Richter's *Codex,* 116 (7): 378–379; *Album Benary* ... possibly the most elaborate and comprehensive vegetable catalogue ever, 116 (8): 427–429; Fungus foray [illustrations of fungi], 116 (9): 496–498; Josiah Conder: An Englishman in Japan, 116 (10): 547–549; Nicolas Fatio de Duillier: A mathematician in the fruit garden, 116 (11): 595–599; The Thomas Edgerley books: Illustrations by a gardener cum self–taught artist, 116 (12): 652–654; **P. Stageman.** 1984. Treasures of the Lindley Library. *Country life* 176 (25 Oct.): 1180–1182.

Bc402. **(Royal) Pharmaceutical Society of Great Britain.** 1911. *Catalogue of the library of the Pharmaceutical Society of Great Britain.* 11th ed. London. 560 pp.
Replaces Several eds., e.g. ___. 1853[–64]. ___. London. [ ]* pp.; ___. 1880. ___. 3rd ed. By J. W. Knapman. London. 445 pp.; ___. 1889. ___. 6th ed. By J. W. Knapman. *Appended is the Catalogue of the North British Branch.* London. 580 pp.; ___. 1894. ___. 7th ed. London. 657 pp.
[*Not examined – pagination not ascertained.]

Bc403. **Smithsonian Institution, United States National Museum.** (No general catalogue.)
For special list *see* ___. 1908. Catalogue of the botanical library of John Donnell Smith presented in 1905 to the Smithsonian Institution. Compiled by A. C. Atwood. *Contributions from the United States National Museum* 12(1): [1–3],1–94.
"some 1,600 bound volumes ... chiefly ... works ... relating to Mexico and Central America."

Bc404. **Société Nationale d'Horticulture de France.** 1900. *Catalogue de la bibliothèque de la Société nationale d'Horticulture de France [etc.].* Classé par G. Gibault. Paris. viii+155 pp.; *Continued* ___. 1905. ___*: Supplément.* Paris. 39 pp.; ___. 1910. ___*: 2e supplément.* Paris. 53 pp.

Bc405. **Society of Apothecaries.** 1913. *Catalogue of the library of the Society of Apothecaries.* By W. B. Taylor. Ed. with an intro. by J. E. Harting. London. viii+106 pp.
Ca.1,750 titles.

Bc406. **Society of Herbalists.** n.d. *A selection from the library.* London. 67 pp.
Classified list of 555 items. *Also* ___. 1953. *An exhibition of books from the library of the Society of Herbalists.* By W. T. Stearn. London, Arts Council of Great Britain. 48 pp.; ___. 1958. *Catalogue of the celebrated collection of botanical books with coloured plates ... the property of The Society of Herbalists formed by their late Chairman, Mrs C. F. Leyel. ... Which will be sold by auction ... on ... the 7th July, 1958 [etc.].* London, Sotheby & Co. 43 pp. & 8 pls. Alphabetical catalogue of 236 lots.

Bc407. **Société Nationale d'Horticulture de France.** 1900. *Catalogue de la bibliothèque de la Société Nationale d'Horticulture de France.* Classé par M. G. Gibault [etc.]. Paris. viii+155 pp.; *Continued* ___. 1905. *Supplément au catalogue [etc.].* Paris. iv+39 pp.; ___. 1910. *2e supplément au catalogue [etc.].* Paris. iv+53 pp.

Bc407A. **South African Public Library, Cape Town.** (No relevant general catalogue.)
For special list *see* **W. Tyrrell-Glynn & M. R. Levyns.** 1963. *Flora Africana: South African botanical books 1600–1963: A selective bibliography based on the collections of the South African Public Library [etc.].* Cape Town [etc.]; *see* Cb467.

Bc408. **Staatsbibliothek Preussischer Kulturbesitz, Berlin.** (No general catalogue.)
For special list *see* ___. 1987. *100 botanische Juwelen. 100 botanical jewels. Catalogue of the exhibition of the Staatsbibliothek Preussischer Kulturbesitz Berlin (West).* Berlin. 216 pp.

Bc409. **Sterling Morton Arboretum.** (No general catalogue.)
For special lists *see* ___. 1972 *Hortus botanicus: The botanic garden and the book. Fifty books from the Sterling Morton Library exhibited at the Newberry Library for the 50th anniversary of the Morton Arboretum.* By I. MacPhail. Introductory essay by J. Ewan. Lisle, IL, Morton Arboretum. 119 pp., illus.
I. MacPhail. 1990. The Sterling Morton Library collection of herbals. *Herbarist* 56: 48–58. Re-pub. in *Morton Arboretum quarterly* 25(4): 49–54, 1989.

Bc410. **Taylor Institution, Oxford.** (No general catalogue.)
For special list *see* **D. J. Gilson.** 1973. Herbals in the Fry collection at the Taylor Institution. *Book collector* 22: 44–62. Collection of 95 works formed by Miss Caroline Sophia Fry (1861–1946).

Bc411. **Tiflisskii Botanicheskii Sad.** 1908. Katalog biblioteki Tiflisskogo Botanicheskogo sada k 1–mu yanvarya 1908 goda. *Trudy Tiflisskogo Botanicheskogo Sada* 8(3): 90+37 pp.

Bc412. **United States Department of Agriculture Library** (*later* **National Agricultural Library**). 1902. *Catalogue of publications relating to botany in the library of the United States Department of Agriculture.* By J. A. Clark. Washington, DC. 242 pp. *Bulletin* no. 42.
Author list, list of serials, subject index.

Bc413. **United States Department of Agriculture Library** (*later* **National Agricultural Library**). 1912. *Catalogue of publications relating to forestry in the library of the United States Department of Agriculture, prepared in the department with the co-operation of the Forest Service.* By E. B. Hawks. Washington, DC. 302 pp. *Bulletin* no. 76. *Continued* **United States Department of Agriculture, Forest Service Library.** 1908+. *Current literature, Forest Service Library, United States Department of Agriculture.* Washington, DC. In progress.

Bc414. ___. 1919–42. *Current author entries: Bureau of Plant Industry Library.* Washington, DC. Nos. 1–165, 1919–25; Continued as *Current botanical literature: Bureau of Plant Industry Library.* Washington, DC. Nos. 166–183, 1925–26; Continued as *Botany current literature. Additions to the botanical catalogue, Bureau of Plant Industry.* Compiled from publications received in the Library of the U. S. Department of Agriculture. Washington, DC, United States Department of Agriculture, vols. 9–16, 1926–34; Continued as *Plant science literature. Selected references [etc.].* Compiled by the Library staff ... (mainly for the botanical catalog) from publications received in the U. S. Department of Agriculture Library. Washington, DC, United States Department of Agriculture, 1935–42. 15 vols.
Continued as *Bibliography of agriculture.* Washington, DC, United States Department of Agriculture, vol. 1+, 1942+.

Bc415. ___. 1958. *Plant science catalog: Botany subject index compiled by [etc.].* Boston, MA, Microphotography Co. 15 vols.
Reproduces 315,000 catalogue cards in the alphabetical subject catalogue, commenced in 1903 and discontinued in July 1952. *See also* **A. C. Atwood.** 1911. Description of the comprehensive catalogue of botanical literature in the libraries of Washington. *Circular, Bureau of Plant Industry, U. S. Department of Agriculture* 87: 7 pp.; **C. R. Barnett.** 1937. Frederick Vernon Coville – friend of the library, his contributions to the early development of the library's collections and bibliographical equipment in the field of botany. *Agricultural library notes* 12: 95–103.

Bc416. ___. 1968. *Historic books and manuscripts concerning horticulture and forestry in the collection of the National Agricultural Library.* Beltsville, MD. 106 pp. *Library list* no. 90.

Bc417. **Universidad de Cordoba.** 1942. *Catálogo sistemático de los libros de la biblioteca. Universidad de Cordoba: Facultad de Ciencias Exactas, Físicas y Naturales.* Bibliotecario: E. Sparn. Cordoba. xii+194 pp.
Ca.2,500 titles.

Bc418. **Universidad Nacional de Buenos Aires.** 1930–31. *Catálogo de la biblioteca. Universidad Nacional de Buenos Aires: Facultad de Ciencias Exactas, Físicas y Naturales.* Buenos Aires. xxi+1,359 pp. *Publicación* ser. A, no. 6.
Classified catalogue of ca.20,000 titles. *Replaces* ___. 1921. *Catálogo de las obras existentes en la biblioteca [etc.].* Buenos Aires. 226 pp.

Bc419. **Universitaire Moretus Plantin, Bibliothèque.** (No general catalogue.)
For special list *see* ___. 1984. *Livres de fleurs du XVIe au XXe siècle dans les collections de la Bibliothèque Universitaire Moretus Plantin,* [by] A.–M. Bogaert–Damin & J. A. Piron. Namur. [ii]+147 pp., illus. Exhibition catalogue of 62 items interspersed within an extensive, illustrated text; *Also* ___. 1992. *Livres de fruits du XVIe au XXe siècle dans les collections de la Bibliothèque Universitaire Moretus Plantin,* [by] A.–M. Bogaert–Damin & J. A. Piron. Namur. [iii]+264 pp., illus.

Bc420. **Universitní Knihovny v Praze.** (No general catalogue.)
For special list *see* **R. Hendrych & V. Skalicky.** 1967–68. Soupis starsí botanické literatury Státní knihovny CSSR – Universitní knihovny v Praze podle starych oborvych svazkovych katalogu. (Verzeichnis der in den alten Fachkatalogen der Staatsbibliothek der CSSR – Universitätsbibliothek in Prag enthalten älteren botanischen Literarur). *Zprávy*

Ceskoslovenské Botanické Spolecnosti pri Ceskoslovenské Akademii Ved 2: 264–721, 1967; *Ibid.* 3: 42–52, 1968.

Bc421. **University of Kansas Libraries.** (No general catalogue.)
For special lists *see* ___. 1985. *Herbals and closely related medico–botanical works, 1472–1753, in the Department of Special Collections, Kenneth Spencer Research Library, and the History of Medicine Collection, Clendening Medical Library.* By S. H. Hocker. viii+94 pp. *University of Kansas publications, library series* 50.
Also **G. S. T. Cavanagh.** 1956. Medical botany. *Books & libraries at the University of Kansas* 1 (Nov.): 5–14; ___. 1959. Florula Darlingtonia. *Ibid.* 21 (May): 3–7.**J. C. Shipman.** 1965. One hundred herbals, 1472–1671: A chapter in the history of botany. Selections from printed works in the University of Kansas Libraries and the Linda Hall Library. *Ibid.* 2: 9–13.

Bc422. **University of Rochester.** (No general catalogue.)
For special information *see* ___. 1950. Early herbals. By M. K. Toth. *University of Rochester Library bulletin* 15: 1–12; ___. 1968: Creatures so very diminutive. By A. Creighton. *Ibid.* 23: 25–33. Descriptions of rare books relating to microscopy in the Medical Library.

Bc423. **University of Texas at Austin, Lundell Rare Book Library.**
See **A. C. Lewis.** 1982. Botany from Linnaeus to Lamarck. *Wrightia* 7 (2): 26–37. Catalogue of an exhibition held in the Leeds Gallery, 1981/82.

Bc424. **Verein zur Beförderung des Gartenbaues in den Königlich Preussischen Staaten, Bibliothek.** 1897. *Katalog der Bibliothek des Vereins zur Beförderung des Gartenbaues in den Königlich Preussischen Staaten ... Nebst einem Anhange enthaltend das Verzeichnis der Obstnachbildungen.* 6. Aufl. Berlin. 139 pp.
Replaces 4. Aufl., 1852. iv+46 pp.; 5. Aufl., 1875. [iii]+77 pp.

Bc425. **Victoria University, Toronto, Centre for Reformation & Renaissance Studies.** (No general catalogue.)
For special list *see* ___. 1985. *Published books (1499 to 1700) on science, medicine, and natural history at the Centre for Reformation and Renaissance Studies, Victoria University, Toronto.* Toronto. ix+37 pp. *Occasional publications, Centre for Reformation and Renaissance Studies* no. 4.

Bc425A. **Yale University, Library.** (No general catalogue.)
For special list *see* ___. 1968. *The art of botanical illustration: Checklist of an exhibition.* By D. R. Roylance. *Yale University Library gazette* 43: 98–122. 72 works, 15th–20th centuries.

Bc426. **___, School of Forestry.** 1962. *Dictionary catalogue of the Yale Forestry Library, Henry S. Graves Memorial Library, Yale University.* Boston, MA, G. K. Hall. 12 vols.
Catalogue of ca.250,000 entries of a collection of ca.90,000 vols.

Cryptogamic botany

Bc427. **British Mycological Society.** 1966. *Catalogue of the library of the British Mycological Society.* [Kew]. 10 pp.

Bc428. **Buller Memorial Library, Canada Department of Agriculture.** 1965. *Catalogue of the Buller Memorial Library. Catalogue de la Bibliothèque commémorative Buller.* By K. D. Oliver. Ottawa, Canada Department of Agriculture, Research Branch. 84 pp., port. *Publication* no. 1226.

Bc429. **Harvard University, Farlow Reference Library of Cryptogamic Botany.** 1979. *Catalog of the Farlow Reference Library of Cryptogamic Botany.* Boston, MA, G. K. Hall. 6 vols.
Vols. 1–5 is an author–title catalogue of ca.60,000 items; vols. 5–6 is a subject catalogue, "not consistent or complete."

Bc430. **Nederlandsche Mycologische Vereeniging.** 1936. *Catalogus van de bibliotheek der Landbouwhoogeschool. Bruikleen van de Nederlandsche Mycologische Vereeniging.* Wageningen. [iii]+57+[6] ff.; *Continued* ___. 1936. ___*: Supplement. Anwinsten gedurende Oct. 1936.* [Wageningen.]
Ca.500 titles in the main work.

Bc431. **University of Michigan.** (No general catalogue.)
For special list *see* **L. C. C. Krieger.** 1924. *Catalogue of the mycological library of Howard A. Kelly.* Baltimore, MD, privately pub. v+260 pp.
Author catalogue of ca.6,500 books and offprints with a separate list of periodicals. The library was presented to the University of Michigan in 1928.

Bc432. **Veterinary Institute, Bogor.** (No general catalogue.)
For special list *see* ___. 1954. *Catalogue of publications on anaerobic bacteria in the library of the Veterinary Institute, Bogor (Indonesia) collected and donated ... by F. C. Kraneveld.* Bogor. 194 pp.
Ca.4,000 titles.

Zoological — General

Bc433. **American Museum of Natural History.** 1977–78. *Research catalog ...*; *Continued* ___. 1983. *New catalog ...*; *see* Bc208.

Bc434. **Australian Museum, Sydney.** 1893 & 1905. *Catalogue of the library of the Australian Museum.* 2nd ed. Part III: *Pamphlets.* By S. Sinclair. Sydney, N.S.W. [iv]+87 pp. *Catalogue* no. VI.
Pt. I, A–Z list of books; Pt. II, A more detailed list of periodical literature, magazines, proceedings of societies, etc. [ever published?] *Replaces* ___. 1883. *Catalogue of the library of the Australian Museum.* By T. Fielding. Sydney, N.S.W. 178 pp. with supplements for 1884, 1885, etc.

Bc435. **Harvard University, Museum of Comparative Zoology.** 1967. *Catalogue of the Museum of Comparative Zoology.* Boston, MA, G. K. Hall. 8 vols. *Continued* ___. 1976. ___*: First supplement.* Boston, MA, [etc.]. 770 pp.
Main work lists their collection of ca.250,000 works on zoology and palaeontology.

Bc436. **Indian Museum.** 1889. *Catalogue of the books in the library of the Indian Museum. Corrected to August 1887.* By R. L. Chapman. Calcutta. 332 pp.; *Continued* ___. 1894. ___*: Supplement I. September, 1887 to August, 1891.* By H. W. Fleming. Calcutta. [iii]+31+69 pp.; ___. 1894. ___*: Supplement II. September, 1891 to August 1895.* By H. B. Perie. Calcutta. [iii]+130 pp.; ___. 1894. ___*: Supplement III. September 1895 to August 1899.* By C. O. Bateman. Calcutta. [iii]+74+54 pp.; ___. 1905. ___*: Supplement IV. September 1899 to August 1903.* By C. O. Bateman. Calcutta. [ii]+83+57 pp.
Replaces ___. 1879. *Catalogue of the library of the Indian Museum.* Calcutta. [ii]+93 pp.

Bc437. **Königliches Zoologisches und Anthropologisch–Ethnographisches Museum in Dresden.** 1898. *Catalog der Handbibliothek des Königlichen Zoologischen und Anthropologisch–Ethnographischen Museums in Dresden.* Berlin. xxv+287 pp.

Bc438. **Koninklijk Zoölogisch Genootschap "Natura Artis Magistra" te Amsterdam.** 1881. *Catalogus der bibliotheek van het Koninklijk Zoölogisch Genootschap "Natura Artis Magistra" te Amsterdam.* By G. Janse. Amsterdam, Scheltema & Holkema. xii+563 pp.
4,381 entries. *Also* ___. 1878. *Linnaeana in Nederland aanwezig. Tentoongesteld op 10 Januari 1878 in hen Koninklijk Zoölogisch Genootschap "Natura Artis Magistra" te Amsterdam.* By D. J. Coster. Amsterdam, Scheltema & Holkema. 58 pp. 272 items.; ___. 1939. *Tentoonstelling van boeken, teekeningen en andere voorwerpen uit de bibliotheek van het Kon.*

Zoölogisch Gen. "Natura Artis Magistra," t. g. v. het 100 j. bestaan van het Genootschap (Mei 1938). Door Jon[a]. Scheffer. Leiden. First pub. in *Bijdragen tot de dierkunde* vol. 27.

Bc439. **London University, London School of Hygiene & Tropical Medicine.** 1965. *Dictionary catalogue of the London School of Hygiene and Tropical Medicine.* Boston, MA, G. K. Hall. 7 vols; *Continued* ___. 1971. ___*: First supplement.* Boston, MA, [etc.]. 441 pp.

Includes much literature on tropical parasites. *Replaces* ___. 1934–35. *Classified catalogue of books in the library, including department libraries [etc.].* London. 3 vols.

Bc440. **McGill University, Montreal, Blacker–Wood Library of Zoology [etc.].** 1931. *An introduction to the literature of vertebrate zoology, based chiefly on the titles in the Blacker Library of Zoology, the Emma Shearer Wood Library of Ornithology, the Bibliotheca Osleriana and other libraries of McGill University Montreal.* By C. A. Wood. London, Oxford Univ. Pr. xix+643 pp. *Re-pub.* Hildesheim, Olms Verlag, 1974.

Bc441. ___**, Blacker–Wood Library of Zoology & Ornithology.** 1966. *A dictionary catalogue of the Blacker–Wood Library of Zoology and Ornithology.* Boston, MA, G. K. Hall. 9 vols.

Lists their collection of over 60,000 vols.

Bc442. **Nederlandsche Dierkundige Vereeniging.** 1924. *Katalogus van der bibliotheek.* 6e verkorte uitg. By H. C. Redeke. Helder, C. de Boer Jr. vi+168 pp.; *Continued* ___. 1926. ___*: 2e verv.* Helder, [etc.]. [ ]* pp.

Replaces ___. 1878. *Catalogus van de bibliotheek [etc.].* By P. P. C. Hoek. Leiden. [ ]* pp.; ___. 1879–81. ___. 3e uitgave. 1e verv. Leiden, E. J. Brill. 58 pp.; ___. 1884–92. ___. 3e uitg. Leiden, E. J. Brill. 2 vols.; ___. 1897–1904. ___. 4e uitg. Helder, C. de Boer, Jr. 2 vols.; ___. 1907. ___. 5e uitg. By H. C. Redeke. Helder, [etc.]. x+205 pp.; ___. 1914. ___*: Eerste vervolg (1 Januari 1907—1 Juli 1914).* Helder, [etc.]. [ii]+99 pp. [*Not examined – pagination not ascertained.]

Bc443. **Pierpont Morgan Library.** (No general catalogue.)

For special list *see* ___. 1940. *The animal kingdom. Illustrated catalogue of an exhibition of manuscripts, illuminations, book illustrations, drawings, cylinder seals, and bindings.* November 19, 1940 – February 28, 1941. (Ed. by H. M. Franc.) New York. 70 pp. & 11 pl.

Bc444. **Société Royale de Zoologie, Antwerp.** (No general catalogue.)

For special information *see* **F. Schrevens.** 1983–86. La precieuse collection de livres de notre bibliothèque dans sa perspective historique: Le 16ième siècle. *Zoo, Antwerp* 49–51: [A series of articles].

Bc445. **Statione Zoologica, Naples** (Zoologische Station in Neapel). 1874. *Die Bibliothek der Zoologischen Station zu Neapel. Verzeichniss der daselbst bis zum Ende des Jahres 1873 vorhandenen Bücher.* By A. Dohrn. Leipzig, W. Engelmann. iv+91 pp.; *Continued* ___. 1879. *Zweiter Nachtrag zum Bibliothekskatalog der Zologischen [etc.].* By A. Dohrn. Leipzig, [etc.]. 51 pp.; ___. 1881. *Dritter Nachtrag [etc.].* By A. Dohrn. Leipzig, [etc.]. 41 pp.

Bc446. **United States Department of the Interior.** 1967. *Dictionary catalog of the Department library.* Boston, MA, G. K. Hall. 37 vols; *Continued* ___. 1968. ___*: First supplement.* Boston, MA, [etc.]. 4 vols; ___. 1971. ___*: Second supplement.* Boston, MA, [etc.]. 2 vols; ___. 1973. ___*: Third supplement.* Boston, MA, [etc.]. 4 vols; ___. 1975. ___*: Fourth supplement.* Boston, MA, [etc.]. 8 vols.

Incorporates, among others, the libraries of the former Bureau of Biological Survey and the Fish & Wildlife Service.

Bc447. **Universitaire Moretus Plantin, Bibliothèque.** (No general catalogue.)

For special list *see* ___. 1987. *Livres d'animaux du XVIe au XXe siècle dans les collections de la Bibliothèque Universitaire Moretus Plantin,* [by] A.–M. Bogaert–Damin & J. A. Piron. Namur, Bibliothèque Universitaire Moretus Plantin. [iv]+223 pp. *Publication* no. 3. Exhibition catalogue interspersed within an extensive, illustrated text.

Bc448. **University of Reading, Cole Library of Early Medicine & Zoology.** 1969. *The Cole Library of Early Medicine and Zoology: Catalogue of books and pamphlets*. Part 1: *1472 to 1800*. By N. B. Eales. Reading. xiv+425 pp. *Continued* ___. 1975. ___. Part 2: *1800 to the present day and supplement to part 1*. By N. B. Eales. Reading. xviii+428 pp.
See also **N. B. Eales.** 1969. On the provenance of some early medical and zoological books. *Journal of the history of medicine* 24: 183–192.

Bc449. **Zoological Society of London.** 1902. *Catalogue of the library of the Zoological Society*. 5th ed. Ed. by F. H. Waterhouse. London. iv+856 pp.
Ca.12,500 entries. *Replaces* ___. 1854. *Catalogue of the library, [etc.]*. London. 44 pp.; ___. 1864. *Catalogue, [etc.]. Supplement*. London. [ ]* pp.; ___. 1872. *Catalogue, [etc.]*. [2nd ed.] London. 155 pp.; ___. 1880. *Catalogue, [etc.]*. [3rd ed.] By F. H. Waterhouse. London. [iv]+236 pp.; ___. 1883. *Catalogue, [etc.]. Supplement: Additions to Aug. 30. 1883*. London. 71 pp.; ___. 1887. *Catalogue, [etc.]*. 4th ed. By F. H. Waterhouse. London. [iv]+515 pp. [*Not examined – pagination not ascertained.]

Ornithological

Bc449A. **Auckland Institute & Museum.** (No general catalogue.)
For special list *see* ___. 1955. *The bird book*. Auckland War Memorial Museum, Exhibition March 24th to April 10th, 1955. By E. G. Turbott. Auckland. 44 pp.

Bc450. **Bibliothèque Royale Albert 1er, Brussels.** (No relevant general catalogue.)
For special list *see* ___. 1969. *Merveilleux plumages: Dix siècles de livres d'oiseaux. Catalogue d'exposition*. Par J. Balis. [Brussels.] viii+165 pp. & 32 pl.
Exhibition held June–August, 1969; 146 items. For Flemish language version *see* ___. 1968. *Van diverse pluimage*. Antwerp, Kon. Mij. Dierk.

Bc451. **Field Museum of Natural History, Ayer Library.** 1926. *Catalogue of the Edward E. Ayer ornithological library*. By J. T. Zimmer. Chicago, IL, Field Museum of Natural History. x+706 pp. & 12 pl. *Publication, Field Museum of Natural History, zoological series* vol. 16. *Re-pub*. New York, Arno Pr., 1974.
Ca.2,000 entries.

Bc451A. **Harvard University.** (No relevant general catalogue.)
For special list *see* **V. Keyes.** 1913. *Catalogue of a collection of books on ornithology in the library of John E. Thayer*. Boston, MA, privately pub. 187 pp. Ca.1,100 entries. Library later presented to Harvard University.

Bc452. **Københavns Universitet, Bibliotek.** (No general catalogue.)
For special list *see* **Anker, J.** 1938. *Bird books and bird art: An outline of the literary history and iconography of descriptive ornithology, based principally on the collection of books containing plates with figures of birds and their eggs now in the University Library at Copenhagen and including a catalogue of these works*. Copenhagen, Levin & Munksgaard. xviii+251 pp. & 12 pl. *Re-pub*. The Hague, W. Junk, 1973.

Bc453. **Library Company of Philadelphia.** (No general catalogue.)
For special information *see* ___. 1977. *A flock of beautiful birds: The ornithological collection of Louise Elkins Sinkler*. By E. Wolf 2nd. Philadelphia, PA, Library Company of Philadelphia. 43 pp.

Bc454. **Manchester University, John Rylands Library.** (No general catalogue.)
For special list *see* ___. 1969. Illustrated bird books [Catalogue of an exhibition]. *Bulletin, John Rylands Library* 51: 263–265.

Bc455. **National Library of Australia, Mathews Ornithological Collection.** 1966. *Checklist to the Mathews ornithological collection*. Canberra, A.C.T. [ii]+v+309 pp. & port.

Bc456. **Naturhistorisches Museum, Bern.** (No general catalogue.)
For special list *see* ___. 1954. *Der Vogel in Buch und Bild: Führer durch eine Ausstellung schöner und seltener ornithologischer Bücher und Bilderwerke aus der Alten und Neuen Welt von Gesner über Naumann bis zur Gegenwart.* Bern. 68 pp. & 28 pl.

Bc457. **New York Public Library.**
See Bc298. For special lists *see* ___. 1988. *The bird illustrated 1550–1900: From the collections of the New York Public Library,* text by J. Kastner; with commentaries by M. T. Gross. Introduction by R. T. Peterson. New York, H. N. Abrams. 128 pp., 100 illus. (45 col.). 96 items.
___. 1991. *The animal illustrated 1550–1900: From the collections of the New York Public Library,* text by J. Kastner; with commentaries by M. T. Gross. New York, H. N. Abrams. 128 pp. 100 items.

Bc458. **Trinity College Library, Hartford.** (No general catalogue.)
For special list *see* ___. 1913. The [Gurdon Wadsworth] Russell collection of books on natural history in Trinity College Library. *Trinity College bulletin* n.s. 10 (2): 28 pp.

Bc459. [Not used.]

Bc460. ___. 1983. *Ornithology books in the library of Trinity College, Hartford, including the library of Ostrom Enders.* Prepared by V. Breit and K. B. Clarke. Hartford, CT, Trinity College Association. 270 pp.
Also ___. 1966. American ornithology before and after Audubon: A descriptive catalogue of an exhibition held in the Trumbull Rare Book Room of the Watkinson Library, Trinity College. *Trinity College Library gazette* 3 (April): 11–16.

Bc461. **University of Kansas Libraries, Ellis Collection.** 1972+. *A catalogue of the Ellis Collection of Ornithological Books in the University of Kansas Libraries.* By R. M. Mengel. Lawrence, KS. Vol. 1+. In progress.
Vol. 1: A–B. 1972; vol. 2: C–D. 1983. Entries #1-735. *Also* **C. Chittenden.** 1976. On birds and bird men: Gouldiana at the University of Kansas. *Ibid.* 13(2): 1–5; **R. M. Mengel.** 1954. [Thomas] Bewick, birds, and bibliography. *Books & libraries at the University of Kansas* 1 (Feb.): 7–8; ___. 1955. More ornithologia. *Ibid.* 1 (Nov.): 6–7; ___.. 1957. *A catalog of an exhibition of landmarks in the development of ornithology from the Ralph N. Ellis collection of ornithology in the University of Kansas libraries.* Lawrence, KS. 33 pp. 100 items; ___. 1962. News from the Ellis. *Books & libraries at the University of Kansas* 2:(3): 1–4; **G. C. Sauer.** 1956. Gouldiana. *Ibid.* 1 (May): 2–3; **R. Vosper.** 1961. A pair of bibliomanes for Kansas: Ralph Ellis and Thomas Jefferson Fitzpatrick. *Papers of the Bibliographical Society of America* 55: 207–225.

Bc462. **Yale University Library.** (No general catalogue.)
For special list *see* ___. 1961. *Ornithological books in the Yale University Library, including the library of William Robertson Coe.* By S. D. Ripley & L. L. Scribner. New Haven, CT, Yale Univ. Pr. [ix]+338 pp. *Re-pub.* Mansfield, CT, Maurizio Martino, [1993]. Ca.5,000 entries. *Also* **S. D. Ripley.** 1952. The Coe ornithological collection. *Yale University library gazette* 27 (2): 66–70; **M. G. Wynne.** 1977. The world of birds. *Ibid.* 51: 10–29. Annotated catalogue of 90 rare books.

See also Bc338A.

Herpetological & Ichthyological

Bc462A. **Brooklyn Museum.** (No general catalogue.)
For special list *see* **E. W. Gudger.** 1925. Notes on certain books of unusual interest in the Blackford Collection of the Brooklyn Museum. *Brooklyn Museum science bulletin* 3: 167–172. Thirty significant books from this collection on fish and fishing.

Bc463. **Cheltenham Public Library.** (No general catalogue.)
For special list *see* ___. 1890. *Catalogue of the lending and reference departments including the 'Day Library of Natural History.'* By W. Jones. Cheltenham. 4+296 pp.

Francis Day's library of ca.1,200 books on natural history, chiefly ichthyology, presented by his daughters.

Bc464. **Directie van de Visserijen, Bibliotheek.** 1951. *Catalogus.* The Hague. [ix]+263 pp.
Ca.5,000 entries.

Bc464A. **New York Public Library.** (No relevant general catalogue.)
For special lists *see* ___. 1899. Check list of works on fish and fisheries in the New York Public Library. *Bulletin of the New York Public Library* 3: 296–312, 334–348; ___. 1909. List of works in the New York Public Library relating to fishing and fish culture. *Ibid.* 13: 259–307. Also pub. separately, New York. 49 pp. The former list excludes literature on fishing as a sport. The latter is confined to works on fishing as a sport and includes books from the library of Thomas Westwood (for details see **J. W. Bouton**, firm. 1873. *Bibliotheca piscatoria ... Catalogue of the library of Thomas Westwood.* For sale by [etc.]. New York. 46 pp. which is now in the NYPL).

Bc465. **University of Kansas Libraries.** (No general catalogue.)
For special information *see* **J. T. Collins.** 1970. The Edward Harison Taylor herpetological library. *Books & libraries at the University of Kansas* 8(2): 2–5.

Bc466. **University of Michigan, William L. Clements Library.** (No general catalogue.)
For special lists *see* ___. 1936. *Ichthyologica et herpetologia Americana: A guide to an exhibition in the William L. Clements Library illustrating the development of knowledge of American fishes, amphibians and reptiles.* By H. H. Peckham, H. T. Gaige & C. L. Hubbs. Ann Arbor, MI. ii+22 pp. *Bulletin of the William L. Clements library* no. 25. 46 items covering 1551–1817.

Entomological

Bc467. **American Entomological Society.** 1868. *Catalogue of works in the library of the American Entomological Society.* Philadelphia, PA. ii+32 pp.
Ca.600 entries. Issued with the Society's *Transactions*, 1867/68.

Bc468. **Deutsche Entomologische Museum.** 1913. *Katalog der Bibliothek des Deutschen Entomologischen Museums.* Teil 1: *Einzelwerke und Separata.* Berlin. [ii]+313 pp.
All published.

Bc469. **Entomologischer Verein zu Stettin.** 1885. *Katalog der Bibliothek des Entomologischen Vereins zu Stettin.* Stettin. 106 pp. *Stettiner entomologische Zeitung* 1885.
Replaces ___. [1862?] *Verzeichniss der am Schlusse des Jahres 1862 in der Bibliothek des Entomologischen Vereins zu Stettin vorhandenen Bücher.* Stettin. 59 pp.

Bc470. **Nederlandsche Entomologische Vereeniging.** 1911. *Catalogus ... der Bibliotheek.* 3e uitg. By C. L. Reuvens [etc.]. [Leiden.] xvi+440+[24] pp.; *Continued* ___. 1915 ___*: Supplement I, [etc.].* The Hague. 38 pp
Replaces ___. 1873. *Catalogus der bibliotheek.* By C. Ritsema Cz. The Hague. [ ]* pp. ___. 1899–[1907?]. *Catalogus A. ... der Bibliotheek.* 2e uitg., [etc.]. ([and] *Supplement I–X.*) Leiden. 11 pts.; ___. 1898–[1907?]. *Catalogus B. ... der Bibliotheek.* 2e uitg., [etc.]. ([and] *Supplement I–X.*) Leiden. 11 pts. [*Not examined – pagination not ascertained.]

Bc471. **Royal Entomological Society of London.** 1980. *Catalogue of the library of the Royal Entomological Society of London.* Boston, MA, G. K. Hall. 5 vols.
Replaces **Entomological Society of London.** 1893. *Catalogue of the library of the Entomological Society of London.* By G. C. Champion, R. McLachlan & D. Sharp. London. [iv]+312 pp. *Continued* ___. 1900. ___*: Supplementary catalogue.* Ed. by G. C. Champion,

W. F. H. Blandford & R. McLachlan. London. [iv]+147 pp.; ___. 1861. *Catalogue [etc.]*. London. 52 pp. Pub. with *Transactions* vol. 5.

Bc472. **Russkoe Entomologicheskoe Obshchestvo v S.–Peterburgh.** 1903–08. *Katalog biblioteki Russkago Entomologicheskago Obshchesta*. [By N. N. Zubovskii & I. Ya. Shevyrev.] St. Petersburg. 273 pp.
Vyp 1: *Inostrannyi otdel'. Izdaniya neperiodicheskie*; Vyp 2: *Russkii otdel. Izdaniya neperiodicheskie*.

Bc473. **Société Entomologique de Belgique.** 1871–81. *Catalogue de la bibliothèque de la Société Entomologique de Belgique*. Brussels. 12 pts., [548] pp.
Ca.10,000 entries.

Bc474. **Société Entomologique de France.** 1867. *Catalogue de la bibliothèque, 1832–1866*. By H. de Bonvouloir. Paris. 103 pp.

Bc475. **South London Entomological & Natural History Society.** 1970. A catalogue of books in the library of the South London Entomological & Natural History Society. *Proceedings of the South London Entomological & Natural History Society* 1959: 104–175. *Also pub. separately* Arbroath, T. Buncle & Co., 1960.

Bc476. **United States Department of Agriculture Library** (*later* **National Agricultural Library**). 1906. *Catalogue of publications relating to entomology in the library of the United States Department of Agriculture*. By J. A. Clark. Washington, DC. 562 pp. *Library bulletin* no. 55.
Classified list of 5,600 entries with indexes.

Bc477. **Universität Hamburg, Zoologisches Museum.** (No general catalogue.)
For special list *see* **H. Weidner.** 1980. Entomologische Schriften der Zeitgenossen und Schüler von Carolus Linnaeus im Archiv der Entomologischen Sammlungen des Zoologischen Instituts und Zoologischen Museums der Universität Hamburg. *Entomologische Mitteilungen aus dem Zoologischen Museum Hamburg* 6: 310–379.

Conchological

Bc478. **Malacological Society.** 1927. *Library catalogue (Radley bequest)*. London. 23 pp.

Bc479. **Nederlandsche Malacologische Vereniging.** 1968. Catalogus van de bibliotheek van de Nederlandse Malacologische Vereniging. Ed. by L. J. M. Butot. *Correspondentieblad van de Nederlandse Malacologische Vereniging, Supplement Bibliotheek* SB 1: 13 pp.

BIBLIOGRAPHIES OF LIBRARY HISTORY & CATALOGUES

Bc480. **Collison, R.** 1973. *Published library catalogues: An introduction to their contents and use*. London, Mansell Information/Publishing. viii+184 pp. ••Part 1. Published library catalogues in various subject fields; part 2. Key to published library catalogues.

Bc481. **Davis, D. G., & J. M. Tucker.** 1989. *American library history: A comprehensive guide to the literature*. Santa Barbara, CA, ABC–Clio. xxi+471 pp. ••*Replaces* **M. H. Harris & D. G. Davis, Jr.** 1978. *American library history: A bibliography*. Austin, TX, Univ. of Texas Pr. xix+260 pp.

Bc482. **Nelson, B. R.** 1982. *A guide to published library catalogs*. Metuchen, NJ, Scarecrow Pr. xvi+342 pp. ••Ca.430 catalogues "published in the last 20 years."

Bc483. **Ranz, J.** 1964. *The printed book catalogue in American libraries, 1723–1900*. Chicago, IL, American Library Association. vii+144 pp. *A C R L monograph* no. 26.

Bc484. **Titley, J.** 1964. Printed catalogues of American medical libraries before 1850: A check list. *Journal of the history of medicine* 19: 61–65.

See also Bc183–185.

Personal libraries

Natural history (including a few medical)

Arents, G.

See New York Public Library, Arents Collection, Bc298A.

Baldinger, E. G.

Bc485. **Conrad, J. G. H.** 1805. *Catalogus bibliothecae medico–physicae Ern. Godofr. Baldingeri.* Curavit, praefatus est, notas rarioribus libris adjecit Jo. Guil. Henric. Conradi. Marburg, typ. Kriegeri. 2 vols. ••Tom. I: *Bibliotheca medica,* x+738 pp.; Tom. II: *Bibliotheca physica,* iv+729–1167 pp.

Banks, Sir J.

Bc486. **Dryander, J.** 1798–1800. *Catalogus bibliothecae historico–naturalis Josephi Banks.* London. 5 vols. *Re-pub.* New York, Johnson Reprint Corporation; Amsterdam, Asher, 1966. ••A minutely carefully classified catalogue. Tom. I: Scriptores generales; Tom. III: Botanici; Tom. V: Supplementum et Index auctorum. The collection was bequeathed to the British Museum, London. *Also* ___. [17 ?] *Catalogus librorum qui desiderantur in bibliotheca historico–naturali Josephi Banks.* [6th ed.] London. 27 pp. (Previous eds. Ed. alt., 1782. [32] pp.; 3rd ed., [1790]. 27 pp.; 4th ed., n.d. 32 pp.; 5th ed., n.d. 32 pp.)

Breyne, J. P.

Bc487. **[Breyne, J. P.]** 1765. *Bibliothecae Breynianae pars prior; sive catalogus librorum philologico–philosophico–historicorum, itinerariorum, imprimis autem medicorum, botanicorum & historiae naturalis scriptorum ... quam ... comparavit Johan. Philipp. Breynius [Auct. 15 Jul. 1765].* Danzig, typ. Bartels. 176 pp. ••Auction catalogue.

Childs, J. L.

Bc488. **Childs, J. L.** 1917. *Catalogue of the North American natural history library of John Lewis Childs.* Floral Park, NY, privately published. [v]+150 pp. ••The library was later auctioned: **American Art Association.** 1923. *Catalogue of the distinguished collection of the late [etc.].* New York. [137] pp.

Cobres, J. P.

Bc489. **Cobres, J. P.** 1781–82. *Deliciae Cobresianae. J. P. Cobres Büchersammlung zur Naturgeschichte.* Augsburg, the Author. 2 pts., xxviii+957 pp. *Re-pub.* Amsterdam, Asher, 1966. 2 vols. ••Classified catalogue of ca.2,500 entries with detailed collations, notes and references to reviews. Author/title index. For background information *see* **F. A. Stafleu.** 1968. Deliciae bibliographicae. *Taxon* 17: 218–220. *See also* **D. E. Johnson.** 1979. A concordance of abbreviations used in references to reviews in botanical bibliographies by Cobres (1782), Böhmer (1787) and Miltitz (1829). *Huntia* 3: 103–114. The augmented library was later sold: **W. Birett,** firm. 1827. *Verzeichniss der Bücher der berühmten naturhistorischen Bibliothek des weiland Herrn J. R. Ritters v. Cobres, deren Verkauf um billige, sehr verminderte Preise besorgt wird von Wilhelm Birett. ... in Augsburg.* Augsburg. iv+134+[2] pp. 3,370 items.

Darwin, C. R.

Bc490. **Cambridge University, Botany School.** 1908. *Catalogue of the library of Charles Darwin now in the Botany School, Cambridge.* By H. W. Rutherford. Cambridge, Univ. Pr. xiv+91 pp.

Bc491. ___, **Library.** 1961. *List of books received in the University Library, March–May 1961.* Cambridge. 29 ff. ••A list of books from Charles Darwin's library that were transferred from Down House, Kent.

Deby, J.

Bc492. **Deby, J.** 1889. *Bibliotheca Debyana being a catalogue of books & abstracts relating to natural science, with special reference to microscopy, in the library of Julien Deby*. Vol. 1. London. [iv]+151 pp. ••*Also* Part III. The Diatomaceae, Bc536.

Dibner, B.

See Smithsonian Institution, Bc319.

Falconet, C.

Bc493. **Barrois, –.**, firm. 1763. *Catalogue de la bibliothèque de feu M. Falconet, médecin consultant du roi et doyen des médecins de la Faculté de Paris, de l'Académie royale des inscriptions et belles–lettres [etc.]*. Paris. 2 vols. ••Natural history, vol. 1, pp. 173–248, vol. 2, pp. 481–829.

Gardner, W.

See National Museum of Wales, Bc290.

Geoffroy, C.–J. & C.–F.

Bc494. **Martin, G.**, firm. 1754. *Catalogue des livres et estampes de défunts Mrs. Geoffroy, de l'Académie royale des sciences, dont la vente de fera en détail le 5 février 1754 et jours suivans [etc.]*. Paris. 90+18 pp. ••Histoire naturelle, pp. 13–32.

Geoffroy, E.–F.

Bc495. **Martin, G.**, firm. 1731. *Catalogus librorum viri cl. D. Stephani–Francisci Geoffroy ... Regiae scientiarum academiae parisiensis necnon Societatis londinensis socii*. Paris. [viii]+156 pp. ••Historia naturalis, pp. 13–34.

Gronovius, L. T.

Bc496. **Haak & Co., T., & J. Meerburg**, [Firms]. 1778. *Catalogus librorum exquisitissimorum, inter quos praecipue eminent rei botanicae & historiae naturalis scriptores & miscellanei. ... Accedit: Museum Gronovianum, ... Quae omnia magna cura sibi comparavit vir, dum in vivis erat, amplissimus Laurent. Theod. Gronovius, ... Quorum omnium publica fiet auctio ... die 5. Octobris & seqq. diebus. 1778*. Leiden, T. Haak & Socios. 42 pp.

Hammer, C.

Bc497. **Lindal, P. J., et al.** 1886–88. *Catalogue de la bibliothèque Hammer à Stockholm. Division étrangère*. Stockholm & Berlin. 9 vols. ••Vol. 7, *includes* Sciences mathématiques et naturelles, #88273–89922; Appendice: Collection Linnéenne, #1–283.

Heurck, H. van.

Bc498. **Koninklijke Maatschappij voor Dierkunde.** 1975. *Henri van Heurck Museum: Bibliotheca van Heurckiana. Catalogue of books collected by Dr. Henri van Heurck (1838–1909) as well as some related publications which have been added after his death*. By F. Schrevens. Antwerp. xii+332 pp. ••The books are scattered in three institutions, the Plantin–Moretus Museum, the Koninklijke Maatschappij voor Dierkunde, and the Stadtsbibliotheek van Antwerpen.

Hooke, R.

Bc499. **Millington, E.**, firm. 1703. *Bibliotheca Hookiana. Sive catalogus diversorum librorum: Viz. Mathematic., Philosophic., Medicorum, Philologicor., Hist. Natural., Nasvigat. &c. plurimis facultatibus linguisque insignium quos Doct. R. Hooke, ... Quorum auctio habenda est Londini, ... the 29th of April, 1703*. Per Edoardum Millington, Bibliop. Londin. London. [iv]+56+[i]+17 pp. *Re-pub. in* **L. Rostenberg.** 1989. *The library of Robert Hooke: The scientific book trade of Restoration England*. Pt. III, pp. 141–221. Santa Monica, CA, Modoc Pr., Inc.; and in: **A. N. L. Munby**, ed. 1975. *Sale catalogues of libraries of eminent persons*. Vol. 11: *Scientists*, ed. with an introduction by H. A. Feisenberger, pp. 57–116. London, Mansell/Sotheby Parke Bernet.

Hulthem, C. J. E. van .

Bc500. **Voisin, A.** 1836–37. *Bibliotheca Hulthemiana ou catalogue méthodique de la riche et précieuse collection de livres et des manuscrits* *délaissés par M. Ch. van Hulthem, [etc.]*. Ghent, imp. J. Poelman. 5 vols. ••"Histoire naturelle," vols. 1,2 & 5 (#5519–7272 & 29606–29667).

Humboldt, A. von.

Bc501. **Stevens, H.** 1863. *The Humboldt library: A catalogue of the library of Alexander von Humboldt, with a bibliographical and biographical memoir*. London. 791 pp. *Re-pub*. Leipzig, Zentralantiquariaat der DDR, 1966.

Huzard, J.–B.

Bc502. **Le Blanc, P.** 1842. *Catalogue des livres, dessins et estampes de la bibliothèque de feu M. J.–B. Huzard. Mis en ordre et rédigé par P. LeBlanc.* Paris, Mme Ve Bouchard–Huzard. 3 vols. ••Classified catalogue of 16,370 lots. Pt. 1. Histoire naturelle et sciences accessoires, (#1323–#4,617); pt. 2. Agriculture [etc.], (#1–#2,480).

Jefferson, T.

Bc503. **Jefferson, T., J. Gilreath, & D. L. Wilson.** 1989. *Thomas Jefferson's library: A catalog with the entries in his own order*. Ed. by J. Gilreath & D. L. Wilson. Washington, DC, Library of Congress. x+149 pp.

Bc504. **Sowerby, E. M.** 1952–53. *Catalogue of the library of Thomas Jefferson*. Washington, DC, U. S. G. P. O. 3 vols.

Keynes, Sir G.

Bc505. **Keynes,** Sir **G.** 1964. *Bibliotheca bibliographici: A catalogue of the library formed by Geoffrey Keynes*. London, Trianon Pr. xxiii+444 pp. & 45 pl.

Lamarck, J. B. P. A. de M.

Bc506. **Barrois l'ainé,** firm. 1830. *Catalogue des livres de la bibliothèque de feu M. le Chevalier J. B. de Lamarck, dont la vente se fera le lundi 19 Avril 1830, et jours suivans [–24 Avril]*. Paris, Barrois, Snr; Benou. 2ff.+36 pp.

Lettsom, J. C.

Bc507. **Leigh & Sotheby,** firm. 1811. *A catalogue of the greater portion of the library of John Coakley Lettsom, M.D. F.R.S. ... which will be sold by auction, ... on Monday, March 26, 1811 ... and seven following days [etc.]*. London. 48 pp. ••1,347 lots.

Lindell, E.

Bc508. **Lindell, E.** 1932. *Bibliotheca Linnæana – works by or relating to Carolus Linnaeus, his predecessors, contemporaries and pupils, with sequels – from the library of Emil Lindell.* Växjö, privately pub. 123 pp. ••Now part of the Strandell collection at the Hunt Institute for Botanical Documentation.

Bc509. [Not used.]

Linnaeus, C.

See Linnean Society of London, Bc261.

Logan, J.

See Library Company of Philadelphia, Bc260

McIlhenny, E. A.

See Louisiana State University, McIlhenny Collection, Bc265.

MacKenzie, K. K.

See Kansas State University Library, Bc252.

Malesherbes, C.–G. L. de.

Bc510. **Nyon, J.–L.,** ed. 1797. *Catalogue des livres de la bibliothèque de feu Chrétien–Guillaume Lamoignon–Malesherbes ... avec une table alphabétique des matières et des auteurs [Auct. 1 May 1797 OS]*. Paris. 32 pp. & unpaged. ••Natural history, items #708–#1957.

Mead, R.

Bc511. **Baker, S.,** firm. 1754–55. *Bibliotheca Meadiana, sive catalogus librorum Richardi Mead, M.D., qui prostabunt venales sub hasta, apud Samuelem Baker ... Die Lunae, 18° Novembris, M.DCC.LIV. iterumque Die Lunae 7° Aprilis, M.DCC.LV [etc.]*. London. [iv]+242 pp. ••Included a considerable natural history content.

Oken, L.

Bc512. **Orell, Füssli & Co.,** firm. 1852. *Catalog der Bibliothek von Laurentius Oken, welche am 17. Mai 1853 in Zürich versteigert werden soll*. Zürich, J. J. Ulrich. 1 f.+viii+204 pp. ••Auction catalogue ed. by Orell, Füssli & Co. in accordance with plan projected by Oken himself. An excellent catalogue of an unusually rich collection.

Orbigny, A. D. D'.

Bc513. **Baillière & Fils, J. B.**, firm. 1858. *Catalogue des livres d'histoire naturelle ouvrages et manuscrits relatifs a l'Amérique composant la bibliothèque de feu Alcide d'Orbigny [Auct. 6 Jan. 1858]*. Paris. 55 p.

Osler, W.

See McGill University, Montreal, Bibliotheca Osleriana, Bc267.

Ray, J.

Bc514. **Ballard, T.**, firm. 1708. *Bibliotheca Rayana: or, A catalogue of the library of Mr. John Ray, ... Which will be sold by auction, ... on ...... the 11th of this instant March, 1707/8 [etc.]*. London. *Re-pub. in* A. N. L. Munby, ed. 1975. *Sale catalogues of libraries of eminent persons*. Vol. 11: *Scientists*, ed. with an introduction by H. A. Feisenberger, pp. 117–148. London, Mansell/Sotheby Parke Bernet. ••1,350 vols.

Rivinus, A. Q.

Bc515. **Hermann, G. S.** 1727. *Bibliotheca Riviniana sive catalogus librorum philologico–philosophico–historicorum, itinerariorum, imprimis autem medicorum, botanicorum et historiae naturalis sciptorum &c. rariorum, quam magno studio et sumptu sibi comparavit D. Augustus Quirinus Rivinus, ... vendenda in vaporario collegii rubri a die XXVII. Octobr. M D XXVII. more auctionis consueto*. Leipzig. [xiv]+740+[107] pp. *Re-pub.* (with an introduction by R. Zaunick) Amsterdam, A. Asher, 1966. ••A classified catalogue of 7,968 lots (natural history #6,277–#6,317; Musea #6,318–6,350; Historiæ plantarum scriptores #6,351–#7,132; Historiæ animalium scriptores #7,133–#7,242) with author index. For background information *see* **F. A. Stafleu.** 1968. Deliciae bibliographicae. *Taxon* 17: 218–220.

Roper, F. C. S.

Bc516. **Roper, F. C. S.** 1865. *Catalogue of works on the microscope, and of those referring to microscopical subjects, in the library of Freeman C. S. Roper*. London, privately pub. 102 pp. *Re-pub.* Bronxville, NY, N. T. Smith, [1979?]. ••The library was later auctioned: **Sotheby**, firm. 1897. *[Auction sale catalogue, 2 & 3 April 1897.]* London. [ ]* pp. [*Not examined – pagination not ascertained.]

Strandell, B.

See Carnegie Mellon University, Hunt Institute for Botanical Documentation, Bc66 & Bc68.

Sturm, J. H. C. F.

Bc517. [**Sturm, J. H. C. F.**] 1863. *Verzeichniss der hinterlassenen sehr werthvollen naturhistorischen Bibliothek Kupferstiche, Radirungen und Lithographieen des Johann Heinrich Christian Friedrich Sturm [Auct. 15 Apr. 1863]*. Nuremberg, Druck d. Sebald'schen Officin. 2 l.+58 pp.

Varenne (de Béost), C.

Bc518. **Hérissant, –.**, firm. 1773. *Notice contenant und grande quantité de très belles estampes ... différentes suites d'histoires, d'animaux, plantes, coquilles, etc. supérieurement colorés, en feuilles et en volumes, provenant du cabinet de M***, dont la vente se fera dans une des salles des Grands Augustins, mercredi 24, jeudi 25, vendredi 26, samedi 27, lundi 29 novembre; mercredi premier et jeudi 2 décembre 1773 [etc.]*. Paris. [2]+24+8 pp.

Bc519. ___, firm. 1773. *Notice des livres du cabinet de M***, dont la vente se fera dans une des salles des Grands Augustins, le vendredi 3 décembre 1773, de relevée et jours suivans aussi de relevée sans interruption. Elle consiste en livres rares, curieux, et ornés d'estampes enluminées, concernant l'histoire naturelle ... plantes, arbres, arbrisseaux, végétaux, minéraux, animaux de tous genres et espèces; quadrupèdes, oiseaux, poissons ...*

Voyages, relations ... et autres livres très intéressans [etc.]. Paris. [4]+10+140 pp.

Wallace, A. R.

See Linnean Society of London, Bc261.

Waller, E.

See Kungliga Universitetet, Uppsala, Bc259.

Botanical

Adanson, M.

Bc520. **Margadant, W. D.** 1963. The Adanson collection of botanical books and manuscripts. In: G. H. M. Lawrence, ed. *Adanson: The bicentennial of Michel Adanson's "Familles des plantes,"* vol. 1: 265–368. Pittsburgh, PA, Hunt Botanical Library.

Barton, B. S.

Bc521. **Ewan, J.** 1986. One professor's chief joy: A catalog of books belonging to Benjamin Smith Barton. In: R. S. Klein, ed. *Science and society in early America: Essays in honor of Whitfield J. Bell, Jr.*, pp. 311–344. Philadelphia, PA, American Philosophical Society.

Bartram, W.

Bc522. **Peck, R. McC.** 1983. Books from the Bartram library. In: Society for the Bibliography of Natural History. *Contributions to the history of North American natural history: Papers from the first North American conference of the Society for the Bibliography of Natural History held at the Academy of Natural Sciences, Philadelphia, 21–23 October 1981*, pp. 46–50. London. *Special publication* no. 2. ••Some books in United States libraries with identified Bartram provenance.

Becher, K.

Bc523. **L'Art Ancien**, firm. 1925. *A catalogue of early herbals, mostly from the well-known library of Dr. Karl Becher, Karlsbad.* With an introduction by Dr. Arnold C. Klebs, Nyon. Lugano, L'Art Ancien S.A. xxiv+61 pp.

Bergius, B. & P. J.

Bc524. **Holmberg, A.** 1945. *I Bergianskabiblioteket representerade svenska bokägare. Alfabetisk förteckning jämte inledning.* Stockholm. xviii+46 pp. *K. Svenska Vetenskapsakademiens årsbok* 1945 Bil. ••Swedish book owners represented in the Bergius library.

Berkeley, M. J.

Bc525. **Wheldon, John**, firm. 1890. Catalogue no. 125. *Catalogue of botanical works.* Part I. *Fungology, including the library of the Rev. M. J. Berkeley.* London. 29 pp. ••411 items.

Boerhaave, H.

Bc526. **Luchtman, S.**, firm. 1739. *Bibliotheca Boerhaaviana, sive catalogus librorum instructissimæ bibliothecæ Hermanni Boerhaave, ... quorum publica fiet auctio ... [8–16 Jun.] 1739.* Leiden. [iv]+88+68 pp. ••Ca.3,311 books.

Bromel, O.

Bc527. **Bromel, Olaus.** 1694. *Catalogus librorum botanicorum ex bibliotheca Bromeliana.* Gothenberg. 10 ff. ••Pub. with his: *Chloris gothica seu catalogus stiurpium circa Gothoburgum nascentium*, Gothenberg, 1694.

Brown, R.

Bc528. **Stevens**, firm. 1859. *A catalogue of the scientific portion (with a few miscellaneous books) of the library, of the late Robert Brown, ... Which will be sold by auction ... on Monday, the 27th day of June, 1859, and the two following days.* London. 34 pp.

Buc'hoz, P.-J.

Bc529. **Bure, G. de.** 1778. *Catalogue des livres de la bibliothèque de M. [Pierre-Joseph] Buc'hoz, disposé et mis en ordre.* Paris, G. de Bure. xvi+212 pp. ••Catalogue of 1,786 lots with author index.

Buller, A. H. R.

See Buller Memorial Library, Canada Department of Agriculture, Bc428.

Earl of Bute.

See Stuart, J.

Caspary, R.

Bc530. **Raabe's Nachfolger, Ferd. (Eug. Heinrich)**, firm. 1888. Catalogue no. 82. *Botanik. Die Bibliothek des Herrn Professor Dr. Robert Caspary*. Königsberg. 121 pp. ••4010 items.

Clifford, G.

Bc531. **Linnaeus, Carl.** 1737. Bibliotheca botanica Cliffortiana. In: *Hortus Cliffortianus, plantas exhibens quas in hortis tam vivis quam siccis Hartecampi in Hollandia, coluit ... Georgius Clifford*, [19 pp.]. Amsterdam. ••295 items. For background information *see* **J. L. Heller.** 1968. "Bibliotheca Cliffortiana." In: Linnaeus's Hortus Cliffortianus. *Taxon* 17: 663–719.

Cavendish family, Dukes of Devonshire.

Bc532. **Pearman, M. A.** [1987.] *Great flower books at Chatsworth: A bibliography*. By M. A. Pearman. Chatsworth, Derbyshire, privately pub. 32 pp.

Danty d'Isnard, A.

Bc533. **Martin, G.**, firm. 1744. *Catalogue des livres de feu M. [Antoine] Danty d'Isnard, Medicin, Ancien Professeur Royal des Plantes au Jardin du Roy; & de l'Académie Royale des Sciences. Dont la Vente se sera en détail le lundy 20. juillet 1744. & jours suivans [etc.]*. Paris, G. Martin. xi+368 pp. ••Catalogue of 3,842 lots, incl. 1,280 lots of natural history.

Davall, E.

Bc534. **Leigh, Sotheby, & Son**, firm. 1803. *A catalogue of the entire and very valuable library of natural history formed by ... Edmund Davall ... Which will be sold by auction ... on Friday, June 17, 1803, and two following days*. London. [ii]+19 pp. ••Davall's library auctioned in 568 + 2 bis lots.

De Belder, R.

Bc535. **Sothebys**, firm. 1987. *A magnificent collection of botanical books being the finest colour-plate books from the celebrated library formed by Robert De Belder*. [Sale cat.] London, Sotheby's. 428 pp. ••388 lots with numerous illus. (some col.).

Deby, J.

Bc536. **Deby, J., & F. Kitton.** 1882. *A bibliography of the microscope and micrographic studies, being a catalogue of books and papers in the library of J. Deby ...* Part III. *The Diatomaceae*. Compiled with the co-operation of F. Kitton. London, Bogue. [vi] +67 p. ••*Also* Part I. Microscopy, Bc492.

Decaisne, J.

Bc537. **Vesque, J.** 1883. *Catalogue de la bibliothèque de feu M. J. Decaisne ... Botanique, horticulture, floriculture, agriculture, sciences naturelles et physiques, ouvrages divers, classé par M. J. Vesque*. Avec une notice biographique par ... Ed. Bornet. Paris, Vve A. Labitte. [viii]+xxi+485 pp. ••Classified catalogue of 3,264 lots, incl: 1. Botanique générale. Botanique descriptive, Flores; 2. Horticulture. Agriculture; 3. Sciences naturelles (autres que la botanique) et physiques. No index.

Delessert, B.

Bc538. **Lasègue, A.** 1845. Bibliothèque botanique. In: *Musée botanique de M. Benjamin Delessert*, pp. 509–552. Paris & Leipzig. vi+588 pp. *Re-pub.* (with an introduction by F. A. Stafleu) Lehre, Cramer, 1970. *Historiae naturalis classica* vol. 81. ••A narrative account of the collection.

Devonshire, Dukes of

See Cavendish family, Dukes of Devonshire.

Fabricius, P. K.

Bc539. **Fabricius, P. K.** 1776. Supellex mea libraria botanica. In: *Enumeratio methodica plantarum horti medici Helmstadiensis, [etc.]*, ed. tertia, [22 pp.]. Helmstadt, Drimborn. ••A list of ca.200 botanical works printed with the prelims.

Flatt von Alföld, K.

Bc540. **Flatt, C. von.** 1892. *Bibliotheca botanica. Verzeichniss der botanischen Bibliothek [C. von] Flatts*. Székesfehérvárott. 26 pp.

Bc541. **Weg, Max**, firm. 1908. (Issued in June. 1908.) Catalogue no. 112. *Bibliotheca botanica. Supplementum I. Hierin die Bibliothek des Linné–forschers Carl von Flatt–Budapest.* Leipzig. 42 pp. ••1275 items.

Fountaine, A.

Bc542. **Sotheby & Co.**, firm. 1971. *Catalogue of mycological books and drawings, the property of Commander A. Fountaine, which will be sold by auction by Sotheby & Co. ... 2nd June 1971.* London. 32 pp., illus. ••Catalogue of 179 lots, especially strong in fine illustrated works.

Fry, C. S.

See Taylor Institution, Oxford, Bc410.

Gaertner, J. & K. F.

Bc543. **Moser, A.**, firm. 1862. Catalogue no. 20. *Verzeichniss einer naturwissenschaftlichen Bibliothek.* Tübingen, Druck von Laupp. 1 f.+94 pp. ••Sale catalogue of library of Joseph Gaertner & his son Karl F. Gärtner.

Gessner, J.

Bc544. **Füssli, J. H.**, firm. 1798. *Catalogus librorum bibliothecae Joannis Gessneri, qui venales prostant apud J. H. Füssli filium.* Zurich, J. H. Fuessli filium. 133 pp.

Goodyer, J.

Bc545. **Gunther, R. T.** 1922. Catalogue of the [John] Goodyer library. In: *Early British botanists and their gardens, based on unpublished writings of Goodyer, Tradescant, and others*, pp. 197–232. Oxford, privately published.

Holmskiold [*formerly* Holm], T.

Bc546. **Ekkard, F.** [1794.] *Catalogus bibliothecæ exquis. atque nitidissimae ... Theodori Holmskiold, quæ publica auctionis lege a. 1794 divendetur.* Copenhagen. 406 pp.

Hunger, F. W. T.

Bc547. **Internationaal Antiquariaat** (Menno Hertzberger), firm. 1951. *Early herbals from the library of De. F. W. T. Hunger, Vooirschoten. To be sold by auction November 7 1951 [etc.].* Amsterdam. 52 pp.; ___. 1952. *Herbals, early books on botany, history of sciences, etc. from the library of the late Dr. F. W. T. Hunger. To be sold by auction November 4 1952.* Amsterdam, International Antiquariaat (M. Hertzberger). 56 pp., illus.; ___. 1952. *Early books ... forming the final portion of the library of the late ..., 24–27 November 1952.* Amsterdam. 20 pp. ••Ca.700 lots. *Also* **International Botanical Congress (6th), Amsterdam.** 1935. *Catalogue of the exhibition of books ... [at] the Royal Zoological Institution "Natura Artis Magistra." Comprising: I. Botanical incunabula (Collection Dr. F. W. T. Hunger, Voorschoten); [etc.].* Amsterdam. 45 pp. "Botanical incunabula and post–incunabula," 45 items.

Hunt, R. M. McM.

See Carnegie Mellon University, Hunt Institute for Botanical Documentation, Bc368.

Insch, J.

See **Linnean Society of London**, Bc261.

Jordan, A.

Bc548. **Klincksieck, Paul**, firm. 1903. *Catalogue de la bibliothèque botanique de feu Alexis Jordan (Lyon 1814–1897) dont la vente aux enchères public aura lieu à Paris les 4, 5, 6, 7, 8 et 9 mai 1903 ...* Paris. 144 pp. ••1,098 lots.

Jussieu family.

Bc549. [**Decaisne, J.**] 1858. *Catalogue de la bibliothèque scientifique de MM. de Jussieu, dont la vente aura lieu le lundi 11 janvier 1858 et jours suivants [etc.].* Paris, H. Labitte. xv+464 pp. ••The collection of three generations of the family from Antoine (1686–1758) to Adrien H. L. (1797–1853). Classified catalogue of 4,069 lots, incl: 5. Sciences naturelles (#285–#3,305). No index.

Kelemen, P.

See New York Public Library, Bc389.

Kelly, H. A.

See University of Michigan, Bc431.

Kjellman.

Bc550. **Weg, Max**, firm. 1908. Catalogue no. 114. *Bibliotheca* *botanica*. Supplementum II. *Hierin die Bibliothek des Algologen Professor F. R. Kjellman Upsala*. Leipzig. 53 pp. ••2152 items.

Kraneveld, F. C.

See Veterinary Institute, Bogor, Bc432.

Krelage, E. H.

Bc551. **Internationaal Antiquariaat** (Menno Hertzberger), firm. 1948. *Catalogue of the botanical library of Dr. E. H. Krelage, Haarlem ... which will be sold by auction on March 30th 1948 [etc.]*. Amsterdam, Internationaal Antiquariaat (Menno Hertzberger). [ii]+46 pp. ••337 lots, incl. "... the exceedingly scarce tulip books and pamphlets of the tulip mania in 1636–1637, the watercolourdrawings [sic] of Hyacinths, Tulips and other bulbous plants, chiefly from the unique collection of the renowned Haarlem florists' family of the Voorhelms ..."

Lambert, A. B.

Bc552. **Sotheby**, firm. 1842. *Catalogue of the valuable botanical library of the late Aylmer Bourke Lambert, which ... will be sold by auction ... on Monday April 18th [–20th], 1842*. London. 42 pp. ••689 lots.

L'Héritier, C. L.

Bc553. **Bure, G. de**, firm. 1805. *Catalogue des livres de la bibliothèque de feu C. L. L'Héritier de Brutelle ... La vente se sera le premier fructidor an XIII et jours suivans [etc.]*. Nouv. ed. Paris, G. Debure, J. G. Mérigot, Thierry. xxxii+359 pp. ••Classified catalogue of 2,957 lots with author and title index. Originally sold en bloc to Mérigot in 1802. The 1802 catalogue was re-issued for the 1805s sale with a new title page and the addition of a three page schedule of the sessions.

Maksimovich, C. J.

Bc554. **Koehler, K. F., Antiquarium**, firm. 1893. Catalogue no. 519. *Botanik. Hierin die Bibliothek des Herrn Dr. C. J. Maximowicz in St. Petersburg*. Leipzig. 76 pp. ••1,947 items.

Marshall, G. G.

See S. H. Johnston Jr., Bc189A and Cleveland Medical Library, Bc370.

Martius, C. F. P. von.

Bc555. **Martius, Carl F. P. von.** 1848. *Catalogue de la bibliothèque Américaine [etc.]*. Munich. 66 pp. ••478 items.

Bc556. **Weigel, T. O.**, firm. 1869. *Katalog der vom Geheimrathe Dr. C. F. Ph. von Martius hinterlassenen Bibliothek ... Catalogue de la bibliothèque de feu [etc.]*. Munich, F. Straub. 180 pp. ••3,180 lots.

Maximowicz, C. J

See Maksimovich, K. I.

Mellon, Mrs Paul.

Bc557. **Raphael, S.** 1989. *An Oak Spring sylva: A selection of the rare books on trees in the Oak Spring Garden Library*. Upperville, VA, Oak Spring Garden Library. xxvii+147 pp., illus. (some col.). ••46 works are described, illustrated and discussed.

Bc558. **___**. 1990. *An Oak Spring pomona: A selection of the rare books on trees in the Oak Spring Garden Library*. Upperville, VA, Oak Spring Garden Library. xxxv+276 pp., illus. (some col.). ••100 works are described, illustrated and discussed.

Meyer, E. H. F.

Bc559. **Weigel, T. O.**, firm. 1859. *Verzeichniss der hinterlassenen Bibliotheken des K. A. Varnhagen von Ense, sowie des Ernst Heinr. Friedrich Meyer, nebst mehreren anderen Büchersammlungen*. Leipzig. 1 f.+ii+326 pp.

Moquin-Tandon, C. H. B. A.

Bc560. **Labitte, A.**, firm. 1874. *Catalogue des livres composant la bibliothèque de ... M. Moquin-Tandon [etc.]*. Paris, A. Labitte. 88 pp. ••*Includes* 1 ptie. Sciences – botanique; 2. ptie. Patois françois.

Morren, C. J. E.

Bc561. **Brockhaus, F. A.**, firm. 1889. *Botanische Bibliothek von Carl und Eduard Morren in Lüttich*. Leipzig. 184 pp. ••4,816 items.

Pearson, A.

See Linnean Society of London, Bc261.

Petch, T.

Bc562. **Dawson**, firm. 1949. *A catalogue of rare and important works on mycology and cryptogamic botany from the library of the late Mr. Tom Petch [etc.]*. London. 40 pp.

Plesch, A.

Bc563. **Gourry, H.-P.** 1973. *Mille et un livres botaniques: Répertoire bibliographique de la Bibliothèque Arpad Plesch.* Brussels, Arcade. 517 pp., illus. (some col.). ••*Replaces* **Bibliothèque Arpad Plesch "La Leonina."** 1954. I. *Botanique [catalogue]*. (Établi par J. Pley). Monte-Carlo. 257 pp.

Bc564. **Sotheby** (*later* **Sotheby Parke Bernet**) **& Co.**, firm. 1975–76. *The magnificent botanical library of the Stiftung für Botanik, Vaduz, Liechtenstein, collected by the late Arpad Plesch. ... sale June, 1975 [and] November, 1975 [and] March, 1976 [etc.]*. [By J. Collins.] London. 3 vols., illus. ••862 lots, index of provenance.

Prantl, K.

Bc565. **Jordan, Richard**, firm. 1893. Catalogue no. 1893 Nos. 3–5. *Botanik. Die Bibliothek des Herrn Professor Dr. Karl Prantl.* Munich. ••Abth. I (Cat. 3): *Annales et Acta. Scripta miscellanea. Phaerogamae*; Abth. II (Cat. 4): *Florae. Geographia plantarum*; Abth. III (Cat. 5): *Cryptogamae. Anatomia et physiologia plantarum.*

Schliemann, J.

Bc566. **Uellner, W.** 1976. *Fungorum libri bibliothecae Joachim Schliemann. Books and prints of four centuries [etc.]*. W. Uellner, ed. Vaduz, Cramer. 287 pp. ••Author catalogue of 1995 items. *Replaces* ___. 1970. *Catalogue bibliographicus librorum botanicorum praesertim mycologicorum quos collegit Joachim Schliemann.* Hamburg, J. Schliemann. [ii]+164 pp.

Smith, J. D.

See Smithsonian Institution, United States National Museum, Bc403.

Sprengel, C.

Bc567. **Deichmann, C.**, firm. 1834. *Verzeichniss der von dem allhier verstorbenen Hrn. Prof. Dr. Curt Sprengel ... nachgelassenen ausgezeichneten Bibliothek; aus allen Fächern der Wissenschaften, ganz besonders aber aus der Botanik, Medicin und Naturwissenschaft ..., welche am 8ten Juli d. J. Nachm. von 2–6 Uhr und folgenden Tagen in dem auf dem grossen Berlin sub N. 334 belegenen Auctionslocale gegen gleich baare Bezahlung öffentlich versteigert werden soll.* Halle. xviii+144 pp.

Stuart, J., 3rd Earl of Bute.

Bc568. **Leigh & Sotheby**, firm. 1794. *A catalogue of the botanical and natural history part of ther library of the late John, Earl of Bute ... which will be sold by auction ... on Thursday, May 8, 1794, and the following days [etc.]*. London. [iv]+64 pp. ••The Earl of Bute's library of books and drawings auctioned in 1,257 + 28 bis lots. Included drawings by Van Rozin, Edwards, De Heer, Withoos, Ehret, Merian, Jeweel, Van Huysum, Taylor, Ditchie, King, Marcous and Moninck (lots. #981–1246.

Bc569. **King**, firm. 1801. *[Auction sale catalogue, 25 April 1801.]* London. 6 pp. ••The Earl of Bute's collection of botanical and shell drawings auctioned in 73 lots.

Sturtevant, E. L.

See Missouri Botanical Garden, Bc386.

Sutherland, J.

Bc569A. **Kelly, W. A.** 1987. Catalogue of James Sutherland's library. *Bibliotheck* 14: 30–106. 303 entries. *See also* **M. V. Mathew.** James Sutherland (1638(?)–1719): Botanist, numismatist and bibliophile, *Ibid*: 1–29.

Symons, G. J.

See Jackson, B. D., Cb564.

Trail, J. W. H.

See Aberdeen University, Bc360.

Trew, C. J.

Bc570. **Trew, C. J.** 1752. *Librorum botanicorum catalogi duo quorum* *prior recentiores quosdam posterior plerosque antiquos ad annum MDL usque excusos ad ductum propriae collectionis breviter recenset.* Nuremberg. 56 pp.; ___. 1757. *Librorum ... catalogus tertius inquo recentiores quosdam ad ductum propriae collectionis [etc.].* Nuremberg. 26 pp.

Ventenat, É. P.

Bc571. **Tilliard Frères**, firm. 1808. *Catalogue des livres de la bibliothèque de feu M. É. P. Ventenat. Suivi de la description de différens herbiers, graines, fruits étrangers, etc. et objets de curiosité.* Paris, Tilliard Frères. xii+83 pp. ••612 lots.

Vogler, H.

Bc572. **Vogler, H.** 1818. *Verzeichniss seiner auserlesnen Sammlung botanischer Werke, auch solcher, welche den Gartenbau, die Obstbaumzucht und Forstwissenschaft betreffen.* Braunschweig, Waisenhausdruckerei. viii+144 pp.

Walpers, G. W.

Bc573. **Friedländer, R. & Sohn**, firm. 1854. *G. W. Walpers' botanische Bibliothek welche am 1 Jun. 1854 ... öffentlich versteigert werden soll.* Berlin, Druck von Trowitsch u. Sohn. 63 pp.+2 ff. ••2,100 lots.

Willett, R.

Bc574. **Leigh & Sotheby**, firm. 1813. *A catalogue of the well known and celebrated library of the late Ralph Willett, Esq., brought from his Seat at Merly in the County of Dorset. Comprising ... a ... selection of botanical drawings by Van Huysun [sic], Taylor, Brown, Lee, ... which will be sold by auction, by Leigh and Sotheby ... on ... December 6, 1813, and 16 following days [etc.].* London. [iv]+119 pp. ••The botanical drawings formed the second part of the sale, the first part consisted of printed books.

Zoological

Audouin, J. V.

Bc575. [**Audouin, J. V.**] 1842. *Catalogue des livres d'histoire naturelle et principalement d'entomologie, composant la bibliothèque de feu M. Victor Audouin.* Paris, Merlin. 169 pp.

Ayer, E. E.

See Field Museum of Natural History, Bc451.

Blainville, H. M. D.

Bc576. **Labitte, H.**, firm. 1850. *Catalogue des livres de géologie, botanique, zoologie, médecine, anatomie, physiologie ... de feu M. H. D. de Blainville [etc.].* Paris. 148 pp.

Bonnier, J. J.

Bc577. **Hermann, A.**, firm. 1908. *Catalogues des ouvrages d'histoire naturelle composant la bibliothèque de feu ... Sous–Director du Laboratoire d'Evolution des Etres Organisés et de la Station Zoologique de Wimereux [etc.].* Paris. 18 pp. ••377 lots.

Broderip, W. J.

Bc578. **Leigh, Sotheby, & Wilkinson**, firm. 1859 *Catalogue of the very valuable library of the late W. J. Broderip, [etc.].* London. 2+53 pp. ••702 + 3 bis lots.

Brünnich, M. T.

Bc579. **Brünnich, M. T.** 1793. *Catalogus bibliothecae historiæ naturalis, libris præsertim ad regnum animale et lapideum spectantibus instructæ, qui auctione publica 1793 vendentur.* Copenhagen. 78 pp.

Buffon, G. L. L.

Bc580. **Rozet & Belin**, firm. 1791. *Catalogue des livres de la bibliothèque de feu M. Mirabeau l'ainé, ... , dont la vente de fera ... le lundi 9 Janvier 1792 et jours suivans [etc.].* Paris. xxvi+440[=434] pp. ••Following Buffon's death his library was sold to Honoré Gabriel Riquetti, Comte de Mirabeau. 2,854 lots including Buffon's library (see pp. 160–212).

Coe, W. R.

See Yale University Library, Bc462.

Cole, F. J.

See University of Reading, Cole Library of Early Medicine & Zoology, Bc448.

Dahl, F.

Bc581. **Dultz, A.**, firm. 1930. *Arthropoda. Alte und neue Werke zur*

Morphologie, Biologie, ... Hierin die Bibliothek des Herrn Professors Dr. F. Dahl, wieland Abetilungsleiter am Zoologischen Museum in Berlin. Munich. 48 pp. ••1,368 items.

Day, F.

See Cheltenham Public Library, Bc463.

Démidoff, P. de.

Bc582. **Fischer von Waldheim, G.** 1806. *Catalogue systématique des livres de la bibliothèque de Paul de Démidoff (à Moscou).* Moscow. lxxiii+275 pp. *Muséum–Démidoff* vol. 1.

Ellis, R. N.

See University of Kansas Libraries, Ellis Collection, Bc461.

Enders, D. F.

See Trinity College Library, Hartford, Bc460.

Evans, E. M.

Bc583. **Parke–Bernet**, firm. 1955. *The important ornithological library formed by the late Dr. Evan Morton Evans.* New York. 105 pp. ••540 lots.

Eschricht, D. F.

Bc584. **List & Francke**, firm. 1863. Catalogue no. 13. *Verzeichnis der Bibliothek des Daniel Fried. Eschricht.* Th. I: *Werke aus allen Zwiegen der Naturwissenschaften.* Leipzig. viii+158 pp.; and: ___. 1864. Catalogue no. 14. ___ Th. II: *Medizinische Wissenschaften.* Leipzig. 73 pp. ••4,566 & 2,629 lots.

Gallatin, F.

Bc585. **Gallatin, D, F., Jr.** 1908. *Catalogue of a collection of books on ornithology in the library of Frederic Gallatin, jr.* New York, privately pub. 179 pp. ••Ca.1,000 entries. The library was later auctioned: **Anderson Galleries.** 1921. *The Frederic Gallatin, Jr. collection of books on ornithology.* New York. 52 pp. 465 lots.

Geoffroy Saint–Hilaire, E.

Bc586. **[Geoffroy Saint–Hilaire, E.]** 1845. *Catalogue des livres de sciences, particulièrement de zoologie, d'anatomie comparée et d'anatomie philosophique, composant la bibliothèque de feu Étienne Geoffroy Saint–Hilaire.* Paris, Delion. 130 pp.

Goldfuss, A.

Bc587. **[Goldfuss, A.]** 1850. *Auctions–Catalog einer ausgezeichneten zoologischen und petrefactologischen Bücher-Sammlung, enthaltend die Bibliothek des verst. A. Goldfuss.* Berlin, gedr. bei Trowitzsch & Sohn. 1 f.+86 pp.

Gould, J.

Bc588. **Puttick & Simpson**, firm. 1881. *Catalogue of the ornithological library of the late John Gould Esq. ... Which will be sold by auction ... on Wednesday, May 4th 1881 [etc.].* London. 2+18 pp. ••301 lots.

Graff, L. von.

Bc589. **Graff, L. von.** 1891. *Bibliothek des Professors de Zoologie und vergl. Anatomie Dr. Ludwig von Graff in Graz.* [Leipzig.] xv+337 pp. ••12,235 entries.

Hoeven, J. van der.

Bc590. **Brill, E. J.**, firm. 1869. *Catalogue de la Bibliothèque délaissé par Mr. J. van der Hoeven [etc.].* Leiden. 215 pp. ••Zoological & palaeontological library.

Kölliker, R. A. von.

Bc591. **Fock, G.**, firm. [1907.] Catalogue no. 328. *Bibliothek A. v Kölliker: normale und vergleichende Anatomie und Physiologie, Entwicklungsgeschichte, Anthropologie, Zoologie.* Leipzig. 1,201 pp. ••5,402 items.

Le Conte, J. L.

Bc592. **Henkels, Stan V.**, firm. 1884. Catalogue no. 133. *Catalogue of the valuable entomological library of the late John L. Le Conte, M.D. of Philadelphia. Being the most important library on entomology ever offered at public sale either in Europe or America. To be sold ... May 6th, 7th and 8th 1884.* Philadelphia, PA, Stan V. Henkels. 52 pp. ••737 lots.

Martin, H. Bradley.

Bc593. **Sotheby's**, firm. 1989. *The library of H. Bradley Martin.* New York. 9 vols. ••Parts 1–3 & 5 contained the ornithological portion: Pt. 1. *John James Audubon: Magnificent books and manuscripts.* Auction June 6. 38 lots; Pt. 2. *Magnificent color–plate ornithology.* Auction June 7. Lots 39–242 lots; Pt. 3. *The original watercolors for Selby's Illustrations of British Ornithology.* Auction, June 8. Lots 243–519; Pt. 5. *Highly important illustrated and scientific ornithology.* Auction Dec. 12 & 13. Lots 1323–1977. *Also* **Antiquariaat Julius Steiner–Asher/Maggs Bros.**, [Firms]. 1991. *Books on palaearctic ornithology from the Bradley Martin collection with additions.* Amsterdam, London. 136 pp. 1258 items.

Mathews, G. M.

See National Library of Australia, Mathews Ornithological Collection, Bc455.

Menke, K. T.

Bc594. **List & Francke**, firm. 1862. Catalogue no. 4. *Verzeichniss einer ausgewählten Sammlung von Werken aus allen Zweigen der Naturwissenschaften aus dem Nachlass des K. Th. Menke ... in Pyrmont.* Leipzig, druck von Melzer. 1 f.+64 pp.

Milne–Edwards, A.

Bc595. **Deyrolle, Les Fils d'Émile**, firm. 1901–04. *Premier (–quatrième et dernier) catalogue des livres d'histoire naturelle de la bibliothèque de feu Alphonse Milne–Edwards ... dont la vente aux enchères publiques ... 1901 (–1904).* Paris. 4 vols. ••3,359 lots, chiefly zoological literature.

Mullens, W. H.

Bc596. **Mullens, W. H.** 1911. *A list of books relating to British birds, published before ... 1815. From the library of W. H. Mullens ... exhibited at the Corporation Museum Hastings, during June 1906 ... Reprinted with additions and corrections.* Hastings. 34 pp. & 7 pl. *Occasional publication, Hastings & St. Leonards Natural History Society* no. 3.

Pennant, T.

Bc597. **Christie, Manson & Woods**, firm. 1938. *Catalogue of the valuable library of books, manuscripts, drawings and engravings sold by order of the executives of the Right Hon. Viscount Fielding ... Including the famous collections of Thomas Pennant (at one time forming a portion of the library at Downing, Flintshire) ... sold by auction ... 4–5 July, 1938.* London. 59 pp. ••342 lots, chiefly Pennant's copies of his own works, some interleaved and annotated.

Radley, P. E.

See Malacological Society, Bc478.

Russell, G. W.

See Trinity College Library, Hartford, Bc458.

Shipley, A. E.

Bc598. **Cutling & Son**, firm. 1927 *[Library sale catalogue ... 29th Nov. 1927.]* Cambridge. 27 pp. ••612 lots.

Sinkler, L. E.

See Library Company of Philadelphia, Bc453.

Smith, G. M.

See Yale University Library, Bc181A.

Stephens, J. F.

Bc599. **Stainton, H. T.** 1853. *Bibliotheca Stephensiana; being a catalogue of the entomological library of ... J. F. Stephens, [etc.].* [With a memoir.] London, privayely pub. 48 pp.

Thayer, J. E.

See Harvard University, Bc451A.

Bc600. [Not used.]

Thienemann, F. A. L.

Bc601. **Wiegel, T. O.**, firm. 1858. *Verzeichniss der hinterlassenen Bibliothek des berühmten Ornithologen F. A. L. Thienemann, nebst mehreren andered bedeutenden Büchersammlungen sowie einem Anhange von Musikalien.* Leipzig. 2 ff.+304 pp.

Vrolik, W.

Bc602. **[Vrolik, W.]** 1865. *Catalogue de la superbe bibliothèque d'ethnographie, de zoologie, d'anatomie comparée, etc. formée par Mr. W. Vrolik. La vente se fera ... 1865, [etc.].* Amsterdam. xii+105 pp.

BIBLIOGRAPHIES OF PERSONAL LIBRARY & BOOK-SALE CATALOGUES

Bc603. **Bay, J. C.** 1910. "Auction and Sales Catalogues of Private Botanical Libraries," and "Booksellers' Catalogues and Auction Catalogues (Trade Lists)." In: Bibliographies of botany. A contribution toward a bibliotheca bibliographica. *Progressus rei botanicae* 3: 442–446, 446–455.

Bc604. **Bibliothèque Royale Albert 1er.** 1982+. *Repertoire des catalogues de ventes de livres imprimés.* By J. Blogie. [Antwerp?], Fl. Tulkens. Vol. 1+. In progress. ••Vol. 1: *Catalogues Belges appartenant a la Bibliothèqie Royale Albert 1er.* 1982. ix pp. +890 cols. Covers 1636–1980; Vol. 2: *Catalogues Français [etc.].* 1985. vi pp. +985 cols. Covers 1643–1980; Vol. 3: *Catalogues Britanniques [etc.].* 1988. vi pp. +1018 cols. Covers 1694–1980; Vol. 4: *Catalogues Néerlandais [etc.].* 1992. vii pp. +570 cols. Covers 1599–1980; etc.

Bc605. **British Museum.** 1915. *List of catalogues of English book sales, 1676–1900, now in the British Museum.* London. xv+523 pp. ••8,000 items chronologically arranged with a name index.

Bc606. **Hunt, J. M. Chalmers–.** 1976. *Natural history auctions, 1700–1972.* London, Sotheby Parke Bernet. xii+189 pp.

Bc607. **Laissus, Y.** 1964. "Description de quelques catalogues de livres et de curiosités d'histoire naturelle intéressant la France du XVIIIe siècle [chronologically arranged]" In: R. Taton, ed. *Enseignement et diffusion des sciences en France au XVIIIe siècle,* pp. 671–679. Paris, Hermann. *Histoire de la pensée* vol. 11.

Bc608. **McKay, G. L.** 1935–36. American book auction catalogues 1713–1934. *Bulletin of the New York Public Library* 39 & 40; *Continued* ___. 1946. ___: Additions. *Ibid.* 50: 177–184; ___. 1948. ___: Supplement no. 2. *Ibid.* 52: 401–412.

Bc609. **Petzholdt, J.** 1866. *Bibliotheca bibliographica: Kritisches Verzeichniss der das Gesammtgebiet der Bibliographie betreffenden Litteratur des In– und Auslandes in systematischer Ordnung. Mit alphabetischem Namen– und Sachregister.* Leipzig, W. Engelmann. xiii+939 pp. Reprinted Nieuwkoop, De Graaf, 1961.

Bc610. **Pollard, H. G., & A. Ehrman.** 1965. *The distribution of books by catalogue from the invention of printing to A.D. 1800, based on material in the Broxbourne library.* Cambridge, for presentation to the Roxburghe Club. xxiii+426 pp. ••*Includes* XI. Auction sales; XII. Institutional library catalogues; XIII. The private library open to scholars; XIV. The personal library.

Bc611. **Taylor, A.** 1987. *Book catalogues: Their varieties and uses.* 2nd ed. Revised by W. P. Barlow. Winchester, St. Paul's Bibliographies. xxxvii+284 pp. ••Reproduces the 1st ed., Chicago, Il, 1957. xii+284 pp. with an introductory account of corrigenda and some addenda. *Includes* A list of catalogues of private libraries that have been recommended for reference use, pp. 223–266.

Bc612. **Vereeniging ter Bevordering van de Belangen des Boekhandels te Amsterdam.** 1920+. *Catalogus der bibliotheek van de [etc.].* Vol. 1+. The Hague, M. Nijhoff; (vol. 8 Amsterdam, Universiteitsbibliotheek). In progress (vol. 8 appeared in 1979.) ••Vols. 4 & 8 are entirely devoted to bookseller's and auction catalogues.

Bc613. **Wells, E. B.** 1983. Scientists' libraries: A handlist of printed sources. *Annals of science* 40: 317–389. ••1,200 titles relating to the libraries of about 880 scientists.

PERIODICALS CATALOGUES

Union catalogues

• From a large number of local, regional and national union catalogues in existence, these few titles will serve as examples of valuable finding aids for older natural–history periodicals:

Bc614. **Bibliothèque Nationale, Paris, Département des Périodiques.** 1967–77. *Catalogue collectif des périodiques du début du XVIIe siècle à 1939, conservées dans les bibliothèques de Paris et dans les bibliothèques universitaires des départements.* Paris. 4 vols.; *Continued* ___. 1982. ___*: Additions et corrections, table des collectivités citées.* Paris. 978 pp.
Holdings of over 75,000 titles in 73 libraries.

Bc615. **Brown, P., & G. B. Stratton.** 1963–65. *World list of scientific periodicals published in the years 1900–1960.* 4th ed. London, Butterworth. 3 vols
Over 60,000 titles in many British libraries; excludes periodicals which terminated before 1900.

Bc616. **Commonwealth Scientific, Industrial & Research Organisation (CSIRO), Australia.** 1975. *Scientific serials in Australian libraries.* Ed. by J. E. Conochie. Melbourne, Vic. 3 vols.
Includes ca.60,000 periodicals in 180 libraries.

Bc617. **Stewart, J.**, ed. 1955–60. *British union catalogue of periodicals: a record of the periodicals of the world from the seventeenth century to the present day in British libraries.* London, Butterworth. 4 vols.
Over 140,000 titles held by 441 libraries.

Bc618. **Titus, E. B.**, ed. 1965. *Union list of serials in libraries of the United States and Canada.* 3rd ed. New York, Wilson. 5 vols.
Over 156,000 titles held by 956 libraries listing titles down to 1960.

Individual catalogues

General

Bc619. [**British Library** *formerly*] **British Museum.** 1963. Periodical publications. In: *General catalogue of printed books to 1955,* vols.184–186. London.
Arranged by place of publication. *Also* microprint edition: ___. [1963] 1967. *General catalogue of printed books to 1955.* Compact ed. Vol.19, pp. 833–1177. New York, Readex Microprint Corporation.

Bc620. **British Library, Document Supply Centre.** 1980+. *Keyword index to serial titles.* Boston Spa. Microfiche.
Quarterly publication. Based on a master list of over 400,000 non-current and current titles; includes serials in the British Library Science Reference & Information Service, Science Museum Library and Cambridge University Library.

Bc621. **New York Public Library.** 1942. *A check list of cumulative indexes to individual periodicals in the New York Public Library.* By D. C. Haskell. New York. 370 pp. *Re-pub.* Detroit, MI, Gale Research Co., 1969.
Ca.6,000 titles.

Natural history

Bc622. **British Museum (Natural History).** 1975. *List of serial publications in the British Museum (Natural History) Library.* 2nd ed. London. 3 vols.
Alphabetical list of ca.17,000 titles. From 1975 onwards various revised eds. have been pub. in microfiche.

Bc622A. **Deutsche Akademie der Wissenschaften, Berlin, Hauptbibliothek.** 1970. *Bestandsverzeichnis der in den Bibliotheken der Deutschen Akademie der Wissenschaften zu Berlin vorhandenen Zeitschriften (BVZ).* Berlin. 2 vols in 3.

Bc623. **Kaiser Leopoldinisch–Carolinische Deutsche Akademie der Naturforscher.** 1937. *Verzeichnis der periodischen Schriften der Bibliothek der Kaiserlich Leopoldinischen–Carolinisch Deutschen Akademie der Naturforscher sitz in Halle.* Bearb. von R. Rauch. Halle. 186 pp.

Bc624. **Koninklijke Nederlandsche Akademie van Wetenschappen, Amsterdam.** 1938. *Catalogus van tijdschriften en serienwerken.* By R. O. Spanier. Amsterdam. [v]+372 pp.
Ca.4,000 titles. *Replaces* **Koninklijke Akademie van Wetenschappen, Amsterdam.** 1855. *Catalogus der boekerij, [etc.].* Eerste aflevering. By W. Vrolik. Amsterdam. viii+83 pp.

Bc625. **Linda Hall Library.** 1989. *Serials holdings in the Linda Hall Library 1989.* Kansas City, MO. 800 pp.
Ca.24,000 titles. *Replaces* ___. 1983. ___ *1983.* Kansas City, MO. ix+739 pp.

Bc626. **Royal Society of London.** 1912. *Catalogue of the periodical publications in the library of the Royal Society of London.* [By L. Newcombe & L. Ellston.] London. viii+455 pp.
An exceptionally well–prepared catalogue of the periodical collection from which the Royal Society *Catalogue of scientific papers* was compiled. *Replaces* ___. 1881. *Catalogue of the scientific books in the library of the Royal Society.* Pt. I. *Transactions – journals – observations – and reports – surveys – museums.* London. [iii]+262 pp.

Bc627. **Smithsonian Institution.** 1866. *Catalogue of publications of societies and of periodical works belonging to the Smithsonian Institution, January 1, 1866, deposited in the Library of Congress.* Washington, DC. 591 pp. *Smithsonian miscellaneous collections* vol. 9(2).
Replaces ___. 1859. ___ *1858.* Washington, DC. 259 pp. *Smithsonian miscellaneous collections* vol. 3(4).

Botanical

Bc628. **Botanical Survey of India.** 1976. *Union catalogue of serials in the libraries of the Botanical Survey of India.* By D. G. Dutton & others. Howrah. 179 pp.

Bc629. **British Library, Science Reference Library.** 1977. *Periodicals on botany held by the SRL.* London. 49 pp.
Ca.700 titles with an appended list of 38 abstracting and indexing services.

Bc630. **Canada Department of Agriculture.** 1977. *Union list of serials in Canada Department of Agriculture libraries.* 2nd ed. Ottawa. v+745 pp.
30,000 entries for periodicals held in 25 libraries.

Bc631. **Commonwealth Forestry Institute, Oxford (CFI).** 1968. *List of periodicals in the Commonwealth Forestry Institute.* 3rd ed., rev. and ed. by J. S. Howse. Oxford. xvii+187 pp.

Bc632. **Conservatoire et Jardin Botanique de Génève.** 1980. *Catalogue des périodiques de la bibliothèque des Conservatoire et Jardin Botaniques de Génève.* Compilé par H. M. Burdet [et al.]. Geneva. vii+271 pp.; *Continued* ___. 1988. *Catalogue des périodiques de la Bibliothèque des Conservatoire et Jardin Botaniques de la Ville de Genève. Supplément 1980–1987.* Genève. 107pp.

Bc633. **Harvard University, Library of Arnold Arboretum.** 1914. Periodicals, reports, and other serial publications. In: *Catalogue of the library*, Vol. 1, pp. 1–26. Cambridge, MA. *Continued* ___. 1917. Serial publications. In: *Catalogue of the library*

[Supplement], Vol. 2, pp. i–iii. Cambridge, MA; ___. 1933. Periodicals, reports, and other serial publications. In: *Catalogue of the library [Supplement, 1917–33]*, Vol. 3, pp. 1–16. Cambridge, MA.

Bc634. **Jardin Botanique de l'État à Bruxelles.** 1911. Catalogue de la bibliothèque collective réunie au Jardin Botanique de l'État à Bruxelles. 1: Publications périodiques ou occasionelles d'académies, de jardins, et d'instituts botaniques et de sociétés savantes. Par P. van Aerschot. *Bulletin, Jardin Botanique de l'État, Bruxelles* 3: 252+xxxiii pp.
Ca.1,750 titles, geographically arranged with an index of titles and editors and a subject index.

Bc635. **Lloyd Library, Cincinnati.** 1936. Catalogue of the periodical literature in the Lloyd Library. By W. H. Aiken & S. Waldbott. *Bulletin of the Lloyd Library and Museum* 34: 1–103.
Ca.2,500 titles. Earlier eds., 1911, 79 pp.; [1914], 123 pp.

Bc636. **London University, Library.** 1954. *Botanical periodicals in London libraries.* London. 25 pp.
Lists 471 titles in 32 libraries.

Bc637. **Rijksuniversiteit, Utrecht, Bibliotheek.** 1935. *Lijst von botanische tijdschriften in de universiteitsbibliotheek, in het Botanisch laboratorium "Willie Commelin Scholten" (te Baarn), zoomede op enkele andere plaatsen te Utrecht.* Utrecht, Rijksuniversiteit. [ii]+38 pp.
Ca.400 titles.

Bc638. **Royal Botanic Garden, Edinburgh.** 1969. *Periodicals list.* Edinburgh. ii+108 pp.

Bc639. **Royal Botanic Gardens, Kew.** 1978. *List of periodical publications in the library.* London, Her Majesty's Stationery Office. iii+90 pp.
Over 2,000 titles. *Replaces* Earlier eds., 1968 & 1971.

Bc640. **United States Department of Agriculture, Forest Service Library.** [1936]. *A union checklist of forestry serials.* Washington, DC. 146 pp.

Zoological

Bc641. **British Museum (Natural History).** 1967. *List of serial publications in the libraries of the Departments of Zoology and Entomology.* London. 281 pp.

Bc642. **Polska Akademija Nauk, Instytut Zoologiczny.** 1958. *Wydawnictwa ciagle w bibliotece Instytutu Zoologicznego Polskiej Akademii Nauk: Katalog i przypisy bibliograficzne. ... Periodicals and serial publications in the Library of the Institute of Zoology of the Polish Academy of Sciences: Catalogue and bibliographical notes.* By J. Roznowska–Feliksiakowa. Warsaw. 709 pp.
Alphabetical index of periodicals; appendix of multi–volume treatises, faunas, floras, etc.; index of corporate authors.

Bc643. **Zoological Society of London.** 1949. *List of the periodicals and serials in the library of the Zoological Society of London.* London. 186 pp.

BIBLIOGRAPHIES OF PERIODICALS CATALOGUES

Bc644. **Besterman, T.** 1971. *Periodical publications: A bibliography of bibliographies.* Totowa, NJ, Rowman & Littlefield, 1971. 2 vols. ••Extracted from the 4th ed. of his *World bibliography of bibliographies*, 1965–66.

Bc645. **Boalch, D. H.** 1961. Draft list of published catalogues of serials in agriculture. *Quarterly Bulletin, International Association of Agricultural Librarians* 6: 100–105; *Ibid.* 6: 229–237.

Bc646. **Fowler, M. J.** 1966. *Guides to scientific periodicals: Annotated bibliography*. London, Library Association. 318 pp. ••Limited to current periodical publications. Well annotated.

Bc647. **Freitag, R. S.** 1964. *Union lists of serials: A bibliography*. Washington, DC, Library of Congress, General Reference & Bibliography Division. xiii+150 pp. *Re-pub*. Boston, MA, Gregg Pr., 1972. *Library reference series; Basic reference sources*. [iv]+xiii+150 pp. ••Geographically arranged list of 1,218 items. *Replaces* **D. C. Haskell & K. Born.** 1943. A bibliography of union lists of serials. In: W. Gregory, ed. *Union list of serials in libraries of the United States and Canada*. 3rd ed. pp. 3053–3065. New York, Wilson.

Bc648. **Krehl, H.** 1966. Verzeichnis der deutschen wissenschaftlichen Zeitschriften. Printed catalogues of German learned periodicals. *Dokumentation, Fackbibliothek, Werksbücherei* 14: 91–92.

Bc649. **Kronick, D. A.** 1974. Studies of the early scientific journal. 1: The source lists. *Texas Reports on Biology & Medicine* 32: 61–74.

Bc650. **Maichel, K.** 1958. Bibliographical guides to Russian periodical publications, 1901–1956. *College & Research Libraries* 19: 209–218, 262. ••Includes lists published outside the Soviet Union. *See also* **E. Beyerly.** 1959. Lists of Soviet periodicals: A review. *Journal of Documentation* 15: 100–105.

MANUSCRIPT RESOURCES & CATALOGUES

(including drawings, theses, etc.)

Guides to repositories

Bc651. **Association of Canadian Archivists.** 1977. *Directory of Canadian records and manuscript repositories*. Ottawa. iii+115 pp.

Bc652. **Foster, J., & J. Sheppard.** 1989. *British archives: A guide to archive resources in the United Kingdom*. 2nd ed. London, Macmillan Publishers. lviii+834 pp.
1,048 entries. *Replaces* 1st ed., 1982. xxv+533 pp.

Bc653. **Grimsted, P. K.** 1972. *Archives and manuscript repositories in the USSR: Moscow and Leningrad*. Princeton, NJ, Princeton Univ. Pr. xxx+436 pp.; *Continued* ___. 1976. ___*: Bibliographical addenda*. Zug, Interdocumentation Co. 203 pp.

Bc654. ___. 1981. *Archives and manuscript repositories in the USSR: Estonia, Latvia, Lithuania, and Belorussia*. Princeton, NJ, Princeton Univ. Pr. 929 pp.

Bc655. ___. 1988+. *Archives and manuscript repositories in the USSR: Ukraine and Moldavia*. Princeton, NJ, Princeton Univ. Pr. Vol. 1+.

Bc656. **International Council on Archives.** 1990. International bibliography of directories and guides to archival repositories. *Archivum* 36: 240 pp.

Bc657. **National Historical Publications & Records Commissions.** 1988. *Directory of archives and manuscript repositories in the United States*. 2nd ed. Phoenix, NY, Oryx Pr. xv+853 pp.
4,225 entries. Includes references to NUCMC (*see* Bc706). *Replaces* ___. 1978. *Directory of archives and manuscript repositories in the United States*. Washington, DC, National Archives & Records Service, General Services Administration. 905 pp.

Bc658. **Society of American Archivists.** 1980. *Directory of college and university archives in the United States and Canada*. Chicago, IL. 80 pp.

Bc659. **Vanrie, A.**, ed. 1988. International directory of archives. *Archivum* 33: 351 pp.

Bc660. **Weil, G. E.** 1978+. *The international directory of manuscripts collections, libraries, private collections, repositories and archives.* Paris, Berger–Levrault.
Vol. 1, pt. 1: Europe: The manuscript collections. 1978. 301 pp.; all published to date.

Surveys of resources

Bc661. **Alton, J.**, ed. 1979. *The catalogues of the archives of scientists, compiled by the Contemporary Scientific Archives Centre, 1973–1978.* Oxford, Oxford Microfilm Publications. 17 microfiche & guide.; *Continued* ___. 1981. ___*: First supplement, 1979–1980.* Oxford, [etc.]; ___. 1985. ___*: Second supplement, 1981–1984.* Oxford, [etc.]
Information on the papers of modern British scientists documented by the Centre and usually then deposited in an appropriate library. In later years the Contemporary Scientific Archives Centre (CSAC) moved from Oxford to Bath and changed its name to the National Cataloguing Unit for Archives of Contemporary Scientists (NCUACS).

Bc662. **Appleby, J. H.** 1986. A survey of some Anglo–Russian medical and natural history material in the British archives, from the 17th century to the beginning of the 19th century. In: J. M. Hartley, ed. *The study of Russian history from British archive sources,* pp. 107–131. London, Mansell.

Bc663. **Australian Science Archives Project.** [1991.] *The guide to the archives of science in Australia: Records of individuals.* Compiled by G. McCarthy. Port Melbourne, Vic., D. W. Thorpe. xi+291 pp.
Information from the ASAP's national database "The Register of the Archives of Science in Australia," available online through the OZLINE network from 1 January 1992.

Bc664. **Bearman, D., & J. T. Edsall**, eds. 1980. *Archival sources for the history of biochemistry and molecular biology: A reference guide and report.* Lists and indexes of archival collections by M. Miller. With a selected bibliography of published sources by M. Konopka. Boston, MA, American Academy of Arts and Sciences; Philadelphia, PA, American Philosophical Society. xii+338 pp. & 6 microfiche.
For background information *see* **J. T. Edsall & D. Bearman.** 1979. Historical records of scientific activity: The survey of sources for the history of biochemistry and molecular biology. *Proceedings of the American Philosophical Society* 123: 279–292.

Bc665. **Bell, W. J., Jr.** 1970. Rattlesnakes and hummingbirds: Philadelphia's resources for the history of science. *Papers of the Bibliographical Society of America* 64: 13–27.

Bc666. **Bridson, G. D. R., V. C. Phillips, & A. P. Harvey.** 1980. *Natural history manuscript resources in the British Isles.* London, Mansell; New York, R. R. Bowker. xxxiv+473 pp.
Arranged by place and repository, with name, place and subject indexes and an annotated bibliography of relevant literature. Excludes the Public Record Office; *see* Bc805 for information on that collection.

Bc667. **Chadwyck–Healey Inc.** 1983+. *The national inventory of documentary sources [NIDS] in the United States.* Alexandria, VA. Microfiche.
A compilation of "finding aids" to manuscript resources reproducing inventories of: Pt. 1. Federal records (1,911 fiche); pt. 2. the Manuscript Division of the Library of Congress (889 fiche); pt. 3. State archives, state libraries, state historical societies, academic libraries and other repositories, mostly hitherto unpublished. To be accompanied by a cumulative index (pt. 3 in progress). Pt. 4 will cover Academic libraries and other repositories.

Bc668. ___. 1991+. *The national inventory of documentary sources [NIDS] in Canada. Inventaire national des sources documentaires au Canada.* Alexandria, VA. Microfiche.
A compilation of "finding aids" to manuscript resources reproducing inventories of collections in the National Archives of Canada, National Library of Canada and several other major

university and research collections, mostly hitherto unpublished. To be published by units, each accompanied by a bilingual index.

Bc669. **Chadwyck Healey Ltd.** 1985+. *National inventory of documentary sources [NIDS] in the United Kingdom and Ireland.* Cambridge. Microfiche.
A compilation of "finding aids" to manuscript resources reproducing inventories of collections in over 60 repositories of every kind at national, university, county, and local level, mostly hitherto unpublished. Being published in units, each accompanied by a list of finding aids and with a cumulating Names and subjects index that is renewed each year.

Bc670. **Chadwyck–Healey France S.A.R.L.** 1989+. *Les inventaires des Archives Nationales de Paris.* Paris. Microfiche (ca.6,500).
An integrated compilation of "finding aids" to manuscript resources reproducing ca.750 inventories of collections in the Archives Nationales and many other public and private sources, many hitherto unpublished. Accompanied by a printed index and guide.

Bc671. **Collins, T., & A. Hamer.** 1981. Archive programs for zoological parks and aquariums. In: American Association of Zoological Parks & Aquariums (AAZPA). *Regional conference proceedings, 1981*, pp. 99–102.

Bc672. **Davis, R. C.** 1977. *North American forest history: A guide to archives and manuscripts in the United States and Canada.* Santa Barbara, CA, Clio Books for The Forest History Society, Inc. xxi+376 pp.
Surveys 3,830 groups of material in 108 archive repositories. Documents a much wider field of nature conservation than just forest history. *Replaces* **C. M. Neiderheiser.** 1956. *Forest history sources of the United States and Canada.* Saint Paul, MN, Forest History Foundation, Inc. xiii+140 pp.

Bc673. **Forbes, E. G.** 1979. Scientific archives in the Federal Republic of Germany. *Berichte zur Wissenschaftsgeschichte* 2: 197–201.

Bc674. **France – Ministère de l'Éducation Nationale.** 1849–85. *Catalogue générale des manuscrits des bibliothèques publiques des départements de France.* Paris. 7 vols. *Repub.* Westmead, Gregg International Publishers, 1968. *Continued* ___. 1885+. *Catalogue générale des manuscrits des bibliothèques publiques de France: Départements [etc.]. [New series.]* Paris, E. Plon, Nourrit & Cie. Vol. 1+.
Vol. 48 of the new series appeared in 1933.

Bc675. ___. 1892. *Catalogue des manuscrits conservés aux archives nationales.* Paris, E. Plon, Nourrit & Cie. ii+532 pp. *Catalogue général des manuscrits des bibliothèques publiques de France.*

Bc676. **___.** 1909–14. *Catalogue générale des manuscrits des bibliothèques publiques de France.* Paris, E. Plon, Nourrit & Cie. 2 vols.
Vol. 1 incl. Faculté de Médecine, Académie de Médecine; École Supérieure de Pharmacie; etc. Vol. 2 incl. Muséum d'Histoire Naturelle; etc.

Bc677. **___.** 1918. *Catalogue générale des manuscrits des bibliothèques publiques de France. Université de Paris et universités des départements.* Paris, E. Plon, Nourrit & Cie. x+801 pp.

Bc678. **___.** 1928. *Catalogue générale des manuscrits des bibliothèques publiques de France. Paris. Bibliothèque de l'Institut. Ancien et nouveau fonds.* Paris, Plon & Nourrit. xv+758 pp.

Bc679. **___.** 1931+. *Catalogue générale des manuscrits des bibliothèques publiques de France. Société savantes [etc.].* Paris, Plon. Vol. 1+.

Bc680. **Great Britain – Royal Commission on Historical Manuscripts.** 1983. *The manuscript papers of British scientists 1600–1940.* Ed. by G. R. David. London. viii+109 pp. *Guides to sources for British history based on the National Register of Archives* vol. 2.

Bc681. **Gundry, L. D.** 1970. *Guide to science and history of science resources in the Philadelphia area.* Philadelphia, PA, American Philosophical Society. [iii]+63 pp. *Library publication* no. 3.

Bc682. **Hamer, P. M.**, ed. 1961. *A guide to archives and manuscripts in the United States.* Compiled for the National Historical Publications Commission. New Haven, CT, Yale Univ. Pr. xxiii+775 pp.
Still a basic source of information for resource searching.

Bc683. **Jameson, J.** 1980. The national park system in the United States: An overview with a survey of selected government documents and archival materials. *Government publications review* 7A(2): 145–158.

Bc684. **Kohlstedt, S. G.** 1984. Historical records in Australian museums of natural history. *Australian historical bibliography* 10: 61–82.

Bc685. **___.** 1985. Historical resources in natural history museums. *History of science in America news & views* 3 (4): 5–6.

Bc686. **Layton, E. T., Jr.**, ed. 1971. *A regional union catalog of manuscripts relating to the history of science and technology located in Indiana, Michigan, and Ohio.* Cleveland, OH, Case Western Reserve Univ. xv+215 pp. *Program in the History of Science & Technology, publication* no. 1.
1,177 entries; subject and name indexes.

Bc687. **Lurie, E.** 1953. Some manuscript resources in the history of nineteenth century American natural science. *Isis* 44: 363–370.

Bc687A. **McConnell, A.** 1990. *Directory of source materials for the history of oceanography.* Paris, Unesco. iii+40 pp. *Unesco technical paper in marine science* no. 58.

Bc688. **MacLeod, R. M., & J. R. Friday.** 1972. *Archives of British men of science.* London, Mansell. 40 pp. & microfiche; *Continued* **R. M. McLeod.** 1973. *___: Supplement to the first edition.* Brighton, Univ. of Sussex. 6 microfiche.
See also **R. M. MacLeod & J. R. Friday.** 1973. The quest for archives of British men of science. *History of science* 11: 8–20.

Bc689. **MacLeod, R. M., J. R. Friday, & C. Gregor.** 1975. *The corresponding societies of the British Association for the Advancement of Science 1883–1929: A survey of historical records, archives and publications.* London, Mansell. xxiv+147 pp.

Bc690. **Mount, E.**, ed. 1989. *Sci–tech archives and manuscript collections.* New York, Haworth Pr. x+144 pp. A re-issue of *Science & technology libraries* vol. 9, no. 4 as a separate book.
Articles by several authors. *Includes* **H. W. Samuels.** Sci–Tech archives and manuscripts: An overview, pp. 3–6; **L. Rohr.** A survey of American zoo and aquarium archives, pp. 75–84; **J. Warnow–Blewett.** The role of a discipline history center. Part II: Promoting archives and research in science and technology, pp. 85–104. References to some others are entered in the relevant sections herein.

Bc691. **Mozley, A.** 1966. *A guide to the manuscript records of Australian science.* Canberra, A.C.T., Australian Academy of Science in association with Australian National Univ. Pr. xxiv+127 pp.

Bc692. **Nelson, E. C.** 1990. "It's a long way to Tipperary" – finding natural history archives in Ireland, with an appendix listing archives. *Archives of natural history* 17: 325–347.
Appendix. Irish archives and their natural history materials, pp. 334–347, arranged alphabetically by place.

Bc693. **Nunn, G. R.**, ed. 1985. *Asia and Oceania: A guide to archival and manuscript sources in the United States.* ... with contributions from A. Freidus, W. Pierson and the

Center for Asian & Pacific Studies, University of Hawaii. London, Mansell Publishing. 5 vols. (2,456 pp.).
State by state survey with index.

Bc694. **Raimo, A. R., & H. J. Mayo.** 1978. A guide to history of medicine and science collections in the Baltimore–Washington area. *American Society for Microbiology news* 44: 458–463.

Bc695. **Rider, R. E., & H. E. Lowood.** 1985. *Guide to sources in Northern California for history of science and technology.* Berkeley, CA, Univ. of California, Office for History of Science & Technology. iv+194 pp. *Berkeley papers in history of science* no. 10.

Bc696. **Rohr, L.** 1989. A survey of American zoo and aquarium archives. *Science & technology libraries* (9) 4: 75–84.

Bc697. **Rookmaaker, L. C.** 1989. *The zoological exploration of southern Africa 1650–1790.* Rotterdam, Brookfield, VT, A. A. Balkema. xii+368 pp.
Includes descriptions and lists of all known manuscript sources in South African, Dutch, British, German, and American libraries, including the H. Claudius drawings, pp. 22–26, the J. G. A. Forster drawings and associated manuscripts, pp. 50–58, the R. J. Gordon drawings and associated manuscripts, pp. 67–111, the F. Levaillant drawings, pp. 210–248, the F. Masson drawings, pp. 130–132, the W. Paterson drawings, pp. 167–173.

Bc698. **Roux, C.** 1906. *Notice bibliographie sur plus de deux cents manuscrits inédits ou peu connus concernant pour la plupart l'histoire naturelle de la région lyonnaise.* Lyon, Société Linnéenne de Lyons. 26 pp.

Bc699. ___. 1908. *Bibliographie méthodique des principaux manuscrits français relatifs aux sciences naturelles.* Lyon, Société d'Agriculture, Sciences et Industrie. 112 pp.
1015 entries.

Bc700. **Royal Commission on Historical Manuscripts, National Register of Archives.** 1955+. List of accessions to repositories (1954). *Bulletin, National Register of Archives* 6, 1955; *Continued* ___. 1956. List ... (1956). *Ibid.* 8; Further continued separately as: ___. 1957+. *List of accessions to repositories* (1956 onwards). London, Her Majesty's Stationery Office.
Annual lists of manuscripts newly acquired by a wide range of institutions, and including much of natural history interest.

Bc701. **Sherman, S. C.** 1986. *Whaling logbooks and journals 1613–1927: An inventory of manuscript records in public collections.* Ed. by J. M. Downey & V. Adams. New York, Garland Publishing. 469 pp. *Garland reference library of the humanities* vol. 682.
Survey of materials in 82 libraries worldwide, ca.5,000 entries.

Bc702. **Stanton, W.** 1991. *American scientific exploration, 1803–1860: Manuscripts in four Philadelphia libraries.* Philadelphia, PA, American Philosophical Society Library. xviii+140 pp. *Library publication* no. 15.

Bc703. **Stieber, M. T.** 1982. Register of collectors' field notes held in North American institutions. *Huntia* 4: 151–202.
List of collectors; list of repositories with details of each set of notes.

Bc704. **United States Library of Congress, Division of Manuscripts, Reference Department.** 1918. *Check list of collections of personal papers in historical societies, university and other learned institutions in the United States.* Washington, DC. 87 pp.

Bc705. ___. 1924. *Manuscripts in public and private collections in the United States.* Washington, DC. 97 pp.

Bc706. ___. 1962. *The national union catalog of manuscript collections 1959–61. Based on reports from American repositories of manuscripts.* Ann Arbor, MI, Edwards. 1,061 pp. *Also* ___. 1964. ___ *Index, 1959–62.* Hamden, CT, Shoe String Pr. 732 pp.; Continued by regular supplements, e.g.: ___. 1964. *The national union catalog of manuscript collections 1962.* Hamden, CT, Shoe String Pr. 532 pp., with periodic cumulative indexes, e.g. for 1963–66, 1967–69, 1970–74, 1975–79, 1980–84, etc.
Records manuscript groups, rather than individual pieces.

Bc707. ___. 1988. *Index to personal names in the National union catalog of manuscript collections 1959–1984.* Ed. under the supervision of H. Ostroff. Alexandria, VA, Chadwyck–Healey. 2 vols. (1,202 pp.)
Indexes about 2 million names.

See also *History of Science Society newsletter* 1972+ (*see* A77) and the *History of science in America news & views* 1981+ (*see* A76) for occasional reports on newly available archive and research collections.

Individual collections

Bc708. **Académie des Sciences, Paris.** 1978. The archives of the Académie des Sciences. *French historical studies* 10: 696–702.
Also see Réaumur correspondence, Ba1414.

Bc709. **Académie Nationale des Sciences, Belles–Lettres et Arts, Bordeaux.** 1879. *Académie des Sciences ... Table historique et méthodique (1712–1875). ... Catalogue des manuscrits de l'ancienne Académie (1712–93).* Bordeaux. 386 pp.

Bc710. **Academy of Natural Sciences of Philadelphia.** 1963. *Guide to the manuscript collections in the Academy.* By V. T. & M. E. Phillips. Philadelphia, PA. xxvi+533 pp. *Special publication* no. 5.
Also see Microfilm of minutes and correspondence, Bb572.

Bc711. **Accademia Nazionale delle Science detta dei XL.** 1984. *Guida all'Archivio storico dell'Accademia Nazionale delle Scienze detta dei XL.* By G. Paoloni & M. T. Croce. Rome. 78 pp.

Bc712. ___. 1984. The archives of the Accademia Nazionale delle Scienze detta dei XL. By G. B. Marini–Bettolo. *Rivista di storia della scienze* 1: 443–452.

Bc713. **Akademiya Nauk SSSR.** 1933–46. *Arkhiv Akademii Nauk SSSR. Obozrenie arkhivnykh materialov. [Archives of the Academy of Sciences of the USSR. Survey of archives materials].* Ed. by G. A. Knyazev & L. B. Modzalevskii. Leningrad. 2 vols. *Trudy arkhiva Akademiya Nauk SSSR* vol. 1 (1 & 2).

Bc714. ___. 1937. *Uchenaya korrespondentsiya Akademii Nauk XVIII veka. Nauchnoe opisanie, 1766–1782. [Scientific correspondence of the Academy of Sciences in the 18th century. Description for the years 1766–82].* Ed. by D. S. Rozhdestvenskii, G. A. Knyazev, & L. B. Modzalevskii. Moscow/Leningrad. 606 pp., illus. *Trudy arkhiva Akademiya Nauk SSSR* vol. 2.

Bc715. ___. 1940. 1940. *Materialy dlya istorii ekspeditsii Akademii Nauk v XVIII i XIX vekakh; khronologicheskie obzory i opisanie arkhivnykh materialov. [Materials for a history of Academy of Science expeditions in the 18th and 19th centuries; chronological survey and description of archiva materials].* By V. F. Gnucheva. Moscow/Leningrad. 310 pp., illus. *Trudy arkhiva Akademiya Nauk SSSR* vol. 4.
Includes descriptive reviews of archival materials relating to expeditions within the present USSR territory between 1717 and 1903.

Bc716. **Akademiya Nauk SSSR.** 1946. *Geograficheskii departament Akademii Nauk XVIII v.* By V. F. Gnucheva. Moscow/Leningrad. 446 pp., illus. *Trudy arkhiva Akademiya Nauk SSSR* vol. 6.

Bc717. ___. 1950. *Arkhiv Akademii Nauk SSSR. Obozrenie arkhivnykh materialov.* Vol. 3. *[Archives of the Academy of Sciences of the USSR. Survey of archives materials].* Leningrad. *Trudy arkhiva Akademiya Nauk SSSR* vol. 9.
IV: Biologicheskie nauk. A. P. Bogdanov (1834–1896); A. O. Kovalevskii (1840–1901); N. V. Nasonov (1855–1939); N. M. Kulagii (1860–1940); N. M. Knipovich (1862–1939); A. A. Kulyabko (1866–1930); A. P. Semenov–Tyan–Shanskii (1866–1942); A. N. Severtsov (1866–1936); A. K. Mordvilko (1867–1938); P. P. Sushkin (1868–1928); V. L. Komarov (1869–1945); S. A. Zernov (1871–1945); V. V. Talanov (1871–1936); B. M. Zhitkov (1872–1943); M. K. Petrova (1874–1948); B. A. Keller (1874–1945); A. A. Ukhtomskii (1875–1942); E. V. Vul'f (1885–1941); K. E. Vorob'eva (1890–1947); I. K. Shishkin (1897–1934); G. I. Petrov (1903–1942); pp. 81–110.

Bc718. ___. 1956–58. *Istoricheskii ocherki i obzor fondov Rukopisnogo Otdela Biblioteki Akademii Nauk. [Historical sketch and synopsis of the collections of the Manuscripts Section of the Library of the Academy of Sciences].* Moscow/Leningrad. 2 vols.
Historical sketch and synopsis of the manuscript collections: Vyp. I. *XVIII vek.* Moscow/Leningrad. 483 pp.; Vyp. II. *XIX–XX veka.* 398 pp.

Bc719. ___. 1966. *Materialy i soobshcheniya po fondam Otdela Rukopisnoi i Redkoi Knigi Biblioteki Akademii Nauk SSSR. [Materials and reports about the collections of the Manuscript and Rare Book Section of the Library of the Academy of Sciences of the USSR.].* Ed. by A. I. Kopanev. Moscow/Leningrad. 204 pp.

Bc720. ___. 1976. *Rukopisnye i redkie pechatnye knigi v fondakh biblioteki AN SSSR: Sbornik nauchnykh trudov. [Manuscripts and rare printed books in the collection of the library of the Academy of Sciences of the USSR: Record of scientific works.]* By S. P. Luppov & A. A. Moiseeva. Leningrad. 157 pp.
Also see Pavlov papers, Ba1355.

Bc721. **Alexander Turnbull Library.** 1976. Turnbull Library holdings in the history of New Zealand sciences. By M. E. Hoare. *Turnbull Library record* 9(2): 4–19.

Bc722. ___. 1982. *A guide to manuscripts relating to science in the Alexander Turnbull Library, Wellington.* By M. Henry & M. E. Hoare. Wellington, N.Z. 30 pp.

Bc723. ___. 1983–84. French explorers in New Zealand, 1769–1840: A list of manuscript material; part I (–II). By I. Ollivier. *Turnbull Library record* 16: 5–19, 95–110.

Bc724. **American Museum of Natural History.** 1987. *Catalog of the American Museum of Natural History film archives.* N. J. Root, ed. New York, Garland Publishing. xxiv+410 pp., illus. *Garland reference library in the humanities* vol. 723.
Lists 291 films produced between 1908 & 1981.
Also **V. Wheat & M. E. Genett.** From curiosities in cabinets to compact shelving: Special collections — American Museum of Natural History Library. In: E. Mount, ed. *Sci–tech archives and manuscript collections*, pp. 23–29. New York, Haworth Pr. A re-issue of *Science & technology libraries* vol. 9, no. 4 as a separate book.
Also see Osborn papers, Ba1327.

Bc725. **American Philosophical Society.** 1970. *Catalog of manuscripts in the American Philosophical Society Library, including the archival shelflist.* Westport, CT, Greenwood. 10 vols.
Also **W. J. Bell Jr.** 1959. Archives and autographs in the American Philosophical Society Library. *Proceedings of the American Philosophical Society* 103: 761–767; ___. 1967. Manuscripts on microfilm: Papers for the history of science. *Library of Congress quarterly journal* 24: 162–170. Microfilmed collections of papers in the Society; **M. D. Smith.** 1989. Manuscript collecting at the American Philosophical Society: The first 101 years. *Journal of the history of collections* 1: 197–206.

Also see Barton papers, Ba502; Darwin letters, Ba693; Dobzhansky papers, Ba738; Dunn papers, Ba761; J. E. LeConte drawings, Ba1111; Morgan papers, Ba1273–1274.

Bc726. ___. 1984. *Historical American sketches: An illustrated guide to sketches in the manuscript collections of the American Philosophical Society.* By M. D. Smith. Boston, MA, G. K. Hall. xiii+275 pp.

Bc727. ___. 1987. *A new guide to the collections in the Library of the American Philosophical Society.* Ed. by J. S. Catlett. Philadelphia, PA. 414 pp.
Replaces ___. 1966. *Guide to the archives and manuscript collections of the American Philosophical Society.* Compiled by W. J. Bell Jr. and M. D. Smith. Philadelphia, PA. ix+182 pp. *Memoirs of the American Philosophical Society* vol. 66.

Bc728. ___. 1987. *A guide to manuscript sources in the history of medicine in the American Philosophical Society.* By J. S. Catlett. Philadelphia, PA. 36 pp.

Bc729. ___. 1988. *A guide to the genetics collection of the American Philosophical Society.* By B. Glass. Boston, MA, G. K. Hall. xxi+122 pp. *Library publication* no. 13.
Describes 41 manuscript collections. *Also* **W. J. Bell, Jr.** 1970. American Philosophical Society: Geneticists' letters and papers. *Mendel newsletter* 5: 1; **B. Glass.** 1983. The preservation of historical materials in genetics. *Bulletin of the history of medicine* 57: 98–105.

Bc730. ___. 1989. *Molecules, cells, and life: An annotated bibliography of manuscript sources on physiology, biochemistry, and biophysics, 1900–1960, in the library of the American Philosophical Society.* By L. E. Kay. Philadelphia, PA. viii+95 pp. *Library publication* no. 14.
Provides notes on some 40 scientists' papers.

Bc731. **Archives Nationales de Paris.** (No relevant general catalogue.)
For special information *see* **G. Beaujouan.** 1950. L'histoire des sciences aux Archives Nationales de Paris. *Archives internationales d'histoire des sciences* 29: 874–881. **M.-J. Imbault-Huart.** 1972. Sources de l'histoire de la médecine aux Archives Nationales de 1750 à 1822. *Revue d'histoire des sciences* 25: 45–53.

Bc732. **Archivo Nacional, Chile.** (No relevant general catalogue.)
For special list *see* Gay papers, Ba835.

Bc733. **Asiatic Society of Bengal.** 1926. On the manuscript drawings of fish in the library of the Asiatic Society of Bengal. By S. L. Hora. *Journal and proceedings of the Asiatic Society of Bengal* n.s. 22: 93–125.
1. Fish drawings in the [Colin] Mackenzie collection, pp. 93–97; 2. Fish drawings in [Francis] Buchanan-Hamilton's zoological drawings, pp. 99–115; 3. Fish drawings among zoological drawings in the collection of Lieut. Col. Sir Alexander Burnes by Dr. P. B. Lord, pp. 117–125.

Bc733A. **Association of American Geographers.** (No relevant general catalogue.)
For special information *see* **R. E. Ehrenberg.** 1976. The archives of the Association of American Geographers. *Professional geographer* 28: 181–185.

Bc734. **Australian Museum, Sydney.** 1988. *A guide to the Australian Museum archives.* By M. Buddin, M. Organ & L. Trott. 2nd ed., rev. Sydney, N.S.W. xiv+91 pp.

Bc735. **Biblioteca Lancisiana, Rome.** 1970. A survey of some of the manuscripts of the Bibliotheca Lancisiana in Rome. *Medical history* 14: 289–294.

Bc736. **Biblioteca Nacional de Madrid.** 1968–70. Indice de los manuscritos existentes en la Biblioteca Nacional de Madrid, de interés a la historia de la farmacia y ciencias afines, con breves comentarios de su contenido. By M. del Pilar Hitos Natera. *Boletin de la Sociedad Española de Historia de la Farmacia* 19: 49–51, 117–128, 1968; *Ibid.* 20: 17–37, 74–82, 103–126, 154–181, 1969; 21: 63–91, 1970.

Bc737. **Biblioteca Nazionale Braidense di Milano.** (No relevant general catalogue.)
For special list see Haller papers, Ba898A. *Also* **F. Bazzi.** 1961. *Catalogo dei manoscritti e*

degli incunabili di interesse medico–naturalistico dell'Ambrosiana e della Braidense. Milan. 191 pp. *Castalia; rivista di storia della medicina* no. 7.

Bc738. **Biblioteka Jagiellonska, Kraków.** 1989. The Clusius and other natural history pictures in the Jagellonian Library, Kraków, by P. J. P. Whitehead, G. van Vliet & W. T. Stearn. *Archives of natural history* 16: 15–32.

Previously held by the Preussischer Staatsbibliothek, Berlin, this series of volumes includes several collections: The "Clusius volumes" (nearly 1,400 drawings, mainly plants, associated with Charles de L'Éscluse); the "Brazilian collection" (805 paintings by Albert Eckhout and others, 1637–44); the "Carli flower album" (97 paintings by Antonio Carli, 1690s); "Conrad–Johrenius: Floreas vivum depictae" (8 volumes of paintings by H. Kribbing); "Garden flowers" (possibly by Johann S. Elsholz); the "Elsholz tulip volume" (127 paintings of tulips by Johann S. Elsholz, 1661); the "Wildenberger birds and plants" (paintings by Jacob Wildenberger, post 1629); the "Herbst butterfly volume" (48 paintings by J. Fr. Schuppel, C18th); "Plantae hortorum Vindoboniensium" (1,453 coloured drawings of garden plants, ca.1800s?); the "Commerson fish drawings" (97 drawings of fishes copied from originals made in Mauritius for Philibert Commerson); the "Wildenberger bird volume" (137 paintings by Jacob Wildenberger, 1629); the "Wagner plant and flower volume" (309 paintings of garden flowers by Johann G. Wagner, 1725); the "Bauer Passiflora volumes" (paintings of 40 Passiflora by Ferdinand Bauer, early C19th).

Bc739. **Biblioteka Zakladu Narodowego im Ossolinskich, Wroclaw.** 1962. *Indeks do inventarza rekopisow Biblioteki Zakladu Narodowego im. Ossolinskich we Wroclawiu. [An index to the inventory of manuscripts in the Ossolineim Library, Wroclaw.]* By J. Turska. Wroclaw, Polish Academy of Sciences. 2 vols., 576 pp.

Bc740. **Bibliothèque Nationale, Paris, Département des Manuscrits.** 1868+. *Catalogue générale des manuscrits français [etc.].* Paris. Vol. 1+.

Two series: Serie in 4° Catalogue des manuscrits français. Anciens fonds, nos. 1–6170; Serie in 8° Nouveau fonds, nos. 6171–33,364, nouvelle acquisitions no. 1+.

___. 1931–48. *Catalogue générale des manuscrits français. Table alphabétique des ancien et nouveaux fonds (nos. 1–33,264) et des nouvelles acquisitions (nos. 1–1,000).* By A. Vidier & P. Perrier. Paris. 6 vols.

___. 1933. *Catalogue alphabétique des livres imprimés mis à la disposition des lecteurs dans la Salle de Travail, suivi de la liste des catalogues usuels de Département des manuscrits.* 4th ed., by E.–A. Van Moe. Paris. 142 pp.

Also see Pasteur papers, Ba1351.

Bc740A. **Bibliothèque Publique et Universitaire, Genève.** (No relevant general catalogue.)

For special list see Bonnet correspondence, Ba565A.

Bc741. **Bibliothèque Royale Albert 1er, Brussels.** 1965–75. *Inventaire des manuscripts scientifiques de la Bibliothèque Royale de Belgique.* By R. Calcoen. Brussels, Centre National d'Histoire des Sciences. 3 vols. *Publications du Centre National d'Histoire des Sciences* no. 1.

Bc742. **Boston Public Library.** 1983. *Botanico–horticultural manuscripts in the Boston Public Library: A descriptive catalog.* Compiled by E. R. Taylor. Boston, MA, Trustees of the Public Library. 34 pp.

Also ___. 1966. *Waiting for Spring: A catalog of an exhibition of books and manuscripts for the gardener, in the Boston Public Library.* Boston, MA. 28 pp.

Bc743. **British Museum.** 1782. *A catalogue of the manuscripts preserved in the British Museum hitherto undescribed: Consisting of five thousand volumes; including the collections of Sir Hans Sloane, Bart., the Rev. Thomas Birch, D.D. and about five thousand volumes bequeathed or purchased at various times.* By S. Ayscough. London. 2 vols.

A classified catalogue. *Also* ___. 1904. *Index to the Sloane manuscripts in the British*

Museum. By E. J. L. Scott. London. viii+583 pp. *Re-pub.* 1971; ___. 1953. The Sloane collection of manuscripts. *British Museum quarterly* 18: 6–10.

Bc744. ___ (*later* **British Library**). 1849+. *Catalogue of additions to the manuscripts in the British Museum.* London.

Lists covering the acquisitions of spans of years were issued at periodic intervals, i.e. for 1783–1835, in 1849; 1836–40, 1843; 1841–45, 1850; 1846–47, 1864; 1848–53, 1868; 1854–60, 1875; 1861–75, 1877; 1854–75, 1880; 1876–81, 1882; 1882–87, 1889; 1888–93, 1894; 1894–99, 1903; 1900–05, 1907; 1906–10, 1912; 1911–15, 1925; 1916–20, 1933; 1921–25, 1950; 1926–30, 1959; 1931–35, 1967; 1936–45, 1970; 1946–50, 1970; ?; New series 1986–1990, 1993+. In progress. Each is separately indexed. Most earlier vols. have been re-pub. in recent decades.)

Bc745. **British Library.** 1984–86. *Index to manuscripts in the British Library.* Cambridge, Chadwyck–Healey. 10 vols.

A comprehensive name and place index to the whole collection, but not subjects.
Also **R. G. C. Desmond.** 1986. *Wonders of creation: Natural history drawings in the British Library.* London, British Library; Wolfeboro, NH, Longwood Publishing Group Inc. 248 pp., including 49 pl. (col.). (Including natural history images in all kinds of manuscripts from eastern and western cultures. Introductory essay followed by 49 individually described and illustrated items); **A. M. Lysaght.** 1959. Some 18th century bird paintings in the library of Sir Joseph Banks (1743–1820). *Bulletin, British Museum (Natural History), historical series* 1: 251–371.; **M. A. E. Nickson.** 1982. *The British Library: Guide to the catalogues and indexes of the Department of Manuscripts.* 2nd ed. London, British Library. 24 pp.; **A. Summers.** 1990. Sources on medical history in the Manuscript Collections of the British Library: A preliminary guide to researchers. *Social history of medicine* 3: 519–523.

Bc746. **British Museum (Natural History).** 1971. A short history of the libraries and list of mss. and original drawings in the British Museum (Natural History). By F. C. Sawyer. *Bulletin of the British Museum (Natural History), Historical* 4: 77–204.

Based on the list pub. in: ___. 1904. *History of the collections contained in the natural history departments in the British Museum.* Vol. 1, pp. 23–53. London. *Also* ___. 1903–15. *Catalogue of the books, manuscripts, maps and drawings in the British Museum (Natural History).* London. 5 vols; *Continued* ___. 1922–40. ___*: Supplement.* London. 3 vols., which lists many manuscripts among the thousands of entries in its 8 vols.

Bc747. ___. 1979. A catalogue of the Richard Owen collection of palaeontological and zoological drawings in the British Museum (Natural History). By J. M. Ingles & F. C. Sawyer. *Bulletin, British Museum (Natural History), historical series* 6: 109–197.

Bc748. ___. 1984–87. Catalogue of the natural history drawings commissioned by Joseph Banks on the Endeavour voyage 1768–1771, held in the British Museum (Natural History). Pt. 1: Botany: Australia. By J. A. Diment et al.; Pt. 2: Botany: Brazil, Java, Madeira, New Zealand, Society Islands, Tierra del Fuego. By J. A. Diment et al. Pt. 3: Zoology. By A. C. Wheeler. *Bulletin, British Museum (Natural History), historical series* 11: 183 pp., 1984; *Ibid.* 12: 200 pp., 1987; *Ibid.* 13: 171 pp., 1987.

Botanical field drawings by S. Parkinson, together with finished full–colour drawings by Parkinson, J. F. Miller, J. Miller, J. Cleveley, T. Burgis and F. P. Nodder; Zoological drawings by S. Parkinson, A. Buchan and H. D. Spöring.
Also **R. E. R. Banks.** 1988. Resources for the history of science in the libraries of the British Museum (Natural History). *British journal for the history of science* 21: 91–97; **J. Diment & L. Newington.** 1985. Botanical prints and drawings at the British Museum (Natural History). *Art libraries journal* 10: 5–27. (Including a selected bibliography on botanical illustration); **B. Smith & A. C. Wheeler**, eds. 1988. *The art of the First Fleet and other early Australian drawings.* Melbourne, Vic., Oxford Univ. Pr. (Australia) in association with the Australian Academy of the Humanities & the British Museum (Natural History). 256 pp., illus. (*Includes* Ch. 4: The natural history drawings, by J. H. Calaby, pp. 140–197. Based on collections in the British Museum (Natural History), especially drawings by T. Watling, the Port Jackson Painter and G. Raper); **P. J. P. Whitehead.** 1969. The [John]

Reeves collection of Chinese fish drawings. *Bulletin, British Museum (Natural History), historical series* 3: 191–233 & 29 pl.; ___, **& P. I. Edwards.** 1974. *Chinese natural history drawings selected from the [John] Reeves collection in the British Museum (Natural History).* London. 109 pp. & 20 pl.
Also see Banks' correspondence, Ba497; Bartram drawings, Ba509; Bauer drawings, Ba517–518; Blake drawings, Ba547; Botanical Society of London, Bb711; Drury letters, Ba748; Forster drawings, Ba805 & Ba809; Hodgson drawings, Ba936–937; Kønig manuscripts, Ba1061; Sowerby papers, Ba1509; Stephens letters, Ba1525; Watling drawings, Ba1596.

Bc749. **Cambridge University Library.** (No relevant general catalogue.)
For special lists *see* Bateson papers, Ba514; Darwin papers (& exhib.), Ba692 & Ba697.

Bc750. **Carnegie Mellon University, Hunt Institute for Botanical Documentation.** 1981+. *Guide to the botanical records and papers in the Archives of the Hunt Institute.* Parts. 1+. Compiled by M. T. Stieber & A. L. Karg. Pittsburgh, PA.
First pub. in *Huntia* 4: 1–89, 1981; 5: 91–210, 1984; [etc.]; 8: 1–88, 1988; In progress.

Bc751. ___. 1985+. *Catalogue of the Botanical Art Collection at the Hunt Institute.* Part 1 [etc.]. Compiled By J. J. White with the assistance of E. R. Smith. Pittsburgh, PA.
Pt. 1: Plant portraits, artists A–D, 1985; pt. 2: ... E–H, 1987; pt. 3: ... H, 1988; pt. 4: ...I–), 1991; ... In progress. A–O includes 11,741 items.
Also **Carnegie Mellon University, Hunt Botanical Library.** 1964. (Prints and paintings). Program and organization of the collection. By J. V. Brindle. *Huntia* 1: 148–155; ___. 1964. Recent acquisitions of botanical art and illustration. By J. V. Brindle. *Ibid.* 1: 156–160; ___, **Hunt Institute for Botanical Documentation.** 1985. *Flora portrayed: Classics of botanical art from the Hunt Institute collection.* J. V. Brindle and J. J. White with the assistance of D. E. Wendel. Pittsburgh, PA. 93 pp., illus. (some col.). 88 items from the 17th to the 20th centuries.
Also see Adanson's papers, Ba431; Arber papers, Ba466; Popenoe papers, Ba1379.

Bc752. **College of Physicians of Philadelphia.** 1983. *A catalogue of the manuscripts and archives of the Library of the College of Physicians of Philadelphia.* Ed. R. Hirsch, T. L. Bratton, & E. T. Miller. Philadelphia, PA, Univ. of Pennsylvania Pr. & the College of Physicians of Philadelphia. xx+259 pp.

Bc753. **Conservatoire et Jardin Botaniques de Genève.** (No relevant general catalogue.)
For special list *see* D75.

Bc754. **Deutsche Akademie der Wissenschaften zu Berlin.** 1960. *Übersicht über die Bestande des Archivs der Deutschen Akademie der Wissenschaften zu Berlin.* Ed. by C. Kirsten. Berlin, Akademie–Verlag. 55 pp.
For special list *see* Forster correspondence, Ba811.

Bc755. **Dublin Philosophical Society.** (No relevant general catalogue.)
For special list *see* Microfilmed papers, Bb216.

Bc756. **Facultad de Ciencias, Instituto de Botánica y Farmacologia, Buenos Aires.** (No relevant general catalogue.)
For special list *see* Bonpland papers, Ba572.

Bc757. **Farlow Reference Library of Cryptogamic Botany, Harvard University.** (No relevant general catalogue.)
For special information *see* **D. H. Pfister.** 1982. The W. G. Farlow correspondence. *Taxon* 31: 607–608. Notes the completion of an unpublished index to this collection.

Bc758. **Fitzwilliam Museum, Broughton Collection.** (No relevant general catalogue.)
However, quite a number of artworks have been described in a series of exhibition catalogues, viz: **Fitzwilliam Museum.** 1973. *A first selection of flower drawings from the Broughton collection. An exhibition in the Fitzwilliam Museum, University of Cambridge.* 20 December 1973 – 27 January 1974. Cambridge. [26] pp. (38 items); ___. 1974. *Botanical drawings from*

the Broughton collection. An exhibition in the Fitzwilliam Museum, University of Cambridge. 28 June – 29 September 1974. Cambridge. [36] pp. (58 items); ___. 1976. *Gardener's pride: A third selection of drawings from the Broughton collection. An exhibition in the Fitzwilliam Museum, University of Cambridge.* 17 July – 25 September 1976. Cambridge. [24] pp. (61 items); ___. 1977. *Roses, roses ...: A fourth selection of drawings from the Broughton collection. An exhibition in the Fitzwilliam Museum, University of Cambridge.* 1 July – 18 September 1977. Cambridge. [26] pp. (52 items); ___. 1978. *The pick of the bunch: A fifth selection of drawings from the Broughton collection. An exhibition in the Fitzwilliam Museum, University of Cambridge.* 11 July – 24 September 1978. Cambridge. [31] pp. (55 items); ___. 1979. *Men versus women: A sixth selection of drawings from the Broughton collection. An exhibition in the Fitzwilliam Museum, University of Cambridge.* 17 July – 7 October 1979. Cambridge. [25] pp. (50 items); ___. 1980. *Dutch and Flemish artists: A seventh selection of drawings from the Broughton collection. An exhibition in memory of John Raven (13 December 1914 – 5 March 1980), Honorary Keeper of the Broughton collection. Fitzwilliam Museum, University of Cambridge.* 15 July – 28 September 1980. Cambridge. [38] pp. (64 items); **D. Scrase.** 1983. *Flowers of three centuries. One hundred drawings and watercolors from the Broughton Collection.* Washington, DC, International Exhibitions Foundation. xv+79 pp., illus. (Catalogue of a choice selection exhibited in various United States cities, 1983–84).

Bc759. **France, Ministère de la France d'Outre–Mer.** 1959. Inventaire du fonds océanique (Polynésie française) conservé aux Archives du Ministère de la France d'Outre–Mer. *Journal de la Société des Oceanistes* 15: 267–320.

Bc760. **Gabinetto Disegni e Stampe degli Uffizi, Florence.** 1984. *Immagini anatomiche e naturalistiche nei disegni degli Uffizi. Secc. XVI. e XVII.* Ed. R. P. Ciardi & L. Tongiorgi Tomasi. Florence, L. Olschki. 152 pp. & 142 pl. *Gabinetto disegni e stampe degli Uffizi* 60.
Includes **L. Tongiorgi–Tomasi.** L'immagine naturalistica a Firenza tra XVI e XVII secolo. Contributo al rapporto 'arte–natura' nel manierismo e prima età barocca, pp. 37–67, 100–134 & figs. 68–142.

Bc761. **Geograficheskoe Obshchestva SSSR, Akademiya Nauk SSSR.** 1971. *Russkie geografy i puteshestvenniki, fondy arkhiva Geograficheskogo Obshchestva. [Russian geographers and voyagers: The archival collections of the Geographical Society of the USSR].* By T. P. Matveeva, T. S. Filonovich, & L. I. Yarukova. Leningrad, Izd. "Nauka." 175 pp.

Bc762. **Harvard University.** 1990. *Guide to manuscripts and archives at Harvard and Radcliffe.* Dec. 1988. Alexandria, VA, Chadwyck–Healey. 17 microfiche & printed guide, 56 pp.
See also ___. 1983. *Directory of archives and manuscript resources at Harvard University and Radcliffe College.* Cambridge, MA. 45 pp. Guide to 53 repositories. For special information *see* **C. A. Elliott.** 1974. Sources for the history of science in the Harvard University Archives. *Harvard Library bulletin* 22: 49–71. A synoptic guide.

Bc762A. ___, **Museum of Comparative Zoology.** (No relevant general catalogue.)
For special list *see* J. L. R. & A. Agassiz letters, Ba434.

Bc762B. **Henry E. Huntington Library.** (No relevant general catalogue.)
For special information *see* **R. B. Haselden.** 1939. Manuscripts in the Huntington Library. *Archives & libraries* 1939: 71–79. *Also* **N. Reingold.** 1964. The anatomy of a collection: The [W. J.] Rhees papers. *American archivist* 27: 251–259. Papers spanning 1744-1906, including National Institution for the Promotion of Science 1816–58, Smithsonian Institution, 1831–1906, etc.

Bc763. **Hudson's Bay Company.** (No relevant general catalogue.)
For special information *see* **S. A. Smith.** 1983. Hudson's Bay Company archives, 1670–1870: A reference source for natural historians. In: Society for the Bibliography of Natural History. *Contributions to the history of North American natural history: Papers from the first North American conference of the Society for the Bibliography of Natural History held at the*

Academy of Natural Sciences, Philadelphia, 21–23 October 1981, pp. 85–89. London. *Special publication* no. 2.

Bc764. **Hunterian Museum, University of Glasgow.** 1908. *A catalogue of the manuscripts in the Library of the Hunterian Museum in the University of Glasgow.* By J. Young & P. Aitken. Glasgow. xi+566 pp.
Also see W. Hunter's MSS, Ba994.

Bc765. **Imperatorskii Sanktpeterburghskii Botanicheskii Sad (Jardin Impérial de Botanique à St. Petersburg).** 1857. *Catalogue alphabétique et méthodique des dessins de plantes, exécutés et conservés au Jardin Impérial de Botanique à St. Pétersbourg.* St. Petersburg. 1 f.+36 pp.

Bc766. **Imperial College of Science & Technology.** 1976. Natural history manuscripts and related material in the archives of Imperial College. By J. Pingree. *Journal of the Society for the Bibliography of Natural History* 7: 371–374.
Also see T. H. Huxley papers, Ba1006, Ba1010–1011.

Bc767. **India Office Library.** 1962. *Natural history drawings in the India Office Library.* By M. Archer. London, Her Majesty's Stationery Office. ix+116 pp. & 35 pl.
Also ___. 1970. *The East India Company and natural history: An exhibition of water–colours, manuscripts, rare books and other naterial illustrating the interests of the East India Company and its servants in the natural history of India and the East Indies.* London. 22 pp. Exhibition catalogue.

Bc768. **Institut de France, Académie des Sciences.** (No relevant general catalogue.)
For special lists *see* Cuvier papers, Ba675; St. Hilaire papers, Ba841.

Bc769. **John Crerar Library.** 1991. *A guide to the historical records of the John Crerar Library, 1828–1984.* By P. A. Carnahan. Ed. by R. L. Popp. Chicago, IL, Univ. of Chicago Library, Department of Special Collections. xvii+128 pp.

Bc769A. **Koninklijke Akademie van Wetenschappen, Amsterdam.** 1937. *Catalogus van der Handschriften der Koninklijke Akademie der Weyenschappen, in bruikleen ep de Koninklijke Bibliotheek.* By D. J. H. Horst. The Hague. 66 pp.

Bc770. **Kungliga Vetenskapsakademiens Bibliotek, Stockholm.** (No relevant general catalogue.)
For special list *see* **A. J. Kohn.** 1991. Paintings of paralectotypes?: *Conus* of the *Museum Ludovicæ Ulricæ. Journal of conchology* 34: 7–18. Catalogue of the paintings commissioned by C. Linnaeus.
Also see Bergius correspondence, Ba529.

Bc771. **Landesbibliothek Gotha.** (No relevant general catalogue.)
For special list *see* Breyne correspondence, Ba587.

Bc772. **Linnean Society of London.** (No relevant general catalogue.)
For special information *see* **G. L. Douglas.** 1988. History of science and technology resources at the Linnean Society of London. *British journal for the history of science* 21: 489–493.
Also see British Ecological Society archives, Bb169; Ellis papers, Ba772; Holmes correspondence, Ba940; Linnean papers, Ba1146 & Ba1175; Pulteney correspondence, Ba1384; Smith papers, Ba1499; Swainson correspondence, Ba1534; Winch correspondence, Ba1628.

Bc773. **McGill University, Montreal.** 1931. *An introduction to the literature of vertebrate zoology, based chiefly on the titles in the Blacker Library of Zoology, the Emma Shearer Wood Library of Ornithology, the Bibliotheca Osleriana and other libraries of McGill University Montreal.* By C. A. Wood. London, Oxford Univ. Pr. xix+643 pp. *Re-pub.* Hildesheim, Olms Verlag, 1974.
Includes a significant number of zoological manuscripts; see especially "Original drawings," pp. 501–505.

Bc774. ___. 1966. [Appendix listing original drawings, paintings, and manuscript letters.] In: *A dictionary catalogue of the Blacker–Wood Library of Zoology and Ornithology.* Boston, MA, Hall. 9 vols.

Also **D. E. Allen.** 1987. The Blacker–Wood letter collection. *Archives of natural history* 14: 112–113; **S. Sheets–Pyenson.** 1982. Geological communication in the nineteenth century: The Ellen S. Woodward autograph collection at McGill University. *Bulletin of the British Museum (Natural History), historical series* 10: 179–226.

Bc775. **Manchester University, John Rylands Library.** (No relevant general catalogue.)

For special information *see* **J. V. Pickstone.** 1989. Some Manchester sources for the history of science, technology, and medicine, with special reference to the John Rylands University Library of Manchester. *Bulletin of the John Rylands University Library of Manchester* 71: 141–157.

Bc776. **Marine Biological Laboratory, Woods Hole.** (No relevant general catalogue.)

For special information *see* **G. E. Allen.** 1968. The MBL library. *Mendel newsletter* 2: 3–4. **R. Davis.** 1984. Resources at the Marine Biological Laboratory. *Ibid.* 24: 4.
Also see Lillie papers, Ba1143; Whitman collection, Ba1618.

Bc777. **Maryland Historical Society, Baltimore.** (No relevant general catalogue.)

For special information *see* **F. Haber.** 1951. Sidelights on American science as revealed in the Hyatt Autograph Collection. *Maryland historical magazine* 46: 233–256.

Bc778. **Massachusetts Institute of Technology Libraries.** 1989. *Selective guide to the collections, Institute Archives and Special Collections, the Libraries, Massachusetts Institute of Technology.* By H. W. Samuels et al. Cambridge, MA. 135 pp.

Replaces ___. 1988. *Selective guide to the collections, Institute Archives and Special Collections, Massachusetts Institute of Technology.* Cambridge, MA. 108 pp. *Also* **M. A. Vargas.** 1989. Archival and manuscript holdings in the MIT institute archives and special collections. In: E. Mount, ed. *Sci–tech archives and manuscript collections*, pp. 41–56. New York, Haworth Pr. A re-issue of *Science & technology libraries* vol. 9, no. 4 as a separate book.

Bc779. **Max–Planck–Gesellschaft.** (No relevant general catalogue.)

For special list *see* Correns papers, Ba664.

Bc780. **Musée de Louvre, Cabinet des Dessins.** (No relevant general catalogue.)

For special list *see* ___. 1984. *Dessins et sciences, XVIe – XVIIIe siècle; 82e exposition du Cabinet des Dessins, Musée du Louvre, 22 juin – 24 septembre 1984.* Paris, Ministère de la Culture, Editions de la Réunion des Musées Nationaux. 127 pp. "Catalogue et exposition par M. Pinault ... avec la collaboration de M. Jaoul ... et D. Cordellier."

Bc781. [Not used.]

Bc782. **Muséum d'Histoire Naturelle du Havre.** (No relevant general catalogue.)

For special list *see* Lesueur drawings, Ba1124, Ba1125 & Ba1128.

Bc783. **Muséum National d'Histoire Naturelle, Paris.** 1889. Vélins conservés à la Bibliothèque du Jardin des Plantes. By H. Stein. In: *Inventaire général des richesses d'art de la France.* Série II: *Paris, monuments civils,* vol. 2 (6): 116–324. Paris, E. Plon, Nourrit & Cie.

Bc784. ___. [1927.] *Les velins du Muséum d'Histoire Naturelle à Paris: Fleurs exotiques.* Ed. by L. Bultingaire. Paris. [xvi] pp. & 88 pl. (24 col.).

Reproductions of work by notable French botanical artists from N. Robert to P. J. Redouté. *Also* **Bibliothèque Royale Albert 1er.** 1974. *Vélins du Muséum: Peintures sur vélin de la collection du Muséum National d'Histoire Naturelle de Paris. Catalogue.* Brussels. xxiv+32 pp. & 38 pl. (35 items by 15 artists); **E. Bonnet.** 1903. Documents pour servir à l'histoire de la collection de miniatures d'histoire naturelle, connue sous la dénomination de Vélins du Muséum. In: *Compte rendu, Association Française pour l'Avancement des Sciences*, 31ème session, Montauban, 1902. 2e partie: *Notes et memoires*, pp. 660–672. Paris; **L. Bultingaire.** 1926.

Les origines de la collection des Vélins du Muséum et ses premiers peintres. *Archives du Muséum d'Histoire Naturelle* (6) 1: 129–149.; ___. 1926. Les richesses d'art du Muséum: La collection des vélins. *Revue scientifique* 64: 545–553.; ___. 1928. Les peintres du Jardin du Roi au XVIIIe siècle. *Archives du Muséum d'Histoire Naturelle* (6) 3: 19–36.; ___. 1930. Les peintres du Muséum à l'époque de Lamarck. *Ibid.* (6) 6: 49–62.; ___. 1932. L'art au Jardin des Plantes. *Ibid.* (6) 12: 665–678.; ___. 1932. Les insectes dans la collection des vélins du Muséum. In: Société Entomologique de France. *Livre du centenaire*, pp. 87–94 & 2 pl. Paris.; **E.–T. Hamy.** 1898. Les peintures de Michel Garnier au Muséum d'Histoire Naturelle. *Bulletin, Muséum d'Histoire Naturelle* 1898 (8): 336–344.; **Y. Laissus.** 1969. Les vélins du Muséum et l'Expédition d'Egypte (1798–1801). *Comptes rendus, 92e congrès national des sociétés savantes, Section des Sciences*, tom.1: *Histoire des Sciences*, pp. 267–300. Paris.
___. 1980. *Nicholas Robert et les vélins du Museum National d'Histoure Naturelle.* Paris, H. Scrépel. [22] pp. & [35] leaves of illus.; **Muséum National d'Histoire Naturelle, Paris.** 1967. *Les vélins du Muséum.* Conférence donnée au Palais de la Découverte le 7 Janvier 1967. Paris. 28 pp.; **H. Stein.** 1890. La collection des vélins du Muséum d'Histoire Naturelle. *Gazette des beaux–arts* 1890: 290–306.

Bc785. ___. 1934. Inventaire des archives du Muséum National d'Histoire Naturelle. Première partie. Série A: Archives du Jardin du Roi. By A.–M. Bidal. *Archives du Muséum National d'Histoire Naturelle* (6) 11: 175–230.
Now preserved in the French national archives.

Bc786. ___. *See* **France, Ministère de l'Instruction Publique, Direction des Bibliothèques de France.** 1965. *Catalogue général des manuscrits des bibliothèques publiques de France.* Tom. 55: *Paris, Bibliothèque Centrale du Muséum National d'Histoire Naturelle (supplement).* By Y. Laissus. Paris. xvii+199 pp.
Also **G. Duprat.** 1957. Essai sur les sources manuscrites conservées au Muséum National d'Histoire Naturelle. In: **Centre National de la Recherche Scientifique (CNRS), Paris.** *Les botanistes français en Amérique du nord avant 1850*, pp. 231–252 & pl. XXIV–XXXI. J. Leroy Ed. Paris; *see* Cb226.
Also see Commerson papers, Ba652; Lamarck papers, Ba1088; St. Hilaire papers, Ba841; Tournefort manuscripts, Ba1556.

Bc787. **National Agricultural Library, United States Department of Agriculture.** 1979. *Guide to manuscripts in the National Agricultural Library.* By A. E. Fusonie. Beltsville, MD. 39 pp. *Miscellaneous publication, United States Department of Agriculture* no. 1374.
Also **A. Fusonie.** 1976. *The Prince family manuscript collection: A register of their papers in the National Agricultural Library.* Beltsville, MD. vi+37 pp. *Library list* no. 101; **M. L. Naftalin.** 1967. *Historic books and manuscripts concerning general agriculture in the collection of the National Agricultural Library.* Washington, DC. 94 pp. *Library list* no. 86; ___. 1968. *Historic books and manuscripts concerning horticulture and forestry in the collection of the National Agricultural Library.* Beltsville, MD. 106 pp. *Library list* no. 90. *Also see* Riley papers, Ba1435.

Bc788. **National Botanic Gardens, Glasnevin.** (No relevant general catalogue.)
For special lists *see* **D. M. Moore & M. J. P. Scannell.** 1986. Bartholomew Sulivan and early watercolours of Falkland Islands vegetation in the National Botanic Gardens, Dublin. *Archives of natural history* 13: 155–163; **B. D. Morley.** 1978. Lydia Shackleton's paintings in the National Botanic Gardens, Glasnevin. *Glasra* 2: 25–32 & 4 pl.; **M. J. P. Scannell & C. I. Houston.** 1980. George V. Du Noyer. A catalogue of paintings at the National Botanic Gardens, Glasnevin, with aspects of his scientific life. *Journal of life sciences, Royal Dublin Society* 2: 1–13; ___. 1985. *Wild flowers from nature painted by Katharine Plunket and Frederica Plunket at the National Botanic Gardens, Glasnevin.* Glasnevin. 55 ff. (Catalogue of 1,200 paintings).

Bc789. **[National Library of Medicine, United States Department of Health.] Surgeon–General's Office, United States Army.** [1950.] *A catalogue of incunabula and manuscripts in the Army Medical Library.* By D. M. Schullian and F. E.

Sommer. New York, Schuman. xiii+361 pp.
Early western manuscripts [before 1501], pp. 213–262; Oriental manuscripts [from 1094 to ca.1825], pp. 293–354.

Bc790. **National Maritime Museum, London.** 1977–80. *A guide to the manuscripts in the National Maritime Museum.* London, Mansell. 2 vols.
Vol. 1: The personal collections. Ed. by R. J. B. Knight.; vol. 2: Public records, business records and artificial collections.

Bc791. **National Zoological Park, Washington, DC.** 1988. *Guide to animal-related records at the National Zoological Park, 1887–1985.* By J. Hamman. Washington, DC, Smithsonian Institution. 11 pp. *Occasional papers, Archives & Special Collections of the Smithsonian Institution* no. 1.

Bc792. ___. 1989. *Guide to the records of the National Zoological Park, 1887–1965, and undated.* By A. L. Bain. Washington, DC, Smithsonian Institution. viii+145+[40] pp., illus. *Occasional papers, Archives & Special Collections of the Smithsonian Institution* no. 10.

Bc793. **New York Botanical Garden.** 1973. *Catalog of the manuscript and archival collections and index to the correspondence of John Torrey.* By S. Lenley et al. Boston, MA, G. K. Hall. viii+473 pp.

Bc794. ___. 1983. *Guide to the archives and manuscripts of the New York Botanical Gardens.* By S. P. Johnson & T. Collins. New York. 94 ff.
Also **C. R. Long.** 1978. Natural history manuscripts and related materials in the archives of the New York Botanical Garden. *Journal of the Society for the Bibliography of Natural History* 8: 343–349; **New York Botanical Garden.** 1988. *The art of botany: Works on paper from the collection of the library of the New York Botanical Garden.* Curated by B. Angell. New York. 28 pp., illus. (An exhibition catalogue of 89 items.)

Bc795. **New York Zoological Society.** 1982. *Guide to the archives of the New York Zoological Society.* Compiled by T. Collins & S. P. Johnson. Bronx, NY, the Society. x+46 pp.

Bc796. **Newcastle upon Tyne Public Libraries Committee.** (No relevant general catalogue.)
For special list *see* Bewick collection, Ba544.

Bc797. **Oxford University.** 1977. Natural history books and manuscripts in Oxford libraries. By S. Raphael. *Journal of the Society for the Bibliography of Natural History* 8: 108–119.

Bc798. ___, **Department of Botany.** 1958. *Catalogue of manuscripts belonging to ... deposited in the Bodleian Library, Oxford, 1957, and referenced MSS. Sherard 1–476.* Oxford. 48 pp.

Bc799. **Oxford University Museum.** 1986. *A history of the Hope Entomological Collections in the University Museum, Oxford. With lists of archives and collections.* By A. Z. Smith. Oxford, Clarendon Pr. xiii+172 pp. & 17 pl.
Appendix A. Contents of the archives, pp. 67–96.

Bc800. **Palacio Nacional de Madrid.** 1971. *Colleccion documental historico–farmaceutica.* I. *Catálogo de documentos de interés histórico–farmacéutico conservados en el Archivo del Palacio Nacional de Madrid.* Ed. by J. L. Valverde & M. D. Vidal. Granada, Universidad. 266 pp. *Cuadernos de historia de la farmacia* Año 1 (2).

Bc801. **Palazzo Pitti, Florence.** (No relevant general catalogue.)
For special list *see* ___. 1988. *Floralia: Florilegio dalle collezioni fiorentine del sei–settecento.* Ed. Marilena Mosco & Milena Rizzotto. Firenze, Palazzo Pitti, Andito degli Angliolini 10 gennnaio – 10 aprile 1988. Florence, Centro Di. 149 pp.
Catalogue of an exhibition organized by Ufficio Documentazione di Palazzo Pitti with the

collaboration of the Dipartimento di Biologia Vegetale dell'Università di Firenze, under the patronage of the Società Botanica Italiana. "Flora spontanea in Toscana: Diseggi a tempera di Jacopo Ligozzi," items #1–14; "Pittori fiamminghi ed olandesi," items #15–27; "Scuola romana e napoletana," items #28–38; "Fioranti attivi a Firenze," items #39–67; "Arti decorativa," items #69–76. With introductory essays: **M. Rizzotto.** Premessa allo studio botanico dei dipinti, p. 10; **M. Mosco.** Flora medico in Florentia, pp. 11–18; **E. Pacini.** Dalla parte del botanico, pp. 19–22.

Bc802. **Polska Akademija Nauk.** (No relevant general catalogue.)
For special information *see* **S. Chankowski.** 1965. Les archives de l'Académie Polonaise des Sciences. *Organon (Warszawa)* 1965 (2): 261–266.

Bc803. **Preussische Staatsbibliothek.** 1938. Die wichtigsten naturwissenschaftlichen Bilderhandschriften nach 1500 in der Preussischen Staatsbibliothek. By H. Wegener. *Zentralblatt für Bibliothekswesen* 55: 1–110.

Bc804. **Princeton University.** (No relevant general catalogue.)
For special information *see* **A. P. Clark.** 1960. The manuscript collections in the Princeton University Library: An introductory survey. [Rev. ed.] Princeton, NJ. 36 pp. First pub. in *Princeton University Library chronicle* 19: 159–190.
Also see J. J. Audubon MSS, Ba477A; E. G. Conklin papers, Ba656.

Bc805. **Public Record Office, London.** 1984. *Guide to the records relating to science and technology in the British Public Record Office.* By M. Jubb. Paris, General Information Programme & UNISIST, United Nations Educational, Scientific & Cultural Organization. v+310 pp.
Replaces ___. 1963–68. *Guide to the contents of the Public Record Office.* London, Her Majesty's Stationery Office. 3 vols. *Also* **C. D. Chalmers.** 1990. Historical and biographical sources for natural history in the Public Record Office. *Archives of natural history* 17: 315–324; **M. Jubb.** 1985. Scientific records in the British Public Record Office. *History of science* 23: 379–389.

Bc806. **Rice University, Woodson Research Center.** (No relevant general catalogue.)
For special list *see* J. S. Huxley papers, Ba1002.

Bc807. **Royal Botanic Gardens, Kew.** 1899. §4: *Manuscripts.* In: *Catalogue of the library [etc.]*, pp. 776–790. *Bulletin of miscellaneous information, Royal Botanic Gardens, Kew* Additional series III; *Continued* ___. 1919. §4: *Manuscripts.* In: ___: *(Supplement.) Additions received or incorporated in the years 1898–1915*, pp. 421–433. London. *Bulletin of miscellaneous information, Royal Botanic Gardens, Kew* Additional series III–2.

Bc808. ___. 1906. Collection of drawings of orchids by the late John Day. By W. B. H[emsley]. *Ibid.* 1906: 177–179.
17 vols containing ca.3,000 coloured drawings, with ca.500 in sepia.

Bc809. ___. 1916. Sir Arthur Church's collection of botanical drawings. *Ibid* 1916: 162–168.
Exhibition of 67 drawings presented to Kew Gardens by Lady Church and 20 others selected from the Kew collection.

Bc810. ___. 1956. The Roxburgh Flora Indica drawings at Kew. By J. R. Sealy. *Kew bulletin* 11: 297–399.

Bc811. ___. 1974. Purchase of a collection of fungus drawings including originals of Persoon's plates. By R. W. G. Dennis. *Ibid.* 29: 711–715.
Including work by Besemann, Guibal, Lamouroux, Letellier, Léveillé, etc.

Bc812. ___. 1988. *Flowers from the Royal Gardens of Kew: Two centuries of Curtis's Botanical Magazine.* By R. L. A. Stiff. Hanover, NH, & London, Univ. Pr. of New England. 80 pp., incl. 47 pl. (col.).

Catalogue of an exhibition of work by 15 artists prepared for the *Botanical magazine*. With essays by R. G. C. Desmond and C. Grey–Wilson.

Bc813. ___. 1988. *The flowers of Kew: 350 years of flower paintings from the Royal Botanic Gardens*. By R. Mabey. London, Century Hutchinson; New York, Atheneum. xvi+208 pp., illus.

Illustrated with approximately 175 coloured reproductions of original artwork from the Gardens' collections.

Also see Marianne North paintings, Ba1318.

Bc814. **Royal College of Surgeons of England.** (No relevant general catalogue.)
For special information *see* John Hunter drawings, Ba989.

Bc815. **Royal Geographical Society.** 1975–77. The R. G. S. archives. By C. Kelly. *Geographical journal* 141: 99–107, 1975; *Ibid.* 142: 117–130, 287–301, 1976; *Ibid.* 143: 73–85, 266–278, 1977.

Bc816. **Royal Horticultural Society, Lindley Library.** 1927. *Catalogue of books, pamphlets, manuscripts and drawings*. London. viii+488 pp.

Also **E. A. Bunyard.** 1927. The Hooker and Lindley drawings. *Journal, Royal Horticultural Society* 52: 218–224. (With a catalogue of this 10–vol. collection of drawings of fruit and flowers, executed by order of the Society from 1815 to 1929, which includes the work of W. Hooker, J. Lindley, C. J. Robertson, A. J. Withers, B. Cotton and E. D. Smith); **Elliott, B.** 1993. Explorations in the Lindley Library: Pictures at an exhibition. *Garden, journal of the Royal Horticultural Society* 118 (3): 108–112 (About the exhibition "Treasures of the Royal Horticultural Society" at Kew Gallery, R.B.G. Kew, 1993); **Royal Horticulturtural Society.** 1954. *Exhibition of manuscripts, books, drawings, portraits, medals and congratulatory addresses on the occasion of the Society's 150th anniversary celebrations*. London. 64 pp. & 5 pl. (186 items); **Synge, P.** 1953. Chinese flower paintings [collected by John Reeves]; important purchase by the Royal Horticultural Society. *Journal, Royal Horticultural Society* 78: 209–213.

Bc817. **Royal Institution of Great Britain.** (No relevant general catalogue.)
For special information *see* **I. M. McCabe & F. A. J. L. James.** 1984. History of science and technology resources at the Royal Institution of Great Britain. *British journal for the history of science* 17: 205–209. "Collections 10."

Bc818. **Royal Scottish Museum.** 1974. *Manuscripts, drawings, etc. in the Natural History Department of the Royal Scottish Museum*. By L. Soane. Edinburgh. 8 pp.

Bc819. ___. 1981+. *Manuscripts in the Royal Scottish Museum*. Edinburgh. Pt. 1+.
Also see Jardine papers, Ba1023; 2: Bruce papers, Ba598; 3: J. A. Harvie–Brown papers, Ba595.

Bc820. **Royal Society of London.** 1907. *Some account of the 'Classified papers' in the archives, with an index of authors*. By A. H. Church. Oxford, privately pub. 38 pp.

Bc821. ___. 1908. *Some account of the 'Letters and Papers' of the period 1741–1806 in the Archives with an index of authors*. By A. H. Church. Oxford, privately pub. 73 pp.

Bc822. ___. 1956. Guide to the archives of the Royal Society and to other documents in its possession. By R. K. Bluhm. *Notes & records of the Royal Society* 12: 21–39.
Replaces ___. 1840. *Catalogue of the miscellaneous manuscripts preserved in the library of the Royal Society*. London. 179 pp.
Appendix: Catalogue of manuscript letters in the possession of the Royal Society.

Bc823. ___. 1985. *The archives of the Royal Society*. Pt. III: *Miscellaneous manuscripts*. Bethesda, MD, Univ. Publications of America. Microfilm, 10 reels with printed guide.

Bc824. ___. 1990. *The early letters and classified papers, 1660–1740*. Bethesda, MD, Univ. Publications of America. Microfilm, 23 reels with printed guide.
Also see Microfilm copies of the journal books & Council minutes, Bb437 & Bb437.

Bc825. **Schiller–Nationalmuseum zu Marbach am Neckar.** (No relevant general catalogue.)
For special information *see* **D. Kuhn.** 1970. Das Cotta–Archive im Schiller–Nationalmuseum zu Marbach am Neckar: Korrespondenz mit deutschen Naturforschen im 19. Jahrhundert. *Isis* 61: 103–104.

Bc826. **Scott Polar Research Institute.** 1982. *Manuscripts in the Scott Polar Research Institute, Cambridge, England: A catalogue.* Ed. by C. Holland. New York, Garland Publishing. xii+815 pp.
Replaces ___. 1959. The manuscript collection of the Scott Polar Research Institute, Cambridge. By A. Savours. *Archives* 4: 102–108.

Bc827. **Scripps Institution of Oceanography Archives.** 1989. *Selected list of collections. Archives of the Scripps Institution of Oceanography, University of California, San Diego.* July 1989. La Jolla, CA, University of California. 53 pp.
Replaces ___. 1986. ___. La Jolla, CA. 31 pp.

Bc828. **Sheffield University Library.** (No relevant general catalogue.)
For special lists *see* Hartlib papers, Ba907, Krebs papers, Ba1064.

Bc829. **Smithsonian Institution.** 1983. *Guide to the Smithsonian archives.* Washington, DC, Smithsonian Institution Pr. xxxiv+431 pp. *Archives and special collections of the Smithsonian Institution* no. 4.
Also ___. 1985. *Manuscripts of the Dibner Collection, in the Dibner Library of the History of Science and Technology of the Smithsonian Institution Libraries.* By R. Maloy et al. Washington, DC, Smithsonian Institution Libraries; Canton, MA, Watson Pub. International. xiii+145 pp. *Smithsonian Institution Libraries research guide* no. 5; **M. T. Stieber.** 1979. Manuscripts produced and/or annotated by Agnes Chase pertinent to grass collections at the Smithsonian Institution. *Huntia* 3: 117–125; **E. B. Wells.** 1983. Charting the history of science. *A B bookman* 1983(Sept. 12): 1491–1492, 1494, 1496. (On MSS in Dibner Library). *Also see* Myers papers, Ba1302; Schmitt papers, Ba1470; Wetmore papers, Ba1605; and National Zoological Park, Washington, DC, archives, Bc791 & Bc792.

Bc829A. **___, National Museum of History & Technology.** 1978. *Guide to manuscript collections in the National Museum of History and Technology.* Washington, DC, Smithsonian Institution Pr. xiv+143 pp. *Archives and special collections of the Smithsonian Institution* no. 3.
Dept. of History of Science, pp. 19–43; Dibner Library, pp. 117–119; Other collections, pp. 121–123. *See also* Guide to the Museum's photographic collections Bc870.

Bc830. **Société de Géographie, Paris.** 1984. *Inventaire des manuscrits de la Société de Géographie.* By A. Fierro. Paris, Bibliothèque Nationale, Département des Cartes et Plans. 305 pp.

Bc831. **Staatsbibliothek, Handschtiftenabteilung, West Berlin.** (No relevant general catalogue.)
For special information *see* **F. B. Churchill.** 1969. Berlin: The Darmstaedter collection. *Mendel newsletter* 3: 1–2. Letters of famous scientists.

Bc832. **Stanford University Archives, Stanford University Libraries.** 1979. *A guide to the manuscript and archival collections of the Stanford University Archives.* By R.-L. Nilan. Stanford, CA. 72 ff.
Also ___. 1988. *Stanford University manuscript collections: History of science and technology.* By H. Lowood. Stanford, CA.
For special list *see* Jordan papers, Ba1040.

Bc832A. **Stazione Zoologica di Napoli** (Zoologische Station in Neapel). (No relevant general catalogue.)
For special information *see* **C. Groeben & K. R. Benson.** 1992. Sources in the history of American biology at the archives of the Stazione Zoologica "Anton Dohrn." *Mendel newsletter*

N.S.2: 5–10. A brief synopsis of the archives and a list of American scientists at the Stazione Zoologiuca (1873–1943) represented in the correspondence.

Bc833. **Stephen F. Austin State College.** (No relevant general catalogue.)
For special information *see* **R. S. Maxwell.** 1978. Forest history collections at Stephen F. Austin. *Texas libraries* 40: 88–94.

Bc834. **United States Library of Congress, Division of Manuscripts, Reference Department.** 1918. *Handbook of manuscripts in the Library of Congress.* Washington, DC. 750 pp.

Bc835. ___. 1922–26. *Accessions of manuscripts, broadsides and British transcripts, July 1920–21 – 1925.* Washington, DC. 5 vols.

Bc836. ___. 1975. *Manuscripts on microfilm: A checklist of the holdings in the Manuscript Division.* By R. B. Bickel. Washington, DC. 82 pp.

Bc837. ___. 1946+. *Library of Congress quarterly journal of current acquisitions.* Washington, DC. In progress.
Also Annual reports on acquisitions: Manuscripts, *Ibid.* vol. 3: 36–49, 1946; 4: 80–94, 1947; 6: 80–94, 1949; 7: 25–31, 1950; 8: 21–30, 1951; 9: 139–154, 1952; 10: 149–169, 1953; 11: 157–179, 1954; 12: 115–143, 1955; 13: 145–167, 1956; 14: 111–130, 1957; 15: 179–196, 1958; 16: 133–151, 1959; 17: 177–191, 1960; 18: 129–145, 1961; 19: 133–145, 1962; 20: 177–192, 1963; 21: 181–201, 1964; 22: 319–335, 1965; 23: 274–289, 1966; 24: 260–285, 1967; 25: 328–361, 1968; 26: 234–269, 1969; Continued as: Recent acquisitions of the Manuscript Division, *Ibid.* vol. 27: 332–375, 1970; 28: 281–325, 1971; 29: 327–349, 1972; 30: 295–337, 1973; 31: 235–276, 1974; 32: 172–206, 1975; 33: 332–382, 1976; 34: 339–379, 1977; 35: 244–281, 1978; 36: 401–446, 1979; *Continued* ___. 1979+. *Library of Congress acquisitions. Manuscript Division 1990.* Washington, DC.

Bc838. ___. 1958+. *Registers of papers in the Manuscript Division of the Library of Congress.* Washington, DC. No. 1+. In progress.

Bc839. ___. 1983. *National inventory of documentary sources in the United States.* Part 2: *Manuscript Division, Library of Congress.* Teaneck, NJ & Cambridge, Chadwyck–Healey Inc. 900 positive microfiche & 219 pp. index.
The published and unpublished registers of 762 manuscript collections.
Also **C. W. Garrison.** 1932. List of manuscript collections in the Library of Congress to July, 1931. *Annual report of the American Historical Association for 1930* I: 123–249. (*Also pub. separately* Washington, DC, 1932); **A. Melville.** 1980. *Special collections in the Library of Congress: A selective guide.* Washington, DC, Library of Congress. xv+464 pp.; **C. P. Powell.** 1939. List of manuscript collections received in the Library of Congress, July 1931 to July 1938. *Annual report of the American Historical Association for 1937* I: 113–145. (Also pub. separately: Washington, DC, 1939. 33 pp.); **N. Reingold.** 1960. Manuscript resources for the history of science and technology in the Library of Congress. *Quarterly journal of the Library of Congress* 17: 161–169.
Also see Palmer papers, Ba1338.

Bc840. ___, **Map Division, Reference Department.** 1962. *Maps showing explorers' routes, trails and early roads in the United States: An annotated list, [etc.].* By R. S. Ladd. Washington, DC, Library of Congress. vi+137 pp., map.

Bc841. **United States National Arboretum.** (No relevant general catalogue.)
For special list *see* **J. J. White & E. A. Neumann.** 1982. The collection of pomological watercolors at the U. S. National Arboretum. *Huntia* 4: 103–123.

Bc842. **United States National Archives & Records Service.** 1987. *Guide to the National Archives of the United States.* New ed. Washington, DC, National Archives Publications. xxxii+896 pp.
Federal governmental agencies records and papers, e.g. (with Record group no.) A: Bur. of Anim. Ind. records (RG17); B: Bur. of Entom. & Pl. Quarant. (RG7); C: Bur. of Pl. Ind., Soils, & Agric. Engin. (RG54); D: Civilian Conservation Corps (RG35); E: Coast & Geodetic Survey records

(RG 23); F: Dept. of Agric., incl: Agric. Section of the Patent Off., Off. of the Civilian Conserv. Corps, Cartog./audiovis. (incl. E. F. Smith colln of port. photos of scientists); G: Fish & Wildlife Serv. (RG22); H: Forest Serv. (RG95); I: Geol. Surv. (RG57), incl. Surveys, viz. U.S. Geol. & Geogr. Surv. of the Territories (Hayden Surv.) lett., U.S. Geogr. & Geol. Surv. of the Rocky Mt. Reg. (Powell Surv.) lett., the U.S. Geogr. Surv. W. of the 100th Merid. (Wheeler Surv.) commun. and lett., the Geol. Explor. of the 40th Parallel (King Surv.) G. M. Wheeler's report (1872); J: Gift collection of materials relating to polar regions (RG401); K: Natl Acad. of Sci. (RG189); L: Natl Park Serv. (RG79); M: Natl Sci. Fdn. (RG307); N: Off. of Exper. Stns (RG164); O: Off. of Sci. & Technol. (RG359); P: Smithsonian Inst. (RG106); Q: U.S. Coast Guard (RG26). A re-pub. of ___. 1974. ___. Washington, DC. xxv+884 pp. with a new preface and an appendix of *Record groups assigned between 1970 and 1977*, pp. 885–896. *See also* **Szucs, L. D., & S. H. Luebking.** 1988. *The archives: A guide to the National Archives field branches.* Salt Lake City, UT, Ancestry Publishing. xvii+340 pp., illus. The National Archives field branches, pp. 11–38; Record groups, pp. 51–321. Same groups as in above *Guide*, plus Environm. Prot. Agency (RG 412).

Bc842A. **United States National Archives & Records Service.** 1940–67. *National Archives accessions.* Washington, DC; *Continued as* ___. 1969+. *Prologue: The journal of the National Archives.* Washington, DC.

First title initially appeared quarterly but then irregularly from 1954–67; *Prologue* thrice yearly from 1969–72, now quarterly. *See also* various relevant lists and articles:

___. 1947. *Preliminary checklist of records of the Soil Conservation Service, 1928–43 (Record Group 114).* By G. A. Lee & F. F. Penney. Washington, DC. v+45 pp. *Publication* no. 47–17. *Preliminary checklist* no. 52.

___. 1952. *Preliminary inventory of the climatological and hydrological records of the Weather Bureau (Record Group 27).* By H. T. Pinkett, H. T. Finneran & K. H. Davidson. Washington, DC. iii+76 pp. *Publication* no. 52–12. *Preliminary inventory* no. 38.

___. 1954. *Preliminary inventory of the records of the Bureau of Plant Industry, Soils and Agricultural Engineering.* Compiled by H. T. Pinkett. Washington, DC, United States National Archives. v+49 pp. *Preliminary inventories* no. 66.

___. 1956. *Preliminary inventory of the records of the Bureau of Entomology and Plant Quarantine.* By H. T. Pinkett. Washington, DC. v+24 pp. *Publication* no. 57–2. *Preliminary inventories* no. 94.

___. 1958. *Preliminary inventory of the records of the Coast & Geodetic Survey (Record Group 23).* By N. Reingold. Washington, DC. v+83 pp. *Publication* no. 59–03. *Preliminary inventories* no. 105.

___. 1958. *Preliminary inventory of the records of the Bureau of Animal Industry (Record Group 17).* By H. T. Pinkett. Washington, DC, United States National Archives. v+18 pp. *Preliminary inventories* no. 106. *Publication* no. 59–04.

___. 1959. *List of photographs of irrigation projects of the Bureau of Reclamation.* By E. B. Haas, A. H. Henry & T. W. Ray. Washington, DC. vi+33 pp. *Publication* no. 60–12. *Special list* no. 15.

___. 1965. *Records of the Agricultural Research Service.* By W. F. Sherman. Washington, DC. ? pp.

___. 1965. *Preliminary inventory of the records of the Fish & Wildlife Service (Record Group 22).* By E. E. Hill. Washington, DC. iii+28 pp.

___. 1966. *Preliminary inventory of the records of the National Park Service (Record Group 79).* By E. E. Hill. Washington, DC. vii+52 pp. *Publication* no. 67–02. *Prelininary inventory* no. 166.

___. 1967. *Preliminary inventory of the cartographic records of the Forest Service (Record Group 95).* By C. M. Ashby. Washington, DC. vii+71p. *Publication* no. 67–5.

___. 1969. *Preliminary inventory of the records of the Forest Service (Record Group 95).* By H. T. Pinkett. Revised by T. W. Good. Washington, DC. vii+23 pp. *Publication* no. 69–10: *Preliminary inventories* no. 18 rev. *Replaces* ___. 1949. ___. Washington, DC. 17 pp. *Preliminary inventories* no. 18.

___. 1971. *Inventory of records of the Hydrographic Office (Record Group 37).* By M. Johnson & W. J. Heynen. Washington, DC. vii+28 pp. *Inventory* no. 4.

___. 1971. *Guide to cartographic records in the National Archives.* By C. M. Ashby et al.

Washington, DC. xi+444 pp. *Publication* no. 71–76.

___. 1973. *Geographical exploration and mapping in the 19th century: A survey of the records in the National Archives.* By R. E. Ehrenberg. Washington, DC. 22 pp. *Reference information paper* no. 66.

___. 1973. *Aerial photographs in the National Archives.* By C. E. Taylor & R. E. Spurr. Rev. ed. Washington, DC. vii+106 pp. *Special list* no. 25.

___. 1974. *Guide to materials on Latin America in the National Archives of the United States.* By G. S. Ulibarri & J. P. Harrison. Washington, DC. xii+489 pp. *Replaces* ___. 1961. ___. By J. P. Harrison. Washington, DC. *Publication* no. 62–3.

___. 1976. *List of record groups of the National Archives and Records Service.* Washington, DC. 64 pp.

___. 1976. *Agricultural maps in the National Archives of the United States ca.1860–1930.* By W. J. Heynen. Washington, DC. vii+25 pp. *Reference information papers* no. 75.

___. 1978. *United States Hydrographic Office manuscript charts in the National Archives.* By W. J. Heynen. Washington, DC. xi+250 pp. *Special list* no. 43.

___. 1979. *Preliminary inventory of the records of the Office of the Secretary of Agriculture (Record Group 16).* By H. F. Ulibarri. Washington, DC. vii+59 pp. *Preliminary inventories* no. 191. *Replaces* ___. 1962. *Preliminary inventory of the records of the Bureau of Agricultural and Industrial Chemistry.* Washington, DC. *Preliminary inventories* no. 149.

___. 1980. *Preliminary inventory of the records of the Civilian Conservation Corps (Record Group 35).* By D. Helms. Washington, DC. vii+23 pp. *Preliminary inventories* no. 11 (Rev.). *Replaces* ___. 1948. ___. By H. T. Pinkett. Washington, DC. iii+16 pp. *Publication no. 49–6. Preliminary inventories* no. 11.

___. 1990. *Guide to the holdings of the Still Picture Branch of the National Archives.* By B. L. Burger. Washington, DC. x+166 pp., illus.

Pt. I. Federal records; pt. II. Donated materials.

Also **S. J. Buck.** 1936. The National Archives and the advancement of science. *Science* 83: 379–385; **S. H. Karren.** Building a better mousetrap: records in the National Archives relating to science and technology. In: E. Mount, ed. *Sci–tech archives and manuscript collections*, pp. 57–67. New York, Haworth Pr. A re-issue of *Science & technology libraries* vol. 9, no. 4 as a separate book; **T. H. Peterson**, ed. 1980. *Farmers, bureaucrats and middlemen: Historical perspectives on American agriculture.* Washington, DC, Howard Univ. Pr. xvii+357 pp. *National Archives conferences* vol. 17. (Papers and proceedings of the Conference on American Agriculture, April 28–29, 1977, jointly sponsored by the National Archives & Records Administration & Agricultural History Society. Also explores the resources available at the National Archives for the study of agricultural history); **H. T. Pinkett.** 1947. Records of research units of the United States Forest Service in the National Archives. *Journal of forestry* 45: 272–275; ___. 1955. Records of the first century of interest of the United States Government in plant industries. *Agricultural history* 29: 38–45; **N. Reingold.** 1955. The National Archives and the history of science in America. *Isis* 46: 22–28.

Bc843. **United States National Park Service, Morristown National Historical Park.** 1967. *A guide to the manuscript collection.* Ed. by B. W. Stewart & J. Reilly. Morristown, NJ. 142 pp.

Bc844. **Università di Torino.** 1986. *Erbari e iconografia botanica: Storia delle collezioni dell' Orto Botanico dell' Università di Torino.* F. Montacchini, ed. Con la collaborazione di G. Forneris e C. Martoglio. Appendice a cura di S. Pettenati. Turin, Allemandi; Milan, Messaggerie. 158 pp. *Archivi di arte e cultura piemontesi* no. 7.

Bc844A. **Universitätsbibliothek Erlangen.** 1928+. *Katalog der Handschriften der Universitätsbibliothek Erlangen.* Erlangen. Vol. 1+.1928+.

For special information *see Katalog* ... vol. 5, Trew papers, Ba1563.

Bc845. **Universitätsbibliothek Uppsala.** (No relevant general catalogue.)

For special list *see* **H. W. Lack & C.–O. Von Sydow.** 1983–84. Dorflers Sammlung von Botanikerbriefen in der Universitatsbibliothek Uppsala. I(–III). Verzeichnis der Briefschreiber A. Verzeichnis der Briefempfunger. Miscellanea. *Willdenowia* 13: 397–428, 1983; *Ibid.* 14:

203–225, 435–456, 1984.
Also see Linnaean MSS, Ba1167.

Bc846. **Université de Liége.** 1959. *Catalogue des manuscripts scientifiques de la Bibliothèque de l'Université de Liége. Acquisition faites de 1886 à 1950.* By J. Hoyoux. Liége, Centre National d'Histoire des Sciences. iv+267 pp.

Bc847. **University of California, Berkeley.** (No relevant general catalogue.)
For special information *see* **A. E. Phillips.** 1973. *Guide to the special collections, University of California, Berkeley Library.* Metuchen, NJ, Scarecrow Pr. vi+157pp.
Also see Goldschmidt papers, Ba859.

Bc848. **University College, London.** (No relevant general catalogue.)
For special information *see* **J. Percival & W. A. Smeaton.** 1978. Library and archive resources in the history of science at University College. *British journal for the history of science* 11: 191–195. "Collections 7."
Also see Galton papers, Ba833.

Bc849. **University of Illinois at Urbana-Champaign.** (No relevant general catalogue.)
For special information *see* Shelford papers, Ba1485.

Bc850. **University of Kansas.** (No relevant general catalogue.)
For special information *see* Herrick papers, Ba931.

Bc851. **University of Leeds.** (No relevant general catalogue.)
For special information *see* **P. B. Wood & J. V. Golinski.** 1981. Library and archives resources in the history of science and medicine at the University of Leeds. *British journal for the history of science* 14: 263–281. "Collections 8."

Bc852. **University of Nebraska.** (No relevant general catalogue.)
For special list *see* Bessey papers, Ba540–541.

Bc853. **University of Strathclyde.** (No relevant general catalogue.)
For special list *see* Geddes papers, Ba836.

Bc854. **Victoria & Albert Museum.** (No relevant general catalogue.)
For special list *see* Bewick relics, Ba546.

Bc855. **Washington State University, Library.** 1976. *North–west botanical manuscripts. An indexed register of the papers, 1867–1957, of Wilhelm Nikolaus Suksdorf, William Conklin Cusick, Charles Vancouver Piper, Rolla Kent Beattie, and Harold St. John in the Washington State University Library.* Pullman, WA. 64 pp.

Bc856. **[Wellcome Institute for the History of Medicine] Wellcome Historical Medical Library.** 1962–76. *Catalogue of western manuscripts on medicine and science in the library.* By S. A. J. Moorat. London. 3 vols.
Vol. 1, Manuscripts written before AD 1650; vol. 2–3, Manuscripts written after AD 1650. Vol. 2 includes a supplement to vol. 1 and an index of writers of autograph letters in the library's collection.

Bc857. **Wellcome Institute for the History of Medicine.** 1989. *A supplementary catalogue of western manuscripts on medicine and science in the library of the Wellcome Institute for the History of medicine. First cumulative supplement.* By R. Palmer. London. 113.

Bc858. **Yale University Library.** (No relevant general catalogue.)
For special information *see* J. J. Audubon MSS, Ba477B; R. G. Harrison papers, Ba906; R. M. Yerkes papers, Ba1636.

Bc859. **Zentralbibliothek Zürich.** (No relevant general catalogue.)
For special list *see* Gesner papers, Ba849.

Bc860. **Zoological Society of London.** (No relevant general catalogue.)
For special list *see* **G. C. Low et al.** 1930. A classification of the original watercolour

paintings of birds of India by B. H. Hodgson ... C. F. Sharpe ... S. R. Tickell [etc.]. *Proceedings of the Zoological Society of London* 1930: 548–625.
Also see Jones drawings, Ba1034–1035.

BIBLIOGRAPHIES OF MANUSCRIPTS CATALOGUES

Bc861. **Evans, F. B.** 1975. *Modern archives and manuscripts: A select bibliography*. Chicago, IL, Society of American Archivists. xiii+209 pp.

Bc862. **Jayawardene, S. A.** 1978. Western scientific manuscripts before 1600: A checklist of published catalogues. *Annals of science* 35: 143–172.

Bc863. **Richardson, E. C.** 1935. *A list of printed catalogues of manuscript books*. New York, Wilson. v+386 pp. *A union world catalogue of manuscript books*, vol. 3. *Re-pub*. New York, Burt Franklin. 1972. ••*Also* **L. Bieler.** 1949. Les catalogues de manuscrits, premier supplément aux listes de Weinberger et de Richardson. *Scriptorium* 3: 303–327.

Bc864. **University of London Library.** 1968. *The palaeography collection in the Library, an author and subject catalogue*. Boston, MA, G. K. Hall. 2 vols.

PHOTOGRAPH COLLECTIONS

Guides to resources

Bc865. **Canada Public Archives, Manuscripts Division, National Photography Collection.** 1979. *Guide to Canadian photographic archives*. A. Clavet, coordinator. Ottawa. 228 pp.
See also ___. 1978. *Researchers' guide*. By A. Clavet. Ottawa. 31 pp.

Bc866. **Evans, M., & A. Nelki.** 1979. *The picture researcher's handbook: An international guide to picture sources and how to use them*. 2nd ed. London, Saturday ventures. 328 pp.

Bc867. **Forest Research Institute, Dehra Dun.** 1912. *Catalogue of the photographic collection at the ... Institute [etc.]*. Calcutta. ii+245 pp.

Bc868. **International Museum of Photography.** 1990. *Index to American photographic collections*. 2nd ed. Ed. by A. H. Eskind & G. Drake. Boston, MA, G. K. Hall. 701 pp.
Replaces ___. 1982. *An index to American photographic collections*. Ed. by J. McQuaid & P. P. Wilson. Boston, MA, G. K. Hall. xiii+407 pp.

Bc869. **Robl, E.**, ed. 1983. *Picture sources 4*. [4th ed.] New York, Special Libraries Association. xi+180 pp. & 8 pp. of illus.
Surveys over 900 collections. Collections, geographic, and subject indexes. *Replaces* **A. Novotny & R. Eakins**, eds. 1975. *Picture sources, collections of prints and photographs in the U. S. and Canada: A project of Picture Division, SLA, and American Society of Picture Professionals*. 3rd ed. New York, Special Libraries Association. xx+387 pp.

Bc870. **Smithsonian Institution.** 1989+. *Guide to photographic collections at the Smithsonian Institution*. Ed. by D. V. O'Connor. Vol. 1+. Washington, DC, Smithsonian Institution Pr.
To be completed in 5 vols. Vol. 1: *National Museum of American History*. 1989. (*Includes* Agriculture & natural resources, pp. 13–17; Medical sciences, pp. 193–204); vol. 2: *National Museum of Natural History, National Zoological Park, Smithsonian Astrophysycal Observatory, Smithsonian Tropical Research Institute*. 1991. x+245 pp. (*Includes* NH. Dept. of Anthropology, Dept. of Botany, Dept. of Entomology, Dept. of Invertebrate Zoology, Dept. of Mineral Sciences, Dept. of Paleobiology, Dept. of Vertebrate Zoology, Off. of the Director, Off. of Education, Off. of Exhibits, Smithsonian Oceanographic Sorting Center, U.S. Biological

Survey, Research Off. of the Assistant Secretary for Research; ZP. Dept. of Animal Health, Dept. of Conservation, Dept. of Pathology, Off. of Design & Exhibit Planning, Off. of Education; SA. ...; ST. Scientific Staff Collections, Off. of Facilities Management, Off. of Photographic Services, Library).

Bc871. **Wall, J.**, ed. 1977. *National photographic record. Directory of British photographic collections.* London, Heinemann. 266 pp.
Ca.1,600 collections listed. Classified arrangement with subject, photographer, owner and location indexes.

PART C — CORE BIBLIOGRAPHY

Section Ca. HISTORY OF NATURAL HISTORY

HISTORIES OF NATURAL HISTORY

General

Ca1. **Academy of Natural Sciences of Philadelphia.** 1977. *Changing scenes in the natural sciences, 1776–1976.* Ed. by C. E. Goulden. Philadelphia, PA. 362 pp. *Special publication* no. 12.
Sixteen articles by various authors under the following heads: Evolution and systematics, Population genetics, Terrestrial ecology, Aquatic ecology, Behavior and sociobiology. *Includes* **G. E. Hutchinson.** The influence of the New World on the study of natural history, pp. 13–34; **E. Mayr.** The study of evolution, historically viewed, pp. 39–58; **P. H. Raven.** The systematics and evolution of higher plants, pp. 59–83; **R. Patrick.** The changing scene in aquatic ecology, pp. 205–222.

Ca2. **Adams, A. B.** 1969. *Eternal quest: The story of the great naturalists.* New York, G. P. Putnam's Sons. 509 pp. & 16 pl.

Ca3. **Atran, S.** 1990. *Cognitive foundations of natural history: Towards an anthropology of science.* Cambridge, Cambridge Univ. Pr. xii+360 pp.

Ca4. **Ballauff, T.** 1954. *Die wissenschaft vom Leben. Eine Geschichte der Biologie.* Vol. 1: *Vom Altertum bis zur Romantik.* Freiburg, K. Alber. xii+432 pp. & 24 pl. *Orbis academicus* vol. II/7.
See E. Ungerer (below) for vol. 3.

Ca5. **Bäumer, Ä.** 1988. *Die Geschichte der Biologie von der Antike bis zur Renaissance.* Mainz, Johannes Gutenberg–Universität. iii+258 pp. *Geschichte der Biologie* vol. 1.

Ca6. **Bodenheimer, F. S.** 1958. *The history of biology: An introduction.* London, Dawson. 465 pp. & 20 pl.

Ca7. **Callot, E.** 1986. *Les étapes de la biologie: Histoire de la biologie illustrée par la vie et les textes des plus grands naturalistes.* Paris, Champion; Geneva, Slatkine. 440 pp.

Ca8. **Candolle, A. L. P. P. de.** 1885. *Histoire des sciences et des savants depuis deux siècles, [etc.].* 2. éd. Geneva, H. Georg. xvi+594 pp. *Re-pub.* [Paris], Fayard, 1987. 313 pp. *Corpus des oeuvres de philosophie en langue française.*
Replaces **___.** 1873. ___. Geneva, [etc.]. vii+482 pp. For German version *see* **___.** 1911. *Zur Geschichte der Wissenschaften und der Gelehrten seit zwei Jahrhunderten.* Transl. by W. Ostwald. Leipzig. xx+466 pp. & port. *Grosser Männer.*

Ca9. **Cuvier, G. L. C. F. D.** 1841–45. *Histoire des sciences naturelles, depuis leur origine jusqu'à nos jours, chez tous les peuples connus, ... complétée, rédigée, annotée et publiée par MM. de Saint–Agy.* Paris, Fortin, Masson & Cie. 3 vols. in 5. *Re-pub.* Brussels, Culture et Civilisation, 1969; Farnborough, Gregg International, 1970.

Ca10. **Gardner, E. J.** 1972. *History of biology.* 3rd ed. Minneapolis, MN, Burgess Publishing Co. vii+469 pp.

Replaces ___. 1960. ___. 2nd ed. v+376 pp.; ___. 1960. *History of life science: Outlines and references*. Minneapolis [etc.]. vi+180 pp.

Ca11. **Giordan, A.**, ed. 1987. *Histoire de la biologie*. Paris, Technique et Documentation — Lavoisier. 2 vols. *Petite collection d'histoire des sciences*.

Ca12. **Glass, B.** 1979. Milestones and rates of growth in the development of biology. *Quarterly review of biology* 54: 31–53.

Ca13. **Gould, S. J.** 1977. *Ever since Darwin: Reflections in natural history*. New York, Norton. 285 pp.; ___. 1980. *The panda's thumb: more reflections in natural history*. New York, [etc.]. 343 pp.; ___. 1983. *Hen's teeth and horse's toes*. New York, [etc.]. 413 pp.; ___. 1985. *the flamingo's smile: [etc.]*. New York, [etc.]. 476 pp.; ___. 1991. *Bully for Brontosaurus: [etc.]*. New York, [etc.]. 540 pp.
Collections of essays, including many historical, first pub. in *Natural history* magazine.

Ca14. **Hoppe, B.** 1976. *Biologie: Wissenschaft von der belebten Materie von der Antike zur Neuzeit. Biologische Methodologie und Lehren von der stofflichen Zusammensetzung der Organismen*. Wiesbaden, Steiner. vii+368 pp. *Sudhoffs Archiv, Beihefte* no. 17.

Ca15. **Jahn, I.** 1990. *Grundzüge der Biologiegeschichte*. Jena, G. Fischer. 507 pp.

Ca16. **___, R. Löther, & K. Senglaub**, eds. 1985. *Geschichte der Biologie: Theorien, Methoden, Institutionen, Kurzbiographien*. 2nd ed. Jena, G. Fischer. 864 pp., illus. (140 port.)
Chapters by various authors. *Replaces* 1st ed., Jena, 1982. 859 pp.

Ca17. **Jenkins, A. C.** 1978. *The naturalists: Pioneers of natural history*. London, Hamish Hamilton. 200 pp., illus.

Ca18. **Magner, L. N.** 1979. *A history of the life sciences*. New York, M. Dekker. xi+489 pp.

Ca19. **Mayr, E.** 1982. *The growth of biological thought: Diversity, evolution, and inheritance*. Cambridge, MA, Belknap Pr. of Harvard Univ. xiii+974 pp.
For German version *see* ___. 1984. *Die Entwicklung der biologischen Gedankenwelt. Vielfalt, Evolution und Vererbung*. Berlin, Springer. xxi+766 pp.

Ca20. **Mikulinskii, S. R.**, ed. 1972. *Istoriya biologii s dreveneishikh vremen do nachala XX veka. [History of biology from the beginning till the commencement of the 20th century.]* Moscow, Izd. "Nauka." 562 pp. *Istoriya biologii s drevneishikh vremen do nashikh dnei*.

Ca21. **Moore, R.** 1967. *The coil of life: The story of the great discoveries on the life science*. New York, A. Knopf; London, Constable. xv+418+8 pp.

Ca22. **Nordenskiold, E.** 1928. *The history of biology: A survey*. Translated from the Swedish by L. B. Eyre. New York, A. Knopf. xii+629+xv pp. *Re-pub*. New York, Tudor Publishing Co., 1935, 1942, 1946, 1949 & 1960.
Translation of *Biologins historia: En överblick*. Stockholm, Björck & Börjesson, 1920–24. 3 vols. Another ed: *Geschichte der Biologie: Ein Überblick*. Deutsch von G. Schneider. Jena, G. Fischer, 1926. vii+648 pp.

Ca23. **Peattie, D. C.** 1936. *Green laurels: The lives and achievements of the great naturalists*. New York, Simon & Schuster. xxiii+368 pp. & 30 pl.

Ca24. **Rádl. E.** 1905–09. *Geschichte der biologischen Theorien*. Leipzig, W. Engelmann. 2 vols.
Pt 1: Geschichte der biologischen Theorien seit dem Ende des siebzehnten Jahrhunderts; pt. 2: Geschichte der Entwicklungstheorien in der Biologie des neunzehnten Jahrhunderts.

Ca25. ___. 1913. *Geschichte der biologischen Theorien in der Neuzeit*. 1 Teil. 2nd ed., Leipzig, W. Engelmann, 1913. xiii+351 pp.

For English version *see* ___. 1930. *The history of biological theories.* Translated & adapted from the German by E. J. Hatfield. Oxford, Oxford Univ. Pr. xii+408 pp.

Ca26. **Schiller, J.** 1978. *La notion d'organisation dans l'histoire de la biologie.* Paris, Maloine. vi+133 pp.

Ca27. **Schöler, W.** 1970. *Geschichte des naturwissenschaftlichen Unterrichts im 17. bis 19. Jahrhundert: Erziehungs–theoretische Grundlegung und schulgeschichtliche Entwicklung.* Berlin, W. de Gruyter. 373 pp.

Ca28. **Singer, C. J.** 1959. *A history of biology to about the year 1900: A general introduction to the study of living things.* 3rd rev. ed. London & New York, Abelard–Schuman. xxxvi+580 pp. *The life of science library* no. 38. *Re-pub.* Ames, IA, Iowa State Univ. Pr., 1989. *History of science & technology reprint series.*
Replaces ___. 1950. ___. 2nd ed. New York, [etc.]. For French version *see* ___. 1934. *Histoire de la biologie.* Éd. française par F. Gidon. Paris, Payot. 613 pp. *Bibliothèque scientifique.*

Ca29. **Sirks, M. J., & C. Zirkle.** 1964. *The evolution of modern biology.* New York, Ronald Pr. v+376 pp.

Ca30. **Smit, P., & R. J. Ch. V. ter Laage.** 1970. *Essays in biohistory.* Utrecht, International Association for Plant Taxonomy. xxii+426 pp. *Regnum vegetabile* vol. 71.
27 articles by various authors. References to some are entered in the relevant sections herein.

Ca31. **Snyder, E. E.** 1940. *Biology in the making.* New York, McGraw Hill Book Co. xii+539 pp.

Ca32. **Society for the Bibliography of Natural History.** 1981. *History in the service of systematics: Papers from the conference to celebrate the centenary of the British Museum (Natural History) 13–16 April, 1981.* Ed. by A. C. Wheeler & J. Price. London. v+164 pp. *Special publication* no. 1.
Twenty–one articles by various authors. References to some are entered in the relevant sections herein.

Ca33. **Taton, R.**, ed. 1963–66. *History of science.* Translated by A. J. Pomerans. New York, Basic Books. 4 vols.
Contributions by various authors. Vol. 1: *Ancient & medieval science from the beginnings to 1450.* 1963; vol. 2: *The beginnings of modern science: From 1450 to 1800.* 1964. (Pt. I. The Renaissance. The descriptive sciences. Ch. 4. Zoology, Ch. 5. Botany, pp. 149–166; Pt. II. The 17th century. The descriptive sciences. Ch. 1. Human & animal biology, Ch. 3. Botany, pp. 333–350, 363–374; Pt. III. The 18th century. Biological & geological sciences. Ch. 1. General biology, Ch. 2. Animal physiology, Ch. 4. Botany, pp. 511–539, 554–564); vol. 3: *Science in the nineteenth century.* 1965. (Pt. V. The biological sciences. Structure & function. Ch. 1. Cytology & histology, Ch. 2. Zoology, Ch. 3. Botany, Ch. 4. Pasteur & microbiology; Ch. 5. Plant physiology, Ch. 6. Animal physiology; The origin of forms. Ch. 1. Comparative vertebrate anatomy, Ch. 2. Vertebrate paleontology, Ch. 3. Theories of animal generation, Ch. 4. Sexuality & reproduction, Ch. 5. Theories of evolution, Ch. 6. The origins of genetics, pp. 348–547); vol. 4: *Science in the twentieth century.* 1966. (Pt. IV. Biology. Introduction, Ch. 1. Elementary life, Ch. 2. Some major problems in animal biology, Ch. 3. Zoology, Ch. 4. Genetics & evolution, Ch. 5. Comparative vertebrate anatomy & paleontology, Ch. 6. Prehistory, Ch. 7. Plant physiology, Ch. 8. Botany, pp. 375–499). Translation of *Histoire générale des sciences.* Paris, Presses Universitaires de France, 1957–61. 4 vols.

Ca34. ___. 1966–83. *Histoire générale des sciences.* 2ème éd. Paris, Presses Universitaires de France. 3 vols.
Vol. 1: *La science antique et médiévale: Des origines à 1450.* By R. Arnaldez, et al. 1966; vol. 2: *La science moderne: De 1450 à 1800.* By E. Bauer, et al. 1969; vol. 3, pt. 1: *La science contemporaine*, pt. 1: *Le XIXe siècle*, pt. 2: *Le XXe siècle.* By F. Aubelès, et al. 1981–83. 4 vols.

Ca35. **Taylor, G. R.** 1963. *The science of life: A picture history of biology.* London, Thames & Hudson. 368 pp. & 16 pl. *Also* New York, McGraw–Hill Book Co. *Re-pub.*

London, Panther Books, 1967.
For French version *see* ___. 1963. *Histoire illustrée de la biologie*. Paris, Hachette. 368 pp.; ___. 1963. *Das Wissen vom Leben. Eine Bildgeschichte der Biologie*. Munich, Zürich, Droemer–Knaur. 368 pp.

Ca36. **Terres, J. K.**, ed. 1961. *Discovery: Great moments in the lives of outstanding naturalists*. Philadelphia, PA, Lippincott. xiii+338 pp.

Ca37. **Ungerer, E.** 1966. *Die Wissenschaft vom Leben*. Vol. 3: *Der Wandel der Problemlage der Biologie in den letzten Jahrzehnten*. Freiburg & Munich, Alber. 405 pp.
See T. Ballauff (above) for vol. 1.

See also Ba1–59.

15th to 18th centuries

Ca38. **Callot, E.** 1951. *La renaissance des sciences de la vie au XVIe siècle*. Paris, Presses Universitaires de France. v+204 pp. *Bibliothèque de philosophie contemporaine: Logique et philosophie des sciences*.

Ca39. ___. 1965. *La philosophie de la vie au XVIIIe siècle étudiée chez Fontenelle, Montesquieu, Maupertuis, La Mettrie, Diderot, d'Holbach, Linné*. Paris, M. Rivière & Cie. 440 pp. *Bibliothèque philosophique*.

Ca40. ___. 1966. *Histoire de la philosophie biologique par les textes*. Paris, Éditions Doin. 436 pp.

Ca41. **Crump, P. E.** 1928. *Nature in the age of Louis XIV*. London, G. Routledge & Sons. xv+224 pp.

Ca42. **Eisnerová, V.** 1972. Research methods in biology in the late 18th century. *Dejiny ved a techniky* 5: 185–201.

Ca43. **Guyénot, E.** 1941. *Les sciences de la vie aux XVIIe et XVIIIe siècles: L'idée d'évolution*. Paris, A. Michel. xxi+462 pp. *L'Évolution de l'humanité, synthèse collective* 3. sect., no. 68. *Re-pub.* 1957.

Ca44. **Lepenies, W.** 1976. *Das Ende der Naturgeschichte: Wandel kultureller Selbstverständlichkeiten in den Wissenschaften des 18. und 19. Jahrhunderts*. Munich, C. Hanser. 277 pp.

Ca45. **Lloyd, C.** 1985. *The travelling naturalists*. London, Croom Helm. 156 pp. *Also* Seattle, VA, Univ. of Washington Pr.
Eight 19th–century naturalists.

Ca46. **Meyer–Abich, A.** 1949. *Biologie der Goethe–Zeit. Klassische Abhandlungen über die Grundlagen und Hauptprobleme der Biologie von Goethe und den grossen Naturforschern seiner Zeit: Georg Forster, Alexander von Humboldt, Lorenz Oken, Carl Gustav Carus, Karl Ernst von Baer und Johannes Müller. Herausgegeben, geistesgeschichtlich eingeleitet und erläutert [etc.]*. Stuttgart, Marquardt. 302 pp.

Ca47. **Miall, L. C.** 1912. *The early naturalists: Their lives and work, 1530–1789*. London, Macmillan & Co. xi+396 pp. *Re-pub.* New York, Hafner Publishing Co., 1969.

Ca48. **Omodeo, P.** [1964.] La biologia nei secoli XVII e XVIII. Milan, Marzorati. [40] pp. *Nuove questioni di storia moderna* vol. 12: 895–935.

Ca49. **Porter, R.** 1988. The new taste for nature in the 18th century. *Linnean* 4(1): 14–30.

Ca50. **Ritterbush, P. C.** 1964. *Overtures to biology: The speculations of eighteenth–century naturalists*. New Haven, CT, Yale Univ. Pr. ix+287 pp. & 8 pl.

Ca51. **Roger, J.** 1980. The living world. In: G. S. Rousseau & R. Porter, eds. *The ferment of knowledge: Studies in the historiography of eighteenth–century science*, pp. 255–283.

Cambridge, Cambridge Univ. Pr. xiii+500 pp.
With eleven other chapters on different disciplines by various authors.

Ca52. **Sarton, G.** 1955. *The appreciation of ancient and medieval science during the Renaissance (1450–1600).* Philadelphia, PA, Univ. of Philadelphia Pr. xvii+233 pp. *Re-pub.* New York, A. S. Barnes & Co., 1961. *A Perpetua book* P-4039.
Expanded versions of Rosenbach Fellowship lectures, January 1953. Lecture II. Natural history, pp. 52–132.

Ca53. ___. 1957. *Six wings: Men of science in the Renaissance.* Bloomington, IN, Indiana Univ. Pr. xiv+318 pp. & 8 pl.
Includes 1st wing. *The frame of the Renaissance: Exploration and education*; 2nd ... *Mathematics and astronomy*; 3rd ... *Physics, chemistry, technology*; 4th ... *Natural history*; 5th ... *Anatomy and medicine*; 6th ... *Leonardo da Vinci: Art and science.*

Ca54. **Schiller, J.** 1974. Queries, answers and unsolved problems in 18th century biology. *History of science* 12: 184–199.

Ca55. **Union Internationale d'Histoire et de Philosophie des Sciences, Division d'Histoire des Sciences.** 1960. *La science au XVIe siècle. Colloque international de Royaumont, Paris, 1–4 Juillet 1957.* Paris, Hermann. 348 pp. *Histoire de la pensée* vol. 2.
Includes **E. Wickersheimer.** Livre des quadrupèdes de Michel Herr, médecin strasbourgeois (1546), pp. 265–283; **F. S. Bodenheimer.** Towards the history of zoology and botany in the XVIth century, pp. 284–296.

19th & 20th centuries

Ca56. **Abir–Am, P. G., & D. Outram**, eds. 1987. *Uneasy careers and intimate lives: Women in science, 1789–1979.* New Brunswick, NJ, Rutgers Univ. Pr. xiii+365 pp.
Articles by various authors. *Includes* **A. B. Shteir.** Botany in the breakfast room: Women and early 19th–century British plant study, pp. 31–44; **M. G. Ainley.** Field work and family: North American women ornithologists, 1900–1950, pp. 60–76; **N. G. Slack.** Nineteenth–century American women botanists: Wives, widows, and work, pp. 77–103.

Ca57. **Allen, G. E.** 1975. *Life science in the twentieth century.* New York, J. Wiley & Sons. xxv+258 pp., illus. *Re-pub.* (with corrections) Cambridge, Cambridge Univ. Pr., 1978.

Ca58. **Blyakher, L. Ya.**, ed. 1975. *Istoriya biologii s nachala XX veka do nashikh dnei. [History of biology from the beginning of the 20th century to our day].* Moscow, Izd. "Nauka." 658 pp.

Ca59. **Borell, M.** 1989. *Album of science: The biological sciences in the twentieth century.* New York, C. Scribner's Sons. xiv+306 pp., illus.
A pictorial survey. 1: The new century; 2: The experimental method in twentieth–century biology; 3: Transformation of the natural history tradition; 4: Investigating life processes; 5: The human animal; 6: Applying biology to social needs; 7: The nature of the organism.

Ca60. **Coleman, W.** 1971. *Biology in the XIXth century: Problems of form, function and transformation.* New York, J. Wiley & Sons. vii+187 pp. *Re-pub.* Cambridge, Cambridge Univ. Pr., 1977.

Ca61. **Cunningham, A., & N. Jardine**, eds. 1990. *Romanticism and the sciences.* Cambridge, Cambridge Univ. Pr. xxii+345 pp.
Includes **D. M. Knight.** Romanticism and the sciences; **T. Lenoir.** Morphotypes and the historical–genetic method in romantic biology; **E. Richards.** "Metaphorical mystifications": The romantic gestation of nature in British biology; **P. F. Rehbock.** Transcendental anatomy; **L. S. Jacyna.** Romantic thought and the origin of the cell theory; **M. Nicolson.**

Alexander von Humboldt and the geography of vegetation; **N. A. Rupke.** Caves, fossils and the history of the earth; etc.

Ca62. **Cuvier, G. L. C. F. D.** 1826–31. *Histoire des sciences naturelles, depuis 1789 jusqu'à ce jour.* Paris, Baudouin Frères. 4 vols. *Oeuvres complètes de Buffon. Complément.* Tome I (–IV). Another ed. Brussels, Société Belge de Librairie, 1837–38, 2 vols.

Ca63. **Dawes, B.** 1952. *A hundred years of biology.* London, G. Duckworth & Co. 429 pp.

Ca64. **Farber, P. L.** 1982. Discussion paper: The transformation of natural history in the nineteenth century. *Journal of the history of biology* 15: 145–152.

Ca65. **Leisewitz, A.** 1982. *Von der Darwinschen Evolutionstheorie zur Molekularbiologie: Wissenschaftshistorische und soziologische Studien zu einer materialistichen Geschichte der Biologie.* Cologne, Pahl–Rugenstein. 474 pp.

Ca66. **McKenzie, J. M.**, ed. 1990. *Imperialism and the natural world.* Manchester, Manchester Univ. Pr. viii+216 pp. *Studies in imperialism.*
Includes **R. H. Grove.** Colonial conservation, ecological hegemony and popular resistance: Towards a global synthesis; **D. Kumar.** The evolution of colonial science in India: Natural history and the East India Company; **M. Worboys.** The Imperial Institute: The state and the development of the natural resources of the Colonial Empire, 1887–1923.

Ca67. **Mendelsohn, E.** 1964. The biological sciences in the nineteenth century: Some problems and sources. *History of science* 3: 39–59.

Ca68. **Merrill, L. L.** 1989. *The romance of Victorian natural history.* New York, Oxford Univ. Pr. xi+288 pp.
Chiefly concerned with the literary study of Victorian natural history writings.

Ca69. **Müller, F. C.** 1902. *Geschichte der organischen Naturwissenschaften im neunzehnten Jahrhundert. Medizin und deren Hilfswissenschaften, Zoologie und Botanik.* Berlin, G. Bondi. xv+714 pp. *Das neunzehnte Jahrh. in Deutschlands Entwickl.* vol. 6.

Ca70. **Olby, R. C., & M. J. S. Hodge**, eds. 1985. The development of the life sciences in the twentieth century. *Annals of science* 42 (3): 187–345.
Special issue. *Includes* **O. G. Meijer.** Hudo de Vries no mendelian?, pp. 189–232; **L. Darden.** Hugo de Vries's lecture plates and the discovery of segregation, pp. 233–242; **D. R. Weiner.** The roots of 'Michurinism'; Transformist biology and acclimatization as currents in the Russian life sciences, pp. 243–260; **N. Roll–Hansen.** A new perspective on Lysenko?, pp. 261–278; **J. Harwood.** Geneticists and the evolutionary synthesis on interwar Germany, pp. 279–301; **P. Weindling.** Weimar eugenics: The Kaiser Wilhelm Institute for Anthropology, Human Heredity and Eugenics in social context, pp. 303–318; **D. L. Hull.** Bias and commitment in science: Phenetics and cladistics, pp. 319–338.

Ca71. **Sander, K.**, ed. 1985. August Weismann (1834–1914) und die theoretische Biologie des 19. Jahrhunderts: Urkunden, Berichte und Analysen. *Freiburger Universitätsblätter* 24: 20–203.
Eighteen articles by various authors on the work and influence of Weisman and his contemporaries.

Ca72. **Shelishch, P. B.** 1982. A quantitative study of biologists in the 18th and 19th centuries. *Scientometrics* 4: 317–329.

Ca73. **Society for the History of Natural History.** 1985. *From Linnaeus to Darwin: Commentaries on the history of biology and geology. Papers from the fifth Easter meeting of the Society for the History of Natural History 28–31 March, 1983, "Natural history in the early nineteenth century."* Ed. by A. Wheeler & J. H. Price. London. v+214 pp. *Special publication* no. 3.
16 papers by various authors. References to some are entered in the relevant sections herein.

Chronologies

Ca74. **Asimov, I.** 1989. *Chronology of science and discovery.* New York, Harper & Row. [xxv]+707 pp.
Covers prehistory to 1988 with name and subject indexes.

Ca75. **Darmstaedter, L.** 1908. *Handbuch zur Geschichte der Naturwissenschaften und der Technik.* 2. Aufl. Berlin, Springer. 1,262 pp. *Re-pub.* n.p., Kraus Reprint Corp., 1960.

Ca76. **Gascoigne, R. M.** 1987. *A chronology of the history of science, 1450–1900.* New York, Garland Publishing. xi+585 pp.

Ca77. **Hellemans, A., & B. Bunch.** 1988. *The timetables of science: A chronology of the most important people and events in the history of science.* New York, Simon & Schuster. x+656 pp.
Covers prehistory to 1988 with name and subject indexes. Includes a column for "Life science" or "Biology."

Ca78. **Power, D'A., S. & C. J. S. Thompson.** 1923. *Chronologia medica: A handlist of persons, periods and events in the history of medicine.* London, Bale. 278 pp.

Ca79. **Thompson, S. J.** 1988. *A chronology of geological thinking from antiquity to 1899.* Metuchen, NJ, Scarecrow Pr. 328 pp.

Ca80. **Valentin, H.** 1950. *Geschichte der Pharmazie und Chemie in Form von Zeittafeln. Unter besonderer Berücksichtigung der Verhältnisse in Deutschland.* 3. Aufl. Stuttgart, Wissenschaftliche Verlagsgesellschaft. 126 pp.

Encyclopaedias & dictionaries

Ca81. **Bynum, W. F., E. J. Browne, & R. Porter**, eds. 1981. *Dictionary of the history of science.* Princeton, NJ, Princeton Univ. Pr. [xxxiv]+494 pp. Also pub. as *Macmillan dictionary of the history of science.* London, Macmillan, 1983.
A dictionary of some 700 key ideas in science. Does not include biographies but has a biographical index to articles.

Ca82. **Mayerhöfer, J., et al.** 1959–[75?]. *Lexikon der Geschichte des Naturwissenschaften: Biographien, Sachwörter und Bibliographien.* Vols. 1–2(8). Vienna, Gebrüder Hollinek.
Later continued in: *Archiv der Geschichte der Naturwissenschaften [etc.].* Ed. by R. Hink. Vienna, Gebrüder Hollinek, Heft 1+, 1980/81+; *see* A93.
In progress. The *Lexikon* had reached the article "Edelsteinkunde" when its slow progress was terminated. It was replaced by the *Archiv*, a loose-leaf periodical issued in numbered Hefte with continuous pagination. The *Archiv* now publishes articles from subsequent parts of the alphabetical sequence of the *Lexikon* as and when they are completed. (The pagination unfortunately becomes disorganized when the loose leaves are filed in alphabetical order.) The *Archiv* also includes some original articles and notices of current publications and events.

BIBLIOGRAPHIES OF NATURAL HISTORY — GENERAL

• For fuller lists of bibliographies see:

Ca83. **Besterman, T.** 1965–66. *A world bibliography of bibliographies and of bibliographical catalogues, calendars, abstracts, digests, indexes and the like.* 4th ed. Lausanne, Societas Bibliographica. 5 vols.; *Continued* **A. F. Toomey.** 1977. *A world bibliography of bibliographies 1964–1974: A list of works represented by the Library of Congress printed catalog cards.* A decennial supplement to Theodore Besterman, *A world bibliography of bibliographies.* Totowa, NJ, Rowman & Littlefield. 2 vols. ••For separate thematic extracts *see* **T. Besterman.** 1971. *Biological sciences: A bibliography of bibliographies.* Totowa, NJ, Rowman & Littlefield. [xiv]+471 pp.; ___.

1971. *Agriculture: A bibliography of bibliographies*. Totowa, NJ, Rowman & Littlefield. [xiv]+303 pp.

Ca84. **Bridson, G. D. R., & A. P. Harvey.** 1971. A checklist of natural history bibliographies and bibliographical scholarship, 1966–1970. *Journal of the Society for the Bibliography of Natural History* 5: 428–467. *Continued* ___. 1973. ___, 1970–1971. *Ibid.* 6: 263–292.

Ca85. **Bronshtein, M. P., & V. V. Gnucheva.** 1956. *Bibliografiya estestvennonauchnoi literatury, [etc.]. [Bibliography of the literature of natural science, etc.]*. Moscow, Gosudarstvennoe Izdatel'stvo Kul'turno–Prosvetitel–noi Literatury. 183+[ix] pp.

Ca86. **Inke, G.** 1968. Bibliographie der Bibliographien der Medizin und der in der Medizin angewandten Naturwissenschaften. *Wissenschaftliche Zeitschrift der Martin–Luther Universität Halle–Wittenberg, Beilage* 1968: xiii+658 pp.

Ca87. **Levin, V. L., V. G. Levina, & D. V. Lebedev.** 1978. *Putevoditel' dlya biologov po bibliograficheskim izdaniyam: Spravochnoe posobie. [Guide to bibliographical works for biologists: Reference manual]*. Leningrad, Izd. "Nauka." 480 pp. ••*Replaces* **V. L. Levin.** 1960. *Spravochnoe posobie po bibliogragii dlya biologov*. Moscow/Leningrad, Akademii Nauk SSSR. 406 pp.

Ca88. **Petzholdt, J.** 1866. *Bibliotheca bibliographica: Kritisches Verzeichniss der das Gesammtgebiet der Bibliographie betreffenden Litteratur des In– und Auslandes in systematischer Ordnung. Mit alphabetischem Namen– und Sachregister*. Leipzig, W. Engelmann. xiii+939 pp. *Re-pub.* Nieuwkoop, De Graaf, 1961.

Ca89. **Simon, H.–R.** 1977. *Die Bibliographie der Biologie: Eine analytische Darstellung unter wissenschaftshistorischen und informationstheoretischen Gesichtspunkten*. Stuttgart, Hiersemann. x+315 pp. *Bibliothek des Buchwesens* vol. 4.

Ca90. **Smith, R. C., W. M. Reid, & A. E. Luchsinger.** 1980. *Smith's guide to the literature of the life sciences*. 9th ed. Minneapolis, MN, Burgess Publishing Co. xi+223 pp. ••*Replaces* 8th ed., 1972.

• A selection of major works:

Ca91. **Beckman, J.** 1770–1806. *Physikalisch–ökonomische Bibliothek worinn von den neuesten Büchern, welche die Naturgeschichte, Naturlehre, und die Land und Stadtwirtschaft betreffen, zuverlässige und volständige Nachrichten ertheilet werden*. Gottingen, Vandenhoeck. & Ruprecht. 23 vols.

Ca92. **Böhmer, G. R.** 1785–89 *Bibliotheca scriptorum historiae naturalis, oeconomiae aliarumque artium ac scientiarum ad illam pertinentium, realis systematica. Systematisch–literaerisches Handbuch der Naturgeschichte Oeconomie und anderer damit verwandten Wissenschaften und Künste*. Leipzig, Junius. 5 parts in 9 vols. ••Ca. 65,000 books and articles. Pars I: Scriptores generales (2 vols.); Pars II: Zoologi (2 vols.); Pars III: Phytologi (2 vols.); Pars IV: Mineralogi (2 vols.); Pars V: Hydrologi acc. index universalis. Index nominalis: A–Z [alphabetical by author], Anonymi [arranged in sections conforming to the sections of the bibliography]. *See also* **D. E. Johnson.** 1979. A concordance of abbreviations used in references to reviews in botanical bibliographies by Cobres (1782), Böhmer (1787) and Miltitz (1829). *Huntia* 3: 103–114.

Ca93. **Bourlière, F.** 1940. *Élémens d'un guide bibliographique du naturaliste*. Macon, Protat Frères. ix+302. *Continued* ___. 1941. ___: *Suppléments I et II*. Paris, P. Lechevalier, pp. 305–368. ••Covers zoology, botany, geology and ethnology. Supplément I – Additions; supplément II – Les grandes expéditions zoologiques.

Ca94. **Bowker, R. R.** [Pub.]. 1982–87. *Pure and applied science books 1876–1982*. New York. 6 vols. (x+7,784 pp.). ••Over 220,000 entries for works published or distributed in the USA. Vol. 6 name & title indexes.

Ca95. **Engelmann,W.** 1846. *Bibliotheca historico–naturalis. Verzeichniss der Bücher über Naturgeschichte welche in Deutschland, Scandinavien, Holland, England, Frankreich, Italien und Spanien in den Jahren 1700–1846 erschienen sind.* 1. Band. *Bücherkunde, Hülfsmittel, allgemeine Schriften, vergleichende Anatomie und Physiologie, Zoologie, Palaeontologie.* Leipzig, W. Engelmann. ix+786 pp. ••Comprehensive coverage of general natural history, pp. 1–202, 627–637. *Continued* **J. V. Carus & W. Engelmann.** 1861. *Bibliotheca zoologica. The literature of zoology, which has appeared in periodicals, transactions etc. and of the books published from 1846 to 1860.* Leipzig, W. Engelmann; London, Williams & Norgate. 2 vols. ••Comprehensive coverage of general natural history, pp. 1–202, 1606–1627; **O. W. Taschenberg.** 1886–1923. *Bibliotheca zoologica II: Verzeichniss der Schriften über Zoologie welche in den periodischen Werken enthalten und vom Jahre 1861–1880 selbständig erscheinen sind, mit Einschluss der allgemein–naturgeschichtlichen periodischen und palaeontologischen Schriften.* Leipzig, W. Engelmann. 8 vols. ••Comprehensive coverage of general natural history, vols. 1 and part of vol. 7.

Ca96. **Fibig, J., & B. S. Nau.** 1789–91. *Bibliothek der gesammten Naturgeschichte.* Frankfurt am Main. 2 vols.

Ca97. **Friedländer & Sohn, R.,** [Pub.]. 1886. *Bibliotheca historico-naturalis et mathematica. Lager–Catalog von R. Friedländer & Sohn … Naturwissenschaften – Exacte Wissenschaften.* Berlin. [1,491] pp. ••Over 50,000 entries.

Ca98. **Junk, W.,** [Pub.]. 1900–39. *Rara historico–naturalia et mathematica.* Berlin, The Hague. 2 vols. & supplement. *Re-pub.* Amsterdam, Asher, 1979. 1 vol., 294 pp. ••Bibliographical notes and, in many instances, good collations. Each vol. has its own index. The supplement comprises *Die historico–Bibliophile* by W. Junk, 1939, pp. 243–294. The 21 nos. of vol. 1 were first pub. in *Laboratorium et Museum* and *Bibliographische Zeitschrift.*

Ca99. **Müller, J. T.** 1780–84. *Einleitung in die oekonomische und physikalische Bücherkunde und in die damit verbundenen Wissenschaften bis auf die neuesten Zeiten.* Leipzig, Schwickert. 2 vols. ••Bd. 1, Klassification der Bücher, Bd. 2, Literar. und krit. Kentn. der Bücher. Includes natural history, agriculture, economic and scientific literature.

Ca100. **Münchhausen, O. von,** Baron. 1774–75. Bibliotheca botanico–physico–oeconomica. Des Hausvaters botanische, physikalische und oekonomische Bibliothek. In: *Der Hausvater*, 2. Aufl., 2. Theil, 1. Stuck. Hanover, Helwing. xlviii+976 pp

Ca101. **Prange, C. F.** 1784. *Systematisches Verzeichniss aller derjenigen Schriften welche die Naturgeschichte betreffen, von den ältesten bis auf die neuesten Zeiten.* Halle, Hendel. 446 pp.

Ca102. **Reuss, J. D.** 1801–21. *Repertorium commentationum a societatibus litterariis editarum.* Göttingen, Dieterich. 16 vols. *Re-pub.* New York, B. Franklin, 1961. *Burt Franklin bibliography & reference series* no. 29. ••Vol. 1, tom. I: *Historia naturalis, generalis et zoologia*, 1801.

Ca103. **Roe, K. E., & R. G. Frederick.** 1981. *Dictionary of theoretical concepts in biology.* Metuchen, NJ, Scarecrow Pr. xli+267 pp. ••Each heading followed by a chronologically arranged literature list.

Ca104. **Royal Society of London.** 1867–1925. *Catalogue of scientific papers. [First series.] (1800–1863).* Vols. 1–6. London, Her Majesty's Stationery Office. *Continued* ___. 1877–79. ___*: [Second series.] (1864–1873).* Vols. 7–8. London, [etc.]; ___. 1891–96. ___*: [Third series.] (1874–1883).* Vols. 9–10. London, Cambridge Univ. Pr; ___.

1902. ___: *(1800–1883). Supplementary volume.* Vol. 12. London [etc.]; ___. 1914–25. ___: *Fourth series. (1884–1900).* Vols. 13–19. Cambridge, Univ. Pr. ••68,577 authors represented. Besides the supplement, vol. 12, there are lists of additions in vol. 6, pp. 560–763 and vol. 19, pp. 867–868 and a list of additional cross–references in vol. 19, pp. 869–877. The separately published subject indexes are incomplete and do not cover the life sciences content. For background information *see* **D. de B. Beaver.** 1972. The Smithsonian origin of the Royal Society Catalogue of Scientific Papers. *Science studies* 2: 385–393; **C. A. Elliott.** 1970. The Royal Society Catalogue as an index to nineteenth century American science. *Journal of the American Society for Information Science* 21: 396–401.

Ca105. **Scheuchzer, J. J.** 1716. *Bibliotheca scriptorum historiae naturalis omnium terrae regionum inservientium. Historiae naturalis Helvetiae prodromus. Accessit Jacobi Le Long ... De scriptoribus historiae naturalis Galliae.* Zurich, H. Bodmer. [xiii]+251 pp. *Re-pub.* Zurich, 1751. [xiv]+241 pp.

See also Library cats. Bc186–357.

Incunabula

Ca106. **Klebs, A. C.** 1937. Incunabula scientifica et medica: Short title list. *Osiris* 4: 1–359. *Re-pub.* Hildesheim, Olms, 1963. *Also* **G. Sarton.** 1938. The scientific literature transmitted through the incunabula. (An analysis and discussion with sixty facsimiles). *Osiris* 5: 41–245, 60 facsimiles. A. Scientific incunabula; B. Iconography; C. Appendices.

Ca107. **Stillwell, M. B.** 1970. *The awakening interest in science during the first century of printing 1450–1550: An annotated checklist of first editions viewed from the angle of their subject content. Astronomy. Mathematics. Medicine. Natural science. Physics. Technology.* New York, Bibliographical Society of America. xxix+355 pp.

Encyclopædias & Dictionaries

Ca108. **Collison, R.** 1966. *Encyclopaedias: Their history throughout the ages. A bibliographical guide with extensive historical notes to the general encyclopaedias issued throughout the world from 350 BC to the present day.* New York, Hafner. xvi+334 pp.

Ca109. **Hughes, A.** 1951. Science in English encyclopædias 1704–1875 — I. *Annals of science* 7: 340–370. ••*Includes* A description of British encyclopædias, pp. 341–350.

Ca110. **Kéler, S. von.** 1963. Übersicht der lexikalischen und der Handbuchliteratur, 1567–1958. A: Lexikalische Werke. In: *Entomologisches wörterbuch mit besonderer Berücksichtigung der morphologisches Terminologie.* 3. Aufl., pp. 1–15, 25–29. Berlin, Akademie–Verlagsgesellschaft. ••1. Entomologische Wörterbücher; 2. Zoologische Wörterbücher; 3. Nomenklatorische Wörterbücher; 4. Naturwissenschaftliche Wörterbücher; Autorenregister. Chronologically arranged lists.

Ca111. **Newberry Library.** 1968. *The circle of knowledge: Encyclopedias past and present. An exhibition to commemorate the 200th anniversary of the Encyclopedia Britannica, on view at the Newberry Library April 9 – May 31, 1968.* Compiled, with an introductory essay by, J. M. Wells. Chicago, IL. 56 pp.

Ca112. **Tonelli, G.** 1971. *A short–title list of subject dictionaries of the 16th, 17th and 18th centuries as aids to the history of ideas.* London, Warburg Institute. [v]+64 pp. *Warburg Institute surveys* no. 4.

Ca113. **Walsh, S. P.** 1968. *Anglo–American general encyclopedias: A historical bibliography 1703–1967.* New York, R. R. Bowker Co. xx+270 pp.

Dissertations

• For a fuller list of dissertation bibliographies see:

Ca114. **Reynolds, M. M.**, ed. 1985. *A guide to theses and dissertations: An* *annotated bibliography of bibliographies*. Rev. ed. Phoenix, AZ, Oryx Pr. vii+263 pp. ••By nation and by subject with institution, name & title, and specific subject indexes.

• A selection of major bibliographies:

Ca115. **Bilboul, R. R., & F. L. Kent**, eds. 1975–77. *The retrospective index to theses of Great Britain and Ireland, 1716–1950*. Oxford, European Bibliographical Center–Clio Pr. 5 vols. & supp. ••Vol. 2: Applied sciences and technology (including agriculture). 1976. xi+159 pp.; Vol. 3: Life sciences, including biological and medical sciences. 1977. xi+327 pp.

Ca116. **Hjelt, O. E. A.** 1911. *Sveriges biologiska disputations– och program–litteratur, 1700–varterminen 1910*. Helsinki. viii+210 pp. *Bidrag till kännedom af Finlands natur och folk* H.70 no. 1. ••I. Disputationslitteratur, 1. Medizin, 2. Zoologi, 3. Botanik; II. Programlitteratur, 1. Medicinska program, 2. Naturalhistoriska program.

Ca117. **Lavaud, S.** 1955. *Catalogue des thèses de doctorat ès sciences naturelles soutenues à Paris de 1891 à 1954*. Paris, Faculté de Pharmacie, Bibliothèque. 257 pp. ••Supplements **A. Maire.** 1892; *see* Ca118.

Ca118. **Maire, A.** 1892. *Catalogue des thèses de sciences soutenues en France de 1810 à 1890 inclusivement*. Paris, Welter. xi+224 pp. *Bibliographie des thèses* no. 2. ••Chronologically arranged with author and subject indexes.

Ca119. **Mundt, H.** 1936–42. *Bio-bibliographisches Verzeichnis von Universitäts– und Hochschuldrucken (Dissertationen) vom Ausgang des 16. bis Ende des 19. Jahrhunderts*. Leipzig, Carlsohn. 2 vols. Reprinted New York, Johnson New York, Johnson Reprint Corporation, 1965. ••A–R unfinished (terminates at "Jo. Detlef Ritter"); *Continued* ___. 1977–80. ___. Vol. 3: *Ritter–Z*; vol. 4: *Personenregister*. Ed. by K. Wickert. Munich, K. G. Saur. 2 vols.

Ca120. **Storey, M.** 1957. Stanford University theses and dissertations of interest to researchers in natural history (systematics, morphology, ecology, etc.). *Circulars, Natural History Museum, Stanford University* 7: 1–58.

Ca121. **Stuart, M. M.** 1973. *A bibliography of master's theses in geography: American and Canadian universities*. Tualatin, OR, Geographic & Area Studies Publications. 275 pp.

Ca122. **Vallinkoski, J.** 1962–63. *Die Dissertationen der alten Universität Turku (Akademia Aboënsis) 1642–1828*. Helsinki. 272 pp. *Helsingfors Universitetsbiblioteks skrifter* no. 30.

Ca123. **Xerox University Microfilms.** 1973. *Comprehensive dissertation index (CDI), 1861–1972*. Ann Arbor, MI. 37 vols; Continued in annual supplements. ••Attempts to document all dissertations accepted at American universities from 1861 onwards, with an incomplete coverage of Canadian and overseas dissertations. Vols.11–13: Biological sciences; vol. 15: Agriculture; vols. 33–37; vol. 16: Geography and geology; Author index.

• The current bibliographic record for dissertations is:

Ca124. ***Dissertation abstracts*** (*later* ***Dissertation abstracts international***). Ann Arbor, MI, University Microfilms, 1952–69, etc. Preceded by: ***Microfilm abstracts.*** Ann Arbor, MI, vols. 1–11, 1938–51. (*See* A177).

Ca125. [Not used.]

Periodical bibliographies

• For a fuller list of periodical bibliographies see:

Ca126. **Owen, D. B.** 1985. *Abstracts and indexes in science and technology: A descriptive guide*. 2nd ed. Metuchen, NJ. xv+235 pp.

• A selection of major bibliographies:

Ca127. ***Bibliographia biotheoretica.*** *[Geschriften van de Prof. Dr. Jan der Hoeven–Stichting voor theoretische Biologie van dier en mensch. Series C.]* <1925–64>. Leiden, Rijksuniversiteit, E. J. Brill, vols. 1–8, 1938–74. ••Pub. irregularly covering spans of several years, i.e. 1925–29, 1930–34, 1935–39, 1940–44, 1945–49, 1950–54, 1955–59, 1960–64.

Ca128. ***Bibliotheca historico-naturalis et physico-chemica*** *oder systematisch geordnete Übersicht der in Deutschland und dem Auslande auf dem Gebiete der gesammten Naturwissenschaften neu erscheinen Bücher.* <1851–87> Herausgegeben von E. A. Zuchold [et al.]. Gottingen, Vandenhoeck & Ruprecht, vols. 1–37, 1851–87.

Ca129. ***Biological abstracts:*** *A comprehensive abstracting and indexing journal of the world's literature in theoretical and applied biology, exclusive of clinical medicine.* <1927+>. Menasha, WI, Philadelphia, PA, 1927+. ••For background information *see* **W. C. Steere.** 1976. *Biological Abstracts/BIOSIS: The first fifty years. The evolution of a major information service.* New York, Plenum Pr. xvii+233 pp.

Ca130. ***Bulletin analytique.*** <1940–55>. Paris, Centre National de la Recherche Scientifique (CNRS), vols. 1–16, 1940–55. (Vols. 3–16 issued in 2 parts. Part II: Sciences biologiques. (Sciences pharmacologiques). Industries alimentaires. Agriculture.) Continued as *Bulletin signalétique.*

Ca131. ***Bulletin signalétique.*** <1956+>. Paris, Centre National de la Recherche Scientifique (CNRS), vols. 17+, 1956+.

Ca132. ***International catalogue of scientific literature.*** *[Section:] L. General biology.* <1901–14>. London, Royal Society of London for the International Council, 1902–19. 14 vols. *Re-pub.* New York, Johnson Reprint Corporation, 1968, 14 vols in 1.

Ca133. ***Naturae novitates:*** *Bibliographie neuer Erscheinungen aller Länder auf dem Gebiete der Naturgeschichte und der exacten Wissenschaften.* <1879–1944>. Berlin, Friedländer. 1879–1944. 66 vols.

Ca134. ***Referativnyi zhurnal. Biologiya.*** <1953+> Moscow, Akademiya Nauk SSSR, Institut Nauchnoi Informatsii (*later* Vsesoyuzni Institut Nauchnoi i Tekhnicheskoi Informatsii <VINITI>), 1954+.

Ca135. ***Science citation index,*** *an interdisciplinary index to the literature of science, medicine, engineering, technology, and the behavioral sciences.* <1961+> Philadelphia, PA, Institute for Scientific Information, 1963+.

Periodicals

• A small selection of major retrospective bibliographies:

Ca136. **Bolton, H. C.** 1898. A catalogue of scientific and technical periodicals (1665 to 1895). Together with chronological tables and a library checklist. 2nd ed. *Smithsonian miscellaneous collections* 40: vii+1,247 pp. *Re-pub.* New York, Johnson Reprint Corporation, 1965. ••Index of subjects, pp. 1141–1154. Excludes learned society periodicals. Arranged alphabetically by the earliest form of the title. *Also* **W. J. Fox.** 1908. A list of American books omitted from Bolton's "Catalogue of Scientific and Technical Periodicals, 1665–1895." *Bulletin of bibliography* 5 (April): 82–85.) *Replaces* ___. 1885. ___ (1665 to 1882). *Smithsonian miscellaneous collections* 29: x+773pp.

Ca137. **Garrison, F. H.** 1934. The medical and scientific periodicals of the 17th and 18th centuries, with a revised catalogue and checklist. *Bulletin of the Institute of the History of Medicine* 2: 285–343. ••Separate lists of medical and scientific titles from the 17th century followed by a classified list of 18th century titles, decade by decade, in 21 sections; with an alphabetical list of medical titles.) *Also* **D. A. Kronick.** 1958. The Fielding H. Garrison list of medical and scientific periodicals of the 17th and 18th centuries: Addenda et corrigenda. *Bulletin of the history of medicine* 32: 456–474.

Ca138. **Kronick, D. A.** 1991. *Scientific and technical periodicals of the seventeenth and eighteenth centuries: A guide.* Metuchen, NJ,

Scarecrow Pr. xix+332 pp. ••Ca. 1860 titles with subject, personal name, institutional name and title indexes.

Ca139. **Lawrence, G. H. M., et al.** 1968. *B–P–H: Botanico–Periodicum–Huntianum.* Pittsburgh, PA, Hunt Botanical Library. 1,063 pp.; *Continued* **G. D. R. Bridson & E. R. Smith.** 1991. *B–P–H/S: Botanico–Periodicum–Huntianum/Supplementum.* Pittsburgh, PA, Hunt Institute for Botanical Documentation. [ii]+1068 pp. ••Over 25,000 titles of periodical (serial) publications, from the 17th century to 1990 that have included articles on the plant sciences, and historical, bibliographical and biographical articles on the plant and life sciences.

Ca140. **Scudder, S. H.** 1897. *Catalogue of scientific serials of all countries, including the transactions of learned societies in the natural, physical and mathematical sciences, 1633–1876.* Cambridge, MA, Library of Harvard Univ. xii+358 pp. *Special publication, Harvard University Library* no.1. Reprint: New York, Kraus, 1965. ••Arranged by country.

See also Library cats. Bc622–643 and:

Ca141. **Kronick, D. A.** 1987. Bibliographic dispersion of early periodicals. *Serials librarian* 12: 55–59.

Chronological lists of periodicals

Ca142. **Bolton, H. C.** 1898. Chronological tables giving the year of publication of 550 periodicals [1725–1895], with an alphabetical index. In: A catalogue of scientific and technical periodicals (1665 to 1895). 2nd ed. *Smithsonian miscellaneous collections* 40: 1017–1140. *Re-pub.* New York, Johnson Reprint Corporation, 1965.

Ca143. **Schmidt, C. E. H.** 1895. Synchronistische Tabellen über die naturwissenschaftliche Journalliteratur von 1650–1893. *Schriften herausgegeben von der Universität Jurjew (Dorpat)* 8: x+63 pp.

REGIONAL HISTORIES OF NATURAL HISTORY

See also Natural–history exploration Ca629–823

European

Ca144. **Akademiya Nauk SSSR, Institut Istorii Estestvozneniya i Tekhniki.** 1967. *Razvitie biologii v SSSR, 1917–1967. [Development of biology in the USSR, 1917–1967].* Ed. by B. E. Bykhovskii. Moscow, Izd. "Nauka." 762 pp. & 12 port. *Sovetskaya nauka i tekhnika za 50 let, 1917–1967* series.

Ca145. **Allen, D. E.** 1976. *The naturalist in Britain: A social history.* London, Allen Lane. xii+292 pp.

Ca146. **___.** 1985. The early professionals in British natural history. In: Society for the History of Natural History. *From Linnaeus to Darwin ... Papers from the fifth Easter meeting of the ... 28–31 March, 1983*, pp. 1–12. London. *Special publication* no. 3.

Ca147. **Barber, L.** 1980. *The heyday of natural history 1820–1870.* London, Johnathan Cape. 320 pp.

Ca148. **Berry, R. J.** 1983. The evolution of British biology. Presidential address to the Linnean Society delivered at the Anniversary Meeting, 24 May 1983. *Biological journal of the Linnean Society* 79: 327–352.

Ca149. **Buican, D.** 1982. Milestones in the philosophic and historical framework of the development of biology in France. *Scientia* 117: 609–621.

Ca150. **Caullery, M.** 1925. *Histoire des sciences biologiques en France.* Paris, Plon. 296 pp. *Histoire de la nation française* vol. 15.

Ca151. **Congrès des Sociétés Savantes.** 1962. Colloque sur les médecins et naturalistes de l'École de Montpellier. *Comptes rendus du 86e Congrès des Sociétés Savantes, Section des Sciences*, pp. 151–239.

Ca152. **Dahl, S.** 1941. *Den Danske plante- og dyreverdens udforskning*. Copenhagen, Udvalget for Folkeoplysnings Fremme. 340 pp.
Explorations of the Danish flora and fauna from pre-Linnaean times.

Ca153. **Ehrard, J.** 1963. *L'idée de nature en France dans la première moitié du XVIIIe siècle*. Paris, S.E.V.P.E.N. 2 vols.

Ca154. **Evans, D.** 1991. *A history of nature conservation in Britain*. London, Routledge. xxv+272 pp., illus.

Ca155. **Figurovskii, N. A.**, ed. 1957–60. *Istoriya estestvoznaniya v Rossi. [History of natural science in Russia]*. Moscow, AN SSSR Institut Estestvoznaniya i Tekhniki. 2 vols. in 3.
Vol. 1, ch. 1: *Razvitie estestvoznaniya v Rossii do XVIII veka. Razvitie estestvoznaniya v Rossii v XVIII veke, Includes* **S. L. Sobol'.** Biologiya, pp. 148–181, 450–492; Vol. 1, ch. 2: *Razvitie estestvoznaniya v Rossii v period polovine XIX veka, Includes* **___.** Biologiya, pp. 275–338; Vol. 3: *Geologo–geograficheskie i biologicheskie nauki.*

Ca156. **Freeman, R. B.** 1976. Children's natural history books before Queen Victoria. *History of Education Society bulletin* 17: 7–21; **___.** 1976. ___: A handlist of texts. *Ibid.* 18: 6–34.
Includes a bibliography of nearly 300 titles.

Ca157. **Fries, T. M.** 1894. *Naturalhistorien i Sverige intill medlet af 1600–talet.* Uppsala, Upsala Universitet. [ii]+75 pp. *Uppsala Universitets årsskrift* 1894, Progr. 2.

Ca158. **___.** 1911. Naturalhistorien i Uppsala under senare hälften af 1600–talet. *Nordisk tidsskrift for vetenskap, konst och industri* 1911: 445–463, 571–585.

Ca159. **___.** 1912. Naturalhistorien i Uppsala under de fyra första decennierna af 1700–talet. *Nordisk tidsskrift for vetenskap, konst och industri* 1912: 343–365, 417–436.

Ca160. **Gesellschaft Deutscher Naturforscher und Ärzte.** 1928. *Naturforschung und Naturlehre im alten Hamburg. Erinnerungsblätter zu Ehren der 90. Versammlung der Gesellschaft Deutscher Naturforscher in Hamburg, September 1928.* Hrsg. von der Hamburger Staats– und Universitäts–Bibliothek. Hamburg. viii+99 pp. & 8 pl.

Ca161. **Hoeniger, F. D. & J. F. M.** 1969. *The growth of natural history in Tudor England.* Charlottesville, VA, Univ. of Virginia Pr. for the Folger Shakespeare Library. 54 pp. & 19 pl. *Folger booklets on Tudor & Stuart civilization.*

Ca162. **___.** 1969. *The growth of natural history in Stuart England from Gerard to the Royal Society.* Washington, DC, Folger Shakespeare Library. 54 pp. & 19 pl. *Folger booklets on Tudor & Stuart civilization.*

Ca162A. **Hrabovec, I.** 1990. *Z dejín botaniky a zoológie na Slovensku do polovice 19. storocia.* Bratislava, Veda. 120 pp., illus.

Ca163. **Jaeger, F. M.** 1919. *Historische studiën. Bijdragen tot de kennis van de geschiedenis der natuurwetenschappen in de Nederlanden gedurende de 16e en 17e eeuw.* Groningen, J. B. Walter. [viii]+275 pp.

Ca164. **Jeandet, A.** 1892. *Recherches bio–bibliographiques pour servir à l'histoire des sciences naturelles en Bourgogne et particulièrement dans le département de Saône–et–Loire depuis le XVIe siècle jusqu'à nos jours.* Macon, Protat Frères. [6]+133 pp.

Ca165. **Jenkins, E. W.** 1979. The development of school biology, 1918–1945. *History of education* 8: 59–73.

Ca166. **Jenkins, E. W.** 1981. Science, sentimentalism or social control? The nature study movement in England and Wales, 1899–1914. *History of education* 10: 33–43.

Ca167. **Jones, W. P.** 1937. The vogue of natural history in England, 1750–1770. *Annals of science* 2: 345–352.

Ca168. **Kaiserlich–Koenigliche Zoologisch–Botanische Gesellschaft in Wien.** 1901. *Botanik und Zoologie in Österreich in den Jahren 1850 bis 1900. Festschrift herausgegeben von K. K. Zool.–Bot. Gesellschaft in Wien ... anlässlich der Feier ihres fünfzigjährigen Bestandes.* Vienna, A. Hölder. ix+620 pp., 5 pl. & 33 port.

Ca169. **Knight, D. M.** 1972. *Natural science books in English 1600–1900.* London, B. T. Batsford. x+262 pp. *Re-pub.* (with corrections and additional bibliography) London, Portman Books, 1989.
Contents include chapters on: The scientific movement; The Royal Society; The rise and fall of natural theology; Science in the eighteenth century; The decline of science in England; Biology and geology in the nineteenth century; The diffusion of science in the nineteenth century; Scientific publications in the nineteenth century. The "additional bibliography" is minimal.

Ca170. **Kongo, L.** 1987. *Istoriya issledovatel'skikh rabot po estestvennym naukam v Estonii v period do 1917g. [History of research work on natural history in Estonia since 1917].* Tallin, Akademiya Nauk Estonskii SSR. 259 pp.

Ca171. [**Krause, E.**] 1901. Geschichte der biologischen Wissenschaften im neunzehnten Jahrhundert, von Carus Sterne. In: *Das Deutsche Jahrhundert.* 2. Bd., Abt. 12, pp. 561–734. Berlin, F. Schneider & Co.

Ca172. **Lenoir, T.** 1981. The Göttingen School and the development of transcendental Naturphilosophie in the romantic era. *Studies in the history of biology* 5: 111–205.

Ca173. **___.** 1982. *The strategy of life: Teleology and mechanics in nineteenth century German biology.* Dordrecht, [etc.], D. Riedel. xii+314 pp. *Studies in the history of modern science* no. 13. *Re-pub.* Chicago, IL, Univ. of Chicago Pr., 1989.

Ca174. **Mikulinskii, S. R.** 1961. *Razvitie obshchikh problem biologii v Rossii: Pervaya polovina XIX veka. [The development of a general biological problematic in Russia in the first half of the 19th century].* Moscow, Akademiya Nauk SSSR. 450 pp.

Ca175. **___.** 1989. The development of biology in Russia in the first half of the 19th century. *Archives internationales d'histoire des sciences* 39: 249–275.

Ca176. **Mornet, D.** 1911. *Les sciences de la nature en France au XVIIIe siècle. Un chapitre de l'histoire des idées.* Paris, A. Colin. x+290 pp. *Re-pub.* New York, Franklin, 1971.

Ca177. **Nicholson, H. A.** 1886. *Natural history: Its rise and progress in Britain as developed in the life and labours of leading naturalists.* London & Edinburgh, W. & R. Chambers. viii+312 pp.

Ca178. **Paul, H. W.** 1985. *From knowledge to power: The rise of the science empire in France, 1860–1939.* Cambridge, Cambridge Univ. Pr. ix+415 pp.
Includes Ch. 2. Fata morgana. Positivism in nineteenth–century French biology, pp. 60–92, 362–367; ch. 3. Biology in the University. The success of marine biology leads but to paradigm lost, pp. 93–133, 367–372; ch. 7. Scientific publication. The flood of monographs, books, and journals unleashed by the new research imperative in schools and societies, pp. 251–287, 389–393; etc.

Ca178A. **Pilet, P.–E.** 1991. *Naturalistes et biologistes à Lausanne: Recherches, enseignements et sociétés savantes en pays Vaudois de 1537 à nos jours.* Lausanne, Payot. 205 pp., illus.

Ca179. **Rapaics, R.** 1953. *A Magyar biologia története.* Budapest, Akadémiai Kiadó. 304 pp. & 32 pl.

Ca180. **Raven, C.** 1947. *English naturalists from Neckam to Ray: A study of the making of the modern world.* Cambridge, Cambridge Univ. Pr. x+379 pp. *Re-pub.* Kraus Reprint, 1968.
A companion to the same author's life of John Ray; *see* Ba1412.

Ca181. **Rehbock, P. F.** 1983. *The philosophical naturalists: Themes in early nineteenth-century British biology.* Madison, WI, Univ. of Wisconsin Pr. xv+281 pp. *Wisconsin publications in the history of science & medicine* no. 3.

Ca182. **Rodolico, F.** 1964. *L'esplorazione naturalistica dell'Appenino.* Florence, Le Monnier. 403 pp.

Ca183. **Ross, H. C. G., & R. Nash.** 1985. The development of natural history in early nineteenth century Ireland. In: Society for the History of Natural History. *From Linnaeus to Darwin ... Papers from the fifth Easter meeting of the ... 28–31 March, 1983*, pp. 13–27. London. *Special publication* no. 3.

Ca184. **Royal Society of Edinburgh.** 1983. Two hundred years of the biological sciences in Scotland. *Proceedings of the Royal Society of Edinburgh* B84: 371 pp.
Includes **K. Blaxter.** Agricultural science and practice, pp. 1–20; **R. Y. Thomson & R. M. S. Smellie.** Biochemistry, pp. 21–35; **A. M. MacLeod.** Brewing, pp. 37–64; **A. G. Morton & M. Noble.** Botany and mycology, pp. 65–83; **C. H. Gimingham, D. H. N. Spence, & A. Watson.** Ecology, pp. 85–118; **J. Hawthorn.** Food science, pp. 119–140; **J. D. Matthews.** Forestry, pp. 141–170; **P. S. Maitland.** Freshwater science, pp. 171–210; **F. W. Robertson.** Genetics, pp. 211–229; **R. I. Currie.** Marine science, pp. 231–250; **G. L. Montgomery.** Medicine, pp. 251–266; **S. Selwyn & A. C. Wardlaw.** Microbiology, including virology, pp. 267–293; **J. M. Boyd.** Nature conservation, pp. 295–336; **D. R. R. Burt.** Parasitology, pp. 337–351; **M. S. Laverack.** Zoology, pp. 353–371.

Ca185. **Sala Catalá, J.** 1981. El evolucionismo en la práctica científica de los biólogos españoles del siglo XIX. *Asclepio* 33: 81–125.

Ca186. **___.** 1982. Cambio de paradigma y polemica cientifica entre los biologos Españoles (1860–1922). *Asclepio* 34: 239–263.

Ca187. **___.** 1984. Los biólogos Españoles entre 1860 y 1922: Una sociedad científica en Cambio. Su descripción. *Quipu* 1: 83–108.

Ca188. **Sellés, M., J. L. Peset, & A. Lafuente**, eds. 1988. *Carlos III y la ciencia de la Ilustración.* Madrid, Alianza. 402 pp.
Includes **F. J. P. Sarmiento.** El Real Jardín Botánico de Madrid durante el reinado de Carlos III; **M. de los A. Calatayud.** El Real Gabinete de Historia Natural de Madrid; **E. Estrella.** Expediciones botánicas; **A. Galeria.** La expedición Malaspina; **C. Minguet.** La obra de Humboldt; etc.

Ca189. **Shapiro, B.** 1979. History and natural history in 16th– and 17th–century England: An essay on the relationship between humanism and science. In: Clark Memorial Library, Los Angeles. *English scientific virtuosi in the 16th and 17th centuries. Papers read at a Clark Library seminar*, pp. 1–55. Los Angeles, CA, Clark Memorial Library, Univ. of California.

Ca190. **Sheail, J.** 1976. *Nature in trust: The history of nature conservation in Britain.* Glasgow, Blackie. xiv+270 pp.

Ca191. **Smit, P.** 1973. De situatie in de Nederlandse biologie rond 1875. *Scientiarum historia* 15: 39–48.

Ca191A. **Società Italiana di Scienze Naturali.** [1939–40.] *Un secolo di progresso scientifico italiano, 1838–1939: Opera storica redatta da circa 200 collaboratori sotto la direzione del comitato scientifico della S.I.P.S.* Pubblicazione a cura del segretario generale Prof. Lucio Silla. Rome. 7 vols.

Vol. 4, *includes* Zoologia; Anatomia; Botanica; Fisiologia; [etc.]. Vol. 5, *includes* Scienze agrarie e forestali.

Ca192. **Taton, R.**, ed. 1964. *Enseignement et diffusion des sciences en France au XVIIIe siècle.* Paris, Hermann. 780 pp. & 16 pl. *Histoire de la pensée* vol. 11.
By various authors. *Includes* 1re partie: L'Enseignement classique. Collèges et universités, pp. 13–168; 2me partie: L'Enseignement de la médecine et de la pharmacie, pp. 169–257; 3me partie: Le Collège Royal et le Jardin du Roi, pp. 259–341; 6me partie: Cabinets scientifiques et observatoires, pp. 617–712. Each part is provided with good bibliographies.

Ca193. **Urteaga, L.** 1987. *La Tierra esquilmada: Las ideas sobre la conservación de la naturaleza en la cultura Español del siglo XVIII.* Barcelona, Serbal; Madrid, C.S.I.C. 221 pp.

Ca194. **Vucinich, A.** 1963. *Science in Russian culture: A history to 1860.* Stanford, CA, Stanford Univ. Pr. xv+463 pp.
Pt. 1: The roots of Russian science (1. The origins of the scientific attitude in old Russia, 2. Peter the Great: Science by decree, 3. The Academy: Dilemmas and triumphs); pt. 2: Science in the age of reason (Science, enlightenment, and absolutism, 5. The rise and decline of humanistic rationalism); pt. 3: Autocratic ideology and science (6. Official nationalism and countervailing thought, 7. The widening horizons of natural science, 8. The end of the formative age).

Ca195. **___.** 1970. *Science in Russian culture 1861–1917.* Stanford, CA, Stanford Univ. Pr. xvi+575 pp.
Ch. 4: The life sciences; ch. 9: Biological evolution: Facts and controversies; ch. 10: The triumphs of experimental biology.

See also Ba60–75 Bb2 Bb13–21 Bb41–568.

North American

Ca196. **Alden, R. H., & J. D. Ifft.** 1943. *Early naturalists of the far West.* San Francisco, CA, California Academy of Sciences. 59 pp. *Occasional papers of the California Academy of Sciences* no. 20.

Ca197. **Baatz, S.** 1988. Philadelphia patronage: The institutional structure in natural history in the new republic, 1800–1833. *Journal of the early republic* 8: 111–138.

Ca198. **Berger, C.** 1983. *Science, God, and nature in Victorian Canada.* Toronto, Univ. of Toronto Pr. xiv+92 pp. *Joanne Goodman lectures* 1982.

Ca199. **Brooks, P.** 1980. *Speaking for nature: How literary naturalists from Henry Thoreau to Rachel Carson have shaped America.* Boston, MA, Houghton Mifflin. 304 pp.

Ca200. **Clepper, H. E.**, ed. 1966. *Origins of American conservation.* New York Ronald Pr. x+193 pp.

Ca201. **Elman, R.** 1982. *America's pioneering naturalists.* [Rev. ed.] Tulsa, OK, Winchester Pr. xx+231 pp., illus.
Replaces ___. 1977. *First in the field: America's pioneering naturalists.* New York, Mason/Charter. xx+231 pp., illus.

Ca202. **Ewan, J.** 1961. The scientist on the frontier. In: J. F. McDermott, ed. *Research opportunities in American cultural history*, pp. 81–101. Lexington, KY, Univ. of Kentucky Pr.

Ca203. **___.** 1965. French naturalists in the Mississippi Valley. In: J. F. McDermott, ed. *The French in the Mississippi Valley*, pp. 159–174. Urbana, IL, Univ. of Illinois Pr.

Ca204. **Fortuine, R.** 1984. Ships' surgeons and naturalists in the early history of Alaska. Pt. 1: Voyages of discovery, 1741–1786; Pt. 2: Voyages of trade and science, 1786–1807; Pt. 3: Voyages for empire and the North West Passage, 1814–1854. *Alaska medicine* 26: 57–61, 82–86, 117–121.

Ca205. **Gándara, G.** 1936. Historia de la enseñanza de las ciencias biológicas en México nacional y prenacional. *Memorias, Academia Nacional de Ciencias "Antonia Alzate"* 53: 387–410.

Ca206. **Genong, W. F.** 1910. The identity of the animals and plants mentioned by the early voyagers to eastern Canada and Newfoundland. *Proceedings & transactions of the Royal Society of Canada* (3) 3 Sect. 2: 197–242.

Ca207. **Goode, G. B.** 1897. The beginnings of natural history in America. *Report of the United States National Museum* 1897 (2): 355–406. *Re-pub. in* ___. 1991. *The origins of natural science in America: The essays of George Brown Goode*, pp. 23–89. Ed. by S. G. Kohlstedt. Washington, DC, Smithsonian Instition Pr. xii+411 pp.
First pub. in: *Proceedings of the Biological Society of Washington* 3: 35–105, 1886.

Ca208. ___. 1898. The beginnings of American science: The third century. *Report of the United States National Museum* 1897 (2): 407–466. *Re-pub. in* ___. 1991. *The origins of natural science in America: The essays of George Brown Goode*, pp. 91–171. Ed. by S. G. Kohlstedt. Washington, DC, Smithsonian Instition Pr. xii+411 pp.
First pub. in: *Proceedings of the Biological Society of Washington* 4: 9–94, 1888.

Ca209. **Greene, J. C.** 1984. *American science in the age of Jefferson.* Ames, IA, Iowa State Univ. Pr. xiv+484 pp.
Includes Ch. 10. Natural history in a new world: Botany, pp. 253–276; Ch. 11. Natural history in a new world: Zoology, pp. 277–319.

Ca210. **Hanley, W.** 1977. *Natural history in America: From Mark Catesby to Rachel Carson.* New York, Quadrangle. xii+339 pp.

Ca211. **Herbst, J.** 1954. *New green world: John Bartram and the early naturalists.* New York, Hastings House Publishers. 272 pp.

Ca212. **Hindle, B.** 1956. *The pursuit of science in revolutionary America, 1735–1789.* Chapel Hill, NC, Univ. of North Carolina Pr. for The Institute of early American History & Culture, Williamsburg, Virginia. xi+410 pp.
Much information on natural history.

Ca213. **Hughes, A. F. W.** 1982. *The American biologist through four centuries.* Springfield, IL, Thomas. xvii+386 pp.

Ca214. **Hurd, P. DeH.** 1961. *Biological education in American secondary schools 1890–1960.* Washington, DC, American Institute of Biological Sciences. ix+263 pp. *Biological Sciences Curriculum Study bulletin* no. 1.

Ca215. **Huth, H.** 1957. *Nature and the American: Three centuries of changing attitudes.* Berkeley & Los Angeles, CA, Univ. of California Pr. xvii+250 pp.

Ca216. **Kastner, J.** 1977. *A species of eternity.* New York, Knopf. xiv+350 pp., illus.
English edition as: *A world of naturalists.* London, John Murray, 1978.

Ca217. **Kohlstedt, S. G.** 1990. Parlors, primers, and public schooling: Education for science in nineteenth–century America. *Isis* 81: 425–445.

Ca218. **Ledesma Mateos, I.** 1990. Esbozo del desarrollo histórico de la biología en Puebla. *Quipu* 7: 93–125.

Ca219. **Merrill, G. P.**, ed. 1920. Contributions to a history of American state geological and natural history surveys. *Bulletin of the United States National Museum* 109: xviii+549 pp. & 37 pl. *Re-pub.* New York, Arno Pr., 1978.
Provides a state by state record of 36 which had surveys before 1885. A modern revision (viz. **A. A. Socolow**, ed. 1988. *The state geological surveys: A history.* Association of American State Geologists. 499 pp.) omits most of the information on natural history and publications.

Ca220. **Nash, R.** 1982. *Wilderness and the American mind.* 3rd ed. New Haven, CT, Yale Univ. Pr. xvii+425 pp.

Ca220A. **Norwood, V.** 1993. *Made from this earth: American women and nature.* Chapel Hill, NC, Univ. of North Carolina Pr. xxiii+368 pp., illus.

Ca221. **Novak, B.** 1980. *Nature and culture: American landscape painting, 1825–1875.* New York, Oxford Univ. Pr. xi+323 pp., illus.

Ca222. **Pauly, P. J.** 1984. The appearance of academic biology in late nineteenth–century America. *Journal of the history of biology* 17: 369–397.
Especially within university biological departments at Harvard, Johns Hopkins, Chicago, Columbia, & University of Pennsylvania.

Ca223. ___. 1991. The development of high school biology: New York City, 1900–1925. *Isis* 82: 662–688.

Ca224. **Porter, C. M.** 1979. "Subsilentio": Discouraged works of early 19th–century American natural history. *Journal of the Society for the Bibliography of Natural History* 9: 109–119.
I.e. Contributions of C. S. Rafinesque, T. Say, J. K. Townsend, and T. R. Peale.

Ca225. ___. 1987. *The eagle's nest: Natural history and American ideas, 1812–1842.* N.p., Univ. of Alabama Pr. xii+[2]+251 pp. *History of American science & technology series.*

Ca226. **Savage, H.** 1970. *Lost heritage.* New York, W. Morrow & Co. 329 pp.
An account of seven pre-Audubon naturalists.

Ca227. **Smallwood, W. M. & M. S.** 1941. *Natural history and the American mind.* New York, Columbia Univ. Pr. xvii+445 pp. *Columbia studies in American culture* no. 8. *Re-pub.* New York, AMS Pr., 1967.

Ca228. **Smith, M. L.** 1987. *Pacific visions: Californian scientists and the environment, 1850–1915.* New Haven, CT, Yale Univ. Pr. ix+243 pp.

Ca229. **Society for the Bibliography of Natural History.** 1983. *Contributions to the history of North American natural history: Papers from the first North American conference of the Society for the Bibliography of Natural History held at the Academy of Natural Sciences, Philadelphia, 21–23 October 1981.* Ed. by A. C. Wheeler. London. 97 pp. *Special publication* no. 2.
Nine papers by various authors. References to some are entered in the relevant sections herein.

Ca230. **Stearns, R. P.** 1932. *Science in the British colonies of America.* Urbana, IL, Univ. of Illinois Pr. xx+760 pp.

Ca231. **Strong, D. H.** 1988. *Dreamers and defenders: American conservationists.* 2nd ed. Lincoln, NE, Univ. of Nebraska Pr. x+282 pp., illus.
Replaces ___. 1971. *The conservationists.* Menlo Park, CA, Addison–Wesley Pub. Co. 196 pp., illus.

Ca232. **Stroud, R. H.**, ed. 1985. *National leaders of American conservation.* Washington, DC, Smithsonian Institution. 432 pp.
Replaces **H. E. Clepper**, ed. 1971. *Leaders of American conservation.* New York, Ronald Pr. vii+353 pp.

Ca233. **Tracy, H. C.** 1931. *American naturists.* New York, E. P. Dutton. viii+282 pp.

Ca234. **Wild, P.** 1979. *Pioneer conservationists of western America.* Missoula, MT; and ___. 1986. ___ *eastern America.* Missoula, MT, [etc.]; *see* Ba98–99.

Ca235. **Wilson, D. S.** 1978. *In the presence of nature.* Amherst, MA, Univ. of Massachusetts Pr. xix+234 pp.
Naturalists and natural history in colonial America.

See also Ba76–99 Bb32–33 Bb569–670 Ca1(Hutchinson) Ca1050 and:

Ca236. **Kohlstedt, S. G., & M. W. Rossiter**, eds. 1986. *Historical writing on American science: Perspectives and prospects.* Baltimore, MD, Johns Hopkins Univ. Pr.

321 pp.
First pub. as *Osiris* n.s. 1, 1985. Essays by various authors. *Includes* **S. G. Kohlstedt.** Institutional history, pp. 17–36; **J. Maienschein.** History of biology, pp. 147–162; **C. A. Elliott.** Bibliographies, reference works, and archives, pp. 295–310.

South & Central American (& Caribbean)

Ca237. **Babini, J.** 1986. *Historia de la ciencia en la Argentina.* Buenos Aires, Ediciones Solar. 273 pp. *Biblioteca "Dimension Argentina."*
Replaces ___. [1949.] ___. Mexico, Fondo de Cultura Económica. 219 pp. *Coleccion terra firme* no. 46. Ch. XXII & XXIII treat biological sciences and natural history.

Ca238. **Chardon, C. E.** 1947. Longevity and casualties among naturalists in tropical America. *Scientific monthly* 64: 108–207.

Ca239. ___. 1949. *Los naturalistas en la América Latina.* Vol. I: *Los siglos XVI, XVII y XVIII, Alejandro Humboldt, Carlos Darwin, La Española, Cuba y Puerto Rico.* Cuidad Trujillo, República Dominicana. vii+386 pp., 27 pl.
All published.

Ca240. **Ewan, J.** 1898. Five American naturalists on the Andes and Amazon: A study in personality. *Archives of natural history* 16: 1–14.

Ca241. ___. 1992. Through the jungle of Amazon travel narratives of naturalists. *Archives of natural history* 19: 185–207.

Ca242. **Furlong Cárdiff, G.** 1948. *Naturalistas Argentinos durante la dominación hispánica.* Buenos Aires, Huarpes. 437 pp. *Cultura colonial Argentina* no 7.

Ca242A. **Isenburg, T.** 1989. *Viaggiatori naturalisti in Brasile nell'ottocento.* Milan, Angeli. 119 pp.

Ca242B. ___. 1989. *Naturalistas italianos no Brasil.* São Paulo, Editora Icone, Arquivo da Estado. 389 pp., illus.
Includes translations of original descriptions by seven travellers.

Ca243. **Lutz, B.** 1941. *British naturalists in Brazil.* Rio de Janeiro, Rodrigues & Cia, Sociedade Brasileira de Cultura Inglesa. 37 pp.

Ca244. ___. 1989. *Esboço historico sobre a botanica e zoologia no Brasil.* Brasilia, Editora UnB. 44 pp.

Ca244A. **O'Brien, M. J. Pires-.** 1993. An essay on the history of natural history in Brazil, 1500–1900. *Archives of natural history* 20: 37–48.

Ca245. **Whitehead, P. J. P., & M. Boeseman.** 1989. *A portrait of Dutch 17th century Brazil: Animals, plants and people by the artists of Johan Maurits of Nassau.* Amsterdam, New York, North-Holland Publishing Co. 359 pp. *Verhandelingen, Koninklijke Nederlandse Akademie van Wetenschappen, Afdeeling Naturkunde*, 2. reeks, no. 87.

See also Ba100–101.

Asian

Ca246. **Azat'yan, A. A.** 1960. *Vydayushchiesya issledovateli prirody srednei Azii: Vtoraya polovina 19 v. [Distinguished investigators of nature in central Asia: second half of the 19th century].* Tashkent, Srednaya i Vyssaya Skola. 170 pp.

Ca247. **Bruun, A. F.** 1961. Danish naturalists in Thailand 1900–1960. *Natural history bulletin of the Siam Society* 20: 71–80, & 3 pl.

Ca248. **Cherepanov, A. I.**, ed. 1968. *Razvitie biologicheskoi nauki v Sibiri za 50 let. [Development of biological science in Siberia during 50 years]*. Novosibirsk, Izd. "Nauka." 266 pp.

Ca249. **Huan, N. T.** 1957. Esquisse d'une histoire de la biologie chinoise des origines jusqu'au 4e siècle. *Revue d'histoire des sciences* 10: 1–37.

Ca250. **Jain, B. L.** 1982. Development of life sciences in India in the eighteenth–nineteenth century. *Indian journal of history of science* 17: 114–131.

Ca251. **Shirai, M.** 1908. *Sotei. Nippon hakubutsu–gaku nenpyo. [A chronological table of natural history in Japan]*. Tokyo. vii+36 pp. & 4 pl.
In Japanese.

Ca252. **Smit, P.** 1978. International influences on the development of natural history in the Netherlands and its East Indian colonies between 1750 and 1850. *Janus* 65: 45–65.

See also Ba102 Bb671–689 & Ca66(Kumar).

Australasian (& Pacific)

Ca253. **Finney, C. M.** 1984. *To sail beyond the sunset: Natural history in Australia, 1699–1829*. Adelaide, S.A., Rigby International. x+206 pp.

Ca254. **Hoare, M. E., & L. G. Bell**, eds. 1984. In search of New Zealand's scientific heritage. *Bulletin of the Royal Society of New Zealand* 21: 123 pp.
Ten papers by various authors. References to some are entered in the relevant sections herein.

Ca255. **Home, R. W.**, ed. 1989. *Australian science in the making: A bicentennial history*. Cambridge, Cambridge Univ. Pr. in association with the Australian Academy of Science. xxx+413 pp.
Articles by various authors. *Includes* **M. J. Hughes.** Philosophical travellers at the ends of the earth: Baudin, Péron and the Tasmanians; **R. A. Stafford.** The long arm of London: Sir Roderick Murchison and imperial science in Australia; **I. Inkster & J. Todd.** Support for the scientific enterprise, 1850–1900; **A. M. Lucas.** Baron von Mueller: Protégé turned patron; **B. W. Butcher.** Gorilla warfare in Melbourne: Halford, Huxley and "Man's Place in Nature"; **R. L. Burt & W. T. Williams.** Plant introduction in Australia; etc.

Ca256. **Home, R. W., & S. G. Kohlstedt**, eds. 1991. *International science and national scientific identity: Australia between Britain and America*. Dordrecht, Kluwer Academic Publishers. x+318 pp. *Australasian studies in history & philosophy of science* vol. 9.
Articles by eleven authors. *Includes* **D. M. Knight.** Tyrannies of distance in British science; **E. D. Newland.** Dr George Bennett and Sir Richard Owen: A case study of the colonization of early Australian science; **S. G. Kohlstedt.** International exchange in the natural history enterprise: Museums in Australia and the United States; **J. Sapp.** Geographic isolation and the origin of species; The migrations of Michael White; etc.

Ca257. **McLeod, R. M., & P. F. Rehbock**, eds. 1988. *Nature in its greatest extent: Western science in the Pacific*. Honolulu, HI, Univ. of Hawaii Pr. xi+288 pp.
Ten articles by various authors. *Includes* **O. H. K. Spate.** Seamen and scientists: The literature of the Pacific, 1697–1798; **A. Frost.** Science for political purposes: European explorations of the Pacific Ocean, 1764–1806; **D. G. Frodin.** The natural world of New Guinea: Hopes, realities and legacies; **R. McLeod.** Imperial reflections in the Southern Seas: The Funafuti expeditions, 1896–1904; **P. F. Rehbock.** Organizing Pacific science: Local and international origins of the Pacific science association; **R. H. Randolph & J. E. Bardach.** Soviet science in the Pacific: The case of marine biology.

Ca258. **Moyal, A.** 1986. *'A bright and savage land': Scientists in colonial Australia*. Sydney, N.S.W., W. Collins. 192 pp.

Ca259. **Pizzey, G.** [1985?] *A separate creation: Discovery of wild Australia by explorers and naturalists.* London, Croom Helm. [vii]+150 pp.

Ca260. **Schedvin, C. B.** 1984. Environment, economy, and Australian biology, 1890–1939. *Historical studies, University of Melbourne* 28: 11–28.

Ca261. ___. 1987. Environment, economy, and Australian biology, 1890–1939. In: N. Reingold & M. Rothenberg, eds. *Scientific colonialism: A cross–national comparison,* pp. 101–126. Washington, DC, Smithsonian Institution Pr.

See also Ba103 Bb690–706.

African

Ca262. **Brown, A. C.**, ed. 1977. *A history of scientific endeavour in South Africa: A collection of essays published on the occasion of the centenary of the Royal Society of South Africa.* Cape Town, Royal Society of South Africa. xii+516 pp., illus.
Includes Essays on marine biology in South Africa, the history and status of nature conservation in South Africa, botanical research in South Africa in the twentieth century, advances in South African vertebrate palaeontology, and the history of the Royal Society of South Africa.

Ca263. **Dekeyser, P. L.** 1963. Principes et historique de la conservation de la nature (1933–1963). *Notes africaines* 99: 65–107.

See also Bb707.

BIBLIOGRAPHIES OF REGIONAL NATURAL HISTORY

European (& Mediterranean)

Ca264. **Boerlage, J. G., & P. P. C. Hoeck.** 1975. *Bibliographie van de flora en fauna Nederland, 1753–1886.* Amsterdam, Backhuys & Meesters. ix+150 pp. ••"Being a partial reprint of *Algemeene Aardrijkskundige Bibliographie van Nederland* ... Uitgegeven door de Afdeeling: Nederland van het Aardeijkskundig Genootschap, Tweede deel. Natuurkundige Toestand, Leiden, 1888."

Ca265. **Brünnich, M. T.** 1783. *Litteraria Danica scientiarum naturalium.* I. *Les progrès de l'histoire naturelle en Dannemarc & en Norvège.* II. *Bibliotheca patria auctorum & scriptorum, scientias naturales tractantium.* Copenhagen & Leipzig. [ii]+124+242+xiv+[xvi] pp.

Ca266. **Buhse, F.** 1854. Uebersicht der naturhistorische Literatur von Liv–, Kur– und Estland, entworfen unter Mitwirkung ... A. Buchholtz. *Correspondenzblatt der Naturforschenden Vereins zu Riga* 7: 49–58, 81–89, 115–122, 129–140.

Ca267. **Deniker, J.** 1895–1922. *Bibliographie des travaux scientifiques (sciences mathématiques, physiques et naturelles publiés par les sociétés savantes de la France depuis l'origine jusqu'en 1888, [etc.].* Dressée sous les auspices du Ministère de l'Instruction Publique. Aine–Orne [–Sarthe]. Vols. 1–2(1). ••No more published.

Ca268. **Desio, A., et al.** 1953. [Bibliografia geologica, paleontologica, zoologica e botanica della Libia dal 1941–1952.] *Annali del Museo Libico si Storia Naturale* 4: 195–214.

Ca269. **Ersch, J. S.** 1828. *Handbuch der deutschen Literatur seit der Mitte des achtzehnten Jahrhunderts.* 2. Bd, 1. Abtheil. *Die Literatur der Mathematik, Natur– und Gewerbskunde enthaltend.* Neue Ausg. von F. W. Schweigger–Seiden. Leipzig. [x] pp. & 1740 columns.

Ca270. **Faudel, –?.** 1874–78. Bibliographie alsatique comprénant l'histoire naturelle, l'agriculture et la médecine, la biographie des hommes de science et les institutions scientifiques de l'Alsace. *Bulletin de la Société d'Histoire Naturelle de Colmar* 14/15: 1–282, 1874; *Ibid.* 16/17: 385–428, 1876; *Ibid.* 18/19: 243–257, 1878.

Ca271. **Feliksiakowa, J.** 1981. Stand der biologischen Bibliographie in Polen in den Jahren 1945–1970. *History & philosophy of the life sciences* 3: 115–131.

Ca272. **Freeman, R. B.** 1980. *British natural history books 1495–1900: A handlist.* Folkestone, Dawson. 437 pp.

Ca273. **Golovachov, A. F.** 1870. *Bibliograficheskii ukazatel' po estestvennym naukam v Rossii c 1856 po 1869 god.* Moscow. 116 pp.

Ca274. **Gomis, A.**, et al. 1988. *Historia natural catálogo ilustrado siglos XVIII y XIX.* [Madrid], Ministerio de Educación y Ciencia. xv+223 pp. ••Illustrated catalogue of 398 works, pub. to accompany an exhibition in the Real Jardin Botánico, Madrid, 25 May to 30 June, 1988. Includes Spanish language titles only.

Ca275. **Haller, G. E. von.** 1777. *Catalogue raisonné des auteurs qui ont écrit sur l'histoire naturelle de la Suisse.* [Berne]. 150 pp.

Ca276. **Harris, C. D.** 1975. *Guide to geographical bibliographies and reference works in Russian or on the Soviet Union.* Chicago, IL, Univ. of Chicago, Department of Geography. xviii+478 pp.

Ca277. **Hérissant, L. A. P.** 1771. *Bibliothèque physique de la France, ou liste de tous les ouvrages, tant imprimés que manuscrits, qui traitent de l'histoire naturelle de ce royaume [etc.].* Paris, [C. J. L. Coquereau]. 40+496 pp.

Ca278. **Iterson, W. J. D., et al.** 1888. *Natuurkundige toestand.* In: Nederlandsch Aardrijkskundige Genootschap. *Algemeene aardrijkskundige bibliographie van Nederland.* 2. deel. Leiden. x+249 pp.

Ca279. **Jeandet, A.** 1892. *Recherches bio-bibliographiques pour servir à l'histoire des sciences naturelles en Bourgogne et particuliérement dans le Département de Saône–et–Loire depuis le XVIe siècle jusqu'a nos jours.* Mâcon. [v]+133 pp.

Ca280. **Krebel, R.** 1847. *Russlands naturhistorische und medizinische Literatur.* I Abteilung: *Schriften und Abhandlungen in nicht–Russischer Sprache.* Jena, F. Mauke. vi+220 pp.

Ca281. **Kutscher, F., H. Reichert & M. Niehuis.** 1980. Bibliographie der naturwissenschaftlichen Literatur über den Hunsrück. *Pollichia* 1: 1–206. ••Teil 2: Pflanzenwelt, Tierwelt, Naturschutz.

Ca282. **Müller, J.** 1883–87 & 1917. *Die wissenschaftlichen Vereine und Gesellschaften Deutschlands in neunzehnten Jahrhundert: Bibliographie ihrer Veröffentlichungen seit ihrer Begründung bis auf die Gegenwart. (Zweiter Band Fortgeführt bis 1914).* Berlin, Asher & Co; Behrend & Co. 2 vols. *Re-pub.* Hildesheim, 1965.

Ca283. **Picatoste y Rodriguez, F.** 1891. *Apuntes para una biblioteca científica española del siglo XVI; estudios biográficos y bibliográqficos de ciencias exactas, físicas y naturales.* Madrid, Imp. M. Tello. 416 pp.

Ca284. **Polska Akademija Nauk, Komisja Nauk Biologicznych.** 1975. *Katalog Polskiej literatury biologicznej. Catalogue of Polish biological literature.* Tom. (XX): *Za lata 1935–1939.* Warsaw. 226 pp.

Ca285. ___. 1963. *Katalog Polskiej literatury biologicznej. Catalogue of Polish biological literature.* Tom. I (XXII): *Za lata 1950–1954.* Warsaw. Continued periodically, e.g.: ___. 1969. *Katalog Polskiej literatury biologicznej. Catalogue of Polish biological literature.* Tom. II (XXIII): *Za lata 1955–1959.* Warsaw. ••A partial continuation of: Polska Akademija Umiejetnosci. *Katalog Polskiej literatury matematyczno–przyrodniczej* (and its predecessors) vols. I–XXI, 1901–49.

Ca286. **Rudescu, L., & A. C. Banu.** 1969. *Bibliographie Roumaine du Danube inférieur.* Bucharest, Commission Nationale de la République Socialiste de Roumanie pour l'UNESCO. 271 pp. ••Section 4: Biologie – Flore, faune, hydrobiologie, etc.

Ca287. **Scheele, M., & G. Natalis**, eds. 1981–82. *Biologie-Dokumentation: Bibliographie der Deutschen biologischen Zeitschriftenliteratur 1796–1965.* Munich [etc.], K. G. Saur. 24 vols. ••Author list of 274,752 articles in 152 periodicals with systematic and keyword indexes.

Ca288. **Sims, R. W., P. Freeman, & D. L. Hawksworth**, eds. 1988. *Key works to the fauna and flora of the British Isles and Northwestern Europe.* 5th ed. Oxford, Oxford Univ. Pr. 312 pp. *Systematics Association special volume* no. 33. ••*Replaces* **G. J. Kerrich, D. L. Hawksworth & R. W. Sims**, eds. 1978. *Key works to the fauna and flora of the British Isles and Northwestern Europe.* 4th ed. London, Academic Pr. for the Systematics Association. xii+178 pp. *Systematics Association special volume* no. 9.

Ca289. **Szinnyei, J. & J. [jr.]** 1878. *Magyarország természettudományi és mathematikai könyvészete, 1472–1875. Bibliotheca hungarica historiae naturalis, 1472–1875.* Budapest, Kir. Magyar Természettudományi Társulat. viii pp. & 1008 columns.

Ca290. **Vereeniging ter Bevordering van de Belangen des Boekhandels te Amsterdam.** 1911. *Catalogus van boeken in Noord-Nederland verschenen van de vroegsten tijd tot op heden.* [Pt. IX. Wis- en natuurkunde.] Samengesteld door de Tentoonstellings-Commissie der Nationale Tentoonstelling van het Boek ... 1910. The Hague. [ii] pp., 96 cols., [viii] pp.

Ca291. **Winther, M.** 1829. *Literaturae scientiae rerum naturalium in Dania, Norvegia & Holsatia usque ad annum MDCCCXXIX enchiridion in usum physicorum & medicorum.* Copenhagen. xvii+233+[xv] pp.

Ca292. **Zemanek-Targoszowa, J.** 1954–58. Bibliografia ochrony przyrody w Polsce ... Bibliography of the protection of nature in Poland for the years 1948–1955. *Ochrona przyrody* 22: 287–354, 1954; *Ibid.* 25: 268–304, 1958. Continued periodically, e.g.: ___. 1964. *Bibliografia ochrony przyrody w Polsce ... Bibliography of the protection of nature in Poland covering the period 1956–1960.* Cracow, Polska Akademija Nauk. [ii]+298 pp.

North American

Ca293. **Batschelet, M. W.** 1990. *Early American scientific and technical literature: An annotated bibliography of books, pamphlets, and broadsides.* Metuchen, NJ, Scarecrow Pr. xii+136 pp. ••Arranged in three classes, Medical titles (#1–54), Technical science titles (#380–722), and Natural and physical science titles (#723–833), and chronologically within each class.

Ca294. **Ellis, M.** 1903. Index to publications of the New York State Natural History Survey and New York State Museum, 1837–1902, also including other New York publications on related subjects. *Bulletin, New York State Museum* 66: 239–653.

Ca295. **Isfort, L. C.** 1949. A partial bibliography of natural history in the Chicago region. *American midland naturalist* 42: 406–472.

Ca296. **Laird, M.**, ed. 1980. *Bibliography of the natural history of Newfoundland and Labrador.* London, Academic Pr. lxxi+376 pp.

Ca297. **Meisel, M.** 1924–29. *A bibliography of American natural history, the pioneer century, 1769–1865: The rôle played by the scientific societies, scientific journals; natural history museums and botanic gardens; state geological and* *natural history surveys; federal exploring expeditions in the rise and progress of American botany, geology, mineralogy, paleontology and zoology*. Brooklyn, NY, Premier Publishing Co. 3 vols. *Re-pub.* New York, Hafner, 1967.

Ca298. **Moore, J. A.** 1989. Nature in the New World: An exhibition on the natural history of North and South America on the occasion of the centennial of the American Society of Zoologists. *Harvard Library bulletin* 36 (2): 95–212.

Ca299. **Owings, L. C.** 1976. *Environmental values, 1860–1972: A guide to information sources.* Detroit, MI, Gale Research Co. 324 pp. ••*Includes* Travel reports, national parks, conservation, nature study, etc.

Ca300. **Richardson, J. A., & B. H. Macdonald.** 1987. *Science and technology in Canadian history: A bibliography of primary sources to 1914*. Thornhill, H S T C Publication. 105 microfiches & 17 pp. guide. *Research tools for the history of Canadian science and technology* no. 3.

Ca301. **Spamer, E. E.** 1981. *Bibliography of the Grand Canyon and the lower Colorado River, 1540–1980*. Grand Canyon, AZ, Grand Canyon Natural History Association. 119 pp. *Monograph* no. 12.

Ca302. **Trübner & Co.** 1858. *A catalogue of books in all branches of natural history published during the last forty years in the United States of America, and constantly on sale by Trübner & Co.* London. 48 pp.

Ca303. **Van Oosten, J.** 1957. *Great Lakes fauna, flora and their environment: A bibliography*. Ann Arbor, MI, Great Lakes Commission. x+86 pp.

South & Central American (& Caribbean)

Ca304. **Gazin, J.** 1926. *Eléments de bibliographie générale, méthodique et historique de la Martinique (Antilles françaises)*. Paris, Leroux. iv+348 pp.

Ca305. **Gillis, W. T., R. Byrne, & W. Harrison.** 1975. Bibliography of the natural history of the Bahama Islands. *Atoll research bulletin* 191: 123 pp.

Ca306. **Instituto Brasileiro de Bibliografia e Documentaçao.** 1963–77. *Amazônia: Bibliografia.* Rio de Janeiro. 4 vols. *Re-pub.* Koenigstein, Otto Koeltz Science Publishers, 1975–77. ••Vol. 1: 1614–1962; vol. 2: 1601–1970; vol. 3: 1971–72; vol. 4: 1973–74. Systematically arranged bibliography of books and articles on the region including botany, zoology, geology, palaeontology, geography.

Ca307. **Trelles y Govin, C. M.** 1918–19. *Biblioteca científica Cubana*. Matanzas, Imp. J. F. Oliver. 2 vols. ••*Includes* Vol. 1. Matemáticas, astronomía, ciencias militares, físicas y naturales, biología, antropología, agricultura; vol. 2. Ciencias médicas, ingeniería.

Ca308. **Weibezahn, F. H., & E. B. Janssen–Weibezahn.** 1990. El Territorio Federal Amazonas, Venezuela: Una bibliografía. The Amazonas Federal Territory, Venezuela: A bibliography. *Scientia Guaianae* 1: 294 pp.

Asian (& Indian Ocean)

Ca309. **Agakhanyants, O. E., & A. S. Sin'kovskaya.** 1968. *Bibliografiya Pamira.* Vyp. 1. *Priroda: Ukazatel' literatury 1920–1964gg. [Bibliography of the Pamirs.* Vol. 1. *Nature: Guide to the literature, 1920–1964.*] Dushanbe, Akademiya Nauk Tadzhikskoi SSR, Tsentral'naya Nauchnaya Biblioteka. 266 pp.

Ca310. **Akademiya Nauk SSSR.** 1962. *Bibliografiya Yakutskoi ASSR (1931–1959).* Vol. 2: *Prirodnye usloviya, resursy i narodnoe khozaistvo. [Bibliography of the Yakutian ASSR (1931–1959).* Vol. 2: *Natural conditions, resources and national economy].* Akademiya Nauk SSSR, Sektor Seti Spetsial'nykh Bibliotek, Gosekonomsovet SSSR, Biblioteka Sops, Yakutskaya Respublikanskaya Biblioteka im. A. S. Pushkina. Moscow. 254 pp.

Ca311. **Alwis, N. A. W. A. T.** 1978. *Bibliography of scientific* *publications relating to Sri Lanka 1960–1976*. Colombo, National Science Council of Sri Lanka. v+244 pp.

Ca312. **Asatkin, A. N., & V. A. Samoilov**, eds. 1935. *Bibliografiya Dal'nevostochnogo kraya. 1890–1931*. Moscow, Vsesoyuznaya Assotsiatsiya Sel'skokhozyaistvennoi Bibliografii. 2(3?) vols. ••Vol. 1: Fizicheskaya geografiya; vol. 2: Geologiya, poleznye iskopaemye, paleontologiya; vol. 3: [Flora & fauna. Ever published?]

Ca313. **Ataur-Rahim, M.** 1979. *Biology of the Arabian Peninsula: A bibliographic study*. Riyadh, Biological Society of Saudi Arabia. xxiv+180 pp. *Publication* no. 3.

Ca314. ___. 1987. *Natural history of Pakistan: A bibliographic survey from 1684–1947*. Islamabad, Pakistan Scientific & Technological Information Centre, Pakistan Science Foundation. iii+494 pp.

Ca315. **Batanouny, K. H.** 1978. *Natural history of Saudi Arabia: A bibliography*. Jeddah, King Abdulaziz Univ. xii+121 pp. *Publication of King Abdulaziz University, biology* no. 1. ••*Includes* I: Environment; II: Plant life (Flora and vegetation, Lower plants); III: Fauna; IV: Land use (Range resources, etc.); V: Explorations and expeditions; VI: Miscellaneous and bibliographies.

Ca316. **Burgess, R. L., A. Mokhtarzadeh & L. Cornwallis.** 1966. A preliminary bibliography of the natural history of Iran. *Science bulletin, Pahlavi University* 1: ii+220+[144] pp. ••English with a supplement in Farsi.

Ca317. **Buryat–Mongol'skii Gos. Nauchno–Issledovatel'skii Universitet Kul'tury i Ekonomiki.** 1939–46. *Bibliografiya Buryat–Mongolii za 1890–1936 gg*. By N. V. Zdobnova, et al. Leningrad, Izd. AN SSSR. 4 vols. ••Vol. 1: Estestvoznanie; vol. 2: Sel'skoe, lesnoe, pushno–zverovoe i rybnoe zhozyaistvo; vol. 3: Zdravookhranenie.

Ca318. **Centre National de la Recherche Scientifique, Paris.** 1972. *Bibliographie du Népal*. Vol. 3: *Sciences naturelles*. Tome 1: *Cartes du Népal dans les bibliothèques de Paris et de Londres*. By L. Boulnois. Paris. 117 pp. & 11 pl. *Cahiers Népalais*.

Ca319. **Cordier, H.** 1904 08. *Bibliotheca sinica: Dictionnaire bibliographique des ouvrages relatifs a l'empire chinois*. 2nd ed. Paris, Guilmoto. 4 vols. (xvi+3,252 cols.); *Continued* ___. 1922–24, 1932. ___: *Supplement et index[*]*. Paris. *Re-pub*. Taipei, Ch'eng–wen Publishing Co., 1966. 5 vols; New York, B. Franklin, [1968]. 6 vols. *Burt Franklin bibliography & reference series* no. 250; ___. 1958. *T'ung–li yüan. China in western literature: A continuation of Cordier's Bibliotheca sinica*. New Haven, CT, Yale Univ. xix+802 pp. ••Including sections on natural history. *The index was never published, but see: *Author index*. Compiled, issued and distributed by the East Asiatic Library, Columbia Universities Libraries. New York, Columbia Univ., East Asiatic Library, 1953. iv+84 ff.

Ca320. ___. 1912. Bibliotheca Japonica: Dictionnaire bibliographique des ouvrages relatifs à l'empire Japonais ranges part ordre chronologique jusqu'a 1870. Paris. xii pp. & 762 cols. *Re-pub*. Hildesheim, G. Olms, 1969.

Ca321. ___. 1912–15. *Bibliotheca indosinica: Dictionnaire bibliographique des ouvrages relatifs à ls péninsule Indochinoise*. Paris, E. Leraux. 4 vols.; *Continued* ___. 1932. ___: *Supplément et index*. Paris, École Française d'Extrême–Orient. *Re-pub*. New York, B. Franklin, [1967]. 5 vols. in 3. *Burt Franklin bibliography & reference series* no. 106; New York, G. Olms, 1972. 5 vols. in 2.

Ca322. **Field, H.** 1953–62. *Bibliography on southwestern Asia. I–VII*. Coral Gables, FL, Univ. of Miami Pr. 7 vols; ___. 1960–64. *Subject index to bibliographies on southwestern Asia: I–V (Vi–VII) Part*

II: Coral Gables, FL, Univ. of Miami Pr. 4 pts., covering zoology & botany, by B. J. Clifton & R. C. Foster; *Continued* ___, **E. M. Laird, & B. J. Clifton.** 1968–72. *Bibliographies on southwestern Asia: Supplement 1(–8).* Coconut Grove, FL, Field Research Projects. 8 vols.

Ca323. **Gorodetskii, V. D.** 1913. *Bibliografiya Turkestana: Geografiya, istoriya, mineralogiya, paleontologiya, botanika, zoologiya, meteorologiya, astronomiya, etnografiya, antropologiya, meditsina, veterinariya, sel'skoe khozyaistvo.* Vol. 1. Tashkent. 6+ii+150 pp.

Ca324. **Gribanovskii, N. N.** 1932. Bibliografiya Yakutii. Ch. 1. Prirodnye resursy i naselenie Yakutskogo kraya. [Bibliography of the Yakutsk region. Part I. Natural resources and population of Yakutsk region]. *Trudy soveta po izucheniyu prirodnykh resursov. Seriya Yakutskaya* 9: [3]+viii+127 pp.; *Continued* **Akademiya Nasuk SSSR, Sovet po Izucheniyu Proizvoditel'nykh Sil.** 1958–62. *Bibliografiya Yakutskoi ASSR, 1931–1955 (–1959).* Ed. by L. V. Pustovalov & V. V. Obruchev. Moscow. 2 vols. Vol. 2: *Prirodnye usloviya resursy i narodnoe khozyaistvo.*

Ca325. **Jones, S., & W. McL. Chapman.** 1971. *Bibliography of the Indian Ocean.* Issued on the occasion of the symposium on Indian Ocean and adjacent seas — their origin and resources, held at Cochin (India), January 12–18, 1971. Mandapam Camp, Marine Biological Association of India. 573 pp. ••Including biology.

Ca326. **Koba, K.** 1935–36. A bibliography of zoogeography and phytogeography of Japan. *Bulletin of the Biogeographical Society of Japan* 5 suppl: 1–52, 1935; *Ibid.* 6 suppl.: 1–23, 1936.

Ca327. **Lal, C., & A. S. Ahsan.** 1978. *Reference sources in Indian agriculture and biology.* New Delhi, Ess Ess Publications. vii+213 pp. ••*Includes* Botany, zoology, general reference sources, agriculture, etc. with author, subject, systematic and geographic index.

Ca328. **Lal Mohan, R. S., D. B. James, & K. K. Appukuttan.** 1968. *Bibliography of the Indian Ocean: 1931–1961.* [Mandapam Camp,] Central Marine Fisheries Research Institute. 176 pp. *Bulletin, Central Marine Fisheries Research Institute* no. 11.

Ca329. **Maslova, O. V.** 1955–62. *Obzor russkikh puteshestvii i ekspeditsii v Srednyuyu Aziyu. Materialy k istorii izucheniya Srednei Azii.* Tashkent, Sredneaziatskii Gos. Universitet im V. I. Lenina. 3 vols. ••Vol. 1 covers 1715–1856, vol. 2 covers 1856–1869, vol. 3 covers 1869–1880.

Ca330. **Mezhov, V. I.** 1891–92. *Sibirskii bibliografiya. Ukazatel' knig i statei o Sibiri na russkom yazyke i odnekh tol'ko knig na inostrannykh yazykakh za ves' period knigopechataniya.* St. Petersburg, I. M. Sibiryakov. 3 vols. ••Vol. 3 includes science.

Ca331. **Nemenzo, C. A.** 1969. The flora and fauna of the Philippines, 1851–1966: An annotated bibliography. Part 1: Plants. *Natural and applied science bulletin, University of the Philippines* 21: xv+307 pp.

Ca332. **Peters, A. J.** 1957. Bibliography of published work bearing on the natural history of the Seychelles and neighbouring archipelagoes. *Journal of the Society for the Bibliography of Natural History* 3: 238–262.

Ca333. **___, & J. F. G. Lionnet.** 1973. Central western Indian Ocean bibliography. *Atoll research bulletin* 165: 322 pp.

Ca334. **Skvortzov, B. V.** 1919–21. The bibliography on the botany, zoology and the rural economy of Manchuria. 1(–2). *Journal of the North China Branch, Royal Asiatic Society* 50: 104–107, 1919; *Ibid.* 52: 104–111, 1921.

Ca335. **Zubashova, L. E.** 1956. *Bibliografiya Irkutskoi oblasti:*

Biologiya. Irkutsk, Irkutskii Univeresitet im. A. A. Zhdanova. 247 pp. *Tr. nauchnoi b–ki* no. 15.

Australalasian (& Pacific)

Ca336. **Aubert de la Rue, E.** 1966. Les sciences naturelles dans les îles françaises, de leur découverte à 1945. Bibliographie. *Comité National Français des Recherches Antarctiques, C N F R A* 16: 1–28. *Continued* **R. Delèpine.** 1964. La biologie depuis 1945 dans les Iles Australes Françaises. Bibliographie. *Ibid.* 10: 173–204; ___. 1966. La biologie dans les Iles Australes Françaises. Supplément bibliographique 1964–65. *Ibid.* 16: 41–61.

Ca337. **Cammack, F. M., & S. Saito.** 1962. *Pacific island bibliography.* New York, Scarecrow. 421 pp.

Ca338. **Edridge, S.** 1985. *Solomon Islands bibliography to 1980.* Suva, Institute of Pacific Studies, Univ. of the South Pacific; Wellington, N.Z., Alexander Turnbull Library; Honiara, Solomon Islands, The National Library. xvi+476 pp.

Ca339. **Reilly, P.** 1985. *Solomon Islands: Bibliography of agriculture and forestry.* Surbiton, Land Resources Development Centre. v+331 pp.

Ca340. ___**, & D. Reitman**, eds. 1967. *Bibliographie de Tahiti et de la Polynésie Française.* Paris, Musée de l'Homme. xvi+1,046 pp. *Publication, Société des Océanistes* no. 14. ••Including botany, zoology, geology, geography, etc.

Ca341. **Snow, P. A.** 1969. *Bibliography of Fiji, Tonga and Rotuma.* Preliminary working edition. Coral Gables, FL, Univ. of Miami Pr. xliii+418 pp. ••Ca.10,000 titles, including botany, biology, forestry, geography, maps.

Ca342. **Utinomi, H.** [1952.] *Bibliography of Micronesia (bibliographia micronesia: Scientiae naturalis et cultus).* Ed. ... by O. A. Bushnell. Honolulu, HI, Pacific Area Bibliographies. xiv+157 pp.

African

Ca343. **Asamani, J. O.** 1975. *Index Africanus.* Stanford, CA, Hoover Institution Pr., Stanford Univ. xiii+659 pp. *Hoover Institution bibliographies* no. 53. ••Catalogue of articles published from 1885 to 1965 in periodicals, Festschriften or memorial volumes, symposia, and proceedings of congresses and conferences. Covers Africa general, six major regions and individual countries. Includes sciences.

Ca344. **Bruel, G.** 1914. *Bibliographie de l'Afrique Equatoriale Française.* Paris, Larose. iv+326 pp.

Ca345. **Grandidier, G.** 1905–06. *Bibliographie de Madagascar.* Paris, Comité de Madagascar. 2 vols; *Continued* ___. 1935. ___, *1904–1933.* Paris, Société d'Éditions Géographiques, Maritimes et Coloniales. viii+759–1,350 pp.

Ca346. **Janvier, G.** 1972. Bibliographie de la Côte–d'Ivoire, sciences de la vie. *Annales de l'Université d'Abidjan* vol. hors série: 354 pp.

Ca347. **Joucla, E.** 1937. *Bibliographie de l'Afrique Occidentale Française.* Paris, Société d'Édit. Géogr., Mar. et Colon. 704 pp.

Ca348. **Mary, G. T.**, ed. 1966. *Afrika Schrifttum: Bibliographie deutschsprachiger wissenschaftlicher Veröffentlichungen über Afrika südlich der Sahara. Literature on Africa: Bibliography of scientific publications on Africa South of the Sahara in the German language.* Wiesbaden, F. Steiner Verlag. [xi]+688 pp. ••*Includes* Geographie (embracing general natural history), (by N. Becker), pp. 1–146; Zoologie, (by J. Niethammer), pp. 497–601; Botanik, (by R. Seelbach), pp. 603–657.

Ca349. **Watkins, B. P., J. Cooper, & I. P. Newton.** 1984. Research into the natural sciences at the Tristan da Cunha islands, 1719–1983: A bibliography. *South African journal of Antarctic research* 14: 40–47.

Polar

Ca350. **Arctic Institute of North America.** 1953–75. *Arctic bibliography.* Washington, DC, United States Government Printing Office (from vol. 13, 1967 Montreal, McGill Univ.). 26 vols. ••Annual list covering literature from pre 1949.

Ca351. **Arnaud, P., & J.–C. Hureau.** 1967. Bibliographie générale de biologie antarctique et subantarctique (cetacés exceptés). *Comité National Français des Recherches Antarctiques, C N F R A* 18: xvi+180 pp.

Ca352. **Denucé, J.** 1913. Bibliographie Antarctique. In: Commission Polaire Internationale. *Procès–verbal de la session tenue a Rome en 1913*, pp. 25–293. Brussels.

Ca353. **United States Library of Congress.** 1970. *Antarctic bibliography, 1951–1961.* Washington, DC, Library of Congress. [x]+349 pp.; *Continued* **National Science Foundation, Office of Antarctic** (*later* **Polar**) **Programs.** 1965+. *Antarctic bibliography.* Washington, DC, Library of Congress. ••Annual list covering literature from 1962+; e.g. vol. 8, 1976. *Also* ___. 1977. *Antarctic bibliography: Indexes to volumes 1–7.* Ed. by G. T. Thuronyi. Washington, DC, Library of Congress.

See also Library cats. Bc212 & Bc316.

SPECIAL TOPICS

Biogeography

Ca354. **Browne, J.** 1983. *The secular ark: Studies in the history of biogeography.* New Haven, CT, & London, Yale Univ. Pr. x+273 pp.

Ca355. **Frankel, H.** 1985. The biogeographical aspect of the debate over continental drift. *Earth sciences history* 4: 160–181.

Ca356. **Hofsten, N. von.** 1916. Zur älteren Geschichte des Diskontinuationsproblems in der Biogeographie. *Zoologische Annalen* 7: 197–353.

Ca357. **Kinch, M. P.** 1980. Geographical distribution and the origin of life: The development of early nineteenth–century British explanations. *Journal of the history of biology* 13: 91–119.

Ca358. **Larson, J. L.** 1986. Not without a plan: Geography and natural history in the late eighteenth century. *Journal of the history of biology* 19: 447–488.

Ca359. **Naumov, G. V.** 1969. *Kratkii istoriya biogeografii.* Moscow, Akademiya Nauk SSSR, Institut Istorii Estestvoznaniya i Tekhniki. 200 pp.

Ca360. **Nelson, G.** 1978. From Candolle to Croizat: Comments on the history of biogeography. *Journal of the history of biology* 11: 269–305.

Ca361. **Robinson, A. H.** 1982. *Early thematic mapping in the history of cartography.* Chicago, IL, Univ. of Chicago Pr. 266 pp.
Ch. 4: Maps of the physical world, pp. 68–108 (Vegetation and animal life, pp. 10–108).

Ca362. **Stott, P.** 1984. History of biogeography. In: J. A. Taylor, ed. *Themes in biogeography*, pp. 1–24. London, Croom Helm.

Ecology

Ca363. **Acot, P.** 1988. *Histoire de l'écologie.* Paris, Presses Universitaires de France. 285 pp. *La politique éclatée.*

Ca364. **Barr, T. C.** 1966. Evolution of cave biology in the United States, 1822–1925. *Bulletin of the National Speleological Society* 28: 15–21.

Ca365. **Dajos, R.** 1984. Éléments pour une histoire de l'écologie: La naissance de l'écologie moderne au XIXe siècle. *Histoire et nature* 24/25: 5–111.

Ca365A. **Deléage, J.–P.** 1992. *Histoire de l'écologie: Une science de l'homme et de la nature*. Paris, Éditions La Découverte. 330 pp.

Ca366. **Egerton, F. N.** 1973. Changing concepts of the balance of nature. *Quarterly review of biology* 48: 322–350.

Ca367. ___. 1976. Ecological studies and observations before 1900. In: B. J. Taylor & T. J. White, eds. *Issues and ideas in America*, pp. 311–351. Norman, OK, Oklahoma Univ. Pr.

Ca368. ___, ed. 1977. *History of American ecology*. New York, Arno Pr. iii+270 pp.
Re-pub. of seven older articles by various authors with two new chapters, i.e. **R. L. Burgess.** History of the Ecological Society of America, 24 pp.; **R. C. Tobey.** American grassland ecology, 1895–1955: The life cycle of a professional research community, 56 pp.

Ca369. ___. 1977. A bibliographical guide to the history of general ecology and population ecology. *History of science* 15: 189–215.

Ca370. ___. 1983–85. The history of ecology: Achievements and opportunities. Part one (–two). *Journal of the history of biology* 16: 259–310, 1983; *Ibid.* 18: 103–143, 1985.

Ca370A. **Hagen, J. B.** 1992. *An entangled bank: The origins of ecosystem ecology*. New Brunswick, NJ, Rutgers Univ. Pr. xi+245 pp.

Ca371. **Harant, H.** 1977. L'écologie: Naissance, évolution et pathologie d'un mot. *Histoire et nature* 11: 95–97.

Ca372. **Kingsland, S. E.** 1985. *Modelling nature: Episodes in the history of population ecology*. Chicago, IL, Univ. of Chicago Pr. [ix]+268 pp.

Ca373. **Klaauw, C. J. van der.** 1937. Zur Geschichte der Definition der Ökologie. *Sudhoffs Archiv für Geschichte der Medizin und der Naturwissenschaften* 29: 136–177.

Ca374. **McIntosh, R. P.** 1976. Ecology since 1900. In: B. J. Taylor & T. J. White, eds. *Issues and ideas in America*, pp. 353–372. Norman, OK, Oklahoma Univ. Pr.

Ca375. ___. 1985. *The background of ecology: Concept and theory*. Cambridge, Cambridge Univ. Pr. xiv+ 383 pp. *Cambridge studies in ecology*.

Ca376. ___. 1989. Citation classics of ecology. *Quarterly review of biology* 64: 31–49.
Eighty "classic" titles, published between 1947–77, derived from computer-based citation statistics.

Ca377. **Maienschein, J., J. P. Collins, & J. Beatty**, eds. 1986. Reflections on ecology and evolution. *Journal of the history of biology* 19(2): 167–312.
J. P. Collins, J. Beatty, & J. Maienschein. Introduction: Between ecology and evolutionary biology, pp. 169–180; **W. Coleman.** Evolution into ecology? The strategy of Warming's *Ecological Plant Geography*, pp. 181–196; **J. B. Hagen.** Ecologists and taxonomists: Divergent traditions in twentieth–century plant geography, pp. 197–214; **W. C. Kimler.** Advantage, adaptiveness, and evolutionary ecology, pp. 215–233; **S. E. Kingsland.** Mathematical figments, biological facts: Population ecology in the thirties, pp. 235–256; **J. P. Collins.** Evolutionary ecology and the use of natural selection in ecological theory, pp. 257–288; **R. E. Michod.** On fitness and adaptedness and their role in evolutionary explanation, pp. 289–302; **D. J. Futuyama.** Reflections on reflections: Ecology and evolutionary biology, pp. 303–312.

Ca378. **Mitman, G.** 1992. *The state of nature: Ecology, community, and American social thought, 1900–1950*. Chicago, IL, Univ. of Chicago Pr. xiv+290 pp., illus. *Science & its conceptual foundations*.

Ca379. **Palladino, P.** 1991. Defining ecology: Ecological theories, mathematical models, and applied biology in the 1960s and 1970s. *Journal of the history of biology* 24: 223–243.

Ca380. **Söderqvist, T.** 1986. *The ecologists: From merry naturalists to saviours of the nation: A sociologically informed narrative survey of the ecologization of Sweden 1895–1975*. Stockholm, Almqvist & Wiksell International. vii+330 pp.

Ca381. **Trepl, L.** 1987. *Geschichte der Ökologie, vom 17. Jahrhundert bis zur Gegenwart: Zehn Vorlesungen.* Frankfurt am Main, Athenäum Verlag. 280 pp.

Ca382. **Weiner, D. R.** 1988. *Models of nature: Ecology, conservation, and cultural revolution in Soviet Russia.* Bloomington & Indianapolis, IN, Indiana Univ. Pr. xiv+312 pp. *Indiana–Michigan Series in Russian & East European Studies.*

Ca383. **Worster, D.** 1977. *Nature's economy: The roots of ecology.* San Francisco, CA, Sierra Club Books. xii+404 pp. *Re-pub. as* ___. 1985. *Nature's economy: A history of ecological ideas.* Cambridge, Cambridge Univ. Pr.

Ca384. **Worthington, E. B.** 1983. *The ecological century: A personal appraisal.* Oxford, Clarendon Pr. xv+206 pp. *Oxford science publications.*

See also Biographies Ba627–628 Ba642 Ba997 Ba1484–1485; Societies Bb169–171 Bb615–616; Misc. Ca66 Ca154 Ca184(Gimingham et al., Boyd) Ca190 Ca200, Ca228 Ca231 Ca232 Ca234 Ca263 & Ca497.

Reproduction

Ca385. **Cafiso, L. M.** 1981. Storia della teoria sulla generazione spontanea: Redi, Spallanzani, Pasteur. *Cultura e scuola* 20: 236–243.

Ca386. **Churchill, F. B.** 1979. Sex and the single organism: Biological theories of sexuality in mid–nineteenth century. *Studies in the history of biology* 3: 139–177.

Ca387. **Farley, J.** 1972. The spontaneous generation controversy (1859–1880): British and German reactions to the problem of abiogenesis. *Journal of the history of biology* 5: 285–319.

Ca388. ___. 1977. *The spontaneous generation controversy from Descartes to Oparin.* Baltimore, MD, Johns Hopkins Univ. Pr. x+225 pp.

Ca389. ___. 1982. *Gametes and spores: Ideas about sexual reproduction, 1750–1914.* Baltimore, MD, Johns Hopkins Univ. Pr. x+299 pp.

Ca390. ___, **& G. L. Geison.** 1974. Science, politics and spontaneous generation in nineteenth–century France: The Pasteur–Pouchet debate. *Bulletin of the history of medicine* 48: 161–198.

Ca391. **Geus, A.** 1972. Der Generationswechsel: Die Geschichte eines biologischen Problem. *Medizinhistorisches Journal* 7: 159–173.

Ca392. **Hertwig, W. A. O.** 1918. *Dokumente zur Geschichte der Zeugungslehre: Eine historische Studie.* Bonn, F. Cohen. 163 pp.

Ca393. **Moeschlin–Krieg, B.** 1953. *Zur Geschichte der Regenerationforschung im 18. Jahrhundert.* Basle, B. Schwabe & Co. [ii]+87 pp. & 5 pl. *Basler Veröffentlichungen zur Geschichte der Medizin und der Biologie.*

Genetics & heredity

Ca394. **Blacher, L. I.** 1982. *The problem of the inheritance of acquired characters: A history of* a priori *and empirical methods used to find a solution.* English translation ed. by F. B. Churchill. New Delhi, Amerind Publishing Co. for the Smithsonian Institution Libraries & the National Science Foundation, Washington, DC. xvi+267 pp.
Translation of **L. I. Blyakher.** *Problema nasledovaniyapriobremrnnykh priznakov: Istoriya aprirnoykh i empiricheskikh popitok ee resheniya.* Moscow, Izd. "Nauka," 1971.

Ca395. **Bowler, P. J.** 1989. *The Mendelian revolution: The emergence of hereditarian concepts in modern science and society.* London, Athlone Pr. viii+207 pp. *Also* Baltimore, MD, Johns Hopkins Univ. Pr.

Ca396. **Buican, D.** 1984. *Histoire de la génétique et de l'évolutionnisme en France.* Paris, Presses Universitaires de France. 421 pp.

Ca397. **Burian, R. M., J. Gayon, & D. Zallen.** 1988. The singular fate of genetics in the history of French biology, 1900–1940. *Journal of the history of biology* 21: 357–402.

Ca398. **Campbell, M.** 1983. The concepts of dormancy, latency and dominance in nineteenth century biology. *Journal of the history of biology* 16: 409–431.

Ca399. **Carlson, E. A.** 1966. *The gene: A critical history.* Philadelphia, PA, Saunders. xi+301 pp., illus. *Re-pub.* Ames, IA, Iowa State Univ. Pr., 1989.

Ca400. **Dunn, L. C.** 1966. *A short history of genetics: The development of some of the main lines of thought, 1864–1939.* New York, McGraw–Hill Book Co. xxii+261 pp., illus. *Re-pub.* Ames, IA, Iowa State Univ. Pr., 1991.

Ca401. **Ford, E. B.** 1969. Ecological genetics. In: R. Harré, ed. *Scientific thought 1900–1960: A selective survey*, pp. 173–195. Oxford, Clarendon Pr.

Ca402. **Gaisinovich, A. E.** 1971. *Problems of variation and heredity in Russian biology at the end of the 19th and the beginning of the 20th centuries.* [From: *Proceedings of the 13th International Congress of the History of Science, Moscow, 1971.*] Moscow, Izd. "Nauka." 23 pp.

Ca403. **___.** 1973. Problems of variation and heredity in Russian biology in the late 19th century. *Journal of the history of biology* 6: 97–123.

Ca404. **Jacob, F.** 1982. *The logic of life: A history of heredity.* Translated by B. E. Spillman. New York, Pantheon. viii+337 pp.
Translation of *La logique du vivant: Une histoire de l'hérédité.* Paris, Éditions Gallimard, 1970. 345 pp. *Bibliothèque des sciences humaines.*

Ca405. **Judson, H.** 1979. *The eighth day of creation: Makers of the revolution in biology.* New York, Simon & Schuster. 686 pp., illus. *Re-pub.* 1979.

Ca406. **Kevles, D. J.** 1980. Genetics in the United States and Great Britain, 1890–1930: A review with speculations. *Isis* 71: 441–455.

Ca407. **Mayr, E.** 1973. The recent historiography of genetics. *Journal of the history of biology* 6: 125–154.

Ca408. **Olby, R. C.** 1966. *Origins of Mendelism.* 2nd ed. Chicago, IL, Univ. of Chicago Pr. viii+310 pp. *Re-pub.* Chicago, [etc.], 1985.
Replaces 1st ed., London, Constable, 1966. 204 pp.

Ca409. **___.** 1984. Historiographical problems in the history of genetics. *Rivista di storia della scienza* 1: 25–38.

Ca410. **Orel, V., & A. Matalova**, eds. 1983. *Gregor Mendel and the foundation of genetics.* Brno, Mendelianum of the Moravian Museum. 368 pp.
Proceedings of pt. 1 of the symposium "The past, present and future of genetics," Kuparovice, Czechoslovakia, August 1982.

Ca411. **Robinson, G.** 1979. *A prelude to genetics: Theories of a material substance of heredity, Darwin to Weisman.* Lawrence, KS, Coronado Pr. xvi+260 pp.

Ca412. **Sapp, J.** 1983. The struggle for authority in the field of heredity, 1900–1932: New perspectives on the rise of genetics. *Journal of the history of biology* 16: 311–342.

Ca413. **Stubbe, H.** 1972. *History of genetics: From prehistoric times to the rediscovery of Mendel's laws.* Translated by T. R. W. Waters. Cambridge, MA, M I T Pr. xi+356 pp.
Translation of *Kurze Geschichte der Genetik bis zur Wiederentdeckung der Vererbungsregeln Gregor Mendels*, 2. Aufl., Jena, G. Fischer, 1965. xiv+272 pp. *Genetik–Grundlagen, Ergebnisse und probleme in Einzeldarstellungen* Bd. 1.

Ca414. **Sturtevant, A. H.** 1965. *A history of genetics.* New York, Harper & Row. viii+154 pp.

Ca415. **Zirkle, C.** 1964. Some oddities in the delayed discovery of Mendelism. *Journal of heredity* 55: 65–72.

See also Biographies Ba513–516 Ba830–833 Ba1234–1247 Ba1598–1599; Misc. Ca70–71 Ca184(Robertson) Ca441(Glass, Glass) Ca459(Lewontin, Darlington, Weinstein, Carson, Dobzhansky) Ca1044 & Cc519(Brown).

Evolution

Ca416. **Allen, G. E.** 1979. Naturalists and experimentalists: The genotype and the phenotype. *Studies in the history of biology* 3: 179–209.

Ca417. **Aulie, R. P.** 1983. Evolution and special creation: Historical aspects of the controversy. *Proceedings of the American Philosophical Society* 30: 418–462.

Ca418. **Beddall, B. G.** 1968. Wallace, Darwin, and the story of natural selection. *Journal of the history of biology* 1: 261–324.

Ca419. **___.** 1972. Wallace, Darwin, and Edward Blyth: Further notes on the development of evolution theory. *Journal of the history of biology* 5: 153–158.

Ca420. **Benasso, G.** 1976. Materiali per una storia dell'evoluzionismo italiano: Da Bonelli a De Filippi: 1811–1864. *Atti della Accademia Roveretana degli Agiati, contributi della classe di scienze matematiche, fisiche e naturali* 14–15, B: 3–106; *Continued* **___.** 1978. ___: Un approccio al Darwinismo, 1864–1900. *Ibid.* 16–17, B: 73–151.

Ca421. **Berman, Z. I.** 1966. *Istoriya evolyutsionnykh uchenii v biologii. [History of evolutionary studies in biology].* Moscow, Akademiya Nauk SSSR, Otdelenie Obshchei Biologii. 323 pp.

Ca422. **Boesiger, E.** 1971. *Evolution des tendencies neoLamarckiennes et neoDarwinistes au vingtième siècle en France.* [From: *Proceedings of the 13th International Congress of the History of Science, Moscow, 1971.*] Moscow, Izd. "Nauka." 35 pp.

Ca423. **Bourdier, F.** 1960. Trois siècles d'hypothèses sur l'origine et la transformation des êtres vivants (1550–1859). *Revue d'histoire des sciences* 13: 1–44.

Ca424. **Bowler, P. J.** 1975. The changing meaning of "evolution." *Journal of the history of ideas* 36: 95–114.

Ca425. **___.** 1983. *The eclipse of Darwinism: Anti-Darwinian evolution theories in the decades around 1900.* Baltimore, MD, Johns Hopkins Pr. xi+291 pp.

Ca426. **___.** 1988. *The non-Darwinian revolution: Reinterpreting a historical myth.* Baltimore, MD, Johns Hopkins Univ. Pr. x+238 pp.

Ca427. **___.** 1989. *Evolution: The history of an idea.* Rev. ed. Berkeley, CA, Univ. of California Pr. xxi+432 pp.
Replaces 1st ed., 1984. xiv+412 pp.

Ca428. **Buican, D.** 1984. *Histoire de la génétique et de l'évolutionisme en France.* Paris, Presses Universitaires de France. 421 pp.

Ca429. **___.** 1989. *La révolution de l'évolution: L'Évolution de l'évolutionisme.* Paris, Presses Universitaires de France. 339 pp.

Ca430. **Centre d'Histoire et d'Épistémologie des Théories Scientifiques et des Doctrines Philosophique de l'Université de Picardie.** 1982. *De Darwin au darwinisme: Science et idéologie. Congrès international pour le centenaire de la mort de Darwin, Paris–Chantilly 13–16 septembre 1982.* Ed. par Y. Conry. Paris, J. Vrin. 284 pp. *L'Histoire des textes; Textes et études.*

Ca431. **Conry, Y.** 1974. *L'Introduction du Darwinisme en France en XIXe siècle.* Paris, J. Vrin. 480 pp.

Ca432. **Corsi, P.** 1988. *The age of Lamarck: Evolutionary theories in France, 1790–1830.* Rev. & updated. Translated by J. Mandelbaum. Berkeley, CA, Univ. of California Pr. xiii+360 pp.

Ca433. **Cravens, H.** 1978. *The triumph of evolution: American scientists and the heredity–environment controversy 1900–1941.* Philadelphia, PA, Univ. of Pennsylvania Pr. xvi+351 pp.

Ca434. **Daniels, G.**, ed. 1972. *Darwinism comes to America.* Waltham, MA, Blaisdell. xix+137 pp.

Ca435. **Danielsson, U.** 1965–67. Darwinismus inträngande i Sverige, I[–II]. *Lychnos* 1963/64: 157–210, 1965; *Ibid.* 1965/66: 261–334, 1967.

Ca436. **Davitashvili, L. Sh.** 1956. *Ocherki po istorii ucheniya ob evolyutsionnom progresse. [Essays on the history of the study of evolutionary progress].* Moscow, Izdalel'stvo Akademii Nauk SSSR. 228 pp.

Ca437. **Ellegård, A.** 1958. *Darwin and the general reader: The reception of Darwin's theory of evolution in the British periodical press, 1859–1872.* Gothenberg, Elanders Boktrycken Akt.; Stockholm, Almqvist & Wiksell. 396 pp. *Göteborgs Universitets årsskrift* vol. 64. *Gothenburg studies in English* no. 8. *Re-pub.* (with a new foreword by D. L. Hull) Chicago, IL, Univ. of Chicago Pr., 1990. [vi]+394 pp.

Ca438. **Farley, J.** 1974. The initial reactions of French biologists to Darwin's *Origin of species. Journal of the history of biology* 7: 275–300.

Ca439. **Ghiselin, M. T.** 1969. *The triumph of the Darwinian method.* Berkeley, CA, Univ. of California Pr. vii+287 pp. *Re-pub.* (with a new preface) Chicago, IL, Univ. of Chicago Pr., 1984. xxii+288 pp.

Ca440. **Giacobini, G., & G. L. Panattoni**, eds. 1983. *Il Darwinismo in Italia: Testi di Filippo de Filippi, Michele Lessona, Paolo Mantegazza, Giovanni Canestrini.* Turin, Unione Tipografico–Editrice Torinese. 127 pp.

Ca441. **Glass, B., O. Temkin & W. L. Straus**, eds. 1959. *Forerunners of Darwin, 1745–1859.* Baltimore, MD, Johns Hopkins Univ. Pr. xx+471 pp.

Articles by various authors. *Includes* **F. C. Haber.** Fossils and early cosmology, pp. 3–29; **B. Glass.** The germination of the idea of biological species, pp. 30–48; **B. Glass.** Maupertuis, pioneer of genetics and evolution, pp. 51–83; **A. O. Lovejoy.** Buffon and the problem of species, pp. 84–; **L. G. Crocker.** Diderot and eighteenth century French transformism, pp. 114–143; **B. Glass.** Heredity and variation in the eighteenth century concept of the species, pp. 144–172; **A. O. Lovejoy.** Kant and evolution, pp. 173–206; ___. Herder: Progressionism without transformism, pp. 207–221; **F. C. Haber.** Fossils and the idea of a process of time in natural history, pp. 222–261; **C. C. Gillispie.** Lamarck and Darwin in the history of science, pp. 265–291; **J. M. Oppenheimer.** An embryological enigma in the *Origin of Species*, pp. 292–322; **O. Temkin.** The idea of descent in post-romantic German biology, pp. 323–355; **A. O. Lovejoy.** The argument for organic evolution before the *Origin of Species*, 1830–1858, pp. 356–414; ___. Schopenhauer as an evolutionist, p. 415–437; ___. Recent criticism of the Darwinian theory of recapitulation: Its grounds and its initiator, pp. 438–458.

Ca442. **Glick, T. F.**, ed. 1974. *The comparative reception of Darwinism.* Austin, TX, Univ. of Texas Pr. ix+505 pp. *Re-pub.* Chicago, IL, Univ. of Chicago Pr., 1988. xxviii+506 pp.
Articles by various authors. *Includes* **M. J. S. Hodge.** England, pp. 3–31; **F. Burkhardt.** England and Scotland: The learned societies, pp. 32–74; **M. J. S. Hodge.** England: Bibliographical essay, pp. 75–80; **W. M. Montgomery.** Germany. [&] German editions of important works on evolution, pp. 81–115–116; **R. E. Stebbins.** France [&] France: Bibliographical essay, pp. 117–164–167; **E. J. Pfeifer.** United States, pp. 168–206; **M. L. Aldrich.** United States: Bibliographic essay, pp. 207–226; **A. Vucinich.** Russia: Biological sciences, pp. 227–255; **J. A. Rogers.** Russia: Social sciences, pp. 256–268; **I. N. Bulhof.** The Netherlands, pp. 269–306; **T. F. Glick.** Spain, pp. 307–345; **R. Moreno.** Mexico, pp. 346–374; **N. A. Bezirgan.** The Islamic world, pp. 375–387; **D. L. Hull.** Darwinism and historiography, pp. 388–402; **H. W. Paul.** Religion and Darwinism: Varieties of Catholic reaction, pp. 403–436; **A. Leeds.** Darwinian and "Darwinian" evolutionism in the study of society and culture, pp. 437–485. The re-pub. ed. includes **T. F. Glick.** Preface, 1988: Reception studies since 1974, pp. xi–xxviii.

Ca443. ___. 1989. *Darwin y el darwinismo en el Uruguay y en America Latina.* Montevideo, Universidad de la República, Facultad de Humanidades y Ciencias. 136 pp.

Ca444. **Gould, S. J.** 1977. *Ontogeny and phylogeny.* Cambridge, MA, Belknap Pr. of Harvard Univ. xv+501 pp

Ca445. **Greene, J. C.** 1959. *The death of Adam: Evolution and its impact on western thought.* Ames, IA, Iowa State Univ. Pr. [xii]+388 pp.
In the interval between John Ray's T*he wisdom of God manifested in the works of the Creation* (1691) and Charles Darwin's *Origin of species* (1859) and *Descent of man* (1871).

Ca446. **Grene, M.**, ed. 1983. *Dimensions of Darwinism: Themes and counterthemes in twentieth–century evolutionary theory.* Cambridge, Cambridge Univ. Pr. ix+336 pp.
Twelve articles by various authors. *Includes* **B. Norton.** Fisher's entrance into evolutionary science: The role of eugenics, pp. 19–29; **B. Rensch.** The abandonment of Lamarckian explanations: The case of climatic parallelism an animal characteristics, pp. 31–42; **W. B. Provine.** The development of Wright's theory of evolution: Systematics, adaptation, and drift, pp. 43–70; **S. J. Gould.** The hardening of the modern synthesis. pp. 71–93; **W. C. Kimler.** Mimicry: Views of naturalists and ecologists before the modern synthesis, pp. 97–127; **J. R. G. Turner.** "The hypothesis that explains mimetic resemblance explains evolution": The gradualist–saltationist schism, pp. 129–169; **W.–E. Reif.** Evolutionary theory in German paleontology, pp. 173–203; **R. Riedl.** The role of morphology in the theory of evolution, pp. 205–238.

Ca447. **Hull, D. L.** 1973. *Darwin and his critics: The reception of Darwin's theory of evolution by the scientific community.* Cambridge, MA, Harvard Univ. Pr. xii+473 pp.
Re-pub. of sixteen contemporary reviews, with commentary.

Ca448. **Kahane, E.** 1977. *Evolution des idées sur l'origine de la vie.* Paris, Centre National de la Recherche Scientifique. 124 pp. *Cahiers d'histoire et de philosophie des sciences* no. 4.

Ca449. **Kelly, A.** 1981. *The descent of Darwin: The popularization of Darwinism in Germany, 1860–1914.* Chapel Hill, NC, Univ. Pr. of North Carolina. 185 pp.

Ca450. **Kohn, D.**, ed. 1985. *The Darwinian heritage* ... With bibliographical assistance from M. J. Kottler. Princeton, NJ, Princeton Univ. Pr., Nova Pacifica. xii+1,138 pp.
Papers by various authors read at the Charles Darwin Centenary Conference. *Includes* **H. E. Gruber.** Going the limit: Toward the construction of Darwin's theory (1832–1839), pp. 9–34; **S. S. Schweber.** The wider British concept in Darwin's theorizing, pp. 35–69; **P. R. Sloan.** Darwin's invertebrate program, 1826–1836: Preconditions for transformism, pp. 71–120; **F. J. Sulloway.** Darwin's early intellectual development: An overview of the *Beagle* voyage (1831–1836), pp. 121–154; **S. P. Rachootin.** Owen and Darwin reading a fossil: Macrauchenia in a boney light, pp. 155–183; **M. J. S. Hodge & D. Kohn.** The

immediate origins of natural selection, pp. 185–206; **M. J. S. Hodge.** Darwin as a lifelong generation theorist, pp. 207–243; **D. Kohn.** Darwin's principle of divergence as internal dialogue, pp. 245–257; **J. Beatty.** Speaking of species: Darwin's strategy, pp. 265 281; **R. W. Burkhardt.** Darwin on animal behavior and evolution, pp. 327–365; **M. J. Kottler.** Charles Darwin and Alfred Russel Wallace: Two decades of debate over natural selection, pp. 367–432; **J. R. Moore.** Darwin of Down: The evolutionist as squarson–naturalist, pp. 435–481; **R. M. Young.** Darwinism *is* social, pp. 609–638; **P. J. Bowler.** Scientific attitudes to Darwinism in Britain and America, pp. 641–681; **P. Corsi & P. J. Weindling.** Darwinism in Germany, France and Italy, pp. 683–729; **F. M. Scudo & M. Acanfora.** Darwin and Russian evolutionary biology, pp. 731–752; **E. Mayr.** Darwin's five theories of evolution, pp. 753–772; **D. L. Hull.** Darwinism as a historical entity: A historiographic proposal, pp. 773–812; **W. B. Provine.** Adaptation and mechanisms of evolution after Darwin: A study in persistent controversies, pp. 825–866; **P. A. Sober.** Darwin on natural selection: A philosophical perspective, pp. 867–899; **A. La Vergata.** Images of Darwin: A historiographic overview, pp. 901–972; **D. M. Porter.** The *Beagle* collector and his collections, pp. 973–1019; Bibliography, pp. 1021–1099.

Ca450A. **Laurent, G.** 1987. *Paléontologie et evolution en France 1800–1860: Une histoire des idées de Cuvier et Lamarck à Darwin.* Paris, Éditions du Comité des Travaux Historiques et Scientifiques. xiv+553 pp.

Ca451. **Lefèvre, W.** 1984. *Die Entstehung der biologischen Evolutionstheorie.* Frankfurt am Main, Ullstein. 290 pp., illus. *Ullstein Buch* nr. 35186.

Ca452. **Lie, T.** 1985. The reception of Darwinism in Norway: The early years 1861–1900. *Archives of natural history* 12: 153–159.

Ca453. **Lovejoy, A. O.** 1904. Some eighteenth century evolutionists. *Popular science monthly* 65: 238–251, 323–340.

Ca454. ___. 1909. The argument for organic evolution before the "Origin of Species." *Popular science monthly* 75: 499–514, 537–549.

Ca455. **Mañé Garzon, F.** 1990. *Un siglo de darwinismo: Un ensayo sobre la historia del pensamiento biologico en el Uruguay.* Montevideo, Facultad de Medicina, Seccion Historia de la Medicina. xxii+347 pp.

Ca456. **Matouskova, B.** 1959. Charles Darwin, 1809–1882. "Origin of species," 1859–1959. The beginnings of Darwinism in Bohemia. *Folia biologica (Prague)* 5: 169–185, & 2 pl.

Ca457. **___, & O. Matousek.** 1959. Darvinism v Chekhoslovakii. [History of Darwinism in Czechoslovakia.] *Annaly biologii* 1: 35–50.

Ca458. **Mayr, E.** 1991. *One long argument: Charles Darwin and the genesis of modern evolutionary thought.* Cambridge, MA, Harvard Univ. Pr. xvi+195 pp. *Questions of science.*

Ca459. **___, & W. B. Provine**, eds. 1980. *The evolutionary synthesis: Perspectives on the unification of biology.* Cambridge, MA, Harvard Univ. Pr. xiii+487 pp.
Articles by various authors. *Includes* **E. Mayr.** Prologue: Some thoughts on the history of the evolutionary synthesis, pp. 1–48; **R. C. Lewontin.** Theoretical population genetics in the evolutionary synthesis, pp. 58–68; **C. D. Darlington.** The evolution of genetic systems: Contributions of cytology to evolutionary theory, pp. 70–80; **A. Weinstein.** Cytology in the T. H. Morgan school, pp. 80–86; **H. L. Carson.** Cytogenetics and the neo–Darwinian synthesis, pp. 86–95; **V. Hamburger.** Embryology and the modern synthesis in evolutionary theory, pp. 97–112; **F. B. Churchill.** The modern evolutionary synthesis and the biogenetic law, pp. 112–122; **E. Mayr.** The role of systematics in the evolutionary synthesis, pp. 123–136; **G. L. Stebbins.** Botany and the synthetic theory of evolution, pp. 139–152; **S. G. Gould.** G. G. Simpson, paleontology and the modern synthesis, pp. 153–172; **W. Coleman.** Morphology in the evolutionary synthesis, pp. 174–180; **M. T. Ghiselin.** The failure of morphology to assimilate Darwinism, pp. 180–193; **M. B.**

Adams. Severtsov and Schmalhausen: Russian morphology and the evolutionary synthesis, pp. 193–225; **T. Dobzhansky.** The birth of the genetic theory of evolution in the Soviet Union in the 1920s, pp. 229–242; **M. B. Adams.** Sergei Chetverikov, the Kol'tsov Institute, and the evolutionary synthesis, pp. 242–278; **B. Rensch.** Historical development of the present synthetic neo-Darwinism in Germany, pp. 284–303; **V. Hamburger.** Evolutionary theory in Germany: A comment, pp. 303–308; **E. Boesiger.** Evolutionary biology in France at the time of the evolutionary synthesis, pp. 309–321; **C. Limoges.** A second glance at evolutionary biology in France, pp. 322–328; **E. B. Ford.** Some recollections pertaining to the evolutionary synthesis, pp. 334–342; **R. W. Burkhardt Jr.** Lamarckism in Britain and the United States, pp. 343–352; **G. E. Allen.** The evolutionary synthesis: Morgan and natural selection revisited, pp. 356–382; Biographical essays: **E. Mayr.** How I became a Darwinian, pp. 413–423; **___.** Curt Stern, pp. 424–429; **A. Weinstein.** Morgan and the theory of natural selection, pp. 432–445; **T. Dobzhansky.** Morgan and his school in the 1930s, pp. 445–452, **E. Mayr.** G. G. Simpson, pp. 452–463.

Ca460. **Mikulinskii, S. R., & Yu. I. Polyanskii**, eds. 1983. *Razvitie evolyutsionnoi teorii v SSSR, 1917–1970–e gody. [The development of the theory of evolution in the USSR (1917–1970)].* Leningrad, Izd. "Nauka." 613 pp., illus.

Ca461. **Milner, R.** 1990. *The encyclopedia of evolution: Humanity's search for its origins.* New York, Oxford, Facts on File. xii+481 pp.
Contents include numerous historical articles.

Ca462. **Nuñez, D.** 1977. *El Darwinismo en España.* Madrid, Castalia. 464 pp. *Biblioteca de pensamiento* no. 5.

Ca463. **Osborn, H. F.** 1929. *From the Greeks to Darwin: The development of the evolution idea through twenty-four centuries.* 2nd ed. New York, C. Scribner's Sons. xvi+398 pp.
Replaces 1st ed., 1894, and New York, Macmillan, 1905 (*Re-pub.* 1913). x+259 pp. *Columbia University biological series* no. 1.

Ca464. **Ospovat, D.** 1981. *The development of Darwin's theory: Natural history, natural theology, and natural selection, 1838–1859.* Cambridge, Cambridge Univ. Pr. xii+301 pp.

Ca465. **Ostoya, P.** 1951. *Les théories de l'évolution: Origine et histoire du transformisme et des idées qui s'y rattachent.* Paris, Payot. 319 pp.

Ca466. **Pancaldi, G.** 1991. *Darwin in Italy: Science across cultural frontiers.* Translated by R. B. Morelli. Bloomington, IN, Indiana Univ. Pr. xv+222 pp.
Translation of *Darwin in Italia: Impresa scientifica e frontiere culturali.* Bologna, Società Editrice il Milano, 1983.

Ca467. **Pfeifer, E. J.** 1965. The genesis of American neo-Lamarckism. *Isis* 56: 156–167.

Ca468. **Pop, E.** 1957. Les débuts du Darwinisme dans notre pays (jusqu'à 1880). *Studii si cercetari de biologie. Academia Republicii Populare Romine. Filiala Cluj* 8: 7–42.
Rumanian text with French summary.

Ca469. **___.** 1959. Rasprostranenie darvinizma v Rumynii. [The spreading of Darwinism in Rumania.] *Annaly biologii* 1: 53–61.

Ca470. **Pruna, P. M.** 1984. La recepción de las ideas de Darwin en Cuba, durante el siglo XIX. *Quipu* 1: 369–389.

Ca471. **___, & A. G. González.** 1989. *Darwinismo y sociedad en Cuba. Siglo XIX.* Madrid, Consejo Superior de Investigaciones Cientificas. ix+204 pp. *Estudios sobre la ciencia* no. 7.

Ca472. **Quatrefages, J. L. A. de.** 1870. *Charles Darwin et ses précurseurs français. Étude sur le transformisme.* Paris, G. Baillière. [iv]+378 pp. *Bibliothèque de philosophie contemporaine.*

Ca473. **Raikov, B. E.** 1947. *Ocherki po istorii evolyutsionnoi idei v Rossii do Darvina. [Essays on the history of evolutyionary ideas in Russia up to Darwin].* Vol. 1. Moscow/Leningrad, AN SSSR. 191 pp.

Ca474. ___, ed. 1951–59. *Russkie biologi–evolyutsionisty do Darvina: Materialy k istorii evolyutzionnoi idei v Rossi. [Russian evolutionary biologists up to Darwin: Materials for the history of evolutionary ideas in Russia].* Moscow & Leningrad, AN SSSR, Institut Istorii Estestvoznaniya i Tekhniki. 4 vols.
Articles by various authors.

Ca475. ___. 1957. Iz istorii Darwinizma v Rossii. [Concerning the history of Darwinism in Russia.] *Trudy Institut Istorii Estestvoznaniya i Tekhniki* 31: 17–81.

Ca476. **Rogers, J. A.** 1973. The reception of Darwin's *Origin of species* by Russian scientists. *Isis* 64: 484–503.

Ca477. ___. 1974. Russian opposition to Darwinism in the 19th century. *Isis* 65: 487–505.

Ca478. **Rostand, J.** 1960. Les précurseurs français de Charles Darwin. *Revue d'histoire des sciences* 13: 45–58.

Ca479. **Ruse, M.** 1979. *The Darwinian revolution: Science red in tooth and claw.* Chicago, IL, Univ. of Chicago Pr. xvi+320 pp.

Ca480. **Russett, C. E.** 1976. *Darwin in America: The intellectual response, 1865–1912.* San Francisco, CA, W. H. Freeman. xii+228 pp.

Ca481. **Sala Catalá, J.** 1981. El evolucionismo en la practica cientifica de los biologos españoles del siglo XIX (1860–1907). *Asclepio* 33: 81–125.

Ca482. **Schmidt, H.** 1918. *Geschichte der Entwicklungslehre.* Leipzig, A. Kröner. ix+549 pp.

Ca483. **Shimao, E.** 1981. Darwinism in Japan, 1877–1927. *Annals of science* 38: 93–102.
Includes A bibliographical survey of Japanese versions of Darwin's works, 1896–1974, pp. 97–102.

Ca484. **Todes, D. P.** 1989. *Darwin without Malthus: The struggle for existence in Russian evolutionary thought.* Oxford, Oxford Univ. Pr. x+221 pp. *Monographs on the history & philosophy of biology.*

Ca485. **Vorzimmer, P. J.** 1970. *Charles Darwin: The years of controversy. The "Origin of species" and its critics 1859–82.* Philadelphia, PA, Temple Univ. Pr. xix+300 pp. *Also* London, Univ. of London Pr., 1972.

Ca486. **Vucinich, A.** 1988. *Darwin in Russian thought.* Los Angeles, CA, Univ. of California Pr. x+468 pp.

Ca487. **Zimmermann, W.** 1953. *Evolution: Die Geschichte ihrer Probleme und Erkenntnisse.* Freiburg, Munich, K. Alber. ix+623 pp. *Orbis academicus, Problemgeschichten der Wissenschaft in Dokumenten und Darstellungen, Naturwissenschaftliche Abteilung.*

Ca488. **Zirkle, C.** 1941. Natural selection before the "Origin of Species." *Proceedings of the American Philosophical Society* 84: 71–123.

Ca489. ___. 1946. The early history of the ideas of the inheritance of acquired characters and of pangenesis. *Transactions of the American Philosophical Society* 35: 91–151.

See also Biographies Ba686–714 Ba885–892 Ba1003–1015 Ba1074–1089 Ba1490 Ba1581–1585; Societies Bb71(Grassé) Bb229(Querner); Misc. Ca1(Mayr) Ca498–500 Ca1035 & D98.

Miscellaneous

Ca490. **Fondation Singer–Polignac.** 1987. *Histoire du concept d'espèce dans les sciences de la vie.* Colloque international (mai 1985) [etc.]. Paris. xiii+324 pp.
15 papers by various authors. *Includes* **A. Leikola.** The development of the species concept in the thinking of Linnaeus; **G. H. Müller.** "Distinguer les uns des autres": Le concept de l'espèce chez Réaumur: Pragmatisme et utilité; **R. Rey.** L'espèce entre science et philosophie chez C. Bonnet; **P. R. Sloan.** From logical intervals to historical individuals: Buffon's idea of biological species; **R. W. Burkhardt Jr.** Lamarck and species; **P. Corsi.** Julien Joseph Virey, le premier critique de Lamarck; **A. La Vergata.** Au nom de l'espèce: Classification et nomenclature au XIXe siècle; **M. J. S. Hodge.** Darwin, species and the theory of natural selection; **J.–L. Fischer.** Espèce et hybrides: À propos des léporides; **A. Diara.** Les espèces sont–elles filles de la nature ou du naturaliste?; **B. Fantini.** L'entrée de la biologie moléculaire dans la définition de l'espèce; **E. Mayr.** The species as category, taxon and population.

Ca491. **Gasking, E. B.** 1970. *The rise of experimental biology.* New York, Random House. viii+178 pp.

Ca492. **Harvey, E. N.** 1957. *A history of luminescence, from the earliest times until 1900.* Philadelphia, PA, American Philosophical Society. xxiii+692 pp. & 44 pl. *Memoirs of the American Philosophical Society* vol. 44.

Ca493. **Paclt, J.** 1983. A chronology of color charts and color terminology for naturalists. *Taxon* 32: 393–405.
Some 30 works, some in more than one edition, from 1665 to 1978.

Ca494. **Rostand, J.** 1959. *Les origines de la biologie expérimentale.* Paris, Université de Paris. 28 pp. *Les Conférences du Palais de la Découverte*, série D, no. 63.

Ca495. **Stearn, W. T.** 1962. The origin of the male and female symbols in biology. *Taxon* 11: 109–113.

Ca496. **Tucker, A. O., M. J. Maciarello, & S. S. Tucker.** 1991. A survey of color charts for biological descriptions. *Taxon* 40: 201–214.
Chronological and critical survey of 46 works, some in more than one edition, from 1776 to 1984.

BIBLIOGRAPHIES OF VARIOUS NATURAL-HISTORY TOPICS

Ca497. **Akademiya Nauk SSSR, Biblioteka.** 1984. *Okhrana okruzhayushchei sredy: Bibliograficheskii ukazatel' osnovnykh otechestvennykh izdanii za 1965–1982 gg. [Protection of the natural environment: Bibliographical index of national editions 1965–1982].* Leningrad. 337 pp.

Ca497A. **Kish, G.,** ed. 1979. *Bibliography of international geographical congresses 1871–1976.* Boston, MA, G. K. Hall. viii+540 pp. ••6,969 items covering 23 congresses.

Ca498. **McIver, T.** 1992. *Anti–evolution: A reader's guide to writings before and after Darwin.* Baltimore, MD, Johns Hopkins Univ. Pr. xv+385 pp. First pub. Jefferson, NC, McFarland, 1988.

Ca499. **Sothebys,** [Firm]. 1992. *Darwin's century: The Jeremy Norman Collection of rare books, manuscripts and portraits concerning the theory of evolution in the nineteenth century ... 11th December 1992 [etc.].* London, Sotheby's. 163 pp., illus. ••431 lots. Especially strong on C. R. Darwin, A. R. Wallace, T. H. Huxley, R. Owen, and F. Galton.

Ca500. **Spengel, J. W.** 1872. *Die Darwinsche Theorie: Verzeichniss der über dieselbe in Deutschland, England, Amerika, Frankreich, Italian, Holland, Belgien und den Skandinavischen Reichen erschienenen Schriften und Aufsätze.* 2. vermehrte Aufl. Berlin, Wiegandt & Hempel. 36 pp. ••I. Deutsche Literatur; II. Englische und Amerikanische Literatur; III.

Französische Literatur; IV. Holländisch–Belgische Literatur; V. Italienische Literatur; Autoren–Verzeichniss.

MICROSCOPY & MICROBIOLOGY

General

Ca501. **Aschoff, L., E. Küster, & W. J. Schmidt.** 1938. *Hundert Jahre Zellforschung.* Berlin, Gebrüder Borntraeger. x+285 pp. *Protoplasma–Monographien* vol. 17.
Three separate contributions, Küster on Plant cytology, pp. 1–64, Schmidt on animal cytology, pp. 65–168, and Aschoff on cyto–pathology, pp. 169–267.

Ca502. **Baker, J. R.** 1948–55. The cell–theory: A restatement, history, and critique. Pt. 1–5. *Quarterly journal of microscopical science* 89: 103–125, 1948; *Ibid.* 90: 87–108, 1949; *Ibid.* 93: 157–190, 1952; *Ibid.* 94: 407–440, 1953; *Ibid.* 96: 449–481, 1955.

Ca503. **Bechtel, W.** 1984. The evolution of our understanding of the cell: A study in the dynamics of scientific process. *Studies in history and philosophy of science* 15: 309–356.

Ca504. **Belloni, L.** 1960. La nascita dell'anatomia microscopica. *Collection de travaux de l'Académie Internationale d'Histoire des Sciences* 1960 (11): 147–161.

Ca505. **Bracegirdle, B.** 1977. The history of histology: A historiographical contribution. *History of science* 15: 77–101.
Bibliography. Pt. I: Survey and secondary sources; pt. II: Chronological list of primary sources, pp. 92–101.

Ca506. **Cole, F. J.** 1938. Microscopic science in Holland in the seventeenth century. *Journal of the Quekett Microscopical Club* (4) 1: 59–78 & 5 pl.

Ca507. **Coleman, W.** 1965. Cell, nucleus and inheritance. *Proceedings of the American Philosophical Society* 109: 124–158.

Ca508. **Collard, P.** 1976. *The development of microbiology.* Cambridge, Cambridge Univ. Pr. vii+201 pp.

Ca509. **Congrès des Sociétés Savantes.** 1961. Colloque sur la microscopie et la micrographie aux XVIIIe et XIXe siècles. *Comptes rendus du 85e Congrès des Sociétés Savantes, Section des Sciences,* pp. 531–640.

Ca510. **Duchesneau, F.** 1987. *Genèse de la théorie cellulaire.* Montreal, Bellarmin; Paris, Vrin. 388+22 pp. *Collection analytiques* no. 1.

Ca511. **Fenner, F.**, ed. 1990. *History of microbiology in Australia.* Canberra, A.C.T., Brolga Pr. for the Australian Society for Microbiology. xiv+610 pp., illus.

Ca512. **Flemming, W.** 1965. Historical paper. Contributions to the knowledge of the cell and its vital processes. *Journal of cell biology* 25 (April) supp: 1–69.

Ca513. **Freund, H., & A. Berg**, eds. 1963–64. *Geschichte der Mikroskopie: Leben und Werk grosser Naturforscher.* Band I: *Biologie.* (Band 2: *Medizin.)* Frankfurt am Main, Umschau Verlag. 3 vols.
See Ba21.

Ca514. **Frey–Wyssling, A.** 1974. Ultrastructure research in biology before the introduction of the electron microscope. *Journal of microscopy* 100: 21–34.

Ca515. **Gooday, G.** 1991. 'Nature' in the laboratory: Domestication and discipline with the microscope in Victorian scientific life. *British journal for the history of science* 24: 307–342.

Ca516. **Gutina, V. N.** 1988. *Ocherki po istorii fiziologii mikroorganizmov. [Essays on the history of the physiology of microorganisms.]* Moscow, Izd. "Nauka." 198 pp.

Ca517. **Hague, E. B.**, ed. 1961. Symposium: History of microscopy, 1490 to the present. *New York State journal of medicine* 61: 430–454.

Ca518. **Harrison, J. Heslop–.** 1979. Botanical microscopy, 1979: Retrospect and prospect. *Proceedings of the Royal Microscopical Society* 14: 256–266.

Ca519. **Hughes, A.** 1959. *History of cytology.* New York & London, Abelard–Schuman. x+158 pp. *Re-pub.* Ames, IA, Iowa State Univ. Pr., 1989. *History of science & technology reprint series.*

Ca520. **Kaiser, H. E.** 1985. Functional comparative histology. 5, Communication: History of histology. *Gegenbaurs morphologisches Jahrbuch* 131: 815–862.

Ca520A. ___. 1987. A concise history of histology and microscopical science. *Zoologische Jahrbücher, Abt. für Anatomie und Ontogenie der Tiere* 115: 9–75.

Ca521. **Klein, M.** 1936. *Histoire des origines de la théorie cellulaire.* Paris, Hermann. 72 pp. *Actualités scientifiques et industrielles* no. 328.

Ca522. **Lechevalier, H. A., & M. Soltorovsky.** 1974. *Three centuries of microbiology.* New York, Dover Publications. vii+536 pp.
Corrected reprint, with additional bibliography, of the 1st ed., New York, McGraw–Hill Book Co., 1965. 536 pp.

Ca523. **Maienschein, J.** 1991. Cytology in 1924: Expansion and collaboration. In: K. R. Benson, R. Rainger, & J. Maienschein, eds. *The expansion of American biology*, pp. 23–51. New Brunswick, NJ, Rutgers Univ. Pr.

Ca524. **Manton, I.** 1983. Some neglected microscopical landmarks (1702–1820). *Proceedings of the Royal Microscopical Society* 18: 157–163.

Ca525. **Möbius, M.** 1938. Hundert Jahre Zellenlehre. *Jenaische Zeitschrift für Naturwissenschaft* 71: 313–326.

Ca526. **Palade, G. E.** 1984. A short history of cell biology. *Noesis* 9: 115–122.

Ca527. **Pas, P. W. van der.** 1964. The earliest microscopical observations in Japan. *Archives internationales d'histoire des sciences* 68/69: 223–238 & 8 pl.

Ca528. ___. 1971. The early history of Brownian motion. *Actes de XIIe [1968] Congrès International d'Histoire des Sciences* 8: 143–158.
See also ___. 1971. The discovery of Brownian motion. *Scientiarum historia* 13: 27–35. Includes a translation of a paper by J. Ingen–Housz, pub. in 1784.

Ca529. **Singer, C. J.** 1914. Notes on the early history of microscopy. *Proceedings of the Royal Society of Medicine, 7 Section History of Medicine* 1913–14: 247–279.

Ca530. ___. 1915. The dawn of microscopical discovery. *Journal of the Royal Microscopical Society* (3) 35: 317–340.

Ca531. ___. 1953. The earliest figures of microscopic subjects. *Endeavour* 12: 197–202.

Ca532. **Terrada Ferrandis, M. L.** 1969. *La anatomía microscópica en España: La doctrina de la fibra y la utilización del microscópico en España durante el Barroco y la ilustración.* Salamanca, Cuadernos de Historia de la Medicina Española. 75 pp. *Monografía* no. 10.

Ca533. **Vandervliet, G.** 1971. *Microbiology and the spontaneous generation debate during the 1870s.* Lawrence, KS, Coronado Pr. 134+12 pp., illus. *Coronado publications in history of science.*

Ca534. **Warner, J. H.** 1982. "Exploring the inner labyrinths of creation": Popular microscopy in 19th century America. *Journal of the history of medicine* 37: 7–33.

Ca535. **Woodruff, L. L.** 1939. Microscopy before the 19th century. *American naturalist* 73: 485–516.

See also Biographies Ba51 Ba270–271 Ba933 Ba1114–1122 Ba1343–1353; Societies Bb376–379 Bb404–408 Bb542 Bb584–587; Misc. Ca184(Selwyn & Wardlaw) & Cb180(Isachenko).

Microscopes & microtechniques

Ca536. **Baker, J. R.** 1945. *The discovery of the uses of colouring agents in biological microtechnique.* London, Williams & Norgate. 22 pp. *Monograph of the Quekett Microscopical Club.*
First pub. in *Journal of the Quekett Microscopical Club* (4) 1 (6).

Ca537. **Bracegirdle, B.** 1978. The performance of seventeenth– and eighteenth–century microscopes. *Medical history* 22: 187–195.

Ca538. ___. 1986. *A history of microtechnique: The evolution of the microtome and the development of tissue preparation.* 2nd ed. Lincolnwood, IL, Science Heritage Ltd. xv+393 pp., illus. (some col.)
Replaces 1st ed., London, Heinemann, (also pub. New York, Cornell Univ. Pr.), 1978. xv+359 pp.

Ca539. **Bradbury, S.** 1967. *The evolution of the microscope.* Oxford, Pergamon Pr. x+357 pp.

Ca540. ___, **& G. L'E. Turner**, eds. 1967. Historical aspects of microscopy. Papers read at a one–day conference, held at Oxford, 18 March, 1966. *Proceedings of the Royal Microscopical Society* 2 (1). *Also pub. separately* Cambridge, Heffer, 1967. vi+227 pp.
Includes **S. Bradbury.** The quality of the image produced by the compound microscope: 1700–1840, pp. 151–173; **G. L'E. Turner.** The microscope as a technical frontier in science, pp. 175–199; **T. Mulvey.** The history of the electron microscope, pp. 201–207.

Ca541. **Clark, G., & F. H. Kasten**, ed. 1983. *History of staining.* 3rd ed. Baltimore, MD, Williams & Wilkins. x+304 pp.
Replaces **H. J. Conn**, ed. 1948. *The history of staining.* 2nd ed. Geneva, NY, Biotech Publications. 141 pp.; 1st ed., Geneva, NY, Biological Stain Commission, 1933. 141 pp.

Ca542. **Clay, R. S., & T. H. Court.** 1932. *History of the microscope up to the introduction of the achromatic microscope.* London, C. Griffin & Co. xiv+266 pp., illus. *Re-pub.* Boston, MA, Longwood Pr., 1978.

Ca543. **Disney, A. N.,** with **C. F. Hill & W. E. W. Baker**, eds. 1928. *Origin and development of the microscope, as illustrated by catalogues of the instruments and accessories, in the collection of the Royal Microscopical Society, together with bibliographies of original authorities.* London, Royal Microscopical Society. xi+303 pp. & 30 pl.
Pt. I: **A. N. Disney.** Historical introduction, (Annotated bibliography, pp. 116–152); pt. II: Catalogues, (Bibliography, pp. 284–297).

Ca544. **Ford, B. J.** 1973. *The revealing lens: Microscopy and mankind.* London, Harrap. 208 pp.

Ca545. ___. 1985. *Single lens: The story of the simple microscope.* London, Heinemann; New York, Harper & Row. x+182 pp.

Ca546. **Gage, S. H.** 1964. Microscopy in America (1830–1945). *Transactions of the American Microscopical Society* 83 (4) supp: 125 pp.

Ca547. **Koninklijke Maatschappij voor Dierkunde.** [1968.] *Henri van Heurck Museum; geillustreerde inventaris van de historische microscopen onderdelen en uitrusting.* By E. Frison. Antwerp. vii+386 pp.
Text in Dutch, French, and English.

Ca548. **Palmer, F. W., & A. B. Sahiar.** 1971. *Microscopes to the end of the 19th century.* London, Her Majesty's Stationery Office. [50] pp.

Ca549. **Pas, P. W. van der.** 1970. A note on the origin of staining techniques for microscopical preparations. *Scientiarum historia* 12: 63–72.

Ca550. **Rijksmuseum voor de Geschiedenis der Naturwetenschappen.** 1954. *Catalogus van vergrootglas tot oog der wetenschap—2 1/2 eeuw microscopie. 25 Juni–19 September 1954.* By M. Rooseboom. Leiden. 53 pp. *Communication* no. 93.

Ca551. **Rooseboom, M.** 1956. *Microscopium.* Leiden, Rijksmuseum voor de Geschiedenis der Naturwetenschappen. 60 pp., illus. *Communication* no. 95.
English and Dutch language editions.

Ca552. ___. 1967. The history of the microscope. *Proceedings of the Royal Microscopical Society* 2: 266–293.

Ca553. **Sobol', S. L.** 1949. *Istoriya mikroskopa i mikroskopicheskikh issledovanii v Rossii v XVIII veke. [History of the microscope and of microscopical investigations in Russia in the 18th century].* Moscow/Leningrad, AN SSSR, Instituta Istorii Estestvoznaniya. 606 pp. *Itogi i problemy sovremennoi nauki.*

Ca554. **Turner, G. L'E.** 1969. The history of optical instruments: A brief survey of sources and modern studies. *History of science* 8: 53–93.
Including microscopy. Bibliography, pp. 75–93.

Ca555. ___. 1972. Micrographia historica: The study of the history of the microscope. Quekett lecture, 11 November 1971. *Proceedings of the Royal Microscopical Society* 7: 120–149.

Ca556. ___. 1980. *Essays on the history of the microscope.* Oxford, Senecio. 245 pp.
Re-pub. of twelve contributions.

Ca557. ___. 1989. *The great age of the microscope: The collection of the Royal Microscopical Society through 150 years.* Bristol, A. Hilger. ix+379 pp.
Includes a catalogue of 453 items.

Ca558. **Van der Star, P.** 1953. *Descriptive catalogue of the simple microscopes in the Rijksmuseum voor de Geschiedenis der Natuurwetenschappen at Leyden.* Leiden, Rijksmuseum [etc.]. 86 pp. & 8 pl. *Communication* no. 87.

See also Cc750.

BIBLIOGRAPHIES OF MICROSCOPY

Ca559. **Grainger, T. H.** [1958.] *A guide to the history of bacteriology.* New York, Ronald Pr. xi+210 pp. *Chronica botanica* vol. 18. ••Pt. I. Selective guide to the literature of bacteriology; pt. II. History of bacteriology – with special reference to specific areas; pt. III. Biographical references – bacteriologists; pt. IV. Selective guide to biographies of [60] selected bacteriologists.

Ca560. **Gray, F. & P.** 1956. *Annotated bibliography of works in Latin alphabet languages on biological microtechnique.* Dubuque, IA, Brown. vii+116 pp.

Ca561. **Lehmann-Haupt, H.** 1973. The microscope and the book: In: E. Geck & G. Pressler, eds. *Festschrift für Claus Nissen*, pp. 471–502. Wiesbaden, Pressler.

Ca562. **Martin, H. de.** 1974. *Vademecum mikroskopischer Literatur: Ein Verzeichnis selbständig erschienener Werke über Mikroskopie von 1590 bis 1970.* Vienna, Höhere Graphische Bundeslehr- und Versuchsanstalt. 242 ff.

Ca563. **Nuttall, R. H.** 1985. Microscopes and popular microscopy: A bibliography for the period 1798–1861. *Microscopy* 35: 177–183.

Ca564. **Sackmann, W.** 1985. *Biographische und bibliographische Materialien zur Geschichte der Mikrobiologie und zur bakteriologischen Nomenklatur.* Frankfurt am Main, P. Lang. 412 pp., illus. *Marburger Schriften zur Medizingeschichte* no. 16.

Ca565. **Stehli, G.** 1925. *Das mikroskopische Schrifttum: Eine Bibliographie der für den mikroskopischer Litteratur der In– und Auslandes.* Stuttgart. 70 pp.

Ca566. **Velde, A. J. J. van de.** 1927–29. Bijdrage tot de bibliographische geschiedenis van het microscoop. 1(–4). *Verslagen en mededeelingen, Koninklijke Vlaamse Academie voor Taal en Letterkunde* 1927: 575–611, 1157–1196; *Ibid.* 1928: 691–735 ; *Ibid.* 1929: 203–267.

See also Library cats. Bc358–359 Bc492 & Bc516.

AQUATIC BIOLOGY & EXPLORATION

Marine

Ca567. **Colloque International sur l'Histoire de la Biologie Marine.** 1965. *Les grandes expéditions scientifiques et la création des laboratoires maritimes. Banyuls–sur–Mer, 2–6 septembre 1963.* Paris. 370 pp. *Vie et milieu supplément* no. 19.
28 papers. *Includes* **M. Wong.** Contribution à l'histoire de la biologie chinoise, pp. 21–34; **P. Huard & M. Wong.** 1965. Bio-bibliographie de quelques médecins naturalistes voyageurs de la marine au début du XIXe siècle, pp. 163–217 (*Includes* Les grandes voyages maritimes du début du XIXe siècle, pp. 164–182); **R.–M. May.** La contribution de l'embryologie des animaux marins à la théorie de l'évolution au XIXe siècle, pp. 239–257; **L. J. Blacher.** Esquisse de l'histoire des stations biologiques maritimes russes, pp. 261–265; **M. J. S. Rudwick.** L'histoire de la paléobiologie marine, pp. 315–321; **F. L. Holmes.** Contributions of marine biology to the development of the concept of the milieu intéreur, pp. 321–335; **J. Lorch.** The history of theories on the nature of corals, pp. 337–347; **J. Theodor.** Aperçu historique sur la plongée et la photographie sous–marines au service de la biologie, pp. 347–351; **M. Bacescu.** La contribution roumaine à l'étude des mers et surtoit de la Mer Noire, pp. 351–361; **F. Garcia del Cid.** L'Institut scientifique des pêches maritimes. Son histoire, pp. 361–365; etc.

Ca568. **De Buen, R.** 1934. *Liste chronologique des croisières océanographiques.* Venice, Ferrari. 73 pp. Commission Internationale pour l'Exploration de la Mer Méditerranée. *Manuel pour les observations océanographiques à la mer,* vol. 1 (2).
Also ___. 1930. Lista cronológica de las campañas y navegaciones a las que se deben observaciones científicas de carácter oceanográfico. *Memorias del Consejo Oceanográfico Ibero–Americano* 5: 1–62.

Ca569. **Deacon, M.** 1971. *Scientists and the sea, 1650–1900: A study of marine science.* London, Academic Pr. xvi+445 pp.

Ca570. **Dean, J. R.** 1966. *Down to the sea: A century of oceanography, [etc.].* Glasgow, Brown, Son & Ferguson. xviii+128 pp. & 16 pl.

Ca571. **Dexter, R. W.** 1988. History of American marine biology and marine biology institutions. Introduction: Origins of American marine biology. *American zoologist* 28: 3–6.

Ca572. **Guberlet, M. L.** 1964. *Explorers of the sea: Famous oceanographic expeditions.* New York, Ronald Pr. vii+226 pp.

Ca573. **Herdman, W. A.** 1923. *The founders of oceanography and their work: An introduction to the science of the sea.* London, E. Arnold. xii+340 pp., illus.

Ca574. **Idyll, C. P.,** ed. 1969. *Exploring the ocean world: A history of oceanography.* New York, Crowell. viii+280 pp.

Ca575. **Institut Océanographique, Monaco.** 1968. Communications, premier Congrès International d'Histoire de l'Océanographie, Monaco, 12–17 décembre 1966. *Bulletin de l'Institut Océanographique, Monaco* numéro spécial 2: xlii+807 pp.
Several sessions, e.g. "Océanographique régionale," "Océanographie biologique et biologie marine," "Grandes expéditions," and "Sources documentaires." Papers by various authors. *Includes* **V. F. Burkhanov.** Russian researches on the Arctic Ocean, pp. 259–268; **G. E. R. Deacon.** Early scientific studies of the Antarctic Ocean, pp. 269–279; **J. B. Tait.** Oceanography in Scotland during the XIXth and early XXth centuries, pp. 281–292; **R. H. Charlier & E. Leloup.** Brief summary of some oceanographic contributions in Belgium until 1922, pp. 293–310; **M. Picotti.** L'océanographie méditerranéenne aux XVIIIe et XIXe siècles, pp. 317–326; **M. Buljan & M. Zore–Armanda.** Aperçu historique sur les recherches hydrographiques en mer Adriatique, en particulier des navires *Ciclope, Najade* et *Vila Velebita*, pp. 337–349; **D. Morovic.** Aperçu historique sur les recherches biologiques effectuées avec le *Vila Velebita* en Adriatique (1913–1914), pp. 351–357; **B. Hoppe.** Influence de la biologie marine sur l'évolution de la pensée écologique au XIXe siècle, pp. 407–416; **G. Trégouboff.** Les précurseurs dans le domaine de la biologie marine dans les eaux des baies de Nice et de Villefranche–sur–Mer, pp. 467–480; **L. A. Zenkevitch.** Histoire des recherches biologiques quantitatives dans les mers et les océans avant la seconde guerre mondiale, pp. 491–504; **T. Wolff.** The Danish expedition to "Arabia Feliz" (1761–1767), pp. 581–601; **H. L. Burstyn.** Science and government in the 19th century: The *Challenger* expedition and its report, pp. 603–611; **C. A. Ronan.** Some illustrations in the history of oceanography, pp. 639–649; **H. L. Burstyn.** The historian of science and oceanography, pp. 665–675. References to other articles are entered in the relevant sections herein.

Ca576. **Kohl, J. G.** 1885. History of the discovery and exploration on the coasts of the United States. *Report, U.S. Coast & Geodetic Survey* 1884, appx. 19: 495–617.

Ca577. **Latil, P. de, & J. Rivoire.** 1954. *A la recherche du monde marin.* Paris, Plon. 382 pp.

Ca578. **Lavbier, L.** 1984. Discovery and study of the deep–sea ecosystem: From zoology to ecology and physiology. *Oceanis* 10: 595–604.

Ca579. **Lenz, W., & M. Deacon,** eds. 1990. *Ocean sciences: Their history and relation to man. Proceedings of the 4th International Congress on the History of Oceanography Hamburg 23.–29. 9. 1987.* Hamburg, Bundesamt für Seeschiffahrt und Hydrographie. xv+603 pp., illus. *Deutsche hydrographische Zeitschrift*, Ergänzungsheft, Reihe B, no. 22.
Seventy–four papers by various authors arranged in 5 sessions: 1. National contributions; 2. International institutions and co-operation in ocean research; 3. Examples of and experience in interdisciplinary research; 4. Economic aspects and the influence on marine research; 5. Scientific and technical assistance in marine science. *Includes* **H. U. Roll.** On the roots of oceanography in Germany, pp. 3–19; **E. L. Mills.** The ocean regarded as pasture: Kiel, Plymouth and the explanation of the marine plankton cycle, 1887 to 1935, pp. 20–29; **G. Hempel.** Marine polar research in the Federal Republic of Germany, pp. 48–61; **C. Reinke-Kunze.** The history and development of German research vessels, pp. 62–69; **H.–P. Bulnheim.** 1990. A century of marine zoological and ecological research around Helgoland Island, pp. 84–93; **A. A. Aleem.** German contributions to marine biology in the Red Sea during the 19th century, pp. 109–113; **W. Lenz.** German marine research in the Atlantic Ocean between World War I and II, pp. 114–131; **H. Rumohr.** A brief history of benthos research in Kiel Bay and in the Baltic, pp. 153–160; **J. Jónsson.** The development of marine knowledge and oceanographic research in Icelandic waters before 1900, pp. 161–165; **A. Majewski.** Oceanography in Poland, pp. 184–196; **K. W. Siudzinski.** Additional material on early Polish biological oceanography, pp. 191–192; **A. Marinescu.** The first Romanian scientific exploration of the Black Sea waters (1893), pp. 203–208; **J. Stiller–Rüdiger & D. Zavodnik.** Hungarian research in the Adriatic Sea, pp. 209–213; **C. Groeben.** The Vettor Pisani circumnavigation (1882 to 1885), pp. 220–234; **G. H. Müller.** Friedrich Alfred Krupp (1854 to 1902), promoter of natural sciences, and his deep–sea researches on board the *Maja* and *Puritan* in the Gulf of Naples, pp. 235–240; **A.**

Boussoulengas. 80 years of oceanographic research in the Aegean Sea, pp. 245–249; **B. Kimor.** Development of oceanographic research in Israel, pp. 250–256; **K. R. Benson.** H. H. Gran and the development of phytoplankton research on the American west coast, pp. 375–377; **M. D. Ehrlich & R. P. Sánchez.** Lights and shadows in biological oceanography research in Argentina: A historical review, pp. 481–498. References to several other articles are entered in the relevant sections herein.

Ca580. **Maienschein, J.** 1988. History of American marine laboratories: Why do research at the seashore? *American zoologist* 28: 15–26.

Ca581. **Maldonado–Koerdell, W.** 1958. Breve historia de los estudios oceanograficos en México (hasta 1957). *Revista de la Sociedad Mexicana de Historia Natural* 19: 173–199.

Ca582. **Mills, E. L.**, ed. 1975. *One hundred years of oceanography.* Halifax, Dalhousie Univ. Pr. ii+89 pp.

Ca583. ___. 1978. Edward Forbes, John Gwyn Jeffreys, and British dredging before the Challenger Expedition. *Journal of the Society for the Bibliography of Natural History* 8: 507–536.

Ca584. ___. 1983. Problems of deep–sea biology: An historical perspective. In: G. T. Rowe, ed. *The sea: Deep sea biology*, pp. 1–79. New York, Wiley. *Ideas and observations on progress in the study of the seas*, vol. 8.

Ca585. ___. 1989. *Biological oceanography: An early history, 1870–1960.* Ithaca, NY, Cornell Univ. Pr. xvii+378 pp.

Ca586. **Musée Océanographique de Monaco.** 1991. Catalogue des appareils d'océanographie en collection au Musée Océanographique de Monaco. 3: Appareils de prélèvement biologique. By C. Carpine. *Bulletin de l'Institut Océanographique de Monaco* 74: 159 pp.

Ca587. **Petit, G.** 1961. *L'Histoire de la biologie marine en France et la création des laboratoires maritimes.* Paris, Université de Paris. 32 pp. *Les Conférences du Palais de la Découverte*, série D, no. 80.

Ca588. **Pirozhnikov, P. L., & L. A. Erman.** 1985. Gidrobiolocheskie issledovaniya v SSSR: Kratkii istoricheskii obzor. [Hydrobiological research in the USSR: A brief historical outline]. *Trudy Vsesoyuznoe Gidrobiologicheskoe Obshenestra* 28: 1–127.
Articles by various authors. *Includes* **L. A. Kuderskii.** Istoriya izucheniya vnutrennikh vodoemov Karelii prileshashchikh territori, pp. 4–24; **A. I. Grigelis & Yu. L. Manyukas.** Gidrobiologicheskie issledovaniya v Litvi, pp. 24–29; **I. V. Kozlova, M. P. Koval'kova & T. S. Lyubimova.** Gidrobiolocheskie issledovaniya na ozerakh srednego i yuzhnogo Urala, pp. 30–41; **M. F. Yaroshenko & F. P. Chorik.** Gidrobiologicheskie issledovaniya v Moldavii, pp. 41–50;
A. Ya. Aldakimova & V. P. Zakutskii. Gidrobiologicheskie issledovaniya v basseine Azovskaga Morya, pp. 51–69; **A. A. Kosova & K. V. Gorbunov.** Ocherk istorii gidrobiolocheskikh issledovanii v del'tovoi oblasti Volgi, pp. 69–85; **P. L. Pirozhnikov.** K istorii izucheniya kormovoi vasy i pitaniya ryb, pp. 85–124.

Ca589. **Ponko, V., Jr.** 1974. *Ships, seas, and scientists: U. S. Naval exploration and discovery in the nineteenth century.* Annapolis, MD, Naval Institute Pr. xiii+283 pp.
Appendix: Ships' data [A–Z by name], pp. 270–273.

Ca590. **Rehbock, P. F.** 1979. The early dredgers: "Naturalizing" in British seas, 1830–1850. *Journal of the history of biology* 12: 293–368.

Ca591. **Rice, A. L.** 1986. *British oceanographic vessels 1800–1950.* London, Ray Society. 193 pp. & 74 pl. *Ray Society publication* no. 157.

Ca592. **Riedl, R.** 1980. Marine ecology – a century of change. *Marine ecology* 1: 3–46.

Ca593. **Royal Society of Edinburgh.** 1972. Proceedings of the second International Congress on the History of Oceanography: *Challenger* expedition centenary, Edinburgh

September 12–20, 1972. Pts. 1 & 2. *Proceedings of the Royal Society of Edinburgh* B 72 & 73. 462+435 pp.

Seventy-four articles by various authors. Sessions on: Founders of oceanography; The sea floor; Deep–sea exploration and research; Deep–sea biology – life in the deep oceans; Life in the upper layers; Fisheries research; etc. *Includes* **C. M. Yonge.** John Graham Dalyell and some predecessors in Scottish marine biology, vol. B 72: 89–97; **J. W. Hedgpeth.** De mirabili maris: Thoughts on the flowering of seashore books, vol. B 72: 107–114; **R. S. & A. H. Scheltema.** Deep–sea biological studies in America, 1846 to 1872 – their contribution to the *Challenger* expedition, vol. B 72: 133–144; **P. S. Meadows & J. I. Campbell.** Habitat selection and animal distribution in the sea: The evolution of a concept, vol. B 73: 145–157. References to several other articles are entered in the relevant sections herein.

Ca594. **Schlee, S.** 1973. *The edge of an unfamiliar world: A history of oceanography.* New York, E. P. Dutton. 398 pp.

Ca595. **Schott, W.** 1987. *Early German oceanographic institutions, expeditions and oceanographers.* Hamburg, Deutsche Gesellschaft für Meeresforschung. 50 pp.

Ca596. **Sears, M., & D. Merriman**, eds. 1980. *Oceanography: The past.* (Proceedings of the third International Congress on the History of Oceanography held September 22–26, 1980, at the Woods Hole Oceanographic Institution, Massachusetts, USA on the occasion of the fiftieth anniversary of the founding of the Institution.) New York, Springer–Verlag. xx+812 pp.

69 articles by various authors. *Includes* **G. S. Ritchie.** Some aspects of the history of oceanography as seen through the publications of the International Hydrographic Bureau 1919–1939, pp. 148–156; **M. Fontaine.** From the physiology of marine organisms to oceanographic physiology or physiological oceanography, pp. 353–359; **E. L. Mills.** Alexander Agassiz, Carl Chun and the problem of the intermediate fauna, pp. 360–372; **A. L. Rice & J. B. Wilson.** The British Association Dredging Committee: A brief history, pp. 373–385; **C. M. Yonge.** The Royal Society and the study of coral reefs, pp. 438–447; **K. W. Siudzinski.** History of Polish biological oceanographic research, pp. 488–495; **F. J. R. Taylor.** Phytoplankton ecology before 1900: Supplementary notes to the "Depths of the Ocean," pp. 509–521; **P. F. Rehbock.** The Victorian aquarium in ecological and social perspective, pp. 522–539; **T. R. Parsons.** The development of biological studies in the ocean environment, pp. 540–550. References to several other articles are entered in the relevant sections herein.

Ca597. **United Nations Educational, Scientific & Cultural Organization.** 1972. Liste des principales expeditions faites en Mediterranée dans le domaine des études hydrographiques, hydrologiques et biologiques de 1810 à 1970. *Newsletter of the cooperative investigation of theMediterranean* 3: 44–53.

Ca598. **Van Oye, P.** 1967. *Geschiedenis van de ontwickkeling der hydrobiologie in Belgie.* Brussels, Paleis der Academien. 175 pp. & 18 port. *Verhandelingen van de Koninklijke Vlaamse Academie voor Wetenschappen, klasse der wetenschappen* Jaarg. 29, no. 93.

Ca599. **Vinogradov, K. A.** 1958. *Ocherki po istorii otechestvennikh gidrobiologicheskikh issledovanii na Chernom More. [Essays on the history of native hydrobiological investigations in the Black Sea].* Kiev, AN USSR. 155 pp. & 6 pl.

Ca600. **Wüst, G.** 1964. The major deep–sea expeditions and research vessels 1873–1960. *Progress in oceanography* 2: 1–52.

Appx. A: Representative vessels in oceanographic deep–sea research, pp. 22–26; Appx. B: Charts of tracks of 20 major deep–sea expeditions, pp. 27–47; Appx. C: Literature on deep–sea research and its history, pp. 48–51. *See also* **M. Tomczak Jr.** 1980. A review of Wüst's classification of the major deep–sea expeditions 1873–1960 and its extension to recent oceanographic research programs. In: M. Sears & D. Merriman, eds. *Oceanography: The past.* (Proceedings of the third International Congress on the History of Oceanography held September 22–26, 1980, [etc.].), pp. 188–194. New York, Springer–Verlag.

Ca601. **Yonge, C. M.** 1956. Development of marine biological laboratories. *Science progress* 44: 1–15.

See also Biographies Ba740–742 Ba927 Ba1214; Societies Bb39 Bb181 Bb289–290 Bb450 Bb849–850; Libr. resources Bc687A; Misc. Ca1(Patrick) Ca178 Ca184(Currie) Ca257(Randolph & Bardach) Ca864–865 Ca992 Ca1010A Ca1018 Ca1011–1013 Ca1020–1022 Ca1028–1029 Ca1055–1062 Ca1065–1067 Ca1069–1070 Ca1081 Ca1090–1091 Ca1093 Cc902 Cc907–919 Cc930 & Cc934 –935.

Freshwater

Ca602. **Beckel, A. L.** 1987. *Breaking new waters: A century of limnology at the University of Wisconsin.* With a contributing chapter by F. Egerton. Madison, Wisconsin Academy of Sciences, Arts & Letters. xiii+122 pp. *Transactions of the Wisconsin Academy of Sciences, Arts, and Letters*, special issue.

Ca603. **Elster, H.–J.** 1974. History of limnology. *Mitteilungen, Internationale Vereiniging für Theoretische und Angewandte Limnologie* 20: 7–30.

Ca604. **Luther, H.** 1982. The first half century of limnology in Finland. *Hydrobiologia* 86: 1–7.

Ca605. **Sladecek, V., & J. Lellak.** 1982. Vyvoj orientace limnologie v Ceskoslovensku. [On the history of limnology in Czechoslovakia]. *Dejiny vedy techniky* 15: 77–87.

Ca606. **Steleanu, A.** 1989. *Geschichte der Limnologie und ihrer Grundlagen.* Frankfurt am Main, Haag & Herchen. 441 pp.

See also Bb218 Bb601 Ca184(Maitland) & Cc94(Reed).

BIBLIOGRAPHIES OF AQUATIC–BIOLOGY & EXPLORATION

Ca607. **Carpine–Lancre, J.** 1965. *Introduction à la bibliographie des sciences de la mer ... (version préliminaire).* Monaco, Musée Océanographique. 55 ff.

Ca608. **___.** 1968. Les expéditions océanographiques et la publication de leurs résultats (étude bibliographique). In: Institut Océanographique, Monaco. Communications, premier Congrès International d'Histoire de l'Océanographie, Monaco, 12–17 décembre 1966. *Bulletin de l'Institut Océanographique, Monaco* numéro spécial 2: 651–664.

Ca609. **Christensen, E. M.** 1956. Bibliography of Utah aquatic biology. *Proceedings of the Utah Academy of Sciences* 33: 91–100.

Ca610. **Curra, R. A.** 1964. Bibliografia e indice de las ciencias marinas de Venezuela I: Biologias marine y pesquera. *Lagena* 4: 2–62.

Ca611. **Grier, M. C.** 1941. *Oceanography of the North Pacific Ocean, Bering Sea and Bering Strait: A contribution toward a bibliography.* Seattle, WA, Univ. of Washington. 312 pp. *University of Washington publications, library series* vol. 2. Re-pub. New York, Greenwood Pr., 1969. ••Includes literature on marine biology.

Ca612. **Hurlbert, S. H.**, ed. 1977. *Aquatic biota of southern South America: Being a compilation of taxonomic bibliographies for the fauna and flora of inland waters of southern South America.* [San Diego, CA, San Diego State Univ.] xiv+342 pp.

Ca613. **___, et al**, eds. 1981. *Aquatic biota of tropical South America: Being a compilation of taxonomic bibliographies for the fauna and flora of inland waters of the tropical portion of South America.* Part 1. *Arthropoda.* San Diego, CA, San Diego State Univ. xii+323 pp.; ___. 1981. *Aquatic biota of tropical South America: Being a compilation* *of taxonomic bibliographies for the fauna and flora of inland waters of the tropical portion of South America.* Part 2. *Anarthropoda.* San Diego, CA, San Diego State Univ. xi+298 pp. ••Covers essentially all of South America except for Chile, Argentina and Uruguay.

Ca614. **Hurlbert, S. H., & A. Villalobos–Figueroa**, eds. 1981. *Aquatic biota of Mexico, Central America and the West Indies.* San Diego, CA, San Diego State Univ. xx+529 pp.

Ca615. **Krebs, B. N.** 1964. A bibliography of the oceanography of the Tasman and Coral Seas 1860–1960. *Memoirs of the New Zealand Oceanographic Institute* 24: 27 pp. & 1 pl.

Ca616. **Kurzikova, T. P.** 1976. *Biologiya Belogo Morya: Bibliograficheskii ukazatel' otechestvennoi literatury 1881–1971. [Biology of the White Sea: Bibliographical index of national literature 1981–1971].* Leningrad, Akademiya Nauk SSSR, Zoologicheskii Institut. 157 pp.; *Continued* **Z. P. Staroverova.** 1983. *Biologiya Belogo morya: Bibliograficheskii ukazatel' otechestvennoi literatury 1972–1981. [Biology of the White Sea: Bibliographical index of national literature 1972–1981].* Leningrad, Akademiya Nauk SSSR, Zoologicheskii Institut. 237 pp.

Ca617. **Laking, P. N.** 1970. *Bibliography of scientific literature on the Black Sea.* Woods Hole, MA, Oceanographic Institution. 280 pp. *Technical report* no. 70–32.

Ca618. **Nikitina, V. R.** 1940. *Spisok literatury po faune i flore Chernogo i Azovskogo morei s 1793 po 1937 gg. (za isklyucheniem Pisces i Mammalia).* Tbilisi, Izd. Gruzinsk. Fil. Akademiya Nauk SSSR. 64 pp.

Ca619. **Prokudina, L. A.** 1952. Spisok literatury po faune i flore Chernogo morya raiona Karadagskoi biologicheskoi stantsii. *Trudy Karadagskoi Biologicheskoi Stantsii* 2: 128–134.

Ca620. **Russell, F. S.** 1928. List of publications recording the results of researches carried out under the auspices of the Marine Biological Association of the U. K. in their laboratory at Plymouth or on the North Sea coast from 1886–1927. *Journal of the Marine Biological Association* 15: 753–828; *Continued* ___. 1952. ___: Supplementary list 1928–1950. *Ibid.* 30: 589–672.

Ca621. **Schwartzlose, R. A., & J. R. Hendrickson.** 1983. *Bibliografía del Golfo de California: Ciencias marinas (Hasta el final de 1981). Bibliography of the Gulf of California: Marine sciences (through 1981).* Mexico, Instituto de Ciencias del Mar y Limnología. [xii]+212 pp. *Publicaciones especiales* no. 7.

Ca622. **United Nations Educational, Scientific & Cultural Organization, Japanese National Commission.** 1957. *Oceanographic papers in Japan (annotated bibliography) (1873–1938).* [Tokyo.] [iv]+235 pp.

Ca623. [Not used.]

Ca624. **Vaissière, R., P. Perno, & C. Charreau.** 1977. Bibliographie commentée des travaux sur les bruits sous–marine d'origine biologique (à l'exception des Mammifères). *Bulletin de l'Institut Océanographique de Monaco* 72 (1432): 62 pp.

Ca625. **Vereshchagin, G. Yu.** 1927. Opyt svoda literatury po Bailalu i ego poberezh'yu. [A contribution to the bibliography of the works concerning the Baikal Lake and its shores]. *Trudy Komissii po Izucheniyu Ozera Baikala* 2: 187–222; *Continued* ___. 1933. Literatury o Baikale za period 1927–1931 gg. i dopolneniya za prezhnie gody. *Trudy Baikal'skoi Limnologicheskoi Stantsii* 5: 163–178.

Ca626. **Watermann, B.** 1987. *Bibliography of the history of oceanography in Germany: Chronological list of titles (1557–1986) and index. Bibliographie zur Geschichte der Deutschen Meeresforschung chronologische Titelaufzahlung (1557–1986) und*

Register. Compiled at the occasion of the 4th International Congress on the History of Oceanography, 23rd–29th September 1987. Hamburg, Deutsche Gesellschaft für Meeresforschung (German Society for Marine Research). xiii+199 pp.

Ca627. **Watermann, B., & O. J. Wrzesinski.** 1989. *Bibliographie zur Geschichte der deutschen Meeresforschung*. Hamburg, Deutsche Gesellschaft für Meeresforschung. 248 pp.

Ca628. **Yentsch, A. E., et al.** 1962. *A partial bibliography of the Indian Ocean: A compilation*. Woods Hole, MA, Woods Hole Oceanographic Institution. 395+iii pp.

See also Library cats. Bc231 Bc270–271 Bc299 Bc317 & Bc333.

NATURAL HISTORY EXPLORATION, VOYAGES & TRAVEL

General

Ca629. **Baker, J. N. L.** 1948. *A history of geographical discovery and exploration*. New ed. London, Harrap & Co. 553 pp. & 68 maps. *Harrap's new geographical series.*
Replaces 1st ed., 1931. 544 pp., etc.

Ca630. **Berg, L. S.**, ed. 1949. *Ocherki po istorii russkikh geograficheskikh otkrytii. [Essays on the history of Russian geographical discoveries]*. New ed. Moscow, Leningrad, Akademiya Nauk SSSR. 467 pp. & 11 pp. of ports. *Seriya itogi problemy sovremennoi nauki.*
Replaces 1st ed., 1926. 358 pp.; *Re-pub.* 1946.

Ca631. **Bériot, A.** 1962. *Grands voiliers autour du monde: Les voyages scientifiques, 1760–1850*. Paris, Editions du Pont-Royal. 296 pp.
Concise bibliography of manuscript and printed sources, pp. 289–295.

Ca632. **Broc, N.** 1969. Voyages et géographie au XVIIIème siècle. *Revue d'histoire des sciences* 22: 137–154.

Ca633. ___. [1975.] *La géographie des philosophes; Géographes et voyageurs français au XVIIIe siècle*. Paris, Ophrys; Association des Publications près les Universités de Strasbourg, Fondation Baulig. 600 pp. & 16 pp. of illus.

Ca634. ___. 1981. Les grandes missions scientifiques françaises au XIXe siècle (Morée, Algérie, Mexique) et leurs travaux géographiques. *Revue d'histoire des sciences* 34: 319–358.

Ca635. ___. 1982. Les explorateurs français du XIXe siècle reconsidérés. *Revue française d'histoire d'outre-mer* 69: 237–272, 323–359.

Ca636. **Brosse, J.** 1985. *The great voyages of discovery: Circumnavigators and scientists, 1764–1843*. Translated by S. Hochman. New York, Facts on File. 228 pp.
Translation of *Les tours du monde des explorateurs: Les grands voyages maritimes, 1764–1843*. Paris, Bordas, 1983.

Ca637. **Cameron, I.** 1980. *To the farthest ends of the earth: 150 years of world exploration by the Royal Geographical Society*. New York, E. P. Dutton. 288 pp.

Ca638. **Delpar, H.**, ed. 1980. *The explorers: An encyclopedia of explorers and exploration*. New York, McGraw-Hill Book Co. 471 pp.

Ca639. **Embacher, F.** 1882. Entdeckungsgeschichte der einzelnen Erdteile. In: *Lexikon der Reisen und Entdeckungen*, Abt. II, pp. 303–393. Leipzig. vi+394 pp. *Re-pub.* Amsterdam, Meridian Publishing Co., 1968.

Ca640. **Fournier, P.** 1932. *Voyages et découvertes scientifiques des missionaires naturalistes français à travers le monde pendant cinq siècles, XVe à XXe siècles*. Paris, P. Lechevalier. 258 pp. & 30 port. *Encyclopédie biologique* no. 10.

Ca641. **Glavnoe Upravlenie Geodezii i Kartografii, USSR.** 1959. *Atlas istorii geograficheskikh otkrytii i issledovanii.* Moscow. 198 pp.
Atlas of the general history of geographical discovery and exploration but emphasizing the Soviet Union and Soviet explorers.

Ca642. **Gnucheva, V. F.** 1940. Materialy dlya istorii ekspeditsii Akademii Nauk v XVIII i XIX vekakh; khronologicheskie obzory i opisanie arkhivnykh materialov. *Trudy Arkhiva Akademiya Nauk SSSR* 4: viii+310 pp.
Covers expeditions within the present USSR territory between 1717 and 1903.

Ca643. **Goetzmann, W. H.** 1986. *New lands, new men: America and the second great age of discovery.* New York, Viking. xiii+528 pp.

Ca644. **Gvozdetskii, N. A.** 1974. *Soviet geographical explorations and discoveries.* Translated by A. Bratov. Moscow, Progress. 342 pp.
Artic and antarctic exploration. Translation of *Sovetskie geograficheskie issledovaniya i otkrytiya.*

Ca645. **Heawood, E.** 1912. *A history of geographical discovery in the seventeenth and eighteenth centuries.* Cambridge, Cambridge Univ. Pr. xii+475 pp. *Cambridge geographical series.*

Ca646. **Huard, P., & M. Wong.** 1964. Les enquêtes scientifiques françaises et l'exploration du monde exotique aux XVIIe et XVIIIe siècles. *Bulletin de l'École Française d'Extrême-Orient* 52: 143–155.

Ca647. **Roberts, G.** 1973. *Atlas of discovery.* Text by Gail Roberts; maps by Geographical Projects. New York, Crown Publications. 192 pp., maps.

Ca648. **Sykes, Sir P.** 1949. *A history of exploration from the earliest times to the present day.* 3rd ed., with Appendix. London, Routledge & Kegan Paul. xiv+426 pp. & 25 pl.

Ca649. **Taranovich, V. P.** 1933. Ekspeditsii Akademii Nauk XVIII veka i ikh rol' v dele razvitiya lesnykh znanii. (Materialy k voprosy ob istorii deyatel'nosti Akademii Nauk v oblasti izucheniya lesa). *Trudy Instituta Istorii Nauki i Tekhniki.* Ser. 1, *arkhiv istorii nauki i tekhniki* 1: 177–204.

Ca649A. **Waldman, C., & A. Wexler.** 1992. *Who was who in world exploration.* New York, Oxford, Facts on File. viii+712 pp.
Information on over 800 explorers, with index of "Explorers by region of exploration."

Ca650. **Wolff, T.** 1967. *Danske ekspeditioner pa verdenshavene. Danish expeditions on the seven seas: 200 years of Danish marine exploration.* Copenhagen, Rhodos. 325 pp.

See also Biographies Ba15 Ba27 Ba31 Ba52 Ba54 Ba58 Ba657–659; Cb817A.

By region

European

Ca651. **Posselt, D.** 1976. Forschungsreisen in Russland im 18. Jahrhundert und ihre Bedeutung für die Entwicklung der Biologie. Überblick. *Wissenschaftliche Zeitschrift der Friedrich-Schiller-Universität, mathematisch-naturwissenschaftliche Reihe* 25: 181–201.

Ca652. ___. 1976. Ergebnisse von Forschungsreisen in Russland im 18. Jahrhundert und ihre Bedeutung für die Geschichte der Evolutionstheorien. *Ibid.* 25: 203–212.

North American

Ca653. **Bartlett, R. A.** 1962. *Great surveys of the American West.* Norman, OK, Univ. of Oklahoma Pr. xxiii+408 pp.
Mid 19th century surveys.

Ca654. **Brebner, J. B.** 1964. *The explorers of North America, 1492–1806.* 2nd ed. London, A. & C. Black. [xiv]+431 pp.

Ca655. **Cooke, A., & C. Holland.** 1970. Chronological list of expeditions and historical events in northern Canada. I: Early times to 1670. *Polar record* 15: 169–183.

Ca656. ___. 1978. *The exploration of northern Canada, 500 to 1920.* Toronto, Arctic Historical Pr. 549+25 pp.

Ca657. **Cumming, W. P., & R. A. Skelton.** 1971. *The discovery of North America.* London, P. Elek. 304 pp., illus.

Ca658. ___**, et al.** 1974. *The exploration of North America 1630–1776.* London, P. Elek. 272 pp., illus.

Ca659. **Goetzmann, W. H.** 1959. *Army exploration in the American West, 1803–1863.* New Haven, CT, Yale Univ. Pr. xx+509 pp.

Ca660. ___. 1966. *Exploration and empire: The explorer and the scientist in the winning of the American west.* New York, A. A. Knopf. xxvi+656+xviii pp.
Covers the 19th century explorations.

Ca660A. **Levere, T. R.** 1993. *Science and the Canadian Arctic: A century of exploration 1818–1918.* Cambridge, Cambridge Univ. Pr. xiv+438 pp.

Ca661. **Sherwood, M.** 1965. *Exploration of Alaska, 1865–1900.* New Haven, CT, Yale Univ. Pr. 207 pp.

Ca662. **Thompson, L. S.** 1985. *Montana's explorers: The pioneer naturalists.* Helena, MT, Montana Magazine. 109 pp. *Montana geographic series.*

Ca663. **Thwaites, R. G.**, ed. 1904–07 *Early western travels: 1748–1846; a series of annotated reprints of some of the best and rarest contemporary volumes of travel, descriptive of the aborigines and social and economic conditions in the Middle and far West, during the period of early American settlement.* Cleveland, OH, The A. H. Clark Co. 32 vols., maps.
Edited with notes, indexes, etc.; vols. 31 & 32: Analytical index.

Ca663A. **Viola, H. J.** 1987. *Exploring the West.* Washington, DC, Smithsonian Books. 256 pp., illus.

Ca664. **Wheeler, G. M.** 1889. Memoir upon the voyages, discoveries, explorations, and surveys to and at the west coast of North America and interior of the United States west of the Mississippi River, between 1500 and 1880 [etc.] In: United States. *Geological and geographical explorations and surveys west of the one hundredth meridian.* Vol. 1: *Geographical report,* appx. F: 481–745. Washington, DC.

See also Bc840.

South & Central American (& Caribbean)

Ca665. **Acosta-Solís, M.** 1976+. *Investigadores de la geografía y la naturaleza de America tropical: Viajeros cronistas e investigadores con especial referencia al Ecuador.* Parte I. Quito, Instituto Panamericano de Geografia y Historia, Seccion Nacional del Ecuador. In progress. *Biblioteca Ecuador.*
Vol. 1. To the end of the nineteenth century. 201 pp.; vol. 2. Twentieth century.

Ca666. **Calatayud Arinero, M. de los A. C.** 1984. *Catálogo de las expediciones y viajes científicos Españoles a América y Filipinas (siglos XVIII y XIX): Fondos del Archivo del Museo Nacional de Ciencias Naturales.* Madrid, Consejo Superior de Investigación Científica, Museo Nacional Ciencias Naturales. 433 pp. *Colección tierra nueva e cielo nuevo* no. 12.

Ca667. **Duviols, J.–P.** 1978. *Voyageurs français en Amérique: Colonies espagnoles et portugaises.* Paris, Bordas. 272 pp.

Ca668. **Engstrand, I. H. W.** 1981. *Spanish scientists in the New World: The eighteenth century expeditions.* Seattle, WA, Univ. of Washington Pr. xiv+220 pp.

Ca669. **Goodman, E. J.** 1972. *The explorers of South America.* New York, Macmillan. viii+408 pp.

Ca670. **Isenberg, T.** 1989. *Viaggiatori naturalisti Italiani in Brasile nell'ottocento.* Milan, Angeli 119 pp. *Geografia umana* no. 51.

Ca671. **Laissus, Y.** 1982. Les naturalistes français en Amérique du Sud au XVIIIe siècle: Les conditions et les résultats. In: Centre National de la Recherche Scientifique, Paris (CNRS). *L'Importance de l'exploration au siècle des lumières*, pp. 65–78. Paris.

Ca672. **Rohl, E.** 1948. *Explorades famosos de la naturaleza Venezolana.* Caracas, Comite Ejecutivo de la Tercera, Conferencia Interamericana de Agriculture. x+220 pp.

Ca673. **Slevin, J. R.** 1959. The Galapagos Islands: A history of their exploration. *Occasional papers, California Academy of Sciences* 25: 1–150.

Ca674. **Smith, A.** 1990. *Explorers of the Amazon.* New York, Viking Penguin. vii+344 pp., illus.

Ca675. **Stehlé, H.** 1970. La contribution des savants français à l'étude des sciences naturelles aux Antilles. In: *Comptes rendus du 94e Congrès National des Sociétés Savantes, Section des Sciences,* 1: 81–98.

Ca676. **Von Hagen, V.** 1945. *South America called them: Explorations of the great naturalists La Condamine, Humboldt, Darwin, Spruce.* New York, Knopf. xii+3ii+ix pp. & 18 pl.
Also ___, ed. 1948. *The green world of the naturalists: A treasury of five centuries of natural history in South America.* Selected and annotated with biographical sketches and introduction by [etc.]. New York, Greenburg. xix+392 pp.

Asian

Ca677. **Clifford, H.** 1904. *Further India: Being the story of exploration from the earliest times in Burma, Malaya, Siam, and Indo–China.* London, Lawrence & Bullen. xi+378 pp., illus.
Appendix: List of the principal publications relative to further India, pp. 347–367.

Ca678. **Efimov, A. V.**, ed. 1964. *Atlas geograficheskikh otkrytii v Sibiri i v Severo–Zapadnoi Amerike XVII–XVIII vv. Atlas of geographical discoveries and north–western America, XVII–XVIII centuries.* Moscow, Izd. "Nauka." xv+134+[166] pp. (incl. 194 maps).

Ca679. **Freshfield, D. W.** 1902. *The exploration of the Caucasus.* 2nd ed. London, E. Arnold. 2 vols.

Ca680. **Keay, J.** 1979. *The Gilgit game: The explorers of the western Himalayas.* London, John Murray. xii+277 pp., illus. *Also* Hamden, CT, Archon Books, 1979. *Re-pub.* Karachi, Oxford Univ. Pr., 1990.

Ca681. **Mekhtiev, Ch. F., & Z. A. Bounyat–Zadé.** 1982. Les recherches de naturalistes des voyageurs et des géologues français en Azerbaidjan (avant 1832). *Histoire et nature* 19/20: 101–106.

Ca682. **Severin, T.** 1976. *The oriental adventure: Explorers of the East.* London, Angus & Robertson. 240 pp. *Also* Boston, MA, Little, Brown, 1976.

Australasian (& Pacific)

Ca683. **Akademiya Nauk SSSR.** 1926. *The Pacific. Russian scientific investigations.* Ed. by A. Fersman. Leningrad. iv+190 pp., maps.
Prepared for the 3rd Pan–Pacific Science Congress, Tokyo, 1926. Articles by V. Akhmatov, A. Belobrov, L. Berg, V. Komarov, A. Kryshtofovich, P. Nikiforov, P. Schmidt, L. Sternberg, & W. Wiese.

Ca684. **Beaglehole, J. C.** 1966. *The exploration of the Pacific.* 3rd ed. London, A. & C. Black. xiii+346 pp. & 4 maps.
Replaces 1st ed., 1934. xv+410 pp.; 2nd ed., 1947.

Ca685. **Buck, P. H.** 1953. *Explorers of the Pacific: European and American discoveries in Polynesia, [etc.].* Honolulu, HI, Bernice Pauahi Bishop Museum. viii+125 pp. *Special publication* no. 43.

Ca686. **Dunmore, J.** 1965–69. *French explorers in the Pacific.* Oxford, Clarendon Pr. 2 vols.

Ca686A. ___. 1991. *Who's who in Pacific navigation.* Honolulu, HI, Univ. of Hawaii Pr. xvi+312 pp.
Selected bibliography, pp. 273–300.

Ca687. **Favenc, E.** 1888. *The history of Australian exploration from 1788 to 1888. Compiled from state documents, private papers, and the most authentic sources of information.* Sydney, Turner & Henderson. xv+474 pp.
Includes Index of names, dates, and incidents, pp. 429–464; Chronological summary, pp. 465–474.

Ca688. **Feeken, E. J. H., G. E. E. Feeken, & O. H. K. Spate.** 1970. *The discovery and exploration of Australia.* Melbourne & Sydney, T. Nelson. viii+318 pp. & 23 maps.
Especially useful series of maps. Gazetteer of Australian place names, pp. 217–261; Key to the maps, pp. 263–273.

Ca689. **Friis, H.**, ed. 1967. *The Pacific Basin: A history of its geographical exploration.* New York, American Geographical Society. xi+457 pp. *Special publication* no. 38.
Papers by various authors.

Ca690. **McClymont, W. G.** 1959. *The exploration of New Zealand.* 2nd ed. London, Oxford Univ. Pr. 125 pp.
Replaces ___. 1940. ___. Wellington, N.Z., Department of Internal Affairs. vii+202 pp., illus. *New Zealand centennial surveys* no. 3.

Ca691. **Makarova, R. V.** 1968. *Russkie na Tikhom Okeane vo vtoroi polovine XVIIIv. [The Russians on the Pacific Ocean in the second half of the 18th century].* Moscow, Moskovskii Gosudarstvennyi Istoriko–Arkhivnii Institut. 200 pp.

Ca692. **Pascoe, J.** 1983. *Explorers and travellers: Early expeditions in New Zealand.* Rev. ed. Wellington, N.Z., Reed. x+158 pp.
Replaces 1st ed. entitled *Exploration New Zealand*, 1971.

Ca693. **Sharp, A.** 1963. *The discovery of Australia.* Oxford, Clarendon Pr. xii+338 pp. & 17 pl.

Ca694. **Stanbury, P.**, ed. 1975. *One hundred years of Australian scientific explorations.* Sydney, Holt, Rinehart & Winston. 124 pp., illus.

Ca695. **Steven, M.** 1988. *First impressions: The British discovery of Australia.* London, British Museum (Natural History). 96 pp., illus.
Prepared to accompany an exhibition.

See also Ba16 & Ca257(Frost).

African

Ca696. **Becker, P.** 1987. *The pathfinders: The saga of exploration in southern Africa.* Harmondsworth, Penguin Books. 282 pp., illus.

Ca697. **Broc, N.** 1988. *Dictionnaire illustré des explorateurs et des grands voyageurs français du XIXe siècle.* 1: *Afrique.* Paris, Editions du C.T.M.S. xxxii+350 pp., illus.

Ca698. **Brown, R.** 1892–95. *The story of Africa and its explorers.* London, Cassell & Co. 4 vols.

Ca699. **Forbes, V. S.** 1965. *Pioneer travellers of South Africa: A geographical commentary, upon routes, records, observations and opinions of travellers at the Cape 1750–1800.* Cape Town & Amsterdam, A. A. Balkema. [vi]+177 pp. & 36 pp. of illus. Incl. 21 maps.

Ca700. **Hallett, R.** 1965+. *The penetration of Africa: European enterprise and exploration, principally in northern and western Africa up to 1830.* London, Routledge & Kegan Paul. Vol. 1+. Vol. 1 also pub. as: *The penetration of Africa: European exploration in north and west Africa to 1815.* New York, Praeger. xxii+458 pp.

Ca701. **Lebrun, J.** 1961. Les biologistes belges au Congo. *Bulletin de l'Académie Royale de Belgique, classe des sciences* 47: 1183–1218.

Ca702. **Santos, M. E. M. dos.** 1978. *Viagems de exploraçao terrestre dos portugueses em Africa.* Lisbon, Centro de Estudos de Cartografia Antiga. 414 pp.

Polar

Ca703. **Efimov, A. V.** 1950. *Iz istorii velikikh russkikh geograficheskikh otkrytii v severnom ledovitom i tikhom okeanakh 17 – pervaya polovina 18 v. [On the history of the great Russian geographical discoveries in the north of the Arctic and Pacific oceans in the 17th and early 18th centuries].* Moscow, Gos. Izd. Geogr. Lit–ry. 317 pp.

Ca703A. **Fogg, G. E.** 1992. *A history of Antarctic science.* Cambridge, Cambridge Univ. Pr. xxi+483 pp., illus. *Studies in polar research.*

Ca704. **Headland, R. K.** 1989. *Chronological list of Antarctic expeditions and related historical events.* Cambridge, Cambridge Univ. Pr. viii+730 pp. *Studies in polar research.* 3,342 entries covering the period from 700 B.C. to 1988. *Replaces* **B. Roberts.** 1958. Chronological list of Antarctic expeditions. *Polar record* 59: 97–134, 191–239.

Ca705. **Mill, H. R.** 1905. *The siege of the South Pole: The story of Antarctic exploration [etc.].* London, Lawrence & Bullen. xvi+455 pp., maps. *Also* New York, F. S. Stokes, 1905.

Ca706. **Mirsky, J.** 1934. *To the North! The story of Arctic exploration from earliest times to the present [etc.].* New York, Viking Pr. xxi+344 pp., illus. *Also* London, Allan Wingate, 1934. *Re-pub.* Chicago, IL, Univ. of Chicago Pr., 1970. xxv+334+xviii pp.

See also Ca660A.

Individual expeditions & voyages

• This selection of expeditions is arranged chronologically by dates of commencement:

1701–02 Tournefort's voyage

Ca707. **Tournefort, J. P. de.** 1982. *Voyage d'un botaniste.* Introduction, notes et bibliographie de S. Yérasimos. Paris, F. Maspéro. 2 vols. *La découverte* nos. 55 & 56.

1732 Linnaeus's Lapland travels

Ca708. **Linnaeus, C.** 1913. *Iter Lapponicum.* Andra upplagan med bilagor och noter ombesörjd af Th. M. Fries. Stockholm, Kungliga Svenska Vetenskaps–Akademien. xi+268 pp. & 3 pl. *Skrifter af Carl von Linné* vol. 5.

1748–50 Kalm's American travels

Ca709. **Kalm, P.** 1937. *The America of 1750. Peter Kalm's travels in North America.* The English version of 1770. Revised from the original Swedish and edited by A. B. Benson ... with a translation of new material from Kalm's diary notes. New York, Wilson–Erickson. 2 vols. *Re-pub.* New York, Dover Publications, 1966. 2 vols.
A bibliography of Peter Kalm's writings on America, pp. 770–776.

Ca710. ___. 1966–70. *Resejournal över resan till Norra Amerika.* Utgiven av M. Kerkkonen & J. E. Roos. Helsinki, Svenska Litteratursällskapet i Finland. Skrifter nos. 419 & 436.

1765–78 Bartrams' American travels

Ca711. **Bartram, J.** 1942. Diary of a journal through the Carolinas, Georgia, and Florida from July 1, 1765, to April 10, 1766. Annotated by F. Harper. *Transactions of the American Philosophical Society* n.s. 33(1): iv+129 pp. & 23 pl.
Geographical, historical, and other comments, pp. 56–78; Annotated index, pp. 79–107.

Ca712. ___, **W.** 1943. Travels in Georgia and Florida, 1773–74: A report to Dr. John Fothergill. Annotated by F. Harper. *Transactions of the American Philosophical Society* n.s. 33(2): [ii]+121–242. & 27 pl.
Geographical, historical, and other comments, pp. 172–193; Annotated index, pp. 194–227

Ca713. ___. 1958. *The travels of William Bartram.* Naturalist's ed. Ed. with commentary and an annotated index by F. Harper. New Haven, CT, Yale Univ. Pr. lxi+727+[4] pp., 14 pl., 15 pp. of illus. & map.
Reprints the text of *Travels through North and South Carolina, Georgia, East & West Florida [etc.].* Philadelphia, PA, James & Johnson, 1791. Commentary, pp. 335–424; Annotated index, pp. 427–667.

Ca714. ___. 1968. *William Bartram: Botanical and zoological drawings, 1756–1788.* Reproduced from the Fothergill Album in the British Museum (Natural History). Ed. with an introduction and commentary by Joseph Ewan. Philadelphia, PA, American Philosophical Society. x+180 pp. & 59 pl. *Memoirs of the American Philosophical Society* vol. 74.

1766 Banks' Newfoundland travels

Ca715. **Lysaght, A. M.** 1971. *Joseph Banks in Newfoundland and Labrador, 1766: His diary, manuscripts and collections.* Berkeley & Los Angeles, CA, Univ. of California Pr. 512 pp., illus.
Includes Biographical sketches of Banks and others, pp. 41–89; The natural history paintings and painters, pp. 101–107; Banks's scientific collection and manuscripts, pp. 291–421. From the manuscript held in the South Australian Branch, Adelaide, of the Royal Geographical Society of Australia; other sections from manuscripts held by McGill University Library and the British Museum (Natural History).

1766–69 Bougainville's voyage

Ca716. **Centre National de la Recherche Scientifique (CNRS).** 1982. *L'importance de l'exploration maritime au siècle de Lumières: À propos du voyage de Bougainville. Centre National de la Recherche Scientifique, Laboratoire associé no. 211 ... Table rond ... Paris 8 et 9 décembre 1978.* Organisée par M. Mollat et E. Taillemite. Paris. 189 pp.

Ca717. **Taillemite, E.**, ed. 1977. *Bougainville et ses compagnons autour du monde, 1766–1769: Journaux de navigation*, établis et commentés par E. Taillemite. Paris. 2 vols. *Voyages et découvertes*, no. 1.

1768–71 "Endeavour" voyage

Ca718. **Banks,** Sir **J.** 1962. *The "Endeavour" journal of Joseph Banks, 1768–1771.* Ed. by J. C. Beaglehole. Sydney, N.S.W., Trustees of the Public Library of New South Wales in Association with Angus & Robertson. 2 vols., 102 pl. (16 col.).
Pub. from the manuscript held in the Mitchell Library.

Ca719. ___. 1980. *The journal of Joseph Banks in the "Endeavour."* With a commentary by A. M. Lysaght. Guildford, Genesis. 2 vols.
A facsimile of the original manuscript.

Ca720. **Carr, D. J.**, et al. 1983. *Sydney Parkinson: Artist of Cook's "Endeavour" voyage.* London & Canberra, A.C.T., British Museum (Natural History) in association with Australian National Univ. Pr. xvi+300 pp., 253 illus.
Includes **W. Blunt.** Sydney Parkinson and his fellow artists, pp. 14–45; **P. Edwards.** Plants of Madeira, Brazil and Tierra del Fuego, pp. 46–75; **F. R. Fosberg & M.–H. Sachet.** Plants of the Society Islands, pp. 76–107; **E. J. Godley.** Plants of New Zealand, pp. 108–127; **R. J. Henderson.** Plants of Australia, pp. 128–177; **W. T. Stearn.** Plants of Java, pp. 178–194; **A. Wheeler.** Animals, pp. 195–241; etc.

Ca721. **Cook, J.** 1955. *The journals of Captain James Cook on his voyages of discovery* ... Ed. by J. C. Beaglehole, [and others]. Vol. 1: *The voyage of the Endeavour, 1768–1771 [etc.].* Cambridge, Cambridge Univ. Pr. for the Hakluyt Society.
Calendar of documents, pp. 602–641. For a bibliography of MS and printed literature on the "Endeavour" voyage *see* **Library of New South Wales.** 1970. First voyage, 1768–1771. In: *Bibliography of Captain James Cook ... circumnavigator.* 2nd ed., pp. 107–203. Ed. K. M. Beddie. Sydney, N.S.W.

Ca722. **Dance, S. P.** 1971. The Cook voyages and conchology. *Journal of conchology* 26: 354–379; ___. 1972. ___: A supplementary note. *Ibid.* 27: 357–358.

Ca723. **Diment, J. A., et al.** 1984–87. Catalogue of the natural history drawings commissioned by Joseph Banks on the *Endeavour* voyage 1768–1771, held in the British Museum (Natural History). Part 1: Botany: Australia. *Bulletin, British Museum (Natural History), historical series* 11: 183 pp.; ___. 1987. ___: Part 2: Botany: Brazil, Java, Madeira, New Zealand, Society Islands, Tierra del Fuego. *Bulletin, British Museum (Natural History), historical series* 12: 200 pp.
Botanical field drawings by Sidney Parkinson, together with finished full–colour drawings by Parkinson, John F. Miller, James Miller, John Cleveley, Thomas Burgis and Frederick P. Nodder.

Ca724. **Lysaght, A. M.** 1979. Banks's artists and his *Endeavour* collections. In: British Museum. *Captain Cook and the South Pacific*, pp. 9–80. Ed. T. C. Mitchell. London, British Museum Publications. *British Museum yearbook* vol. 3.

Ca725. **Merrill, E. D.** 1954. The botany of Cook's voyages and its unexpected significance in relation to anthropology, biogeography and history. *Chronica botanica* 14(5/6): iv+161–384 pp., & pl. 80–93.

Ca726. **Stearn, W. T.** 1969. A Royal Society appointment with Venus in 1769: The voyage of Cook and Banks in the *Endeavour* in 1768–1771 and its botanical results. *Notes & records of the Royal Society of London* 24: 64–90.
Also ___. 1968. The botanical results of the *Endeavour* voyage. *Endeavour* 27: 3–10.

Ca727. **Wheeler, A. C.** 1987. Catalogue of the natural history drawings commissioned by Joseph Banks on the *Endeavour* voyage 1768–1771, held in the British Museum (Natural History). Part 3: Zoology. *Bulletin, British Museum (Natural History), historical series*

13: 171 pp.
Zoological drawings by Sidney Parkinson, Alexander Buchan and Herman Diedrich Spöring.

Ca728. **Whitehead, P. J. P.** 1969. Zoological specimens from Captain Cook's voyages. *Journal of the Society for the Bibliography of Natural History* 5: 161–201.

Ca729. ___. 1978. A guide to the dispersal of zoological material from Captain Cook's voyages. *Pacific studies* 1978 (Fall): 53–93.

1777–79 Paterson's Cape travels

Ca730. **Paterson, W.** 1980. *Paterson's Cape travels 1777 to 1779.* Ed. and with biographical and other chapters by V. S. Forbes and J. Rourke. Johannesburg, Brenthurst Pr. 202 pp., illus. (col.).
Based on the author's manuscript *An account of four journies into the southern parts of Africa* and three albums of water–colour drawings (now in the Brenthurst Library). *Includes* The life and career of William Paterson, pp. 17–32; Writings and drawing–collection of William Paterson, pp. 43–58; Appendix A: Drawings in the three Paterson albums in the Brenthurst Library, Johannesburg, pp. 173–177.

1777–88 Ruiz & Pavon botanical expedition to Peru

Ca731. **Expedición Botánica al Virreinato del Peru.** 1988. *La expedición botánica al Virreinato del Peru (1777–1788).* Barcelona, Lunwerg. 2 vols.

Ca732. **Lang, C.** 1985. Joseph Dombey et l'éxpedition de Ruiz et Pavón: Étude des itineraires (1778–1784). *Bulletin, Société Botanique de France* 132: 259–275.

Ca733. ___. 1988. Joseph Dombey (1742–1794), un botaniste au Pérou et au Chili: Présentation des sources. *Revue d' histoire moderne et contemporaine* 35: 262–274.

Ca734. **Ruiz, H.** 1940. *Travels of Ruiz, Pavón and Dombey in Peru and Chile (1777–1788).* With an epiloque and official documents added by A. J. Barreiro. Translated by B. E. Dahlgren. *Field museum of natural history. Botanical series* 21: 372 pp.

Ca735. **Steele, A. R.** 1964. *Flowers for the king: The expedition of Ruiz and Pavón and the flora of Peru.* Durham, NC, Duke Univ. Pr. xv+378 pp. & 8 pl.

1783–1816 Royal Botanical Expedition to Nueva Granada [Colombia & Ecuador]

Ca736. **Amaya, J. A.** 1983. *Bibliografia de la Real Expedicion Botanica del Nuevo Reyno de Granada.* Bogota, Instituto Colombiano de Cultura Hispanica. 184 pp., illus. *Ediciones del segunda centenario de la Real Expedicion Botanica.*

Ca737. **Díaz–Piedrahíta, S.** 1984. José Celestino Mutis y la Real Expedicíon Botánica del Nuevo Reino de Granada. *Revista de la Academia Colombiana de Ciencias Exactas, Fisicas y Naturales* 15 (59): 19–29.

Ca738. **Jara, V. E.** 1984. *La expedición botánica de Mutis y la cultura Hispanica.* 2a ed. Bogota, Editorial Kelly. 243 pp.

Ca739. **Pineros Corpas, J.** 1973. *Real Expedición Botánica del Nuevo Reino de Granada (1783–1816). Patria naturaleza; documentos y mensajes de la Expedición Botánica.* Seleccion y notas introductivas de J. P. Corpas. [Bogota,] Caja de Credito Agrario. 356 pp., illus. *Biblioteca caja agraria* no. 3.

Ca740. **Hernández de Alba, G.** 1986. *Historia documental de la Real Expedicíon Botánica del Nuevo Reino de Granada después de la muerte de su director Don José Celestino Mutis, 1808–1952.* Bogota, Fundacion Segunda Expedición Botánica, Istituto Colombiano de Cultura Hispanica. 489 pp., illus.

Ca741. **Mendoza, D.** 1909. *Expedición botánica de José Celestino Mutis al Nuevo Reino de Granada y memorias ineditas dr Francisco José de Caldas.* Madrid, Libreria General de V. Suarez. 2 vols.

Ca742. **Silva, R.** 1984. *Contribucion a una bibliografia especializada de la Real Expedicion Botanica en el Nuevo Reino de Granada.* Bogota, Instituto Colombiano de Cultura Hispanica. 88 pp. *Ediciones de la segunda centenario de la Real Expedicion Botanica.*

Ca743. **Uribe-Uribe, L.** 1953. La Expedición Botánica del Nuevo Reino de Granada: Su obra y sus pintores. *Revista de la Academia Colombiana de Ciencias Exactas, Fisicas y Naturales* 9 (33/34): 1–13.

1785–87 Castiglioni's American travels

Ca744. **Castiglioni, L.** 1983. *Luigi Castiglioni's Viaggio: Travels in the United States of North America 1785–87.* With natural history commentary and Luigi Castiglioni's Botanical observations, translated by A. Pace and ed. by J. & N. Ewan. Syracuse, NY, Syracuse Univ. Pr. xliii+487 pp., illus.
Translation of *Viaggio negli Stati Uniti dell'America Settentrionale fatto negli anni 1785, 1786, e 1787*, Milano, Marelli, 1790.

1785–88 La Pérouse's voyage

Ca745. **Association Lapérouse.** 1988. *Bicentenaire du voyage de Lapérouse: 1785–1788. Actes du Colloque Lapérouse, Albi, mars 1985.* Albi. 463+55 pp. & 28 pp. of illus.

Ca746. **Dunmore, J., & M. de Brossard**, eds. 1985. *Le voyage de Lapérouse 1785–1788: Récit et documents originaux.* Préséntés par J. Dunmore ... M. de Brossard [etc.]. Paris, Imprimerie Nationale. 2 vols. *Voyage et découvertes.*

Ca747. **Gaziello, C.** 1984. *L'Éxpedition de Lapérouse 1785–1788: Réplique française aux voyages de Cook.* Paris, Comité des Travaux Historiques et Scientifiques. 323 pp. *Mémoires de la section d'histoire des sciences et des techniques* no. 1.

Ca748. **Girault de Coursac, P. & P.** 1985. *Le voyage de Louis XVI autour du monde: L'expedition La Pérouse.* Paris, La Table Ronde. 251 pp.

Ca749. **La Pérouse, J.–F. de.** 1980. *Voyage autour du monde sur "L'Astrolabe" et "La Boussole" 1785–1788.* Jean–François de Lapérouse. Choix de textes, introduction et notes de H. Minguet. Paris, F. Maspéro. 414 pp. *La découverte* no. 31.

1787–1803 Royal botanical expedition to New Spain (Sessé & Mocino)

Ca750. **Arias Divito, J. C.** 1968. *Las expediciones científicas Españolas durante el siglo XVIII: Expedición botánica de Nueva España.* Madrid, Ediciones Cultura Hispanica. 427 pp. & 59 pl.

Ca751. **McVaugh, R.** 1977. Botanical results of the Sessé & Mocino expedition (1787–1803). 1: Summary of excursions and travels. *Contributions from the University of Michigan Herbarium* 11: 97–195.

Ca752. **Real Jardín Botánico de Madrid.** 1987. *La real expedicion botanica a Nueva España 1787–1803.* Madrid. xvi+358 pp.
Includes **F. de Solano.** El virreinato de la Nueva España en el último tercio del siglo XVIII; **A. G. de Vierna.** Clasificación de las expediciones españoles a América durante el siglo XVIII según su finalidad y disciplina científica; **M. D. Higueras.** Marina y ciencia en el siglo XVIII; **A. Galera.** La expedición alrededor del mundo de Alejandro Malaspina; **M. A. Puig–Samper.** La enseñanza de la botánica en la España ilustrada: El Jardín Botánico de Madrid; **F. J. P. Sarmiento.** Casimiro Gómez Ortega y la organización de las expediciones botánicas ultramarinas; **J. L. Peset.** Las polémicas de la nueva botánica; **M. E. Rodríguez.** La medicina y su enseñanza en el México ilustrado; **M. Peset.** La ilustración y la Universidad de México; **R. A. Peláez.** La obra de Hernández y su recuperación ilustrada; **J. L. M. Polo.** El Jardín Botánico de México y la Cátedra de Botánica; **S. B. Albert.** 1792. La expedición botánica en el noroeste de América: Los viajes a California y Nutka; **J. de la S. Rius.** La expedición en Centroamérica; **A. G. Blanco.** Sessé y la expedición de Mopox a Cuba; **M. L. M. Calvo.** La aventura española de Martín Sessé y José Mariano Mociño; **M. de los A. C. Arinero.** Los fondos dosumentales de la expedición de Nueva España en el Museo Nacional de

Ciencias Naturales; **P. B. F. de Caleya.** Los herbarios de Sessé y Mociño; **J. F. Pérez & F. P. López.** La contribución faunística de la expedición a Nueva España; **A. Z. Mouriz & B. S. Gómez.** La imagen gráfica de la expedición.

Ca753. **Rickett, H. W.** 1947. The royal botanical expedition to New Spain 1788–1820 as described in documents in the Archivo General de la Nación [Mexico] now translated and collated. *Chronica botanica* 11(1): 1–86.

See also Ca188(Galeria).

1789–94 Malaspina's voyage

Ca754. **González Claverán, V.** 1988. *La expedición científica de Malaspina en Nueva España (1789–1794).* Mexico City, Centro de Estudios Históricos, Colegio de México. 528 pp.

Ca755. **Real Jardín Botánico, Madrid.** 1989. *La botanica en la Expedición Malaspina, 1789–1794: Pabellón Villanueva, Real Jardin Botanica, octubre–noviembre 1989.* Madrid, Turner. 218 pp., illus. *Coleccion encuentros, serie catalogos.*
Includes **M. D. Higueras.** El marino ilustrado y las expediciones científicas; **J. L. Peset.** Las riquezas de los nuevos mares; **A. G. Gómez.** La botánica en el proyecto científico de Antonio Pineda; **M. V. Ibáñez.** Nuevas aportaciones a la investigación Haenkeana; **F. M. Garmendia.** Algo más sobre Neé y sus aportaciones a la botánica; **C. S. Serrano.** La botánica y el dibujo en el siglo XVIII; **O. Kaspar.** Tadeo Haenke y sus fondos documentales en Bohemia; **M. de los A. C. Arinero.** Documentación científica y colecciones de la expedición Malaspina en el Museo Nacional de Ciencias Naturales; **M. V. Ibáñez.** Primer inventario del fondo documental "Tadeo Haenke" en el Real Jardín Botánico de Madrid.

1791–93 d'Entrecasteaux' voyage

Ca756. **Hamy, E. T.** 1895. Notice sur une collection de dessins provenant de l'expédition d'Entrecasteaux. *Bulletin de la Société de Géographie* 1895: 1–18.

Ca757. **Richard, H.** 1986. *Une grande expédition scientifique au temps de la Révolution française: Le voyage de d'Entrecasteaux à la recherche de Lapérouse.* Paris, Comité des Travaux Historiques et Scientifiques. 376 pp. *Mémoires de la Section d'Histoire des Sciences et des Techniques* no. 3.

1798–1801 French Egyptian expedition

Ca758. **Gillispie, C. C.** 1989. Scientific aspects of the French Egyptian expedition. *Proceedings of the American Philosophical Society* 133: 447–474.

Ca759. **Laurens, H.** 1989. *L'expedition d'Égypte, 1798–1801.* Paris, Colin. 519 pp., illus.
Contents include articles by C. C. Gillispie, J.–C. Golvin, and C. Traunecker.

Ca760. **Muséum National d'Histoire Naturelle, Paris.** 1969. Les vélins du Muséum et l'Expédition d'Egypte (1798–1801). By Y. Laissus. *Comptes rendus, 92e Congrès National des Sociétés Savantes, Section des Sciences,* tom.1: *Histoire des Sciences,* pp. 267–300. Paris.

1799–1804 Humboldt & Bonpland's American travels

Ca761. **Meyer–Abich, A., & H. Beck.** 1971. *Alexander von Humboldts grosses amerikanische Reisewerk: Eine bibliographische Einleitung.* New York, Da Capo Pr; Amsterdam, Theatrum Orbis Terrarum. 22 pp.

Ca762. **Minguet, C.** 1969. *Alexandre de Humboldt: Historien et géographe de l'Amérique Espagnole 1799–1804.* Paris, F. Maspero. 693 pp., illus.

Ca763. **Stearn, W. T.**, ed. 1968. *Humboldt, Bonpland, Kunth and tropical American botany: A miscellany on the "Nova genera et species plantarum."* Lehre, J. Cramer. [viii]+159 pp.

A collection of articles, including: VI. **W. T. Stearn.** Humboldt and Bonpland's travels in America, pp. 44–48; VII. **H. Beck.** Das Ziel der großen reise Alexander von Humboldts, pp. 49–68; VIII–XII. papers by **N. Y. Sandwith, W. T. Stearn, & T. A. Sprague** on itineraries in Venezuela, Cuba, Colombia, Ecuador & Peru, and Mexico, pp. 69–98; XIII. **W. T. Stearn.** Collective index of Humboldt and Bonpland's localities, pp. 99–115.

See also Ca188(Minguet).

1800–04 Baudin's Australian expedition

Ca764. **Bonnemains, J., E. Forsyth, & B. Smith**, eds. 1988. *Baudin in Australian waters: The artwork of the French voyage of discovery to the Southern Lands, 1800–1804.* With a complete descriptive catalogue of drawings and paintings of Australian subjects by C.–A. Lesueur and N.–M. Petit from the Lesueur Collection at the Muséum d'Histoire Naturelle, Le Havre, France. Melbourne, Vic., Oxford Univ. Pr. in association with the Australian Academy of the Humanities. xi+347 pp.

French ships "Le Naturaliste" and "Le Geographe" under the command of T. N. Baudin.

1804–06 Lewis & Clark's expedition

Ca765. **Burroughs, R. D.**, ed. 1961. *The natural history of the Lewis and Clark expedition.* East Lansing, MI, Michigan State Univ. Pr. xii+340 pp.

Ca766. **Clarke, C. G.** 1970. *The men of the Lewis and Clark expedition: A biographical roster of the fifty–one members and a composite diary of their activities from all known sources.* Glendale, CA, A. H. Clark Co. 351 pp.

Ca767. **Coues, E.** 1875. An account of the various publications relating to the travels of Lewis and Clarke, with a commentary on the zoological results of their expedition. *Bulletin, U.S. Geological & Geographical Survey of the Territories* (2) 1: 417–444.

Ca768. ___. 1893. Description of the original manuscript journals and field notebooks of Lewis and Clark, on which was based Biddle's History of the expedition of 1804–06, and which are now in the possession of the American Philosophical Society. *Proceedings of the American Philosophical Society* 31 (140): 17–33.

Ca769. **Jackson, D.**, ed. 1978. *Letters of the Lewis and Clark Expedition, with related documents, 1783–1854.* 2nd ed., with additional documents and notes. Urbana, IL, Univ. of Illinois Pr. 2 vols.

Ca770. **Lewis, M., & W. Clark.** 1893. *History of the expedition under the command of Lewis and Clark, to the sources of the Missouri River ... during ... 1804–5–6, [etc.].* New ed. faithfully reprinted from the only authorized edition of 1814, with copious critical commentary ... together with a new biographical and bibliographical introduction, ... and a complete index, by E. Coues. New York, F. P. Harper. 4 vols. *Re-pub.* New York, Dover Publications, 1965.

Ca771. ___. 1904–05. *Original journals of the Lewis and Clark expedition 1804–1806.* Printed from the original manuscripts in the library of the American Philosophical Society ... together with manuscript material of Lewis and Clark from other sources including note–books, letters, maps, etc. ... Ed., with introduction, notes, and index by R. G. Thwaites. New York, Dodd, Mead & Co. 7 vols. & atlas. *Re-pub.* New York, Arno Pr., 1990.

Ca772. **Meisel, M.** 1926. Lewis and Clark Expedition (1804). In: *A bibliography of American natural history, the pioneer century, 1769–1865, [etc.]*, vol. 2: 88–103. Brooklyn, NY, Premier Publishing Co. *Re-pub.* New York, Hafner, 1967.

Historical note and bibliography.

Ca773. **Moulton, G. E., et al**, ed. 1983+. *The journals of the Lewis and Clark expedition.* Lincoln, NE, Univ. of Nebraska Pr. Vol. 1+.

Vol. 1, 1983, entitled *Atlas of the Lewis & Clark expedition*; vols. 2+, 1986+, entitled *Journals [etc.]*. Latest seen, vol. 8, 1993.

Ca774. **Rudd, V.** 1954. Botanical contributions of the Lewis and Clark expedition. *Journal of the Washington Academy of Science* 44: 351–356.

See also Cc94(Quinn).

1821–29 Langsdorff's expedition

Ca775. **Akademiya Nauk SSSR, Arkhiv.** 1973. *Materialy ekspeditsii Akademika Grigoriya Ivanovicha Langsdorfa v Braziliyu v 1821–1829 gg. Nauchnoe opisanie.* Izdanie podgotovili D. E. Bertel's, B. N. Komissarov, T. I. Lysenko. Pod red. L. A. Shura. Leningrad, Izd. "Nauka." 228 pp. *Trudy Arkhiva Akademiya Nauk SSSR* vol. 25.

Ca776. **Monteiro, S., & L. Kaz**, eds. 1988. *Expedição Langsdorff a Brasil 1821–1829.* Rio de Janeiro, Ed. Alumbramento. 3 vols.

1832–36 "Beagle" voyage

Ca777. **Darwin, C. R.** 1979. *The journal of a voyage in H. M. S. "Beagle."* Guildford, Genesis. 843 pp. , illus.
A facsimile of the original manuscript.

Ca778. ___. 1985. Appendix II: Darwin's *Beagle* records. In: C. R. Darwin. *Correspondence of Charles Darwin.* Ed. by F. Burkhardt, S. Smith, et al., vol. 1: 545–548. Cambridge, Cambridge Univ. Pr.

Ca779. ___. 1988. *Charles Darwin's* Beagle *diary.* Ed. by R. D. Keynes. Cambridge, Cambridge Univ. Pr. xxix+464 pp.
Biographical register pp. 448–455.

Ca780. **Keynes, R. D.**, ed. 1979. *The Beagle record. Selections from the original pictorial records and written accounts of the voyage of H.M.S. Beagle.* Cambridge, Cambridge Univ. Pr. xiv+409 pp., illus.
Illustrations reproduced from originals by Conrad Martens.

Ca781. **Moorehead, A.** 1969. *Darwin and the* Beagle. London, Hamish Hamilton. 224 pp., illus.

1838–42 United States Exploring Expedition

Ca782. **American Philosophical Society.** 1940. Centenary celebration: The Wilkes Exploring Expedition of the United States Navy, 1838–1842, and symposium on American Polar exploration, February 23–24, 1940. *Proceedings of the American Philosophical Society* 82: 519–950.
Includes **E. G. Conklin.** Connection of the American Philosophical Society with our first national exploring expedition, pp. 519–541; **J. A. G. Rehn.** Connection of the Academy of Natural Sciences of Philadelphia with our first exploring expedition, pp. 543–549; **G. S. Bryan.** The purpose, equipment and personnel of the Wilkes Expedition, pp. 551–560; **H. H. Bartlett.** The reports of the Wilkes expedition, and the work of specialists in science, pp. 601–705; **M. E. Cooley.** The Exploring Expedition in the Pacific, pp. 707–719.

Ca783. **Collins, F. S.** 1912. The botanical and other papers of the Wilkes Exploring Expedition. *Rhodora* 14: 57–68.

Ca784. **Haskell, D. C.** 1942. *The United States Exploring Expedition, 1838–1842, and its publications, 1844–1874: A bibliography.* New York, New York Public Library. xii+188 pp. *Bulletin of the New York Public Library* 45–46, 1941–42. *Re-pub.* New York, Greenwood Pr., 1968.
Publications of the expedition, #1–87; Works relating to the expedition and its results, #88–385; Manuscripts, #386–533.

Ca785. **Meisel, M.** 1926. Wilkes United States Exploring Expedition (1838). In: *A bibliography of American natural history, the pioneer century, 1769–1865, [etc.]*, vol. 2: 650–673. Brooklyn, NY, Premier Publishing Co. *Re-pub.* New York, Hafner, 1967.
Historical note and bibliography.

Ca786. **Poesch, J.** 1961. *Titian Ramsay Peale, 1799–1885, and his journals of the Wilkes Expedition.* Philadelphia, PA, American Philosophical Society. x+214 pp. *Memoirs of the American Philosophical Society* vol. 52.

Ca787. **Stanton, W.** 1975. *The great United States Exploring Expedition of 1838–1842.* Berkeley, CA, Univ. of California Pr. x+433 pp.

Ca788. **Tyler, D. B.** 1968. *The Wilkes Expedition: The first United States Exploring Expedition (1838–1842).* Philadelphia, PA, American Philosophical Society. xviii+435 pp. *Memoirs of the American Philosophical Society* vol. 73.

Ca789. **Viola, H. J., & C. J. Margolis, et al.**, eds. 1985. *Magnificent voyagers: The U. S. Exploring Expedition, 1838–1842.* Washington, DC, Smithsonian Institution Pr. 304 pp.
Chapters by various authors. *Includes* **R. H. Eyde.** Expedition botany: The making of a new profession, pp. 24–41; **G. E. Watson.** Vertebrate collections: Lost opportunities, pp. 42–69; **F. M. Bayer.** The invertebrates of the U S Exploring Expedition, pp. 70–87; **N. Reingold & M. Rothenberg.** The exploring Expedition and the Smithsonian Institution, pp. 242–253.

<u>1848–55 United States & Mexican Boundary Survey</u>

Ca790. **Meisel, M.** 1929. United States & Mexican Boundary Survey. In: *A bibliography of American natural history, the pioneer century, 1769–1865, [etc.]*, vol. 3: 99–105. Brooklyn, NY, Premier Publishing Co. *Re-pub.* New York, Hafner, 1967.
Historical note and bibliography.

Ca791. Not used.

<u>1853–56 North Pacific Exploring Expedition</u>

Ca792. **Meisel, M.** 1929. Rodgers & Ringgold North Pacific Exploring Expedition (1853). In: *A bibliography of American natural history, the pioneer century, 1769–1865, [etc.]*, vol. 3: 221–228. Brooklyn, NY, Premier Publishing Co. *Re-pub.* New York, Hafner, 1967.
Historical note and bibliography (including list of natural history publications 1855–1907).

See also Cc779.

<u>1853–60s Pacific railroad surveys</u>

Ca793. **Meisel, M.** 1929. The Pacific Railroad Surveys (1853). In: *A bibliography of American natural history, the pioneer century, 1769–1865, [etc.]*, vol. 3: 189–220. Brooklyn, NY, Premier Publishing Co. *Re-pub.* New York, Hafner, 1967.
Historical note and bibliography (including list of natural history publications in the *Reports* 1855–60).

Ca794. **Moore, J. A.** 1986. Zoology of the Pacific Railroad Surveys. *American zoologist* 26: 331–341.

<u>1866–69 Father David's travels</u>

Ca795. **David, A.** 1949. *Abbé David's diary: Being an account of the French naturalist's journeys and observations in China in the years 1866 to 1869.* Translated & ed. by H. M. Fox. With a foreword by E. D. Merrill. Cambridge, MA, Harvard Univ. Pr. xxxii+302 pp., illus.

See also Ba723.

1872–76 "Challenger" voyage

Ca796. **Linklater, E.** 1972. *The voyage of the Challenger.* London, J. Murray; Garden City, NY, Doubleday. 288 pp. & 28 pl.

Ca797. **Royal Society of Edinburgh.** 1972. Proceedings of the second International Congress on the History of Oceanography: *Challenger* expedition centenary, Edinburgh September 12–20, 1972. Pts. 1 & 2. *Proceedings of the Royal Society of Edinburgh* B 72 & 73. 462+435 pp.
Numerous articles on the *Challenger* voyage and associated research, including: **C. M. Yonge.** The inception and significance of the *Challenger* expedition, vol. B 72: 1–13; **D. Merriman.** Challengers of Neptune: The 'philosophers,' vol. B 72: 15–45; **H. L. Burstyn.** Pioneering in large–scale scientific organisation: The *Challenger* and its report. I. Launching the expedition, vol. B 72: 47–61.

See also Ca575(Burstyn) Ca849 & Cc666.

1899 Harriman Alaskan expedition

Ca798. **Goetzmann, W. H., & K. Sloan.** 1982. *Looking far north: The Harriman Expedition to Alaska, 1899.* New York, Viking–Overlook Pr. xxv+244 pp., illus.

1933–34 John Murray expedition

Ca799. **Rice, A. L.**, ed. 1986. *Deep–sea challenge: The John Murray/Mabahiss expedition to the Indian Ocean, 1933–34.* [Paris,] UNESCO. 336 pp., illus.
Includes **A. L. Rice.** The John Murray bequest and the organization of the expedition; **R. B. Seymour Sewell.** Narrative of the John Murray expedition to the Arabian Sea; **A. L. Rice.** Biographical notes on the principal participants; **A. L. Rice.** Mabahiss, the story of the ship; **G. E. R. Deacon & A. L. Rice.** The significance of the expedition.

BIBLIOGRAPHIES OF EXPLORATION & TRAVEL

Ca800. **Arias Divito, J. C.** 1983. *Expediciones científicas Españoles a América en el siglo XVIII: Indice documental.* Buenos Aires, Instituto Bibliográfico "Antonio Zinny." xi+96 pp. *Indices documentales* no. 1.

Ca801. **Bay, J. C.** 1950. *Journeys and voyages to nature: A survey of one hundred books.* Chicago, IL, John Crerar Library. 67 pp. & 2 pl.

Ca802. **Bourlière, F.** 1941. Les grandes expéditions zoologiques. In: *Élémens d'un guide bibliographique du naturaliste.* Supplement II, pp. 335–353. Paris, P. Lechevalier.

Ca803. **Casada, J. A.** 1993? *An annotated bibliography of exploration in Africa.* Oxford, Zell. Ca.400 pp. ••Forthcoming publication announced.

Ca804. **Clark, T. D.** 1956–59. *Travels in the Old South: A bibliography.* Norman, OK, Univ. of Oklahoma Pr. 3 vols.

Ca805. **Cox, E. G.** 1935–49. *A reference guide to the literature of travel, including voyages, geographical descriptions, adventures, shipwrecks and expeditions [etc.].* Seattle, WA, Univ. of Washington. 3 vols. *University of Washington publications, language & literature* vols. 9, 10 & 12. *Re-pub.* New York, Greenwood Pr., 1969; Cambridge, MA, Maurizio Martino, 1992. ••Primary literature up to 1800. Vol. 1, Old World; vol. 2, New World; vol. 3, Great Britain.

Ca806. **Davidson, R.** 1970. *A book collector's notes on items relating to the discovery of Australia, the first settlement and the early coastal exploration of the continent.* North Melbourne, Vic., Cassell Australia Ltd. [ix]+138 pp.

Ca807. **Day, A. E.** 1980+. *Discovery and exploration: A reference handbook.* [Vol. 1.] *The old world.* London, Bingley; Munich, K. G. Saur. 295 pp.

Ca808. **Decourbe, S.** 1979. *Essai de bibliographie analytique des sources imprimées concernant les* *explorations françaises en Afrique équatoriale de 1839 à 1914*. Paris, École Pratique des Hautes Études. 2 vols. *Mémoire, sciences sociales, École Pratique des Hautes Études* 1979.

Ca809. **Engelmann, W.** 1857–58. *Bibliotheca geographica: Verzeichniss der seit der Mitte des vorigen Jahrhunderts bis zu Ende des Jahres 1856 in Deutschland erscheinenen Werke über Geographie und Reisen, [etc.]*. Leipzig, E. Engelmann. iv+1,225 pp. *Re-pub.* Amsterdam, Meridian Publishing Co., 1965.

Ca810. **Goodman, E. J.** 1983. *The exploration of South America: An annotated bibliography*. New York, Garland. xx+174 pp. *Themes in European expansion* no. 4.

Ca811. **Hasse, A. R.** 1899. *Reports of explorations printed in the documents of the United States government (Contribution towards a bibliography)*. Washington, DC, Government Printing Office. 90 pp. ••554 reports, including botany & zoology.

Ca812. **Hess, R. L., & D. M. Coger.** 1972. *A bibliography of primary source for nineteenth-century tropical Africa as recorded by explorers, missionaries, traders, travelers, administrators, military men, adventurers, and others.* Stanford, CA, Hoover Institution Pr., Leland Stanford Junior Univ. xxv+800 pp. *Hoover Institution bibliographical series* no. 47. ••Excludes North Africa and the Republic of South Africa. Geographically arranged list of books and articles published in the nineteenth century.

Ca813. **Hulth, J. M.** 1910. *Swedish arctic and antarctic explorations 1758–1910: Bibliography*. Uppsala & Stockholm. 189 pp. *Kongliga Svenska Vetenskapsakademiens Årsbok* 1910, bilaga 2.

Ca814. **Monaghan, F., S. J. Marino.** 1961. *French travellers in the United States 1765–1932: A bibliography*. New York, Antiquarian Pr. Ltd. xxiii+130 pp., illus.

Ca815. **Muséum National d'Histoire Naturelle.** 1981. *Voyages et découvertes: Des voyageurs naturalistes aux chercheurs scientifiques: Jardin des Plantes [Exposition 4 février – 31 décembre 1981]*. Paris. [112] pp. ••Exhibition catalogue.

Ca816. **Robert, W. C. H.** 1970. *Contributions to a bibliography of Australia and the South Sea Islands.* Vol. 3: *Printed material relating to discovery, exploration and travel issued in Europe, except the Netherlands, to 1853 [etc.]*. 2nd ed. Amsterdam, Philo Pr. x+401 pp. ••*Replaces* 1st ed., 1967.

Ca817. **Royal Geographical Society.** 1893. *Review of British geographical work during the hundred years, 1789–1889: Bibliography.* London. viii+257 pp.

Ca818. **Schmeckebier, L. F.** 1904. *Catalogue and index of the publications of the Hayden, King, Powell, and Wheeler surveys, namely: Geological and Geographical Survey of the Territories, Geological Exploration of the Fortieth Parallel, Geographical & Geological Surveys of the Rocky Mountain Region, Geographical Surveys West of the One Hundredth Meridian.* Washington, DC, Government Printing Office. 208 pp. *Bulletin of the United States Geological Survey* no. 222. *Re-pub.* Portland, OR, Northwest Books, 1970. *Reprints in geology.*

Ca819. **Tikhomirov, G. S.** 1968. *Bibliograficheskii ocherk istorii geografii v Rossii XVIII veka. [Bibliographical study of the history of geography in Russia in the 18th century]*. Moscow, Akademiya Nauk SSSR. 133 pp.

Ca820. **United States Library of Congress, Map Division, Reference Department.** 1962. *Maps showing explorers' routes, trails and early roads in the United States: An annotated list, [etc.]*. By R. S. Ladd. Washington, DC, Library of Congress. vi+137 pp., map.

Ca821. **Wagner, H. R., & C. L. Camp.** 1982. *The plains and the Rockies: A critical bibliography of* *exploration, adventure and travel in the American West, 1800–1965*. 4th rev. ed. San Francisco, CA, J. Howell. 745 pp.

Ca822. **Waterston, E.**, et al. 1989. *The travellers – Canada to 1900: An annotated bibliography of works published in English from 1577*. Guelph, Univ. of Guelph. [ii]+viii+321 pp.

Ca823. **Welch, T. L., & M. Figueras.** 1982. *Travel accounts and descriptions of Latin America and the Caribbean, 1800–1920: A selected bibliography*. Washington, DC, Columbus Memorial Library, Organization of American States. [vi]+v+293 pp. *Documentation & information series* no. 6. ••Country by country lists with general sections for Central America, Latin America, South America and West Indies.

See also Library cats. Bc189 Bc207 Bc212 Bc235 Bc289 Bc302 Bc305–308 Bc316 & Bc332.

NATURAL HISTORY CABINETS & MUSEUMS

General

Ca824. **Alexander, E. P.** 1983. *Museum masters: Their museums and their influence*. Nashville, TN, American Association for State & Local History. x+428 pp.
Includes Sir Hans Sloane and the British Museum, Charles Wilson Peale and his Philadelphia Museum, William Jackson Hooker and the Royal Botanic Garden of Kew, Carl Hagenbeck and his Stellingen Tierpark, Oskar von Miller and the Deutsches Museum.

Ca825. **Bacescu, M.** 1972. L'évolution des musées de sciences naturelles. *Cahiers d'histoire mondiale* 14: 74–102.

Ca826. **Cahn, M.** 1987. Raritätenkabinette und Florilegien: Vom Wissen des Sammlers und einer nervösen Lektüre. *Tumult; Zeitschrift für Verkehrswissenschaft* 9: 49–62.

Ca827. **Cohen, D. M., & R. F. Cressey.** 1969. Papers presented at a symposium on natural history collections past, present and future. *Proceedings of the Biological Society of Washington* 82: 559–762.

Ca828. **Farrington, O. C.** 1915. The rise of natural history museums. *Proceedings of the American Association of Museums* 9: 36–53.

Ca829. **Impey, O., & A. MacGregor**, eds. 1986. *The origins of museums: The cabinet of curiosities in sixteenth and seventeenth century Europe*. Oxford, Clarendon Pr. xiii+335+[5] pp.
Chapters by various authors. *Includes* **G. Olmi.** Science – honour – metaphor: Italian cabinets of the sixteenth and seventeenth centuries, pp. 5–16; **A. Aimi, V. de Michele, & A. Morandotti.** Towards a history of collecting in Milan in the late Renaissance and baroque periods, pp. 24–28; **E. Scheicher.** The collection of Archduke Ferdinand II at Schloss Ambras: Its purpose, composition and evolution, pp. 29–38; **R. Distelberger.** The Habsburg collections in Vienna during the seventeenth century, pp. 39–46; **E. Fucikova.** The collection of Rudolf II at Prague: Cabinet of curiosities or scientific museum?, pp. 47–53; **O. Neverov.** 'His Majesty's Cabinet' and Peter I's *Kunstkammer*, pp. 54–61; **H. C. Ackermann.** The Basle cabinets of art and curiosities in the sixteenth and seventeenth centuries, pp. 62–68; **J. Menzhausen.** Elector Augustus's *Kunstkammer*: An analysis of the inventory of 1587, pp. 69–75; **L. Seelig.** The Munich *Kunstkammer*, 1565–1807, pp. 76–89; **H.-O. Boström.** Philipp Hainhofer and Gustavus Adolphus's *Kunstschrank* in Uppsala, pp. 90–101; **F. A. Dreier.** The *Kunstkammer* of the Hessian Landgraves in Kassel, pp. 102–109; **C. Theuerkauff.** The Brandenburg *Kunstkammer* in Berlin, pp. 110–114; **T. H. L. Scheurleer.** Early Dutch cabinets of curiosities, pp. 115–120; **H. D. Schepelern.** Natural philosophers and princely collectors: Worm, Paludanus, and the Gottorp and Copenhagen collections, pp. 121–127; **B. Gundestrup & M. Bencard.** From the royal *Kunstkammer* to the modern museums of Copenhagen, pp. 128–135; **R. Lightbown.** Some notes on Spanish baroque collectors, pp. 136–146; **A. MacGregor.** The cabinet of curiosities in seventeenth

century Britain, pp. 147–158; **M. Hunter.** The cabinet institutionalized: The Royal Society's 'Repository' and its background, pp. 159–168; **W. Schupbach.** Some cabinets of curiosities in European academic institutions, pp. 169–178; **W. George.** Alive or dead: Zoological collections in the seventeenth century, pp. 179–187; **H. E. Coomans.** Conchology before Linnaeus, pp. 188–192; **J. D. Hunt.** '*Curiosities* to adorn *Cabinets* and *Gardens*', pp. 193–203; **H. Torrens.** Early collecting in the field of geology, pp. 204–213; **G. L'E. Turner.** The cabinet of experimental philosophy, pp. 214–222; etc.

Ca830. **Macgregor, A.** 1983. Collectors and collections of rarities in the sixteenth and seventeenth centuries. In: A. Macgregor, ed. *Tradescant's rarities: Essays on the foundations of the Ashmolean Museum, 1683*, pp. 70–97. Oxford, Clarendon Pr.

Ca831. **Pyenson, S. Sheets–.** 1987. Cathedrals of science: The development of colonial natural history museums during the late 19th century. *History of science* 25: 279–300.

Ca832. **___.** 1987. Civilizing by nature's example: The development of colonial museums of natural history, 1850–1900. In: N. Reingold & M. Rothenberg, eds. *Scientific colonialism: A cross–national comparison*, pp. 351–377. Washington, DC, Smithsonian Institution Pr.

Ca833. **___.** 1988. How to "grow" a natural history museum: The building of colonial collections, 1850–1900. *Archives of natural history* 15: 121–147.

Ca834. **___.** 1988. *Cathedrals of science: The development of colonial natural history museums during the late 19th century.* Kingston & Montreal, McGill–Queen's Univ. Pr. xii+144 pp., illus.

Ca835. **Stearn, W. T.** 1959. Museums and the eighteenth century: Natural history. *Museums journal* 59: 44–48.

See also Ca256(Kohlstedt).

By region

Ca836. **Allingham, E. G.** 1924. *A romance of the rostrum, being the business life of Henry Stevens and the history of thirty–eight King Street, together with some account of famous sales held there during the last hundred years.* London, Witherby. 333 pp., illus.
Stevens' auction rooms handled the sales of several important collections of natural history specimens.

Ca837. **Carter, H. B., et al.** 1981. The Banksian natural history collections of the *Endeavour* voyage and their relevance to modern taxonomy. In: Society for the Bibliography of Natural History. *History in the service of systematics: Papers from the conference to celebrate the centenary of the British Museum (Natural History) 13–16 April, 1981*, pp. 61–70. London. *Special publication* no. 1.

Ca838. **Danilov, V. J.** 1990. *America's science museums.* Westport, CT, Greenwood Pr. x+483 pp.
Brief descriptions of 476 museums providing brief information on history, description of collections, etc. Sections on Aquariums, marine museums, and marinelife parks; Botanical gardens, conservatories, and arboretums; Natural history museums; Zoos and wildlife parks.

Ca839. **Goode, G. B.** 1890. The origin of the national scientific and educational institutions of the United States. *Report of the United States National Museum* 1897 (2): 263–354. *Re-pub. in* ___. 1991. *The origins of natural science in America: The essays of George Brown Goode*, pp. 173–296. Ed. by S. G. Kohlstedt. Washington, DC, Smithsonian Institution Pr. xii+411 pp.

Ca840. **Gunther, R. T.** 1925. *Early science in Oxford.* Vol. 3. i: *The biological sciences.* Vol. 3. ii: *Biological collections.* Oxford, privately pub. xii+564 pp.

Also ___. 1926. *Early medical and biological science. Extracted from "Early science in Oxford."* London, Oxford Univ. Pr. 246 pp.

Ca841. **Hunt, J. M. Chalmers–.** 1976. *Natural history auctions 1700–1972: A register of sales in the British Isles.* With articles by S. P. Dance, Peter G. Embrey, W. D. Ian Rolfe, C. Stimson, W. T. Stearn, J. M. Chalmers–Hunt, A. Wheeler. London, Sotheby Parke Bernet. xiii+189 pp.

Ca842. **Kohlstedt, S. G.** 1980. Henry A. Ward: The merchant naturalist and American museum development. *Journal of the Society for the Bibliography of Natural History* 9: 647–661.

Ca843. ___. 1983. Australian museums of natural history: Public priorities and scientific initiatives in the 19th century. *Historical records of Australian science* 5(4): 1–29.

Ca844. ___. 1987. International exchange and national style: A view of natural history museums in the United States, 1860–1900. In: N. Reingold & M. Rothenberg, eds. *Scientific colonialism: A cross–national comparison*, pp. 167–190. Washington, DC, Smithsonian Institution Pr.

Ca845. ___. 1988. Curiosities and cabinets: Natural history museums and education on the antebellum campus. *Isis* 79: 405–426.

Ca846. ___. 1988. Museums on campus: A tradition of inquiry and teaching. In: R. Rainger, K. R. Benson, & J. Maienschein, eds. 1988. *The American development of biology*, pp. 15–47. Philadelphia, PA, Univ. of Pennsylvania Pr.

Ca847. **Laissus, Y.** 1964. Les cabinets d'histoire naturelle. In: R. Taton, ed. *Enseignement et diffusion des sciences en France au XVIIIe siècle*, pp. 659–712. Paris, Hermann. *Histoire de la pensée* vol. 11.
Includes Annexes. I. Description de quelques catalogues de livres et de curiosités d'histoire naturelle intéressant la France du XVIIIe siècle [chronologically arranged], pp. 671–679; II. Essai de répertoire des cabinets d'histoire naturelle Français au XVIIIe siècle [Parisian and provincial lists], pp. 679–712; *See* Ca192.

Ca848. **Lamy, E.** 1930. *Les cabinets d'histoire naturelle en France au XVIIIe siècle et le Cabinet du Roi (1635–1793).* Paris, privately pub. 58 pp.

Ca849. **Lingwood, P. F.** 1981. The dispersal of the collections of H.M.S. *Challenger*; an example of the importance of historical research in tracing a systematically important collection. In: Society for the Bibliography of Natural History. *History in the service of systematics: Papers from the conference to celebrate the centenary of the British Museum (Natural History) 13–16 April, 1981*, pp. 71–77. London. *Special publication* no. 1.

Ca850. **Olmi, G.** 1982. Ordine e fama: Il museo naturalistico in Italia nei secoli XVI e XVII. *Annali dell'Istituto Storico Italo–Germanico in Trento* 8: 225–274.

Ca851. **Orosz, J. J.** 1990. *Curators and culture: The museum movement in America, 1740–1870.* Tuscaloosa, AL, Univ. of Alabama Pr. xiv+304 pp. *History of American science & technology series.*

Ca852. **Pomian, K.** 1987. *Collectors and curiosities: Paris and Venice, 1500–1800.* Translated by W. Wiles–Portier. Cambridge, Polity Pr. v+348 pp.
Translation of *Collectionneurs, amateurs et curieux*, Paris, Editions Gallimard, 1987. 367 pp.

Ca853. **Rudner, I.** 1982. The earliest natural history museums and collectors in South Africa. *South African journal of science* 78: 434–437.

Ca854. **Szabadváry, F.** 1973. Geschichte der Sammlung naturwissenschaftlicher und technischer Objekte in Ungarn. *Medizinhistorisches Journal* 8(2–3): 328–332.

See also Ba1589.

Individual museums

European

Ashmolean Museum, Oxford

Ca855. **Jahn, M. E.** 1966. The Old Ashmolean Museum and the Lhwyd collections. *Journal of the Society for the Bibliography of Natural History* 4: 237–275.

Ca856. **MacGregor, A.**, ed. 1983. *Tradescant's rarities: Essays on the foundations of the Ashmolean Museum, 1683, with a catalogue of the surviving early collections.* Oxford, Clarendon Pr. xiii+382 pp. & 186 pl. & 5 microfiches.
Chapters by various authors.

Ca857. ___. 1983. *Ark to Ashmolean: The story of the Tradescants, Ashmoles and the Ashmolean Museum.* Oxford, Ashmolean Museum. 24 pp.

Ca858. **Ovenell, R. F.** 1986. *The Ashmolean Museum 1683–1894.* Oxford, Clarendon Pr. xii+276 pp.

Ca859. **Simcock, A. V.** 1984. *The Ashmolean Museum and Oxford Science 1683–1983.* Oxford, Museum of the History of Science. vii+48 pp.

Ca860. ___. 1985. *Robert T. Gunther and the old Ashmolean Museum.* Oxford, Museum of the History of Science. xii+99 pp.

Bergens Museum

Ca861. **Bergens Museum.** 1926. *Register til Bergens Museums Aarbok, 1883–1925.* Bergen. 24 pp.

Ca862. ___. 1900. *Bergens Museum, 1825–1900: En historisk fremstilling.* By J. Brunchorst. Bergen. 340+106 pp., illus.

Ca863. ___**, Library.** 1955. *Bergens Museums publikasjoner 1825–1948. Generalregister utarbeidet ved Universitetsbiblioteket.* Bergen. 108 pp.

Ca864. **Brattström, H.** 1967. The biological stations of the Bergens Museum and the University of Bergen, 1892–1967. *Sarsia* 29: 7–80.

Ca865. **Høisaeter, T.** 1967. Publications from the Biological Stations of the Bergen Museum and the University of Bergen 1892–1967. *Sarsia* 29: 97–136.

British Museum (Natural History), London

Ca866. **British Museum (Natural History).** 1904–06. *History of the collections contained in the natural history departments in the British Museum.* London. 2 vols. in 3.
Vol. 1: Libraries, Botany, Geology, Minerals; vol. 2: Separate historical accounts of the several collections included in the Department of Zoology; vol. 2 appendix: General history of the Department of Zoology from 1856 to 1895, By A. Gunther.

Ca867. ___. 1911. *List of works relating to the Natural History Departments of the British Museum, to the end of the year 1910.* By B. B. Woodward. London. 71 pp.

Ca868. ___. 1931. *A short history of the collections.* London. xi+62 pp. *Special guide* no. 9.

Ca869. ___. 1982. The British Museum (Natural History): Celebrating one hundred years at South Kensington, 1881–1981. Ed. by A. P. Harvey. *Bulletin of the British Museum (Natural History), historical series* 10: 117–140.

Ca870. ___. 1987. Fifty years of oceanography in the Department of Mineralogy, British Museum (Natural History). By D. R. C. Kempe & H. A. Buckley. *Bulletin of the British Museum (Natural History), historical series* 15: 59–97.

Ca871. **Gunther, A. E.** 1980. *The founders of science at the British Museum 1753–1900.* Halesworth, Halesworth Pr. ix+219 pp.

Ca872. **Stearn, W. T.** 1981. *The Natural History Museum at South Kensington: A history of the British Museum (Natural History) 1753–1980.* London, Heinemann. xxiii+414 pp.

See also Ca824.

Erfurter Naturkundemuseum

Ca873. **Erfurter Naturkundmuseum.** 1983. 60 Jahre Erfurter Naturkundmuseum (Teil 2.). By H. Pontius. *Veroffentlichengen des Naturkundmuseums Erfurt* 1983: 3 17.

Florence — Museums

Ca874. **Berzi, A., C. Cipriani, & M. Poggesi.** 1980. Florentine scientific museums. *Journal of the Society for the Bibliography of Natural History* 9: 413–425.

Goulandris Museum, Kifissia

Ca875. **Anon.** 1983. The Goulandris natural history museum, Kifissia, Greece, 1963–1983. *Annales Musei Goulandris* 6: xi–xxii.

Humboldt Universität zu Berlin (*formerly* Friedrich–Wilhelms–Universität in Berlin), Museum für Naturkunde

Ca876. **Humboldt Universität zu Berlin, Museum für Naturkunde.** 1970. Museum für Naturkunde an der Humboldt Universität zu Berlin — 200 Jahre. By G. Hoppe [et al.]. *Wissenschaftliche Zeitschrift der Humboldt–Universität zu Berlin, mathematisch–naturwissenschaftliche Reihe* 19: 123–328.

Ca877. **___.** 1971. Zur Frühgeschichte der wissenschaftlichen Sammlungen im Museum für Naturkunde an der Humboldt–Universität zu Berlin (1770–1810). By R. Daber. *Ibid.* 20: 165–173.

Ca878. **___.** 1989. Beiträge zur Geschichte des Museums für Naturkunde an der Humboldt Universität zu Berlin und zu seinen aktuellen Forschungs– und Bildungsaufgaben. *Ibid.* 38: 277–425.

Eighteen articles by various authors. *Includes* **G. Natho & H. Schmidt.** Zur Geschichte der Systematischen Botanik am Museum für Naturkunde; **R. Angermann.** Historische und aktuelle Bedeutung wissenschaftlicher Sammlungen am Beispiel der Säugetiersammlung des Museums für Naturkunde; **F. Hieke.** Die Laufkäferfauna Bulgariens; **E. Kohler & B. Mory.** Die Flora–Cuba–Forschungen und Sammlungen des Bereichs Botanik und Arboretum; **D. Benkert.** Laienforscher und die regionale botanische Arbeit in den Brandenburgischen Bezirken; **H.–J. Paepke.** Schutzmassnahmen für einheimische Wildfische, eine Aufgabe für Berufs– und Freizeitforscher; and **D. H. H. Kühlmann.** Korallenforschung im Museum für Naturkunde.

Ca879. **Jahn, I.** 1972. Über die Einwirkung Alexander von Humboldts auf die Entwicklung der Naturwissenschaften an der Berliner Universität. *Wissenschaftliche Zeitschrift der Humboldt–Universität zu Berlin, mathematisch–naturwissenschaftliche Reihe* 21: 131–144.

Imperatorskaya Akademiya Nauk, St. Petersburg

Ca880. **Imperatorskaya Akademiya Nauk.** 1865. *Ocherk istorii muzeev Imperatorskoi Akademii Nauk.* St. Petersburg. 88 pp.

Imperial Institute, London

See Ca66 (Worboys).

India Museum, London

Ca881. **Desmond, R. G. C.** 1982. *The India Museum 1801–1879.* London, Her Majesty's Stationery Office. xv+215 pp.

Kaiserlichen Leopoldinisch–Carolinischen Deutschen Akademie der Naturforscher, Halle (Saale)

Ca882. **Kaiserliche Leopoldinisch–Carolinische Deutsche Akademie der Naturforscher.** 1894. *Geschichte der Bibliothek und Naturaliensammlung der Kaiserlichen Leopoldinisch–Carolinischen Deutschen Akademie der Naturforscher.* By O. Grulich. Halle. ix+300 pp. & 1 pl.

Liverpool — Museums

Ca883. **Greenwood, E. F.** 1980. A history of Liverpool natural history collections. *Journal of the Society for the Bibliography of Natural History* 9: 375–382.
Origin of collections presently held by the Merseyside County Museums and the University of Liverpool.

Musée d'Histoire Naturelle de Dijon

Ca884. **Chabeuf, M., & J. Philibert.** 1980. Le Musée d'Histoire Naturelle de Dijon de 1836 à 1976. *Bulletin scientifique de Bourgogne* 33: 1–12.

Musée d'Histoire Naturelle de Marseille

Ca884A. **Musée d'Histoire Naturelle de Marseille.** 1925. Index alphabétique par noms d'auteurs des travaux contenus dans les vingt premiers tomes des Annales (1882 à 1925) et dans l'unique tome du Bulletin (2me série, t. 1, 1898–1899). *Annales du Musée d'Histoire Naturelle de Marseille* 20: 123–142.

Musée d'Histoire Naturelle de Neuchâtel

Ca885. **Dufour, C., & J.–P. Haenni.** 1985. *Le Musée d'Histoire Naturelle de Neuchâtel, à travers les richesses de ses collections: Deux siècles d'histoire et de vie scientifique.* Hauterive, Attinger. 84 pp.

Musée Royale d'Histoire Naturelle de Belgique

Ca885A. **Musée Royal d'Histoire Naturelle de Belgique.** 1943. *Table analytique des tomes 1 à 15 du Bulletin du Musée Royale d'Histoire Naturelle de Belgique 1882–1888 et 1930–1939.* By M. Glibert. Brussels. 119 pp.; ___. 1950. *Table analytique des tomes 16 à 24 du Bulletin du Musée Royale d'Histoire Naturelle de Belgique 1940–1948.* By M. Glibert. Brussels. 175 pp.; ___. 1971. *Table analytique des tomes 25 à 44 du Bulletin du Musée Royale d'Histoire Naturelle de Belgique 1959–1968.* By M. Glibert. Brussels. 122 pp.

Ca885B. ___. 1944. *Table analytique des Mémoires 1 à 100 du Musée Royale d'Histoire Naturelle de Belgique (1900–1943).* By M. Glibert. Brussels. 72 pp.; ___. 1952. *Table analytique des Mémoires du Musée Royale d'Histoire Naturelle de Belgique. Mémoires 101 à 110 (1943–1948) et Mémoires 2éme série, 1 à 19 (1935–1947).* By M. Glibert. Brussels. 62 pp.

Museo Civico di Storia Naturale (di Genova) "Giacomo Doria," Genoa

Ca886. **Capocaccia, L., & R. Poggi.** 1982. Short history of the Museo Civico di Storia Naturale "Giacomo Doria" in Genoa, Italy. *Archives of natural history* 11: 107–122.

Ca887. **Museo Civico di Storia Naturale (di Genova) Giacomo Doria.** 1901. *Annali del Museo Civico di Storia Naturale (di Genova) Giacomo Doria. Indice generale sistematico delle due prime serie, vols. 1–40, (1870–1901).* Genoa. 48 pp.

Museo Civico di Storia Naturale, Milan

Ca888. **Museo Civico di Storia Naturale.** 1989. *I 150 anni del Museo Civico di Storia Naturale di Milano (1838–1988).* By A. Visconti. Milan. 51 pp.

Museo Civico de Storia Naturale, Trieste

Ca889. **Pilleri, G.** [1971.] *Zur Geschichte des Naturhistorischen Museums in Triest. H. Koch, S. A. de Syrski, C. de Marchesetti, G. Müller, biographische Skizzen.* Waldau, Hirnanatomisches Institut. 16 pp. & 11 pl.

Muséum d'Histoire Naturelle, d'Ethnographie et de Préhistoire, Rouen

Ca890. **Fouray, M., & M. Lerond**, eds. 1978. *150e anniversaire du Muséum de Rouen.* Rouen, Centre de Documentation du Muséum. 181 pp.

Muséum d'Histoire Naturelle, Geneva

Ca891. **Aellen, V.** 1970. *150 ans du Muséum.* Geneva, Muséum d'Histoire Naturelle. 39 pp.

Muséum d'Histoire Naturelle de La Rochelle

Ca892. **Chanudet, F., et al.** 1982. *Bicentenaire du Muséum d'Histoire Naturelle de la Rochelle 1782–1982.* La Rochelle, Société des Sciences Naturelles de la Charente-Maritime. 48 pp. *Annales* supp. Sept. 1982.

Muséum d'Histoire Naturelle de Poitiers

Ca893. **Richard, H.** 1987. Les débuts du Muséum d'Histoire Naturelle de Poitiers. In: Comité des Travaux Historiques et Scientifiques. *Sciences et techniques en Poitou-Charentes*, pp. 195–207. Paris.

Muséum National d'Histoire Naturelle, Paris

Ca894. **Bourdier, F.** 1962. Le cabinet d'histoire naturelle du muséum, 1635–1935. *Sciences (Paris)* 18: 35–50.

Ca895. **Brygoo, E. R.** 1988. Du Jardin et du Cabinet du Roi au Muséum d'Histoire Naturelle, en 1793: La continuité par les hommes. *Histoire et nature* 28/29: 47–63.

Ca896. **Denise, L.** 1903. *Bibliographie historique et iconographique du Jardin des Plantes: Jardin Royal des Plantes Médicinales et Muséum d'Histoire Naturelle.* Paris, Daragon. iii+268 pp. *Bibliothèque du Vieux-Paris.*
Includes 1. Histoire générale; 5. Le Muséum d'Histoire Naturelle; 6. Descriptions générales; 7. Déscriptions particulières. Author and subject indexes.

Ca897. **Laissus, Y.** 1981. Les voyageurs naturalistes du Jardin du Roi et du Muséum d'Histoire Naturelle: Essai de portrait–robot. *Revue d'histoire des sciences* 34: 259–317.

Ca898. **Limoges, C.** 1980. The development of the Muséum d'Histoire Naturelle of Paris, c.1800–1914. In: Fox, R., & G. Weisz, eds. *The organization of science and technology in France: 1808–1914*, pp. 211–240. Cambridge, Cambridge Univ. Pr.

Ca899. **Muséum d'Histoire Haturelle, Paris.** 1888. Table générale des Archives et Nouvelles Archives du Muséum d'histoire naturelle. 1839 à 1888. *Nouvelles archives du muséum d'histoire naturelle, Paris* 2 sér., 10: xxiv pp.; ___. [1899.] Table générale de la troisième série des Nouvelles archives ... 1889 à 1898. *Ibid.* (3) 10: xv–xxiii.; ___. 1908. Table générale de la quatrième série des Nouvelles archives ... 1899 à 1908. *Ibid.* (4) 10: vii pp.

Ca900. **___.** 1893. *Centenaire de la fondation du Muséum d'Histoire Naturelle, 10 Juin 1793 — 10 Juin 1893: Volume commémoratif publié par les professeurs du Muséum.* Paris, Imprimerie Nationale. vii+575 pp. & 15 pl.
Includes **E.-T. Hamy.** Les derniers jours du Jardin du Roi et la fondation du Muséum d'Histoire Naturelle, 162 pp.

Ca901. **Muséum National d'Histoire Naturelle.** 1935. Le Muséum National d'Histoire Naturelle: Son histoire – son état actuel. *Archives du Muséum National d'Histoire Naturelle* (6) 12: 1–79.

Introduction to a Museum Festschrift: 1. partie – Histoire général; 2. partie – Historique des chaires.

See also Inventories of the Museum's archives Bc785–786.

Muzeul de Istorie Naturala, Iasi

Ca902. **Muzeul de Istorie Naturala, Iasi.** 1963. *Muzeul de Istorie Naturala – Iasi 1833–1963: Album jubiliar.* Iasi, Universitatea "Al. I. Cuza". 76 pp.

Ca903. ___. 1984. *Volum festiv: 150 de ani de la infiintarea Muzeului.* Iasi, Universitatea "Al. I. Cuza". 414 pp., illus.
Includes **C. V. Mandru.** 150 de ani de la infiintarea Muzeului de Istorie Naturala din Iasi, 4 Februarie 1834 – 4 Februarie 1984, pp. 11–54; **T. I. Stefureac.** Botanisti Romani explorati ai florei din peninsula Balcanica, pp. 55–58.

Muzeul National de Istorie Naturala "Grigore Antipa," Bucharest

Ca904. **Muzeul National de Istorie Naturala "Grigore Antipa."** 1957. A l'occasion de l'anniversaire du cinquantenaire du Musée d'Histoire Naturelle "Grigore Antipa" de Bucarest. By M. A. Ionescu & B. Schnapp. *Travaux du Museum d'Histoire Naturelle "Grigore Antipa"* 1: 7–19.

Ca905. ___. 1961. *Le Musée d'histoire naturelle de Bucharest au cours des années (1831–1961).* Bucharest, [the Museum]. By M. A. Ionescu & B. Schnapp. 120 pp.

Ca906. ___. 1985. Le Muséum National d'Histoire Naturelle de Bucharest (1834–1984) – aperçu chronologique. By A. Marinescu & A. Ionescu. *Travaux du Muséum d'Histoire Naturelle "Grigore Antipa"* 27: 373–417.

Naturhistorisches Landes–Museum von Kärnten

Ca907. **Naturhistorisches Landes–Museum von Kärnten.** 1898. *Festschrift zum 50–jährigen Bestehen des Kärntnerischen naturhistorischen Landes–Museums in Klagenfurt, 1848–1898: Seine Gründung und Entwickelung.* Klagenfurt, F. von Kleinmayr. 75 pp. & 1 pl.

Naturhistorisches Museum, Bern

Ca908. **Stalder, H. A.** 1982. 150 Jahre Naturhistorisches Museum Bern. *Schweizer Strahler* 6 (2): 41–92.

Naturhistorisches Museum, Vienna

Ca909. **Naturhistorisches Museum in Wien.** 1976. *100 Jahre Naturhistorisches Museum in Wien: Jubilaumsfestaustellung.* Ed. by F. Bachmayer & O. Schultz. Vienna, the Museum. 105 pp. *Veröffentlichungen aus dem Naturhistorischen Museum* no. 12.

Ca910. ___. 1976. *Die Geschichte der Wiener naturhistorischen Sammlungen bis zum Ende der Monarchie. Unter Verwendung älterer Arbeiten von Leopold Joseph Fitzinger und Hubert Scholler.* By G. Hamann. Mit einem Kapital über die Zeit nach 1919 von M. Fischer, I. Moschner, R. Schönmann. Vienna, Verlag Naturhistorisches Museum. 98 pp. *Veröffentlichungen aus dem Naturhistorischen Museum* no. 13.
Replaces **L. J. F. J. Fitzinger.** 1856–80. Geschichte des Kais. Kön. Hof–Naturalien–Cabinet zu Wien. *Sitzungsberichte der Kaiserlichen Akademie der Wissenschaften. Mathematisch–naturwissenschaftliche Classe* 21: 433–479, 1856; *Ibid.* Abt. 1 57: 1013–1092, 1868; *Ibid.* Abt. 1 58: 35–120, 1868; *Ibid.* Abt. 1 81: 267–329, 1880; *Ibid.* Abt. 1 82: 279–333, 1881; and **H. Scholler.** 1958. *Die Geschichte der Wiener naturhistorischer Sammlungen.* Vienna.

Naturhistorisches Museum zu Wiesbaden

Ca911. **Geisthardt, M.** 1985. Die zoologisch–botanischen Sammlungen der Naturwissenschaftlichen Sammlung des Museums Wiesbaden: Ein Rückblick auf die Entwicklung seit 1829. *Jahrbücher des Nassauischen Vereins für Naturkunde* 108: 48–61.

Ca912. **Thomä, C.** 1842. *Geschichte des Vereins für Naturkunde im Herzogthum Nassau und des Naturhistorischen Museums zu Wiesbaden.* Wiesbaden, Friedrich. viii+196 pp.

Naturhistoriska Riksmuseum, Stockholm

Ca913. **Broberg, G.** 1989. The Swedish Museum of Natural History. In: T. Frängsmyr, ed. *Science in Sweden: The Royal Swedish Academy of Sciences, 1739–1989.* Canton, MA, Science History Publications.

Ca914. **Kungliga Svenska Vetenskapsakademien.** 1916. *Naturhistoriska Riksmuseets historia: Dess uppkomst om utveckling, [etc.].* Stockholm. 290 pp. & 9 pl.

Ca915. **Lönnberg, E.** 1933. The Natural History Museum (Naturhistoriska Riksmuseum) Stockholm. *Natural history magazine* 4: 77–93.

Naturkunde–Museum, Bamberg

Ca916. **Anon.** 1983. 180 Jahre Naturkunde–Museum Bamberg (Linder'sche Stiftung). *Kultur & Technik* 7: 94–102.

Oberösterreichisches Landesmuseum, Linz

Ca917. **Kohn, H.**, ed. 1983. *150 Jahre Oberösterreichisches Landesmuseum.* Linz, Oberösterreichisches Landesmuseum. 293 pp.
Chapters by various authors.

Ca918. [Not used.]

Real Gabinete de Historia Natural de Madrid

See Ca188 (Calatayud).

Royal Society of London

See Ca829 (Hunter).

Senckenberg–Museum, Frankfurt am Main

Ca919. **Senckenberg–Museum.** 1967. *Geschichte des Senckenberg–Museums in Grundriss*, pp. 7–167. By W. Schäfer. In: *Aus der Geschichte des Senckenberg–Museums* 1–2. Frankfurt am Main, Kramer. 571 pp. *Senckenberg–Buch* 46.

Ca920. **Schäfer, W.** 1967. Forschungsanstalt für Meeresgeologie und Meeresbiologie "Senckenberg" in Wilhelmshaven. *Senckenbergiana lethaea* 48: 191–217.

Slovenského Národného Múzea v Bratislave

Ca921. **Slovenského Národného Múzea v Bratislave.** 1974. Príspevok k dejinám Prirodovedného Ustavu Slovenského Národného Múzea v Bratislave: 1. cast. (50 rokov prírodovednych zbierok v Bratislave.) (Beitrag zur Geschichte des Naturhistorischen Instituts des Slowakischen Nationalmuseums in Bratislava: 1. Teil. (50 Jahre der Naturhistorischen Sammlungen in Bratislava).) By B. Matousek & I. Okáli. *Zbornik Slovenského Narodného Muzea. Prírodné vedy* 20: 5–24.

Ca922. **___.** 1980. Príspevok k dejinám Prirodovedného Ustavu Slovenského Národného Múzea v Bratislave: 7. cast. (55 rokov prírodovednych zbierok v Bratislave.) (Beitrag zur Geschichte des Naturhistorischen Instituts des Slowakischen Nationalmuseums in Bratislava: 7. Teil. (55 Jahre der Naturhistorischen Sammlungen in Bratislava).) By B. Matousek & I. Okáli [et al.]. *Ibid.* 26: 231–243.

Staatliches Museum für Naturkunde, Stuttgart

Ca923. **Staatliches Museum für Naturkunde in Stuttgart.** 1967. *175 Jahre Staatliches Museum für Naturkunde in Stuttgart: Aus Geschichte und Gegenwart einer Bildungs– und Forschungsstatte.* Stuttgart. 40 pp.
Issued with *Jahreshefte des Vereins für vaterländische Naturkunde in Wurttemberg* vol. 122.

Stavanger Museum

Ca924. **Stavanger Museums Bibliotek.** 1977. *Stavanger museums publikasjoner 1877–1977: En bibliografi.* T. Wisham, ed. Stavanger. 96 pp.

Teyler's Stichting, Haarlem

Ca925. **Teylers Stichting.** 1978. *Teyler: 1778–1978: Studies en bijdragen over Teylers Stichting naar aanleiding van het tweede eeuwfeest.* Haarlem, Antwerp, Schuyt. 328 pp.

Tromsø Museum

Ca926. **Tromsø Museum.** 1924. *Tromsø Museum i 50 aar. Beretning om Virksomheten i aarene 1872–1922.* Ved Rektor J. Qvigstad. Med bistand av ... O. Nicolaissen, ... O. T. Grønlie, ... O. Klykken og ... T. Soot–Ryen. *(Tillegg til Tromsø Museums 50 aars–beretning. general–register over de av Tromsø Museum 1872–1922 utgitte avhandlinger).* Tromsø. 22 pp. & 4 pl.

Ca927. ___. 1947. *Tromsø Museum gjennom 75 år, 1872–1947.* Trondheim. 96 pp., illus.

Ca928. ___. 1973. *Tromsø Museum publikasjoner 1872–1972.* Tromsø. 61 pp.

North American

Albany Institute, Albany, NY

Ca929. **Meads, O.** 1872. Annual address [etc.]. *Transactions of the Albany Institute* 7: 1–34.

Short history of the Institute and its predecessors.

Ca930. **Meisel, M.** 1926. Albany Institute, Albany, N.Y. (1824). In: *A bibliography of American natural history, the pioneer century, 1769–1865, [etc.]*, vol. 2: 432–436. Brooklyn, NY, Premier Publishing Co. *Re-pub.* New York, Hafner, 1967.

Historical note and bibliography (including list of natural history publications in its *Transactions* 1824–63).

American Museum, New York

Ca931. **Goler, R. I.** 1986. "Here the book of nature is unfolded": The American Museum and the diffusion of scientific knowledge in the early republic. *Museum studies journal* 2(2): 10–21.

The American Museum flourished in New York city in the 1790s.

American Philosophical Society, Philadelphia, PA

Ca932. **Bell, W. J., Jr.** The cabinet of the American Philosophical Society. In: W. M. Whitehead, ed. *A cabinet of curiosities: Five episodes in the evolution of American museums*, pp. 1–34. Charlottesville, VA, Univ. of Virginia Pr.

Bernice Pauahi Bishop Museum, Honolulu, HI

Ca933. **Rose, R. G.** 1980. *A museum to instruct and delight: William T. Brigham and the founding of Bernice Pauahi Bishop Museum.* Honolulu, HI, Bishop Museum Pr. viii+77 pp. *Special publication* no. 68.

Some historical information is also contained in: **F. Radovsky.** 1979. Featured institution – Bernice P. Bishop Museum. *A S C newsletter* 7 (5): 41-47.

Chester County Cabinet of Natural Science, West Chester, PA

Ca934. **Meisel, M.** 1926. Chester County Cabinet of Natural Science, West Chester, PA, (1826). In: *A bibliography of American natural history, the pioneer century, 1769–1865, [etc.]*, vol. 2: 445–449. Brooklyn, NY, Premier Publishing Co. *Re-pub.* New York, Hafner, 1967.

Historical note and bibliography, 1828–44.

Cleveland Academy of Natural Science, Cleveland, OH

Ca935. **Meisel, M.** 1929. Cleveland Academy of Natural Science (1845). In: *A bibliography of American natural history, the pioneer century, 1769–1865, [etc.]*, vol. 3: 5–9. Brooklyn, NY, Premier Publishing Co. *Re-pub.* New York, Hafner, 1967.
Historical note and bibliography (including list of natural history papers 1851–58).

Columbian Institute for the Promotion of Arts & Sciences, Washington, DC

Ca936. **Meisel, M.** 1926. Columbian Institute for the Promotion of Arts and Sciences, Washington, D.C. (1816). In: *A bibliography of American natural history, the pioneer century, 1769–1865, [etc.]*, vol. 2: 228–230. Brooklyn, NY, Premier Publishing Co. *Re-pub.* New York, Hafner, 1967.
Historical note and bibliography.

Ca937. **Rathbun, R.** 1917. The Columbian Institute for the Promotion of Arts and Sciences: A Washington society of 1816–1838, which established a museum and botanic garden under government patronage. *Bulletin of the United States National Museum* 101: iii+87 pp.

Field Museum of Natural History, Chicago, IL

Ca938. **Collier, D.** 1969. Chicago comes of age: The world's Columbian Exposition and the birth of the Field Museum. *Bulletin, Field Museum of Natural History* 40: 3–7.

Ca939. **Dexter, R. W.** 1970. The role of F. W. Putnam in founding the Field Museum. *Curator* 13 (1): 21–26.
Some historical information is also contained in: **M. H. Nitecki.** 1980. Featured institution – Field Museum of Natural History. *Association of Systematics Collections newsletter* 8 (5): 61–70.

Harvard University, Cambridge, MA

Ca940. **Cohen, I. B.** 1950. *Some early tools of American science: An account of the early scientific instruments and mineralogical and biological collections in Harvard University.* Cambridge, MA, Harvard Univ. Pr. xxi+201 pp.

Illinois State Museum of Natural History, Springfield, IL

Ca941. **Crook, A. R.** 1927. *Illinois State Museum, 1877–1927: Its fifty years of service.* Springfield, IL. 18 pp. & 17 pl.
Some historical information is also contained in: **Anon.** 1979. Featured institution – Illinois State Museum. *Association of Systematics Collections newsletter* 7 (4): 33-36.

Maclurian Lyceum of Philadelphia, PA

Ca942. **Meisel, M.** 1926. Maclurian Lyceum of Philadelphia (1826). In: *A bibliography of American natural history, the pioneer century, 1769–1865, [etc.]*, vol. 2: 449–451. Brooklyn, NY, Premier Publishing Co. *Re-pub.* New York, Hafner, 1967.
Historical note and bibliography (including list of natural history publications in its *Contributions* 1827–29).

Ca943–945. [Not used.]

Museu Nacional de Historia Natural, Mexico City

Ca946. **Museo Nacional de Mexico.** 1923. *Reseña histórico e indices de los Anales del Museo nacional correspondientes á las tres primeras épocas de éstos (1877–1913).* Mexico City, Museo Nacional. xiv+63 pp.

National Institution for the Promotion of Science, Washington, DC

Ca947. **Meisel, M.** 1926. National Institution for the Promotion of Science, Washingdon, D.C. (1840). In: *A bibliography of American natural history, the pioneer century, 1769–1865, [etc.]*, vol. 2: 698–704. Brooklyn, NY, Premier Publishing Co. *Re-pub.* New

York, Hafner, 1967.
Historical note and bibliography (including list of natural history publications in its *Bulletin of the proceedings* and *Proceedings* 1840–57).

National Museum of Natural History, Washington, DC
See Smithsonian Institution, Washington, DC

New York State Museum (*formerly* New York State Cabinet of Natural History), Albany, NY

Ca948. **Meisel, M.** 1926. New York State Cabinet of Natural History, Albany, N.Y. (1843). In: *A bibliography of American natural history, the pioneer century, 1769–1865, [etc.]*, vol. 2: 716–726. Brooklyn, NY, Premier Publishing Co. *Re-pub.* New York, Hafner, 1967.
Historical note and bibliography (including list of natural history publications in its *Reports* 1847–66).

Ca949. **New York State Museum of Natural History.** 1903. Index to publications of the New York State Natural History Survey and New York State Museum, 1837–1902; also including other New York publications on related subjects. By M. Ellis. *Bulletin of the New York State Museum of Natural History, miscellaneous* 66: 239–653.

Peabody Museum, Salem, MA

Ca950. **Parker, F.** 1967. George Peabody and the Peabody Museum of Salem. *Curator* 10: 137–153.

Smithsonian Institution, Washington, DC

Ca951. **Hellmann, G. T.** 1967. *The Smithsonian: Octopus on the mall.* Philadelphia, PA, J. B. Lippincott. 224 pp. *Re-pub.* Westport, CT, Greenwood Pr., 1978.

Ca952. **Meisel, M.** 1929. Smithsonian Institution, Washington, D.C. (1846). In: *A bibliography of American natural history, the pioneer century, 1769–1865, [etc.]*, vol. 3: 19–41. Brooklyn, NY, Premier Publishing Co. *Re-pub.* New York, Hafner, 1967.
Historical note and bibliography (including list of natural history publications in *Smithsonian contributions, Smithsonian miscellaneous collections* and *Annual reports* 1847–66).

Ca953. **Oehser, P. H.** 1949. *Sons of science: The story of the Smithsonian Institution.* New York, H. Schuman. xvii+220 pp., illus. *The life of science library.*

Ca954. **___, & L. Heskett.** 1983. *The Smithsonian Institution.* 2nd ed. Boulder, CO, Westview. xiv+223 pp. *Westview library of federal departments, agencies & systems.*
Replaces **P. H. Oehser, P. H.** 1970. *The Smithsonian institution.* New York, Praeger. xiii+275 pp. *Praeger library of U. S. government departments and agencies.*

Ca955. **Smithsonian Institution.** 1889. Catalogue of publications of the Smithsonian Institution [1846 to 1st July, 1886]. By W. J. Rhees. *Report (Annual) of the Board of Regents of the Smithsonian Institution* 1885/86 (1): 485–867.

Ca956. **___.** 1897. *The Smithsonian Institution 1846–1896: The history of its first half century.* Ed. by G. B. Goode. Washington, DC, Smithsonian Institution. x+856 pp., 4 pl. 21 ports.
Includes **W. G. Farlow.** Botany, pp. 697–710; **T. Gill.** Zoölogy, pp. 711–744; **H. C. Bolton.** Bibliography, pp. 785–804.

Ca957. **___.** 1901. The Smithsonian Institution: Documents relative to its origin and history, 1835–1899. By W. J. Rhees. Washington, DC. *Smithsonian miscellaneous collections* vol. 42, 1901 & vol. 43, 1901.

Ca958. **___.** 1903. List of publications of the Smithsonian Institution, 1846–1903 [etc.]. By W. J. Rhees. *Smithsonian miscellaneous collections* 44 (1376): [vi]+99 pp.

Ca959. ___. 1963. *Author–subject index to articles in Smithsonian annual reports, 1849–1961.* By R. Stemple. Washington, DC, Smithsonian Institution. v+200 pp. *Smithsonian publication* no. 4503.

See also Rhees papers Bc762B; guide to the Institution's archives Bc829.

Smithsonian Institution, United States National Museum, Washington, DC

Ca960. **Kopper, P.** 1982. *The National Museum of Natural History.* New York, H. N. Abrams in association with the Smithsonian Institution. 496 pp.

Ca961. **United States National Museum.** 1892. The genesis of the United States National Museum. By G. B. Goode. *Report, United States National Museum, Smithsonian Institution* 1891: 273–380.

Ca962. ___ 1905. The United States National Museum: An account of the buildings occupied by the national collections. By R. Rathbun. *Report, United States National Museum, Smithsonian Institution* 1903: 177–309 & 29 pl.

Ca963. ___. 1947. A list and index of the publications of the United States National Museum. By I. G. Geare. *Bulletin of the United States National Museum* 193: iii+306 pp.

Ca964. ___. 1985. *The National Museum of Natural History: 75 years in the Natural History Building.* By E. I. Yochelson. Ed. by M. Jarrett. Washington, DC, Smithsonian Institution Pr. for the National Museum of Natural History. 216 pp.

South American

Museo de La Plata

Ca965. **Museo de La Plata.** 1928. *Revista del Museo de La Plata. [Author index to vols. 1–30, 1890–1927.]* By R. Lehman-Nitsche. Buenos Aires. 23 pp.

Museo Nacional de Historia Natural de Buenos Aires

Ca966. **Museo Nacional de Historia Natural de Buenos Aires.** 1914. *Anales del Museo Nacional de Historia Natural de Buenos Aires. Indices de los tomos 1–20, (1864–1911).* Buenos Aires. 30 pp.

Museu Nacional do Rio de Janeiro

Ca967. **Museu Nacional do Rio de Janeiro.** 1920. Indice dos Archivos do Museu Nacional organizado por B. M. J. Lutz. vols. I–XXII, 1876–1919. *Arquivos do Museu Nacional do Rio de Janeiro* 22: 275–290.

Ca968. ___. 1934. Indice dos Boletins do Museu Nacional. Vol. I a X [1923–34]. By M. Gikovate. *Boletim do Museu National de Rio de Janeiro* 10: 117–126.

Museum Paraense Emílio Goeldi, Belém, Pará

Ca969. **Museu Goeldi (Museu Paraense) de Historia Natural e Ethnographia.** [1906.] *Relação das publicações scientificas feitas pelo Museu Goeldi ... durante ... 1894–1904, sendo im retrospecto bibliografico sobre o primeiro decennio da existencia do mesmo Museu contendo tambem uma enumeraçao completa dos trabalhos sobre historia natural publicados pelo Director ... E. A. Goeldi de 1879–1904, [etc.].* Bern. 58 pp.

Ca969A. **Museu Paraense Emílio Goeldi.** 1989. *Talento e atitude: Estudos biográficos do Museu Emílio Goeldi.* By O. R. da Cunha. Belem, PR/SCT/CNPq, Museu [etc.]. 159 pp., illus.

Universidade de São Paulo, Museu Paulista

Ca970. **Museu Paulista.** 1931. Indice geral das materias contidas na "Revista do Museu Paulista" de vol. 1 ao vol. 16 [1895–1929]. By O. M. de Oliveira Pinto. *Revista do*

Museu Paulista 17 (1): 571–626; *Continued* ___. 1958. Revista do Museu Paulista. Nove série, volumes I–X. Indice de matérias [and] ... colaboradores. *Ibid.* N.S. 10: 359–370.

Australasian

Auckland Institute & Museum

Ca971. **Auckland Institute & Museum.** 1967. *The centennial history of the Auckland Institute and Museum.* Ed. by A. W. B. Powell. Auckland, N.Z. 88 pp., illus.

Queensland Museum, Brisbane, Qld.

Ca972. **Queensland Museum.** 1956. The Queensland Museum, 1855–1955. By G. Mack. *Memoirs of the Queensland Museum* 13: 107–124.

Ca973. ___. 1986. *A time for a museum: The history of the Queensland Museum 1862–1986.* By P. Mather et al. South Brisbane, Qld. ix+366 pp., illus. *Memoirs of the Queensland Museum* no. 24.

South Australian Museum, Adelaide, S.A.

Ca974. **South Australian Museum.** 1956. *The first hundred years of the museum — 1856–1956.* By E. N. Hale. Adelaide, S.A., South Australian Museum. xi+225 pp. & 22 port. *Records of the South Australian Museum* vol. 12.

Personal collections

Bullock's London Museum, etc.

Ca975. **Edmondson, J.** 1989. The Regency exhibitionists: A fresh look at the Bullocks. *Linnean* 5 (1): 17–26.

Ca976. **Mullens, W. H.** 1917. Some museums of old London. 1: William Bullock's London Museum. *Museums journal* 17: 51–56 & 1 pl., 132–137, 180–187 & 1 pl.

Ca977. **Sweet, J. M.** 1970. William Bullock's collection and the University of Edinburgh. *Annals of science* 26: 23–32.

Cassiano Dal Pozzo Museo Cartaceo

Ca978. **Haskell, F., et al.** 1989. *Il Museu Cartaceo di Cassiano dal Pozzo. Cassiano naturalista.* [Ivrea,] Olivetti. viii+106 pp., illus. *Quaderni puteani* #1.
Chapters by various authors. *Includes* **H. McBurney.** A brief history of the Museo Cartaceo, pp. 5–9; **D. Freedberg.** Cassiano, natural historian, pp. 10–15; ___. Cassiano dal Pozzo's drawings of citrus fruits, pp. 16–36; **H. McBurney.** Cassiano dal Pozzo's drawings of birds, pp. 37–47; **F. Solinas.** The oryx, pp. 48–51; ___. L'erbario miniato e altri fogli di iconografia botanica appartenuti a Cassiano dal Pozzo, pp. 52–76; ___. Il primo erbario Azteco e la copia Romana di Cassiano dal Pozzo, pp. 77–83; ___. Il trattato del legno fossile minerale di Francesco Stelluti e i quattro volumi della Natural History of Fossils nelle raccolte della Biblioteca Reale di Windsor, pp. 84–94; **C. Napoleone.** Osservazioni sul Natural History of Fossils V, pp. 95–102. Part of the collection is now in the Royal Library at Windsor, other parts dispersed.

Chasseur collection

Ca979. **Duchesne, R.** 1983. Magasin de curiosités ou musée scientifique? Le musée d'histoire naturelle de Pierre Chasseur à Québec (1824–1854). *H S T C bulletin, journal of the history of Canadian sciences, technology, and medicine* 24: 59–79.

Leverian Museum, London

Ca980. **Mullens, W. H.** 1915. Some museums of old London. 1: The Leverian Museum. *Museums journal* 15: 123–129, 162–172 & 1 pl.

Linnaeus collection

Ca981. **Jackson, B. D.** 1890. History of the Linnean collections. *Proceedings of the Linnean Society of London* 1887/88: 18–34.

Peale's Museum, Philadelphia, PA

Ca982. **Appel, T. A.** 1980. Science, popular culture and profit: Peale's Philadelphia Museum. *Journal of the Society for the Bibliography of Natural History* 9: 619–634.

Ca983. **Ellis, R. P.** 1966. The founding, history, and significance of Peale's Museum in Philadelphia, 1785–1841. *Curator* 9: 235–258.

Ca984. **Faxon, W.** 1915. Relics of Peale's Museum. *Bulletin of the Museum of Comparative Zoology* 59: 117–148.

Ca985. **Greene, J. C.** 1966. The founding of Peale's Museum. In: T. R. Buckman, ed. *Bibliography and natural history*, pp. 66–72. Lawrence, KS, Univ. of Kansas Libraries.

Ca986. **Hart, S., & D. C. Ward.** 1988. The waning of an enlightenment ideal: Charles Willson Peale's Philadelphia Museum, 1790–1820. *Journal of the early republic* 8: 389–418.
Re-pub. in **L. B. Miller & D. C. Ward**, eds. 1991. *New perspectives on Charles Willson Peale: A 250th anniversary collection*, pp. 219–235. Pittsburgh, PA, Univ. of Pittsburgh Pr. for the Smithsonian Institution. xviii+317 pp. & 88 pp. of illus.

Ca987. **Sellers, C. C.** 1980. *Mr. Peale's Museum: Charles Willson Peale and the first popular museum of natural science and art.* New York, W. W. Norton & Co. xiv+370 pp.

See also Ca824.

Spallanzani's museum

Ca988. **Spallanzani, M. F.** 1985. *La collezione naturalistica di Lazzaro Spallanzani: I modi e i tempi della sua formazione.* Reggio Emillia, Civici Musei. 147 pp. *Documenti per la storia delle arti, dell'archeologia e delle scienze a Reggio Emilia* no. 3.

Worm's museum

Ca989. **Schepelern, H. D.** 1990. The Museum Wormianum reconstructed: A note on the illustration of 1655. *Journal of the history of collections* 2: 81–85.

BIBLIOGRAPHIES OF MUSEUMS

Ca990. **Meisel, M.** 1926–29. The institutions which have contributed to the rise and progress of American natural history, which were founded or organized between 1769–1865, arranged chronologically by their respective dates of formation or organization. In: *A bibliography of American natural history, the pioneer century, 1769–1865 [etc.].* Vol. 2: 1–741, vol. 3: 1–329. Brooklyn, NY, Premier Publishing Co. 3 vols. *Re-pub.* New York, Hafner, 1967.

Ca991. **Murray, D.** 1904. *Museums: Their history and their use, with a bibliography and list of museums in the United Kingdom.* Glasgow, J. MacLehose & Sons. 3 vols.

See also Individual museums, e.g. Ca867 Ca958–959

NATURAL HISTORY EDUCATION & RESEARCH INSTITUTIONS

General

Ca992. **Benson, K. R.** 1988. Laboratories on the New England shore: The 'Somewhat different direction' of American marine biology. *New England quarterly* 61: 55–78.

Ca993. **Fusonie, A. E., & M. Jacobs.** 1983. *The land–grant colleges; a selective historical bibliography and legislative chronology.* Beltsville, MD, National Agricultural Library, Historical Program. [ii]+26 pp.

Ca994. **Hendrickson, W. B.** 1973. Science and culture in the American middle West. *Isis* 64: 326–340.
Thirteen academies of science of the antebellum period are examined.

Ca995. **Irsay, S. d'.** 1933–35. *Histoire des universités françaises et étrangères des origines à nos jours.* Paris. 2 vols.

Ca996. **Jack, H. A.** 1945. Biological field stations of the world. *Chronica botanica* 9(1): 1–73.
Deals with the purpose, history, location, administration, equipment, living facilities, instruction, and research at biological field stations. Directory, pp. 39–73, arranged alphabetically by country. The descriptive accounts given are corrected up to 1940.

Ca997. **Kaavere, V.**, ed. 1983. *Teaduse ajaloo lehekulgi Eestist.* 4 kogumik: *Teadusorganisatsioonide arengust Eestis. [Pages from the history of science in Estonia.* No. 4: *On the history of scientific institutions].* Tallinn, Valgus. 126 pp. & 8 pp. of illus.
Chapters by various authors.

Ca998. **Kofoid, C. A.** 1910. *The biological stations of Europe.* Washington, DC, United States Bureau of Education. xiii+360 pp. & 55 pl. *Bulletin* no. 440.
Includes a brief historical note on each station.

Ca999. **Kohlstedt, S. G.** 1986. Institutional history. In: S. G. Kohlstedt & M. W. Rossiter, eds. *Historical writing on American science: Perspectives and prospects*, pp. 17–36. Baltimore, MD, Johns Hopkins Univ. Pr. 321 pp.
First pub. as *Osiris* n.s. 1, 1985.

Ca1000. **Lindholm, T.** 1982. The history of biological stations in Finland. *Eidema* 1(3): 112–123.

See also Ca192 Ca219 Ca222–223 Ca580 & Cc94(Spencer).

Individual institutions

European

Berliner Universität

Ca1001. **Various authors.** 1985. Beiträge zur Entwicklung der Biologie in der Geschichte der Berliner Universität. *Wissenschaftliche Zeitschrift der Humboldt Universität zu Berlin, mathematisch–naturwissenschaftliche Reihe* 34: 183–406.
19 articles by various authors.

Cambridge University

Ca1002. **Gunther, R. T.** 1937. *Early science in Cambridge.* Oxford, privately pub. xii+513 pp. *Re-pub.* London, Dawsons, 1969.
Includes chapters on Medicine, Anatomy, Physiology, Apothecaries and materia medica, Zoology, Botany, Husbandry and horticulture, etc.

Collège de France, Paris

Ca1003. **Collège de France, Paris.** 1932. *Le Collège de France (1530–1930): Livre jubilaire composé à l'occasion de son quatrième centenaire.* By A. Lefranc et al. Paris, Presses Universitaires de France. 417 pp., illus.
Includes **A. Mayer.** L'histoire naturelle et la physiologie au Collège de France, pp. 167–190.

Ca1004. ___. 1981. *450e anniversaire de la fondation du Collège de France (1530–1980): Compte–rendu de la réunion extraordinaire des professeurs le 26 octobre 1980 [etc.]*. Paris. 50 pp.

Ca1005. **Torlais, J.** 1964. Le collége royal. In: R. Taton, ed. *Enseignement et diffusion des sciences en France au XVIIIe siècle*, pp. 261–286. Paris, Hermann. *Histoire de la pensée* vol. 11.

Re-pub. in **Y. Laissus & J. Torlais.** 1986. Le Jardin du Roi et le Collège royal dans l'enseignement des sciences au XVIIIe siècle. Paris, Herman. 261–341. *Enseignement et diffusion des sciences en France au XVIIIe siècle.*

Debreceni Tudományegyetem, Budapest

Ca1006. **Debreceni Tudományegyetem.** 1956. *Bibliographia Universitatis Debreceniensis.* Pars 1. *Facultas scientiarum naturalium 1914–1955. Debreceni egyetemi bibliográfia. Elsö rész. Természettudományi kar 1914–1955.* By L. Módis. *Acta universitatis debreceniensis de Ludovico Kossuth nominatae. Series biologica* III(3): 119–245. Also pub. separately as *A Debreceni Kossuth Lajos Tudományegyetem Könyvtárának Közleményei* no. 5.

Faculté des Sciences, Paris

Ca1007. **Charle, C., & E. Telkes.** 1989. *Les professeurs de la Faculté des Sciences de Paris: Dictionnaire biographique (1901–1939).* Paris, Institut National de Recherche Pédagogique, Éditions du C N R S. 271 pp. *Histoire biographique de l'enseignement.*

Friedrich–Schiller Universität, Jena

Ca1008. **Friedrich–Schiller Universität.** 1959. *Beiträge zur Geschichte der Mathematisch–Naturwissenschaftlichen Fakultät der Friedrich–Schiller Universität Jena anlässlich der 400 Jahr–Feier.* Jena, G. Fischer. 109 pp. & 11 pl.

Georg–August–Universität zu Göttingen

Ca1009. **Grossbach, U.** 1988. Der Weg zur allgemeinen Biologie: Entwicklungsbiologie und Genetik in Göttingen. In: H.–H. Voigt, ed. *Naturwissenschaften in Göttingen: Eine Vortragsreihe*, pp. 85–97. Göttingen, Vandenhoeck u. Ruprecht. *Göttinger Universitätsschriften* Serie A: *Schriften*, 13.

Imperatorskii Yur'evskii Universitet (*formerly* Kaiserliche Universität, Dorpat; *later* Tartu Riiklik Ulikool)

Ca1010. **Levitskii, G. V.** 1902–03. Biograficheskii slovar' professorov i prepodavatelei Imperatorskago Iur'evskago, byvshago Derptskago, Universiteta za sto liet ego sushchestvovaniya. *[Biographical dictionary of the Professors, Imperial University of Yur'yev, formerly Dorpat, during one hundred years of its existence, 1802–1902.]* Yur'ev. 2 vols.

Including biographies of biological faculty members.

Institut Océanographique de Monaco

Ca1010A. **Institut Océanographique de Monaco.** 1953. Bulletin de l'Institut Océanographique. *Table générale (1904–1953). Vol. 1 à 50, — No. 1 à 1034 bis. Table des Bulletins, index des auteurs, index des matières.* Monaco. 80 pp.

International Council for the Exploration of the Sea

Ca1011. **Food & Agriculture Organization of the United Nations, Fisheries Division.** 1968. *North Atlantic bibliography and citation index: An index of the publications of the International Council for the Exploration of the Sea and the International Commission for the North Atlantic Fisheries.* Rome. [246] pp. *F A O fisheries technical paper* no. 54.

Ca1012. **Went, A. E. J.** 1972. The history of the International Council for the Exploration of the Sea. In: Royal Society of Edinburgh. Proceedings of the second International Congress on the History of Oceanography: *Challenger* expedition centenary, Edinburgh September 12–20, 1972. *Proceedings of the Royal Society of Edinburgh* B 73: 351–360.

Ca1013. ___. 1972. Seventy years a-growing: A history of the International Council for the Exploration of the Sea, 1902–1972. *Rapport et procès–verbaux des réunions de la Commission Internationale pour l'Exploration Scientifique de la Mer* 165: ii+252 pp.

Karl–Ferdinands–Universität, Prague

Ca1014. **Ludy, W.** 1970. *Personalbibliographien von Professoren der Philosophie, Zoologie und Botanik an der Philosophischen Fakultät der Karl–Ferdinands–Universität in Prag im ungefähren Zeitraum von 1860 bis 1918. Mit kurzen biographischen Angaben und Ueberblick über die Tatigkeitsberichte.* Diss. Erlangen–Nürnberg. 227 pp.

Ca1015. **Vogel, H.** 1972. *Personalbibliographien von Professoren und Dozenten der Botanik, Chemie, Diätetik, Pharmazeutischen Warenkunde, Physiologie, Psychiatrie, Tierheilkunde und Zoologie an der Medizinischen Fakultät der Karl–Ferdinands–Universität in Prag im ungefähren Zeitraum von 1800 bis 1860. Mit kurzen biographischen Angaben und Ueberblick über die Tatigkeitsberichte.* Diss. Erlangen–Nürnberg. 159 pp.

Kievskii Gosudarstvennii Universitet, Kiev

Ca1016. **Kievskii Gosudarstvennii Universitet.** 1935. *Rozvitok nauki v Kiyivs'komy Universiteti za sto rokiv. 1835–1936.* Kiev. xviii+293 pp.
Includes **O. U. Sokolov'skii.** Naris roboti kiyivs'kogo universitetu v galuzi sistematiki roslin ta vivchennya mistsevoi flori za minule storichchya, pp. 130–148; **M. M. Voskoboinikov & O. P. Markevich.** Morfologiya tvarin v Kiyivskomy universiteti, pp. 149–170; **D. E. Beling.** Storichna rabota Kiyivs'kogo universitetu v galuzi gidrofaunistichnikh i gidrobiologichnikh doslidzhen', pp. 183–204; **V. M. Artobolevs'kii.** Zoologichnii musei Kiyiv'skogo universitetu za pershe storichchya iogo isnuvannya, pp. 205–225.

Kongeligt Norsk Frederiks Universitet, Christiania (Oslo)

Ca1017. **Kongeligt Norsk Frederiks Universitet.** 1911. *Det Kongelige Frederiks Universitet, 1811–1911. Festskrift.* Christiania. 2 vols.
Includes **H. H. Gran.** Botaniken og zoologien, pp. 539–575.

Laboratoire Arago, Université de Paris, Banyuls–sur–Mer

Ca1018. **Laboratoire Arago.** 1982. Centenaire du Laboratoire Arago (1882–1982): 12 et 13 octobre 1982. *Vie et milieu* 32(4): 199–285.
Includes **P.–P. Grasse.** Le Laboratoire Arago et quelques–uns de ses naturalistes, pp. 211–218; **J. Théodoridès.** Le Laboratoire Arago il y a 35 ans: Souvenirs des années 1947–1954, pp. 219–224; **P. Drach.** Premier developpements de l'océanographie au Laboratoire Arago, pp. 225–233 ; **A. Guille.** L'Océanographie biologique au Laboratoire Arago les temps modernes, pp. 235–240; **C. Delamare Deboutteville.** L'Écologie terrestre au Laboratoire Arago passe, present et futur, pp. 241–247 ; **G. Poljansky.** Un siècle de recherches protistologiques au Laboratoire Arago, pp. 249–253; **P. Tardent.** La Suisse, la mer et le Laboratoire Arago, pp. 259–261; **R. Reinboth.** Recherche et enseignement des biologistes allemands au Laboratoire Arago, pp. 263–264; **M. Bacescu & A. Marinescu.** Emile Racovitza, un nom Roumain entre dans l'histoire du Laboratoire Arago, pp. 267–274; **W. L. M. Geilenkirchen.** La Labora toire Arago et le Benelux, pp. 279–282.

Leningradskii Gosudarstvennyi Ordena Lenina Universitet, St. Petersburg (Leningrad)

Ca1019. **Dogel', V. A.** 1946. Istoriya razvitiya biologicheskikh nauk v Leningradskom Gosudarstvennom Universitete. In: Leningradskii Gosudarstvennyi Ordena Lenina Universitet. *Trudy yubileinoi nauchnoi sessii*, pp. 3–18. Leningrad.

Marine Biological Association of the United Kingdom, Plymouth

Ca1020. **Southward, A. J., & E. K. Roberts.** 1984. The Marine Biological Association, 1884–1984: One hundred years of marine research. *Report & transactions of the Devonshire Association* 116: 155–199. *Also pub. separately* Plymouth, Marine Biological Association of the United Kingdom, 1984. *Occasional publication* no. 3.

Ca1021. ___. 1987. One hundred years of marine research at Plymouth. *Journal of the Marine Biological Association of the United Kingdom* 67: 465–506.

Marine Biological Station, Millport, Scottish Marine Biological Association

Ca1022. **Marshall, S. M.** 1987. *An account of the Marine Station at Millport.* Glasgow. 133 pp. *Occasional publication* no. 4.

Oxford University

Ca1023. **Gunther, R. T.** 1925. *Early science in Oxford.* Vol. 3. i: *The biological sciences.* Vol. 3. ii: *Biological collections.* Oxford, privately pub. xii+564 pp.
Also ___. 1926. *Early medical and biological science. Extracted from "Early science in Oxford."* London, Oxford Univ. Pr. 246 pp.

Philipps–Universität Marburg

Ca1024. **Schmitz, R.** 1978. *Die Naturwissenschaften an der Philipps–Universität Marburg: 1527–1977.* Marburg, Elwert. xxxiv+540 pp., illus.

Rijksuniversiteit te Gent

Ca1025. **Oye, P. van.** 1968. *Anderhalve eeuw biologie aan de Rijksuniverseteit te Gent.* Brussels, Palais der Academien. 194 pp. *Vlaamse Academie voor Wetenschappen, Letteren en Schone Kunsten van Belgie, Klasse der wetenschappen* vol. 30, no. 104.

Rostovskii na Donu Gosudarstvennyi Universitet im. V. M. Molotova, Rostov on Don

Ca1026. **Rostovskii na Donu Gosudarstvennyi Universitet im. V. M. Molotova.** 1941. *Yubil. sbornik 1915–XXV–1940.* Pod red. S. E. Belozerova. Rostov on Don. 136 pp.
Includes **I. V. Novopokrovskii.** Botanicheskaya spetsial'nost' v RGU, pp. 68–84; **S. Ya. Veisig.** Zoologicheskie kafedry, pp. 85–90.

Sevastopol'skaya biologicheskaya stantsiya

See Bb124.

Turun Yliopisto (Universitas Turkuensis), Turku (Åbo)

Ca1027. **Hjelt, O. E. A.** 1896. *Naturhistoriens studium vid Åbo Universitet.* Helsinki. [v]+iv+446 pp. *Skrifter utgifna af Svenska Literatursällskapet i Finland* vol. 32.

Universität Kiel, Institut für Meereskunde

Ca1028. **Krauss, W.** 1990. The Institute of Marine research in Kiel. In: W. Lenz & M. Deacon, eds. *Ocean sciences: Their history and relation to man. Proceedings of the 4th International Congress on the History of Oceanography Hamburg [etc.]*, pp. 131–140. Hamburg, Bundesamt für Seeschiffahrt und Hydrographie. *Deutsche hydrographische Zeitschrift*, Ergänzungsheft, Reihe B, no. 22.

Ca1029. **Wüst, G., et al.** 1956. Das Institut für Meereskunde der Universität Kiel nach seinem Wiederaufbau. *Kieler Meeresforschungen* 12: 127–153.

Universität zu Köln

Ca1030. **Schwarzbach, M.**, ed. 1985. *Naturwissenschaften und Naturwissenschaftler in Köln zwischen der alten und der neuen Universität, 1798–1919.* Cologne, Böhlau Verlag. xi+291 pp. *Studien zur Geschichte der Universität zu Köln* vol. 2.
Includes **G. Quarg.** F. F. Wallraf (1748–1824) und die Naturgeschichte an der alten Köln

Universität; **K. Nappzinn.** Die "Kölner Botanik" zwischen alter und neuer Universität; **H. Englander.** Die Zoologische Garten in Köln; ___. Das Naturkundmuseum in Köln (1892–1944) und seine Vorgänger; **M. Schwarzbach.** Naturwissenschaftliche Vereine; **K. Pabst.** Emil Du Bois–Reymond und der Verein für wissenschaftliche Vorlesungen in Köln; **M. Schwarzbach.** Die Gesellschaft deutscher Naturforscher und Aerzte; **G. Quarg & M. Schwarzbach.** Die naturwissenschaftlichen Lehrstühle and der Handelshochschule; etc.

Université de Liège

Ca1031. **Godeaux, J.** 1990. Aperçu de la biologie marine à l'Université de Liège au cours de ces cent vingt dernières années. *Tijdschrift voor de Geschiedenis der Geneeskunde Natuurwetenschappen, Wiskunde en Techniek* 13: 56–65.

Universitet im. Petera Stuckas, Riga

Ca1032. **Universitet im. Petera Stuckas.** 1971. *Petera Stuckas Latvijas valsts universitates macibu speku publikacijas 1945–1965. Bibliografija … [Publikatsii prepodavatelei Latviiskogo gosudarstvennogo universita im. P. Stucki 1945–1965. Bibliografiya.* 1: *Khimiya, geografiya, biologiya.* Riga. 124 pp.

University of Edinburgh

Ca1033. **Morrell, J. B.** 1971. The University of Edinburgh in the late eighteenth century: Its scientific eminence and academic structure. *Isis* 62: 158–171.

See also Ca977.

Uniwersytecie Wilenskim, Vilnius

Ca1034. **Fedorowicz, Z.** 1957. *Organizacja studiów przyrodniczych na Wszechnicy Wilenskiej w latach 1781–1832; Katedra historii materialnej w dawnej Wszechnicy Wilenskiej; Katedra zoologii i anatomii porównawcej w Uniwersytecie Wilenskim. [Organisation des études des sciences naturelles à l'Université de Wilno dans les années 1781–1832; La chaire d'histoire naturelle à l'ancienne Université de Wilno; La chaire de zoologie et d'anatomie comparée à l'Université de Wilno].* Warsaw, Akademia nauk, Komitet Historii Nauki, Panstwowe Wydawnictwo Naukowe, Warszawa. 203 pp. *Studia i materialy z dziejów nauki polskiej* vol. 5.

Ca1035. ___. 1959. Ewolucjonizm na Uniwersytecie Wilenskim przed Darwinem. (L'Evolutionnisme à l'Université de Wilno avant Darwin). *Memorabilia zoologica* 4: 123 pp.
Summary in French.

North American

Academy of Science of St. Louis, MO

Ca1036. **Academy of Science of St. Louis.** 1906. Celebration of the fiftieth anniversary of the first meeting, March 10, 1906. *Transactions of the Academy of Science at St. Louis* 16: 15–99.
Includes **H. M. Whelpley.** A sketch of the history of the Academy; **G. Hambach.** A history of the museum and library.

Ca1037. **Meisel, M.** 1929. Academy of Science of St. Louis (1856). In: *A bibliography of American natural history, the pioneer century, 1769–1865, [etc.]*, vol. 3: 247–255. Brooklyn, NY, Premier Publishing Co. *Re-pub.* New York, Hafner, 1967.
Historical note and bibliography (including list of natural history publications in its *Transactions* 1856–65).

California Academy of (Natural) Sciences, San Francisco, CA

Ca1038. **Meisel, M.** 1929. California Academy of Natural Sciences (1853). In: *A bibliography of American natural history, the pioneer century, 1769–1865, [etc.]*, vol. 3:

148–157. Brooklyn, NY, Premier Publishing Co. *Re-pub.* New York, Hafner, 1967.
Historical note and bibliography (including list of natural history publications in its *Proceedings* 1854–65).

Canada Department of Agriculture, Ottawa

Ca1039. **Canada Department of Agriculture, Biosystematics Research Centre.** 1986. *One hundred harvests: Research Branch, Agriculture Canada, 1886–1986.* By T. Anstey. Ottawa, Biosystematics Research Centre. xvi+432 pp. *Historical series* no. 27.

Ca1040. **___.** 1986. *Systematics in Agriculture Canada at Ottawa, 1886–1986. La recherche en systématique à Agriculture Canada Ottawa, 1886–1986.* By W. J. Cody, D. B. O. Savile & M. J. Sarazin. Ottawa, Biosystematics Research Centre. [vi]+81 & [vi]+83 pp. *Historical series* no. 28.
Histories of the Division of Entomology & Botany 1887–1909, Division of Botany (& Plant Pathology) 1909–59, Plant Research Institute 1959–72, Division of Entomology 1910–13, Entomological Branch 1914–38, Entomology Division 1939–59, Entomology Research Institute 1959–72, and Biosystematics Research Institute 1973. English and French versions of the text.

Ca1041. **___, Headquarters Library.** 1975. *Publications of the Canada Department of Agriculture 1867–1974.* 2nd rev. ed. Ottawa. 341+136 pp.
Replaces ___. 1963. *Publications of the Canadian Department of Agriculture, 1867–1959.* By E. S. G. Minter. Ottawa, Canadian Department of Agriculture, Information Division. 387 pp.

Carnegie Institution of Washington

Ca1042. **Carnegie Institution of Washington.** 1948. *Catalogue of publications and depositories of the Institution.* Washington, DC. vii+388 pp.

Carnegie Institution of Washington, Department of Marine Biology
See Tortugas Marine Laboratory

Cold Spring Harbor Laboratory, NY

Ca1043. **Ebert, J. D.** 1985. Evolving institutional patterns for excellence: A brief comparison of the organization and management of the Cold-Spring Harbor Laboratory and the Marine Biological Laboratory. *Biological bulletin* 168 supp: 183–186.

Ca1043A. **Roberts, L.** 1990. Cold Spring Harbor turns 100. *Science* 250 (4980): 496–498.

Ca1044. **Watson, E. L.** 1991. *Houses for science: A pictorial history of Cold Spring Harbor Laboratory.* (With *Landmarks in twentieth century genetics: A series of essays* by J. D. Watson.) Plainview, NY, Cold Spring Harbor Pr. xvi+351 pp., illus.

Cornell University, Ithaca, NY

Ca1045. **Wright, A. H.** 1953. Biology at Cornell University. *Bios* 24: 123–145.

Harvard University, Cambridge, MA

Ca1046. **Elliott, C. A., & M. W. Rossiter**, eds. 1992. *Science at Harvard University: Historical perspectives.* Bethlehem, PA, Lehigh Univ. Pr. 380 pp., illus.
Chapters by various authors. *Includes* **T. Appel.** A scientific career in the age of character: Jeffries Wyman and natural history at Harvard; **C. A. Elliott.** The further study of Harvard science (1636–1945): Sources and suggestions; Appendix 1. Bibliography on the history of the sciences at Harvard (1636–1945); Appendix 2. Chronological overview of Harvard science (1636–1945).

Ca1047. **Sax, K.** 1966. The Bussey Institute: Harvard University Graduate School of Applied Biology, 1908–1936. *Journal of heredity* 57: 175–178.

Ca1048. **Wheeler, W. M.** 1930. The Bussey Institution. In: S. E. Morison. 1930. *The development of Harvard University since the inauguration of President Eliot*, pp. 508–517. Cambridge, MA. *The tercentenary history of Harvard College and University, 1636–1936.*

Ca1049. **Williams, C. M.** 1982. *The jubilee of the Harvard Biological Laboratories. Retrospections.* [Cambridge, MA.] 24 pp.

Ca1050. [Not used.]

Illinois Natural History Survey

Ca1051. **Hays, R. G.** 1980. *State science in Illinois: The scientific surveys, 1850–1978.* Carbondale, IL, Southern Illinois Univ. Pr. for The Board of Natural Resources & Conservation of the Illinois Institute of Natural Resources; London, Feffer & Simons. xii+257 pp.

Ca1052. **Illinois, Natural History Survey.** 1931. *Serial list of publications of the Illinois State Natural History Survey and its predecessors, the State Laboratory of Natural History and the State Entomologist's Office.* Urbana, IL. 30 pp.

John Carroll University, Biology Department, Cleveland, OH

Ca1053. **Gavin, D. P.** 1986. History of the Biology Department of John Carroll University, Biology Department. *Ohio journal of science* 86 (3): 51–60.

Johns Hopkins University, Baltimore, MD

Ca1054. **Ebert, J. D., et al.** 1987. The role of Johns Hopkins University in the development of experimental and quantitative biology in America. *American zoologist* 27: 749–817.

Six papers by Ebert and others.

Marine Biological Laboratory, Oceanographic Institution, Woods Hole, MA

Ca1055. **Dexter, R. W.** 1974. From Penikese to the Marine Biological Laboratory at Woods Hole – The role of Agassiz's students. *Essex Institute historical collections* 110: 151–161.

Ca1056. **Galtsoff, P.** 1962. *The story of the Bureau of Commercial Fisheries Biological Laboratory, Woods Hole, Massachusetts.* Washington, DC, United States Department of the Interior. iv+121 pp.

Ca1057. **Groeben, C.** 1984. History: The Naples Zoological Station and Woods Hole. *Oceanus* 27: 60–69.

Ca1058. **Lillie, F. R.** 1944. *The Woods Hole Marine Biological Laboratory.* Chicago, IL, Univ. of Chicago Pr. 284 pp.

Ca1059. **Maienschein, J.** 1989. *100 years of exploring life, 1888–1988: The Marine Biological Laboratory at Woods Hole.* Selection and arrangement of photographs by R. Davis. Boston, MA, Jones & Bartlett. xvi+192 pp.

Ca1060. **Pauly, P. J.** 1988. Summer resort and scientific discipline: Woods Hole and the structure of American biology, 1882–1925. In: R. Rainger, K. R. Benson, & J. Maienschein, eds. 1988. *The American development of biology*, pp. 121–150. Philadelphia, PA, Univ. of Pennsylvania Pr.

Ca1061. **Sears, M., & D. Merriman**, eds. 1980. *Oceanography: The past.* (Proceedings of the third International Congress on the History of Oceanography held September 22–26, 1980, at the Woods Hole Oceanographic Institution, Massachusetts, USA on the occasion of the fiftieth anniversary of the founding of the Institution.) New York, Springer–Verlag. xx+812 pp.

Includes **R. Revelle.** The Oceanographic and how it grew, pp. 10–23; **G. E. R. Deacon.** The Woods Hole Oceanographic Institution: An expanding influence, pp. 25–31; **H. L.**

Burstyn. Reviving American oceanography, pp. 57–66; **R. L. Haedrich & K. O. Emery.** Growth of an oceanographic institution, pp. 67–82; **R. W. Dexter.** The Annisquam sea–side laboratory of Alpheus Hyatt, predecessor of the Marine Biological Laboratory at Woods Hole, 1880–1886, pp. 94–160; and other articles on this institution.

Ca1062. **Various authors.** 1985. The Naples Zoological Station and the Marine Biological Laboratory: One hundred years of biology. *Biological bulletin* 168 supp: 207 pp.
Twenty–two articles by various authors. *Includes* **J. Maienschein.** Agassiz, Hyatt, Whitman, and the birth of the Marine Biological Laboratory, pp. 26–34; **N. Reingold & J. N. Bodansky.** The sciences, 1850–1900, a North Atlantic perspective, 44–61; **P. R. Gross.** Laying the ghost: Embryonic development, in plain words, pp. 62–79; **J. D. Ebert.** Cell interactions: The roots of a century of research, pp. 80–87; **W. D. Russell–Hunter.** An evolutionary century at Woods Hole: Instruction in invertebrate zoology, pp. 88–98; **B. Fantini.** The sea urchin and the fruit fly: Cell biology and heredity, 1900–1910, pp. 99–106; **G. E. Allen.** Heredity under an embryological paradigm: The case of genetics and embryology, pp. 107–121; **S. S. Cohen.** Some struggles of Jacques Loeb, Albert Mathews, and Ernest Just at the Marine Biological Laboratory, pp. 127–136; **J. D. Ebert.** Evolving institutional patterns for excellence: A brief comparison of the organization and management of the Cold–Spring Harbor Laboratory and the Marine Biological Laboratory, pp. 183–186; **J. Maienschein.** Early struggles at the Marine Biological Laboratory over mission and money, pp. 192–196; **W. D. Russell–Hunter.** The Woods–Hole laboratory site: History and future ecology, pp. 197–199; **W. D. Russell–Hunter.** From Woods Hole to the world: The *Biological bulletin*, pp. 200–202; *See* Cc916 for note of other contributions.

See also Ba1143.

Natural History Survey of New York

Ca1063. **Meisel, M.** 1926. The Natural History Survey of New York (1836). In: *A bibliography of American natural history, the pioneer century, 1769–1865, [etc.]*, vol. 2: 606–618. Brooklyn, NY, Premier Publishing Co. *Re-pub.* New York, Hafner, 1967.
Historical note and bibliography (including list of natural history papers 1836–47). Some historical information is also contained in: **N. G. Miller & J. K. Barnes.** 1986. Featured institution – New York State Museum, sesquicentennial of the Biological and Geological Surveys. *Association of Systematics Collections newsletter* 14 (3): 25-29.

Princeton University, NJ

Ca1064. **Conklin, E. G.** 1948. Biology at Princeton. *Bios* 19: 151–171.

Queen's University, Kingston, Ont.

Ca1064A. **Queen's University, Department of Biology.** 1991. *Queen's biology: An academic history of innocence lost and fame gained, 1858-1965.* By B. N. Smallman, H. M. Good & A. S. West. Kingston, Ont. xiv+215 pp., illus.

Scripps Institution of Oceanography, La Jolla, CA

Ca1065. **Raitt, H., & B. Moulton.** 1967. *Scripps Institution of Oceanography: First fifty years.* La Jolla, CA, Ward Ritchie Pr. xix+217 pp.

Ca1066. **Shor, E. N.** 1978. *Scripps Institution of Oceanography: Probing the ocean, 1936 to 1976.* San Diego, CA, Tofua Pr. x+502 pp.

Ca1067. **___.** 1980. The role of T. Wayland Vaughan in American oceanography. In: M. Sears & D. Merriman, eds. *Oceanography: The past.* (Proceedings of the third International Congress on the History of Oceanography held September 22–26, 1980, [etc.].), pp. 127–137. New York, Springer–Verlag.

See also Selected list of the Institution's archives Bc827.

Stanford University, CA

Ca1068. **Baumberger, J. P.** 1954. A history of biology at Stanford University. *Bios* 22: 123–147.

Tortugas Marine Laboratory, Dry Tortugas, FL

Ca1069. **Colin, P. L.** 1980. A brief history of the Tortugas Marine Laboratory and the Department of Marine Biology, Carnegie Institution of Washington. In: M. Sears & D. Merriman, eds. *Oceanography: The past.* (Proceedings of the third International Congress on the History of Oceanography held September 22–26, 1980, [etc.].), pp. 138–147. New York, Springer–Verlag.

Ca1070. **Ebert, J. D.** 1985. Carnegie Institution of Washington and marine biology: Naples, Woods Hole, and Tortugas. *Biological bulletin* 168 supp: 172–182.

Transylvania University, Lexington, KY

Ca1071. **Gobar, A., & J. Hamon.** 1982. *A lamp in the forest: Natural philosophy in Transylvania University 1799–1859.* Lexington, KY, Transylvania Univ. Pr. 198 pp.

United States Department of Agriculture

Ca1072. **Beal, W. H., & E. L. Ogden.** 1908. The contributions of the U. S. Department of Agriculture to the bibliography of science. *Proceedings & papers, Bibliographical Society of America* 2: 135–152.

Ca1073. **Harding, T. S.** 1947. *Two blades of grass: A history of scientific development in the U.S. Department of Agriculture.* Norman, OK, Oklahoma Univ. Pr. xv+352 pp. *Repub.* New York, Arno Pr., 1980.

Ca1074. **Smith, M. S.** 1974. *A list of references for the history of the United States Department of Agriculture.* Davis, CA, Univ. of California, Agricultural History center. 88 pp.

Ca1075. **United States Department of Agriculture.** 1937. *A history of agricultural experimentation and research in the United States, 1607–1925, including a history of the United States Department of Agriculture.* By A. C. True. Washington, DC. vi+321 pp. *Miscellaneous publication* no. 251.

Ca1076. ___. 1962. After a hundred years. *The Yearbook of agriculture* 1962: xv+688 pp.
Articles by various authors arranged under the following heads: Backgrounds, Plants, Conservation, Forests, Animals, Insects, Technologies, Markets, Economics, So people may know, Homes.

See also Cb50.

Universidad Nacional Autonoma de Mexico, Instituto de Biología

Ca1077. **Universidad Nacional Autonoma de Mexico, Instituto de Biología.** 1961. *Historia bibliográfica del Istituto Médico Nacional de México (1888–1915). Antecesor del Instituto de Biología de la UNAM.* By F. Fernández del Castillo. Mexico City. 207 pp.

University of Chicago, IL

Ca1078. **Hall, D. L.** 1979. Academics, bluestockings, and biologists: Women at the University of Chicago, 1892–1932. In: A. M. Briscoe & S. M. Pfafflin, eds. *Expanding the role of women in the sciences*, pp. 300–320. New York, New York Academy of Sciences. *Annals of the New York Academy of Sciences* vol. 323.

University of Pennsylvania, Philadelphia, PA

Ca1079. **Wenrich, D. H.** 1951. Biology at the University of Pennsylvania. *Bios* 22: 151–190.

University of Vermont, College of Natural History, Burlington, VT

Ca1080. **Dann, K. T.** 1985. The College of Natural History at the University of Vermont. *Vermont history* 53: 77–94.

Virginia Institute of Marine Science, Gloucester Point, VA

Ca1081. **Hargis, J., & T. M. Armitage.** 1983. Products and productivity of the men and women of the Virginia Institute of Marine Science from 1940 to June 1983 – A comprehensive annotated inventory. *Special scientific report, Virginia Institute of Marine Science* 114: 457 pp.

Wake Forest College, Department of Biology, Winston–Salem, NC

Ca1082. **Wake Forest College, Department of Biology.** 1969. *A short history of the Department of Biology of Wake Forest College, 1834–1967.* By E. C. Cocke. Winston–Salem, NC. ix+84 pp.

Western Academy of Natural Sciences, Cincinnati, OH

Ca1083. **Shapiro, H. D.** 1976. The Western Academy of Natural Sciences of Cincinnati and the structure of science in the Ohio Valley, 1810–1850. In: A. Oleson & S. C. Brown, eds. *The pursuit of knowledge in the early American republic: American scientific and learned societies from colonial times to the civil war*, pp. 219–247. Baltimore, MD, Johns Hopkins Univ. Pr.

Yale University, New Haven, CT

Ca1084. **Coe, W. R.** 1950. The development of the biological sciences at Yale. *Bios* 21: 147–170.

South American

Instituto Biológico de São Paulo

Ca1085. **Reis, J.** 1977. Instituto biológico de São Paulo. *Arquivos do Instituto Biológico (São Paulo)* 44(3): 119–144.

Universidad de Buenos Aires

Ca1086. **Universidad de Buenos Aires.** 1971. *Las ciencias naturales en la Universidad de Buenos Aires: Estudio histórico.* By H. H. Camacho. Buenos Aires. xvi+150 pp.

Universidad Central de Venezuela, Escuela de Biologia, Caracas

Ca1087. **Texera Arnal, Y.** 1986. La biologia en un contexto periférico: La Escuela de Biologia de la Universidad Central de Venezuela. *Quipu* 3: 29–52.

Asian

Biologo–Geograficheskii Nauchno–Issledovatel'skii Institut, (Gosudarstvennii Irkutskii Universitet', Irkutsk

Ca1088. **Egorov, A. G., & P. P. Khoroshkikh.** 1950. Dvadtsat' pyat' let deyatel'nosti Biologi–Geograficheskogo Nauchno–Issledovatelakogo Instituta (1923–1948). [Twenty–five years of activity of the Biologo–Geographical Institute of Scientific Research (1923–48.] *Izvestiya Biologo–Geograficheskogo Nauchno Issledovatel'skogo Instituta pri Irkutskom Gosudarstvennom Universitete imeni A. A. Zdanova* 11: 15–68 & 3 pl.

Tomskii Gosudarstvennii Universitet im. V. V. Kuibysheva

Ca1089. **Maslov, D. P., & M. R. Filimonov.** 1954. Bibliografiya nauchnykh rabot, opublikovannykh v izdaniyakh Tomskogo Gosudarstvennogo Universiteta im. V. V. Kuibysheva za 1889–1950 gg. Sistematicheskii ukazatel'. 3: [*Includes* Biologicheskie nauki; meditisina; sel'skoe khozyaistvo.] *Uchenye zapiski Tomskogo Gosudarstvennogo Universiteta im. V. V. Kuibysheva* 21: 113–154.

Otsu Hydrobiological Station, Kyoto Imperial University

Ca1090. **Otsu Hydrobiological Station.** 1964. *Otsu Hydrobiological Station: Fifty years*. Otsu. [i]+32 pp., illus.

Australasian

Great Barrier Reef Committee

Ca1091. **Hill, D.** 1985. The Great Barrier Reef Committee, 1922–82. Part 2: The last three decades. *Historical records of Australian science* 6: 195–221.

Office de la Recherche Scientifique et Technique d'Outre Mer (ORSTOM)

Ca1091A. **Fages, J.** 1991. Les centres ORSTOM dans le Pacifique Sud (1946–1990)/ORSTOM centres in the South Pacific (1946–1990). *Iles et Archipels* 14: 21–34 (French text) & 42–49 (English).

University of Auckland

Ca1092. **Foster, B., J. Rattenbury, & J. Marbrook.** 1983. *A history of biology at the University of Auckland, 1883–1983*. Auckland, Univ. of Auckland, Departments of Botany, Cell Biology & Zoology. 111 pp.

African

Institute of Oceanography, Port Sudan

Ca1093. **Nasr, D. H.** 1990. The history of the Institute of Oceanography at Port Sudan and its role in research activities in the Red Sea. In: W. Lenz & M. Deacon, eds. *Ocean sciences: Their history and relation to man. Proceedings of the 4th International Congress on the History of Oceanography Hamburg [etc.]*, pp. 557–561. Hamburg, Bundesamt für Seeschiffahrt und Hydrographie. *Deutsche hydrographische Zeitschrift*, Ergänzungsheft, Reihe B, no. 22.

BIOLOGICAL CLASSIFICATION

Ca1094. **Bather, F. A.** 1927. Biological classification: Past and future. *Quarterly journal of the Geological Society of London* 83: lxii–civ.

Ca1095. **Bauverie, J.** [1932.] *La systématique des formes: Les grandes étapes du progrès de la classification naturelle, ses tendences actuelles*. Paris, Éditions du Cerf. 64 pp.

Ca1096. **Burckhardt, C. R.** 1903. Zur Geschichte der biologischen Systematik. *Verhandlungen der Naturforschenden Gesellschaft in Basel* 16: 388–440.

Ca1097. **Cain, A. J.** 1958. Logic and memory in Linnaeus's system of taxonomy. *Proceedings of the Linnean Society of London* 169: 144–163.

Ca1098. ___. 1959. Deductive and inductive methods in post–Linnaean taxonomy. *Proceedings of the Linnean Society of London* 170: 185–217.

Ca1099. ___. 1959. The post–Linnaean development of taxonomy. *Proceedings of the Linnean Society of London* 170: 234–244.

Ca1100. **Croizat, L.** 1945. History and nomenclature of the higher units of classification. *Bulletin of the Torrey Botanical Club* 72: 52–75.

Ca1101. **Dagognet, F.** 1970. *Le catalogue de la vie: Étude méthodologique sur la taxinomie*. Paris, Presses Universitaires de France. 189 pp. *"Galien," histoire et*

philosophie de la biologie et de la médecine.
Includes Ch. 1: Botanique et linguistique (Linné, de Jussieu, Desfontaines), pp. 13–63; ch. 2: La systématique animale (Vicq d'Azyr, Cuvier, Geoffroy Saint-Hilaire), pp. 65–124, Ch. 3: La nosologie (de Tenon à Pinel et à Laennec), pp. 125–171.

Ca1102. **Daudin, H.** 1926. *Études d'histoire des sciences naturelles.* Paris, Alcan. 2 vols. *Re-pub.* Paris, Editions des Archives Contemporaines, 1983. 2 vols.
I: *De Linné à Jussieu. Méthodes de la classification et idée de série en botanique et en zoologie, (1740–90).* II: *Cuvier et Lamarck: Les classes zoologiques et l'idée de série animale (1790 1830).*

Ca1103. **Hoppe, B.** 1978. Der Ursprung der Diagnosen in der botanischen und zoologischen Systematik. *Sudhoffs Archiv für Geschichte der Medizin und der Naturwissenschaften* 62: 105–130.

Ca1104. **Hünemörder, C.** 1983. Aims and intentions of botanical and zoological classification in the Middle Ages and Renaissance. *History & philosophy of the life sciences* 5: 53–67.

Ca1105. **Knight, D. M.** 1981. *Ordering the world: A history of classifying man.* London, Burnett Books. 215 pp.

Ca1106. **Larson, J. L.** 1971. *Reason and experience: The representation of natural order in the work of Carl von Linné.* Berkeley, CA, Univ. of California Pr. vii+171 pp.
Ch. I. The System of Nature and the natural method; ch. II. Classes and orders; ch. III. Genera; ch. IV. Species and varieties; ch. V. Nomenclature.

Ca1107. **Lewis, J. E.** 1963. A short history of taxonomy from Aristotle to Linnaeus. *Medical arts & sciences* 17: 106–123.

Ca1108. **Linnean Society of London.** 1950. *Lectures on the development of taxonomy delivered in the rooms of the Linnean Society during the session 1948–49.* London. [iv]+83 pp.

Ca1109. **Lovejoy, A.** 1936. *The great chain of being: A study of the history of an idea.* Cambridge, MA, Harvard Univ. Pr. ix+382 pp. *Re-pub.* New York, Harper, 1960.

Ca1109A. **Panchen, A. L.** 1992. *Classification, evolution, and the nature of biology.* Cambridge, Cambridge Univ. Pr. xi+403 pp.

Ca1110. **Rossi, P.** 1984. Universal languages, classifications and nomenclatures in the seventeenth century. *History & philosophy of the life sciences* 6: 119–131.

Ca1111. **Schiller, J.** 1980. *Physiology and classification: Historical relations.* Paris, Maloine S.A. xiii+194 pp. *Series Université de Compiegne.*

Ca1112. **Slaughter, M. M.** 1982. *Universal languages and scientific taxonomy in the seventeenth century.* Cambridge, Cambridge Univ. Pr. x+277 pp.

Ca1113. **Sloan, P. R.** 1972. John Locke, John Ray and the problem of the natural system. *Journal of the history of biology* 5: 1–53.

Ca1114. **___.** 1979. Buffon, German biology, and the historical interpretation of biological species. *British journal for the history of science* 12: 109–153.

Ca1115. **Sneath, P. H. A.** 1965, Key works in the history of biological classification. *Classification Society bulletin* 1: 5–13.

Ca1116. **Stearn, W. T.** 1959. The background of Linnaeus's contributions to the nomenclature and methods of systematic biology. *Systematic zoology* 8: 4–22.

Ca1117. **Vernon, K.** 1988. The founding of numerical taxonomy. *British journal for the history of science* 21: 143–159.

Ca1118. **Voss, E. G.** 1952. The history of keys and phylogenetic trees in systematic biology. *Journal of the Science Laboratories, Denison University* 43: 1–25.

CLASSIFICATION SCHEDULE

Ca1119. **Parker, S. P.**, ed. 1982. *Synopsis and classification of living organisms*. New York, McGraw–Hill Book Co. 2 vols. (2,400 pp.)

Natural–history terms, etymology

Ca1120. ***Documents pour l'histoire du vocabulaire scientifiques.*** Besançon, Institut de la Langue Française, vol. 1+, 1980+. ••Including etymological studies of natural–history terms, e.g.: **J.–L. Fischer.** 1981. Sens, contre sens et synonymie dans l'emploi des termes "mulet," "métis" et "hybride" en zoologie de 1749 à 1860. *Ibid.* 2: 23–35; **P. Acot.** 1982. Éléments de réflexion sur l'origine du mot "écologie." *Ibid.* 3: 103–109; **J.–L. Fischer & R. Rey.** 1983. De l'origine et de l'usage des termes "Taxinomie" et "taxonomie." *Ibid.* 5: 97–113; **R. Olry.** 1989. Histoire des nomenclatures anatomiques. *Ibid.* 9: 91–98; etc.

Ca1121. **Heller, J. L.** 1945. Classical mythology in the "Systema Naturae" of Linnaeus. *Transactions of the American Philological Association* 76: 333–357.

Ca1122. **Hogben, L. T., & M. Cartwright.** 1969. *The vocabulary of science*. London, Heinemann. [vii]+184 pp. *Also* New York, Stein & Day, 1970. ••The Latin legacy; The reinstatement of Greek; Background to the French reform; Spelling conventions as tools of diagnosis; Basic Greek and Latin vocabularies; Medical and biological Greek and Latin terms.

Ca1123. **Jaeger, E. C.** [1962.] *A source–book of the biological names and terms*. 3rd ed., rev. 3rd printing Springfield, IL, C. C. Thomas. xxxv+323 pp. ••Ca. 15,000 word elements with derivations. *Replaces* ___. 1950. ___. 2nd ed. Springfield, [etc.]. xxxv+287 pp.

Ca1124. **Melander, A. L.** 1940. *Source book of biological terms*. 2nd ed. New York, College of the City of New York, Department of Biology. iii+157 pp. ••*Replaces* ___. 1937. ___. New York, [etc.]. iii+157 pp.

Ca1125. **Oxford University Press** [Pub.]. 1989. *The Oxford English dictionary*. 2nd edn. Ed. by J. A. Simpson & E. S. C. Weiner. Oxford, Clarendon Pr. 20 vols. ••Integrates the material from the original O E D and the Supplement with the addition of over 5,000 new words. Over 464,000 words defined, with ca. 2,400,000 illustrative quotations. Special attention was given to natural history terms in the preparation of the *Supplement* (1972–86) by the Senior Editor (Natural History), Sandra Raphael. *Replaces* ___. 1933. *The Oxford English dictionary: Being a corrected re-issue, with an introduction, supplement, and bibliography of A new English dictionary on historical principles* ... Ed. by J. A. H. Murray [et al.] Oxford, Clarendon Pr. 12 vols. & suppl. Reprinted 1961, 1970, 1975 & 1978; *Continued* ___. 1972–86. *A supplement to the Oxford English dictionary*. Ed. by R. W. Burchfield. Oxford, Clarendon Pr. 4 vols. *See also* **S. Raphael.** 1973. Natural history and the *Oxford English Dictionary*. *Journal of the Society for the Bibliography of Natural History* 6: 229–235.

Ca1126. **Potter, S., & L. Sargent.** 1974. *Pedigree: Essays on the etymology of words from nature*. London, Collins. 320 pp. *New naturalist* no. 56. American ed. as: *Pedigree: The origins of words from nature*. New York, Taplinger.

Ca1127. **Werner, C. F.** 1956. *Wortelemente lateinisch–griechischer Fachausdrücke in der Biologie, Zoologie und vergleichenden Anatomie*. Leipzig, Akademische Verlagsges. Geest & Portig. 397 pp.

Ca1128. **Yancey, P. H.** 1945. Origins from mythology of biological names and terms. *Bios* 16: 3–47.

See also Bibliographies of dictionaries; *see* Ca110 & Ca112

NATURAL HISTORY ILLUSTRATION

Ca1129. **Biblioteche Estense e Universitaria, Modena.** [1984.] *Immagine e natura. L'immagine naturalistica nei codici e libri a stampa delle Biblioteche Estense e Universitaria. Secolo XV–XVII. La scoperta della natura, tra fantasia e realta, nei secoli della rivoluzione scientifica.* Modena, Edizioni Pannini. 234 pp., illus.
Exhibition held 21 Mar. – 15 May, 1984. Including 15 essays by M. G. Tavoni, L. Tongiorgi Tomasi & P. Tongiorgi inserted between sections of the catalogue, with 45 coloured and 253 monochrome illustrations.

Ca1130. **Bridson, G. D. R.** 1989. From xylography to holography: Five centuries of natural history illustration. *Archives of natural history* 16: 121–141.

Ca1131. **___, & J. J. White.** 1990. *Plant, animal and anatomical illustration in art and science: A bibliographical guide from the 16th century to the present day.* Winchester, St. Paul's Bibliographies; Detroit, MI, Omnigraphics Inc. xxxix+451 pp.

Ca1132. **Choulant, J. L.** 1858. *Graphische Incunabeln für Naturgeschichte und Medicin: Enthaltend Geschichte und Bibliographie der ersten naturhistorischen und medicinischen Drucke des 15. und 16. Jahrhunderts, welche mit illustrierenden Abbildungen versehen sind.* Leipzig, Weigel. xx+168 pp. *Re-pub.* Munich, Münchner Drucke, 1924; Hildesheim, Olms, 1963.

Ca1133. **Kemp, M.** 1990. Taking it on trust: Form and meaning in naturalistic representation. *Archives of natural history* 17: 127–188, incl. 54 illus.

Ca1134. **Landwehr, J.** 1976. *Studies in Dutch books with coloured plates published 1662–1875: Natural history, topography and travel, costumes and uniforms.* The Hague, W. Junk. ix+136, xvii+605 pp. & 16 pl. (col.)
Includes a chronological list and alphabetical catalogue of colour-plate natural history books.

Ca1135. **Randall, L. M. C.** 1966. *Images in the margins of gothic manuscripts.* Los Angeles & Berkeley, CA, Univ. of California Pr. xi+235 pp. & 158 pl.
Alphabetical index of subjects, including plants, pp. 45–235; list of manuscripts indexed, pp. 27–38.

Ca1136. **Rosen, F.** 1903. *Die Natur in der Kunst: Studien eines Naturforschers zur Geschichte der Malerei, mit 120 Abbildungen nach Zeichnungen von Erwin Süss und Photographien des Verfassers.* Leipzig, B. G. Teubner. xii+344 pp., illus.

Ca1137. **Smith, B.** 1985. *European vision and the South Pacific.* 2nd ed. New Haven, CT, Yale Univ. Pr., 1985. [xiii]+370 pp., illus.
Including plants and animals in art. *Replaces* ___. 1960. *European vision and the South Pacific, 1768–1850.* Oxford, [etc.]. xix+287 pp. & 60 pp. of illus.

Ca1138. **Uppsala Universitet** (Departments of Art History & History of Science). 1985. *The natural sciences and the arts: Aspects of interaction from the Renaissance to the 20th century. An international symposium.* Ed. by A. Ellenius. Uppsala, Almqvist & Wiksell. vii+177 pp. *Acta Universitatis Upsaliensis, figura* n.s. 22.
Includes **J. S. Ackerman.** Early Renaissance 'naturalism' and scientific illustration, pp. 1–17; **P. Hulton.** Realism and tradition in ethnological and natural history imagery of the 16th century, pp. 18–31; **D. E. Allen.** Natural history and visual taste: Some parallel tendencies, pp. 32–45; **W. Ashworth Jr.** The persistent beast: Recurring images in early zoological illustration, pp. 46–66; **W. Harms.** On natural history and emblematics in the 16th century, pp. 67–83; **G. Broberg.** Natural history frontispieces and ecology, pp. 84–97; **M. Kemp.** Yellow, red and blue: The limits of colour science in painting, 1400–1730, pp. 98–105; **D. M. Knight.** Scientific theory and visual language, pp. 106–124; **M. Baxandall.** The bearing of the scientific study of vision on painting in the 18th century: Pieter Camper's De visu (1746), pp. 125–132; **B.–O. Landin.** Entomological illustration and illumination with

special reference to Swedish treatises: A bibliographical survey, pp. 133–146; **A. Ellenius.** Ecological vision and wildlife painting towards the end of the 19th century, pp. 147–165.

See also Biographies Ba773 Ba809 Ba1124–1128 Ba1248–1252 Ba1342 & Ba1596; Ca245 Ca274 & Ca764.

BIBLIOGRAPHIES OF ILLUSTRATION RESOURCES

Ca1139. **Ellis, J. C.** 1949. *Nature and its applications: Over 200,000 selected references to nature forms and illustrations of nature as used in every way*. Boston, Faxon. xii+861 pp. ••Books and periodicals used in index pp. xi–xii. The alphabetically arranged index uses common names with relatively few references from scientific names. There is a small section on landscapes, pp. 854–861. The references are mostly to photographs in books and magazines of the late 19th and 20th century.

Ca1140. **Munz, L. T., & N. G. Slausen.** 1981. *Index to illustrations of living things outside North America: Where to find pictures of flora and fauna*. Hamden, CT, Archon. 441 pp. ••Guide to illustrations published in books since 1963.

Ca1141. **Thompson, J. W.** 1977. *Index to illustrations of the natural world: Where to find pictures of the living things of North America*. Ed. by N. Slauson. Syracuse, NY, Gaylord Professional Publications. 265 pp. ••References are to 178 books, mostly pub. since 1960. The alphabetically arranged index of ca. 6,200 entries uses common names (there is a Scientific name index, pp. 191–258). Bibliography, pp. 259–265.

Section Cb. HISTORY OF BOTANY

HISTORIES OF BOTANY

General

Cb1. **Adanson, M.** 1847 [1864]. Préface istorike sur l'état ancien et actuel de la botanike, & une théorie de cette science. In: *Histoire de la botanique et plan des familles naturelles des plantes.* 2me éd. preparée par l'auteur, publiée sur ses manuscrits par A. Adanson et J. Payer, vol. 1: i–cccxxv. Paris, Masson et fils, Imprimée en 1847, Publiée en 1864.

Cb2. **Bazilevskaya, N. A., I. P. Belokony, & A. A. Shcherbakova.** 1968. Kratkaya istoriya botaniki. [A short history of botany]. *Trudy Moskovskogo Obshchestva Ispytatelei Prirody, Biol.* 31: 311 pp.

Cb3. **Cailleux, A.** 1953. Progression du nombre d'espèces de plantes décrites de 1500 à nos jours. *Revue d'histoire des sciences* 6: 42–49.

Cb4. **Gibson, E. J. Harvey–.** 1919. *Outlines of the history of botany.* London, A. & C. Black. x+274 pp.

Cb5. **Guérin, L.** 1869. *Précis de l'histoire de la botanique pour servir de complément à l'étude du règne végétal.* Paris, L. Guérin, T. Morgand. 533 pp. *La règne végétal* (A. Dupuis et al., eds.) vol. 17.
For background information *see* **J. Brunel.** 1949. Another little–known history of botany. *Nature* 163: 732; **A. Arber.** 1949. Another little–known history of botany. *Ibid.* 163: 882.

Cb6. **Hawks, E.** 1928. *Pioneers of plant study [etc.].* London, Sheldon Pr. x+288 pp. & 17 pl.

Cb7. **Högrell, B.** 1886. *Botanikens historia in öfversigt.* Goteborg, H. L. Bolinder. viii+304 pp.
Botanistförteckning med personalnotiser, pp. 220–287.

Cb8. **Jessen, K. W. F.** 1864. *Botanik der Gegenwart und Vorzeit in culturhistorischer Entwickelung. Ein Beitrag zur Geschichte der abendländischen Völker.* Leipzig, F. A. Brockhaus. xviii+495 pp. *Re-pub.* Waltham, MA, Chronica Botanica, 1948; Walluf, 1972.
"The best concise, all–round history of botany ever attempted." F. Verdoorn.

Cb9. **Mägdefrau, K.** 1992. *Geschichte der Botanik: Leben und Leistung grosser Forscher.* 2. Aufl. Stuttgart, G. Fischer. [vii]+359 pp.
Replaces 1st ed., 1973. [viii]+359 pp.

Cb10. **Meyer, E. H. F.** 1854–57. *Geschichte der Botanik.* Königsberg, Gebrüder Borntraeger. 4 vols. *Re-pub.* (with a preface by F. Verdoorn) Amsterdam, Asher, 1965.

Cb11. **Möbius, M.** 1968. *Geschichte der Botanik von den ersten Anfängen bis zur Gegenwart.* 2., unveränd Aufl. Stuttgart, G. Fischer. viii+458 pp.
Replaces 1st ed., Jena, G. Fischer, 1937.

Cb12. **Morton, A. G.** 1981. *History of botanical science: An account of the development of botany from ancient times to the present day.* London, Academic Pr. xii+474 pp.

Cb13. **Reed, H. S.** 1942. *A short history of the plant sciences.* Waltham, MA, Chronica Botanica. [x]+320 pp. *New series of plant science books* vol. 7.
1: Introduction; 2: Gardeners and herbalists of antiquity; 3: The nascent period; 4: The

retrogressive period; 5: The renascent period; 6: The seventeenth century; 7: The eighteenth century; 8: Gardens and other things; 9: Plant geography in the nineteenth century; 10: Morphology; 11: Cytology; 12: The water economy of plants; 13: The fixation of carbon by plants; 14: The assimilation of nitrogen; 15: The fixation and metabolism of nitrogen; 16: Plant nutrition; 17: Mineral constituents in metabolism; 18: Mycology; 19: Plant pathology; 20: Significant names in the history of botanical science.

Cb14. **Sachs, J. von.** 1890. *History of botany (1530–1860).* Authorised translation by H. F. Garnsey. Revised by I. B. Balfour. Oxford, Clarendon Pr. xvi+568 pp.
Book I. History of morphology and classification; Book II. History of vegetable anatomy; Book III. History of vegetable physiology. Translation of *Geschichte der Botanik vom. 16. Jahrhundert bis 1860.* Munich, R. Oldenbourg, 1875. xii+612 pp. *Geschichte der Wissenschaften in Deutschland* vol. 15. *Re-pub.* New York, Johnson Reprint Corporation, 1965, and Hildesheim, Olms, 1966. Another ed: ___. 1892. *Histoire de la botanique du XVI siècle à 1860.* Traduction française par H. de Varigny. Paris, C. Reinwald & Cie. xvi+584 pp. For a continuation by J. R. Green, *see* Cb38.

Cb15. **Schultes, J. A.** 1817. *Grundriss einer Geschichte und Literatur der Botanik, von Theophrastos Eresios bis auf die neuesten Zeiten; nebst einer Geschichte der botanischen Gärten.* Vienna, C. Schaumberg & Co. xvi+411 pp.
With an added title-page: *Anleitung zum gründlichen Studium der Botanik, zum Gebrauche bey Vorlesungen und zum Selbstunterrichte.*

Cb16. **Sprengel, K. P. J.** 1817–18. *Geschichte der Botanik.* New ed. Altenberg & Leipzig, F. A. Brockhaus. 2 vols.
Replaces ___. 1807–08. *Historia rei herbariae.* Amsterdam, sumtibus Tabernae. 2 vols. For French version *see* ___. 1832. *Histoire de botanique depuis son origine jusqu'au 19ième siècle.* Traduit de l'Allemand sur la seconde édition par A. J. L. Jourdan. Paris. 2 vols.

Cb17. **Troyer, J. R.** 1992. On the history and characteristics of some multiple discoveries in botany, 1648–1965. *American journal of botany* 79: 833–841.

Cb18. **Verdoorn, F.** 1970. De plant in de biohistorie. *Jaarboek, Koninklijke Nederlandse Botanische Vereniging* 1970: 29–84.
Summary and selective table of contents in English, pp. 29–32.

Cb19. **Winckler, E.** 1854. *Geschichte der Botanik.* Frankfurt am Main, Literarische Anstalt. xvi+646 pp.

See also Ba125–156 Bb708–778 Ca33–34 Ca459(Stebbins) & D86.

15th to 18th centuries

Cb20. **Anderson, F. J.** 1977. *An illustrated history of the herbals.* New York, Columbia Univ. Pr. xiv+270 pp.

Cb21. **Arber, A.** 1986. *Herbals: Their origin and evolution. A chapter in the history of botany 1470–1670.* 3rd ed. Cambridge, Cambridge Univ. Pr. xxxii+358 pp.
Re-pub. of the 2nd edition, 1938, with an introduction and annotations by W. T. Stearn. Also includes a reprint of "From medieval herbalism to the birth of modern botany", first pub. in *Science, medicine and history: Essays ... in honour of Charles Singer*, Oxford, Oxford Univ. Pr., 1953.

Cb22. **Barlow, H. M.** 1913. *Old English herbals, 1525–1640.* London. 42 pp.
Details of 40 works.

Cb23. **Burgess, E. S.** 1902. *History of pre-Clusian botany in its relation to* Aster. New York, Torrey Botanical Club. xii+447 pp. *Memoirs of the Torrey Botanical Club* vol. 10.

Cb24. **Callery, B. G.** 1978. Ancient, moderns and reliable friends: Stated sources of some English herbals. *Journal of the Society for the Bibliography of Natural History* 8: 435–444.

Cb25. **Ferrero, P. M., et al.** 1986. *Actas del simposium CCL aniversario nacimiento de José Celestino Mutis.* Cadiz, Paz Martin Ferrero. 457 pp. *Historia* vol. 3.
Many papers by various authors. *Includes* **M. D. Pérez Murillo, J. M. De La Casa Rivasa, & A. De Dueñas Olmo.** El interés por la historia natural en la época de Carlos III según los fondos documentales del Archivo General de Indias; **M. del C. Vidal Casero.** Cultura en el siglo XVIII: Desarrollo de sociedades cientificas y reales academias; **J. L. Valverde & J. Alarcón.** Casimiro Gómez Ortega y las expediciones científicas españolas del siglo XVIII; **J. Pettegui Estrada.** La expedición de Malaspina; **G. Garcia Guardia.** Itinerario de José Quer en su expedición botánica de 1761; **R. M. Ba sante Pol.** Influencias de Casimiro Gómez Ortega en la creación y crecimiento de los jardines botánicos; **M. E. Mansilla Legorburo.** Estudio estadístico de los comisionados del Jardín Botánico de Madrid, nombrados en conformidad del capítulo 26 del reglamento del nuevo Jardín Botânico; **A. González Bueno & S. Muñoz Calvo.** El Jardin Botánico del Colegio de Farmacéuticos de Madrid a lo largo del siglo XVIII; **S. Alba Romero.** Farmacéuticos españoles del siglo XVIII ante la botánica de su época; **S. Muñoz Calvo & A. González Bueno.** Contribución del Colegio de Farmacéuticos de Madrid el desarrollo de la botánica española en el siglo XVIII; **A. Aguilera Guirao.** Noticias sobre la vida del botánico valenciano D. José Cavanilles; etc.

Cb26. **Fischer, H.** 1929. *Mittelalterliche Pflanzenkunde.* Munich, Verlag der Münchner Drucke. viii+326 pp. *Geschichte der Wissenschaften: Geschichte der Botanik* vol. 2.

Cb27. **Fries, R. E.** 1951. De linneanska 'apostlarnas' resor: Kommentar till en karta. *Svenska Linné–Sällskapets årsskrift* 33/34: 31–40.

Cb28. **Greene, E. L.** 1983. *Landmarks of botanical history.* Ed. by F. N. Egerton, with contributions by R. P. McIntosh & R. McVaugh. Stanford, CA, Stanford Univ. Pr. for Hunt Institute for Botanical Documentation. 2 vols.
From antiquity to the eighteenth century, concentrating on some two dozen major figures. Bibliography, pp. 1039–1109. Vol. 1 first pub. as *Smithsonian miscellaneous collections* no. 54, 1909, 392pp.

Cb29. **Heilmann, K. E.** 1966. *Kräuterbücher in Bild und Geschichte.* Munich, K. Kölbl. 425 pp.

Cb30. **Lawrence, G. H. M.** 1965. Herbals: Their history and significance. In: Clark Memorial Library, University of California, Los Angeles. *History of botany: Two papers presented at a symposium held at the William Andrews Clark Memorial Library, December 7, 1963*, pp. 1–35. Los Angeles, CA; Pittsburgh, PA.
References to the literature on herbals, pp. 21–35.

Cb31. **Reeds, K. M.** 1976. Renaissance humanism and botany. *Annals of science* 33: 519–542

Cb32. ___. 1991. *Botany in medieval and renaissance universities.* New York, Garland Publishing. xix+316+[52] pp., illus. *Harvard dissertations in the history of science.*
An "Annex" includes two reprinted articles by the same author: Renaissance humanism and botany. *Annals of botany* 33: 519–542, 1976, & Publishing scholarly books in the sixteenth century. *Scholarly publishing* 1983 (April): 259–274, 1983.

Cb33. **Rohde, E. S.** 1922. *The old English herbals.* London, Longmans Green. xii+243 pp. *Re-pub.* New York, Dover Publications, 1971; London, 1974.

Cb34. **Scharffenberg, J.** 1902. Bidrag til botanikens historie i Norge i det 17de aarhundrede. *Nyt magazin for naturvidenskaberne* 40: 193–202.

Cb35. **Wein, K.** 1933. Die Geschichte der Floristik des 17. Jahrhunderts. *Repertorium specierum novarum regni vegetabilis, Beihefte* 71–D: 1–67.

Cb36. **Wellisch, H. H.** 1978. Early multilingual and multiscript indexes in herbals. *Indexer* 11: 81–104.

19th & 20th centuries

Cb37. **Costantin, J.** 1934. Aperçu historique des progrès de la botanique depuis cent ans (1834–1934). *Annales des sciences naturelles, botanique* (10) 16 (fasc. sp.): cciii pp.
As recorded in the *Annales des sciences naturelles.*

Cb38. **Green, J. R.** 1909. *A history of botany 1860–1900: Being a continuation of Sachs 'History of botany, 1530–1860.'* Oxford, Clarendon Pr. 543 pp. *Re-pub.* New York, Russell & Russell, 1967.
Book I. Morphology; Book II. The anatomy of plants; Book III. The physiology of plants. Bibliography, pp. 507–538.

Cb39. **Stuckey, R. L., & E. D. Rudolph.** 1974. 25 years of botany: History of botany with a bibliographic appendix. *Annals of the Missouri Botanic Garden* 61: 237–261.

Chronologies

Cb40. [Not used.]

See Cb311A Cb634 Cb654 & Cb1246.

Plant exploration

• A list of general histories of plant hunting. N. B. Histories relating to the botanical exploration of individual countries are included in "Histories of botany, regional" below.

Cb41. **Castellanos, A.** 1943. Las exploraciones botánicas en la época colonial. *Cursos y conferencias, revista del Colegio Libre de Estudios Superiores* 136/137: 411–453.

Cb42. **Coats, A. M.** 1969. *The quest for plants: A history of the horticultural explorers.* London, Studio Vista. 400 pp. & 26 pl.
American ed. as: *The plant hunters: Being a history of the horticultural pioneers, their quests and their discoveries, from the Renaissance to the twentieth century.* New York, McGraw–Hill Book Co., 1969.

Cb43. **Duval, M.** 1982. *The King's garden.* Translated by A. Tomarken & C. Cowen. Charlottesville, VA, Univ. Pr. of Virginia. xii+214 pp., illus.
French plant exploration in the 16th to 19th centuries. Translation of *La planète des fleurs.* Paris, R. Laffont, 1977. 286 pp.

Cb44. **Ewan, J.** 1991. Who conquered the New World?; or, Four centuries of exploration in an indehiscent capsule. *Annals of the Missouri Botanical Garden* 78: 57–64.

Cb45. **Healey, B. J.** 1975. *The plant hunters.* New York, Scribner. vii+214 pp.

Cb46. **Lasègue, A.** 1845. Expéditions et voyages dont les collections botaniques sont conservées dans l'Herbier de M. Delessert. [and] Voyages botaniques. In: *Musée botanique de M. Benjamin Delessert*, pp. 74–287, & 359–506. Paris & Leipzig. vi+588 pp. *Re-pub.* (with an introduction by F. A. Stafleu) Lehre, J. Cramer; Codicote, Wheldon & Wesley, 1970. *Historiae naturalis classica* vol. 81.
Both sections are geographically arranged, the second has a "Liste générale des expéditions et des voyageurs–botanistes dont les itinéraires sont tracés dans les chapitres qui précèdent."

Cb47. **Lemmon, K.** 1968. *The golden age of plant hunters.* London, Phoenix House. x+229 pp. & 20 pl.
I.e. Ca. 1760–1860.

Cb48. **Maguire, B.** 1958. Highlights of botanical exploration in the New World. In: W. C. Steere, ed. *Fifty years of botany: Golden jubilee volume of the Botanical Society of America*, pp. 209–246. New York, McGraw–Hill Book Co.

Cb49. **Maule, A. F.** 1974. Danish botanical expeditions and collections in foreign continents. *Botanisk tidsskrift* 69: 167–205.

Cb50. **Ryerson, K. A.** 1967. The history of plant exploration and introduction in the United States Department of Agriculture. In: Escuela Agricola Panamericana. *Proceedings of the International Symposium on Plant Introduction ... November 30 – December 2, 1966*, pp. 1–19. Tegucigalpa.
Appendix: Principal expeditions of the USDA, pp. 16–19.

Cb51. **Schildt, G.** 1960. *Resenärer i floras rike, [etc.]*. Stockholm, Biblioteksförlaget. 316 pp.
Kalm, Hasselquist, Thunberg, etc.

Cb52. **Spongberg, S. A.** 1990. *A reunion of trees: The discovery of exotic plants and their introduction into North American and European landscapes*. Cambridge, MA, Harvard Univ. Pr. xiv+270 pp.

Cb53. **Stearn, W. T.** 1958. Botanical exploration to the time of Linnaeus. *Proceedings of the Linnean Society of London* 169: 173–196.

Cb54. **Veitch, J. H.** 1906. *Hortus Veitchii: A history of the rise and progress of the nurseries of Messrs James Veitch and Sons, together with an account of the botanical collectors and hybridists employed by them and a list of the most remarkable of their introductions*. London, J. Veitch & Sons. 542 pp. & 50 pls.

Cb55. **Whittle, M. S. Tyler–.** 1988. *The plant hunters: Being an examination of collecting with an account of the careers and methods of those who have searched the world for wild plants*. New York, P A J Publ. xii+281 pp. & 22 pp. of illus. First pub. 1970.

See also Biographies Ba745–746 Ba759 Ba786–787 Ba1587–1588 Ba1627; Ca188(Estrella) Cb134(Davy de Virville & Schnell) Cb190A.

BIBLIOGRAPHIES OF BOTANY — GENERAL

• For fuller lists of bibliographies see:

Cb56. **Bay, J. C.** 1910. Bibliographies of botany. A contribution toward a bibliotheca bibliographica. *Progressus rei botanicae* 3: 331–456.

Cb57. **Cole, G.** 1952. *A contribution toward a bibliography of botanical bibliographies*. Madison, WI. 59 pp.

Cb58. **Garnier, G.** 1958. Botanique et physiologie végétale. In: L.–N. Malclès, ed. *Les sources du travail bibliographique*, vol. 3: Bibliographies spécialisées (sciences exactes et techniques), pp. 391–425. Geneva, E. Droz; Paris, Minard. *Re-pub.* 1965.

Cb59. **Swift, L. H.** 1970. *Botanical bibliographies: A guide to bibliographic materials applicable to botany*. Minneapolis, MN, Burgess. xxxvii+804 pp. *Re-pub.* Koenigstein, Koeltz, 1974.

• For a bibliography of earlier histories see:

Cb60. **Johnson, D. E.** 1985. Literature on the history of botany and botanic gardens, 1730–1840: A bibliography. *Huntia* 6: 3–121.

• A selection of major works:

Cb61. **Blake, S. F., & A. C. Atwood.** 1942. *Geographical guide to floras of the world: An annotated list prepared with special reference to useful plants and common plant names*. Part 1, *Africa, Australia, North America, South America, and Islands of the Atlantic, Pacific and Indian Oceans*. Washington, DC, United States Department of Agriculture. 336 pp. *Miscellaneous publication* no. 401. *Re-pub.* Koenigstein, Otto Koeltz Science Publishers, 1974.

Cb62. **Blake, S. F.** 1961. *Geographical guide to floras of the world: An annotated list prepared with special reference to useful plants and common plant names*. Part II, *Western Europe. Finland, Sweden, Norway, Denmark, Iceland, Great Britain with Ireland, Netherlands, Belgium,*

Luxenbourg, France, Spain, Portugal, Andorra, Monaco, Italy, San Marino, and Switzerland. Washington, DC, United States Department of Agriculture. 742 pp. *Miscellaneous publication* no. 797. *Re-pub.* Koenigstein, Otto Koeltz Science Publishers, 1974.

Cb63. **Davis, S. D., et al.** 1986. *Plants in danger: What do we know?* Cambridge, I U C N Conservation Monitoring Centre. 488 pp. ••Geographical survey with extensive bibliographies.

Cb64. **Dierbach, J. H.** 1831. *Repertorium botanicum oder Versuch einer systematischen Darstellung der neuesten Leistungen im ganzen Umfange der Pflanzenkunde.* Lemgo, Meyer. xi+268 pp.

Cb65. **Frodin, D. G.** 1984. *Guide to standard floras of the world: An annotated, geographically arranged systematic bibliography of the principal floras, enumerations, checklists and chronological atlases of different areas.* Cambridge, Cambridge Univ. Pr. xx+619 pp. ••*Replaces* ___. 1964. *Guide to the standard floras of the world (arranged geographically).* Knoxville, TN, Univ. of Tennessee, Department of Botany. iv+59 pp.

Cb66. **Haller, A. von.** 1771–72. *Bibliotheca botanica. Qua scripta ad rem herbariam facienta a rerum initiis recensentur.* Zurich, Orell, Gessner, Fuessli, et Socc. 2 vols. *Re-pub.* New York, Johnson Reprint Corporation, 1967; Hildesheim, G. Olms, 1969. *Also* **C. T. de Murr.** 1805. *Adnotationes ad Bibliothecas Hallerianas botanicam, anatomicam, chirurgicam et medicinae practicae, cum variis ad scripta M. Serveti pertinentibus.* Erlangen. [iii]+67 pp.; **J. C. Bay.** 1908. *Bibliotheca botanica auctore Alberto de Haller. Index emendatus.* Bern, Benteli & Son. [v]+57 pp. Bay's index reprinted with 1967 Johnson Reprint Corporation edition of Haller. ••Ca. 15,000 entries; Murr added ca. 300 titles.

Cb67. **Jackson, B. D.** 1881. *Guide to the literature of botany; being a classified selection of botanical works, including nearly 6000 titles not given in Pritzel's 'Thesaurus.'* London, Index Society. xl+626 pp. *Index society publication* no. 8. *Re-pub.* New York, Hafner, 1964; Koenigstein, Koeltz, 1974. ••For background information *see* ___. 1880. Remarks on botanical bibliography. *Journal of botany* n.s. 9:167–177; for background information on the compilation of the *Guide.*

Cb68. **Junk, W.** 1909. *Bibliographia botanica.* Berlin, Junk. xviii+268 pp. *Continued* ___. 1916. *Bibliographiae botanicae supplementum.* Berlin, [etc.]. pp. 289–1052. ••First vol. includes: Die botanische Literatur vom bibliographische Standpunkte, pp. ii–xv.

Cb69. **Krüger, M. S.** 1841. *Bibliographia botanica. Handbuch der botanischen Literatur in systematischer Ordnung nebst kurzen biographischen Notizen über die botanischen Schriftsteller. Zum Gebrauche für Freunde und Lehrer der Pflanzenkunde.* Berlin, Haude & Spener. vi+464 pp.

Cb70. **Lipshits, S. Yu.** 1974. Bibliografiya trudov russkikh po floram zarubezhnykh stran (1725–1973). Index scriptorum auctorum rossicarum floram regionum extranearum spectantium (1725–1973). *Byulleten' Moskovskogo Obshchestva Ispytatelei Prirody, otdel biologicheskii* 79: 145–154.

Cb71. **Linnaeus, C.** 1751. *Bibliotheca botanica recensens libros plus quam mille de plantis huc usque editos.* Editio altera, priori longe auctior & emendatior. Amsterdam, Schouten. [xvi]+220+[13] pp. *Continued* ___, Praeses. 1759. *Auctores botanici, in dissertatione propositi.* A. Loo, Resp. Uppsala. [ii]+20 pp. *Re-pub.* Stockholm, Rediviva, 1970. ••Ca. 2,000 selected works, a "... classification of books according to their presumed purpose and the varying degrees to which they attained it ..." (Heller). Index auctorum. Supplemented by the 1759 dissertation which is an alphabetical list of over 350 authors and their works, with geographical and classified indexes. *Replaces* 1st ed., 1736, [xvi]+153+[19]+36 pp. *Re-pub.* Munich, Fritsch, 1968. *Historiae scientiarum elementa* fasc. 4; and Ed. nova, 1747, [xvi]+220+[13] pp. For background information *see* **J. L. Heller.** 1970. Linnaeus's Bibliotheca Botanica. *Taxon* 19: 363–411.

Cb72. **Lloyd Library, Cincinnati.** 1914–18. Bibliography relating to botany, exclusive of floras. *Bibliographical contributions from the Lloyd Library* 2(2)–(12): 125–795; 3(1)–(7): 1–338.

Cb73. **Miltitz, F. A.** 1829. *Bibliotheca botanica secundum botanices partes, locos, chronologiam, formam, auctores, volumen, titulos, pretium et recensiones. Handbuch der botanischen Literatur für Botaniker, Bibliothekare, Buchhändler und Auctionatoren, [etc.].* Berlin, A. Rücker. vii pp.+543 coll.

Cb74. **Montalbani, O.** 1657. *Bibliotheca botanica, seu herbaristarum scriptorum promota synodia; cui accessit infividualis gramium omnium ab auctoribus hucusque observatorum numerosissima nomenclatura, Ia. Antonio Bumaldo ... collectore.* Bologna, heirs of V. Bennacci. 188 pp. ••The first general botanical bibliography.

Cb75. **Pritzel, G. A.** 1872 [1871–77]. *Thesaurus literaturae botanicae omnium gentium inde a rerum botanicarum initiis ad nostra usque tempora. Quindecim millia operum recensens.* Edit. novam reformatam. Leipzig, Brockhaus. v+577 pp. *Re-pub.* Milan, Görlich, 1950 and Koenigstein, Koeltz Antiquariat, 1972. ••The standard bibliography; 10,871 entries. *Also* **E. A. Zuchold.** 1853–66. Additamenta ad Georgii Augusti Pritzelii Thesaurum literaturae botanicae. *Jahresbericht des Naturwissenschaftlichen Vereines in Halle* 5: 505–582, 1853; ___. 1866. ___: Fasc. 2. *Zeitschrift für die gesammten Naturwissenschaften* 27: 122–134, 1866. *Also pub. separately* Halle. 59 pp.; & Leipzig. 16 pp.; **E. von Berg.** 1859. *Additamenta ad Thesaurum Literaturae Botanicae. Index librorum botanicorum Bibliothecae Horti Imperialis Botanici Petropolitani quorum inscriptiones in G. A. Pritzelii Thesauro Literaturae Botanicae et in Additamentis ad Thesaurum illum ab E. A. Zuchold editis desiderantur.* Halle. 40 pp. According to Jackson (1881, p.3) "Although Pritzel acknowledges the help he had received from this author (Thes. Ed. 2, p.23), he does not enumerate these additions."; **O. Kuntze.** 1891–93. Notizen zu Pritzel's Thesaurus literaturae botanicae. In: *Revisio generum plantarum,* Pars I, pp. cxxii–cxlvi, and Weitere Notizen zu Pritzel's Thesaurus literaturae botanicae. In: *Ibid.*, Pars III, pp. 153–162. Leipzig, [etc.], Felix. For background information *see* **F. A. Stafleu.** 1973. Pritzel and his Thesaurus. *Taxon* 22: 119–130. *Replaces* ___. 1851. *Thesaurus literaturae botanicae* ... Leipzig, [etc.]. viii+547 pp.

Cb76. **Rehder, A.** 1911–18. *The Bradley bibliography: A guide to the literature of the woody plants of the world published before the beginning of the twentieth century.* Cambridge, MA, Riverside Pr. 5 vols. *Publication of the Arnold Arboretum* no. 3. ••Vols. 1–2: Dendrology; vol. 3: Arboriculture; vol. 4: Forestry; vol. 5.

Cb77. **Reuss, J. D.** 1801–21. *Repertorium commentationum a societatibus litterariis editarum.* Göttingen, Dieterich. 16 vols. *Re-pub.* New York, B. Franklin, 1961. *Burt Franklin bibliography & reference series* no. 29. ••Vol. 1, tom II: Botanica et mineralogia, 1802.

Cb78. **Royal Botanic Gardens, Kew.** 1889. Guide to the botanical literature of the British Empire. *Bulletin of miscellaneous information, Royal Gardens, Kew* 31: 153–189.

Cb79. **Séguier, J. F.** 1740 [i.e. 1760]. *Bibliotheca botanica, sive catalogus auctorum et librorum omnium, qui de re botanica, de medicamentis ex vegetabilibus paratis, de re rustica, et de horticultura tractant* ... Accessit Bibliotheca botanica Jo. Ant. Bumaldi, seu potius Ovidii Montalbani Bononiensis. Nec non auctuarium in Bibliothecam botanicam cl. Seguierii opera Laur. Theod. Gronov. Leiden, C. Haak. xvi+450+66[i.e. 74] pp. & [ii]+65+[2] pp. ••A re-publication of the ed. of 1740 with the addition of Gronovius's Auctuarium. The title-page incorrectly preserves the date of the first ed., 1740.

Cb80. **Stafleu, F. A., & R. S. Cowan.** 1976–88. *Taxonomic literature: A selective guide to botanical publications and collections with dates, commentaries and types.* 2nd edn. Utrecht & Antwerp, Bohn, Scheltema & Holkema; The Hague, Boston, MA, W. Junk, for the International Association for Plant Taxonomy. 7 vols. *Regnum vegetabile* vols 94, 98, 105, 110, 112, 115, & 116; *Continued* **F. A. Stafleu & E. A. Mennega.** 1992+. ___*: Supplement.* Koenigstein, Koeltz. In progress (vol. 1, A–Ba. 1992). *Regnum vegetabile* vols 125, [etc.]. ••Known as "TL–2." A bibliographical catalogue of pre-1939 books, pamphlets, series, and other bibliographically independent and limited entities. For background information *see* **F. A. Stafleu & R. S. Cowan.** 1979. The making of a book: An interim report on TL–2. *Taxon* 28: 77–86. *Replaces* **F. A. Stafleu.** 1967. *Taxonomic literature* ... Utrecht, International Bureau for Plant Taxonomy and Nomenclature. xx+556 pp. *Regnum vegetabile* vol. 52. *Also* **J. A. Mears**, ed. 1989. *Plant taxonomic literature: Bibliographic guide.* Cambridge, Chadwyck–Healey. [iii]+177 pp. Catalogue of the microfiche editions of 4,679 items selected from the "TL–2" guide and offered for sale by Chadwyck–Healey.

Cb81. **Warner, M. F., & J. S. Brown.** 1939. *Early horticultural literature: A check list of works published in the 16th, 17th and 18th centuries, with other materials for the history and bibliography of horticulture.* Washington, DC. [ii]+15+911+143 ff. ••Typescript of an author catalogue with title index, the latter including full entries for anonymous works. (Original document in the United States Department of Agriculture, National Agricultural Library, Washington, DC. Copy in Hunt Institute for Botanical Documentation, Pittsburgh, PA.

Cb82. **World Conservation Monitoring Unit, Royal Botanic Gardens, Kew.** 1990. *World plant conservation bibliography.* Kew. 645 pp.

See also Cb876, Library cats. Bc360–426 & (Periodicals lists) Bc628–640.

Herbals

Cb83. **Nissen, C.** 1956. *Kräuterbücher aus fünf Jahrhunderten. Medizinhistorischer und bibliographischer Beitrag.* Zurich, L'Art Ancien; Munich, R. Wölfle; Olten, Weiss–Hesse. x+83 pp. & 50 leaves. ••With leaves from 50 German, French, Dutch, English, Italian, and Swiss herbals. *Also* ___. 1958. *Herbals of five centuries. A contribution to medical history and bibliography.* (Translated by W. Bodenheimer & A. Rosenthal). Los Angeles, CA, Dawson. x+91 pp.

Cb84. **Rohde, E. S.** 1922. [A bibliography of manuscript and printed herbals.] In: *The old English herbals*, pp. 189–235. London, Longmans Green. xii+243 pp. *Re-pub.* New York, Dover Publications, 1971.

Cb85. **Schmid, A.** 1939. *Ueber die alte Kräuterbücher.* Berne & Leipzig, P. Haupt. [viii]+76 pp. & 5 pl. ••Ca. 100 entries.

Cb86. **Schreiber, W. L.** 1982. *Die Krauterbücher des XV und XVI Jahrhunderts. Unveranderter Nachdruck der Ausgabe von 1924 zum Gedenken an den Autor anlasslich* *seines 50 Todestages herausgegeben und mit einem Nachwort versehen von Reimar W. Fuchs.* Stuttgart. iv+74 pp. ••125 entries. First pub. in Munich in 1924 with a facsimile of the *Hortus sanitatis*, Mainz 1485.

See also Library cats. Bc523 & Bc547.

Plant groups

• A selection of major works:

Cb87. **Dietrich, H.** 1980–81. *Bibliographia Orchidacearum: Aussereuropäische Arten.* Band 2: *Literaturauswahl der Jahre 1900–1978.* Teil 1: A–K; Teil 2: L–Z. Jena 62+158 pp.; 320 pp. *Bibliographische Mitteilungen der Universitätsbibliothek Jena* no. 30: 1 & 2.; Updated by periodic supplements, e.g. Bande 3–6, pub. 1985–90, covering the literature of 1979–81, 1982–84, 1984–86, & 1987–1990. Each includes addenda to the previously pub. lists.

Cb88. **Farjon, A.** 1990. *A bibliography of conifers: Selected literature on taxonomy and related disciplines of the Coniferales, and especially of the families Cupressaceae (with Taxodiaceae) and Pinaceae.* Königstein, Koeltz Scientific Books. 124 pp. *Regnum vegetabile* vol. 122.

Cb89. **Kostermans, A. J. G. H.** 1964. *Bibliographia Lauracearum.* Bogor, Ministry of National Research. xvi+1,450 pp.

Cb90. **Rollet, B.** 1981. *Bibliography on mangrove research 1600–1975.* Paris, UNESCO. xxviii+479 pp.

Cb91. **Schulze, W., & H. Vogt.** 1989. *Liliifloren: Internationale Auswahlbibliographie, 1898–1986.* Jena, Universitätsbibliothek Jena. 181 pp. *Bibliographische Mitteilungen der Universitätsbibliothek Jena* no. 48.

Cb92. **Stock, K. L.** 1984. *Rose books: A bibliography of books and important articles in journals on the genus* Rosa, *in English, French, German and Latin, 1550–1975.* Codicote & Milton Keynes, Wheldon & Wesley. x+533 pp.

Cb93. **Weinert, M.** 1989. *Bibliographie über Vermehrung, Kultur und Mykorrhiza terrestrischer Orchideen (insbesondere der europäischen Arten) 1804–1988.* Koenigstein, Koeltz Scientific Books. 223 pp.

Cb94. **Willing, B. & E.** 1977. Bibliographie über die Orchideen Europas und der Mittelmeerländer 1744–1976: Spezielle Orchideen–Literatur ohne Bercksichtigung allgemeiner floristischer Werke. *Willdenowia Beiheft* 11: 325 pp.; *Continued* **B. Willing.** 1985. ___: Supplement. Spezielle Orchideen–Literatur ohne Berücksichtigung allgemeiner floristischer Werke. *Englera* 5: 280 pp.

Dissertations

Cb95. **Inter–University Board of India.** 1975. *Botany: A bibliography of doctoral dissertations accepted by Indian universities, 1857–1970.* New Delhi. x+146 pp.

Cb96. **Johnson, D.** 1983. Palm literature: A bibliography of graduate theses on palms. *Principes* 27: 85–88.

Cb97. **Kleppa, P.** 1982. Hovedoppgaver i botanikk ved norske universiteter 1909–1979. *Blyttia* 40: 1–40. ••Cand. real and mag. scient. theses in botany at the Norwegian universities 1909–1979.

Cb98. **Watt, A. D.** 1979. *North American paleobotanical dissertations, 1857–1977.* Missoula, MT, Botanical Society of America, Paleobotanical Section. 38 pp.

Periodical bibliographies

A selection of major works:

Cb99. ***Archiv für Naturgeschichte.*** *[Section:] Bericht über die Leistungen in der Pflanzengeographie (1845+: ... und systematischen Botanik; 1848–53+: ... geographischen und systematischen Botanik).* <1840–53>. Berlin, 1841–55. Re-pub. with separate title–pages, 1843–53.

Cb100. ***Års–berättelse om botaniska arbeten och***

upptäckter. <1828–44> Stockholm, Kongliga Svenska Vetenskaps–Akademien, 1828–44. 13 vols.

Cb101. ***Årsberättelse om framstegen uti botanik.*** <1826–27> Stockholm, Kongliga Svenska Vetenskaps–Akademien, 1826–27. 2 vols. Continued as *Års–berättelse om botaniska arbeten och upptäckter.*

Cb102. ***Årsberättelser om vetenskapernas framsteg.*** <1821–25> Stockholm, Kongliga Svenska Vetenskaps–Akademien, 1822–25. 5 vols. Continued as *Öfversigt af botaniska arbeten och upptäckter.*

Cb103. ***Botanical abstracts:*** *A monthly serial furnishing abstracts and citations of publications in the international field of botany in its broadest sense.* <1918–26>. Baltimore, MD, Williams & Wilkins, vols. 1–15, 1918–26. Continued as *Biological abstracts.*

Cb104. ***Botanische Jahrbücher für Systematik, Pflanzengeschichte und Pflanzengeographie.*** *[Section:] Übersicht der wichtigeren und umfassenden, im Jahre ... über Systematik, Pflanzengeschichte und Pflanzengeographie erschienen Arbeiten.* <1880–1959?>. Leipzig, W. Engelmann (*later* Stuttgart, Schweizerbart), vols. 1–78(4), 1884–1959.

Cb105. ***Botanischer Jahresbericht,*** *systematisch geordnetes Repertorium der botanischen Literatur aller Länder.* <1873–82> Berlin, Gebrüder Borntrăger, vols. 1–10, 1873–82. Continued as *Just's botanischer Jahresbericht.*

Cb106. ***Botanisches Centralblatt (Zentralblatt):*** *Referiendes Organ für das Gesamtgebiet der Botanik des In– und Auslandes* (subtitle varies). Cassel, Jena, G. Fischer, vols. 1–179(8), 1880–1945. (Section: Neue Literatur. <1902–44>, vols. 91–179(8) [pub. suspended 1920–21]). Continued as *Excerpta botanica. Also* ___. *General–Register über Band 1 bis 60.* Cassel, [etc.], 1903. 667 pp.

Cb107. ***Excerpta botanica.*** Sectio A: *Taxonomica et chorologica.* <1959+>. Stuttgart, G. Fischer, vol. 1+, 1959+. *Also* **Boivin, B.** 1976. *Biography index to* Excerpta botanica, Sectio A, *volumes 1–20, 1959–1973.* Ottawa. 38 pp.

Cb108. ***Excerpta botanica.*** Sectio B: *Sociologica.* <1959+>. Stuttgart, G. Fischer, 1959+.

Cb109. ***Flora:*** *Oder allgemeine botanische Zeitung. [Section:] Repertorium der periodischen botanischen Literatur. Als Beiblatt zur Flora.* Pts. 1–8.) <1864–71>. Jena, Regensburg, Königl. Botanischen Gesellschaft in Regensburg, pts. 1–8, 1864–71.

Cb110. ***Fortschritte der Botanik.*** <1931+> Berlin [etc.], Springer–Verlag, vol. 1+, 1932+.

Cb111. ***International catalogue of scientific literature.*** *[Section:] M. Botany.* <1901–14>. London, Royal Society of London for the International Council, 1902–21. 14 vols. Reprinted New York, Johnson Reprint Corporation, 1968, 14 vols in 3.

Cb112. ***Just's botanischer Jahresbericht:*** *Systematisch geordnetes Repertorium der botanischen Litteratur aller Länder.* <1886–1935>. Berlin, Leipzig, Gebrüder Bornträger, vols. 11–67 [vol. 67 incomplete], 1886–1944. ••Divided into two parts with vol. 6, 1879; divided into three parts with vol. 33, 1905. Part 2 terminated 1935; part 3 terminated 1927.

Cb113. ***Kew record of taxonomic literature relating to vascular plants for 1971*** [+]. London, H.M.S.O. for Royal Botanic Gardens, Kew, 1974+.

Cb114. ***Öfversigt af botaniska arbeten och upptäckter.*** <1825> Stockholm, Kongliga Svenska Vetenskaps–Akademien, 1825. 1 vols. Continued as *Årsberättelse om framstegen uti botanik.*

Cb115. ***Repertorium annuum literaturae botanicae periodicae.*** <1872–79>. Haarlem, Loosjes, vols. 1–8, 1873–86.

REGIONAL HISTORIES OF BOTANY

European (& Mediterranean)

Cb116. **Amigo, J.–J.** 1985. Les débuts de l'histoire de la botanique dans les Pyrénées–Orientales (1570–1800). *Comptes rendus du 110 Congrès National des Sociétés Savantes, Section d'Histoire des Sciences [etc.]*, tom. 1: 9–20.

Cb117. **Borzì, A.** 1906. Botanica e botanici in Sicilia nel secolo XVIII. *Bollettino del R[eale] Orto Botanico di Palermo* 5: 3–21.

Cb118. ___. 1912. *Elenco alfabetico degli autori che si occupano della Libia sotto l'aspetto botanico ed agrario e delle loro pubblicazioni.* Rome, Ministerio degli Affari Esteri. 32 pp. *Monografie* no. 9. *Continued* ___. 1912. ___*: Secundo elenco.* Rome, [etc.] 30 pp.

Cb119. **Botanical Society of the British Isles.** 1986. The long tradition. The botanical exploration of the British Isles. Ed. by H. J. Noltie. *Scottish naturalist* 98: 191 pp. *Botanical Society of the British Isles conference report* no. 20.
Includes **G. T. L. Chapman.** William Turner of Morpeth, Northumberland (1508–1568), pp. 11–27; **F. H. Brightman.** Thomas Johnnson's "Iter plantarum investigationis ... in agrum cantianum", pp. 29–41; **W. T. Stearn.** John Ray's natural history trends in Britain, pp. 43–57; **A. A. Slack.** Lightfoot and the exploration of the Scottish flora, pp. 59–76; **S. R. J. Woodell.** Oxfordshire botanists and the discovery of the British flora, pp. 77–96; **J. G. Roger.** George Don, 1764–1814, pp. 97–108; **R. G. Ellis.** Welsh botanists of the 16th and 17th centuries, pp. 109–122; **D. A. Webb.** The hey–day of Irish botany, 1866–1916, pp. 123–134; **A. J. Byfield.** The Lizard flora: A history of discovery, pp. 135–145; **G. Halliday.** The discovery of lakeland flora, pp. 147–156; **M. A. Taylor.** Francis Buchanan White (1842–1894) and Scottish botany, pp. 157–173; **D. E. Allen.** The discoveries of Druce, pp. 175–189. *Also* **E. C. Nelson.** 1987. Scottish connections in Irish botany and horticulture. *Scottish naturalist* 1987: 3–31. Supplementary to the above work.

Cb120. **Bower, F. O.** 1938. *Sixty years of botany in Britain (1875–1935): Impressions of an eye–witness.* London, Macmillan. 112 pp.

Cb121. **Brandes, D.** 1988. *400 Jahre Erforschung der Flora von Niedersachsen.* Braunschweig, Universitätsbibliothek der Technischen Universität. 61 pp. *Veröffentlichungen der Universitätsbibliothek Braunschweig* no. 2.

Cb122. **[Brito, A.]** 1883. *Historia de botanica em Portugal.* Lisbon, D. Corazzi 63 pp. *Bibliotheca do povo e das escolas* no. 65.

Cb123. **Broberg, G.** 1991. Fruntimmersbotaniken. [Botany for women.]. *Svenska Linnésällskapets årsskrift* 1991–1991: 177–231.
With a brief English summary. "The focus is the influence of Linnaeus, ... This study, ending at the middle of the nineteenth century, finally, puts the question whether botany for women has liberated women and in what way botany for women might have affected the development of botany during the Linnean period."

Cb124. **Brown, P. S.** 1982. Herbalists and medical botanists in mid-nineteenth century Britain with special reference to Bristol. *Medical history* 26: 405–420.

Cb125. **Camarasa, J. M.** 1989. *Botanica i botanics dels paisos Catalans.* Barcelona, Enciclopedia Catalana. 268 pp. *Biblioteca universitaria* no. 10.
In Catalan.

Cb126. **Candolle, A. P. de.** 1832. Histoire de la botanique genévoise. *Mémoires de la Société de Physique et d'Histoire Naturelle de Genève* 5: 1–61.

Cb127. **Christ, H.** 1915. Die ersten erforscher der schweizerischen Alpenflora im XVI Jahrhundert: C. Gesner, B. Aretius, Joh. Fabricius, etc. und ihre Ergebnisse. *Schweizerische Apotheker–Zeitung* 53: 344–349, 357–361.

Cb128. **Christensen, C.** 1924–26. *Den Danske botaniks historie med tilhørende bibliografi.* Copenhagen, Hagerups. 3 vols.
1. Den danske botaniks historie fra de ældste Tider til 1912; 2. Bibliografi.

Cb129. **Collander, R.** 1965. *The history of botany in Finland 1828–1918.* With an appendix on forest science by Y. Ilvessalo. Translated by D. Barrett. Helsinki, Societas Scientiarum Fennica. 159 pp. *The history of learning and science in Finland 1828–1918* vol. 8.

Cb130. **Colmeiro, M.** 1858. *La botánico y los botánicos de la península Hispano–Lusitana. Estudios bibliográficos y biográficos.* Madrid, M. Rivadeneyra. [xi]+216 pp.

Cb131. ___. 1870. *Exámen histórico de los trabajos concernientes á la flora Hispano–Lusitana. (Fragmenta que alcanza hasta fines del siglo XVI).* Madrid, T. Frey. 86 pp.

Cb132. **Combes, R.** 1933. *Histoire de la biologie végétale en France.* Paris, F. Alcan. [iii]+172 pp. *Bibliothèque de philosophie contemporaine.*

Cb133. **Cosson, E. St.–C.** 1881. Notice sur les voyages et les explorations de l'Algérie, de la Tunisie et du Maroc. In: *Compendium florae Atlanticae ... ou flore des états barbaresques Algérie, Tunisie et Maroc,* vol. 1: 7–101. Paris, Imprimerie Nationale.

Cb134. **Davy de Virville, A., et al.** 1954. *Histoire de la botanique en France.* Publié par le Comité Français du VIIIe Congrès International de Botanique, Paris–Nice, 1954. Paris, Société d'Edition d'Enseignement Supérieur. 395 pp.
1e partie: **A. Davy de Virville.** Botanique des temps anciens; 2e partie: Botanique des temps modernes – **M. Hocquette.** Morphologie, anatomie, cytologie; **J. Carles.** Physiologie végétale; **A. Guillaumin.** Phanérogamie; Cryptogamie – **R. Gaume.** Bryologie; **J. Feldmann.** Phycologie; **M. Chadefaud.** Mycologie; **H. des Abbayes.** Lichénologie; **P. Jovet.** Flore et phytogéographie de la France; **J. Bustarret.** Botanique agronomique; **P. Guinier.** Botanique forestière; **G. Viennot-Bourgin.** Pathologie végétale; **R. Paris.** Botanique médicale; **E. Rivalier.** Mycologie médicale; **A. Davy de Virville.** La notion d'éspèce et d'évolution; **E. Boureau.** Paléobotanique; Mme **Van Campo.** Palynologie; **A. Davy de Virville & R. Schnell.** XIXe siècle: Voyages et explorations botaniques.

Cb135. **Dobat, K., et al.** 1975. 300 Jahre Botanik in Tübingen. *Attempto* 55/56: 8–49.

Cb136. **Dokhman, G. I.** 1973. *Istoriya geobotaniki v Rossii.* Moscow, Izd. "Nauka." 286 pp. *Materialy k poznaniyu fauny i flory SSSR, otdel botanicheskii* no. 17.

Cb137. **Egle, K., & G. Rosenstock.** 1966. *Die Geschichte der Botanik in Frankfurt am Main: Ein Vermächtnis von Dr. Johann Christian Senckenberg.* Frankfurt am Main, Umschau. 100 pp.

Cb138. **Eisnerová–Spudliová, V. J. B.** 1969. *Botanika v Ceskych zemích 2. poloviny 19. století. [Botanical research in Bohemia in the second half of the 19th century].* Prague, Oddeleni pro Dejiny Prírodních Ved a Techniky, Historického Ustavu C S A V. 136 pp. *Práce z dejin prírodních ved* no. 1.

Cb139. **Emberger, L., & R. Maire.** 1934. L'exploration botanique du Maroc du XVIIe au XIXe siècle. *Archeion* 16: 168–180.

Cb140. ___. 1934. Histoire de nos connaissances botaniques sur le Maroc. *Mémoires de la Société des Sciences Naturelles du Maroc* 38: 5–35.

Cb141. **Eriksson, G.** 1969. *Botanikens historia i Sverige intill år 1800.* Stockholm, Almqvist & Wiksell. 370 pp. *Naturvetenskapskapernas historia i Sverige intill år 1800* vol. 3. *Lychnos bibliotek* 17: 3.

Cb142. **Fernandes, A.** 1986. História da botânica em Portugal até finais do século XIX. In: Academia das Ciências de Lisboa. *Colóquio sobre a história e desenvolvimento da ciência em Portugal. I Colóquio – até ao século XX. Lisboa, 15 a 19 de Abril de 1985.* Lisbon. 2 vols. (1,407 pp.).

Cb143. **Fischer, E.** 1914. Botanik und Botaniker in Bern. *Verhandlungen der schweizerischen naturforschenden Gesellschaft* 2: 3–28.

Cb144. **Fitting, H.** 1896. Geschichte der hallischen Floristik. *Zeitschrift für die gesammten Naturwissenschaften* 69: 289–386.

Cb145. **Flerov, A. F.** 1907. Okskaya flora. *Trudy Imperatorskago S.–Peterburgskago Botanicheskago Sada* 27 (1): xvi+286 pp.
Includes Pt. 2. Istoriya issledovaniya Okskoi flory, pp. 153–286.

Cb146. **Fries, R. E.** 1950. *A short history of botany in Sweden.* Uppsala, Almqvist & Wiksell for the Seventh International Botanical Congress, Stockholm. 162 pp.
With contributions by K. V. O. Dahlgren, A. Müntzing, B. Åberg and H. Osvald.

Cb147. **Froment, P.** 1953. La phytogéographie et la phytosociologie en France du début du XIXe siècle à nos jours. *Bulletin de la Société Botanique de France* B100: 362–387.

Cb148. **Funke, H.** 1965. Beiträge zur Geschichte der Floristik in Sachsen. *Berichte der Arbeitsgemeinschaft sächsischer Botaniker* 5–6: 447–472.

Cb149. **Gérard, M.** 1896. *La botanique à Lyon avant la révolution et l'histoire du jardin municipal de cette ville.* Paris, Masson & Cie. [v]+96 pp. *Annales de l'Université de Lyon* no. 23.

Cb150. **Gerber, C.** 1925. Les jardins botaniques toulousains et l'étude de la flore pyrénéenne, sous l'ancien régime et la révolution, d'après des documents inédits. *Bulletin de la société Botanique de France* 71: 788–842.

Cb151. **Green, J. R.** 1914. *A history of botany in the United Kingdom from the earliest times to the end of the 19th century.* London, Dent. xii+648 pp.

Cb152. **Gunther, R. T.** 1922. *Early British botanists and their gardens, based on unpublished writings of Goodyer, Tradescant, and others.* Oxford, privately pub. viii+417 pp., illus. *Re-pub.* New York, Kraus Reprint, 1971.

Cb153. **Henrey, B.** 1975. *British botanical and horticultural literature before 1800. Comprising a history and bibliography of botanical and horticultural books printed in England, Scotland, and Ireland from the earliest times until 1800.* London, Oxford Univ. Pr. 3 vols.
Detailed, plentifully illustrated, literary history covering herbals, early scientific and floristic works, gardening books, floras (of Britain and foreign areas), works associated with botanical and other gardens, horticultural trades, garden design and some minor subjects.

Cb154. **Hryniewiecki, B.** 1948. *Rozwój botaniki w Polsce.* Cracow, Polska Akademia Umiejetnosci. 53 pp. *Polska Akademia Umiejetnosci, historia nauki Polskiej w monografiach* no. 8.
Replaces ___. 1931. *Zarys historji botaniki w Polsce. Précis de l'histoire de la botanique en Pologne.* Warsaw, Société Botanique en Pologne. 45 pp.

Cb155. ___. 1948. Pierwsze flory okolic Warszawy. [First floristic monographs on the environments of Warsaw]. *Monographiae botanicae* 2: 76 pp.

Cb156. **Junker, T.** 1989. *Darwinismus und Botanik: Rezeption, Kritik und theoretische Alternativen im Deutschland des 19. Jahrhunderts.* Mit einem Geleitwort von R. Schmitz. Stuttgart, Deutscher Apotheker–Verlag. x+367 pp. *Quellen und Studien zur Geschichte der Pharmazie* vol. 54.

Cb157. **Kanitz, A.** 1863. *Geschichte der Botanik in Ungarn. (Skizzen).* Hannover, W. Riemschneider; Pest, Mor. Rath. iii+199 pp.

Cb158. ___. 1865. Versuch einer Geschichte der Ungarischen Botanik. *Linnaea* 33: 401–664. *Also pub. separately* Halle, Gebauer Schwetschké'sche Buchdruck. iv+264+1 pp.

Cb159. **Knuth, P. E. O. W.** 1890. *Geschichte der Botanik in Schleswig–Holstein.* Kiel & Leipzig, Lipsius & Tischer. 216 pp.

Cb160. **Kobrle, A.** 1965. *Dejiny botanického vyzkumu vychodních Cech. [The history of botanical works of eastern Czechs].* Pardubice, Vychodoceske Muzeum. 34 pp.

Cb161. **Kosiek, Z.** 1971. Badania nad historia botaniki w Polsce w latach 1945–1969. [Research concerning the history of botany in Poland, 1945–1969]. *Kwartalnik historii nauki i techniki* 16: 577–594.

Cb162. **Lacroix, A.** 1941. *Notice historique sur les membres et correspondants de l'Académie des Sciences ayant travaillé dans l'Afrique du Nord française depuis le XVIIIe siècle. Les botanistes.* Paris, Gauthier–Villars. 58 pp. *Mémoires de l'Académie des Sciences de l'Institut de France* vol. 64.

Cb163. **Lefort, F. L.** 1949. Contribution à l'histoire botanique du Luxembourg. *Bulletin de la Société des Naturalistes Luxembourgeois* n.s. 43: 31–160 & 18 pl.

Cb164. **Legré, L.** 1899. *La botanique en Provence au XVIe siècle. Pierre Pena et Mathias de Lobel.* Marseille, Aubertin & Rolle. xii+263 pp.

Cb165. **___.** 1899. *La botanique en Provence au XVIe siècle. Hugues de Solier.* Marseille, Aubertin & Rolle. 45 pp.

Cb166. **___.** 1900. *La botanique en Provence au XVIe siècle. Félix et Thomas Platter. Avec extraits relatifs à la Provence, des mémoires de Félix et de Thomas Platter traduits de l'allemand par M. Kieffer.* Marseille, Aubertin & Rolle. viii+91 pp.

Cb167. **___.** 1900. *La botanique en Provence au XVIe siècle. Léonard Rauwolff, Jacques Raynaudet.* Marseille, Aubertin & Rolle. x+147+[2] pp.

Cb168. **___.** 1901. *La botanique en Provence au XVIe siècle. Louis Anguillara, Pierre Belon, Charles de l'Escluse, Antoine Constantin.* Marseille, Aubertin & Rolle. 193+[2] pp.
With "Appendice: Lettres de N. C. Fabri de Peirese à Charles de l'Escluse."

Cb169. **___.** 1904. *La botanique en Provence au XVIe siècle. Les deux Bauhin, Jean Henri Cherler, et Valerand Dourez.* Marseille, Aubertin & Rolle. xi+117 pp.

Cb170. **Lerond, M.** 1978. Eléments pour une histoire de la botanique en Normandie. *Histoire et nature* 12/13: 111–132.

Cb171. **Lorenzen, H.**, ed. 1988. *Beiträge zur neueren Geschichte der Botanik in Deutschland.* Stuttgart, Fischer. ix+419 pp.
First pub. as *Berichte der Deutschen Botanischen Gesellschaft* vol. 100. Articles by various authors. *Includes* **U. Willerding**, Die Palaeo–Ethnobotanik und ihre Entwicklung im deutschsprachigen Raum; **A. Hüttermann**, History of forest botany (Forstbotanik) in Germany from the beginning in 1800 until 1940 [etc.]; **F. K. Timler & B. Zepernick**, German colonial botany; **H. J. Conert**, Geschichte der Botanik in Naturmuseum und Forschungsinstitut Senckenberg; **C. Hünemörder**, Zur Geschichte der Botanik in Hamburg bis 1945; **L. Kies**, Phykologie in Hamburg, etc.

Cb172. **Lundqvist, Å.** 1981. Från ölandsfloristikens ungdomsår. [Early years of floristic studies in Öland]. *Svenska Linnésällskapets årsskrift* 1975/77: 29–49.

Cb173. **Macdonald, J. A.** 1984. *Plant science and scientists in St. Andrews up to the middle of the 20th century.* [St. Andrews], privately pub. 57 pp.

Cb174. **Maire, R.** 1931. *Les progrès des connaissances botaniques en Algérie depuis 1830.* Paris, Masson & Cie. 253 pp. & 8 pl. *Collection du centenaire de l'Algérie.*

Cb175. **Maiwald, V.** 1901–02. Die Opizische Periode in der floristischen Erforschung Böhmens. *Jahresbericht d. öffentl. Stifts–Obergymnasium d. Benedictiner zu Braunau in Böhmen an Schlusse des Schuljahres* 1901 & 1902: 184 pp.

Cb176. **Maiwald, V.** 1904. *Geschichte der Botanik in Böhmen.* Vienna & Leipzig, C. Fromme. viii+297 pp.

Cb177. **Masalles, R. M.** 1984. La cartografia botanica a Catalunya. *Butlleti de la Institucio Catalana d'Historia Natural* 50: 249–254.

Cb178. **Meshkov, A. P.** 1948. Ocherk istorii izucheniya flory i rastitel'nosti Sredne–Russkoi vozvyshennosti Voronezh. [Sketch of the history of the study of the flora and vegetation of central Russian elevation Voronezh.] *Izvestiya Voronezhskoi Gosudarstvennyi Pedagogicheskii Institut* 9 (2): 24–39.

Cb179. **___.** 1949. Ocherk istorii izucheniya flory i rastitel'nosti Oksko–Donskoi ravniny. [Sketch of the history of the study of the flora and vegetation of the Oksk–Donsk plain.] *Izvestiya Voronezhskoi Gosudarstvennyi Pedagogicheskii Institut* 11 (1): 3–37.

Cb180. **Moskovskoe Obshchestvo Ispytatelei Prirody.** 1947. *Ocherki po istorii Russkoi botaniki. [Essays on the history of botany in Russia].* Ed. by N. A. Komarnitskii. Moscow/Leningrad. 320 pp. & 92 pl.
Includes **S. Yu. Lipshits.** Sistematika, floristikai geografiya rastenii, pp. 6–114; **N. A. Komarnitskii.** Morfologiya rastenii, pp. 115–174; **N. A. Maksimov.** Fiziologiya rastenii, pp. 211–274; **B. L. Isachenko.** Mikrobiologiya, pp. 275–318.

Cb181. **___.** 1972. Kratkaya istoriya izucheniya rastitel'nogo pokrova Moskovskoi oblasti. [Brief history of the study of the vegetation of the Moscow region.] In: *Flora i rastitel'nost Moskovskoi oblasti: Istoriya izucheniya i annotirovannaya bibliografiya*, pp. 10–53. By I. A. Gubanov, B. A. Starostin, & V. N. Tikhomirov. Moscow, Izd. Moskovskogo Universiteta.

Cb182. **Muñoz Medina, D. J. M.** 1969. *Historia del desarrollo de la botanica en España.* Pamplona. 34 pp. Universidad de Novara dissertation.

Cb183. **Nekrasova, V. L.** 1958. *Istoriya izucheniya dikorastushchikh syr'evykh rastenii v SSSR. Historia investigationis plantarum usualium spontanearum florae URSS.* Moscow/Leningrad, AN SSSR. 275 pp. & 9 pl.
History down to 1917. Bibliography, pp. 241–273.

Cb184. **Nelson, E. C.** 1982. The influence of Leiden on botany in Dublin in the early eighteenth century. *Huntia* 4: 133–146.

Cb185. **Nuet, J., & J. M. Panareda.** 1984. Notes historiques sobre l'estudi de la flora i vegetacio de Montserrat. *Butlleti de la Institucio Catalana d'Historia Natural* 50: 115–124.

Cb186. **Opsomer, J. E.** 1972. Cinq cents ans de botanique exotique en belgique (XIIIe–XVIIe siècles). *Bulletin des séances, Académie Royal des Sciences d'Outre–Mer* 1972: 446–479.

Cb187. **Oudemans, C. A. J. A.** 1877–81. De ontwikkeling onzer kennis aangaande de flora van Nederland, mit de bronnen geschetst toegelicht I–IV. *Nederlandsch kruidkundig archief* (2) 2: 214–277, 279–391, 1877; *Ibid.* (2) 3: 1–75, 325–398, 1878–81.

Cb188. **Pedrotti, F.**, ed. 1988. *100 anni di ricerche botaniche in Italia, 1888–1988.* Florence, Società Botanica Italiana. viii+1,124 pp.

Cb189. **Pée–Laby, C. E.** 1895–96. Les botanistes pyrénéennes. *Explorations pyrénéennes* 30: 77–117, 309–320, 1895; *Ibid.* 31: 33–49, 1896.

Cb190. **Pop, E.** 1971. La période Linnéenne chez les botanistes en Roumaine. *Actes de XIIe [1968] Congrès International d'Histoire des Sciences* 8: 159–193.

Cb190A. **Priszter, Sz.** 1955. Ausländische Forschungstätigkeit der ungarischenn Botaniker. *Acta botanica Academiae Scientiarum Hungaricae* 1: 267–300.

Cb191. **Puerto Sarmiento, F. J.** 1988. *La ilusión quebrada: Botánica, sanidad y política científica en la España ilustrada.* Barcelona & Madrid, Ediciones del Serbal y Consejo Superior de Investigaciones Científicas. 315 pp. *Libros del Buen Andar* no. 22.

Cb192. **Pulteney, R.** 1790. *Historical and biographical sketches of the progress of botany in England, from its origin to the introduction of the Linnaean system.* London, T. Cadell. 2 vols.
Other eds: ___. 1798. *Geschichte der Botanik bis auf die neuern Zeiten, mit besonderer Rücksicht auf England [etc.].* Translated by C. G. Kühn. Leipzig, Weygand. 2 vols; ___. 1809. *Esquisses historiques et biographiques des progrès de la botanique en Angleterre, depuis son origine jusqu'à l'adoption du système de Linnée [etc.]* Translated by A. M. H. Boulard. Paris, Maradan. 2 vols.

Cb193. **Reess, M.** 1884. *Ueber die Pflege der Botanik in Franken von der Mitte des 16. bis zur Mitte des 19. Jahrhunderts, nebst einigen Bemerkungen über gegenwärtige Zustände. Rede beim Antritt des Prorectorats [etc.].* Erlangen, Junge. 56 pp.

Cb194. **Richter, R., H. von Guttenberg, & E. Libbert.** 1968. Die Entwicklung der Botanik in Rostock. *Wissenschaftliche Zeitschrift der Universität Rostock, Mathematisch–naturwissenschaftliche Reihe* 17: 263–275.

Cb195. **Saccardo, P. A.** 1895–1901. *La botanica in Italia: Materiali per la storia di questa scienza.* Venice, C. Ferrari. 2 vols.
First pub. in *Memorie del R. Istituto Veneto di Scienze, Lettere ed Arte* vol. 25 (4) & vol. 36 (6).

Cb196. **___.** 1909. *Cronologia della flora Italiana ossia repertorio sistematico delle piu antiche date ed autori del rinvenimento delle piante (Fanerogame e Pteridofite) indigene, naturalizzate e aventizie d'Italia e della introduzione di quelle esotiche piu communemente coltivate fra noi.* Padua. xxxviii+390 pp. *Re-pub.* Bologna, Edagricole, 1971.

Cb196A. **Schnarrenberger, C., & H. Scholz**, eds. 1990. *Geschichte der Botanik in Berlin.* Berlin, Colloquium Verlag. 382 pp. *Wissenschaft und Stadt* no. 15.
Includes **C. Schnarrenberger.** Botanik an der Friedrich–Wilhelms–Universität; **___.** Botanik an Westberliner Universitäten; **___.** Botanik an den Kaiser–Wilhelm–Instituten (cf. Cb1067); **W. Plarre.** Zur Geschichte der Vererbungsforschung in Berlin (cf. Cb548); **A. Lang.** Elisabeth Schiemann: Leben und Laufbahn einer Wissenschaften in Berlin; **D. Mollenhauer.** Zur Geschichte der Phykologie in Berlin (cf. Cb709); **W. Laux.** Zur Geschichte der Phytomedizin in Berlin (cf. Cb197); **H. Sukopp.** Zur Geschichte der pflanzengeographisch–pflanzenökologischen Forschung in Berlin (cf. Cb480); **H. W. Lack.** Opera magna der Berliner systematischen Botanik (cf. Cb197); **P. Hiepko.** Die Sammlungen des Botanischen Museums Berlin–Dahlem und ihre Geschichte (cf. Cb984); **B. Zepernick & F. K. Timler.** Beiträge zur botanischen Erforschung aussereuropäischer Länder; **H. Scholz.** Phylogenetik und Evolutionismus: Berliner Alternitaven und Kritik; **___.** Wenig Behanntes aus der Berliner Botanik–Geschichte; **___.** Botanik und Nationalsozialismus in Berlin.

Cb197. **Scholz, H.**, ed. 1987. Botany in Berlin. On the occasion of the XIV International Botanical Congress. *Englera* 7: 288 pp.
Includes **W. Laux.** Contributions to the development of phytomedicine in Berlin, pp. 51–84; **H. W. Lack.** Opera magna in Berlin plant taxonomy, pp. 253–281. Other articles are entered at relevant places herein. For German versions, see Cb196A.

Cb198. **Schröter, C.** 1918. Vierhundert Jahre Botanik in Zürich. *Verhandlungen der Schweizerischen Naturforschenden Gesellschaft* 99: 3–28.

Cb199. **Schube, T.** 1890. Zur Geschichte der schlesischen Floren–Erforschung bis zum Beginn des siebzehnten Jahrhunderts. *Jahresbericht der Schlesischen Gesellschaft für Vaterländische Kultur* 68 Erganzungsheft: 48 pp.

Cb200. **Scourse, N.** 1983. *The Victorians and their flowers.* London, Croom Helm; Portland, OR, Timber Pr. [xi]+195 pp.

Cb201. **Shcherbakova, A. A.** 1961. Razvitie botanicheskikh znanii v Rossii do XVIII veka. [Development of botanical knowledge in Russia before the eighteenth century.] *Trudy Institut Istorii Estestvoznaniya i Tekhniki* 36: 136–175.

Cb202. ___. 1979. *Istoriya botaniki v Rossii do 60–x gg XIX v. (dodarvinovskii period). [History of botany in Russia in the 19th century.]* Novosibirsk, Izd. "Nauka." 365 pp.

Cb203. **Shteir, A. B.** 1984. Linnaeus's daughters: Women and British botany. In: B. J. Harris & J. K. McNamara, eds. *Women and the structure of society*, pp. 67–73. Durham, NC, Duke Univ. Pr.

Cb204. ___. 1987. Botany in the breakfast room: Women and early nineteenth–century British plant study. In: P. G. Abir–Am & D. Outram, eds. *Uneasy careers and intimate lives: Women in science, 1789–1979*, pp. 31–43. New Brunswick, NJ, Rutgers Univ. Pr.

Cb205. **Sirks, M. J.**, ed. 1935. *Botany in the Netherlands.* Ed. for the organizing committee of the VI International Botanical Congress. Leiden, Brill. vii+140 pp.

Cb206. **Stankov, S. S.** 1940. *80–letnie itogi izucheniya flory i rastitel'nosti Kryma. [Results of 80 years of study of the flora and vegetation of Crimea.]* Moscow, Moskovskoe Obshchestvo Ispytatelei Prirody. 28 pp.

Cb207. **Stawinski, W.** 1966. Dzieje botaniki w Polsce w latach 1795–1860. (The history of botany in Poland, 1795–1860). *Studia i materialy z dziejow nauki Polskiej* B 10: 97–103.

Cb208. **Stearn, W. T.** 1961. *The influence of Leyden on botany in the seventeenth and eighteenth centuries.* Assen, Van Gorcum & Co. 63 pp.
Also ___. 1962. The influence of Leyden on botany in the seventeenth and eighteenth centuries. *British journal for the history of science* 1: 137–158.

Cb209. ___. 1966. Early Marburg botany. In: K. Moench. *Methodus plantas horti botanici et agri Marburgensis* [Marburg, 1794 & 1802, facsimile edition], pp. i–xix. Koenigstein, Otto Koeltz Antiquariat.

Cb210. ___. 1975. History of the British contribution to the study of the European flora. In: Botanical Society of the British Isles. *European floristic and taxonomic studies*, pp. 1–17. Ed. by S. M. Walters. Faringdon, E. W. Classey.

Cb211. **Stefanoff, B.** 1930. Istoricheskii pregled" na izsledvaniyata v"rkhu florata na B"lgariya. Historische Übersicht der Untersuchungen über die Flora Bulgariens. *Izvestiya na Tsarskite Prirodonauchni Instituti v Sofiya* 3: 61–112, illus.
Text in Bulgarian and German.

Cb212. **Sulma, T., & Z. Schwarz.** 1968. Z dziejow botaniki na Pomorzu Gdanskim. (A contribution to the history of botany in Pomorze Gdanskie). *Studia i materialy z dziejow nauki Polskej* B 14: 25–74.

Cb213. **Szafer, W.** 1970. Outline of the history of botany in Poland from the middle ages to the year 1918. In: P. Smit & R. J. Ch. V. ter Laage, eds. *Essays in biohistory*, pp. 381–388. Utrecht, International Association for Plant Taxonomy. *Regnum vegetabile* no. 71.

Cb214. **Trautvetter, E. R. von.** 1837. *Grundriss einer Geschichte der Botanik in Bezug auf Russland.* Petrograd, K. Akademie der Wissenschaften. [ii]+145 pp.
pp. 55–136 is a bibliography.

Cb215. **Vallot, J. N.** 1828. Histoire de la botanique en Bourgogne. *Séance publique, Académie des Sciences, Arts et Belles–Lettres de Dijon* 1827: 112–160.

Cb216. **Voss, W.** 1884–85. Versuch einer Geschichte der Botanik in Krain (1754 bis 1883). *Jahresbericht der Staats– Oberrealschule in Laubach* 1884: 1–59; *Ibid.* 1885: 1–41.
Biographical sketches of 12 distinguished botanists.

Cb217. **White, J.** 1982. A history of Irish vegetation studies. *Journal of life sciences, Royal Dublin Society* 3: 15–42.

Cb218. **Wildeman, E. de.** 1950. Notes pour l'histoire de la botanique et d'horticulture en Belgique. *Mémoires de l'Académie Royale des Sciences de Belgique, classe des sciences* (2) 25: 1–832.
I: Introduction, pp. 5–37; II: Le Jardin de Jean Hermans, Bruxelles 1652–1653, pp. 38–638; III: Les plantes en culture à Dileghem (Bruxelles, 1633) chez B. Wynhouts; à Lille (1644) chez P. Ricart et chez J. Hermans (Bruxelles 1652–1653), pp. 639–692; IV: Enumération des plantes cultivées à Lille par P. Ricart, en 1644, pp. 693–727; V: Quelques conclusions, pp. 728–769; VI: Table alphabétique des noms utilisés dans le texte, auteurs et végétaux, pp. 770–832.

Cb219. **Zhukovskii, P. M.** 1945. Russkoe klassiki botaniki. *Vestnik, Akademiya Nauk SSSR* 1945(5/6): 126–141.

See also Ba157–190 Bb51(Morren) Bb71(Plantefol) Bb86 Bb710–711 Bb713 Bb715–717 Bb720–729 Bb732–734 Bb738–749 Bb751–752 Bb756–768 Bb771–774 & Ca184(Morton & Noble).

North American

Cb220. **American Philosophical Society.** 1957. [Historical papers on French botanists in North America.] *Proceedings of the American Philosophical Society* 101: 336–368.
Includes **J. R. Schramm.** Influence — past and present — of François-André Michaux on forestry and forest research in America, pp. 336–343; **G. Chinard.** André and François-André Michaux and their predecessors: An essay on early botanical exchanges between America and France, pp. 344–361; **W. J. Robbins.** French botanists and the flora of the northeastern United States: J. G. Gilbert and Elias Durand, pp. 362–368. First presented at the *Colloque international du Centre National de la Recherche Scientifique*, 63, Paris 11–14 Septembre, 1956, and published in French language versions, see Cb226.

Cb221. **Bailey, D. L.** 1949. Botany. In: Royal Canadian Institute. *The Royal Canadian Institute centennial volume, 1849–1949*, pp. 25–35. Ed. W. S. Wallace. Toronto.

Cb222. **Blankinship, J. W.** 1904. A century of botanical exploration in Montana, 1805–1905: Collectors, herbaria and bibliography. *Montana College of Agriculture & Mechanic Arts. Science studies, botany* 1: 1–31.

Cb223. **Boone, W. W.** 1965. *A history of botany in West Virginia.* Parsons, WV, McClain Printing Co. [xi]+196 pp. & 1 pl.

Cb224. **Brendel, F.** 1879. Historical sketch of the science of botany in North America from 1635 to 1840. *American naturalist* 13: 754–771; *Continued* ___. 1880. Historical sketch of the science of botany in North America from 1840 to 1858. *Ibid.* 14: 25–38.
Both articles re-pub. in **R. L. Stuckey**, ed. 1978. *Development of botany in selected regions of North America before 1900.* New York, Arno Pr.; *see* Cb263.

Cb225. **Butler, J. R.** 1938. America: A hunting ground for eighteenth–century naturalists with special reference to their publications about trees. *Papers of the Bibliographical Society of America* 32: 1–16.

Cb226. **Centre National de la Recherche Scientifique (CNRS), Paris.** 1957. *Les botanistes français en Amérique du nord avant 1850.* J. Leroy Ed. Paris. 360 pp. *Colloques internationaux du Centre National de la Recherche Scientifique*, 63, Paris 11–14 Septembre, 1956.
29 papers by various authors. *Includes* **J. Ewan.** L'Activité des premiers explorateurs français dans le S.E. des Etats–Unis, pp. 17–40; **J. Robbins.** Les botanistes français et la flore du N.E. des Etats–Unis: J. G. Milbert et Elias Durand, pp. 41–51; **A. Camus.** La contribution française à l'étude des Graminées de l'Amérique du Nord au XVIIIe siècle et dans la première

moitié du XIXe siècle, pp. 107–121; **A. Guillaumin & V. Chaudun.** L'Introduction en France des plantes horticoles originaires d'Amérique du Nord avant 1850, pp. 123–135; **C. Baehni.** Les relations de Rafinesque et A.–P. de Candolle, pp. 137–147; **J. Rousseau.** Michel Sarrazin, J. F. Gaultier et l'étude prélinnéenne de la flore canadienne, pp. 149–157; **C. Skottsberg.** Linné, Kalm et l'étude de la flore nord–américaine au XVIIIe siècle, pp. 179–187; **J. F. Leroy.** Le premier explorateur de Terre–Neuve et des îles Saint–Pierre et Miquelon: Bachelot de la Pylaie (1786–1856), pp. 203–230 & pl. xvi–xxii; **G. Duprat.** Essai sur les sources manuscrites conservés au Muséum National d'Histoire Naturelle, pp. 231–252 & pl. xxiv–xxxi; **G. Chinard.** Les Michaux et leurs précurseurs, pp. 263–284; **J. R. Schramm.** Influence actuelle et passée de F. A. Michaux sur la foresterie et la recherche forestière en Amérique, pp. 287–300 (**P. Guinier.** A propos de la communication de M. Schramm, p.301); **R. F. Wood & J. D. Matthews.** Arbres nord–américains et sylviculture anglaise, pp. 303–313; **A. Galoux.** Les grandes étapes de l'introduction des arbres nord–américains en Belgique, pp. 315–328; **P. Guinier.** Ce que les jardins et les forêts de France doivent aux forêts américaines, pp. 329–337; **J. Pourtet.** Rôle et importance des essences forestières américaines dans les techniques actuelles de la foresterie française, pp. 339–348.

Cb227. **Christensen, E. M.** 1967. The growth of botany in Utah determined from botanical literature (1945–1964). *Proceedings of the Utah Academy of Science* 44: 367–396.

Cb228. **Cittadino, E.** 1980. Ecology and the professionalization of botany in America, 1890–1905. *Studies in the history of biology* 4: 171–198.

Cb229. **Core, E. L.** 1970. The botanical exploration of the southern Appalachians. In: P. C. Holt & R. A. Patterson, eds. *The distributional history of the biota of the southern Appalachians.* Pt. II: *Flora*, pp. 1–65. Blacksburg, VA, Virginia Polytechnic Institute & State Univ. *Research division monograph* no. 2.

Cb230. **Coulter, J. M.** 1917. A century of botany in Indiana. *Proceedings of the Indiana Academy of Science* 1916: 236–260.

Cb231. **Dickerman, C.** 1985. Mid-19th century botanical exploration in New Mexico. *New Mexico historical review* 60: 159–171.

Cb232. **Eifert, V. L.** 1965. *Tall trees and far horizons: Adventures and discoveries of early botanists in America.* New York, Dodd Mead & Co. xvii+301 pp.

Cb233. **Ewan, J.**, ed. 1969. *A short history of botany in the United States.* New York, Hafner Publishing for the XI International Botanical Congress, Seattle. xiii+174 pp.
Contributions by various authors: **J. Ewan.** Calendar of events, pp. 1–25; ___. Early history, pp. 27–48; **S. Carlquist.** Morphology and anatomy, pp. 49–57; **C. Zirkle.** Plant genetics and cytology, pp. 58–66; **S. B. Hendricks.** Plant physiology, pp. 67–73; **W. R. Taylor.** Phycology, pp. 74–81; **K. F. Baker.** Plant pathology and mycology, pp. 82–88; **E. D. Rudolph.** Bryology and lichenology, pp. 89–96; **R. Tryon.** Pteridology, pp. 97–102; **C. A. Arnold.** Paleobotany, pp. 103–108; **C. B. Heiser Jr.** Taxonomy, pp. 110–114; **J. Ewan.** Plant geography, pp. 115–123; **P. B. Sears.** Plant ecology, pp. 124–131; **G. H. M. Lawrence.** Horticulture, pp. 132–145; **J. Stannard.** Medical botany, pp. 146–152.

Cb234. **Ford, C. E.** 1964. Botany texts: A survey of their development in American higher education, 1643–1906. *History of education quarterly* 4: 59–70.

Cb235. **Frick, G. F., et al.** 1987. Botanical explorations and discoveries in colonial Maryland, 1688 to 1753. *Huntia* 7: 5–59.
Accompanied by these related articles: **C. R. Broome et al.** A 1698 Maryland florula by the London apothecary James Petiver (ca.1663–1718), pp. 61–90; **J. L. Reveal et al.** The identification of pre-1753 polynomials and collections of vascular plants from the British colony of Maryland, pp. 91-208; ___. On the identities of Maryland plants mentioned in ther first two editions of Linnaeus's Species plantarum, pp. 209–245; **M. L. Brown et al.** Comments on the vegetation of colonial Maryland, pp. 247–283.

Cb236. **Friesner, R. C.** 1952. The pioneer period in the study of the Indiana flora. *Butler University botanical studies* 10: 144–152.

Cb237. **Harshberger, J. W.** 1911. History and literature of the botanic works and explorations of the North American continent. In: *Phytogeographic survey of North America* pp. 1–92. Leipzig, W. Engelmann. E. Engler & O. Drude, eds. *Die vegetation der Erde*, vol. 13. *Re-pub.* Weinheim, H. R. Engelmann (J. Cramer); New York, Hafner Publishing Co., 1958.
Ch. I. History. 1. Floristic work, 2. Plant geography, pp. 1–45; Ch. II. Bibliography. A. Gener works on North American botany and plant geography, B. Special work on the territories, pp. 46–92.

Cb238. **Hartung, S. A.** 1957. A history of botanical exploration of South Dakota. *Proceedings of the South Dakota Academy of Sciences* 35: 111–115.

Cb239. **Hultén, E.** 1940. History of botanical explorations in Alaska and Yukon territories from the time of their discovery to 1940. *Botaniska notiser* 1940: 289–346.
With chronological list of ca. 350 collectors.

Cb240. **Izquierdo, J. J.** 1955. La botanique Aztèque et la botanique mexicaine moderne. *Archives internationales d' histoire des sciences* 32: 227–240.

Cb241. **Juel, H. O.** 1920. Early investigations of north American flora, with special reference to Linnaeus and Kalm. *Svenska Linné–Sällskapets årsskrift* 3: 61–79.
Contents include a list of: North American plants first described by Linnaeus, pp. 72–76.

Cb241A. **Keeney, E. B.** 1992. *The botanizers: Amateur scientists in nineteenth–century America.* Chapel Hill, NC, Univ. of North Carolina Pr. xii+206 pp.

Cb242. **Langman, I. K.** 1960. Works prior to 1800 useful for studies in Mexican botany. *Revista interamericana de bibliografía* 10: 219–243.

Cb243. **López, E. A.** 1945. Las plantas de America en la botánica europea del siglo XVI. *Revista de Indias* 6: 221–288.

Cb244. **McClintock, E.** 1967. Early plant explorers in the west. *Journal of the California Horticultural Society* 28: 114–121, 152–160; ___. 1990. ___. Part II. *Fremontia* 18: 11–16.

Cb245. **McKelvey, S. D.** 1955. *Botanical exploration of the Trans–Mississipi West 1790–1850.* Jamaica Plain, NY, Arnold Arboretum of Harvard Univ. xl+1,144 pp. & 10 maps. *Re-pub.* Corvallis, OR, Oregon State Univ. Pr., 1991.
A laborious, well–documented, decade–by–decade account of travels. The 1991 ed. includes an introduction by J. Ewan and an additional map.

Cb246. **McVaugh, R.** 1972. Botanical exploration in Nueva Galicia, Mexico from 1790 to the present time. *Contributions from the University of Michigan Herbarium* 9: 207–357.
Includes over 300 biographical sketches.

Cb247. **Mathias, M. E.** 1989. The fascinating history of the early botanical exploration and investigations in southern California. *Aliso* 12: 407–434.

Cb248. **Murrill, W. A.** 1945. *Historic foundations of botany in Florida (and America).* Gainesville, FL, the Author. 51 pp.
Largely based on personal recollections.

Cb249. **Penhallow, D. P.** 1888–95. A review of Canadian botany from the first settlement of New France to the nineteenth century. Pt. I. A review of Canadian botany from 1880 to 1895. Pt. II. *Proceedings & transactions of the Royal Society of Canada* 5 Sect. 4: 45–61, 1888; *Ibid.* (2) 3 Sect. 4: 3–56, 1897.

Cb250. **Pringle, J. S.** 1989. *Botanical exploration of the Canadian watershed of Lake Huron during the nineteenth century.* Hamilton, Ont., Royal Botanical Gardens, Centre for Canadian Historical Horticultural Studies. 88 pp. *Canadian horticultural history* 2.

Cb251. **Pringle, J. S.** 1991. Contributions by Moravian missionaries to the knowledge of the flora of Labrador. *Canadian horticultural history* 2: 187–222.

Cb252. **Reveal, J. L.** 1972. Botanical explorations in the intermountain region. In: A. Cronquist et al. *Intermountain flora*, vol. 1: 40–72. New York, Hafner Publishing Co. for the New York Botanical Garden.

Cb253. ___. 1983. The significance of pre-1753 botanical explorations in temperate North America on Linnaeus's first edition of *Species Plantarum. Phytologia* 53: 1–96.

Cb254. ___. 1991. Botanical explorations in the American west — 1889–1989: An essay on the last century of a floristic frontier. *Annals of the Missouri Botanical Garden* 78: 65–80.

Cb255. ___. 1992. *Gentle conquest: The botanical discovery of North America.* With illustrations from the Library of Congress. Washington, DC, Starwood Publishing. 160 pp., illus.

Cb256. **Rodgers, A. D.** 1944. *American botany, 1873–1893: Decades of transition.* Princeton, NJ, Princeton Univ. Pr. vii+340 pp. *Re-pub.* New York, Hafner Publishing Co., 1968.
Focussed on Asa Gray and his contemporaries.

Cb257. **Rousseau, J.** 1937. La botanique canadienne à l'époque de Jacques Cartier, [etc.]. *Contributions de Laboratoire de Botanique de l'Université de Montréal* 28: 86 pp.

Cb258. **Rudolph, E. D.** 1982. The introduction of the natural system of classification of plants to 19th century American students. *Archives of natural history* 10: 461–468.

Cb259. ___. 1982. Women in 19th–century American botany: A generally unrecognized constituency. *American journal of botany* 69: 1346–1355.

Cb260. **Rydberg, P. A.** 1907. Scandinavians who have contributed to the knowledge of the flora of North America. A memoir prepared for the celebration, at Augustana College and Theological Seminary, of the 200th anniversary of the birth of Linné. *Augustana Library publications* 6: ii+49 pp.

Cb261. **Stannard, J.** 1966. Early American botany and its sources. In: T. R. Buckman, ed. *Bibliography and natural history*, pp. 73–102. Lawrence, KS, Univ. of Kansas Libraries.

Cb262. **Steere, W. C.**, ed. 1958. *Fifty years of botany.* Golden jubilee volume of the Botanical Society of America. New York, McGraw–Hill Book Co. xiii+638 pp.
34 contributions by various authors.

Cb263. **Stuckey, R. L.**, ed. 1978. *Development of botany in selected regions of North America before 1900.* Ed. with an introduction by [etc.]. New York, Arno Pr. [268] pp.
Re-pub. of several historical articles. *Includes* **F. Brendel.** Historical sketch of the science of botany in North America from 1635 to 1840. *American naturalist* 13: 754–771, 1879; ___. Historical sketch of the science of botany in North America from 1840 to 1858. *Ibid.* 14: 25–38, 1880; **J. W. Harshberger.** Introduction. From: *The botanists of Philadelphia and their work*, pp. 1–40. Philadelphia, PA, 1899; **J. E. Humphrey.** Botany and botanists in New England. *New England magazine* N.S. 14: 27–44, 1896; **H. H. Rushby.** A historical sketch of the development of botany in New York City. *Torreya* 6: 101–111, 133–145, 1906; **F. V. Coville.** Early botanical activity in the District of Columbia. *Records of the Columbia Historical Society* 5: 176–194, 1902; **F. Lamson–Scribner.** Southern botanists. *Bulletin of the Torrey Botanical Club* 20: 315–334, 1893; **R. S. Cocks.** Historical sketch of the botany of Louisiana. *Proceedings of the Louisiana Society of Naturalists* 1897/99: 69–74, 1900; **P. Spaulding.** A biographical history of botany at St. Louis, Missouri. *Popular science monthly* 73: 488–499, 1908; *Ibid.* 74: 48–57, 124–133, 240–258, 1909.

Cb264. **Thomas, J. H.** 1961. The history of botanical collecting in the Santa Cruz Mountains of central California. *Contributions from the Dudley Herbarium of Stanford*

University 5: 147–168.
Includes biographical notes on 95 collectors.

Cb265. ___. 1979. Botanical exploration in Washington, Oregon, California and adjacent regions. *Huntia* 3: 5–66.

Cb266. **Voss, E. G.** 1978. *Botanical beachcombers and explorers: Pioneers of the 19th century in the Upper Great Lakes.* Ann Arbor, MI, Univ. of Michigan Herbarium. viii+100 pp. *Contributions from the University of Michigan Herbarium* no. 13.

Cb267. **Zeller, S.** 1987. Part 3. Botany. In: *Inventing Canada: Early Victorian science and the idea of a transcontinental nation*, pp. 181–268, (pp. 313–335 notes). Toronto, Univ. of Toronto Pr.

See also Ba191–215 Bb592(Coulter) Bb593(Fernald) Bb712 Bb714 Bb718–719 Bb735–737 Bb750 Bb753–755 Bb769–770 Bb775 & Ca56(Slack).

South & Central American (& Caribbean)

Cb268. **Acosta-Solís, M.** 1968. *Naturalistas y viajeros científicos que han contribuido al conocimiento florístico y fitigeográfico del Ecuador.* Quito, Casa de la Cultura Ecuatoriana. 138 pp. *Contribución, Instituto Ecuatoriano de Ciencias Naturales* no. 65.

Cb269. **Beltrán, E.** 1968. Las reales expediciones botánicas del siglo XVIII a Hispanoamérica. *Revista de la Sociedad Mexicana de Historia Natural* 28: 179–249.

Cb270. **Colmeiro, M.** 1892. *Primeras noticias acerca de la vegetación Americana suministradas por el almirante colón y los inmediatos continuadores de las investigaciones dirigidas al conocimiento de las plantas con un resumen de las expediciones botánicas de los españoles.* Madrid, Sucesores de Rivadeneyra. 59 pp.

Cb271. **Conde, J. A.** 1958. *Historia de la botánica en Cuba.* Havana, Junta Nacional de Arqueologia y Etnologia. xvi+353 pp. & 2 pl.

Cb271A. **Diaz-Piedrahita, A.** 1991. La botanica en Colombia: Hechos notables en su desarrollo. *Colecçión Enrique Perez-Arbeleaez, Académia Colombiana de Ciencias Exactas* 6: xii+118 pp.

Cb272. **Herrera, F. L.** 1936. El Perú botanica. Del itinerario de viajes del professor don Antonio Raimondi. *Revista del Museo Nacional, Lima* 5: 89–119.

Cb273. ___. 1937. Botánistas de fines del siglo XVIII: Primeros escritos de vulgarización científica en el Perú. *Revista del Museo Nacional, Lima* 6: 95–124.

Cb274. ___. 1937. Apuntes para la historia de la botánica en el Perú. (La expedición botánica de Ruiz y Pavón (1777).) *Revista del Museo Nacional, Lima* 6: 151–158.

Cb275. ___. 1937. Exploraciones botánicas en el Perú. *Revista del Museo Nacional, Lima* 6: 291–358.

Cb276. ___. 1938. Precursores de los estudios botánicos en el Departamento del Cuzco (apuntes sueltos). *Revista del Museo Nacional, Lima* 7: 53–130.

Cb277. **Howard, R. A.** 1979. Early botanical records from the West Indies, particularly Barbados: Ligon (1657) to Lord Seaforth (1806). *Botanical journal of the Linnean Society* 79: 65–96.

Cb278. **Huber, O., & J. J. Wurdack.** 1984. History of botanical exploration in Territorio Federal Amazonas, Venezuela. *Smithsonian contributions to botany* 56: iii+83 pp.

Cb279. **Léon, J. S. S.** 1918. Los exploraciones botanicas de Cuba. *Memorias de la Sociedad Cubana de Historia Natural "Felipe Poey"* 3: 178–224.

Cb280. **Moscosa, R. M.** 1974. Las exploraciones botánicas en Santo Domingo. *Eme* 3 (13): 57–83.

Cb281. **Parodi, L. R.** 1961. Ciento cincuenta años de botanica en la Republica Argentina. *Boletín de la Sociedad Argentina de Botánica* 9: 1–68.
Includes biographical information on native botanists.

Cb282. **Standley, P. C.** 1928. History of botanical exploration in Panama. *Contributions from the United States National Herbarium* 27: 40–49.

Cb283. **Texera Arnal, Y.** 1987. Exploradores botánicos europeos en Venezuela durante el signo XIX. *Quipu* 4: 185–211.

Cb284. ___. 1991. *La exploración botánica en Venezuela, 1754–1950*. Caracas, Fondo Editorial Acta Cientifica Venezolana. 187 pp. *Colección de estudios sociales de la ciencia.*

Cb285. **Turrill, W. B.** 1920. Botanical exploration in Chile and Argentina. *Kew bulletin* 1920: 57–66.

Cb286. **Weberbauer, A.** 1936. Historia de la exploración botánica del Perú. *Revista de la Universidad Católica del Perú* 14: 293–305, 368–377, 509–525.

See also Ba217–225.

Asian

Cb287. **Bhattacharyya, P. K.** 1982. Beginning of modern botany in India by Dutch in 16th–18th century (basic features and characteristics). *Indian journal of history of science* 17: 365–376.

Cb288. **Boulger, G. S.** 1900. Appx. IV: History of Ceylon botany. In: H. Trimen. *Handbook to the flora of Ceylon*, pt. 5: 366–381. London, Dulau & Co.

Cb289. **Blatter, E.** 1933. Flora arabica, the botanical exploration of Arabia. *Records of the Botanical Survey of India* 8: 451–501.

Cb290. **Bretschneider, E.** 1882–85. Botanicon Sinicum. *Journal of the North China branch of the Royal Asiatic Society* n.s. 16: 18–230, 1882; *Ibid.* n.s. 25: ii+468 pp., 1893; *Ibid.* n.s. 29: 623 pp., 1896. Separately pub. London, Trübner & Co., 1882–95.
I: *Notes on Chinese botany from native and western sources*; II: *The botany of the Chinese classics.* With annotations, appendix and index by E. Faber; III: *Botanical investigations into the materia medica of the ancient Chinese.*

Cb291. ___. 1880. Early European researches into the flora of China. *Journal of the North-China branch of the Royal Asiatic Society* n.s. 15: 1–192. *Also pub. separately* Shanghai, American Presbyterian Mission Pr., 1881.

Cb292. ___. 1898. *History of European botanical discoveries in China.* London, Sampson Low. xv+1,167 pp. (in 2 vols.) *Re-pub.* Leipzig, Koehler, 1935; Leipzig, Zentral–Antiquariat der Deutschen Demokratischen Republik, 1962.

Cb293. **Burkill, I. H.** 1927. Botanical collectors, collections and collecting places in the Malay Peninsula: A review of the work so far done towards a knowledge of the plant geography of Malaya. *Gardens' bulletin. Straits Settlements* 4: 113–202.

Cb293A. ___. 1965. *Chapters on the history of botany in India.* Calcutta, Botanical Survey of India. xi+245+[1] pp., 4 pl. & 2 maps.
First pub. in: *Journal of the Bombay Natural History Society* 51: 846–878, 1953; *Ibid.* 54: 42–86, 1956; *Ibid.* 58: 678–706, 1961; *Ibid.* 59: 335–359, 747–777, 1962; *Ibid.* 60: 49–83, 356–370, 1963.

Cb294. **Cherepnin, L. M.** 1954. Istoriya issledovanii rastitel'nogo pokrova yuzhnoi chasti Krasnoyarskogo kraya. *Uchenye Zapiski, Krasnoyarskii gosudarstvennyi pedagogicheskii instituta* 3 (1): 80 pp.

Cb295. **Cox, E. H. M.** 1945. *Plant hunting in China: A history of botanical exploration in China and the Tibetan marches.* London, Collins. 230 pp., illus. *Re-pub.* London, Oldbourne Book Co., 1961.

Cb296. **Desmond, R. G. C.** 1992. *The European discovery of the Indian flora.* Oxford, Royal Botanic Gardens & Oxford Univ. Pr. xii+355 pp., illus.

Cb297. **Galazii, G. I., & V. N. Moloznikov.** 1982. *Istoriya botanicheskikh issledovanii na Baikale (itogi i perspektivy ekologo–botanicheskikh rabot).* Novosibirsk, Akademiya Nauk SSR, Sibirskoe Otdelenie, Limnologicheskii Institut. 152 pp.

Cb298. **Govorukhina, V. A.** 1977. *Iz istorii botanicheskikh issledovanii na territorii sovremennogo Turkmenistana v dosovetskii period.* Ashkhabad, Ylym. 163 pp. *Continued* ___. 1982. *Iz istorii botanicheskikh issledovanii v Turkmenistane. Chast 2.* Ashkhabad, Ylym. 194 pp.

Cb299. **Honig, P., & F. Verdoorn**, eds. 1945. *Science and scientists in the Netherlands Indies.* New York, Board for Netherlands Indies, Surinam and Curaçao. xxiv+491 pp.
Includes Chapters in the history of Cinchona [5 papers by various authors], pp. 181–207; **F. A. F. C. Went.** A short history of general botany in the Netherlands Indies, pp. 390–402.

Cb300. **Indian Botanical Society.** 1958+. *History of botanical researches in India, Burma and Ceylon.* Bangalore City. Various paging.
Part 1: **S. N. Das Gupta.** 1958. *Mycology and plant pathology.* iv+118 pp. (Bibliography, pp. 50–118); Part 2: **H. Santapau.** 1958. *Systematic botany of angiosperms.* vii+77 pp. (Bibliography, pp. 11–77); Part 3: **A. R. Rao.** 1958. *Palaeobotany.* viii+57 pp. (Bibliography, pp. 39–57); Part 4: **V. Puri.** 1961. *Floral anatomy.* [v]+31 pp. (Bibliography, pp. 23–31).

Cb300A. **Kerr, A.** 1939. Early botanists in Thailand. *Journal of the Thailand research society. Natural history supplement* 12: 1–27.

Cb301. **Koidzumi, G.** 1936. A historical sketch of herbal and botanical investigations on the flora of Nippon. *Acta phytotaxonomica et geobotanica* 5: 1–26.

Cb302. **Krylov, G. V., & N. G. Salatova.** 1969. *Istoriya botanicheskikh i lesnykh issledovanii v Sibiri i na Dal'nem Vostoke. [The history of botanical and forestry investigations in Siberia and the Far East].* Novosibirsk, Akademiya Nauk SSSR, Sibirskoe Otdelenie, Biologicheskii Institut. 276 pp. & 12 pl. of ports.
Bibliography, pp. 162–229.

Cb303. **Lipskii, V. I.** 1899. *Flora Kavkaza. Svod svedenii o flore Kavkaza za dvukhsotletnii period ee issledovaniya nachinaya ot Turnefora i konchaya XIX v. [Flora of the Caucasus. Collected information on the flora of the Caucasus in two hundred years of investigation, beginning with Tournefort and ending with the nineteenth century].* St. Petersburg. xv+584 pp. *Trudy Tiflisskogo Botanicheskogo Sada* vol. 4; ___. 1902. ___*: Dopolnenie 1. [* ___*: Supplement 1.]* St. Petersburg. 100 pp. *Trudy Tiflisskogo Botanicheskogo Sada* vol. 6 (1).

Cb304. ___. 1903. Istoriya botanicheskogo issledovaniya. Historiae florae exploratae. In: Flora Srednei Azii ... Flora Asiae mediae ... [etc]. *Trudy Tiflisskogo Botanicheskogo Sada* 7 (2): 246–337.

Cb305. **Maheshwari, P., & R. N. Kapil.** 1963. *Fifty years of science in India. Progress of botany.* Calcutta, Indian Science Congress Association. vii+178 pp.
Separate chapters on: Algae; Fungi and allied organisms; Bryophytes; Pteridophytes; Gymnosperms; Taxonomy; Morphology and anatomy; Embryology; Cytogenetics; Plant breeding; Ecology; Physiology; Palaeobotany. Literature cited (817 titles).

Cb306. **Markov, M. V.** 1952. Iz istoriya Kazanskoi geo-botanicheskoi shkoly. *Uchenye zapiski Kazakhskogo Gosudarstvennogo Universiteta im S. M. Kirova* 112 (7): 3–46.

Cb307. **Matthew, K. M.** 1982. Botany and its technologies in peninsula India in the eighteenth and nineteenth centuries. *Indian journal of the history of science* 17: 353–364.

Cb308. **Needham, J., et al.** 1986+. *Science and civilisation in China*. Vol. 6: *Biology and biological technology*. Part 1: *Botany*. By J. Needham, with the collaboration of Lu Gwei–Djen & Huang Hsing–Tsung. Cambridge, Cambridge Univ. Pr. xxxii+718 pp.
Section 38 (d) The literature and its content, pp. 182–471; Bibliographies. A. Chinese and Japanes books before +1800; B. Chinese and Japanese books and journal articles since +1800; C. Books and journal articles in western languages, pp. 555–662. *Also* **J. Needham & W. Ling.** 1954. *Science and civilisation in China*. Vol. 1: *Introductory orientations*. Cambridge, Cambridge Univ. Pr. xxxviii+318 pp.

Cb309. **Quisumbing, E.** 1964. Botanical explorations in the Philippines. *Philippine geographical journal* 8: 21–38.

Cb309A. **Rajbhandri, K. R.** 1976. History of botanical explorations in Nepal. *Journal of the Bombay Natural History Society* 73: 468–481.

Cb310. **Rechinger, K. H.** 1989. Fifty years of botanical research in the *Flora Iranica* area (1937–1987). In: Tan, K., ed. *Plant taxonomy, phytogeography and related subjects: The Davis and Hedge Festschrift [etc.]*, pp. 301–349. Edinburgh, Edinburgh Univ. Pr.

Cb310A. **Robinson, C. B.** 1906. The history of botany in the Philippine Islands. *Journal of the New York Botanical Garden* 7: 104–112.

Cb311. **Rowell, M.** 1973. Early Russian botanical exploration in the north Pacific. *Episteme* 7: 165–185.

Cb311A. **Steenis–Kruseman, M. J. van.** 1950. Ch. IV: Chronology of the collections; being a key to the history of botanical exploration of Malaysia. In: C. G. G. J. van Steenis, ed. *Flora malesiana*. Series 1, *Spermatophyta*. Vol. 1, pp. lxxxvi–cvi. Leyden, Noordhof; *Continued* ___. 1958. ___: Supplement 1. In: *Ibid.* Series 1. Vol. 5(4), pp. ccxxxvii–ccxlv. Leyden, Noordhoff.

Cb312. **Stewart, R. R.** 1979. The first plant collectors in Kashmir and the Punjab. *Taxon* 28: 5–11.

Cb313. ___. 1982. Missionaries and clergymen as botanists in India and Pakistan. *Taxon* 31: 57–64.

Cb314. ___. 1982. *History and exploration of plants in Pakistan and adjoining areas*. Islamabad, [University of Karachi, Department of Botany]. v+186 pp. Forms part of: E. Nasir & S. I. Ali, eds. *Flora of Pakistan* [unnumbered part].

Cb314A. **Wit, H. C. D. de.** 1949. Short history of the phytography of the Malaysian vascular plants. In: C. G. G. J. van Steenis, ed. *Flora malesiana*, Series 1: *Spermatophyta*, vol. 4, pp. lxx–clxi. Djakarta, Republic of Indonesia, Ministry of Agriculture.

See also Ba226–232 Bb708–709.

Australasian (& Pacific)

Cb315. **Carr, D. J., & S. G. M.**, eds. 1981. *People and plants in Australia*. Sydney, Academic Pr. xxi+416 pp.
A collection of 18 articles by various authors.

Cb316. ___, eds. 1981. *Plants and man in Australia*. Sydney, Academic Pr. xiv+313 pp.
A collection of 13 articles by various authors.

Cb317. **Gilbert, L. A.** 1982. Plants and parsons in nineteenth century New South Wales. *Historical records of Australian science* 5(3): 17–32.

Cb318. **Glenn, R.** 1950. *The botanical explorers of New Zealand.* Wellington, N.Z., A. H. & A. W. Reed. 176 pp. & 14 pl.
28 biographies.

Cb319. **Godley, E. J.** 1965. Botany of the Southern zone, exploration to 1843. *Tuatara* 13: 140–181.

Cb320. ___. 1967. A century of botany in Canterbury. *Transactions of the Royal Society of New Zealand* 1: 243–266.

Cb321. **McCaughey, V.** 1918–19. History of botanical exploration in Hawaii. *Hawaiian forester & agriculturist* 15: 388–396, 417–429, 508–510, 1918; *Ibid.* 16: 25–28, 49–54, 1919.

Cb322. **Oliver, W. R. B.** 1921. Botanical discovery in New Zealand: The visiting botanists. *Post-primary school bulletin* 5: 17–48.

Cb323. **Short, P. S.**, ed. 1990. *History of systematic botany in Australia: Proceedings of a symposium held at University of Melbourne, 25–27 May 1988.* South Yarra, Vic., Australian Systematic Botany Society. v+326 pp.

Cb324. **Smith, G. G.** 1982. Pioneer seashore botanists in Western Australia. *Early days* 8: 86–102.

Cb325. **Thomson, A. D.** 1988. Some sources for a history of New Zealand botany. *Journal, Canterbury Botanical Society* 22: 34–37.

See also Ba233–242.

African

Cb326. **Fernandes, A.**, ed. 1962. Histoire de l'exploration botanique de l'Afrique au sud du Sahara. *Comptes rendus, IVe réunion plénière, Association pour l'Étude Taxonomique de la Flore d'Afrique Tropicale* (2): 47–248.
Includes **A. W. Exell.** Pre-Linnean collections in the Sloane herbarium from Africa south of the Sahara, pp. 47–49; **A. Aubréville.** L'exploration botanique de l'Afrique occidentale Française, pp. 51–54; **R. W. J. Keay.** Botanical collectors in West Africa prior to 1860, pp. 55–68; **F. N. Hepper.** Botanical collectors in West Africa, except French territories, since 1860, pp. 69–75; **L. A. G. Barbosa.** Les botanistes dans l'archipel du Cap Vert, pp. 77–94; **A. W. Exell.** Botanical collecting on the islands of the Gulf of Guinea, pp. 95–102; **J. B. Teixeira.** La naturaliste Joaquim José da Silva et les itinéraires des expéditions qu'il a effectuées en Angola, de 1783 à 1804, pp. 103–109; **F. A. Mendonça.** Botanical collectors in Angola, pp. 111–121; **A. D. J. Meeuse.** A short history of botanical exploration in South West Africa, pp. 123–126; **H. Humbert.** Histoire de l'exploration botanique à Madagascar, pp. 127–144; **F. A. Mendonça.** Botanical collectors in Mozambique, pp.145–152; **A. & R. Fernandes.** Les voyages de Manoel Galvão da Sylva au Mozambique, pp. 153–159; **H. Wild.** Biographical notes on botanical collectors in the Federation of Rhodesia and Nyasaland, pp. 161–173; **F. White.** A history of plant collecting in Northern Rhodesia, pp. 175–203; **J. B. Gillett.** The history of the botanical exploration of the area of The Flora of Tropical east Africa (Uganda, Kenya, Tanganyika and Zanzibar), pp. 205–229; **G. Cufodontis.** A preliminary contribution to the knowledge of the botanical exploration of north-east tropical Africa, pp. 233–248.

Cb327. **Gossweiler, J.** [1939?] Elementos para a história da exploração botânica de Angola: Itinerário e relação dos viajantes e exploradores que fizeram colecções botânicas en Angola. *Boletim da Sociedade Broteriana* (2) 13: 283–305.

Cb328. **Gunn, M., & L. E. Codd.** 1981. Historical outline up to about 1750. In: *Botanical exploration of southern Africa: An illustrated history of early botanical literature on the Cape flora. Biographical accounts of the leading plant collectors and their activities in southern Africa from the days of the East India Company until modern times*, pp.1–74.

Cape Town, Balkema for the Botanical Research Institute. xv+400 pp. Introductory volume to the Flora of Southern Africa.
Also **L. E. Codd & M. Gunn.** 1985. More early illustrations of Cape plants. *Veld & flora* 71 (3): 86–88.

Cb329. **Hepper, F. N.** 1991. The Niger and the Nile: Botanical exploration around two African rivers. *Annals of the Missouri Botanical Garden* 78: 81–86.

Cb330. **Hutchinson, J.** 1946. Notes on the history of botanical exploration in South Africa up to the time of Harvey and Sonder's Flora Capensis [1859–65]. In: *A botanist in South Africa*, pp. 609–647. London, P. R. Gawthorn.
Also **G. H. Spaulding.** 1953. *Index: A botanist in South Africa.* Arcadia, CA, California Arboretum Foundation. [iv]+53 ff. *Lasca miscellanea* vol. 1. An index to plants mentioned.

Cb330A. **McCracken, D. P. & P. A.** 1990. *Natal: The garden colony: Victorian Natal and the Royal Botanic Gardens, Kew.* Sandton, Frandsen Publishers. xxi+99 pp.

Cb331. **Phillips, E. P.** 1930. A brief historical sketch of the development of botanical science in South Africa and the contribution of South Africa to botany. *South African journal of science* 27: 39–80.

Cb332. **Reynolds, G. W.** 1969. Introduction. In: *The Aloes of South Africa*, pp. 1–96. Johannesburg, Aloes of South Africa Book Fund.
Illustrated historical account of botanical exploration. Bibliographical notes (i.e. "Bibliographical notes – pre-Linnean," and "Bibliographical notes from 1753 onwards," pp. 71–96).

Cb333. **Vallot, J.** 1882. Notice sur les voyages botaniques accomplis dans l'Afrique tropicale. *Bulletin de la Société Botanique de France* 29: 171–205.

See also Ba243–250 & Cb470.

BIBLIOGRAPHIES OF REGIONAL BOTANY

European (& Mediterranean)

Cb334. **Akchurina, R. M.** 1963. *Flora i rastitel'nost' Yugo–Vostoka Evropeiskoi chasti SSSR, chast' vtoraya (1949–1960). Bibliograficheskii ukazatel'; botanicheskoi literatury.* Ed. by R. M. Akchurina. Saratov, Nauchnaya Biblioteka Saratovskogo Universiteta im N. G. Chernyshevskogo. 219 pp. *Bibliografiya Saratovskoi oblasti* vyp. 6.

Cb335. **Aytug, N. & A. Cakman.** 1972. *Türkiye flora biblyografyasi, 1843–1968.* Ankara, Türdok. 38 pp. *Türdok biblyografya serisi* no. 5.

Cb336. **Baruisheva, M. D., et al.** 1959. *Oboshchevodstvo i kartofelevodstvo na Krainem severe: Bibliograficheskii ukazatel', 1932–1957gg.* Leningrad, Tsentral'naya Nauchnaya Sel'skokhozyaistvennaya Biblioteka [etc.]. 51 pp.

Cb337. **Boltova, V. M.** 1957. *Flora i rastitel'nost' Komi ASSR. (1814–1955 gg): Bibliograficheskii ukazatel'.* Moscow, Syktyvkar. 179 pp.

Cb338. **Buhse, F.** 1854. Uebersicht der naturhistorische Literatur von Liv–, Kur– und Estland, entworfen unter Mitwirkung ... A. Buchholtz. *Correspondenzblatt der Naturforschenden Vereins zu Riga* 7: 49–58, 81–89, 115–122, 129–140.

Cb339. **Bychkova, E. Kh.** 1950. *Flora i rastitel'nost' Yugo–Vostoka Evropeiskoi chasti SSSR.* Saratov, Nauchnaya Biblioteka Saratovskii Gosod. Universiteta im N. G. Chernyshevskogo. 202 pp. *Bibliografiya Saratovskoi oblasti* vyp. 2.

Cb340. **Chmielinska, M.** 1954. *Polska bibliografia zielarstwa, za okres od poczatku XVI wieku do roku 1940.* Warsaw. xxxii+516 pp.

Cb341. **Christensen, C. F. A.** 1924–26. Bibliografi: Kronologisk med korte biografier og systematisk. In: *Den Danske botaniks historie med* *tilhørende bibliografi*, vol. 2. Copenhagen, Hagerups. xix+680 pp.

Cb342. ___. 1913. *Den Danske botaniske Litteratur 1880–1911, sammenstillet, [etc.]. Bibliographia botanica danica 1880–1911 [etc.].* Copenhagen, Munksgaard. xxi+279 pp.; *Continued* ___. 1940. *Den Danske botaniske Litteratur. Bibliographia botanica danica, 1912–1939. Med portræter af 78 danske botanikere.* Udgivet i anledning af Dansk Botanisk Forenings 100–Aars Jubilæum. Copenhagen, Munksgaard. [viii]+350 pp.; **A. Hansen.** 1963. II. Bibliografi. In: Den Danske botaniske litteratur: Bibliographia botanica Danica 1940–1959 med tilhørende forfatter–biografier. *Dansk botanisk arkiv* 21 (1): 73–318.

Cb343. **Christiansen, W. & W.**, eds. 1932–36. *Das botanische Schrifttum von Schleswig–Holstein, Hamburg und Lübeck.* Kiel, 'Heimat und Erbe.' [iv]+323 pp.

Cb344. **Colmeiro, M.** 1858. Estudios bibliográficos. Biblioteca botánica Hispano–Lusitana, ó enumeracion metódica y analítica de los escritos Españoles y Portugueses, tanto impresos como inéditos, que interesan á los botánicos bajo diferentes aspectos, y de los publicados en otras naciones sobre la vegetacion de la península Hispano–Lusitana. In: *La botánico y los botánicos de la península Hispano–Lusitana*, Parte primera, pp. 1–139. Madrid, M. Rivadeneyra.

Cb345. **Deleuil, G.** 1954–58. Bibliographie botanique Provençale (végétaux vasculaires), [etc.]. *Bulletin du Muséum d'Histoire Naturelle, Marseille* 14: 1–208, 1954; Ibid. 18: 105–148, 1958. ••Covers the Départements des Alpes Maritimes, Basses–Alpes, Bouches-du–Rhône, Var et Valeuse.

Cb346. **Dillemann, G.** 1939–48. Bibliographie des flores régionales de la France. *Monde des plantes* 40(237): 22–23, (238): 27–31, (239/240): 36–38, 1939; 41(241): 3, (242): 6–8, 1940; 42(244): 12–14, (245): 21–22, (246): 28–29, (247): 36–38, 1947; 43(249): 3–6, (250/251): 13–15, 1948.

Cb347. **Dussaussois, G.** 1978. *Documents de bibliographie botanique Pyrénéenne.* 1: *Bibliographie botanique des Pyrénées centrales et occidentales de France et d'Europe.* [Toulouse,] Université Paul Sabatier, Laboratoire de Botanique et Biogéographie. ii+114 pp.

Cb348. **Fischer, E.** 1901. *Flora helvetica, 1530–1900.* Berne, Centralkommission für Schweizerische Landeskunde. xviii+242 pp. *Bibliographie der Schweizerischen Landeskunde* IV.5. *Continued* ___. 1922. ___: *Nachträge.* Berne. ix+40 pp. *Ibid.* IV.5.

Cb349. **Flerov, A. F.** 1907. Okskaya flora. *Trudy imperatorskago S.–Peterburgskago botanicheskago sada* 27 (1): xvi+286 pp. ••*Includes* Pt. 1. Obzor literatury po flore okskikh gubernii, pp. 1–151, illus.

Cb350. ___. 1907. *Flora Kaluzhskoi gubernii.* Kaluga, [Gouvern. Zemstvo Statist. Abt.]. 3 vols. ••*Includes* Pt. 1. Literatura po flore Kaluzhskoi gubernii, pp. 1–61.

Cb351. **Futák, J. & K. Domin.** 1960. *Bibliografia k flóre CSR do r. 1952.* Bratislava, Slovenská Akadémia Vied. 883 pp.

Cb351A. **Gattefosse, J., & E. Jahandiez.** 1922. Essai de bibliographie botanique marocaine. *Bulletin de la Société des Sciences Naturelles du Maroc* 2: 71–86.

Cb352. **Gombocz, E.** 1939. *A Magyar növénytani irodalom bibliográfiája 1578–1900. Bibliographie der ungarischen botanischen Literatur 1578–1900.* Budapest, Bethlen. 360 pp. *Az Országos Magyar Természettudományi Múzeum Növénytára.*

Cb353. **Hamann, U., & G. Wagenitz.** 1977. *Bibliographie zur Flora von Mitteleuropa.* 2. Aufl. Berlin, Hamburg, P. Parey. 376 pp. ••*Replaces* 1st ed., Munich, Hanser Verlag, 1970. 328 pp.

Cb354. **Henrey, B.** 1975. *British botanical and horticultural literature* *before 1800. Comprising a history and bibliography of botanical and horticultural books printed in England, Scotland, and Ireland from the earliest times until 1800.* London, Oxford Univ. Pr. 3 vols.

Cb355. **Hylander, N.** 1970. Prima loca plantarum vascularium Sueciae. Plantae subspontaneae vel in tempore recentiore adventitiae. Första litteraturuppgift för Sveriges vildväxende kärlväxter jämte uppgifter om förstu svenska fynd. Fövildade eller senare tid in komma växter. *Svensk botanisk tidskrift* 64 supplement: 1–322.

Cb356. **Jäger, E. J., & C. Müller-Uri.** 1981–82. *Wuchsform und Lebensgeschichte dere Gefässpflanzen Zentral-europas: Bibliographie.* Halle. Teile 1 & 2. iv+423 pp. *Terrestrische Ökologie,* sonderhefte 1.

Cb357. **Kent, D. H.** 1963. Progress in the study of the British flora since World War II. *Webbia* 18: 129–150; *Continued* ___. 1974. Progress in the study of the British flora, 1961–1971. *Memórias da Sociedade Broteriana* 24: 353–375.

Cb358. ___. 1967. *Index to botanical monographs: A guide to monographs and taxonomic papers relating to phanerogams and vascular cryptogams found growing wild in the British Isles.* London, Academic Pr. for the Botanical Society of the British Isles. xi+163 pp.

Cb359. **Kleppa, P.** 1973. *Norsk botanisk bibliografi 1814–1964.* Oslo–Bergen–Tromsø, Universitetsforlaget. [xi]+334 pp. *Universitetsbiblioteket i Oslo, skrifter* no. 2.

Cb360. **Kobrle, A.** 1967. *Bibliografie botanické literatury vychodních Cech. Bibliographie der botanischen Litteratur Ostböhmens.* Pardubice, Vychodoceské Muzeum. 163 pp.

Cb361. **Krause, K.** 1927. Die botanische Litteratur über die Turkei. *Repertorium specierum novarum regni vegetabilis* 24: 113–126.

Cb362. **Krok, T. O. B. N.** 1925. *Bibliotheca botanica suecana ab antiquissimis temporibus ad finem anni MCMXVIII. Svensk botanisk litteratur från äldsta tider t. o. m. 1918.* Uppsala, Almqvist & Wiksell. xvi+799 pp. ••*Replaces* **J. E. Wikström.** 1831. *Conspectus litteraturae botanicae in Suecia ab antiquissimis temporibus usque ad finem anni 1831, notis bibliographicis et biographiis auctorum adjectis.* Stockholm. [2]+xlix+342 pp.

Cb363. **Lebedev, D. V.** 1956. *Vvedenie v botanicheskuyu literaturu SSSR: Posobie dla geobotanikov. [Introduction to the literature of the botany of the U. S. S. R.]* Moscow & Leningrad, Akademii Nauk SSSR. 382 pp.

Cb364. **Lipshits, S.** 1975. *Florae URSS fontes. Literaturnye istochniki po flore SSSR. [Literature sources on the flora of the USSR.]* Leningrad, Izd. "Nauka." 231 pp.

Cb365. **Lloyd Library, Cincinnati.** 1911–14. Bibliography relating to the floras. *Bibliographical contributions from the Lloyd Library* 2–13: 1–513.

Cb366. **L'vina, I. N., & G. E. Kushnirenko.** 1970. *Flora i rastitel'nost' Moldavii bibliografia, 1715–1965. [Bibliography of the flora and vegetation of Moldavia, 1715–1965].* Kishinev, Kishinevskii Gosudarstvennyi Universitet im. V. I. Lenina, Kafedra Botaniki, Nauchnaya Biblioteka. 206 pp.

Cb366A. **Matuskiewicz, A., & J. B. Falinski.** 1964–72 *Bibliografia fitosocjologiczna Polski.* Warsaw. 4 pts. *Materialy zakladu fitosocjologii stosowanej Universitetu Warzawskiego* nos. 1, 5, 19 & 28; *Continued* ___. 1974–75. Bibliographia phytosociologica Polonica. Pars 2. *Excerpta botanica* B 14: 252–320; *Ibid.* 15: 1–139. ••Covering literature of 1959 onwards.

Cb367. **Mikhailova, M. G.** 1938. *Flora ta roslinnist' URSR. Bibliografiya. Pokazhchik knizhok i zhurnal'nikh statei.* Kiev, Akad. Nauk URSR, In-t. Botaniki. 63 pp.

Cb368. ___. 1941. *Flora i rastitel'nost' URSR. Ukazatel' knig i zhurnal'nykh statei 1747–1938 gg. Dopolnenie 1.* Kiev, Akad. Nauk URSR, In–t. Botaniki. 41 pp.

Cb369. **Moskovskoe Obshchestvo Ispytatelei Prirody.** 1972. Bibliografiya. In: *Flora i rastitel'nost Moskovskoi oblasti: Istoriya izucheniya i annotirovannaya bibliografiya,* pp. 54–288. [*Flora and vegetation of Moscow region: History of research and annotated bibliography.*] By I. A. Gubanov, B. A. Starostin, & V. N. Tikhomirov. Moscow, Izd. Moskovskogo Universiteta. 288 pp. ••The bibliography is chronologically arranged with an author index.

Cb369A. **Nowak, M.** 1991. *Bibliografia flory Polskiej — Rosliny naczyniowe za lata 1971–1980. Bibliography of Polish flora — vascular plants for the years 1971–1980.* Cracow, Polska Akademia Nauk, Instytut Botaniki im. W. Szafera. 272 pp. *Bibliografie botaniczne* no. 5.

Cb370. **Osipchik, L. A. & V. S. Gel'tman.** 1970. *Sovetskaya literatura po flore i rastitel'nosti Belorussii: Bibliografiya 1919–1968gg.* [*Soviet literature on the flora and vegetation of Belorussia: A bibliography for 1919–1968.*] Minsk, Akademiya Nauk Belorusskoi SSR, Fundamental'naya Biblioteka im Ya. Kolasa. ii+433 pp. *Continued* ___. 1971. ___*: Bibliograficheskii ukazatel' za 1970 god.* Minsk. 90 pp.

Cb371. **Pax, F.** 1929. *Bibliographie der schlesischen Botanik.* Breslau, Historische Kommission für Schlesien. [viii]+167 pp. *Schlesische Bibliographie* vol. 4.

Cb372. **Pole, R.** 1915. Ukazatel' lesovodstvennoi i botaniko–geograficheskoi literatury severnoi Rossii i Finlyandii. *Lesnoi zhurnal* 45 (3): viii+268.

Cb373. **Saccardo, P. A.** 1895–1901. Repertorio biografico e bibliografico dei botanici Italiani aggiuntivi gli stranieri che trattarono della flora Italiana and Repertorio biografico e bibliografico supplementare ... In: *La botanica in Italia: Materiali per la storia di questa scienza,* Vol. 1, pp. 7–180, 220–233, Vol. 2, pp. 1–132. Venice. 2 vols.

Cb374. **Sælan, T.** 1916. Finlands botaniske litteratur till och med år 1900. *Acta Societatis pro Fauna et Flora Fennica* 43: 1–633; *Continued* **R. Collander, V. Erkamo & P. Lehtonen.** 1973. Bibliographia botanica Fenniae 1901–1950. *Acta Societatis pro Fauna et Flora Fennica* 81: 1–646.

Cb375. **Salanon, R.** 1970. Bibliographie botanique et l'Auvergne et des contrées limitrophes. *Arvernia biologica,* n.s. *botanique* (2) 10: xv+144 pp.

Cb376. **Sapiraite, S.** 1971. *Lietuvos botanikos bibliografija 1800–1965.* Mokslinis redaktorius K. Brundza. Vilnius, Lietuvos T S R Mokslu Akademijos, Centrine Biblioteka. 528 pp.; *Continued* **S. Norkuniene.** 1984. *Lietuvos botanika, 1966–1970: Literaturos rodykle.* Vilnius, [etc.]. 534 pp.

Cb377. **Simpson, N. D.** 1960. *A bibliographical index to the British flora, including floras, herbals, periodicals, societies and references relating to the identification, distribution and occurrence of phanerogams, vascular cryptogams and charophytes in the British Isles.* Bournemouth, Privately. xix+429 pp.

Cb378. **Soó, R. v.**, ed. 1978. *Bibliographia synoecologica scientifica Hungarica, 1900–1972.* Budapest, Akademiae Kiado. 500 pp.

Cb379. **Szymkiewicz, D.** 1925. *Bibliografia flory Polskiej.* Cracow, Polska Akademia Umiejetnosci. [i]+159 pp. *Prace monograficzne Komisji Fizjograficznej* vol. 2.

Cb380. **Trautvetter, E. R. von.** 1880. Florae Rossicae fontes. *Trudy imperatorskago S.–Peterburgskago botanicheskago sada* 7(1): 341 pp.; *Continued* **F. E. Herder.** 1881. Fontes florae rossicae. Continuatio, 1846–1879. *Botanisches Centralblatt* 5: 155–158, 185–188, 220–222, 281–284, 316–318, 385–388, 406–408, 1881; *Ibid.* 6: 31–34, 63–67, 137–138, 176–178, 245–247, 277–296, 324–329, 1881.

Cb381. **Utkin, L. A., A. F. Gammerman & V. A. Nevskii.** 1957. *Bibliografiya po lekarstvennym rasteniyam. Ukazatel' otechestvennoi literatury: Rukopisi XVII–XIX vv., pechatnye izdaniya 1732–1954 gg.* [*Bibliography of medicinal plants. A guide to national literature: MSS of the 17th–19th centuries, printed publications of 1732–1954.*] Moscow/Leningrad, Akademiya Nauk SSSR. 725 pp.

Cb382. **Warming, E.** 1880. Den Danske botaniske Literatur fra de aeldeste tider til 1880. *Botanisk tidskrift* 12: 42–131, 153–247; *Also* **J. C. Bay.** 1890–92. Tillaeg til "Den Danske botaniske literatur dra de aeldste tider til 1880," sammenstillet af Eug. Warming. *Botanisk tidskrift* 17: 315–329, 1890; *Ibid.* 18: 95–103, 1892.

Cb383. **Winkler, K.** 1877. Literatur und Pflanzenverzeichniss der Flora Baltica. *Archiv für die Naturkunde Liv–, Ehst– und Kurlands* Ser. 2, 7 (1): 387–490.

Cb384. **Zatikian, S. S., & A. M. Barsegian.** 1980. *Flora, rastitel'nost, rastitel'nie resursy Armenii: Anatomiya, fiziologiya i biokhimiya rastenii. Bibliografiya (1900–1975).* Erevan. 426 pp.

North American

Cb385. **Adams, J.** 1928–30. A bibliography of Canadian plant geography I (–IV). *Transactions of the Royal Canadian Institute* 16: 293–355, 1928; 17: 103–145, 1929; 17: 227–265, 267–295, 1930; *Continued* **J. Adams & M. H. Norend.** 1932–36. ___. V (–VI). *Ibid.* 18: 343–373, 1932; 21: 95–135, 1936; **H. A. Senn.** 1951. ___. VII. Indices. (–VIII & corrections). *Ibid.* 26: 9–151, 1946; 26: 153–344, 1947; 28: 161–163, 1950; ___. 1951. ___ *IX, 1941–45.* Ottawa, Canada Department of Agriculture. 183 pp. *Publication* no. 863.

Cb386. **Allison, E. M.** 1908. Bibliography and history of Colorado botany. *University of Colorado studies* 6: 51–76.

Cb387. **Blake, S. F.** 1954. Guide to popular floras of the United States and Alaska: An annotated, selected list of nontechnical works for the identification of flowers, ferns and trees. *Bibliographical bulletin, U. S. Department of Agriculture* 23: 56 pp.

Cb388. **Bourgeron, P. S., et al.** 1988. Bibliography of Montana vegetation description. *Great Basin naturalist* 48(3): 301–323.

Cb389. **Bowers, J. E.** 1981. Local floras of Arizona: An annotated bibliography. *Madroño* 28: 193–209.

Cb390. ___. 1982. Local floras of the southwest, 1920–1980: An annotated bibliography. *Great Basin naturalist* 42: 105–112.

Cb391. **Catling, P. M., et al.** 1986. Bibliography of vascular plant floristics for New Brunswick, Newfoundland (insular), and Nova Scotia. *Technical bulletin, Research Branch, Agriculture Canada* 1986–3E: 28 pp.

Cb392. **Christensen, E. M.** 1967. Bibliography of Utah botany and wildland conservation. *Science bulletin, Brigham Young University, biological series* 9: 1–136. *Continued* ___. 1967. ___, no. 2. *Proceedings of the Utah Academy of Sciences* 44: 545–566.

Cb393. **Core, E. L., W. H. & B. J. Gillespie.** [1962.] *Bibliography of West Virginia plant life.* New York. [v]+46 pp. *SL bibliography series.*

Cb394. **Dayton, W. A.** 1952. United States tree books: A bibliography of tree identification. *Bibliographical bulletin, United States Department of Agriculture* 20: 32 pp.

Cb395. **Douglas, G. W., A. Ceska, & G. G. Ruyle.** [1983.] *A floristic bibliography for British Columbia.* Victoria, B.C., Province of British Columbia, Ministry of Forests. 143 pp. *Land management report* no. 15.

Cb396. **Egler, F. E.** 1948. Regional vegetation literature. I: Connecticut. *Phytologia* 3: 1–26; ___. 1949. ___ II: Rhode Island. *Ibid.* 3: 49–56; ___. 1950. ___ III: Massachusetts. *Ibid.* 3: 193–237.

Cb397. **Ewan, J.** 1967. A bibliography of Louisiana botany. *Southwestern Louisiana journal* 7: 1–83; *Continued* **E. Sundell.** 1979. A bibliography of Louisiana botany, 1951–1975. *Tulane studies in zoology and botany* 21: 1–66. ••Ewan covers the literature from 1728–1950.

Cb398. **Fairbrothers, D. E.** 1964. An annotated, chronological bibliography of the floristic publications of New Jersey from 1753–1961. *Bulletin of the Torrey Botanical Club* 91: 47–58, 141–151; *Continued* ___. 1966. An annotated bibliography of the floristic publications of New Jersey from 1962–1965. *Ibid.* 93: 352–359.

Cb399. **Greene, H. C., & J. T. Curtis.** 1955. *A bibliography of Wisconsin vegetation.* Milwaukee, WI, Milwaukee Public Museum. 84 pp. *Publications in botany* no. 1.

Cb400. **Habeck, J. R. & E. Hartley.** 1965. The vegetation of Montana: A bibliography. *Northwest science* 39: 60–72.

Cb401. **Hodgins, J. L.** 1978. *A guide to the literature on the herbaceous vascular flora of Ontario.* Toronto, Botany Pr. 73 pp.

Cb402. **House, H. D.** 1941–42. Bibliography of botany of New York state, 1751–1940. *Bulletin, New York State Museum* 328: 1–174, 1941; *Ibid.* 329, 177–233, 1942.

Cb403. **Johnson, F. L., & T. H. Milby.** 1989. *Oklahoma botanical literature.* Norman, OK, Univ. of Oklahoma Pr. 162 pp.

Cb404. **Kellerman, W. A.** 1893. Bibliography of Ohio botany. *Bulletin of technical services, Ohio Agricultural Experiment Station* 1 (3): 180–201.

Cb405. **Langman, I. K.** 1964. *A selected guide to the literature on the flowering plants of Mexico.* Philadelphia, PA, Univ. of Pennsylvania Pr. 1,015 pp. *A Morris Arboretum monograph.* ••*Replaces* **N. León.** 1895. *Biblioteca botánico-Mexicana: Catálogo bibliográfico, biográfico y crítico de autores y escritos referentes a vegetales de México y sus aplicaciones, desde la conquista hasta el presente.* Suplemento a la Materia Médica Mexicana publicada por el Instituto Médico Nacional. México, Secretaría de Fomento. 372 pp.

Cb406. **Little, E. C. S., & B. H. Honkala.** 1976. *Trees and shrubs of the United States: A bibliography for identification.* Washington, DC, United States Department of Agriculture, Forest Service. 56 pp. *Miscellaneous publication, United States Department of Agriculture* no. 1336.

Cb407. **McKay, A. H.** 1901–17. Botanical bibliography of Canada (*later* Bibliography of Canadian botany), 1900 (–1915). *Proceedings and Transactions of the Royal Society of Canada* (2) 7 (sect. IV): 141–142, 1901; 8 (IV): 157–160, 1902; 9 (IV): 169–172, 1903; 10 (IV): 153–160, 1905; 11 (IV): 143–152, 1906; 12 (IV): 33–48, 1906; (3) 4 (IV): 121–153, 1911; 5 (IV): 177–189, 1912; 6 (IV): 129–137, 1913; 7 (IV): 175–185, 1914; *Transactions of the Royal Society of Canada* (3) 8 (sect. IV): 25–35, 1914; 9 (IV): 251–261, 1916; 10 (IV): 189–199, 1917.

Cb408. **Peet, R. K.** 1979. A bibliography of the vegetation of the Carolinas. *Veröffentlichungen des Geobotanischen Institutes der Eidgenossosche Technische Hochschule, Stiftung Rübel* 68: 263–297.

Cb409. **Schmutz, E. M.** 1978. *Classified bibliography on native plants of Arizona.* Project in cooperation with the Natural Vegetation Committee, Arizona Chapter, Soil Conservation Society of America. Tucson, AZ, Univ. of Arizona Pr. xii+160 pp.

Cb410. **Standley, P. C.** 1910. The type localities of plants first described from New Mexico: A bibliography of New Mexican botany. *Contributions from the United States National Herbarium* 13: 143–246. ••Bibliography, pp. 229–246.

Cb411. **Torrey Botanical Club.** 1969. *Index to American botanical literature, 1886–1966.* Boston, MA,

G. K. Hall. 4 vols. *Continued* ___. 1977. ___*: 1967–1976*. Boston, MA, [etc.] 740 pp. ••An author catalogue. Cumulation of the entries first pub. in the *Bulletin of the Torrey Botanical Club*.

Cb412. **Watson, S.** 1878. Bibliographical index to north American botany: Or citations of authorities for all the recorded indigenous and naturalized species of the flora of North America ... Part 1: Polypetalae. *Smithsonian miscellaneous collections* 258: [v]+476 pp. ••All published.

South & Central American (& Caribbean)

Cb413. **Angely, J.** 1982. *The South America botanical bibliography*. São Paulo, Privately. 5 vols.

Cb414. **Bertoni, B. S., J. M. Mascherpa & R. Spichiger.** 1982. Datos bibliográficos para el estudio de la vegetación y de la flora del Paraguay. (Eléments bibliographiques pour l'étude de la végétation et de la flore du Paraguay). *Candollea* 37: 277–313.

Cb415. **Cabrero, A. L., & M. Ferrario.** 1970. *Bibliografia botanica de la Provincia de Buenos Aires: Plantas vasculares*. La Plata, Comision de Investigaciones Cientificas – Provincia de Buenos Aires. 96 pp.

Cb416. **Castellanos, A., & R. A. Pérez-Moreau.** 1940–41. Contribución a la bibliografía botánica Argentina: I, [parts 1–2]. *Lilloa* 6: 5–161, 1940; 7: 5–549, 1941. *Also pub. separately* Buenos Aires, Imprenta y Casa Editori Coni, 1941. 555 pp.

Cb417. **Coimbra Filho, A. F., & A. Magnanini.** 1964. Bibliografia florestal Brasileira. 1.a, contribuição. *Publicações, Instituto Nacional de Pesquisas da Amazônia* 20: 1–93.

Cb418. **Duek, J. J., M. Ramirez & O. Leon.** 1972. Sistema automatizado de informacion bibliographica. Bibliografía botánica con especial enfasis en la botanica sobre Chile. Parte 1. Indice alfabetico de autores. *Miscelánia botanica Gayana* 2: 1–167; ___. 1973. ___. Parte 2. Indice dual de descriptores. *Ibid.* 3: 1–114.

Cb419. **Estrada, J., J. Fuertes, & J. M. Cardiel**, eds. 1991. *Bibliografía botánica para Colombia*. Bogota, Agencia Española de Cooperación Internacional AECI–ICI Oficina Técnica de Cooperación en Colombia. xvii+400+[1] pp.

Cb420. **Graham, A.** 1973. Literature on vegetational history in Latin America. In: ___, ed. *Vegetation and vegetational history of northern Latin America*, pp. 315–360. Amsterdam, Elsevier. *Continued* ___. 1979. ___: Supplement 1. *Review of palaeobotany and palynology* 27: 29–52.

Cb421. **Hampshire, R. J., & D. A. Sutton.** 1988. *Flora Mesoamericana: A preliminary bibliography of the Mesoamerican flora*. St. Louis, MO, Missouri Botanical Garden; Mexico, D.F., Instituto de Biología, U.N.A.M.; London, British Museum (Natural History). [vi]+194 pp.

Cb422. **Hicken, C. M.** 1927. Contribución al conocimiento de la bibliografía botánica Argentina. *Darwiniana* 1: 191–318, 1927. *Continued* ___. 1929. ___: Segunda contribucion. *Ibid.* 2: 319–430; ___. 1930. ___: Tercera contribucion. *Ibid.* 3: 431–539.

Cb423. **Kurtz, F.** 1913–15. Essai d'une bibliographie botanique de l'Argentine. [Ed. 2.] *Boletín de Academia Nacional de Ciencias, Córdoba* 19: 221–376, 1913; 20: 369–467, 1915. *Continued* **L. Hauman & A. Castellanos.** 1922. Bibliografia botánica Argentina, especialmente para los años 1914–1921: Primer suplemento a la obra bibliografica de F. Kurtz. *Physis* 5: 263–291; ___. 1927. ___ para los años 1922–1926: Segundo suplemento [etc.] *Ibid.* 8: 436–467. ••Supplement covers literature of 1914–21; supplement 2 covers 1922–26. *Replaces* **F. Kurtz.** 1899. ___. [1st ed.] *Boletín de Academia Nacional de Ciencias, Córdoba* 16: 117–205; *Also pub. separately*

Buenos Aires, Coni Frères, 1900. 90 pp. Supplemented by C. M. Hicken, 1927; *see* Cb422.

Cb424. **Looser, G.** 1927. Bibliographía botánica Chilena. *Revista de bibliografía Chilena* 1927: 212–230, 364–390.

Cb425. **Marticorena, C.** 1992. *Bibligrafia botánica taxonómica de Chile*. St. Louis, MO, Missouri Botanical Garden. iii+587 pp. *Monographs in systematic botany from the Missouri Botanical Garden* no. 41.

Cb426. **Reiche, K.** 1934. Bibliografia de la flora Chilena: Literatura, mapas, colecciones. In: *Geografía botánica de Chile*, Trad. del alemán de G. Looser, vol. 1, pp. 44–85. Santiago de Chile, Imprenta Universitaria. ••*Also* ___. 1929. Bibliografia de la flora Chilena: Literatura, mapas, colecciones. *Revista chilena de historia y geografía* 62: 164–205.

Cb427. **Sparn, E.** 1938. Bibliografía de la flora fanerogámica de la Provincia de Córdoba, Argentina. *Revista del Museo Provincial de Ciencias Naturales de Córdoba* 3: 3–108.

Cb428. **Urban, I.** 1898. Bibliographia India occidentalis botanica. In: *Symbolae antillanae*, vol. 1: 1–195; ___. 1900. Continuatio I. *Ibid.* 2: 1–7; ___. 1902. Continuatio II. *Ibid.* 3: 1–13; ___. 1904. Continuatio III. *Ibid.* 5: 1–16. *Re-pub.* with original pagination: Amsterdam, Asher, 1964.

Cb429. **Verdoorn, F.** 1945. A selected list of travel books of botanical interest. In: *Plants and plant science in Latin America*, pp. xxiii–xxviii. Waltham, MA, Chronica Botanica Co.

Asian

Cb430. **Akademiya Nauk Kazakskoi SSR, Tsentral'naya Nauchnaya Biblioteka.** 1971. *Botanicheskaya literatura Kazakhstana, 1937–1965gg.* [*Kazakhstan botanical literature, 1937–1965.*] Alma–Ata. [iv]+426pp.

Cb431. **Akademiya Nauk Kirgizskoi SSR, Institut Biologii, Tsentral'naya Nauchnaya Biblioteka.** 1982–88. *Flora i rastitel'nost' Kirgizii: Bibliograficheskii ukazatel'. [Flora and vegetation of Kirgizia; Bibliographical index]*. By A. F. Zakachurina. Frunze, Izd. "Ilim." 2 vols. ••Vol. 1: 1918–65. 1988 ; vol. 2: 1966–75. 1982.

Cb432. **Aksenova, N. N.** 1966. *Flora zapadnoi Sibiri: Bibliograficheskii ukazatel'.* [*The flora of western Siberia: A bibliographical index.*] Tomsk, Tomskii Gosudarstvennii Universitet im Kuibysheva, Nauchnaya Biblioteka. 105pp.

Cb433. **Blatter, E.** 1911. A bibliography of the botany of British India and Ceylon. *Journal of the Bombay Natural History Society* 20: lxxix–clxxvi; *Continued* **H. Santapau.** 1952. Contributions to the bibliography of Indian botany. *Journal of the Bombay Natural History Society* 50: 520–548, 1952; 51: 205–259, 1952.

Cb434. **Centre National de la Recherche Scientifique, Paris.** 1972. *Bibliographie du Népal.* Vol. 3: *Sciences naturelles*. Tome 2: *Botanique*. By J.–F. Dobremez, F. Vigny & L. H. J. Williams. Paris. 126 pp. & 4 pl. *Cahiers Népalais*.

Cb435. **Devyatkina, A. V.** 1966. *Rastitel'nyi i zhovotnyi mir Uzbekistana: Bibliograficheskii ukazatel' (1917–1952)*. Tashkent, Akademii Nauk Uzbekskoi SSR, Fundamental'naya Biblioteka. 468 pp. *Priroda i prirodnye resursy Uzbekistana* vyp. 3.

Cb436. **Frey, W. & H.–J. Meyer.** 1971. Botanische Literatur über den Iran. *Botanische Jahrbücher* 91: 348–382.

Cb437. **Gorovoi, P. G.**, ed. 1973. *Flora, rastitel'nost' i rastitel'nye resursy dal'nego vostoka: Ukazatel' literatury (1928–1969 gg). [Flora, vegetation and plant resources of the Far east: Bibliography (1928–1969).]* Vladivostok, Akademiya Nauk SSSR, Dal'nevostochnyi Nauchnyi Tsentr im V. L. Komarova, [etc.]. 552 pp.

Cb438. **Karthikeyan, S., M. P. Nayar & R. S. Raghavan.** 1981. *Annotated bibliography of taxonomic* *botany of peninsular India, 1959–1978.* Howrah, Botanical Survey of India. 201 pp.

Cb439. **Kazmi, S. M. A.** 1970–77. *Bibliography on the botany of West Pakistan and Kashmir.* 1–5. Coconut Grove, FL, Field Research Projects. 5 pts. 603 pp.

Cb440. **Komarov, V. L.** 1928. *Bibliografiya k flore i opisaniyu rastitel'nosti dal'nego vostoka. [Bibliography of the flora and vegetation of the far East].* Vladivostok, Gosudarstvennoe Russkoi Geograficheskoe Obshchestvo. 278 pp. *Zapiski Yuzhno Ussuriiskogo Otdela* vol. 2.

Cb441. **Kurata, S., et al.**, eds. 1966. *A bibliography of forest botany in Japan (1940–1963).* Tokyo, Univ. of Tokyo Pr. xiii+146 pp.

Cb442. **Lebedev, D. V.** 1956. *Vvedenie v botanicheskuyu literaturu SSSR: Posobie dla geobotanikov. [Introduction to the literature of the botany of the U. S. S. R.]* Moscow & Leningrad, Akademii Nauk SSSR. 382 pp.

Cb443. **Lipskii, V. I.** 1902. Literatura po flore Srednai Azii. Literatura florae Asiae Mediae. In: Flora Srednei Azii ... Flora Asiae mediae ... [etc]. *Trudy Tiflisskogo Botanicheskogo Sada* 7 (1): 1–245.

Cb444. **Litvinov, D. I.** 1909. *Bibliografiya flory Sibiri.* St. Petersburg, Izd. Imperatorskoi Akademii Nauk. ix+460 pp. *Trudy Botanicheskago Muzeya Imperatorskoi Akademii Nauk* vol. 5. ••Supplemented by T. A. Vorob'eva et al. 1964; *see* Cb458.

Cb445. **Margolina, D. L.** 1941. Flora i rastitel'nost' Tadzhikistana: Bibliografiya. *Trudy Tadzhikskii Filial, Akademii Nauk SSSR* 15: 338 pp.

Cb446. **Merrill, E. D.** 1915. A contribution to the bibliography of the botany of Borneo. *Sarawak Museum journal* 2: 99–136. *Also* ___. 1921. A bibliographical enumeration of Bornean plants. *Journal of the Straits Branch of the Royal Asiatic Society* special no: 2–6.

Cb447. ___. 1926. A discussion and bibliography of Philippine flowering plants. *Popular bulletin, Bureau of Science, Philippines* 2: 1–239.

Cb448. **Merrill, E. D., & E. H. Walker.** 1938. *A bibliography of Eastern Asiatic botany.* Jamaica Plain, NY, Arnold Arboretum. xlii+719 pp. *Continued* **E. H. Walker.** 1960. ___*: Supplement 1.* Washington, DC, American Institute of Biological Sciences. xl+552 pp. ••*See also* **L. C. Goodrich.** 1940. Some bibliographical notes on Eastern Asiatic botany. *Journal of the Amererican Oriental Society* 60: 258–260; **Y. Tsiang.** 1977. (A commentary on the part of appendix (older Chinese works) in Merrill and Walker *A bibliography of eastern asiatic botany.) Acta phytotaxonomica sinica* 15: 96–106[in Chinese].

Cb449. **Narayanaswami, V.** 1961–65. *A bibliography of Indology (enumerating basic publications on all aspects of Indian culture).* Vol. 2: *Indian botany.* Calcutta, Government of India National Library. 2 vols, xxx+782 pp.

Cb450. **Nayar, M. P.** 1984. *Key works to the taxonomy of flowering plants of India.* Calcutta, Botanical Society of India. 5 vols. ••Vol. 1: Acanthaceae to Crypteroniaceae; vol. 2: Cucurbitaceae to Juncaginaceae; vol. 3: ; Vol. 4: Magnoliaceae to Orchidaceae; vol. 5: Gymnosperms; General index.

Cb451. ___, **& G. S. Giri.** 1988+. *Key works of floristics of India.* Calcutta, Botanical Society of India. Vol. 1+. *Flora of India* series 4. ••Bibliography by states. Vol. 1: Up to Madhya Pradesh.

Cb452. **Okuyama, S.** 1948–51. A list of botanical literature on the local flora of Japan. *Miscellaneous reports, National Science Museum, Tokyo* 3: 1–48, 1948; 4: 1–75, 1951.

Cb453. **Pavlov, N. V.** 1940. Rastitel'nost Kazakhstana. Vol. 2: Literaturnye istochniki po flore i rastitel'nosti Kazakhstana. *Trudy Kazakhstanskii Filial, Akademii Nauk SSSR* 19: 182 pp.

Cb454. **Petch, T.** 1925. *Bibliography of books and papers relating to agriculture and botany to the end of the year 1915*. Colombo. [iii]+256 pp. *Peradeniya manuals* vol. 3.

Cb455. **Pételot, A.** 1955. Bibliographie botanique de l'Indochinc. *Archives des recherches agronomiques et pastorales au Viêtnam* 24: 1–108. *Continued* **J. E. Vidal.** 1972. Bibliographie botanique Indochine de 1955–1969. *Bulletin de la Société des Études Indo-Chinoises de Saigon* 47: 657–748; **J. E. & Y. Vidal, & Pham Hoang Hô.** 1988. *Bibliographie botanique Indochinoise de 1970 à 1985: Documents pour la flore du Cambodge, du Laos et du Viêtnam*. Paris, Muséum National d'Histoire Naturelle, Laboratoire de Phanerogamie. [ii]+132 pp. ••Name, subject, taxonomic and serials indexes. *Replaces* La botanique en Indochine: Bibliographie. *Bulletin de l'Agence Économique de l'Indochine* 32: 587–632, 1929.

Cb456. **Reed, C. F.** 1969. *Bibliography to floras of southeast Asia: Burma, Laos, Thailand (Siam), Cambodia, Viet Nam (Tonkin, Annam, Cochinchina), Malay peninsula and Singapore*. Baltimore, MD, P. M. Harrod. [ii]+191 pp. & map.

Cb457. **Savich, V. P., & A. A. Elenkin.** 1950. Vvedenie k flore lishainikov aziatskoi chasti SSSR. *Trudy Botanicheskogo Instituta im. V. L. Komarova, Akademii Nauk SSSR* (2) 6: 181–343.

Cb458. **Vorob'eva, T. A., V. P. Sokolova & A. A. Konograi.** 1964. *Rastitel'nost' i rastitel'nye resursy zapanoi Sibiri: Bibliografiya, 1909–1962 gg. [The vegetation and plant resources of Western Siberia: Bibliography (1909–1962)]*. Moscow, Izd. "Nauka." 151 pp. ••Supplements D. I. Litvinov. 1909; *see* Cb444.

Cb459. **Walker, E. H.** 1952. Contribution toward a bibliography of Thai botany. *Natural history bulletin, Siam Society* 15: 27–88; *Continued* **B. Hansen.** 1973. Bibliography of Thai botany. *Natural history bulletin, Siam Society* 24: 319–408.

Cb460. **Wang, Z.-X., et al.** 1983. *Bibliography of Chinese botany*. Ed. by the Chinese Botanical Society. 3 vols. Beijing, Science Pr. xxx+1,793 pp. ••Text mainly Chinese with part in English.

Australasian (& Pacific)

Cb461. **Merrill, E. D.** 1947. A botanical bibliography of the islands of the Pacific. *Contributions from the United States National Herbarium* 30: 1–322, and **E. H. Walker**, A subject index to Elmer D. Merrill's [etc.]. *Ibid.* : 323–404.

Cb462. **Mill, S. W., et al.** 1988. *Indexed bibliography on the flowering plants of Hawai'i*. Honolulu, HI, Univ. of Hawaii Pr. 214 pp. *Special publication, Bishop Museum* no. 82.

Cb462A. **Morat, P.** 1991. Publications botaniques faites sur le Pacifique par des chercheurs français et associés de 1960 à 1991. *Iles et archipels* 14: 305–328.

Cb463. **Sachet, M.-H., & F. R. Fosberg**, eds. 1955. *Island bibliographies: Micronesian botany, land environment and ecology of coral atolls: Vegetation of tropical Pacific islands*. Washington, DC, National Academy of Sciences, National Research Council. [3]+v+577 pp. *N R C publication* no. 335. *Continued* **___.** 1971. ___*: Supplement*. Washington, DC, National Academy of Sciences, National Research Council. ix+427 pp.

Cb464. **Schofield, E. K.** 1973. Annotated bibliography of Galapagos botany 1836–1971. *Annals of the Missouri Botanic Garden* 60: 461–477; *Continued* **___.** 1980. ___: Supplement 1. *Brittonia* 32: 537–547.

Cb465. **State Library of New South Wales.** 1981. *A fine field for botanizing: An exhibition arranged in association with XIII International*

Botanical Congress, Sydney, 21–28 August 1981. 7 August 1981 – 21 March 1982, Mitchell and Dixson Galleries, State Library of New South Wales [etc.] Sydney, N.S.W. [ii]+42 pp. ••218 items. Introduction by M. G. Pitman.

See also Library cat. Bc193.

African (& Indian Ocean)

Cb466. **Bullock, A. A.** 1978. *Flora of southern Africa: Bibliography of South African botany (up to 1951)*. Ed. by O. Leistner. Pretoria, Government Printer. iii+194 pp.

Cb467. **Glynn, W. Tyrrell–, & M. R. Levyns.** 1963. *Flora Africana: South African botanical books 1600–1963: A selective bibliography based on the collections of the South African Public Library (with the addition of items from the Bolus Herbarium of the University of Cape Town)*. Pub. in commemoration of the Kirstenbosch Jubilee. Cape Town, South African Public Library. [x]+77 pp. & 8 pl. *Grey bibliographies* no. 8.

Cb468. **Hutchinson, J.** 1946. Review of South African botanical literature. In: *A botanist in South Africa*, pp. 554–608. London, P. R. Gawthorn.

Cb469. **Judy, T. M.** 1971. *Plant sciences in east Africa: A bibliography*. Morgantown, WV, West Virginia Univ. Library. 5 vols. ••Printout of a computer bibliography.

Cb470. **Kerkham, A. S.** 1988. *Southern African botanical literature 1600–1988 (SABLIT)*. Cape Town, South African Library. 256 pp. & 16 pl. *Grey bibliographies* vol. 16. ••*Includes* "South African botany: Historical background and botanical illustration," pp. 13–30.

Cb471. **Lebrun, J.-P.** 1969. Nouvelle contribution à la connaissance de la flore de la République du Sénégal et bibliographie botanique sénégalaise 1941–1969. *Bulletin de la Société Botanique de France* 116: 249–277.

Cb472. **Lorence, D. H., & R. E. Vaughan.** 1992. *Annotated bibliography of Mascarene plant life, including the useful and ornamental plants of the region, covering the period 1609–1990*. Lawai, Kauai, HI, National Tropical Botanical Garden. [vi]+274 pp.

Cb473. **Thirgood, J. V.** 1966. An annotated bibliography of Liberian forestry and botany, 1849–1964. 2nd ed. *Excerpta botanica* B 7: 75–91. ••*Replaces* 1st ed., 1849–63. 1964. 11 pp.

SPECIAL TOPICS

Plant geography

Cb474. **Del Tredici, P.** 1990. The new USDA plant hardiness zone map. *Arnoldia (Jamaica Plain)* 50(3): 16–20.
"The publication of this map provides an opportunity to review the history of hardiness zone maps."

Cb475. **Engler, A.** 1899. *Die Entwicklung der Pflanzengeographie in den letzten hundert Jahren und weitere Aufgaben derselben*. In: Gesellschaft für Erdkunde. *Wissenschaftliche Beiträge zum Gedächtniss der hundertjährigen Wiederkehr des Antritts von Alexander von Humboldt's Reise nach Amerika am 5. Juni 1799*. Berlin, W. H. Kühl. 247 pp.

Cb476. **Good, R.** 1955. Plant geography. In: California Academy of Sciences. *A century of progress in the natural sciences 1853–1953*, pp. 747–765. San Francisco, CA.

Cb477. **Hagen, J. B.** 1986. Ecologists and taxonomists: Divergent traditions in twentieth-century plant geography. *Journal of the history of biology* 19: 197–214.

Cb478. **Mennema, J.** 1985. The first plant distribution map. *Taxon* 34: 115–117.
I.e. by J. Schouw in his *Plantegeographisk Atlas*, Copenhagen, 1824.

Cb479. **Sampaio, A. J. de.** 1943. Histórico da geografia botânica. *Arquivos do Museu Nacional do Rio de Janeiro* 37: 465–486.

Cb480. **Sukopp, H.** 1987. On the history of plant geography and plant ecology in Berlin. *Englera* 7: 85–103.
For a German version, *see* Cb196A.

Cb481. **Turrill, W. B.** 1959. Plant geography. *Vistas in botany* 1: 172–229.

See also Ba1462–1463 Ca377(Hagen) & Cb233(Ewan).

Ecology (& forest botany)

Cb482. **Allen, D. E.** 1987. Changing attitudes to nature conservation: The botanical perspective. *Biological journal of the Linnean Society* 32: 203–212.

Cb483. **Boerker, R. H.** 1916. A historical study of forest ecology: Its development in the fields of botany and forestry. *Forestry quarterly* 14: 380–432.

Cb484. **Carron, L. T.** 1985. *A history of forestry in Australia.* Canberra, A.C.T., Australian National Univ. Pr. 372 pp.

Cb485. **Cittadino, E.** 1990. *Nature as the laboratory: Darwinian plant ecology in the German empire, 1880–1900.* Cambridge, Cambridge Univ. Pr. xi+199 pp.

Cb486. **Cox, T. R.** 1985. Americans and their forests: Romanticism, progress, and science in the 19th century. *Journal of forest history* 29: 156–168.

Cb487. **Clepper, H. E., & A. B. Meyer**, eds. 1960. *American forestry: Six decades of growth.* Washington, DC, Society of American Foresters. x+319 pp.
Includes **A. B. Meyer.** The literature of American forestry, pp. 50–64.

Cb488. **Davis, R. C.**, ed. 1983. *Encyclopedia of American forest and conservation history.* New York, Macmillan; London, Collier Macmillan. 2 vols, (xxxii+820 pp.).
Sponsored by the Forest History Society this comprises the work of over 200 contributors.

Cb489. **Fabricius, L.** 1905. *Geschichte der Naturwissenschaften in der Forstwissenschaft bis zum Jahre 1830.* Stuttgart. vii+137 pp. *Naturwissenschaftliche Zeitschrift für Land- und Forstwirtschaft, Stuttgart*, Beiheft 2.

Cb490. **Langlet, O.** 1971. Two hundred years' genecology. *Taxon* 20: 653–721.

Cb491. **Lowe, P. D.** 1976. Amateurs and professionals: The institutional emergence of British plant ecology. *Journal of the Society for the Bibliography of Natural History* 7: 517–535.

Cb492. **Nicolson, M.** 1989. National styles, divergent classifications: A comparative case study from the history of French and American plant ecology. *Knowledge & society* 8: 139–186.

Cb493. **Reed, H. S.** 1905. A brief history of ecological work in botany. *Plant world* 8: 163–170, 198–208.

Cb494. **Seddon, G.** 1974. Xerophytes, xeromorphs and sclerophylls: The history of some concepts in ecology. *Biological journal of the Linnean Society of London* 6: 65–87.

Cb495. **Sheail, J.** 1986. Grassland management and the early development of British ecology. *British journal for the history of science* 19: 283–299.

Cb496. **Stebbing, E. P.** 1922–62. *The forests of India.* London, J. Lane. 4 vols. *Re-pub.* New Delhi, A J Reprints Agency, 1982–83; *Continued* Sir H. Champion & F. S. Osmaston, eds. 1962. *E. P. Stebbing's "The forests of India" vol. IV. Being the history from 1925 to 1947 of the forests now in Burma, India, and Pakistan.* Oxford, Oxford Univ. Pr., 1962. xvi+485 pp. & 66 pl.
Vol. 1: 1796 to 1864. 1921; vol. 2: The development of the Indian Forest Service, 1865 to

1900. 1923; vol. 3: The progress of conservancy and the development of research in forestry, 1901 to 1920, including brief reviews of the progress of conservancy in the several presidencies and provinces between 1871–1900. 1926.

Cb497. **Tansley, A. G.** 1947. The early history of modern plant ecology in Britain. *Journal of ecology* 35: 130–137.

Cb498. **Tobey, R. C.** 1981. *Saving the prairies: The life cycle of the founding school of American plant ecology, 1895–1955.* Berkeley, CA, Univ. of California Pr. x+315 pp.

Cb499. **Whittaker, R. H.** 1958. Recent evolution of ecological concepts in relation to the eastern forests of North America. In: W. C. Steere, ed. *Fifty years of botany: Golden jubilee volume of the Botanical Society of America*, pp. 340–358. New York, McGraw–Hill Book Co.

See also Bb596 Ca368(Tobey) Ca377(Coleman, Hagen) Cb134(Guinier) Cb171(Hüttermann) Cb226 Cb228 & Cb233(Sears).

Morphology, anatomy & embryology

Cb500. **Arber, A.** 1950. *The natural philosophy of plant form.* Cambridge, Cambridge Univ. Pr. 260 pp. *Re-pub.* Darien, CT, Hafner, 1970.

Cb501. **Baranov, P. A.** 1955. *Istoriya embriologii rastenii v svyazi s razvitiem predstavlenii o zarozhedenii organizmov. [History of plant embryology in connection with the development of ideas on the origin of organisms.]* Moscow, Akademiya Nauk SSSR. 439 pp., illus.

Cb502. **Cheadle, V. I.** 1956. Research on xylem and phloem – progress in fifty years. *American journal of botany* 43: 719–731.

Cb503. **Eames, A. J.** 1958. Some aspects of progress in plant morphology during the past fifty years. In: W. C. Steere, ed. *Fifty years of botany: Golden jubilee volume of the Botanical Society of America*, pp. 606–614. New York, McGraw–Hill Book Co.

Cb504. **Eisnerová–Spudliová, V. J. B.** 1968. Rostlinná anatomie v Cecháh ve 2. polovine 19. století. (Plant anatomy in Bohemia in the 2nd half of the 19th century). *Dejiny ved a techniky* 1968: 150–157.

Cb505. ___. 1971. The anatomy of plants and its contribution to the origin of cellular theory in the early 19th century. *Acta historiae rerum naturalium nec non technicorum* 5: 269–333.

Cb506. **Eyde, R. H.** 1975. The foliar theory of the flower. *American scientist* 63: 430–437.

Cb507. **Guédès, M.** 1969. La théorie de la métamorphose en morphologie végètale: Des origines à Goethe et Batsch. *Revue de l'histoire des sciences* 22: 323–363.

Cb508. ___. 1972. La théorie de la métamorphose en morphologie végètale: A.–P. de Candolle et P. J. F. Turpin. *Revue de l'histoire des sciences* 25: 253–270.

Cb509. ___. 1973. La théorie de la métamorphose en morphologie végètale: La métamorphose et l'idée d'evolution chez Alexandre Braun. *Episteme* 7 (1): 32–51.

Cb510. **Hanstein, A. von.** 1886. *Ueber die Begründung der Pflanzenanatomie durch Nehemia Grew und Marcello Malpighi. Inaug.–Dissertation [etc.].* Bonn, Univ.–Buchdruckerei von C. Georgi. [iv]+91 pp.

Cb511. **Lorch, J.** 1968. The elusive cambium. *Archives internationales d'histoire des sciences* 80: 253–283.

Cb512. **Maheshwari, P.** 1950. Ch. 1. Historical sketch. In: *An introduction to the embryology of angiosperms*, pp. 1–27. New York, McGraw–Hill Book Co.

Cb513. **Maltzahn, K. E. von.** 1971. New formation of organs in plants — the foundation of plant morphogenesis. *Journal of the history of biology* 4: 307–317.

Cb514. **Metcalfe, C. R.** 1959. A vista in plant anatomy. *Vistas in botany* 1: 76–99.

Cb515. **___, L. Chalk, & D. F. Cutler.** 1979. History of systematic anatomy. In: C. R. Metclafe & L. Chalk. *Anatomy of the dicotyledons*, 2nd ed., vol. 1: 1–11. Oxford, Clarendon Pr.

Cb516. **Modilevskii, Ya. S.** 1956. *Istoriya otechestvennoi embriologii vysshikh rastennii.* Kiev, AN USSR. 202 pp.

Cb517. **Montgomery, W. M.** 1970. The origins of the spiral theory of phyllotaxis. *Journal of the history of biology* 3: 299–323.

Cb518. **Poddubnaya–Arnol'di, V. A.** 1973. K istorii tsitoembriologii rastenii. [On the history of cytoembryology]. *Iz istorii biologii* 4: 3–40.

Cb519. **Potonié, H.** 1903. *Ein Blick in die Geschichte der botanischen Morphologie und die Pericaulom–Theorie.* Jena, G. Fischer. [iii]+45 pp.
First pub. in *Naturwissenschaftliche Wochenschrift* 18: 3–8, 13–15, 25–28.

Cb520. **Schmucker, T., & G. Linnemann.** 1951. Geschichte der Anatomie des Holzes. In: H. Freund, ed. *Handbuch der Mikroskopie in der Technik*, Bd. 5, Teil 1, pp. 1–78. Frankfurt am Main, Umschau Verlag.

Cb521. **Schopfer, W.–H.** 1948. Les débuts de l'anatomie végétale moderne. L'Histoire du cambium. *Archives internationales d'histoire des sciences* 2: 270–279.
Also ___. 1949. Remarque bibliographique sur l'histoire du terme "Cambium." *Ibid.* 6: 457–458.

Cb522. **Souèges, R.** 1934. *L'embryologie végétale. Résumé historique.* Paris, Hermann & Cie. 2 vols. *Actualités scientifiques et industrielles* nos. 142 & 175.
1. *1ere époque: Des origines à Hanstein (1870).* 58 pp.; 2. *De Hanstein (1870) à nos jours.* 57 pp.

Cb523. **Studhalter, R. A.** 1955. Tree growth. I. Some historical chapters. *Botanical review* 21: 1–72.

See also Cb134(Hocquette) Cb180(Komarnitskii) Cb233(Carlquist) Cb1044(Aleksandrov & Yakovlev).

Physiology

Cb524. **Delaporte, F.** 1982. *Nature's second kingdom: Explorations of vegetality in the eighteenth century.* Translated by A. Goldhammer. Cambridge, MA, M I T Pr. xii+266 pp.
1. The central problems of botany; 2. Nutrition; 3. Generation; 4. Movement. Bibliography, pp. 233–252; Biographical glossary, pp. 253–261. Translation of *La second règne de la nature: Essai sur les questions de la végétalité au XVIIIe siècle.* Paris, Flammarion, 1978. 242 pp.

Cb525. **Evenari, M.** 1984. Seed physiology: Its history from antiquity to the beginning of the 20th century. *Botanical review* 50: 119–142.

Cb526. **Ivanov, L. A.** 1947. Istoriya sozdaniya ucheniya o fotosinteze. K 175–letiyu ego otkrytiya. [History of the creation of study of photosynthesis. On the 175th anniversary of its discovery]. *Izvestiya Akademii Nauk SSSR, seriya biologicheskaya* 1947(3): 329–340.

Cb527. **Künkele, W.** 1971. *Zur Entwicklungsgeschichte der Pflanzenchemie. Beginn der chemischen Pflanzenanalyse unter besonderer Berücksichtigung der Forschungen an der Akademie der Wissenschaften in Paris vom Ende des 17. bis Mitte des 18. Jahrhunderts.* Diss. rer. nat. Marburg. 276 pp.

Cb528. **Lyubimenko, V. N.** 1928. Poltorasta let izycheniya fotosinteza. [One hundred and fifty years of the study of photosynthesis]. *Prirody* 1928(5): 467–482.

Cb529. **Maksimov, N. A.** 1947. Ocherk istorii fiziologii rastenii v Rossii. [Sketch of the history of plant physiology in Russia]. *Trudy Instituta Istorii Estestvoznaniya* 1: 21–80.

Cb530. **Nash, L. K.** 1952. *Plants and the atmosphere.* In: J. B. Conant, ed. *Harvard case histories in experimental science. Case 5*, pp. 323–436. Cambridge, MA, Harvard Univ. Pr.

Cb531. **Sanders, A. P. M.** 1973. Physiologische Erklärungen der Phyllotaxis und ihre Vorgeschichte. *Janus* 60: 115–127.

Cb532. **Weevers, T.** 1949. *Fifty years of plant physiology.* With an introduction by F. W. Went. Amsterdam, Scheltema & Holkema. xi+308 pp.
A study of the researches of 1895–1945.

Cb533. **Went, F. W.** 1958. Fifty years of plant physiology. In: W. C. Steere, ed. *Fifty years of botany: Golden jubilee volume of the Botanical Society of America*, pp. 615–628. New York, McGraw–Hill Book Co.

See also Ba1016–1019 Cb134(Carles) Cb233(Hendricks) Cb1044(Brilliant et al.).

Floral biology

Cb534. **Ducker, S., & R. B. Knox.** 1985. Pollen and pollination: A historical review. *Taxon* 34: 401–419.

Cb535. **Johnson, D. S.** 1914. The history of the discovery of sexuality in plants. *Science* 39: 299–319. Also pub. in *Annual report of the Smithsonian Institution* for 1914: 383–406.

Cb536. **Knuth, P.** 1906. Historical development of flower pollination. Bibliography of flower pollination. In: *Handbook of flower pollination [etc.].* Translated by J. A. Davis. Vol. 1: 1–28, 212–372. Oxford, Clarendon Pr.
The bibliography runs to 3,748 entries.

Cb537. **Leroy, J.–F. P.** 1959. *Histoire de la notion de sexe chez les plantes.* Paris, Université de Paris. 40 pp. *Les Conférences du Palais de la Découverte* no. 68.

Cb538. **Loew, E.** 1895. *Einführung in die Blütenbiologie auf historischer Grundlage.* Berlin, Dümmler. 432+xii pp.

Cb539. ___. 1966. The discovery of sexuality and fertilization in higher plants. *Janus* 53: 212–235.

Cb540. ___. 1978. The discovery of nectar and nectaries and its relation to views on flowers and insects. *Isis* 69: 514–533.

Cb541. **Manten, A. A.** 1968. A short history of palynology in diagrams. *Review of palaeobotany & palynology* 6: 177–188.

Cb542. ___. 1969. On the earliest microscopical observation of pollen grains. *Review of palaeobotany & palynology* 9: 5–16.

Cb543. **Neishtadt, M. I.** 1952. *Sporovo–pyl'tsevoi metod v SSSR: Istoriya i bibliografiya.* Moscow, AN SSSR. 223 pp.
Bibliografiya po sporovo–pyl'tsevomu analizu v SSSR (1906–1951 gg.), pp. 49–199.

Cb544. **Schmid, R.** 1975. Two hundred years of pollination biology: An overview. *Biologist* 57: 26–35.

Cb545. **Tröndle, A.** 1925. *Geschichte des Atmungs– und Ernährungsproblems bei den Pflanzen.* Zurich & Leipzig, Orell Füssli. 111 pp. & 4 pl. *Veröffentlichungen der*

Schweizerischen Gesellschaft für Geschichte der Medizin und der Naturwissenschaften no. 4.

Cb546. **Wodehouse, R. P.** 1935. [Historical introduction.] In: *Pollen grains: Their structure, identification and significance in science and medicine.* New York, McGraw–Hill Book Co. *Re-pub.* New York, Hafner Publishing Co., 1959.

See also Cb134(Van Campo).

Genetics & heredity

Cb547. **Chouard, P.** 1951. *Éléments de génétique et d'amélioration des plantes.* I: *Les mécanismes de l'hérédité élémentaire ou factorielle.* Paris, Centre de Documentation Universitaire. 137 pp.

Cb548. **Plarre, W.** 1987. A contribution to the history of science of heredity in Berlin. *Englera* 7: 147–217.
For a German version, *see* Cb196A.

Cb549. **Roberts, H. F.** 1929. *Plant hybridization before Mendel.* Princeton, NJ, Princeton Univ. Pr. 374 pp.

Cb550. **Zirkle, C.** 1935. *The beginnings of plant hybridization.* Philadelphia, PA, Pennsylvania Univ. Pr.; London, Oxford Univ. Pr. xiii+231 pp. *Morris Arboretum monographs* no. 1.

Cb551. ___. 1969. Plant hybridization and plant breeding in eighteenth–century American agriculture. *Agricultural history* 43: 25–38.

See also Cb233(Zirkle).

Palaeobotany

Cb552. **Andrews, H. N.** 1980. *The fossil hunters: In search of ancient plants.* Ithaca, NY, Cornell Univ. Pr. 421 pp.

Cb553. **Arber, E. A. N.** 1921. A sketch of the history of palaeobotany with special reference to the fossil flora of the British coal measures. In: C. J. Singer, ed. *Studies in the history and method of science*, vol. 2: 472–489. Oxford, Clarendon Pr.

Cb554. **Gothan, W.** 1948. *Die Probleme der Paläobotanik und ihre geschichtliche Entwicklung.* Berlin, Wissenschaftliche Editionsgesellschaft MBH. 92 pp. *Probleme der Wissenschaft in Veregangenheit und Gegenwart* no. 10.

Cb555. ___. 1948. Die Paläobotanik in Deutschland in den letzten hundert Jahren. *Zeitschrift der Deutschen Geologischen Gesellschaft* 100: 94–105.

Cb556. **Just, T.** 1958. Fifty years of paleobotany 1906–1956. In: W. C. Steere, ed. *Fifty years of botany: Golden jubilee volume of the Botanical Society of America*, pp. 590–605. New York, McGraw–Hill Book Co.

Cb557. ___. 1959. Progress in paleobotany, 1908–1958. *Journal of paleontology* 33: 500–510.

Cb558. **Krishtofovich, A. N.** 1956. *Istoriya paleobotaniki v SSSR.* Moscow, AN SSSR. 111 pp.

Cb559. **Rao, A. R.** 1958. Palaeobotany. In: Indian Botanical Society. *History of botanical researches in India, Burma and Ceylon.* Part 3. Bangalore City. viii+57 pp.
Bibliography, pp. 39–57.

Cb560. **Ward, L. F.** 1884. Sketch of palaeobotany. *Report of the United States Geological Survey* 5th 1883/84: 357–452.

See also Cb134(Boureau) Cb233(Arnold) Cb1005 Cb1006(Conert) & Cb1044(Krishtofovich).

BIBLIOGRAPHIES OF SPECIAL TOPICS

• A selection of major works:

Cb561. **Barton, L. V.** 1967. *Bibliography of seeds*. New York, Columbia Univ. Pr. 858 pp.

Cb562. **Fahl, R. J.** 1977. *North American forest and conservation history: A bibliography*. Santa Barbara, CA, A.B.C.–Clio Pr. for the Forest History Society. iv+408 pp. ••Annotated list of over 8,000 entries.

Cb563. **Gel'man, N. S., & G. D. Zenkevich.** 1956. *Biokhimiya rastenii. Bibliograficheskii ukazatel' otechestvennoi literatury 1738–1952*. Moscow, AN SSSR. 394 pp.

Cb564. **Jackson, B. D.** 1882. *Vegetable technology: A contribution towards a bibliography of economic botany with a comprehensive subject index*. Founded upon the collection of George James Symons. London, Index Society. xii+356 pp. *Index Society publication* no. 11.

Cb565. **Kovaleva, A. N.** 1972–77. *Kornevaya sistema drevesnykh i kustarnikovykh rastenii: Ukazatel' otechestvennoi i inostrannoi literatury. [Root systems of woody plants: Guide to Russian and foreign literature.]* Moscow, Moskovskoi Gosudarstvennyi Universitet im M. B. Lomonosova, Nauchnaya Biblioteka im A. M. Gor'kogo. 2 vols. ••Covers literature from 1900–65 and 1965–72.

Cb566. **Manten, A. A.** 1969. Bibliography of palaeopalynology 1836–1966. *Review of palaeobotany & palynology* 8: 1–570.

Cb567. **Matsuura, H.** 1933. *A bibliographical monograph on plant genetics (genic analyses), 1900–1929*. 2nd ed. Sapporo, Hokkaido Imperial Univ. xx+787 pp. ••"Bibliography on genic analysis in plants," pp. 529–743. *Replaces* 1st ed., Tokyo, 1929. xii+499 pp.

Cb568. **Rehder, A.** 1949. *Bibliography of cultivated trees and shrubs hardy in the cooler temperate regions of the northern hemisphere*. Jamaica Plain, NY, Arnold Arboretum of Harvard Univ. xl+825 pp. *Re-pub.* Koenigstein, Otto Koeltz Science Publishers, 1978. *Collectanea bibliographica* vol. 10.

Cb569. **Solbrig, O., & T. W. J. Gadella.** 1970. *Biosystematic literature: Contributions to a biosystematic literature index (1945–1964)*. Utrecht, International Bureau of Plant Taxonomy and Nomenclature. v+566p. *Regnum vegetabile* vol. 69.

Cb570. **Tsitsin, N. V.**, ed. 1970. *Otdalennaya gibridizatsiya rastenii: Bibliograficheskii ukazatel' otechestvennaya literatura 1759–1967gg. [Hybridization of plants. Bibliographical index of domestic literature 1759–1967]*. Moscow, Izd. "Nauka." 263 pp.

Cb571. **Warner, M. F., M. A. Sherman, & E. M. Colvin.** 1934. *A bibliography of plant genetics*. Washington, DC, United States Department of Agriculture. 552 pp. *Miscellaneous publication* no. 164.

See also Cb543 & Cb1063(Gregory).

CRYPTOGAMIC BOTANY

General

Cb572. **Komarnitskii, N. A.** 1948. Ocherk istorii izucheniya nizshikh rastenii v Rossii i SSSR. *Trudy Botanicheskogo Sada, Moskovskii Ordena Lenina Gosudarstvennii Universitet imeni M. V. Lomonosova* 6: 71–119.

Cb573. **Leroy, J.–F.** 1977. L'alternance des générations chez les plantes: Notes sur la genèse d'une théorie. *Histoire et nature* 11: 3–44.

See also Ba251 & Cb994; Library cat. Bc429.

Pteridology

Cb574. **Allen, D. E.** 1969. *The Victorian fern craze: A history of pteridomania*. London, Hutchinson. xii+85 pp.

Cb575. **Manton, I.** 1955. Pteridology. In: California Academy of Sciences. *A century of progress in the natural sciences 1853–1953*, pp. 301–321. San Francisco, CA.

Cb576. **Pichi-Sermolli, R. E. G.** 1959. Pteridophyta. *Vistas in botany* 1: 421–490.

Cb577. **___.** 1973. Historical review of the higher classification of the Filicopsida. In: A. C. Jermy et al., eds. *The phylogeny and classification of the ferns*, pp. 11–40 & 19 pl. London, Academic Pr. for the Linnean Society of London.

Cb578. **Tryon, R. M., Jr.** 1952. A sketch of the history of fern classification. *Annals of the Missouri Botanic Garden* 39: 255–262.

Cb579. **___.** 1969. Pteridology. In: J. Ewan, ed. *A short history of botany in the United States*, pp. 97–102. New York, Hafner Publishing.

Cb580. **Wagner, W. H., Jr.** 1974. Pteridology 1947–1972. *Annals of the Missouri Botanical Garden* 61: 86–111. ••Bibliography, pp. 104–111.

See also Ba252 & Cb233(Tryon).

BIBLIOGRAPHIES OF PTERIDOLOGY

• A selection of major works:

Cb581. **Blake, S. F.** 1941. State and local fern floras of the United States. *American fern journal* 31: 81–90, 131–143. *Continued* ___. 1950. ___: Supplement 1. *Ibid.* 40: 148–165; **M. A. Miasek.** [1977?] *Regional, state and local fern floras, manuals, checklists and new fern and fern–ally records for the U. S. and adjacent Canada, since 1950*. Bronx, NY, New York Botanical Garden. 21 pp.

Cb582. **Christensen, C. F. A.** 1906. Catalogus literaturae. In: *Index filicum sive enumeratio omnium generum specierumque filicum et hydropteridum ab anno 1753 ad finem anni 1905 descriptorum*, pp. 671–744. Copenhagen, H. Hagerup; *Continued* ___. 1913. Supplementum. In: ___*: Supplementum, 1906–1912*, pp. 75–86 & 132. Copenhagen, [etc.]; ___. 1917. Supplementum 1913–1916. In: ___*: Supplément préliminaire pour les années 1913, 1914, 1915, 1916*, pp. 34–39. Copenhagen, Triers; ___. 1934. Supplementum 1917–1933. In: ___*: Supplementum tertium pro annis 1917–1933*, pp. 200–219. Copenhagen, [etc.]. (*All above re-pub.* Koenigstein, O. Koeltz Antiquariat, 1973); **R. E. G. Pichi–Sermolli.** 1965. Catalogus litteraturae 1934–1960. In: ___*: Supplementum quartum pro annis 1934–1960*, pp. 311–349. Utrecht, International Bureau for Plant Taxonomy & Nomenclature. *Regnum vegetabile* vol. 37; **F. M. Jarrett et al.** 1985. Catalogus litteraturae. In: ___*: Supplementum quintum pro annis 1961–1975*, pp. 219–245. Oxford, Oxford Univ. Pr.

Cb583. **Jones, G. N.** 1967. *An annotated bibliography of Mexican ferns*. Urbana, IL, Univ. of Illinois Pr. xxxiii+297 pp.

Cb584. **Kaur, S., & F. Raza.** 1983. *Bibliography on Indian Pteridology*. Lucknow, Economic Botany Information Service, National Botanical Research Institute. iv+83 pp.

Cb585. **Reed, C. F.** 1953. Bibliography of the Isoëtales. In: Index Isoëtales. *Boletim da Sociedade Broteriana* (2)27: 58–72.

Cb586. **___.** 1954. Bibliography. In: Index Marsileata et Salviniata. *Boletim da Sociedade Broteriana* (2)28: 47–61. *Continued* ___. 1965.

Bibliography. In: ___: Supplement. *Ibid.* (2)39: 279–302.

Cb587. **Reed, C. F.** 1965–66. Bibliography to Index Seliginellarum. In: Index Selaginellarum. *Memórias de Sociedade Broteriana* 18: 246–279.

Cb588. ___. 1966. Bibliography to the Psilotales. In: Index Psilotales. *Boletim da Sociedade Broteriana* 40: 86–96.

Cb589. ___. 1968. Bibliography. In: Index Thelypteridis. *Phytologia* 17: 325–328 & 466.

Cb590. ___. 1971. Bibliography to extant Equisetaceae. In: *Index to Equisetophyta.* Part II: *Extantes, Index Equisetorum*, pp. 100–128. Baltimore, MD, Reed. *Contribution of the Reed Herbarium* no. 19.

Cb591. **Hall, N.** 1983. Collecting Victorian fern books. *Pteridologist* 1: 17–21.

Cb592. ___. 1983. Collecting Victorian fern books. *Antiquarian book monthly review* 10: 170–173, 175.

Cb593. **Schollick, H. L.** 1971. An anatomy of pteridological bibliomania. *British Pteridological Society newsletter* 9: 20–23.

See also Cb580.

Bryology

Cb594. **Anderson, L. E.** 1974. Bryology 1947–1972. *Annals of the Missouri Botanical Garden* 61: 56–85.
Extensive bibliography, pp. 79–85.

Cb595. **Hagen, I.** 1897. Norges bryologi i det 18de århundrede. *Kongelige Norske Videnskabers Selskabs skrifter* 1897 (3): 195 pp. & 1 pl.

Cb596. **Lindberg, S. O.** 1877. *Hepaticologiens utveckling från äldsta tider till och med Linné. Installationsprogram.* Helsinki, J. C. Frenckell & Sons. 54 pp.

Cb597. **Mägdefrau, K.** 1975. Die Geographie der Moose, ihre Begründung und Entwickung. In: K. Mothes & J.-H. Scharf, eds. *Beiträge zur Geschichte der Naturwissenschaften und der Medizin; Festschrift für Georg Uschmann ... zum 60. Geburtstag am 18. Oktober 1873*, pp. 95–111. Halle, Deutsche Akademie der Naturforscher Leopoldina. *Acta historica Leopoldina* no. 9.

Cb598. **Margadant, W. D.** 1973. The development of the opinions concerning the sexuality of Bryophytes. *Janus* 60: 193–198.

Cb599. **Petkov, S.** 1939. Prouchvaniyata na nizshata flora (talosnitie i mukhovitie rasteniya) na Bulgariya prez poslednitie 50 godini. [History of investigations on thallophyta and bryophytes (Muscineae) of Bulgaria. *Izvestiya na Bulgarskoto Botanichesko Druzhestvo* 8: xix–xxxiv.

Cb600. **Richards, P. W.** 1959. Bryophyta. *Vistas in botany* 1: 387–420.

Cb601. **Rodgers, A. D., 3rd.** 1940. *Noble fellow: William Starling Sullivant.* New York, Putnam. xii+366 pp.
A history of American bryology and its makers.

Cb602. **Rudolph, E. D.** 1969. Bryology and lichenology. In: J. Ewan, ed. *A short history of botany in the United States*, pp. 89–96. New York, Hafner Publishing.

Cb603. **Steere, W. C.** 1955. Bryology. In: California Academy of Sciences. *A century of progress in the natural sciences, 1853–1953*, pp. 267–299. San Francisco, CA.

Cb604. ___. 1977. North American muscology and muscologists: A brief history. *Botanical review* 43: 285–343.
Provides a list of North American muscologists and collectors, pp. 288–294, and biographical sketches of 12 leading workers.

See also Ba253–258 Bb779–782A Bb786 Cb134(Gaume) & Cb233(Rudolph).

BIBLIOGRAPHIES OF BRYOLOGY

• For a fuller list of bibliographics see:

Cb605. **Rückert, D., & W. Korth.** 1990. Bibliographie bryologischer Bibliographien. Bibliography on bryological bibliographies. *Courier Forschungsinstitut Senckenberg* 132: 144 pp.

• A selection of major works:

Cb606. **Greene, S. W., & A. J. Harrington.** 1988. *The conspectus of bryological taxonomic literature.* Part 1: *Index to monographs and regional reviews.* Berlin, Stuttgart, J. Cramer in der Gebrüder Bornträger Verlagsbuchhandlung. 272 pp. *Bryophytorum bibliotheca* vol. 35.

Cb607. ___. 1989. ___. Part 2: *Guide to national and regional literature.* Berlin, Stuttgart, J. Cramer in der Gebrüder Bornträger Verlagsbuchhandlung. 321 pp. *Bryophytorum bibliotheca* vol. 37.

Cb608. **Lane, D.** 1978. A geographical guide to bryophyte floras of the world. 1 & 2. *Miscellanea bryologica et lichenologica* 8: 15–23, 29–34.

Cb609. **Margadant, W. D.** 1968. *Early bryological literature: A descriptive bibliography of selected publications treating musci during the first decades of the nineteenth century and especially of the years 1825, 1826 and 1827.* Pittsburgh, PA, Hunt Botanical Library. xi+278 pp. Also pub. as *Mededelingen van het Botanisch Museum en Herbarium van de Rijksuniversiteit de Utrecht* no. 283.

Cb610. **Sayre, G.** 1959. *Dates of publications describing Musci, 1801–1821.* Troy, NY, Privately. [v]+102 pp.

Cb611. **Tsvetkova, N. N., & Z. F. Kryukova.** 1976. *Katalog literatury po mokhoobraznym (Bryophyta) 1946–1960 po fondam otdeleniya biblioteki AN SSSR pri botanicheskom instituta im V. L. Komarova AN SSSR.* [*Catalogue of publications on Bryophyta, 1946–1960, based on the collections of the library of the Institute of Botany of the Academy of Sciences of the USSR.*] Leningrad, Biblioteka Akademiya Nauk SSSR. 412 pp.; *Continued* ___, **& M. N. Smidt.** 1975. *Katalog literatury po mokhoobraznym (Bryophyta) 1961–1970 po fondam otdeleniya biblioteki AN SSSR pri botanicheskom instituta im V. L. Komarova AN SSSR.* [*Catalogue of works on Bryophytes, 1961–1970, based on the collections of the library of the Institute of Botany of the Academy of Sciences of the USSR.*] Leningrad, Biblioteka Akademiya Nauk SSSR. 200 pp.; ___. 1977. *Katalog literatury po mokhoobraznym (Bryophyta) 1971–1975 po fondam otdeleniya biblioteki AN SSSR pri botanicheskom instituta im V. L. Komarova AN SSSR.* [*Catalogue of works on Bryophytes, 1971–1975, based on the collections of the library of the Institute of Botany of the Academy of Sciences of the USSR.*] Leningrad, Biblioteka Akademiya Nauk SSSR. 331 pp. This and the two preceding parts reprinted Koenigstein, Otto Koeltz Science Publishers, 1978. 849 pp.; **Akademii Nauk SSSR, Biblioteka.** 1983. *Katalog literatury po mokhoobraznym (Bryophyta) 1976–1980 po fondam otdela Biblioteki AN SSR pri Botanicheskom Institute im V. L. Komarova AN SSR ...* [*Catalogue of works on Bryophytes, 1976–1980, based on the collections of the library of the Institute of Botany of the Academy of Sciences of the USSR.*] Leningrad. 231 pp.

Cb612. **Underwood, L. M.** 1892. Index hepaticarum. Part 1: Bibliography. *Memoirs of the Torrey Botanical Club* 4(1): 1–93.

See also Cb594.

Lichenology

Cb613. **Doidge, E. M.** 1950. History of observations on the fungi and lichens of South Africa. *Bothalia* 5: 18–44.

Cb614. **Fink, B.** 1904. Two centuries of North American lichenology. *Proceedings of the Iowa Academy of Science* 11: 11–38.

Cb615. **Galloway, D. J.** 1981. Erik Acharius, Olof Swartz and the evolution of generic concepts in lichenology. In: Society for the Bibliography of Natural History. *History in the service of systematics: Papers from the conference to celebrate the centenary of the British Museum (Natural History) 13–16 April, 1981*, pp. 119–127. London. *Special publication* no. 1.

Cb616. ___. 1985. Lichenology in the South Pacific, 1790–1840. In: Society for the History of Natural History. *From Linnaeus to Darwin ... Papers from the fifth Easter meeting of the ... 28–31 March, 1983*, pp. 205–214. London. *Special publication* no. 3.

Cb617. **Godínez, J. L., & M. M. Ortega.** 1989. Liquenológia de México: Historia y bibliografía. *Cuadernos del Instituto de Biología* 3: 46 pp.

Cb618. **Hawksworth, D. L., & M. R. D. Seaward.** 1977. *Lichenology in the British Isles 1568–1975: A historical and bibliographical survey.* Richmond, Sy., Richmond Publishing. xi+231 pp.
History, pp. 1–41.

Cb619. **Kantvilas, G.** 1983. A brief history of lichenology in Tasmania. *Papers & proceedings of the Royal Society of Tasmania* 117: 41–51.

Cb620. **Krempelhuber, A. von.** 1867–72. *Geschichte und Litteratur der Lichenologie von den ältesten Zeiten bis zum Schlusse des Jahres 1865*. Munich, Wolf. 3 vols.
Vol. 1: Geschichte und Litteratur; vol. 2: Die Flechten-Systeme und Flechten-Spezies; vol. 3: Die Fortschritte und die Litteratur der Lichenologie in dem Zeitraume von 1866–1870 incl. nebst Nachträgen zur den früheren Perioden.

Cb621. **Llimona, X.** 1984. La recerca Micologica i liquenologica als Paisos Catalans. *Butlleti de la Institucio Catalana d'Historia Natural* 50: 97–114.

Cb622. **Rudolph, E. D.** 1988. Lichens in U S introductory botany textbooks, 1836–1986. *Bioscience* 38(6): 412–416.

Cb623. **Thomson, J. W.** 1974. Lichenology in North America 1947–1972. *Annals of the Missouri Botanical Garden* 61: 45–55.
Bibliography pp. 52–55.

See also Ba259–263 Cb134(des Abbayes) Cb233(Rudolph) & Cb1013.

BIBLIOGRAPHIES OF LICHENOLOGY

• For a fuller list of bibliographies see:

Cb624. **Korth, W., & D. Rückert.** 1988. Bibliographie lichenologischer Bibliographien: Flechten – Lichenes. Bibliography on lichenological bibliographies. *Courier Forschungsinstitut Senckenberg* 104: 153 pp.

• A selection of major works:

Cb625. **Culberson, W. L.** 1955. *A guide to the literature on the lichen flora and vegetation of the United States.* Beltsville, MD, U. S. Department of Agriculture, Agricultural Research Service, Plant Industry Station. 54 pp. *Special publication* no. 7.

Cb626. **Delzenne-van Haluwyn, C.** 1976. *Bibliographia societatum lichenorum.* Vaduz, Cramer. ii+177 pp. *Bibliographia phytosociologica syntaxonomica suppl.* 1.

Cb627. **Des Abbayes, H.** 1953. Travaux sur les lichens parus de 1939 à 1952. *Bulletin de la Société Botanique de France* 100: 83–123.

Cb628. **Faltynowicz, W.** 1983. *Polska bibliografia lichenologiczna. (The Polish lichenological bibliography).* Cracow, Polska Akademia Nauk, Instytut Botaniki 184 pp. *Bibliografie botaniczne* vol. 1.

Cb629. **Hawksworth, D. L., & T. Ahti.** 1990. A bibliographic guide to the lichen floras of the world. *Lichenologist* 22: 78 pp.

Cb630. **Pluntke, M.** 1984. *Die Flechten Kubas (Flora lichenum Cubensis) Bibliographie.* Halle (Saale), Universitäts– und Landesbibliothek Sachsen–Anhalt. 157 pp. *Terrestrische Ökologie, Sonderheft* 4.

Cb631. **Scholz, F.** 1986. Die Flechtenflora der DDR – Bibliographie. Halle (Saale), Universitäts– und Landesbibliothek Sachsen–Anhalt. 175 pp. *Terrestrische Ökologie, Sonderheft* 7.

Cb632. **Tuxen, R.** 1964. Bibliographia phytosociologica cryptogamica. Pars 4: Lichenes (sine epiphyta). *Excerpta botanica* B 6: 208–244.

See also Cb673 & Cb680.

Mycology (including plant pathology)

Cb633. **Ainsworth, G. C.** 1965, Historical introduction to mycology. In: G. C. Ainsworth, F. K. Sparrow & A. S. Sussman, eds. *The fungi: An advanced treatise*, vol. 1: 3–20. New York, Academic Pr.

Cb634. ___. 1976. *Introduction to the history of mycology.* Cambridge, Cambridge Univ. Pr. xi+359 pp., illus.
Includes Chronology and bibliography (from 1491 to 1974), pp. 308–337, and name and subject indexes.

Cb635. ___. 1981. *Introduction to the history of plant pathology.* Cambridge, Cambridge Univ. Pr. xii+315 pp., illus.
Includes Biographical references, pp. 252–258; Bibliography, pp. 259–293 (chronologically arranged from "Before 1500" to 1979, with introductory lists of "History" and "Representative phytopathological textbooks and handbooks"), and name and subject indexes.

Cb636. ___. 1986. *Introduction to the history of medical and veterinary mycology.* Cambridge, Cambridge Univ. Pr. xi+228 pp., illus.
Includes bibliography, pp. 181–214. *Also* **G. C. Ainsworth & P. M. Stockdale.** 1983. Landmarks in medical and veterinary mycology. *Review of medical and veterinary mycology* 18: 1–18.

Cb637. **Baker, K. F.** 1969. Plant pathology and mycology. In: J. Ewan, ed. *A short history of botany in the United States*, pp. 82–88. New York, Hafner Publishing.

Cb638. **Baldwin, R. S.** 1981. *The fungus fighters: Two women scientists and their discovery.* Ithaca, Cornell Univ. Pr. 212 pp.
Re. the work of Rachel Brown (1898–1980) and Elizabeth Lee Hazen (1888–?).

Cb639. **Bessey, E. A.** 1955. Mycology. In: California Academy of Sciences. *A century of progress in the natural sciences, 1853–1953*, pp. 225–264. San Francisco, CA.

Cb640. **Bresinsky, A.** 1973. 200 Jahre Mykologie in Bayern. *Zeitschrift für Pilzkunde* 39: 15–38.

Cb641. **Conners, I. L.**, ed. 1972. *Plant pathology in Canada.* Winnipeg, Canadian Phytopathological Society. 251 pp., illus.

Cb642. **Das Gupta, S. N.** 1958. Mycology and plant pathology. In: Indian Botanical Society. *History of botanical researches in India, Burma and Ceylon.* Part 1. Bangalore City. iv+118 pp.
Bibliography, pp. 50–118.

Cb643. **Doidge, E. M.** 1950. History of observations on the fungi and lichens of South Africa. *Bothalia* 5: 18–44.

Cb644. **Guétrot, (Dr.).** 1934. La mycologie avant la fondation de la Société Mycologique de France. In: *Le quarantenaire de la Société Mycologique de France (1884–1924)*, pp. 1–18. Paris, Société Mycologique de France.

Cb645. **Hassell, F. C.** 1957. The early Irish mycologists, 1726–1900. *Irish naturalists' journal* 12: 116–120.

Cb646. **Ingold, C. T.** 1959. Fungi. *Vistas in botany* 1: 348–386.

Cb647. ___. 1972. The advance of mycology. *Transactions of the British Mycological Society* 58 (2) supp: 5–14.

Cb648. **Keitt, G. W.** 1959. History of plant pathology. In: J. G. Horsfall & A. E. Dimond, eds. *Plant pathology: An advanced treatise*, vol. 1: 61–97. New York, Academic Pr.

Cb649. **Large, E. C.** 1940. *The advance of the fungi.* London, J. Cape. 488 pp. *Re-pub.* New York, Dover Publications, 1962.

Cb650. **Lazzari, G.** 1973. *Storia della micologia Italiana contributo dei botanici Italiani allo sviluppo delle scienze micologiche.* Trento, Arti Grafiche Saturnia. 351 pp.

Cb651. **Llimona, X.** 1984. La recerca Micologica i liquenologica als Paisos Catalans. *Butlleti de la Institucio Catalana d'Historia Natural* 50: 97–114.

Cb652. **Lütjeharms, W. J.** 1936. *Zur Geschichte der Mykologie. Das XVIII Jahrhundert.* Gouda, Koch & Knuttel. xxi+262 pp. *Mededelingen Nederlandische Mycologische Vereeniging* no. 23.

Cb653. **McKenzie, E. H. C.** 1983. Mycological history and exploration in New Zealand. *Archives of natural history* 11: 233–247.

Cb654. **Parris, G. K.** 1968. *A chronology of plant pathology.* Starkville, MS, Johnson & Sons. 167 pp.

Cb655. **Pfister, D. H.** 1986. Mycology, universities and 19th century American science. *Mycotaxon* 26: 65–74.

Cb656. **Raab, H.** 1965–75. Aus der Geschichte der Mykologie. *Schweizerische Zeitschrift für Pilzkunde* 443: 81–84, 1965; *Ibid.* 44: 149–154, 1966; *Ibid.* 46: 17–19, 105–110, 1968; *Ibid.* 48: 13–16, 1970; *Ibid.* 49: 153–158, 1971; *Ibid.* 50: 41–45, 1972; *Ibid.* 51: 53–59, 1973; *Ibid.* 53: 22–28, 1975.

Cb657. **Ramsbottom, J. R.** 1941. The expanding knowledge of mycology since Linnaeus. *Proceedings of the Linnean Society of London* 151: 280–367 & pl. 5–7.

Cb658. ___. 1963. History of Scottish mycology. *Transactions of the British Mycological Society* 46: 161–178.

Cb659. **Reed, H. S.** 1942. Ch. 18: Mycology; 19: Plant pathology. In: *A short history of the plant sciences*, pp. 266–284, 285–308. Waltham, MA, Chronica Botanica.

Cb660. **Shape, P. F.** 1929. History of mycological collectors in Colorado. *Mycologia* 21: 292–296.

Cb661. **Smith, A. L.** 1908. *Presidential address. Microfungi: A historical sketch.* Worcester, British Mycological Society. 8 pp. & 1 pl.

Cb662. **Smith, E. F.** 1929. Fifty years of pathology. In: *Proceedings of the International Congress of Plant Science, Ithaca, 1926*, vol. 1: 13–46, & 34 pl. Ithaca, NY.
The pls. provide dozens of portraits of plant pathologists.

Cb663. **Société Mycologique de France.** 1984. Le livre d'or du centenaire: Documents historiques sur la mycologie, Société Mycologique de France: 1884–1984. *Bulletin* *trimestriel de la Société Mycologique de France* 100 (3): CCXXVIII pp., illus. *Includes* **Y. Mottez.** La mycologie et les militaires, pp. I–XIV; **G. Bertram.** La mycologie et les ecclésiastiques, pp. XV–XXXI; **J. Mornand.** Histoire de la mycologie en Anjou, pp. XXXII–LXVI, (& Addenda, *Ibid.* Atlas, pl. 238, p. 33); **G. Dillemann.** La mycologie et les pharmaciens, pp. LXX–CIV; **G. Garnier.** La mycologie et les enseignants, pp. CV–CXXXVII; **D. Lamy.** À propos de la correspondance de E. Boudier à P. V. A. Feuilleaubois, pp. CXXXIX–CLXI; **S. Jovet-Ast.** La mycologie au Jardin du Roy et au Muséum National d'Histoire Naturelle, pp. CLXII–CCVI; **F. Guyon.** La mycologie et les médecins, pp. CCX–CCXVIII; **Anon.** La mycologie et les vétérinaires, pp. CCXVIII–CCXX; **G. Segretain.** La mycologie de l'Institut Pasteur, pp. CCXXI–CCXXVIII; *also* **P. Joly.** Les Vélins de la collection "ancienne" de champignons et la "méthode nouvelle" de classification des plantes, Atlas, pl. 235, 10 pp.; **J. Perreau.** Les peintures d'espèces "cueillies en Italie" dans la collection "ancienne" des vélins de champignons. Une représentation du *Boletus erythropus*, Atlas, pl. 237, 14 pp.

Cb664. **Sparrow, F. K.** 1974. Mycology, 1947–1972. *Annals of the Missouri Botanical Garden* 61: 3–13.
Classified bibliography, pp. 10–13.

Cb665. **Walker, J. C.** 1969. History of plant pathology. In: *Plant pathology*, 3rd ed., pp. 14–46. New York, McGraw–Hill Book Co.

Cb666. ___. 1975. Some highlights in plant pathology in the United States. *Annual review of phytopathology* 13: 15–29.

Cb667. **Wasson, V. P., & R. G.** 1957. *Mushrooms, Russia and history.* New York, Pantheon. 2 vols.

Cb668. **Wehnelt, B.** 1943. *Die Pflanzenpathologie der deutschen Romantik als Lehre vom kranken Leben und Bilden der Pflanzen, ihre Ideenwelt und ihre Beziehungen zu Medizin, Biologie und Naturphilosophie historisch–romantischer Zeit.* Bonn, Scheur. viii+237 pp. & 8 pl.

Cb669. **Weston, W. H.** 1958. Mycology during the past fifty years. In: W. C. Steere, ed. *Fifty years of botany: Golden jubilee volume of the Botanical Society of America*, pp. 570–580. New York, McGraw–Hill Book Co.

Cb670. **Whetzel, H. H.** 1918. *Outline of the history of phytopathology.* Philadelphia, PA, Saunders. 130 pp.

See also Biographies Ba264–269 Ba528 Ba660 Ba1498; Societies Bb783–785 Bb787–788; Misc. Cb134(Chadefaud, Rivalier) Cb233(Baker) Cb1056 Cb1060 Cb1095 Cb1098 Cb1205 Cb1207 & Cb1210.

BIBLIOGRAPHIES OF MYCOLOGY — GENERAL

Cb671. **Arnold, G. R. W.** 1976. Internationale Bibliographie der Hypomycetaceae (*Mycophyta, Ascomycotina*). Jena, Universitätsbibliothek. 129 pp. *Bibliographische Mitteilungen der Universitätsbibliothek Jena* no. 25.

Cb672. **Bessey, E. A.** 1950. Guide to the literature for the identification of fungi. In: *Morphology and taxonomy of fungi*, pp. 660–756. New York, McGraw–Hill Book Co. *Re-pub.* New York, Hafner Publishing Co., 1964.
••19th & 20th century taxonomic works arranged in 51 lists.

Cb673. **Ciferri, R.** 1957–60. *Thesaurus literaturae mycologicae et lichenologicae (G. Lindau et P. Sydow). Supplementum 1911–1930.* Pavia, Cortina. 4 vols.

Cb674. ___, **& P. Redaelli.** 1957–58. *Bibliographia mycopathologica (1800–1940).* Florence, Sanson. 2 vols. *Bibliotheca bibliographica italica* vols. 18, 20.

Cb675. **Dörfelt, H., & C. Müller-Uri.** 1986. *Geastrales–Bibliographie.* Halle. 254 pp. *Terrestrische Oekologie, Sonderheft* 6.

Cb676. **Fischer, G. W.** 1951. *The smut fungi: A guide to the literature with bibliography*. New York, Ronald Pr. x+387 pp.

Cb677. **Flück–Wirth, F.** 1982. *Mykologie, Phytopathologie: Spezialkatalog Nr. 2. September 1982*. Teufen, F. Flück–Wirth. Krypto. [vi]+198 pp. ••Supplemented by *Neues von Krypto*. Teufen, F. Flück–Wirth. Krypto, no. 1+, 1978+.

Cb678. **Guba, E. F., & P. A. Young.** 1924. Check list of important references dealing with taxonomy of fungi. *Transactions of the American Microscopical Society* 43: 17–67. *Continued* ___. 1925. Literature on fungus flora in North America and an addition to ... check list. *Ibid.* 44: 109–118.

Cb679. **Hacskaylo, E., & C. M. Tomkins.** 1973. *World literature on Mycorrizae*. Baltimore, MD, Reed Herbarium. 142 pp. *Contributions from the Reed Herbarium* no. 22.

Cb680. **Lindau, G., & P. Sydow.** 1908–17. *Thesaurus litteraturae mycologicae et lichenologicae ratione habita praecipue omnium quae adhuc scripta sunt de mycologia applicata quem congresserunt ...* Leipzig, Borntraeger. 5 vols. *Re-pub.* New York, Johnson Reprint Corporation, 1966. ••Literature through 1910.

Cb681. **Pfister, D. H., J. R. Boise, & M. A. Eifler.** 1990. *A bibliography of taxonomic mycological literature, 1753–1821*. Berlin, Stuttgart, J. Cramer in der Gebrüder Borntrāger Verlagsbuchhandlung. 161 pp. *Mycologia memoirs* no. 17.

Cb682. **Sturgis, W. C.** 1912. A guide to the botanical literature of the myxomycetes from 1875 to 1912. *Colorado College publications, science series* 12: 385–433.

Cb683. **Tamiya, H., & S. Morita.** 1929. Bibliographie von *Aspergillus* 1729–1928. *Botanical magazine, Tokyo* 43: 60–71, 145–156, 179–189, 237–249, 281–291, 321–332, 371–381, 427–438, 501–515.

Cb684. **Watling, R., & A. E.** 1980. *A literature guide for identifying mushrooms*. Eureka, CA, Mad River Pr. [vi]+121 pp.

See also Library cats. Bc427–428 Bc430–431 Bc525 Bc542 Bc562 & Bc566.

Regional

• A selection of major works:

Cb685. **Burk, W. R.** 1980. *A bibliography of North American Gasteromycetes. 1: Phallales*. Vaduz, Cramer. xv+216 pp. *Bibliotheca mycologica* vol. 73.

Cb686. **Das Gupta, S. N.** 1958. Mycology and plant pathology. In: Indian Botanical Society. *History of botanical researches in India, Burma and Ceylon*. Part 1. Bangalore City. iv+118 pp. ••Bibliography, pp. 50–118.

Cb687. **Farlow, W. G., & W. Trelease.** 1887. A list of works on North American fungi. *Bibliographical contributions, Harvard University library* 25: 36 pp. *Continued* **W. G. Farlow.** 1888. A supplemental list of works on North American fungi. *Ibid.* 31: 9 pp.

Cb688. **Farlow, W. G.** 1905. *Bibliographic index of North American fungi*. Vol. 1, part 1. *Abrothallus to Badhamia*. Washington, DC, Carnegie Institution of Washington. xxxv+312 pp. *Publication, Carnegie Institution of Washington* no. 8. ••All published.

Cb689. **Firman, I. D.** 1978. *Bibliography of plant pathology and mycology in the area of the South Pacific Commission*, 1820–1976. Noumea, South Pacific Commission. iv+78 pp. *Technical paper, South Pacific Commission* no. 176.

Cb690. **Guba, E. F., & P. A. Young.** 1925. Literature on fungus flora in North America [etc.] *Transactions of the American Microscopical Society* 44: 109–118.

Cb691. **Holden, M.**, ed. 1975. Guide to the literature for the identification of British fungi. 3rd ed. *Bulletin of the British Mycological Society* 9:

67–106. ••Key literature on Myxomycetes, Phycomycetes, Ascomycetes, and Basidiomycetes. *Replaces* ___. 1969. ___. 2nd ed. *Bulletin of the British Mycological Society* 3: 19–54; **British Mycological Society**. 1952. *Guide [etc.]*. London, Lewis.

Cb692. **Hsiang, Wang-nien.** 1957. [Bibliographies of fungi of China and botanical diseases.] Beijing, K'o Hsueh Ch'u Pan She. x+323 pp. ••In Chinese.

Cb693. **Lepik, E.** 1930. Bibliographische Beiträge zur ostbaltischen Pilzflora. *Tartu ülikooli juures oleva loodusuurijate seltsi aruanded* 36: 27–88.

Cb694. **Ostero, J. I., & M. T. Cook.** 1937. Bibliography of mycology and phytopathology of Central and South America, Mexico, and the West Indies. *Journal of agriculture of the University of Puerto Rico* 23: 249–486.

Cb695. **Palmer, J. T.** 1968. A chronological catalogue of the literature of the British Gasteromycetes. *Nova hedwigia* 15: 65–178. *Re-pub.* 1968. 114 pp.

Cb696. **Skirgiello, A.** 1988. *Polska bibliografia mikologiczna. The Polish mycological bibliography.* Warsaw, Polska Akademia Nauk, Instytut Botaniki im. W. Szafera. xliii+891 pp. *Bibliografie botaniczne* no. 2.

Cb697. **United States Department of Agriculture, Library.** 1919. *A check list of the publications of the Department of Agriculture on the subject of plant pathology, 1837–1918.* Washington, DC. 38 pp. *Bibliographical contribution* no. 1.

Cb698. ___. 1925. *Author and subject index to the publications on plant pathology issued by the U. S. Department of Agriculture up tp January 1, 1925.* By J. M. Allen. Washington, DC. 158 pp. *Bibliographical contribution* no. 8.

Cb699. ___. 1928. *Author and subject index to the publications on plant pathology issued by the state agricultural experiment stations up to December 1, 1927.* By J. M. Allen. Washington, DC. xiii+251 pp. *Bibliographical contribution* no. 16.

Cb700. **Vasil'kov, B. P.** 1953. *Izuchenie shlyapochnykh gribov v SSSR: Istoriko-bibliograficheskii obzor.* Moscow/Leningrad, Akademiya Nauk SSSR. 192 pp.

Phycology

Cb701. **Bold, H. C.** 1974. Phycology 1947–1972. *Annals of the Missouri Botanical Garden* 61: 14–44.
With an extensive bibliography pp. 32–44.

Cb702. **Brunel, J.** 1944. *Les grandes étapes de l'algologie Américaine.* Montreal, Institut Botanique de l'Université de Montréal. 32 pp. *Contributions* no. 52.

Cb703. **Cassie, V.** 1983. A history of freshwater diatom research in New Zealand. *Archives of natural history* 11: 223–231.

Cb704. **Ducker, S. C.** 1979. History of Australian phycology: The significance of early French exploration. *Brunonia* 2: 19–42.

Cb705. ___. 1981. History of Australian phycology: Early German collectors and botanists. In: Society for the Bibliography of Natural History. *History in the service of systematics: Papers from the conference to celebrate the centenary of the British Museum (Natural History) 13–16 April, 1981*, pp. 43–51. London. *Special publication* no. 1.

Cb706. **Hansen, J. B.** 1985. *An outline of phycology in Denmark.* Copenhagen, 2nd International Phycological Congress. 24 pp., illus.

Cb707. **Hoppe, B.** 1990. Motives and results of early German researches on marine plants; Their relationship to the development of botany and the role of international collaboration. In: W. Lenz & M. Deacon, eds. *Ocean sciences: Their history and relation to man.*

Proceedings of the 4th International Congress on the History of Oceanography Hamburg [etc.], pp. 94–101. Hamburg, Bundesamt für Seeschiffahrt und Hydrographie. *Deutsche hydrographische Zeitschrift*, Ergänzungsheft, Reihe B, no. 22.

Cb708. **Lorch, J.** 1968. The history of the sexuality of marine algae. In: Institut Océanographique, Monaco. Communications, premier Congrès International d'Histoire de l'Océanographie, Monaco, 12–17 décembre 1966. *Bulletin de l'Institut Océanographique, Monaco* numéro spécial 2: 397–406.

Cb709. **Mollenhauer, D.** 1987. History of phycology in Berlin. *Englera* 7: 29–50.
For a German version, *see* Cb196A.

Cb710. **Papenfuss, G. F.** 1955. Classification of the algae. In: California Academy of Sciences. *A century of progress in the natural sciences 1853–1953*, pp. 115–224. San Francisco, CA.

Cb711. ___. 1958. Progress and outstanding achievements in phycology during the last fifty years. In: W. C. Steere, ed. *Fifty years of botany: Golden jubilee volume of the Botanical Society of America*, pp. 554–569. New York, McGraw–Hill Book Co.

Cb712. **Setchell, W. A.** 1931. Some early algal confusions. *University of California publications in botany* 16: 351–366, 31 pl.
Includes valuable notes on the early phycologists.

Cb713. **Smith, G. M.** 1947. The historical significance of names applied to reproductive structures of algae. *Farlowia* 3: 217–224.

Cb714. ___. 1958. The role of study of algae in the development of botany. In: W. C. Steere, ed. *Fifty years of botany: Golden jubilee volume of the Botanical Society of America*, pp. 471–483. New York, McGraw–Hill Book Co.

Cb715. **Taylor, W. R.** 1959. Phycology in retrospect and anticipation. *Vistas in botany* 1: 328–347.

Cb716. ___. 1969. Phycology. In: J. Ewan, ed. *A short history of botany in the United States*, pp. 74–81. New York, Hafner Publishing.

Cb717. **Whitford, L. A.** 1968. Notes on the history of freshwater phycology. *Journal of phycology* 4: 169–173.

See also Ba293 Ba933 Bb850(Reinke) Ca579(Benson) Ca596(Taylor) Cb134(Feldmann) Cb171(Kies) Cb233(Taylor) & Cc916(Thomas).

BIBLIOGRAPHIES OF PHYCOLOGY

• A selection of major works:

Cb718. **Cassie, V.** 1980. Bibliography of the freshwater algae of New Zealand. *New Zealand journal of botany* 18: 433–447. *Continued* ___. 1981. ___: Additions [etc.]. *Ibid.* 19: 389–391.

Cb719. **Cesati, V.** 1882. Saggio di una bibliografis algologica Italiana. *Memorie della Società Italiana delle Scienze a Napoli* 1882 (Sept.): 1–77. *Also pub. separately* Naples, Accademia Reale delle Scienze, 1882. [ii]+77 pp.

Cb720. **Dawson, E. Y.** 1946. A guide to the literature and distributions of the marine algae of the Pacific coast of North America. *Memoirs of the Southern California Academy of Sciences* 3 (1): 1–134.

Cb721. **De Toni, G. B.** 1931–32. *Bibliographia algologica universalis, seu repertorium totius litteraturae physologicae hucusque editae quam digessit J. de Toni. Ad syllogem algarum omnium hucusque cognitarum, auctore J. B. de Toni additamentum.* Padua. ix+436 pp.

Cb722. **Dixon, P. S., D. E. G. Irvine & J. H. Price.** 1966. The distribution of benthic marine algae: A bibliography for the British Isles. *British phycological bulletin* 3: 87–

142. Reprinted Port Erin, British Phycological Society, 1966. 56 pp.

Cb723. **Gaidukov, N. M., et al.** 1901. *Literaturnye istochniki k' russkoi flore vodoroslei. Literatur-Quellen zur Algenflora Russlands. Literature resources to the algal flora of Russia.* St. Petersburg. 126 pp. *Botanicheskiya zapiski, St Petersburg* no. 27; *Continued* **A. A. Elenkin & L. A. Ol'.** 1929. Bibliografiya al'gologicheskikh trudov v predelakh SSSR s 1900 po 1925gg. Bibliographie des ouvrages algologiques en USSR parus depuis 1900 jusqu'à 1925 inclusivement [#453–1308]. *Trudy Glavnogo Botanicheskago Sada* 42(1): iv+139 pp. In Russian with French summary. *Continued* ___. 1934. ___ s 1926 po 1930g vklyuchitel'no [#1309–1906]. *Trudy Botanicheskago Instituta Akademii Nauk SSSR*, ser. 2 *sporovye rasteniya* 2: 171–256; ___. 1950. ___ s 1931 po 1935g [etc.] [#1907–2329]. *Trudy Botanicheskago Instituta imeni Akademika V. L. Komarova, Akademii Nauk SSSR*, ser. 2 *sporovye rasteniya* 5: 95–115; **M. M. Gollerbakh et al.** 1966. *Bibliografiya Sovetskoi literatury po vodoroslyam za 1936–1940gg. S dopolneniem za predydushchie gody. (Bibliographia algologica Sovietica annorum 1936–1940. Cum supplemento annorum praeteritorum.) [#2330–3821].* Leningrad, Biblioteka Akademiya Nauk SSSR. 169 pp.; **L. K. Krasavina & M. M. Gollerbakh.** 1968. *Bibliografiya ... 1941–1960gg. [Bibliography of Soviet literature on algae, 1941–1960] [#3822–7037.]* Leningrad, Biblioteka Akademiya Nauk SSSR. 342 pp.; ___. 1983. *Vodorosli: Ukazatel' k "Bibliografiya Sovetskoi literatury po vodorosliam 1961–1970." [Index to algological bibliography of the USSR 1961–1970] [#7038+].* Leningrad, Botanicheskii Institut im. V. L. Komarova, Akademii Nauk SSSR. 460 pp.; **L. K. Krasavina, N. N. Tsvetkova & I. N. Matveevskaya.** 1978. *Bibliografiya Sovetskoi literatury po vodoroslyam, 1971–1975. [Bibliography of Soviet literature on algae, 1971–1975.]* Leningrad, [etc.]. 515 pp. ••*Also* **A. A. Elenkin et al.** 1950. Dopolnitel'nye dannye po bibliografii al'gologicheskikh trudov v predelakh SSSR. Dopolnenie II. *Trudy botanicheskago instituta imeni akademika V. L. Komarova, akademii nauk SSSR*, ser. 2 *sporovye rasteniya* 6: 5–22. Additions to Elenkin's bibliographies listed above. *Re-pub. as* **Gaidukov, N. et al.** 1976. *Algological bibliography of the U. S. S. R., from the beginning up to 1960.* Koenigstein, Otto Koeltz Science Publishers. xxix+1,334 pp. *Collectanea bibliographica* vol. 3. Facsimile of the first six bibliographies, 1901–68, listed above.

Cb724. **Gollerbakh, M. M., & L. K. Krasavina.** 1971. *Vodorosil: Svodniai ukazatel' k otechestvennym bibliografiyam po vodorosliam za 1737–1960gg. [Cumulated index to the national bibliography on algae for 1737–1960 incl.]* Leningrad, Akademii Nauk SSSR. 623 pp. Reprinted Koenigstein, Otto Koeltz Science Publishers, 1977. ••Provides a systematic key to the bibliographies of Gaidukov et al., Elenkin & Ol', and Krasavina listed above.

Cb725. **Prescott, G. W.** 1956. A guide to the literature on ecology and life histories of the Algae. *Botanical review (Lancaster)* 22: 167–240.

Cb726. ___. 1983–84. *Bibliographia desmidiacearum universalis. (A contribution to a bibliography of Desmid systematics, biology and ecology from 1744 to 1982).* Koenigstein, Koeltz Scientific Books. viii+612 pp.

Cb727. **Royal Microscopical Society.** 1928. *Complete index to the articles on, and references to the Diatomaceæ in the Transactions and Journals of the Royal Microscopical Society, 1853–1915, including subject index and indices to authors, genera, species and plates.* Compiled by Miss A. M. Mainland. London. 42 pp.

Cb728. **Sieminska, J.** 1990. *Polska bibliografia fykologiczna. The Polish phycological bibliography*. Cracow, Wroclaw, Polska Akademia Nauk, Instytut Botaniki im. W. Szafera. 464 pp. *Bibliografie botaniczne* no. 3.

Cb729. **Tilden, J. E.** 1895. A contribution to the bibliography of American algae. *Minnesota botanical studies* 1(23): 295–421.

Cb730. ___. 1920. *Bibliography of the literature relating to the Pacific Ocean algae and to the freshwater algae of the countries bordering on the Pacific Ocean*. Minneapolia, MN. 58 pp. ••Reprinted from *Index algarum universalis: A universal card index of the marine, freshwater, epiphytic, and parasitic algae of the world.*

Cb731. **Tsuda, R. T., & F. O. Wray.** 1977. Bibliography of marine benthic algae in Micronesia. *Micronesica* 13: 85–120. *Continued* **R. T. Tsuda.** 1981. ___: Addenda. *Ibid.* 17: 213–218.

See also Cb701; Library cat. Bc536.

BOTANICAL GARDENS

General

Cb732. **Born, W.** 1949. Early botanical gardens. *Ciba symposia* 11: 1093–1124.

Cb733. **Brockway, L. H.** 1979. *Science and colonial expansion: The role of the British royal botanic gardens.* New York, Academic Pr. xiv+215 pp.

Cb734. **Chiarugi, A.** 1953. Le date di fondazione dei primi orti botanici del mondo: Pisa (estate 1543); Padova (7 Iuglio 1545); Firenze (1 dicembre 1545). *Nuovo giornale botanico Italiano* 60: 785–839.

Cb735. **Gager, C. S.** 1937–38. Botanic gardens of the world: Materials for a history. 2nd ed. *Brooklyn Botanic Garden record* 26: 149–353, 1937; 27: 151–406, 1938.

Cb736. **Guéguen, E.** 1976. Les jardins botaniques de Bretagne aux XVIIIe et XIXe siècles. In: *Comptes rendus du 97e Congrès National des Sociétés Savantes, Section des Sciences,* tom. 2: 127–136.

Cb736A. **Hill, A. W.** 1915. The history and functions of botanic gardens. *Annals of the Missouri Botanical Garden* 2: 185–240 & pl. 4–12.

Cb737. **Holttum, R. E.** 1970. The historical significance of botanic gardens in S. E. Asia. *Taxon* 19: 707–714.

Cb738. **Hyams, E., & W. MacQuitty.** 1969. *Great botanical gardens of the world.* London, Nelson. 288 pp. *Re-pub.* London, Bloomsbury Books, 1985.

Cb739. **Leiva, A. T.** 1981. Notas acerca de la historia y organizaci6n cientifica de los jardines botánicos del mundo. *Revista del Jardin Botanico Nacional, Universidad de la Habana* 2: 69–115.

Cb740. **McCracken, D. P. & E. M.** 1988. The way to Kirstenbosch. *Annals of Kirstenbosch Botanic Gardens* 18: 125 pp., illus.
History of botanical gardens in South Africa from 1652.

Cb741. **Morris, Sir D.** 1906. Botanical institutions of Jamaica. *Kew bulletin* 1906: 61–68.
History of various botanical gardens.

Cb742. **Prest, J.** 1981. *The Garden of Eden: The botanic garden and the re-creation of Paradise.* New Haven, CT, Yale Univ. Pr. 122 pp.

Cb743. **Purseglove, J. W.** 1957. History and functions of botanic gardens with special reference to Singapore. *Tropical agriculture* 34: 165–189.

Cb744. **Randhawa, G. S., K. L. Chadha, & D. Singh**, eds. 1971. *The famous gardens of India.* New Delhi, Malhotra Pub. House. 168 pp. *Re-pub.* 1978.
Includes several botanical gardens.

Cb745. **Schultes, J. A.** 1817. *Grundriss einer Geschichte und Literatur der Botanik, von Theophrastos Eresios bis auf die neuesten Zeiten; nebst einer Geschichte der botanischen Gärten.* Vienna, C. Schaumburg & Co. xvi+411 pp.
With an added title–page: *Anleitung zum gründlichen Studium der Botanik, zum Gebrauche bey Vorlesungen und zum Selbstunterrichte.*

Cb746. **Stafleu, F. A.** 1969. Botanical gardens before 1818. *Boissiera* 14: 31–46.

Cb747. **Stearn, W. T.** 1961. Botanical gardens and botanical literature in the eighteenth century. In: Hunt Botanical Library. *Catalogue of botanical books in the collection of Rachel McMasters Miller Hunt.* Vol. 2, part i: *Introduction to printed books 1701–1800*, pp. xli–cxl. Pittsburgh, PA.

Cb748. **Verdoorn, F.** 1953. Botanical gardens and arboretums of the past and their reconstruction. In: International Union of Biological Sciences. *Sur l'organisation scientifique des jardins botaniques*, pp. 23–30. Paris. 226 pp. *Compte rendu des colloques* Série B, no. 13.

Also

Cb749. **Morren, C. J. É.** 1874. *Correspondance botanique: Liste des jardins botaniques du monde, des chaires de botanique et de quelques établissements de botanique.* Liège, Fédération des Sociétés d'Horticulture de Belgique. 32 pp. ••This directory went through ten editions, e.g: ___. 1874. *Correspondance botanique. Liste des jardins, des chaires, des musées, des revues, et des sociétés de botanique du Monde.* 2me éd. [Etc.]. 40 pp.; ___. 1875. ___. 3me éd. [Etc.]. 65 pp.; ___. 1876. ___. 4me éd. [Etc.]. 80 pp.; ___. 1877. ___. 5me éd. [Etc.]. 92 pp.; ___. 1878. ___. 6me éd. [Etc.]. 149 pp.; ___. 1879. ___. 7me éd. [Etc.]. 154 pp.; ___. 1880. ___. 8me éd. [Etc.]. 169 pp.; ___. 1881. ___. 9me éd. [Etc.]. [ ]* pp.; ___. 1884. ___. 10me éd. [Etc.]. 196 pp. [*Not examined – pagination not ascertained.]

Individual gardens

European (& Mediterranean)

Akademiya Nauk BSSR, Tsentral'nii Botanicheskii Sad, Minsk

Cb750. **Petrovskii, P. Ya.** 1984. Tsentral'nomu Botanicheskomu Sadu Akademii Nauk BSSR: 50 let. *Byulleten' Glavnogo Botanicheskogo Sada, Moscow* 132: 87–89.

Cb751. **Sidorovich, E. A.** 1982. Tsentral'namu Batanichnamu Sadu AN BSSR: 50 hadow. *Vestsi Akademii Navuk Belaruskai SSR, seryya biyalagichnykh i sel'skagaspadarchykh navuk* 1982(5): 3–14.

Akademiya Nauk Gruzinskoi SSR, Botanicheskii Sad, Tbilisi

Cb752. **Akademiya Nauk Gruzinskoi SSR, Botanicheskii Sad.** 1986. *Tsentral'nyi (Tbilisskii) Botanicheskii Sad AN Gruzinskoi SSR: Kratkaya istoriya.* By M. A. Gogolishvili & V. S. Skhiereli. Tbilisi, Metsniereba. 75 pp.

Akademiya Nauk Kazakhskoi SSR, Glavnyi Botanicheskii Sad, Alma Ata

Cb753. **Akademiya Nauk Kazakhskoi SSR, Glavnyi Botanicheskii Sad.** 1982. *Glavnyi Botanicheskii Sad: Istoriya obrazovaniya, struktura, itogi nauchnoi deyatel'nosti Akademiya Nauk Kazakhskoi SSR.* Ed. by I. O. Baitulin. Alma–Ata, Izd. "Nauka" Kazakskoi SSR. 94 pp.

Akademiya Nauk SSSR, Botanicheskii Sad, Botanicheskii Institut im. V. L. Komarova, St. Petersburg (Leningrad)

Cb754. **Botanicheskii Sad Akademii Nauk SSSR.** 1964. *Annotirovannyi bibliograficheskii spravochnikh k "Izvestiyam S.–Peterburgskogo Botanicheskogo Sada" — "Izvestiyam Botanicheskogo Sada Akademii Nauk SSSR" (1901–1932 gg.).* By Z. V. Naumova. Moscow, Leningrad, Nauka. 238 pp.

Cb754A. **Rodionenko, G. I.** 1974. Botanicheskii Sad Botanicheskogo Instituta im. V. L. Komarova AN SSSR za 150 let ego sushchestvovaniya (1823–1973). [The Botanical Garden of the V. L. Komarov Botanical Institute of the Academy of Sciences of the U.S.S.R. through 150 years of its existence (1823–1973)]. *Botanicheskii zhurnal* 59: 1699–1709.

See also Cb1044(Bobrov, Sochava, Sokolov).

Akademiya Nauk SSSR, Glavnyi Botanicheskii Sad, Moscow

Cb755. **Akademiya Nauk SSSR, Glavnyi Botanicheskii Sad.** 1957. *Izdaniya Glavnogo Botanicheskogo Sada SSSR (1947–1956 gg.). Bibliograficheskii ukazatel'.* By F. M. Portnaya. Moscow, AN SSSR, Biblioteka Glavnogo Botanichesogo Sada. 105 pp.

Cb756. ___. 1958. Alfabitnyi ukazatel' statei, opublikovannykh v "Byulletene Glavnogo Botanicheskogo Sada, Akademiya nauk SSSR". Vyp. 1–30. *Byulleten' Glavnogo Botanicheskogo Sada, Akademiya Nauk SSSR* 30: 109–129.

Cb757. **Andreev, L. N.** 1985. Glavnyi Botanicheskii Sad AN SSSR: K 40–letiyu osnovaniya. [The Central Botanical gardens: Towards the 40 years anniversary of organization]. *Vestnik Akademii Nauk SSSR* 1985 (7): 104–110.

Akademiya Nauk SSSR, Ural'skii Nauchnii Tsentr, Botanicheskii Sad, Sverdlovsk

Cb758. **Mamaev, S. A.** 1988. 50 let Botanicheskomu Sadu Ural'skogo Nauchnogo Tsentra AN SSSR. *Byulleten' Glavnogo Botanicheskogo Sada, Moscow* 147: 90–94.

Bergianska Botaniska Trädgården, Stockholm

Cb759. **Fries, R. E.** 1918. Några drag ur den Bergianska trädgårdens historia 1885–1914. *Acta Horti Bergiani* 6 (1): 24 pp. & 54 pl.

Cb759A. **Jonsell, B., et al.** 1991. *Bergianska botanister.* Stockholm, Atlantis. 141 pp. *Bidrag till Kungliga Svenska Vetenskapsakademiens historia* no. 24.
Includes **B. Jonsell.** Olof Swartz; **B. Nordenstam.** Johan Emanuel Wikström; **G. Eriksson.** Nils Johan Anderson; **G. Broberg.** Veit Brecher Wittrock; **B. Jonsell.** Bergianska stiftelsen under vart sekel; **B. Jonsell.** Bergianska stiftelsen, P. Bergius.

Cb760. **Swederus, M. B.** 1877. *Botaniska Trädgården i Upsala, 1655–1807. Ett bidrag till den Svenska Naturforskningens historia.* Falun. iii+141 pp.

Berliner Botanischer Garten

Cb761. **Timler, F. K., & B. Zepernick.** 1978. *Der Berliner Botanische Garten, seine 300 jährige Geschichte vom Hof– und Künchengarten des Grossen Kurfürsten zur wissenschaften Forschungsstätte.* Berlin, Presse– und Informationsamt des Landes Berlin. 95 pp.

Cb762. **Urban, I.** 1881. Geschichte des Königl. Botanischen Gartens und des Königl. Herbariums zu Berlin nebst einer Darstellung des augenblicklichen Zustandes dieser Institute. *Jahrbuch des Königlichen Botanischen Gartens und des Botanischen Museums zu Berlin* 1: 1–164.

Cb763. **Zepernick, B.** 1983. Der Botanische Garten Berlin in Geschichte und Gegenwart. *Sitzungsberichte der Gesellschaft Naturforschender Freunde zu Berlin* n.s. 23: 30–38.

Cb764. ___, **E.–M. Karlsson.** 1979. *Berlins Botanischer Garten.* Berlin, Haude & Spener. 124 pp. *Berliner Reminiszenzen* no. 51.

Cb765. ___, **& F. K. Timler.** 1979. *Grundlagen zur 300 jährigen Geschichte des Berliner Botanischen Gartens.* Berlin, Botanischer Garten und Botanisches Museum Berlin–Dahlem. 303 pp. *Englera* vol. 1.
Includes Die hauptamtlichen Mitarbeiter. Kurzbiographien, pp. 28–120; Bibliographie, pp. 141–194.

Birmingham Botanical & Horticultural Society's Garden

Cb766. **Ballard, P.** 1980. John Claudius Loudon and the Birmingham Botanical and Horticultural Society's gardens at Edgbaston, 1831–1845. *Garden history* 8(2): 66–74.

Cb767. ___. 1983. *An oasis of delight: The history of the Birmingham Botanical Gardens.* London, Duckworth. 152 pp., 12 pp. of illus.

Botanicheskii Sad Akademii Nauk Moldavskoi SSR, Kishinev

Cb768. **Chebotar', A. A.**, ed. 1978. *Issledovaniya Botanicheskogo Sada Akademii Nauk Moldavskoi SSR: 1947–1977gg.* Kishinev, "Shtiintsa." 150 pp.

Cb769. ___, **et al.** 1986. *Bibliograficheskii ukazatel' rabot sotrudnikov Botanicheskogo Sada (Instituta) AN MSSR, 1950–1984.* Kishinev, "Shtiintsa." 337 pp.

Botanicheskite Gradini na Negovo Velichestvo Tsaria na Bulgarite, Sofia

Cb770. **Buresh, I.** 1939. Iokhan Kelerer in Botanicheskite Gradini na Negovo Velichestvo Tsaria na Bulgarite. [Johann Keller und die Botanischen Gärten S. M. des Königs der Bulgaren]. *Izvestiya na Tsarskite Prirodonauchni Instituti v Sofiya* 12: 123–180.

Botanikus Kert, Budapest

Cb771. **Gombocz, E.** 1914. *A Budapesti egyetemi botanikus kert és tanszék történéte. Historia horti botanici nec non cathedrae botanicae Regiae Scientiarum Universitatis Hungaricae Budapestinensis. 1770–1866. Egy fejezet a magyar botanika történetéből.* Budapest, A. Fritz. 200 pp.

Cb772. **Priszter, Sz.** 1971. Kurze Chronologie des 200–Jährigen Botanischen Gartens der Budapester Universität. *Annales Universitatis Scientiarum Budapestinensis de Rolando Eötvös nominatae. Sectio biologica* 13: 137–141.

Cb773. ___. 1971. *A Budapesti egyetemi botanikus kert 1771–1971.* Budapest. 102 pp.

Botanischer Garten, Frankfurt am Main

Cb774. **Möbius, M.** 1903. Geschichte und Beschreibung des botanischen Gartens zu Frankfurt am Main. *Bericht der Senckenbergischer Naturforschenden Gesellschaft.* 1903: 117–154 & 2 pl.

Botanischer Garten, Göttingen

Cb774A. **Peter, A.** 1901. *Geschichte der Gründung und Entwicklung des botanischen Gartens zu Göttingen.* In: Königliche Gesellschaft der Wissenschaften, Göttingen. *Festschrift zur Feier des hundertfünfzigjährigen Bestehens der Königlichen Gesellschaft der Wissenschaften zu Göttingen. Abhandlungen der mathematisch–physikalischen Klasse*, pp. 263–389. Berlin.

Botanischer Garten, Hamburg

Cb775. **Voigt, J. F.** 1869. *Geschichtliches über die Gärten um Hamburg.* Hamburg. 46 pp.

Botanischer Garten, Munchen–Nymphenburg

Cb776. **Friedrich, H. C.** 1990. 75 Jahre Botanischer Garten Munchen–Nymphenburg. *Jahrbesbericht der Generaldirektion der Staatlichen Naturwissenschaftlichen Sammlungen Bayerns* 1989: 9–49.

Botanischer Garten, Munich

Cb777. **Schmid, G.** 1938. Aus der Frühgeschichte des botanischen Gartens in München. *Archiv für die Geschichte der Medizin* 31: 148–164.

Botanischer Garten der Universität Kiel

Cb778. **Universität Kiel.** 1912. *Festschrift der Universität Kiel zur feier des Geburtsfestes Seiner Majestät des Kaisers und Königs Wilhelm II. Der älteste botanische Garten Kiels: Urkundliche Darstellung der Begründung eines Universitäts–Instituts im siebzehnten Jahrhundert.* By J. Reinke. Kiel, Lipsius & Tischer. 84 pp.

Botaniska Trädgården i Upsala

Cb779. **Förberg, E.** 1923. Hortus Upsaliensis några anteckningar om dess öden under 180 år. *Svenska Linné–Sällskapets årsskrift* 6: 120–135.

Cb780. **Juel, H. O.** 1929. Några drag ur Uppsala botaniska institutions historia. *Svenska Linné–Sällskapets årsskrift* 12: 67–87.

Cb781. **Swederus, M. B.** 1877. *Botaniska trädgården i Upsala, 1655–1807. Ett bidrag till den Svenska Naturforskningens historia. (Minnet af dess fyrahundraåriga Verksamhet.)* Falun, Falu Boktryckeri–Aktiebolags Förlag. iii+141 pp.

Chelsea Physic Garden, London

Cb782. **Field, M.** 1878. *Memoirs of the Botanic Garden at Chelsea belonging to the Society of Apothecaries at London.* Revised, corrected, and continued to the present time by R. H. Semple. London, Gilbert & Rivington. [iii]+272 pp.
Replaces Memoirs, historical and illustrative, of the Botanick Garden at Chelsea. London, R. Gilbert, 1820. [vii]+111 pp.

Cb783. **Drewitt, F. D.** 1928. *The romance of the apothecaries' garden at Chelsea.* 3rd ed. London, Cambridge Univ. Pr. xii+175 pp.
Replaces 2nd ed., London, Chapman & Dodd, 1924. x+136 pp.

Cb784. **Hollman, A.** 1973. The Chelsea Physic Garden. *Journal of the Royal College of Physicians* 8: 87–94.

Cb785. **Stearn, W. T.** 1972. Philip Miller and the plants from the Chelsea Physic Garden presented to the Royal Society of London, 1723–1796. *Transactions of the Botanical Society of Edinburgh* 41: 293–307.

Cb786. ___. 1975. The Chelsea Physic Garden 1673–1973: Three centuries of triumph in crises: A tercentenary address. *Garden history* 3(2): 68–73.

Conservatoire et Jardin Botanique de Genève

Cb787. **Miège, J.** 1968. Le jardin botanique de Genève: 150 ans d'histoire. *Candollea* 23: 1–16.

Eesti N S V Teaduste Akadeemia, Tallinna Botaanikaaed (Akademiya Nauk Estonskoi SSR, Tallinskii Botanicheskii Sad)

Cb788. **Akademiya Nauk Estonskoi SSR, Botanicheskii Sad.** 1986. *Tallinnskii botanicheskii sad Akademii Nauk Estonskoi SSR, 1961–1986.* Ed. by R. Ratas. Tallinn, "Valgus." 86 pp., illus.

Forst–Botanische Garten, Tharandt

Cb789. **Tesche, M.** 1988. 175 Jahre Forschung für die Forstwirtschaft im Forst-botanischen Garten Tharandt. *Wissenschaftliche Zeitschrift der Friedrich–Schiller–Universität Jena/Thüringen. Beiträge zur Ökologie und Landeskultur* 37: 93–98.

Friedrich–Alexander–Universität Erlangen–Nürnberg, Botanische Garten

Cb790. **Roehrich, H.** 1965. Der Botanische Garten der Friedrich–Alexander–Universität Erlangen–Nürnberg (1743–1965). *Erlanger Bausteine zur fränkischen Heimatforschung* 12: 43–55.

Glasgow University Physic Garden

Cb791. **Boney, A. D.** 1989. *The lost gardens of Glasgow University*. Bromley, Christopher Helm Publishers Ltd. vi+298 pp., illus.
Replaces ___. 1986. A history of the Glasgow University Physic Garden. *Transactions of the Botanical Society of Edinburgh* 45: 43–62.

Gorenskii Botanicheskii Sad [Count Razumovsky's], Gorenky

Cb792. **Nekrasova, V. L.** 1949. Gorenskii botanicheskii sad. (Materialy k istorii russkikh botanicheskikh sadov). *Trudy Instituta Istorii Estestvoznaniya* 3: 330–350.

Gosudarstvennyi Nikitskii Botanicheskii Sad, Yalta

Cb793. **Gosudarstvennyi Nikitskii Botanicheskii Sad.** 1938. *Kratkie itogi rabot Nikitskogo Botanicheskogo Sada im. Molotova (1812–1938). [Short account of the work of Nikita Botanical Garden ...]*. By G. V. Verbenko. Yalta. 86 pp.

Cb794. ___. 1962. *Bibliograficheskii ukazatel' rabot Gosudarstvennogo Nikitskogo Botanicheskogo Sada s 1812 po 1960 god. [Bibliographical index of works of the State Nikita Botanical garden from 1812 to 1960]*. By A. M. Torbek et al. Yalta. 163 pp.; *Continued* ___. 1967. *Bibliograficheskii ukazatel' rabot Gosudarstvennogo Nikitskogo Botanicheskogo Sada. Dopolnenie pervoe 1961–1965 gg*. By N. P. Onokhova, Z. S. Chernobai, & D. D. Mikhailenko. Simferopol'. 51 pp.
Sections on "General information and history of the garden" and "Personalia" also include extra–mural publications.

Cb795. ___. 1964. 150 let Gosudarstvennomu Nikitskomu Botanicheskomu Sadu. *Sbornik nauchnykh trudov, Gosudarstvennyi Ordena Trudovogo Krasnogo Znameni Nikitski Botanicheskii Sad* 37: 616 pp.

Cb796. **Kochin, M. A.** 1972. K 160–petiyu Gosudarstvennogo Nikitskogo Botanicheskogo Sada. [To the 160th anniversary of the State Nikita Botanical gardens]. *Byulleten' Gosudarstvennogo Nikitskogo Botanicheskogo Sada* 2: 5–10.

Cb797. **Maleev, O. F.** 1931. Nikitskii sad pri Stevene (1812–1824 gg.) Ocherk po istorii Gosudarstvennyi Botanicheskogo Sada. [The Nikita Botanical garden under Steven ...]. *Zapiski Gosudarstvennogo Nikitskogo Opytnogo Botanicheskogo Sada* 17 (1): 34 pp.

Cb798. **Vavilov, N. I., & I. L. Nikitina**, eds. 1939. *Nikitskii Botanicheskii Sad imeni V. M. Molotova za 125 let deyatel'nosti (1812–1937 gg.). Materialy v yubileinogo plenuma sektsii subtropicheskikh kul'tur (Yalta, 25–27 Oktyabrya 1938 g). [Nikita Botanical Garden ... on the 125th anniversary ...]*. Moscow, Vsesoyuznaya Akademiya Sel'sko–Khozyaistvennykh Nauk im. V. I. Lenina. 192 pp.
Papers by various authors.

Hortus Botanicus Academicus, Rijksuniversiteit Leiden

Cb799. **Karstens, W. K. H., & H. Kleibrink.** 1982. *De Leidse Hortus: Een botanische erfenis*. Zwolle, Waanders. 192 pp.

Cb800. **Veendorp, H., & L.–G. M. Baas Beckling.** 1938. *Hortus Academicus Lugduno Batavus 1587–1937: The development of the gardens of Leyden University.* Haarlem, Ex. Typ. Enschedaiana. 218 pp., illus. *Re-pub.* Leiden, Rijksherbarium/Hortus Botanicus, 1990. [x]+218 pp. With a new introduction by C. Kalkman.

Hortus Groningianus, Rijks–Universiteit, Groningen

Cb801. **Andreas, C. H.** 1976. *In en om een botanische tuin: Hortus Groningianus 1626–1966.* Groningen, Van der Kamp. 199 pp., illus.

Imperatorskii Sanktpeterburghskii Botanicheskii Sad, St. Petersburg (Leningrad)

Cb802. **Fischer von Waldheim, A. A.**, ed. 1913–15. *Imperatorskii S.–Peterburgskii Botanicheskii Sad za 200 let ego sushchestvovaniya (1713–1913). [The Imperial St. Petersburg Botanical Garden during 200 years of its existence (1713–1913].* Yubileinoe izdanie, sostavlennoe chlenami sada pod gl. red. A. A. Fishera–fon–Val'dgeima. St. Petersburg. 3 vols.

Includes Vol. 1: **V. I. Lipskii.** Istoricheskii ocherk Imperatorskago S.–Peterburgskago Botanicheskago Sada (1713–1913), pp. 1–378 & 5 pl.; vol. 2: **S. V. Liubimov.** Lichnyisostav i administrativnaya organizatsiya. Istoriya stroenii Imperatorskago Botanicheskago Sada, pp. 241–321; vol. 3: **V. I. Lipskii.** Biografii i literaturnaya deyatel'nost' botanikov i lits, soprikasavshikhsya s imp. Botanicheskim sadom. [Biographies and literary activity of the botanists and the persons connected with the Imperial Botanical Garden], pp. 1–536.

Cb803. **Lukina, T. A.** 1966. Botanicheskii sad Peterburgskoi Akademii Nauk vo vtoroi polovine XVIII veka. [The botanical garden of the St. Petersburg Academy of Science in the second half of the 18th century]. *Iz istorii biologicheskikh nauk* 1: 117–141.

Cb804. **Nekrasova, V. L.** 1945. K istorii Botanicheskogo Sada Akademii Nauk. (Na Vasil'evskom ostrove, (1735–1812). [Contribution to the history of the botanical garden of the Academy of Sciences (1735–1812]. *Sovetskaya botanika* 13: 13–37.

Cb805. **Stapf, O.** 1913. The Imperial Botanic Garden of Peter the Great at St. Petersburg. *Kew bulletin* 1913: 243–255 & 2 pl.

Jardin Botanique d'Anvers

Cb806. **Frison, E., & R. Aernouts.** 1973. Le Jardin Botanique d'Anvers, 1797–1926. *Janus* 60: 149–192.

Jardin Botanique de Bruxelles

Cb807. **Jardin Botanique de Bruxelles.** 1970. *Histoire des jardins botaniques de Bruxelles: Brochure éditée à l'occasion du centième anniversaire du Jardin Botanique National de Belgique.* Brussels, Crédit Communal de Belgique. 32 pp.

Includes **J. Balis.** Le premier jardin botanique de Bruxelles; **E. Witte.** Le jardin botanique de la S. A. "Société Royale d'Horticulture des Pays-Bas" (1826–1870); **R. Tournay.** Le Jardin Botanique National de Belgique. First pub. in: *Bulletin trimestriel du credit communal de Belgique* 25 (95): 107–138.

Cb808. ___. 1970. *Le Jardin Botanique de Belgique. De Nationale Plantentuin van Belgie, 1870–1970.* Brussels. 78 pp.

Text in French and Flemish.

Jardin Botánico, Tenerife

Cb809. **Rodríguez Garcia, V.** 1980. *El Jardin Botánico de Tenerife: esquema de su historia.* Las Palmas de Gran Canaria, Mancomunidad de Cabildos. 48 pp.

Jardin des Plantes, Caen

Cb810. **Lignier, O.** 1904. Essai sur l'histoire du Jardin des Plantes de Caen. *Bulletin de la Société Linnéenne de Normandie* 8: 27–175 & 4 pl.

Jardin des Plantes de Montpellier

Cb811. **Association des Amis du Jardin des Plantes de Montpellier.** 1983. *Le jardin des plantes de Montpellier: Le plus ancien jardin de France, 1593.* Montpellier. 60 pp.

Cb812. **Bonaventure-Laurens, J. J.** 1964. Le jardin des Plantes de Montpellier au milieu du XIXe siècle. *Monspeliensis Hippocrates* 7: 8–10.

Cb813. **Emberger, L., & H. Harant.** 1959. *La botanique à Montpellier. L'histoire de la botanique à Montpellier. Ouvrage édité à l'occasion de l'inauguration du nouvel Institut de Botanique et de la restauration du Jardin des Plantes de l'Université de Montpellier.* [Montpellier.] 67 pp., illus.

Cb814. **Harant, H.** 1954. The Montpellier botanical garden. *Endeavour* 13: 97–100.

Jardin des Plantes, Paris (Previously Jardin du Roi)

Cb815. **Barthélemy, G.** 1979. *Les jardiniers du roy: Petite histoire du Jardin des Plantes de Paris.* Paris, diffusion Librairie du Muséum. 293 pp.

Cb816. **Brygoo, E.** 1979. Les Médecins de Montpellier et le Jardin du Roi à Paris. *Histoire et nature* 14: 3–29.

Cb817. ___. 1988. Du Jardin et du Cabinet du Roi au Muséum National d'Histoire Naturelle, en 1793. La continuité par les hommes. *Histoire et nature* 28/29: 47–63.

Cb817A. **Chevalier, A.** 1930. L'œuvre des voyageurs–naturalistes et du Jardin des Plantes depuis sa fondation jusqu'à la révolution. *Revue de botanique appliquée et d'agriculture coloniale* 10: 463–498.

Cb818. **Crestois, P.** 1953. *Contribution à l'histoire de l'enseignement de la pharmacie: L'Enseignement de la botanique au Jardin Royal des Plantes de Paris.* Cahors, Coueslant. 132 pp.

Cb819. **Denise, L.** 1903. *Bibliographie historique et iconographique du Jardin des Plantes: Jardin Royal des Plantes Médicinales et Muséum d'Histoire Naturelle.* Paris, Daragon. iii+268 pp. *Bibliothèque du Vieux–Paris.*
Includes 1. Histoire générale; 2. La création du Jardin Royal des Plantes Médicinales; 3 & 4. Le Jardin du Roi. Author and subject indexes.

Cb820. **Falls, W.** 1933. Buffon et l'agrandissement du Jardin du Roy à Paris. *Archives du Muséum* 1933: 131–198.

Cb821. **Hamy, E.-T.** 1893. *Les derniers jours du Jardin du Roi et la fondation du Muséum d'Histoire Naturelle.* Paris, Imprimerie Nationale. [i]+162 pp. Extrait du *Volume commémoratif du centenaire de la fondation du Muséum d'Histoire Naturelle.*

Cb822. **Howard, R.** 1974. The founding of the Jardin des Plantes in Paris. *Proceedings of the annual meeting of the Western Society for French History* 2: 138–150.

Cb823. ___. 1980. Medical politics and the founding of the Jardin des Plantes in Paris. *Journal of the Society for the Bibliography of Natural History* 9: 395–402.

Cb824. ___. 1983. The garden that grew a revolution. *Garden, New York* 7(5): 20–23, 30–32.

Cb825. ___. 1983. *La bibliothèque et le laboratoire de Guy de La Brosse au Jardin des Plantes à Paris.* Geneva, Droz. 133 pp.

Cb826. [Not used.]

Cb827. **Laissus, Y.** 1967. La succession de Le Monnier au Jardin du Roi: Antoine-Laurent de Jussieu et René Louiche-Desfontaines. In: *Comptes rendus du 91e Congrès National des Sociétés Savantes, Section des Sciences,* tom. 1: 137–152.

Cb828. **Laissus, Y.** 1981. Les voyageurs naturalistes du Jardin du Roi et du Muséum d'Histoire Naturelle: Essai de portrait–robot. *Revue d'histoire des sciences* 34: 259–317.

Cb829. ___, **& J. Torlais.** 1964. Le Jardin du Roi et le Collège royal dans l'enseignement des sciences au XVIIIe siècle. Paris, Hermann. pp. 261–341. *Enseignement et diffusion des sciences en France au XVIIIe siècle.*
Includes Annexe. Le personnel scientifique du Jardin du Roi de 1635 a 1793, pp. 319–341. First pub. as part of: R. Taton, ed. 1964. *Enseignement et diffusion des sciences en France au XVIIIe siècle.* Paris, Hermann. 780 pp. & 16 pl. *Histoire de la pensée* vol. 11; *see* Ca192.

Cb830. **Leroy, J.–F.** 1971. La botanique au Jardin des Plantes (1626–1970). *Adansonia* 11: 225–250.

Cb831. **Letouzey, Y.** 1989. *Le Jardin des Plantes a la croisée des chemins avec André Thouin 1747–1824.* Paris, Muséum National d'Histoire Naturelle. 678 pp. & 23 pl. (col.).

Cb832. **Perret, J.** 1984. *Le Jardin des Plantes.* Paris, Julliard. 258+[5] pp., illus.

See also Inventory of the archives of the Jardin du Roi Bc785.

Jardin Royal des Plantes de Perpignan

Cb833. **Amigo, J.–J.** 1985. Le Jardin Royal des Plantes de Perpignan. *Comptes rendus du 110 Congrès National des Sociétés Savantes, Section d'Histoire des Sciences [etc.],* tom. 1: 21–32.

Jenaer Botanische Garten

Cb834. **Jahn, I.** 1988. Zur Grundungs– und Entwicklungsgeschichte der Jenaer Botanischen Garten (von 1586 bis 1864). *Wissenschaftliche Zeitschrift der Friedrich–Schiller–Universität Jena/Thüringen. Beiträge zur Ökologie und Landeskultur* 37: 17–25.

Cb835. **Klotz, G.** 1988. Der Hortus Botanicus Jenensis. *Wissenschaftliche Zeitschrift der Friedrich–Schiller–Universität Jena/Thüringen. Beiträge zur Ökologie und Landeskultur* 37: 26–33.

Cb836. ___. 1988. Die Direktorenn des Botanischen Gartens der Universität Jena 1794–1986. *Wissenschaftliche Zeitschrift der Friedrich–Schiller–Universität Jena/Thüringen. Beiträge zur Ökologie und Landeskultur* 37: 34–35.

Kjøbenhavns Universitet, Botanisk Have

Cb837. **Jensen, J. N.** 1974. The Botanical Garden from 1924–1974. *Botanisk tidsskrift* 69: 75–87.

Cb838. **Lange, J. M. C.** 1876. Erindringer fra Universitets Botaniske Have ved Charlottenborg, 1778–1874. *Botanisk tidsskrift* 9: 1–68.

Cb839. **Olsen, O.** 1974. The centenary of the Copenhagen Botanical Garden. *Botanisk tidsskrift* 69: 73–74.

Cb840. ___. 1974. Der Botanische Garten Kopenhagens 1879–1974. *Gärtnerisch–botanische Briefe* 44: 1464–1475.

Cb841. **Ostenfeld, C. H.** 1924. *Botanisk Have gennem 50 aar. 1874–1924.* Copenhagen, G. E. C. Gads Forlag. 101 pp.

Latvijas P S R Zinatnu Akademija, Botaniskais Darzs, Riga (Akademiya Nauk Latviiskoi SSR, Botanicheskii Sad, Riga)

Cb842. **Akademiya Nauk Latviiskoi SSR, Botanicheskii Sad.** 1983. *Botanicheskii sad Akademii Nauk Latviiskoi SSR: 1956–1981. The Botanical Garden of the Latvian SSR Academy of Sciences 1956–1981.* Ed. by R. Tsinovskis. Riga, Zinatne. 320 pp.

Leicester University, Botanic Garden

Cb843. **Gornall, R. J.** 1983. A history of Leicester University Botanic Garden. *Transactions of the Leicester Literary & Philosophical Society* 77: 42–48.

Martin–Luther Universität Halle–Wittenberg, Botanische Garten

Cb844. **Ebel, F., & F. Kümmel.** 1973. Zur Entwicklung des Botanischen Gartens der Martin–Luther Universität Halle–Wittenberg in den vergangenen 25 Jahren (1947–1972). Ein Beitrag zu seiner 275 jährigen Geschichte. *Hercynia* 10: 193–233.

Cb845. **Kraus, G. K. M., & K. P. J. Sprengel.** 1888–94. *Der botanische Garten den Universität Halle*. Leipzig, W. Engelmann. 2 vols.

Moskovskii Universitet, Botanicheskii Sad

Cb846. **Kropotova, I. I., V. S. Novikov & V. N. Tikhomirov.** 1980. Nachal'nyi period istorii botanicheskogo sada Moskovskogo Universiteta. [First period of history of Moscow State University Botanical Garden]. *Vestnik Moskovskogo Universiteta*, Seriya 16, *biologiya* 1980 (1): 3–8.

Cb847. **___.** 1981. *Botanicheskii Sad Moskovskogo Universiteta 1706–1981 (bibliografiya)*. Moscow, Izdatel'stvo Moskovskogo Universiteta. 180 pp.

Cb848. **Moskovskii Universitet.** 1954. Kratkaya istoriya sada. In: *Botanicheskii Sad Moskovskogo Universiteta: Putevoditel'*, pp. 9–24. Moscow. 180 pp.

See also Cb1073(Meier).

Mortola (La), Hanbury Botanical Gardens

Cb849. **Forbes, D. Hanbury–.** 1960. La Mortola: The story of the Hanbury Botanical Gardens in Italy, since the days of Daniel Hanbury, F.R.S., M.P.S., F.L.S., and Sir Thomas Hanbury, F.L.S. *Pharmaceutical journal* 185: 149–151.

National Botanic Gardens, Glasnevin

Cb850. **Nelson, E. C.** 1980. A select annotated bibliography of the National Botanic Gardens, Glasnevin, Dublin. *Glasra* 5: 1–20.

Cb851. **___, & E. M. McCracken.** 1987. *The brightest jewel: A history of the National Botanic Gardens Glasnevin, Dublin*. Kilkenny, Boethius Pr. [xiv]+275 pp.

Ogród Botaniczny w Krzemiencu

Cb852. **Oleszakowa, J.** 1971. Ogród botaniczny w Krzemiencu i jego katalogi (1805–1832). [The Botanical Garden at Krzemieniec and its catalogs (1805–1832)]. *Studia i materialy z dziejow nauki Polskiej, ser. B* 21: 59–82.

Orto Botanico di Firenze

Cb853. **Mattirolo, O.** 1899. *Cenni cronologici sugli orti botanici di Firenze*. Florence. 27 pp. *Pubblicazioni del R. Istituto di Studi Superiori Pratici e di Perfezionamento in Firenze, sezione di scienze fisiche e naturali* [24].

Orto Botanico di Napoli

Cb854. **Catalano, G.** 1958. Storia dell'Orto Botanico di Napoli. *Delpinoa* 11: 170 pp.

Cb855. **Cavara, F.** 1913. Celebrazione del centenario del R. Orto Botanico e inaugurazione del monumento a Michele Tenore. *Bulletino dell' Orto Botanico della Regia Università di Napoli* 3: I–LXX.

Cb856. **___.** 1913. Cenni sul R. Orto Botanico di Napoli. *Bulletino dell' Orto Botanico della Regia Università di Napoli* 3: 1–35.

Orto Botanico di Padova

Cb857. **Azzi Visentini, M.** 1984. *L'orto botanico di Padova e il giardino del Rinascimento.* (Saggi, 3). Milan, Il Polifilo. 288 pp.

Cb858. **Christopher, T. A.** 1981. The seeds of botany. *Natural history* 90(3): 50–57.

Cb859. **Dupuigrenet–Desroussilles, F.** 1989. Regards et savoirs: Images du jardin botanique de l'Université de Padoue au XVIe siècle. *Revue de l'histoire des sciences* 42: 281–291.

Cb860. **Rhodes, D. E.** 1984. The botanical garden of Padua: The first hundred years. *Journal of garden history* 4: 327–331.

Orto Botanico di Pisa

Cb861. **Garbari, F., L. Tongiorgi Tomasi, & A. Tosi.** 1991. *Giardino dei Semplici: L'orto botanico di Pisa dal XVI al XX secolo.* Pisa, Casa di Risparmio. 397 pp., illus. (col.).
Includes **F. Garbari & L. Tongiorgi Tomasi.** Le origini del Giardino del Semplici: Dall' Orto dell'Arsenale all'"Orto novo" di via Santa Maria, pp. 15–26; **F. Garbari.** I "Prefetti" del Giardino, delle origini, pp. 27–114; **A. Tosi.** Arte e scienza tra neoclassicismo e romanticismo: Il giardino en età moderna, pp. 213–274; **L. Tongiorgi Tomasi.** Arte e natura nel Giardino dei Semplici: Dalle origini alla fine dell'età Medicea, pp. 115–212; Appendici documentarie, pp. 277–349; Bibliografia, pp. 353–378.

Oxford Botanic Garden

Cb862. **Druce, G. C.** 1924. The foundation of the Oxford Botanic Garden and its tercentenary. *Report, Botanical Society and Exchange Club of the British Isles* 7: 335–366.

Cb863. **Osborn, T. G. B.** 1951. The Oxford Botanic Garden. *Endeavour* 10: 70–77.

Plantentuin der Universiteit te Gent

Cb864. **Thiery, L. M.** 1947. *Bij de 150e verjaring van de Plantentuin der Universiteit te Gent (1797–1947).* Gent, Schnoeck–Ducaju. 461 pp.

Real Jardín Botánico de Madrid

Cb865. **Añón, C.** 1987. *Real Jardín Botánico de Madrid: Sus orígenes, 1755–1781.* Madrid, Real Jardín Botánico. 182 pp., illus.

Cb866. **Colmeiro, M.** 1875. Bosquejo histórico y estadístico del Jardin Botánico de Madrid. *Anales de la Sociedad Española de Historia Natural* 4: 241–345.

See also Ca188(Sarmiento).

Regia Università degli Studi, Modena, Orto Botanico

Cb867. **De Toni, G. B.** 1906. Il R. Orto botanico di Modena dal 1772 al 1906. Notizie raccolte dal direttore prof. G. B. De Toni. *Malpighia* 20: 272–283.

Royal Botanic Garden, Edinburgh

Cb868. **Cowan, J. M.** 1933. The history of the Royal Botanic Garden, Edinburgh. *Notes from the Royal Botanic Garden, Edinburgh* 19: 1–134.

Cb869. **Fletcher, H. R., & W. H. Brown.** 1970. *The Royal Botanic Garden, Edinburgh, 1670–1970.* Edinburgh, Her Majesty's Stationery Office. xvi+309 pp. & 24 pl.

Cb870. **Royal Botanic Garden, Edinburgh.** 1925. *Notes from the Royal Botanic Garden, Edinburgh: Index to volumes 1–14, 1900–1924.* Edinburgh. 54 pp.; ___. 1956. ___: General index, volumes 15–21, 1925–56. *Notes from the Royal Botanic Garden, Edinburgh* 21: 305–343.; ___. 1971. ___: *General index, volumes 22–30, 1955–70.*

Edinburgh. 141 pp.; ___. 1983. ___: *General index, volumes 31–38, 1971–80.* Edinburgh. 136 pp.

Royal Botanic Gardens, Kew

Cb871. **Allan, M.** 1967. *The Hookers of Kew, 1785–1911.* London, M. Joseph. 273 pp. & 28 pl.

Cb872. **Blunt, W.** 1978. *In for a penny: A prospect of Kew Gardens: Their flora, fauna, and falballas.* London, Hamilton. xxi+218 pp.

Cb873. **Desmond, R. G. C.** 1975. The Hookers and the development of the Royal Botanic Gardens, Kew. *Biological journal of the Linnean Society* 7: 173–182.

Cb873A. **___., & F. N. Hepper.** 1993. *A century of Kew plantsmen: A celebration of the Kew Guild.* Kew, Kew Guild. vii+188 pp., illus.
Includes Distinguished Kewites: An anthology of biographies in the Journal of the Kew Guild, pp. 129–183. Ca. 50 people with portraits.

Cb874. **Dyer, W. T. Thiselton–.** 1891. Historical account of Kew to 1841. *Bulletin of miscellaneous information, Royal Gardens, Kew* 60: 279–327.

Cb875. **Hepper, F. N.** 1982. *Kew: Gardens for science and pleasure.* London, Her Majesty's Stationery Office. 195 pp.

Cb876. **Jackson, B. D.** 1897. [List of Kew publications, 1841–1895]. *Bulletin of miscellaneous information, Royal Botanic Gardens, Kew* 1897: 1–84, 238–240; *Continued* **Anon.** 1907. List of Kew publications, 1896–1906. *Ibid.* 1907 Appx. V:95–152.
The earlier list is chronologically arranged, the later one is arranged by authors; continued by annual lists of staff publications in the *bulletin*.

Cb877. **Simmons, J.** 1981. The history and development of the arboretum at the Royal Botanic Gardens, Kew. *Arboricultural journal* 5: 173–188.

Cb878. **Stearn, W. T.** 1965. The self–taught botanists who saved the Kew botanic garden. *Taxon* 14: 293–298.
I.e. Duke of Bedford, W. J. Hooker, J. Lindley & J. Paxton.

Cb879. **Turrill, W. B.** 1959. *The Royal Botanic Gardens, Kew, past and present.* London, H. Jenkins. 256 pp. & 16 pl.

See also Ba945–946 Ba948 Ca824 Cb330A, Cb1063 & Cb1206.

Royal Botanic Society, London

Cb880. **Meynell, G.** 1980. The Royal Botanic Society's Garden, Regent's Park. *London journal* 6: 135–146.

Royal Horticultural Society's Garden, Wisley

Cb881. **Chittenden, F. J.** 1933. The Royal Horticultural Society's gardens, Wisley: The first twenty–five years. *Journal of the Royal Horticultural Society* 58: 258–287.

Cb882. **Royal Horticultural Society.** 1978. *Wisley, the first 100 years.* London. [40] pp.

Trinity College Dublin, Botanic Garden

Cb883. **Jackson, P., & S. Wyse.** 1987. The Botanic Garden of Trinity College Dublin, 1687 to 1987. *Botanical journal of the Linnean Society* 95: 301–311.

Cb884. **McCracken, E.** 1979. The Botanic Garden, Trinity College, Dublin. *Garden history* 7(1): 86–90.

Università di Pavia, Orto Botanico

Cb885. **Giacomini, V.** 1959. *Alle origini della Lettura dei Semplici (1546) dell' Orto dei Semplici (1558) e dell' Orto Botanico (1773) nell' Università di Pavia.* Pavia. 167 pp.

Cb886. **Tomaselli, R.** 1973. Ricorrenze storiche dell' Orto Botanica dell' Università di Pavia negli anni settanta. *Atti dell' Istituto Botanico dell' Università di Pavia* 9: 3–129.

Università di Torino, Orto Botanico

Cb887. **Mattirolo, O.** 1929. *1729–1929: Cronistoria dell' Orto Botanico (Valentino) della R. Università di Torino.* Turin, L. Checchini. cxli pp.
Biographical sketches of "I professori di botanica" and "Assistenti, giardinieri, erborizzatori, conservatori," pp. xxxii–xc.

Cb888. **Montacchini, F., G. Forneris & C. Martoglio.** 1986. *Erbari e iconografia botanica: Storia delle collezioni dell'Orto Botanico dell'Università di Torino.* Turin, Um Allemandi. 158 pp., illus. *Archivi di arte e cultura piemontesi* no. 7.

Universität Basel, Botanische Garten

Cb889. **Burckhardt, F.** 1905. Geschichte der botanische Anstalt. *Verhandlungen der Naturforschenden Gesellschaft in Basel* 18 (1): 83–118.

University of Helsinki, Botanic Garden

Cb890. **Uotila, P., & P. Lehtonen.** 1983. Helsingin Yliopiston Kasvitieteellisen Puutarhan bibliografia vuoteen 1982. Bibliography of the Botanical Garden of the University of Helsinki, up to the year 1982. *Norrlinia* 1: 1–39.

Uniwersytet Jagiellonski, Ogrodu Botanicznego, Cracow

Cb891. **Piekielko, A.** 1980. Poczatki Krakowskiego Ogrodu Botanicznego (1783–1809). *Kwartalnik historii nauki i techniki* 25: 67–87.

Cb892. ___. 1981. Dzieje Krakowskiego Ogrodu Botanicznego w latach 1809–1917. *Kwartalnik historii nauki i techniki* 26: 111–136.

Cb893. ___. 1983. *Historia Ogrodu Botanicznego Uniwersytetu Jagiellonskiego w Krakowie.* Cracow, Uniwersytet Jagiellonski. 113 pp.
With an English summary.

Uniwersytet Warszawski, Ogrodu Botanicznego

Cb894. **Karpowiczowa, L.** 1967. *150 lat Ogrodu Botanicznogo Uniwersytetu Warszawskiego.* Warsaw, Panstwowe Wydawnictwo Naukowe. 72 pp.

Voronezhskii Gosudarstvennii Universitet, Botanicheskii Sad

Cb895. **Mashkin, S. I.** 1954. *Botanicheskii sad V[oronezhskogo] G[osudarstvennogo] U[niversitet].* Voronezh, Knizhn. Izd. 104 pp., illus.
Bibliografiya nauchnykh rabot [etc.], pp. 95–103.

North American

Arnold Arboretum of Harvard University, Jamaica Plain, NY

Cb896. **Arnold Arboretum of Harvard University.** 1973. *Journal of the Arnold Arboretum: Index to authors and titles, vols. 1 through 50, 1919–1969.* By E. B. Bernstein. Cambridge, MA. 74 pp.; *Continued* ___. 1985. Journal of the Arnold Arboretum: Index to authors and titles, vols. 51–65 (1970–1984). By E. B. Schmidt. *Journal of the Arnold Arboretum* 66: 39–72.

Cb897. **Howard, R. A.** 1972. Scientists and scientific contributions of the Arnold Arboretum, the first century. *Arnoldia* 32: 49–58.
Details of 80 past and present staff members.

Cb898. **Sargent, C. S.** 1922. The first fifty years of the Arnold Arboretum. *Journal of the Arnold Arboretum* 3: 127–171.

Cb899. **Spongberg, S. A.** 1990. *A reunion of trees: The discovery of exotic plants and their introduction into North American and European landscapes.* Cambridge, MA, Harvard Univ. Pr. xiv+270 pp.
Focusses on the work of the Arnold Arboretum.

Cb900. **Stearn, W. T.** 1972. From medieval park to modern arboretum: The Arnold Arboretum and its historical background. *Arnoldia* 32: 173–197.

Cb901. **Sutton, S. B.** 1970. *Charles Sprague Sargent and the Arnold Arboretum.* Cambridge, MA, Harvard Univ. Pr. xvii+382 pp.

Cb902. ___. 1971. *The Arnold Arboretum: The first century.* Foreword by W. H. Whitehill and an afterword by R. A. Howard. Jamaica Plain, NY, Arnold Arboretum of Harvard Univ. 72 pp.

Cb903. **Wilson, E. H.** 1925. *America's greatest garden: The Arnold Arboretum.* Boston, MA, Stratford Co. [vi]+ix+123 pp., illus.

Bartram's garden, Philadelphia, PA

Cb904. **Philadelphia Botanical Club.** 1931. An account of the two hundredth anniversary of the founding of the first botanic garden in the American colonies by John Bartram. *Bartonia* 12 supp: 68 pp.
Seven articles by various authors.

Cb905. [Not used.]

Brooklyn Botanic Garden, New York

Cb906. **Brooklyn Botanic Garden.** 1944. C. Stuart Gager and the Brooklyn Botanic Garden. *Brooklyn Botanic Garden record* 33(2): 69–178.

Cb907. ___. [1961.] *Brooklyn Botanic Garden: The first fifty years [etc.].* Brooklyn, NY. 81 pp.

Cambridge Botanic Garden, Harvard University, Cambridge, MA

Cb908. **Ingersoll, E.** 1886. Harvard's botanic garden and its botanists. *Century magazine* 32: 236–248.

Elgin Botanic Garden, New York

Cb909. **Brown, A.** 1908. The Elgin Botanic Garden: Its later history, and relation to Columbia College and the Vermont land controversy. *Bulletin of the New York Botanical Garden* 5: 319–372.

Cb910. **Meisel, M.** 1926. The Elgin Botanic Garden, New York City (1801). In: *A bibliography of American natural history, the pioneer century, 1769–1865, [etc.]*, vol. 2: 81–84. Brooklyn, NY, Premier Publishing Co. *Re-pub.* New York, Hafner, 1967.
Historical note and bibliography.

Fairchild Tropical Garden, Coral Gables, FL

Cb911. **Zuckerman, B.** 1988. *The dream lives on: A history of the Fairchild Tropical Garden 1938–1988.* Miami, FL, Fairchild Tropical Garden. 127 pp.

Huntington Botanical Gardens, San Marino, CA

Cb912. **Hertrich, W.** 1949. *The Huntington Botanical Gardens, 1905–1949. Personal recollections [etc.].* San Marino, CA, Huntington Library. [vi]+167 pp., illus.

Missouri Botanical Garden, St. Louis, MO

Cb913. **Various authors.** 1984. Missouri Botanical Garden: 125th anniversary. *Gateway heritage* 5(1): 1–48.
Five articles by various authors.

Cb914. **Bry, C.** 1989. Henry Shaw and the early foundations of the Missouri Botanic Garden. In: C. Bry et al. *A world of plants: The Missouri Botanic Garden*, pp. 41–65. New York, Abrams.

Cb915. **Faherty, W. B.** 1989. *A gift to glory in: The first hundred years of the Missouri Botanical Garden (1859–1959)*. Ocean Park, WA, Harris & Friedrich. xxi+234 pp.
A sequel to: ___. 1987. *Henry Shaw: His life and legacies*. Columbia, MO, Univ. of Missouri Pr. xi+228 pp.

Cb916. **Missouri Botanical Garden.** 1898. Report (Annual) of the Missouri Botanical Garden. Indexes to volumes 1–10 [1889–98]: – a. General index; b. Index to Latin names. *Report (Annual) of the Missouri Botanical Garden* 10: 162–211.; ___. 1908. ___: Indexes to volumes 11–20 [1889–1908]: – a. General index; b. Index to Latin names. By C. J. Hogan. *Report (Annual) of the Missouri Botanical Garden* 20: 152–223.

Cb917. ___. 1973. *Annals of the Missouri Botanical Garden. Index. An author and title index to volumes 1–55, (1914–1968).* By C. Lange. St. Louis, MO. 54 pp.

Cb918. **Rudolph, E. D.** 1991. One hundred years of the Missouri Botanical Garden. *Annals of the Missouri Botanical Garden* 78: 1–18.

New York Botanical Garden

Cb919. **Howe, M. A.** 1929. New York's first botanical garden. *Journal of the New York Botanical Garden* 30: 49–58.

Cb920. **New York Botanical Garden.** 1914. Index to volumes 1–15 [1900–14]. *Journal of the New York Botanical Garden* 15 (180): 233–261.; ___. 1929. Index to volumes 16–30 [1915–1929]. *Ibid.* 30 (360): 289–477.

Cb921. ___. 1918. Students and investigators at the New York Botanical Garden, 1897–1918. *Journal of the New York Botanical Garden* 19: 231–277.

See also Catalogue of the Garden's archives Bc793.

Rancho Santa Ana Botanic Garden, Claremont, CA

Cb922. **Lenz, L. W.** 1977. Rancho Santa Ana Botanic Garden – the first fifty years, 1927–1977. *Aliso* 9(1): 1–156.
Some historical information is also contained in: **K. L. Tomlinson.** 1987. Featured institution – Rancho Santa Ana Botanic Garden: Sixty years of botany in California. *Association of Systematics Collections newsletter* 15 (2): 13–15.

Transylvania Botanic Garden, Lexington, KY

Cb923. **Harrison, I. W.** 1913. The Transylvania Botanic garden: A little–known American enterprise of great historic, scientific, and educational interest; [etc.]. *Journal of American history* 1: 901–919.

South & Central American (& Caribbean)

Jardim Botánico, Rio de Janeiro

Cb924. **Laszlo, H.** 1990. Der botanische Garten in Rio de Janeiro. *Archiv der Geschichte der Naturwissenschaften* 25: 1219–1230.

Cb925. **Niemeyer de Lavor, J. C.** 1979. Jardim Botanico do Rio de Janeiro, do seu inicio aos nossos dias. *Rodriguesia* 31: 275–283.

Cb926. ___. 1983. Historiografia do Jardim Botánico do Rio de Janeiro, no contexto da fazenda real da Logoa Rodrigo de Freitas e seus desdobramentos. *Rodriguesia* 35: 51–99.

Cb927. **Rio de Janeiro, Jardim Botánico.** 1948. Lista bibliográfica dos naturalistas do jardim botánico. *Archivos do Jardim Botânico do Rio de Janeiro* 8: 517–567.

Jardin Botánico, São Paulo

Cb928. **Hoehne, F. C., M. Kuhlmann & O. Handro.** 1941. *O Jardim Botánico de Sao Paulo: Precedido de prólogo histórico e notas bio–bibliograficas de naturalistas botânicos que trabalharam para o progresso do conhecimento de flora do Brasil*, por ... e catálogo das espécies ali cultivadas; em colaboração com: M. Kuhlmann e O. Handro. [São Paulo,] Secretaria de Agricultura, Indústria e Comércio de São Paulo. 656 pp.

Asian

Botanic Garden, Manila

Cb929. **Rodriguez, L.** 1962. The botanical garden of Manila and Sebastian Vidal y Soler. *Unitas* 35: 258–277.

Botanic Gardens, Singapore

Cb930. **Botanic Gardens, Singapore.** 1959. An issue [of the *Gardens bulletin*] to mark the occasion of the hundredth anniversary of the founding of the Botanic Gardens, Singapore 1859–1959. *Gardens bulletin* 17(2): 125–340.
Articles various authors.

Cb931. **Tinsley, B.** 1983. *Singapore green: A history and guide to the botanic gardens.* Singapore, Times Books International. 139 pp., illus.

Cb932. **___.** 1989. *Visions of delight: The Singapore Botanic Gardens through the ages.* Singapore, Singapore Botanic Gardens. 95 pp., illus.

Kebun Raya Indonesia, Bogor (*formerly* Buitenzorg)

Cb933. **Bernard, C. J.** 1945. Le Jardin Botanique de Buitenzorg et les institutions de botanique appliquée aux Index Néerlandaises. In: P. Honig & F. Verdoorn, eds. *Science and scientists in the Netherlands Indies*, pp. 10–15. New York, Board for Netherlands Indies, Surinam and Curaçao.

Cb934. **Demmerman, K. W.** 1935. The quinquagenary of the foreigners' laboratory at Buitenzorg, 1884–1934. *Annales du Jardin Botanique de Buitenzorg* 45: 1–60, & 5 pls.
Also ___. 1945. A history of the visitors' laboratory ("Treub Laboratorium") of the Botanic Gardens, Buitenzorg, 1884–1934. In: P. Honig & F. Verdoorn, eds. *Science and scientists in the Netherlands Indies*, pp. 59–75. New York, Board for Netherlands Indies, Surinam and Curaçao.

Cb935. **Jardin Botanique de Buitenzorg.** 1918. Tables du contenu des volumes 1–30 [et] suppléments 1–4 [1876–1918]. *Annales du Jardin Botanique de Buitenzorg supp.* 4: 18–62.

Cb936. **'s Lands Plantentuin te Buitenzorg.** 1893. *Der botanische Garten "'s Lands Plantentuin" zu Buitenzorg auf Java. Festschrift zur Feier seines 75 jährigen Bestehens (1817–1892).* Leipzig. vi+426 pp. & 12 pl.

Cb936A. **Rijnberg, T. F.** 1992. *'s Lands Plantentuin, Buitenzorg, 1817–1992.* Bogor, Kebun Raya Indonesia. ix+212 pp., illus.
Text in Dutch, English summary.

Cb937. **Steenis, C. G. G. J. van.** 1935. Survey of botanical researches carried on by visitors to the Treub Laboratory 1884–1934, morphology, physiology, anatomy and cryptogams excluded. *Annales du Jardin Botanique de Buitenzorg* 45: 61–104.

Peradeniya Botanic Garden, Colombo

Cb938. **Stockdale, F. A., T. Petch, & H. F. Macmillan.** 1922. *The Royal Botanic Gardens, Peradeniya, Ceylon, 1822–1922.* Colombo. [iv]+65 pp., illus.
First pub. in *Tropical agriculturist*, vol. 58, no. 3, 1922.

Royal Botanic Garden, Calcutta

Cb939. **Biswas, K.**, ed. 1942. *The 150th anniversary volume of the Royal Botanic Garden, Calcutta.* Alipore, Bengal Government Pr. 301 pp.
Chapters by various authors.

Cb940. ___. 1950. *The original correspondence of Sir Joseph Banks relating to the foundation of the Royal Botanic Garden, Calcutta and the summary of the 150th anniversary volume of the Royal Botanic Garden, Calcutta.* Calcutta, Royal Asiatic Society of Bengal. [vi]+62 pp. & 16 pl.

Turkmenskii Botanicheskii Sad, Ashkhabad

Cb941. **Gosteva, A. N.** 1956. Bibliografiya rabot Turkmenskogo botanicheskogo sada (1929–1954). *Trudy Turkmenskogo Botanicheskogo Sada* 2: 158–179.

Australasian

Adelaide Botanic Garden

Cb942. **Adelaide, Botanic Garden.** [1955.] *The Botanic Garden Adelaide, South Australia. Centenary volume 1855–1955: History, guides and catalogues of plants.* Adelaide. 412 pp.

Botanic Garden, Wellington, N.Z.

Cb943. **Shepherd, W., & W. Cook.** 1988. *The Botanic Garden, Wellington: A New Zealand history 1840–1987.* Wellington, N.Z., Millwood Pr. [iv]+396 pp. & 22 pl. (col.).

Christchurch Botanic Gardens

Cb944. **Christchurch City Council.** [1963?] *A garden century: The Christchurch Botanic Gardens 1863–1963.* Ed. by M. J. Barnett, H. G. Gilpin & L. J. Metcalf. Christchurch, N.Z. 182 pp.

Royal Botanic Gardens, Melbourne, Vic.

Cb945. **Morrison, C.** 1946. *Melbourne's garden: A description and pictorial record of the Botanic Garden, Melbourne.* Melbourne, Vic., Melbourne Univ. Pr. 158 pp.

Cb946. **Prescott, R. T. M.** 1982. *The Royal Botanic Gardens, Melbourne: A history from 1845 to 1970.* Melbourne, Vic., Oxford Univ. Pr. xii+212 pp.

Cb947. **Smith, J. Law–, & E. Herbert.** 1984. *The Royal Botanic Gardens, Melbourne.* Melbourne, Vic., Maud Gibson Trust. 84 pp.

Royal Botanic Gardens, Sydney, N.S.W.

Cb948. **Gilbert, L. A.** 1986. *The Royal Botanic Gardens, Sydney: A history 1816–1986.* Melbourne, Vic., Oxford Univ. Pr. xiv+210 pp.

Cb949. **Maiden, J. H.** 1906. History of Sydney Botanic Gardens. *Kew bulletin* 1906: 205–218.

Royal Tasmanian Botanical Gardens, Hobart, Tas.

Cb950. **Hurburgh, M.** 1987. *The Royal Tasmanian Botanical Gardens 1818–1986: A history in stone, soil and superintendents.* Sandy Bay, Tas., Shearwater Pr. 88 pp.

African (& Indian Ocean)

Durban Botanic Gardens

Cb951. **McCracken, D. P.** 1987. Durban Botanic Gardens, Natal, 1851–1913. *Garden history* 15: 64–73.

Dutch East India Company garden, Cape Town

Cb952. **Karsten, M. C.** 1951. *The Old Company's garden at the Cape and its superintendents; involving an historical account of early Cape botany.* Cape Town, Maskew Miller. xvi+188 pp.

Natal Botanic Garden, Pietermaritzburg

Cb953. **Law, P. A. S.** 1971. Natal Botanic Garden, Pietermaritzburg: A brief history. *Journal of the Botanical Society of South Africa* 57: 46–49.

National Botanic Gardens of South Africa (Nasionale Botaniese Tuine van Suid Afrika), Kirstenbosch

Cb954. **National Botanic Gardens of South Africa.** 1965. *Kirstenbosch: Garden for a nation. Being the story of the first 50 years of the National Botanic Gardens of South Africa 1913–1963.* By R. H. Compton, with an historical introduction by M. A. Cook, [etc.]. Cape Town, Tafelberg–Uitgewers. 168 pp.

Royal Botanic Gardens, Pamplemousses, Mauritius

Cb955. **Schöller, M.** 1992. Zur Geschichte tropischer botanischer Gärten im Kolonialismus am Beispiel des ehemalig Königlichen Botanischen Garten von Pamplemousses (Mauritius). *Palmengarten* 56 (2): 92–96 .

BIBLIOGRAPHIES OF BOTANICAL GARDENS

Cb956. **Anon.** 1949. Literatura o botanicheskikh sadakh SSSR. *Byulleten' Glavnogo botanicheskogo sada* 2: 103–107. ••Literature covering 1781 to 1949.

Cb957. **Ebel, F., et al.** 1983. Bibliographie der Botanischen Gärten Europas. *Terrestrische ökologie* Sonderheft 3: xxx+364 pp. ••Covers history, development of collections, etc In 3 parts, the first part covers German gardens (East & West), the other two parts list countries in alphabetical order and the town within each likewise.

Cb958. **Johnson, D. E.** 1985. Literature on the history of botany and botanic gardens, 1730–1840: A bibliography. *Huntia* 6: 3–121.

Cb959. **Kuijlen, J., C. S. Oldenburger–Ebbers, & D. O. Wijnands.** 1983. *Paradisus Batavus: Bibliografie van plantencatalogi van onderwijstuinen, particuliere tuinen en kwerkerscollecties in de Noordelijke en Zuidelijke Nederlanden (1550–1839).* Wageningen, Pudoc. 232 pp. ••A bibliography of plant catalogues of educational and private gardens and plant breeders' collections in the Low Countries; historical introduction.

Cb960. **Saccardo, P. A.** 1895–1901. Cenni storici e bibliografici degli orti botanici pubblici e privati d'Italia. In: *La botanica in Italia: Materiali per la storia di questa scienza,* Vol. 1, pp. 187–216; Vol. 2, pp. 133–152. Venice. 2 vols.

Cb961. **Stearn, W. T.** 1971. Sources of information about botanic gardens and herbaria. *Biological journal of the Linnean Society* 3: 225–233.

HERBARIA & MUSEUMS

General

Cb962. **Allen, D. E.** 1959. The history of the vasculum. *Proceedings of the Botanical Society of the British Isles* 3: 135–150.

Cb963. **Camus, J.** 1895. Historique des premiers herbiers. *Malpighia* 9: 283–314.

Cb964. **Flatt von Alföld, K.** 1902–03. Zur Geschichte der herbare. *Magyar botanikai lapok* 1: 61–75, 115–117, 147–151, 178–185, 210–215, 249–252, 272–279, 315–317,

348–349, 382–387, 1902; *Ibid.* 2: 30–37, 86–94, 123–127, 150–157, 184–194, 213–217, 1903.

Cb965. **Matouschek, F.** 1901. Über alte Herbarien, insbesondere über die ältesten in Österreich angelegten: Eine Skizze. *Mitteilungen aus dem Vereine der Naturfreunde in Reichenberg* 32: 1–23.

Cb966. **Opsomer, J. E.** 1976. Sur quelques herbiers du XVIIème siècle conservés à la Bibliothèque Royale Albert 1er. *Bulletin des séances de l'Académie Royale des Sciences d'Outre–Mer, Bruxelles* 1976: 142–166.

Cb967. **Saint–Lager, J. B.** 1866. Histoire des herbiers. *Annales de la société botanique de Lyon* 13: 1–120.

Cb968. **___.** 1886. *Recherches sur les anciens herbaria.* Paris, J.–B. Baillière. 45 pp.

Cb969. **Schmid, G.** 1936. Geschichte der Botanisierbüchse. *Oesterreichische botanische Zeitschrift* 85: 140–150, illus.

Cb970. **Shetler, S. G.** 1969. The herbarium: Past, present and future. *Proceedings of the Biological Society of Washington* 82: 687–758.

Cb971. **Stuckey, R. L.** 1971. The first public auction of an American herbarium, including an account of the fate of the Baldwin, Collins and Rafinesque herbaria. *Taxon* 20: 443–459.
i.e. William Baldwin, Zaccheus Collins & C. S. Rafinesque.

Cryptogamic herbaria

Cb972. **Culberson, W. L.** 1959. "Lichenes exsiccati" in herbariis Americae septentrionalis asservati. *Bryologist* 62: 45–62.

Cb973. **Lynge, B.** 1913. On the world's "Lichenes exsiccati." *Nyt magazin for naturvidenskaberne* 51: 95–122.

Cb974. **___.** 1915–22. Index specierum et varietatum "Lichenum exsiccatorum." *Nyt magazin for naturvidenskaberne* 53: 1–112, 1916; 54:113–304, 1916; 55: 305–384, 1917; 56: 385–464, 1918; 57: 465–559, 1919; 58: 1–96, 1920; 59: 97–192, 1921; 60: 193–316, 1922. *Also pub. separately* Pars I. 1916–1919; Pars II. 1920–22. *Continued* **___.** 1939. ___: Supplementum I. *Ibid.* 79: 233–323.

Cb975. **Morton, C. V.** 1968–71. The fern collections in some European herbaria. *American fern journal* 58: [London & Paris.] 158–168, 1968; 59: [Berlin.] 11–22, [Geneva.] 48–55, [Kew.] 89–106, [Florence.] 37–149, 1969; 60: [Hamburg.] 49–61, 1970; 61: [Brussels.] 59–75, 1971.
Series never completed.

Cb976. **Pfister, D. H.** 1985. A bibliographic account of exsiccatae containing fungi. *Mycotaxon* 23: 1–139.

Cb977. **Sayre, G.** 1969. Cryptogamae exsiccatae – An annotated bibliography of published exsiccatae of algae, lichenes, hepaticae, and musci. *Memoirs of the New York Botanical Garden* 19: 1–423.
Introduction, pp. 1–4; I. General cryptogams, pp. 5–56; V. Unpublished exsiccatae, i. Collectors, pp. 277–423.

Cb978. **Stevenson, J. A.** 1971. An account of fungus exsiccati containing material from the Americas. *Beihefte zur Nova Hedwigia* 36: vii+563 pp.
Alphabetically arranged by "author," with bibliographies of secondary references to each.

Cb979. **Underwood, L. M.** 1903–05. The early writers on ferns and their collections. *Torreya* 3: 145–150, 1903; 4: 49–52, 145–150, 1904; 5: 37–41, 1905.

Individual herbaria & museums

European

Akademiya Nauk SSSR, Botanicheskii Institut im. V. L. Komarova, St. Petersburg (Leningrad) (LE)

Cb980. **Budantsev, L. Yu.** 1974. Muzei Botanicheskogo Instituta im. V. L. Komarova Akademii Nauk SSSR (k 150–letiyu so dnya osnovaniya). [The Museum of the V. L. Komarov Botanical Institute, Academy of Sciences of the U.S.S.R. (to the 150th anniversary of foundation)]. *Botanicheskii zhurnal* 59: 906–914.

Cb981. **Lipshits, S., & I. T. Vasil'chenko.** 1968. *Tsentral'nyi Gerbarii SSSR: Istoricheskii ocherk. Herbarium Centrale URSS (Herbarii Instituti Botanici nomine V. L. Komarovii, Academiae Scientiarum URSS: Historia ab initiis ad tempora nostra).* Leningrad, Izd. "Nauka." 141 pp.
Obraztsy avtografov otechestvennykh botanikov i kollektorov rastenii. Autographa botanicorum et collectorum Rossicorum illustrata, pp. 83–136.

Cb982. **Lipskii, V. I.** 1908. *Gerbarii Imperatorskogo S.–Peterburgskago Botanicheskago Sada (1823–1908). Herbarium Horti Botanici Imperialis Petropolitani (1823–1908).* 2nd ed. Yur'yev, K. Mattisen. 238 pp.
Replaces 1st ed., St. Petersburg, 1898. 128+vi pp.

See also Bb124 Cb1044(Baranov, Shishkin, Sochava, Polyanskii).

Botanisches Museum, Berlin–Dahlem (B)

Cb983. **Eckhardt, T.** 1966. 150 Jahre Botanisches Museum, Berlin (1815–1965). *Willdenowia* (4) 2: 151–182 & 15 pl.

Cb984. **Hiepko, P.** 1987. The collections of the Botanical Museum Berlin–Dahlem (B) and their history. *Englera* 7: 219–249.
Appx. A: Selective list of collections acquired by the Botanical Museum Berlin–Dahlem (B) between 1914 and 1986, pp. 240–249; Appx. B: Extant collections of the Botanical Museum Berlin–Dahlem (B) from the time before 1943, pp. 250–252. For a German version, *see* Cb196A.

Cb985. **Urban, I.** 1917. Geschichte des Königlichen Botanischen Museums zu Berlin–Dahlem (1815–1913), nebst Aufzählung seiner Sammlungen. *Beihefte zum botanischen Zentralblatt* 34 (1): 457 pp.

British Museum (Natural History), London (BM)

Cb986. **Murray, G., & J. Britten.** 1904. The Department of Botany. In: British Museum (Natural History). *The history of the collections contained in the Natural History Departments of the British Museum.* Vol. 1, pp. 77–193. London, British Museum.

Conservatoire et Jardin Botaniques de Genève (G)

Cb987. **Neuwly, R., & H. M. Burdet.** 1981. *Index Candollea – Boissiera – Annuaire du Conservatoire et du Jardin Botaniques de Genève.* Geneva, Conservatoire et du Jardin Botaniques de Genève. 180pp.

Eidgenossische Technische Hochschule, Zurich (ETH)

Cb988. **Baltisberger, M.** 1985. 125 Jahre Herbarium an de ETH Zürich. *Berichte des Geobotanischen Institutes der Eidgenossosche Technische Hochschule Stiftung Rubel* 52: 35–40.

Gosudarstvennyi Nikitskii Botanicheskii Sad, Yalta (YALT)

Cb989. **Kosykh, V. M.** 1976. Gerbariyu Gosudarstvennogo Nikitskogo Botanicheskogop Sada 60 let. [Sixty years of the herbarium of the State Nikitsk Botanical Garden]. *Botanicheskii zhurnal* 61: 585–589.

Cb990. **Popov, K. P.** 1964. Kratkaya istoriya i sovremenoe sostayine gerbariya Nikitskogo Botanicheskoao Sada. [The brief history and present condition of the Nikitsky Botanical Garden herbarium]. *Sbornik nauchnykh trudov, Gosudarstvennyi Ordena Trudovogo Krasnogo Znameni Nikitski Botanicheskii Sad* 37: 606–614.

Humboldt Universität zu Berlin, Museum für Naturkunde (BHU)

Cb991. **Hoffmann, P., U. Günther, & D. Blohm.** 1965. Die Botanik und ihre Vertreter an der Berliner Universität (von 1810 bis 1945). *Wissenschaftliche Zeitschrift der Humboldt–Universität zu Berlin, mathematisch–naturwissenschaftliche Reihe* 14: 799–816.

See also Ca878(Natho & Schmidt).

Kazanskii Gosudarstvennyi Universitet imeni V. I. Ul'yanova–Lenina (KAZ)

Cb992. **Gordyagin, A. Ya.** 1933. Iz istorii botanicheskogo kabineta. *Uchenye zapiski, Kazanskii Gosudarstvennyi Universitet imeni V. I. Ul'yanova–Lenina* 6(1): 46–65.

Kjøbenhavns Universitet, Botanisk Museum (C)

Cb993. **Skovsted, A.** 1974. Botanical Museum 1924–1974. *Botanisk tidsskrift* 69: 88–93.

Laboratorio Crittogamico dell' Università di Pavia (PAV)

Cb994. **Picco, D., & R. Tomaselli.** 1971. Il centenario del Laboratorio Crittogamico dell' Università di Pavia. *Atti del'Istituto Botanico e del Laboratorio Crittogamico dell'Università di Pavia* 7: 3–28.

Moskovskii Gosudarstvennyi Universitet im. M. V. Lomonosova, Gerbariya

See Cb1073 (Smirnov).

Muséum Nationale d'Histoire Naturelle, Paris (P)

Cb995. **Lecomte, H.** 1907. Historique des collections de botanique du Muséum d'Histoire Naturelle de Paris. *Revue scientifique* (5) 7: 673–680, 716–721.

Naturhistorisches Museum, Vienna (W)

Cb996. **Beck von Mannagetta, G.** 1888. Geschichte des Wiener Herbariums. *Botanisches Centralblatt* 33: 249–251, 280–283, 312–314, 378–380; 34: 28–31, 86–87, 147–151 [16 pp.]

Orto Botanico di Pisa, Istituto Botanico dell'Università, herbarium (PI)

Cb997. **Chiarugi, A.** 1950. L'Herbarium Horti Botanici Pisani nel ventennio 1930–1950. *Nuovo giornale botanico Italiano* n.s. 57: 640–645.

Oxford University, Department of Botany (OXF)

Cb998. **Clokie, H. N.** 1964. *An account of the herbaria of the Department of Botany in the University of Oxford.* Oxford, Oxford Univ. Pr. 280 pp.

Rijksherbarium, Leiden (L)

Cb999. **Bas, C., et al.** 1979. *Voorjaar 1979. 150 jaar Rijksherbarium.* Leiden. 65 pp.

Cb1000. **Goddijn, W. A.** 1931. Rijks herbarium 1830–1930. *Mededeelingen van's Rijks–Herbarium* 62A: 1–6; *Ibid.* 62b: 1–53.

Cb1001. **Kalkman, C., & P. Smit**, eds. 1979. *Rijksherbarium 1829–1979: A jubilee volume.* Leiden. 140 pp. First pub. in *Blumea* vol. 25.
Eleven articles by various authors.

Cb1002. **Rijksuniversiteit, Leiden.** 1979. *Honderd vijftig jaar Rijksherbarium.* Leiden. 64 pp.

Cb1003. **Veldkamp, J. F.** 1980. The Rijksherbarium at 150 years. *Taxon* 29: 101–104.

Royal Pharmaceutical Society of Great Britain, London

Cb1004. **Forster, K. Arnold–.** 1991. The museum of the Royal Pharmaceutical Society of Great Britain. *Pharmaceutical historian* 33: 11–18.

Cb1005. **Shellard, E. J.** 1981. A history of British pharmacognosy: part 4. The materia medica and herbaria of the Pharmaceutical Society. *Pharmaceutical journal* 227: 201–209.

Senckenberg–Museum, Frankfurt am Main (FR)

Cb1006. **Senckenberg–Museum.** 1967. Aus der Geschichte der Senckenberg–Museums. *Senckenbergiana biologica* 48 Sonderheft A & B.
Includes **H. J. Conert.** Die Geschichte der Botanisch– Paläobotanischen Abteilung.

See also Ca171(Conert).

Städtisches Museum für Natur– Völker– und Handelskunde, Bremen

Cb1007. **Buchenau, F. G. P.** 1885, 1897, 1907. Aus dem Städtischen museum für natur– völker– und handelskunde. Geschichte der botanischen Sammlungen. *Abhandlungen herausgegeben vom Naturwissenschaftlichen Vereine zu Bremen* 9: 245–256, 1885; *Ibid.* 15: 116–132, 1897; *Ibid.* 19: 25–44, 1907.

Természettudományi Múzeum, Budapest (BP)

Cb1008. **Szujkó Lacza, J.** 1973. A 100 éves Növénytár herbáriumainak története. 1: Altalános rész. (Ungarische Herbarien: 10. Die Geschichte der Herbarien der 100–jährigen Botanischen Abteilung des Museums: 1. Allgemeiner Teil). *Botanikai közlemények* 60: 49–52.

Cb1009. ___. 1973. The history of the Centennial Botanic Department of the Hungarian Natural History Museum. *Acta agronomica Academiae Scientiarum Hungaricae* 22: 479–487.

Trinity College, Dublin (TCD)

Cb1010. **Webb, D. A.** 1991. The herbarium of Trinity College, Dublin: Its history and contents. *Botanical journal of the Linnean Society* 106: 295–327.

Cb1011. **Wright, E. P.** 1896. The herbarium of Trinity College: A retrospect. *Notes from the Botanical School of Trinity College, Dublin* 1: 1–14.

Universitatea din Bucuresti, Gradina Botanica, Bucharest (BUC)

Cb1012. **Nedelcu, G.** 1981. Hundert Jahre seit dem Bestehen des Herbariums im Botanischen Garten von Bukarest. *Lucrarile grarile Gradinii Botanice din Bucuresti* 1981/82: 21–27.

Uniwersytet Wroclawski im Boleslawa Bieruta, Instytut Botaniczny (WRSL)

Cb1013. **Seaward, M. R. D.** 1984. Lichen herbarium of Wroclaw University. Historical background. *Acta Universitatis Wratislaviensis, prace botaniczne* 29: 149–161.
With a Polish summary.

University of Edinburgh

Cb1013A. **Balfour, I. B.** 1913. A sketch of the professors of botany in Edinburgh from 1670 until 1887. In: F. W. Oliver, ed. *Makers of British botany: A collection of biographies of living botanists*, pp. 280–301. Cambridge, Cambridge Univ. Pr.

North American

Academy of Natural Sciences of Philadelphia,PA. (PH)

Cb1014. **Pennell, F. W.** 1950. Historic botanical collections of the American Philosophical Society and the Academy of Natural Sciences of Philadelphia. *Proceedings of the American Philosophical Society* 94: 137–151.

American Philosophical Society, Philadelphia, PA

Cb1015. **Pennell, F. W.** 1950. Historic botanical collections of the American Philosophical Society and the Academy of Natural Sciences of Philadelphia. *Proceedings of the American Philosophical Society* 94: 137–151.

Harvard University, Botanical Museum, Cambridge, MA (AMES, ECON)

Cb1016. **Harvard University, Botanical Museum.** 1966. *A quarter century of publications of the Botanical Museum of Harvard University 1940–1965*. Cambridge, MA. iv+66 pp.

Cb1016A. **Schultes, R. A., et al.** 1984. The Botanical Museum of Harvard University in its 125th year 1858–1983. *Botanical Museum leaflets, Harvard University* 30: 1–62.

Herbario Nacional de Mexico (MEXU)

Cb1017. **Universidad Nacional Autonoma de Mexico, Instituto de Biología.** 1991. *Herbario Nacional de Mexico*. By P. D. Dávila Aranda & M. T. Germán Ramírez. Mexico. xi+123 pp., illus.
1a parte: Historia, pp. 1–42.

Missouri Botanical Garden, St. Louis, MO

Cb1018. [**Greenman, J. M.**] 1919. Early collections in the garden herbarium. *Missouri Botanical Garden bulletin* 7: 29–35, pl. 8–11.

Ohio State Herbarium, Ohio State University, Columbus, OH (OS)

Cb1019. **Stuckey, R. L.** 1984. Early Ohio botanical collections and the development of the state herbarium. *Ohio journal of science* 84: 148–174.

Smithsonian Institution, Washington, DC, United States National Herbarium (US)

See Ca951–959 *passim*.

University of North Carolina, Raleigh, NC (NCSC)

Cb1020. **Radford, L. S.** 1984–85. The history of the herbarium of the University of North Carolina. Part 1. [–5]. *Friends of the UNC Herbarium newsletter* 2(1): 10–16, 1984; *Ibid.* 2(2): 4–11, 1984; *Ibid.* 3(1): 1–8, 1985; *Ibid.* 3(2): 7–14, 1985; *Ibid.* 4 (1): 9–15, 1986.

University of Washington University Herbaria (WTU)

Cb1021. **Hitchcock, C. L., et al.** [1973?] *The University herbaria, Department of Botany, University of Washington, Seattle: A history, summary of activities and future needs*. Seattle, WA. 27 pp.

West Virginia University Herbarium, Morgantown, WV (WVA)

Cb1022. **Clarkson, R. B., & L. Rader.** 1991. *The West Virginia University Herbarium — the first hundred years*. N.p. 20+[16] ff.

Asian

Tomskii Gosudarstvennyi Universitet im. V. V. Kuibysheva, Gerbarii im P. N. Krylova, Tomsk (TK)

Cb1023. **Polozhiy, A. V.** 1986. *Gerbarii imeni P. N. Krylova v Tomskom Universitete: K 100–letiyu so vremeni osnovaniya.* Tomsk, Izd. Universita. 87 pp.

Cb1024. **Sergeiskaya, L. P.** 1961. *Gerbarii imeni P. N. Krylova pri Tomskoi Gosudarstvennom Universitete imeni V. V. Kuibysheva. K 75 letiyu so dnya osnovaniya.* Tomsk, Tomskii Gosudarstvennyi Universitet. 56 pp.

Australasian

Queensland Herbarium, Brisbane, Qld. (BRI)

Cb1025. **Everist, S. L.** 1982. History of the Queensland Herbarium and botanical library, 1855 to 1976. *Austrobaileya* 1: 429–445.

Personal herbaria

• A selection of some representative collections:

Bauhin herbarium

Cb1026. **Candolle, A. P. de.** 1904. L'herbier de Gaspard Bauhin. *Bulletin de l'Herbier Boissier* 4: 201–216, 297–312, 459–474, 721–754.
Now at Botanisches Institut der Universität Basel (BAS).

Delessert herbarium

Cb1027. **Lasègue, A.** 1845. *Musée botanique de M. Benjamin Delessert. Notices sur les collections de plantes et la bibliothèque qui le composent; contenant en outre des documents sur les principaux herbiers d'Europe.* Paris & Leipzig. vi+588 pp. *Re-pub.* (with an introduction by F. A. Stafleu) Lehre, J. Cramer; Codicote, Wheldon & Wesley, 1970. *Historiae naturalis classica* vol. 81.
Now at Conservatoire et Jardin Botaniques de Genève (G).

Dillenius herbarium

Cb1028. **Druce, G. C.** 1907. *The Dillenian herbaria: An account of the Dillenian collections in the herbarium of the University of Oxford, together with a biographical sketch of Dillenius, selections from his correspondence, notes, ...* Ed. with an introduction by S. H. Vines. Oxford, Clarendon Pr. cxii+258 pp.

Haller herbarium

Cb1029. **Zoller, H.** 1958. Albrecht von Hallers Pflanzensammlungen in Göttingen, sein botanisches Werk und sein Verhältnis zu Carl von Linné. *Nachrichten der Akademie der Wissenschaften in Göttingen, mathematisch–physikalische Klasse* 10: 217–251.
Now at the Systematisch–Geobotanisches Institut, Universität Göttingen (GOET).

Lamarck herbarium

Cb1030. **Aymonin, G. G.** 1981. L'herbier de Lamarck. *Revue de l'histoire des sciences* 34: 25–58.
Now at Muséum Nationale d'Histoire Naturelle, Paris (P).

Lambert herbarium

Cb1031. **Miller, H. S.** 1970. The herbarium of Aylmer Bourke Lambert; notes on its acquisition, dispersal, and present whereabouts. *Taxon* 19: 489–553.

Lindemann herbarium

Cb1032. **Lindemann, E.** 1863–85. Kurze Nachrichten über den Bestand meines Herbariums. *Bulletin de la Société Impériale des Naturalistes de Moscou* 36: 233–253, 1866; *Continued Ibid.* 46: 56–101, 1872; *Ibid.* 60: 265–312, 1884; *Ibid.* 61: 37–92, 1885.
Includes details of the collectors, their travels, and the collections they contributed.

Linnaean herbarium

Cb1033. **Jackson, B. D.** 1922. Notes on a catalogue of the Linnaean herbarium. *Proceedings of the Linnean Society of London* 134 supp: 38 pp.
Now at the Linnean Society of London (LINN). Includes details of ca. 150 collectors.

Cb1034. **Lindman, C. A. M.** 1907–09. A Linnæan herbarium in the Natural History Museum in Stockholm. *Arkiv för botanik* 7(3): 57 pp., 1907; *Ibid.* 9(6): 1–50 pp., 1909.

Cb1035. **Savage, S.** 1937. Sir James E. Smith and the Linnaean herbarium. *Proceedings of the Linnean Society of London* 149: 6–10.

Cb1035A. ___. 1945. *A catalogue of the Linnean herbarium.* London. 225 pp.
With illus. of 72 examples of handwriting of contributors to the herbarium.

Cb1036. **Stearn, W. T.** 1957. Linnaean herbaria. In: C. Linnaeus. *Species Plantarum. A facsimile of the first edition 1753*, (An introduction to the *Species Plantarum* and cognate botanical works of Carl Linnaeus) vol. 1, pp. 103–124.

Morison herbarium

Cb1037. **Vines, S. H., & G. C. Druce.** 1914. *An account of the Morisonian herbarium in the possession of the University of Oxford, together with biographical and critical sketches of Morison and the two Bobarts and their works in the early history of the Physic Garden 1619–1720, [etc.].* Oxford, Clarendon Pr. lxviii+350 pp. & 5 pl.

Platter herbarium

Cb1038. **Rytz, W.** 1933. Das Herbarium Felix Platters: Ein Beitrag zur Geschichte der Botanik des XVI. Jahrhunderts. *Verhandlungen der Naturforschenden Gesellschaft in Basel* 44: 222 pp.

Ratzenberger herbarium

Cb1039. **Kessler, H. F.** 1870. *Das älteste und erste Herbarium Deutschlands, im Jahr 1592 von ... C. Ratzenberger angelegt, gegenwärtig noch im Königlichen Museum zu Cassel befinlich, [etc.].* Cassel, Freyschmidt. 92 pp.

Sloane herbarium

Cb1040. **Britten, J., & J. E. Dandy.** 1958. *The Sloane herbarium: An annotated list of the* horti sicci *composing it; with biographical accounts of the principal contributors.* Based on records compiled by the late James Britten. With an introduction by Spencer Savage. Rev. & ed. by J. E. Dandy. London, British Museum (Natural History). 246 pp. & 35 pl.
Now at the British Museum (Natural History) (BM). Nearly 100 illustrations of specimens of handwriting.

Steven herbarium

Cb1041. **Kukkonen, I., & V. Kiljamae.** 1971. Herbarium of Christian Steven. *Helsingin Yliopiston Kasvimuseon monisteita* 4: 1–109, illus.
Now at the Botanical Museum, University of Helsinki (H). With a list of ca.400 collectors.

Webb herbarium

Cb1042. **Steinberg, C. H.** 1977. The collectors and collections in Herbarium Webb. *Webbia* 32: 1–49.
Now at Herbarium Universitatis Florentinae, Instituto Botanico (FI).

BOTANICAL EDUCATION & RESEARCH INSTITUTIONS

Individual institutions

European

Académie des Sciences, Paris, Laboratoire d'Anatomie et de Botanique

Cb1043. **Schiller, J.** 1964. Les laboratoires d'anatomie et de botanique à l'académie des sciences au XVIIe siècle. *Revue d'histoire des sciences* 17: 97–114.

Akademiya Nauk SSSR, Botanicheskii Institut, St. Petersburg (Leningrad)

Cb1044. **Baranov, P. A., & E. G. Bobrov**, eds. 1957. *Ot aptekarskogo ogoroda do Botanicheskogo Instituta. Ocherk po istorii Botanicheskogo Instituta Akademii Nauk SSSR.* Moscow/Leningrad, AN SSSR. 308 pp.

Includes **P. A. Baranov.** Sad v Goronkakh i poslednie gody sada Akademii Nauk, pp. 25–31; **E. G. Bobrov.** Botanicheskii sad (1801–1916), pp. 32–71; ___. Botanicheskii Muzei Akademii Nauk (1823–1916), pp. 72–80; **B. K. Shishkin.** Botanicheskii Muzei Akademii Nauk i gerbarii Glavnogo Botanicheskogo Sada v gody 1917–1931, pp. 81–86; ___. Otdel sistematiki i geografii vysshikh rastenii, pp. 87–97; **A. N. Krishtofovich.** Rabota v oblasti paleobotaniki, pp. 98–114; **V. P. Savich.** Otdel sporobykh rastenii, pp. 115–133; **V. B. Sochava.** Geobotanika v Peterburgskom botanicheskom sadu i Botanicheskom Muzee Akademii Nauk (XVIII–XIX–XX vv. po 1916 g.), pp. 134–142; **E. M. Lavrenko & V. D. Aleksandrova.** Otdel geobotaniki i ego rol' v razvitii sovetskoi geobotaniki, pp. 143–173; **V. L. Nekrasova.** Izuchenie rastitel'nogo syr'ya za 1917–1957 gg., pp. 174–188; **V. G. Aleksandrov & M. S. Yakovlev.** Raboty po morfologii, anatomii i embriologii rastenii (s 40–kh g. XIX v. do 50–kh g. XX v.), pp. 189–201; **V. A. Brilliant, I. N. Konovalov & A. P. Shennikov.** Voprosy fiziologii i ekologii rastenii v rabotakh instituta (s 40–kh g. XIX v. do 50–kh g. XX v.), pp. 202–215; **S. Ya. Sokolov.** Botanicheskii sad i ego introduktsionnaya i kul'turno–prosvetitel'naya rabota s XVIII v. do 50–kh g. XX v.), pp. 216–240; **V. I. Polyanskii.** Kratkii istoricheskii ocherk muzeya (s 1823 g. – data osnovaniya do 50–kh g. XX v.), pp. 241–262; **I. T. Vassil'chenko.** Raboty po semenovedeniyu v gody 1877–1931, pp. 263–270; **D. V. Lebedev.** Biblioteka Botanicheskogo Instituta (s 1824 g. (god osnovaniya) do 50–kh g. XX v.), pp. 271–277; ___. Literaturno–izdatel'skaya deyatel'nost' Botanicheskogo Instituta (XIX–XX vv. do 50–kh gg.), pp. 278–296; ___. Khronologicheskaya tablitsa, pp. 297–299.

Cb1045. **Levina, F. Ya.** 1971. *Geobotanika v Botanicheskom Institute im V. L. Komarova AN SSSR 1922–1964. (Geobotany in the Komarov Botanical Institute of the Academy of Sciences of the USSR (1922–1964).* Leningrad, Izd. "Nauka." 319 pp., illus. (ports.)

Pt. 1: Ekspeditsionnye marshrutnye i statsionarnye geobotanicheskie issledovaniya i publikatsii. Proizvodstvennye konferentsii, teoreticheskie diskussii, pp. 15–102; Pt. 2: Problemno–tematicheskie gryppy publikatsii, pp. 103–160; Pt. 3: Bibliografiya, pp. 161–297.

Cb1046. **Lipshits, S. Yu.** 1945. K istorii sistematiki, floristiki i geografii rastenii v Akademii Nauk. *Botanicheskii zhurnal SSSR* 30 (3): 99–114.

Cb1047. **Shetler, S.** 1967. *The Komarov Botanical Institute: 250 years of Russian research.* Washington, DC, Smithsonian Institution Pr. xiv+240 pp.

Cb1048. **Sokolov, V. S., & A. Fedorov.** 1947. *Botanicheskii Institut imeni V. L. Komarova Akademii Nauk SSSR.* Leningrad. 72 pp.

See also Bb124.

Akademiya Nauk Ukrainskoi RSR, Instytut Botaniky, Kiev

Cb1049. **Akademiya Nauk Ukrainskoi SSR.** 1971. *Institut Botaniki im N. G. Kholognogo.* By K. K. Sytnik [et al.]. Kiev, Izd. "Naukova Dumka." 104 pp.

Cb1050. **Dobrochaeva, D. N., B. V. Zaverukha, & L. M. Sipailova.** 1972. Botanicheskii Muzei Instituta Botaniki im. N. G. Kholodnogo AN USSR (k 50–letiyu Instituta Botanuku AN USSR). [Botanical Museum of the N. G. Kholodny Institute of Botany of the Academy of Sciences of the Ukrainian SSR (to the 50th anniversary of the Institute ...]. *Botanicheskii zhurnal* 57: 416–423.

Cb1051. **Sytnik, K. M.** 1971. Institut botaniki im. N. G. Kholodnogo Akademii Nauk Ukrainskoi SSR (k 50–letiyu so dnya osnovaniya). [50th anniversary of the Kholodny Institute of Botany of the Academy of Sciences of the Ukrainian S. S. R.]. *Botanicheskii zhurnal* 56: 1714–1724.

Cb1052. ___. 1971. Pidsumky pivvikovoyi diyal'nosti Instytutu botaniky im M. H. Kholodnogo Akademiyi Nauk Ukrayins'koyi SSR. [Results of the activity of the N. G. Kholodny Institute of Botany of the Academy of Sciences of the Ukrainina SSR for 50 years]. *Ukrayins'kyi botanichnyi zhurnal* 28: 545–559.

Akademiya Navuk Belaruskai SSR, Instytutu Eksperymental'nai Bataniki im. V. F. Kuprevicha, Minsk

Cb1053. **Parfenov, V. I.** 1981. Instytutu Eksperimental'noi Bataniki imya V. F. Kuprevicha AN BSSR — 50 hadow. [The 50th anniversary of the V. F. Kuprevich Institute of Experimental Botany, Academy of Sciences of the BSSR]. *Vestsi Akademii Navuk Belaruskai SSR, seryya biyalagichnykh i sel'skagaspadarchykh navuk* 1981 (2): 1–22.

Berliner Universität

Cb1054. **Hoffmann, P., U. Günther, & D. Blohm.** 1965. Die Botanik und ihre Vertreter an der Berliner Universität (von 1810 bis 1945). *Wissenschaftliche Zeitschrift der Humboldt Universität zu Berlin, mathematisch–naturwissenschaftliche Reihe* 14: 799–816.

Cb1055. **Höxtermann, E., et al.** 1985. Kurzbiographien und Porträts von Botanikern in der Geschichte der Berliner Universität. *Wissenschaftliche Zeitschrift der Humboldt Universität zu Berlin, mathematisch–naturwissenschaftliche Reihe* 34: 360–406.

Biologische Zentralanstalt für Land– und Forstwirtschaft

Cb1056. **Biologische Zentralanstalt für Land– und Forstwirtschaft.** 1949. *50 Jahre Deutsche Pflanzenschutzforschung. Festschrift zum fünfzigjähren Bestehen der Biologischen Zentralanstalt für Land– und Forstwirtschaft in Berlin–Dahlem.* Berlin, Deutscher Zentralverlag. 231 pp.

Botanische Staatssammlung München

Cb1057. **Hertel, H., & A. Schreiber.** 1988. Die Botanische Staatssammlung München 1813–1988. (Eine Übersicht über die Sammlungsbestande). *Mitteilungen der Botanischen Staatssammlung München* 26: 81–512.

Cambridge University

Cb1058. **Walters, S. M.** 1981. *The shaping of Cambridge botany: A short history of whole–plant botany in Cambridge from the time of Ray to the present century.* Pub. on the occasion of the sesquicentenary of Henslow's New Botanic Garden, 1831–1981. Cambridge, Cambridge Univ. Pr. xv+121 pp.

Ceskoslovenska Akademie Ved, Botanicky Ustav, Prague

Cb1059. **Hejny, S., J. Holub, & J. Kvet.** 1981. The Botanical Institute of the Czechoslovak Academy of Sciences: Twenty five years. *Folia geobotanica et phytotaxonomica* 16: 113–123.

Commonwealth Mycological Institute, Kew

Cb1060. **Ainsworth, G. C.** 1980. CMI 1920–1980. *Review of plant pathology* 59(6): 249–255.

Conservatoire Botanique de Genève

Cb1061. **Miège, J.** 1975. Le Conservatoire Botanique de Genève: 150 ans d'histoire. *Saussurea* 6: 113–134.

Deutsche Akademie der Wissenschaften zu Berlin, Institut für Kulturpflanzenforschung Gatersleben

Cb1062. **Deutsche Akademie der Wissenschaften zu Berlin, Institut für Kulturpflanzenforschung Gatersleben.** 1982. *Geschichte des Instituts für Kulturpflanzenforschung Gatersleben der Deutsche Akademie der Wissenschaften zu Berlin, 1943–1968.* By H. Stubbe. Berlin, Akademie–Verlag. 428 pp. *Studien zur Geschichte der Akademie der Wissenschaften der DDR* vol. 10.

Friedrich–Wilhelms–Universität in Berlin (*later* Humboldt Universität zu Berlin)

See Cb196A (Schnarrenberger).

Jodrell Laboratory, Royal Botanic Gardens, Kew

Cb1063. **Metcalfe, C. R.** 1976. History of the Jodrell Laboratory as a centre for systematic anatomy. *Leiden botanical series* 3: 1–19.
Also **M. Gregory.** 1976. Jodrell Laboratory publications 1877–1975. *Notes from the Jodrell Laboratory* 1976 (April): 38 pp.

John Innes Institute (John Innes Horticultural Institution), Norwich (*formerly* Bayfordbury)

Cb1064. **John Innes Horticultural Institution.** [1935?] *The John Innes Horticultural Institution, 1910–1935.* Cambridge, Cambridge Univ. Pr. 58 pp.

Cb1065. **Olby, R.** 1989. Scientists and bureaucrats in the establishment of the John Innes Horticultural Institution under William Bateson. *Annals of science* 46: 497–510.

Jószef Attila Tudományegyetem, (Universitas Szegediensis), Szeged

Cb1066. **Végh, M.** 1971. History of botany in the Attila József University (1921–1970). *Acta biologica, Szeged* 17: 7–16.

Kaiser Wilhelm–Institut, Berlin–Dahlem

Cb1067. **Schnarrenberger, C.** 1987. Botany at the Kaiser Wilhelm Institutes. *Englera* 7: 105–146.
For a German version, *see* Cb196A.

Khar'kovskii Universitet, Kharkov

Cb1068. **Shatalova–Zalesskaya, E. O.** 1955. K istorii kafedry fiziologii rastenii v Khar'kovskom Universitete. *Uchenye zapiski, Khar'kovskii Universiteta* 59 (*Trudy Nauchno–Issledovatel'skogo Instituta Biologii i Biologicheskogo Fakul'teta* 22): 227–234.

Cb1069. **Shkorbatov, L. A.** 1955. K istorii al'gologii i gidrobiologii v Khar'kovskom Universitete. *Uchenye zapiski, Khar'kovskii Universiteta* 59 (*Trudy Nauchno–Issledovatel'skogo Instituta Biologii i Biologicheskogo Fakul'teta* 22): 123–160.

Kjøbenhavns Universitet

Cb1070. **Jakobsen, K.** 1974. The Botanical Laboratory 1924–1959. *Botanisk tidsskrift* 69: 99–105.

Lunds Universitet, Botaniska Institution

Cb1071. **Almborn, O.** 1980. Three centuries of botany at the University of Lund, a short survey. *Botaniska notiser* 133: 451–474.

Cb1072. **Törje, A.** 1968. *De Botaniska Institutionerna vid Lunds Universitet.* Ur *Lunds universitets historia* no. 3. Lund, Gleerup. 136 pp. *Acta Universitatis Lundensis* sectio 1, 6.

Moskovskii Gosudarstvennyi Universitet im. M. V. Lomonosova

Cb1073. **Moskovskii Gosudarstvennyi Universitet.** 1940. [Papers on the history of botany at Moscow University]. *Uchenye zapiski, Moskovskii Gosudarstvennyi Universitet, yubil. seriya* 54: 259–339.
Includes **V. A. Petrov.** Pervye etapy razvitiya botaniki v MU, pp. 259–268; **V. V. Alexin.** Floristika i sistematika rastenii, botanicheskaya geografiya i fitotsenologiya v MU, pp. 269–287; **K. I. Meier.** Sravnitel'no–morfologicheskoe napravlenie v izuchenii vysshikh rastenii v MU, pp. 288–293; **L. I. Kursanov.** Izuchenie nizshikh rastenii v MU, pp. 294–304; **V. N. Shaposhnikov & D. A. Sabinin.** Fiziologiya rastenii v MU, pp. 305–314; **V. N. Shaposhnikov.** Mikrobiologiya v MU, pp. 322–324; **P. A. Smirnov.** Gerbarii MU, pp. 325–332; **K. I. Meier.** Botanicheskii sad, pp. 332–339.

Odesskii Universitet

Cb1074. **Potapenko, G. I.** 1960. Materialy po istorii botaniki v Odesskom Universitete. (Iz vospominanii). *Trudy Institut Istorii Estestvoznaniya i Tekhniki* 41: 44–62.

Rijksuniversiteit, Leiden

Cb1075. **Smit, P.** 1973. Carolus Clusius and the beginnings of botany in Leiden University. *Janus* 60: 87–92.

Statione Zoologica di Napoli (Zoologische Station in Neapel)

Cb1076. **Funk, G.** 1897. Zur Geschichte botanischer Forschung an der Zoologischen Station zu Neapel. *Berichte der Deutschen Botanischen Gesellschaft* 40: 221–241.

Sveriges Utsädesforening, Svalöf [Plant Breeding Institute]

Cb1077. **Åkerman, Å.**, ed. 1948. *Svalöf, 1886–1946: History and present problems.* Lund, C. Blom. 389 pp.

Tbilisskii Botanicheskii Institut, Akademiya Nauk Gruzinskoi SSR

Cb1078. **Gvaladze, T. N.** 1960. *Bibliografiya pechatnikh rabot Tbiliisskogo Botanicheskogo Instituta, 1933–1958.* Tbilisi, AN GruzSSR. 157 pp.

Cb1079. **Tbilisskii Botanicheskii Institut.** 1956. *Ukazatel' statei periodicheskikh izdanii Tbilisskogo Botanicheskogo Instituta Akademii Nauk Gruzinskoi SSR.* Tbilisi. 86 pp.

Universität Greifswald

Cb1080. **Borriss, H.** 1956. Die Entwicklung der Botanik und der botanischen Einrichtungen an der Universität Greifswald. In: Festschrift zur 500–Jahrfeier der Universität Greifswald 17.10.1956, Bd. 2, pp. 515–540. *Wissenschaftliche Zeitschrift der Ernst–Mortiz–Arndt–Universität Greifswald.*

Universität Innsbruck, Institut für Botanik

Cb1081. **Universität Innsbruck, Institut für Botanik.** 1988. *75 Jahre, Institut für Botanik in Innsbruck–Hötting, 1913–1988.* Innsbruck, Universität Innsbruck. 50 pp. *Veröffentlichungen der Universität Innsbruck* no. 164.

Universität Jena

Cb1082. **Chemnitius, F.** 1930. *Die Botaniker an der Universität Jena.* Jena, Frommann. 22 pp. & 1 pl.

Cb1083. **Renner, O.** 1947. 150 Jahre Botanische Anstalt zu Jena. *Jenaische Zeitschrift für Medizin und Naturwissenschaft* 78 (2): 131–162.

Universität Tübingen, Mathematische–Naturwissenschaftliche Fakultät

Cb1084. **Mägdefrau, K.** 1965. Die Entwicklung der neueren Botanik. *Naturwissenschaftliche Rundschau* 18: 219–227.
From Hugo von Mohl onwards.

Université de Genève, Département de Biologie Végétale

Cb1085. **Miège, J.** 1975. Le Département de Biologie Végétale de l'Universitaire de Genève. *Saussurea* 6: 61–67.

University of Manchester

Cb1086. **Weiss, F. E.** 1929. Fifty years' growth of the Botanical Department of Owens College and the University of Manchester. [Manchester.] 8 pp. & 2 pl.
Reprinted from the *Old Owensian journal* Dec. 1929.

Universytet Stefana Batorego, Vilnius

Cb1087. **Mowszowicz, J.** 1966. Botanika i botanicy w Uniwersytecie Stefana Batorego w Wilnie w latach 1919–1939. (The chair of botany and its representatives at the University of Stefan Batory at Wilna [i.e. Vilnius] during 1919–1939). *Studia i materialy z dziejow nauki Polskej* B 12: 99–124.

Uniwersytet Jagiellonski, Cracow, Zaklad Botanicznego im. Janczewskiego

Cb1088. **Szafer, W.** 1969. *Concise history of botany in Cracow against the background of six centuries of the Jagiellonian University [etc.].* Translated from Polish [by W. Kulerski & H. Markiewicz]. Warsaw, for the Smithsonian Institution and the National Science Foundation by the Central Institute for Scientific, Technical & Economic Information, Scientific Publications Foreign Co-operation Center. 171 pp. & 18 pl.
Replaces ___. 1964. *Zary historii botaniki w Krakowie na tle szescíu wieków Uniwersytetu Jagiellonskiego.* Cracow, Uniwersytet Jagiellonski. 169 pp. & 12 pl.

Cb1089. **Zablocki, J.** 1968. Historia i dzialalnosc Zakladu Botanicznego im. Janczewskiego, Uniwersytetu Jagiellonskiego w Krakowie. (The history and activity of the Janczewski Botanical Institute of the Jagiellonian University, Cracow). *Studia i materialy z dziejow nauki Polskej* B 14: 187–209.

Cb1090. **Zemanek, A.** 1988. *Bibliografia botaniki w Uniwersytecie Jagiellonskim (1532–1917).* Cracow, Uniwersytet Jagiellonski. 126 pp. *Varia, Uniwersitetu Jagiellienskiego* no. 251.

Cb1091. ___. 1989. *Historia botaniki w Uniwersytecie Jagiellonskim, 1780–1917. History of botany at the Jagiellonian University, (1780–1917).* Cracow, Uniwersytet Jagiellonski. 167 pp. *Rozprawy habilitacyjne* no. 164.

Cb1091A. ___. 1991. *Dzieje nauczania botaniki u Uniwersytecie Jagiellonskim. History of teaching the botany at the Jagiellonian University (1783–1917).* Cracow, Uniwersytet Jagiellonski. 88 pp., illus. *Varia, Uniwersitetu Jagiellienskiego* no. 290.
English summary.

Wiener Universität

Cb1092. **Linsbauer, K., & L. von Portheim.** 1903. *Wiesner und seine Schule: Ein Beitrag zur Geschichte der Botanik.* Festschrift anläszlich des dreiszigjährigen Bestandes des Pflanzenphysiologischen Institutes der Wiener Universität. Vienna, A. Hölder. xviii+259 pp. *Continued* ___. 1910. ___*: [supplement].* Vienna. 72 pp.

North American

Butler University, Indianapolis, IN

Cb1093. **Dailey, F. K.** 1959. Botanists of the Butler University Faculty (1920–1955). *Proceedings of the Indiana Academy of Science* 69: 237–272.

California Academy of Sciences, Department of Botany, San Francisco, CA

Cb1094. **California Botanical Club.** 1953. Our first hundred years: Botany at the Academy from the days of gold ... A symposium in the centennial year of the California Academy of Sciences in honor of its Department of Botany. *Leaflets of western botany* 7(3): 41–104 & 8 pl.

Canada Department of Agriculture, Ottawa

Cb1095. **Canada, Department of Agriculture, Biosystematics Research Centre.** 1986. *Systematics in Agriculture Canada at Ottawa, 1886–1986. La recherche en systématique à Agriculture Canada Ottawa, 1886–1986.* By W. J. Cody, D. B. O. Savile & M. J. Sarazin. Ottawa, Biosystematics Research Centre. [vi]+81 & [vi]+83 pp. *Historical series* no. 28.
Ch. 2. Botany; ch. 3. Mycology. Includes histories of the Division of Entomology & Botany 1887–1909, Division of Botany (& Plant Pathology) 1909–59, Plant Research Institute 1959–72, and Biosystematics Research Institute 1973. English and French versions of the text.

Harvard University, Cambridge, MA

Cb1096. **Robinson, B. L.** 1930. Botany. In: S. E. Morison, ed. 1930. *The development of Harvard University since the inauguration of President Eliot*, pp. 338–378. Cambridge, MA. *The tercentenary history of Harvard College and University, 1636–1936.*

Ohio State University, Columbus, OH

Cb1097. **Meyer, B. S.** 1983. *Botany at Ohio State University: The first 100 years.* Columbus, OH, Ohio State University, College of Botanical Sciences. vi+177 pp. *Bulletin of the Ohio Biological Survey* n.s. vol. 6, no. 2.

United States Department of Agriculture, Bureau of Plant Industry

Cb1098. **United States Department of Agriculture, Library.** 1919. *A check list of the publications of the Department of Agriculture on the subject of plant pathology, 1837–1918.* Washington, DC. 38 pp. *Bibliographical contribution* no. 1.

Cb1099. ___. 1921. *Check list of publications issued by the Bureau of Plant Industry ... 1901–20 and by the Divisions and Offices which combined to form this Bureau 1862–1901.* By J. M. Allen. Washington, DC. 127 pp. *Bibliographical contribution* no. 3.

United States Forest Service

Cb1100. **Ogden, G.** 1976. *The United States Forest Service: A historical bibliography, 1876–1972.* Davis, CA, Univ. of California, Agricultural History Center. 439 pp.

See also Guide to the Service's archives Bc842A.

Universidad Nacional Autonoma de Mexico, Instituto de Biología, Departamento de Botánica

Cb1101. **Universidad Nacional Autonoma de Mexico, Instituto de Biología.** 1967. Historia del Departamento de Botánica del Instituto de Biología de la Universidad Nacional Autónoma de Mexico. By T. Herrera. *Anales del Instituto de Biológia de la Universidad Nacional Autónoma de México, série botánica* 38A: 193–201.

Cb1102. Not used.

Université de Montréal, Institut Botanique

Cb1103. **Université de Montréal, Institut Botanique.** 1941. *Histoire de l'Institut Botanique de l'Université de Montréal 1920–1940.* By Frère Marie-Victorin. Montreal. 70 pp. *Contributions* no. 40.

University of California, Berkeley, Department of Botany, Berkeley, CA

Cb1104. **Constance, L.** 1978. *Botany at Berkeley: The first hundred years.* [Berkeley, CA.] 22 pp. & 36 pp. of ports.

University of Chicago

Cb1105. **University of Chicago.** [1916.] *A record of the doctors in botany of the University of Chicago 1897–1916, presented to John Merle Coulter, professor and head of the Department of Botany, by the doctors in botany, at the quarter-centennial of the university, June 1916.* Chicago, IL. viii+87 pp.

University of Wisconsin, Madison, WI

Cb1106. **Bryan, G. S.** 1950. A brief history of the development of botany and of the Department of Botany at the University of Wisconsin. *Transactions of the Wisconsin Academy of Sciences, Arts and Letters* 40: 1–27.

South American

Instituto Biológico, Secção de Botanico Agronomia, São Paulo

Cb1107. **Hoehne, F. C.** 1937. *Resenha historica para a commemoração do vigesimo anniversario da Secção de Botanica e Agronomia annexa ao Instituto Biologico de São Paulo; sua producção scientifica, a Estação Biologico do Alto da Serra, o Parque e o Jardim Botanico do estado.* Indice das especies novas. São Paulo, Secretaria da Agricultura, Industria e Commercio do Estado de São Paulo. 166 pp.

Asian

Academia Sinica, Kunming Institute of Botany, Kunming

Cb1108. **Academia Sinica, Kunming Institute of Botany.** 1984. *A list of the works and monographs of Kunming Institute of Botany, Academia Sinica (1958–1983).* Kunming. [ix]+86 pp., illus.

Akademiya Nauk Kazakhskoi SSR, Institut Botaniki, Alma Ata

Cb1109. **Akademiya Nauk Kazakhskoi SSR, Institut Botaniki.** 1982. *Institut botaniki (istoriya obrazovaniya, struktura, itogi nauchnoi deyatel'nosti).* Alma–Ata, Izd. "Nauka" Kazakhskoi SSR. 89 pp.

Botanical Survey of India, Howrah

Cb1110. **Basak, R. K.** 1983. B S I: *An account of its establishment, development, and activities.* Howrah, Botanical Survey of India. viii+300 pp.

Cb1111. **Santapau, H.** 1964. The Botanical Survey of India. *Science & culture* 30: 2–11.
Chiefly a history of its museum.

Indian Forest Service

Cb1112. **Stebbing, E. P.** 1923. The development of the Indian Forest Service, 1865 to 1900. In: *The forests of India*, vol. 2. London, J. Lane. *Re-pub.* New Delhi, A. J. Reprints Agency, 1982.

Turkmenskii Institut Rastenievodstva, Ashkhabad

Cb1113. **Muradov, K. M., & V. A. Govorukhina.** 1984. Nauchnyi tsentr rastenievodstva v Turkmenistane (k 50–letiyu sozdaniya Turkmenskogo Instituta Rastenievodstva). *Izvestiya Akademiya Nauk Turkmenskoi SSR, seriya biologicheskikh nauk* 3: 59–64.

BOTANICAL CLASSIFICATION & NOMENCLATURE

General

Cb1114. **Atran, S.** 1987. Origin of the species and genus concepts: An anthropological perspective. *Journal of the history of biology* 20: 195–279.
In man's study of plants.

Cb1115. **Baehni, C.** 1957. Les grands systèmes botaniques depuis Linné. *Gesnerus* 14: 83–93.

Cb1116. **Bartlett, H. H., et al.** 1940. History of the generic concept in botany. *Bulletin of the Torrey Botanical Club* 67: 319–362.

Cb1117. **Briggs, B. G.** 1991. One hundred years of plant taxonomy, 1889–1989. *Annals of the Missouri Botanical Garden* 78: 19–32.

Cb1118. **Brizicky, George K.** 1969. Subgeneric and sectional names: Their starting points and early sources. *Taxon* 18: 643–660.

Cb1119. **Constance, L.** 1955. The systematics of the angiosperms. In: California Academy of Sciences. *A century of progress in the natural sciences 1853–1953*, pp. 405–483. San Francisco, CA.

Cb1120. **Croizat, L.** 1945. History and nomenclature of the higher units of classification. *Bulletin of the Torrey Botanical Club* 72: 52–75.

Cb1121. **Florin, R.** 1955. The systematics of the Gymnosperms. In: California Academy of Sciences. *A century of progress in the natural sciences 1853–1953*, pp. 323–403. San Francisco, CA.

Cb1122. **Gibbs, R. D.** 1963. History of chemical taxonomy. In: T. Swain, ed. *Chemical plant taxonomy*, pp. 41–88. London, Academic Pr.

Cb1123. **Guédès, M.** 1973. Duchesne, Bruisson, Durand: Early followers of the natural method of the Jussieus. *Taxon* 22: 211–219.
I.e. A.–N. Duchesne, J. P. Bruisson, & J. F. Durand.

Cb1124. **Hagen, J. B.** 1983. The development of experimental methods in plant taxonomy, 1920–1950. *Taxon* 32: 406–416.

Cb1125. ___. 1984. Experimentalists and naturalists in 20th–century botany: Experimental taxonomy, 1920–1950. *Journal of the history of biology* 17: 249–270.

Cb1126. **Hegnauer, R.** 1965. Chemotaxonomy — past and present. *Lloydia* 28: 267–278.

Cb1127. **Heywood, V. H.** 1980. The impact of Linnaeus on botanical taxonomy — past, present and future. *Veröffentlichungen, Joachim Jungius–Gesellschaft der Wissenschaften Hamburg* 43: 97–115.

Cb1128. **Rytz, W.** 1947. Beiträge zur Geschichte der botanischen Systematik. Wege zum Artbegriff. Von den Kräuterbüchern bis zu C. von Linné. *Gesnerus* 4: 121–127; **C. Baehni.** 1947. Beiträge zur Geschichte der botanischen Systematik. 2. Naissance et développement de la systématique moderne de Linné aux temps actuels. *Ibid.* 4: 127–145.

Cb1129. **Sprague, T. A.** 1950. The evolution of botanical taxonomy from Theophrastus to Linnaeus. In: Linnean Society of London. *Lectures on the development of taxonomy*, pp. 1–23. London.

Cb1130. **Stafleu, F. A.** 1971. *Linnaeus and the Linnaeans: The spreading of their ideas in systematic botany, 1735–1789*. Utrecht, Oostheok for the International Association for Plant Taxonomy. xvi+386 pp. *Regnum vegetabile* vol. 79.

Cb1131. **Starostin, B. A.** 1970. Razvitie biokhimicheskoi sistematiki vysshikh rastenii. [The development of biochemical taxonomy of higher plants]. *Iz istorii biologicheskikh nauk* 2: 145–163.

Cb1132. **Stearn, W. T.** 1957. An introduction to the *Species Plantarum* and cognate botanical works of Carl Linnaeus. In: C. Linnaeus. *Species plantarum. A facsimile of the first edition 1753*. Vol. 1, pp. 1–176. London, Ray Society.
Includes IV. Linnaeus's sexual system of classification; V. The *Genera Plantarum* and the typification of Linnaean genera; VI. The *Hortus Cliffortianus*; VII. The *Amoenitates academicae* and the authorship of Linnaean dissertations; VIII. The preparation of the *Species Plantarum* and the introduction of binomial nomenclature; IX. The reception of the *Species Plantarum* in England and its influence on British botany; XVII. The species–concept of Linnaeus.

Cb1133. **Stevens, P.** 1984. Metaphors and typology in the development of botanical systematics 1690–1960, or the art of putting new wine in old bottles. *Taxon* 33: 169–211.

Cb1134. **___.** 1984. Haüy and A.-P. de Candolle: Crystallography, botanical systematics, and comparative morphology. *Journal of the history of biology* 17: 49–82.

Cb1135. **Ulbrich, E.** 1919–20. *Pflanzenkunde*. Leipzig, Reclam. 2 vols.
Vol. 1: Geschichte des Pflanzensystems. Die niederen Pflanzen; vol. 2: Die Blütenpflanzen.

Cb1136. **Walters, S. M.** 1961. The shaping of modern angiosperm taxonomy. *New phytologist* 60: 70–84.

Cb1137. **Zunck, H. L.** 1840. *Die natürlichen Pflanzensysteme geschichtlich Entwickelt*. Leipzig, Hinrich. 208 pp.

See also Ca1(Raven) Cb134(Davy de Virville) Cb233(Heiser) Cb258 & Cb577–578.

CLASSIFICATION SCHEDULES

Cb1138. **Becker, K. M.** 1973. A comparison of the angiosperm systems. *Taxon* 22: 19–50. ••A tabular comparison of the classification of the orders and families of flowering plants by Cronquist, Takhtajan, Thorne, Engler, Hutchinson, and Bentham & Hooker arranged according to the sequence of Cronquist.

Cb1139. **Swift, L. H.** 1974. *Botanical classifications: A comparison of eight systems of angiosperm classification*. Hamden, CT, Archon Books. xiv+374 pp.

Botanical nomenclature

Cb1140. **Buchanan, R. E.** 1959. The International Code of Nomenclature of the Bacteria and Viruses. *Systematic zoology* 8 (1): 27–39.

Cb1141. **Fryxell, P. A.** 1983. Very personal generic names (Nomina perpropria): A contribution to whimsical botany. *Sida* 10 (2): 95–102.

Cb1142. **Green, M. L.** 1927. History of plant nomenclature. *Bulletin of miscellaneous information, Royal Botanic Gardens, Kew* 1927 (10): 403–415.

Cb1143. **Heller, J. W.** 1964. The early history of binomial nomenclature. *Huntia* 1: 33–70.

Cb1144. **Nicholson, D. H.** 1991. A history of botanical nomenclature. *Annals of the Missouri Botanical Garden* 78: 33–56.
"Chronologic bibliography on botanical nomenclature, with some annotations," covering 1737–1989, pp. 42–56.

Cb1145. **Parkinson, P. G.** 1975. The international code of botanical nomenclature: An historical review and bibliography. *Tane* 21: 153–173.

Cb1146. **Zeebroek, R.** 1986. Histoire savante et "pensée sauvage" dans les nomenclatures botaniques en Europe. *Civilisations* 36: 349–363.

Cb1147–1148. [Not used.]

CODES OF NOMENCLATURE & INDEXES

Cb1149. **Brickell, C. D., et al.**, eds. 1980. *International code of nomenclature for cultivated plants—1980: Formulated and adopted by the International Commission for the Nomenclature of Cultivated Plants of the I U B S.* Utrecht, Bohn, Scheltema & Holkema. 32 pp. *Regnum vegetabile* vol. 104.

Cb1150. **Greuter, W., et al**, eds. 1988. *International code of botanical nomenclature adopted by the fourteenth International Botanical Congress, Berlin, July–August 1987.* Königstein, Koeltz Scientific Books. xiv+328 pp. *Regnum vegetabile* vol. 118.

Spermatophyta

Cb1151. **Andrews, H. N.** 1970. Index of generic names of fossil plants, 1820–1965. *Bulletin, U. S. Geological Survey* 1300: 354 pp. ••Alphabetical list. *Replaces* ___. 1955. Index of generic names of fossil plants, 1820–1950. *Bulletin, U. S. Geological Survey* no.1013: 262 pp.

Cb1152. **Dalla Torre, C. G., & H. Harms.** 1907. *Genera siphonogamarum ad systems Englerianum conscripta.* Leipzig, W. Engelmann. *Re-pub.* Wiesbaden, Verlag für Wissenschaftliche Neudrucke GMBH, 1963. ••Taxa above the rank of species arranged by 280 families, with sub-families, tribes, etc. Includes a supplement, appendix and index of genera. The *Index nominum* was revised, updated to include names in the supplement and appendix to the main work and re-issued as *Register zu de Dalla Torre et Harms Genera Siphonogamarum ad systema Englerianum conscripta.* Weinheim, H. R. Engelmann (Cramer), 1958. 568 pp.

Cb1153. **Dandy, J. E.** 1967. *Index of generic names of vascular plants 1753–1774.* Utrecht, International Bureau for Plant Taxonomy and Nomenclature. 130 pp. *Regnum vegetabile* vol. 51. ••Includes an annotated bibliography, index of 2,998 names and systematic arrangement of the names.

Cb1154. **Farr, E. R., J. A. Leussink & F. A. Stafleu**, eds. 1979. *Index nominum genericorum (plantarum).* Utrecht, Bohn, Scheltema & Holkema; The Hague, W. Junk, for the International Association for Plant Taxonomy. 3 vols. *Regnum vegetabile* vols. 100–102. *Continued* **E. R. Farr, J. A. Leussink & G. Zijlstra**, eds. 1986+. ___*: Supplementum I [etc.].* Utrecht & Antwerp, Bohn, Scheltema & Holkema; The Hague, Boston, MA, W. Junk, for the International Association for Plant Taxonomy. *Regnum vegetabile* vols. 113, [etc.]. ••A list of validly published generic names of all plant genera, recent and fossil (but not bacteria). For background information *see* **R. S. Cowan.** 1970. The Index Nominum Genericorum project – past, present, and future. *Taxon* 19: 52–54, 1970.

Cb1155. ***Fossilium catalogus II. Plantae.*** Berlin, The Hague, W. Junk, (*later* Amsterdam, Kugler Publications), pt. 1+, 1913+.

Cb1156. **Harvard University, Gray Herbarium.** 1968. *Gray Herbarium index.* Boston, MA, G. K. Hall. 10

vols. ••An index of all names of vascular plant taxa attributable to the Western Hemisphere beginning with the literature of 1886, including pteridophytes. Intended to supplement the original volumes of the *Index Kewensis*. For background information *see* **E. A. Shaw.** 1971. The Gray Herbarium index. *Taxon* 20: 333–336.

Cb1157. **Jackson, B. D.**, ed. 1895. *Index Kewensis plantarum phanerogamarum. Nomina et synonyma omnium generum et specierum a Linnaeo usque ad annum MDCCCLXXXV complectens nomina recepto auctore patria unicuique plantae subjectis.* Oxford, Clarendon Pr. 2 vols in 4. *Continued* **T. Durand & B. D. Jackson.** 1902–06. *Index Kewensis plantarum phanerogamarum. Supplementum primum. Nomina et synonyma omnium generum et specierum ab initio anni MDCCCLXXXVI usque ad finem anni MDCCCXCV complectens.* Brussels. 519 pp.; **Royal Botanic Gardens, Kew.** 1904+. ___*: Supplementum secundum [etc.].* Oxford, [etc.] 18 supplements to date. *Some vols. re-pub.* Koenigstein, Otto Koeltz Science Publishers, 1978–79. {Consolidated microfiche edition (cumulation of 4 original vols. & 16 supplements). London (*later* Bath), Mindata Ltd., 1983.} Continued by: *Kew Index [etc.]*; *see* Cb1158. For background information *see* **W. Greuter.** 1985. The "Index Kewensis" as a source of validation of new specific names. *Candollea* 40: 211–213; **B. D. Jackson.** 1924. The history of the compilation of the "Index Kewensis." *Journal of the Royal Horticultural Society* 49: 224–229; **R. D. Meikle.** 1971. The history of the Index Kewensis. *Biological journal of the Linnean Society* 3: 295–299; **E. Rouleau.** 1970. *Guide to Index Kewensis and its supplements I–XIV.* Montreal. vi+370 pp. (Includes a list of misprints and other corrections to Index Kewensis); ___. 1981. *Répertoire des nom génériques mentionnés dans l'Index Kewensis et ses quinze premiers suppléments.* Lac de Brome, Chatelain. Unpaged; **F. A. Stafleu.** 1966. The Index Kewensis. *Taxon* 15: 270–274.

Cb1158. ***Kew index for 1986** [etc.]: Names of seed–bearing plants, ferns, and fern allies at the rank of family and below published during 1986 [etc.].* Oxford, 1987+. ••Annual publication supplementing *Index kewensis*, see Cb1157.

Cb1159. **Pfeiffer, L.** 1873–74. *Nomenclator botanicus: Nominum ad finem anni 1858 publici juris factorum, classes, ordines, familias, divisiones, genera, subgenera vel sectiones designantium enumeratio alphabetica. Adjectis auctoribus, temporibus, locis systematicis apud varios, notis literariis atque etymologicis et synonymis.* Cassel, T. Fischer. 2 vols in 4 pts.

Cb1160. **Post, T. E. von & C. E. O. Kuntze.** 1904. *Lexicon generum Phanerogamarum inde ab anno MDCCXXXVII, cum nomenclatura legitima internationali et systemate inter recentia medio.* Stuttgart, Deutsche Verlags–Anstalt. xlvii+714 pp.

Cb1161. **Richter, H. E.** 1840. *Caroli Linnæi Systema, Genera, Species Plantarum uno volumine. Editio critica, adstricta, conferta sive Codex botanicus Linnæanus. Textum Linnæanum integrum ex omnibus Systematis, Generum, Specierum Plantarum, editionibus, Mantissis, additamentis selectumque ex ceteris ejus botanicis libris digestum, collatum, contractum, cum plena editionum discrepantia exhibens.* In usum Botanicorum practicum edidit brevique adnotatione explicavit H. E. Richter. [etc.]. Leipzig, sumpt. O. Wigand. xxxii+1,102 pp.; ___. 1840. ___ *In codicem botanicum Linnaeanum Index alphabeticus generum, specierum ac synonymorum omnium completissimus.* Composuit atque edidit Dr. G. L. Petermann, [etc.]. Leipzig, [etc.]. iv+202 pp.

Cb1162. **Steudel, E. T.** 1840–41. *Nomenclator botanicus, seu synonymia plantarum universalis, enumerans ordine alphabetico nomina atque synonyma, tum generica tum* *specifica et a Linnaeo et a recentioribus de re botanica scriptoribus plantis phanerogamis imposita.* Ed. 2. Stuttgart & Tübingen, J. G. Cotta. 2 parts. ••Nomina autorum citatorum, pt. 2: 803–810.

Pteridophyta

Cb1163. **Christensen, C. F. A.** 1906. *Index filicum sive enumeratio omnium generum specierumque filicum et hydropteridum ab anno 1753 ad finem anni 1905 descriptorum.* Copenhagen, H. Hagerup. lx+744 pp.; *Continued* ___. 1913. ___: *Supplementum, 1906–1912.* Copenhagen, [etc.]. iv+132 pp.; ___. 1917. ___: *Supplément préliminaire pour les années 1913, 1914, 1915, 1916.* Copenhagen, Triers. iv+60 pp.; ___. 1934. ___: *Supplementum tertium pro annis 1917–1933.* Copenhagen, [etc.]. 219 pp. All reprinted: Koenigstein, O. Koeltz Antiquariat, 1973; **Pichi-Sermolli, R. E. G.** 1965. ___: *Supplementum quartum pro annis 1934–1960.* Utrecht, International Bureau for Plant Taxonomy & Nomenclature. xiv+370 pp. *Regnum vegetabile* vol. 37; **F. M. Jarrett et al.** 1985. ___: *Supplementum quintum pro annis 1961–1975.* Oxford, Oxford Univ. Pr. vi+246 pp.

Musci

Cb1164. **Van der Wijk, R., W. D. Margadant & P. A. Florschütz.** 1959–69. *Index muscorum.* Utrecht, International Bureau for Plant Taxonomy & Nomenclature. 5 vols. *Regnum vegetabile* vols. 17, 26, 33, 48 & 65. *Continued* **M. R. Crosby.** 1977. Index muscorum supplementum: 1974–1975. *Taxon* 26: 285–307; ___. 1979. ___: 1976–1977. *Ibid.* 28: 237–264.

Hepaticae

Cb1165. **Bonner, C. E. B., et al.** 1962+. *Index hepaticarum.* Parts 1–6, 7A, 8, 10, 12. Weinheim, Vaduz, Braunschweig, J. Cramer. *Continued* **J. J. Engel.** 1978. Index hepaticarum supplementum: 1974–1975. *Taxon* 27: 393–418; ___. 1989. ___: 1980–1981. *Taxon* 38: 414–439; **P. Geissler & H. Bischler**, eds. 1987. *Index hepaticarum.* Vol. 8/9. 2nd ed. Berlin, Borntraeger. 310 pp.

Cb1166. **Grolle, R.** 1983. *Nomina generica hepaticarum: references, types and synonymies.* Helsinki, Finnish Botanical Publishing Board. 62 pp. *Acta botanica fennica* no. 121.

Lichenes

Cb1167. **Commonwealth Mycological Institute.** 1972. *Index of fungi. Supplement. Lichens 1961–1969.* By D. L. Hawksworth. Kew. ix+78 pp.

Cb1168. **Lamb, I. M.** 1968. *Index nominum lichenum inter annos 1932 et 1960 divulgatorum.* New York, Ronald Pr. 809 pp.

Cb1169. **Zahlbruckner, A.** 1921–40. *Catalogue lichenum universalis.* Leipzig, Borntraeger. 10 vols. Reprinted New York, Johnson Reprint Corporation, 1951. ••Vol. 1: Pyrenocarpeae. Gymnocarpeae, Coniocarpineae; vol. 2: Graphidineae – Pyrenopsidaceae; vol. 3: Lichinaceae – Lecideaceae, genus Lecidia; vol. 4: Lecideaceae, genus Mycoblastus – Umbilicariaceae; vol. 5: Acarosporaceae – Lecanoraceae; vol. 6: Parmeliaceae – Usneaceae; vol. 7: Caloplacaceae – Lichens imperfecti vel morbosi; vol. 8: Supplementum. Addenda et corrigenda; vol. 9: Index; vol. 10: Ergänzungsband (Opus posthumum).

Fungi

Cb1170. **Commonwealth Mycological Institute.** 1940+. *Index of fungi: Comprising a list of new genera, species, and varieties of fungi, new combinations, and new names published since the beginning of 1940.* Kew. ••Pub. as supplements to the *Review of applied mycology* from 1940–47. A compilation from world literature of new names of fungi and lichens with citations and typifications. Continues F. Petrak lists; *see* Cb1171. Cumulated host and subject index, 1940–49.

Cb1171. **Petrak, F.** 1930–44. Verzeichniss der neuern Arten, Varietäten, Formen, Namen und wichtigsten Synonyme der Pilze 1920 (–35). *Just's botanischer Jahresbericht* 48(2): 184–256, 1930. *Re-pub.* Kew, 1953; *Ibid.* 56(2): 291–697, 1937. *Re-pub.* Kew, 1953; *Ibid.* 57(2): 592–631, 1938. *Re-pub.* Kew 1952; *Ibid.* 58(1): 447–570, 1938. *Re-pub.* Kew, 1952; *Ibid.* 60(1): 449–514, 1939. *Re-pub.* Kew, 1953; *Ibid.* 63(2): 805–1216, 1935. *Re-pub.* Kew, 1953. *Continued* ___. 1950. *Index of fungi 1936–39: List of new species and varieties of fungi, new combinations and new names published 1936–1939.* Kew, Commonwealth Mycological Institute. 115 pp.; **Commonwealth Mycological Institute.** 1956–57. *Index of fungi. Cumulative index of Petrak's lists, 1920–1939.* London. 307 pp.; ___. 1969. *Index of fungi. Supplement. A supplement to Petrak's lists 1920–1939. Fungi.* Kew. 236 pp. ••Spans the gap between Saccardo and the *Index of fungi.* The CMI supplements comprise indexes of generic names and specific epithets, hosts, host genera, and other substrata, and corrections to Petrak's lists #1–8 and to the cumulative index, supplementary list of fungus names and supplementary host index.

Cb1172. **Saccardo, P. A., et al.** 1882–1972. *Sylloge fungorum omnium hucusque cognitorum.* Pavia, Saccardo et al., Berlin, Gebrüder Bornträger; Avellino, [vols. 23–25]; New York, [vol. 26]. *Re-pub.* New York, Johnson Reprint Corporation, 1967. 25 vols. in 28. ••Vols. 9–11, 1891–95: *Supplementum universale* ... pars I (–III); vols. 12–13, 1897–98: *Index universalis et locupletissimus generum, specierum, subspecierum et varietatum in toto opere expositorum* (i.e. vols. 1–11), *Index universalis et locupletissimus nominum plantarum hospitum*; vol. 14, 1899: *Supplementum universale* ... pars IV; vol. 15, 1901: *Synonymia generum, specierum subspecierumque in vol. 1–14 descriptorum*; vols. 16–18, 1902–06: *Supplementum universale pars 5–7. Index universalis ... in toto opere* (i.e. vols. 10–18); vols. 21–26, 1912–72: Supplementum universale pars 8–11. *Also* **Commonwealth Mycological Institute.** 1985. *Index of fungi. Supplement. Saccardo's omissions.* Ed. by P. M. Kirk. Kew. 101 pp. 2,600 names which were published before 1920.

Algae

Cb1173. **Norstedt, C. F. O.** 1896–1908. *Index Desmidiacearum citationibus locupletissimus atque bibliographia. (& Suppl.)* Berlin, Borntraeger. ii+310+149 pp. Reprinted 1978. ••20,000 references.

Cb1174. **Silva, P. C.** 1980. *Names of classes and families of living algae with special reference to their use in the Index Nominum Genericorum (Plantarum).* Utrecht, Bohn, Scheltema & Holkema for the International Association for Plant Taxonomy. [v]+156 pp. *Regnum vegetabile* vol. 103. ••Incl. an extensive bibliography, pp. 113–142.

Cb1175. **Vanlandingham, S. L.** 1967–79. *Catalogue of the fossil and recent genera and species of diatoms and their synonyms. A revision of F. W. Mills' "An index to the genera and species of the Diatomaceae and their synonyms."* Lehre, Cramer. 8 vols. ••Covers 1786–1964. Includes a complete bibliographical references to each entry. A–Z vols. 1–7; vol. 8, Supplementary taxa (through 1964), supplementary references, synonymy addendum, corrections, additions.

Botanical terms, etymology

Cb1176. **Backer, C. A.** 1936. *Verklarend Woordenboek der wetenschappelijke namen van de in Nederland en Nederlandsch–Indië in het wild groeiende en in tuinen en parken gekweekte varens en hoogere planten.* Groningen, P. Noordhoff. xii+664 pp. ••Etymological dictionary of plant generic and specific names with biographical notes on the people for whom they were named.

Cb1177. **Dean, D. R.** 1972. Erasmus Darwin's botanic terms. *English language notes* 10: 27–29.

Cb1178. **Genaust, H.** 1976. *Etymologisches Wörterbuch der botanischen Pflanzennamen.* Basle, Birkhäuser. 390 pp.

Cb1179. **Gentil, A.** 1923. *Dictionnaire étymologique de la flore française.* Paris, P. Lechevalier. xiv+241 pp. *Encyclopédie pratique du naturaliste* no. 13.

Cb1180. **Jackson, B. D.** 1928. *A glossary of botanic terms with their derivation and accent.* 4th edn. London, Duckworth. xii+481 pp. Various reprints, e.g. New York, Hafner, 1965.

Cb1181. **Rydén, M.** 1978. *Shakespearean plant names: Identifications and interpretations.* Stockholm, Almqvist & Wiksell. 117 pp. *Acta Universitatis Stockholmiensis, Stockholm studies in English* no. 43. ••*Also* ___. 1984. The contextual significance of Shakespeare's plant names. *Studia neophilogica* 56: 155–162.

Cb1182. ___. 1984. *The English plant names in the Grete Herball (1526): A contribution to the historical study of English plant–name usage.* Stockholm, Almqvist & Wiksell. 110 pp. *Acta Universitatis Stockholmiensis, Stockholm studies in English* no. 61.

Cb1183. **Smith, A. W.** 1972. *A gardener's dictionary of plant names: A handbook on the origin and meaning of some plant names.* Edit., rev. & enl. by W. T. Stearn, I. L. L. Smith. London, Cassell. xii+391 pp. ••*Replaces* ___. 1963. *A gardener's book of plant names.* New York, Harper & Row. 428 pp.

Cb1184. **Stearn, W. T.** 1993. *Botanical Latin: History, grammar, syntax, terminology and vocabulary.* 4th ed. Newton Abbot, David & Charles. xiv+546 pp. ••*Replaces* 3rd rev. ed., Newton Abbot, David & Charles, 1983. xiv+566 pp.

Cb1185. **Wittstein, G. C.** 1967. *Etymologisch–botanisches Handwörterbuch.* 2. Aufl. Wiesbaden, Sändig. 952 pp.

BOTANICAL ILLUSTRATION

General

Cb1186. **Anderson, F. J.**, ed. 1983. *Herbals through 1500.* New York. 2 vols., illus. *The Illustrated Bartsch* no. 90.
Vol. 1 is a "Commentary" volume, vol. 2 is a "supplement" providing reproductions of every woodcut illustration from 9 early herbals.

Cb1187. **Arber, A.** 1986. Ch. VII: The evolution of the art of botanical illustration. In: *Herbals: Their origin and evolution. A chapter in the history of botany 1470–1670.* 3rd ed., pp. 185–246. Cambridge, Cambridge Univ. Pr. xxxii+358 pp., illus.
Reprint of the 2nd edition, 1938, with an introduction and annotations by W. T. Stearn. Also includes a reprint of "The colouring of 16th century herbals", first pub. in *Nature* 165, 1940.

Cb1188. **Bartlett, H. H., & H. Shohara.** 1961. *Japanese botany during the period of wood–block printing. I. An essay on the development of natural history, especially botany, in Japan; on the influence of early Chinese and western contacts; on Japanese books and wood–block illustration. II. An exhibition of Japanese books and manuscripts, mostly botanical, held at the Clements Library [Ann Arbor].* Los Angeles, CA, Dawson's Book Shop. vi+271 pp., illus.
Bibliographical description of 112 items. First pub. in *Asa Gray bulletin* n.s. 3: 289–561, 1961.

Cb1189. **Blunt, W.** 1967. *The art of botanical ilustration.* With the assistance of W. T. Stearn. (4th ed.) London, Collins. xxxii+304 pp., illus. *New naturalist* no. 14. *Re-pub.*

1971.
Actually a corrected reprint of the 1st ed., pub. 1950, with some addenda.

Cb1190. ___, **& S. Raphael.** [1979.] *The illustrated herbal.* London, Frances Lincoln. 192 pp., illus. *Also* New York, Thames & Hudson.

Cb1191. **Cave, R., & G. Wakeman.** 1967. *Typographia naturalis.* Wymondham, Brewhouse Pr. 36 pp. & 4 pl.

Cb1192. **Coats, A. M.** 1973. *The book of flowers: Four centuries of flower illustration.* London, Phaidon Pr. 208 pp., illus. *Also* New York, McGraw–Hill Book Co., [1973]. 2nd ed., 1974.

Cb1193. ___. 1975. *The treasury of flowers.* London, Phaidon Pr. 190 pp., illus. *Also* New York, McGraw–Hill Book Co., [1975].

Cb1194. **Desmond, R. G. C.** 1987. *A celebration of flowers: Two hundred years of Curtis's Botanical Magazine.* London, Royal Botanic Gardens, Kew, in association with Collingridge. 208 pp., illus.
Appendix: Artists, engravers and lithographers in 'Curtis's Botanical Magazine,' pp. 197–202.

Cb1195. **Dunthorne, G.** 1938. *Flower and fruit prints of the 18th and early 19th centuries: Their history, makers and uses, with a catalogue raisonné of the works in which they are found.* Washington, DC, privately; London, Dulau. xiv+275 pp., illus. *Re-pub.* New York, Da Capo Pr., 1968; London, Holland Pr., 1970.
The "Catalogue raisonné" of 335 works provides "complete descriptions of the characteristics of each print, artist, engraver, publisher and other details necessary for their identification."

Cb1196. **Eisendrath, E. R.** 1961. Portraits of plants: A limited study of the "Icones." *Annals, Missouri Botanical Garden* 48: 291–327 & 14 pl.

Cb1197. **Ewing, W. A.** 1991. *Flora photographica: Masterpieces of flower photography, 1835 to the present.* London, Thames & Hudson. 224 pp., illus. *Also* New York, Simon & Schuster, 1991.
180 works by over 100 photographers.

Cb1198. **Fischer, E.** 1933. 200 Jahre Naturselbstdruck. *Gutenberg Jahrbuch* 1933: 186–213.

Cb1199. **Heilmann, P. N.** 1982. *Die Natur als Drucker. Naturselbstdrucke der k. u. k. Hof- und Staatsdruckerei, Wien.* Dortmund; Harenberg. 144 pp., illus. *Die Bibliophilen Taschenbücher* no. 330.

Cb1200. ___. 1988. Typennaturselbstdruck aus heutiger Sicht: Ist die Entdeckung des Naturdrucks eine Wiener Erfindung? *Journal für Druckgeshichte* 1: 17–25, 2 pl. & a specimen of nature printing.

Cb1201. **Hulton, P., & L. Smith.** 1979. *Flowers in art from east and west.* London, British Museum Publications. x+150 pp., illus.
1: Naturalism in Sung China; 2: The rebirth of naturalism in Europe; 3: The herbal tradition; 4: Artists of discovery; 5: Flowers for pleasure in India and Islam; 6: A world of colour – floral painting in China and Japan; 7: Ink painting in China and Japan; 8: Woodblock prints in East Asia; 9: Printing techniques in Europe; 10: Art and botany.

Cb1202. **Hunt Institute for Botanical Documentation, Carnegie Mellon University.** 1980. *A selection of late 18th and early 19th century Indian botanical paintings recording the indigenous and introduced flora of the subcontinent, commissioned by the Honourable East India Company and executed in water colour by native artists, variously lent from the collections of the British Museum (Natural History), India Office Library, Linnean Society of London, & Royal Botanic Gardens, Kew:* Illustrated, with accompanying catalogue by P. I. Edwards, J. J. White, D. H. Nicholson & R. W. Kiger,

preface by J. V. Brindle, and introductory essays by S. C. Welch and P. I. Edwards. Pittsburgh, PA. 72 pp., illus.
72 items by various artists.

Cb1203. ___. 1982. *Talking in flowers: Japanese botanical art.* Catalogue of an exhibition 5 April to 16 July 1982. J. V. Brindle & J. J. White. Pittsburgh, PA. 95 pp., illus.
Approximately 85 items, dating from the 17th century to the present time.

Cb1204. ___. 1986. *Printmaking in the service of botany.* 21 April to 31 July 1986. Catalogue of an exhibition. By G. D. R. Bridson & D. E. Wendel with the assistance of J. J. White. Pittsburgh, PA. 166 pp., illus.
72 items covering five centuries of illustrations and prints.

Cb1205. **Krieger, L. C. C.** 1922. A sketch of the history of mycological illustration. *Mycologia* 14: 311–331.

Cb1206. **Mabey, R.** 1988. *The flowers of Kew: 350 years of flower paintings from the Royal Botanic Gardens.* London, Century Hutchinson; New York, Atheneum. xvi+208 pp., illus.
Illustrated with approximately 175 coloured reproductions of original artwork from the Gardens' collections.

Cb1207. **Middelhoek, A.** 1946. De mycologische iconographie van Clusius tot Lange. *Mededelingen, Nederlandsche Mycologische Vereenining* 1946 (Mei): 55–103.

Cb1208. **Nissen, C.** 1933. *Botanische Practwerke. Die Blütezeit der Pflanzenillustration von 1740 bis 1840.* Vienna, H. Reichner, 1933. 44 pp.
First pub. as: Botanische Practwerke. Die Blütezeit der Pflanzenillustration von 1740 bis 1840. *Philobiblon* 6: 243–256, 287–303, 336–348, 1933.

Cb1209. ___. 1951. *Die botanische Buchillustration: Ihre Geschichte und Bibliographie.* Stuttgart, Hiersemann. 2 vols. *Continued* ___. 1966. ___*: Supplement.* Stuttgart, Hiersemann. vii+97 pp.
Vol I: Geschichte – Einleitung; Altertum und Mittelalter, 1. Die Handschriften, 2. Die Frühdrucke; Renaissance; Barock, 1. Die Blumen– und Gartenbücher, 2. Die wissenschaftliche Botanikmalerei, 3. Frankreich; Neuzeit, 1. England, 2. Frankreich (seit Aubriet), 3. Italien, 4. Spanien und Portugal, 5. Deutschland, 6. Schweiz, 7. Russland, 8. Skandinavien, 9. Niederlande, 10. Belgien; Die Botanikmalerei ausserhalb Europas, 1. Nordamerika, 2. Südafrika, 3. Australien, 4. Asien; Anhang, 1. Die primären Dokumentationsmittel der Botanik, 2. Die Paläobotanische Abbildung, 3. Die Photographie in der botanischen Illustration; Allgemeines Schrifttum. Vol II: Bibliographie – Titel–Verzeichnis, 1. Verfasser–Teil (Titel Nr. 1–2207), 2. Zeitschriften, anonyme und Serienwerke (Titel Nr. 2208–2387); Register zu Band I und II, Künstler – Pflanzen – Länder – Verfasser. Supplement incl: Nachträge und Ergänzungen zu Band II; Register zu Band III, Künstler – Pflanzen – Länder – Verfasser.
Bände I–II & Supplement were also issued as: 2. Aufl. Durchgesehener und verbesserter Abdruck der zweibänden Erstauflage ergänzt durch ein Supplement. Stuttgart, 1966. 3 vols. in one.

Cb1210. **Rea, E. A.** 1916. Notes on fungus illustrations. *Transactions, British Mycological Society* 5: 211–228.

Cb1211. **Rix, M.** 1981. *The art of the botanist.* Guildford & London, Lutterworth Pr. 224 pp., illus.
American ed. entitled: The art of the plant world: The great botanical illustrators and their work. Woodstock, NY, Overlook Pr., 1981. *Re-pub. as* ___. 1989. *The art of botanical illustration.* London, Bracken Books.

Cb1212. **Sampson, F. B.** 1985. *Early New Zealand botanical art.* Birkenhead, Auckland, Reed Methuen. 142 pp., illus.
Including chapters on: Parkinson, Banks and Solander; Johann & George Forster; Dumont D'Urville's voyages; Etienne Raoul; William Curtis's Botanical Magazine; Sir William & Sir Joseph Hooker; Martha King – first resident botanical artist; Fanny Osborne of Great Barrier Island; John Buchanan and nature printing; Mr & Mrs Fenton's Art Album; Emily Harris's

Flowers, berries and ferns; Mrs Hetley's Native flowers; Henry Field's Ferns of New Zealand; Thomas Kirk Forest flora of New Zealand; and Thomas Cheeseman Illustrations of the New Zealand flora.

Cb1213. **Sitwell, S., & W. Blunt.** 1956. *Great flower books 1700–1900: A bibliographical record of finely-illustrated flower books.* ... The bibliography ed. by P. M. Synge. London, Collins. x+94 pp. & 36 pl. *Re-pub.* (in reduced format with new illustrations) New York, The Atlantic Monthly Pr., 1990. xii+189 pp., illus.
The bibliography, by W. T. Stearn, S. Wilson & H. Buchanan, pp. 47–85.

Cb1214. **Steenis, C. G. G. J. van.** 1950. List of works principally containing illustrations of Malaysian plants, and of collections of drawings and photographs. In: ___, ed. *Flora malesiana*, Series 1: *Spermatophyta*, vol. 1, pp. xxx–xxxiii. Djakarta, Republic of Indonesia, Ministry of Agriculture. *Continued* ___. 1958. ___: Supplement 1. In: *Ibid.* Series 1. Vol. 5(4), pp. ccxxxvii. Leyden, Noordhoff; ___. 1974. ___: Supplement 2. In: *Ibid.* Series 1. Vol. 8(1), pp. i. Leyden, Noordhoff.

See also Biographies Ba375–378 Ba380 Ba517–518 Ba768 Ba794–795 Ba1123 Ba1315–1318 Ba1342 Ba1382 Ba1423–1425; & Cb663.

Plant images in art

Cb1215. **Baldwin, M. F.** 1933. Plant forms in ornament: A selective list of references in the New York Public Library and other libraries of New York City. *Bulletin, New York Public Library* 37: 511–525, 613–632, 715–741. *Also pub. separately* New York, 1933. 59 pp.
A classified list.

Cb1216. **D'Ancona, M. L.** 1977. *The garden of the Renaissance: Botanical symbolism in Italian painting.* Florence, Olschki. 603 pp., illus. *Arte e archeologia, Studi e documenti* no. 10.
Main text arranged alphabetically by English name of plant, with notes on symbolism, and lists works of art in which each plant is depicted. Bibliography; Dictionary [of plant names, French/English]; Glossary [of plant names, Latin/English]; Index of artists; Iconographical index; Index of paintings by place; Index of symbols, attributes, meanings and legends.

Cb1217. **Foshay, E. M.** 1980. Charles Darwin and the development of American flower imagery. *Winterthur portfolio: A journal of American material culture* 15 (Winter): 299–314.

See also Ca221 Ca720 & D48.

BIBLIOGRAPHIES OF ILLUSTRATION RESOURCES

Cb1218. **Bamps, P., J.-P. Lebrun & A. L. Stork.** 1981–88. *Index iconographique des plantes vasculaires d'Afrique, 1935–1980.* Maisons-Alfort, Institut d'Élevage et de Médecine Vétérinaire des Pays Tropicaux. 4 vols. *Étude botanique* nos. 9, 11, 12 & 14. ••Supplements *Index Londinensis.*

Cb1219. **Ellis, J. C.** 1949. *Nature and its applications: Over 200,000 selected references to nature forms and illustrations of nature as used in every way.* Boston, Faxon. xii+861 pp. ••Books and periodicals used in index, pp. xi–xii. The alphabetically arranged index uses common names with relatively few references from scientific names. There is a small section on landscapes, pp. 854–861. The references are mostly to photographs in books and magazines of the late 19th and 20th century.

Cb1220. **Fritsch, F. E.** 1964. *The Fritsch collection of algae illustrations.* Zug, Inter Documentation. Microfiche. ••Reproduction of the collection of over 25,000 sheets of illustrations and a card index, collected and systematized by Fritsch and now preserved at the Freshwater Biological Association library at Ambleside, England.

Cb1221. **Hamilton, R. M.** 1967. *Orchid flower index: A world list of reproductions in colour in books and periodicals 1736 to 1966*. Richmond, B.C., privately. x+124 pp.; *Continued* **G. E. Woolfson.** 1978. *Ten year index to plant illustrations in 10 influential orchid publications (with a few extras thrown-in) 1968–1977*. Lowell, MI, Twin Oaks Books. xii+68 pp. ••Supplement is an index to: *Orchid review*, *Orchid digest*, *Australian orchid review*, *Orquidea* (Mexico), *Orquideologia* (Colombia), *Florida orchidist*, *Orchid advocate*, *Cymbidium Society news*, *Hawaii orchid journal*, *Paphiopedilum world*, *Orchadian* (Australia), *Orchid journal*, *Harvard botanical leaflets*.

Cb1222. **Isaacson, R. T.** 1979. *Flowering plant index of illustrations and information*. [Sponsored by] The Garden Center of Greater Cleveland. Boston, MA, G. K. Hall. 2 vols. *Continued* ___. 1982. ___*: 1979–81*. Boston, MA, [etc.]. 2 vols. ••The 4 vols. include over 100,00 entries arranged A–Z by scientific name. Works indexed are generally post 1935 (the close of the *Index Londinensis* coverage).

Cb1223. ___. 1991. *FPI: Flowering plant index*. Chanhassen, MN, , Univ. of Minnesota Libraries. 2 vols.

Cb1224. **Nissen, C.** 1951. *Die botanische Buchillustration: Ihre Geschichte und Bibliographie*. Vol. 2: *Bibliographie*. Stuttgart, Hiersemann. *Continued* ___. 1966. ___*: Supplement*. Stuttgart, [etc.]. vii+97 pp. •• Band 2. *Bibliographie* and Band 3. *Supplement* include classified indexes to "Pflanzen" and "Länder" that enable the user to survey five hundred years of published illustration resources for both of these themes.

Cb1225. **Pam, A.** 1942. Check–list of Amaryllid coloured plates chiefly latter part 18th and 1st half 19th centuries. *Herbertia* 9: 85–98.

Cb1226. **Pritzel, G. A.** 1866. *Iconum botanicarum index locupletissimus. An alphabetic register of phanerogamic plants and ferns, compiled from botanical and horticultural publications of the XVIIIth and XIXth centuries, [etc.]*. Berlin, Nicolaische Verlagsbuchhandlung. xxxi pp.+1184 coll. *Continued* ___. 1866. ___*: Zweiter Theil fortgeführt bis zu Ende des Jahres1865*. Berlin, Nicolaische Verlagsbuchhandlung. [7] pp.+298 coll.

Cb1227. **Saccardo, P. A., & J. B. Traverso.** 1910–11. Index iconum fungorum enumerans eorundem figuras omnes hucusque editas ab auctoribus sive antiquis sive recentioribus. Ductu et consilio P. A. Saccardo congessit J. B. Traverso. In: *Sylloge fungorum omnium hucusque cognitorum*, vols. 19–20. Padua, Saccardo et al. *Re-pub.* New York, Johnson Reprint Corp., 1967.

Cb1228. **Stapf, O.** 1929–31. *Index Londinensis to illustrations of flowering plants, ferns and fern allies; Being an amended and enlarged edition, continued up to the end of the year 1920, of Pritzel's 'Alphabetical register of representations of flowering plants and ferns' ...*; prepared ... by O. Stapf. Oxford, Clarendon Pr. 6 vols. *Continued* ___. 1941. ___*: Supplement for 1921–35*, prepared by W. C. Worsdell. Oxford, Clarendon Pr. 2 vols. 8 vols. *Re-pub.* Koenigstein, Otto Koeltz Science Publishers, 1979–80. ••Not continued. "Although no further Supplements are to be issued, it has been arranged that references to new illustrations shall be given in the entries of new names included in future Supplements of the *Index Kewensis* from 1936 onwards. Such entries will be prefixed by an asterisk."

Cb1229. **Sternberg, C. von.** 1821. *Catalogus plantarum ad septem varias editiones commentariorum Mathioli in Dioscoridem*. Prague, F. Tempsky. v+30 pp. ••Index secundus "Nomina plantarum botanica, nec non iconum in septem variis Mathioli commentariorum editionibus citationes et annotationes continens" is a tabulated listing of the contents of the 1565, 1562, 1558, 1554, 1611, 1596 & 1598 eds., in that order, arranged by Linnaean names.

Cb1230. **White, J. J., & R. F. Schallert.** 1986. Illustrations in the "Contributions from the U. S. National Herbarium," 1890–1974. *Huntia* 6: 147–164.

See also Ca245 Ca714 Ca720 Ca723 Ca760 & Ca979.

PLANTS & MAN

General

A selection of some representative works:

Cb1231. **Bennett, J.** 1991. *Lilies of the hearth: The historical relationship between women and plants.* Camden East, Ont., Camden House. 191 pp.

Cb1232. **Berrall, J. S.** 1953. *A history of flower arrangement.* London & New York, Studio Publications. 160 pp., illus.

Cb1233. **Bertsch, K. & F.** 1949. *Geschichte unserer Kulturpflanzen.* 2nd ed. Stuttgart, Wissenschaftliche Verlagsgesellschaft. 275 pp.

Cb1234. **Candolle, A. L. P. P. de.** 1886. *The origin of cultivated plants.* 2nd ed. London, Routledge. viii+468 pp. *International scientific series* vol. 49. *Re-pub.* New York, Hafner, 1959 & 1964.
Translation of *Origine des plantes cultivées.* Paris, G. Baillière & Cie, 1883. viii+377 pp. *Bibliothéque internationale scientifique.*

Cb1235. **Coats, A. M.** 1956. *Flowers and their histories.* London, A. & C. Black. xii+347 pp. & 14 pl.

Cb1236. **___.** 1992. *Garden shrubs and their histories.* New illustrated ed., with notes by J. L. Creech. New York, Simon & Schuster, 1992. 222 pp., illus.
Replaces ___. 1963. ___. London, Vista Books. 416 pp. *Also* New York, E. P. Dutton, 1965. The new ed. has supplementary notes to "fill in what may appear as gaps from an American standpoint" and 112 colour reproductions of 18th & 19th century illustrations.

Cb1237. **___, P.** 1970. *Flowers in history.* London, Weidenfeld & Nicholson. 264 pp.

Cb1238. **Dodge, B. S.** 1962. *Plants that changed the world.* London, Phoenix House. vii+174 pp.

Cb1239. **Gibault, G.** 1912. *Histoire des légumes.* Paris, Libraire Horticole. viii+404 pp., illus.

Cb1240. **Griggs, B.** 1981. *Green pharmacy: A history of herbal medicine.* London, Jill Norman & Hobhouse. xiii+379 pp.

Cb1241. **Hedrick, U. P.** 1988. *A history of horticulture in America to 1860.* With an addendum of books published from 1861–1920, by E. Woodburn. Portland, OR, Timber Pr. xv+634 pp., illus.
Replaces Earlier ed. New York, Oxford Univ. Pr., 1950. 551 pp.

Cb1242. **Jackson, J. R.** 1890. *Commercial botany of the nineteenth century: A record of progress in the utilisation of vegetable products in the United Kingdom, and the introduction of economic plants to British colonies, during the present century.* London, Cassell & Co. [vi]+168 pp.

Cb1243. **Kraus, G. K. M.** 1894. *Geschichte der Pflanzeneinführungen in die europäischen botanischen Gärten.* Leipzig, W. Engelmann. [ii]+73 pp.
Forms part of *Der botanische Garten der Universität Halle.* Leipzig, W. Engelmann, 1888–94.

Cb1244. **Leighton, A.** 1970. *Early English gardens in New England. "For Meate or Medicine."* London, Cassell & Co. xviii+441 pp., illus.; *Continued* ___. 1976. *American gardens in the eighteenth century. "For Use or for Delight."* Boston, MA, Houghton Mifflin Co. xxiii+514 pp., illus.; ___. 1987. *American gardens of the nineteenth century. "For comfort and affluence."* Amherst, MA, Univ. of Massachusetts Pr. xviii+395 pp., illus.

Cb1245. **Nelson, K.** 1950. *America's crop heritage: The history of foreign plant introduction by the federal government.* Ames, IA, Iowa State College Pr. xii+156 pp.

Cb1246. **Pickering, C.** 1879. *Chronological history of plants: Man's record of his own existence illustrated through their names, uses and companionship.* Boston, MA. xvi+1,222 pp. *Re-pub.* Koenigstein, 1975; Dehra Dun, 1986.

Cb1247. **Roach, F. A.** 1985. *Cultivated fruits of Britain: Their origin and history.* Oxford, Basil Blackwell. [viii]+349 pp., illus.

Cb1248. **Salaman, R. N.** 1985. *The history and social influence of the potato.* Rev. impr. ed. by J. G. Hawkes. Cambridge, Cambridge Univ. Pr. xliv+685 pp., illus.
Replaces 1949 ed.

Cb1249. **Simmonds, N. W.**, ed. 1976. *Evolution of crop plants.* London, Longman Group Ltd. xii+339 pp.
Accounts of individual crops by 77 authors. Arranged by plant family, each account includes a brief history of each crop's early and recent history.

Cb1250. **Sturtevant, E. L.** 1919. *Sturtevant's notes on edible plants.* Ed. by U. P. Hedrick. Albany, NY, New York Department of Agricyulture. vii+686 pp. *Report, New York Agricultural Experiment Station* 1919, pt. 2. *Re-pub. as* ___. [1972.] *Sturtevant's edible plants of the world.* New York, Dover Publications.

Cb1251. **Tolkowsky, S.** 1938. *Hesperides: History of the culture and use of citrus fruits.* London, J. Bale, Sons & Curnow. xx+371 pp., illus.

Cb1252. **Williams, M.** 1988. *Americans and their forests: A historical geography.* Cambridge, Cambridge Univ. Pr. 640 pp., illus. *Studies in environment and history.*

Plant mythology & folklore

Cb1253. **Dyer, T. F. Thiselton–.** 1889. *The folk–lore of plants.* London, Chatto & Windus. [iii]+328 pp.

Cb1254. **Friend, H.** 1884. *Flowers and flower lore.* 2nd ed. London, Swan, Sonnenschein & Co. xvi+704 pp.

Cb1255. **Gubernatis, A. de.** 1872. *La mythologie des plantes ou les légendes du règne végétal; botanique générale et spéciale.* Paris, Reinwald. 2 vols.

Cb1256. **Gupta, S. M.** 1971. *Plant myths and traditions in India.* Leiden, E. J. Brill. xii+117 pp. & 33 pl.

Section Cc. HISTORY OF ZOOLOGY

HISTORIES OF ZOOLOGY

General

Cc1. **Brooks, W. K.** 1899. *The foundations of zoölogy.* New York, Macmillan Co. viii+339 pp. *Columbia University biological series* no. 5.

Cc2. **Burckhardt, C. R.** 1921. *Geschichte der Zoologie und ihrer wissenschaftlichen Probleme.* 2nd ed. Neu bearb. von H. Erhard. Berlin, De Gruyter. 2 vols. *Sammlung Goschen* nos. 357 & 823.
Replaces 1st ed., Leipzig, G. J. Goschen, 1907. 156 pp. *Sammlung Goschen* [357].

Cc3. **Carus, J. V.** 1872. *Geschichte der Zoologie bis auf Joh. Müller und Charl. Darwin.* Munich, R. Oldenbourg. xii+739 pp. *Geschichte der Wissenschaften in Deutschland, neuere Zeit,* 12. *Re-pub.* New York, Johnson Reprint Corporation, 1964.
For French version *see* ___. 1880. *Histoire de la zoologie depuis l'antiquité jusqu'au XIXe siècle.* Traduction française par P.–O. Hagenmüller et notes par A. Schneider. Paris, J. B. Baillière. viii+623 pp.

Cc4. **Fedorowicz, Z.** 1962. *Zarys historii zoologii.* Warsaw, Panstwowe Wydawnictwo Naukowe. 349 pp.

Cc5. **Hagberg, K.** 1941. *Från Aristoteles till Brehm: Några blad ur djurskildringens historia.* Stockholm, Bonnier. 99 pp.
I.e. From Aristotle to Brehm: Some pages from the history of animal description.

Cc6. **Hays, H. R.** 1972. *Birds, beasts and men: A humanistic history of zoology.* New York, Putnam. 383 pp. *Also* London, J. M. Dent & Sons, 1973.
For German version *see* ___. 1972. *Das Abenteuer Biologie: Die Erforschung der Tierwelt von Aristoteles bis Konrad Lorenz.* Dusseldorf, Cologne, Diedrichs. 326 pp.

Cc7. **Hoefer, F.** 1890. *Histoire de la zoologie depuis les temps les plus reculés jusqu'à nos jours.* Nouv. ed. Paris, Hachette & Cie. [iv]+412 pp. *Histoire universelle publiée par une Société de Professeurs et de Savants sous la direction de M. V. Duruy.*
Replaces 1st ed., 1873. [iv]+412 pp.

Cc7A. **Lankester, E. R.** 1911. [Article] Zoology. In: *Encyclopaedia Britannica*, 11th ed., vol. 28, pp. 1022–1039. Cambridge, Cambridge Univ. Pr.
Concise history. *Replaces* [Article] Zoology. In: *Encyclopaedia Britannica*, 9th ed., 1889, also pub. in: ___. 1890. *The advancement of science, [etc.]*, pp. 287–387. London, Macmillan & Co.

Cc8. **Ley, W.** 1948. *The lungfish, the dodo, and the unicorn: An excursion into romantic zoology*. 2nd ed., rev. & enl. New York, Viking. xi+361 pp.
Replaces 1st ed. New York, Modern Age, 1941.

Cc9. **Locy, W. A.** 1915. *Biology and its makers*. 3rd ed., rev. New York, H. Holt & Co. xxvi+477 pp.
Replaces 2nd ed., 1910. xxvi+469 pp. Re-pub. several times, e.g. New York, Holt, Rinehart & Winston, 1962. For German version *see* ___. 1915. *Die Biologie und ihre Schöpfer*. Autorisierte Übersetzung der zweiten amerikanischen Aufl. von E. Nitardy. Jena, G. Fischer. xii+415 pp.

Cc10. ___. 1918. *The main currents of zoology*. New York, H. Holt & Co. vii+216 pp.

Cc11. **Plavil'shchikov, P. N.** 1941. *Ocherki po istorii zoologii. [Essays on the history of zoology]*. Moscow, Uchpedgiz. 296 pp.

Cc12. **Spix, J. B. von.** 1811. *Geschichte und Beurtheilung aller Systeme in der Zoologie nach ihrer Entwicklungsfolge von Aristoteles bis auf gegenwärtige Zeit*. Nuremberg, Schrag'sche Buchhandlung. xxvi+710 pp.

Cc13. **Wendt, H.** 1959. *Out of Noah's Ark: The story of man's discovery of the animal kingdom*. Translated from German by M. Bullock. London, Weidenfeld & Nicholson. xii+464 pp.

Cc14. ___. 1968. *Before the deluge*. Translated from German. London, Gollancz. xii+419 pp.

See also Ba303 Ba306 Ba309 Ba314 Ba316 Bb789–822 & Ca33–34.

15th to 18th centuries

Cc15. **Delaunay, P.** 1955. La divulgation des "œuvres de nature" dans la zoologie du XVIe siècle. *Archives internationales d'histoire des sciences* 31: 134–144.

Cc16. ___. 1962. *La zoologie au seizième siècle*. Paris, Hermann. xi+338 pp. & 9 pl. *Histoire de la pensée* no. 7.

Cc17. **Ley, W.** 1968. *Dawn of zoology*. Englewood Cliffs, NJ, Prentice–Hall. viii+280 pp.
Translation of *Eine Geschichte der Tiere, ihr Einfluss auf Zivilisation und Kultur*. Hamburg, Rowohlt, 1952.

Cc18. **Petit, G., & J. Théodoridès.** 1962. *Histoire de la zoologie des origines à Linné*. Paris, Hermann. [xii]+360 pp. & 22 pl. *Histoire de la pensée* no. 8.

Chronologies

Cc19. **Koller, G.** 1949. *Daten zur Geschichte der Zoologie: Zeittafel, Forscherliste, Artentabelle*. Bonn, Athenäum–Verlag. 64 pp.
Covers 1550 to 1915.

Cc20. **Rothschuh, K. E.** 1952. *Entwicklungsgeschichte physiologischer Probleme in Tabellenform*. Munich, Urban & Schwarzenberg. 122 pp.
Covers ca. 1600 to 1920.

BIBLIOGRAPHIES OF ZOOLOGY — GENERAL

For fuller lists of bibliographies see:

Cc21. **Madier, G.** 1958. Zoologie. In: L.-N. Malclès, ed. *Les sources du travail bibliographique*, vol. 3: Bibliographies spécialisées (sciences exactes et techniques), pp. 321–390. Geneva, E. Droz; Paris, Minard. *Re-pub.* 1965.

Cc22. **Sawyer, F. C.** 1955. Books of reference in zoology, chiefly bibliographical. *Journal of the Society for the Bibliography of Natural History* 3: 72–91.

Cc23. **Simon, H. R.** 1983. Research and publication trends in systematic zoology 1758–1970. *Berichte der Ökologische Aussenstelle Schluchtern* 11: 403 pp.

Cc24. **Smith, R. C., & R. H. Painter.** 1967. *Guide to the literature of the zoological sciences.* 7th ed. Minneapolis, MN, Burgess Publishing Co. [ii]+xiv+238 pp. ••For later editions *see* Ca90.

• A selection of major works:

Cc25. **Agassiz, J. L. R.** 1848–54. *Bibliographia zoologiae et geologiae: A general catalogue of all books, tracts, and memoirs on zoology and geology.* Corrected, enlarged and ed. by H. E. Strickland. London, Ray Society. 4 vols. Ray Society publications nos. 13, 18, 22 & 26. *Re-pub.* New York, Johnson Reprint Co., 1968. ••Author list, no index. List of periodicals and serials indexed.

Cc26. **Brückmann, F. E.** 1743. *Bibliotheca animalis oder Verzeichnis derer meisten Schriften, so von Thieren und deren Theilen handeln, [etc.].* Wolfenbüttel, Meissner. 277+[56] pp. *Continued* ___. 1747. ___*: Bibliotheca animalis continuatio [etc.].* Wolfenbüttel. 178 pp. ••The first general zoological bibliography.

Cc27. **Engelmann,W.** 1846. *Bibliotheca historico-naturalis. Verzeichniss der Bücher über Naturgeschichte welche in Deutschland, Scandinavien, Holland, England, Frankreich, Italien und Spanien in den Jahren 1700 – 1846 erschienen sind.* 1. Band. *Bücherkunde, Hülfsmittel, allgemeine Schriften, vergleichende Anatomie und Physiologie, Zoologie, Palaeontologie.* Leipzig, W. Engelmann. ix+786 pp. *Re-pub.* Weinheim, J. Cramer; Codicote, Wheldon & Wesley, 1960. *Historiae naturalis classica* vol. 14. ••*Continued* **J. V. Carus & W. Engelmann.** 1861. *Bibliotheca zoologica. The literature of zoology, which has appeared in periodicals, transactions etc. and of the books published from 1846 to 1860.* Leipzig, W. Engelmann; London, Williams & Norgate. 2 vols.; **O. W. Taschenberg.** 1886–1923. *Bibliotheca zoologica II: Verzeichniss der Schriften über Zoologie welche in den periodischen Werken enthalten und vom Jahre 1861–1880 selbständig erscheinen sind, mit Einschluss der allgemein-naturgeschichtlichen periodischen und palaeontologischen Schriften.* Leipzig, W. Engelmann. 7 vols. These three works include ca. 10,000, 40,000 & 160,000 entries respectively. Vol. 7 forms a supplement to the first 6 vols.

Cc28. **Fischer de Waldheim, G.** 1834. *Bibliographia palaeonthologica animalium systematica.* Moscow, Typis Universitatis Caes. viii+414 pp.

Cc29. **Gronovius, L. T.** 1760. *Bibliotheca regni animalis atque lapidei, seu recensio auctorum et librorum, qui de regno animali et lapides ... in usum naturalis tractant, historiae studiosorum.* Leiden, the Author. [viii]+326 pp.

Cc30. **Kingsley, J. S.** 1882. *The naturalist's assistant: A handbook for the collector and student. With a bibliography of fifteen hundred works necessary for the systematic zoologist.* Boston, MA, S. E. Cassino. ix+228+(8) pp.

Cc31. **Reuss, J. D.** 1801–21. *Repertorium commentationum a societatibus litterariis editarum.* Göttingen, Dieterich. 16 vols. *Re-pub.* New York, B. Franklin, 1961. *Burt Franklin bibliography & reference series* no. 29. ••Vol. 1, tom. I: Historia naturalis, generalis et zoologia, 1801. 574 pp.

Cc32. **Rensch, B.** 1930. Tiergeographie (Für die Literatur von 1908–1930). *Geographisches Jahrbuch* 45: 51–132. *Continued* ___. 1932–37. [Literature of 1931–34]. *Zoogeographica* vols. 1–3. Further continued in *Fortschritte der Zoologie.*

Cc33. **Sherborn, C. D.** 1902. Bibliography, being a rough list of books referred to in the compilation of this index. In: *Index animalium sive index nominum quae ab A. D. MDCCLVIII (1758) generibus et speciebus animalium imposta sunt ... sectio prima a kalendis Januariis MDCCLVIII (1758) usque ad finem Decembris MDCCC (1800)*, pp. xi–lvi. Cambridge. lix+1195 pp.

Cc34. ___. 1922–32. Bibliography, being a rough list of books referred to in the compilation of this index. (And: Additions to bibliography.) In: *Index animalium sive index nominum quae ab A. D. MDCCLVIII (1758) generibus et speciebus animalium imposta sunt ... sectio secunda a kalendis Ianuariis, MDCCCI usque ad finem Decembris, MDCCCL*, pt. 1: xv–cxxxi; pt. 29: cxxxiii–cxlvii. London, British Museum (Natural History). ••Both these lists include annotations on correct dates of publication.

Cc35. **Sims, R. W., & D. Hollis**, eds. 1981. *Animal identification: A reference guide.* London, British Museum (Natural History); Chichester, New York, J. Wiley & Sons. 3 vols. ••Vol. 1: Marine and brackish water animals; vol. 2: Land and freshwater animals (not insects); vol. 3: Insects.

See also Library cats. Bc433–449 & Periodicals lists Bc641–643.

Dissertations

Cc36. **Moore, J. L.** 1970. *Bibliography of wildlife theses, 1900–1968.* Los Angeles, CA, Biological Information Service. xii+[599] pp.

Periodical bibliographies

Cc37. ***Anatomischer Anzeiger;*** *Litteratur* [Section: Internationale Bibliographie der Anatomie des Menschen und Wirbeltiere – title varies]. <1895+> Jena, G. Fischer, 1896–1927.

Cc38. ***Archiv für Naturgeschichte.*** <1834–1919> Berlin, Nicolai'sche Buchhandlung, vols. 1–89, 1835–1923. Vol. 1 has no bibliography.

Cc39. ***Årsberättelse (later Berättelse) om framstegen i insekternas, myriapodernas och arachnidernas naturalhistoria for ...*** <1847–56> Stockholm, Kongliga Svenska Vetenskaps–Akademien, 1847–56. 5 vols.

Cc40. ***Årsberättelse om zoologens framsteg under åren*** ... <1840–46> Stockholm, Kongliga Svenska Vetenskaps–Akademien, 1840–46. 3 vols. Continued as *Årsberättelse (later Berättelse) om framstegen i insekternas, myriapodernas och arachnidernas naturalhistoria for ...* , *Berättelse om framstegen i molluskernas, crustaceernas och de lägre skelettlösa djurens naturhistoria under åren ...*; and *Berättelse om framstegen i vertebrerade djurens naturalhistoria och ethnografien under åren ...*

Cc41. ***Årsberättelser om nyare zoologiska Arbeten och upptäckter.*** <1826–40> Stockholm, Kongliga Svenska Vetenskaps–Akademien, 1826–42. 12 vols. Continued as *Årsberättelse om zoologens framsteg under åren ...*

Cc42. ***Årsberättelser om vetenskapernas framsteg.*** <1821–25> Stockholm, Kongliga Svenska Vetenskaps–Akademien, 1822–25. 5 vols. Continued as *Årsberättelser om nyare zoologiska Arbeten och upptäckter.*

Cc43. ***Berättelse om framstegen i molluskernas, crustaceernas och de lägre skelettlösa djurens naturhistoria under åren*** *1845–1849 [etc.].* <1845–49> Stockholm, Kongliga Svenska Vetenskaps–Akademien, 1852.

Cc44. ***Berättelse om framstegen i vertebrerade djurens naturalhistoria och ethnografien under åren*** *1845–50 [etc.].* <1845–50> Stockholm, Kongliga Svenska Vetenskaps–Akademien, 1853.

Cc45. ***Bibliographia zoologica*** *(diario 'Zoologischer Anzeiger' adnexa).* <1896–1934> Leipzig, W. Engelmann (*later* Zurich, Concilium bibliographicum), vols. 1–43, 1896–1934. ••For background information *see* **H. H. Field.** 1905. A bibliography of publications relating to the Concilium Bibliographicum 1893–1904. *Annotationes Concilii Bibliographici* 1: 19–27. The *Annotationes* (vols. 1–8, 1905–12), contains various original and reprinted articles about the Concilium.

Cc46. ***Biological abstracts,*** *a comprehensive abstracting and indexing journal of the world's literature in pure and applied biology, [etc.].* <1927+> Philadelphia, PA, Union of American Biological Societies (etc.), vol. 1+, 1927+.

Cc47. ***Fortschritte der Zoologie, neue Folge.*** <1935+> Jena, G. Fischer, vol. 1+, 1937+.

Cc48. ***Records of zoological literature.*** <1864–69> London, J. Van Voorst, vols. 1–6, 1864–69. Continued as *Zoological record.*

Cc49. ***Referativnyi zhurnal. Biologiya.*** <1954+> Moscow, Akademiya Nauk SSSR, Institut Nauchnoi Informatsii, 1954+.

Cc50. ***Zoological record,*** *[etc.].* <1870+> London, Zoological Society of London, vol. 7+, 1870+. Comprises an annual set of bibliographies [from 1876+ separately paginated] devoted to the major animal groups, viz: *General subjects* (later *Comprehensive zoology*) <1886+>, *Protozoa* <1865+>, *Spongozoa* (later *Porifera*) <1876+> [previously contained in *Protozoa], Coelenterata* <1865+>, *Echinodermata* <1864+>, *Vermes* <1870+>, *Molluscoida* <1864–83>, *Brachiopoda* <1884+> [previously contained in *Molluscoida], Polyzoa* (later *Bryozoa*) <1884–1905, 1937+> [previously contained in *Molluscoida,* from 1906–36 contained in *Brachiopoda*], *Mollusca* <1864+>, *Arthropoda* <1906–20>, *Crustacea* <1864+>, *Trilobita* <1938+> [previously contained in *Arachnida*], *Arachnida* <1864+>, *Myriapoda* <1864–1905> [from 1906+ contained in *Arachnida*], *Insecta* <1964+> [later divided into subsections], *Tunicata* (later *Protochordata*) <1884+> [formerly contained in *Molluscoida*], *Vertebrata* <1906–20>, *Pisces* <1864+>, *Amphibia* <1953+> [previously contained in *Reptilia*], *Reptilia* <1864+>, *Aves* <1864+>, *Mammalia* <1864+>, *List of new genera, etc.* <1870+>. For background information *see* **G. D. R. Bridson.** 1968. The Zoological record – a centenary appraisal. *Journal of the Society for the Bibliography of Natural History* 5: 23–34.

Cc51. ***Zoologischer Anzeiger,*** *[Litteratur].* <1891–95> Leipzig, W. Engelmann, vols. 14–18, 1891–95. Continued as *Bibliographia zoologica (diario 'Zoologischer Anzeiger' adnexa).*

Cc52. ***Zoologischer Berichte*** *im Auftrage der Deutschen Zoologischen Gesellschaft.* <1922–44> Jena, G. Fischer, vols. 1–55, 1922–44.

Cc53. ***Zoologischer Jahresbericht,*** *herausgegeben von der Zoologischen Station zu Neapel.* <1879–1913> Leipzig, W. Engelmann (*later* Berlin, R. Friedländer & Sohn), 1879–1913, 35 vols.

Cc54. ***Zoologisches Centralblatt.*** <1894–1911> Leipzig, W. Engelmann, vols. 1–18, 1894–1911.

REGIONAL HISTORIES OF ZOOLOGY

European (& Mediterranean)

Cc55. **Appel, T. A.** 1987. *The Cuvier–Geoffroy debate: French biology in the decades before Darwin*. New York, Oxford Univ. Pr. 305 pp. *Monographs on the history & philosophy of biology*.

Cc56. **Bogdanov, A. P.** 1888–92. Materialy dlya istorii nauchnoi i prikladnoi deyatel'nosti v Rossii po zoologii i soprikasayushchimsya s neyu otraslyami znaniya, preimushchestvenno za poslednee tridtsatipyatiletie (1850–1888). (Materials for the history of scientific and applied activities in Russia in zoology and related branches of knowledge especially during the last thirty-five years etc.) *Izvestiya Imperatorskogo Obshchestva Lyubitelei Estestvoznaniya, Antropologii i Etnografii pri Imperatorskom Moskovskom Universitete* 55(3): iv+222 pp. & 12 pl., 1888; *Ibid.* 57(4): 304 pp. & pl. 13–24, 1889; *Ibid.* 70(6): 310 pp. & pl. 25–37, 1891; *Ibid.* 71(bis): 126 pp. & pl. 33–41, 1892.

Cc57. **Broch, H.** 1954. *Zoologiens historie i Norge til annen verderskrig*. Oslo, Akademisk Vorlag. 158 pp. & 12 pl.
The history of zoology in Norway up to the second World War.

Cc58. **Brzek, G.** 1947–55. Historia zoologii w Polsce do r. 1918. [History of zoology in Poland till 1918]. Part I. Poczatki zoologi w Polsce i rozwoj tej nauki do konca XVIII wieku. *Annales Universitatis Mariae Curie–Sklodowska*, Sect. C, supp. 2: vii+112; ___. Part 2. Osrodet Wilenski. *Ibid.*, Sect. C, supp. 2a: 113–253, 1947; ___. Part III. Materialy do historii osrodka Warszawskiego. *Ibid.*, Sect. C, supp. 7: xii+455 pp.
Text in Polish with English summaries.

Cc59. ___. 1970. Materialy do historii zoologii w Polsce w latach 1860–1918. [Materials for the history of zoology in Poland in the years 1860–1918]. *Studia i materialy z dziejow nauki Polskiej* B 19: 13–80.

Cc60. **Fedorowicz, Z.** 1966. Zarys historii zoologii w Polsce do 1860 r. (History of zoology in Poland to 1860). *Ibid.* B 10: 9–43.

Cc61. ___. 1968. Zoologia w Gdansku w stuleciach XVII i XVIII. (La zoologie à Gdansk aux XVII et XVIII siècles). *Memorabilia zoologica* 19: 119 pp.
Summary in French.

Cc62. ___. 1970. Materialy do historii zoologii w Polsce w latach 1914–1939. [Materials for the history of zoology in Poland in the years 1914–1939]. *Studia i materialy z dziejow nauki Polskiej* B 19: 81–145.

Cc63. ___. 1970. Dotychczasowe badania nad historia zoologii in Polsca. (The present state of research on the history of zoology in Poland). *Kwartalnik historii nauki i techniki* 15: 579–592.

Cc64. **Imperatorskaya Akademiya Nauk (Académie Impériale des Sciences de St. Pétersbourg).** 1879. *Bericht über die Fortschritte, welche die zoologischen Wissenschaften den von der Kaiserlichen Akademie der Wissenschaften zu St. Petersburg von 1831 bis 1879 herausgegebenen Schriften verdanken. Ein Beitrag zur neuern Geschichte der zoologischen Wissenschaftlichen in Russland.* By J. F. Brandt. St. Petersburg. 213 pp. *Beiträge zur Kenntniss des Russischen Reiches und der angräzendern Länder Asiens [etc.]* Folg. 1, Bd. 1.

Cc65. **Kozuchowski, J.** 1963. Zarys historiografii zoologii na ziemiach polskich. (Précis de l'historiographie de la zoologie en territoire polonais.] *Memorabilia zoologica* 11: 1–112.
Summary in French.

Cc66. **Leclercq, J.** 1959. *Perspectives de la zoologie européenne: Histoire problèmes contemporains.* Gembloux, Duculot; Paris, Libraire Agricole de la Maison Rustique. 163 pp.

Cc67. ___, **& P. Dagnellie.** 1966. *Perspectives de la zoologie européenne: Une sondage d'opinions des zoologistes.* Gembloux, J. Duculot. 215 pp.

Cc68. **Levin, V. L.** 1957. Zoologiya v Russkikh zhurnalakh 18 v. [Zoology in Russian journals of the 18th century]. *Trudy Institut Istorii Estestvoznaniya i Tekhniki* 16: 201–238.

Cc69. ___. 1960. Materialy po istorii zoologii v Russkikh universitetakh dosovetskogo perioda. [Materials towards the history of zoology in Russian universities of the pre-Soviet period]. *Ibid.* 31: 403–434.

Cc70. ___. 1960. Materialy po istorii zoologii v Russkikh universitetakh. Zoologiya v trudakh obshchestv estestvoispytatelei. [Materials towards the history of zoology in Russian universities. Zoology in the work of the naturalists' society]. *Ibid.* 41: 391–413.

Cc71. **Novikov, P. A.** 1961. Faunisticheskie i zoogeograficheskie issledovaniya v Rossii v pervoi polovine XIX veka. [Faunistic and zoogeographical investigations in Russia in the second half of the 19th century]. *Ibid.* 36: 294–304.

Cc72. **Ritvo, H.** 1985. Learning from animals: Natural history for children in the eighteenth and nineteenth centuries. *Children's literature* 13: 72–93.

Cc73. ___. 1985. Animal pleasures: Popular zoology in eighteenth– and nineteenth–century England. *Harvard Library bulletin* 33: 239–279.

Cc74. **Sacarrão, G. da F.** 1953. As origens dos estudos zoológicos portugueses. *Naturália* 4 (1): 1–40.

Cc75. **Saldanha, L.** 1990. The Forbes' azoic theory and the Portuguese zoologists of the 19th century. In: W. Lenz & M. Deacon, eds. *Ocean sciences: Their history and relation to man. Proceedings of the 4th International Congress on the History of Oceanography Hamburg [etc.]*, pp. 166–173. Hamburg, Bundesamt für Seeschiffahrt und Hydrographie. *Deutsche hydrographische Zeitschrift*, Ergänzungsheft, Reihe B, no. 22.

Cc76. **Scheele, I., & C. Hünemörder.** 1990. On the contribution of zoologists from Hamburg to marine biology. In: W. Lenz & M. Deacon, eds. *Ocean sciences: Their history and relation to man. Proceedings of the 4th International Congress on the History of Oceanography Hamburg [etc.]*, pp. 102–108. Hamburg, Bundesamt für Seeschiffahrt und Hydrographie. *Deutsche hydrographische Zeitschrift*, Ergänzungsheft, Reihe B, no. 22.

Cc77. **Seurat, L.–G.** 1930. *L'éxploration zoologique de l'Algérie de 1830 à 1930.* Paris, Masson & Cie. 708 pp. & 16 pl. *Collection centenaire de l'Algérie, 1830–1930: Études scientifiques.*
Notice sur les zoologistes et les explorateurs qui ont contribué à faire connaître les animaux de l'Algérie, pp. 689–694.

Cc78. **Szilady, Z.** 1927. *Die Geschichte der Zoologie in Ungarn.* Debrecen, Kertesz. 116 pp.

Cc79. **Tembrock, O.** 1959. Zur Geschichte der Zoologie in Berlin. *Wissenschaftliche Zeitschrift der Humboldt Universität zu Berlin, mathematisch–naturwissenschaftliche Reihe* 8: 185–196.

Cc80. **Uschmann, G.** 1960. *Geschichte der Zoologie und der zoologischen Anstalten in Jena 1779–1919.* Jena, G. Fischer. xv+250 pp. & 69 pl.

See also Ba295 Ba300 Bb51(van Beneden) Bb797–802 Bb806 Bb808–811 Bb813–821 Ca184(Laverack) & Cc1030–1032.

North American

Cc81. **Benson, K. R., R. Rainger, & J. Maienschein**, eds. 1991. *The expansion of American biology*. New Brunswick, NJ, Rutgers Univ. Pr. xiii+357 pp.
Commissioned by the American Society of Zoologists to celebrate its centennial (1889–1989). Twelve articles on twentieth–century developments by various authors. *Includes* **R. Rainger.** Introduction, pp. 1–22; **M. B. Ogilvie.** The "New Look" women and the expansion of American zoology: Nettie Maria Stevens (1861–1912) and Alice Middleton Boring (1883–1955), pp. 52–79; **L. F. Laporte.** George G. Simpson, paleontology, and the expansion of biology, pp. 80–106. References to some others are entered in the relevant sections herein.

Cc82. **Borland, H., & the National Wildlife Federation.** 1975. *The history of wildlife in America*. New York, Arch Cape Pr. 208 pp.

Cc83. **Coe, W. R.** 1918. A century of zoology in America. *American journal of science* (4) 46: 355–398. *Also pub. in* E. S. Dana et al. 1918. *A century of science in America with special reference to the American Journal of Science 1818–1918 [etc.]*, pp. 391–438. New Haven, CT, Yale Univ. Pr.

Cc84. **Davis, G. P., N. B. Carmony, & D. E. Brown**, eds. 1982. *Man and wildlife in Arizona: The American exploration period, 1824–1865*. Phoenix, AZ, Arizona Game & Fish Department. xvii+235 pp.

Cc85. **Dexter, R. W.** 1979. The impact of evolutionary theories on the Salem group of Agassiz zoologists (E. W. Morse, A. Hyatt, A. S. Packard, F. W. Putnam). *Essex Institute historical collections* 115: 144–175.

Cc86. **Dunlap, T. R.** 1988. *Saving America's wildlife: Ecology and the American mind, 1850–1990*. Princeton, NJ, Princeton Univ. Pr. xvi+222 pp. *Re-pub.* 1991.

Cc87. **Dymond, J. R.** 1949. Zoölogy. In: Royal Canadian Institute. *The Royal Canadian Institute centennial volume, 1849–1949*, pp. 108–120. Ed. W. S. Wallace. Toronto.

Cc88. **Evermann, B. W.** 1917. A century of zoology in Indiana, 1816–1916. *Proceedings of the Indiana Academy of Science* 1916: 189–224.

Cc89. **Maienschein, J.** 1991. *Transforming traditions in American biology, 1880–1915*. Baltimore, MD, Johns Hopkins Univ. Pr. xii+366 pp.
Studies especially the work of E. G. Conklin, E. B. Wilson, T. H. Morgan and R. Harrison.

Cc90. **Matthiessen, P.** 1987. *Wildlife in America*. Rev. ed. New York, Elisabeth Sifton Books, Viking. 332 pp. & 24 pl.

Cc91. **May, R.–M.** 1963. *Les zoologistes français aux Etats–Unis, au début du XIXe siècle*. Paris, Université de Paris. 32 pp. *Les Conférences du Palais de la Découverte* série D, no. 90.

Cc92. **Packard, A. S.** 1876. A century's progress in American zoology. *American naturalist* 10: 591–598.

Cc93. **Rainger, R., K. R. Benson, & J. Maienschein**, eds. 1988. *The American development of biology*. Philadelphia, PA, Univ. of Pennsylvania Pr. xii+380 pp.
Commissioned by the American Society of Zoologists to celebrate its centennial (1889–1989). Ten articles by various authors. *Includes* **K. R. Benson.** From museum research to laboratory research: The transformation of natural history into academic biology, pp. 49–83. References to others are entered in the relevant sections herein.

Cc94. **Spencer, L. T., et al.** 1986. Naturalists and natural history institutions of the American West. *American zoologist* 26: 295–384.
Includes **L. T. Spencer.** A preview, pp. 295–297; **C. E. Quinn.** A zoologist's view of the Lewis and Clark expedition, pp. 299–306; **R. G. Beidleman.** The 1820 Long expedition, pp. 307–313; **E. B. Reed.** Limnological aspects of some early explorations in the western United States, pp. 315–320; **L. T. Spencer.** Naturalists of the Pacific shore: Early

explorer/naturalists of California, pp. 321–329; **J. A. Moore.** Zoology of the Pacific railroad surveys, pp. 331–341; **M. Foster.** Ferdinand Vandeveer Hayden as naturalist, pp. 343–349; **K. R. Benson.** The Young Naturalists' Society and natural history in the Northwest, pp. 351–361; **D. M. Armstrong.** Edward Royal Warren (1860–1942) and the development of Coloradan mammalogy, pp. 363–370; **L. T. Spencer.** Filling in the gaps: A survey of 19th century institutions associated with the exploration and natural history of the American West, pp. 371–380; **K. R. Benson.** Concluding remarks: American natural history and biology in the 19th century, pp. 381–384.

Cc95. **Wenrich, D. H.** 1955. Some early Philadelphia zoologists. *Proceedings of the Pennsylvania Academy of Science* 29: 22–35.

Cc96. **Wetherbee, D. K.** 1985. *Early zoological exploration of Central America for new vertebrate species.* Boston, MA, the Author. 69 pp.

Cc97. ___. 1985. *Zoological exploration of Haiti for endemic species.* Shelburne, MA, the Author. 557 ff.

Cc98. ___. 1985. *Zoological exploration of Jamaica for endemic species.* Shelburne, MA, the Author. 214 ff.

Cc99. ___. 1985. *Zoological exploration of Panama for endemic species.* Boston, MA, the Author. 246 pp.

Cc100. ___. 1985. *The historical development of comparative zoology in the West Indies.* Santo Domingo, Dominican Republic. 75 pp.

Cc101. ___. 1985. *Zoological exploration of the Bahamas for endemic species.* Shelburne, MA, the author. i+54 ff. *Re-pub.* Cologne, Bohlau Verlag, 1986. *Universität zu Köln, Studien in Geschichte* no. 3.

Cc102. ___. 1987. *Further contributions to the history of zoology in Hispaniola.* Restauracion, Dominican Republic, the Author. 114 pp.

See also Bb789 Bb792–795 Bb803–805 Bb822.

South & Central American (& Caribbean)

Cc103. **Cutright, P. R.** 1940. *The great naturalists explore South America.* New York, Macmillan. xii+340 pp.
A brief general introduction is followed by chapters on individual animals and the story of their discovery.

Cc104. **MacDonagh, E. J.** 1957. Para una historia de la zoologia Argentina: II. Nuevos datos sobre Charles Darwin en su viaje Argentino. *Ciencia e investigación* 13(2): 51–69.

Asian

Cc105. **Cranbrook, Earl of, & C. M. U. Leh.** 1983. A history of zoology in Sarawak. *Sarawak Museum journal* 32: 15–33, & pl. vi–x.

Cc106. **Holthuis, L. B., & T. Sakai.** 1970. *Ph. F. von Siebold and Fauna Japonica: A history of early Japanese zoology.* Tokyo, Academic Pr. of Japan. [xx]+323 pp. & 32 pl.

Cc107. **Kolosov, A. M.** 1938. Istoriya faunisticheskikh issledovanii Altaya. (Obzor literatury po pozvonochnym). *Trudy Altaiskogo Gosudarstvennogo Zapovednika* 1: 327–390.

Cc108. **Narzikulov, M. N.** 1985. *Zoologicheskaya nauka Tadzhikistana za 60 let. [Zoological science in Tadzhikistan over the last 60 years].* Dushanbe, Izd. Donish. 245 pp.

Cc109. **Srinivasa Rao, H.** 1957. History of our knowledge of the Indian fauna through the ages. *Journal of the Bombay Natural History Society* 54: 251–280, & 2 pl.

Cc110. **Sultanova, G. S., & L. V. Persianovoi.** 1982. *Zoologicheskie issledovaniya u srednei Asii (1820–1975). [Zoological research in central Asia (1820–1975)]*. Tashkent, Fan. 240 pp.

Australasian

Cc111. **Andrews, J. R. H.** 1987. *The southern ark: Zoological discovery in New Zealand, 1769–1900.* London, Century. xii+237 pp. Also Honolulu, HI, Univ. of Hawaii Pr.

Cc112. **Dugan, K. G.** 1987. The zoological exploration of the Australasian region and its impact on biological theory. In: N. Reingold & M. Rothenberg, eds. *Scientific colonialism: A cross–national comparison*, pp. 79–100. Washington, DC, Smithsonian Institution Pr.

Cc113. **Fell, H. B., et al.** 1953. *The first century of New Zealand zoology 1769–1868, comprising abstracts and extracts from the early work on the New Zealand fauna.* Wellington, N.Z., Victoria Univ. College. 88 pp.

Cc114. **Fletcher, J. J.** 1901. On the rise and early progress of our knowledge of the Australian fauna. *Report, Australian Association for the Advancement of Science* 7: 69–104.

Cc115. **Whitley, G. P.** 1970. *Early history of Australian zoology.* Sydney, N.S.W., Royal Zoological Society of New South Wales. 75pp.; *Continued* ___. 1975. *More early history of Australian zoology.* Sydney, N.S.W., [etc.]. 91 pp.

See also Bb807.

African

Cc116. **Brink, P.** 1955. Swedish exploration of South African animal life during 200 years. In: B. Hanström, P. Brink, & G. Rudebeck, eds. *South African animal life: Results of the Lund University expedition in 1950–51*, vol. 1: 11–61. Stockholm, Almqvist & Wiksell.

Cc117. **Hoppe, B.** 1990. Naturwissenschaftliche und zoologische Forschungen in Afrika während der deutschen Kolonialbewegung bis 1914. *Berichte der Wissenschaftsgeschichte* 13: 193–206.

Cc118. **Rookmaaker, L. C.** 1989. *The zoological exploration of southern Africa 1650–1790.* Rotterdam, Brookfield, VT, A. A. Balkema. xii+368 pp.

Part 1: Reports on South African animals; Part 2: The main explorers; Part 3: Species accounts; Part 4: Sources.

See also Bb812 Cc1028.

BIBLIOGRAPHIES OF REGIONAL ZOOLOGY

European (& Mediterranean)

Cc119. **Allodiatoris, I.** 1966. *Bibliographie der Zoologie im Karpatenbechen 1900–1925.* Budapest, Akadémiai Kiadó. 574 pp.

Cc120. **Aytug, N., & A. Cakman.** 1972. *Türkiye fauna biblyografyasi, 1843–1968.* Ankara, Türdok. 59 pp. *Türdok biblyografya serisi* no. 6.

Cc121. **Bianconi, G. G.** 1853–54. *Repertorio italiano per la storia naturale. Repertorium italicum complectens zoologiam, mineralogiam, geologiam et palaeontologiam.* Bononiae, the Author. 2 vols.

Cc122. **Blasius, W.** 1891. *Die faunistische Litteratur Braunschweigs und der Nachbargebiete mit Einschluss des ganzen Harzes.* Braunschweig. 239 pp. *Jahresbericht des Vereins für Naturwissenschaft zu Braunschweig* no. 6.

Cc123. [Not used.]

Cc124. **Cavanna, G.** 1880. *Elementi per una bibliografia italiana intorno all'idrofauna, agli allevamenti degli animali acquatici e alla pesca.* Florence, Coi Tipi dell'Arte della Stampa. viii+170 pp.

Cc125. **Feliksiaka, S., et al.** 1954. *Przewodnik dla poslugujacych sie pismiennictwem do oznaczania zwierzat karjowych. [Literature guide to Polish animals].* Warsaw, Panstwowe Wydawnictwo Naukowe. 199 pp.

Cc126. **Feliksiakowa, J.** 1969. *Polska bibliografia zoologiczna za lata 1945–1954.* Warsaw, Polskiej Akademii Nauk, Komitet Zoologiczny. xxii+558 pp. ••*See also* ___. 1970. Na marginesie Polskiej Bibliografii Zoologicznej za lata 1945–1954. [On the margin of Polish Zoological Bibliography for the years 1945–1954.] *Przeglad zoologiczny* 14: 178–180.

Cc127. **Gosch, C. C. A.** 1870–78. *Udsigt over Danmarks zoologiske Litteratur med en inledende fremstilling af de videnskabelige grundsætninger for naturvidenskabens især zoologiens studium.* Copenhagen, O. Schwarz's efterfölger (E. Jespersen). 3 vols in 4; *Continued* **S. Dahl.** 1910. *Bibliotheca zoologica Danica, 1876–1906.* Copenhagen, Lybeckers. xxii+262 pp.; **P. & S. Borella.** 1986 [1987]. *Dansk zoologisk bibliografi ... 1907–1945.* Copenhagen, Univ. Library, Scientific & Medical Department. 2 vols. *Acta historica scientiarum naturalium et medicinalium* vols. 36–37.

Cc128. **Gusenleitner, F.** 1983. Systematische Aufstellung der Evertebraten–Literatur Oberosterreichs 1781–1982. *Linzer biologische Beiträge* 15: 3–266.

Cc129. **Jakubski, A., & M. Dyrdowskiej.** 1927–28. *Bibljografja fauny Polskej do roku 1880.* Cracow, Polska Akademija Umiejetnosci. 2 vols. *Prace monograficzne Komisji Fizjograficznej, Cracovie* vols. 3 & 4.

Cc130. **Kanellis, A., & Ch. Hatzissarantos.** 1950. Bibliographia faunae Graecae (1800–1950). *To Vouno, journal of the Hellenic Alpine Club* 1949/50: 78 pp.; *Continued* ___. 1960. Bibliographia faunae graecae, 1950–1960 und Nachträge. *Verhandlungen der Zoologisch–Botanischen Gesellschaft in Wien* 100: 96–105.

Cc131. **Köppen, F. T.** 1905–08. *Bibliotheca zoologica Rossica. Litteratur über die Thierwelt Gesammtrusslands bis zum Jahre 1885 incl.* Band 1, 2(1–2): Allgemeiner Theil. St. Petersburg, Imperatorskaya Akademiya Nauk. 2 vols. ••All published.

Cc132. **Kumerlöwe, H.** 1986. *Bibliographie der Saügetiere und Vögel der Türkei (rezente Fauna) unter Berücksichtigung der benachbarten Gebiete und mit Hinweisen auf weiterführendes Schrifttum.* Bonn, Zoologisches Forschungsinstitut und Museum Alexander Koenig. 132 pp. *Bonner zoologische Monographien* no. 21.

Cc133. **Nardo, G. D.** 1875–76. Bibliografia cronologica della fauna delle provincie venete e del mare Adriatico. *Atti del Reale Istituto Veneto di Scienze, Lettere ed Arti* 1874/75: 199–210, 305–317, 459–468, 539–567, 711–730; 1876/77: 169–172. *Also pub. separately* Venice, 1877. 86 pp.

Cc134. **Neu, W., & H. Kumerlöwe.** 1939. *Bibliographie der zoologischen Arbeiten über die Türkei und ihre Grenzgebiete.* Leipzig, Harrasowitz. xii+62 pp. *Continued* ___. 1940. ___: Supplement. *Folia zoologica et hydrobiologica* 10: 390–397.

Cc135. **Pax, F., & H. Tischbierek.** 1930. *Bibliographie der Schlesischen Zoologie.* Breslau, Priebatsch. xii+520 pp. *Schlesische Bibliographie* no. 5; *Continued* **F. Pax.** [1957.] *Bibliografia zoologi Slaska: Czesc (1935–1950). Bibliography of the Silesian zoology: Part III (1928–1935).* Breslau, [etc.]. x+178 pp.; **Panstwowe Wydawnictwo Naukowe.** 1957. *Bibliografia zoologii*

Slaska: Czesc III (1935–1950). Wroclaw. xvi+184 pp.

Cc136. **Polska Akademija Nauk, Oddzial w Krakowie, Komisja Nauk Biologicznych.** 1975. *Katalog Polskiej literatury biologicznej. Catalogue of Polish biological literature. Tom. XX za lata 1935–1939*. Wroclaw, [etc.]. x+226 pp.; ___. 1963. *Katalog Polskiej literatury biologicznej. Catalogue of Polish biological literature.* Tom. I (XXII): *Za lata 1950–1954*. Cracow. 324 pp.; ___. 1970. *Katalog Polskiej literatury biologicznej. Catalogue of Polish biological literature.* Tom. II (XXIII): *Za lata 1955–1959*. Cracow. 2 vols.

Cc137. **Reuter, O. M., & A. Luther.** 1909. Bibliotheca zoologica Fenniae. *Acta Societatis pro Fauna et Flora Fennica* 24: ixii+1–144, ix+145–361 pp. *Continued* **H. Lindberg.** 1937. Bibliotheca zoologica Fenniae. Opera annorum 1901–1930. *Ibid.* 59: xi+438 pp.; ___. 1953. ___, 1931–1940. *Ibid.* 70: ix+284 pp.

Cc138. **Rikhter, A. A.** 1948. Bibliografiya nauchno-issledovatel'skikh rabot po zoologii v Armenii za 25 let (1920–1945). *Zoologicheskii Sbornik, Zoologicheskii Institut, Akademiya Nauk Armyanskoi SSR* 5: 189–212.

Cc139. **Spuris, E.** 1958. Faunisticheskaya i zooekologicheskaya literatura po Latviiskoi SSR s 1945 po 1956 god. *Trudy Instituta Biologii, Akademiya Nauk Latviiskoi SSR* 5: 291–323.

Cc140. **Urbanski, J.** 1958. Materialy do bibliografii zoologicznej ziemi Lubuskiej i Pomorza zachodniego oraz pogranicznych terenow zachodnich. Czesc 1. *Badania fizjograficzne nad Polska zachodnia, zool.* 4: 293–407.

Cc141. **Zavattari, E.** 1939. Le nostre conoscenze zoologiche sulla Libia sino al 1938. *Annali del Museo Libico di Storia Naturale* 1: 55–88; *Continued* ___. 1953. Bibliografica zoologica della Libia per il decennio 1941–1951. *Ibid.* 4: 203–206.

Cc142. **Zschokke, F.** 1897. *Seenfauna*. Berne, Centralkommission für Schweizerische Landeskunde. [x]+24 pp. *Bibliographie der schweizerischen Landeskunde* IV.6.2.

North American

Cc143. **Bick, G. H.** 1954. A bibliography of the zoology of Louisiana. *Proceedings of the Louisiana Academy of Sciences* 27: 5–48.

Cc144. **Gerrick, D.** 1968. A contribution toward a bibliography of Ohio zoology. *Bulletin of the Ohio Biological Survey* n. s. 2 (4): ix+273 pp.

Cc145. **Hotchkiss, N.** 1954. *Wildlife abstracts 1935–51: An annotated bibliography of the publications abstracted in the* Wildlife review, *nos. 1–66*. Washington, DC, United States Department of the Interior, Fish & Wildlife Service. iv+434 pp.

Cc146. **Jackson, H. H. T., et al.** 1949. Literature on the natural history of the Arctic region, with special reference to Alaska and Canada. *Wildlife leaflet, U. S. Fish & Wildlife Service* 317: 48 pp.

Cc147. **Langdon, F. W.** 1883. Bibliography of the Cincinnati fauna. *Journal of the Cincinnati Natural History Society* 6: 5–39.

Cc148. **McKinley, D.** 1960. A chronology and bibliography of wildlife in Missouri. Columbia, MO, Univ. of Missouri Library. 128 pp. *Bulletin of the University of Missouri bulletin* vol. 61 (13), *Library series* no. 26.

Cc149. **Osborn, H.** 1930. Bibliography of Ohio zoology. *Ohio Biological Survey bulletin* 4 (23): 351–410.

Cc150. **Shoup, C. S.** 1939. An annotated bibliography of the zoology of Tennessee and the Tennessee Valley region. *American midland naturalist* 21: 583–635.

Cc151. **Walker, E. M.** 1917. Bibliography of Canadian zoology (exclusive of entomology). *Transactions of the Royal Society of Canada* 10: 201–215.

South & Central American (& Caribbean)

Cc152. **Argentina, Consejo Federal de Inversiones.** 1963. *Fauna silvestre*. Vol. 1. By J. C. Godoy. Buenos Aires. *Evaluación de los recursos naturales de la Argentina*, primera etapa vol. 8. ••*Includes* Bibliografía sobre la fauna argentina, pp. 63–74; Notas y citas bibliográficas (sobre explotación de la fauna argentina), pp. 180–189; Apéndice bibliográfico por autores, pp. 263–299.

Asian

Cc153. **Chauhan, B. S.** 1954. A list of references to Indian zoology (excluding Insecta, fishes and helminths) published during the years 1938–1950. *Records of the Indian Museum* 51: 427–480.

Cc154. **Elera, C. de.** 1895–96. *Catálogo sistemático de toda la fauna de Filipinas.* Manila, Imprenta del Colegio de Santo Thomas. 3 vols.

Cc155. **Fotedar, D. N., & R. C. Bhagat.** 1984. *Jammu, Kashmir, and Ladakh, zoological studies: An annotated bibliography, 1848–1983.* New Delhi, International Books & Periodicals Supply Service. 133 pp.

Cc156. **Ivanova-Berg, M. N.** 1927. *Ukazatel' literatury po zhivotnomu miru Turkestana. Zhivotnyi mir, vrediteli zemledeliya, rybolovstvo, okhota i zhivotnovodstvo. [Bibliography of the literature of the animal kingdom of Turkestan].* Leningrad, Akademiya Nauk SSSR. iv+235 pp.

Cc157. **Nahif, A. A.** 1986. Bibliographie der zoologischen Literatur über Afghanistan. *Bonner zoologische Beiträge* 37: 311–339.

Cc158. **Pételot, L.** 1927. Complément au chapitre de la Bibliotheca indosinica relatif à la zoologie de l'Indochine Française. *Bulletin de l'École Française d'Extrême-Orient, Hanoi* 1927: 239–282.

Cc159. **Suvatti, C.** 1950. Bibliography. In: *Fauna of Thailand*, pp. 655–929. Bangkok, Department of Fisheries.

Australasian

Cc160. **Freed, D.** 1963. Bibliography of New Zealand marine zoology 1769–1899. *Bulletin, New Zealand Department of Scientific & Industrial Research* 148: 46 pp. & 12 pl.

African

Cc161. **Ghidini, G. M.** 1939–43. *Materiali per una bibliografia zoologica dell'Africa orientale italiana.* Rome. 8 pts. ••Originally pub. in *Rivista di biologia coloniale* vols. 2–6.

Cc162. **Goodwin, S. M.** 1972. *South African animal life: A bibliography.* Johannesburg, Univ. of Witwatersrand, Department of Bibliography. iii+66 pp.

Cc163. **Sprenger, D.** 1968. *Animal distribution in southern Africa: A bibliography.* Johannesburg, Univ. of Witwatersrand, Department of Bibliography, Librarianship & Typography. vii+31 ff.

SPECIAL TOPICS

Animal geography

Cc164. **Fichman, M.** 1977. Wallace: Zoogeography and the problem of land bridges. *Journal of the history of biology* 10: 45–63.

Cc165. **George, W.** 1969. *Animals and maps.* London, Secker & Warburg. 239 pp.

Cc166. ___. 1980. Sources and background to discoveries of new animals in the 16th and 17th centuries. *History of science* 18: 79–104.

Cc167. **Janssens, E.** 1952. Esquisse d'une histoire critique des théories zoogéographiques. *Bulletin de la Société Royale Belge de Géographie* 76: 33–50.

Cc168. **Schmidt, K. P.** 1955. Animal geography. In: California Academy of Sciences. *A century of progress in the natural sciences 1853–1953*, pp. 767–794. San Francisco, CA.

Ecology

Cc169. **Allee, W. C., & T. Park.** 1949. Ch. 2. Ecological background and growth before 1900. Ch. 3. First four decades of the twentieth century. In: W. C. Allee et al. *Principles of animal ecology*, pp. 13–43, 43–72. Philadelphia, PA, Saunders.

Cc170. **Chesnova, L. V.** 1970. *Ocherki po istorii ekologii. [Essays on the history of ecology].* Moscow, Izd. "Nauka." 291 pp., illus.

Cc171. **Colwell, R. K.** 1985. The evolution of ecology. *American zoologist* 25: 771–777.

Cc172. **Egerton, F. N.** 1968. Studies of animal populations from Lamarck to Darwin. *Journal of the history of biology* 1: 225–260.

Cc173. **Klaauw, C. J. van der.** 1936. Zur Geschichte der Definitionen der Ökologie, besonders auf Grund der Systeme der zoologischen Disziplinen. *Archiv für Geschichte des Medizin* 29: 136–177.

Cc174. **Novikov, G. A.** 1957. Istorii otechestvennoi ekologii nazemnykh pozvonochnykh zhivotnykh. [On the history of investigations into the ecology of terrestrial vertebrates in Russia]. *Trudy Institut Istorii Estestvoznaniya i Tekhniki* 16: 146–200.

Cc175. ___. 1980. *Ocherk istorii ekologii zhivotnykh. [Essay on the history of animal ecology].* Leningrad, Izd. "Nauka." 287 pp.

Cc176. **Zenkevich, L. A.** 1948. Russkie issledovaniya fauny morei. *Trudy Instituta Istorii Estestvoznaniya* 2: 170–196.

See also Cc86 Cc519(Andrewartha & Birch) Cc530 Cc684 Cc723–733 & Cc905.

Morphology & anatomy

N.B. For histology *see* Ca501–Ca535.

Cc177. **Balan, B.**1979. *L'ordre et le temps: l'Anatomie comparée et l'histoire des vivants au XIXe siècle.* Paris, J. Vrin. 610 pp. *L'Histoire des sciences: Textes et études.*

Cc178. **Chaine, J.** 1922–25. *Histoire de l'anatomie comparative.* Bordeaux, E. Daguerre. 2 vols.

Cc179. **Cole, F. J.** 1944. *A history of comparative anatomy from Aristotle to the eighteenth century.* London, Macmillan. 524 pp. *Re-pub.* 1949.

Cc180. ___**, & N. B. Eales.** 1917. The history of comparative anatomy. Part 1. — A statistical analysis of the literature. *Science progress* 11: 578–596.

Cc181. **Desmond, A.** 1989. *The politics of evolution: Morphology, medicine and reform in radical London.* Chicago, IL, Univ. of Chicago Pr. x+504 pp. *Science & its conceptual foundations.*
Grant and others in the pre-Darwinian London scene.

Cc182. **Faller, A.** 1948. *Die Entwicklung der makroscopisch–anatomischen Präparierkunst von Galen bis zur Neuzeit.* Basle, S. Karger. [iii]+115 pp., illus.

Cc183. **Franklin, K. J.** 1927. Valves in veins: An historical survey. *Proceedings of the Royal Society of Medicine* 21 *Section History of medicine*: 1–33.

Cc184. **Lubosch, W.** 1931. Geschichte der vergleichenden Anatomie. In: L. Bolk et al. *Handbuch der vergleichenden Anatomie der Wirbeltiere*, vol. 1, pp. 3–76. Berlin, Urban & Schwarzenberg.

Cc185. **Maienschein, J., R. Rainger, & K. R. Benson.** 1981. American morphology at the turn of the century. *Journal of the history of biology* 14: 83–191.
Papers read at a session of the History of Science Society meeting, New York, December 1979, entitled "Morphology and the emergence of American biology." *Includes* **J. Maienschein.** Shifting assumptions in American biology: Embryology, 1890–1910, pp. 89–113; **K. R. Benson.** Problems of individual development: Descriptive embryological morphology in America at the turn of the century, pp. 115–128; **G. E. Allen.** Morphology and the twentieth century: A response, pp. 159–176.

Cc186. **Nyhart, L.** 1987. The disciplinary breakdown of German morphology, 1870–1900. *Isis* 78: 365–389.

Cc187. **Russell, E. S.** 1916. *Form and function: A contribution to the history of animal morphology*. London, J. Murray. ix+383 pp. *Re-pub*. Farnborough, Gregg, 1973; and, With an introduction by G. V. Lauder. Chicago, IL, Univ. of Chicago Pr., 1982. xlvi+384 pp.

Cc188. **Swinton, W. E.** 1960. Early history of comparative anatomy. *Endeavour* 19: 209–214.

Cc189. **Zanetti, C., & U. Wimmer-Aeschlimann.** 1968. *Eine Geschichte der Anatomie und Physiologie von Albrecht von Haller*. Berne & Stuttgart, Huber. 157 pp. *Berner Beiträge zur Geschichte der Medizin und Naturwissenschaften* N.F. vol. 1.

See also Biographies Ba299 Ba302 Ba304–305 Ba312 Ba433–446 Ba674–683 Ba885–892 Ba1332 Ba1567; Societies Bb71(Millot); Misc. Ca53 Ca459(Coleman, Ghiselin, Adams) Cc519(Richards) Cc939 & Cc1022–1026.

Embryology

Cc190. **Adelmann, H. D.** 1966. The main currents of embryological thought before Malpighi [and] Malpighi's contribution to embryology. In: *Marcello Malpighi and the evolution of embryology*, vol. 2, pp. 729–816, 819–886. Ithaca, NY, Cornell Univ. Pr.

Cc191. **Blacher, L. I.** 1982. *History of embryology in Russia, from the middle of the eighteenth century to the middle of the nineteenth century*. Translated by H. I. Youssef & B. A. Malek. Introduction by J. Maienschein. Washington, DC, Smithsonian Institution Pr. v+673 pp.
Translation of **L. I. Blyakher**, *Istoriya embriologii v Rossii*, vol. 1, 1955, see below.

Cc192. **Blyakher, L. I.** 1955–59. *Istoriya embriologii v Rossii. [The history of embryology in Russia.* Moscow, Akademiya Nauk SSSR. 2 vols.
Vol. 1: *S srediny XVIII do serediny XIX veka. [From the mid-18th century to the mid-19th century]*; vol. 2: *S srediny XIX do serediny XX veka. Bespozvonochnye. [From the mid-19th to the mid-20th century. Invertebrates]*.

Cc193. **Bloch, B.** 1904. Die geschichtlichen Grundlagen der Embryologie bis auf Harvey. *Nova acta Academiae Caesareae Leopoldino–Carolinae Germanicae Naturae Curiosorum* 82: 217–334.

Cc194. **Bodemer, C. W.** 1964. Regeneration and the decline of preformationism in eighteenth century embryology. *Bulletin of the history of medicine* 38: 20–31.

Cc195. **Bowler, P. J.** 1971. Preformation and pre-existence in the seventeenth century: A brief analysis. *Journal of the history of biology* 4: 221–244.

Cc196. **Churchill, F. B.** 1970. The history of embryology as intellectual history. *Journal of the history of biology* 3: 155–181.

Cc197. **Clarke, A. E.** 1991. Embryology and the rise of American reproductive sciences, circa 1910–1940. In: K. R. Benson, R. Rainger, & J. Maienschein, eds. *The expansion of American biology*, pp. 107–132. New Brunswick, NJ, Rutgers Univ. Pr.

Cc198. **Fischer, J. L., & J. Smith.** 1984. French embryology and the 'Mechanics of development' from 1887 to 1910: L. Chabry, Y. Delage & E. Bataillon. *History and philosophy of the life sciences* 6: 25–39.

Cc199. **Gilbert, S. F.**, ed. 1991. *A conceptual history of modern embryology*. New York, Plenum Pr. xiv+266 pp. *Developmental biology* vol. 7.
Chapters by eleven authors. *Includes* **F. B. Churchill.** The rise of classical descriptive embryology, pp. 1–29; **J. Maienschein.** The origins of Entwicklungsmechanik, pp. 43–

61; **G. B. Grunwald.** The conceptual and experimental foundations of vertebrate embryonic cell adhesion research, pp. 129–158; **S. F. Gilbert.** Induction and the origins of developmental genetics, pp. 181–206; **J. Sapp.** Concepts of organization: The leverage of ciliate protozoa, pp. 229–258; etc.

Cc200. **Horder, T. J., J. A. Witkowski, & C. C. Wylie**, eds. 1986. *History of embryology. The eighth symposium of the British Society for Developmental Biology.* Cambridge, Cambridge Univ. Pr. xxiv+477 pp.
Fifteen papers by various authors and a list of "Selected references to sources relating to the history of embryology." *Includes* **F. B. Churchill.** Weismann, hydromedusae, and the biogenetic imperative: A reconsideration, pp. 7–33; **M. Ridley.** Embryology and classical zoology in Great Britain, pp. 35–67; **J. Maienschein.** Preformation or new formation — or neither or both, pp. 73–108; **J. A. Witkowski.** Ross Harrison and the experimental analysis of nerve growth: The origins of tissue culture, pp. 149–177; **T. J. Horder & P. J. Weindling.** Hans Spemann and the organiser, pp. 183–242; **J. Brachet.** Early interactions between embryology and biochemistry, pp. 246–259; **M. Borell.** Instruments and an independent physiology: The Harvard Physiological Laboratory, 1871–1906, pp. 293–321; **H. Wallace.** Regeneration, pp. 331–346; **L. Wolpert.** Gradients, position and pattern: A history, pp. 347–362; etc.

Cc201. [Not used.]

Cc202. **Maienschein, J.** 1978. Cell lineage, ancestral reminiscence, and the biogenetic law. *Journal of the history of biology* 11: 129–158.

Cc203. ___. 1984. What determines sex? A study of converging approaches, 1880–1916. *Isis* 75: 457–480.

Cc204. **Meyer, A. W.** 1939. *The rise of embryology.* Stanford, CA, Univ. of Stanford Pr. xv+367 pp.

Cc205. **Needham, J.** 1959. *A history of embryology.* 2nd ed., rev. with the assistance of A. Hughes. Cambridge, Cambridge Univ. Pr. 304 pp. & 19 pl.
History to the end of the 18th century. *Replaces* 1st ed., 1934. xviii+274 pp. & 16 pl. For Russian version *see* ___. 1947. *Istoriya embriologii.* Per. c angl. A. V. Yudinov. Moscow, Gos. Izd. 342 pp.

Cc206. **Oppenheimer, J. M.** 1967. *Essays in the history of embryology and biology.* Cambridge, MA, M I T Pr. x+374 pp.
Re-publication of thirteen essays: Embryological concepts in the twentieth century; Questions posed by classical descriptive and experimental embryology; Ross Harrison's contributions to experimental embryology; Analysis of development: Problems, concepts and their history; Analysis of development: Methods and techniques; Embryology and evolution: Nineteenth century hopes and twentieth century realities; An embryological enigma in the *Origin of species*; The non-specificity of the germ–layers; K. E. von Baer's beginning insights into causal–analytical relationships during development; John and William Hunter and some eighteenth century scientific methods; John Hunter, Sir Thomas Browne and the experimental method; William Harvey and historical change; William Gilbert: Plant grafting and the grand analogy; Postscript: Additional references.

Cc207. **Roe, S.** 1981. *Matter, life, and generation: Eighteenth–century embryology and the Haller–Wolff debate.* Cambridge, Cambridge Univ. Pr. x+214 pp.

See also Ba480–484 Ba1631 Bb71(Wolff) Ca459(Hamburger) Ca567(May) Ca1062(Gross, Allen) & Cc185(Maienschein, Benson) & Cc916(Monroy & Groeben).

Physiology & reproduction

Cc208. **Andreoli, A.** 1961. *Zur geschichtlichen Entwicklung der Neuronentheorie.* Basle, B. Schwabe & Co. 88 pp. *Basler Veröffentlichungen zur Geschichte der Medizin und der Biologie* no. 10.

Cc209. **Bastholm, E.** 1950. *The history of muscle physiology from the natural philosophers to Albrecht von Haller.* Copenhagen, E. Munksgaard. 256 pp. *Acta historica scientiarum naturalium et medicinalium* vol. 7.

Cc210. **Belloni, L.** 1965. Leeuwenhoek, Boerhaave und Bleyswyk über Spermatozoën. *Janus* 52: 193–217.

Cc211. **Canguilhem, G.** 1955. *La formation du concept de réflexe aux XVIIe et XVIIIe siècles.* Paris, Presses Universitaires de France. 206 pp. *Bibliothèque de philosophie contemporaine.*

Cc212. **Castellani, C.** 1965. *La storia della generazione: Idee e teorie dal diciasettesimo al diciottesimo secolo.* Milan, Longanesi. 414 pp. & 42 pl. *I panorami scientifici* vol. 28.

Cc213. **___.** 1973. Spermatozoan biology from Leeuwenhoek to Spallanzani. *Journal of the history of biology* 6: 637–668.

Cc214. **Cole, F. J.** 1930. *Early theories of sexual generation.* Oxford, Clarendon Pr. x+230 pp.

Cc215. **Corsi, P.**, ed. 1991. *The enchanted loom: Chapters in the history of neuroscience.* New York, Oxford Univ. Pr. (in cooperation with the Fidia Research Foundation). 383 pp., illus. *History of neuroscience* no. 4.
Expanded edition of an exhibition catalogue, interspersed with thirteen essays by various authors, in three sections: Part 1: The art of memory; Part II: The discovery of the brain: From Descartes to Gall; Part III: Birth and frontiers of neuroscience. Translation of *La fabbrica del pensiero: Dall'arte della memorie alle neuroscience.* Milan, Electa, 1989.

Cc216. **Delaporte, F.** 1983. Theories of osteogenesis in the eighteenth century. *Journal of the history of biology* 16: 343–360.

Cc217. **Dinsmore, C. E.**, ed. 1991. *A history of regeneration research: Milestones in the evolution of a science.* Cambridge, Cambridge Univ. Pr. x+228 pp.
Includes **R. J. Goss.** The natural history (and mystery) of regeneration, pp. 7–23; **D. M. Skinner & J. S. Cook.** New limbs for old: Some highlights in the history of regeneration in Crustacea, pp. 25–45; **H. M. & S. G. Lenhoff.** Abraham Trembley and the origins of research on regeneration in animals, pp. 47–66; **C. E. Dinsmore.** Lazzaro Spallanzani: Concepts of generation and regeneration, pp. 67–89; **K. R. Benson.** Observation versus philosophical committment in eighteenth–century ideas of regeneration and generation, pp. 91–100; **M. Singer & J. Géraudie.** The neurotrophic phenomenon: Its history during limb regeneration in the newt, pp. 101–112; **F. B. Churchill.** Regeneration, 1885–1901, pp. 113–131; **J. Maienschein.** T. H. Morgan's regeneration epigenesis and (w)holism, pp. 133–149; **J. W. Vanable Jr.** A history of bioelectricity in development and regeneration, pp. 151–178; **R. A. Liversage.** Origin of the blastema cells in epimorphic regeneration of urodele appendages: A history of ideas, pp. 179–199; **L. Wolpert.** Morgan's ambivalence: A history of gradients and regeneration, pp. 201–217.

Cc218. **Foster,** Sir **M.** 1901. *Lectures on the history of physiology during the sixteenth, seventeenth and eighteenth centuries.* Cambridge, Cambridge Univ. Pr. [viii]+310 pp. *Repub.* 1924.

Cc219. **Franklin, K. J.** 1949. *A short history of physiology.* 2nd ed. London, Staples Pr. 147 pp. & 16 pl.
Replaces 1st ed., London, Bale, Sons & Danielsson, 1933. iv+122 pp.

Cc220. **Fulton, J. F.** 1931. *Physiology.* New York, P. B. Hoeber. xv+141 pp. *Clio medica* no. 5.

Cc221. **Garrison, F. H.** 1969. *Garrison's history of neurology.* Rev. and enl. with a bibliography of classical, original and standard works in neurology by L. C. McHenry, Jr. Springfield, IL, C. C. Thomas. xv+552 pp.

Cc222. **Gasking, E. B.** 1967. *Investigations into generation, 1651–1828.* London, Hutchinson & Co. 192 pp.

Cc223. **Geison, G. L.** 1978. *Michael Foster and the Cambridge school of physiology: The scientific enterprise in late Victorian society.* Princeton, NJ, Princeton Univ. Pr. xix+401 pp.

Cc224. **Goodfield, G. J.** 1960. *The growth of scientific physiology: Physiological method and the mechanist–vitalist controversy, illustrated by the problems of respiration and animal heat.* London, Hutchinson. 174 pp.
Biographical bibliography, pp. 165–174.

Cc225. **Graubard, M. A.** 1984. *Circulation and respiration: The evolution of an idea.* New York, Harcourt Brace. x+278 pp.

Cc226. **Hall, D. L.** 1981. *Why do animals breathe? Physiological problems and iatromechanical research in the early eighteenth century.* New York, Arno Pr. xviii+296 pp. *The development of science: Sources for the history of science.*

Cc227. **Hall, T. S.** 1969. *Ideas of life and matter: Studies in the history of general physiology, 600 B.C. – 1900 A.D.* Chicago, IL, Univ. of Chicago Pr. 2 vols. *Re-pub.* 1975 as: *History of general physiology, 600 B.C. to A.D. 1900.* Chicago, IL, [etc.]. 2 vols.

Cc228. **Hoff, H. E., & P. Kellaway.** 1952. The early history of the reflex. *Journal of the history of medicine* 7: 211–249.

Cc229. **Koshtoyants, Kh. S.** 1964. *Essays on the history of physiology in Russia.* Translated by D. P. Boder, K. Hanes, & N. O'Brian. Washington, DC, American Institute of Biological Sciences (AIBS). xiv+321 pp. *Russian monographs on brain and behavior* no. 1.
Translation of *Ocherki po istorii fiziologii v Rossii.* Moscow/Leningrad, Akademiya Nauk SSSR, 1946. 494 pp., illus.

Cc230. **Lesch, J. E.** 1984. *Science and medicine in France: The emergence of experimental physiology, 1790–1855.* Cambridge, MA, Harvard Univ. Pr. xi+276 pp.

Cc231. **Liddell, E. G. T.** 1960. *The discovery of reflexes.* Oxford, Clarendon Pr. vi+174 pp.

Cc232. **Mani, N.** 1959–67. *Die historischen Grundlagen der Leberforschung.* Basle, Schwabe Verlag. 2 vols. *Basler Veröffentlichungen zur Geschichte der Medizin* fasc. 9 & 21.
1 Teil: *Die Vorstellungen über Anatomie, Physiologie und Pathologie der Leber in der Antike*; 2 Teil: *Die Geschichte der Leberforschung von Galen bis Claude Bernard.*

Cc233. **Mendelsohn, E.** 1964. *Heat and life: The development of the theory of animal heat.* Cambridge, MA, Harvard Univ. Pr; London, Oxford Univ. Pr. xii+208 pp.

Cc234. **___.** 1965. Physical models and physiological concepts: Explanation in nineteenth century biology. *British journal for the history of science* 2: 201–219.

Cc235. **Neuberger, M.** 1981. *The historical development of experimental brain and spinal chord physiology before Flourens.* Translated & ed. with additional material by E. Clarke. Baltimore, MD, John Hopkins Pr. xxvii+391 pp.

Cc236. **Polyak, S.** 1957. Part 1: History of investigations of the structure and function of the eye and of the visual pathways and centers of the brain. In: *The vertebrate visual system [etc.]*, pp. 9–203. Chicago, IL, Univ. of Chicago Pr.

Cc237. **Premuda, L.** 1966. *Storia della fisiologia: Problemi e figure.* Udine, Del Bianco Editore. 328 pp.

Cc238. **Rapp, D.** 1970. *Die Entwicklung der physiologischen Methodik von 1784 bis 1911: Eine quantitative Untersuchung.* Munster, Institut für Geschichte der Medizin der Universität Münster. v+233 pp. *Münstersche Beiträge zur Geschichte und Theorie der Medizin* no. 2.

Cc239. **Riese, W.** 1959. *A history of neurology.* New York, M.D. Publications. 223 pp.

Cc240. **Roger, J.** 1971. *Les sciences de la vie dans la pensée française de XVIIIe siècle: La génération des animaux de Descartes à l'Encyclopédie.* 2nd ed. complétée. Paris, A. Colin. 848 pp.
Bibliographie (905 references), pp. 781–816. *Replaces* 1st ed., Paris, [etc.], 1963. 842 pp.

Cc241. **Rostand, J.** 1930. *La formation de l'être: Histoire des idées sur la génération.* Paris, Hachette. 222 pp.

Cc242. **Rothschuh, K. E.** 1952. *Entwicklungsgeschichte physiologischer Probleme in Tabellenform.* Munich, Urban & Schwarzenberg. 122 pp.
Covers ca. 1600 to 1920. Designed as a companion to his *Geschichte der Physiologie.* Berlin, Springer, 1953.

Cc243. ___. 1959. Wer ist gross und was ist bedeutend in der Geschichte der Wissenschaft? Ein Beitrag mit spezieller Hinsicht auf die Geschichte der Physiologie. *Naturwissenschaftliche Rundschau* 1959 (2): 45–52.

Cc244. ___. 1966. Zur Geschichte der physiologischen Reizmethodik im 17. und 18. Jahrhundert. *Gesnerus* 23: 147–160.

Cc245. ___. 1968. *Physiologie: Der Wandel ihrer Konzepte, Probleme und Methoden vom 16. bis 20. Jahrhundert.* Freiburg, Munich, K. Alber. 407 pp. *Orbis academicus* vol. II/15.

Cc246. ___. 1969. *Physiologie im Werden.* Stuttgart, G. Fischer. xi+188 pp. *Medizin in Geschichte und Kultur* vol. 9.
Re-publication of several of this author's essays from various sources.

Cc247. ___. 1973. *History of physiology.* Translated & ed., with a new English bibliography, by G. B. Risse. Huntington, NY, R. E. Krieger. xxi+379 pp.
Translation of *Geschichte der Physiologie.* Berlin, Springer, 1953. xi+249 pp.

Cc248. ___**, & A. Schäfer.** 1955. Quantitative Untersuchungen über die Entwicklung des physiologischen Fachschrifttums (Periodica) in den letzten 150 Jahren. *Centaurus* 4: 63–66.

Cc249. **Schiller, J.** 1968. Physiology's struggle for independence in the first half of the nineteenth century. *History of science* 7: 64–89.

Cc250. **Singer, C. J.** 1922. *The discovery of the circulation of the blood.* London, G. Bell & Sons. x+80 pp. *Re-pub.* London, Dawson, 1956.

Cc251. **Smith, A. H. T. Robb–.** 1962. Unravelling the functions of the blood. *Medical history* 6: 1–21.

Cc252. **Taschenberg, O.** 1892. *Historische Entwickelung der Lehre von der Parthenogenesis.* Halle, M. Niemeyer. 89 pp.

Cc253. **Willius, F. A., & T. J. Dry.** 1948. *A history of the heart and circulation.* Philadelphia, PA, W. B. Saunders. xvii+456 pp., illus.
Includes General history, followed by biographies of selected researchers, and a summary chronology arranged under eighteen separate themes.

See also Biographies Ba296 Ba301 Ba308 Ba310 Ba315 Ba530–537 Ba813 Ba908–918 Ba1208 Ba1221–1222; Misc. Cc519(Wigglesworth) & Cc916(Ghiretti, Florey, Young, Bennett).

Behaviour

Cc254. **Bodemer, C. W.** 1967. Overtures to behavioral science: Eighteenth– and nineteenth–century ideas relating to light and animal behavior. *Episteme* 1: 135–152.

Cc255. **Burkhardt, R. W., Jr.** 1988. Charles Otis Whitman, Wallace Craig, and the biological study of animal behavior in the United States, 1898–1925. In: R. Rainger, K. R. Benson, & J. Maienschein, eds. *The American development of biology*, pp. 185–218. Philadelphia, PA, Univ. of Pennsylvania Pr.

Cc256. **Chevroton, D.** 1978. La querelle de l'âme des bêtes: Une entrave à la naissance de la psychologie animale ou une incitation à la recherche de nouveaux points de vue? *Histoire et nature* 12/13: 43–82.
Appendix: Courtes notices biographiques concernant les [22] auteurs cites, pp. 80–81.

Cc257. **Dewsbury, D. A.**, ed. 1985. *Studying animal behaviour: Autobiographies of the founders.* Chicago, IL, Univ. of Chicago Pr. 519 pp. *Re-pub.* Chicago, [etc.], 1989. *See* Ba298.

Cc258. ___. 1989. A brief history of the study of animal behavior in North America. *Perspectives in ethology* 8: 85–122.

Cc259. **Durant, J. R.** 1981. Innate character in animals and man: A perspective on the origins of ethology. In: C. Webster, ed. *Biology, medicine and society*, pp. 157–192. Cambridge, Cambridge Univ. Pr.

Cc260. **Gray, P. H.** 1968. The early animal behaviorists: Prolegomenon to ethology. *Isis* 59: 372–383.

Cc261. **Mitman, G., & R. W. Burkhardt, Jr.** 1991. Struggling for identity: The study of animal behavior in America, 1930–1945. In: K. R. Benson, R. Rainger, & J. Maienschein, eds. *The expansion of American biology*, pp. 164–194. New Brunswick, NJ, Rutgers Univ. Pr.

Cc262. **Richards, R. J.** 1982. The emergence of evolutionary biology of behaviour in the early nineteenth century. *British journal for the history of science* 15: 241–280.

Cc263. ___. 1989. *Darwin and the emergence of evolutionary theories of mind and behaviour.* Chicago, IL, Univ. of Chicago Pr. xviii+700 pp.

Cc264. **Schurig, V., & S. van Mourik.** 1986. Die Begründung der zoologischen Verhaltensforschung als "Ethologie". *Biologische Rundschau* 24: 197–208.

Cc265. **Sparks, J.** 1982. *The discovery of animal behaviour.* London, W. Collins Sons & Co. & British Broadcasting Corporation. 288 pp. *Also* Boston, MA, Little, Brown & Co.

Cc266. **Stamm, R. A.**, ed. 1984. *Tierpsychologie: Die biologische Erforschung tierischen und menschlichen Verhaltens.* Weinheim, Beltz Verlag. xvii+581 pp. *Kindlers "Psychologie des 20. Jahrhunderts."*
Includes **R. A. Stamm.** Zeittafel zur Entwicklung der Ethologie; **G. M. Burghardt.** Die Geschichte der Tierpsychologie, mit besonderer Berücksichtigung von England und Amerika; **M. Meyer–Holzapfel.** Die Geburt der Ethologie; **A. Portmann.** Jean–Henri Fabre (1823–1915); **M. Lindauer.** Karl von Frisch (1886–1982); **I. Eibl–Eibesfeldt.** Konrad Lorenz (*1903); **C. G. Beer.** Niko Tinbergen (*1907); **B. Hassenstein.** Erich von Holst (1908–1962), etc.

Cc267. **Tinbergen, N.** 1958. *Curious naturalists.* London, Country Life. 280 pp. *Re-pub.* Amherst, MA, Univ. of Massachusetts Pr. 268 pp.

Cc268. ___. 1969. Ethology. In: R. Harré, ed. *Scientific thought 1900–1960: A selective survey*, pp. 238–268. Oxford, Clarendon Pr.

Cc269. **Thorpe, W. H.** 1979. *The origins and rise of ethology: The science of the natural behaviour of animals.* London, Heinemann Educational. xi+174 pp.

Cc270. **Young, R. M.** 1966. Scholarship and the history of the behavioural sciences. *History of science* 5: 1–51.
Bibliography, pp. 41–51.

Cc271. ___. 1990. *Mind, brain and adaptation in the nineteenth century: Cerebral localization and its biological context from Gall to Ferrier.* Oxford, Oxford Univ. Pr. 278 pp. *History of neuroscience* no. 3.
First pub. 1970, here re-pub. with a new preface.

See also Ba298 Ba925 Ba1201–1204 Ba1355–1356 Ba1551 Bb796 Cc349(Nice) & Cc519(Richard).

Palaeontology

Cc272. **Bowler, P. J.** 1976. *Fossils and progress: Paleontology and the idea of progressive evolution in the nineteenth century.* New York, Science History Publications. viii+191 pp.

Cc273. **Buffetaut, E.** 1987. *A short history of vertebrate palaeontology.* London, Croom Helm. 223 pp., illus.

Cc274. **Colbert, E. H.** 1968. *Men and dinosaurs: The search in field and laboratory.* New York, E. P. Dutton. xviii+283 pp., illus. *Re-pub. as* ___. 1984. *The great dinosaur hunters and their discoveries.* New York, Dover Publications. xxii+283 pp., illus.

Cc275. **Desmond, A.** 1982. *Archetypes and ancestors: Palaeontology in Victorian London 1850–1875.* London, F. Muller. 287 pp. *Also* Chicago, IL, Univ. of Chicago Pr., 1984.

Cc276. ___. 1984. Interpreting the origin of mammals: New approaches to the history of palaeontology. *Zoological journal of the Linnean Society* 82: 7–16.

Cc277. **Gregory, J. T.** 1979. North American vertebrate paleontology, 1776–1976. In: Schneer, C. J. *Two hundred years of geology in America. Proceedings of the New Hampshire Bicentennial conference on the history of geology*, pp. 305–335. Hanover, NH, Univ. Pr. of New England for the Univ. of New Hampshire.

Cc278. **Howard, R. W.** 1975. *The dawnseekers: The first history of American paleontology.* New York, Harcourt Brace Jovanovich. xiii+314 pp., illus.

Cc279. **Howe, H. V.** 1959. Fifty years of micropaleontology. *Journal of paleontology* 33: 511–517.

Cc280. **Lanham, U.** 1973. *The bone hunters: The heroic age of paleontology in the American west.* New York, Columbia Univ. Pr. x+285 pp. *Re-pub. as* ___. 1991. *The bone hunters: The heroic age of paleontology in the American west.* New York, Dover Publications.

Cc280A. **Osborn, H. F.** 1911. [Article] Palaeontology. In: *Encyclopaedia Britannica*, 11th ed., vol. 20, pp. 579–591. Cambridge, Cambridge Univ. Pr.
Concise history.

Cc281. **Romer, A. S.** 1959. Vertebrate paleontology1908–1958. *Journal of paleontology* 33: 915–925.

Cc282. **Rudwick, M. J. S.** 1985. *The meaning of fossils: Episodes in the history of palaeontology.* 2nd ed. Chicago, IL, Univ. of Chicago Pr. xvi+288 pp.

Cc283. **Shor, E. N.** 1974. *The fossil feud between E. D. Cope and O. C. Marsh.* Hicksville, NY, Exposition Pr. xi+340 pp., illus.

Cc284. **Simpson, G. G.** 1942. The beginnings of vertebrate paleontology in North America. *Proceedings of the American Philosophical Society* 86: 130–188.

Cc285. ___. 1984. *Discoverers of the lost world: An account of some of those who brought back to life South American mammals long buried in the abyss of time.* New Haven, CT, Yale Univ. Pr. viii+222 pp., illus.

See also Biographies Ba297 Ba307 Ba313 Ba317 Ba435–446 Ba662 Ba674–683 Ba1327–1329 Ba1332 Ba1526; Societies Bb576A Bb592(Scott); Misc. Ca441(Haber, Haber) Ca446(Reif) Ca450A Ca459(Gould) Cc81(Laporte) Cc349(Wetmore) Cc701 & Cc925.

VERTEBRATE ZOOLOGY

General

BIBLIOGRAPHIES OF VERTEBRATES

Cc286. **Blackwelder, R. E.** 1972. *Guide to the taxonomic literature of vertebrates.* Ames, IA, Iowa State Univ. Pr. xix+259 pp.

Cc287. **Hay, O. P.** 1902. *Bibliography and catalogue of the fossil vertebrates of North America.* Washington, DC, United States Geological Survey. 868+iii pp. *Bulletin* no. 179. *Re-pub.* New York, Arno Pr., 1974; *Continued* ___. 1929–30. *Second bibliography [etc.].* Washington, DC, Carnegie Institution of Washington. 2 vols. *Publication* no. 390; Further continued by periodic bibliographies, viz. **C. L. Camp et al.** 1940, 1942, 1949, 1953, 1961, 1964, 1968 & 1972. *Bibliography of fossil vertebrates, 1928–1933 (... 1934–1938. 1939–1943. 1944–1948. 1949–1953. 1954–1958. 1959-1963. 1964-1968).* New York, Geological Society of America. *Special papers* no. 27, & 42, and *Memoirs* no. 37, 57, 84, 92, 117 & 134; **J. T. Gregory et al.** 1973. *Bibliography ... 1969-1972.* New York, [etc.]. xlvi+733 pp. *Memoir* no. 141; **M. Green et al.** 1979. *Bibliography ... 1973–1977.* Rapid City, SD, South Dakota School of Mines & Technology. v+311+165+144 pp.; Continued annually *1978+*, Falls Church, VA, American Geological Institute, 1981+

Cc288. **Junk, W.** 1918. *Vertebratorum bibliographia.* Berlin, W. Junk. 662 pp.

Cc289. **Minot, C. S.** 1893. A bibliography of vertebrate embryology. *Memoirs of the Boston Society of Natural History* 4: 487–618.

Cc290. **Romer, A. S., et al.** 1962. *Bibliography of fossil vertebrates, exclusive of North America, 1509–1927.* New York, Geological Society of America. 2 vols. (lxxxix+1,544 pp.) *Memoir* no. 87.

Cc291. **Wood, C. A.** 1931. *An introduction to the literature of vertebrate zoology, based chiefly on the titles in the Blacker Library of Zoology, the Emma Shearer Wood Library of Ornithology, the Bibliotheca Osleriana, and other libraries of McGill University, Montreal.* London, Oxford Univ. Pr., 1931. xix+643 pp. *Re-pub.* Hildesheim, Olms, 1974. ••Ca. 15,000 entries.

Mammalogy

Cc291A. **Almaça, C.** 1991. *The beginning of the Portuguese mammalogy.* Lisbon, Museu Nacional de Historia Natural. 44 pp., illus.

Cc292. **Arvy, L., & G. Pilleri.** 1983. Quatre cents ans d'observations sur l'hyperoodon, le cachelot et l'orque. *Investigations on cetacea* 15 supp: 265 pp.

Cc293. **Barsanti, G.** 1990. Storia naturale delle scimmie, 1600–1800. *Nuncius* 5 (2): 99–165.

With a bibliography.

Cc294. **Broberg, G.** 1972. Järven – Filfrassen – Frossaren: En studie i zoologisk exoticism. [A study in zoological exoticism: The wolverine]. *Lychnos* 1971–72: 181–216.

Cc295. **Brodhead, M. J.** 1991. Of mice and mastadons: Contributions to the literature of mammalogy by officers and men of the United States Army in the nineteenth century. *Archives of natural history* 18: 363–374.

Cc296. **Gregory, W. K.** 1910. [Historical survey of mammalian classification]. In: The orders of mammals. *Bulletin of the American Museum of Natural History* 27: 1–102.

Cc297. **Hamilton, W. J., Jr.** 1955. Mammalogy in North America. In: California Academy of Sciences. *A century of progress in the natural sciences 1853–1953*, pp. 661–687. San Francisco, CA.

Cc298. **Hershkovitz, P.** 1987. A history of the recent mammalogy of the neotropical region from 1492–1850. *Fieldiana, zoology* n.s. 39: 11–98.

Cc299. **Janson, H. W.** 1952. *Apes and ape lore in the Middle Ages and the Renaissance.* London, Warburg Institute. 384 pp. & 56 pl.

Cc300. **Kinnear, N. B.** 1952. The history of Indian mammalogy and ornithology. I: Mammals. *Journal of the Bombay Natural History Society* 50: 766–778.

Cc301. **Krumbiegel, I.** 1933. Artkenntnis und –Erkenntnis in der Säugetierkunde: Ein Beitrag zur Geschichte zer zoologischen Systematik. *Sitzungsberichte der Gesellschaft Naturforschender Freunde zu Berlin* 1933: 110–125.

Cc302. **Lever,** Sir **C.** 1985. *Naturalized mammals of the world.* London, Longmans. xvii+487 pp.
When, where, why, how, and by whom non-indigenous mammals living today in the wild were first introduced.

Cc303. **Querner, H.** 1986. Zur Geschichte der Anthropologie. *Anthropologische Anzeiger* 44: 281–297.

Cc304. **Schultz, A. H.** 1976. The rise of primatology in the twentieth century. *Folia primatologica* 26: 5–23.

Cc305. **Spencer, F.**, ed. 1982. *A history of American physical anthropology, 1920–1980.* New York, Academic Pr. xvi+495 pp.
Includes **R. Ribnick.** A short history of primate field studies: Old World monkeys and apes, pp. 49–73; **J. G. Fleagle & W. L. Jungers.** Fifty years of higher primate phylogeny, pp. 187–230; **G. J. Armelagos, D. S. Carlson, & D. P. Van Gerven.** The theoretical foundations and development of skeletal biology, pp. 305–328; **C. C. Lovejoy, R. P. Mensforth, & G. J. Armelagos.** Five decades of skeletal biology as reflected in the American journal of physical anthropology, pp. 329–336; **M. A. Little.** The development of ideas on human ecology and adaptation, pp. 405–433; etc.

Cc306. **Sterling, K. B.**, ed. 1987+. *An international history of mammalogy.* Bel Air, MD, One World Pr. Vol. 1+.
A multi–volume series by various authors that is intended to cover the development of mammalogy in every country of the world. Vol. 1. *Eastern Europe and Fennoscandia.* xxvi+198 pp.

Cc306A **Stevenson, J. C.** 1991. *Dictionary of concepts in physical anthropology.* New York, Greenwood Press. xiv+432 pp. *Reference sources for the social sciences & humanities* no. 10.
Essay length articles, good bibliographies.

See also A45 Ba311 Ba1636 Cc94(Armstrong) Cc772(Brambell & Mathews, Manton) Cc829(Felten) Cc949 Cc1038 Cc1040 Cc1050 Cc1054 & D85.

BIBLIOGRAPHIES OF MAMMALOGY

Cc307. **Allen, J. A.** 1882. Preliminary list of works and papers relating to the mammalian order Cete and Sirenia. *Bulletin, U. S. Geological & Geographical Survey of the Territories* 6(3): 399–562. *Re-pub.* (with an author index) Hythe, A. D. Lilly, 1977. [ii]+399–562+[12] pp. ••Chronological bibliography, 1495–1840. Describes almost all general mammalogical literature for the period covered.

Cc308. **Angress, S., & C. A. Reed.** 1962. An annotated bibliography on the origin and descent of domestic animals. *Fieldiana, anthropology* 54: 1–143.

Cc309. **Bekierz, F. W., & H. Brasseler.** 1985. Bibliographie über Fledermaus Bibliographien Chiroptera. Bibliography on bat bibliographies. *Courier Forschungsinstitut Senckenberg* 77: 139 pp.

Cc310. **Elliot, O.** 1971. Bibliography of the tree shrews 1780–1969. *Primates* 12; 323–414.

Cc311. **Fletcher, J. J.** 1885. Catalogue of papers and works relating to the mammalian orders Marsupialia and Monotremata. *Proceedings of the Linnaean Society of New South Wales* 9: 809–863.

Cc312. **Ruch, T. C.** 1941. *Bibliographia primatologica: A classified bibliography of primates other than man. Part 1: Anatomy, embryology and quantitative morphology; physiology, pharmacology and psychobiology; primate phylogeny and miscellanea.* Springfield, IL, Thomas. xxviii+241 pp. *Yale Medical Library, Hist. Libr. publ.* no. 4. ••All pub. *See also* **T. C. Ruch & J. F. Fulton.** 1942. Growth of primate literature since 1800. *Science* 95 (Jan. 9): 47–48.

Cc313. **Saint-Girons, M.-C.** 1986. *Bibliographie des mammifères du Paléarctique occidental (1850–1980).* Paris, Muséum National d'Histoire Naturelle, Secrétariat de la Faune et de la Flore. 440 pp. *Inventaire de faune et de flore* no. 28.

Cc314. **Voss, H.** 1955. *Bibliographie der Menschenaffen (Schimpanse, Orang, Gorilla), [etc.].* Jena, G. Fischer Verlag. viii+163 pp.

Cc315. **Walker, E. P., et al.** 1964. *Mammals of the world.* Vol. 3: *A classified bibliography.* Baltimore, MD, Johns Hopkins Univ. Pr. xiii+769 pp.

Cc316. **Wolfe, L. D.** 1987. *Field primatology: A guide to research.* New York, Garland Publishing. x+288 pp. *Garland reference library of social science* vol. 356. ••Covering the literature of the last 25 years.

Regional

Cc317. **Allouse, E.** 1954. *A bibliography of the vertebrate fauna of Iraq and neighbouring countries:* I. *Mammals.* Baghdad, Iraq Natural History Museum. 30 pp. *Publications* no. 4: *Continued* ___. 1968. *A bibliography of the vertebrate fauna of Iraq and neighbouring countries: Mammals (second compilation).* Baghdad, [etc.] 36 pp. *Ibid.* no. 25.

Cc318. **Ealey, D. M.** 1987. *A bibliography of Alberta mammalogy.* Edmonton, Alberta, Alberta Culture, Historical Resources Div. 400 pp.

Cc319. **Fischer-Sigwart, H.** 1900. *Säugethiere.* Berne, Centralkommission für Schweizerische Landeskunde. xv+104 pp. *Bibliographie der schweizerischen Landeskunde* IV.6.3.

Cc320. **Gill, T. N., & E. Coues.** 1877, Material for a bibliograpy of North American mammals. *Report, U. S. Geological & Geographical Survey of the Territories* 11, appx. B: 951–1081.

Cc321. **Marelli, C. A.** 1936. *Bibliographía eurística de las mammíferos, de caza e caza marítima. Contribuciones al estuio de la fauna Argentina.* La Plata, Ministerio de Obras Públicas de la Provincia de Buenos Aires. 189 pp. *Memorias del Jardin Zoológico de La Plata* vol. 3, pt. 2.

Cc322. **Ol'kova, N. V., & K. A. Bashanov.** 1970. *Bibliografiya po gryzunam i zaitseobraznym Sibiri i dal'nego vostoka (1786–1967 gg). [Bibliography of rodents and lagomorphs of Siberia and the far-East (1786–1967).]* Kyzyl, Irkutskii Gosudarstvennyi Nauchno–Isledovatel'skii Protivochumnoi

Institu Sibiri i Dal'nego Vostoka. 382 pp.

Cc323. **Owen, R. D., & G. D. Schnell.** 1989. *Oklahoma mammalogy: An annotated bibliography and checklist.* Norman, OK, Univ. of Oklahoma Pr. x+240 pp.

Cc324. **Petrov, A. I.** 1935. *Ukazatel' russkoi literatury po prikladnoi entomologii, fitopatologii i mammaliologii Srednei Azii. Vrediteli sel'skokhozyaistvennykh kul'tur i lesov (1870–1933).* Tashkent, Gosizdat UzSSR. 160 pp.

Cc325. **Phillips, J. C.** 1930. *American game mammals and birds: A catalogue of books, 1582–1925, sport, natural history and conservation.* Boston, MA, Houghton Mifflin & Co. 638 pp. *Re-pub.* New York, Arno Pr., 1978.

Cc326. **Raun, G. G.** 1962. A bibliograpy of the recent mammals of Texas. *Bulletin of the Texas Memorial Museum* 3: 1–81.

Cc327. **Verheyen, W. N., & E. Van Der Straeten.** 1969. An annotated bibliography of the African Muridae (Rodentia). *Documentation zoologique, Musée Royale de l'Afrique Centale* 16: 1–237.

Periodical bibliographies

See periodical bibliographies listed at Cc37–54 (especially Zoological record; *see* Cc50).

Ornithology — General

Cc328. **Beddall, B. G.** 1957. Historical notes on avian classification. *Systematic zoology* 6: 129–136.

Cc329. **Bock, W. J.** 1990. A special review: Peters' "Check–list of birds of the world" and a history of avian checklists. *Auk* 107: 629–648.

Cc330. **Boubier, M.** 1932. *L'evolution de l'ornithologie.* [New impression.] Paris, Alcan. ii+308 pp. *Nouvelle collection scientifique.*
First pub., 1925.

Cc331. **British Ornithologists' Club.** 1980. [Various papers.] *Bulletin of the British Ornithologists' Club* 100.
Special issue. *Includes* **E. M. Nicholson.** Co-operative ornithology and conservation in western Europe, pp. 44–47; **E. Bezzel.** Ornithological advances in western Europe during the last 50 years, pp. 47–50; **Z. Bochenski.** Some trends in ornithology in east European countries during the last fifty years, pp. 50–55; **Y. A. Isakov.** Some of the results of ornithological investigations in the Soviet Union for the past fifty years, pp. 55–61; **J. D. R. Vernon.** 50 years of ornithology in north–west Africa 1930–1980, pp. 61–66; **G. J. Morel.** Fifty years or ornithology in west Africa, pp. 66–68; **P. L. Britton.** Ornithological progress in eastern Africa during the last 50 years, pp. 68–73; **R. K. Brooke.** Ornithology in southern Africa, 1930–1980, pp. 73–76; **C. W. Benson.** Fifty years of ornithology in the Malagasy faunal region, pp. 76–80; **S. Ali.** Indian ornithology: The current trends, pp. 80–83; The **Earl of Cranbrook.** The state of ornithology in eastern Asia, pp. 84–89; **D. L. Serventy.** Developments in Australian ornithology, pp. 89–93; **J. A. Gibb.** New Zealand ornithology during the past 50 years, pp. 93–96; **J. Warham.** Recent trends in sub–Antarctic ornithology, pp. 96–102; **E. C. Young.** The present status of Antarctic ornithology, pp. 102–115; **H. Ouellet & W. E. Godfrey.** Ornithology in Canada in the 20th century: A capsule overview, pp. 115–118; **R. E. Ricklefs & F. B. Gill.** Fifty years of American ornithology, pp. 118–122; **D. W. Snow.** Ornithological research in tropical America — the last 35 years, pp. 123–131.

Cc332. **Cooke, C. H.** 1959–61. The historical background to the study of bird migration. *Country–side* n.s. 18: 366–369, 428–431, 491–493, 539–541, 1959; *Ibid.* 19: 11–13, 67–69, 107–109, 143–144, 185–187, 1961.

Cc333. **Farber, P. L.** 1982. *The emergence of ornithology as a scientific discipline: 1760–1850.* Dordrecht, D. Riedel Publishing Co. xxi+191 pp. *Studies in the history of modern science* no. 12.

Cc334. ___, ed. 1988. The historical impact of ornithology on the biological sciences. In: H. Ouellet, ed. *Acta XIX Congressus Internationalis Ornithologici, Ottawa, 1986*, vol. 2: 2176–2759. Ottawa, Univ. of Ottawa Pr. for the National Museum of Natural Sciences.
Includes **E. Mayr.** The contributions of birds to evolutionary theory; **K. B. Sterling.** Three 19th–century American ornithologists: J. A. Allen, Elliott Coues, and C. Hart Merriam – and their leadership in related fields; **M. G. Ainley.** William Rowan and the experimental approach in ornithology; **R. J. O'Hara.** Diagrammatic classifications of birds, 1819–1901: Views of the natural system in 19th–century British ornithology.

Cc335. **Gurney, J. H.** 1921. *Early annals of ornithology.* London, Witherby. iv+240 pp. *Re-pub.* Chicheley, Minet, 1972.

Cc336. **Hall, J. J.** 1991. The classification of birds, in Aristotle and early modern naturalists. *History of science* 29: 111–151, 223–243.

Cc337. **Lever,** Sir **C.** 1987. *Naturalized birds of the world.* Harlow, Longman Scientific & Technical; New York, J. Wiley. xx+615 pp.
When, where, why, how, and by whom non-indigenous birds living today in the wild were first introduced.

Cc338. **McClymont, J. R.** 1920. *Essays on early ornithology and kindred subjects.* London, B. Quaritch. 35 pp. & 3 pl.
Includes The rukh of Marco Polo. The penguins and the seals of the Angra de Sam Brás. The Banda Islands and the Bandan birds. The etymology of the name 'Emu.' Australian birds in 1697. New Zealand birds in 1772.

Cc339. **Newton, A., & H. Gadow.** 1893–96. Introduction [historical]. In: *A dictionary of birds*, pp. 1–120. London, A. & C. Black.

Cc340. **___, & P. Chalmers–Mitchell.** 1911. [Article] Ornithology. In: *Encyclopaedia Britannica*, 11th ed., vol. 20, pp. 299–326. Cambridge, Cambridge Univ. Pr.
Detailed historical account.

Cc341. **Rydzewski, W.** 1951. A historical review of bird marking. *Dansk Ornithologisk Forenings tidsskrift* 45: 61–95.

Cc342. **Stresemann, E.** 1975. *Ornithology: From Aristotle to the present.* Translated by H. J. & C. Epstein; ed. by G. W. Cottrell. With a foreword and an epilogue on American ornithology by E. Mayr. Cambridge, MA, Harvard Univ. Pr. xii+432 pp.
Mayr's epilogue entitled "Materials for a history of American ornithology," pp. 365–396. Translation of *Die Entwicklung der Ornithologie von Aristoteles bis zur Gegenwart.* Berlin, F. W. Peters, 1951. xv+431 pp.

Cc343. **Strom, D.**, ed. 1986. *Birdwatching with American women: A selection of nature writings.* New York, W. W. Norton & Co. xxviii+286 pp.
Eighteen authors are included with biographical memoirs and some portraits.

Cc344. **Wood, C. T.** [1835.] *The ornithological guide, in which are discussed several interesting points in ornithology.* London, Whittaker. xviii+236+[4] pp.
Critical reviews of books, pp. 79–176.

Cc345. **___, H. B.** 1945. The history of bird banding. *Auk* 62: 256–265.

Cc346. **___, N.** 1836. *The ornithologist's text–book. Being reviews of ornithological works, with an appendix containing discussions on various topics of interest.* London, J. W. Parker. [iv]+232 pp.
Critical reviews of a large number books, pp. 3–98 & 181–203; Synopsis of systems, (Willughby 1678 to Fleming 1828), pp. 99–152.

See also Biographies Ba323–324 Ba327–328 Ba456–457 Ba470–477 Ba595 Ba637 Ba729 Ba867–873 Ba935–938 Ba1057 Ba1140 Ba1232–1233 Ba1305–1308 Ba1338 Ba1363 Ba1367 Ba1434 Ba1533 Ba1586 Ba1605 Ba1621–1626; Societies Bb823–845; Misc. Cc772(Olney) Cc780–781 Cc1005–1012 Cc1014 Cc1033 Cc1040 Cc1042 Cc1044 & D87(Mengel).

Ornithology — Regional

Cc347. **Ainley, M. G.** 1980. The contribution of the amateur to North American ornithology: A historical perspective. *Living bird* 18: 161–177.

Cc348. **Allen, E. G.** 1951. The history of American ornithology before Audubon. *Transactions of the American Philosophical Society* n.s. 41 (3): 386–591. Re–pub. New York, Russell & Russell, 1969.

Cc349. **American Ornithologists' Union.** 1933. *Fifty years progress of American ornithology, 1883–1933.* On the occasion of its semi–centennial anniversary. New York, November 13–16, 1933. (Ed. by F. M. Chapman & T. S. Palmer.) Lancaster, PA. 249 pp.
Includes **T. S. Palmer.** A brief history of the American Ornithologists' Union; **W. Stone.** American ornithological literature—1833–1933; **W. Rowan.** Fifty years of bird migration; **F. C. Lincoln.** Bird banding; **M. M. Nice.** The theory of territorialism and its development; **H. Friedmann.** Advances in life history work; **W. L. McAtee.** Economic ornithology; **J. L. Peters.** Collections of birds in the United States and Canada: Study collections; **F. M. Chapman.** Collections of birds in the United States and Canada: Exhibition collections; **A. O. Gross.** History and progress of bird photography in America; **G. M. Sutton.** Fifty years progress in American bird–art; **T. G. Pearson.** Fifty years of bird protection; **A. A. Allen.** Ornithological education in America; **A. Wetmore.** Fossil birds. *Whole vol. re-pub. in* **K. B. Sterling**, ed. 1974. *Contributions to the history of American ornithology.* New York, Arno Pr.; *see* Cc370.

Cc350. **Brooke, R. K.** 1986. Historical sketch of Afrotropical ornithology. *Revue de zoologie Africaine* 100 (1): 7–12.

Cc351. **Brouwer, G. A.** 1954. *Historische gegevens over onze vroegere ornithologen en over de avifauna van Nederland.* Leiden, Brill. 226 pp. *Ardea* jubilee number.

Cc352. **Brzck, G.** 1959. Zloty wiek ornitologii polskiej. (The golden age of Polish ornithology). *Memorabilia zoologica* 3: 1–175.
Summary in English.

Cc353. **Eifert, V. S.** 1962. *Men, birds, and adventure ... the discovery of American birds.* New York, Dodd, Mead & Co. xiv+273 pp.

Cc354. **Ferrer, X.** 1984. L'ornitologia a Catalunya. *Butlleti de la Institucio Catalonia d'Historia Natural* 50: 187–197.

Cc355. **Fisher, J.** 1966. *The Shell bird book.* London, Ebury Pr. & Michael Joseph. 344 pp., illus.
Historical matter relating to ornithology in Britain and to the development of the British list of birds.

Cc356. **Gibbons, F., & D. Strom.** 1988. *Neighbours to the birds: A history of birdwatching in America.* New York, W. W. Norton. 364 pp., illus.

Cc357. **Harting, J. E.** 1871. *The birds of Shakespeare, or the ornithology of Shapespeare critically examined, explained and illustrated.* London, J. Van Voorst. xxii+321 pp. *Re-pub.* Chicago, IL, Argonaut Inc., 1965. xxxviii+321 pp.
1965 edition *includes* **G. Steiner.** Of men and birds: Prolegomena to the Birds of Shakespeare, pp. v–xxi.

Cc358. **Johansen, H.** 1952. Ornithology in Russia. *Ibis* 94: 1–48.

Cc358A. **Junge, G. C. A.** 1954. Ornithologisch onderzoek in de Indische Archipel. *Ardea* 41: 301–336.

Cc359. **Kastner, J.** 1986. *A world of watchers.* New York, A. A. Knopf. x+241 pp., illus.
"An informal history of the American passion for birds — from its scientific beginnings to the great birding boom of today." (blurb).

Cc360. **Kinnear, N. B.** 1952. The history of Indian mammalogy and ornithology. II: Birds. *Journal of the Bombay Natural History Society* 51: 104–110.

Cc361. **Leedy, D. L.** 1961. Some federal contributions to bird conservation during the period 1885–1960. *Auk* 78: 167–175.

Cc362. **McAtee, W. L.** 1953. North American bird records in the *Philosophical Transactions*, 1665–1800. *Journal of the Society for the Bibliography of Natural History* 3: 56–60; ___. 1955. North American birds of Virginia chroniclers, 1588–1686. *Ibid.* 3: 92–101; ___. 1957. The North American birds of Mark Catesby and Eleazar Albin. *Ibid.* 3: 177–194; ___. 1957. The North American birds of Linnaeus. *Ibid.* 3: 291–300; ___. 1963. The North American birds of Thomas Pennant: A review. *Ibid.* 4: 100–124.

Cc363. **McNamara, K. R.** 1990. The feathered scribe: The discourses of American ornithology before 1800. *William & Mary quarterly* 47: 210–234.

Cc364. **Medway, D. G.** 1979. Some ornithological results of Cook's third voyage. *Journal of the Society for the Bibliography of Natural History* 9: 315–351.

Cc365. ___. 1981. The contribution of Cook's third voyage to the ornithology of the Hawaiian Islands. *Pacific science* 35: 105–175, & 1 pl.

Cc366. **Pearse, T.** 1968. *Birds of the early explorers in the Northern Pacific.* Comox, B.C., the Author. 274 pp.

Cc367. **Pinto, O. M. de O.** 1979. *A ornitologia do Brasil através das idades (século XVI a século XIX).* São Paulo. 117 pp. *Brasiliensia documenta* no. 13.

Cc368. **Rounds, R. S.** 1990. *Men and birds in South America, 1492–1900.* Fort Bragg, Q.E.D. Pr. 190 pp., illus.

Cc369. **Sibley, C. G.** 1955. Ornithology. In: California Academy of Sciences. *A century of progress in the natural sciences 1853–1953*, pp. 629–659. San Francisco, CA.

Cc370. **Sterling, K. B.**, ed. 1974. *Contributions to the history of American ornithology.* Introduction by K. B. Sterling. New York, Arno Pr. 328–373+249+109 pp.
Includes **S. P. Fowler.** Ornithology of the United States: Its past and present history; **E. Coues.** Progress in American ornithology; **American Ornithologists' Union.** *Fifty years progress of American ornithology, 1883–1933.* On the occasion of its semi–centennial anniversary. New York, November 13–16, 1933. (Ed. by F. M. Chapman & T. S. Palmer.) Lancaster, PA (*see* Cc349); **C. F. Batchelder.** An account of the Nuttall Ornithological Club, 1873–1919.

Cc371. **Stimson, C.** 1976. Ornithology. In: J. M. Chalmers–Hunt. *Natural history auctions 1700–1972: A register of sales in the British Isles*, pp. 23–31. London, Sotheby Parke Bernet. xiii+189 pp.

Cc372. **Tate, P.** 1986. *Birds, men and books: A literary history of ornithology.* London, H. Sotheran. [8]+193 pp., illus.
British and American.

Cc373. **Welker, R. H.** 1955. *Birds and men: American birds in science, art, literature and conservation, 1800–1900.* Cambridge, MA, Belknap Pr. of Harvard Univ. Pr. viii+230 pp. *Re-pub.* New York, Atheneum, 1966.

Cc374. **Whittell, H. M.** 1954. [History]. In: *The literature of Australian birds: A history and a bibliography of Australian ornithology*, pp. 1–116. Perth, W.A., Brokenshaw. xi+116+788 pp.

See also Ba318–322 Ba325–326 Bb803–805 Ca56(Ainley) Cc787–788 Cc807 Cc816 Cc829(Steinbacher) Cc842 Cc867 Cc889 Cc951 Cc1013 & Cc1055.

BIBLIOGRAPHIES OF ORNITHOLOGY — GENERAL

• For a fuller list of bibliographies see:

Cc375. **Brasseler, H., & F. W. Bekierz.** 1984. Bibliographie ornithologischer Bibliographien. Bibliography on ornithological bibliographies. *Courier Forschungsinstitut Senckenberg* 72: 158 pp.; *Continued* **H. Brasseler.** 1987. Bibliographie ornithologischer Bibliographien II. Aves. Bibliography on ornithological bibliographies II. Aves. *Courier Forschungsinstitut Senckenberg* 99: 214 pp.

• A selection of major works:

Cc376. **Berndt, R., & W. Meise.** 1966. *Naturgeschichte der Vögel.* Vol. 3: *Bibliografien* [etc.]. Stuttgart, Franckh. vii+474 pp.

Cc377. **Giebel, C. A.** 1872–77. *Thesaurus ornithologiae. Repertorium der gesammten ornithologischen Literatur und Nomenclator sämmtlicher Gattungen und Art der Vögel nebst Synonymen und geographischer Verbreitung.* Leipzig, F. A. Brockhaus. 3 vols. ••Repertorium ornithologicum, vol. 1: 1–252.

Cc378. **Low, G. C.** 1931. *The literature of the Charadriiformes from 1894–1928.* 2nd ed. London, H. F. & G. Witherby. xiii+637 pp. ••*Replaces* 1st ed., 1924. xi+220 pp.

Cc379. **Miller, M. A.** 1986. *Birds: A guide to the literature.* New York, London, Garland Publishing. xii+887 pp. *Garland reference library of the humanities* vol. 680.

Cc380. **Olendorff, R. R. & S. E.** 1968. *An extensive bibliography on falconry, eagles, hawks, falcons and other diurnal birds of prey.* Fort Collins, CO, privately pub. 3 vols. iv+244 pp.

Cc381. **Pettingill, O. S.** 1956. Bibliography of life history studies. In: *Laboratory and field manual of ornithology*, 3rd ed., pp. 335–345. Minneapolia, MN, Burgess.

Cc382. **Strong, R. M.** 1939–46. *A bibliography of birds with special reference to anatomy, behavior, biochemistry, embryology, pathology, physiology, genetics, ecology, aviculture, economic ornithology, poultry culture, evolution.* Chicago, IL, Field Museum of Natural History. 3 vols. *Publication, Field Museum of Natural History, zoological series* vol. 25. ••Ca. 20,000 entries.

Cc383. **Tate, P.** 1985. *A century of bird books.* 2nd ed. London, H. F. & G. Witherby. 272 pp. ••*Replaces* ___. 1979. ___. London, [etc.]. 264 pp. Ornithological books published in the British Isles.

See also Library cats. Bc450–462 Bc583 Bc585 Bc588 Bc593 Bc596 & Bc601.

Regional

European (& Mediterranean)

Cc384. **Boano, G.** 1985. *Bibliografia ornitologica piemontese. Aggiornata al 1983.* Carmagnola. 193 pp.

Cc385. **Catuneanu, I. I., et al.** 1971. *Bibliografia Romana si straina asupra pasarilor din Romania. (Bibliographie Roumaine et etrangère concernant les oiseaux de Roumanie).* Bucharest, Comitetul de Stat Pentru Cultura si Arta, Directia Muzeelor. 321 pp.

Cc386. **Hulth, J. M.** 1899. *Öfversikt af faunistiskt och biologiskt vigtigare litteratur rörande nordens fåglar.* Stockholm. [ii]+16.

Cc387. **Irwin, R.** 1951. *British bird books: An index to British ornithology A.D. 1481 to A.D.* 1948. London, Grafton. xix+398 pp.

Cc388. **Karasek, J.** 1929. Bibliographiae ornithologicae

republicae Bohemo–Slovenicae conspectus. *Sbornik Klubu Prirodoveckeho v Brne* 11: 61–128.

Cc389. **Knorre, D. von.** 1984. *Ornithologische Veröffentlichungen aus Thüringen 1945–1981*. Jena, Friedrich–Schiller–Universität. 159 pp. *Bibliographische Mitteilungen der Universitätsbibliothek Jena* no. 40.

Cc390. **La Cour, T.** 1949. *Fuglebøger: Danske ornithologiske skrifter, 1648–1948*. Copenhagen, Om Bøger. 47 pp.

Cc391. **Legendre, M.** 1936. Bibliographie des faunes ornithologiques des regions françaises. *Memoires de la Société Ornithologique et Mammalogique de France* 4: 1–127. *Continued* ___. 1941. ___: Supplement. *L'Oiseau* n.s. 11: 49–61, 119–123.

Cc392. **Mayer, G. T.** 1982. Ornithologische Bibliographie Oberösterreichs 1850–1980. *Linzer biologische Beiträge* 14: 53–92.

Cc393. **Mosansky, A.** 1968. Ornitologicka bibliografia vychodneho Slovenska do roku 1965. *Zborník Vychodoslovenského Múzea v Kosiciach. Séria B, zoologia, botanika* 9B: 33–75.

Cc394. **Mullens, W. H., & H. K. Swann.** [1916–] 1917. *A bibliography of British ornithology from the earliest times to the end of 1912; including biographical accounts of the principal writers and bibliographies of their published works*. London, Macmillan & Co. xx+691 pp. Reprint Codicote, Wheldon & Wesley Ltd., 1986.

Cc395. ___ **& F. C. R. Jourdain.** 1919–20. *A geographical bibliography of British ornithology from the earliest times to the end of 1918, arranged under counties, being a record of printed books, published articles, notes, and records, relating to local avifauna*. London, Witherby & Co. viii+558 pp. *Continued* ___. 1923. ___*: Supplement. A chronological list of British birds*. London, [etc.]. xvii+42 pp.

Cc396. **Nilsson, R.** 1920. *Förteckning över Sveriges ornitologiska litteratur, rörande Svenska fågelfaunen (1729–1918)*. Lund, Gleerup. 160 pp.

Cc397. **Normand, N.** 1976. *Bibliographie ornithologique de la région parisienne*. Paris, Association Parisienne Ornithologique. [v]+119 ff. *L'Epeiche* no. spécial.

Cc398. **Ronsil, R.** 1948–49. *Bibliographie ornithologique française. Travaux publiés en langue française et en Latin en France et dans les Colonies Françaises de 1473 à 1944*. Paris, P. Lechevalier. 2 vols. *Encyclopédie ornithologique* vols. 8 & 9.

Cc399. **Rootsmae, I. & L.** 1983. *Eesti ornitoloogiline kirjandus 1656–1975*. Tartu, Eesti Looduseurijate. xxvi+317 pp.

Cc400. **Schaanning, H. T. L.** 1924. Bibliotheca ornithologica Norvegica, 1591–1924. *Norsk ornitologisk Tidsskrift* 2 (5): 57–131.

Cc401. **Schalow, H.** 1906. Beiträge zu einer ornithologischen Bibliographie des Atlas Gebietes. *Journal für Ornithologie* 54: 100–143.

Cc402. **Schiøler, E. L.** 1925–26. Kortfattet oversigt over Danmarks ornithologiske Litteratur [and] Kortfattet oversigt over ornithologiske Skrifter, behandlende Grønlands Fuglefauna. In: *Danmarks fugle med henblik paa de i Grønland, [etc.]*, vol. 1: 11–197 & vol. 2: 11–30. Copenhagen, Gyldendal.

Cc403. **Studer, T.** 1895. *Vögel*. Berne, Centralkommission für Schweizerische Landeskunde. xiv+43 pp. *Bibliographie der schweizerischen Landeskunde* IV.6.4.

Cc404. **Tschusi–Schnidhoffen, V. von.** 1879. Bibliographia ornithologica. Verzeichniss der gesammten ornithologischen Kiteratur der Österreich–Ungarischen Monarchi. *Verhandlungen der kaiserlich–königlichen zoologisch–botanischen Gesellschaft in Wien* 28: 491–544.

Cc405. **Wüst, W.** 1973. Bibliographie zur Avifauna Bayerns. *Abhandlungen, Bayerische Akademie der Wissenschaften, mathematisch–naturwissenschaftliche Klasse* 153:

203 pp.; *Continued* ___. 1985. ___: Nachtrag I. *Ibid*. 164: 92 pp.

North American

Cc406. **Anderson, A. H.** 1971. *A bibliography of Arizona ornithology*. Tucson, AZ. 272 pp.

Cc407. **Banks, R. C.** 1968. Annotated bibliography of Nevada ornithology since 1951. *Great Basin naturalist* 28: 49–60.

Cc408. **Brooks, E. A.** 1938. *A descriptive bibliography of West Virginia ornithology*. Newton Highlands, MA, the Author. iii+28 pp.

Cc409. **Coues, E.** 1878–80. (American ornithological bibliography. Pts. 1–4). Bibliographical appendix: List of faunal publications relating to North American ornithology [etc.]. In: Birds of the Colorado Valley [etc.] Vol. 1: Passeres to Laniidae. *Miscellaneous publications of the U. S. Geological Survey of the Territories* 11: 567–784, 1878. *Continued* ___. 1879. ___: Second instalment of American ornithological bibliography, being the titles of "Faunal Publications" relating to the Ornithology of the rest of America. *And* Third instalment ..., being those titles belonging to the Systematic Department of the whole bibliography. *Bulletin of the U. S. Geological Survey of the Territories* 5(2): 239–330, 1879; *Ibid*. 5(4): 521–1066, 1879; ___. 1880. Fourth instalment ..., being a list of faunal publications relating to British birds. *Smithsonian miscellaneous collections* 19: 359–482, 1880.

Cc410. **Greene, E. R., et al.** 1945. Birds of Georgia: A preliminary check-list and bibliography. *Occasional papers, Georgia Ornithological Society* 2: 111 pp.

Cc411. **Grinnell, J.** 1909. *A bibliography of California ornithology [1797–1907]*. Santa Clara, CA, Cooper Ornithological Club. 166 pp. *Pacific coast avifauna* no. 5; *Continued* ___. 1924. ___: *Second installment to end of 1923*. Berkeley, CA, [etc.] 191 pp. *Pacific coast avifauna* no. 16; ___. 1939. ___: *Third installment to end of 1938*. Ed. by A. H. Miller & H. W. Grinnell. Berkeley, CA, [etc.]. 235 pp. *Pacific coast avifauna* no. 26.

Cc412. **Grzybowski, J. A., & G. D. Schnell.** 1984. *Oklahoma ornithology: An annotated bibliography*. Norman, OK, Univ. of Oklahoma Pr. 192 pp. *A Stovall Museum publication.*

Cc413. **Lovell, H. B. & M.** 1949. A bibliography of Kentucky ornithology. *Occasional publications, Kentucky Ornithological Society* 1: 50 pp.

Cc414. **McAtee, W. L.** 1913. *Index to papers relating to the food of birds by members of the Biological Survey in publications of the United States Department of Agriculture, 1885–1911*. Washington, DC, U. S. Department of Agriculture, Bureau of Biological Survey. 69 pp. *Bulletin* no. 43.

Cc415. **Phillips, J. C.** 1930. *American game mammals and birds: A catalogue of books, 1582–1925, sport, natural history and conservation*. Boston, MA, Houghton. 638 pp.

Cc416. **Rylander, K., & B. White.** 1968. A bibliography of Texas birds. *Bulletin, Texas Ornithological Society* 2: 42–51.

Cc417. **Sage, J. H., L. B. Bishop, & W. P. Bliss.** 1913. Bibliography of the birds of Connecticut. *Bulletin, Connecticut State Geological & Natural History Survey* 20: 200–257.

Cc418. **Shaver, J. M.** 1931. A bibliography of Tennessee ornithology. *Journal of the Tennessee Academy of Science* 6: 179–190.

Cc419. **Stephens, T. C.** 1957. An annotated bibliography of South Dakota ornithology. *Occasional papers, Nebraska Ornithologists' Union* 2: 22 pp.

Cc420. ___. 1957. An annotated bibliography of Iowa ornithology. *Occasional papers, Nebraska Ornithologists' Union* 4: vi+114 pp.

Cc421. **Wheaton, J. M.** 1882. Bibliography of Ohio ornithology. *Report, Geological Survey of Ohio* 4 (1) Zoology & botany D: 594–612.

South & Central American (& Caribbean)

Cc422. **Marelli, C. A.** 1934. Contribuciones al estudio de la fauna Argentina. Bibliografía relativa a la ornitología. *Memorias del Jardin Zoológico, La Plata* 5: 37–106.

Cc423. **Rohl, E.** 1932. Apuntes para la historia y la bibliografía de la ornitología Venezolana. *Boletín de la Sociedad Venezolana de Ciencias Naturales* 1932 (6): 201–248.

Asian

Cc424. **Allouse, E.** 1954. *A bibliography of the vertebrate fauna of Iraq and neighbouring countries:* II. *Birds.* Baghdad, Iraq Natural History Museum. 44+[4] pp.

Cc425. **Hachisuka, M.** 1952. Bibliography of Chinese birds. *Quarterly journal of the Taiwan Museum* 5: 71–209.

Cc426. **Hoogerswerf, A.** 1953. An ornithological bibliography having particular reference to the study of the birds of Java. *Bulletin, Organisatie voor Natuurwetenschappelijk Onderzoek in Indonesië* 13, 14, 15 & 16: 168 pp.

Cc427. **Riley, J. H., & C. W. Richmond.** 1922. Partial bibliography of Chinese birds. *Journal of the North China branch of the Royal Asiatic Society* 53: 196–237.

Australasian (& Pacific)

Cc428. **Fisher, H. I.** 1947. Bibliography of Hawaiian birds since 1890. *Auk* 64: 78–97.

Cc429. **Mathews, G. M.** 1925. *Bibliography of the birds of Australia.* London, H. F. & G. Witherby. viii+149 pp. ••Supplements 4–5 of his *Birds of Australia.*

Cc430. **Oliver, H. C.** 1968. *Annotated index to some early New Zealand bird literature.* Wellington, N.Z., Department of Internal Affairs. x+222 pp. *Wildlife publication* no. 106.

Cc431. **Whittell, H. M.** 1954. *The literature of Australian birds: A history and a bibliography of Australian ornithology.* Perth, W.A., Brokenshaw. xi+116+788 pp.

African

Cc432. **Reichenow, A.** 1894. Bibliographia ornithologiae Aethiopicae. *Journal für Ornithologie* 42: 172–226.

Polar

Cc433. **Roberts, B.** 1941. A bibliography of Antarctic ornithology. In: British Museum (Natural History). *British Graham Land expedition 1934–37: Scientific reports*, vol. 1(9): 337–367. London.

Periodical bibliographies

See periodical bibliographies listed at Cc37–54 (especially Zoological record; *see* Cc50).

Herpetology

Cc434. **Adler, K.** 1979. A brief history of herpetology in North America before 1900. *Herpetological circular, Society for the Study of Amphibians & Reptiles* 8: 1–40.

Cc435. **Banta, B. H., & W. W. Tanner.** 1964. A brief historical resumé of herpetological studies in the Great Basin of the western United States. Part 1: The reptiles. *Great Basin naturalist* 24: 37–57.

Cc436. **Best, A. E.** 1968. The discovery of the mechanism of colour–changes in the chameleon. *Annals of science* 2: 147–167.

Cc437. **Bodson, L.** 1984. Living reptiles in captivity: A historical survey from the origins to the end of the XVIIth century. *Acta zoologica et pathologica Antverpiensia* 78: 15–32.

Cc438. **Gunther, A. C. L. G., & A. S. Woodward.** 1911. [Article] Reptiles. In: *Encyclopaedia Britannica.* 11th ed., vol. 23: 136–176. Cambridge, Cambridge Univ. Pr. History of herpetology, pp. 136–141.

Cc439. **Montori, A., J. Maluquer, & X. Pascual.** 1984. Sintesi historica dels estudis herpetologics als Paisos Catalans. *Butlletí de la Institucio Catalana d'Historia Historia* 50: 173–185.

Cc440. **Morris, R. & D.** 1965. *Men and snakes.* London, Hutchinson & Co. 224 pp.

Cc441. **Schmidt, K. P.** 1955. Herpetology. In: California Academy of Sciences. *A century of progress in the natural sciences 1853–1953*, pp. 591–627. San Francisco, CA.

Cc442. **Terent'ev, P. V.** 1957. Materialy k istorii otechestvennoi gerpetologii. [Materials for the history of Russian herpetology]. *Trudy Institut Istorii Estestvoznaniya i Tekhniki* 16: 97–122.

See also Biographies Ba329–331 Ba580 Ba756 Ba1020 Ba1541; Societies Bb846–848 Bb851; Misc. Cc772(Bellairs & Ball) Cc829(Mertens) & Cc871–872.

BIBLIOGRAPHIES OF HERPETOLOGY — GENERAL

• For a fuller list of bibliographies see:

Cc443. **Brasseler, H.** 1989. Bibliographie herpetologischer Bibliographien I. Amphibia + Reptilia. Bibliography on herpetological bibliographies I. *Courier Forschungsinstitut Senckenberg* 116: 214 pp.

• A selection of major works:

Cc444. **Gans, C., & T. S. Parsons.** 1970. Taxonomic literature on reptiles. In: C. Gans, ed. *Biology of the Reptilia.* Vol. 2: *Morphology B*, pp. 315–333. London, Academic Pr.

Cc445. **Russell, F. E., & R. S. Scharffenberg.** 1964. *Bibliography of snake venoms and venemous snakes.* West Covina, CA, Bibliographic Associates. viii+220 pp.

Regional

Cc446. **Allouse, E.** 1955. *A bibliography of the vertebrate fauna of Iraq and neighbouring countries:* III. *Reptiles and amphibians.* Baghdad, Iraq Natural History Museum. 21+[4] pp.

Cc447. **Banta, B. H.** 1965. An annotated chronological bibliography of the herpetology of the State of Nevada. *Wasmann journal of biology* 23 (1/2): 224 pp.

Cc448. **Boring, A. M.** 1945. *Chinese amphibians, living and fossil forms: A complete bibliography analysed, tabulated, annotated and indexed.* Ed. by P. Leroy. Peking, Inst. Geobiol. viii+152 pp. *Publication* no. 13.

Cc449. **Carpenter, C. C., & J. J. Krupa.** 1989. *Oklahoma herpetology: An annotated bibliography.* Norman, OK, Univ. of Oklahoma Pr. 258 pp.

Cc450. **Enge, K. N., & K. Dodd.** 1986. A bibliography of the herpetofauna of Florida. *Smithsonian Herpetological Information Service* 72: 68 pp.

Cc451. **Fischer–Sigwart, H.** 1898. *Reptilien und Amphibien.* Berne, Centralkommission für Schweizerische Landeskunde. xii+27 pp. *Bibliographie der schweizerischen Landeskunde* IV.6.5.

Cc452. **Gans, C.** 1949. A bibliography of the herpetology of Japan. *Bulletin of the American Museum of Natural History* 93: 389–496.

Cc453. **Jones, G. S.** 1969. A list of references to the Amphibia and Reptilia of southeast Asia. *Quarterly journal of the Taiwan Museum* 22: 91–120.

Cc454. **Marelli, C. A.** 1935. Contribuciones al estudio de la fauna Argentina. Bibliografía de reptiles y batracios. *Memorias del Jardin Zoologico de La Plata* 4(2): 1–52. *Also pub. separately* Buenos Aires, Coni, 1935. 52 pp.

Cc455. **Medem, F.** 1965. Bibliografia comentada de reptiles colombianos. *Revista de la Academia Colombiana de Ciencias Exactas, Físicas y Naturales correspondiente de la Española* 12: 299–346.

Cc456. **Parent, G. H.** 1982. Bibliographie de l'herpétofaune française. Paris, Secrétariat de la Faune et de la Flore. 431 pp. *Bibliographie des inventaires faunistiques de France: 1758–1979. Inventaire de la faune et de la flore* no. 17 & 18.

Cc457. **Vanzolini, P. E.** 1977–78. *An annotated bibliography of the land and fresh-water reptiles of South America (1758–1975).* São Paulo, Universidade de São Paulo, Museu de Zoologia. 2 vols. ••Vol. 1: 1758–1900; vol. 2: 1901–1975.

Cc458. **Welch, K. G.** 1982. *Herpetology of Africa: A checklist and bibliography of the orders Amphisbaenia, Sauria and Serpentes.* Malabar, FL, Robert E. Krieger. x+293 pp.

Cc459. **___.** 1983. *Herpetology of Europe and southwest Asia: A checklist and bibliography of the orders Amphisbaenia, Sauria and Serpentes.* Malabar, FL, Robert E. Krieger. viii+135 pp. ••*See* Addenda provided by **S. C. Anderson** in his review. *A S C newsletter* 12 (2): 19-21, 1984.

Cc460. **Wright, A. H.** 1949. Scientific and popular writers on American snakes (1517–1944): A check list and short bibliography. *Herpetologica* 5 supp. 1: 1–55.

Cc461. **___ & A. A.** 1962. *Handbook of snakes of the United States and Canada.* Vol. 3: *Bibliography.* Ithaca, NY, privately pub. ii+179 pp.

Periodical bibliographies

See periodical bibliographies listed at Cc37–54 (especially Zoological record; *see* Cc50).

Ichthyology

Cc462. **Bauchot, M.-L., J. Daget & R. Bauchot.** 1990. L'ichthyologie en France au début du XIXe siècle: L'histoire naturelle des poissons de Cuvier et Valenciennes. *Bulletin du Muséum National d'Histoire Naturelle* (4) 12A Supp. 1: 142 pp., illus.
Includes biographical essays on Cuvier and Valenciennes.

Cc462A. **Bornbusch, A. H.** 1989. Lacépède and Cuvier: A comparative case study of goals and methods in late 18th and early 19th century fish classification. *Journal of the history of biology* 22: 141–162.

Cc463. **Dey, S. C.** 1984. Ichthyological developments in Assam during the nineteenth century. *Indian journal for the history of science* 19: 297–313.

Cc464. **Dymond, J. R.** 1964. A history of ichthyology in Canada. *Copeia* 1964: 2–33.

Cc465. **Gudger, E. W.** 1934. The five great naturalists of the 16th century: Belon, Rondelet, Salviani, Gesner, and Aldrovandi. A chapter in the history of ichthyology. *Isis* 22: 21–40.

Cc466. **___.** 1936. Beginnings of fish teratology, 1555–1642. Belon, Rondelet, Gesner and Aldrovandi, the fathers of ichthyology, the first to figure abnormal fishes. *Scientific monthly* 43: 252–261.

Cc467. **Gunther, A. C. L. G., & G. A. Boulenger.** 1911. [Article] Ichthyology. In: *Encyclopaedia Britannica.* 11th ed., vol. 14: 243–270. Cambridge, Cambridge Univ. Pr.
See: History and literature down to 1880 *and* ___ from 1880, pp. 243–250.

Cc468. **Hubbs, C. L.** 1964. History of ichthyology in the United States after 1850. *Copeia* 1964: 42–60.

Cc469. **Hünemörder, C.** 1975. Die Geschichte der Fischbücher von Aristoteles bis zum Ende des 17. Jahrhunderts. *Schriften des Deutschen Schiffahrtsmuseums* 5: 185–200.

Cc470. **Jordan, D. S.** 1902. The history of ichthyology: An address. *Proceedings of the American Association for the Advancement of Science* 51: 427–456. Also pub. in *Science* n.s. 16: 241–258, 1902.

Cc471. **___.** 1905. The history of ichthyology. In: *Guide to the study of fishes*, vol. 1: 387–428. New York, H. Holt & Co.

Cc472. **Marston, R. B.** 1894. *Walton and some earlier writers on fish and fishing.* London, E. Stock. xxvii+264 pp.

Cc473. **Morovic, D.** 1971. Histoire des recherches ichtyologiques sur les côtes Yougoslaves de l'Adriatique au cours du XIXe siècle. *Actes de XIIe [1968] Congrès International d'Histoire des Sciences* 7: 81–86.

Cc474. **Myers, G. S.** 1964. A brief sketch of the history of ichthyology in America to the year 1850. *Copeia* 1964: 33–41.

Cc475. **Oppenheimer, J. M.** 1936. Historical introduction to the study of teleostean development. *Osiris* 2: 124–148.

Cc476. **Thomazi, A.** 1947. *Histoire de la pêche.* Paris, Payot. 645 pp.

See also Biographies Ba330 Ba549–550 Ba818 Ba883–884 Ba1037–1040; Ba1441 Ba1569–1571; Societies Bb846–850; Misc. Cc829(Klausewitz) Cc876 Cc882–883 Cc934–935 Cc945–947 Cc952 Cc1015 & D94.

BIBLIOGRAPHIES OF ICHTHYOLOGY — GENERAL

• A selection of major works:

Cc477. **Artedi, P.** [1788–] 1789–93. *Bibliotheca ichthyologica, seu historia litteraria ichthyologiæ, [etc.].* Pars I. Emendata et aucta a J. J. Walbaumi. Greifswald. 230 pp. ••*Replaces* ___. 1738. *Bibliotheca ichthyologica [etc.].* Leiden. 66 pp.

Cc478. **Bosgoed, D. M.** 1874. *Bibliotheca ichthyologicae et piscatoria – catalogus van boeken en geschriften over de natuurlijke geschiedenis van de visschen en walvisschen, de kunstmatige vischteelt, de visscherijen, de wetgeving op de visscherijen [etc.]. ... Catalogue de livres d'écrits sur l'histoire naturelle des poissons et des cétacés, la pisciculture, les pêches, la législation des pêches [etc.].* Haarlem, E. Loosjes. xxvi+474 pp. ••*Replaces* ___. 1871. *Proeve van eene ichthyologische bibliographie; catalogus [etc.].* Haarlem. 247 pp.

Cc479. **Dean, B.** 1916–23. *A bibliography of fishes.* Vols. 1–2 enlarged and ed. by C. R. Eastman; vol. 3 extended and ed. by E. W. Gudger, with the cooperation of A. W. Henn. New York, American Museum of Natural History. 3 vols. *Re-pub.* New York, Russell & Russell, 1962. ••Ca. 50,000 entries. *See also* **E. W. Gudger.** 1924. 1925. The classified continuation card catalogue of the Bibliography of fishes. *Science* 61: 516–517.

Cc480. **Healey, P. J., & F. W. H. Beamish.** 1984. *The Cyclostomata: An annotated bibliography.* Ann Arbor, MI, Great Lakes Fishery Commission. 320 pp.

Cc481. **Romanov, N. S.** 1968. *Annotated bibliography on population dynamics, behavior and distribution of fish, marine mammals, commercial invertebrates and algae (covering publications of 1956).* Jerusalem, Israel Program for Scientific Translations. [iv]+168 pp.

Cc482. **Wheeler, A. C.** 1979. The sources of Linnaeus's knowledge of fishes. *Svenska Linné–Sällskapets årsskrift* 1978 commemorative vol: 156–211. ••Bibliography of sources used for fish taxa in the 10th, & 12th editions of *Systema naturae*, pp. 186–205.

See also Library cats. Bc181A Bc463–464 & Bc466.

Regional

Cc483. **Allouse, E.** 1955. *A bibliography of the vertebrate fauna of Iraq and neighbouring countries:* IV. *Fishes.* Baghdad, Iraq Natural History Museum. ii+34 pp.

Cc484. **Coad, B., & L. al–Hussan.** 1988. *A bibliography of the fishes of the Tigris–Euphrates basin. Bibliographie der Fische des Euphrat–Tigris–Beckens.* Heidelberg, M. Kasparek Verlag. 54 pp.

Cc485. **Diaz, E. L.** 1965. Bibliographic material on the fishes of Colombia and northwestern South America. *F A O fisheries technical paper* 53: [iv]+72 pp.

Cc486. **Fischer–Sigwart, H.** 1900. *Fische.* Berne, Centralkommission für Schweizerische Landeskunde. xiv+85 pp. *Bibliographie der schweizerischen Landeskunde* IV.6.5.

Cc487. **Kasparek, M.**, ed. 1986. *Zoologische Bibliographie der Turkei. Pisces, Amphibia, Reptilia.* Heidelberg, M. Kasparek Verlag. 118 pp.

Cc488. **McDowall, R. M.** 1964. A bibliography of the indigenous freshwater fishes of New Zealand. *Transactions of the Royal Society of New Zealand, Zoology* 5: 1–38. ••*See also* **D. Bassett.** 1961. *A bibliography of the freshwater fishes of New Zealand.* Wellington, N.Z., National Library Service Library School. iv+46 pp. *Library School bibliographical series* no. 5.

Cc489. **Marelli, C. A.** 1930. Bibliografía de ictiología marina, fluvial y lacustre, pesca y piscicultura de la República Argentina. *Memorias del Jardin Zoologico de La Plata* 3: 1–62.

Cc490. **Matthes, H.** 1973. *A bibliography of African freshwater fish. Bibliographie des poissons d'eau douce de l'Afrique.* Rome, Food & Agriculture Organization of United Nations. xv+299 pp.

Cc491. **Morovic, D.** 1950. Prilog bibliografiji Jadranskog ribarstva. (Contributo alla bibliografia della pesca nell'Adriatico (1864–1949). *Posebna izdanja, Institut za Oceanografiju i Ribarstvo u Splitu* 1: 143 pp.

Cc492. **Murty, V. S., et al.** 1968. Bibliography of marine fisheries and oceanography of the Indian Ocean. *Bulletin, Central Marine Fisheries Research Institute* 1: 1–208.

Cc493. **Okada, Y.** 1960. Bibliography. In: Studies on the freshwater fishes of Japan. *Journal of the Faculty of Fisheries of the University of Mie* 4: 860 pp. ••Historical bibliography, pp. 3–230, general bibliography, pp. 763–846.

Cc494. **___, & K. Matsubara.** 1953. *Bibliography of fishes in Japan (1612–1950).* Mie, Prefectural Univ., Faculty of Fisheries. iv+228 pp.

Cc495. **Phillips, W. J.** 1927. Bibliography of New Zealand fishes. *Fish bulletin, New Zealand Marine Department* 1: 1–68.

Cc496. **Romanov, N. S.** 1955. *Ukazatel' literatury po rybnomu khozyaistvu yuzhnikh basseinov SSSR za 1918–1953gg. Syr'evaya baza i vosproizvodstvo zapasov promyslovykh ryb.* Moscow, Akademii Nauk SSSR, Otdelenie Biologicheskikh Nauk, Ikhtiologicheskaya Komissiya. 294 pp. ••English translation: *Annotated bibliography on fisheries of the southern basins of the U. S. S. R. (1918–1953): Fish resources and replenishment of commercial fish stocks.* Jerusalem, Israel Program for Scientific Translations, 1968. iii+400 pp.

Cc497. **___.** 1959. *Ukazatel' literatury po rybnomu khozyaistvu Dal'nego Vostoka za 1923–1956gg.* Moscow, Akademii Nauk SSSR. 291 pp. ••3,960 references. English translation: *Annotated bibliography on Far Eastern aquatic fauna, flora and fisheries.* Jerusalem, Israel Program for Scientific Translations, 1964. [iii]+391 pp.

Cc498. **Whitley, G. P.** 1964. A survey of Australian ichthyology. *Proceedings of the Linnaean Society of New South Wales* 89: 11–127. ••Chiefly bibliographical.

Periodical bibliographies

See periodical bibliographies listed at Cc37–54 (especially Zoological record; *see* Cc50).

INVERTEBRATE ZOOLOGY

BIBLIOGRAPHIES OF INVERTEBRATES — GENERAL

Cc499. **Allouse, E.** 1956. *A bibliography of the invertebrate fauna of Iraq and neighbouring countries.* Baghdad, Iraq Natural History Museum. [ii]+46 pp. *Publications* no. 8.

Cc500. **Davies, B. R., et al.** 1982. *A bibliography of African inland water invertebrates (to 1980).* Pretoria, CSIR. 418 pp. *South African National Scientific programmes report* no. 58.

Entomology — General

Cc501. **Allen, G. E.** 1975. The introduction of *Drosophila* into the study of heredity and evolution: 1900–1910. *Isis* 66: 322–333.

Cc502. **Bodenheimer, F. S.** 1928–29. *Tabellen des vor–Linné'schen Insekten–Bestandes.* Berlin, W. Junk. 176 pp.

Cc503. ___. 1928–29. *Materialien zur Geschichte der Entomologie bis Linné.* Berlin, W. Junk. 2 vols.
1: Die orientalische Urzeit. Das europäische Altertum. Das Mitelalter. Die Neuzeit; 2: Die Neuzeit (Fortsetz.).

Cc504. **Brémond, J., & J. Lessertisseur.** 1973. Lamarck et l'entomologie. *Revue d'histoire des sciences et leurs applications* 26: 231–250.

Cc505. **Chesnova, L. V.** 1980. *Preemstvennost' nauchnykh shkol v entomologii. [The succession of scientific schools in entomology.]* Moscow, Izd. "Nauka." 172 pp., illus.

Cc506. **Cushing, E. C.** 1957. *History of entomology in World War II.* Washington, DC, Smithsonian Institution. vi+117 pp. & 9 pl.

Cc507. **Dumortier, B.** 1966. La stridulation et l'audition chez les insectes orthoptères: Aperçu historique sur les idées et les decouvertes jusqu'au début du XXe siècle. *Revue d'histoire des sciences et de leurs applications* 19: 1–28.

Cc508. **Eiselt, J. N.** 1836. *Geschichte, Systematik und Literatur der Insektenkunde, von den ältesten Zeiten bis auf die gegenwart.* Leipzig, C. H. F. Hartmann. viii+250 pp.

Cc509. **Essig, E. O.** 1931. *A history of entomology.* New York, Macmillan. x+1,039 pp. *Re-pub.* New York, Hafner Publishing Co., 1965.
Strongly biassed towards American work.

Cc510. ___. 1936. Sketch history of entomology. *Osiris* 2: 80–123.

Cc511. [Not used.]

Cc512. **Marlatt, C. L.** 1898. A brief historical survey of the science of entomology. *Proceedings of the Entomological Society of Washington* 4: 83–120.

Cc513. **Mickel, C. E.** 1955. Some milestones in the history of insect classification. *Canadian entomologist* 87: 57–66.

Cc514. **Müller, G. H.** 1985. The development of thought on the respiration of insects. *History & philosophy of the life sciences* 7: 301–314.

Cc515. **Osborn, H.** 1937. *Fragments of entomological history including some personal recollections of men and events.* Columbus, OH, privately pub. vii+394 pp. & 47 pl.
Strongly biassed towards American work. Many portraits.

Cc516. ___. 1952. *A brief history of entomology, including time of Demosthenes and Aristotle to modern times.* With over 500 portraits. Columbus, OH, Spahr & Glenn Co. iii+303 pp. & 58 pl.

Cc517. **Prete, F. R., & M. M. Wolfe.** 1992. Religious supplicant, seductive cannibal, or reflex machine? In search of the praying mantis. *Journal of the history of biology* 25: 91–136.

Cc518. **Ross, E., et al.** 1955. Systematic entomology. In: California Academy of Sciences. *A century of progress in the natural sciences 1853–1953*, pp. 485–590. San Francisco, CA.

Cc519. **Smith, R. F., T. E. Mittler, & C. N. Smith**, eds. 1973. *History of entomology*. Palo Alto, CA, Annual Reviews Inc. in cooperation with the Entomological Society of America. vii+517 pp.
20 contributions by various authors. *Includes* **M. Beier.** The early naturalists and anatomists during the Renaissance and seventeenth century, pp. 81–94; **S. L. Tuxen.** Entomology systematizes and describes: 1700–1815, pp. 95–118; **C. H. Lindroth.** Systematics specializes between Fabricius and Darwin: 1800–1859, pp. 119–154; **H. H. Ross.** Evolution and phylogeny, pp. 171–184; **A. G. Richards.** Anatomy and morphology, pp. 185–202; **V. B. Wigglesworth.** The history of insect physiology, pp. 203–228; **H. G. Andrewartha & L. C. Birch.** The history of insect ecology, pp. 229–266; **S. W. Brown.** Genetics — the long story, pp. 407–432; **G. Richard.** The historical development of nineteenth and twentieth century studies on the behavior of insects, pp. 477–502.

Cc520. **Weidner, H.** 1983. Schriften von Liebhaber–Entomologen im ausgehenden 18. und beginnenden 19. Jahrhundert. *Entomologische Mitteilungen aus dem Zoologischen Museum Hamburg* 7 (117): 279–342.

Cc521. **Weiss, H. B.** 1915. Some old classifications of insects. *Canadian entomologist* 47: 369–376.

Cc522. **Wilson, H. F., & M. H. Dorner.** 1937. *The historical development of insect classification*. Madison, WI, privately pub. 133 pp.

Cc523. **Winsor, M. P.** 1976. The development of Linnaean insect classification. *Taxon* 25: 57–67.

See also Biographies Ba333 Ba339 Ba458 Ba646 Ba671 Ba744 Ba778–784 Ba793 Ba850A Ba905 Ba952–953 Ba1093–1094 Ba1310 Ba1325 Ba1330 Ba1333 Ba1414–1417 Ba1435 Ba1475 Ba1525 Ba1607 Ba1616; Organzations Bb22 Bb852–894; Misc. Cc782 Cc1016–1018 Cc1020 Cc1053 Cc1063 D90 D95–96 & D99.

Entomology — Regional

Cc524. **Allan, P. B. M.** 1943. *Talking of moths*. Newtown, Montgomery. 340 pp.
Includes historical information on insect collecting.

Cc525. **Baeta Neves, C. M. L.** 1986. A entomologia em Portugal até aos fins do século XIX. In: Academia das Ciências de Lisboa. *Colóquio sobre a história e desenvolvimento da ciência em Portugal. I Colóquio – até ao século XX. Lisboa, 15 a 19 de Abril de 1985*. Lisbon. 2 vols. (1,407 pp.).

Cc526. **Barnes, J. K.** 1985. Insects in the new nation: A cultural context for the emergence of American entomology. *Bulletin, Entomological Society of America* 31: 21–30.

Cc527. **Beirne, B. P.** 1985. Irish entomology: The first hundred years. *Irish naturalists journal, special entomological supplement*: 40 pp.

Cc528. **Calvert, P. P.** 1928. A contribution to the history of entomology in North America. *Transactions of the American Entomological Society* 52 (supp.): i–lxiii.

Cc529. **Chesnova, L. V.** 1968. *Essays from the history of applied entomology in Russia*. [Translated by B. Bojarski. Ed. by E. Donahaye.] Jerusalem, Israel Program for Scientific Translations. 90 pp., illus.

Translation of *Ocherki iz istorii prikladnoi entomologii v Rossi.* Moscow, Akademiya Nauk SSSR, 1962.

Cc530. ___. 1988. *Osnovnie etapi razvitiya ekologii nasekomykh v SSSR.\ [Fundamental stages of the development of insect ecology in the USSR].* Moscow, Izd. "Nauka." 175 pp.

Cc531. **Deutsche Entomologische Gesellschaft.** 1906. *Ein Beitrag zur Geschichte der systematischen Entomologie: Dem Gründer der Deutschen Entomologischen Gesellschaft, des Deutschen Entomologischen National–Museums und des Entomologischen Vereins in Berlin zum 50–jährigen Jubilaeum als entomologischer Präsident und Redakteur und zur Feier des goldenen Doktor–Jubilaeums gewidmet von der Deutschen Entomologischen Gesellschaft. Eine Festgabe für alle Mitglieder.* By G. Kraatz. Berlin. 164 pp. & 5 pl.

Cc532. **Entomological Society of Canada.** 1964. *Centennial of entomology in Canada, 1863–1963. Proceedings of a commemorative joint meeting of the Entomological Society of Canada and the Entomological Society of Ontario, Carleton University, Ottawa, 3–6 September 1963.* Ottawa. 475 pp. *Canadian entomologist* vol. 96(1/2).

Cc533. **Entomological Society of India.** 1964. *Entomology in India, 1938–1963, [etc.].* [New Delhi.] [xii]+529 pp. *Indian journal of entomology* silver jubilee number; *Continued* ___. 1965. ___*: Supplement.* [New Delhi.] 147 pp.

Cc534. **Essig, E. O.** 1926. *Outlines of the history of American entomology.* Berkeley, CA. 72 pp.
Mimeographied publication.

Cc535. **Everly, R. T., et al.** 1955. Symposium on a century of entomology in Indiana. *Proceedings of the Indiana Academy of Science* 64: 140–172.

Cc536. **Glen, R.** 1956. Entomology in Canada up to 1956: A review of developments and accomplishments. *Canadian entomologist* 88: 290–371.

Cc537. **Goidanich, A.** 1975. *Italian men, histories and insects in the science of then past. Uomini, storie e insetti italiani nelle scienza del passato.* Florence, Istituto Sperimentale per la Zoologia Agraria. 1,060 pp. *Redia* vols. 57 & 58.

Cc537A. **Hatch, M. H.** 1949. *A century of entomology in the Pacific Northwest.* Seattle, WA, Univ. of Washington Pr. v+42 pp., illus.

Cc538. **Henriksen, K. L.** 1921–37. Oversigt over Dansk entomologisk historie. *Entomologiske meddelelser* 15: 578 pp., ports.

Cc539. **Highley, E., & R. W. Taylor**, eds. 1983. *Australian systematic entomology: A bicentenary perspective.* Melbourne, Vic., Commonwealth Scientific & Industrial Research Organization. [vii]+147 pp.
Includes **E. N. Marks.** *Dramatis personae*: An anecdotal account of some historical figures; **T. R. New.** Systematics and ecology: Reflections from interface; **I. F. B. Common.** Landmarks in the taxonomy of Lepidoptera; **J. A. L. Watson & F. J. Gay.** Taxonomy and applied entomology of Australian termites: A small order in perspective; **R. H. Crozier.** Genetics and insect systematics: Retrospect and prospect; **R. W. Taylor.** Descriptive taxonomy; past, present and future.

Cc540. **Howard, L. O.** 1930. A history of applied entomology (somewhat anecdotal). *Smithsonian miscellaneous collections* 84: viii+545 pp. & 51 pl.
Strongly biassed towards American work. *Also* ___. 1931. The rise of applied entomology in the United States. *Annual report, Smithsonian Institution* 1930: 387–393.

Cc541. **Hunt, J. M. Chalmers–.** 1976. Entomological sales. In: *Natural history auctions 1700–1972: A register of sales in the British Isles*, pp. 3–14. London, Sotheby Parke Bernet. xiii+189 pp.

Cc542. **Illingworth, J. F.** 1923. *Early references to Hawaiian entomology*. Honolulu, HI, Bernice P. Bishop Museum. 63 pp. *Bulletin* no. 2.
Chronologically arranged bibliography, pp. 19–50.

Cc543. **Kevan, D. K. McE.** 1981. The terrestrial arthropods of the Bermudas: An historical review of our knowledge. *Archives of natural history* 10: 1–29.

Cc544. **Lamas, G. & J. M.** 1980. Introducción a la historia dela entomología en el Peru. *Revista Peruana de entomología* 23: 17–37.

Cc545. **Liu, G.** 1939. Some extracts from the history of entomology in China. *Psyche* 46: 23–28.

Cc546. **Maso, A., F. Vallhonrat, & J. J. Perez.** 1984. La lepidopterologia a Catalunya. *Butlleti de la Institucio Catalana d'Historia Natural* 50: 147–154.

Cc547. **Montalenti, G.** 1983. L'evoluzionismo negli entomologi Piemontesi dell' 800. *Atti Congresso Nazionale Italiano di Entomologia* 13: 15–28.

Cc548. **Musgrave, A.** 1930. The history of Australian entomological research. *Australian zoologist* 6: 189–203.

Cc549. **Nash, R.** 1983. A brief summary of the development of entomology in Ireland during the years 1790–1870. *Irish naturalists' journal* 21: 145–192.

Cc550. **Papavero, N.** 1971–76. *Essays on the history of neotropical dipterology, with special reference to collectors (1750–1905)*. São Paulo, Universidad de São Paulo, Museu de Zoologia. 3 pts.

Cc551. **Riegert, P. W.** 1980. *From arsenic to DDT: A history of entomology in western Canada*. Toronto, Univ. of Toronto Pr. xii+357 pp., illus.

Cc551A. **Roepke, W. K. J.** 1951. History of entomology in the Dutch East Indies. In: International Congress of Entomology, 9th, Amsterdam. *Entomology in the Netherlands and their overseas territories*, pp. 1–20. Amsterdam.

Cc552. **Silva Figueroa, C.** 1914. Reseña histórica y bibliográfica de la entomología Chilena. *Boletín del Museo Nacional de Chile* 7: 166–193.

Cc553. **Société Entomologique de France.** 1983. Premier Congrès Internationale des Entomologistes d'Expression Française, Paris, 1982. Compte rendus des travaux ... Groupe 1. *Bulletin de la Société Entomologique de France* 88(1–2): 1–166.
Includes **C. Dupuis.** La volunte d'etre entomologiste ..., pp. 18–38; **R. Codreanu.** La signification evolutive trans-disciplinaire de l'oeuvre René Jeannel, pp. 48–57; **M. Leon-Dufour.** Quelques aspects de la personnalité de Leon Dufour (1780–1865), pp. 58–65; **M. Sartori & D. Cherix.** Histoire de l'étude des insectes sociaux en Suisse à travers l'oeuvre d'Auguste Forel, pp. 66–74; **L. Botosaneanu.** Histoire de la trichoptérologie, pp. 74–81; **J. Leclercq.** Succes, mais aussi obstacles epistémiologiques rencontres dans la "Cartographie des Invertebres Europeens," pp. 81–97; **C. Dufour, W. Geiger, & J. P. Haeni.** Les méthodes mises en oeuvre par la diptérologie helvétique, pp. 98–109; **J. C. & J. Léonide.** La rigueur méthodologique en biotaxonomie: L'Exemple des *Blaesoxipha*, pp. 110–121; **A. Orian.** Histoire de l'entomologie d'expression française dans les pays de l'Ocean Indien, p.122; **C. Verstraeten.** Les grand noms de l'entomologie francophone en Belgique, pp. 122–134; **M. Farley.** L'institutionalisation de l'entomologie française, pp. 134–143; **J. Wuest.** l'Entomologie génévoise, du XVIIIe siècle à nos jours, pp. 143–154; **C. Herbulot.** Cinq grands lepidoptéristes français du siècle dernier, pp. 154–157; **M. Ghilarov.** Le *Referativnyi Zhurnal, Entomologia*, pp. 163–165.

Cc554. **Sorensen, C.** 1988. The rise of government sponsored applied entomology, 1840–1870. *Agricultural history* 62(2): 98–115.

Cc555. **Spencer, G. J.** 1964. A century of entomology in Canada. *Canadian entomologist* 96: 33–59.

Cc556. **Weidner, H.** 1967. Geschichte der Entomologie im Hamburg. *Abhandlungen und Verhandlungen des Naturwissenschaftlichen Vereins in Hamburg* 9 Supp: 387 pp.

Cc557. **Weiss, H. B.** 1936. *The pioneer century of American entomology.* New Brunswick, NJ, privately published. 320 pp. *Re-pub.* Lemont, PA, Lion Books, 1981.

Cc558. **Wiggins, G. B.**, ed. 1966. *Centennial of entomology in Canada, 1863–1963: A tribute to E. M. Walker.* Toronto, Univ. of Ontario Pr. 94 pp. & 9 pl. *Life sciences contribution, Royal Ontario Museum* no. 69.

Cc559. **Wilkinson, R. S.** 1966–68. English entomological methods in the seventeenth and eighteenth centuries. Part 1: To 1720. *Entomologists' record* 78: 143–151, 1966; ___. Part 2: Wilkes and Dutfield. *Ibid.* 78: 285–292, 1966; ___. Part 3: Moses Harris' *The Aurelian. Ibid.* 80: 193–200, 1968.

Cc560. **Willink, A.** 1969. Contribucion a la historia de la entomologia Argentina. *Miscelanea, Fundacion e Instituto Miguel Lillo* 28: 30 pp.

Cc561. **Yepez, F. F.** 1978. Contribución a la historia de la entomología en Venezuela. *Revista de la Facultad de Agronomia, Universidad Central de Venezuela* 26: 11–27.

See also Ba332 Ba334–338 Ba340 Ba1568A Bb593(Calvert) Cc821 Cc824 Cc829(Franz) Cc849 Cc861 Cc880 Cc931–932 Cc941–944 Cc956 Cc1019 & Cc1021.

BIBLIOGRAPHIES OF ENTOMOLOGY — GENERAL

• For fuller lists of bibliographies see:

Cc562. **Chamberlin, W. J.** 1952. *Entomological nomenclature and literature.* 3nd ed., rev. & enl. Dubuque, IA, Brown. vii+141 pp.

Cc563. **Gilbert, P., & C. J. Hamilton.** 1990. *Entomology: A guide to information sources.* 2nd ed. Poole, Mansell. 272 pp. ••*Replaces* ___. 1983. ___. London, Mansell. viii+237 pp. List of reference sources, entomological suppliers, photo sources, libraries and societies.

Cc564. **Rohlfien, K.** 1977. Bibliographie entomologischer Bibliographien (1920–1970). *Beiträge zur Entomologie* 27: 313–379. ••*Replaces* **C. Dupuis.** 1955. Les sources bibliographiques de l'entomologiste. 1re partie: Bibliographies de bibliographies, bibliographies rétrospectives et courantes. Sources diverses. Bibliographies des ordres d'insectes. *Cahiers des naturalistes* n.s. 10 supp.: 77–112.

• A selection of major works:

Cc565. **Derksen, W., & U. Scheiding.** 1963–75. *Index litteraturae entomologicae.* Serie II: *Die Welt-Literatur über die gesamte Entomologie von 1864 bis 1900.* Berlin, Deutsche Akademie der Landwirtschaftswissenschaften zu Berlin. 5 vols. ••For background information *see* **H. Sachtleben.** 1949. Über Bibliographie und "Index Litteraturae Entomologicae." *Entomon* 1: 2–4.

Cc566. **Hagen, H. A.** 1862–63. *Bibliotheca entomologica. Die Litteratur über das ganze gebiet der Entomologie bis zum Jahre 1862.* Leipzig, W. Engelmann. 2 vols. *Re-pub.* Weinheim, J. Cramer; Codicote, Wheldon & Wesley, 1960. 2 vols in 1. *Historiae naturalis classica* vol. 13. For addenda see: **G. Kraatz.** 1874. Ergänzungen und Nachtrage. *Berliner entomologische Zeitschrift* 18: 209–226; **H. M. Schmidt-Gäbel.** 1876. Zusätze und Berichtigungen. *Deutsche entomologische Zeitschrift* 20: 145–160; **K. W. von Dalla Torre.** 1878–81. Addenda und Corrigenda. *Entomologische Nachrichten* 4: 324–330; *Ibid.* 6: 125–129, 137–140, 168–171, 261–267, 1880; *Ibid.* 7: 45–48, 163–170, 1881; **A. Müller.** 1873. Contributions to entomological bibliography up to 1862 [No. 1]. *Transactions of the Entomological Society of London* 1873: 207–217; ___. 1873. ___. *No. 2 [–3].* London, E. W. Janson. 15 & 16 pp.

Cc567. **Horn, W., & S. Schenkling.** 1928–29. *Index litteraturae entomologicae.* Serie I: *Die Welt–Literatur über die gesamte Entomologie bis inklusive 1863.* Berlin–Dahlem, privately pub. 4 vols. *Also* **R. Gaedike & O. Smetana.** 1978–84. Ergänzungen und Berichtigungen zu Walter Horn und Sigmund Schenkling *Index litteraturae entomologicae. Beiträge zur Entomologie* 28: 329–436, 1978 & *Ibid.* 34: 167–291, 1984.

Cc568. **Kéler, S. von.** 1963. Übersicht der lexikalischen und der Handbuchliteratur. B: Hand– und Lehrbücher der Entomologie. In: *Entomologisches wörterbuch mit besonderer Berücksichtigung der morphologisches Terminologie.* 3. Aufl., pp. 15–29. Berlin, Akademie–Verlagsgesellschaft. ••1. Allgemeine Entomologie; 2. Forst– und landwirtschaftliche Entomologie; 3. Medizinische und veterinärmedizinische Entomologie. Chronologically arranged lists.

Cc569. **Nodier, J. C. E.** 1793. *Bibliographie entomologique; ou catalogue raisonné des ouvrages relatifs à l'entomologie et aux insectes, avec des notes critiques et l'exposition des méthodes.* Paris, Moutardier. viii+64 pp.

Cc570. **Percheron, A.** 1837. *Bibliographie entomologique, comprenant l'indication par ordre alphabétique de noms d'auteurs 1° des ouvrages entomologiques publiés en France et à l'étranger depuis les temps le plus reculés jusque et y compris l'année 1834; 2° des monographies et mémoires contenus dans les recueils, journaux et collections académiques françaises et étrangères, accompagnée de notices sur les ouvrages périodiques, les dictionnaires et les mémoires des sociétés savantes; suivie d'une table méthodique et chronologique des matières.* Paris, J. B. Baillière. 2 vols.

Cc571. **Rautenberg, L. E.** 1953–57. Beiträge zur Kenntniss alter naturwissenschaftlicher Werke. *Beiträge zur Entomologie* 3: 105–111, 618–626, 1953; 4: 99–104, 231–233, 656–662, 1954; 5: 60–69, 432–437, 640–649, 1955; 6: 180–190, 337–353, 1956; 7: 429–444, 609–631, 1957. ••Chiefly entomological works are described.

See also Library cats. Bc467–477 Bc581 Bc592 & Bc599.

Regional

Cc572. **Banks, N.** 1910. A list of works on North American entomology. *Bulletin, Bureau of Entomology, U. S. Department of Agriculture* 81: 1–120. ••*Replaces* ___. 1900. List of works on North American entomology. *Bulletin of the Division of Entomology, United States Department of Agriculture* n.s. 24: 1–95.

Cc573. **Bogdanov–Kat'kov, N. N.** 1924. *Russkaya literatura po prikladnoi entomologii (preimushchestvenno sel'sko-khozyaistvennaya).* Leningrad, Gosizdat'. xii+224 pp.

Cc574. **Conci, C.** 1975. Repertorio delle biografie e bibliografia degli scrittori e cultori Italiani di entomologia. *Memorie della Societa Entomologica Italiana* 48: 817–1069.

Cc575. **Costa Lima, A. M. da.** 1936. Bibliographia entomologica (trabalhos publicados no Brazil). In: *Terceiro catalogo dos insectos que vivem nas plantas do Brazil*, pp. 3–104. Rio de Janeiro, Directoria de Estatistica da Producção, Seçcão de Publicidade.

Cc576. **D'Aubigny, J. Le D.** 1980. *Insectes I.* Étude réalisée sour la direction de J.–C. Lefeuvre. Paris, Secrétariat de la Faune et de la Flore. 91 ff. *Bibliographie des inventaires faunistiques de France: 1758–1979. Inventaire de la faune et de la flore* no. 3–8.

Cc577. **Göllner–Scheiding, U.** 1967–71. Bibliographie der Bestimmungstabellen europäischer Insecten (1880–1963). Teil I: Apterygota bis Siphonaptera. Beiträge zur Entomologie 17: 697–958, 1967; Teil II: Hymenoptera. *Mitteilungen aus d. zool mus Berlin* 45: 1–156, 1969; Teil III: Coleoptera & Strepsiptera. *Deutsche entomologische Zeitschrift* n.s. 17: 33–118, 433–476, 1970; *Ibid.* 18: 1–84, 287–360, 1971.

Continued *Beiträge zur Entomologie* 26: 49–166, 1976; *Ibid.* 31: 235–304, 1981; *Ibid.* 36: 261–319, 1986; *Ibid.* 38: 239–276, 1988.

Cc578. **Gressitt, J. L., & J. J. H. Szent-Ivany.** 1968. Bibliography of New Guinea entomology. *Pacific insects monograph* 18: 674 pp.

Cc579. **Hobby, B. M.** 1932–35. A bibliography of entomological notes and papers contained in the serial publications issued by local scientific societies in the British Isles – part 1 (–3). *Transactions of the Entomological Society of the South of England* 8 (2): 117–139, 1932; *Transactions of the Society for British Entomology* 1 (1): 85–102, 1934; *Ibid.* 2 (2): 167–233, 1935.

Cc580. **Hunt, J. M. Chalmers–.** 1989. *Local lists of Lepidoptera, or, a bibliographic catalogue of local lists and regional accounts of the butterflies and moths of the British Isles.* Uffington, Hedera Pr. [vi]+247 pp.

Cc581. **Illingworth, J. F.** 1923. Early references to Hawaiian entomology. *Bulletin of the Bernice P. Bishop Museum* 2: 63 pp.

Cc582. **Jiron, L. F.** 1985. *Indice de publicaciones entomologicas de Costa Rica.* San Jose, Costa Rica, Universidad de Costa Rica. 433 pp.

Cc583. **Lisney, A. A.** 1960. *A bibliography of British lepidoptera 1608–1799.* London, Chiswick Pr. xviii+315 pp. & 40 pl. (ports.)

Cc584. **Lizer, C.** 1919. *Primer ensayo bibliográfico de entomología Argentina.* Buenos Aires, "Coni." 380 pp. ••*Also* ___. 1927. Apuntaciones para la bibliografía entomológica Argentina. *Physis, Buenos Aires* 8: 505–535.

Cc585. **Lomholdt, O., P. Nielsen, & K. Schnack,** eds. 1984. Entomologisk Litteratur — en hjoelp til studiet af den danske insektfauna. *Entomologiske meddelelser* 51: 1–85.

Cc586. **Miller, D.** 1956. Bibliography of New Zealand entomology, 1775–1952 (with annotations). Nelson, N.Z., Cawthron Institute. xxiv+492 pp. *Bulletin, New Zealand Department of Scientific & Industrial Research* no. 120.

Cc587. **Musgrave, A.** 1932. *Bibliography of Australian entomology 1775–1930, with biographical notes on authors and collectors.* Sydney, Royal Zoological Society of New South Wales. viii+380 pp.

Cc588. **Ouchi, Y.** 1934. Bibliographical introduction to the study of Chinese insects. *Journal of the Shanghai Institute* Sect. 3, 2: [iv]+533 pp. *Continued* ___. 1935–38. ___: Supplement 1934 (–1936). *Ibid.* Sect. 3, 3: 17–88, 125–220, 233–327.

Cc589. **Petrov, A. I.** 1935. *Ukazatel' russkoi literatury po prikladnoi entomologii, fitopatologii i mammaliologii Srednei Azii. Vrediteli sel'skokhozyaistvennykh kul'tur i lesov (1870–1933).* Tashkent, Gosizdat UzSSR. 160 pp.

Cc590. **Pfeffer, A.** 1950. Bibliographie entomologique de la Bohême, de la Moravie et de la Sinésie pour les années 1800–1940. *Casopis Ceske Spolecnosti Entomologické* 47: 231–356.

Cc591. **Piran, A. A.** 1954–57. *Curso de entomologia* no. 10. Bibliografia entomologica Argentina, 1847–1952. In: *Publicaciones de extension cultural y didactica, Museo Argentino de Ciencias Naturales "Bernardino Rivadavia"* no. 1: 621–804. *Also* ___. 1961. *Bibliografía entomólogica Argentina: Erratas, addenda y supplemento I.* Buenos Aires, Instituto Nacional de Tecnología Agropecuaria, Instituto de Patología Vegetal. 40 pp. *Publ. tec.* no. 76.

Cc592. **Roonwal, M. L.** 1954. A brief review of the progress of entomology in India during the period 1938–1950, together with a bibliography. *Memoirs of the Entomological Society of India* 3: 1–119.

Cc593. **Ryan, J. G., J. P. O'Connor & B. P. Beirne.** 1984. *A bibliography of Irish entomology.* Dublin, Flyleaf Pr. vii+363 pp.

Cc594. **Schwarz, E. A.** 1888. An enumeration of the published synopses, catalogues, and lists of

North American insects; together with other information intended to assist the student of American entomology. *Bulletin of the Division of Entomology, United States Department of Agriculture* (1) 19: 77 pp.

Cc595. **Shui-Chen, C.** 1958. *Bibliography of entomology in Taiwan (1684–1957).* Taipei, Taiwan Agricultural Research Institute. viii+vi+246 pp. *Special publication* no. 1.

Cc596. **Smetana, O.** 1969. *Entomologicka bibliografie Ceskoslovenska, 1951–1960.* Prague, Akademia. 329 pp.

Cc597. **Steck, T.** 1926. *Insects 1634–1900.* Berne, Centralkommission für Schweizerische Landeskunde. vi+292 pp. *Bibliographie der schweizerischen Landeskunde* IV.6.7.

Insect groups

Cc598. **Arnett, R. H.**, ed. 1978. *Bibliography of Coleoptera of North America north of Mexico, 1758–1948.* Baltimore, MD, World Natural History Publications for the Biological Research Institute of America. Re-pub. of bibliographies from C. W. Leng, ed., *Catalog of the Coleoptera of America, north of Mexico* and the 5 supplements, Mount Vernon, NY, J. D. Sherman, 1920–48.

Cc599. **Blackwelder, R.** 1957. Part 6: Bibliography. In: Checklist of the coleopterous insects of Mexico, Central America, the West Indies and South America. *Bulletin of the United States National Museum* 185: 927–1344. *Re-pub.* Washington, DC, Smithsonian Institution Pr., 1982.

Cc600. **Bridges, C. A.** 1989. *Bibliography (Lepidoptera: Rhopalocera).* Urbana, IL, the Author. [587] pp. ••Pt. 1. Bibliography; pt. 2. Index to journals and serials; pt. 3. Index to bibliography by year.

Cc601. **Clay, T., & G. H. E. Hopkins.** 1950–60. The early literature on Mallophaga. Pt. 1. *Bulletin of the British Museum (Natural History), Entomology* 1: 221–272, 1950; ___: Part II: 1763–1775. *Ibid.* 2: 1–37, 1951; ___: Pt. III: 1776–1786. *Ibid.* 3: 223–265, 1954; ___: Pt. IV: 1787–1818. *Ibid.* 9: 1–61, 1960.

Cc602. **Cotti, G.** 1963. *Bibliografia ragionata 1930–1961 del gruppo* Formica rufa, *in Italiano, Deutsch, English.* Rome, Ministero dell'Agricoltura delle Foreste. iv+413 pp. *Collana verde* no. 8.

Cc603. **Evenhuis, N. L.** 1983. *An indexed bibliography of Bombyliidae (Insecta: Diptera).* Koenigstein, Koeltz Scientific Books. 494 pp. & 4 pl. *Theses zoologicae* vol. 4.

Cc604. **Field, W. D., C. F. Dos Passos, & J. H. Masters.** 1974. A bibliography of the catalogs, lists, faunal and other papers on the butterflies of North America north of Mexico, arranged by state and province (Lepidoptera/Rhopalocera). *Smithsonian contributions to zoology* 157: 104 pp.

Cc605. **Gerini, V.** 1984. *Bibliografia sulle termiti dal 1758 al 1980.* Florence, Istituto Agronomico per l'Oltremare. 214 pp. *Relazioni e monografie agrarie subtropicali e tropicali* n.s. no. 93.

Cc606. **Junk, W.** 1912. *Bibliographia coleopterologica.* Berlin, 1912. xvi+146 pp.; *Continued* ___. 1935. *Bibliographia coleopterologica.* Neue Aufl. The Hague, Junk. xvi+149–323 pp.

Cc607. ___. 1913. *Bibliographie lepidopterologica.* Berlin, Junk. xxvi+142 pp.

Cc608. ___. 1933. *Bibliographie dipterologica.* Berlin, Junk. 120 pp.

Cc609. **Kéler, S. von.** 1960. Bibliographie der Mallophagen. *Mitteilungen aus der Zoologischen Museum Berlin* 36: 145–403.

Cc610. **Kudrna, O.**, ed. 1985. *Concise bibliography of European butterflies.* Wiesbaden, AULA Verlag. viii+447 pp. *Butterflies of Europe.* Vol. 1. ••Literature of 1901–83.

Cc611. **Metcalf, Z. P.** 1945. *A bibliography of the Homoptera (Auchenorhyncha).* Raleigh, NC, North Carolina State College of Agriculture & Engineering, Department of Zoology & Entomology. 2 vols.

Cc612. **Metcalf, Z. P.** 1960. A bibliography of the Cercopoidea (Homoptera: Auchenorhyncha). *Papers of the Department of Entomology, North Carolina Agriculture Experiment Station* 1135: iv+262 pp.

Cc613. ___. 1962. *A bibliography of the Cicadoidea (Homoptera: Auchenorhyncha).* Raleigh, NC, North Carolina State College of Agriculture & Engineering. iv+229 pp. *General catalogue of the Homoptera* fasc. 8.

Cc614. **Mocsáry, A.** 1883. Literatura hymenopterorum. *Természetrajzi füzetek* 6: 3–122.

Cc615. **Moulds, M. S.** 1977. *Bibliography of the Australian butterflies (Lepidoptera: Hesperioidea and Papilionoidea) 1773–1973.* Greenwich, N.S.W., Australian Entomological Pr. 239 pp.

Cc616. **Parshley, H. M.** 1925. *A bibliography of the North American Hemiptera–Heteroptera.* Northampton, MA, Smith College. ix+252 pp. *Smith College 50th anniversary publications* no. 4.

Cc617. **Roonwal, M. L.** 1958. Bibliographia acrididiorum: A bibliography of the orthopterous insects of the family Acrididae (comprising the short–horned grasshoppers and locusts) from the earliest times to the end of 1954 (with some additions for 1955–57). *Records of the Indian Museum* 56: ix+611 pp.

Cc618. **Sachtleben, H.** 1962. Bibliographie der paläarktischen Ichneumoniden. *Beiträge zur Entomologie* 12: 1–242, 720–731, 916–939.

Cc619. **Schenkling, S.** 1938–39. *Faunistischer Führer durch die Coleopteren–Literatur: Die wichtigste Käfer–Literatur nach geographischen Gebieten geordnet.* Neubrandenburg, Feller. 320 pp. ••All published; deals with European literature.

Cc620. **Schmidt, E.** 1933. *Bibliographia odonatologica.* Vienna, Wagner. 116 pp.

Cc621. **Sharma, M. L.** 1969–77. *Bibliographie des Aphidoidea.* Sherbrooke, Quebec, Éditions Paulines. 4 vols. *Publications de l'Université de Sherbrooke, Faculté des Sciences, Département de Biologie* no. 1.

Cc622. **Smith, C. F.** 1972. Bibliography of the Aphididae of the world. *Technical bulletin, North Carolina Experiment Station* 216: 717 pp.

Cc623. **Snyder, T.** 1956. Annotated, subject heading–bibliography of termites 1350 B.C. to A. D. 1954. *Smithsonian miscellaneous collections* 130: 305 pp. *Continued* ___. 1961. Supplement to the annotated, subject–heading bibliography of termites 1955 to 1960. *Ibid.* 143 (3): 137 pp.; ___. 1968. Second supplement to the annotated, subject–heading bibliography of termites, 1961–1965. *Ibid.* 152(3): iii+188 pp.; **E. Ernst et al.** 1986. *A bibliography of termite literature, 1966–1978.* New York, Chichester, J. Wiley & Sons. 903 pp.

Cc624. **Tietz, H. M.** 1972. *An index to the described life histories, early stages and hosts of the macrolepidoptera of the continental United States and Canada.* Sarasota, FL, A. C. Allyn for the Allyn Museum of Entomology. iv+1041 pp.

Cc625. **West, L. S., & O. B. Peters.** 1973. *An annotated bibliography of* Musca domestica *Linnaeus.* Marquette, MI, Northern Michigan Univ. xiii+743 pp. *Also* Folkestone & London, Dawson.

See also Cc580 & Cc583.

Periodical bibliographies

• See periodical bibliographies listed at Cc37–54 (especially *Zoological record*) and:

Cc626. ***Abstracts of entomology.*** <1970+> Philadelphia, PA, Biosciences Information Service of Biological Abstracts, vol. 1+, 1970+.

Cc627. ***Bibliography of the more important contributions to American economic entomology.*** <1889–1904> Washington, DC, United States Department of Agriculture, pts. 1–8, 1889–1905. ••*Includes* Pts 1–3: S.

Henshaw. The more important writings of B. D. Walsh & C. V. Riley. 1889–90. 454 pp.; 4–5: ___. The more important writings of government and state entomologists, and of other contributors to the literature of American economic entomology. 1895–96. 167+179 pp.; 6: **N. Banks.** The more important writings published between June 30, 1888, and Dec. 30, 1896. 1898; 7: ___ Dec. 31, 1896, and Jan. 1, 1900. 1901; 8. ___ Dec. 31, 1899, and Jan. 1, 1905. 1905. Continued as *Index to the literature of American economic entomology*.

Cc628. ***Entomologische Litteraturblätter***, repertorium der neuesten Arbeiten auf dem Gesamtgebiet der Entomologie. <1901–14> Berlin, R. Friedländer & Sohn, vols. 1–14, 1901–14. ••Continued as *Repertorium entomologicum*.

Cc629. ***Entomology abstracts.*** <1969+> London, Information Retrieval Inc. (*later*, <1982+> Bethesda, MD, Cambridge Scientific Abstracts), 1969+.

Cc630. ***Entomology current literature.*** <1932–42> Washington, DC, U. S. Department of Agriculture, Bureau of Entomology, vols. 1–11, 1932–42.

Cc631. ***Index to the literature of American economic entomology.*** <1905–59> Washington, DC, Entomological Society of America (*later* American Association of Economic Entomologists), Vols. 1–18, 1917–62.

Cc632. ***Repertorium entomologicum***, *fortsetzung der "Entomologischen Literaturblätter."* Edit. Deutschen Entomologischen Gesellschaft. <1924–33> Berlin, R. Friedländer & Sohn, vols. 1–10, 1924–35.

Conchology

Cc633. **Altaba, C. R., & A. Travesset.** 1984. La malacologia als paisos Catalans. *Butlleti de la Institucio Catalana d'Historia Natural* 50: 155–171.

Cc634. **Burnay, L. P., & A. A. Monteiro.** 1988. *Historia da malacologia em Portugal.* Lisbon, Sociedade Portuguesa de Malacologua. 123 pp., illus. *Publicações ocasionais* no. 12.

Cc635. **Dance, S. P.** 1971. The Cook voyages and conchology. *Journal of conchology* 26: 354–379; ___. 1972. ___: A supplementary note. *Ibid.* 27: 357–358.

Cc636. ___. 1986. *A history of shell collecting.* Rev. ed. Leiden, E. J. Brill & Dr. W. Backhuys. xv+265 pp. & 32 pl.
Replaces First ed., London, Faber, 1966. 344 pp., illus.

Cc637. ___. 1989. *Rare shells.* London, Faber & Faber. 128 pp.

Cc638. ___. 1976. Shells. In: J. M. Chalmers–Hunt. *Natural history auctions 1700–1972: A register of sales in the British Isles*, pp. 45–51. London, Sotheby Parke Bernet. xiii+189 pp.

Cc639. **Dean, J. D.** 1936. Conchological cabinets of the last century. *Journal of conchology* 20: 225–252.

Cc640. **Dexter, R. W.** 1974. Historical sketch of Indian Ocean malacology. *Journal of the Marine Biological Association of India* 16: 512–519, 1974 [i.e. 1977].

Cc641. **Gillespie, N. C.** 1984. Preparing for Darwin: Conchology and natural theology in Anglo–American natural history. *Studies in the history of biology* 7: 93–145.

Cc642. **Sykes, E. R.** 1901. Conchology at the dawn and close of the nineteenth century. *Journal of conchology* 10: 35–42.

Cc643. **Tryon, G. W., Jr.** 1862. A sketch of the history of conchology in the United States. *American journal of science & arts* (2) 33: 161–180.

Cc644. **Zilch, A.** 196 ?–72*. Zur Geschichte der Deutschen malakozoologie. *Mitteilungen der Deutschen Malakozoologischen Gesellschaft* 2: 308–312.
A series of eleven [?] articles, *includes* VI: Briefe von O. Boettger an V. M. Gredler. Ibid. 2: 156–187, 1970; VIII: Die conchylien–Sammlung von Siegfried H. Jaeckel. *Ibid.* 2: 308–312, 1971; IX: Zur Konchylien Sammlung von K. L. Pfeiffer. *Ibid.* 2: 324–326, 1972; X: Zur Konchylien Sammlung von P. Ehrmann. *Ibid.* 2: 327–329, 1972; XI: Zur Konchylien Sammlung von H. Kaltenboschr. *Ibid.* 2: 330–331, 1972. [*Earlier articles not examined.]

See also Ba341–344 Ca829(Coomans) Bb895–900 Cc789 Cc810 Cc818 Cc865 Cc874–875 & Cc888.

BIBLIOGRAPHIES OF CONCHOLOGY

• A selection of major works:

Cc645. **Bieler, R., & A. R. Kabat.** 1991. Malacological journals and newsletters, 1773–1990. *Nautilus* 105 (2): 39–61.

Cc646. **Binney, W. G.** 1863–64. Bibliography of North American conchology, previous to the year 1860, prepared for the Smithsonian Institution. *Smithsonian miscellaneous collections* 5: 650 pp.; *Ibid.* 9 (article 1): 306 pp. ••Pt. 1: American authors; pt. 2: Foreign authors.

Cc647. **Bouchet, P., & V. Heros.** 1981. *Mollusques.* Paris, Secrétariat de la Faune et de la Flore. 100 ff. *Bibliographie des inventaires faunistiques de France: 1758–1979. Inventaire de la faune et de la flore* no. 14.

Cc648. **Burk, R., & J. H. Jungbluth.** 1985. *140 Jahre Molluskunde im deutschsprachigen Raum 1844–1984.* Budapest, Termeszettudomanyi Muzeum (Ungarischen Naturwissenschaftlichen Museum). 348 pp., illus. ••Index to a dozen German malacological periodicals published 1844–1984.

Cc649. **Clarke, A. H.** 1962. Annotated list and bibliography of the abyssal marine molluscs of the world. *Bulletin of the National Museums of Canada* 181: 1–114.

Cc650. **Hidalgo y Rodriguez, J. G.** 1890–1913. Bibliografía crítica. In: Obras malacólogicas. Pte. II. Estudios preliminares sobre los moluscos terrestres y marinos de España, Portugal y las Baleares. *Memorias, Real Academia de Ciencias Exactas, Físicas y Naturales* 15: 273–2431.

Cc651. **Kanamaru, T.** 1928–29. Bibliography of the conchological classics of Japan. *Venus* 1: 32–38, 55–58, 82–87, 132–136.

Cc652. **McClelland, H.** [1926.] *General index of all families, genera, species and varieties described and noted in the Journal of conchology, vols. I–XVI, 1874–1922, the Proceedings of the Malacological Society vols. I–XV, 1893–1923, the Conchologist vols. I & II, 1891–93, continued as Journal of malacology vols. III–XII, 1894–1905.* Birmingham., Birbeck & Sons. 165 pp.

Cc653. **Maton, W. G., & T. Rackett.** 1804. An historical account of testaceological writers. *Transactions of the Linnean Society of London* 7: 119–244. ••*Includes* Syllabus classium in systemate testaceologorum, pp. 215–244. (*Also* ___. 1811. *Bibliothèque chronologique et systématique ou méthodique des auteurs testacéologiques.* Traduit de l'Anglois par A. M. H. Boulard. Paris, Garnery. Cited in W. Engelmann, 1846, *Bibl. hist.–nat.* and J.–C. Brunet, 1860–66, *Manuel du libraire,* Table méthodique #31725, but I am unable to trace a copy).

Cc654. **Radwin, G. E., & E. V. Coan.** 1976. A catalogue of collations of works of malacological importance. *Occasional paper, Western Society of Malacologists* no. 2: 34 pp.

Cc655. **Russell, H. D.** 1971. *Index Nudibranchia (1554–1965).* Greenville, DE, Delaware Museum of Natural History. iv+141 pp.

Cc656. **Studer, T., G. Arnstein, & A. Brot.** 1896. *Mollusken.* Berne, Centralkommission für Schweizerische Landeskunde. xi+12 pp.
Bibliographie der schweizerischen Landeskunde IV.6.6.

Cc657. **Taylor, D. W.** 1970. West American freshwater mollusca. i: Bibliography of Pleistocecene and recent species. *Memoirs of the San Diego Society of Natural History* 4: 73 pp.

Cc658. **Tryon, G. W., Jr.** 1861. *List of American writers on conchology; With titles of their memoirs and dates of publication.* New York, Baillière Bros. 68 pp.

Cc659. **Vokes, H. E.** 1967. Genera of the bivalvia: A systematic and bibliographic catalog. *Bulletins of American paleontology* 51 (232): 394 pp.

See also Library cats. Bc478 & Bc479.

Periodical bibliographies

See periodical bibliographies listed at Cc37–54 (especially Zoological record; *see* Cc50).

Arachnology, crustaceology, etc.

Cc660. **Allen, E. Heron–.** 1928. *Barnacles in nature and myth.* Oxford, Oxford Univ. Pr. xv+180 pp.

Cc661. **Barnard, J. L.** 1958. The question of decline in systematic activity measured in the marine Amphipoda. *Systematic zoology* 7: 123–125.

Cc662. **Damkaer, D. M., & T. Mrozek–Dahl.** 1980. The Plankton–Expedition and the copepod studies of Friedrich and Maria Dahl. In: M. Sears & D. Merriman, eds. *Oceanography: The past.* (Proceedings of the third International Congress on the History of Oceanography held September 22–26, 1980, [etc.].), pp. 462–473. New York, Springer–Verlag.

Cc663. **Geus, A.** 1975. Die Anfänge der Tardigradenforschung. In: K. Mothes & J.–H. Scharf, eds. *Beiträge zur Geschichte der Naturwissenschaften und der Medizin; Festschrift für Georg Uschmann ... zum 60. Geburtstag am 18. Oktober 1873*, pp. 403–410. Halle, Deutsche Akademie der Naturforscher Leopoldina. *Acta historica Leopoldina* no. 9.

Cc664. **Hennings, C. J. A. C.** 1908. Sie systematische Stellung und Einteilung der Myriopoden: Anschauungen und Erfahrungen 1758–1905. *Zoologische Annalen* 2: 47–70.

Cc665. **Holthuis, L. B.** 1959. Notes on pre-Linnaean carcinology (including the study of Xiphosura). In: Wit, H. C. D. de *Rumphius memorial volume*, pp. 63–125. Baarn, Hollandia.
Includes Chronological survey of carcinological literature, pp. 64–95.

Cc665A. **Jones, D. S.** 1991. A history of the discovery and description of Australian barnacles (Cirripedia: Thoracica), including a bibliography of reference works. *Archives of natural history* 18: 149–178.

Cc666. **Mills, E. L.** 1972. T. R. R. Stebbing, the Challenger and knowledge of deep–sea Amphipoda. In: Royal Society of Edinburgh. Proceedings of the second International Congress on the History of Oceanography: *Challenger* expedition centenary, Edinburgh September 12–20, 1972. *Proceedings of the Royal Society of Edinburgh* B 72: 69–87.

Cc667. **Oudemans, A. C.** 1926–37. *Kritisch–historisch Overzicht der Acarologie (Critico–historical survey of acarology from 850 B.C. to 1850).* Leiden, E. J. Brill. 3 vols. (in 8). Vols. 1–2, *Tijdschrift voor Entomologie* 69 (supp.): viii+500 pp., 1926; Ibid. & 72 (supp): xvii+1097 pp., 1929; vol. 3 (in 6), xx+3379 pp., 1936–37.
1: 850 v. Chr. tot 1758; 2: 1759 to 1804; 3: 1805 to 1850.

Cc668. **Prasad, V.** 1982. *History of acarology*. Oak Park, MI, Indira. vii+472 pp.

Cc669. **Savory, T. H.** 1961. *Spiders, men and scorpions: Being the history of arachnology*. London, Univ. of London Pr. vi+191 pp. & 15 pl.

Cc670. **Strand, E.** 1916. Collectanea arachnologica: Beiträge zur Bibliographie und Geschichte der Arachnologie. *Archiv für Naturgeschichte* 82A: 42–69.

Cc671. **Winsor, M. P.** 1969. Barnacle larvae in the nineteenth century: A case study in taxonomic theory. *Journal of the history of medicine* 24: 294–309.

See also Cc217(Skinner & Cook) & Cc798A.

BIBLIOGRAPHIES OF ARACHNOLOGY & CRUSTACEOLOGY

• A selection of major works:

Cc672. **Bonnet, P.** 1945–61. *Bibliographia araneorum: Analyse méthodique de toute la littérature aranéologique jusqu'en 1939*. Toulouse, Douladoure. 3 vols in 7 ••Ca. 8,000 entries. 1: Introduction, liste alphabétique par auteurs des ouvrages et articles, étude bibliographique par sujets; 2: Systématique des araignées; 3: Index alphabétique. For background information *see* ___. 1961. Bibliographia araneorum: Résultats numérique et considerations diverses. *Bulletin de la Société d'Histoire Naturelle de Toulouse* 96: 255–258. Also pub. in *Bollettino delle sedute accademia Gioenia di scienze naturali* (3) 73: 205–210, 1961; and *Annals and magazine of natural history* (13) 4: 749–752, 1962.

Cc673. **Brignoli, P. M.** 1983. Bibliography. In: *A catalogue of the Araneae described between 1940 and 1981*. Ed. by P. Merret, pp. 13–108. Manchester, Manchester Univ. Pr. 755 pp.; *Continued* **N. I. Platnick.** 1989. *Advances in spider taxonomy, 1981–1987: A supplement to Brignoli [etc.]*. Manchester, [etc.]. 673 pp. ••Supplements Bonnet; *see* Cc672.

Cc674. **D'Aubigny, J. Le D.** 1980. *Myriapodes*. Étude réalisée sour la direction de J.–C. Lefeuvre. Paris, Secrétariat de la Faune et de la Flore. [iii]+xi+20 ff. *Bibliographie des inventaires faunistiques de France: 1758–1979. Inventaire de la faune et de la flore* no. 2.

Cc675. ___. 1980. *Arachnides sauf acariens*. Étude réalisée sour la direction de J.–C. Lefeuvre. Paris, Secrétariat de la Faune et de la Flore. 95 ff. *Bibliographie des inventaires faunistiques de France: 1758–1979. Inventaire de la faune et de la flore* no. 9–13.

Cc676. **Gordan, J.** 1956. A bibliography of pagurid crabs, exclusive of Alcock, 1905. *Bulletin of the American Museum of Natural History* 108: 253–352. ••I.e. **A. Alcock.** 1905. *Catalogue of Indian decapod crustacea in the collection of the Indian Museum*. Part 2. Anomura. Fasc. 1. Pagurides. Calcutta. xi+197 pp.

Cc677. ___. 1957. A bibliography of the order Mysidacea. *Bulletin of the American Museum of Natural History* 112: 283–393.

Cc678. **Gurney, R.** 1939. *Bibliography of the larvae of decapod crustacea*. London, Ray Society. vii+123 pp. ••Supplement included in his *Larvae of* ... London, Ray Society, 1942, pp. 288–299. *Both re-pub. together* Weinheim, J. Cramer, Codicote; Wheldon & Wesley, 1960. *Historiae naturalis classica* vol. 8.

Cc679. **Hart, C. W., Jr., & J. Clark.** 1987. *An interdisciplinary bibliography of freshwater crayfishes (Astacoidea and Parastacoidea) from Aristotle through 1985. Updated through 1987*. Washington, DC, Smithsonian Institution Pr. 497 pp. ••First pub. in: *Smithsonian contributions to zoology* 455: 437 pp.

Cc680. **Hoogstraal, H.** 1970–74. *Bibliography of ticks and tickborne diseases, from Homer (about 800 B.C.) to 31 December 1973*. Cairo, United States Naval Medical Research Unit. 5 vols.

Cc681. **Nuttall, G. F. H., L. E. Robinson & W. F. Cooper.** 1911–15. *Bibliography of the Ixodoidea.* Cambridge, Cambridge Univ. Pr. 2 vols.

Cc682. **Vervoort, W.** 1986–88. *Bibliography of Copepoda, up to and including 1980.* Leiden, E. J. Brill. 3 vols. *Crustaceana supplement* vols. 10–12.

Cc683. **Viets, K.** 1955. *Die Milben des Süsswassers und des Meeres. Hydrachnellae et Halacaridae (Acari).* Teil 1: *Bibliographie.* Jena, G. Fischer. [iv]+476 pp.

See also Cc720

Periodical bibliographies

See periodical bibliographies listed at Cc37–54 (especially Zoological record; *see* Cc50).

Helminthology

Cc684. **Chesnova, L. V.** 1983. To the history of development of helminth ecology. *Helminthologia* 20: 161–173.

Cc685. **Enigk, K.** 1986. *Geschichte der Helminthologie in deutschsprachigen Raum.* Stuttgart, New York, G. Fischer. xviii+356 pp.

Cc686. **Farley, J.** 1972. The spontaneous generation controversy (1700–1860): The origin of parasitic worms. *Journal of the history of biology* 5: 95–125.

Cc687. **Grove, D. I.** 1990. *A history of human helminthology.* Wallingford, C A B International. viii+848 pp.

Cc688. **Gupta, P. D.** 1984. Helminthology in India in 18th–19th centuries with some remarks on its recent progress. *Indian journal of history of science* 19: 109–117.

Cc689. **Skryabin, K. I., et al.** 1962–72. *Stroitel'stvo gel'mintologicheskoi nauki i praktiki v SSSR. [Development of the helminthological science and practice in the USSR].* Moscow, Izdatel'stvo Akademii Nauk SSSR. 5 vols.

Cc690. **Smit, P., & R. P. Visser.** 1984. Ideas concerning the generation and development of intestinal works, 1650–1850. In: A. H. Kerkhoff, A. M. Lugendijk-Elshout, & M. J. Poulissen, eds. *De novis inventis*, pp. 401–411. Amsterdam, A P A Holland Univ. Pr.

See also Ba1491–1492.

BIBLIOGRAPHIES OF HELMINTHOLOGY

• A selection of major works:

Cc691. **Baker, A. D., M. Oostenbrink & J. A. Van Berkum.** 1967–71. *Bibliography on plant, soil and freshwater nematodes.* Leiden, E. J. Brill. Card index, pub. in 8 instalments. *Continued* ___. 1967–71. ___*: Supplement.* Leiden, E. J. Brill. Card index, pub. in 6 instalments.

Cc691A. **Gavin, A. M., & V. I. Macdonald.** 1977. Annotated bibliography of the Polychaeta of the Pacific Northwest. Sidney, B.C., Institute of Ocean Sciences. 18+136+93+125 pp. *Contractor report series* 1977 no. 6.

Cc692. **Hargis, W. J., et al.** 1969. *Bibliography of the monogenetic trematode literature of the world, 1758 to 1969.* Gloucester Point, VA, Institute of Marine Science. v+195 pp. *Special scientific report* no. 55. *Continued* ___. 1970. ___*: Supplement 1.* Gloucester Point, VA, [etc.]. iv+13 pp.; ___. 1971. ___*: Supplement 2.* Gloucester Point, VA, [etc.]. v+25 pp.

Cc693. **Hartman, O.** 1951. *Literature of the polychaetous annelids.* Vol. 1: *Bibliography.* Los Angeles, CA, the Author. vi+290 pp.

Cc694. **Levashov, M. M.** 1949. Opyt bibliografii russkoi gel'minto–

faunisticheskoi literatury za period s 1771 po 1947 g. *Trudy Gel'mintologicheskoi Laboratorii* 2: 143–204.

Cc695. **Modeer, A.** 1786. *Bibliotheca helminthologica seu enumeratio auctorum qui de vermibus scilicet eryptozoois, gymnodelis, testaceis atque phytozoois tam vivis quam petrifactis scripserunt.* Erlangen, Palm. viii+223 pp.

Cc696. **Wiszniewski, J.** 1954. Bibliographie des travaux sur les rotifères, à partir de 1913. *Polskie archiwum hydrobiologii* (2) 15: 164–249. ••Covers 1915–1939.

Cc697. **Zschokke, F.** 1902. *Parasitische Würmer.* Berne, Centralkommission für Schweizerische Landeskunde. xii+39 pp. *Bibliographie der schweizerischen Landeskunde* IV.6.8.

See also Cc720.

Periodical bibliographies

See periodical bibliographies listed at Cc37–54 (especially *Zoological record*) and:

Cc698. ***Helminthological abstracts.*** <1932+> St. Albans, Farnham Royal, Imperial (*later* Commonwealth) Agricultural Bureaux, vols. 1–38, 1932–69; divided into Series A, Animal & human helminthology, vol. 39–58, 1970–89; Series B, Plant nematology, vol. 39–58, 1970–89; re-united vol. 59+, 1990+.

Other invertebrate zoology

Cc699. **Dawson, V. P.** 1987. *Nature's enigma: The problem of the polyp in the letters of Bonnet, Trembley, and Réaumur.* Philadelphia, PA, American Philosophical Society. ix+266 pp. *Memoirs of the American Philosophical Society* vol. 174.

Cc700. **Edwards, C.** 1972. The history and state of the study of medusae and hydroids. In: Royal Society of Edinburgh. Proceedings of the second International Congress on the History of Oceanography: *Challenger* expedition centenary, Edinburgh September 12–20, 1972. *Proceedings of the Royal Society of Edinburgh* B 73: 247–257.

Cc701. **Gould, S. J.** 1989. *Wonderful life: The Burgess shale and the nature of history.* New York, W. W. Norton. 347 pp.

Cc702. [Not used.]

Cc703. **Lühe, M.** 1905. Geschichte und Ergebnisse der Echinorhynchen–Forschung bis auf Westrumb (1821). *Zoologische Annalen* 1: 139–353.

Cc704. **Schiller, J.** 1968. Controverses autour de certaines structures chez les tuniciers au XIXe siècle. In: Institut Océanographique, Monaco. Communications, premier Congrès International d'Histoire de l'Océanographie, Monaco, 12–17 décembre 1966. *Bulletin de l'Institut Océanographique, Monaco* numéro spécial 2: 387–396.

Cc705. **Schopf, T. J. M.** 1967. The literature of the phylum Ectoprocta, 1555–1963. *Systematic zoology* 16: 318–327.

Cc706. **Winsor, M. P.** 1972. A historical consideration of the Siphonophores. In: Royal Society of Edinburgh. Proceedings of the second International Congress on the History of Oceanography: *Challenger* expedition centenary, Edinburgh September 12–20, 1972. *Proceedings of the Royal Society of Edinburgh* B 73: 315–323.

Cc707. ___. 1976. *Starfish, jellyfish and the order of life: Issues in 19th–century science.* New Haven, CT, Yale Univ. Pr. x+228 pp. *Yale studies in the history of science and medicine* no. 10.

Cc708. **Wood, H. M. Muir–.** 1955. *A history of the classification of the phylum Brachiopoda.* London, British Museum (Natural History). vii+124 pp.

See also Cc200(Churchill) & Cc217(Lenhoff).

BIBLIOGRAPHIES OF OTHER INVERTEBRATE ZOOLOGY

Cc709. **Arndt, W.** 1940. *Bibliographia spongiologica.* 1: *Spongiae utiles.* The Hague, Junk. iv+114 pp.

Cc710. **Clapp, W. F., & R. Kenk.** 1963. *Marine borers: An annotated bibliography, [etc.].* Washington, DC, Office of Naval Research. xii+1136 pp.

Cc711. **Davidson, T., & W. H. Dalton.** 1886. Bibliography of the Brachiopoda. In: *A monograph of the British fossil Brachiopoda,* vol. 6. London, Palaeontographical Society.

Cc712. [Not used.]

Cc713. **Holland, N. D. & L. Z.** 1969. A bibliography of echinoderm biology, continuing Hyman's 1955 bibliography through 1965. *Pubblicazioni della Stazione Zoologica di Napoli* 37: 441–542. ••i.e. **L. H. Hyman.** 1955. *The invertebrates.* Vol. 4: *Echinodermata.* New York, McGraw–Hill Book Co.

Cc714. **Hopkinson, J.** 1913. *A bibliography of the Tunicata, 1469–1910.* London, Ray Society. xii+288 pp.

Cc715. **Musée National des Sciences Naturelles, Ottawa, Division de la Zoologie des Invertébrés.** 1983+. *Bibliographia invertebratorum aquaticorum Canadensium.* Ottawa.
1: **D. R. Laubitz, I. Sutherland, & N. Sharma.** *Index generum.* 1983. v+56 pp.
2: **N. Sharma, D. R. Laubitz, & I. Sutherland.** 1983. *Index bibliographicus.* 1983. v+238 pp.

Cc716. **Neviani, A.** 1896. Materiali bibliografici per servire alla storia degli studi sui briozoi. *Rivista Italiani de scienze naturali* 16(1): 1–7; (2): 25–28; (3): 35–38.

Cc717. ___. 1930. Appunti bibliografici sopra antiche opere relative allo studio di animali marini inferiori. *Memorie dell'Accademia Pontificia dei Nuovi Lincei. Rome* (2) 14: 25–44.

Cc718. **Penney, J. T.** 1960. Distribution and bibliography (1892–1957) of the fresh water sponges. *University of South Carolina publications, biology* (3) 3: 1–97.

Cc719. **Ranson, G., & J. Paretias.** 1958. Coraux et récifs coralliens (bibliographie). *Bulletin de l'Institut Océanographique, Monaco* 1121: 80 pp.

Cc720. **Studer, T., et al.** 1898. *Crustacea, bryozoa, annelida, rotifera, turbellaria, spongien und hydroiden, protozoa.* Berne, Centralkommission für Schweizerische Landeskunde. viii+27 pp. *Bibliographie der schweizerischen Landeskunde* IV.6.9.

Cc721. **Thompson, D'A. W.** 1885. *A bibliography of protozoa, sponges, coelenterata and works including also the polyzoa, brachiopoda and tunicata for the years 1861–1883.* Cambridge, Cambridge Univ. Pr. viii+284 pp.

Cc722. **Vosmaer, G. C. J.** 1928. *Bibliography of sponges, 1551–1913.* Ed. by G. P. Bidder & C. S. Vosmaer–Röell. Cambridge, Cambridge Univ. Pr. xii+234 pp.

Periodical bibliographies

See periodical bibliographies listed at Cc37–54 (especially Zoological record; *see* Cc50).

Parasitology

Cc723. **Aleksandrov, D. A.** 1984. Iz istorii parazitologii v Leningrade. [History of parasitology in Leningrad]. *Parazitologiya* 18: 489–494.

Cc724. **Chesnova, L. V.** 1978. *Evolyutsionnaya kontseptsiya v parazitologii: Ocherki istorii. [Evolution of concepts in parasitology: Historical essays].* Moscow, Izd. "Nauka." 163 pp., ports.

Cc725. **Chesnova, L. V.** 1989. Osnovnye napravleniya i tendentsii razvitiya evolutsionnoi idei v parazitologii. [The main trends and tendencies in the development of evolutionary ideas in parasitology]. *Istoriko–biologicheskie issledovaniya* 10: 51–70.

Cc726. **Foster, W. D.** 1965. *A history of parasitology.* Edinburgh & London, E. & S. Livingstone. vii+210 pp. & 15 pl.

Cc727. **Garnham, P. C.** 1988. History of discoveries of malaria parasites and of their life cycles. *History & philosophy of the life sciences* 10: 93–108.

Cc728. **Harant, H.** 1955. *Histoire de la parasitologie.* Paris, Université de Paris. 26 pp. *Les Conférences du Palais de la Découverte* série D, no. 35.

Cc729. **Hoeppli, R. J. C.** 1959. *Parasites and parasitic infections in early medicine and science.* Singapore, Univ. of Malaya Pr. xvi+528 pp. & 23 pl.

Cc730. **Reinhard, E. G.** 1957. Landmarks of parasitology. 1: The discovery of the life cycle of the liver fluke. *Experimental parasitology* 6: 208–232; *Continued* ___. 1958. ___. Part 2: Demonstrations of the life cycle and pathogenicity of the spiral threadworm. *Ibid.* 7: 108–123.

Cc731. **Rogers, W. P.** 1987. Advances in parasitology: 1886–1986. *International journal for parasitology* 17: 1–13.

Cc732. **Sergiev, P. G., & N. N. Plotnikov.** 1953. *Ocherki po istorii parazitologii. [Essays on the history of parasitology].* Moscow, Medgiz. 148 pp.

Cc733. **Théodoridès, J.** 1966. Les grandes étapes de la parasitologie. *Clio medica* 1: 129–145, 185–208.

See also Bb71(Gallien) Bb790–791 Ca184(Burt) & Cc948.

BIBLIOGRAPHIES OF PARASITOLOGY

• A selection of major works:

Cc734. **Bezukladnikova, N. A.** 1957. *Literatura po parazitologii Kazakhstana. Referirovannaya bibliografiya po parazitam i parazitarnym boleznyam cheloveka, sel'skokhozyaistvennykh i dikikh zhivotnykh.* Alma–Ata, Akad. Nauk KazSSR. 312 pp.

Cc735. **Lim, H.**, ed. 1974. *Bibliography on medical parasitology in Indonesia, 1855–1968.* Taipei, U. S. Naval Medical Research Unit. No. 2. iv+194 pp.

Cc736. **Stiles, C. W., & A. Hassall (et al.).** 1932–52. *Index–catalogue of medical and veterinary zoology. Authors.* [New ed.] by A., Hassall , M. Potter, et al. Washington, DC, United States Department of Agriculture. 18 pts, 5,711 pp.; *Continued* ___. 1953+. *Index–catalogue of medical and veterinary zoology: Supplement series.* Washington, DC, United States Department of Agriculture. I.e.: Nos. 1–5 (A–B, A–C, A–I, A–K, A–Q), by M. A. Doss, J. M. Humphry, & D. B. Segal, 1953–56, with annual lists (A–Z) thereafter, 1956+. ••*Replaces* ___. 1902–12. Index–catalogue of medical and veterinary zoology. Authors. *Bulletin, Bureau of Animal Industry* 39: 36 pts., 2,766 pp. For background information *see* **W. W. Becklund.** 1969. The Index–Catalogue of Medical & Veterinary Zoology at the Beltsville Parasitological Laboratory. *Journal of parasitology* 55: 381–384; **M. A. Doss.** 1953. The Index–Catalogue of Medical and Veterinary Zoology. *Bulletin of the Medical Library Association* 41: 110–113.

Cc737. ___. 1908–20. *Index–catalogue of medical and veterinary zoology. Subject catalogs.* Washington, DC, United States Department of Agriculture.
I.e.: **C. W. Stiles & A. Hassall.** 1908. *Trematoda and trematode diseases.* 401 pp. *Bulletin, United States Hygiene Laboratory* no. 37; ___. 1912. *Cestoda and cestodaria.* 467 pp. *Bulletin, United States Hygiene Laboratory* no. 85; ___.

1920. *Roundworms (Nematoda, Gordiacea and Acanthocephali) and the diseases they cause.* 886 pp. *Bulletin, United States Hygiene Laboratory* no. 114.

Cc738. ___. 1963–69. *Index–catalogue of medical and veterinary zoology. Subject catalogs.* [New ed.] by A., Hassall , M. Potter, et al. Washington, DC, United States Department of Agriculture. I.e. **M. A. Doss et al.** 1963–69. *Trematoda and trematode diseases.* 12 pts.

Cc739. ___. 1925–35. *Index–catalogue of medical and veterinary zoology. Key catalogs.* [New ed.] by A., Hassall , M. Potter, et al. Washington, DC, United States Department of Agriculture. *Bulletin, United States Hygiene Laboratory* nos. 140, 142, 148, 150, 152, 155, 159, & 163.
Key catalogue of the protozoa reported for man. 1925. 67 pp. *Bulletin* no. 140.
Key catalogue of the worms reported for man. 1926. pp. i–iv, 69–196. *Bulletin* no. 142.
Key catalogue of the crustacea and arachnoids of importance in public health. 1927. pp. 197–289 *Bulletin* no. 148.
Key catalogue of insects of importance in public health. 1928. pp. 291–408. *Bulletin* no. 150.
Key catalogue of parasites reported for primates (monkeys and lemurs) with their possible public health importance. By C. W. Stiles & A. Hassall. 1929. pp. 409–491. *Bulletin* no. 152.
Key catalogue of primates for which parasites are reported. By C. W. Stiles & M. O. Nolan. 1929. pp. 492–601. *Bulletin* no. 152.
Key catalogue of parasites reported for Chiroptera (bats) with their possible public health importance and the confused nomenclature of Nycteriba Lat. 1796 and Spinlurmix Heyden 1826. By C. W. Stiles & M. O. Nolan. 1930. pp. 603–789. *Bulletin* no. 155.
Key–catalog of parasites reported for Insectivora (moles, shrews, etc.) with their possible public health importance. By C. W. Stiles & M. O. Nolan. 1932. pp. 791–911. *Bulletin* no. 159.
Key–catalogue of parasites reported for Carnivora (cats, dogs, bears, etc), with their possible public health importance. By C. W. Stiles & C. E. Baker. 1935. pp. 913–1,223. *Bulletin* no. 163.

Protozoology

Cc740. **Belloni, L.** 1966. Appunti per una storia pre-Leeuwenhoekiana degli "animalcula." *Gesnerus* 23: 13–22.

Cc741. **Beltrán, E.** 1948. Notas de historia protozoológica, 1: El descubrimiento de los sarcodarios y los trabajos de F. Dujardin. *Revista de la Sociedad Mexicana de Historia Natural* 9: 341–345; ___. 1950. ___, 2: Cien años de estudio de las amibas parasitas del hombre, 1849–1949. Ibid. 10: 285–307; ___. 1974. ___, 3: Leeuwenhoek y el tricentenario del descubrimiento de los protozoarios. ... 4: Las amibas parasitas. *Anales de la Sociedad Mexicana de Historia de la Ciencia y de la Tecnnologia* 4: 225–308; ___. 1979. ___, 5: El renacimiento de la protozoología: Siete fecundos lustros, 1941–1976. 6: La regeneración en los protozoarios. *Ibid.* 5: 91–114, 115–126.

Cc742. **Brodskii, A. L.** 1937. *Istoriya protozoologii. (Istoriya razvitiya nashikh znanii o prosteishikh zhivotnykh). [History of the development of our knowledge about protozoan animals].* Tashkent/Samarkand, Uchpedgiz. 63 pp.

Cc743. **Churchill, F. B.** 1989. Introduction: Toward the history of protozoology. *Journal of the history of biology* 22: 185–187.

Cc744. ___. 1989. The guts of the matter. Infusoria from Ehrenberg to Bütschli: 1838–1876. *Journal of the history of biology* 22: 189–213.

Cc745. **Cole, F. J.** 1926. *History of protozoology. Two lectures delivered before the University of London at King's College in May 1925*. London, Univ. of London Pr. 64 pp. & 2 pl.

Cc746. **Corliss, J. O.** 1978–79. A salute to fifty–four great microscopists of the past: A pictorial footnote to the history of protozoology. Pt. 1 (–2). *Transactions of the American Microscopical Society* 97: 419–458, 1978; *Ibid.* 98: 26–58, 1979.

Cc747. ___. 1989. The protozoon and the cell: A brief twentieth–century overview. *Journal of the history of biology* 22: 307–323.

Cc748. **Jacobs, N. X.** 1989. From unit to unity: Protozoology, cell theory, and the new concept of life. *Journal of the history of biology* 22: 215–242.

Cc749. **Muller, M.** 1985. The search for cell organelles in protozoa. *Journal of protozoology* 32: 559–563.

Cc750. **Neviani, A.** 1936. Appunti per una storia intorno ai foraminiferi dall' antichità al secolo XVIII. Con appendice sugli antichi microscopi dell'Istituto di Bologna venduti all'esterno nel 1886. *Memorie dell'Accademia Pontificia delle Scienze dei Nuovi Lincei. Rome* (3) 2: 131–210.

Cc751. **Richmond, M. L.** 1989. Protozoa as precursors of metazoa: German cell theory and its critics at the turn of the century. *Journal of the history of biology* 22: 243–276.

Cc752. **Rothschild, L. J.** 1989. Protozoa, Protista, Protoctista: What's in a name? *Journal of the history of biology* 22: 277–305.

Cc753. **Théodoridès, J.** 1972. État des connaissances de la structure des protozoaires avant la formulation de la théorie cellulaire. *Revue d'histoire des sciences et de leurs applications* 25: 27–44.

Cc754. **Wenrich, D. H.** 1956. Some American pioneers in protozoology. *Journal of protozoology* 3: 1–7.

See also Ca1018(Poljansky) Cc199(Sapp).

BIBLIOGRAPHIES OF PROTOZOOLOGY

• A selection of major works:

Cc755. **Cushman, J. A.** 1932. *A bibliography of American foraminifera*. Sharon, MA, Cushman Laboratory for Foraminiferal Research. 40 pp. *Special publication* no. 3.

Cc755A. **Hagenmüller, F.** 1899. Bibliotheca sporozoologica. Bibliographie générale et spéciale des travaux concernant les sporozoaires parus antérieurement au 1er janvier 1899. *Annales de Musée d'Histoire Naturelle de Marseille* (2) 1 supp.: 233 pp.

Cc756. **Paranagua, M. N., et al.** 1981. Bibliografia sobre Rotíferos com enfase em espécies planctonicas. *Trabalhos oceanográficos, laboratorio de ciencias do mar, Universidade Federal de Pernambuco* 16: 297–364.

Cc757. **Sherborn, C. D.** 1888. *A bibliography of the Foraminifera, recent and fossil, from 1565–1888; with notes explanatory of some of the rare and little–known publications.* London, Dulau & Co. vi+152 pp.

Cc758. **Thalmann, H. E., & A. Liebus.** 1931–33. *Bibliographia foraminiferum recentius et fossilium.* The Hague, W. Junk. 3 vols. *Fossilium catalogus* nos. 49, 59 & 60.

Cc759. **Thompson, D'A. W.** 1885. *A bibliography of protozoa, sponges, coelenterata and works including also the polyzoa, brachiopoda and tunicata for the years 1861–1883.* Cambridge, Cambridge Univ. Pr. viii+284 pp.

See also Cc720–721.

Periodical bibliographies

See periodical bibliographies listed at Cc37–54 (especially Zoological record; *see* Cc50).

ZOOLOGICAL GARDENS

General

Cc760. **Loisel, G.** 1912. *Histoire des ménageries de l'antiquité à nos jours*. Paris, H. Laurens. 3 vols.
Vol. 1: Antiquité, Moyen âge, Renaissance; vol. 2: Temps modernes (XVIIe et XVIIIe siècles); vol. 3: Époque contemporaine (XIXe et XXe siècles).

Cc761. **Pieters, F. J. M.** 1978. Diergaarden in de Nederlanden 1750–1850 en hun betekenis voor de zoologie. In: *Acta VIII conventus historiae scientiae medicinae matheseos naturaliumque excolendae, Bergae ad Zomam*, pp. 51–72. Amsterdam.

Cc762. **Smit, P., A. P. M. Sanders & J. P. F. Van der Veer**, eds. 1986. *Hendrik Engel's alphabetical list of Dutch zoological cabinets and menageries*. 2nd enlarged ed. Amsterdam, Editions Rodopi. x+340 pp. *Nieuwe Nederlandse bijdragen tot de geschiedenis der geneeskunde en der natuurwetenschappen* no. 19.
Replaces **H. Engel.** 1939. Alphabetical list of Dutch zoological cabinets and menageries. *Bijdragen tot de dierkunde* 27: 247–346.

Cc763. **Stott, R. J.** 1981. The historical origins of the zoological park in American thought. *Environmental review* 5: 52–65.

Cc764. **Zuckermann, S.**, ed. 1979. *Great zoos of the world: Their origins and significance*. London, Weidenfeld & Nicholson. 231 pp.

Individual zoos

National Zoological Park, Washington, DC

See Guide to the records Bc792.

New York Zoological Park

Cc765. **Bridges, W.** 1974. *Gathering of animals: An unconventional history of the New York Zoological Society*. New York, Harper & Row. viii+518 pp.

Cc766. **Horowitz, H. L.** 1975. Animal and man in the New York Zoological Park. *New York history* 56: 426–455.

See also Guide to the Society's archives Bc795.

Scottish National Zoological Park, Edinburgh

Cc767. **Gillespie, T. H.** 1964. *The story of the Edinburgh Zoo: the Royal Zoological Society of Scotland and the Scottish National Zoological Park. An account of their origin and progress*. Aberdeen, M. Slains. xii+123 pp.

Tierpark, Hamburg

Cc768. **Reichenbach, H.** 1980. Carl Hagenbeck's Tierpark and modern zoological gardens. *Journal of the Society for the Bibliography of Natural History* 9: 573–585.

See also Ca824.

Zoological Gardens, London

Cc769. **Blunt, W.** 1976. *The ark in the park: The zoo in the 19th century*. London, Hamish Hamilton. 256 pp.

Cc770. **Mitchell, P. Chalmers–.** 1929. *Centenary history of the Zoological Society of London*. London, Zoological Society. xi+307 pp. & 33 pl.

Cc771. **Scherren, H.** 1905. *The Zoological Society of London: A sketch of its foundation and development and the story of its farm, museum, gardens, menagerie and library.* London, Cassell. xii+252 pp. & 62 pl.

Cc772. **Zoological Society of London.** 1976. *The Zoological Society of London 1826–1976 and beyond.* Ed. by ... Lord Zuckerman. London. xviii+353 pp. *Symposia of the Zoological Society of London* no. 40.
Includes **H. G. Vevers.** Management of a public aquarium, pp. 105–118; **A. d'A. Bellairs & D. J. Ball.** Reptiles, pp. 119–132; **P. J. S. Olney.** The policy of keeping birds in the Society's collections 1826–1976, pp. 133–145; **M. R. Brambell & S. J. Mathews.** Primates and carnivores at Regent's Park, pp. 147–165; **V. J. A. Manton.** Mammals at Whipsnade, pp. 167–178; **J. W. Toovey.** 150 years of building at London zoo, pp. 179–202; **L. G. Goodwin.** Research, pp. 215–222.

Zoologischer Garten, Hanover

Cc773. **Dittrich, L., & A. Rieke–Müller.** 1990. *Ein Garten für Menschen und Tiere: 125 Jahre Zoo Hannover.* Ronnenberg/Hannover, Grütter. 227 pp., illus.

Zoologisk Have, Copenhagen

Cc774. **Zoologisk Have, Kjøbenhavn.** 1959. *Zoo, gennem hundrede år.* Udgivet ... i anledning af dens hundredårige beståen ved R. Spärck. Copenhagen. 24 pp.

Personal menageries

Cc775. **Woolfall, S. J.** 1990. History of the 13th Earl of Derby's menagerie and aviary at Knowsley Hall, Liverpool (1806–1851). *Archives of natural history* 17: 1–47.
Known as the Knowsley menagerie.

ZOOLOGICAL CABINETS & MUSEUMS

General

Cc776. **Cleevely, R. J.** 1983. *World palaeontological collections.* London, British Museum (Natural History); Mansell Publishing Ltd.
Includes Index of institutions and collection holdings, pp. 324–365, with literature references. Partly based on: **Sherborn, C. D.** 1940. *Where is the ___ collection? [etc.]*; *see* Cc785.

Cc777. **Dance, S. P.** 1986. *A history of shell collecting.* Rev. ed. Leiden, E. J. Brill & Dr. W. Backhuys. 280 pp., illus.
Replaces First ed., London, Faber, 1966. 344 pp., illus.

Cc778. **Dean, J. D.** 1936. Conchological cabinets of the last century. *Journal of conchology* 20: 225–252.

Cc779. **Deiss, W. A., & R. B. Manning.** 1981. The fate of the invertebrate collections of the North Pacific Exploring Expedition. In: Society for the Bibliography of Natural History. *History in the service of systematics: Papers from the conference to celebrate the centenary of the British Museum (Natural History) 13–16 April, 1981*, pp. 79–85. London. *Special publication* no. 1.
See Ca792.

Cc780. **Farber, P. L.** 1977. The development of taxidermy and the history of ornithology. *Isis* 68: 550–566.

Cc781. ___. 1980. The development of ornithological collections in the late eighteenth and early nineteenth centuries and their relationship to the emergence of ornithology as a scientific discipline. *Journal of the Society for the Bibliography of Natural History* 9: 391–394.

Cc782. **Horn, W., & J. Kahle.** 1935–37. *Über entomologische Sammlungen, Entomologen und Entomo–Museologie: Ein Beitrag zur Geschichte der Entomologie.* Berlin. vi+536 pp. & 38 pl. *Entomologische Beihefte aus Berlin–Dahlem* vol. 2: 1–160, 1935; *Ibid.* 3: 161–296, 1936; *Ibid.* 4: 297–535, 1937; *Continued* **H. Sachtleben.** 1961. Nachträge zu Walter Horn und Ilse Kahle: 'Über entomologische Sammlungen.' *Beiträge zur Entomologie* 11: 481–540.
Replaces **W. Horn.** 1926. Ueber den Verbleib der entomologischen Versammlungen der Welt: Ein Beitrag zur geschichte der Entomo–Museologie. Berlin, Deutsches Entomologisches Institut. 133 pp. *Supplementa entomologica* 12.

Cc783. **Löwegren, Y.** 1952. *Naturalienkabinett i Sverige under 1700–talet: Ett bidrag till zoologiens historie.* Lund. 407 pp. *Lychnos bibliotek* no. 13.

Cc784. **Serebryakov, A. E.** 1936. Zoologicheskii kabinet kunstkamery. *Trudy Instituta Istorii Nauki i Tekhniki.* Ser. 1, *arkhiv istorii nauki i tekhniki* 9: 70–128.

Cc785. **Sherborn, C. D.** 1940. *Where is the ___ collection? An account of the various natural history collections which have come under the notice of the compiler ... between 1880 and 1939.* Cambridge, Cambridge Univ. Pr. 149 pp.
Almost entirely zoological and palaeontological.

Cc786. **Smit, P., A. P. M. Sanders & J. P. F. Van der Veer**, eds. 1986. *Hendrik Engel's alphabetical list of Dutch zoological cabinets and menageries.* 2nd enlarged ed. Amsterdam, Editions Rodopi. x+340 pp. *Nieuwe Nederlandse bijdragen tot de geschiedenis der geneeskunde en der natuurwetenschappen* no. 19.
Replaces **H. Engel.** 1939. Alphabetical list of Dutch zoological cabinets and menageries. *Bijdragen tot de dierkunde* 27: 247–346.

Cc787. **Stone, W.** 1899. Some Philadelphia ornithological collections and collectors, 1784–1850. *Auk* 16: 166–177.

Cc788. **Stresemann, E.** 1923. Die Anfänge ornithologischer Sammlungen. *Journal für Ornithologie* 11: 112–127, 1923.

Cc789. **Tomlin, J. R. Le B.** 1941–49. Shell sales. *Proceedings of the Malacological Society of London* 24: 157–160, 1941; *Ibid.* 25: 25–33, 96–99, 1942; *Ibid.* 27: 254–256, 1949.

Cc790. **Whitehead, P. J. P.** 1969. Zoological specimens from Captain Cook's voyages. *Journal of the Society for the Bibliography of Natural History* 5: 161–201.

Cc791. **___.** 1970–71. Museums in the history of zoology. *Museums journal* 70: 49–57, 155–160.

Cc792. **___.** 1978. A guide to the dispersal of zoological material from Captain Cook's voyages. *Pacific studies* 1978 (Fall): 53–93.

See also Ba345 & Ca829(George).

Individual museums

European

Akademiya Nauk SSSR, Zoologicheskii Muzei, St. Petersburg (Leningrad)

Cc793. **Akademiya Nauk SSSR.** 1925. *Zoologicheskii Muzei. Osnovan v 1832 g.* Leningrad. 23 pp.

Cc794. **Byalynitskii–Birulya, A. A.** 1925. Zoologicheskii Muzei Akademii Nauk. *Prirody* 1925(7/9): 171–182.

Cc795. **Imperatorskaya Akademiya Nauk (Académie Impériale des Sciences de St. Pétersbourg).** 1889. *Das Zoologische Museum der Kaiserlichen Akademie der*

Wissenschaften zu St. Petersburg in seinem fünfzigjährigen Bestehen, [etc.]. By A. Strauch. St. Petersburg. iv+372 pp. & 2 pl.

Cc796. **Oshanin, V. F.** 1911. Obshchii ukazatel' rabot, pomeshchenn ikh v pyatnadtsati pervykh tomakh Ezhegodnika Zoologicheskogo Muzeya Akademiya Nauk. St. Petersburg. *Ezhegodnik Zoologicheskii Muzei Akademiya Nauk* 15 prilozhenie: 60 pp.

Cc797. **Serebrovskii, P.** 1925. Zoologicheskii Muzei Vsesoyuznoi Akademii Nauk. *Chelovek i priroda* 1925(9): 13–26.

See also Bb124.

British Museum (Natural History), London

Cc798. **British Museum (Natural History).** 1906. Separate historical accounts of the several collections included in the Department of Zoology. In: *History of the collections contained in the natural history departments in the British Museum*, vol. 2. London. 782 pp.

Cc798A. **___.** 1991. Carcinology in the Natural History Museum, London: The brachyuran crab collections and their curation from 1813–1904 (Leach to Calman). By R. W. Ingle. *Bulletin of the British Museum (Natural History), historical series* 19: 161–224.

Cc799. **Gunther, A. E.** 1975. *A century of zoology at the British Museum through the lives of two keepers 1815–1914.* London, Dawsons. 533 pp.
i.e. J. E. Gray (1800–1875) and A. C. L. G. Günther (1830–1914).

See also Ba1332 & Cc885.

Deutsche Akademie der Landwirtschaftswissenschaften, Berlin, Deutsche Entomologische Institut

Cc800. **Ebert, W., et al.** 1986. Einhundert Jahre Deutsches Entomologisches Institut. *Beiträge zur Entomologie* 36: 5–52.

Edinburgh University, Zoological Museum

Cc801. **Chitnis, A. C.** 1970. The University of Edinburgh's Natural History Museum and the Huttonian–Wernerian controversy. *Annals of science* 26: 85–94.

Göteborgs Naturhistoriska Museum

Cc802. **Göteborgs Naturhistoriska Museum.** 1983. Göteborgs Naturhistoriska Museum, 150 år. Ed. by L. Orrhage. *Arstryck Göteborgs Naturhistoriska Museum* 1983: 120 pp.
Includes **S. Matthiasson.** Zoologen, mångsysslaren och djurskyddsmannen A. W. Malm: Instiftare av "Sällskapet Småfoglarnas Vänner," föregångaren till Göteborgs Djurskyddsförening, pp. 23–31; **___.** Naturhistoriska Museets ryggradsdjur och tjänstemännen kring dem, pp. 47–67; **H. W. Waldén.** Göteborgs Naturhistoriska Museums markfaunainventering 1921–1981, pp. 69–86.

Cc803–805. [Not used.]

Humboldt Universität zu Berlin (*formerly* Friedrich–Wilhelms–Universität in Berlin), Museum für Naturkunde

Cc806. **Braun, A.** 1910. Das zoologische Museum. In: M. Lenz, ed. *Geschichte der königlichen Friedrich–Wilhelms–Universität zu Berlin*, vol. 3: 372–389. Halle, Verlag der Buchhandlung des Waisenhauses.

Cc807. **Stresemann, E.** 1922. Die Entwicklung der Vogelsammlung des Berliner Museums unter Illiger und Lichtenstein. *Journal für Ornithologie* 10: 498–503.

See also Ca878.

Hunter's Museum, Glasgow

Cc808. **Brock, C. H.** 1980. Dr. William Hunter's Museum Glasgow University. *Journal of the Society for the Bibliography of Natural History* 9: 403–412.

Kazanskii Gosudarstvennyi Universitet, Zoologicheskii Kabinet, Kazan

Cc809. **Ostroymov, A. A.** 1926. K istorii Zoologicheskogo kabineta Kazanskogo Gosudarstvennogo Universiteta. *Uchenye zapiski, Kazanskii Gosudarstvennyi Universitet imeni V. I. Ul'yanova–Lenina* 86: 12–21.

Kievskii Gosudarstvennii Universitet, Zoologichnii Musei, Kiev

See Ca1016(Artobolevs'kii).

Koninklijk Zoölogisch Genootschap "Natura Artis Magistra," Museum, Amsterdam

Cc810. **Benthem Jutting, W. S. S. van.** 1939. A brief history of the conchological collections at the Zoological Museum of Amsterdam, with some reflections on eighteenth–century shell cabinets and their proprietors, on the occasion of the centenary of the Royal Zoological Society "Natura Artis Magistra." *Bijdragen tot de dierkunde* 27: 167–246, illus.

Cc811. **Van Pel, P.** 1985. Zoologisch Museum van Amsterdam. *Gloria maris* 24 (4): 81–88.

Lunds Universitet, Zoologiska Museet

Cc812. **Löwegren, Y.** 1968. *Zoologiska museet och institutionen vid Lunds Universitet.* Ur Lunds universitets historia no. 6. Lund, Gleerup. 142 pp. *Acta Universitatis Lundensis* sectio 1, 9.

Moskovskii Gosudarstvennyi Universitet, Zoologicheskii Muzei

Cc813. **Turov, S. S.** 1956. *Zoologicheskii Muzei Moskovskogo Universiteta.* Moscow, Izd. Moskovskogo Universiteta. 43 pp.

Museo Zoologico di Firenze

Cc814. **Giglioli, E. H.** 1909. *Storia della collezione centrale degli animali vertebrati Italiani nel R. Museo Zoologico di Firenze, 1876–1908.* Florence. 21 pp. *Pubblicazioni, Reale Istituto di Studi Superiori, sezione di scienze fisiche e naturali.*

Musée Zoologique, Lausanne

Cc815. **Blanc, H.** 1912. *Le Musée Zoologique de Lausanne: Ses origines, son installation au Palais de Rumine, ses collections.* Lausanne, Imprimeries Réunies. 59 pp.

Muséum National d'Histoire Naturelle, Paris

Cc816. **Berlioz, J.** 1938. Les collections ornithologiques du Muséum de Paris. *L'Oiseau* 1938 (2): 237–260.

Cc817. **Bourdelle, E.** 1927. La chaire de Mammalogie et d'Ornithologie du Muséum National d'Histoire Naturelle de Paris. *Revue scientifique* 65: 577–586.

Cc818. **Fischer–Piette, E.** 1944. Histoire de la chaire de Malacologie. *Bulletin du Muséum National d'Histoire Naturelle, Paris* (2) 16: 385–404.

Cc819. **Fontaine, M.** 1944. La chaire de Physiologie générale du Muséum National d'Histoire Naturelle. *Bulletin du Muséum National d'Histoire Naturelle, Paris* (2) 16: 204–222.

Národní Muzeum, Prague

Cc820. **Hanzák, J.** 1980. History of the Prague National Museum, with special regard to zoology. *Journal of the Society for the Bibliography of Natural History* 9: 369–373.

Naturkundemuseum Görlitz

Cc821. **Franke, R.** 1982. Zur Geschichte der entomologische Sammlungen des Staatlichen Museums für Naturkunde Gorlitz. *Abhandlungen und Berichte des Naturkunde Museums Gorlitz* 56(3): 1–21.

Oxford University Museum

Cc822. **Chancellor, G., et al.** 1988. Charles Darwin's Beagle collections in the Oxford University Museum. *Archives of natural history* 15: 197–231.

Cc823. **Davies, K. C., & J. Hull.** 1976. *The zoological collections of the Oxford University Museum: A historical review and general account, with comprehensive donor index to the year 1975.* Oxford, Oxford Univ. Museum vii+136 pp., illus.

Cc824. **Oxford University Museum.** 1986. *A history of the Hope Entomological Collections in the University Museum, Oxford. With lists of archives and collections.* By A. Z. Smith. Oxford, Clarendon Pr. xiii+172 pp. & 17 pl.

Peterburgskii Kunstkamera

Cc825. **Müller–Dietz, H. E.** 1989. Anatomische Präparate in der Petersburg "Kunstkammer." *Zentralblatt für allgemeine Pathologie und pathologische Anatomie* 135: 757–767.

Cc826. **Novikov, P. A.** 1957. Zoologicheskii otdel Peterburgskoi Kunstkamery v ego istoricheskom razvitii. [Zoological section of the St. Petersburg Kunstkammer in its historical development]. *Trudy Institut Istorii Estestvoznaniya i Tekhniki* 14: 302–352.

Polska Akademija Nauk, Instytut Zoologiczny, Warsaw

Cc827. **Leszckowska, W.** 1975. *Z dziejow ksiegozbioru Warszawskiego Gabinetu Zoologicznego 1819–1915.* Warsaw. 72 pp.

Rijksmuseum van Natuurlijke Historie, Leiden

Cc828. **Gijzen, A.** 1938. *'s Rijks Museum van Natuurlijke Historie 1820–1915.* Rotterdam, W. L. & J. Brusse. xii+335 pp.

Cc828A. **Rijksmuseum van Natuurlijke Historie, Leiden.** 1901. *Notes from the Leyden Museum: Index, vols. 1–20, 1879–1899.* Leiden, E. J. Brill. [iii]+218 pp.; ___. 191 ? *[___: Index, vols. 21–38 , 1899/1900–1914].** Leiden. [ ] pp.
[*Not examined – full details not ascertained.]

Senckenberg–Museum, Frankfurt am Main

Cc829. **Senckenberg–Museum.** 1967. Aus der Geschichte der Senckenberg–Museums. *Senckenbergiana biologica* 48 Sonderheft A & B.
Includes Sonderheft A: **R. Mertens.** Die herpetologische Sektion; Sonderheft B: **A. Zilch.** Zur Geschichte der zoologischen Abteilung; **H. Felten.** Geschichte der Säugetier–Sektion; **J. Steinbacher.** Geschichte der Ornithologischen Sektion; **W. Klausewitz.** Geschichte der Ichthyologischen Sektion; **E. Franz.** Geschichte der entomologischen Sektion; **O. Krause.** Geschichte der Vertebraten Sektion (Crustaceen–Sektion von **R. Bott**).

Spècola (La), Florence

Cc830. **Azzaroli Puccetti, M. L.** 1972. La Spècola, the Zoological Museum of the University of Florence. *Curator* 15: 93–112.

Università di Milano, Istituto di Anatomia degli Animali Domestici

Cc831. **Aureli, G., & B. Cozzi.** 1984. Il Museo Anatomico dell'Istituto di Anatomia degli Animali Domestici dell' Università di Milano. *Natura (Milan)* 74: 129–155.

Università di Padova, Musei Zoologica

Cc832. **Minelli, A., & C. Pasqual.** 1982. Documenti del sec. XIX concernenti il Musei Zoologica dell'Universita di Padova. *Lavori Societa Veneziana di Scienze Naturali* 7: 227–247.

Università di Palermo, Museo di Zoologia

Cc833. **Sara, M.** 1985. La collezione di apporati anatomici del Museo di Zoologia dell'Università di Palermo. *Naturalista Siciliano* 9: 3 18.

Universität Kiel, Zoologisches Museum

Cc834. **Kacker, J.** 1984. Vom Kuriositätenkabinett zum wissenschaftlichen Museum: Die Entwicklung der Zoologischen Sammlungen der Kieler Universität von 1665 bis 1868. *Mitteilungen aus dem Zoologischen Museum der Universität Kiel* 1984 supp. 1: 43 pp.

University of Athens, Zoological Museum

Cc835. **Legakis, A.** 1984. The Zoological Museum of the University of Athens. i. Historical notes. *Biologia Gallo–Hellenica* 11: 79–83.

University of St. Andrews Natural History Museum

Cc836. **McIntosh, W. C.** 1913. *Brief sketch of the Natural History Museum of the University of St. Andrews.* [St. Andrews.] 63 pp. & 17 pl.

Universität Halle, Zoologische Museum

Cc837. **Taschenberg, O.** 1894. Geschichte der Zoologie und der zoologischen Sammlungen an der Universität Halle, 1694–1894. Halle, M. Niemeyer. 177 pp. & 9 pl. *Abhandlungen der Naturforschenden Gesellschaft zu Halle* no. 20.

Zoologische Museum, Hamburg

Cc838. **Panning, A.** 1957. Beiträge zur Geschichte des Zoologischen Staatsinstituts und Zoologischen Museums in Hamburg. 3 teil. *Mitteilungen aus den Hamburgischen Zoologischen Museum und Institut* 55: 1–23.

Cc839. **Weidner, H.** 1968. 125 Jahre Naturhistorisches Museum zu Hamburg. *Abhandlungen und Verhandlungen des Naturwissenschaftlichen Vereins in Hamburg* 13: 5–30.

(Königliches) Zoologisches und Anthropologisch–Ethnographisches Museum zu Dresden

Cc840. **Meyer, A. B.**, ed. 1902. *Hauptverzeichniss zu den 3 Heften Mittheilungen aus dem K. Zoologischen Museum zu Dresden, 1875–78, und den 9 Bänden Abhandlungen und Berichte des Königl. Zoologischen und Anthropologisch–Ethnographisch Museums zu Dresden, 1886–1901.* Berlin, R. Friedländer & Sohn. 65 pp. *Abhandlungen und Berichte des Königlichen Zoologischen und Anthropologisch–Ethnographischen Museums zu Dresden* Bd. 9, Beiheft 2.

Zoologisk Museum i Kjøbenhavn

Cc841. **Spärck, R.** 1945. *Zoologisk Museum i Kjøbenhavn gennen tre aarhundredar.* Copenhagen, Munksgaard. 110 pp.

North American

Academy of Natural Sciences of Philadelphia, PA

Cc842. **Stone, W.** 1899. A study of the type specimens of birds in the collection of the Academy of Natural Sciences of Philadelphia, with a brief history of the collection. *Proceedings of the Academy of Natural Sciences of Philadelphia* 51: 5–62.

American Museum of Natural History, New York

Cc843. **American Museum of Natural History.** 1902. *List of papers published in the Bulletin and Memoirs of the American Museum of Natural History, volumes 1–16, 1881–1902*. New York. 32 pp.

Cc844. ___. 1911. *The American Museum of Natural History: Its origin, its history, the growth of its departments to December 31, 1901*. By H. F. Osborn. Written with the cooperation of members of the administrative and scientific staffs. 2nd ed. New York, Irving Pr. 148 pp., illus.

Cc845. **Hellmann, G. T.** 1969. *Bankers, bones and beetles: The first century of the American Museum of Natural History*. Garden City, NY, Natural History Pr. for the American Museum of Natural History. [xi]+275 pp. & 18 pl.

Cc846. **Osborn, H. F.** 1919. Fifty years of the American Museum of Natural History. *Science* 49: 477–481.

Cc847. **Preston, D. J.** 1986. *Dinosaurs in the attic: An excursion into the American Museum of Natural History*. New York, St. Martin's Pr. 244 pp.

Cc848. **Rainger, R.** 1991. *An agenda for antiquity: Henry Fairfield Osborn and vertebrate paleontology at the American Museum of Natural History, 1890–1935*. Tuscaloosa, AL, Univ. of Alabama Pr. xiii+360 pp. *History of American science & technology series.*

See also Ba1327–1329.

Lyman Entomological Museum, St. Anne de Bellevue, Que

Cc849. **Vickery, V. R., & G. A. Moor.** 1964. The Lyman Entomological Museum 1914–1964. *Canadian entomologist* 96: 1489–1494.

Museum of Comparative Zoology, Harvard University, Cambridge, MA

Cc850. **Museum of Comparative Zoology, Harvard University.** 1913. *Publications of the Museum of Comparative Zoology, Harvard College, Mass.* Cambridge, MA. 32+4 pp.

Cc851. ___. 1936. *Notes concerning the history and contents of the Museum of Comparative Zoology by members of the staff.* Cambridge, MA.

Cc852. **Lurie, E.** 1960. *The founding of the Museum of Comparative Zoology.* Cambridge, MA, Museum of Comparative Zoology. 46 pp. & 3 pl.

Cc853. **Winsor, M. P.** 1990. *Reading the shape of nature: Comparative zoology at the Agassiz museum.* Chicago, IL, Univ. of Chicago Pr. xviii+328 pp. *Science & its conceptual foundations series.*

Peter Redpath Museum, Montreal

Cc854. **Pyenson, S, Sheets.** 1982. "Stones and bones and skeletons": The origins and early development of the Peter Redpath Museum (1882–1912). *McGill journal of education* 17 (Winter 1982): 49–62.

Smithsonian Institution, United States National Museum, Washington, DC

Cc855. **Deiss, W. A.** 1980. Spencer F. Baird and his collectors. *Journal of the Society for the Bibliography of Natural History* 9: 635–645.

See also Ca956(Gill).

University of California, Museum of Vertebrate Zoology, Berkeley

Cc856. **Star, S. L., & J. R. Griesemer.** 1989. Institutional ecology, 'translations' and boundary objects: Amateurs and professionals in Berkeley's Museum of Vertebrate Zoology, 1907–1939. *Social studies of science* 19: 387–420.

Asian

Colombo Museum, Colombo

Cc857. **Colombo Museum.** 1977. *One hundred years 1977, Museum centenary souvenir.* By P. H. D. H. de Silva. Colombo. ix+148 pp.

Indian Museum, Calcutta

Cc858. **Indian Museum.** 1914. *The Indian Museum, 1814–1914.* Calcutta. xii+136+lxxxvii pp. & 9 pl.
Ch. VII. The zoological collections, pp. 68–89; Ch. VIII. Biological work of the Marine Survey of India, pp. 90–102; Ch. IX. Curators and superintendents, pp. 103–122.

Cc859. ___. 1923. *Records of the Indian Museum: General index to vols. I–XX, 1907–1920.* Calcutta. 282 pp.; ___. 1954. ___*: General index to vols. XXI–XL, 1921–1938.* Calcutta. 320 pp.

Australasian

Australian Museum, Sydney, N.S.W

Cc860. **Strahan, R., & D. F. Branagan.** 1979. *Rare and curious specimens: An illustrated history of the Australian Museum, 1827–1979.* Sydney, N.S.W., Australian Museum. ii+173 pp., illus.

See also Guide to the Museum's archives Bc734.

Macleay Museum, Sydney, N.S.W

Cc861. **Horning, D. S.** 1984. The history and significance of the Macleay insect collection. In: D. H. Borchardt, ed. 1984. *Some sources for the history of Australian science*, pp. 23–32. Sydney, N.S.W., History Project Incorporated. 85 pp.

Cc862. **Stanbury, P., & J. Holland**, eds. 1988. *Mr Macleay's celebrated cabinet: The history of the Macleays and their museum.* Sydney, N.S.W., The Macleay Museum. 171 pp.
History of the collection's formation by the Macleays and the Macleay Museum at Sydney University.

African

South African Museum, Cape Town

Cc863. **South African Museum.** 1938. *Annals of the South African Museum. List of papers, vols. 1–30, [1898–1931.]* Cape Town. 56 pp.

Cc864. **Summers, R. F. H.** 1975. *A history of the South African Museum, 1825–1975. Compiled from the Museum records.* Cape Town, A. A. Balkema. ix+245 pp., illus.

Personal collections

• A selection of representative collections:

Cuming collection

Cc865. **Dance, S. P.** 1980. Hugh Cuming (1791–1865): Prince of collectors. *Journal of the Society for the Bibliography of Natural History* 9: 477–501.

Fabricius collection

Cc866. **Zimsen, E.** 1964. *The type material of I. C. Fabricius.* Copenhagen, Munksgaard. 656 pp.
A systematic guide to the whereabouts of type specimens.

Hoffmannsegg collection

Cc867. **Stresemann, E.** 1950. Die Brasilianischen Vogelsammlungen des Grafen [Johann Centurius] von Hoffmannsegg aus den Jahren 1800–1812. *Bonner zoologische Beiträge* 1: 43–51, 126–143.

Hunter's Museum

Cc868. **Dobson, J.** 1959. John Hunter's museum. In: Z. Cope. *The Royal College of Surgeons of England; a history.* London, A. Blond. xii+360 pp.

Cc869. ___. 1962. John Hunter's animals. *Journal of the history of medicine* 17: 479–486.

Cc870. ___. 1970. Curiosities of natural history as illustrated in John Hunter's museum. *Annals of the Royal College of Surgeons of England* 47: 233–242.

Linnaeus's collections

Cc871. **Anderson, L. G.** 1899. *Catalogue of Linnæan type–specimens of snakes in the Royal Museum in Stockholm.* Stockholm. 35 pp. *Bihang til Kongliga Svenska Vetenskaps–Akademiens handlingar* vol. 24, afd. IV, no. 6.

Cc872. ___. 1900. *Catalogue of Linnean type–specimens of Linnaeus's reptilia in the Royal Museum in Stockholm.* Revised by F. A. Smitt and Hj. Théel. Stockholm. 29 pp. *Bihang til Kongliga Svenska Vetenskaps–Akademiens handlingar* vol. 26, afd. IV, no. 1.

Cc873. **Brunton, C. H. C.** 1967. Brachiopods in the Linnean shell collection. *Proceedings of the Linnean Society of London* 178: 161–183.

Cc874. **Dance, S. P.** 1967. Report on the Linnean shell collection. *Proceedings of the Linnean Society of London* 178: 1–24, 10 pl.

Cc875. **Dodge, H.** 1952–59. A historical review of the mollusks of Linnaeus. *Bulletin of the American Museum of Natural History* 100: 1–263, 1952; *Ibid.* 103: 1–134, 1953; *Ibid.* 107: 1–157, 1955; *Ibid.* 111: 153–312, 1956; *Ibid.* 113: 73–224, 1957; *Ibid.* 116: 153–224, 1958.
Also ___. 1959. Evidential factors in the identification of the Linnaean molluscs. *Journal of the Linnean Society (Zoology)* 44: 170–179.

Cc876. **Fernholm, B., & A. Wheeler.** 1983. Linnaean fish specimens in the Swedish Museum of Natural History, Stockholm. *Zoological journal of the Linnean Society* 78: 199–286.

Cc877. **Holm, Å.** 1957. *Specimena Linnaeana. I Uppsala bevarade Zoologiska Samlingar från Linnés tid. Inbjudningar till Doktors–Promotionerna i Uppsala Universitets aula. Fredagen den 31 Maj 1957.* Uppsala. 68 pp. *Uppsala Universitets årsskrift* 1957(6).

Cc878. **Jackson, B. D.** 1913. Catalogue of the Linnean specimens of Amphibia, Insecta and Testacea. *Proceedings of the Linnean Society of London* 125 (supp.): 48 pp.; ___. 1922. ___: Additions and corrections. *Ibid.* 134 supp: 33–38.

Cc879. **Lönnberg, E.** 1896. Linnean type–specimens of birds, reptiles, batrachians and fishes in the Zoological Museum of the R. University in Upsala. *Bihang til Kongliga Svenska Vetenskaps–Akademiens handlingar* 22, Afd. IV (1): 45 pp.

Cc880. **Oomen, H. C. J.** 1983. Ook de sapelden in de insekten collecties van Linnaeus habben diagnostische waarde. *Vakblad voor biologen* 63 (7): 1–131.
See also **K. Mikkola.** 1983. Diagnostic insect pins: Some problems of the Linnaean insect collection solved. *Antenna* 7 (1): 16–17.

Cc881. **Wallin, L.** 1992. Linnaean specimens in the Zoological Museum of Uppsala University. *Archives of natural history* 19: 219–230.
With a bibliography of many previous and more specialised studies on Linnaean specimens.

Cc882. **Wheeler, A. C.** 1985. The Linnaean fish collection in the Linnean Society of London. *Zoological journal of the Linnean Society* 84: 16–76.
Replaces **A. C. L. G. Günther.** 1899. Complete catalogue of Linné's private collection of fishes, now in possession of the Linnean Society. *Proceedings of the Linnean Society of London* 111: 15–38.

Cc883. ___. 1991. The Linnaean fish collection in the Zoological Museum of the University of Uppsala. *Zoological journal of the Linnean Society* 103: 145–195.

Macleay Museum, Sydney, N.S.W.

Cc884. **Stanbury, P., & J. Holland**, eds. 1988. *Mr Macleay's celebrated cabinet: The history of the Macleays and their museum.* Sydney, N.S.W., The Macleay Museum. 171 pp.

Rothschild Museum, Tring

Cc885. **Rothschild, M.** 1983. *Dear Lord Rothschild: Birds, butterflies and history.* Glenside, PA, Balaban; London, Hutchinson. xxiii+398 pp. *Also* Glenside, Balaban, 1983.
Includes details of the growth of his zoological museum at Tring, bequeathed to the British Museum (Natural History).

Seba's cabinet

Cc886. **Boeseman, M.** 1970. The vicissitudes and dispersal of Albertus Seba's zoological specimens. *Zoologische mededelingen* 44: 177–206, pl. 1–4.

Cc887. **Engel, H.** 1961. The sale–catalogue of the cabinets of natural history of Albertus Seba (1752): A curious document from the period of the Naturae curiosi. *Bulletin of the Research Council of Israel* 10B: 119–131.

Swainson collection

Cc888. **McMillan, N. F.** 1980. William Swainson (1789–1855) and his shell collection. *Journal of the Society for the Bibliography of Natural History* 9: 427–434.

Cc889. **Parkinson, P.** 1988. William Swainson's ornithological collections. *Archives of natural history* 15: 77–88.

William IV, Stadholder

Cc890. **Pieters, F. F. J. M.** 1980. Notes on the menagerie and zoological cabinet of Stadholder William V of Holland, directed by Aernout Vosmaer. *Journal of the Society for the Bibliography of Natural History* 9: 539–563.
Remnants now in the Rijksmuseum van Natuurlijke Historie at Leiden.

COLLECTIONS BIBLIOGRAPHIES

Cc891. **Wiktor, J., & W. Rydzewski.** 1991. *Bibliography of catalogues of type specimens in world's zoological and palaeozoological collections.* Wroclaw, Wroclaw Univ. Pr. 308 pp. *Prace zoologiczne* vol. 22.
Replaces **A. W. F. Banfield.** 1968. *Une list préliminaire de catalogues des spécimens types en zoologie et palaéontologie.* Bucharest, Revista Muzeelor. 30 pp.

ZOOLOGICAL EDUCATION & RESEARCH INSTITUTIONS

General

Cc892. **Desmond, A.** 1985. The making of institutional zoology in London 1822–1836. *History of science* 23: 153–185, 223–250.

Individual institutions

European

Akademiya Nauk SSSR, Institut Morphologii Zhivotnykh im. A. N. Severtsova, Moscow

Cc893. **Bobukh, I. I., & M. V. Kuzyatina.** 1957. *Izdaniya Instituta Morfologii Zhivotnykh im A. N. Severtsova (1949–1956). (Bibliograficheskii ukazatel').* Moscow. 36 pp.

Akademiya Nauk SSSR, Zoologicheskii Institut, St. Petersburg (Leningrad)

Cc894. **Skarlato, O. A.**, ed. 1982. *Zoologicheskii Institut, 150 let. [Zoological Institute, 150 years.]* Leningrad, Izd. "Nauka." 244 pp.

See also Bb124.

Cambridge University

Cc895. **Fozzard, J. A.** 1983. *Professors of anatomy in the University of Cambridge: The first two hundred and sixty–one years of the Cambridge University Department of Anatomy, 1707–1968.* Cambridge, FOSSLIA. 87 pp.

(Königliche) Christian–Albrechts–Universität, Kiel

Cc896. **Remane, A.** 1968. Zoologie und Meereskunde. In: Christian–Albrechts–Universität Kiel. *Geschichte der Christian–Albrechts–Universität Kiel, 1665–1965.* Vol. 6: *Geschichte der Mathematik, der Naturwissenschaften und der Landwirtschaftwissenschaften*, K. Jordan, ed., pp. 161–179. Neumünster, K. Wachholtz Verlag.

Friedrich–Schiller Universität, Jena

Cc897. **Friedrich–Schiller Universität.** 1965. *Kleine Festgabe aus Anlass der hundertjährigen Wiederkehr der Gründung des Zoologischen Institutes der Friedrich–Schiller–Universität Jena im Jahre 1865 durch Ernst Haeckel.* Ed. M. Gersch. Jena, G. Fischer. 140 pp.
Four articles by various authors.

Institut "Edouard Van Beneden", Université de Liège

Cc898. **Godeaux, J.** 1986. Cent ans de biologie marine et d'océanographie biologique à l'Institut de Zoologie Edouard Van Beneden. *Revue medicale de Liege* 41: 786–789.

Cc899. **Jeuniaux, C.** 1986. Essai d'un panorama de cent ans de recherche scientifique à l'institut zoologique "Edouard Van Beneden." *Revue medicale de Liege* 41: 790–801.

Khar'kovskii Ordena Trudovogo Krasnogo Znameni Gosudarstvennyi Universitet im. A. M. Gor'kogo, Kharkov

Cc900. **Volchanetskii, I. B., S. I. Medvedev, & M. P. Volovik.** 1955. Ocherk istorii zoologii v Khar'kovskom Universitete. [Sketch of the history of zoology at Kharkov University]. *Uchenye zapiski, Khar'kovskii Universiteta* 59 (*Trudy Nauchno–Issledovatel'skogo Instituta Biologii i Biologicheskogo Fakul'teta* 22): 81–122.

Københavns Universitet

Cc901. **Spärck, R.** 1962. *Undervisningen i zoologi ved Københavns universitet: Et tilbageblik over 300 år. (Teaching of zoology at the University of Copenhagen, 300 years in retrospect).* Copenhagen. 301 pp.

Kommission zur Wissenschaftlichen Untersuchung der Deutschen Meere in Kiel

Cc902. **Brandt, K.** 1921. Die zoologischen Arbeiten der Kieler Kommission, 1870–1920. In: *Festschrift der Preussischen Kommission zur wissenschaftlichen Untersuchung der deutschen Meer zu Kiel aus Anlass ihres 50 jährigen Bestehens*, pp. 76–194. Kiel & Leipzig, Lipsius & Tischer.

Leningradskii Gosudarstvennyi Ordena Lenina Universitet, St. Petersburg (Leningrad)

Cc903. **Raikov, B. E.** 1953. Iz istoriya zoologii v Leningradskom Gosudarstvennom Universitete. *Vestnik Leningradskogo Universiteta* 1953(4): 73–86.

Moskovskii Gosudarstvennyi Universitet im. M. V. Lomonosova

Cc904. **Moskovskii Gosudarstvennyi Universitet.** 1940. [Papers on the history of zoology at Moscow University]. *Uchenye zapiski, Moskovskii Gosudarstvennyi Universitet, yubil. seriya* 54: 3–196.
Includes **S. D. Yudinitsev, & P. P. Bondarenko.** Istoriya Biologicheskogo fakul'teta i ego nauchno–issledovatel'skikh institutov, pp. 3–19; **N. M. Kulagin.** Pervye etapy razvitiya zoologii v MU (1755–1896), pp. 20–31; **L. A. Zenkevich.** Istoriya zoologii bespozvonochnykh v MU, pp.32–37; ___. Istoriya zoologii bespozvonochnykh po osnovnym razdelam, pp. 37–46; ___. Kafedra i laboratoriya zoologii bespozvonochnykh, pp. 47–50; **N. M. Kulagin.** Kafedra i laboratoriya entomologii, pp. 50–51; **V. V. Alpatov.** Laboratoriya ekologii, pp. 51–59; **B. S. Matveev.** Istoriya zoologii pozvonochnykh, spavnitel'noi anatomii i ekologii v MU, pp. 60–61; ___., Spavitel'naya anatomiya v MU (1755–1940), pp. 62–77; **A. N. Formozov.** Ekologiya pozvonochnykh v MU, pp. 77–87; **S. I. Ognev.** Ocherk sistematicheskikh, faunisticheskikh i zoogeograficheskikh issledovanii po pozvonochnym zhivotnym v MU (1896–1940), pp. 87–102; **S. N. Kadovskii et al.** Istorii gidrobiologii i ikhtiologii v MU, pp. 140–156; **S. S. Turov, & F. P. Dement'ev.** Ocherk istorii Zoologicheskogo Muzeya MU, pp. 176–196.

Oxford University, Bureau of Animal Population

Cc905. **Crowcroft, P. W.** 1991. *Elton's ecologists: A history of the Bureau of Animal Population.* Chicago, IL, Univ. of Chicago Pr. xx+177 pp.
I.e. Charles Elton, who founded the Bureau in 1932.

Reale Università di Napoli, Catedra di Zoologia

Cc905A. **Reale Università di Napoli.** (1906) 1907. *Centenario della Catedra di Zoologia nella R. Università di Napoli, 1806–1906.* Naples. 58 pp.

Rijksuniversiteit, Utrecht

Cc906. **Visser, R. P. W.** 1984. De Utrechtse zoologie rond 1900: Van morfologie naar vergelijkende fysiologie. *Tijdschrift voor de Geschiedenis der Geneeskunde Natuurwetenschappen, Wiskunde en Techniek* 7: 49–59.

Station Zoologique, Villefranche–sur–Mer

Cc906A. **Blacher, L. J.** 1968. La période russe dans l'activité de la Station Zoologique de Villefranche–sur–Mer. In: Institut Océanographique, Monaco. Communications, premier Congrès International d'Histoire de l'Océanographie, Monaco, 12–17 décembre 1966. *Bulletin de l'Institut Océanographique, Monaco* numéro spécial 2: 481–490.

Cc907. **Trégouboff, G.** 1983. *Histoire de la Station Zoologique de Villefranche–sur–Mer.* Paris, Comité des Travaux Historiques et Scientifiques. 133 pp. *Bulletin de la section des sciences* no. 4.

Stazione Zoologica di Napoli (Zoologische Station in Neapel)

Cc908. **Benson, K. R.** 1988. Review paper: The Naples Stazione Zoologica and its impact on the emergence of American marine biology. *Journal of the history of biology* 21: 331–341.

Cc909. **Dohrn, F. A.**, ed. [1897.] *Das 25 jährige Jubiläum der Zoologischen Station zu Neapel am 14 April 1897.* Leipzig. 44 pp.

Cc910. **Groeben, C.**, ed. 1975. *The Naples Zoological Station at the time of Anton Dohrn.* Exhibition and catalogue by C. Groeben, in collaboration with I. Müller. English translation by R. & C. Ivell. Naples, Naples Zoological Station. 110 pp.

Cc911. ___. 1984. History: The Naples Zoological Station and Woods Hole. *Oceanus* 27: 60–69.

Cc912. **Muller, G. M., & C. Groeben.** 1984. Die Zoologische Station in Neapel von ihren Anfangen bis heute: Ein "Permanenter Kongress." *Naturwissenschaftliche Rundschau, Stuttgart* 37: 429–437.

Cc913. **Oppenheimer, J. M.** 1980. Some historical backgrounds for the establishment of the Stazione Zoologica at Naples. In: M. Sears & D. Merriman, eds. *Oceanography: The past.* (Proceedings of the third International Congress on the History of Oceanography held September 22–26, 1980, [etc.].), pp. 179–187. New York, Springer–Verlag.

Cc914. **Partsch, K. J.** 1980. *Die Zoologische Station in Neapel: Modell internationaler Wissenschaftszusammenarbeit.* Gottingen, Vandenhoeck & Ruprecht. 369 pp. *Studien zu Naturwissenschaft, Technik und Wirtschaft im 19. Jahrhundert* vol. 11.

Cc915. **Simon, H.–R.** 1980. *Anton Dohrn und die Zoologische Station Neapel.* Frankfurt am Main, Erbrich. 164 pp. *Bibliographie et scientia* no. 1.

Cc916. **Various authors.** 1985. The Naples Zoological Station and the Marine Biological Laboratory: One hundred years of biology. *Biological bulletin* 168 supp: 207 pp. Twenty–two articles by various authors. *Includes* **C. Groeben.** Anton Dohrn – the statesman of Darwinism, pp. 4–25; **A. Monroy & C. Groeben.** The "new" embryology at the Zoological Station and at the Marine Biological Laboratory, pp. 35–43; **F. Ghiretti.** Comparative physiology and biochemistry at the Zoological Station of Naples, pp. 122–126; **E. Florey.** The Zoological Station at Naples and the neuron: Personalities and encounters in a unique institution, pp. 137–152; **J. Z. Young.** Cephalopods and neuroscience, pp. 153–158; **M. V. L. Bennett.** Nicked by Occam's Razor: Unitarianism in the investigation of synaptic transmission, pp. 159–167; **C. R. Thomas.** Marine botany and ecology at Stazione Zoologica, pp. 168–171; **J. D. Ebert.** Carnegie Institution of Washington and marine biology: Naples, Woods Hole, and Tortugas, pp. 172–182; **J. Maienschein.** First impressions: American biologists at Naples, pp. 187–191; *see* Ca1062 for note of other contributions.

Cc917. **Zoologische Station in Neapel.** 1893. Mitteilungen aus der Zoologischen Station zu Neapel. Autorenregister zu Band 1–10 (1879–1893). *Mitteilungen aus der Zoologische Station zu Neapel* 10: 675–680; ___. 1913. Autorenregister zu Band 11–20 (1893–1913). *Ibid.* 20: 711–715.

See also Guide to some of the archives Bc832A.

Stazione Zoologica di Trieste (Zoologische Station in Trieste)

Cc917A. **Specchi, M.** 1990. La Stazione Zoologica di Trieste, centro scientifico all'avanguardia a cavallo di due secoli (1875–1918). In: Istituto Veneto di Scienze, Lettere ed Arti. *La scienze mediche nel Veneto dell'ottocento*, pp. 99–108. Venice.

Sveriges Zoologiska Hafsstation, Kristineberg

Cc918. **Bernhard, C. B.** 1978. *Kristinebergs marinbiologiska station 1877–1977.* [Stockholm.] 55 pp. *Bidrag till Kungliga Svenska Vetenskapsakademiens historia* no. 14.

Cc919. **Theel, Hj.** 1928. *Bidrag till Kristinebergs Zoologiska Station, 1877–1927.* Stockholm. 48 pp. *Skriftserie utgiven av K. Svenska Vetenskapsakademien* no. 1.

Trinity College, Dublin

Cc920. **Gatenby, J. B.** 1960. The history of zoology and comparative anatomy in Trinity College, Dublin. *Irish journal of medical sciience* (6) 1960 (417): 395–407.

Tvärminne Zoologiska Station

Cc921. **Luther, A.** 1957. Tvärminne Zoologiska station. *Acta Societatis pro Fauna et Flora Fennica* 73: 128 pp.

Universität Erlangen

Cc922. **Geus, A.** 1969. *Die Zoologie in Erlangen. Zur 150. Wiederkehr der Errichtung eines Lehrstuhls für Naturgeschichte* ... Ausstellung 26. Juli bis 30. September 1969. Erlangen, Stadtmuseum. 72 pp. & 42 pl.

Universität Greifswald

Cc922A. **Keilbach, R.** 1956. Chronik des Zoologischen Instituts und Museums der Ernst Moritz Arndt-Universität Greifswald 17.10.1956, Bd. 2, pp. 561–571. *Wissenschaftliche Zeitschrift der Ernst–Mortiz–Arndt–Universität Greifswald.*

Universität Halle

Cc923. **Kaiser, W., & W. Piechocki.** 1969. Medizinisch–zoologischer Unterricht im 18. Jahrhundert an der Universität Halle. *Hercynia* 6: 258–284.

Cc924. **Taschenberg, O.** 1894. Geschichte der Zoologie und der zoologischen Sammlungen an der Universität Halle, 1694–1894. Halle, M. Niemeyer. 177 pp. & 9 pl. *Abhandlungen der Naturforschenden Gesellschaft zu Halle* no. 20.

Universität Wien, Paläontologisches Institut

Cc925. **Universität Wien, Paläontologisches Institut.** 1973. *100 Jahre Paläontologisches Institut der Universität Wien, 1873–1973.* By F. Steininger & E. Thennius. Vienna. 68 pp.

University College, London

Cc926. **Freeman, R. B.** 1964. *Notes on Robert E. Grant, M.D. (Edin. 1814); and on the Department of Zoology and Comparative Anatomy, University College, London.* [London, privately pub.] 31+[3] pp.
Xerographed production.

Uniwersytet Jagiellonski, Cracow

Cc927. **Fedorowicz, Z.** 1962. Materialy do historii zoologii na Uniwersytecie Jagiellonskim (1777–1914). (Matériaux à l'histoire de la zoologie à l'Université Jagellonienne à Cracovie (1777–1914).). *Memorabilia zoologica* 9: 1–124.
Summary in French.

Cc928. **___.** 1965. *Dzieje zoologii na uniwersytecie Jagiellonskim w latach 1780–1960.* Cracow, Uniwersytet Jagiellonski. 144 pp.

Wiener Universität

Cc929. **Kühnelt, W.** 1985. Zoologische Forschung im Bereich der Wiener Universität. *Archiv der Geschichte der Naturwissenschaften* 14/15: 663–679.

Zoologische Station des Berliner Aquariums, Rovigno

Cc930. **Zavodnik, D.** 1990. The ships and marine research of Rovinj Zoological Station before 1918. In: W. Lenz & M. Deacon, eds. *Ocean sciences: Their history and relation to man. Proceedings of the 4th International Congress on the History of Oceanography*

Hamburg [etc.], pp. 214–219. Hamburg, Bundesamt für Seeschiffahrt und Hydrographie. *Deutsche hydrographische Zeitschrift*, Ergänzungsheft, Reihe B, no. 22.

North American

Canada Department of Agriculture, Ottawa

Cc931. **Canada Department of Agriculture.** 1938. *Index to entomological publications of the Department of Agriculture for 1884–1936.* By C. E. Petch. Ottawa, Department of Agriculture, Entomological Branch. 410 pp.

Cc932. **___, Biosystematics Research Centre.** 1986. *Systematics in Agriculture Canada at Ottawa, 1886–1986. La recherche en systématique à Agriculture Canada Ottawa, 1886–1986.* By W. J. Cody, D. B. O. Savile & M. J. Sarazin. Ottawa, Biosystematics Research Centre. [vi]+81 & [vi]+83 pp. *Historical series* no. 28.
Ch.4. Entomology. Includes histories of the Division of Entomology & Botany 1887–1909, Division of Entomology 1910–13, Entomological Branch 1914–38, Entomology Division 1939–59, Entomology Research Institute 1959–72, and Biosystematics Research Institute 1973. English and French versions of the text.

Columbia University, New York

Cc933. **Crampton, H. E.** 1950. A history of the Department of Zoology at Columbia University. *Bios* 21: 219–246.

Fisheries Research Board of Canada

Cc934. **Fisheries Research Board of Canada.** 1968. *Index and list of titles, Fisheries Research Board of Canada and associated publications, 1900–1964.* By N. M. Carter. Ottawa. 649 pp. *Bulletin* no. 164; *Continued* ___. 1973. ___*: 1965–1972.* By N. M. Carter. Ottawa. vi+588 pp. *Miscellaneous special publication* no. 18.

Cc935. **Johnstone, K.** 1977. *The aquatic explorers: A history of the Fisheries Research Board of Canada.* Toronto, Univ. of Toronto Pr. xv+342 pp.

Harvard University, Cambridge, MA

Cc936. **Mark, E. L.** 1930. Zoology. In: Morison, S. E. 1930. *The development of Harvard University since the inauguration of President Eliot*, pp. 378–393. Cambridge, MA. *The tercentenary history of Harvard College and University, 1636–1936.*

Cc937. **Romer, A. S.** 1948. Zoology at Harvard. *Bios* 19: 7–20.

Harvard University, Cambridge, MA, Physiological Laboratory

See Cc200(Borell).

Indiana University, Bloomington, IN

Cc938. **Torrey, T. W.** 1949. Zoology and its makers at Indiana University. *Bios* 20: 67–99.

Johns Hopkins University, Biology Department, Baltimore, MD

Cc939. **Benson, K. R.** 1985. American morphology in the late nineteenth century: The Biology Department at Johns Hopkins University. *Journal of the history of biology* 18: 163–205.

Cc940. **Swanson, C. P.** 1951. A history of biology at the John Hopkins University. *Bios* 22: 222–262.

United States Bureau (*later* Department) of Entomology

Cc941. **Dunlap, T. R.** 1980. Farmers, scientists, and insects. *Agricultural history* 54: 93–107.
Development of the Bureau of Entomology, 1900–20.

Cc942. **United States Bureau of Entomology.** 1906–10. *List of the publications of the Bureau of Entomology.* (Revised to February 1, 1910). By M. Colcord. Washington, DC. *Circular* ser. 2, no. 76.

Cc943. **United States Department of Entomology.** 1930. *Check list of publications on entomology issued by the U. S. Department of Agriculture through 1927, with subject index.* By M. Colcord, I. L. Hawes & A. J. Carabelli. Washington, DC. iv+261 pp. *Bibliographical contribution* no. 20.

Cc944. **Weber, G. A.** 1930. *The Bureau of Entomology: Its history, activities and organization.* Washington, DC, Brookings Institute. 177 pp. *Monograph of the U. S. Government* no. 60.

See also Guide to the Bureau's archives Bc842A.

United States Bureau of Sport Fisheries & Wildlife

Cc945. **United States Bureau of Sport Fisheries & Wildlife.** 1974. *Bibliography of research publications of the U. S. Bureau of Sport Fisheries & Wildlife, 1928–1972.* By P. H. Eschmeyer & Van T. Harris. Washington, DC. 154 pp. *Resource publication* no. 120.

United States Commission of Fish & Fisheries (*later* Bureau of Fisheries)

Cc946. **Allard, D. C., Jr.** 1978. *Spencer Fullerton Baird and the U. S. Fish Commission.* New York, Arno Pr. xii+424 pp.

Cc947. **United States Bureau of Fisheries.** 1920. An analytical subject bibliography of the publications of the Bureau of Fisheries, 1871–1920. By R. M. E. MacDonald. *Report of the United States Commissioner of Fisheries* 1920 Appendix 5: 306 pp.

United States Department of Agriculture

Cc948. **Andrews, J. S.** 1986. Animal parasitology in the United States Department of Agriculture: 1886–1984. *Journal of N A L Associates* 11: 113–166.

United States Department of Agriculture, Bureau of Biological Survey

Cc949. **Cameron, J.** 1929. *The Bureau of Biological Survey: Its history, activities and organization.* Baltimore, MD, Johns Hopkins Univ. Pr. x+339 pp. *Re-pub.* New York, Arno Pr., 1974.

Cc950. **Sterling, K. B.** 1989. Builders of the U.S. Biological Survey, 1885–1930. *Journal of forest history* 33: 180–187.

Cc951. **United States Department of Agriculture, Bureau of Biological Survey.** 1913. *Index to papers relating to the food of birds, by members of the Biological Survey, in publications of the United States Department of Agriculture, 1885–1911.* By W. L. McAtee. Washington, DC. 69 pp. *Bulletin* no. 43.

United States Fish & Wildlife Service

Cc952. **United States Fish & Wildlife Service.** [1955.] *Fishery publication index, 1920–54; publications of the Bureau of Fisheries and fishery publications of the Fish and Wildlife Service by series, authors, and subjects.* Washington, DC. 254 pp. *Circular* no. 36; *Continued* ___. 1969. *Fishery publication index, 1955–64; [etc.].* Washington, DC. 240 pp. *Circular* no. 296.

Some historical information is also contained in: **R. D. Fisher.** 1982. Featured institution – Museum Section, U. S. Fish and Wildlife Service. *Association of Systematics Collections newsletter* 10 (3): 29–31.

See also Guide to the Bureau's archives Bc842A.

University of California, Berkeley, CA

Cc953. **Eakin, R. M.** 1956. History of zoology at the University of California, Berkeley. *Bios* 27: 62–90.

University of Chicago, IL

Cc954. **Newman, H. H.** 1948. History of the Department of Zoology in the University of Chicago. *Bios* 19: 215–239.

University of Illinois, Department of Zoology, Urbana, IL

Cc955. **Van Cleave, H. J.** 1947. A history of the Department of Zoology in the University of Illinois. *Bios* 18: 75–97.

University of Maryland, College Park, MD

Cc956. **Bissell, T. L.** 1960. History of entomology at the University of Maryland. *Bulletin of the Entomological Society of America* 6: 80–85.

University of Missouri, Department of Zoology, Columbia, MO

Cc957. **Curtis, W. C.** 1949. History of the Department of Zoology, University of Missouri. *Bios* 20: 147–172.

University of North Carolina, Department of Zoology, Chapel Hill, NC

Cc958. **Coker, R. E.** 1949. The Department of Zoology, University of North Carolina. *Bios* 20: 211–224.

University of Texas, Austin, TX

Cc959. **Oliver, C. P.** 1952. Zoology at the University of Texas. *Bios* 23: 91–108.

University of Toronto, Department of Zoology

Cc960. **Craigie, E. H.** 1966. *A history of the Department of Zoology of the University of Toronto up to 1962. With appendices covering the building and opening of the Ramsay Wright Zoological Laboratories.* Toronto, Univ. of Toronto. 108 pp.

University of Wisconsin, Department of Zoology, Madison, WI

Cc961. **Noland, L. E.** 1950. History of the Department of Zoology, University of Wisconsin. *Bios* 21: 83–109.

Asian

Tomskii Gosudarstvennii Universitet im. V. V. Kuibysheva

Cc962. **Ruzskii, M. D.** 1940. Kratkii ocherk zoologicheskogo izucheniya Sibiri za 50 let sushchestvovaniya Tomskogo Gosudarstvennogo Universita. [Short sketch of zoological studies in Siberia during 50 years existence of Tomsk State University]. *Trudy Biologicheskogo Nauchno–Issledovatel'skogo Instituta Tomskogo Gosudarstvennom Universitete* 7: 5–24.

ZOOLOGICAL CLASSIFICATION & NOMENCLATURE

General

Cc963. **Balss, H.** 1933. Zur Geschichte der Systematik der Wirbeltiere. *Archeion* 15: 16–26.

Cc964. **Carleer, L. H. M.** [1861.] *Question de sciences naturelles. Examen des principales classifications adoptées par les zoologistes.* [Brussels.] 284 pp. *Annales des Universités de Belgique* vol. 1, 1858/59.

Cc965. **Daudin, H.** 1926. *Études d'histoire des sciences naturelles*. Paris, Alcan. 2 vols. *Re-pub.* Paris, Editions des Archives Contemporaines, 1983. 2 vols.
1: De Linné à Lamarck. Méthodes de la classification et idée de série en botanique et en zoologie (1740–1790); 2: Cuvier et Lamarck. Les classes zoologiques et l'idée de série animale (1790–1830).

Cc966. **Di Gregorio, M. A.** 1982. In search of the natural system: Problems of zoological classification in Victorian Britain. *History & philosophy of the life sciences* 4: 225–254.

Cc967. **Farber, P. L.** 1976. The type–concept in zoology during the first half of the nineteenth century. *Journal of the history of biology* 9: 93–119.

Cc968. **Günther, K.** 1956. Systematik und Stammesgeschichte der Tiere, 1939–53. *Fortschritte der Zoologie* 10: 33–278; *Continued* ___. 1962. ___, 1954–1959. *Ibid.* 14: 268–547.
Includes good bibliographies.

Cc969. **Hofsten, N. von.** 1960. Linnés djursystem. *Svenska Linné–Sällskapets årsskrift* 42 (1959): 9–49.
Linnaeus's system of animal classification.

Cc970. **___.** 1963. A system of 'double entries' in the zoological classification of Linnaeus. *Zoologiska bidrag från Uppsala* 35: 603–631.

Cc971. **Hopwood, A. T.** 1950. Animal classification from Linnaeus to Darwin. In: Linnean Society of London. *Lectures on the development of taxonomy*, pp. 46–59. London.

Cc972. **Ritvo, H.** 1990. New presbyter or old priest? Reconsidering zoological taxonomy in Britain, 1750–1840. *History of the human sciences* 3: 259–276.

Cc973. **Winsor, M.** 1976. *Starfish, jellyfish and the order of life: Issues in 19th–century science*. New Haven, CT, Yale Univ. Pr. x+228 pp. *Yale studies in the history of science and medicine* no. 10.

See also Cc296 Cc305(Fleagle & Jungers) Cc328 Cc334(O'Hara) Cc336 Cc462 Cc502–503 Cc513 Cc519(Tuxen, Lindroth) Cc521–523 Cc539 & Cc708.

CLASSIFICATION SCHEDULES

Cc974. **Blackwelder, R. E.** 1963. *Classification of the animal kingdom*. Carbondale, IL, Southern Illinois Univ. Pr. [v]+94 pp.

Cc975. **Rothschild, N. M. V.** (Lord). 1965. *A classification of living animals*. 2nd ed. London, Longmans, Green & Co. 1965. ix+134 pp. ••*Replaces* 1st ed., 1961. vii+106 pp.

Zoological nomenclature

Cc976. **Heppell, D.** 1981. The evolution of the code of zoological nomenclature. In: Society for the Bibliography of Natural History. *History in the service of systematics: Papers from the conference to celebrate the centenary of the British Museum (Natural History) 13–16 April, 1981*, pp. 135–141. London. *Special publication* no. 1.

Cc977. **Linsley, E. G., & R. L. Usinger.** 1959. Linnaeus and the development of the International Code of Zoological Nomenclature. *Systematic zoology* 8 (1): 39–47.

Cc978. **Stejneger, L.** 1924. A chapter in the history of zoological nomenclature. *Smithsonian miscellaneous collections* 77 (1): 1–21.

CODE OF NOMENCLATURE & INDEXES

Cc979. **International Trust for Zoological Nomenclature.** 1985. *International code of zoological nomenclature adopted by the XX General Assembly of the International Union of Biological Sciences, Helsinki, ... 1979.* 3rd ed. London, International Trust for Zoological Nomenclature in association with British Museum (Natural History); Berkeley, CA, Univ. of California Pr. xx+338 pp.

• Indexes:

Cc980. **Agassiz, J. L. R.** 1842–46. *Nomenclator zoologicus: Continens nomina systemica generum animalium tam viventium quam fossilium, secundum ordinen alphabeticum disposita, adjectis auctoribus, libris, in quibus, reperiunter, anno editionis, etymologia et familiis, ad quas pertinent, in singulis classibus ...* Solothurn, Jent & Gassmann. 26 separately paged pts. *Also* ___. 1846. *Nomenclatoris zoologici index universalis.* Soloduri. vii+393 pp. ••Later ed. of the index *Nomenclatoris zoologici index universalis ... secundum ordinem alphabeticum unicum disposita.* Solothurn, 1848. x+1,135 pp. *See also* **D. R. Bowley & H. M. Smith.** 1968. The dates of publication of Louis Agassiz's *Nomenclator zoologicus. Journal of the Society for the Bibliography of Natural History* 5: 35–36; **D. K. McE. Kevan.** 1970. Agassiz's Nomenclator zoologici index universalis – a correction. *Ibid.* 5: 286.

Cc981. ***Fossilium catalogus I. Animalia.*** Berlin, The Hague, W. Junk, (*later* Amsterdam, Kugler Publications), pt. 1+, 1913+.

Cc982. **Marschall, A. F.** 1873. *Nomenclator zoologicus continens nomina systematica generum animalium tam viventium quam fossilium, secundum ordinem alphabeticum disposita.* Vienna, Kaiserlich–Königliche Zoologisch–Botanische Gesellschaft. iv+482 pp.

Cc983. **Neave, S. A.** 1939–40. *Nomenclator zoologicus: A list of the names of genera and subgenera in zoology from the tenth edition of Linnaeus 1758 to the end of 1935.* London, Zoological Society of London. 4 vols. *Continued* ___. 1950. ___*: Vol. 5. (Supplement, 1936–1945).* London, [etc.]. vi+308 pp.; ___. 1966. ___*: Vol. 6. (Supplement, 1946–55).* Ed. by M. A. Edwards & A. T. Hopwood. London, [etc.]. 329 pp.; ___. 1975. ___*: Vol. 7. (Supplement, 1956–65).* London, [etc.]. 374 pp. Supplemented by: ***Zoological record.*** Section 19: "List of new genera and sub–generic names." ••*Replaces* **S. H. Scudder.** 1882. Nomenclator zoologicus — An alphabetical list of all generic names that have been employed by naturalists for recent and fossil animals from the earliest times to the close of the year 1879 ... in two parts; I: Supplemental list; II: Universal index. *Bulletin of the United States National Museum* 19: xix+376+340 pp. I: Supplemental list of genera in zoology [supplementing Agassiz, Marschall, *Zoological record,* etc.]; II: Universal index to genera in zoology [etc.].

Cc983A. **Schulze, F. E., et al.** 1926–54. *Nomenclator animalium generum et subgenerum.* Ed. Preussischen (*later* Deutsche) Akademie der Wissenschaften zu Berlin. 5 vols.

Cc984. **Sherborn, C. D.** 1902. *Index animalium sive index nominum quae ab A. D. MDCCLVIII (1758) generibus et speciebus animalium imposta sunt; societatibus eruditorum adiuvantibus a Carolo Davies Sherborn; sectio prima a kalendis Januariis MDCCLVIII (1758) usque ad finem Decembris MDCCC (1800).* Cambridge. lix+1195 pp.

Cc985. **Sherborn, C. D.** 1922–32. *Index animalium sive index nominum quae ab A. D. MDCCLVIII (1758) generibus et speciebus animalium imposta sunt; societatibus eruditorum adiuvantibus a Carolo Davies Sherborn; sectio secunda a kalendis Ianuariis, MDCCCI usque ad finem Decembris, MDCCCL.* London, British Museum (Natural History). 9 vols. *Also* **F. Poche.** 1939. Supplement zu C. D. Sherborns Index Animalium. In: *Festschrift zum 60. Geburtstage von*

Prof. Dr. Embrik Strand, vol. 5: 477–615. Riga, "Latvija."

Cc986. **Waterhouse, C. O.** 1902. *Index zoologicus, an alphabetical list of names of genera and subgenera proposed for use in zoology as recorded in the "Zoological Record" 1880–1900 together with other names not included in the "Nomenclator Zoologicus" of S. H. Scudder*. London, Zoological Society of London. xii+421 pp. *Continued* ___. 1912. *Index zoologicus no. II, an alphabetical list of names ... in the "Zoological Record" vol. 38–47 inclusive (1901–1910) and the zoology volumes of the "International Catalogue of Scientific Literature" annual issues 1–10 together with other names not included in previous nomenclators*. London, [etc.] vi+324 pp. *See also* **F. Poche.** 1908. Supplement zu C. O. Waterhouses Index Zoologicus. *Zoologische Annalen* 2: 273–343; ___. 1914. Supplement zu C. O. Waterhouses Index Zoologicus II. *Zoologische Annalen* 6: 33–46.

Zoological terms, etymology

Cc987. **Dobson, J.** 1962. *Anatomical eponyms: Being a biographical dictionary of those anatomists whose names have become incorporated into anatomical nomenclature, with definitions of the structures to which their names have been attached and references to the works in which they are described*. 2nd ed. Edinburgh & London, E. & S. Livingstone. [viii]+235 pp. ••*Replaces* 1st ed., London, Baillière, Tindall & Cox, 1946. [x]+280 pp.

Cc988. **Field, E. J., & R. J. Harrison.** 1968. *Anatomical terms: Their origin and derivation*. 3rd ed. Cambridge, Heffer. xvii+312 pp. ••*Replaces* ___. 1957. ___. 2nd ed. Cambridge, Heffer. x+190 pp.

Cc989. **Gruson, E. S.** 1972. *Words for birds — a lexicon of North American birds with biographical notes*. New York, Quadrangle Books. xii+212 pp.

Cc990. **Hahn von Dorsche, H.** 1981. Die Nomenklaturen der Anatomen: Ein historischer Überblick. *Anatomischer Anzeiger* 150: 11–43.

Cc991. **Jobling, J. A.** 1991. *A dictionary of scientific bird names*. Oxford, Oxford Univ. Pr. xxix+272 pp.

Cc992. **Scarborough, J.** 1992. *Medical terminologies: Classical origins*. Norman, OK, Univ. of Oklahoma Pr. ix+303 pp. ••More than the title suggests. *Includes* 1. Special vocabularies in medical and other English: Why there are jargons; 2. Botany; 3. Invertebrates: Protozoa to Molluscs; 4. Arthropods; 5. Bones; 6. Nerves; 7. Muscles; 8. Breathing and how it works; 9. Eating, digestion, elimination; 10 Sex; 11. Vascular matters: Heart and blood; 12. Some more fluids; 13. The human context: Sight, sound, voice.

Cc993. **Singer, C. J.** 1959. The strange histories of some anatomical terms. *Medical history* 3: 1–7.

Cc994. **Suolahti, H.** 1909. *Die deutschen Vogelnamen: Eine wortgeschichtliche Untersuchung*. Strasbourg, K. J. Trübner. xxxiii+540 pp.

Cc995. **Swann, H. K.** 1913. *A dictionary of English and folk-names of British birds, with their history, meaning, and first usage [etc.]*. London, Witherby & Co. xii+266 pp.

Cc996. **Wynne, O. E.** 1969. *Biographical key — names of birds of the world — to authors and those commemorated*. Sandleheath, Hants, privately pub. iv+246 pp. ••*Replaces* Provisional ed., 1966.

ZOOLOGICAL ILLUSTRATION

General

Cc997. **Dal, B.** 1974. Sveriges zoologiska litteratur: Översikt av våra zoologiska tryck fram till år 1900, med spridda anteckningar kring zoologi och zoologer, illustratörer och boktryckare. *Biblis* 1974: 29–162, illus.

English "Swedish zoological literature. A summary," pp. 153–156. A general survey of Swedish zoological prints up to the year 1900. With a name index.

Cc998. **Dance, S. P.** 1978. *The art of natural history: Animal illustrators and their work.* London, Country Life; Woodstock, NY, Overlook Pr. 224 pp., illus.

Cc998A. **Guggisberg, C. A. W.** 1977. *Early wildlife photographers.* Foreword by Eric Hosking. Newton Abbot, David & Charles; New York, Taplinger. 128 pp., illus.

Cc999. **Hutchinson, G. E.** 1978. Zoological iconography in the West after A. D. 1200. *American scientist* 66: 675–684.

Cc1000. **Knight, D. M.** 1977. *Zoological illustration: An essay towards a history of printed zoological pictures.* London, Dawson; Hamden, CT, Archon Books. xii+204 pp.

Cc1001. **Nissen, C.** 1966–78. *Die zoologische Buchillustration. Ihre Bibliographie und Geschichte.* Stuttgart, Hiersemann. 2 vols., illus.
(This work and its ornithological companion (*see* Cc1012) are complementary; the zoological work refers the user to its older partner for most ornithological titles.) Band I: Bibliographie, Teil I: Verfasser; Teil II: Zeitschriften, anonyme und Serienwerke. — Zugleich Titelregister zum Verfasser–Teil; Register — i. Länder, ii. Sachen, iii. Tiere, iv. Verfasser, Herausgeber, Übersetzer etc. Band II: Geschichte — I. Vorgeschichte; II. Antike und Mittelalter; III. Die Renaissance des Naturstudiums; IV. Neuzeit, I. Teil: Die Zeit des Holzschnittes; V. Neuzeit, II. Teil: Die Zeit des Kupferstiches; VI. Neueste Zeit; Einzelfragen der zoologischen Buchillustration: Beiträge verschiedener Verfasser [contributions on various topics by H. Schlegel, C. Nissen, H.–R. Simon, T. A. Wohlfahrt, F. Kinkelin, W. Langer, and H. Walravens]; Nachtrag; Register — i. Länder, Orte und Völker (Band II), ii. Sachen (Band II), iii. Tiere (Band II), iv. Verfasser und einige Anonyma (Band II), v. Künstler (Band I und II).

Cc1002. **___.** 1969. *Tierbücher aus fünf Jahrhunderten: Einführung und Beschreibung von [etc.].* Zurich, L'Art Ancien; Munich, R. Wölfle; Olten, Weiss–Hesse. 115 pp. & 60 original leaves.
With leaves from 60 European illustrated zoological books.

Cc1003. **Norelli, M. R.** 1975. *American wildlife painting.* New York, Watson–Guptill. 224 pp.
Studies of Mark Catesby, Alexander Wilson, John J. Audubon, Martin J. Heade, Abbott H. Thayer, and Louis A. Fuertes.

Cc1004. **Schäfer, W.** 1951. *Das wissenschaftliche Tierbild.* Herausgegeben von der Senckenbergischen Naturforschenden Gesellschaft zu Frankfurt am Main. 2. Aufl. Frankfurt am Main, Waldemar Kramer. 143 pp.
Replaces 1st ed., 1949. 136 pp.

See also Biographies Ba384 Ba542–546 Ba809 Ba1248–1252 & Ba1596.

Ornithological

Cc1005. **Anker, J.** 1938. *Bird books and bird art: An outline of the literary history and iconography of descriptive ornithology, based principally on the collection of books containing plates with figures of birds and their eggs now in the University Library at Copenhagen and including a catalogue of these works.* Copenhagen, Levin & Munksgaard. xviii+251 pp. & 12 pl. *Re-pub.* The Hague, W. Junk, 1973; New York, Maurice Martino, 1990.

Cc1006. **Jackson, C. E.** 1975. *Bird illustrators: Some artists in early lithography.* London, H. F. & G. Witherby. 136 pp. & 14 pl.

Cc1007. **___.** 1978. *Wood engravings of birds.* London, H. F. & G. Witherby. 144 pp., illus.

Cc1008. ___. 1985. *Bird etchings: The illustrators and their books 1655–1855.* Ithaca, NY, Cornell Univ. Pr. 296 pp., illus.

Cc1009. **Lambourne, M.** 1990. *The art of bird illustration.* London, Collins. 192 pp., illus. *Also* Secaucus, NJ, The Wellfleet Pr.

Cc1010. **Lysaght, A. M.** 1975. *The book of birds: Five centuries of bird illustration.* London, Phaidon Pr. 208 pp., illus.

Cc1011. **Nissen, C.** 1936. *Schöne Vogelbücher: Ein Überblick der ornithologischen Illustration nebst Bibliographie.* Vienna, Reichner. 95 pp.
First pub. as: ___. 1935. Die ornithologischen Illustration: Ein Überblick nebst einer Bibliographie schöner Vogelbücher. *Philobiblon* 8: 23–40, 69–89, 123–136, 169–179, 225–234, 436–443.

Cc1012. ___. 1953. *Die illustrierten Vögelbucher: Ihre Geschichte und Bibliographie.* Stuttgart, Hiersemann. 222 pp., illus.
Geschichtlicher Teil: Altertum, Mittelalter, Neuzeit (15./16. Jahrhundert, 17. Jahrhundert, 18. Jahrhundert, 19. Jahrhundert <Frankreich, Deutschland, Skandinavien, Amerika, England>, 20. Jahrhundert <England, Niederlande, Skandinavien, Deutschland, Frankreich, Italien, Amerika>); Bibliographischer Teil; Register. 1. Künstler, 2. Vögel, 3. Länder, 4. Verfasser.

Cc1013. **Ronsil, R.** 1957. *L'art français dans le livre d'oiseaux. (Éléments d'une iconographie ornithologique française.)* Paris, Société Ornithologique de France. [iii]+136 pp. & 20 pl. *Memoires de la Société Ornithologique et Mammalogique de France* no. 6. First pub. in*Mémoires, Muséum National d'Histoire Naturelle* n.s. A 15, 1957.

Cc1014. **Sitwell, S., H. Buchanan & J. Fisher.** 1953. *Fine bird books 1700–1900.* London, Collins. viii+120 pp. *Re-pub.* (in reduced format with new illustrations) New York, The Atlantic Monthly Pr., 1990. xii+180 pp.
The bibliography, by H. Buchanan & J. Fisher, pp. 65–158.

See also Biographies Ba379 Ba470–447 Ba826–828 Ba1034–1035 Ba1051 Ba1099 Ba1104 Ba1141–1142 Ba1198–1199 Ba1478 Ba1630; Misc. Cc349(Gross, Sutton).

Ichthyological

Cc1015. **Nissen, C.** 1951. *Schöne Fischbücher: Kurze Geschichte der ichthyologischen Illustration. Bibliographie fischkundlicher Abbildungswerke.* Stuttgart, L. Hempe. 108 pp.
Also **J. C. Briggs.** 1955. Illustrated works on fishes. *Copeia* 1955: 243–245. Supplements Nissen's 1951 publication.

Entomological

Cc1016. **Ashmolean Museum, McAlpine Gallery.** 1987. *Insects in art.* [Exhibition catalogue.] Oxford. 31 pp. & errata sheet.

Cc1017. **Bayerische Staatsbibliothek.** 1938. *Das Insekt in der Darstellung vom Mittelalter bis auf Linné.* Ausstellung der Bayerischen Staatsbibliothek 1938/39. Munich. 47 pp. & 22 pl.

Cc1018. **Freeman, R. B.** 1962. Illustrations of insect anatomy from the beginning to the time of Cuvier. *Medical & biological illustration* 12: 174–181.

Cc1019. **Landin, B.-O.** 1978. On Swedish insect books with hand-coloured plates. *Biblis* 1977/78: 63–147.

Cc1020. **Pfeiffer, M. A.** 1936. Schmetterlings- und andere Insektenwerke, die künstlerisch und bibliographisch bemerkenswert sind. *Philobiblon* 1936: 51–67.

Cc1021. **Walton, W. R.** 1921. Entomological drawings and draughtsmen: Their relation to the development of economic entomology in the United States. *Proceedings, Entomological Society of Washington* 23: 69–99.

See also Ba1111 & Ba1111A.

Anatomical

Cc1022. **Choulant, J. L.** 1945. *History and bibliography of anatomic illustration.* Translated and annotated by M. Frank. Further essays by F. H. Garrison, M. Frank, E. C. Streeter; with a new historical esssay by C. Singer [etc.]. New York, Schuman. xxvii+435 pp., illus. *Re-pub.* New York, Hafner, 1952; Cambridge, MA, Maurizio Martino, [1993].

Cc1023. **Herrlinger, R.** 1970. *History of medical illustration from antiquity to A. D. 1600.* London, Pitman. iv+178 pp., illus.
Translation of *Geschichte der medizinischen Abbildung. I: Von der Antike bis zum 1600.* Munich, Moos, 1967. 180 pp.

Cc1024. **Putscher, M.** 1972. *Geschichte der medizinischen Abbildung von 1600 bis zur Gegenwart.* 2 Aufl. Munich, H. Moos. 207 pp., illus.
Continues the history commenced in 1967 by R. Herrlinger, bringing it down to date.

Cc1025. **Roberts, K. B., & J. D. W. Tomlinson.** 1992. *The fabric of the body: European traditions of anatomical illustration.* Oxford, Clarendon Pr. xx+638 pp., illus.

Cc1026. **Simpson, I.** 1991. *Anatomy of humans: Including works by Leonardo da Vinci, John Flaxman, Henry Gray and others.* New York, Crescent Books. 144 pp., illus.
Survey of changes that took place in the 18th and 19th centuries.

Animal images in art

Cc1027. **Klingender, F.** 1971. *Animals in art and thought to the end of the Middle Ages.* Ed. by E. Antal & J. Harthan. London, Routledge & Kegan Paul; Cambridge, MA, M I T Pr. xxviii+580 pp., illus.

Cc1028. **Lloyd, J. B–.** 1971. *African animals in renaissance literature and art.* Oxford, Clarendon Pr. xi+145 pp., illus. *Oxford studies in the history of art and architecture.*

Cc1029. **Rawson, J.** 1977. *Animals in art.* London, British Museum Publications. x+150 pp., illus.

Cc1030. **Yale Center for British Art, Paul Mellon Collection.** 1978. *British sporting and animal paintings, 1655–1867.* By J. Egerton. London, Tate Gallery for the Yale Center for British Art. xxiv+382 pp., illus. *Sport in Art and Books.*

Cc1031. ___. 1980. *British sporting and animal drawings, ca.1500–1850.* By J. Egerton and D. Snelgrove. London, Tate Gallery for the Yale Center for British Art. 2 vols., illus. *Sport in Art and Books.*

Cc1032. ___. 1981. *British sporting and animal prints, 1658–1874.* A catalogue compiled by D. Snelgrove. London, Tate Gallery for the Yale Center for British Art. xxii+257 pp., illus. *Sport in Art and Books.*

BIBLIOGRAPHIES OF ILLUSTRATION RESOURCES

Cc1033. **Hopkinson, E.** 1918. Index-list of the coloured plates of birds in "The Ibis" 1859 to 1917. *Ibis* (10) 6: 10–51. ••Classified list, following system used in B.M.(N.H.) Catalogue of Birds.

Cc1034. **Nissen, C.** 1953. *Die illustrierten Vögelbucher: Ihre Geschichte und Bibliographie.*

Stuttgart, Hiersemann. 222 pp. ••Includes classified indexes to "Vögel" and "Länder" that enable the user to survey five hundred years of published illustration resources for both of these themes.

Cc1035. ___. 1966–78. *Die zoologische Buchillustration: Ihre Bibliographie und Geschichte.* Stuttgart, Hiersemann. 2 vols. ••(This work and its ornithological companion are complementary; the zoological work refers the user to its older partner for most ornithological titles.) Band I. *Bibliographie* provides classified indexes to "Länder," "Sachen," and "Tiere" that enable the user to survey five hundred years of published illustration resources for all of these themes.

Cc1036. **Root, N. J., & B. R. Johnson.** 1986. *The Proceedings of the Zoological Society of London: An Index to the artists 1848–1900.* New York & London, Garland Publishing. xxiii+947 pp.

Cc1037. ___. 1986. *The Transactions of the Zoological Society of London: An index to the artists 1835–1936.* New York & London, Garland Publishing. xix+444 pp.

Cc1038. **Rothamsted Experimental Station.** 1958. *Prints and paintings of British farm livestock 1780–1910: A record of the Rothamsted collection,* by D. H. Boalch. Harpenden. xxxvi+127 pp. & 32 pl.

See also Ca245 Ca714 Ca720 Ca727 Ca760 Ca764 & Ca979.

ANIMALS & MAN

General

Cc1039. **Dance, S. P.** 1976. *Animal fakes and frauds.* Maidenhead, Sampson Low. 128 pp.
See also **R. Vane–Wright & P. E. S. Whalley.** 1985. Linnaeus' fabulous butterfly. *Linnean* 1 (5): 19–24.

Cc1040. **Day, D.** 1989. *Vanished species.* London, London Editions; New York, Gallery Books. 288 pp.
Chiefly birds and mammals.

Cc1041. **Delaunay, P.** 1958. Le XVIe siècle et l'exploitation du monde zoologique. *Archives internationales d'histoire des sciences* 42: 3–14.
With a short bibliography of C16th books on falconry, hunting and fishing.

Cc1042. **Doughty, R. W.** 1975. *Feather fashions and bird preservation: A study in nature protection.* Berkeley, CA, Univ. of California Pr. 184 pp.

Cc1043. **Evans, E. P.** 1906. *The criminal prosecution and capital punishment of animals.* London, Heinemann. x+384 pp.

Cc1044. **Fuller, E.** 1987. *Extinct birds.* New York, Oxford, Facts on File Publications. 256 pp.
With contemporary eye–witness accounts of their natural history, illustrations from historic sources, and details of their demise.

Cc1045. **Hastings, H.** 1936. *Man and beast in French thought of the eighteenth century.* Baltimore, MD, Johns Hopkins Univ. Pr. 297 pp.

Cc1046. **Leavitt, E. S.** 1970. *Animals and their legal rights: A survey of American laws from 1641 to 1970.* 2nd ed. New York, Animal Welfare Institute. 165+lii pp.

Cc1047. **Lewinsohn, R.** 1954. *Animals, men and myths: A history of the influence of animals on civilization and culture.* By "Morus." London, V. Gollancz. 374 pp. & 31 pl.
Translation of *Eine Geschichte der Tiere, ihr Einfluss auf Zivilisation und Kultur.* Hamburg, Rowohlt, 1952. 400 pp.

Cc1048. **Lutts, R. H.** 1990. *The nature fakers: Wildlife, science, and sentiment.* Golden, CO, Fulcrum Publishing. xv+255 pp.

Cc1049. **McKenzie, J. M.** 1988. *The empire of nature: Hunting, conservation, and British imperialism.* Manchester, Manchester Univ. Pr. x+340 pp.

Cc1050. **Palmer, T. S.** 1912. Chronology and index of the more important events in American game protection: 1776–1911. *Bulletin of the Bureau of Biological Survey, United States Departmnent of Agriculture* 41: 62 pp.

Cc1051. **Ritvo, H.** 1987. *The animal estate: The English and other creatures in the Victorian age.* Cambridge, MA, Harvard Univ. Pr. viii+347 pp.

Cc1052. **Serpell, J.** 1986. *In the company of animals: A history of human–animal relationships.* New York, Blackwell. vii+215 pp.

Cc1053. **Thompson, J. L. C–.** 1976. *Insects and history.* New York, St. Martin's Pr. 242 pp.

Cc1054. **Zeuner, F. E.** 1963. *A history of domesticated animals.* London, Hutchinson & Co. 560 pp. *Also* New York, Harper & Row, 1963.
For German version *see* ___. 1967. *Geschichte der Haustiere.* Munich, Bayerischer Landwirtschaftsverlag. 448 pp.

Mythology, folklore & cryptozoology

Cc1055. **Armstrong, E. A.** 1958. *The folklore of birds ... some magico–religious traditions.* London, Collins. xvi+272 pp. & 33 pl.

Cc1056. **Barber, R., & A. Riches.** 1971. *A dictionary of fabulous beasts.* London, Macmillan. 167 pp.

Cc1057. **Gubernatis, A. de.** 1872. *Zoological mythology, or the legends of animals.* London, Trübner & Co. 2 vols.

Cc1058. **Heuvelmans, B.** 1958. *On the track of unknown animals.* Translated by R. Garnett. London, R. Hart–Davis. 558 pp. & 30 pl.
Translation of *Sur la piste des bêtes ignorées.* Paris, Librairie Plon, 1955. 2 vols.

Cc1059. ___. 1968. *In the wake of the sea serpents.* Translated by R. Garnett. London, R. Hart–Davis. 645 pp. & 32 pp. of illus.
Translation of *Le grande Serpent–de–Mer.* Paris, Librairie Plon, 1965. 751 pp. & 48 pl. with the addition of a much condensed version of *Le Kraken et le Poulpe colossal.* Paris, [etc.]., 1958.

Cc1060. ___. 1984. The birth and early history of cryptozoology. *Cryptozoology* 3: 1–30.

Cc1061. **Oudemans, A. C.** 1917. *The great sea serpent: An historical and critical treatise.* Leiden, E. J. Brill; London, Luzac & Co. xvi+592 pp.

Cc1062. **Robinson, M. W.** 1961. *Fictitious beasts: A bibliography.* London, Library Association. 76 pp.

Cc1063. **Siganos, A.** 1985. *Les mythologies de l'insecte: Histoire d'une fascination.* Paris, Librairie des Méridiens. 397 pp.

Cc1064. **South, M.** 1987. *Mythical and fabulous creatures: A source book and research guide.* Westport, CT, Greenwood Pr. 422 pp.
Glossary. Pt. 1: Twenty fabulous creatures. Birds and beasts: **M. South.** The Unicorn; **J. D. Evans.** The Dragon; **D. J. McMillan.** The Phoenix; ___. The Roc; **W. Bertscht.** The Griffin; **D. A. Leeming.** The Cimera; **L. A. Breiner.** The Basilisk; Human–animal composites: **D. R. Cheney.** The Manticora; **R. Berman.** Mermaids; ___. Sirens; **B. Rowland.** Harpies; **J. D. Suther.** The Gorgon Medusa; **F. A. Scafella.** The Sphinx; **M. J. Curley.** The Minotaur; **P. Grootkerk.** The Satyr; **J. J. Kollmann.** The Centaur; Creatures of darkness: **J. C. Holte.** The Vampire; **M. Chelik.** The Werewolf; Giants and

fairies: **J. L. Pallister.** Giants; **J. M. Carter.** Fairies. Part 2. A miscellany and taxonomy. A bibliography of works useful for the study of fabulous creatures. Index.

PART D —
HISTORICAL & BIBLIOGRAPHICAL METHODS

HISTORIOGRAPHY & HISTORICAL METHODS

D1. **Allen, D. E.** 1976. Natural history and social history. *Journal of the Society for the Bibliography of Natural History* 7: 509–516.

D2. **___.** 1977. Naturalists in Britain: Some tasks for the historian. *Journal of the Society for the Bibliography of Natural History* 8: 91–107.

D3. **___.** 1990. Arcana ex multidudine: Prosopography as a research technique. *Archives of natural history* 17: 349–359.

D3A. **Altick, R. D, & J. J. Fenstermaker.** 1993. *The art of literary research.* 4th ed. New York, W. W. Norton. xiii+353 pp.
Replaces **Altick, R. D.** 1981. ___. 3rd ed., rev. by J. J. Fenstermaker. New York, [etc.]. xii+318 pp. Includes a valuable discussion on the principles underlying the critical examination of evidence, with advice on bibliographical procedures and note-taking.

D4. **Baron, W.**, ed. 1967. *Beiträge zur Methodik der Wissenschaftsgeschichte.* Wiesbaden, F. Steiner. 80 pp. *Beiträge zur Geschichte derWissenschaft und der Technik* No. 9.
Includes **W. Baron.** Die Entwicklung der Biologie im 19. Jahrhundert und ihre geistesgeschichtlichen Voraussetzungen, pp. 7–11; **A. Meyer–Abich.** Geschichte der Naturwissenschaften oder Geschichte der Naturwissenschaft?, pp. 34–41.

D5. **Bell, W. J., Jr.** 1955. *Early American science: Needs and opportunities for study.* Williamsburg, VA, Institute of Early American History. ix+85 pp. *Re-pub.* New York, Russell & Russell, 1971.
Includes Early American science: a survey; Bibliography (General history of science; History of science in America; Periodicals; Fifty early American scientists: selected bibliographies).

D6. **Callot, É.** 1965. Système et méthode dans l'histoire de la botanique. *Revue d'histoire des sciences et de leurs applications* 18: 45–53.

D7. **Clarke, E.**, ed. 1971. *Modern methods in the history of medicine.* London, Athlone Pr. xiv+389 pp.

Includes **J. B. Blake.** Automation and the control of historical sources: Problems and possibilities, pp. 260–276; **K. E. Rothschuh.** The graphic presentation of data and relationships in the history of medicine, pp. 314–334; other chapters by J. C. Bearn, S. Benison, D. Brothwell, J. Z. Fullmer, B. Gandevia, C. J. Hackett, M. B. Hall, F. L. Holmes, G. M. Howe, E. R. Kerley, P. DeLacy, R. S. Roberts, C. E. Rosenberg, O. Temkin, W. C. Wake, & B. I. Williams.

D8. **Ewan, J.** 1970. Opportunities in botanical history. *Taxon* 19: 825–830.

D9. ___. 1981. What's history to him or he to history that the systematist should bother. In: Society for the Bibliography of Natural History. *History in the service of systematics: Papers from the conference to celebrate the centenary of the British Museum (Natural History) 13–16 April, 1981*, pp. 157–164. London. *Special publication* no. 1.

D10. **Garfield, E.** 1967. Primordial concepts, citation indexing and historio–bibliography. *Journal of library history* 2: 235–249.

D11. ___. 1970. Citation indexing, historio–bibliography and the sociology of science. In: *Proceedings of the third International Congress of Medical Librarianship, Amsterdam, 5–9 May 1969*, pp.187–204. Amsterdam, Excerpta Medica.

D12. **Garfield, E., I. H. Sher, & R. J. Torpie.** 1964. *The use of citation data in writing the history of science.* Philadelphia, PA, Institute for Scientific Information v+75 pp., 6 transparent charts.

D13. **Hankins, T. L.** 1979. In defence of biography: The use of biography in the history of science. *History of science* 17: 1–16.

D14. **Knight, D. M.** 1975. *Sources for the history of science, 1660–1914.* Cambridge, Cambridge Univ. Pr.; Ithaca, NY, Cornell Univ. Pr. 223 pp. *The sources of history: Studies in the uses of historical evidence.*
Includes 1. The history of science; 2. Histories of science; 3. Manuscripts; 4. Journals; 5. Scientific books; 6. Non-scientific books; 7. Surviving physical objects.

D15. **Kragh, H.** 1987. *An introduction to the historiography of science.* Cambridge, Cambridge Univ. Pr. viii+235 pp.

D16. **Layton, D.** 1965. Diction and dictionaries in the diffusion of scientific knowledge: An aspect of the history of the popularization of science in Great Britain. *British journal for the history of science* 2: 221–234.
Includes General English dictionaries, General scientific dictionaries, and Special dictionaries.

D17. **Outram, D.** 1976. Scientific biography and the case of Georges Cuvier: With a critical bibliography. *History of science* 14: 101–137.

D18. **Pang, A. S.–K.** 1989. Oral history and the history of science: A review essay with speculations. *International journal of oral history* 10: 270–285.

D19. **Parker, E. B., W. J. Paisley, & R. Garret.** 1967. *Bibliographic citations as unobtrusive measures of scientific communication.* Palo Alto, CA, Stanford Univ., Institute for Communication Research. iv+125 ff.
"The purpose of the present study was to examine the bibliographic citations of a set of behavioral science journals in which communication researchers publish articles to be read by their colleagues and students, or which are frequently cited by them as sources of relevant information."

D20. **Pyenson, L.** 1977. "Who the guys were": Prosopography in the history of science. *History of science* 15: 155–188.

D21. **Reingold, N.** 1980. Through paradigm–land to a normal history of science. *Social studies of science* 10: 475–496. *Re-pub. in* ___. 1991. *Science, American style*, pp. 389–409. New Brunswick, NJ, Rutgers Univ. Pr. 429 pp.

D22. **Roger, J.** 1964. Réflexions sur l'histoire de la biologie (XVIIe–XVIIIe siècle): problèmes de méthodes. *Revue de l'histoire des sciences* 17: 25–40.

D23. **Rosenberg, C.** 1964. On the study of American biology and medicine: Some justifications. *Bulletin of the history of medicine* 38: 364–376.

D24. **Russell, N.** 1988. Towards a history of biology in the twentieth century: Directed autobiographies as historical sources. *British journal for the history of science* 21: 77–89. Proposal for recorded autobiographical interviewing programme.

D25. **Russo, F.** 1983. *Nature et méthode de l'histoire des sciences.* Paris, A. Blanchard. xxx+503 pp.

D26. **Schofield, R. E.** 1963. Histories of scientific societies: Needs and opportunities for research. *History of science* 2: 70–83.

D27. **Shapin, S., & A. Thackray.** 1974. Prosopography as a research tool in history of science: The British scientific community, 1700–1900. *History of science* 12: 1–28.

D28. **Stephens, L. D.** 1975. *Historiography: A bibliography.* Metuchen, NJ, Scarecrow Pr. vi+271 pp.

D29. **Truesdell, C.** 1976. The scholar: A species threatened by professions. *Critical enquiry* 2: 631–648.

D30. **Verdoorn, F.** 1944. On the aims and methods of biological history and biography. *Chronica botanica* 8(4): 425–448.

D31. ___. 1951. Problems of botanical historiography. *Archives internationales d'histoire des sciences* 15: 448–457.

D32. ___. 1958. Biohistory, its aims and scope. In: *Actes, VIIIe Congrès International d'Histoire des Sciences (Florence 1956)*, pp. 762–769. Paris, Hermann.

D33. ___. 1968. Horizontal versus vertical approaches in the history of the life sciences. In: *Actes, 11th International Congress of the History of Science (Warsaw–Cracow 1965)*, pp. 5–9. Wroclaw.

D34. ___. 1971. Horticultural biographies: Challenges and opportunities. *Chronica horticulturae* 11: 26–27.

D35. **Walford, A. J.** 1986. *Reviews and reviewing: A guide.* London, Mansell Publishing. vii+248 pp.

D36. **Whitehead, P. J. P.** 1978. History in the service of taxonomy: Cultural environment of fishes. *Environmental biology of fishes* 3: 153–161.

D37. **Williams, R. B.** 1990. Who, what, where, when and why: The elements of history. *Archives of natural history* 17: 263–282.

• Appx. Special sources of historical evidence:

D38. **Challenger, S. C.** 1981. Early nursery catalogues: Overlooked archival material. *Archifacts* n.s. 18: 447–453.

D39. **Cheek, M.** 1989. On typifying names validated by plates unaccompanied by text. *Taxon* 38: 636–638.

D39A. **Dodge, H.** 1959. Evidential factors in the identification of the Linnaean molluscs. *Journal of the Linnean Society (Zoology)* 44: 170–179.

D40. **Edwards, P. I.** 1978. The taxonomic importance of the original drawings of published illustrations, with special reference to those of the 17th and 18th centuries relating to the flora of the Cape of Good Hope. *Journal of the Society for the Bibliography of Natural History* 8: 335–341.

D41. **Ewan, J.** 1969. Historical problems for the working taxonomist. *Taxon* 18: 194–203.

D42. **Leussink, J. A.** 1975. Dates in the rules of the International Code of Botanical Nomenclature. *Taxon* 24: 194–199; *Also* ___. 1978. Dates in the International Code of Botanical Nomenclature. *Taxon* 27: 561 562; ___. 1984. Dates in the International Code of Botanical Nomenclature, the Sydney code. *Taxon* 33: 308–309.

D43. **Mikkola, K.** 1983. Diagnostic insect pins: Some problems of the Linnaean insect collection solved. *Antenna* 7 (1): 16–17.

D44. **Spencer, F.** 1990. *Piltdown: A scientific forgery.* Based on research by Ian Langham (1942–1984). London, Natural History Museum Publications. xxvi+272 pp.

D45. ___. 1990. *The Piltdown papers, 1908–1955: The correspondence and other documents relating to the Piltdown forgery.* London, Natural History Museum Publications. xii+282 pp.

D46. **Standley, P. C.** 1927. A counterfeit collection of Mexican plants falsely attributed to Brother G. Arsene. *Science* 65: 130–133.

D47. **Wright, R. Vane-, & P. E. S. Whalley.** 1985. Linnaeus' fabulous butterfly. *Linnean* 1 (5): 19–24.

D48. **Zeven, A. C., & W. A. Brandenburg.** 1986. Use of paintings from the 16th to 19th centuries to study the history of domesticated plants. *Economic botany* 40: 397–408.

ARCHIVAL SOURCES

Significance & use

D49. **Bailey, M. J.** 1972. The laboratory notebook as an R & D record. *Special libraries* 65: 189–194.

D50. **Brichford, M.** 1969. *Scientific and technological documentation: Archival evaluation and processing of university records relating to science and technology.* Urbana, IL, Univ. of Illinois at Urbana–Champaign. 35 pp.

D51. **Brooks, P. C.** 1969. *Research in archives: The use of unpublished and primary sources.* Chicago, IL, Univ. of Chicago Pr. xi+127 pp.

D52. **Burkhardt, F.** 1988. Editing the correspondence of Charles Darwin. *Studies in bibliography* 41: 149–159.
See also N. **Reingold.** 1988. The Darwin industry encounters Tanselle and Bowers. *Documentary editing* 1988 (June): 16–19.

D53. **Catchpole, A. J. W., & D. W. Moodie.** 1978. Archives and the environmental scientist. *Archivia* 6: 113–136.

D54. **Centre International de Synthèse.** 1976. Les correspondances: Leur importance pour l'historien des sciences et de la philosophie; problèmes de leur édition. Journées organisées par le Centre International de Synrhèse les 5, 6 et 7 mai 1975 à Chantilly. *Revue de synthèse* 97: 1–189.
Articles by various authors, including: **R. Taton.** Le rôle et l'importance des correspondances scientifiques aux XVIIe et XVIIIe siècles; **W. Voise.** L'art épistolaire, son passé et son avenir; **P. Dibon.** Les échanges épisatolaires dans l'Europe savante du XVIIe siècle; **J. Roger.** Correspondences de savants et de naturalistes aux XVIIIe et XIXe siècles.

D55. **Cook, F.** 1971. The archivist: Link between scientist and historian. *American archivist* 34: 377–381.

D56. **Doligh, F. I.** 1975. The importance of archives for science and technology. *UNESCO bulletin for libraries* 29: 331–335.

D57. **Elliott, C. A.** 1974. Experimental data as source for the history of science. *American archivist* 37: 27–35.

D58. **Elliott, C. A.** 1981. Citation patterns and documentation for the history of science: Some methodological considerations. *American archivist* 44: 131–142.

D59. **Gruman, G. J.** 1957. Saving our primary source materials. *Isis* 48: 183–185.

D60. **Haas, J. K., H. W. Samuels & H. W. Simmons.** 1985. *Appraising the records of modern science and technology: A guide.* Cambridge, MA, Massachusetts Institute of Technology. 96 pp.

D61. **History of Science Society.** 1962. The conference on science manuscripts. *Isis* 53: 1–155.
These conference proceedings, although thirty years old now, still deserve reading. The following papers are of interest here: **R. H. Shyrock.** The viewpoint of an historian and manuscript librarian, pp. 9–13; **E. Lurie.** Some observations on research in nineteenth–century American science, pp. 21–30; **W. C. Grover.** The role of the archivist in the preservation of scientific records, pp. 55–62; **A. H. Dupree.** What manuscripts the historian wants saved, pp. 62–66; **K. Birr.** What shall we save? pp. 72–79; **N. Reingold.** Subject analysis and description of manuscript collections, pp. 106–112; **S. Benison.** Oral history and manuscript collecting, pp. 113–117; **D. Fleming.** The ends in view of the preservation of the private papers of American scientists, pp. 118–121; **B. Glass.** The scientist and the preservation of science manuscripts, pp. 136–142; **C. Leake.** Responsibility for science archives, pp. 143–148.

D62. **International Congress of the History of Science, XVth.** 1977. Symposium 7. Problems of source materials in the history of science. In: *Human implications of scientific advance.* Proceedings of the XVth International Congress of the History of Science, Edinburgh, 10–15 August 1977. Ed. by E. G. Forbes, pp. 369–440. Edinburgh, Edinburgh Univ. Pr.
Includes reviews of available sources and projects in France, USSR, Italy, Poland, Great Britain and the USA.

D63. **Kline, M.-J.** 1987. *A guide to documentary editing.* Baltimore, MD, Johns Hopkins Univ. Pr. xvii+228 pp.

D64. **Knight, D. M., et al.** 1987. International conference on documentary editing. *British journal for the history of science* 20: 1–79.
Papers by seven authors. *Includes* **D. Knight.** Background and foreground: Getting things in context; **W. Montgomery.** Editing the Darwin correspondence: A quantitative perspective; **N. Reingold.** On not doing the papers of great scientists; **A. J. Meadows.** Changing records and changing realities; **F. G. Burke.** Automation and documentary editing.

D65. **Lawrence, G. H. M.** 1964. The botanical significance of letters and manuscripts. *Huntia* 1: 162–165.

D66. **Levere, T. H.**, ed. 1982. *Editing texts in the history of science and medicine. Papers given at the seventeenth annual conference on editorial problems, University of Toronto, 6–7 November 1981.* New York, Garland Publishing. 190 pp.
Papers by S. Drake, L. E. Voights, B. S. Hall, & I. B. Cohen.

D67. **McAlister, D. T.** 1969. Collecting archives for the history of science. *American archivist* 32: 327–332.

D68. **National Historical Publications & Records Commission, Joint Committee on Archives of Science & Technology.** 1983. *Understanding progress as process: Documentation of the history of post–war science and technology in the United States.* Final report of the Joint Committee on Archives of Science & Technology. C. A. Elliott, ed. Chicago, IL, (distributed by) American Society of Archivists. 64 pp.
Also **C. A. Elliott.** 1984. Joint Committee on Archives of Science and Technology (JCAST): Summary from the final report. *Isis* 75: 158–162.

D69. **Reingold, N.** 1961. Manuscripts of science – Analysis and description. *American archivist* 24: 163–170.

D70. ___. 1968. Confessions of a reformed archivist. *American archivist* 31: 371–377.

D71. ___, ed. 1990. Symposium on documents, interpretations and the history of the sciences. *Proceedings of the American Philosophical Society* 134: 338–441.
Papers by several authors including: **N. Reingold.** Tales from the archives; **F. L. Holmes.** Laboratory notebooks: Can the daily record illuminate the broader picture?

D72. **Sulloway, F. J.** 1982. Darwin's conversion: The Beagle voyage and its aftermath. *Journal of the history of biology* 15: 325–396; ___. 1983. Further remarks on Darwin's spelling habits and the dating of Beagle voyage manuscripts. *Ibid.* 16: 361–390.

D73. **Tanselle, G. T.** 1978. The editing of historical documents. *Studies in bibliography* 31: 451–506. *Re-pub. in* ___. 1979. *Selected studies in bibliography*, pp. 451–506. Charlottesville, VA, Univ. Pr. of Virginia for the Bibliographical Society of the Univ. of Virginia.

Handwriting

D74. **Bowers, F.** 1976. Transcription of manuscripts: The record of variants. *Studies in bibliography* 29: 212–264.
I.e. Transcription for scholarly printed editions, e.g. for the *Correspondence of Charles Darwin* project.

D75. **Burdet, H. M.** 1979. *Auxilium ad botanicorum graphicem.* Geneva, Conservatoire et Jardin Botaniques. 21+[428]+xxiv pp. *Publication hors–série* no. 2.
First pub. as "Cartulae ad botanicorum graphicem" in *Candollea* 27: 307–340, 1972; 28: 137–170, 407–440, 1973; 29: 207–240, 489–522, 1974; 30: 203–234, 379–410, 1975; 31: 127–158, 319–360, 1976; 32: 165–207, 377–418, 1977; 33: 139–180, 365–456, 1978; 34: 167–218, 1979. Provides "... samples of signatures, herbarium labels, and of fragments of letters and other documents preferably showing the different forms which the handwriting of a single botanist may take in space and time" from the archives of the Conservatoire et Jardin Botaniques de Genève.

D76. **Marshall, J. B.** 1978. The handwriting of Joseph Banks, his scientific staff and amanuenses. *Bulletin of the British Museum (Natural History) Botany series* 6: 1–85.

D77. **Nickell, J.** 1990. *Pen, ink, and evidence: A study of writing and writing materials for the penman, collector, and document detective.* Lexington, KY, Univ. Pr. of Kentucky. ix+228 pp.

D78. **Stoddard, R. E.** 1985. *Marks in books, illustrated and explained.* Cambridge, MA, Houghton Library, Harvard Univ. 44 pp.
Large format illustrated catalogue of 64 exhibits that demonstrate the nature and interpretation of manuscript annotations (and certain other marks) found in printed books from all periods.

D79. **Weidemann, H. R.** 1989. *Briefe grosser Naturforscher und Arzte in Handschriften. Mit Textabdruck und Porträts der Autoren.* Lübeck, Verlag Graphische Werkstätten. vii+445 pp.

BIBLIOGRAPHY

Bibliographies of bibliographical literature

D80. ***Annual bibliography of the history of the printed book and libraries (ABHB).*** Ed. by H. D. L. Vervliet. <1970+> The Hague, Nijhoff (*later* Dordrecht, Kluwer Academic), vol. 1+, 1973+.

A comprehensive, classified annual bibliography of the literature. Suffers from being several years late in its literature coverage. For a rival current bibliography see *BBB* (D82).

D81. **Besterman, T.** 1971. *Printing, book collecting and illustrated books: A bibliography of bibliographies.* Totowa, NJ, Rowman & Littlefield. 2 vols. ([xiv]+609 pp.) *The Besterman world bibliographies.*

D82. ***Bibliographie der Buch- und Bibliotheksgeschichte (BBB).*** <1980/81+>. Bad Iburg, Bibliographischer Verlag Dr. Horst Meyer, 1982+.
Initially concentrating on German language literature *BBB* has developed its coverage of international literature and now shares much common ground with*ABHB* (D80)

D83. **Tanselle, G. T.** 1973. The periodical literature of English and American bibliography. *Studies in bibliography* 26: 167–191.

D84. **Vereeniging ter Bevordering van de Belangen des Boekhandels te Amsterdam.** 1920+. *Catalogue der bibliotheek van de [etc.].* The Hague, M. Nijhoff (*later*) Amsterdam, Universiteitsbibliotheek. Vol. 1+. In progress (vol. 8, 1979).
Classified subject catalogue (with periodic author indexes) of a vast international collection of literature with very broad coverage of bibliography, printing–history, and the book trade (vols. 4 & 8 are entirely devoted to publishers', booksellers', auction sale and other catalogues).

Bibliographical methods & problems

D85. **Anderson, S., & R. G. Van Gelder.** 1970. The history and status of the literature of mammalogy. *Bioscience* 20(17): 949–957.

D86. **Bay, J. C.** 1905. Contributions to the theory and history of botanical bibliography. *Proceedings & papers of the Bibliographical Society of America* 1(1): 75–83. Also pub. in *Annotationes Concilii Bibliographici* 2: 15–19, 1906.

D87. **Buckman, T. R.**, ed. 1966. *Bibliography and natural history: Essays presented at a conference convened in June 1964 by Thomas R. Buckman.* Lawrence, KS, Univ. of Kansas Libraries. ix+148 pp.
Nine essays, including: **W. T. Stearn.** The use of bibliography in natural history, pp. 1–26; **R. M. Mengel.** Bibliography and the ornithologist, pp. 121–130. Other essays are entered elsewhere in this work.

D88. **Cole, F. J.** 1939. Bibliographical reflections of a biologist. *Proceedings & papers of the Oxford Bibliographical Society* 5: 169–186. Also pub. in *Osiris* 8: 289–315, 1948.

D89. ___. 1958. Obiter dicta bibliographica. *Journal of the history of medicine* 13: 2–9.

D90. **Derksen, W.** 1965. Entomobibliographisches. *Beiträge zur Entomologie* 15: 101–110.

D91. **Dobell, C.** 1938. Dr. O. Uplavici (1887–1938). *Parasitology* 30: 239–241.
This obituary, a cautionary bibliographical tale, is required reading for bibliographers of all ages!

D92. **Eyles, V. A.** 1955. Bibliography and the history of science. *Journal of the Society for the Bibliography of Natural History* 3: 63–71.

D93. **Greene, E. L.** 1898. Bibliographical difficulties in botany. *Bulletin of the Catholic University* 4: 62–75.

D94. **Gudger, E. W.** 1939. A bibliographer turns detective: How a bibliographer of fish literature "ran down" an obscure citation to its ultimate source. *Scientific monthly* 48: 345–349.

D95. **Horn, W.** 1929. Über die Not der Entomo–Bibliographie. In: *Xe Congres Internationale de Zoologie, Budapest, 4–10 Septembre, 1927. [Report]*, vol. 2: 1043–1052. Budapest.

D95A. **Jackson, B. D.** 1880. Remarks on botanical bibliography. *Journal of botany* n.s. 9:167–177.
Account of methods and problems associated with the compilation of his *Guide to the literature of botany [etc.]*, 1881; *see* Cb67.

D96. **Köppen, H.** 1968. Über bibliographische Arbeit in der Entomologie. *Beiträge zur Entomologie* 18: 265–268.

D97. **Perman, D. H.**, ed. 1967. *Bibliography and the historian: The conference at Belmont of the Joint Committee on Bibliographical Services to History, May 1967.* Santa Barbara, CA, Clio Pr. viii+176 pp.

D98. **Poulton, E. B.** 1913. Presidential address, 1913. A remarkable work on evolution. *Proceedings of the Linnean Society of London* 125: 26–45: *Continued* ___. 1914. Presidential address, 1914 [etc.]. *Ibid.* 126: 23–44.
Apropos George Washington Sleeper's forged pamphlet "Shall we have common sense. Some recent lectures written and delivered by Geo. W. Sleeper, Boston: Wm. Bense, Printer ... 1849," purporting to have contained a lecture given in 1847 that anticipated some of Darwins ideas.

D99. **Sachtleben, H.** 1949. Systematische und bibliographische Entomologie. In: ___. *Festschrift zum 50. Jahr. bestehen der Biologischen Zentralanstalt für Land und Forstwirtschaft*, pp. 97–112.

D100. **Sarton, G.** 1953. La bibliographie de l'histoire des sciences. *Archives internationales d'histoire des sciences* 32: 395–419.

D101. ___. 1920. Synthetic bibliography: With special reference to the history of science. *Isis* 3: 159–170.

D102. **Sherborn, C. D.** 1908. Notes on bibliography, publication and nomenclature. *Naturalist* 1908: 5–9.

D103. **Stearn, W. T.** 1981. Bibliography in the British Museum (Natural History). In: Society for the Bibliography of Natural History. *History in the service of systematics: Papers from the conference to celebrate the centenary of the British Museum (Natural History) 13–16 April, 1981*, pp. 1–6. London. *Special publication* no. 1.

D104. **Stokes, R.** 1982. *The function of bibliography.* 2nd ed. London, Gower Publishing Co. xi+201 pp.
Replaces 1st ed., London, Deutsch, 1969. 174 pp.

D105. **Wells, E. B.** 1986. [Article] Reprint. In: A. Kent & H. Lancour. *Encyclopedia of library and information science*, vol. 40, supp. 5, pp. 385–394. New York, M. Dekker.
A discussion of the bibliographical status of reprints, also known as "extracts," "offprints" or "separates."

D106. **Whitrow, M.** 1964. Classification schemes for the history of science: A comparison. *Journal of documentation* 20: 120–136.

D107. ___. 1976. A classification scheme for the history of science, medicine and technology. In: Isis *cumulative bibliography 1913–1965: A bibliography of the history of science formed from* Isis *critical bibliographies 1–90, indexing literature published from 1913 through 1965.* Vol. 3: *Subjects*, pp. 621–678. London, Mansell.
Includes schedules for "Form and aspect subdivisions" and "Scientific subject fields" and an "Index to subject classmarks."

D108. ___. 1978. A classification scheme for the history of science, medicine and technology. In: E. G. Forbes, ed. *Human implications of scientific advance*, pp. 518–530. Edinburgh, Edinburgh Univ. Pr.

D109. ___. 1983. Interrelationships in the history of science: Bibliography as a guide to subjects for research. *Isis* 74: 81–88.
Briefly explains the organization of and points out the scholarly potential of vols. 3 (Subjects)

& 4–5 (Civilizations and periods) of the Isis *cumulative bibliography 1913–1965*; *see* A11. *Also* ___. 1983. The *Isis* cumulative bibliography, 1913–65. *Indexer* 13(3): 158–165.

D110. **Willoughby, E. E.** 1957. *The uses of bibliography to the students of literature and history.* Hamden, CT, Shoe String Pr. 105 pp.

Analytical & descriptive bibliography

D111. **Bridson, G. D. R.** 1976. The treatment of plates in bibliographical description. *Journal of the Society for the Bibliography of Natural History* 7: 469–488.

D112. **Gascoigne, B.** 1986. *How to identify prints: A complete guide to manual and mechanical processes from woodcut to ink jet.* London, Thames & Hudson. 208 pp.
And, of course, printed book illustrations or "plates." Glossary index, pp. 202–208.

D113. **Harmon, R. B.** 1989. *Elements of bibliography.* Rev. ed. Metuchen, NJ, Scarecrow Pr. xv+288 pp.
"By design, this work is primarily a guide to the literature of bibliography; it is accompanied by basic standards for compiling various kinds of bibliographic instruments." *Replaces* ___. 1981. ___. Metuchen, NJ, [etc.]. viii+253 pp.

D115. **MacPhail, I.** 1964. An introduction to bibliography for botanists. *Huntia* 1: 103–115.

D116. **Margadant, W. D.** 1968. Descriptive bibliography applied to botany. In: *Early bryological literature: A descriptive bibliography of selected publications treating musci during the first decades of the nineteenth century and especially of the years 1825, 1826 and 1827*, pp. 1–33. Pittsburgh, PA, Hunt Botanical Library. xi+278 pp. Also pub. in *Mededelingen van het Botanisch Museum en Herbarium van de Rijksuniversiteit de Utrecht* no. 283.
Includes a novel key to the determination of formats.

D117. **Stevenson, A.** 1961. A bibliographical method for the description of botanical books. In: Hunt Botanical Library. *Catalogue of botanical books in the collection of Rachel McMasters Miller Hunt.* Vol. 2, part i: *Introduction to printed books 1701–1800*, pp. cxli–ccxiv. Pittsburgh, PA.

D118. **Tanselle, G. T.** 1968. Tolerances in bibliographical description. *Library* (5) 23: 1–12. *Re-pub. in* J. B. Jones, ed. 1974. *Readings in descriptive bibliography*, pp. 42–56. Kent, OH, Kent State University Pr.

D119. ___. 1971. The bibliographical description of paper. *Studies in bibliography* 24: 27–67. *Re-pub. in* ___. 1979. *Selected studies in bibliography*, pp. 203–243. Charlottesville, VA, Univ. Pr. of Virginia for the Bibliographical Society of the Univ. of Virginia; *and in* J. B. Jones, ed. 1974. *Readings in descriptive bibliography*, pp. 71–116. Kent, OH, Kent State University Pr.

D120. ___. 1975. The bibliographical concepts of issue and state. *Papers of the Bibliographical Society of America* 69: 17–66.

D121. ___. 1977. Descriptive bibliography and library cataloguing. *Studies in bibliography* 30: 1–56. *Re-pub. in* ___. 1979. *Selected studies in bibliography*, pp. 37–92. Charlottesville, VA, Univ. Pr. of Virginia for the Bibliographical Society of the Univ. of Virginia.

D122. ___. 1982. The description of non-letterpress material in books. *Studies in bibliography* 35: 1–42.

D123. ___. 1985. Title–page transcription and signature collation reconsidered. *Studies in bibliography* 38: 45–81.

D124. ___. 1987. A sample bibliographical description with commentary. *Studies in bibliography* 40: 1–30.

D124A. **United States Library of Congress.** 1981. *Bibliographic description of rare books: Rules formulated under AACR2 and ISBD(A) for the descriptive cataloging of rare books and other special printed materials.* Washington, DC. vii+62 pp.

Historical bibliography

D125. **Adams, F. B., Jr.** 1969. *The uses of provenance.* Berkeley & Los Angeles, CA, Univ. of California. 26 pp. *Ninth annual Zeitlin & Ver Brugge Lecture in Bibliography.*

D126. **Berry, W. T., & H. E. Poole.** 1966. *Annals of printing: A chronological encyclopaedia from the earliest times to 1950.* London, Blandford. 315 pp.

D127. **Callery, B. G.** 1981. The persistence of ephemera: Bibliographical value of part wrappers. In: Society for the Bibliography of Natural History. *History in the service of systematics: Papers from the conference to celebrate the centenary of the British Museum (Natural History) 13–16 April, 1981*, pp. 15–22. London. *Special publication* no. 1.

D128. **Deschamps, P.** 1870 & 1904. *Dictionnaire de géographie ancienne et moderne, contenant: ... 2º Les recherches ... sur les origines de la typographie dans toutes les villes, bourgs, abbayes d'Europe, jusqu'au XIXe siècle exclusivement ... Suivi de l'imprimerie hors l'Europe.* Paris, Firmin–Didot. *Re-pub.* Paris, G.–P. Maisonneuve & Larose, 1964. viii+1591+[v]+208 pp.
Invaluable for information on place names, Latin or otherwise, as found in imprints.

D129. **Farber, E. I., et al.**, E ds. 1979–82. *Combined retrospective index to book reviews in scholarly journals, 1886–1974.* Arlington, VA, Carrollton Pr. 15 vols.
Indexes 459 journals covering history, sociology, etc.

D130. ___, E ds. 1982–84. *Combined retrospective index to book reviews in humanities journals, 1802–1974.* Woodbridge, CT, Research Publications. 10 vols.
Indexes principally English language literary journals.

D131. **Feather, J.** 1986. *A dictionary of book history.* Oxford, Oxford Univ. Pr. [x]+278 pp.

D132. **Gray, R. A.** 1968. *A guide to book review citations: A bibliography of sources.* Columbus, OH, Ohio State Univ. Pr. xv+223 pp.
Includes sources for reviews of 18th and 19th century books; Section 8:0 Science & technology, pp. 189–197.

D133. **Johnson, D. E.** 1979. A concordance of abbreviations used in references to reviews in botanical bibliographies by Cobres (1782), Böhmer (1787) and Miltitz (1829). *Huntia* 3: 103–114.

D134. **Kerzhner, I. M.** 1984. Converting dates from the Julian (old style) or French rebublican (revolutionary) calendars to the Gregorian (new style) calendar. *Taxon* 33: 410–412.

D135. **Peddie, R. A.** 1932. *Place names in imprints: An index to the Latin and other forms used on title pages.* London, Grafton & Co. vii+62 pp.
"... not meant to replace Graesse's *Orbis Latinus*, or Deschamps' *Dictionnaire de Geographie.* It is intended simply to bring together in one alphabet, in a handy volume, the actual forms in Latin and in the various modern languages and dialects, into which place names have been translated and twisted."

D136. **Price, J. H.** 1981. Order in disorder: A summary of some characteristics of part–publication. In: Society for the Bibliography of Natural History. *History in the service of* *systematics: Papers from the conference to celebrate the centenary of the British Museum (Natural History) 13–16 April, 1981*, pp. 23–26. London. *Special publication* no. 1.

D137. **Schmid, R.** 1986. Intercontinental and transcontinental mail transit times from Europe to Berkeley, 1892 to 1941. *Archives of natural history* 13: 141–154.

D138. **Stokes, R.** 1989. *A bibliographical companion.* Metuchen, NJ, & London, Scarecrow Pr. x+298 pp.
A dictionary of bibliography and book history.

D139. **Tanselle, G. T.** 1969. Copyright records and the bibliographer. *Studies in bibliography* 22: 77–124. *Re-pub. in* ___. 1979. *Selected studies in bibliography*, pp. 93–139. Charlottesville, VA, Univ. Pr. of Virginia for the Bibliographical Society of the Univ. of Virginia.

D140. ___. 1971. Book–jackets, blurbs, and bibliographers. *Library* (5) 26: 91–134.

D141. **Wallis, P. J.** 1974. Book subscription lists. *Library* (5) 29: 255–286.

D142. **West, J. L. W., 3rd.** 1983. The bibliographical concept of plating. *Studies in bibliography* 36: 252–266.

D143. **Woodruff, J. F., & H. B. De Groot.** 1978. "This day is published": Advertisements in periodicals as a means of dating eighteenth and nineteenth century publications. *Victorian periodicals newsletter* 11: 104–108.

Publishing history

D144. **Altick, R. D.** 1957. *The English common reader.* Chicago, IL, Univ. of Chicago Pr. ix+430 pp.
Includes Appendix B. Best–sellers, pp. 381–380, (some are natural history books). Further lists appeared separately: ___. 1969. Nineteenth–century English best–sellers: A further list. *Studies in bibliography* 22: 187–206; ___. 1986. Nineteenth–century English best–sellers: A third list. *Ibid.* 39: 235–241.

D145. **Boutry, G.–A.** 1969. L'évolution des publications scientifiques primaires: Études statistiques. *Bulletin des bibliothèques de France* 14: 425–454.

D146. **Desmond, R. G. C.** 1980. Loudon and nineteenth century horticultural journalism. In: Dumbarton Oaks Research Library & Collection. *John Claudius Loudon and the early nineteenth century in Great Britain*, pp. 79–103. Washington, DC.

D147. **Eisnerová, V.** 1973. Botanická periodika v minulosti. *Dejiny ved a techniky* 6: 75–89.
Botanical periodicals in the past, especially the 19th century.

D148. **Kronick, D. A.** 1965. Scientific journal publication in the eighteenth century. *Papers of the Bibliographical Society of America* 59: 28–44.

D149. ___. 1976. *A history of scientific and technical periodicals: The origins and development of the scientific and technical press, 1665–1790.* 2nd ed. Metuchen, N.J, Scarecrow Pr. xvi+336pp.
Replaces 1st ed., New York, Scarecrow, 1972. 274pp.

D150. ___. 1977. Toward a typology of the 17th and 18th century scientific and technical periodical. *Serials librarian* 1: 385–398, 1977; 2: 67–81, 155–166, 1977.

D151. **McNally, P. F.**, ed. 1987. *The advent of printing: Historians of science respond to Elizabeth Eisenstein's* The printing press as an agent of change. Montreal, Graduate School of Library & Information Studies. 36 pp. *Occasional paper* no. 10.
Includes **S. Sheets–Pyenson.** The effects of changes in printing technology on the development of natural history sciences during the 19th century, pp. 21–26.

D152. **Meadows, A. J.**, ed. 180. *Development of science publishing in Europe.* Amsterdam, etc., Elsevier Science. ix+268 pp.
Includes **A. A. Manten.** The growth of European scientific journal publishing before 1850, pp. 1–22; **D. M. Knight.** The growth of European scientific monograph publishing before 1850, pp. 23–41; **J. G. Shaw.** Patterns of journal publication in scientific natural history from 1800–1939, pp. 149–176.

D153. **Noblett, W.** 1985. Dru Drury, his *Illustrations of natural history* (1770–82) and the European market for printed books. *Quaerendo* 15: 83–102.

D154. **Price, J. H.** 1982. Publication in parts: A background to the concept, efficacy and taxonomic complexity. *Archives of natural history* 10: 443–459.

D155. **Pyenson, S. Sheets–.** 1976. Popular science periodicals in Paris and London: The emergence of a low scientific culture, 1820–1875. *Annals of science* 42: 549–572.
Appendix: Popular science periodicals, 1820–1875, pp. 565–572.

D156. ___. 1981. War and peace in natural history publishing: *The Naturalist's Library. Isis* 72: 50–72.
Edited by Sir William Jardine and published by William H. Lizars.

D157. ___. 1981. A measure of success: The publication of natural history journals in early Victorian Britain. *Publishing history* 9: 21–36.

D158. ___. 1981. From the north to Red Lion Court: The creation and early years of the *Annals of natural history. Archives of natural history* 10: 221–249.

D159. **Reeds, K. M.** 1983. Publishing scholarly books in the 16th century. *Scholarly publishing* 14: 259–274.

D160. **Robinson, F. J. G., & P. J. Wallis.** 1975. *Book subscription lists: A revised guide.* Newcastle–upon–Tyne, Book Subscription Lists Project. xxv+112 pp.; *Continued* **C. Wadham.** 1976. ___*: First supplement.* Newcastle–upon–Tyne, [etc.]. 23 pp.

D161. **Smith, A. H. T. Robb–.** 1970. John Hunter's private press. *Journal of the history of medicine* 25: 262–269.

D162. **Snelders, H. A. M.** 1988. Die niederländischen naturwissenschaftlichen Zeitschriften im 18. Jahrhundert. In: A. Völker, ed. *Dixhuitième: Zur Geschichte von Medizin und Naturwissenschaften im 18. Jahrhundert*, pp. 149–159. Halle, Martin–Luther–Universität. *Wissenschaftliche Beiträge* vol. 20.

D163. **Thornton, J. L., & R. I. J. Tully.** 1971. Scientific publishing and bookselling. In: *Scientific books, libraries and collectors: A study of bibliography and the book trade in relation to science.* 3rd rev. ed., pp. 354–366. London, Library Association; ___. 1978. ___*: Supplement 1969–75*, pp. 110–112. London, Library Association.
Replaces 2nd rev. ed., 1962. xiii+406 pp.

D164. **Whalen, M. D., & M. F. Tobin.** 1980. Periodicals and the popularization of science in America, 1860–1910. *Journal of American culture* 3: 195–203.

D165. **Wiles, R. M.** 1957. *Serial publication in England before 1750.* Cambridge, Cambridge Univ. Pr. xv+391 pp. & 9 pl.
Details the birth, growth and methods of publishing in parts, later to become significant as a means of disseminatiing expensive illustrated natural history books. *Includes* Appendix B. Short–title catalogue of books published in fascicules before 1750, pp. 267–356. Early examples of serial publication can be found in: **Arents Collection, New York Public Library.** 1957–64. *The Arents collection of books in parts … checklist*; *see* Bc298.

Some individual publishers

Brill

D166. **Brill, E. J.** 1983. *Tuta sub aegide Pallas: E. J. Brill and the world of learning.* Leiden. 87 pp.

Cramer

D167. **Cramer Verlag.** 1987. *Cramer books on botany 1956–1987.* Berlin & Stuttgart, J. Cramer in der Gebrüder Borntraeger Verlagsbuchhandlung. 172 pp.

Engelmann

D168. **Engelmann, W.** [Firm]. 1911. *Wilhelm Engelmann, 1811–1911: Jubiläums Katalog der Verlagsbuchhandlung.* Leipzig. 447 pp.

D168A. ___. 1935. *Verlagskatalog, 1811–1935.* Leipzig. 5 pts.

Fischer

D169. **Fischer, Gustav** [Firm]. [1937.] *Verzeichnis naturwissenschaftlicher Werke: Botanik.* Zweite Ausgabe. 1878–1937. Jena. 270 pp. & 11 pl.
Classified catalogue with full statements of contents, extracts from contemporary reviews, author and subject indexes. Illustrated with portraits of distinguished authors.

D170. ___. [1928.] *Verzeichnis naturwissenschaftlicher Werke: Zoologie, mit Anhang:Theorie und Geschichte der Biologie, Naturphilosophie.* Jena. 330 pp.

D171. ___. [1935.] *Verzeichnis naturwissenschaftlicher Werke: Zoologie.* 2. Ausg. Jena. 330 pp. & Nachtrag 86 pp.

D172. **Lütge, F. K.** 1928. *Das Verlaghaus Gustav Fischer in Jena: Seine Geschichte und Vorgeschichte; aus Anlass des 50 jährigen Firmenjubiläums ... 1928.* Jena. vii+243 pp. & port.

Friedländer

D173. **Friedländer & Sohn, R.** [Firm]. 1891. *Verlags–Catalog. 1830–1890.* Berlin. xxviii+240 pp.

Hardwicke

D174. **English, M. P.** 1986. Robert Hardwicke (1822–1875), publisher of biological and medical books. *Archives of natural history* 13: 25–37.
Also ___. 1989. Robert Hardwicke and T. H. Huxley. *Ibid.* 16: 245.

Junk

D175. **Junk B.V., Dr. W.** [Firm]. 1939. *Catalogue of publications. Fonds–catalogus. Verlags–catalog. Livres de fonds.* The Hague. 130 pp.
An annotated catalogue of the firm's "more important books."

D176. ___. 1959. *General catalogue 1959.* The Hague. 168 pp.
Publications from 1899–1959.

McGraw Hill

D177. **McGraw–Hill Book Co.** 1959. *Fifty years of publishing, 1909–59.* New York. 569 pp.

Masson

D178. **Masson & Cie.** 1954. *Masson et Cie, 1804–1954: Un siècle et demi d'édition médicale et scientifique.* Paris. 54 pp.

Oxford University Pres

D179. **Barker, N.** 1978. *The Oxford University Press and the spread of learning 1478–1978.* Oxford, Clarendon Pr. xiv+332 pp.

Panckoucke

D180. **Watts, G. B.** 1969. Charles Joseph Panckoucke, "l'Atlas de la librairie Française". *Studies on Voltaire & the 18th century* 68: 67–205.
Publisher of the *Encyclopédie méthodique* etc.

Springer

D181. **Davidis, M.**, ed. 1985. *Wissenschaft und Buchhandel: Der Verlag von Julius Springer end seine Autoren: Briefe und Dokumente aus den Jahren 1880–1946.* Munich, Deutschen Museum. xii+96 pp.

Taylor & Francis

D182. **Brock, W. H., & A. J. Meadows.** 1984. *The lamp of learning: Taylor & Francis and the development of science publishing.* London, Taylor & Francis. xv+240 pp.

Van Voorst

D183. **Williams, R. B.** 1988. John Van Voorst: Patron publisher of Victorian natural history. *Private library* (4) 1 (1): 4–12.

Dates-of-publication literature

D184. **Ardagh, J.** 1926. Index to "Bibliographical notes" in the Journal of botany, 1893–1924. *Journal of botany* 64: 274–280.

D184A. **Browning, M. R., & B. L. Monroe Jr.** 1991. Clarifications and corrections of the dates of issue of some publications containing descriptions of North American birds. *Archives of natural history* 18: 381–405.
Information on twenty-eight works.

D185. **Burdet, H. M.**, ed. 1981. Med-checklist notulae bibliographicae: 1 & 2. *Candollea* 36: 397–408; ___. 1982. ___: 3 à 5. *Ibid.* 37: 589–597; ___. 1983. ___: 6 à 8. *Ibid.* 38: 691–698; ___. 1985. ___: 9 à 13. *Ibid.* 40: 211–216; ___. 1989. ___: 14 à 17. *Ibid.* 44: 555–568;
A series of miscellaneous contributions by various authors on subjects of botanical bibliography studied in the course of the drawing up of Med-Checklist.

D186. **Fortuner, R.** 1984. *Dates of publication of some journals with taxonomic data in nematology.* 2nd ed. Sacramento, CA, California Department of Food & Agriculture, Analysis & Identification – Nematology Laboratory. [22] ff.

D187. **Griffin, F. J., C. D. Sherborn & H. S. Marshall.** 1936. A catalogue of papers concerning the dates of publication of natural history books. *Journal of the Society for the Bibliography of Natural History* 1: 1–30. *Continued* **F. J. Griffin.** 1943. ___: First supplement. *Ibid.* 2: 1–17; **W. T. Stearn & A. C. Townsend.** 1953. ___: Second supplement. *Ibid.* 3: 5–12; **G. H. Goodwin Jr.** 1957. ___: Third supplement. *Ibid.* 3: 165–174; **___, W. T. Stearn & A. C. Townsend.** 1962. ___: Fourth supplement. *Ibid.* 4: 1–19.

D188. **Guédès, M.** 1969. Notules de bibliographie botanique: I–IV. *Journal of the Society for the Bibliography of Natural History* 5: 110–116; ___. 1972. ___: V–VIII. *Ibid.* 6: 174–180; ___. 1978. ___: IX–XV. *Ibid.* 9: 42–52; ___. 1983. ___: XVI–XXI. *Archives of natural history* 11: 195–206; ___. 1987. Notules de bibliographie biologique: XXII–XXV. *Ibid.* 14: 159–167.

i.e. I. Une édition de Bruxelles (1840) du *Nouveau système de physiologie végétale* ... de Raspail; II. Un exemplaire particulier de la *Monographie des Mélastomacées* d'Humboldt et Bonpland; III. Un exemplaire d'épreuves de la Physique des arbres de H. L. Duhamel du Monceau (1758); IV. L'édition française de la *Botanik für Frauenzimmer* ... d'A. J. G. C. Batsch; V. Les premiers états de la *Rariorum historia* de Ch. de L'Écluse (1601); VI. Le tirage a part de l'*Esquisse d'organographie végétale* de P. J. F. Turpin (1837); VII. Un exemplaire de l'*Essai sur l'histoire naturelle des fraisiers* de Duchesne annoté par l'auteur; VIII. Le second volume des *Herbarium vivae eicones* de Brunfels (1531–1532); IX. L'*Appendix altera* à L'"Édition" de 1686 du *Botanicum Monspeliense* de P. Magnol; X. Le *Manuel économique des plantes* de Buc'hoz (1800); XI. Une remise en vente en 1839 de la première édition du *Nouveau dictionnaire* de Déterville; XII. Les plantes du *Jardin du Roy* ... *Louis XIIII* de P. Vallet; XIII. Le format de la *Deutschlands flora* de J. Sturm; XIV. Un exemplaire d'un *Cours de botanique* autographié d'A. P. de Candolle; XV. Une réimpression de deux mémoires de Delile sur la flore d'Égypte; XVI. Les trois émissions de *Discours sur la structure des fleurs* de S. Vaillant (1718, 1727, 1728); XVII. La réimpression (ca 1750) de l'édition de 1719 des *Institutiones rei herbariae* de J. P. de Tournefort; XVIII. Le format et le papier à *Zierrandwasserzeichen* de la *Deutscxhlands flora* de G. F. Hoffmann (1791); XIX. Le quatrième fascicule des *Musci Thuringici* de J. C. Zenker & F. D. Dietrich (1825); XX. La reproduction de 1863 du premier livre des *Campi elysii* de O. J. Rudbeck et O. O. Rudbeck (1702); XXI. Nouvelles remarques sur les tirages des *Plantes du* Roi; XXII. La réémission de 1741 de l'édition de 1729 de l'*Histoire des plantes qui naissent aux environs de Paris* de J. P. de Tournefort; XXIII. Les éditions Lyonnaises de l'*Anatomie de l'homme* de Pierre Dionis; XXIV. L'édition collective de 1764 des *Œuvres philosophiques* de J. O. de la Mettrie; XXV. La puce et le pou de R. Hooke dans la seconde édition du *Philosophical account of the works of nature* de R. Bradley (1739).

D189. **Heppner, J. B.** 1982. Dates of selected Lepidoptera literature for the western hemisphere fauna. *Journal of the Lepidopterists' Society* 36: 87–111.

D190. **Manitz, H.** 1982. Bibliographische Notizen: 2. 1. Briefsammlungen als Hilfsmittel zur Datierung botanischer Publikationen. *Wissenschaftliche Zeitschrift der Friedrich-Schiller Universität Jena, Mathematisch-naturwissenschaftliche Reihe* 31: 259–262.

D191. **Margadant, W. D.** 1968. *Early bryological literature: A descriptive bibliography of selected publications treating musci during the first decades of the nineteenth century and especially of the years 1825, 1826 and 1827.* Pittsburgh, PA, Hunt Botanical Library. xi+278 pp. Also pub. in *Mededelingen van het Botanisch Museum en Herbarium van de Rijksuniversiteit de Utrecht* no. 283.

Introduction, pp. i–viii; Descriptive bibliography applied to botany, pp. 1–34; Special bibliography, pp. 35–239; Chronology, pp. 242–251; Indexes of plant names, artists, engravers, botanists, printers, publishers, and subject index. Includes lists of "Bibliographical studies" for each work treated.

D192. **Mathews, G. M.** 1920. Dates of ornithological works. *Austral avian record* 4: 1–27.

Revision of a list that first appeared in his *Birds of Australia*, vol. 7, appx. B.

D193. **Muir, T. B.** 1981. Dates of publication of Australian pharmacy journals in connection with taxonomy. *Muelleria* 4: 385–387.

D194. **Sayre, G.** 1959. *Dates of publications describing Musci, 1801–1821.* Troy, NY, privately published. [v]+102 pp.

Publications arranged by years, pp. 8–67; Dates of selected periodicals and serial works, pp. 69–81; Literature used as sources of dates, pp. 82–91.

D195. **Sherborn, C. D., & B. B. Woodward.** 1901. Notes on the dates of publication of the natural history portions of some French voyages. — Part I. 'Amerique méridionale'; 'Indes orientales'; 'Pôle Sud ('Astrolabe' and 'Zélée'); 'La Bonite'; 'La Coquille'; and 'L'Uranie et Physicienne.' *Annals & magazine of natural history* (7) 7: 388–392. *Continued* **___.** 1901. ___: Part 2. Ferret and Galinier's 'Voyage en Abyssinie'; Lefebvre's 'Voyage en Abyssinie'; 'Exploration scientifique de l'Algérie'; Castelneau's

'Amérique du Sud'; Dumont d'Urville's 'Voyage de l'Astrolabe'; Laplace's 'Voyage sur la Favorite'; Jacquemont's 'Voyage dans l'Inde'; Tréhouart's 'Commission scientifique d'Islande'; Cailliaud, 'Voyage à Méroë'; 'Expedition scientifique de Morée'; Fabre, 'Commission scientifique du Nord'; Du Petit–Thouars, 'Voyage de la Venus'; and on the dates of the 'Faune Française.' *Ibid.* (7) 8: 161–164, 333–336, 491–494; ___. 1901. Addenda et corrigenda to Part 1. *Ibid.* (7) 8 :494; ___. 1906. Notes on the dates of publication of the natural history portions of some French voyages. — 'Voyage autour du Monde ... sur ... la Coquille pendant ... 1822 25 ... par L. J. Duperry,' &c. — A correction. *Ibid.* (7) 17: 335–336.

D196. **Sherborn, C. D.** 1926. Dates of publication of early catalogues of natural history issued by the British Museum. *Annals & magazine of natural history* (9) 17: 271–272; ___. 1934. Dates of publication of catalogues of natural history (post 1850) issued by the British Museum. *Ibid.* (10) 13: 308–312.

D197. ___. 1938. Dates of books received: Extracted from the Gentleman's Magazine, 1758–1800. *Journal of the Society for the Bibliography of Natural History* 1: 141–142.

D198. ___. 1939. Notices and reviews in the "Gentleman's Magazine" 1801–1833. *Journal of the Society for the Bibliography of Natural History* 1: 197–198.

D199. **Stafleu, F. A.** 1963. Dates of botanical publications 1788–1792. *Taxon* 12: 43–87.

D200. **___, & R. S. Cowan.** 1976–88. *Taxonomic literature: A selective guide to botanical publications and collections with dates, commentaries and types.* 2nd edn. Utrecht & Antwerp, Bohn, Scheltema & Holkema; The Hague, Boston, MA, W. Junk, for the International Association for Plant Taxonomy. 7 vols. *Regnum vegetabile* vols 94, 98, 105, 110, 112, 115, & 116; *Continued* **F. A. Stafleu & E. A. Mennega.** 1992+. ___*: Supplement.* Koenigstein, Koeltz. In progress (vol. 1, A–Ba, 1992). *Regnum vegetabile* vols 125, [etc.].

Familiarly known as "TL–2", the work is a bibliographical catalogue of pre-1939 books, pamphlets, series, and other bibliographically independent and limited entities and lays particular emphasis on providing information on their publication dates, especially those published in parts. *Replaces* **F. A. Stafleu.** 1967. *Taxonomic literature [etc.].* Utrecht, International Bureau for Plant Taxonomy and Nomenclature. xx+556 pp. *Regnum vegetabile* vol. 52.

D201. **Steenis, C. G. G. J. van**, ed. 1947+ (1955+ **M. J. van Steenis–Kruseman**, ed.). Dates of publication of Malaysian phytotaxonomical literature. *Flora Malesiana bulletin* 1(2): 49–51.

The first of a regular series of bibliographical studies by H. C. D. de Wit, M. J. van Steenis–Kruseman, et al. on this subject. e.g. 1(3): 69–70, 1948; 1(4): 96–99, 1948; 1(5), 137–141, 1949; 1(6): 164–168, 1950; 1(7): 193–196, 1950; 2(10): 363, 1953; *2(11): 432–433, 1955; 2(12): 488–491, 1956; 3(13): 568–570, 1957; 3(14): 641–645, 1959; 3(15): 730–743, 1960; 3(16): 821–825, 1961; 4(17): 905–911, 1962; 4(18): 1011–1016, 1963; 4(19): 1141–1149, 1964; 4(20): 1274–1278, 1965; 5(24): 1801, 1969; 6(25): 1906, 1971; 6(26): 2017–2019, 1972; 6(27): 2190–2191, 1974; 6(28): 2364–2365, 1975. (*Later* *Dates of publication and bibliographical notes). Includes lists of "Bibliographical studies" for each work treated.

D202. ___, ed. 1948. List of bibliographic references to publication dates of taxonomic literature of importance to Malaysian taxonomy. *Flora Malesiana bulletin* 1(4): 99–109.

A checklist of bibliographic data sources: General sources; Special sources.

D203. **Steenis–Kruseman, M. J. van, & W. T. Stearn.** 1954. Dates of publication [of works treating Malaysian plants.] In: C. G. G. J. van Steenis, ed. *Flora Malesiana.* Series 4, vol. 5, pp. clxiii–ccxix. Leyden, Noordhof.

D204. **Walker, E. G.** 1961. References to dates of issue and other bibliographic data on entries in the Merrill–Walker: "Bibliography of eastern Asiatic botany" 1938. *Taxon* 10: 45–49.

Natural–history bookselling

D205. **Fock, Gustav**, [Firm]. 1929. *Aus Wissenschaft und Antiquariaat. Festschrift zum 50 Jährigen bestehen der Buchhandlung Gustav Fock.* Leipzig. [xii]+390+[2] pp. & 23 pl.
Historical essays on bibliopegic and bibliographic subjects by various authors. *Includes* **F. Schulze.** Buchhandlung Gustav Fock G.m.b.H. 1879–1929.

D206. **Junk, W.** 1949. *50 Jahre Antiquar.* Ein nachgelassenes Manuskript herausgegeben von A. C. Klooster und W. Weisbach. The Hague, Dr. W. Junk. [viii]+431 pp.
Index nominum, pp. 391–397; Index titulorum, pp. 398–430.

D207. **Swann, C. Kirke.** 1956. Natural history books from a bookseller's point of view. *Journal of the Society for the Bibliography of Natural History* 3: 117–126.

D208. ___. 1972. Natural history bookselling. *Journal of the Society for the Bibliography of Natural History* 6: 118–126.

D209. **Voyles, J.** 1971. Dealers' catalogs. In: A. Kent & H. Lancour. *Encyclopedia of library and information science*, vol. 6, pp. 467–481. New York, M. Dekker.

D210. **Zeitlin, J.** 1966. Natural history books from a rare bookseller's point of view. In: T. R. Buckman, ed. *Bibliography and natural history*, pp. 140–148. Lawrence, KS, Univ. of Kansas Libraries.

Systematic bibliography

D211. **Berger, S. E.** 1991. *The design of bibliographies: Observations, references and examples.* London, Mansell Publishing. x+198 pp. *Also* Westport, CT, Greenwood Pr., 1992. *Bibliographies & indexes in library & information science* no. 6.
Includes an extensive annotated bibliography, pp. 49–103 and Appendix I: "Facsimile examples", pp. 103–189, featuring 35 examples.

D212. **Hazen, R. M., M. H. Hazen, & L. W. Finger.** 1983. Compiling *American geological literature, 1669 to 1850*: A systematic approach to natural history bibliography. In: Society for the Bibliography of Natural History. *Contributions to the history of North American natural history: Papers from the first North American conference of the Society for the Bibliography of Natural History held at the Academy of Natural Sciences, Philadelphia, 21–23 October 1981*, pp. 91–97. London. *Special publication* no. 2.

D213. **Krummel, D. W.** 1984. *Bibliographies: Their aims and methods.* London, Mansell Publishing. x+192 pp.

D214. **National Agricultural Library, United States Department of Agriculture.** 1982. *Guidelines for the preparation of bibliographies.* Beltsville, MD. 64p. *Miscellaneous publication* no. 1414.

D215. **Robinson, A. M. L.** 1979. *Systematic bibliography: A practical guide to the work of compilation.* 4th ed. With an additional chapter by M. Lodder. London, C, Bingley; Munich, K. G. Saur. 135 pp.

D216. **Schneider, G.** 1934. *Theory and history of bibliography.* Translated by R. R. Shaw. New York, Columbia Univ. Pr. xiv+306 pp. *Re-pub.* New York, Scarecrow Pr., 1961.
Translation of "Theoretisch–Geschichtlicher Teil." In: *Handbuch der Bibliographie.* 3rd ed., pp. 3–199. Leipzig, Hiersemann, 1926.

Periodical title abbreviations

D217. **Alkire, L. G.**, ed. 1992. *Periodical title abbreviations.* 8th ed. Detroit, MI, Gale Research Co. 3 vols.
Vol. 1: By abbreviation; vol. 2: By title; vol. 3: New periodical title abbreviations. 130,000 entries. Abbreviations filed letter by letter; titles files word by word.

D218. **Brown, P., & G. B. Stratton.** 1963–65. *World list of scientific periodicals published in the years 1900–1960.* 4th ed. London, Butterworth. 3 vols.
Lists over 60,000 titles with recommended title abbreviations, and provides a "Select list of abbreviations used," vol. 3, pp. xiii–xxii.

D219. **Lawrence, G. H. M., et al.** 1968. *B–P–H: Botanico–Periodicum–Huntianum.* Pittsburgh, PA, Hunt Botanical Library. 1,063 pp.; *Continued* **G. D. R. Bridson & E. R. Smith.** 1991. *B–P–H/S: Botanico–Periodicum–Huntianum/Supplementum.* Pittsburgh, PA, Hunt Institute for Botanical Documentation. [ii]+1068 pp.
Lists over 25,000 titles of periodical (serial) publications with recommended title–abbreviations and A–Z lists of word abbreviations used.

D220. **Leistner, G.** 1981. *Internationale Titelabkürzungen (ITA) von Zeitschriften, Zeitungen, wichtigen Handbüchern, Wörterbüchern, Gesetzen usw.* 3. Aufl. Osnabrück, Biblio Verlag. 2 vols.
A–Z by abbreviation. *Replaces* ___. 1977. ___. 2. Aufl. Osnabrück, [etc.]. 1,137 pp.

D221. **Royal Society of London.** 1867–72. *Catalogue of scientific papers (1800–1900).* London, Her Majesty's Stationery Office (*later* Cambridge, Cambridge Univ. Pr.). 6 vols.
Vol. 1: List of title abbreviations of periodicals cited included in vols. 1, 6, 7, 9, 12, & 13.

D222. **Rust, W.** 1927. *Verzeichnis von unklaren Titelkürzungen deutscher und ausländer Zeitschriften.* Leipzig, Harrassowitz. viii+142 pp.
A–Z list of ca. 4,000 title–initial abbreviations.

HISTORY OF BIBLIOGRAPHY

General

D223. **Altick, R. D.** 1951. *The scholar adventurers.* New York, Macmillan Co. viii+338 pp.
Includes accounts of some bibliographical adventures.

D224. **Beaver, D. de B.** 1972. The Smithsonian origin of the Royal Society Catalogue of scientific papers. *Science studies (London)* 2: 385–393.

D225. **Besterman, T.** 1950. *Les débuts de la bibliographie méthodique.* 3me ed. revue. Traduit de l'Anglais. Paris, La Palme. 95 pp. & 12 pl.
Replaces ___. 1936. *The beginnings of systematic bibliography.* 2nd ed. London. xi+81 pp. & 12 pl. Re–pub. New York, B. Franklin, 1966. *Burt Franklin bibliography & reference series* no. 216.

D226. **Bolton, H. C.** 1897. Bibliography. In: Smithsonian Institution. *The Smithsonian Institution 1846–1896*, pp. 785–804. Ed. by G. B. Goode. Washington, DC.

D227. **Breslauer, B. H., & R. Folter.** 1984. *Bibliography: Its history and development.* New York, Grolier Club. 224 pp.
Annotated catalogue of 169 items exhibited at the Club to celebrate the completion of the *National union catalog, pre-1956 imprints.*

D228. **Elliott, C. A.** 1970. The Royal Society Catalogue as an index to 19th–century American science. *Journal of the American Society for Information Science* 21: 396–401.

D229. **Malclès, L.–N.** 1961. *Bibliography*. Translated [and ed.] by T. C. Hines. New York, Scarecrow Pr. [xii]+152 pp.
A history of systematic bibliography. Translation of *La bibliographie*. Paris, Presses Universitaire de France, 1956. 134 pp. *Que sais–je?*

D230. **Murra, K. O.** 1951. History of some attempts to organize bibliography internationally. In: J. H. Shera & M. E. Egan, eds. *Bibliographic organization: Papers presented before the fifteenth annual conference of the Graduate Library School, July 24–29, 1950*, pp. 24–53. Chicago, IL, Univ. of Chicago Pr. *University of Chicago studies in library science.*
Examines seven experiments in this field, including: the Royal Society *Catalogue of scientific papers*, the *International catalogue of scientific literature*, Concilium bibliographicum, and UNESCO.

D231. **Norris, D. M.** 1939. *A history of cataloguing and cataloguing methods 1100–1850: With an introductory survey of ancient times.* London, Grafton & Co. x+246 pp.

D232. **Sorkin, A. M.** [1963.] Zur Entstehungsgeschichte des "International catalogue of scientific literature." *N T M, Zeitschrift für Geschichte der Naturwissenschaften, Technik und Medizin* 1 (4): 67–84.

D233. **Thornton, J. L., & R. I. J. Tully.** 1971. Scientific bibliographies and bibliographers. In: *Scientific books, libraries and collectors: A study of bibliography and the book trade in relation to science.* 3rd rev. ed., pp. 294–338. London, Library Association; ___. 1978. ___*: Supplement 1969–75*, pp. 91–103. London, Library Association.
Replaces 2nd rev. ed., 1962. xiii+406 pp.

D234. **Townsend, A. C.** 1959. Life and the librarian: Aspects of biological bibliography. *Library Association record* 1959 (Dec.): 319–327.

Bibliographers & historians

D235. **Ewan, J.** 1988. Field notes on five "jackdaws of scholarship." *Taxon* 37: 782–790.
Notes on J. Dryander, G. A. Pritzel, J. Sabin, A. Rehder, & C. Nissen.

D236. **Fulton, J. F.** 1951. *The great medical bibliographers: A study in humanism.* Philadelphia, PA, Univ. of Pennsylvania Pr. xv+107 pp. & 22 pp. of illus. *Rosenberg Fellowship lectures.*

D237. **Guerra, F.** 1962. Some bibliographers of early medical Americana. *Journal of the history of medicine* 17: 94–115, ports.

D238. **Jayawardene, S. A., & J. Lawes.** 1979. Biographical notices of historians of science: A checklist for Kurt Vogel on his 90th birthday (30 September 1978). *Annals of science* 36: 315–394.
Includes 1. Science (general) (#1–109); 6. Natural history, botany, agriculture, pharmacy (#231–252); 7. Medicine (#253–370).

D239. **Sarton, G.** 1955. "Historians and philosophers" of science: Biographies available in *Isis* (1–45) & *Osiris* (1–11). *Isis* 46: 360–366.

Individual bibliographers & historians

Aldrovandi, U.

D240. **Adversi, A.** 1966. Nuovi appunti su Ulisse Aldrovandi, bibliofilo, bibliotecario e bibliografo, a sulla sua inedita "Bibliologia." *Bibliofilia* 68: 51–90.

Anker, J.

D241. **Drachmann, A. G.** 1957. Jean Anker, 1892–1957. *Centaurus* 5: 167–169, port.

Bay, J. C.

D242. **Taylor, K. L.** 1941. *J. Christian Bay at seventy: A review and a bibliography.* Chicago, IL, John Crerar Library. 17 pp.

D243. **Thompson, L. S.** 1962. Jens Christian Bay, bibliologist. *Libri* 12: 320–330.

Billings, J. S.

D244. **Bradway, F.** 1965. Bibliography of the writings of John Shaw Billings. In: F. B. Rogers, ed. *Selected papers of John Shaw Billings*, pp. 285–300. Chicago, IL, Medical Library Association.

D245. **Marson, J.** 1969. John Shaw Billings as a bibliographer. *Bulletin of the Medical Library Association* 57: 379–390.

Bodenheimer, F. S.

D246. **Zaunick, R.** 1960. In memoriam Frederick Simon Bodenheimer (1896–1959). *Physis (Florence)* 2: 270–275, port.

Burckhardt, R.

D247. **Smit, P.** 1985. The Swiss zoologist Rudolf Burckhardt (1866–1908): Pioneer in biohistory. *Gesnerus* 42: 67–83.

Carus, J. V.

D248. **Simon, H–R.** 1975. Julius Victor Carus (1823–1903) als Bibliograph der Zoologie. *Bibliothek und Wissenschaft* 9: 250–275.

Choulant, J. L.

D249. **ter Laage, R. J. Ch. V.** 1970. Reflections on Johann Ludwig Choulant and his medico–historical bibliographies. In: P. Smit & R. J. Ch. V. ter Laage, eds. *Essays in biohistory*, pp. 115–133. Utrecht, International Association for Plant Taxonomy. *Regnum vegetabile* no. 71.

Cole, F. J.

D250. **Eales, N. B.** 1959. Francis Joseph Cole, 1872–1959. *Journal of the history of medicine* 14: 267–272, port.

D251. **Franklin, K. J.** 1959. Francis Joseph Cole 1872–1959. *Biographical memoirs of the Fellows of the Royal Society* 5: 37–47, port.

Dryander, J.

D252. **Uggla, A. Hj.** 1944. Jonas Dryander (1748–1810). *Proceedings of the Linnean Society of London* 156: 99–102.
Also **R. Du Rietz.** 1964. More light on Jonas Dryander. *Svenska Linné–Sällskapets årsskrift* 47: 82–83.

Ewan, J. & N.

D253. **Crotz, K.** 1989. *Ewaniana: The writings of Joe and Nesta Ewan.* Chillicothe, IL, The American Botanist. [64] pp.

Fulton, J.

D254. **LeFanu, W.** 1962. John Fulton's historical and bibliographical work. *Journal of the history of medicine* 17: 38–50.

D255. **S., M. E., & E. H. T.** 1962. Bibliography of John Farquhar Fulton. *Ibid.* 17: 51–71.

Gesner, C.

D256. **Bay, J. C.** 1916. Conrad Gesner (1516–1565) the father of bibliography: An appreciation. *Papers of the Bibliographical Society of America* 10 (2): 53–86.

D257. **Escher, H.** 1934. Die Bibliotheca Universalis Konrad Gesners. *Vierteljahresschrift der Naturforschenden Gesellschaft Zürich* 79: 174–194.

D258. **Fischer, H.** 1966. Conrad Gesner (1516–1565) as bibliographer and encyclopedist. *Library* (5) 21: 269–281.

D259. **Mayerhöfer, J.** 1965. Conrad Gesner als Bibliograph und Enzyklopädist: Der Zusammenbruch der mittelalterlichen artes liberales. *Gesnerus* 22: 176–195.

Goode, G. B.

D260. **Kohlstedt, S. G.** 1988. History in a natural history museum: George Brown Goode and the Smithsonian Institution. *Publishing history* 10(2): 7–26.

Gudger, E. W.

D261. **Sarton, G.** 1951. Bibliography of Dr. E. W. Gudger's contributions to the history of ichthyology (1905–1951). *Isis* 42: 237–242.

Gunther, R. T.

D262. **Gunther, A. E.** 1967. *Robert T. Gunther, a pioneer in the history of science, 1869–1940.* Oxford, privately pub. xiv+520 pp. & 18 pl. *Early science in Oxford* vol. 15.

Haller, A. von.

D263. **Bay, J. C.** 1960. Albrecht von Haller, medical encyclopedist. *Bulletin of the Medical Library Association* 48: 393–403.

D264. **Fulton, J. F.** 1932. Haller and the humanization of bibliography. *New England journal of medicine* 206: 323–328.

Herrlinger, R.

D265. **Feiner, E.** 1970. *Bibliographie Robert Herrlinger (1914–1968), eingeleitet und zusammengestellt von [etc.].* [Kiel, Institut für Geschichte der Medizin und Pharmazie and der Universität Kiel.] 94 pp.
See also **M. Quick.** 1990. Bibliographie Robert Herrlinger (1914–1968). Korrekturen, Ergänzungen und Kommentar zu der Vorarbeiten Edith Feiners. *Wurzburger medizinhistorische Mitteilungen* 8: 307–324.

Junk, W.

D266. **Schwarz, F. H.** 1975. Bibliographia Junkiana: Verzeichnis, der von Dr. rer. nat. h.c. et Dr. phil. h.c. Wilhelm Junk verfassten oder herausgegeben Publikationen (1893–1949). In: Dr. W. Junk B.V., [Pub.]. *General catalogue 1975*, pp. xiii–xlii. The Hague.

Keynes, Sir G.

D267. **Keynes,** Sir **G.** 1953. Religio bibliographici. Presidential address to the Bibliographical Society, March 1953. *Library* (5) 8: 63–76. *Re-pub. in* ___. 1964. *Bibliotheca bibliographici: A catalogue of the library formed by Geoffrey Keynes.* London, Trianon Pr. xxiii+444 pp. & 45 pl.

D268. ___. 1981. *The gates of memory.* Oxford, Clarendon Pr. xi+428 pp.

Klebs, A. C.

D269. **Bungartner, L.** 1943. Arnold C. Klebs, 1870–1943. *Bulletin of the history of medicine* 14: 201–216.
See also **Lang, A.** 1940. Bibliography of the writings of Arnold C. Klebs. *Bulletin of the history of medicine* 8: 523–532.

Linnaeus, C.

D270. **Heller, J. L.** 1970. Linnaeus's Bibliotheca botanica. *Taxon* 19: 363–411.
Includes a discussion of other early botanical bibliographies.

Needham, J.

D271. **Teich, M., & R. Young**, eds. 1973. *Changing perspectives in the history of science: Essays in honour of Joseph Needham.* London, Heinemann. xxi+490 pp.
Includes Bibliography of Needham, pp. 472–477.

Nissen, C.

D272. **Geck, E., & G. Pressler**, eds. 1973. *Festschrift für Claus Nissen zum siebzigsten Geburtstag, 2 September 1971.* Wiesbaden, G. Pressler. 686 pp.
Includes Verzeichnis der Veröffentlichungen von Claus Nissen in den Jahren 1925–1970, pp. 15–24.

Nordenskiöld, E.

D273. **Hofsten, N. von.** 1947. Erik Nordenskiöld (1872–1933). *Isis* 38: 103–106, port.

Pritzel, G. A.

D274. **Helm, J.** 1966. Dr. Georg August Pritzel 1851–1874: Ein Gedenken zum 150. Geburtstag. (Eine Biobibliographie). *Kulturpflanze* 14: 311–357.

D275. **Stafleu, F. A.** 1973. Pritzel and his Thesaurus. *Taxon* 22: 119–130.

Raikov, B. E.

D276. **Lukina, T. A.** 1970. *Boris Evgen'evich Raikov, 1880–1966.* Leningrad, Izd. "Nauka." 208 pp.

D277. ___. 1970. Boris Evgen'evich Raikov, kak istorisk biologii i pedagog. [B. E. Raikov as a historian of biology]. *Iz istorii biologii* 1970 (2): 5–20.
Includes a bibliography.

Reuss, J. D.

D278. **Geiger, P. J. D.** 1905. Reuss und seine Bibliothek. *Zentralblatt für Bibliothekswesens* 22: 465–490.

Rothschuh, K. E.

D279. **Toellner, R.**, ed. 1983. *Karl Eduard Rothschuh: Bibliographie, 1935–1983.* 2nd ed. Tecklenburg, Burgerverlag. ix+54 pp. *Münstersche Beiträge zur Geschichte und Theorie der Medizin* no. 19.
Replaces **G. Poloczek.** 1973. ___. Munster, Institut für Geschichte der Medizin. ix+37 pp. *Münstersche Beiträge zur Geschichte und Theorie der Medizin* no. 8.

Sarton, G.

D280. **Garfield, E.** 1985. The life and career of George Sarton: The father of the history of science. *Journal of the history of the behavioral sciences* 21: 107–117.

D281. **Strelsky, K.** 1957. Bibliography of the publications of George Sarton. *Isis* 48: 336–350.
Forms part of The George Sarton Memorial Issue of *Isis* (vol. 48 no. 3), with contributions by I. Bernard Cohen, Charles & Dorothea Singer, John F. Fulton and others. Sarton's personal library is now at Harvard University Library, *see* **Anon.** 1950. The Sarton Collection in the History of Science. *Harvard Library bulletin* 4: 276–277.

Sherborn, C. D.

D282. **Norman, J. R.** 1944. *Squire: Memories of Charles Davies Sherborn.* London, G. C. Harrap & Co. 202 pp. & 4 pl.

Singer, C. J.

D283. **Hall, A. R.** 1960. Charles Joseph Singer, 1876–1960. *Isis* 51: 558–560.

D284. **Sheppard, J.** 1987. Papers in the Contemporary Medical Archives Center. *Medical history* 31: 466–471.

D285. **Underwood, E. A.** 1953. A bibliography of the published writings of Charles Singer. In: ___, ed. *Science, medicine and history, essays ... written in honour of Charles Singer*, vol. 2: 555–586. Oxford, Oxford Univ. Pr.

D286. **___.** 1960. Charles Singer (1876–1960). *Medical history* 4: 353–358, port.
Includes a bibliography of his publications since 1953.

Stearn, W. T.

D287. **Stearn, W. T.** 1976–92. Publications by William T. Stearn on bibliographical, botanical and horticultural subjects, 1929–1976; a chronological list. *Biological journal of the Linnean Society* 8: 299–318, 1976; *Continued* **___.** 1992. Publications ... 1977–1991; a chronological list. *Botanical journal of the Linnean Society* 109: 443–451, 1992.

Verdoorn, F.

D288. **Sanders, A. P. M.** 1984. In memoriam: Frans Verdoorn, bryologist, publisher and biohistorian. *Janus* 71: 165–179.

Wood, C. A

D289. **Astbury, E. C.** 1981. *Casey A. Wood (1856–1942), ophthalmologist, bookman, ornithologist: A bio-bibliography.* Montreal, McGill Univ., Graduate School of Library Science. 66 pp. *Occasional papers* no. 7.

Zimmer, J. T.

D290. **Murphy, R. C., & D. Amadon.** 1959. In memoriam: John Zimmer. *Auk* 76: 418–423.

SUBJECT INDEX

GEOGRAPHICAL INDEX

This index uses contemporary country names. Therefore, for older geographical entities, always search for the nations that occupy the corresponding territory today. However, a few changes occurred too late for incorporation. "Czechoslovakia" and "Yugoslavia" are still used; Byelorussia, Georgia, Moldavia, and the central Asian republics of the Russian commonwealth, are included under the heading "Russia."

NAME INDEX

This is primarily an index to all authors (personal or corporate), editors, and translators. Also entered are the subjects of biographies and bibliographies, creators of book and specimen collections, etc. — references to these are prefixed "*Subj.*" Title-headings are included for a few entries such as periodicals.